NY LIBRARY

Prion Biology and Diseases

SECOND EDITION

COLD SPRING HARBOR MONOGRAPH SERIES

Prion Biology and Diseases

SECOND EDITION

EDITED BY

Stanley B. Prusiner

University of California, San Francisco

COLD SPRING HARBOR LABORATORY PRESS
Cold Spring Harbor, New York

PRION BIOLOGY AND DISEASES
Second Edition

Monograph 41
© 2004 by Cold Spring Harbor Laboratory Press
All rights reserved
Printed in the United States of America

Publisher	John Inglis
Acquisitions Editor	David Crotty
Production Manager	Denise Weiss
Project Coordinator	Joan Ebert
Production Editors	Melissa Frey and Patricia Barker
Desktop Editor	Susan Schaefer
Interior Book Designer	Denise Weiss and Emily Harste
Cover Designer	Ed Atkeson and Denise Weiss

Front Cover Artwork: In the background, a high-power view of a two-dimensional crystal of PrP 27-30 after image processing. In the foreground, a trimeric model (*left*) for the structure of PrP 27-30 is super-imposed and in scale with the 2D crystal background. A short segment of a model fiber (*right*) is composed of PrP 27-30 trimers. Several of these small fibers might form a typical prion rod. The models show the proposed β-sheet structure in yellow and α-helices in red. (Image courtesy of H. Wille and C. Govaerts.)

Library of Congress Cataloging-in-Publication Data

Prion biology and diseases / edited by Stanley B. Prusiner.-- 2nd ed.
    p. cm.
Includes bibliographical references and index.
 ISBN 0-87969-693-1 (hardcover : alk. paper)
 1. Prions. 2. Prion diseases. I. Prusiner, Stanley B., 1942- II.
Title.
 QR502.P745 2003
 616.8'3--dc22

                   2003019306

10 9 8 7 6 5 4 3 2

All Cold Spring Harbor Laboratory Press publications may be ordered directly from Cold Spring Harbor Laboratory Press, 500 Sunnyside Boulevard, Woodbury, New York 11797-2924. Phone: 1-800-843-4388 in Continental U.S. and Canada. All other locations: (516) 422-4100. FAX: (516) 422-4097. E-mail: cshpress@cshl.edu. For a complete catalog of Cold Spring Harbor Laboratory Press publica-tions, visit our World Wide Web Site http:// www.cshlpress.com/

It often happens, with regard to new inventions, that one part of the general public finds them useless and another part considers them to be impossible. When it becomes clear that the possibility and the usefulness can no longer be denied, most agree that the thing was fairly easy to discover and that they knew about it all along.

—— *Abraham Edelcrantz, 1796*

From *"The Early History of Data Networks"* by G.J. Holzmann and B. Pehrson. 1994. Wiley–IEEE Computer Society Press.

Contents

TRANSGENETICS AND CELL BIOLOGY OF PRIONS

PRION DISEASES IN ANIMALS

HUMAN PRION DISEASES

METHODOLOGY, BIOSAFETY, AND THERAPEUTICS

Preface

FOUR YEARS AGO, THE FIRST EDITION of *Prion Biology and Diseases* was published. Because a large number of recent studies on prions have greatly enhanced our understanding of these novel infectious pathogens, we decided to revise and modestly expand this book when the first edition became a "bestseller" and all the copies were sold.

In general, most scholars have come to accept the existence of prions. They believe that prions are different from all other infectious pathogens, and this view has aroused considerable interest. Bovine prions have spread from cattle with bovine spongiform encephalopathy (BSE) to humans in Great Britain and other parts of Europe and this continues to worry physicians, public health officials, and politicians. Despite the fact that molecular investigations of prions began only 21 years ago with the discovery of the protease-resistant fragment of the prion protein (PrP 27-30), there has been a steady increase in new knowledge, which is quite impressive. With the isolation of PrP 27-30, the application of modern molecular biological approaches, immunologic techniques, and structural biological studies became possible.

Like BSE in Europe, the continuing "spread" of chronic wasting disease (CWD) in free-ranging as well as captive deer and elk in the United States and Canada is of considerable concern. Whether CWD prions can be transmitted to humans either directly or first through sheep or cattle is unknown. The discovery of an 8-year-old cow with BSE in Alberta, Canada, led more than 30 countries to ban Canadian beef. Many questions have been raised by this BSE case. How did the Canadian cow become infected with prions? What should be done to prevent additional cases of mad cow disease in Canada? How can consumers be assured that they are not eating bovine prions? And how can Canada reassure other nations that its beef is safe to eat? Ironically, Canada banned the importation of all Japanese beef when the first BSE case was announced in Japan;

this ban has remained in place even though all slaughtered cattle in Japan are now tested for BSE prions. Recently, a 23-month-old slaughtered Japanese cow was found to have BSE, but the prion strain seems to differ from that seen previously in Japan and Europe. Is this a sporadic case of prion disease in cattle? Certainly, this is a reasonable hypothesis to investigate.

Much scientific investigation will need to be performed in order to answer the questions posed above concerning livestock with prion diseases. We have virtually no data on the frequency of sporadic or spontaneous prion diseases in livestock or free-ranging cervids. Although the vast majority of prion disease in humans is sporadic, we have little understanding of the molecular mechanism responsible for prion propagation in these unfortunate people.

Both editions of this book were assembled with the hope of stimulating young investigators to enter the field of prion biology. These young scientists will need to develop sensitive antemortem diagnostic tests that will detect prion replication many months or years before neurologic dysfunction is evident. Such tests would allow drug intervention long before the brain degenerates. Having stated that, we also need to develop drugs that either block prion replication or stimulate the clearance of prions. Antemortem diagnostic tests could also be used to test all livestock for prions prior to slaughter and thus assure that the human food supply is free of prions.

In addition to enticing young scientists to enter the prion field, we hope that this book will provide an authoritative source for more senior investigators who wish to pursue investigations on prions or who find themselves having to lecture to students on the subject. Because of the size constraints for this book imposed by the publisher, many investigators who have made important contributions to prion biology could not be asked to write chapters: To these individuals, the editor apologizes. Instead, a representative group of authors, who have contributed substantially to the study of prions over the past decade, were selected to participate in this undertaking. To the investigators who have so generously given their time and effort to write the chapters in this book, I express my deep appreciation. To Hang Nguyen at the University of California San Francisco and to John Inglis and his many colleagues at Cold Spring Harbor Laboratory Press, I extend a very special thank you for their dedication and talents.

The six chapters at the beginning of this book were written to provide a proper introduction for those not already schooled in prion biology.

These chapters give an overview of prion biology, record the development of the prion concept, describe bioassays for prions, explain the basic principles of prion replication, and outline some methods for the study of prions. The remaining chapters of this book cover an array of diverse topics, ranging from the cell biology of prions to transgenic and gene-targeted mice, from inherited and infectious prion diseases of humans to bovine spongiform encephalopathy. In addition, scrapie of sheep, the experimental neuropathology of genetically engineered mice, therapeutics, biocontainment and biosafety issues in prion disease, as well as studies of mammalian prion strains and fungal prions, are discussed. Because the literature on prions has grown to be immense, interested readers are urged to consult papers published in refereed journals that are readily retrievable through Internet searches.

Many questions about the atomic structure of prions, the mechanism of replication, and the pathogenesis of prion disease remain unanswered. Perhaps the most fascinating question involves deciphering the molecular language used to encipher prion strain–specific information in the structure of the disease-causing isoform, PrP^{Sc}.

By virtually any measure, the discovery of prions must be considered an extraordinary chapter in the history of biology and medicine. Ritualistic cannibalism causing kuru, industrial cannibalism causing BSE, and prion-tainted pituitary growth hormone causing iatrogenic prion disease are but a few of the astonishing sagas that embellish the history of prions. The difficulties that marked acceptance of the prion concept are not unique. Skepticism is the rule in science, rather than the exception, when new ideas are first introduced. Only the steady accumulation of data has served to change the thinking of most scholars. That a few vocal skeptics remain seems to be a frequent and perhaps constant feature of paradigm shifts in science. Certainly, studies of prions are no exception; fortunately, the science of prion biology has clearly triumphed over prejudice.

STANLEY B. PRUSINER

1

An Introduction to Prion Biology and Diseases

Stanley B. Prusiner

Institute for Neurodegenerative Diseases
Departments of Neurology and Biochemistry and Biophysics
University of California, San Francisco, California 94143

THE HISTORY OF PRIONS IS A FASCINATING SAGA in the annals of biomedical science. For nearly five decades with no clue as to the cause, physicians watched patients with a central nervous system (CNS) degenerative disorder called Creutzfeldt-Jakob disease (CJD) die, often within a few months of its onset (Creutzfeldt 1920; Jakob 1921; Kirschbaum 1968). CJD destroys the brain while the body remains unaware of this process. No febrile response, no leukocytosis or pleocytosis, and no humoral immune response is mounted in response to this devastating disease. Despite its recognition as a distinct clinical entity, CJD remained a rare disease; first, it was in the province of neuropsychiatrists and, later, of neurologists and neuropathologists. Although multiple cases of CJD were recognized in families quite early (Kirschbaum 1924; Meggendorfer 1930; Stender 1930; Davison and Rabiner 1940; Jacob et al. 1950; Friede and DeJong 1964; Rosenthal et al. 1976; Masters et al. 1979, 1981a,b), this observation did little to advance understanding of the disorder.

The unraveling of the etiology of CJD is a wonderful story that has many threads, each representing a distinct piece of the puzzle. An important observation was made by Igor Klatzo, Carleton Gajdusek, and Vincient Zigas in 1959, when they recognized that the neuropathology of kuru resembled that of CJD (Klatzo et al. 1959). That same year, William Hadlow suggested that kuru, a disease of New Guinea highlanders, was similar to scrapie, a hypothesis also based on light microscopic similarities (Hadlow 1959). Hadlow's insight was much more profound that that of Klatzo and his colleagues because Hadlow suggested that kuru is a

Prion Biology and Diseases, 2nd Ed. ©2004 Cold Spring Harbor Laboratory Press 0-87969-693-1/04

1

transmissible disease, like scrapie. Moreover, Hadlow proposed that demonstration of the infectivity of kuru might be best accomplished using chimpanzees because they are so closely related to humans. He also noted that many months or years might be required before clinically recognizable disease would be seen in these inoculated nonhuman primates. He argued that brain tissue from patients dying of kuru should be homogenized and injected intracerebrally into the chimpanzees, similar to the procedures used to demonstrate the transmission of scrapie from an infected sheep to a healthy recipient.

At the time Hadlow set forth his hypothesis, scrapie was thought to be caused by a "slow virus." The term "slow virus" had been coined by Bjorn Sigurdsson in 1954 based on his studies on scrapie and visna of sheep in Iceland (Sigurdsson 1954). Although Hadlow suggested that kuru, like scrapie, was caused by a slow virus, he did not perform the experiments required to demonstrate this phenomenon (Hadlow 1959, 1995). In fact, seven years passed before the transmissibility of kuru was established by Gajdusek, Clarence Gibbs, and Michael Alpers, who passaged kuru to chimpanzees (Gajdusek et al. 1966). Two years later, Gibbs, Gajdusek, and their colleagues reported the transmission of CJD to chimpanzees after intracerebral inoculation (Gibbs et al. 1968).

An early clue to the unusual properties of the scrapie agent emerged from studies of 18,000 sheep that were inadvertently inoculated with the scrapie agent. The sheep had been vaccinated against louping ill virus with a formalin-treated suspension of ovine brain and spleen which was subsequently shown to have been contaminated with the scrapie agent (Gordon 1946). Three different batches of vaccines were administered, and two years later, 1500 sheep developed scrapie. These findings demonstrated that the scrapie agent is resistant to inactivation by formalin, in contrast to most viruses that are readily inactivated by such treatment.

The unusual biological properties of the scrapie agent are no less puzzling than the disease process itself, for the infectious agent causes devastating degeneration of the CNS in the absence of an inflammatory response (Zlotnik 1962; Beck et al. 1964). Although the immune system remains intact, its surveillance mechanism fails to detect a raging infection. The infectious agent that causes scrapie, now generally referred to as a "prion," achieved status as a scientific curiosity when its extreme resistance to ionizing and ultraviolet irradiation was discovered (Alper et al. 1966, 1967; Latarjet et al. 1970). Later, similar resistance to inactivation by UV and ionizing radiation was reported for the CJD agent (Gibbs et al. 1978). Tikvah Alper's radiation resistance data on the scrapie agent evoked a torrent of hypotheses concerning its composition. Suggestions as to the nature of the scrapie agent ranged from small DNA viruses and membrane fragments to polysaccharides and proteins, the last of which eventually proved to be correct (Pattison 1965; Gibbons and Hunter 1967; Griffith 1967; Pattison and Jones 1967; Hunter et al. 1968; Field et al. 1969; Hunter 1972; Chapter 2).

Because the scrapie agent had been passaged from sheep into mice (Chandler 1961), for many years scrapie was the most amenable of these diseases to study experimentally. Although scrapie, CJD, and kuru were thought for many years to be caused by different slow viruses, all these diseases are known now to be caused by prions and, thus, the distinction was artificial.

Studies of prions have wide implications, ranging from basic principles of protein conformation to the development of effective therapies for prion diseases (Prusiner 1998a). In this chapter, the structural biology of prion proteins, as well as the genetics and molecular neurology of prion diseases, is introduced. Furthermore, this chapter describes how information is enciphered within the infectious prion particle.

PRION BIOLOGY AND DISEASES

Our current understanding of prions that cause scrapie, CJD, and related diseases in mammals is described in this book (Table 1). Prions are unprecedented infectious pathogens that cause a group of invariably fatal, neurodegenerative diseases by means of an entirely novel mechanism (Prusiner 1998a). Prion diseases may present as genetic, infectious, or sporadic disorders, all of which involve modification of the prion protein (PrP). CJD generally presents as a progressive dementia, whereas scrapie of sheep and bovine spongiform encephalopathy (BSE) are generally manifest as ataxic illnesses (Wells et al. 1987). Prions have also been reported in yeast and other fungi. These fascinating studies have greatly expanded our thinking about the role of prions in biology (Chapter 7).

PRIONS

Prions are infectious proteins. In mammals, prions reproduce by recruiting the normal, cellular isoform of the prion protein (PrP^C) and stimulating its conversion into the disease-causing isoform (PrP^{Sc}). PrP^C is rich in α-helical content and has little β-sheet structure, whereas PrP^{Sc} has less α-helical content and a high β-sheet structure. Comparisons of secondary structures of PrP^C and PrP^{Sc} were performed on the proteins purified from Syrian hamster (SHa) brains (Pan et al. 1993). On the basis of these

Table 1. The prion diseases

Disease	Host	Mechanism of pathogenesis
A. Kuru	Fore people	infection through ritualistic cannibalism
Iatrogenic CJD	humans	infection from prion-contaminated HGH, dura mater grafts, etc.
Variant CJD	humans	infection from bovine prions
Familial CJD	humans	germ-line mutations in PrP gene
GSS	humans	germ-line mutations in PrP gene
FFI	humans	germ-line mutation in PrP gene (D178N,M129)
Sporadic CJD	humans	somatic mutation or spontaneous conversion of PrP^C into PrP^{Sc}?
sFI	humans	somatic mutation or spontaneous conversion of PrP^C into PrP^{Sc}?
B. Scrapie	sheep	infection in genetically susceptible sheep
BSE	cattle	infection with prion-contaminated MBM
TME	mink	infection with prions from sheep or cattle
CWD	deer, elk	unknown
FSE	cats	infection with prion-contaminated bovine tissues or MBM
Exotic ungulate encephalopathy	greater kudu, nyala, oryx	infection with prion-contaminated MBM

(BSE) Bovine spongiform encephalopathy; (CJD) Creutzfeldt-Jakob disease; (CWD) chronic wasting disease; (FFI) fatal familial insomnia; (FSE) feline spongiform encephalopathy; (sFI) sporadic fatal insomnia; (GSS) Gerstmann-Sträussler-Scheinker disease; (HGH) human growth hormone; (MBM) meat and bone meal; (TME) transmissible mink encephalopathy.

data, structural models for PrP^C and PrP^{Sc} were proposed (Huang et al. 1995). Subsequently, nuclear magnetic resonance (NMR) solution structures of recombinant SHa and mouse (Mo) PrPs produced in bacteria showed that PrP^C likely has three α-helices and not four as predicted by molecular modeling (Riek et al. 1996; Liu et al. 1999). The computational model of PrP^{Sc} is supported by studies with recombinant antibody fragments, which have been used to map the surfaces of PrP^C and PrP^{Sc} (Peretz et al. 1997). This α-to-β transition in PrP structure is the fundamental event underlying prion diseases.

Limited proteolysis of PrP^{Sc} produces a smaller protease-resistant molecule of ~142 amino acids, designated PrP 27-30; under the same conditions, PrP^C is completely hydrolyzed (Fig. 1). In the presence of detergent, PrP 27-30 polymerizes into amyloid (McKinley et al. 1991). Prion rods formed by limited proteolysis and detergent extraction (Fig. 2) are

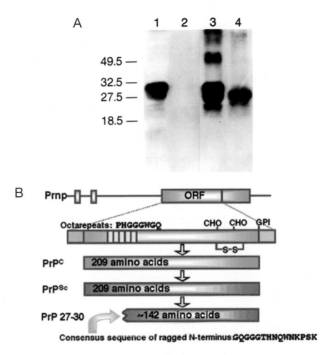

Figure 1. Prion protein isoforms. (*A*) Western immunoblot of brain homogenates from uninfected (lanes *1* and *2*) and prion-infected (lanes *3* and *4*) Syrian hamsters. Samples in lanes *2* and *4* were digested with 50 μg/μl proteinase K for 30 minutes at 37°C. PrP^C in lanes 2 and 4 was completely hydrolyzed under these conditions, whereas ~67 amino acids were digested from the amino terminus of PrP^{Sc} to generate PrP 27-30. After polyacrylamide gel electrophoresis (PAGE) and electrotransfer, the blot was developed with anti-SHaPrP R073 polyclonal rabbit antiserum (Scrban et al. 1990). Molecular-weight markers are depicted in kilodaltons (kD). (*B*) Bar diagram of the *PrnP* gene in Syrian hamsters that encodes a protein of 254 amino acids. After processing of the amino and carboxyl termini, both PrP^C (*green*) and PrP^{Sc} (*red*) consist of 209 residues. After limited proteolysis, the amino terminus of PrP^{Sc} is truncated to form PrP 27-30, which is composed of ~142 amino acids, the amino-terminal sequence of which was determined by Edman degradation.

indistinguishable from the filaments that aggregate to form PrP amyloid plaques in the CNS. Both the rods and the PrP amyloid filaments found in brain tissue exhibit similar ultrastructural morphology and green-gold birefringence after staining with Congo red dye (Prusiner et al. 1983).

Much has been learned about the basic biology of prions using yeast, in which two different proteins, unrelated to PrP, form prions (Wickner

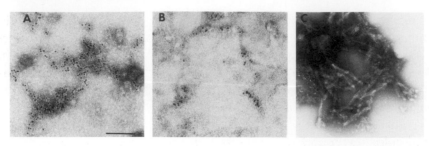

Figure 2. Electron micrographs of negatively stained and immunogold-labeled prion proteins. (*A*) PrP^C and (*B*) PrP^{Sc}. Neither PrP^C nor PrP^{Sc} forms recognizable, ordered polymers. (*C*) Prion rods composed of PrP 27-30 were negatively stained. The prion rods are indistinguishable from many purified amyloids. Bar, 100 nm. (Reprinted, with permission, from Pan et al. 1993 [copyright National Academy of Sciences].)

1994; Sparrer et al. 2000). These studies are described below and in Chapter 7.

PRIONS ARE DISTINCT FROM VIRUSES

The major feature that distinguishes prions from viruses is that both PrP isoforms are encoded by a chromosomal gene, designated *PRNP* in humans and *Prnp* in mice, located on the short arm of chromosome 20 and the syntenic region of chromosome 2, respectively (Prusiner 1998b). Prions differ from viruses and viroids in that they lack a nucleic acid genome that directs the synthesis of their progeny. In contrast, strains of viruses and viroids have distinct nucleic acid sequences that produce pathogens with different properties. Many investigators argued for a nucleic acid genome within the infectious prion particle while others contended for a small, noncoding polynucleotide of either foreign or cellular origin (Dickinson et al. 1968; Kimberlin 1982, 1990; Dickinson and Outram 1988; Bruce et al. 1991; Weissmann 1991b). However, no nucleic acid was found despite intensive searches using a wide variety of techniques and approaches (Kellings et al. 1992, 1994). Based on a wealth of evidence, it is reasonable to assert that such a nucleic acid has not been found because it does not exist. Prions are composed of an alternative isoform of a cellular protein, whereas most viral proteins are encoded by a viral genome, and viroids are devoid of protein.

Many features of prion structure and replication distinguish prions from viruses and other known infectious pathogens. Prions can exist in multiple molecular forms, whereas viruses exist in a single form with a distinct ultrastructural morphology. Prions do not have a consistent structure, in marked contrast to viruses. Infectivity has been detected in fractions containing prion particles with an extremely wide range of sizes (Kimberlin et al. 1971; Prusiner et al. 1978b; Diringer and Kimberlin 1983). Initially, the small size of prions as determined by ionizing radiation inactivation studies (Alper et al. 1966) was confusing because scrapie infectivity was clearly associated with larger particles (Rohwer 1984). Eventually, aggregation due to hydrophobic interactions between PrP^{Sc} molecules was found to be responsible for such anomalous behavior (Prusiner et al. 1978b).

In contrast to viruses, prions are nonimmunogenic. They do not elicit an immune response because the host has been rendered tolerant to PrP^{Sc} by PrP^{C} (Prusiner et al. 1993a; Williamson et al. 1996). In contrast, the foreign proteins of viruses that are encoded by the viral genome often elicit a profound immune response. Thus, it seems unlikely that vaccination, which has been so effective in preventing many viral illnesses, will be a useful strategy for preventing or treating prion diseases.

Despite these distinctions, prions are similar to viruses in some ways. Like viruses, prions are infectious because they stimulate a process by which more of the pathogen is produced. As prions or viruses accumulate in an infected host, they eventually cause disease. Both prions and viruses exist in different varieties or subtypes, called strains. The phenomenon of prion strains posed a profound conundrum with respect to how prions might be composed of only host-encoded PrP^{Sc} molecules and yet exhibit diversity. An enlarging body of data argues that strains of prions are enciphered in the conformation of PrP^{Sc}.

That many of the prion diseases were discovered prior to our current understanding of prion biology has created confusion in an environment in which decisions of great economic, political, and possibly public health consequence are being made (Phillips et al. 2000). For example, scrapie and BSE have different names, yet they are the same disease in two different species. Scrapie and BSE are prion diseases that differ from each other in only two respects: First, the sheep PrP sequence differs from that of cattle at seven or eight positions of 270 amino acids (Goldmann et al. 1990a, 1991b), which results in different PrP^{Sc} molecules. Second, some aspects of each disease are determined by the particular prion strain that infects the respective host.

Understanding prion strains and the "species barrier" is of paramount importance with respect to the BSE epidemic in Britain, in which more than 180,000 cattle have died over the past decade (Wells and Wilesmith 1995; Chapters 11 and 17). Brain extracts from eight cattle with BSE resulted in similar incubation times and patterns of vacuolation in the neuropil when inoculated into a variety of inbred mice (Bruce et al. 1994; Bruce 1996). Incubation times and profiles of neuronal vacuolation have been used for three decades to study prion strains. Brain extracts prepared from three domestic cats, one nyala, and one kudu, all of which died with a neurologic illness, produced incubation times and lesion profiles indistinguishable from those found in the BSE cattle. Cats and exotic ungulates (such as the kudu) presumably developed prion disease from eating food containing bovine (Bo) prions (Jeffrey and Wells 1988; Wyatt et al. 1991; Kirkwood et al. 1993).

NOMENCLATURE

The Glossary has definitions for all of the terms discussed in this section on nomenclature. A listing of the different prion diseases is given in Table 1. Although the prions that cause transmissible mink encephalopathy (TME) and BSE are referred to as "TME prions" and "BSE prions," this may be unjustified because both are thought to originate from the oral consumption of scrapie prions in sheep-derived foodstuffs and because many lines of evidence argue that the only difference among various prions is the sequence of PrP, which is dictated by the host and not the prion itself. Human (Hu) prions present a similar semantic conundrum. Transmission of Hu prions to laboratory animals produces prions carrying PrP molecules with sequences dictated by the PrP gene of the host, not that of the species from which the inoculum was derived.

The "Sc" superscript of PrP^{Sc} was initially derived from the term scrapie because scrapie was the prototypic prion disease. Because all of the known prion diseases (Table 1) of mammals involve aberrant metabolism of PrP similar to that observed in scrapie, the "Sc" superscript was suggested for all abnormal, pathogenic PrP isoforms (Prusiner et al. 2000). In this context, the "Sc" superscript is used to designate the disease-causing isoform of PrP. The development of the conformation-dependent immunoassay (CDI) for PrP^{Sc} led to the discovery of a protease-sensitive form of PrP^{Sc}, designated $sPrP^{Sc}$ (Safar et al. 1998). The CDI is based on measuring antibody binding to an epitope that is exposed in PrP^{C} but buried in PrP^{Sc}. Under conditions of limited proteolysis in which the protease-resistant form of PrP^{Sc}, designated $rPrP^{Sc}$, is converted into PrP 27-30, $sPrP^{Sc}$ is completely hydrolyzed. Whether $sPrP^{Sc}$ is an intermediate in the formation of infectious prions remains to be established.

In the case of mutant PrP, mutations and polymorphisms can be denoted in parentheses following the PrP isoform. For fatal familial

Table 2. Examples of human PrP gene mutations found in the inherited prion diseases

Inherited prion disease	PrP mutation
Gerstmann-Sträussler-Scheinker disease	P102L*
Gerstmann-Sträussler-Scheinker disease	A117V
Familial Creutzfeldt-Jakob disease	D178N, V129
Fatal familial insomnia	D178N, M129*
Gerstmann-Sträussler-Scheinker disease	F198S*
Familial Creutzfeldt-Jakob disease	E200K*
Gerstmann-Sträussler-Scheinker disease	Q217R
Familial Creutzfeldt-Jakob disease	octarepeat insert*

*Signifies genetic linkage between the mutation and the inherited prion disease (Hsiao et al. 1989; Dlouhy et al. 1992; Petersen et al. 1992; Poulter et al. 1992; Gabizon et al. 1993).

insomnia (FFI), in which it might be important to identify the mutation, the prions would be designated HuPrP^{Sc}(D178N,M129) (Table 2).

Parentheses following PrP^{Sc} can also be used to notate a particular prion strain. For example, prions from Syrian hamsters inoculated with Sc237 or 139H prion strains can be designated SHaPrP^{Sc}(Sc237) or SHaPrP^{Sc}(139H), respectively. Similarly, sheep inoculated with scrapie or BSE prions produce OvPrP^{Sc}(Sc) or OvPrP^{Sc}(BSE), respectively.

The terms PrP^{res} and PrP-res were derived to describe the protease-resistance of PrP^{Sc} and have sometimes been used interchangeably with PrP^{Sc}. The use of PrP^{res} and PrP-res became particularly problematic with the discovery of sPrP^{Sc}. Protease-resistance, insolubility, and high β-structure content should be considered only as surrogate markers of PrP^{Sc} infectivity because not all may be present. For example, MoPrP^{Sc}(P101L) from transgenic (Tg) mice that express high levels of MoPrP(P101L) is transmissible to Tg(MoPrP,P101L)196/$Prnp^{0/0}$ mice, but these prions are sensitive to proteolytic digestion at 37°C. When digestions were performed at 4°C or the CDI was used to detect PrP^{Sc}, then MoPrP^{Sc}(P101L) was detected (Tremblay et al. 2003). In contrast, when Tg196 mice were inoculated with mouse RML prions, the resulting MoPrP^{Sc}(P101L) was resistant to proteolytic digestion at 37°C (Manson et al. 1999). These findings emphasize the ambiguities that may arise from simply assessing resistance to proteolytic digestion. Whether PrP^{res} is useful in denoting PrP molecules that have been subjected to procedures modifying resistance to proteolysis, but that neither convey infectivity nor cause disease, remains questionable. Perhaps PrP^{res} is best reserved for PrP molecules that exhibit resistance to limited proteolysis after binding to PrP^{Sc} (Kocisko et al. 1994).

To simplify the terminology, the generic term PrP^{Sc} was suggested in place of such terms as PrP^{CJD}, PrP^{BSE}, and PrP^{res} (Prusiner et al. 2000). To distinguish PrP^{Sc} found in humans and cattle from that found in other animals, "HuPrP^{Sc}" and "BoPrP^{Sc}" are suggested instead of PrP^{CJD} and PrP^{BSE}, respectively. Once human prions, and thus HuPrP^{Sc} molecules, have been passaged into animals, the prions and PrP^{Sc} are no longer of the human species unless they were formed in an animal expressing a HuPrP transgene.

The term "PrP*" has been used in two different ways. First, it has been used to identify a fraction of PrP^{Sc} molecules that are infectious (Weissmann 1991a). Such a designation is thought to be useful because there are ~10^5 PrP^{Sc} molecules per infectious (ID$_{50}$) unit (Prusiner et al. 1982a, 1983). Second, PrP* has been used to designate a metastable intermediate of PrP^C that is bound to a putative conversion cofactor, provisionally designated protein X (Cohen et al. 1994). It is noteworthy that neither a subset of biologically active PrP^{Sc} molecules nor a metastable intermediate of PrP^C has been identified, to date.

In mice, *Prnp* is now known to be identical to two genes, *Sinc* and *Prn-i*, that are known to control the length of the incubation time in mice inoculated with prions (Carlson et al. 1994; Moore et al. 1998). A gene, designated *Pid-1*, on mouse chromosome 17 also appears to influence experimental CJD and scrapie incubation times, but information on this locus is limited.

Distinguishing among CJD, FFI, and Gerstmann-Sträussler-Scheinker syndrome (GSS) has grown increasingly difficult with the recognition that familial (f) CJD, GSS, and FFI are autosomal-dominant diseases caused by mutations in *PRNP* (Table 2). Initially, it was thought that each *PRNP* mutation was associated with a particular cliniconeuropathologic phenotype, but more exceptions are being recognized. Multiple examples of variations in the cliniconeuropathologic phenotype have been recorded within a single family in which all affected members carry the same *PRNP* mutation. Most patients with a *PRNP* mutation at codon 102 present with ataxia and have PrP amyloid plaques; such patients are generally given the diagnosis of GSS, but some individuals present with dementia, a clinical characteristic that is usually associated with CJD. For most inherited prion diseases, the disease is specified by the respective mutation, such as fCJD(E200K) and GSS(P101L). In the case of FFI, describing the D178N mutation and M129 polymorphism seems unnecessary because this is the only known mutation–polymorphism combination that results in the FFI phenotype. The sporadic form of fatal insomnia is denoted sFI (Mastrianni et al. 1997; Gambetti and Parchi 1999).

DISCOVERY OF THE PRION PROTEIN

The discovery of PrP transformed research on scrapie and related diseases (Bolton et al. 1982; Prusiner et al. 1982a). It provided a molecular marker that was subsequently shown to be specific for these illnesses and identified the major, and the only known, component of the prion particle. The protease-resistant fragment of PrP^{Sc}, designated PrP 27-30 because it has an apparent molecular weight (M_r) of 27 kD to 30 kD (Fig. 1), was discovered by enriching fractions from SHa brain for scrapie infectivity (Bolton et al. 1982; Prusiner et al. 1982a).

Over a 20-year period beginning in 1960, there were many unsuccessful attempts to purify the scrapie agent or to identify a biochemical marker that copurified with it (Hunter et al. 1963, 1969, 1971; Hunter and Millson 1964, 1967; Kimberlin et al. 1971; Millson et al. 1971, 1976; Marsh et al. 1974, 1978, 1980; Siakotos et al. 1976; Gibbs and Gajdusek 1978; Millson and Manning 1979). Studies on the sedimentation properties of scrapie infectivity in mouse spleens and brains suggested that hydrophobic interactions were responsible for the non-ideal physical behavior of the scrapie particle (Prusiner 1978; Prusiner et al. 1978a). Indeed, the scrapie agent presented a biochemical nightmare: Infectivity spread from one end of a sucrose gradient to the other and from the void volume of chromatographic columns to fractions eluting at 5 to 10 times the included volume. Such results demanded new approaches and better assays. Only with the development of improved bioassays did the purification of the infectious pathogen that causes scrapie and CJD become possible (Prusiner et al. 1980, 1982b).

Enriching fractions from the brains of scrapie-infected Syrian hamsters for infectivity yielded a single protein, PrP 27-30, as noted above. PrP 27-30 was later found to be the protease-resistant core of PrP^{Sc} (Prusiner et al. 1984; Basler et al. 1986). Copurification of PrP 27-30 and scrapie infectivity demands that the physicochemical properties as well as the antigenicity of these two entities be similar (Gabizon et al. 1988; Chapter 2). The results of a wide array of inactivation experiments demonstrated the similarities in the properties of PrP 27-30 and scrapie infectivity (McKinley et al. 1983; Prusiner et al. 1983). To explain these findings in terms of the virus hypothesis, it is necessary to postulate the existence of either a virus that has a coat protein that is highly homologous with PrP or a virus that binds tightly to PrP^{Sc}. In either case, the PrP-like coat proteins or the PrP^{Sc}–virus complexes must display properties indistinguishable from PrP^{Sc} alone (Chapter 2).

The inability to inactivate preparations highly enriched with scrapie infectivity by procedures that modify nucleic acids was interpreted as evidence against the existence of a scrapie-specific nucleic acid (Alper et al. 1967; Prusiner 1982). To explain the findings in terms of a virus, PrP^{Sc} or an as-yet-undetected PrP-like protein of viral origin must protect the viral genome from being inactivated.

PrP mRNA levels are similar in normal, uninfected and in scrapie-infected tissues (Chesebro et al. 1985; Oesch et al. 1985). This finding produced skepticism about whether or not PrP 27-30 is related to the infectious prion particle. Nevertheless, the search for a protein encoded by PrP mRNA revealed PrP^C, a protease-sensitive protein that is soluble in non-denaturing detergents (Oesch et al. 1985; Meyer et al. 1986).

Determination of the Amino-terminal Sequence of PrP 27-30

The molecular biology and genetics of prions began with the purification of PrP 27-30 to allow determination of its amino-terminal amino acid sequence (Prusiner et al. 1984). Multiple signals in each cycle of the Edman degradation suggested that either multiple proteins were present in these "purified fractions" or a single protein with a ragged amino terminus was present. When the signals in each cycle were grouped according to their strong, intermediate, and weak intensities, it became clear that a single protein with a ragged amino terminus was being sequenced (Fig. 3). Determination of a single, unique sequence for the amino terminus of PrP 27-30 permitted the synthesis of isocoding mixtures of oligonucleotides that were subsequently used to identify incomplete PrP cDNA clones from hamster (Oesch et al. 1985) and mouse (Chesebro et al. 1985). cDNA clones encoding the entire open reading frames (ORFs) of SHaPrP and MoPrP were eventually recovered (Basler et al. 1986; Locht et al. 1986).

PRION PROTEIN ISOFORMS

In Syrian hamsters, PrP^C and PrP^{Sc} are 209-residue proteins that are anchored to the cell surface by a glycosylphosphatidyl inositol (GPI) moiety and have the same covalent structure (Fig. 1). The amino-terminal sequencing, the deduced amino acid sequences from PrP cDNA, and immunoblotting studies argue that PrP 27-30 is a truncated protein of approximately 142 residues, which is derived from PrP^{Sc} by limited proteolysis of the amino terminus (Prusiner et al. 1984; Oesch et al. 1985; Basler et al. 1986; Locht et al. 1986; Meyer et al. 1986).

Relative Amount	Amino Acid Sequence*
1	G-Q-F-F-F-T-H-N-Q-W-N-K-P-S-K
0.4	X-X-X-T-H-N-X-W-X-K-P
0.2	X-X-P-W-X-Q-X-X-X-T-H-X-Q-W

*single-letter amino acid code. X = amino acid not determined at that cycle.

Figure 3. Interpreted sequence, shown by single-letter amino acid codes, of the amino terminus of PrP 27-30. The "ragged ends" of PrP 27-30 are shown. X = amino acid that was not detected at that cycle of the Edman degradation (Prusiner et al. 1984).

In general, $\sim 10^5$ PrP^{Sc} molecules correspond to one ID$_{50}$ unit using the most sensitive bioassay (Prusiner et al. 1982a, 1983). PrP^{Sc} is probably best defined as the alternative or abnormal isoform of PrP, which stimulates conversion of PrP^C into nascent PrP^{Sc}, accumulates, and causes disease (see Chapter 2). Although resistance to limited proteolysis has proved to be a convenient tool for detecting PrP^{Sc}, not all PrP^{Sc} molecules possess protease resistance, as discussed above (Hsiao et al. 1994; Telling et al. 1996a). Furthermore, PrP^{Sc} from different species or prion strains may exhibit different degrees of protease resistance. Some investigators equate protease resistance with PrP^{Sc}, and this erroneous view has been compounded by the use of the term "PrP-res" (Caughey et al. 1990).

Although insolubility as well as protease resistance was used in initial studies to differentiate PrP^{Sc} from PrP^C (Meyer et al. 1986), subsequent investigations showed that these properties are only surrogate markers, as are high β-structure content and polymerization into amyloid (Prusiner et al. 1983; Caughey et al. 1991b; McKinley et al. 1991; Gasset et al. 1993; Safar et al. 1993b; Muramoto et al. 1996; Riesner et al. 1996). When these surrogate markers are present, they are useful, but their absence does not establish a lack of prion infectivity. PrP^{Sc} is usually not detected by western immunoblotting if fewer than 10^5 ID$_{50}$ units/ml are present in a sample (Lasmézas et al. 1996a).

In our experience, the method of sample preparation from scrapie-infected brain also influences the sensitivity of PrP^{Sc} immunodetection, in part because PrP^{Sc} is not uniformly distributed in the brain (DeArmond et al. 1987; Taraboulos et al. 1992). Some experiments in which PrP^{Sc} detection proved to be problematic in partially purified preparations (Czub et al. 1986; Xi et al. 1992) were repeated with crude homogenates, in which PrP^{Sc} was readily measured (Jendroska et al. 1991; McKenzie et al. 1994).

Cell Biology of PrP^{Sc} Formation

In scrapie-infected cells, PrP^C molecules destined to become PrP^{Sc} exit to the cell surface prior to conversion into PrP^{Sc} (Stahl et al. 1987; Borchelt et al. 1990; Caughey and Raymond 1991). Like other GPI-anchored proteins, PrP^C appears to re-enter the cell through a subcellular compartment bounded by cholesterol-rich, detergent-insoluble membranes, which may be caveolae or early endosomes (Gorodinsky and Harris 1995; Taraboulos et al. 1995; Vey et al. 1996; Kaneko et al. 1997a; Naslavsky et al. 1997). Within this cholesterol-rich, nonacidic compartment, GPI-anchored PrP^C can be either converted into PrP^{Sc} or partially degraded (Taraboulos et al. 1995). Subsequently, PrP^{Sc} is trimmed at the amino terminus in an acidic compartment in scrapie-infected cultured cells to form PrP 27-30 (Caughey et al. 1991a). In contrast, amino-terminal trimming of PrP^{Sc} is minimal in the brain, where little PrP 27-30 is found (McKinley et al. 1991).

RODENT MODELS OF PRION DISEASE

Mice and hamsters are commonly used in experimental studies of prion disease (Chapters 4 and 16). The shortest incubation times are achieved with intracerebral inoculation of prions with a sequence identical to that of the host animal; under these conditions, all animals develop prion disease within a narrow interval for a particular dose. When the PrP sequence of the donor prion differs from that of the recipient host, the incubation time is prolonged, and can be quite variable; often, many of the inoculated animals do not develop disease (Carlson et al. 1989; Telling et al. 1994, 1995; Tateishi et al. 1996). This phenomenon is often called the "species barrier" (Pattison 1965).

PrP GENE STRUCTURE AND ORGANIZATION

The PrP gene is a member of the *Prn* gene family. The second member of this family to be identified is the *Prnd* gene that lies approximately 19 kb downstream from the PrP locus and encodes the doppel (Dpl) protein (Moore et al. 1999; Chapter 6). The respective genes that encode PrP and Dpl appear to represent ancient gene duplication that occurred prior to the speciation of mammals. The sequences are only ~25% identical, but the structures of the two proteins are highly conserved (Fig. 4) (Mo et al. 2001). In contrast to PrP, which is expressed in many different tissues, Dpl expression is confined to the testis. Both Dpl and PrP are found on the surface of sperm, but their functions are unknown. In contrast to PrP-

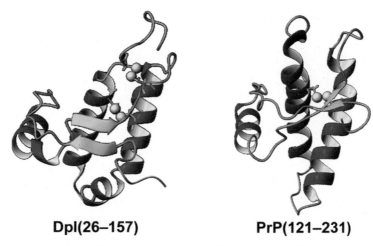

Dpl(26–157)   **PrP(121–231)**

Figure 4. Comparison of the NMR structures of mouse Dpl and PrP. Backbone topology of mouse Dpl(26–157) (Mo et al. 2001) and mouse PrP(121–231) (Riek et al. 1996). (Figure prepared by J. Dyson and P. Wright.)

deficient ($Prnp^{0/0}$) mice, Dpl-deficient ($Prnd^{0/0}$) mice are sterile (Behrens et al. 2002). The knockout of both the PrP and Dpl genes resulted in a sterile phenotype indistinguishable from that of the $Prnd^{0/0}$ mice (D. Paisley and D.W. Melton, pers. comm.).

The entire ORF of all known mammalian and avian PrP genes resides within a single exon (Basler et al. 1986; Westaway et al. 1987; Hsiao et al. 1989; Gabriel et al. 1992), which eliminates the possibility that PrP^{Sc} arises from alternative RNA splicing (Basler et al. 1986; Westaway et al. 1987, 1991). The two exons of the SHaPrP gene are separated by a 10-kb intron; exon 1 encodes a portion of the 5′ untranslated leader sequence, whereas exon 2 encodes the ORF and 3′ untranslated region (Basler et al. 1986). Encoded by the SHaPrP gene, a low-abundance PrP mRNA containing an additional small exon in the 5′ untranslated region was discovered (Li and Bolton 1997). The PrP genes of mouse, sheep, and rat contain three exons, with exon 3 analogous to exon 2 of the hamster PrP gene (Westaway et al. 1991, 1994a,b; Saeki et al. 1996). The promoters of both the SHaPrP and MoPrP genes contain multiple copies of GC-rich repeats and are devoid of TATA boxes. These GC nonamers represent a motif that may function as a canonical binding site for the transcription factor Sp1 (McKnight and Tjian 1986). Mapping of PrP genes to the short arm of human chromosome 20 and to the analogous region of mouse chromosome 2 argues for the existence of PrP genes prior to the speciation of mammals (Robakis et al. 1986; Sparkes et al. 1986).

Like the PrP gene, the entire ORF of the Dpl gene is encoded within a single exon. In some lines of $Prnp^{0/0}$ mice, high levels of Dpl expression were found in the brain (Moore et al. 1999). The expression of Dpl in the CNS was due to intergenic splicing of the nontranslated exons of the PrP gene with the translated exon of Dpl. That Dpl expression is neurotoxic was subsequently demonstrated by construction of Tg mice expressing Dpl in the brain (Moore et al. 2001).

Expression of the PrP Gene

Although PrP mRNA is constitutively expressed in the brains of adult animals (Chesebro et al. 1985; Oesch et al. 1985), it is highly regulated during development. In the septum, levels of PrP mRNA and choline acetyltransferase were found to increase in parallel during development (Mobley et al. 1988). In other brain regions, PrP gene expression occurs at an earlier age. In situ hybridization studies show that the highest levels of PrP mRNA are found in neurons (Kretzschmar et al. 1986).

Because no antibodies are currently available that clearly distinguish between PrP^C and PrP^{Sc}, PrP^C is generally measured in tissues from uninfected control animals, in which no PrP^{Sc} is found. The report that an α-PrP IgM monoclonal antibody (mAb) reacts exclusively with PrP^{Sc} has not been confirmed (Korth et al. 1997), and thus, PrP^{Sc} must be measured in tissues of infected animals after PrP^C has been hydrolyzed by digestion with a proteolytic enzyme. Alternatively, PrP^{Sc} can be measured using the CDI, which does not require proteolysis of PrP^C (Safar et al. 1998). PrP^C expression in brain was defined by standard immunohistochemistry (DeArmond et al. 1987) and by histoblotting, a procedure for the detection of PrP^{Sc} in situ, in the brains of uninfected controls (Taraboulos et al. 1992). Immunostaining of PrP^C in the SHa brain was most intense in the stratum radiatum and stratum oriens of the CA1 region of the hippocampus and was virtually absent from the granule cell layer of the dentate gyrus and the pyramidal cell layer throughout Ammon's horn (Chapter 15).

PrP^{Sc} staining was minimal in the regions that intensely stained for PrP^C. A similar relationship between PrP^C and PrP^{Sc} was found in the amygdala. In contrast, PrP^{Sc} accumulated in the medial habenular nucleus, the medial septal nuclei, and the diagonal band of Broca; these areas were virtually devoid of PrP^C. In the white matter, bundles of myelinated axons contained PrP^{Sc} but were devoid of PrP^C. These findings suggest that prions are transported along axons, which is consistent with earlier findings in which scrapie infectivity migrated in a pattern consistent with retrograde transport (Kimberlin et al. 1983; Fraser and Dickinson 1985; Jendroska et al. 1991). Whereas the rate of PrP^{Sc} synthesis appears to be a function of the level of PrP^C expression in Tg mice, the level of PrP^{Sc} accumulation appears to be independent of PrP^C concentration (Prusiner et al. 1990).

Overexpression of Wild-type PrP Transgenes

Mice expressing different levels of the wild-type (wt) SHaPrP transgene were constructed (Chapter 8). Inoculation of these Tg(SHaPrP) mice with SHa prions demonstrated abrogation of the species barrier, resulting in abbreviated incubation times due to a nonstochastic process (Prusiner et al. 1990). The length of the incubation time after inoculation with SHa prions was inversely proportional to the level of SHaPrP^C in the brains of Tg(SHaPrP) mice (Prusiner et al. 1990). Bioassays of brain extracts from clinically ill Tg(SHaPrP) mice inoculated with Mo prions revealed that Mo prions, but no SHa prions, were produced. Conversely, inoculation of Tg(SHaPrP) mice with SHa prions led to the synthesis of only SHa prions.

During transgenetic studies, uninoculated older mice harboring high copy numbers of wt PrP transgenes derived from Syrian hamsters, sheep, and Prnp^b mice spontaneously developed truncal ataxia, hind-limb paralysis, and tremors (Westaway et al. 1994a). These Tg mice exhibited profound necrotizing myopathy involving skeletal muscle, demyelinating polyneuropathy, and focal vacuolation of the CNS. Development of disease was dependent on transgene dosage. For example, Tg(SHaPrP$^{+/+}$)7 mice homozygous for the SHaPrP transgene array regularly developed disease between 400 and 600 days of age, whereas hemizygous Tg(SHaPrP$^{+/0}$)7 mice developed disease after >650 days. Whether overexpression of wt or somatic mutant PrP^C is responsible for this neuromyopathy remains to be established.

PrP Gene Dosage Controls the Incubation Time

Incubation times have been used to isolate prion strains inoculated into sheep, goats, mice, and hamsters (Dickinson et al. 1968). The *Sinc* gene is a major determinant of incubation periods in mice. Once molecular clones of *Prnp* became available, a study was performed to determine whether control of the length of the incubation time is genetically linked to the PrP gene. Because the availability of VM mice with prolonged incubation times that were used to define *Sinc* were restricted, I/LnJ mice (Kingsbury et al. 1983) were used in the crosses. Indeed, the incubation time locus, designated *Prn-i*, was found to be either congruent with or closely linked to *Prnp* (Carlson et al. 1986).

Although the amino acid substitutions in PrP that distinguish *Prnp*[a] from *Prnp*[b] mice argued for the congruency of *Prnp* and *Prn-i* (Westaway et al. 1987), experiments with *Prnp*[a] mice expressing *Prnp*[b] transgenes demonstrated a "paradoxical" shortening of incubation times (Westaway et al. 1991). These Tg mice were predicted to exhibit a prolongation of the incubation time after inoculation with RML prions based on (*Prnp*[a] × *Prnp*[b]) F$_1$ mice, which exhibit long incubation times. Those findings were described as a "paradoxical" shortening because we and other investigators had believed for many years that long incubation times are dominant traits (Dickinson et al. 1968; Carlson et al. 1986). From studies of congenic and transgenic mice expressing different numbers of the *a* and *b* alleles of *Prnp*, these findings were discovered to be not paradoxical; indeed, they resulted from increased PrP gene dosage (Carlson et al. 1994). When the RML isolate was inoculated into congenic and transgenic mice, increasing the number of copies of the *a* allele was found to be the major determinant in reducing the incubation time; however, increasing the number of copies of the *b* allele also reduced the incubation time, but not to the same extent as that seen with the *a* allele. Gene-targeting studies established that the *Prnp* gene controls the incubation time and, as such, is congruent with both *Prn-i* and *Sinc* (Moore et al. 1998).

PrP-deficient Mice

Ablation of the PrP gene in mice did not affect development of these animals (Chapter 8) (Büeler et al. 1992; Manson et al. 1994). *Prnp*$^{0/0}$ mice remain healthy for more than two years. Although brain slices from *Prnp*$^{0/0}$ mice were reported to show altered synaptic behavior (Collinge et al. 1994; Whittington et al. 1995), these results could not be confirmed by other workers (Herms et al. 1995; Lledo et al. 1996).

In two subsequent *Prnp*$^{0/0}$ lines, Purkinje cell loss was accompanied by ataxia beginning at ~70 weeks of age (Sakaguchi et al. 1996). Crossing one of these *Prnp*$^{0/0}$ lines with Tg mice overexpressing MoPrP rescued the ataxic phenotype (Nishida et al. 1999). With the discovery of Dpl, it became clear, as described above, that Dpl expression in the brains of these *Prnp*$^{0/0}$ mice provoked cerebellar degeneration (Moore et al. 1999).

Prnp$^{0/0}$ mice inoculated with prions are resistant to infection (Büeler et al. 1993; Prusiner et al. 1993a). *Prnp*$^{0/0}$ mice were sacrificed 5 days, 60 days, 120 days, and 315 days after inoculation with RML prions, and brain extracts were bioassayed in Swiss CD-1 mice. Except for residual infectivity from the inoculum detected at 5 days after inoculation, no infectivity was found in the brain extracts, as the Swiss CD-1 mice did not develop disease (Prusiner et al. 1993a). One group of investigators found that $Prnp^{0/0}$ mice inoculated with mouse-passaged scrapie prions and sacrificed 20 weeks later had $10^{3.6}$ ID_{50} units/ml of homogenate by bioassay (Büeler et al. 1993). Another group also found measurable titers of prions many weeks after inoculation of $Prnp^{0/0}$ mice with mouse-passaged CJD prions (Sakaguchi et al. 1995) (Chapter 8). Some investigators have argued that these data imply that prions replicate in the absence of PrP gene expression (Chesebro and Caughey 1993; Caughey and Chesebro 1997; Lasmézas et al. 1997).

$Prnp^{0/0}$ mice crossed with Tg(SHaPrP) mice were rendered susceptible to SHa prions but remained resistant to Mo prions (Büeler et al. 1993; Prusiner et al. 1993a). Because the absence of PrP^C expression does not provoke disease, it is likely that scrapie and other prion diseases are a consequence of PrP^{Sc} accumulation rather than inhibition of PrP^C function (Büeler et al. 1992). Such an interpretation is consistent with the dominant inheritance of familial prion diseases.

Mice heterozygous ($Prnp^{0/+}$) for ablation of the PrP gene had prolonged incubation times when inoculated with Mo prions and developed signs of neurologic dysfunction at 400–460 days after inoculation (Prusiner et al. 1993a; Büeler et al. 1994). These findings are in accord with studies on Tg(SHaPrP) mice in which increased SHaPrP expression was accompanied by shortened incubation times (Prusiner et al. 1990).

Because $Prnp^{0/0}$ mice do not express PrP^C, we reasoned that they might more readily produce α-PrP antibodies. $Prnp^{0/0}$ mice immunized with Mo or SHa prion rods produced α-PrP antisera that bound MoPrP, SHaPrP, and HuPrP (Prusiner et al. 1993a; Williamson et al. 1996). These findings contrast with earlier studies in which α-MoPrP antibodies could not be produced in mice, presumably because the mice had been rendered tolerant by the presence of $MoPrP^C$ (Barry and Prusiner 1986; Kascsak et al. 1987; Rogers et al. 1991). That $Prnp^{0/0}$ mice readily produce α-PrP antibodies is consistent with the hypothesis that the lack of an immune response in prion diseases is due to the fact that PrP^C and PrP^{Sc} share many epitopes.

SPECIES VARIATIONS IN THE PrP SEQUENCE

PrP is posttranslationally processed to remove a 22-amino acid, amino-terminal signal peptide. The carboxy-terminal 120 amino acids contain two conserved disulfide-bonded cysteines and a sequence that marks the addition of a GPI anchor. Twenty-three residues are removed during the

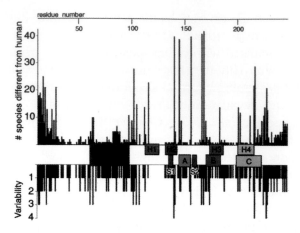

Figure 5. Species variations and mutations of the prion protein gene. The *x*-axis represents the human PrP sequence, with the five octarepeats and H1–H4 regions of putative secondary structure shown, as well as the three α-helices A, B, and C and the two β-strands S1 and S2 as determined by NMR. Vertical bars above the axis indicate the number of species that differ from the human sequence at each position. Below the axis, the length of the bars indicates the number of alternative amino acids at each position in the alignment. (Data compiled by P. Bamborough and F.E. Cohen.)

addition of this GPI moiety, which anchors the protein to the cell membrane (Stahl et al. 1990). Contributing to the mass of the protein are two asparagine side chains linked to large oligosaccharides with multiple structures that have been shown to be complex and diverse (Endo et al. 1989). Although many species variants of PrP have been sequenced (Schatzl et al. 1995), only the chicken sequence has been found to differ greatly from the human sequence (Harris et al. 1989; Gabriel et al. 1992). The alignment of the translated sequences from more than 40 PrP genes shows a striking degree of conservation between the mammalian sequences and is suggestive of the retention of some important function through evolution (Fig. 5, Appendix). Cross-species conservation of PrP sequences makes it difficult to draw conclusions about the functional importance of many of the individual residues in the protein.

Amino-terminal Sequence Repeats

The amino-terminal domain of mammalian PrP contains five copies of a P(H/Q)GGG(G)WGQ octarepeat sequence, occasionally more, as in the case of one sequenced bovine allele, which has six copies (Goldmann et al.

1991b; Prusiner et al. 1993b). These repeats are remarkably conserved between species, which implies a functionally important role. The chicken sequence contains a different repeat, PGYP(H/Q)N (Harris et al. 1989; Gabriel et al. 1992). Although insertions of extra repeats have been found in patients with familial prion disease, naturally occurring deletions of single octarepeats do not appear to cause disease, and deletion of all these repeats does not prevent PrP^C from undergoing a conformational transition into PrP^{Sc} (Rogers et al. 1993; Fischer et al. 1996).

It was suggested that the histidine residues in the octarepeats might bind metal ions and that immobilized metal ion affinity chromatography (IMAC) might facilitate the purification of PrP (Sulkowski 1985); indeed, IMAC did prove to be useful in the purification of PrP^C (Pan et al. 1992). The availability of purified PrP^C allowed studies comparing the secondary structures of PrP^C and PrP^{Sc} (Pan et al. 1993; Hornshaw et al. 1995a). The metal ion–peptide complexes were found to be much more soluble than the metal ions bound to full-length PrP. Using full-length recombinant (rec) PrP, Cu^{++} was found to bind with a much higher avidity than any other metal ion, but the concentration for half-maximal binding for Cu^{++} was 14 μM at pH 6.0, indicating a rather low affinity of PrP for Cu^{++} (Stöckel et al. 1998). At neutral pH, Cu^{++}–PrP complexes tended to form large aggregates and precipitate; therefore, synthetic peptides containing the octarepeats were used to study more extensively the interaction of Cu^{++} with PrP (Hornshaw et al. 1995b). The concentration for half-maximal binding for Cu^{++} to peptide (51–75) containing two octarepeats was 6 μM at pH 7.5. When peptide (58–91) containing four octarepeats was studied, the binding of Cu^{++} was found to be cooperative and highly pH-dependent (Fig. 6) (Viles et al. 1999). The midpoint of the pH-dependence transition was pH 6.7, suggesting that the binding of Cu^{++} occurred through the imidazole nitrogens of histidine residues.

In studies of full-length recPrP, Cu^{++} was found to catalyze the oxidation of the histidine residues within the octarepeats. These findings argue that PrP may function as Cu^{++}-binding protein (Requena et al. 2001). Consistent with studies on recPrP are investigations reporting that membranes isolated from the brains of $Prnp^{0/0}$ mice had substantially lower levels of Cu^{++} than wt mice (Brown et al. 1997). However, attempts to confirm this relationship between brain Cu^{++} and PrP levels were unsuccessful (Waggoner et al. 2000). Interestingly, the copper-chelating reagent cuprizone administered to rodents causes spongiform degeneration resembling that induced by prions (Pattison and Jebbett 1971; Kimberlin et al. 1976; Kimberlin and Millson 1976).

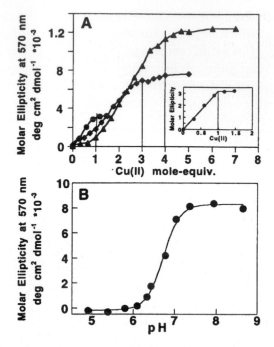

Figure 6. Binding of copper ions to synthetic peptides containing the octarepeats of PrP. (*A*) Cu(II)-binding curves: molar ellipticity at 570 nm with increasing amounts of Cu(II), pH 7.5. (*Filled circles*) 2-His peptide, PrP(51-75) (0.34 mM). (*Filled diamonds*) 3-His peptide, PrP(66-91) (0.021 mM). (*Filled triangles*) 4-His peptide, PrP(58-91) (0.033 mM). (*B*) pH dependence of the ellipticity at 570 nm for PrP(58-91). The pH dependence curve has been fitted to the following equation: $\Delta\varepsilon_{obs} = \{\Delta\varepsilon_{acid}[H^+]^n + \Delta\varepsilon_{base}[H^+] \, K_a\} / [H^+]^n + K_a\}$, in which n = Hill coefficient and K_a = acid dissolution constant for the transition. The midpoint of the transition is pH 6.7. (Reprinted, with permission, from Viles et al. 1999 [copyright National Academy of Sciences].)

Conserved Alanine–Glycine Region

In addition to the octarepeat, the other region of notable conservation is in the sequence at the carboxy-terminal end of the last octarepeat. Here, an unusual glycine- and alanine-rich region from A113 to Y128 is found (Fig. 5). Although no differences between species have been found in this part of the sequence, a single point mutation A117V is linked to GSS (Hsiao et al. 1991a). The conservation of structure suggests an important role in the function of PrP^C; in addition, this region is likely to be important in the conversion of PrP^C into PrP^{Sc}.

STRUCTURES OF PrP ISOFORMS

Mass spectrometry and gas phase sequencing were used to search for post-translational chemical modifications that might explain the differences in the properties of PrP^C and PrP^{Sc}. No modifications differentiating PrP^C from PrP^{Sc} were found (Stahl et al. 1993). These observations forced the consideration of the possibility that a conformational change distinguishes the two PrP isoforms.

When the secondary structures of the PrP isoforms were compared by optical spectroscopy, they were found to be markedly different (Pan et al. 1993). Fourier transform infrared (FTIR) and circular dichroism (CD) spectroscopy studies showed that PrP^C contains ~40% α-helix and little β-sheet, whereas PrP^{Sc} is composed of ~30% α-helix and 45% β-sheet (Pan et al. 1993; Safar et al. 1993a). That the two PrP isoforms have the same amino acid sequence runs counter to the widely accepted view that the amino acid sequence specifies only one biologically active conformation of a protein (Anfinsen 1973). Like PrP^{Sc}, the amino-terminally truncated PrP 27-30 has a high β-sheet content (Caughey et al. 1991b; Gasset et al. 1993), which is consistent with the earlier finding that PrP 27-30 polymerizes into amyloid fibrils (Prusiner et al. 1983). Denaturation of PrP 27-30 under conditions that reduced infectivity resulted in a concomitant diminution of β-sheet content (Gasset et al. 1993; Safar et al. 1993b).

Prior to comparative studies on the structures of PrP^C and PrP^{Sc}, metabolic labeling studies showed that the acquisition of protease resistance in PrP^{Sc} is a posttranslational process (Borchelt et al. 1990). In a search for chemical differences that would distinguish PrP^{Sc} from PrP^C, ethanolamine was identified in hydrolysates of PrP 27-30, which signaled the possibility that PrP might contain a GPI anchor (Stahl et al. 1987). Both PrP isoforms were found to carry GPI anchors. PrP^C was found on the surface of cells, where it could be released by cleavage of the anchor. Subsequent studies showed that PrP^{Sc} formation occurs after PrP^C reaches the cell surface (Caughey and Raymond 1991) and localizes to caveolae-like domains (CLDs) (Gorodinsky and Harris 1995; Taraboulos et al. 1995).

NMR Structures of recPrP

Modeling studies and subsequent NMR investigations of a synthetic PrP peptide containing residues 90–145 suggested that PrP^C might contain an α-helix within this region (Huang et al. 1994). This peptide contains the region (residues 113–128) that is most highly conserved among all species studied and corresponds to a transmembrane segment that was delineated in cell-free translation studies. A transmembrane form of PrP was found in brains of patients with GSS caused by the A117V mutation and of Tg mice overexpressing either mutant or wt PrP (Hegde et al. 1998, 1999). That no evidence for an α-helix in this region has been found in NMR studies of recPrP in an aqueous environment (Riek et al. 1996; Donne et al. 1997; James et al. 1997) suggests that these recPrPs correspond to the secreted form of PrP that was also identified in the cell-free translation studies. This contention is supported by studies with recombinant antibody fragments (recFabs) showing that GPI-anchored PrP^C on the surface of cells exhibits an immunoreactivity profile similar to that of recPrP refolded into α-helical conformation (Williamson et al. 1996; Peretz et al. 1997).

The NMR structure of recSHaPrP(90–231) was determined after the protein was purified and refolded (Fig. 7A). Residues 90–112 are not shown because marked conformational heterogeneity was found in this region, whereas residues 113–126 constitute the conserved hydrophobic region that also displays some structural plasticity (James et al. 1997). Although some features of the structure of recSHaPrP(90–231) are similar to those reported earlier for the smaller recMoPrP(121–231) fragment (Riek et al. 1996), substantial differences were found. For example, the loop at the amino terminus of helix B is defined in recSHaPrP(90–231) but is disordered in recMoPrP(121–231); in addition, helix C is composed of residues 200–227 in recSHaPrP(90–231) but extends only from residues 200 to 217 in recMoPrP(121–231). The loop and the carboxy-terminal portion of helix C are particularly important, as described below. Whether the differences between the two recPrP fragments are due to (1) their different lengths, (2) species-specific differences in sequences, or (3) the conditions used for solving the structures remains to be determined.

NMR studies of MoPrP(23–231) and SHaPrP(29–231) showed that like the shorter fragments described above, full-length PrP is also a three-helix–bundle protein with two short antiparallel β-strands. Whereas the three helices form a globular carboxy-terminal domain, the amino-terminal domain is highly flexible and lacks identifiable secondary structure under the experimental conditions employed (Fig. 7B) (Donne et al. 1997). Studies of SHaPrP(29–231) indicate transient interactions between the carboxy-terminal end of helix B and the highly flexible, amino-terminal random coil containing the octarepeats (residues 29–125) (Donne et al. 1997).

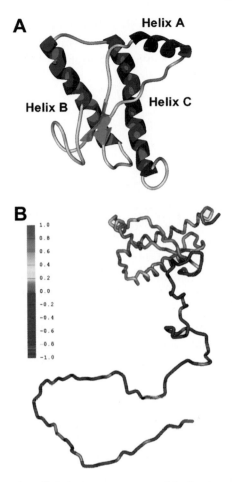

Figure 7. Structures of PrP^C. (*A*) NMR structure of Syrian hamster (SHa) recombinant (rec) PrP(90–231). Presumably, the structure of the α-helical form of recPrP(90–231) resembles that of PrP^C. recPrP(90–231) is viewed from the interface where PrP^{Sc} is thought to bind to PrP^C. The color scheme is: α-helices A (residues 144–157), B (172–193), and C (200–227) in blue; loops in yellow; residues 129–134 encompassing strand S1 and residues 159–165 encompassing strand S2 in green (James et al. 1997). (*B*) Schematic diagram showing the flexibility of the polypeptide chain for PrP(29–231) (Donne et al. 1997). The structure of the portion of the protein representing residues 90–231 was taken from the coordinates of PrP(90–231) (James et al. 1997). The remainder of the sequence was hand-built for illustration purposes only. The color scale corresponds to the heteronuclear {^{1}H}-^{15}N NOE data: red for the lowest (most negative) values, where the polypeptide is most flexible, to blue for the highest (most positive) values in the most structured and rigid regions of the protein (Donne et al. 1997).

The NMR structures of more than 10 PrPs from different species have been determined (Table 3) and are quite similar. Even the structure of MoDpl is similar to that of MoPrP, although the two proteins share only 25% sequence similarity (Fig. 4).

Electron Crystallography of PrP^{Sc}

Because the insolubility of PrP^{Sc} has frustrated structural studies by X-ray crystallography or NMR spectroscopy, electron crystallography was used to characterize the structure of two infectious variants of PrP. Isomorphous, two-dimensional crystals of PrP 27-30 and a miniprion (PrP^{Sc}106) were identified by negative stain electron microscopy. Image processing allowed the extraction of limited structural information to 7-Å resolution. Comparing projection maps of PrP 27-30 and PrP^{Sc}106, the 36-residue internal deletion of the miniprion was visualized and the N-linked sugars were localized (Fig. 8). The dimensions of the monomer and the locations of the deleted segment and sugars were used as constraints in the construction of models for PrP^{Sc}. Only models featuring parallel β-helices as the key element could satisfy the constraints.

Previous endeavors to model the structure of PrP^{Sc} attempted to fit sequence-specific conformational preferences with spectroscopic, antibody binding, and other biological data. Originally, an antiparallel β-sheet formed from residues 90–170 packed against the two carboxy-terminal α-helices was postulated (Huang et al. 1995). The results obtained from the 2D crystals described here, the existence of PrP^{Sc}106 (implying the spatial colocalization of residues 140 and 177), and increasing data pointing to parallel β-sheet structure in amyloid-forming proteins (Benzinger et al. 1998; Antzutkin et al. 2000) caused us to revisit this model (Huang et al. 1995).

A single antiparallel β-sheet was not consistent with the observed densities in the projection maps obtained from the 2D crystals. Specifically, the sheets were far too wide to fit into the observed hexameric arrangement. Efforts to adjust the sheet morphology to fit the density required the use of shorter strands. The amount of β-structure in these altered β-sheets was no longer compatible with the amounts of β-sheet observed by FTIR spectroscopy (Wille et al. 1996; Supattapone et al. 1999b). Furthermore, antiparallel β-sheets typically have a twist of ~20° per strand. A six- to eight-stranded β-sheet would be difficult to accommodate in the electron density. Parallel β-sheets are commonly observed in protein structures as part of planar α/β folds, α/β barrels, and parallel β-helices. Planar α/β

Table 3. Structures of PrP, Dpl, and Ure2p

Species	Protein	Determined by	PDB accession codes	Reference
Mouse	recPrP(121-231)	NMR	1AG2	Riek et al. (1996)
Mouse	recPrP(23-231)	NMR	data not released?	Riek et al. (1997)
Syrian hamster	recPrP(90-231)	NMR	1B10, updated 1999	James et al. (1997)
Syrian hamster	recPrP(29-231)	NMR	data not released?	Donne et al. (1997)
Syrian hamster	synPrP(104-113) in complex with 3F4	X-ray	1CU4 in complex with 3F4	Kanyo et al. (1999)
Syrian hamster	recPrP(90-231)	NMR	used to update 1B10	Liu et al. (1999)
Human	recPrP(23-230)	NMR	1QLX and 1QLZ	Zahn et al. (2000)
Human	recPrP(90-230)	NMR	1QM0 and 1QM1	Zahn et al. (2000)
Human	recPrP(121-230)	NMR	1QM2 and 1QM3	Zahn et al. (2000)
Cattle	recPrP(121-230)	NMR	1DWY and 1DWZ	García et al. (2000)
Cattle	recPrP(23-230)	NMR	1DX0 and 1DX1	García et al. (2000)
Human	recPrP(121-230), M166V	NMR	1E1G and 1E1J	Calzolai et al. (2000)
Human	recPrP(121-230), S170N	NMR	1E1P and 1E1S	Calzolai et al. (2000)
Human	recPrP(121-230), R220K	NMR	1E1U and 1E1W	Calzolai et al. (2000)
Human	recPrP(90-231), E200K	NMR	1QM0, 1FKC, and 1FO7	Zhang et al. (2000)
Human	recPrP(90-231)	X-ray	1I4M	Knaus et al. (2001)
Human	recPrP(121-230), M166C, E221C	NMR	1H0L	Zahn et al. (2003)
Sheep	synPrP(145-169)	NMR	1G04	Kozin et al. (2001)
Sheep	synPrP(145-169)	NMR	1M25	Kozin et al. (2001)
Mouse	recDpl(26-157)	NMR	1I17	Mo et al. (2001)
Human	recDpl(24-152)	NMR	1LG4	Lührs et al. (2003)
S. cerevisiae	recUre2p(95-354)	X-ray	1G6W and 1G6Y	Bousset et al. (2001b)
S. cerevisiae	recUre2p(95-354) glutathione complex	X-ray	1JZR, 1K0B, and 1K0D	Bousset et al. (2001a)
S. cerevisiae	recUre2p(95-354) S-P-nitrobenzyl-glutathione complex	X-ray	1K0C	Bousset et al. (2001a)
S. cerevisiae	recUre2p(95-354) S-hexylgluta-thione complex	X-ray	1K0A	Bousset et al. (2001a)
S. cerevisiae	recUre2p(97-354)	X-ray	1HQO	Umland et al. (2001)

Data compiled by Holger Wille.

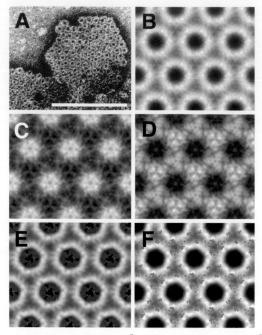

Figure 8. Two-dimensional crystals of PrP^Sc. (*A*) A 2D crystal of PrP^Sc106 stained with uranyl acetate. Bar, 100 nm. (*B*) Image-processing result after correlation-mapping and averaging followed by crystallographic averaging. (*C*, *D*) Subtraction maps between the averages of PrP 27-30 and PrP^Sc106 (panel *B*). (*C*) PrP^Sc106 minus PrP 27-30 and (*D*) PrP 27-30 minus PrP^Sc106, showing major differences in lighter shades. (*E*, *F*) The statistically significant differences between PrP 27-30 and PrP^Sc106 calculated from *C* and *D* in red and blue, respectively, overlaid onto the crystallographic average of PrP 27-30. (Reprinted, with permission, from Wille et al. 2002 [copyright National Academy of Sciences].)

folds encounter the same problems as twisted, antiparallel β-sheets. α/β barrels have alternating α-helices and β-strands, with the fraction of α-helical residues exceeding that of the β-stranded residues. This structure would be in conflict with the FTIR results for both PrP 27-30 and PrP^Sc106. Although a novel protein fold for the structure of PrP^Sc cannot be excluded, the parallel β-helix is the only known fold that provides the necessary β-sheet content, parallel β-architecture, and room to accommodate the α-helices that are expected at the carboxyl terminus of the molecule.

PrP^Sc was modeled as a parallel β-helical fold (Fig. 9), placing the structurally conserved carboxy-terminal α-helices and the glycosylation sites (N181 and N197) on the periphery of the oligomer and with the highly flexible N-linked sugars pointing above and below the plane of the

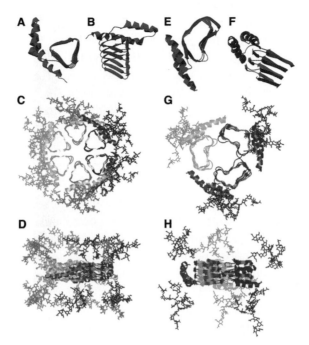

Figure 9. β-Helical models of PrP 27-30. (*A, B*) Top and side views, respectively, of PrP 27-30 modeled with a left-handed β-helix. The β-helical portion of the model is based on the *Methanosarcina thermophila* γ carbonic anhydrase structure. (*C, D*) Top and side views, respectively, of the trimer-of-dimer model of PrP 27-30 with left-handed β-helices. (*E, F*) Top and side views, respectively, of PrP 27-30 modeled with a right-handed β-helix. The β-helical portion of the model is based on the most regular helical turns of *Bordetella pertussis* P.69 pertactin. (*G, H*) Top and side views, respectively, of the trimer model of PrP 27-30 with right-handed β-helices. The structure of the α-helices was derived from the solution structure of recombinant hamster PrP (Donne et al. 1997; James et al. 1997; Liu et al. 1999). In the single molecule images (panels *A, B, E,* and *F*), residues 141–176 that are deleted in PrP106 are colored blue. (Reprinted, with permission, from Wille et al. 2002 [copyright National Academy of Sciences].)

oligomer (Fig. 9). In this conformation, PrP^{Sc} is compact and fits readily into the density observed by electron microscopy (Fig. 8). Because there is very little twist or bend to parallel β-helices, the modeled oligomers have relatively planar faces that permit stacking along the fibril axis. The β-helices also provide flat sheets for lateral assembly into disc-like oligomers and filamentous assemblies (Seckler 1998; Schuler et al. 1999). The deletion of 36 residues in PrP106 correlates favorably with exactly two turns of an average left-handed β-helix. Therefore, the orientation of the α- and β-helices, the sugars, as well as the fold of the oligomeric face, could be

is deletion mutant. This would allow full-length PrP^{Sc} as well
o template the replication of PrP^{Sc}106. Furthermore, the
ation of the β-helical deletions in our models are consistent
with the difference densities observed between the 2D crystals of PrP
27-30 and PrP^{Sc}106. Finally, the PrP sequence can be threaded onto the β-
helical folds in a register that is consistent with secondary structure predictions, mutational information, and the negative electrostatic potential
at the center of the oligomers.

PRION REPLICATION

The mechanism of prion replication is not well understood. The formation of nascent prions requires that PrP^C be converted into PrP^{Sc}. Some
investigators have argued that this process involves a nucleation–polymerization (NP) reaction, whereas others have contended that a more
likely mechanism employs a template-assisted process.

In an uninfected cell, PrP^C with the wt sequence is likely to exist in
equilibrium in its monomeric α-helical, protease-sensitive state or bound
to some other protein, such as protein X (Fig. 10). The conformation of
PrP^C that is bound to protein X is denoted PrP* (Cohen et al. 1994); this
conformation is likely to be different from that determined under aqueous conditions for monomeric recPrP. The PrP*–protein X complex will
bind PrP^{Sc}, thereby creating a replication-competent assembly. Order-of-
addition experiments demonstrate that for PrP^C, protein X binding precedes productive PrP^{Sc} interactions (Kaneko et al. 1997c). A conformational change takes place wherein PrP*, in a shape competent for binding
to protein X and PrP^{Sc}, represents the initial phase in the formation of
infectious PrP^{Sc}.

Several lines of evidence argue that the smallest infectious prion particle is an oligomer of PrP^{Sc}, perhaps as small as a dimer (Bellinger-
Kawahara et al. 1988). Upon purification, PrP^{Sc} tends to aggregate into
insoluble multimers that can be dispersed into liposomes (Gabizon et al.
1988). Insolubility does not seem to be a prerequisite for PrP^{Sc} formation
or prion infectivity, as suggested by some investigators (Gajdusek 1988;
Caughey et al. 1995); a protease-resistant PrP^{Sc} that is soluble in 1%
Sarkosyl was generated in prion-infected neuroblastoma (ScN2a) cells by
expression of PrP106 (Muramoto et al. 1996).

In attempts to form PrP^{Sc} in vitro, PrP^C has been exposed to 3 M
GdnHCl and then diluted 10-fold prior to binding to PrP^{Sc} (Kocisko et al.
1994; Kaneko et al. 1997b). On the basis of these results, exposure of PrP^C
to GdnHCl is presumed to convert PrP^C into a PrP*-like molecule.
Whether this PrP*-like protein is converted into PrP^{Sc} is unclear. Although

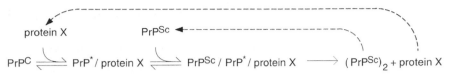

Figure 10. Hypothetical scheme for template-assisted PrP^{Sc} formation. In the initial step, PrP^C binds to protein X to form the PrP*–protein X complex. Next, PrP^{Sc} binds to PrP* that has already formed a complex with protein X. When PrP* is transformed into a nascent molecule of PrP^{Sc}, protein X is released and a dimer of PrP^{Sc} remains. The inactivation target size of an infectious prion suggests that it is composed of a dimer of PrP^{Sc} (Bellinger-Kawahara et al. 1988). In the model depicted here, a fraction of infectious PrP^{Sc} dimers dissociates into uninfectious monomers as the replication cycle proceeds, while a majority of the dimers accumulates in accord with the increase in prion titer that occurs during the incubation period. Another fraction of PrP^{Sc} is cleared presumably by cellular proteases. The precise stoichiometry of the replication process remains uncertain. (Reprinted, with permission, from Prusiner et al. 1998 [copyright Cell Press].)

the PrP*-like protein bound to PrP^{Sc} is protease-resistant and insoluble, it has not been reisolated in order to assess whether or not it was converted into PrP^{Sc}. It is noteworthy that recPrP can be refolded into either α-helical or β-sheet forms but no forms have been found to possess prion infectivity as judged by bioassay.

Mechanism of Prion Propagation

In attempting to probe the mechanism of prion formation, it is reasonable to ask, What is the rate-limiting step in prion formation? First, the impact of the concentration of PrP^{Sc} in the inoculum, which is inversely proportional to the length of the incubation time, must be considered. Second, the sequence of PrP^{Sc} that forms an interface with PrP^C must be taken into account. When the sequences of the two isoforms are identical, the shortest incubation times are observed. Third, the strain-specific conformation of PrP^{Sc} must be considered. Some prion strains exhibit longer incubation times than others; the mechanism underlying this phenomenon is not understood. From these considerations, a set of conditions exists under which initial PrP^{Sc} concentrations can be rate-limiting. These effects presumably relate to the stability of the PrP^{Sc}, its targeting to the correct cells and subcellular compartments, and its ability to be cleared. Once infection in a cell is initiated and endogenous PrP^{Sc} production is operative, then the following discussion of PrP^{Sc} formation seems most applicable. If the assembly of PrP^{Sc} into a specific dimeric or multimeric arrangement is difficult, then a NP formalism would be relevant. In NP processes, nucleation is the rate-limiting step, and elongation, or polymerization, is facile. These conditions are frequently observed in peptide models of aggregation phenomena (Caughey et al. 1995); however, studies in ScN2a cells and with Tg mice expressing foreign PrP genes suggest that a different process is occurring.

Template-assisted Prion Formation

From investigations with mice expressing both the SHaPrP transgene and the endogenous MoPrP gene, designated Tg(SHaPrP) mice, it is clear that PrP^{Sc} provides a template for directing prion replication; a template is defined as a catalyst that leaves its imprint on the product of the reaction (Prusiner et al. 1990). Inoculation of these mice with $SHaPrP^{Sc}$ leads to the production of nascent $SHaPrP^{Sc}$ and not $MoPrP^{Sc}$. Conversely, inoculation of the Tg(SHaPrP) mice with $MoPrP^{Sc}$ results in $MoPrP^{Sc}$ and not $SHaPrP^{Sc}$ formation.

Even stronger evidence for templating has emerged from studies of prion strains passaged in Tg(MHu2M)$Prnp^{0/0}$ mice, which express a chimeric mouse–human gene, as described in more detail below (Telling et al. 1996b; Prusiner 1997). Even though the conformational templates were initially generated with PrP^{Sc} molecules having different sequences from patients with inherited prion diseases, these templates are sufficient to direct replication of distinct PrP^{Sc} molecules when the amino acid sequences of the substrate PrPs are identical.

Another line of evidence for template-assisted prion replication comes from studies of FI. In FFI, the protease-resistant fragment of PrP^{Sc} after deglycosylation has an M_r of 19 kD when measured by immunoblotting of brain extracts from humans with FFI or from Tg(MHuM)$Prnp^{0/0}$ mice that were inoculated with FFI prions. In both humans and inoculated Tg(MHuM)$Prnp^{0/0}$ mice on both first and second passage, PrP^{Sc} is confined largely to the thalamus (Telling et al. 1996b). These findings argue that the conformation of PrP^{Sc} which yields a 19-kD polypeptide after deglycosylation is propagated both in humans and in mice expressing an artificial chimeric transgene; in addition, prion accumulation is confined to the thalamus. Additional evidence supporting these assertions comes from a patient who died after developing a clinical disease similar to FFI. Because both PrP alleles encoded the wt PrP sequence and methionine at position 129, we labeled this sFI. At autopsy, the spongiform degeneration, reactive astrocytic gliosis, and PrP^{Sc} deposition were confined to the thalamus (Mastrianni et al. 1997, 1999; Gambetti and Parchi 1999; Parchi et al. 1999). Moreover, the PrP^{Sc} both in the patient's brain and in the brains of inoculated Tg(MHuM)$Prnp^{0/0}$ mice yielded a 19-kD fragment after deglycosylation. These findings argue that the clinicopathologic phenotype is determined by the conformation and not the amino acid sequence of PrP^{Sc}. Additionally, FI prions can be replicated with mutant human PrP(D178N), wt human PrP, or chimeric human–mouse PrP. With all these different PrPs as substrates, the conformation of PrP^{Sc} must be faithfully copied, and template-assisted prion replication provides such a mechanism.

Evidence for Protein X

Protein X was postulated to explain the results of the transmission of Hu prions to Tg mice (Table 4) (Telling et al. 1994, 1995). Mice expressing both MoPrP^C and HuPrP^C were resistant, whereas those expressing only HuPrP^C were susceptible to Hu prions. These results argue that MoPrP^C inhibited transmission of Hu prions, i.e., the formation of nascent HuPrP^{Sc}. In contrast, mice expressing both MoPrP^C and chimeric MHu2MPrP^C were susceptible to Hu prions, and mice expressing MHu2MPrP^C alone were only slightly more susceptible. These findings contend that MoPrP^C has only a minimal effect on the formation of chimeric MHu2MPrP^{Sc}.

When the data on Hu prion transmission to Tg mice were considered together, they suggested that MoPrP^C prevented the conversion of HuPrP^C into PrP^{Sc} by binding to another Mo protein (protein X) with a higher affinity than does HuPrP^C. It was postulated that MoPrP^C had little effect on the formation of PrP^{Sc} from MHu2MPrP^C (Table 4) because MoPrP^C and MHu2MPrP^C share the same amino acid sequence at the carboxyl terminus. This also suggested that MoPrP^C only weakly inhibited transmission of SHa prions to Tg(SHaPrP) mice because SHaPrP is more closely related to MoPrP than is HuPrP.

The search for protein X has been frustrating because many proteins are known to bind to PrP^C, but none has been shown to participate in PrP^{Sc} formation (Oesch et al. 1990; Kurschner and Morgan 1995; Edenhofer et al. 1996; Tatzelt et al. 1996; Martins et al. 1997; Rieger et al. 1997; Yehiely et al. 1997; Graner et al. 2000; Keshet et al. 2000; Mouillet-Richard et al. 2000; Gauczynski et al. 2001; Hundt et al. 2001; Schmitt-Ulms et al. 2001). Whether protein X is one protein or a complex of proteins remains to be established. It is reasonable to expect that mice defi-

Table 4. Distinct prion strains generated in humans with inherited prion diseases and transmitted to transgenic mice

Inoculum	Host species	Host PrP genotype	Incubation time days ± S.E.M. $(n/n_0)^a$	PrP^{Sc} (kD)
—	human	FFI(D178N,M129)	—	19
FFI	mouse	Tg(MHu2M)	206 ± 7 (7/7)	19
FFI→Tg(MHu2M)	mouse	Tg(MHu2M)	136 ± 1 (6/6)	19
—	human	fCJD(E200K)	—	21
fCJD	mouse	Tg(MHu2M)	170 ± 2 (10/10)	21
fCJD→Tg(MHu2M)	mouse	Tg(MHu2M)	167 ± 3 (15/15)	21

Data from Telling et al. (1996b).

$^a n$, number of ill mice; n_0, number of inoculated mice.

cient for protein X would not replicate prions or would do so extremely slowly, resulting in prolonged incubation times.

Dominant-negative Inhibition

To extend the foregoing findings in Tg mice, an expression vector with an insert encoding a chimeric mouse–hamster PrP, designated MHM2PrP, was transfected into ScN2a cells. MHM2PrP carries two Syrian hamster residues, which create an epitope that is recognized by the anti-SHaPrP 3F4 mAb that does not react with mouse PrP (Scott et al. 1992). Substitution of a Hu residue at position 214 or 218 in MHM2PrP prevented formation of MHM2PrP^{Sc} in ScN2a cells (Kaneko et al. 1997c). The side chains of these residues protrude from the same surface of the carboxy-terminal α-helix, forming a discontinuous epitope with residues 167 and 171 in an adjacent loop. Like MHM2PrP(Q218K), substitution of a basic residue at position 167 or 171 prevented PrP^{Sc} formation. When MHM2PrP and MHM2PrP(Q218K) were coexpressed, the conversion of MHM2PrP^C into PrP^{Sc} was inhibited, arguing that MHM2PrP(Q218K) was acting as a dominant negative. Similar results were obtained when studies were performed with cells expressing MHM2PrP(Q167R).

One interpretation of dominant-negative inhibition of prion propagation is that mutant PrP binds more avidly to an auxiliary factor that participates in prion replication (protein X) than does wt PrP. In this scenario, mutant PrP binds to protein X, which prevents the binding of wt PrP. This explanation is also consistent with the protective effects of basic polymorphic residues in PrP of humans and sheep. The E219K substitution seems to render humans resistant to sCJD (Shibuya et al. 1998), and Q171R renders sheep resistant to scrapie (Hunter et al. 1993; Westaway et al. 1994b). In MHM2, the Q218K mutation corresponds to E219K in humans, and the Q167R mutation corresponds to Q171R in sheep (Fig. 11).

To determine whether dominant-negative inhibition of prion formation occurs in vivo, Tg mice expressing PrP with either the Q167R or Q218K substitutions alone or in combination with wt PrP were produced (Perrier et al. 2002). Tg(MoPrP,Q167R)$Prnp^{0/0}$ mice expressing mutant PrP at levels equal to non-Tg FVB mice were inoculated with prions and remained healthy for more than 550 days, indicating that inoculation did not initiate a process sufficient to cause disease. Immunoblots of brain homogenates and histologic analysis did not reveal abnormalities. Tg(MoPrP,Q167R)$Prnp^{+/+}$ mice expressing both mutant and wt PrP exhibited neurologic dysfunction at ~450 days after inoculation; the brains of three of these mice that were sacrificed at 300 days revealed low levels of PrP^{Sc} as well as numerous vacuoles and severe astrocytic gliosis. Both Tg(MoPrP,Q218K)$Prnp^{0/0}$ and Tg(MoPrP,Q218K)$Prnp^{+/+}$ mice expressing the transgene product at 16X remained healthy for more than

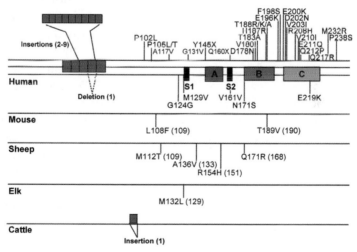

Figure 11. Mutations and polymorphisms of the prion protein gene. PrP mutations causing inherited human prion disease and PrP polymorphisms found in humans, mice, sheep, elk, and cattle. Above the line of the human sequence are mutations that cause inherited prion disease. Below the lines are polymorphisms, some but not all of which are known to influence the onset as well as the phenotype of disease. Residue numbers in parentheses correspond to the human codons. (Many of the data were compiled by J.-L. Laplanche.)

450 days after inoculation. Neither PrP^{Sc} nor neuropathologic changes were found. Tg mice expressing MoPrP(Q218K) at 32x developed spontaneous neurodegeneration at ~450 days of age. These studies demonstrate that although dominant-negative inhibition of wt PrP^{Sc} formation occurs, expression of dominant-negative PrP at the same level as wt PrP does not completely prevent prion formation. However, expression of dominant-negative PrP alone had no deleterious effects on the mice and did not support prion propagation.

SPORADIC PRION DISEASES

In most patients with CJD (Masters et al. 1978), there is neither an infectious nor a heritable etiology. How prions arise in patients with sCJD is unknown; hypotheses include: (1) horizontal transmission from humans or animals (Gajdusek 1977), (2) somatic mutation of the *PRNP* ORF, (3) spontaneous conversion of PrP^{C} into PrP^{Sc} (Prusiner 1989; Hsiao et al. 1991b), or (4) the accumulation of PrP^{Sc}, which is normally present at very low levels (Peretz et al. 2001b; Chapter 2). Numerous attempts to establish an infectious link between sCJD and a preexisting prion disease in animals or humans have been unrewarding (Malmgren et al. 1979; Cousens et al. 1990).

HERITABLE PRION DISEASES

The recognition that 10–15% of CJD cases are familial led to the suspicion that genetics plays a role in this disease (Meggendorfer 1930; Masters et al. 1981b). As with scrapie, the relative contributions of genetic and infectious etiologies in the human prion diseases remain puzzling. More than 30 different mutations of the PrP gene have been shown to segregate with the heritable human prion diseases (Fig. 11; Chapter 14). Five of these mutations have been genetically linked to the inherited human prion diseases (Table 2). Virtually all cases of GSS and FFI appear to be caused by germ-line mutations in the PrP gene. The brains of humans dying of inherited prion disease contain infectious prion particles that have been transmitted to experimental animals.

GSS and Genetic Linkage

The discovery that GSS, which was known to be a familial disease, could be transmitted to apes and monkeys was first reported when many still thought that scrapie, CJD, and related disorders were caused by viruses (Masters et al. 1981b). Only the discovery that a proline-to-leucine mutation at codon 102 of the human PrP gene was genetically linked to GSS permitted the unprecedented conclusion that prion disease can have both genetic and infectious etiologies (Hsiao et al. 1989; Prusiner 1989). In that study, the codon-102 mutation was linked to development of GSS with a logarithm of the odds (LOD) score exceeding 3, demonstrating a tight association between the altered genotype and the disease phenotype. This mutation may be caused by the deamination of a methylated CpG in a germ-line PrP gene, which results in the substitution of a thymine for cytosine. The mutation has been found in many families in numerous countries, including the first family identified to have GSS (Doh-ura et al. 1989; Goldgaber et al. 1989; Kretzschmar et al. 1991).

fCJD Caused by Octarepeat Inserts

An insert of 144 bp containing six octarepeats at codon 53, in addition to the five that are normally present, was described in patients with CJD from four families residing in southern England (Owen et al. 1989; Poulter et al. 1992). Genealogic investigations have shown that all four families are related, arguing for a single founder born more than two centuries ago. The LOD score for this extended pedigree exceeds 11. Studies from several laboratories have demonstrated that inserts of two, four, five, six, seven, eight, or nine octarepeats in addition to the normal five are found in individuals with inherited CJD (Fig. 11) (Owen et al. 1989; Goldfarb et al. 1991b).

fCJD in Libyan Jews

The unusually high incidence of CJD among Israeli Jews of Libyan origin was thought to be due to the consumption of lightly cooked sheep brain or eyeballs (Kahana et al. 1974). Molecular genetic investigations revealed that Libyan and Tunisian Jews with fCJD have a PrP gene point mutation at codon 200, resulting in a glutamic acid–to-lysine substitution (Fig. 11) (Goldfarb et al. 1990a; Hsiao et al. 1991b). The E200K mutation has been genetically linked to the disease, with a LOD score exceeding 3 (Gabizon et al. 1993); the same mutation has also been found in patients from Orava in North Central Slovakia (Goldfarb et al. 1990a), in a cluster of familial cases in Chile (Goldfarb et al. 1991c), and in a large German family living in the United States (Bertoni et al. 1992).

Most patients are heterozygous for the mutation and thus express both mutant and wt PrP^C. In the brains of patients who die of fCJD(E200K), the mutant PrP^{Sc} is both insoluble and protease-resistant, whereas much of wt PrP differs from both PrP^C and PrP^{Sc} in that it is insoluble but readily digested by proteases. Whether this form of PrP is an intermediate in the conversion of PrP^C into PrP^{Sc} remains to be established (Gabizon et al. 1996).

Penetrance of fCJD

Life-table analyses of carriers harboring the codon-200 mutation exhibit complete penetrance (Chapman et al. 1994; Spudich et al. 1995). In other words, if the carriers live long enough, they will all eventually develop prion disease. Some investigators have argued that the inherited prion diseases are not fully penetrant, and thus an environmental factor, such as the ubiquitous "scrapie virus," is required for illness to be manifest, but as reviewed above, no viral pathogen has been found (Goldfarb et al. 1990b, 1991c).

Fatal Familial Insomnia

The D178N mutation has been linked to the development of FFI, with an LOD score exceeding 5 (Petersen et al. 1992). More than 30 families worldwide with FFI have been recorded (Gambetti et al. 1995). Studies of inherited human prion diseases demonstrate that the amino acid at polymorphic residue 129 with the D178N pathogenic mutation alters the clinical and neuropathologic phenotype. The D178N mutation combined with M129 results in FFI (Goldfarb et al. 1992; Medori et al. 1992). In this disease, adults generally over age 50 years present with a progressive sleep disorder and usually die within one year (Lugaresi et al. 1986). In their brains, deposition of PrP^{Sc} is confined largely within the anteroventral and the dorsal medial nuclei of the thalamus. In contrast, the same D178N mutation with V129 produces fCJD, in which the patients present with dementia, and widespread deposition of PrP^{Sc} is evident postmortem (Goldfarb et al. 1991a). The first family to be recognized with CJD was found to carry the D178N mutation (Meggendorfer 1930; Kretzschmar et al. 1995).

Human PrP Gene Polymorphisms

At PrP codon 129, an amino acid polymorphism for methionine→valine has been identified (Fig. 11) (Owen et al. 1990). This polymorphism appears able to influence prion disease expression not only in inherited forms, but also in iatrogenic and sporadic forms of prion disease (Palmer et al. 1991; Goldfarb et al. 1992; Collinge and Palmer 1997). A second polymorphism resulting in an amino acid substitution at codon 219 (glutamic acid→lysine) has been reported to occur with a frequency of about 12% in the Japanese population but not in Caucasians (Kitamoto and Tateishi 1994; Furukawa et al. 1995). In people heterozygous for the lysine substitution, lysine at 219 seems likely to act as a dominant-negative and protect against CJD by binding to protein X. By sequestering protein X, PrP^C(K219) prevents the conversion of PrP^C(E219) into PrP^{Sc} (Kaneko et al. 1997c; Perrier et al. 2002). In people homozygous for the lysine substitution, it seems likely that PrP^C(K219) is not converted into PrP^{Sc} (Perrier et al. 2002). A third polymorphism results in an amino acid substitution at codon 171 (asparagine→serine) (Fink et al. 1994), which lies adjacent to the protein X–binding site. This polymorphism has been found in Caucasians, but it has not been studied extensively and it is not known to influence the binding of PrP^C to protein X (Kaneko et al. 1997c; Chapter 10). A fourth polymorphism is the deletion of a single octarepeat (24 bp), which has been found in 2.5% of Caucasians (Laplanche et al. 1990; Vnencak-Jones and Phillips 1992; Cervenáková et al. 1996). In another study of >700 individuals, this single octarepeat deletion was found in 1.0% of the population (Palmer et al. 1993).

Studies of Caucasian patients with sCJD have shown that most are homozygous for methionine or valine at codon 129 (Palmer et al. 1991). This contrasts with the general Caucasian population, in which frequencies for the codon-129 polymorphism are 12% Val/Val, 37% Met/Met, and 51% Met/Val (Collinge et al. 1991). In contrast, the frequency of the valine allele in the Japanese population is much lower (Doh-ura et al. 1991; Miyazono et al. 1992). Heterozygosity at codon 129 is more frequent in Japanese CJD patients (18%) than in the general population, in which the polymorphism frequencies are 0% Val/Val, 92% Met/Met, and 8% Met/Val (Tateishi and Kitamoto 1993).

Although no specific mutations have been identified in the PrP gene of patients with sCJD (Goldfarb et al. 1990c), homozygosity at codon 129 (Palmer et al. 1991) is consistent with the results of Tg mouse studies. The finding that homozygosity at codon 129 predisposes to sCJD supports a model of prion production that favors PrP interactions between homologous proteins, as appears to occur in Tg(SHaPrP) mice inoculated with either hamster prions or mouse prions (Scott et al. 1989; Prusiner et al. 1990; Prusiner 1991), as well as in Tg(MHM2) mice inoculated with "artificial" prions (Scott et al. 1993).

INFECTIOUS PRION DISEASES

Prions from a variety of different sources have infected humans. Human prions have been transmitted to others both by ritualistic cannibalism and by iatrogenic means. Kuru in the highlands of New Guinea was transmitted by ritualistic cannibalism, as people in the region attempted to immortalize their dead relatives by eating their brains (Glasse 1967; Alpers 1968; Gajdusek 1977). Iatrogenic transmissions include prion-tainted human growth hormone (HGH) and gonadotropin, dura mater grafts, and corneal transplants obtained from people who died of CJD. In addition, CJD cases have been recorded after neurosurgical procedures in which ineffectively sterilized depth electrodes or instruments were used (Chapter 13).

Human Growth Hormone

More than 120 young adults have been diagnosed with iatrogenic (i) CJD between 4 and 30 years after receiving HGH or gonadotropin from cadaveric pituitaries (Koch et al. 1985; Public Health Service 1997). The longest incubation periods (20–30 years) are similar to those associated with more recent cases of kuru (Gajdusek et al. 1977; Klitzman et al. 1984; Chapter 13). Since 1985, recombinant HGH produced in *Escherichia coli* has been used in place of cadaveric HGH. With recombinant HGH, no cases of iCJD have been identified.

Variant Creutzfeldt-Jakob Disease

More than 130 teenagers and young adults have died of variant (v) CJD in Britain, Ireland, Italy, and France. The average age of vCJD patients is 26 years of age; the youngest patient was 12 years old and the oldest was 74 years old (Spencer et al. 2002). The median duration of illness was 13 months, with the range from 6 to 69 months. Both epidemiologic and experimental studies have built a convincing case that vCJD is the result of prions being transmitted from cattle with BSE to humans through consumption of contaminated beef products (Chazot et al. 1996; Will et al. 1996; Cousens et al. 1997; Chapter 12).

The majority of vCJD patients presented with psychiatric symptoms, including dysphoria, withdrawal, anxiety, insomnia, and loss of interest (Spencer et al. 2002). Generally, neurologic deficits did not appear until at least four months later; these neurologic changes consisted of memory loss, paraesthesias, sensory deficits, gait disturbances, and dysarthria. To date, all vCJD cases have been reported from Britain, except for six cases

in France, one in Italy, one in Ireland, and one in the Un
U.S. case is a 23-year-old woman, who is thought to have l
bovine prions while living in Britain during the first 12 y
As discussed elsewhere, the evidence is now quite compelli
due to the transmission of bovine prions to humans through the consumption of tainted beef products (Bruce et al. 1997; Hill et al. 1997; Scott
et al. 1999).

Studies of the prion diseases have taken on new significance with the
identification of vCJD (Bateman et al. 1995; Britton et al. 1995; Chazot et
al. 1996; Will et al. 1996; Chapters 12 and 17). The number of cases of
vCJD caused by bovine prions that will occur in the years ahead is
unknown (Cousens et al. 1997; Ghani et al. 2000; Phillips et al. 2000).
Until more time passes, it will be impossible to assess the magnitude of
this problem. These tragic cases have generated a continuing discourse
concerning mad cows, prions, and the safety of the human and animal
food supplies throughout the world. Untangling politics and economics
from the science of prions seems to have been difficult in disputes
between Great Britain and other European countries over the safety of
beef and lamb products.

DE NOVO GENERATION OF PRIONS

Perhaps the most widely held view of the initial transmissions of prions
from the brains of patients who died of GSS or fCJD to apes and monkeys
was that these individuals carried a mutant gene that rendered them susceptible to the "CJD virus" (Masters et al. 1981b). Once mutations in the
PrP gene of such patients were discovered (Hsiao et al. 1989), these transmissions could be reinterpreted in terms of the prion concept: Patients
carrying a pathogenic mutation of their PrP gene generate PrP^{Sc} and thus,
prion infectivity de novo (Prusiner 1989). Indeed, the prion concept provided, for the first time, an intellectual framework within which to explain
how a single disease process can be manifest as genetic, infectious, and
sporadic illnesses.

The de novo formation of prions from noninfectious materials has
often been suggested as proof of the prion concept. Several approaches
have been taken to develop such experimental data. First, multiple
attempts have been made to demonstrate prion infectivity in preparations
of recPrP. No fractions of purified recPrP produced in E. coli or mammalian cells have exhibited prion infectivity when bioassayed in mice or
hamsters (Baskakov et al. 2001, 2002).

Second, PrP molecules with pathologic mutations responsible for inherited prion diseases have been extensively studied. Mutant PrP^C is transformed into mutant PrP^{Sc} in humans and Tg mice. The dominantly inherited E200K mutation in humans is fully penetrant, producing fCJD in 100% of carriers if they live long enough (Chapman et al. 1994; Spudich et al. 1995). Similar to humans with GSS, Tg mice expressing high levels of PrP(P101L) develop CNS dysfunction and accumulate a modified form of PrP^{Sc} (Hsiao et al. 1990). The CNS degeneration in these Tg mice is indistinguishable from experimental murine scrapie, with neuropathology consisting of widespread spongiform morphology, astrocytic gliosis, and PrP amyloid plaques (Hsiao et al. 1994; Telling et al. 1996a). Prions in the brains of Tg mice expressing high levels of PrP(P101L) can be transmitted to another Tg line, designated Tg196, that expresses low levels of PrP(P101L) (Hsiao et al. 1994; Telling et al. 1996a). A few Tg196 mice develop spontaneous neurodegeneration at ~500 days of age, but most remain healthy. In the brains of Tg mice expressing high levels of PrP(P101L) and of inoculated Tg196 mice, a modified form of PrP^{Sc} was found, based on measurements of resistance to proteolytic digestion at 4°C and on measurements of abnormally folded PrP by the CDI (Tremblay et al. 2003).

These studies, as well as transmission of prions from patients who died of GSS to apes and monkeys (Masters et al. 1981b) and to Tg(MHu2M,P102L) mice (Telling et al. 1995), argue persuasively that prions are generated de novo by mutations in PrP. In contrast to most species-specific variations in the PrP sequence, all of the known point mutations occur either within or adjacent to regions of secondary structure and, as such, are thought to destabilize the structure of PrP (Huang et al. 1994; Zhang et al. 1995; Riek et al. 1996). It is noteworthy that denaturation studies on some recPrP isolated from *E. coli* which carry these pathologic mutations do not support the notion of destabilization (Riek et al. 1998; Liemann and Glockshuber 1999). In other studies, such mutations are associated with altered properties when the mutant PrP is expressed in cultured mammalian cells (Lehmann and Harris 1995, 1996; Daude et al. 1997; Stewart and Harris 2001). Defining the mechanism by which mutations cause inherited prion disease is an important avenue of investigation.

Why mutations of the PrP gene require many decades in humans to be manifest as CNS dysfunction is unknown. In Tg(MoPrP,P101L) mice, the level of expression of the mutant transgene is inversely related to the age of disease onset (Hsiao et al. 1990; Telling et al. 1996a). In addition, the presence of the wt MoPrP gene slows the onset of disease and diminishes the severity of the neuropathologic changes.

Third, a PrP peptide composed of residues 89–143 has been shown to induce disease in 100% of inoculated Tg196 mice (Kaneko et al. 2000). This activity of the PrP peptide requires the P101L substitution and must be folded into a β-rich conformation. Serial transmission of disease in Tg196 mice initiated by the PrP(89–143,P101L) peptide argues that the 55-mer peptide is an artificial prion (Tremblay et al. 2003).

Fourth, experiments with the yeast prion [*PSI*⁺] also argue for the de novo synthesis of prions (Sparrer et al. 2000). The prion domain of Sup35 and a portion of the adjacent M domain were expressed in *E. coli* and, after purification, aggregated into amyloid. The amyloid was then incorporated into liposomes, which were used to transform yeast into the [*PSI*⁺] state. Like the PrP peptide studies described above, these studies argue for de novo synthesis of prions.

Taken together, the results of these studies argue quite persuasively for the prion concept. They contend that prions are infectious proteins. Some investigators would like to see studies in which wt recPrP is converted into PrP^Sc in vitro and then inoculated into mice to produce disease (Aguzzi and Weissmann 1997; Chesebro 1998). As the structure of PrP^Sc becomes clearer and auxiliary proteins involved in the formation of PrP^Sc are identified, forming PrP^Sc in vitro should be possible.

Strains of Prions

That goats with scrapie can manifest two different syndromes, one in which the goats become hyperactive and the other in which they become drowsy, raises the possibility that strains of prions might exist (Pattison and Millson 1961; Chapter 9). Subsequent studies with mice documented the existence of multiple strains through careful measurements of incubation times and the distribution of vacuoles in the CNS (Dickinson et al. 1968; Fraser and Dickinson 1973). Two different groups of prion strains were identified using two strains of mice, C57BL and VM: one group, typified by Me7 prions, exhibited short incubation times in C57BL mice and long durations in VM mice. The other group of prions, including 22A and 87V, showed the converse behavior with respect to the length of the incubation time. Particularly puzzling was the finding that long incubation times were a dominant trait for both groups of prions, as evidenced by studies in F₁ (C57BL × VM) mice.

Molecular genetic studies using NZW and Iln mice later showed genetic linkage in controlling incubation times to the PrP gene (Carlson et al. 1986) that proved to be analogous in the C57BL and VM mice

(Hunter et al. 1987). The PrP sequences of NZW and Iln mice differ at residues 108 and 189 and were termed PrP-A and PrP-B, respectively (Westaway et al. 1987). Using Tg mice expressing different levels of PrP^C-A and PrP^C-B, it was shown that long incubation times in F_1 mice were not a dominant trait, but rather were due to a gene dosage effect (Carlson et al. 1994).

The search for an explanation of how biological information encrypting the disease phenotype could be enciphered within the prion posed a conundrum. Many investigators kept arguing for a small nucleic acid (Bruce and Dickinson 1987; Weissmann 1991b), but none could be found (Chapter 2). Others argued that this information must be enciphered within the structure of PrP^{Sc} (Prusiner 1991; Ridley and Baker 1996). The first evidence supporting the hypothesis that strain-specific information is enciphered in PrP^{Sc} came from studies on prions causing TME, which were passaged into Syrian hamsters. On serial passaging, two strains emerged: one strain (HY) produced hyperactivity in Syrian hamsters and the other (DY) was manifest as a drowsy syndrome, as was seen with scrapie strains in goats (Marsh et al. 1991; Bessen and Marsh 1992). The HY strain was similar to the Sc237 prion strain with respect to protease resistance and sedimentation properties, whereas the DY strain showed minimal protease resistance and was less readily sedimented (Bessen and Marsh 1994).

Because the HY and DY strains emerged after passaging TME prions in Syrian hamsters, concern over the origin of these strains made the conclusions somewhat questionable. A much more convincing study came from the transmission of FFI and fCJD to Tg mice (Telling et al. 1996b). In this study, two different strains of prions were generated de novo in patients with PrP gene mutations and propagated in mice expressing the chimeric MHu2MPrP transgene. Brain homogenates of FFI and fCJD(E200K) patients transmitted disease to Tg(MHu2M) mice ~200 days after inoculation. The FFI inoculum induced formation of a 19-kD PrP^{Sc} fragment, as measured by SDS-PAGE after limited proteolysis and removal of asparagine-linked carbohydrates; in contrast, fCJD(E200K) produced a 21-kD PrP^{Sc} fragment (see Table 4) (Telling et al. 1996b; Korth et al. 2003). On second passage, Tg(MHu2M) mice inoculated with FFI prions showed an incubation time of ~130 days and a 19-kD PrP^{Sc}, whereas those inoculated with fCJD(E200K) prions exhibited an incubation time of ~170 days and a 21-kD PrP^{Sc} (Prusiner 1997). The experimental data demonstrated that MHu2MPrP^{Sc} can exist in two different conformations, based on the sizes of the protease-resistant fragments, within an invariant amino acid sequence. The results of these studies argue that PrP^{Sc} acts as a template for the conversion of PrP^C into nascent PrP^{Sc}. Imparting the size of the protease-resistant fragment of PrP^{Sc} through conformational templating provides a mechanism for both the generation and propagation of prion strains.

In another set of studies, prions from cattle with BSE and humans with vCJD were used to demonstrate distinct prion strains in humans expressing wt PrP, in contrast to the studies described above with prions from patients harboring the E200K or D178N mutation. Both BSE and vCJD prions transmitted to Tg mice expressing BoPrP but not to mice expressing either HuPrP or MHu2MPrP (Hill et al. 1997; Scott et al. 1999). Moreover, sCJD, iCJD, and fCJD(E200K) prions failed to transmit disease to Tg(BoPrP) mice after more than 400 days (M. Scott and S.B. Prusiner, in prep.). These findings demonstrate the prions that cause BSE and vCJD are similar, but distinct from those responsible for sCJD, iCJD, and fCJD(E200K).

Isolation of New Strains

Additional evidence implicating PrP in the phenomenon of prion strains comes from studies on the interspecies transmission of strains (Kimberlin et al. 1989). Such studies were especially revealing when mice expressing chimeric hamster–mouse transgenes were used. Evidence was obtained showing that new strains pathogenic to Syrian hamsters could be obtained from prion strains that had been previously replicated by limiting dilution in mice (Dickinson et al. 1969; Scott et al. 1997a). Both the generation and propagation of prion strains result from interactions between PrP^{Sc} and PrP^C. Additionally, strains that were isolated from different breeds of scrapied sheep and thought to be distinct were shown to have identical properties. Such findings argue for the convergence of some strains and raise the issue of the limits of prion diversity.

Studies with different strains of prions propagated in Syrian hamsters and subsequently in Tg mice expressing MH2MPrP are most illustrative with respect to the isolation of new prion strains. These Tg mice were inoculated with the two prion strains, Sc237 and DY, that had been serially passaged in Syrian hamsters (Peretz et al. 2002). On first passage in Tg(MHM2)$Prnp^{0/0}$ mice, the SHa(Sc237) prions exhibited prolonged incubation times, typical of a species barrier (Chapter 2). On subsequent passage in Tg(MHM2)$Prnp^{0/0}$ mice, the MH2M(Sc237) strain showed a

rtening of incubation time. Moreover, PrP^{Sc} of the
strain possesses different structural properties from PrP^{Sc}
37) strain, as demonstrated by relative conformational stability measurements (Peretz et al. 2001a). Conversely, transmission of
SHa(DY) prions to Tg(MHM2)$Prnp^{0/0}$ mice did not encounter a species
barrier and the MH2M(DY) strain retained the conformational and phenotypic properties of SHa(DY). These results extend the findings
described above for the FFI and fCJD(E200K) strains of prions and contend that a change in PrP^{Sc} conformation is intimately associated with the
emergence of a new prion strain.

Selective Neuronal Targeting

Although the foregoing scenario is unprecedented in biology, considerable experimental data now support these concepts. However, it is not yet
known whether multiple conformers of PrP^C, which serve as precursors
for selective conversion into different PrP^{Sc}, exist. In this light, it is useful
to consider another phenomenon that is not yet understood: the selective
targeting of neuronal populations in the CNS of the host mammal. Data
suggest that variations in glycosylation of the asparagine-linked sugars
may influence the rate at which a particular PrP^C molecule is converted
into PrP^{Sc} (DeArmond et al. 1997). Because asparagine-linked oligosaccharides are known to modify the conformation of some proteins (Otvos
et al. 1991; O'Connor and Imperiali 1996), it seemed reasonable to
assume that variations in complex-type sugars may alter the size of the
energy barrier that must be traversed during formation of PrP^{Sc}. If this is
the case, then regional variations in oligosaccharide structure in the CNS
could account for selective targeting: i.e., formation of PrP^{Sc} in particular
areas of the brain. Such a mechanism could also explain the variations in
the ratio of the various PrP^{Sc} glycoforms observed by some investigators
(Collinge et al. 1996). But such a mechanism, while accounting for specific patterns of PrP^{Sc} distribution, does not seem to influence to any measurable degree the properties of the resulting PrP^{Sc} molecule. In fact, molecular modeling and NMR structural studies may provide an explanation
for such phenomena because the asparagine-linked oligosaccharides
appear to be on the face of PrP opposite to where PrP^C and PrP^{Sc} are
expected to interact during the formation of nascent PrP^{Sc} (Huang et al.
1994, 1995; Zhang et al. 1995; Riek et al. 1996).

With the development of histoblotting (Taraboulos et al. 1992), it
became possible to localize and to quantify PrP^{Sc} as well as to determine

whether or not distinct prion strains produce different, reproducible patterns of PrP^{Sc} accumulation. The patterns of PrP^{Sc} accumulation were found to be different for each prion strain when the genotype of the host was held constant (Hecker et al. 1992; DeArmond et al. 1993; Scott et al. 1999; Peretz et al. 2002). This finding was in accord with earlier studies showing that spongiform degeneration is strain-specific (Fraser and Dickinson 1973) because PrP^{Sc} accumulation precedes vacuolation.

Although studies with both mice and Syrian hamsters established that each strain has a specific signature as defined by a specific pattern of PrP^{Sc} accumulation in the brain (Hecker et al. 1992; DeArmond et al. 1993; Carlson et al. 1994), comparisons must be performed on an isogenic background (Scott et al. 1993; Hsiao et al. 1994). When one strain is inoculated into mice expressing different PrP genes, variations in the patterns of PrP^{Sc} accumulation were found to be as great as those seen between two strains (DeArmond et al. 1997). On the basis of initial studies that were performed in animals of a single genotype, PrP^{Sc} synthesis was suggested to occur in specific populations of cells for a given prion isolate (Prusiner 1989; Hecker et al. 1992). In conclusion, the pattern of PrP^{Sc} deposition is a manifestation of the particular prion strain, but can be modified by the conditions under which it is propagated (DeArmond et al. 1997). The results of studies of three prion strains prepared from three brain regions and spleens of inbred mice support this contention (Carp et al. 1997).

Interplay between the Species and Strains of Prions

The recent advances described above in our understanding of the role of the primary and tertiary structures of PrP in the transmission of disease have given new insights into the pathogenesis of the prion diseases. The amino acid sequence of PrP encodes the species of the prion (Table 4) (Scott et al. 1989; Telling et al. 1995), and the prion derives its PrP^{Sc} sequence from the last mammal in which it was passaged (Scott et al. 1997a). Although the primary structure of PrP is likely to be the most important or even sole determinant of the tertiary structure of PrP^{C}, existing PrP^{Sc} seems to function as a template in determining the tertiary structure of nascent PrP^{Sc} molecules as they are formed from PrP^{C} (Prusiner 1991; Cohen et al. 1994). In turn, prion diversity appears to be enciphered in the conformation of PrP^{Sc}, and prion strains represent different conformers of PrP^{Sc} (Bessen and Marsh 1994; Telling et al. 1996b; Scott et al. 1997a). The total number of prion strains, i.e., different conformations of PrP^{Sc}, in existence remains to be established.

PRION DISEASES OF ANIMALS

The prion diseases of animals include scrapie of sheep and goats, BSE, TME, chronic wasting disease (CWD) of mule deer and elk, feline spongiform encephalopathy, and exotic ungulate encephalopathy (Table 1; Chapters 11 and 12).

PrP Polymorphisms in Sheep, Cattle, and Elk

In 1962, Parry argued that host genes were responsible for the development of scrapie in sheep. He was convinced that natural scrapie is a genetic disease that could be eradicated by proper breeding protocols (Parry 1962, 1983). He considered its transmission by inoculation of importance primarily for laboratory studies and communicable infection of little consequence in nature. Other investigators viewed natural scrapie as an infectious disease and argued that host genetics only modulates susceptibility to an endemic infectious agent (Dickinson et al. 1965).

In sheep, polymorphisms at codons 136, 154, and 171 of the PrP gene that produce amino acid substitutions have been studied with respect to the incidence of scrapie (Fig. 11) (Goldmann et al. 1990a,b; Laplanche et al. 1993; Clousard et al. 1995). Studies of natural scrapie in the U.S. have shown that ~85% of afflicted sheep are of the Suffolk breed. Only those Suffolk sheep homozygous for glutamine (Q) at codon 171 developed scrapie although healthy controls with Glu/Glu, Glu/Arg, and Arg/Arg genotypes were also found (Hunter et al. 1993, 1997a,b; Goldmann et al. 1994; Westaway et al. 1994b; Belt et al. 1995; Clousard et al. 1995; Ikeda et al. 1995; O'Rourke et al. 1997). These results argue that susceptibility in Suffolk sheep is governed by the PrP codon-171 polymorphism. As with the Suffolk breed, the PrP codon-171 polymorphism in Cheviot sheep has a profound influence on susceptibility to scrapie, and codon 136 also modulates susceptibility but less so than codon 171 (Goldmann et al. 1991a; Hunter et al. 1991).

In contrast to sheep, different breeds of cattle have no specific PrP polymorphisms. The only polymorphism recorded in cattle is a variation in the number of octarepeats: Most cattle, like humans, have five octarepeats, but some have six (Goldmann et al. 1991b; Prusiner et al. 1993b). However, the presence of six octarepeats does not seem to be overrepresented in BSE (Goldmann et al. 1991b; Prusiner et al. 1993b; Hunter et al. 1994).

In studies of CWD, the susceptibility of elk, but not deer, seems to be modulated by codon 132, which corresponds to codon 129 in humans (Fig. 11). Elk with CWD consistently express Met/Met at position 132; no

elk with CWD expressing leucine at this residue have been found
(O'Rourke et al. 1999).

Bovine Spongiform Encephalopathy

Prion strains and the species barrier are of paramount importance in
understanding the BSE epidemic in Britain, where it is estimated that
almost one million cattle were infected with prions (Anderson et al. 1996;
Nathanson et al. 1997). The mean incubation time for BSE is ~5 years.
Therefore, most cattle did not manifest disease because they were slaughtered between 2 and 3 years of age (Stekel et al. 1996). Nevertheless, more
than 180,000 cattle, primarily dairy cows, have died of BSE over the past
decade (Anderson et al. 1996; Phillips et al. 2000). BSE is a massive common-source epidemic caused by meat and bone meal (MBM) fed primarily to dairy cows (Wilesmith et al. 1991; Nathanson et al. 1997). MBM was
prepared from the offal of sheep, cattle, pigs, and chickens as a high-protein nutritional supplement. In the late 1970s, the hydrocarbon-solvent
extraction method used in the rendering of offal began to be abandoned,
resulting in MBM with a much higher fat content (Wilesmith et al. 1991).
It is now thought that this change allowed scrapie prions from sheep to
survive the rendering process and to be passed into cattle. Alternatively,
bovine prions may have been present at low levels prior to modification
of the rendering process and with the processing change, survived in sufficient numbers to initiate the BSE epidemic when reintroduced into cattle through ingestion of MBM (Phillips et al. 2000). Perhaps a particular
conformation of BoPrP^{Sc} selected for heat resistance during the rendering
process and then reselected multiple times as cattle infected by ingesting
prion-contaminated MBM were slaughtered and their offal rendered into
more MBM. Against the latter hypothesis is the widespread geographical
distribution throughout England of the initial 17 cases of BSE, which
occurred almost simultaneously (Wilesmith 1991; Kimberlin 1996;
Nathanson et al. 1997). Evidence of a preexisting prion disease of cattle,
either in Great Britain or elsewhere, is scant, but an outbreak of TME in
Wisconsin has been cited as evidence for sporadic BSE (Marsh et al. 1991).
In July 1988, the practice of feeding MBM to sheep and cattle was banned.
Statistical analyses demonstrate that the epidemic is disappearing as a
result of this ruminant feed ban (Anderson et al. 1996; Phillips et al.
2000), reminiscent of the disappearance of kuru in the Fore people of
New Guinea (Gajdusek 1977; Alpers 1987).

    The origin of BSE prions cannot be determined by examining the
amino acid sequence of PrP^{Sc} from cattle with BSE because it has the

bovine sequence regardless of whether the initial prions originated from sheep or cattle. The bovine PrP sequence differs from that of sheep at seven or eight positions (Goldmann et al. 1990a, 1991b; Prusiner et al. 1993b).

Brain extracts from cattle with BSE cause disease in cattle, sheep, mice, pigs, and mink after intracerebral inoculation (Fraser et al. 1988; Dawson et al. 1990a,b; Bruce et al. 1993, 1997), but prions in brain extracts from sheep with scrapie fed to cattle produced an illness substantially different from BSE (Robinson et al. 1995). However, no exhaustive effort has been made to test different strains of sheep prions or to examine the disease following bovine-to-bovine passage.

Monitoring Cattle for BSE Prions

Although many plans have been offered for the culling of older cattle in order to minimize the spread of BSE (Anderson et al. 1996), it seems more important to monitor the frequency of prion disease in cattle as they are slaughtered for human consumption. No reliable, specific test for prion disease in live animals is available, but immunoassays for PrP^{Sc} in the brain stems of cattle might provide a reasonable approach to establishing the incidence of subclinical BSE in cattle entering the human food chain (Hope et al. 1988; Serban et al. 1990; Taraboulos et al. 1992; Prusiner et al. 1993b; Grathwohl et al. 1997; Korth et al. 1997). Determining how early in the incubation period PrP^{Sc} can be detected by immunological methods is now possible because a reliable bioassay is available with the creation of Tg(BoPrP) mice (Scott et al. 1997b, 1999; Buschmann et al. 2000) (Chapter 3).

Prior to development of Tg(BoPrP)$Prnp^{0/0}$ mice, non-Tg mice inoculated intracerebrally with BSE brain extracts required more than 300 days to develop disease (Taylor 1991; Fraser et al. 1992; Bruce et al. 1997; Lasmézas et al. 1997). Depending on the prion titer in the inoculum, and the structures of PrP^C, PrP^{Sc}, and protein X, the number of inoculated animals developing disease can vary over a wide range. Some investigators have stated that transmission of BSE to mice is quite variable, with incubation periods exceeding one year (Lasmézas et al. 1997), whereas others report low prion titers in BSE brain homogenates (Taylor 1991; Fraser et al. 1992) compared to scrapied rodent brain (Hunter et al. 1963; Eklund et al. 1967; Kimberlin and Walker 1977; Prusiner et al. 1982b).

The CDI was calibrated using Tg(BoPrP)$Prnp^{0/0}$ mice. As described in Chapter 3, the sensitivities of the bioassay and the CDI are similar (Safar et al. 2002). These studies suggest that the CDI may be able to detect BSE prions in asymptomatic cattle even when titers are quite low.

Epidemiology of vCJD

In 1994, the first cases of CJD in teenagers and young adults that were eventually labeled vCJD occurred in Great Britain (Will et al. 1996; Chapters 13 and 15). In addition to the young age of these patients (Bateman et al. 1995; Britton et al. 1995), vCJD is characterized by numerous PrP amyloid plaques surrounded by a halo of intense spongiform degeneration in the brain (Ironside 1997). These unusual neuropathologic changes have not been seen in CJD cases in the United States, Australia, or Japan (Centers for Disease Control 1996; Ironside 1997). Both macaque monkeys and marmosets developed neurologic disease several years after inoculation with bovine prions (Baker et al. 1993), but only the macaques exhibited numerous PrP plaques similar to those found in vCJD (Lasmézas et al. 1996b; R. Ridley and H. Baker, unpubl.).

The restricted geographical occurrence and chronology of vCJD raised the possibility that BSE prions have been transmitted to humans. That only ~130 vCJD cases have been recorded and the incidence has remained relatively constant made establishing the origin of vCJD difficult. No set of dietary habits distinguishes vCJD patients from apparently healthy people. Moreover, there is no explanation for the predilection of vCJD for teenagers and young adults. Why have older individuals not developed vCJD-based neuropathologic criteria? It is noteworthy that epidemiologic studies over the past three decades have failed to find evidence for transmission of sheep prions to humans (Malmgren et al. 1979; Brown et al. 1987; Harries-Jones et al. 1988; Cousens et al. 1990). Attempts to predict the future number of cases of vCJD, assuming exposure to bovine prions prior to the offal ban, have been uninformative because so few cases of vCJD have occurred (Collinge et al. 1995; Cousens et al. 1997; Raymond et al. 1997; Ghani et al. 2000). Are we at the beginning of a prion disease epidemic in Britain, as seen for BSE and kuru, or will the number of vCJD cases remain small, as seen with iCJD caused by cadaveric HGH (Billette de Villemeur et al. 1996; Public Health Service 1997)?

Recent studies of PrP^{Sc} from brains of patients who died of vCJD show a PrP glycoform pattern different from those found for sCJD and iCJD (Collinge et al. 1996; Hill et al. 1997). The utility of measuring PrP glycoforms is questionable in trying to relate BSE to vCJD (Parchi et al. 1997; Somerville et al. 1997) because PrP^{Sc} is formed after the protein is glycosylated (Borchelt et al. 1990; Caughey and Raymond 1991) and enzymatic deglycosylation of PrP^{Sc} requires denaturation (Endo et al. 1989; Haraguchi et al. 1989).

Compelling Evidence for Transmission of Bovine Prions to Humans

As described above, laboratory studies have been unconvincing in establishing a relationship between the conformations of PrP^{Sc} from cattle with BSE and those from humans with vCJD. Perhaps slightly more persuasive is the relationship demonstrated between vCJD and BSE based on similar incubation times in non-Tg RIII mice of ~310 days after inoculation with human vCJD or bovine BSE prions (Bruce et al. 1997). However, such studies suffer from transmission of both BSE and vCJD prions to a heterologous host, i.e., non-Tg mice expressing MoPrP^{C}. Using Tg mice, as was done for strains generated in the brains of patients with FFI or fCJD (Telling et al. 1996b; Scott et al. 1997b), compelling evidence for the transmission of bovine prions to humans was found. BSE prions transmitted to all inoculated Tg(BoPrP) mice after ~240 days but not to Tg mice expressing either HuPrP or MHu2MPrP (Scott et al. 1999). On second passage to Tg(BoPrP)$Prnp^{0/0}$ mice, the incubation time was unaltered, demonstrating the complete absence of a species barrier. Similar to BSE prions, vCJD prions transmitted readily to Tg(BoPrP)$Prnp^{0/0}$ mice, with a slightly longer incubation time of ~270 days (Scott et al. 1999), but poorly to Tg(HuPrP) (Hill et al. 1997) and Tg(MHu2M) mice. Moreover, sCJD, iCJD, and fCJD(E200K) prions failed to transmit disease to Tg(BoPrP) mice after more than 400 days. On second passage of vCJD prions to Tg(BoPrP)$Prnp^{0/0}$ mice, the incubation time was reduced to ~225 days, demonstrating a small but expected species barrier. These findings argue that the strain-specific PrP^{Sc} conformations from BSE and vCJD prions are quite similar despite substantial differences in the amino acid sequences of BoPrP and HuPrP. Clearly, the conformation of HuPrP^{Sc}(vCJD) makes these prions much more readily transmittable to Tg(BoPrP)$Prnp^{0/0}$ mice than to either Tg(HuPrP)$Prnp^{0/0}$ or Tg(MHu2M)$Prnp^{0/0}$ mice.

Chronic Wasting Disease

Mule deer, white-tailed deer, and elk have been reported to develop CWD. CWD is unique among the animal prion diseases because it seems to be far more communicable than scrapie, BSE, or TME; moreover, it is the only prion disease known in free-ranging animals. CWD was first described in 1967 and was reported to be a spongiform encephalopathy in 1978 on the basis of histopathology in the brain. CWD has been found in the United States, Canada, and South Korea. In the U.S., CWD has been reported in Colorado, Wyoming, South Dakota, Nebraska, Oklahoma, Montana, New Mexico, Minnesota, and Wisconsin. In captive cervid herds, up to 90% of mule deer have been reported to be positive for prions (Williams and Young 1980), and up to 60% of elk in Colorado and Wyoming develop CWD (Miller et al. 1998; Peters et al. 2000; Williams and Miller 2002). Moreover, the incidence of CWD in cervids living in the wild has been estimated to be as high as 15% (Miller et al. 2000). The mode of transmission of the CWD prion among mule deer, white-tailed deer, and elk is unresolved, but contamination of grass with prions excreted in fecal matter seems to be a likely source (Williams and Miller 2002). The high content of PrP^{Sc} in the intestinal lymphoid tissue of cervids with CWD (Sigurdson et al. 1999) supports such a scenario.

Brain homogenates from mule deer with CWD have transmitted disease to 4 of 13 cattle after intracerebral inoculation (Chapter 11) (Hamir et al. 2001). These findings are particularly important because there is great concern that CWD prions might be transmitted to cattle grazing in contaminated pastures. In addition, CWD has been transmitted to ferrets, mink, squirrel monkeys, goats, and mice after intracerebral inoculation (Williams and Miller 2002); however, only mule deer demonstrate efficient transmission of CWD prions by intracerebral inoculation. To date, endpoint titrations of CWD prions have not been performed in mule deer, and Tg mice susceptible to CWD prions have not been developed, but the CDI has been adapted to measure CWD prions rapidly (Safar et al. 2002).

FUNGAL PRIONS

Although prions were originally defined in the context of an infectious pathogen (Prusiner 1982), it is now becoming widely accepted that prions are elements that impart and propagate variability through multiple conformers of a normal, cellular protein (Chapter 7). It is likely that such a mechanism will not be restricted to a single class of transmissible pathogens. Indeed, it is probable that this original definition will need to be extended to encompass other situations in which a similar mechanism of information transfer occurs. Two notable prion-like determinants, [URE3] and [PSI], have been described in yeast (Wickner 1994), and another prion-like determinant has been reported in other fungi (Deleu et al. 1993; Coustou et al. 1997).

The [URE3] Determinant

In considering what properties yeast prion-like determinants would possess, Wickner and colleagues (Wickner et al. 1995) proposed a series of

1) They will behave as non-Mendelian genetic elements; :d phenotype will be reversible; (3) a maintenance gene rmal protein will manifest as a related, Mendelian genet- overproduction of the maintenance gene product will increase the generation of the non-Mendelian element; and (5) defective, interfering replicons will not be evident. Two non-Mendelian genetic determinants that fulfill these criteria, [URE3] and [PSI], were first described over 25 years ago. The ure2 and [URE3] mutations were isolated by their ability to utilize ureidosuccinate in the medium, thereby overcoming a defect in uracil biosynthesis caused by mutations in aspartate transcarbamylase (Ura2p) (Lacroute 1971). Whereas the behavior of the ure2 mutations was entirely consistent with a normal chromosomal locus, when [URE3] strains were mated with wt strains, an irregular segregation pattern was observed (Lacroute 1971; Aigle and Lacroute 1975). Subsequently, it was shown by cytoduction that [URE3] could be transferred in the absence of nuclear fusion, confirming their non-Mendelian nature, and that [URE3] can be cured by growth of cells on rich medium containing 5 mM GdnHCl (Wickner et al. 1995). Significantly, the cured strains could then be used to generate additional [URE3] mutants, arguing strongly against the participation of a nucleic acid genome, and thus satisfying two of the aforementioned criteria expected of a prion-like determinant. Notably, the ure2 mutations are recessive and display the same phenotype as [URE3] mutants, and a series of genetic arguments showed clearly that the URE2 chromosomal gene is necessary for propagation of the [URE3] phenotype (Wickner 1994). However, because the phenotype of ure2 mutants is the same as that observed in the presence of [URE3], it seems most likely that URE2 encodes the normal, active form of the protein. Conversion to the abnormal, inactive form leads to the [URE3] state (Wickner 1994). Following the introduction of the URE2 gene on a high-copy plasmid, a 50- to 100-fold increase was observed in the frequency with which [URE3] mutants were obtained (Wickner 1994). This is entirely expected for a prion mechanism, because the stochastic event that gives rise to the abnormal conformer will increase in frequency in cells in which the normal "precursor" is overproduced. A similar mechanism may lead to spontaneous prion disease in transgenic mice overexpressing PrP (Westaway et al. 1994c).

The [PSI] Determinant

Another non-Mendelian genetic element in yeast, called the [PSI] factor, exaggerates the effect of a weak chromosomal ochre suppressor, SUQ5

(Cox 1965). Subsequent studies showed that the action of [PSI] is more general, affecting other weak ochre suppressors (Broach et al. 1981), and that strong ochre suppressors become lethal in the presence of [PSI] (Cox 1965), probably due to an intolerably low frequency of correct translational termination. [PSI] also affects the efficiency of suppression of UGA and UAG codons by the aminoglycoside antibiotics (Palmer et al. 1979).

Many lines of evidence suggest that [PSI] is an abnormal, prion-like conformer of the Sup35 protein (Sup35p) (for reviews, see Lindquist et al. 1995; Wickner et al. 1995; Tuite and Lindquist 1996). Like [URE3], [PSI] can be cured by growth on 5 mM GdnHCl (Tuite et al. 1981) as well as hyperosmotic media (Singh et al. 1979). Other characteristics of [PSI] mirror those of [URE3]: The [PSI] phenotype is the same as that of the omnipotent suppressor mutations sup35 and sup45 (Hawthorne and Mortimer 1968), and overproduction of Sup35p leads to a 100-fold increase in the frequency of the occurrence of [PSI] (Chernoff et al. 1993). In addition to these similarities to [URE3], the influence of [PSI] on protein synthesis in vitro provides further evidence for a prion-like mode of propagation (Tuite et al. 1987). When the efficiency of translational readthrough by extracts of yeast cells in the presence of added suppressor tRNAs in vitro was assessed, it was found that substantial readthrough occurred only when the extracts were prepared using strains that contained [PSI] (Tuite et al. 1987).

Yeast Prion Domains

[PSI] and [URE3] share another important characteristic. In both cases, the "functional" determinants have been mapped to the carboxy-terminal region of the protein, distinct from the "prion" domain, which comprises the amino-terminal residues 65 and 114 of Ure2p (Masison and Wickner 1995) and Sup35p, respectively (Doel et al. 1994; Ter-Avanesyan et al. 1994; Derkatch et al. 1996). Although neither of the prion domains displays sequence identity either to each other or to PrP, the amino-terminal regions of both Sup35p and mammalian PrP contain short repeated sequence elements: PQGGYQQYN in Sup35p and PHGGGWGQ in PrP (Tuite and Lindquist 1996). Interestingly, when the prion domains of both Sup35 and Ure2p are expressed in E. coli and purified, they polymerize spontaneously into amyloid-like fibrils (Glover et al. 1997; King et al. 1997). Polymerization of the prion domains of both Sup35 and Ure2p is able to create [PSI] and [URE3] states in yeast, respectively.

Dependence of Yeast Prions on Molecular Chaperones

The intrinsic power of the yeast genetic system has provided striking evidence for the involvement of chaperones in the propagation of yeast [PSI] "prions." A genetic screen for factors that suppress the [PSI] phenotype resulted in the isolation of a single suppressor plasmid, which was found to contain the chaperone Hsp104 (Chernoff et al. 1995). Furthermore, propagation of [PSI] was eliminated by either overproduction or absence of Hsp104, and treatment of cells with guanidine or UV light led to induction of Hsp104 (Chernoff et al. 1995; Patino et al. 1996). The significance of Hsp104 is unclear, because no published data exist to indicate that [URE3] utilizes Hsp104; furthermore, overexpression of Sup35 at high levels can induce [PSI] in the absence of Hsp104 (Glover et al. 1997).

Differences between Yeast and Mammalian Prions

A wealth of studies has established the concept of fungal prions. Whereas fungal prions are produced in the cytoplasm, mammalian prions are produced within cholesterol-rich microdomains on the surface of cells, called rafts or CLDs. Whether the processes of fungal and mammalian prion replication are fundamentally different or quite similar remains to be established. Probably many more similarities than differences are likely to emerge because distinct strains of both fungal and mammalian prions seem to represent different conformations of a particular protein. This finding necessitates a mechanism by which a particular conformation can be templated and reproduced with a high degree of fidelity.

Interestingly, the prion state in yeast is proposed to be functionally inert in the case of both [PSI] and [URE3] and produces the same phenotype as inactivation of the maintenance gene. In contrast, prion diseases in mammals cannot be explained simply by the loss of function of PrP because ablation of the PrP gene has no as-yet-detectable deleterious effect (Büeler et al. 1992).

PRION DISEASES ARE DISORDERS OF PROTEIN CONFORMATION

The study of prions has followed several unexpected directions over the past three decades. The discovery that prion diseases in humans are uniquely genetic and infectious has greatly strengthened and extended the prion concept. To date, more than 30 different mutations in the human

PrP gene, all resulting in nonconservative substitutions, have been found either to be linked genetically to, or to segregate with, the inherited prion diseases (Fig. 11). Yet the transmissible prion particle is composed of abnormal PrP^{Sc} (Prusiner 1991).

Understanding how PrP^C unfolds and refolds into PrP^{Sc} will be of paramount importance in transferring advances in the prion diseases to studies of other degenerative illnesses (Prusiner 2001). The mechanism by which PrP^{Sc} is formed must involve a templating process in which existing PrP^{Sc} directs the refolding of PrP^C into nascent PrP^{Sc} with the same conformation. Undoubtedly, molecular chaperones of some type participate in a process that appears to be associated with CLDs on the cell surface.

Studies of prions in fungi have been extremely helpful in establishing the prion concept (Chapter 2). In the case of yeast, Sup35 and Ure2p fold into alternative conformations that create new metabolic states, [PSI] and [URE3], respectively. Whether PrP^{Sc} in mammals represents a misfolded protein or an alternatively folded protein, as with yeast prions, discussed in Chapter 7, remains to be established.

PREVENTION AND THERAPEUTICS FOR PRION DISEASES

Because people at risk for inherited prion diseases can now be identified decades before neurologic dysfunction is evident, the development of an effective therapy for these fully penetrant disorders is imperative (Chapman et al. 1994; Spudich et al. 1995). Although it is difficult to predict the number of individuals who may develop neurologic dysfunction from bovine prions in the future (Cousens et al. 1997; Ghani et al. 2000), seeking an effective therapy now seems most prudent. Interfering with the conversion of PrP^C into PrP^{Sc} would likely be the most attractive therapeutic target (Cohen et al. 1994).

Defining the pathogenesis of prion disease is an important issue with respect to developing an effective therapy. The issue of whether large aggregates of misprocessed proteins or misfolded monomers (or oligomers) cause CNS degeneration has been addressed in several studies of prion diseases in humans as well as in Tg mice. In humans, the frequency of PrP amyloid plaques varies from 100% in GSS and vCJD (Will et al. 1996) to ~70% in kuru (Klatzo et al. 1959) and ~10% in sCJD, arguing that these plaques are a nonobligatory feature of the disease (DeArmond and Prusiner 1997). In Tg mice expressing both MoPrP and SHaPrP, animals inoculated with hamster prions produced hamster prions and developed amyloid plaques composed of SHaPrP^{Sc} (Prusiner et al. 1990). In contrast, Tg mice inoculated with mouse prions did not develop plaques, even though they produced mouse prions and died of prion disease.

Prion Therapeutics

Various compounds have been proposed as potential therapeutics for treatment of prion diseases; these include polysulfated anions, dextrans, Congo red dye, oligonucleotides, and cyclic tetrapyrroles, all of which have been shown to increase survival time when given prior to prion infection in rodents, but not when administered a month or more after infection has been established (Dickinson et al. 1975; Kimberlin and Walker 1983; Ehlers and Diringer 1984; Diringer 1991; Ingrosso et al. 1995; Priola et al. 2000; Sethi et al. 2002). A more complete discussion of prion therapeutics can be found in Chapter 18.

Besides studies in rodents, ScN2a cells chronically infected with scrapie prions have been used to identify several candidate antiprion drugs (Caughey and Race 1992; Gorodinsky and Harris 1995; Taraboulos et al. 1995; Supattapone et al. 1999a; Perrier et al. 2000; Korth et al. 2001; Proske et al. 2002), but none has been shown to be effective in halting prion diseases in either animals or humans. Structure-based drug design based on dominant-negative inhibition of prion formation has produced several lead compounds (Perrier et al. 2000, 2002). Prion replication depends on protein–protein interactions, and a subset of these interactions gives rise to dominant-negative phenotypes produced by single-residue substitutions (Kaneko et al. 1997c; Zulianello et al. 2000). A particularly interesting set of drugs is the branched polyamines, or dendrimers, which enhance the clearance of PrP^{Sc} from cells (Supattapone et al. 1999a, 2001). Although these compounds cure cultured cells of prion infection, they have not been successfully deployed in mice because of difficulties in delivering such highly charged compounds to the CNS.

Quinacrine and Other Acridine Derivatives

Tricyclic derivatives of acridine exhibit half-maximal inhibition of PrP^{Sc} formation at effective concentrations (IC$_{50}$) between 0.3 μM and 3 μM in cultured ScN2a cells (Doh-ura et al. 2000; Korth et al. 2001). The IC$_{50}$ for chlorpromazine was 3 μM, whereas quinacrine was 10 times more potent. A variety of 9-substituted, acridine-based analogs of quinacrine were synthesized, which demonstrated variable potencies similar to chlorpromazine and emphasized the importance of the side chain in mediating the inhibition of PrP^{Sc} formation (Korth et al. 2001). These studies showed that tricyclic compounds with an aliphatic side chain at the middle ring moiety constitute a new class of antiprion agents. Because quinacrine and chlorpromazine have been used in humans as antimalarial and antipsychotic drugs, respectively, for many years and are known to pass the blood–brain barrier (BBB), these compounds became immediate candidates for the treatment of CJD and other human prion diseases (Miller and Prusiner [2001 FDA Investigational New Drug Application]).

An asymmetric carbon in the side chain of quinacrine creates two stereoisomers. (S)-Quinacrine is two to three times more potent than the (R)-isomer in cultured cells (Ryou et al. 2003). Whether the use of (S)-quinacrine in place of the racemic mixture will allow significantly more drug to be given remains to be established. Studies performed almost six decades ago in humans demonstrated that (S)-quinacrine is selectively metabolized and that (R)-quinacrine remains intact (Hammick and Chambers 1945). Whether (R)-quinacrine or the metabolites of (S)-quinacrine are responsible for the side effects recorded in patients receiving the racemic mixture is unknown. Side effects of quinacrine include liver toxicity, cardiomyopathy, and toxic psychoses (Goodman and Gilman 1970).

Recombinant Antibody Fragments

A panel of recombinant antibody fragments (recFabs) recognizing different epitopes on PrP was studied with respect to inhibition of prion propagation in cultured ScN2a cells (Peretz et al. 2001b). Recombinant Fabs binding to PrP^{C} on the cell surface inhibited PrP^{Sc} formation in a dose-dependent manner. In ScN2a cells treated with the most potent recFab D18, prion replication was completely abolished and preexisting PrP^{Sc} rapidly cleared, suggesting that this antibody may cure established infection. The activity of recFab D18 is associated with its ability to recognize more completely the total population of PrP^{C} molecules on the cell surface than other recFabs, and with the location of its epitope on PrP^{C}. In other studies, a monoclonal antibody, 6H4, which is thought to bind to the same region of PrP as recFab D18, was found to inhibit prion accumulation in ScN2a cells (Enari et al. 2001).

Whether antibodies or Fabs can be effectively administered to humans for the prevention and treatment of prion diseases is unclear. Neither antibodies nor Fabs cross the BBB in high concentration, so that delivery of these proteins to the CNS remains a critical issue.

Dominant-negative Sheep

The production of domestic animals that do not replicate prions may also prove to be a practical way to prevent prion disease. Sheep encoding Arg/Arg at polymorphic position 171 seem resistant to scrapie (Hunter et al. 1993, 1997a,b; Goldmann et al. 1994; Westaway et al. 1994b; Belt et al. 1995; Clousard et al. 1995; Ikeda et al. 1995; O'Rourke et al. 1997); presumably, this was the genetic basis of Parry's scrapie eradication program in Great Britain 30 years ago (Parry 1962, 1983). A more effective approach using dominant negatives for producing prion-resistant domestic animals, including sheep and cattle, is probably the expression of PrP transgenes encoding K219 and/or R171 (Fig. 11). Such an approach has been evaluated in Tg mice and shown to be effective (Perrier et al. 2002). In fact, the replacement of all sheep with Arg/Arg at codon 171 is under way in the Netherlands. Such an approach can be instituted by artificial insemination using sperm from males homozygous for Arg/Arg at residue 171. A less practical approach is the production of PrP-deficient cattle and sheep. Although such animals would not be susceptible to prion disease (Büeler et al. 1993; Prusiner et al. 1993a), they might suffer some deleterious effects from the ablation of the PrP gene (Collinge et al. 1994; Lledo et al. 1996; Sakaguchi et al. 1996; Tobler et al. 1996).

SOME PRINCIPLES OF PRION BIOLOGY

Many principles of prion replication are clearly unprecedented in biology. As such, it is not surprising that some of these principles have been slow to be embraced. Although prion replication resembles viral replication superficially, the underlying principles are quite different. For example, in prion replication, the substrate is a host-encoded protein, PrP^C, which undergoes modification to form PrP^{Sc}, the only known component of the infectious prion particle. In contrast, viruses carry a DNA or RNA genome that is copied and directs the synthesis of most, if not all, of the viral proteins. The mature virion consists of a nucleic acid genome surrounded by a protein coat, whereas a prion appears to be composed of a dimer of PrP^{Sc}.

When viruses pass from one species to another, they often replicate without any structural modification, whereas prions undergo a profound change. The prion adopts a new PrP sequence, which is encoded by the PrP gene of the current host. That change in amino acid sequence can result in a restriction of transmission for some species, while making the new prion permissive to others. In viruses, different properties exhibited by distinct strains are encoded in the viral genome, whereas in prions, strain-specific properties are enciphered in the conformation of PrP^{Sc}.

Implications for Common Neurodegenerative Diseases

Understanding how PrP^{C} unfolds and refolds into PrP^{Sc} may also open new approaches to deciphering the causes of, and to developing effective therapies for, some common neurodegenerative diseases, including Alzheimer's disease, Parkinson's disease, and amyotrophic lateral sclerosis (ALS) (Prusiner 2001). Whether or not therapies designed to prevent the conversion of PrP^{C} into PrP^{Sc} will be effective in these more common neurodegenerative diseases is unknown. Alternatively, developing a therapy for the prion diseases might provide a blueprint for designing somewhat different drugs for these common disorders. Like the inherited prion diseases, subsets of Alzheimer's disease and ALS are caused by mutations that result in nonconservative amino acid substitutions in proteins expressed in the CNS.

As knowledge about prions continues to expand, our understanding of how prions replicate and cause disease will undoubtedly evolve. It is important to add that many of the basic principles of prion biology, as set forth in succeeding chapters, are becoming increasingly well understood.

REFERENCES

Aguzzi A. and Weissmann C. 1997. Prion research: The next frontiers. *Nature* **389:** 795–798.

Aigle M. and Lacroute F. 1975. Genetical aspects of [URE3], a non-Mendelian, cytoplasmically inherited mutation in yeast. *Mol. Gen. Genet.* **136:** 327–335.

Alper T., Haig D.A., and Clarke M.C. 1966. The exceptionally small size of the scrapie agent. *Biochem. Biophys. Res. Commun.* **22:** 278–284.

Alper T., Cramp W.A., Haig D.A., and Clarke M.C. 1967. Does the agent of scrapie replicate without nucleic acid? *Nature* **214:** 764–766.

Alpers M.P. 1968. Kuru: Implications of its transmissibility for the interpretation of its changing epidemiological pattern. In *The central nervous system: Some experimental models of neurological diseases* (ed. O.T. Bailey and D.E. Smith), pp. 234–251. Williams and Wilkins, Baltimore, Maryland.

——— 1987. Epidemiology and clinical aspects of kuru. In *Prions—Novel infectious pathogens causing scrapie and Creutzfeldt-Jakob disease* (ed. S.B. Prusiner and M.P. McKinley), pp. 451–465. Academic Press, Orlando, Florida.

Anderson R.M., Donnelly C.A., Ferguson N.M., Woolhouse M.E.J., Watt C.J., Udy H.J., MaWhinney S., Dunstan S.P., Southwood T.R.E., Wilesmith J.W., Ryan J.B.M., Hoinville L.J., Hillerton J.E., Austin A.R., and Wells G.A.H. 1996. Transmission dynamics and epidemiology of BSE in British cattle. *Nature* **382:** 779–788.

Anfinsen C.B. 1973. Principles that govern the folding of protein chains. *Science* **181:** 223–230.

Antzutkin O.N., Balbach J.J., Leapman R.D., Rizzo N.W., Reed J., and Tycko R. 2000. Multiple quantum solid-state NMR indicates a parallel, not antiparallel, organization of β-sheets in Alzheimer's β-amyloid fibrils. *Proc. Natl. Acad. Sci.* **97:** 13045–13050.

Baker H.F., Ridley R.M., and Wells G.A.H. 1993. Experimental transmission of BSE and scrapie to the common marmoset. *Vet. Rec.* **132:** 403–406.

Barry R.A. and Prusiner S.B. 1986. Monoclonal antibodies to the cellular and scrapie prion proteins. *J. Infect. Dis.* **154:** 518–521.

Baskakov I.V., Legname G., Prusiner S.B., and Cohen F.E. 2001. Folding of prion protein to its native α-helical conformation is under kinetic control. *J. Biol. Chem.* **276:** 19687–19690.

Baskakov I.V., Legname G., Baldwin M.A., Prusiner S.B., and Cohen F.E. 2002. Pathway complexity of prion protein assembly into amyloid. *J. Biol. Chem.* **277:** 21140–21148.

Basler K., Oesch B., Scott M., Westaway D., Wälchli M., Groth D.F., McKinley M.P., Prusiner S.B., and Weissmann C. 1986. Scrapie and cellular PrP isoforms are encoded by the same chromosomal gene. *Cell* **46:** 417–428.

Bateman D., Hilton D., Love S., Zeidler M., Beck J., and Collinge J. 1995. Sporadic Creutzfeldt-Jakob disease in a 18-year-old in the UK (letter). *Lancet* **346:** 1155–1156.

Beck E., Daniel P.M., and Parry H.B. 1964. Degeneration of the cerebellar and hypothalamo-neurohypophysial systems in sheep with scrapie; and its relationship to human system degenerations. *Brain* **87:** 153–176.

Behrens A., Genoud N., Naumann H., Rülicke T., Janett F., Heppner F.L., Ledermann B., and Aguzzi A. 2002. Absence of the prion protein homologue Doppel causes male sterility. *EMBO J.* **21:** 3652–3658.

Bellinger-Kawahara C.G., Kempner E., Groth D.F., Gabizon R., and Prusiner S.B. 1988. Scrapie prion liposomes and rods exhibit target sizes of 55,000 Da. *Virology* **164:** 537–541.

Belt P.B.G.M., Muileman I.H., Schreuder B.E.C., Ruijter J.B., Gielkens A.L.J., and Smits M.A. 1995. Identification of five allelic variants of the sheep PrP gene and their association with natural scrapie. *J. Gen. Virol.* **76:** 509–517.

Benzinger T.L.S., Gregory D.M., Burkoth T.S., Miller-Auer H., Lynn D.G., Botto R.E., and Meredith S.C. 1998. Propagating structure of Alzheimer's β-amyloid$_{(10-35)}$ is parallel β-sheet with residues in exact register. *Proc. Natl. Acad. Sci.* **95:** 13407–13412.

Bertoni J.M., Brown P., Goldfarb L., Gajdusek D., and Omaha N.E. 1992. Familial Creutzfeldt-Jakob disease with the *PRNP* codon 200^{lys} mutation and supranuclear palsy but without myoclonus or periodic EEG complexes. *Neurology* (suppl. 3) **42:** 350. (Abstr.)

Bessen R.A. and Marsh R.F. 1992. Identification of two biologically distinct strains of transmissible mink encephalopathy in hamsters. *J. Gen. Virol.* **73:** 329–334.

———. 1994. Distinct PrP properties suggest the molecular basis of strain variation in transmissible mink encephalopathy. *J. Virol.* **68:** 7859–7868.

Billette de Villemeur T., Deslys J.-P., Pradel A., Soubrié C., Alpérovitch A., Tardieu M., Chaussain J.-L., Hauw J.-J., Dormont D., Ruberg M., and Agid Y. 1996. Creutzfeldt-Jakob disease from contaminated growth hormone extracts in France. *Neurology* **47:** 690–695.

Bolton D.C., McKinley M.P., and Prusiner S.B. 1982. Identification of a protein that purifies with the scrapie prion. *Science* **218:** 1309–1311.

Borchelt D.R., Scott M., Taraboulos A., Stahl N., and Prusiner S.B. 1990. Scrapie and cellular prion proteins differ in their kinetics of synthesis and topology in cultured cells. *J. Cell Biol.* **110:** 743–752.

Bousset L., Belrhali H., Melki R., and Morera S. 2001a. Crystal structures of the yeast prion Ure2p functional region in complex with glutathione and related compounds. *Biochemistry* **40:** 13564–13573.

Bousset L., Belrhali H., Janin J., Melki R., and Morera S. 2001b. Structure of the globular region of the prion protein Ure2 from the yeast *Saccharomyces cerevisiae. Structure* **9:** 39–46.

Britton T.C., Al-Sarraj S., Shaw C., Campbell T., and Collinge J. 1995. Sporadic Creutzfeldt-Jakob disease in a 16-year-old in the UK (letter). *Lancet* **346:** 1155.

Broach J.R., Friedman L.R., and Sherman F. 1981. Correspondence of yeast UAA suppressors to cloned tRNA$^{ser}_{UCA}$. *J. Mol. Biol.* **150:** 375–387.

Brown D.R., Qin K., Herms J.W., Madlung A., Manson J., Strome R., Fraser P.E., Kruck T., von Bohlen A., Schulz-Schaeffer W., Giese A., Westaway D., and Kretzschmar H. 1997. The cellular prion protein binds copper *in vivo. Nature* **390:** 684–687.

Brown P., Cathala F., Raubertas R.F., Gajdusek D.C., and Castaigne P. 1987. The epidemiology of Creutzfeldt-Jakob disease: Conclusion of a 15-year investigation in France and review of the world literature. *Neurology* **37:** 895–904.

Bruce M.E. 1996. Strain typing studies of scrapie and BSE. In *Methods in molecular medicine: Prion diseases* (ed. H.F. Baker and R.M. Ridley), pp. 223–236. Humana Press, Totowa, New Jersey.

Bruce M.E. and Dickinson A.G. 1987. Biological evidence that the scrapie agent has an independent genome. *J. Gen. Virol.* **68:** 79–89.

Bruce M.E., McConnell I., Fraser H., and Dickinson A.G. 1991. The disease characteristics of different strains of scrapie in *Sinc* congenic mouse lines. Implications for the nature of the agent and host control of pathogenesis. *J. Gen. Virol.* **72:** 595–603.

Bruce M., Chree A., McConnell I., Foster J., and Fraser H. 1993. Transmissions of BSE, scrapie and related diseases to mice. In *Abstracts from the Proceedings of the 9th International Congress of Virology*, Glasgow, Scotland, p. 93.

Bruce M., Chree A., McConnell I., Foster J., Pearson G., and Fraser H. 1994. Transmission of bovine spongiform encephalopathy and scrapie to mice: Strain variation and the species barrier. *Philos. Trans. R. Soc. Lond. B* **343:** 405–411.

Bruce M.E., Will R.G., Ironside J.W., McConnell I., Drummond D., Suttie A., McCardle L., Chree A., Hope J., Birkett C., Cousens S., Fraser H., and Bostock C.J. 1997. Transmissions to mice indicate that "new variant" CJD is caused by the BSE agent. *Nature* **389:** 498–501.

Büeler H., Raeber A., Sailer A., Fischer M., Aguzzi A., and Weissmann C. 1994. High prion and PrP^{Sc} levels but delayed onset of disease in scrapie-inoculated mice heterozygous for a disrupted PrP gene. *Mol. Med.* **1:** 19–30.

Büeler H., Aguzzi A., Sailer A., Greiner R.-A., Autenried P., Aguet M., and Weissmann C. 1993. Mice devoid of PrP are resistant to scrapie. *Cell* **73:** 1339–1347.

Büeler H., Fisher M., Lang Y., Bluethmann H., Lipp H.-P., DeArmond S.J., Prusiner S.B., Aguet M., and Weissmann C. 1992. Normal development and behaviour of mice lacking the neuronal cell-surface PrP protein. *Nature* **356:** 577–582.

Buschmann A., Pfaff E., Reifenberg K., Müller H.M., and Groschup M.H. 2000. Detection of cattle-derived BSE prions using transgenic mice overexpressing bovine PrP^C. *Arch. Virol.* (suppl.) **16:** 75–86.

Calzolai L., Lysek D.A., Güntert P., von Schroetter C., Riek R., Zahn R., and Wüthrich K.

2000. NMR structures of three single-residue variants of the human prion protein. *Proc. Natl. Acad. Sci.* **97:** 8340–8345.

Carlson G.A., Westaway D., DeArmond S.J., Peterson-Torchia M., and Prusiner S.B. 1989. Primary structure of prion protein may modify scrapie isolate properties. *Proc. Natl. Acad. Sci.* **86:** 7475–7479.

Carlson G.A., Kingsbury D.T., Goodman P.A., Coleman S., Marshall S.T., DeArmond S., Westaway D., and Prusiner S.B. 1986. Linkage of prion protein and scrapie incubation time genes. *Cell* **46:** 503–511.

Carlson G.A., Ebeling C., Yang S.-L., Telling G., Torchia M., Groth D., Westaway D., DeArmond S.J., and Prusiner S.B. 1994. Prion isolate specified allotypic interactions between the cellular and scrapie prion proteins in congenic and transgenic mice. *Proc. Natl. Acad. Sci.* **91:** 5690–5694.

Carp R.I., Meeker H., and Sersen E. 1997. Scrapie strains retain their distinctive characteristics following passages of homogenates from different brain regions and spleen. *J. Gen. Virol.* **78:** 283–290.

Caughey B. and Chesebro B. 1997. Prion protein and the transmissible spongiform encephalopathies. *Trends Cell Biol.* **7:** 56–62.

Caughey B. and Race R.E. 1992. Potent inhibition of scrapie-associated PrP accumulation by Congo red. *J. Neurochem.* **59:** 768–771.

Caughey B. and Raymond G.J. 1991. The scrapie-associated form of PrP is made from a cell surface precursor that is both protease- and phospholipase-sensitive. *J. Biol. Chem.* **266:** 18217–18223.

Caughey B., Kocisko D.A., Raymond G.J., and Lansbury P.T., Jr. 1995. Aggregates of scrapie-associated prion protein induce the cell-free conversion of protease-sensitive prion protein to the protease-resistant state. *Chem. Biol.* **2:** 807–817.

Caughey B., Raymond G.J., Ernst D., and Race R.E. 1991a. N-terminal truncation of the scrapie-associated form of PrP by lysosomal protease(s): Implications regarding the site of conversion of PrP to the protease-resistant state. *J. Virol.* **65:** 6597–6603.

Caughey B.W., Dong A., Bhat K.S., Ernst D., Hayes S.F., and Caughey W.S. 1991b. Secondary structure analysis of the scrapie-associated protein PrP 27-30 in water by infrared spectroscopy. *Biochemistry* **30:** 7672–7680.

Caughey B., Neary K., Butler R., Ernst D., Perry L., Chesebro B., and Race R.E. 1990. Normal and scrapie-associated forms of prion protein differ in their sensitivities to phospholipase and proteases in intact neuroblastoma cells. *J. Virol.* **64:** 1093–1101.

Centers for Disease Control (CDC). 1996. Surveillance for Creutzfeldt-Jakob disease—United States. *Morb. Mortal. Wkly. Rep.* **45:** 665–668.

Cervenáková L., Brown P., Piccardo P., Cummings J.L., Nagle J., Vinters H.V., Kaur P., Ghetti B., Chapman J., Gajdusek D.C., and Goldfarb L.G. 1996. 24-nucleotide deletion in the *PRNP* gene: Analysis of associated phenotypes. In *Transmissible subacute spongiform encephalopathies: Prion diseases* (ed. L. Court and B. Dodet), pp. 433–444. Elsevier, Paris, France.

Chandler R.L. 1961. Encephalopathy in mice produced by inoculation with scrapie brain material. *Lancet* **1:** 1378–1379.

Chapman J., Ben-Israel J., Goldhammer Y., and Korczyn A.D. 1994. The risk of developing Creutzfeldt-Jakob disease in subjects with the *PRNP* gene codon 200 point mutation. *Neurology* **44:** 1683–1686.

Chazot G., Broussolle E., Lapras C.I., Blättler T., Aguzzi A., and Kopp N. 1996. New variant of Creutzfeldt-Jakob disease in a 26-year-old French man. *Lancet* **347:** 1181.

Chernoff Y.O., Derkach I.L., and Inge-Vechtomov S.G. 1993. Multicopy *SUP35* gene induces *de novo* appearance of *psi*-like factors in the yeast *Saccharomyces cerevisiae*. *Curr. Genet.* **24:** 268–270.

Chernoff Y.O., Lindquist S.L., Ono B., Inge-Vechtomov S.G., and Liebman S.W. 1995. Role of the chaperone protein Hsp104 in propagation of the yeast prion-like factor [*psi⁺*]. *Science* **268:** 880–884.

Chesebro B. 1998. BSE and prions: Uncertainties about the agent. *Science* **279:** 42–43.

Chesebro B. and Caughey B. 1993. Scrapie agent replication without the prion protein? *Curr. Biol.* **3:** 696–698.

Chesebro B., Race R., Wehrly K., Nishio J., Bloom M., Lechner D., Bergstrom S., Robbins K., Mayer L., Keith J.M., Garon C., and Haase A. 1985. Identification of scrapie prion protein-specific mRNA in scrapie-infected and uninfected brain. *Nature* **315:** 331–333.

Clousard C., Beaudry P., Elsen J.M., Milan D., Dussaucy M., Bounneau C., Schelcher F., Chatelain J., Launay J.-M., and Laplanche J.-L. 1995. Different allelic effects of the codons 136 and 171 of the prion protein gene in sheep with natural scrapie. *J. Gen. Virol.* **76:** 2097–2101.

Cohen F.E., Pan K.-M., Huang Z., Baldwin M., Fletterick R.J., and Prusiner S.B. 1994. Structural clues to prion replication. *Science* **264:** 530–531.

Collinge J. and Palmer M.S. 1997. Human prion diseases. In *Prion diseases* (ed. J. Collinge and M.S. Palmer), pp. 18–56. Oxford University Press, Oxford, United Kingdom.

Collinge J., Palmer M.S., and Dryden A.J. 1991. Genetic predisposition to iatrogenic Creutzfeldt-Jakob disease. *Lancet* **337:** 1441–1442.

Collinge J., Sidle K.C.L., Meads J., Ironside J., and Hill A.F. 1996. Molecular analysis of prion strain variation and the aetiology of "new variant" CJD. *Nature* **383:** 685–690.

Collinge J., Whittington M.A., Sidle K.C., Smith C.J., Palmer M.S., Clarke A.R., and Jefferys J.G.R. 1994. Prion protein is necessary for normal synaptic function. *Nature* **370:** 295–297.

Collinge J., Palmer M.S., Sidle K.C., Hill A.F., Gowland I., Meads J., Asante E., Bradley R., Doey L.J., and Lantos P.L. 1995. Unaltered susceptibility to BSE in transgenic mice expressing human prion protein. *Nature* **378:** 779–783.

Cousens S.N., Vynnycky E., Zeidler M., Will R.G., and Smith P.G. 1997. Predicting the CJD epidemic in humans. *Nature* **385:** 197–198.

Cousens S.N., Harries-Jones R., Knight R., Will R.G., Smith P.G., and Matthews W.B. 1990. Geographical distribution of cases of Creutzfeldt-Jakob disease in England and Wales 1970–84. *J. Neurol. Neurosurg. Psychiatry* **53:** 459–465.

Coustou V., Deleu C., Saupe S., and Begueret J. 1997. The protein product of the *het-s* heterokaryon incompatibility gene of the fungus *Podospora anserina* behaves as a prion analog. *Proc. Natl. Acad. Sci.* **94:** 9773–9778.

Cox B.S. 1965. PSI, a cytoplasmic suppressor of super-suppressor in yeast. *Heredity* **20:** 505–521.

Creutzfeldt H.G. 1920. Über eine eigenartige herdförmige Erkrankung des Zentralnervensystems. *Z. Gesamte Neurol. Psychiatrie* **57:** 1–18.

Czub M., Braig H.R., and Diringer H. 1986. Pathogenesis of scrapie: Study of the temporal development of clinical symptoms of infectivity titres and scrapie-associated fibrils in brains of hamsters infected intraperitoneally. *J. Gen. Virol.* **67:** 2005–2009.

Daude N., Lehmann S., and Harris D.A. 1997. Identification of intermediate steps in the conversion of a mutant prion protein to a scrapie-like form in cultured cells. *J. Biol. Chem.* **272:** 11604–11612.

Davison C. and Rabiner A.M. 1940. Spastic pseudosclerosis (disseminated encephalomyelopathy; corticopal-lidospinal degeneration). Familial and nonfamilial incidence (a clinico-pathologic study). *Arch. Neurol. Psychiatry* **44:** 578–598.

Dawson M., Wells G.A.H., and Parker B.N.J. 1990a. Preliminary evidence of the experimental transmissibility of bovine spongiform encephalopathy to cattle. *Vet. Rec.* **126:** 112–113.

Dawson M., Wells G.A.H., Parker B.N.J., and Scott A.C. 1990b. Primary parenteral transmission of bovine spongiform encephalopathy to the pig. *Vet. Rec.* **127:** 338.

DeArmond S.J. and Prusiner S.B. 1997. Prion diseases. In *Greenfield's neuropathology*, 6th edition (ed. P. Lantos and D. Graham), pp. 235–280. Edward Arnold, London, United Kingdom.

DeArmond S.J., Mobley W.C., DeMott D.L., Barry R.A., Beckstead J.H., and Prusiner S.B. 1987. Changes in the localization of brain prion proteins during scrapie infection. *Neurology* **37:** 1271–1280.

DeArmond S.J., Yang S.-L., Lee A., Bowler R., Taraboulos A., Groth D., and Prusiner S.B. 1993. Three scrapie prion isolates exhibit different accumulation patterns of the prion protein scrapie isoform. *Proc. Natl. Acad. Sci.* **90:** 6449–6453.

DeArmond S.J., Sánchez H., Yehiely F., Qiu Y., Ninchak-Casey A., Daggett V., Camerino A.P., Cayetano J., Rogers M., Groth D., Torchia M., Tremblay P., Scott M.R., Cohen F.E., and Prusiner S.B. 1997. Selective neuronal targeting in prion disease. *Neuron* **19:** 1337–1348.

Deleu C., Clavé C., and Bégueret J. 1993. A single amino acid difference is sufficient to elicit vegetative incompatibility in the fungus *Podospora anserina*. *Genetics* **135:** 45–52.

Derkatch I.L., Chernoff Y.O., Kushnirov V.V., Inge-Vechtomov S.G., and Liebman S.W. 1996. Genesis and variability of [*PSI*] prion factors in *Saccharomyces cerevisiae*. *Genetics* **144:** 1375–1386.

Dickinson A.G. and Outram G.W. 1988. Genetic aspects of unconventional virus infections: The basis of the virino hypothesis. *Ciba Found. Symp.* **135:** 63–83.

Dickinson A.G., Fraser H., and Outram G.W. 1975. Scrapie incubation time can exceed natural lifespan. *Nature* **256:** 732–733.

Dickinson A.G., Meikle V.M.H., and Fraser H. 1968. Identification of a gene which controls the incubation period of some strains of scrapie agent in mice. *J. Comp. Pathol.* **78:** 293–299.

———. 1969. Genetical control of the concentration of ME7 scrapie agent in the brain of mice. *J. Comp. Pathol.* **79:** 15–22.

Dickinson A.G., Young G.B., Stamp J.T., and Renwick C.C. 1965. An analysis of natural scrapie in Suffolk sheep. *Heredity* **20:** 485–503.

Diringer H. 1991. Transmissible spongiform encephalopathies (TSE) virus-induced amyloidoses of the central nervous system (CNS). *Eur. J. Epidemiol.* **7:** 562–566.

Diringer H. and Kimberlin R.H. 1983. Infectious scrapie agent is apparently not as small as recent claims suggest. *Biosci. Rep.* **3:** 563–568.

Dlouhy S.R., Hsiao K., Farlow M.R., Foroud T., Conneally P.M., Johnson P., Prusiner S.B., Hodes M.E., and Ghetti B. 1992. Linkage of the Indiana kindred of Gerstmann-Sträussler-Scheinker disease to the prion protein gene. *Nat. Genet.* **1:** 64–67.

Doel S.M., McCready S.J., Nierras C.R., and Cox B.S. 1994. The dominant *PNM2⁻* mutation which eliminates the ψ factor of *Saccharomyces cerevisiae* is the result of a missense mutation in the *SUP35* gene. *Genetics* **137:** 659–670.

Doh-ura K., Iwaki T., and Caughey B. 2000. Lysosomotropic agents and cysteine protease inhibitors inhibit scrapie-associated prion protein accumulation. *J. Virol.* **74:** 4894–4897.

Doh-ura K., Kitamoto T., Sakaki Y., and Tateishi J. 1991. CJD discrepancy. *Nature* **353:** 801–802.

Doh-ura K., Tateishi J., Sasaki H., Kitamoto T., and Sakaki Y. 1989. Pro→Leu change at position 102 of prion protein is the most common but not the sole mutation related to Gerstmann-Sträussler syndrome. *Biochem. Biophys. Res. Commun.* **163:** 974–979.

Donne D.G., Viles J.H., Groth D., Mehlhorn I., James T.L., Cohen F.E., Prusiner S.B., Wright P.E., and Dyson H.J. 1997. Structure of the recombinant full-length hamster prion protein PrP(29–231): The N terminus is highly flexible. *Proc. Natl. Acad. Sci.* **94:** 13452–13457.

Edenhofer F., Rieger R., Famulok M., Wendler W., Weiss S., and Winnacker E.-L. 1996. Prion protein PrP^C interacts with molecular chaperones of the Hsp60 family. *J. Virol.* **70:** 4724–4728.

Ehlers B. and Diringer H. 1984. Dextran sulphate 500 delays and prevents mouse scrapie by impairment of agent replication in spleen. *J. Gen. Virol.* **65:** 1325–1330.

Eklund C.M., Kennedy R.C., and Hadlow W.J. 1967. Pathogenesis of scrapie virus infection in the mouse. *J. Infect. Dis.* **117:** 15–22.

Enari M., Flechsig E., and Weissmann C. 2001. Scrapie prion protein accumulation by scrapie-infected neuroblastoma cells abrogated by exposure to a prion protein antibody. *Proc. Natl. Acad. Sci.* **98:** 9295–9299.

Endo T., Groth D., Prusiner S.B., and Kobata A. 1989. Diversity of oligosaccharide structures linked to asparagines of the scrapie prion protein. *Biochemistry* **28:** 8380–8388.

Field E.J., Farmer F., Caspary E.A., and Joyce G. 1969. Susceptibility of scrapie agent to ionizing radiation. *Nature* **222:** 90–91.

Fink J.K., Peacock M.L., Warren J.T., Roses A.D., and Prusiner S.B. 1994. Detecting prion protein gene mutations by denaturing gradient gel electrophoresis. *Hum. Mutat.* **4:** 42–50.

Fischer M., Rülicke T., Raeber A., Sailer A., Moser M., Oesch B., Brandner S., Aguzzi A., and Weissmann C. 1996. Prion protein (PrP) with amino-proximal deletions restoring susceptibility of PrP knockout mice to scrapie. *EMBO J.* **15:** 1255–1264.

Fraser H. and Dickinson A.G. 1973. Scrapie in mice. Agent-strain differences in the distribution and intensity of grey matter vacuolation. *J. Comp. Pathol.* **83:** 29–40.

———. 1985. Targeting of scrapie lesions and spread of agent via the retino-tectal projection. *Brain Res.* **346:** 32–41.

Fraser H., McConnell I., Wells G.A.H., and Dawson M. 1988. Transmission of bovine spongiform encephalopathy to mice. *Vet. Rec.* **123:** 472.

Fraser H., Bruce M.E., Chree A., McConnell I., and Wells G.A.H. 1992. Transmission of bovine spongiform encephalopathy and scrapie to mice. *J. Gen. Virol.* **73:** 1891–1897.

Friede R.L. and DeJong R.N. 1964. Neuronal enzymatic failure in Creutzfeldt-Jakob disease. A familial study. *Arch. Neurol.* **10:** 181–195.

Furukawa H., Kitamoto T., Tanaka Y., and Tateishi J. 1995. New variant prion protein in a Japanese family with Gerstmann-Sträussler syndrome. *Mol. Brain Res.* **30:** 385–388.

Gabizon R., McKinley M.P., Groth D., and Prusiner S.B. 1988. Immunoaffinity purification and neutralization of scrapie prion infectivity. *Proc. Natl. Acad. Sci.* **85:** 6617–6621.

Gabizon R., Telling G., Meiner Z., Halimi M., Kahana I., and Prusiner S.B. 1996. Insoluble wild-type and protease-resistant mutant prion protein in brains of patients with inherited prion disease. *Nat. Med.* **2:** 59–64.

Gabizon R., Rosenmann H., Meiner Z., Kahana I., Kahana E., Shugart Y., Ott J., and Prusiner S.B. 1993. Mutation and polymorphism of the prion protein gene in Libyan Jews with Creutzfeldt-Jakob disease (CJD). *Am. J. Hum. Genet.* **53:** 828–835.

Gabriel J.-M., Oesch B., Kretzschmar H., Scott M., and Prusiner S.B. 1992. Molecular cloning of a candidate chicken prion protein. *Proc. Natl. Acad. Sci.* **89:** 9097–9101.

Gajdusek D.C. 1977. Unconventional viruses and the origin and disappearance of kuru. *Science* **197:** 943–960.

———. 1988. Transmissible and non-transmissible amyloidoses: Autocatalytic post-translational conversion of host precursor proteins to β-pleated sheet configurations. *J. Neuroimmunol.* **20:** 95–110.

Gajdusek D.C., Gibbs C.J., Jr., and Alpers M. 1966. Experimental transmission of a kuru-like syndrome to chimpanzees. *Nature* **209:** 794–796.

Gajdusek D.C., Gibbs C.J., Jr., Asher D.M., Brown P., Diwan A., Hoffman P., Nemo G., Rohwer R., and White L. 1977. Precautions in medical care of, and in handling materials from, patients with transmissible virus dementia (Creutzfeldt-Jakob disease). *N. Engl. J. Med.* **297:** 1253–1258.

Gambetti P. and Parchi P. 1999. Insomnia in prion diseases: Sporadic and familial. *N. Engl. J. Med.* **340:** 1675–1677.

Gambetti P., Parchi P., Petersen R.B., Chen S.G., and Lugaresi E. 1995. Fatal familial insomnia and familial Creutzfeldt-Jakob disease: Clinical, pathological and molecular features. *Brain Pathol.* **5:** 43–51.

García F.L., Zahn R., Riek R., and Wüthrich K. 2000. NMR structure of the bovine prion protein. *Proc. Natl. Acad. Sci.* **97:** 8334–8339.

Gasset M., Baldwin M.A., Fletterick R.J., and Prusiner S.B. 1993. Perturbation of the secondary structure of the scrapie prion protein under conditions that alter infectivity. *Proc. Natl. Acad. Sci.* **90:** 1–5.

Gauczynski S., Peyrin J.M., Haik S., Leucht C., Hundt C., Rieger R., Krasemann S., Deslys J.P., Dormont D., Lasmezas C.I., and Weiss S. 2001. The 37-kDa/67-kDa laminin receptor acts as the cell-surface receptor for the cellular prion protein. *EMBO J.* **20:** 5863–5875.

Ghani A.C., Ferguson N.M., Donnelly C.A., and Anderson R.M. 2000. Predicted vCJD mortality in Great Britain. *Nature* **406:** 583–584.

Gibbons R.A. and Hunter G.D. 1967. Nature of the scrapie agent. *Nature* **215:** 1041–1043.

Gibbs C.J., Jr. and Gajdusek D.C. 1978. Atypical viruses as the cause of sporadic, epidemic, and familial chronic diseases in man: Slow viruses and human diseases. In *Perspectives in virology* (ed. M. Pollard), pp. 161–198. Raven Press, New York.

Gibbs C.J., Jr., Gajdusek D.C., and Latarjet R. 1978. Unusual resistance to ionizing radiation of the viruses of kuru, Creutzfeldt-Jakob disease. *Proc. Natl. Acad. Sci.* **75:** 6268–6270.

Gibbs C.J., Jr., Gajdusek D.C., Asher D.M., Alpers M.P., Beck E., Daniel P.M., and Matthews W.B. 1968. Creutzfeldt-Jakob disease (spongiform encephalopathy): Transmission to the chimpanzee. *Science* **161:** 388–389.

Glasse R.M. 1967. Cannibalism in the kuru region of New Guinea. *Trans. N.Y. Acad. Sci. (Ser. 2)* **29:** 748–754.

Glover J.R., Kowal A.S., Schirmer E.C., Patino M.M., Liu J.-J., and Lindquist S. 1997. Self-seeded fibers formed by Sup35, the protein determinant of [*PSI⁺*], a heritable prion-like factor of *S. cerevisiae*. *Cell* **89:** 811–819.

Goldfarb L.G., Mitrova E., Brown P., Toh B.H., and Gajdusek D.C. 1990a. Mutation in codon 200 of scrapie amyloid protein gene in two clusters of Creutzfeldt-Jakob disease in Slovakia. *Lancet* **336:** 514–515.

Goldfarb L., Brown P., Goldgaber D., Garruto R., Yanaghiara R., Asher D., and Gajdusek D.C. 1990b. Identical mutation in unrelated patients with Creutzfeldt-Jakob disease.

Lancet **336:** 174–175.

Goldfarb L.G., Haltia M., Brown P., Nieto A., Kovanen J., McCombie W.R., Trapp S., and Gajdusek D.C. 1991a. New mutation in scrapie amyloid precursor gene (at codon 178) in Finnish Creutzfeldt-Jakob kindred. *Lancet* **337:** 425.

Goldfarb L.G., Brown P., Goldgaber D., Asher D.M., Rubenstein R., Brown W.T., Piccardo P., Kascsak R.J., Boellaard J.W., and Gajdusek D.C. 1990c. Creutzfeldt-Jakob disease and kuru patients lack a mutation consistently found in the Gerstmann-Sträussler-Scheinker syndrome. *Exp. Neurol.* **108:** 247–250.

Goldfarb L.G., Brown P., McCombie W.R., Goldgaber D., Swergold G.D., Wills P.R., Cervenakova L., Baron H., Gibbs C.J.J., and Gajdusek D.C. 1991b. Transmissible familial Creutzfeldt-Jakob disease associated with five, seven, and eight extra octapeptide coding repeats in the *PRNP* gene. *Proc. Natl. Acad. Sci.* **88:** 10926–10930.

Goldfarb L.G., Petersen R.B., Tabaton M., Brown P., LeBlanc A.C., Montagna P., Cortelli P., Julien J., Vital C., and Pendelbury W.W. 1992. Fatal familial insomnia and familial Creutzfeldt-Jakob disease: Disease phenotype determined by a DNA polymorphism. *Science* **258:** 806–808.

Goldfarb L.G., Brown P., Mitrova E., Cervenakova L., Goldin L., Korczyn A.D., Chapman J., Galvez S., Cartier L., Rubenstein R., and Gajdusek D.C. 1991c. Creutzfeldt-Jacob disease associated with the *PRNP* codon 200^{Lys} mutation: An analysis of 45 families. *Eur. J. Epidemiol.* **7:** 477–486.

Goldgaber D., Goldfarb L.G., Brown P., Asher D.M., Brown W.T., Lin S., Teener J.W., Feinstone S.M., Rubenstein R., Kascsak R.J., Boellaard J.W., and Gajdusek D.C. 1989. Mutations in familial Creutzfeldt-Jakob disease and Gerstmann-Sträussler-Scheinker's syndrome. *Exp. Neurol.* **106:** 204–206.

Goldmann W., Hunter N., Manson J., and Hope J. 1990a. The PrP gene of the sheep, a nat ural host of scrapie. *Proceedings of the 8th International Congress of Virology*, Berlin, p. 284.

Goldmann W., Hunter N., Benson G., Foster J.D., and Hope J. 1991a. Different scrapie-associated fibril proteins (PrP) are encoded by lines of sheep selected for different alleles of the *Sip* gene. *J. Gen. Virol.* **72:** 2411–2417.

Goldmann W., Hunter N., Martin T., Dawson M., and Hope J. 1991b. Different forms of the bovine PrP gene have five or six copies of a short, G-C-rich element within the protein-coding exon. *J. Gen. Virol.* **72:** 201–204.

Goldmann W., Hunter N., Smith G., Foster J., and Hope J. 1994. PrP genotype and agent effects in scrapie: Change in allelic interaction with different isolates of agent in sheep, a natural host of scrapie. *J. Gen. Virol.* **75:** 989–995.

Goldmann W., Hunter N., Foster J.D., Salbaum J.M., Beyreuther K., and Hope J. 1990b. Two alleles of a neural protein gene linked to scrapie in sheep. *Proc. Natl. Acad. Sci.* **87:** 2476–2480.

Goodman L.S. and Gilman A., Eds. 1970. *The pharmacological basis of therapeutics: A textbook of pharmacology, toxicology, and therapeutics for physicians and medical students*, 4th edition. Macmillan, New York.

Gordon W.S. 1946. Advances in veterinary research. *Vet. Res.* **58:** 516–520.

Gorodinsky A. and Harris D.A. 1995. Glycolipid-anchored proteins in neuroblastoma cells form detergent-resistant complexes without caveolin. *J. Cell Biol.* **129:** 619–627.

Graner E., Mercadante A.F., Zanata S.M., Forlenza O.V., Cabral A.L.B., Veiga S.S., Juliano M.A., Roesler R., Walz R., Mineti A., Izquierdo I., Martins V.R., and Brentani R.R. 2000. Cellular prion protein binds laminin and mediates neuritogenesis. *Mol. Brain Res.* **76:** 85–92.

Grathwohl K.-U.D., Horiuchi M., Ishiguro N., and Shinagawa M. 1997. Sensitive enzyme-linked immunosorbent assay for detection of PrP^{Sc} in crude tissue extracts from scrapie-affected mice. *J. Virol. Methods* **64:** 205–216.

Griffith J.S. 1967. Self-replication and scrapie. *Nature* **215:** 1043–1044.

Hadlow W.J. 1959. Scrapie and kuru. *Lancet* **2:** 289–290.

———. 1995. Neuropathology and the scrapie-kuru connection. *Brain Pathol.* **5:** 27–31.

Hamir A.N., Cutlip R.C., Miller J.M., Williams E.S., Stack M.J., Miller M.W., O'Rourke K.I., and Chaplin M.J. 2001. Preliminary findings on the experimental transmission of chronic wasting disease agent of mule deer to cattle. *J. Vet. Diagn. Invest.* **13:** 91–96.

Hammick D.L. and Chambers W.E. 1945. Optical activity of excreted mepacrine. *Nature* **155:** 141.

Haraguchi T., Fisher S., Olofsson S., Endo T., Groth D., Tarantino A., Borchelt D.R., Teplow D., Hood L., Burlingame A., Lycke E., Kobata A., and Prusiner S.B. 1989. Asparagine-linked glycosylation of the scrapie and cellular prion proteins. *Arch. Biochem. Biophys.* **274:** 1–13.

Harries-Jones R., Knight R., Will R.G., Cousens S., Smith P.G., and Matthews W.B. 1988. Creutzfeldt-Jakob disease in England and Wales, 1980–1984: A case-control study of potential risk factors. *J. Neurol. Neurosurg. Psychiatry* **51:** 1113–1119.

Harris D.A., Falls D.L., Walsh W., and Fischbach G.D. 1989. Molecular cloning of an acetylcholine receptor-inducing protein. *Soc. Neurosci. Abstr.* **15:** 70.7.

Hawthorne D.C. and Mortimer R.K. 1968. Genetic mapping of nonsense suppressors in yeast. *Genetics* **60:** 735–742.

Hecker R., Taraboulos A., Scott M., Pan K.-M., Torchia M., Jendroska K., DeArmond S.J., and Prusiner S.B. 1992. Replication of distinct scrapie prion isolates is region specific in brains of transgenic mice and hamsters. *Genes Dev.* **6:** 1213–1228.

Hegde R.S., Tremblay P., Groth D., Prusiner S.B., and Lingappa V.R. 1999. Transmissible and genetic prion diseases share a common pathway of neurodegeneration. *Nature* **402:** 822–826.

Hegde R.A., Mastrianni J.A., Scott M.R., DeFea K.A., Tremblay P., Torchia M., DeArmond S.J., Prusiner S.B., and Lingappa V.R. 1998. A transmembrane form of the prion protein in neurodegenerative disease. *Science* **279:** 827–834.

Herms J.W., Kretzschmar H.A., Titz S., and Keller B.U. 1995. Patch-clamp analysis of synaptic transmission to cerebellar purkinje cells of prion protein knockout mice. *Eur. J. Neurosci.* **7:** 2508–2512.

Hill A.F., Desbruslais M., Joiner S., Sidle K.C.L., Gowland I., Collinge J., Doey L.J., and Lantos P. 1997. The same prion strain causes vCJD and BSE. *Nature* **389:** 448–450.

Hope J., Reekie L.J.D., Hunter N., Multhaup G., Beyreuther K., White H., Scott A.C., Stack M.J., Dawson M., and Wells G.A.H. 1988. Fibrils from brains of cows with new cattle disease contain scrapie-associated protein. *Nature* **336:** 390–392.

Hornshaw M.P., McDermott J.R., and Candy J.M. 1995a. Copper binding to the N-terminal tandem repeat regions of mammalian and avian prion protein. *Biochem. Biophys. Res. Commun.* **207:** 621–629.

Hornshaw M.P., McDermott J.R., Candy J.M., and Lakey J.H. 1995b. Copper binding to the N-terminal tandem repeat region of mammalian and avian prion protein: Structural studies using synthetic peptides. *Biochem. Biophys. Res. Commun.* **214:** 993–999.

Hsiao K.K., Scott M., Foster D., Groth D.F., DeArmond S.J., and Prusiner S.B. 1990. Spontaneous neurodegeneration in transgenic mice with mutant prion protein. *Science* **250:** 1587–1590.

Hsiao K.K., Cass C., Schellenberg G.D., Bird T., Devine-Gage E., Wisniewski H., and Prusiner S.B. 1991a. A prion protein variant in a family with the telencephalic form of Gerstmann-Sträussler-Scheinker syndrome. *Neurology* **41:** 681–684.

Hsiao K., Baker H.F., Crow T.J., Poulter M., Owen F., Terwilliger J.D., Westaway D., Ott J., and Prusiner S.B. 1989. Linkage of a prion protein missense variant to Gerstmann-Sträussler syndrome. *Nature* **338:** 342–345.

Hsiao K.K., Groth D., Scott M., Yang S.-L., Serban H., Rapp D., Foster D., Torchia M., DeArmond S.J., and Prusiner S.B. 1994. Serial transmission in rodents of neurodegeneration from transgenic mice expressing mutant prion protein. *Proc. Natl. Acad. Sci.* **91:** 9126–9130.

Hsiao K., Meiner Z., Kahana E., Cass C., Kahana I., Avrahami D., Scarlato G., Abramsky O., Prusiner S.B., and Gabizon R. 1991b. Mutation of the prion protein in Libyan Jews with Creutzfeldt-Jakob disease. *N. Engl. J. Med.* **324:** 1091–1097.

Huang Z., Prusiner S.B., and Cohen F.E. 1995. Scrapie prions: A three-dimensional model of an infectious fragment. *Fold. Des.* **1:** 13–19.

Huang Z., Gabriel J.-M., Baldwin M.A., Fletterick R.J., Prusiner S.B., and Cohen F.E. 1994. Proposed three-dimensional structure for the cellular prion protein. *Proc. Natl. Acad. Sci.* **91:** 7139–7143.

Hundt C., Peyrin J.M., Haik S., Gauczynski S., Leucht C., Rieger R., Riley M.L., Deslys J.P., Dormont D., Lasmezas C.I., and Weiss S. 2001. Identification of interaction domains of the prion protein with its 37-kDa/67-kDa laminin receptor. *EMBO J.* **20:** 5876–5886.

Hunter G.D. 1972. Scrapie: A prototype slow infection. *J. Infect. Dis.* **125:** 427–440.

Hunter G.D. and Millson G.C. 1964. Studies on the heat stability and chromatographic behavior of the scrapie agent. *J. Gen. Microbiol.* **37:** 251–258.

———. 1967. Attempts to release the scrapie agent from tissue debris. *J. Comp. Pathol.* **77:** 301–307.

Hunter G.D., Kimberlin R.H., and Gibbons R.A. 1968. Scrapie: A modified membrane hypothesis. *J. Theor. Biol.* **20:** 355–357.

Hunter G.D., Millson G.C., and Chandler R.L. 1963. Observations on the comparative infectivity of cellular fractions derived from homogenates of mouse-scrapie brain. *Res. Vet. Sci.* **4:** 543–549.

Hunter G.D., Gibbons R.A., Kimberlin R.H., and Millson G.C. 1969. Further studies of the infectivity and stability of extracts and homogenates derived from scrapie affected mouse brains. *J. Comp. Pathol.* **79:** 101–108.

Hunter G.D., Kimberlin R.H., Millson G.C., and Gibbons R.A. 1971. An experimental examination of the scrapie agent in cell membrane mixtures. I. Stability and physico-chemical properties of the scrapie agent. *J. Comp. Pathol.* **81:** 23–32.

Hunter N., Foster J.D., Benson G., and Hope J. 1991. Restriction fragment length polymorphisms of the scrapie-associated fibril protein (PrP) gene and their association with susceptiblity to natural scrapie in British sheep. *J. Gen. Virol.* **72:** 1287–1292.

Hunter N., Goldmann W., Smith G., and Hope J. 1994. Frequencies of PrP gene variants in healthy cattle and cattle with BSE in Scotland. *Vet. Rec.* **135:** 400–403.

Hunter N., Hope J., McConnell I., and Dickinson A.G. 1987. Linkage of the scrapie-associated fibril protein (PrP) gene and *Sinc* using congenic mice and restriction fragment length polymorphism analysis. *J. Gen. Virol.* **68:** 2711–2716.

Hunter N., Goldmann W., Benson G., Foster J.D., and Hope J. 1993. Swaledale sheep affected by natural scrapie differ significantly in PrP genotype frequencies from healthy sheep and those selected for reduced incidence of scrapie. *J. Gen. Virol.* **74:** 1025–1031.

Hunter N., Moore L., Hosie B.D., Dingwall W.S., and Greig A. 1997a. Association between natural scrapie and PrP genotype in a flock of Suffolk sheep in Scotland. *Vet. Rec.* **140:** 59–63.

Hunter N., Cairns D., Foster J.D., Smith G., Goldmann W., and Donnelly K. 1997b. Is scrapie solely a genetic disease? *Nature* **386:** 137.

Ikeda T., Horiuchi M., Ishiguro N., Muramatsu Y., Kai-Uwe G.D., and Shinagawa M. 1995. Amino acid polymorphisms of PrP with reference to onset of scrapie in Suffolk and Corriedale sheep in Japan. *J. Gen. Virol.* **76:** 2577–2581.

Ingrosso L., Ladogana A., and Pocchiari M. 1995. Congo red prolongs the incubation period in scrapie-infected hamsters. *J. Virol.* **69:** 506–508.

Ironside J.W. 1997. The new variant form of Creutzfeldt-Jakob disease: A novel prion protein amyloid disorder (editorial). *Amyloid: Int. J. Exp. Clin. Invest.* **4:** 66–69.

Jacob H., Pyrkosch W., and Strube H. 1950. Die erbliche Form der Creutzfeldt-Jakobschen Krankheit. *Arch. Psychiatr. Z. Neurol.* **184:** 653–674.

Jakob A. 1921. Über eigenartige Erkrankungen des Zentralnervensystems mit bemerkenswertem anatomischen Befunde (spastische Pseudosklerose-Encephalomyelopathie mit disseminierten Degenerationsherden). *Z. Gesamte Neurol. Psychiatrie* **64:** 147–228.

James T.L., Liu H., Ulyanov N.B., Farr-Jones S., Zhang H., Donne D.G., Kaneko K., Groth D., Mehlhorn I., Prusiner S.B., and Cohen F.E. 1997. Solution structure of a 142-residue recombinant prion protein corresponding to the infectious fragment of the scrapie isoform. *Proc. Natl. Acad. Sci.* **94:** 10086–10091.

Jeffrey M. and Wells G.A.H. 1988. Spongiform encephalopathy in a nyala (*Tragelaphus angasi*). *Vet. Pathol.* **25:** 398–399.

Jendroska K., Heinzel F.P., Torchia M., Stowring L., Kretzschmar H.A., Kon A., Stern A., Prusiner S.B., and DeArmond S.J. 1991. Proteinase-resistant prion protein accumulation in Syrian hamster brain correlates with regional pathology and scrapie infectivity. *Neurology* **41:** 1482–1490.

Kahana E., Milton A., Braham J., and Sofer D. 1974. Creutzfeldt-Jakob disease: Focus among Libyan Jews in Israel. *Science* **183:** 90–91.

Kaneko K., Vey M., Scott M., Pilkuhn S., Cohen F.E., and Prusiner S.B. 1997a. COOH-terminal sequence of the cellular prion protein directs subcellular trafficking and controls conversion into the scrapie isoform. *Proc. Natl. Acad. Sci.* **94:** 2333–2338.

Kaneko K., Wille H., Mehlhorn I., Zhang H., Ball H., Cohen F.E., Baldwin M.A., and Prusiner S.B. 1997b. Molecular properties of complexes formed between the prion protein and synthetic peptides. *J. Mol. Biol.* **270:** 574–586.

Kaneko K., Zulianello L., Scott M., Cooper C.M., Wallace A.C., James T.L., Cohen F.E., and Prusiner S.B. 1997c. Evidence for protein X binding to a discontinuous epitope on the cellular prion protein during scrapie prion propagation. *Proc. Natl. Acad. Sci.* **94:** 10069–10074.

Kaneko K., Ball H.L., Wille H., Zhang H., Groth D., Torchia M., Tremblay P., Safar J., Prusiner S.B., DeArmond S.J., Baldwin M.A., and Cohen F.E. 2000. A synthetic peptide initiates Gerstmann-Sträussler-Scheinker (GSS) disease in transgenic mice. *J. Mol. Biol.* **295:** 997–1007.

Kanyo Z.F., Pan K.-M., Williamson A., Burton D.R., Prusiner S.B., Fletterick R.J., and Cohen F.E. 1999. Antibody binding defines a structure for an epitope that participates in the PrP^C→PrP^{Sc} conformational change. *J. Mol. Biol.* **293:** 855–863.

Kascsak R.J., Rubenstein R., Merz P.A., Tonna-DeMasi M., Fersko R., Carp R.I., Wisniewski H.M., and Diringer H. 1987. Mouse polyclonal and monoclonal antibody to scrapie-associated fibril proteins. *J. Virol.* **61:** 3688–3693.

Kellings K., Prusiner S.B., and Riesner D. 1994. Nucleic acids in prion preparations: Unspecific background or essential component? *Philos. Trans. R. Soc. Lond. B* **343:** 425–430.

Kellings K., Meyer N., Mirenda C., Prusiner S.B., and Riesner D. 1992. Further analysis of nucleic acids in purified scrapie prion preparations by improved return refocussing gel electrophoresis (RRGE). *J. Gen. Virol.* **73:** 1025–1029.

Keshet G.I., Bar-Peled O., Yaffe D., Nudel U., and Gabizon R. 2000. The cellular prion protein colocalizes with the dystroglycan complex in the brain. *J. Neurochem.* **75:** 1889–1897.

Kimberlin R.H. 1982. Reflections on the nature of the scrapie agent. *Trends Biochem. Sci.* **7:** 392–394.

———. 1990. Scrapie and possible relationships with viroids. *Semin. Virol.* **1:** 153–162.

——— 1996. Speculations on the origin of BSE and the epidemiology of CJD. In *Bovine spongiform encephalopathy: The BSE dilemma* (ed. C.J. Gibbs, Jr.), pp. 155–175. Springer, New York.

Kimberlin R.H. and Millson G.C. 1976. The effects of cuprizone toxicity on the incubation period of scrapie in mice. *J. Comp. Pathol.* **86:** 489–496.

Kimberlin R. and Walker C. 1977. Characteristics of a short incubation model of scrapie in the golden hamster. *J. Gen. Virol.* **34:** 295–304.

———. 1983. The antiviral compound HPA-23 can prevent scrapie when administered at the time of infection. *Arch. Virol.* **78:** 9–18.

Kimberlin R.H., Collis S.C., and Walker C.A. 1976. Profiles of brain glycosidase activity in cuprizone-fed Syrian hamsters and in scrapie-affected mice, rats, Chinese hamsters and Syrian hamsters. *J. Comp. Pathol.* **86:** 135–142.

Kimberlin R.H., Field H.J., and Walker C.A. 1983. Pathogenesis of mouse scrapie: Evidence for spread of infection from central to peripheral nervous system. *J. Gen. Virol.* **64:** 713–716.

Kimberlin R.H., Millson G.C., and Hunter G.D. 1971. An experimental examination of the scrapie agent in cell membrane mixtures. III. Studies of the operational size. *J. Comp. Pathol.* **81:** 383–391.

Kimberlin R.H., Walker C.A., and Fraser H. 1989. The genomic identity of different strains of mouse scrapie is expressed in hamsters and preserved on reisolation in mice. *J. Gen. Virol.* **70:** 2017–2025.

King C.-Y., Tittman P., Gross H., Gebert R., Aebi M., and Wüthrich K. 1997. Prion-inducing domain 2-114 of yeast Sup35 protein transforms *in vitro* into amyloid-like filaments. *Proc. Natl. Acad. Sci.* **94:** 6618–6622.

Kingsbury D.T., Kasper K.C., Stites D.P., Watson J.D., Hogan R.N., and Prusiner S.B. 1983. Genetic control of scrapie and Creutzfeldt-Jakob disease in mice. *J. Immunol.* **131:** 491–496.

Kirkwood J.K., Cunningham A.A., Wells G.A.H., Wilesmith J.W., and Barnett J.E.F. 1993. Spongiform encephalopathy in a herd of greater kudu (*Tragelaphus strepsiceros*): Epidemiological observations. *Vet. Rec.* **133:** 360–364.

Kirschbaum W.R. 1924. Zwei eigenartige Erkrankungen des Zentralnervensystems nach Art der spastischen Pseudosklerose (Jakob). *Z. Gesamte Neurol. Psychiatrie* **92:** 175–220.

———. 1968. *Jakob-Creutzfeldt disease*. Elsevier, Amsterdam.

Kitamoto T. and Tateishi J. 1994. Human prion diseases with variant prion protein. *Philos. Trans. R. Soc. Lond. B* **343:** 391–398.

Klatzo I., Gajdusek D.C., and Zigas V. 1959. Pathology of kuru. *Lab. Invest.* **8:** 799–847.

Klitzman R.L., Alpers M.P., and Gajdusek D.C. 1984. The natural incubation period of kuru and the episodes of transmission in three clusters of patients. *Neuroepidemiology* **3:** 3–20.

Knaus K.J., Morillas M., Swietnicki W., Malone M., Surewicz W.K., and Yee V.C. 2001. Crystal structure of the human prion protein reveals a mechanism for oligomerization. *Nat. Struct. Biol.* **8:** 770–774.

Koch T.K., Berg B.O., DeArmond S.J., and Gravina R.F. 1985. Creutzfeldt-Jakob disease in a young adult with idiopathic hypopituitarism. Possible relation to the administration of cadaveric human growth hormone. *N. Engl. J. Med.* **313:** 731–733.

Kocisko D.A., Come J.H., Priola S.A., Chesebro B., Raymond G.J., Lansbury P.T., Jr., and Caughey B. 1994. Cell-free formation of protease-resistant prion protein. *Nature* **370:** 471–474.

Korth C., May B.C.H., Cohen F.E., and Prusiner S.B. 2001. Acridine and phenothiazine derivatives as pharmacotherapeutics for prion disease. *Proc. Natl. Acad. Sci.* **98:** 9836–9841.

Korth C., Stierli B., Streit P., Moser M., Schaller O., Fischer R., Schulz-Schaeffer W., Kretzschmar H., Raeber A., Braun U., Ehrensperger F., Hornemann S., Glockshuber R., Riek R., Billeter M., Wüthrich K., and Oesch B. 1997. Prion (PrP^{Sc})-specific epitope defined by a monoclonal antibody. *Nature* **389:** 74–77.

Korth C., Kaneko K., Groth D., Heye N., Telling G., Mastrianni J., Parchi P., Gambetti P., Will R., Ironside J., Heinrich C., Tremblay P., DeArmond S.J., and Prusiner S.B. 2003. Abbreviated incubation times for human prions in mice expressing a chimeric mouse–human prionprotein transgene. *Proc. Natl. Acad. Sci.* **100:** 4784–4789.

Kozin S.A., Bertho G., Mazur A.K., Rabesona H., Girault J.P., Haertle T., Takahashi M., Debey P., and Hoa G.H. 2001. Sheep prion protein synthetic peptide spanning helix 1 and beta-strand 2 (residues 142–166) shows beta-hairpin structure in solution. *J. Biol. Chem.* **276:** 46364–46370.

Kretzschmar H.A., Neumann M., and Stavrou D. 1995. Codon 178 mutation of the human prion protein gene in a German family (Backer family): Sequencing data from 72 year-old celloidin-embedded brain tissue. *Acta Neuropathol.* **89:** 96–98.

Kretzschmar H.A., Prusiner S.B., Stowring L.E., and DeArmond S.J. 1986. Scrapie prion proteins are synthesized in neurons. *Am. J. Pathol.* **122:** 1–5.

Kretzschmar H.A., Honold G., Seitelberger F., Feucht M., Wessely P., Mehraein P., and Budka H. 1991. Prion protein mutation in family first reported by Gerstmann, Sträussler, and Scheinker. *Lancet* **337:** 1160.

Kurschner C. and Morgan J.I. 1995. The cellular prion protein (PrP) selectively binds to Bcl-2 in the yeast two-hibrid system. *Mol. Brain Res.* **30:** 165–168.

Lacroute F. 1971. Non-Mendelian mutation allowing ureidosuccinic acid uptake in yeast. *J. Bacteriol.* **106:** 519–522.

Laplanche J.-L., Chatelain J., Launay J.-M., Gazengel C., and Vidaud M. 1990. Deletion in prion protein gene in a Moroccan family. *Nucleic Acids Res.* **18:** 6745.

Laplanche J.-L., Chatelain J., Beaudry P., Dussaucy M., Bounneau C., and Launay J.-M. 1993. French autochthonous scrapied sheep without the 136Val PrP polymorphism. *Mamm. Genome* **4:** 463–464.

Lasmézas C.I., Deslys J.-P., Demaimay R., Adjou K.T., Hauw J.-J., and Dormont D. 1996a.

Strain specific and common pathogenic events in murine models of scrapie and bovine spongiform encephalopathy. *J. Gen. Virol.* **77:** 1601–1609.

Lasmézas C.I., Deslys J.-P., Demaimay R., Adjou K.T., Lamoury F., Dormont D., Robain O., Ironside J., and Hauw J.-J. 1996b. BSE transmission to macaques. *Nature* **381:** 743–744.

Lasmézas C.I., Deslys J.-P., Robain O., Jaegly A., Beringue V., Peyrin J.-M., Fournier J.-G., Hauw J.-J., Rossier J., and Dormont D. 1997. Transmission of the BSE agent to mice in the absence of detectable abnormal prion protein. *Science* **275:** 402–405.

Latarjet R., Muel B., Haig D.A., Clarke M.C., and Alper T. 1970. Inactivation of the scrapie agent by near monochromatic ultraviolet light. *Nature* **227:** 1341–1343.

Lehmann S. and Harris D.A. 1995. A mutant prion protein displays aberrant membrane association when expressed in cultured cells. *J. Biol. Chem.* **270:** 24589–24597.

———. 1996. Two mutant prion proteins expressed in cultured cells acquire biochemical properties reminiscent of the scrapie isoform. *Proc. Natl. Acad. Sci.* **93:** 5610–5614.

Li G. and Bolton D.C. 1997. A novel hamster prion protein mRNA contains an extra exon: Increased expression in scrapie. *Brain Res.* **751:** 265–274.

Liemann S. and Glockshuber R. 1999. Influence of amino acid substitutions related to inherited human prion diseases on the thermodynamic stability of the cellular prion protein. *Biochemistry* **38:** 3258–3267.

Lindquist S., Patino M.M., Chernoff Y.O., Kowal A.S., Singer M.A., Liebman S.W., Lee K.-H., and Blake T. 1995. The role of Hsp104 in stress tolerance and [*PSI*⁺] propagation in *Saccharomyces cerevisiae. Cold Spring Harbor Symp. Quant. Biol.* **60:** 451–460.

Liu H., Farr-Jones S., Ulyanov N.B., Llinas M., Marqusee S., Groth D., Cohen F.E., Prusiner S.B., and James T.I. 1999. Solution structure of Syrian hamster prion protein rPrP(90–231). *Biochemistry* **38:** 5362–5377.

Lledo P.-M., Tremblay P., DeArmond S.J., Prusiner S.B., and Nicoll R.A. 1996. Mice deficient for prion protein exhibit normal neuronal excitability and synaptic transmission in the hippocampus. *Proc. Natl. Acad. Sci.* **93:** 2403–2407.

Locht C., Chesebro B., Race R., and Keith J.M. 1986. Molecular cloning and complete sequence of prion protein cDNA from mouse brain infected with the scrapie agent. *Proc. Natl. Acad. Sci.* **83:** 6372–6376.

Lugaresi E., Medori R., Montagna P., Baruzzi A., Cortelli P., Lugaresi A., Tinuper P., Zucconi M., and Gambetti P. 1986. Fatal familial insomnia and dysautonomia with selective degeneration of thalamic nuclei. *N. Engl. J. Med.* **315:** 997–1003.

Lührs T., Riek R., Guntert P., and Wüthrich K. 2003. NMR structure of the human doppel protein. *J. Mol. Biol.* **326:** 1549–1557.

Malmgren R., Kurland L., Mokri B., and Kurtzke J. 1979. The epidemiology of Creutzfeldt-Jakob disease. In *Slow transmissible diseases of the nervous system* (ed. S.B. Prusiner and W.J. Hadlow), vol. 1, pp. 93–112. Academic Press, New York.

Manson J.C., Clarke A.R., Hooper M.L., Aitchison L., McConnell I., and Hope J. 1994. 129/Ola mice carrying a null mutation in PrP that abolishes mRNA production are developmentally normal. *Mol. Neurobiol.* **8:** 121–127.

Manson J.C., Jameison E., Baybutt H., Tuzi N.L., Barron R., McConnell I., Somerville R., Ironside J., Will R., Sy M.-S., Melton D.W., Hope J., and Bostock C. 1999. A single amino acid alteration (101L) introduced into murine PrP dramatically alters incubation time of transmissible spongiform encephalopathy. *EMBO J.* **18:** 6855–6864.

Marsh R.F., Bessen R.A., Lehmann S., and Hartsough G.R. 1991. Epidemiological and experimental studies on a new incident of transmissible mink encephalopathy. *J. Gen. Virol.* **72:** 589–594.

Marsh R.F., Malone T.G., Semancik J.S., and Hanson R.P. 1980. Studies on the physico-chemical nature of the scrapie agent. In *Search for the cause of multiple sclerosis and other chronic diseases of the central nervous system* (ed. A. Boese), pp. 314–320. Verlag Chemie, Weinheim, Germany.

Marsh R.F., Malone T.G., Semancik J.S., Lancaster W.D., and Hanson R.P. 1978. Evidence for an essential DNA component in the scrapie agent. *Nature* **275:** 146–147.

Marsh R.F., Semancik J.S., Medappa K.C., Hanson R.P., and Rueckert R.R. 1974. Scrapie and transmissible mink encephalopathy: Search for infectious nucleic acid. *J. Virol.* **13:** 993–996.

Martins V.R., Graner E., Garcia-Abreu J., de Souza S.J., Mercadante A.F., Veiga S.S., Zanata S.M., Neto V.M., and Brentani R.R. 1997. Complementary hydropathy identifies a cellular prion protein receptor. *Nat. Med.* **3:** 1376–1382.

Masison D.C. and Wickner R.B. 1995. Prion-inducing domain of yeast Ure2p and protease resistance of Ure2p in prion-containing cells. *Science* **270:** 93–95.

Masters C.L., Gajdusek D.C., and Gibbs C.J., Jr. 1981a. The familial occurrence of Creutzfeldt-Jakob disease and Alzheimer's disease. *Brain* **104:** 535–558.

———. 1981b. Creutzfeldt-Jakob disease virus isolations from the Gerstmann-Sträussler syndrome. *Brain* **104:** 559–588.

Masters C.L., Gajdusek D.C., Gibbs C.J., Jr., Bernouilli C., and Asher D.M. 1979. Familial Creutzfeldt-Jakob disease and other familial dementias: An inquiry into possible models of virus-induced familial diseases. In *Slow transmissible diseases of the nervous system* (ed. S.B. Prusiner and W.J. Hadlow), vol. 1, pp. 143–194. Academic Press, New York.

Masters C.L., Harris J.O., Gajdusek D.C., Gibbs C.J., Jr., Bernouilli C., and Asher D.M. 1978. Creutzfeldt-Jakob disease: Patterns of worldwide occurrence and the significance of familial and sporadic clustering. *Ann. Neurol.* **5:** 177–188.

Mastrianni J., Nixon F., Layzer R., DeArmond S.J., and Prusiner S.B. 1997. Fatal sporadic insomnia: Fatal familial insomnia phenotype without a mutation of the prion protein gene. *Neurology* (suppl.) **48:** A296.

Mastrianni J.A., Nixon R., Layzer R., Telling G.C., Han D., DeArmond S.J., and Prusiner S.B. 1999. Prion protein conformation in a patient with sporadic fatal insomnia. *N. Engl. J. Med.* **340:** 1630–1638.

McKenzie D., Kaczkowski J., Marsh R., and Aiken J. 1994. Amphotericin B delays both scrapie agent replication and PrP-res accumulation early in infection. *J. Virol.* **68:** 7534–7536.

McKinley M.P., Bolton D.C., and Prusiner S.B. 1983. A protease-resistant protein is a structural component of the scrapie prion. *Cell* **35:** 57–62.

McKinley M.P., Meyer R.K., Kenaga L., Rahbar F., Cotter R., Serban A., and Prusiner S.B. 1991. Scrapie prion rod formation *in vitro* requires both detergent extraction and limited proteolysis. *J. Virol.* **65:** 1340–1351.

McKnight S. and Tjian R. 1986. Transcriptional selectivity of viral genes in mammalian cells. *Cell* **46:** 795–805.

Medori R., Montagna P., Tritschler H.J., LeBlanc A., Cortelli P., Tinuper P., Lugaresi E., and Gambetti P. 1992. Fatal familial insomnia: A second kindred with mutation of prion protein gene at codon 178. *Neurology* **42:** 669–670.

Meggendorfer F. 1930. Klinische und genealogische Beobachtungen bei einem Fall von spastischer Pseudosklerose Jakobs. *Z. Gesamte Neurol. Psychiatrie* **128:** 337–341.

Meyer R.K., McKinley M.P., Bowman K.A., Braunfeld M.B., Barry R.A., and Prusiner S.B. 1986. Separation and properties of cellular and scrapie prion proteins. *Proc. Natl. Acad. Sci.* **83:** 2310–2314.

Miller M.W., Wild M.A., and Williams E.S. 1998. Epidemiology of chronic wasting disease in captive Rocky Mountain elk. *J. Wildl. Dis.* **34:** 532–538.

Miller M.W., Williams E.S., McCarty C.W., Spraker T.R., Kreeger T.J., Larsen C.T., and Thorne E.T. 2000. Epizootiology of chronic wasting disease in free-ranging cervids in Colorado and Wyoming. *J. Wildl. Dis.* **36:** 676–690.

Millson G.C. and Manning E.J. 1979. The effect of selected detergents on scrapie infectivity. In *Slow transmissible diseases of the nervous system* (ed. S.B. Pruiner and W.J. Hadlow), vol. 2, pp. 409–424. Academic Press, New York.

Millson G.C., Hunter G.D., and Kimberlin R.H. 1971. An experimental examination of the scrapie agent in cell membrane mixtures. II. The association of scrapie infectivity with membrane fractions. *J. Comp. Pathol.* **81:** 255–265.

———. 1976. The physico-chemical nature of the scrapie agent. In *Slow virus diseases of animals and man* (ed. R.H. Kimberlin), pp. 243–266. Elsevier, New York.

Miyazono M., Kitamoto T., Doh-ura K., Iwaki T., and Tateishi J. 1992. Creutzfeldt-Jakob disease with codon 129 polymorphism (Valine): A comparative study of patients with codon 102 point mutation or without mutations. *Acta Neuropathol.* **84:** 349–354.

Mo H., Moore R.C., Cohen F.E., Westaway D., Pruiner S.B., Wright P.E., and Dyson H.J. 2001. Two different neurodegenerative diseases caused by proteins with similar structures. *Proc. Natl. Acad. Sci.* **98:** 2352–2357.

Mobley W.C., Neve R.L., Prusiner S.B., and McKinley M.P. 1988. Nerve growth factor increases mRNA levels for the prion protein and the β-amyloid protein precursor in developing hamster brain. *Proc. Natl. Acad. Sci.* **85:** 9811–9815.

Moore R.C., Hope J., McBride P.A., McConnell I., Selfridge J., Melton D.W., and Manson J.C. 1998. Mice with gene targetted prion protein alterations show that *Prn-p, Sinc* and *Prni* are congruent. *Nat. Genet.* **18:** 118–125.

Moore R.C., Mastrangelo P., Bouzamondo E., Heinrich C., Legname G., Prusiner S.B., Hood L., Westaway D., DeArmond S.J., and Tremblay P. 2001. Doppel-induced cerebellar degeneration in transgenic mice. *Proc. Natl. Acad. Sci.* **98:** 15288–15293.

Moore R.C., Lee I.Y., Silverman G.L., Harrison P.M., Strome R., Heinrich C., Karunaratne A., Pasternak S.H., Chishti M.A., Liang Y., Mastrangelo P., Wang K., Smit A.F.A., Katamine S., Carlson G.A., Cohen F.E., Prusiner S.B., Melton D.W., Tremblay P., Hood L.E., and Westaway D. 1999. Ataxia in prion protein (PrP) deficient mice is associated with upregulation of the novel PrP-like protein doppel. *J. Mol. Biol.* **292:** 797–817.

Mouillet-Richard S., Ermonval M., Chebassier C., Laplanche J.L., Lehmann S., Launay J.M., and Kellermann O. 2000. Signal transduction through prion protein. *Science* **289:** 1925–1928.

Muramoto T., Scott M., Cohen F.E., and Prusiner S.B. 1996. Recombinant scrapie-like prion protein of 106 amino acids is soluble. *Proc. Natl. Acad. Sci.* **93:** 15457–15462.

Naslavsky N., Stein R., Yanai A., Friedlander G., and Taraboulos A. 1997. Characterization of detergent-insoluble complexes containing the cellular prion protein and its scrapie isoform. *J. Biol. Chem.* **272:** 6324–6331.

Nathanson N., Wilesmith J., and Griot C. 1997. Bovine spongiform encephalopathy (BSE): Cause and consequences of a common source epidemic. *Am. J. Epidemiol.* **145:** 959–969.

Nishida N., Tremblay P., Sugimoto T., Shigematsu K., Shirabe S., Petromilli C., Erpel S.P., Nakaoke R., Atarashi R., Houtani T., Torchia M., Sakaguchi S., DeArmond S.J., Prusiner S.B., and Katamine S. 1999. A mouse prion protein transgene rescues mice deficient for the prion protein gene from Purkinje cell degeneration and demyelination. *Lab. Invest.* **79:** 689–697.

O'Connor S.E. and Imperiali B. 1996. Modulation of protein structure and function by asparagine-linked glycosylation. *Chem. Biol.* **3:** 803–812.

O'Rourke K.I., Holyoak G.R., Clark W.W., Mickelson J.R., Wang S., Melco R.P., Besser T.E., and Foote W.C. 1997. PrP genotypes and experimental scrapie in orally inoculated Suffolk sheep in the United States. *J. Gen. Virol.* **78:** 975–978.

O'Rourke K.I., Besser T.E., Miller M.W., Cline T.F., Spraker T.R., Jenny A.L., Wild M.A., Zebarth G.L., and Williams E.S. 1999. PrP genotypes of captive and free-ranging Rocky Mountain elk (*Cervus elaphus nelsoni*) with chronic wasting disease. *J. Gen. Virol.* **80:** 2765–2769.

Oesch B., Teplow D.B., Stahl N., Serban D., Hood L.E., and Prusiner S.B. 1990. Identification of cellular proteins binding to the scrapie prion protein. *Biochemistry* **29:** 5848–5855.

Oesch B., Westaway D., Wälchli M., McKinley M.P., Kent S.B.H., Aebersold R., Barry R.A., Tempst P., Teplow D.B., Hood L.E., Prusiner S.B., and Weissmann C. 1985. A cellular gene encodes scrapie PrP 27-30 protein. *Cell* **40:** 735–746.

Otvos L., Jr., Thurin J., Kollat E., Urge L., Mantsch H.H., and Hollosi M. 1991. Glycosylation of synthetic peptides breaks helices. *Int. J. Pept. Protein Res.* **38:** 476–482.

Owen F., Poulter M., Collinge J., and Crow T.J. 1990. Codon 129 changes in the prion protein gene in Caucasians. *Am. J. Hum. Genet.* **46:** 1215–1216.

Owen F., Poulter M., Lofthouse R., Collinge J., Crow T.J., Risby D., Baker H.F., Ridley R.M., Hsiao K., and Prusiner S.B. 1989. Insertion in prion protein gene in familial Creutzfeldt-Jakob disease. *Lancet* **1:** 51–52.

Palmer E., Wilhelm J., and Sherman F. 1979. Phenotypic suppression of nonsense mutants in yeast by aminoglycoside antibiotics. *Nature* **277:** 148–150.

Palmer M.S., Dryden A.J., Hughes J.T., and Collinge J. 1991. Homozygous prion protein genotype predisposes to sporadic Creutzfeldt-Jakob disease. *Nature* **352:** 340–342.

Palmer M.S., Mahal S.P., Campbell T.A., Hill A.F., Sidle K.C.L., Laplanche J.-L., and Collinge J. 1993. Deletions in the prion protein gene are not associated with CJD. *Hum. Mol. Genet.* **2:** 541–544.

Pan K.-M., Stahl N., and Prusiner S.B. 1992. Purification and properties of the cellular prion protein from Syrian hamster brain. *Protein Sci.* **1:** 1343–1352.

Pan K.-M., Baldwin M., Nguyen J., Gasset M., Serban A., Groth D., Mehlhorn I., Huang Z., Fletterick R.J., Cohen F.E., and Prusiner S.B. 1993. Conversion of α-helices into β-sheets features in the formation of the scrapie prion proteins. *Proc. Natl. Acad. Sci.* **90:** 10962–10966.

Parchi P., Capellari S., Chin S., Schwarz H.B., Schecter N.P., Butts J.D., Hudkins P., Burns D.K., Powers J.M., and Gambetti P. 1999. A subtype of sporadic prion disease mimicking fatal familial insomnia. *Neurology* **52:** 1757–1763.

Parchi P., Capellari S., Chen S.G., Petersen R.B., Gambetti P., Kopp P., Brown P., Kitamoto T., Tateishi J., Giese A., and Kretzschmar H. 1997. Typing prion isoforms (letter). *Nature* **386:** 232–233.

Parry H.B. 1962. Scrapie: A transmissible and hereditary disease of sheep. *Heredity* **17:** 75–105.

———. 1983. *Scrapie disease in sheep.* Academic Press, New York.

Patino M.M., Liu J.-J., Glover J.R., and Lindquist S. 1996. Support for the prion hypothesis for inheritance of a phenotypic trait in yeast. *Science* **273:** 622–626.

Pattison I.H. 1965. Experiments with scrapie with special reference to the nature of the agent and the pathology of the disease. In *Slow, latent and temperate virus infections* (ed.

D.C. Gajdusek et al.), NINDB Monogr. 2, pp. 249–257. U.S. Government Printing Office, Washington, D.C.

Pattison I.H. and Jebbett J.N. 1971. Clinical and histological observations on cuprizone toxicity and scrapie in mice. *Res. Vet. Sci.* **12:** 378–380.

Pattison I.H. and Jones K.M. 1967. The possible nature of the transmissible agent of scrapie. *Vet. Rec.* **80:** 1–8.

Pattison I.H. and Millson G.C. 1961. Scrapie produced experimentally in goats with special reference to the clinical syndrome. *J. Comp. Pathol.* **71:** 101–108.

Peretz D., Scott M., Groth D., Williamson A., Burton D., Cohen F.E., and Prusiner S.B. 2001a. Strain-specified relative conformational stability of the scrapie prion protein. *Protein Sci.* **10:** 854–863.

Peretz D., Williamson R.A., Legname G., Matsunaga Y., Vergara J., Burton D., DeArmond S.J., Prusiner S.B., and Scott M.R. 2002. A change in the conformation of prions accompanies the emergence of a new prion strain. *Neuron* **34:** 921–932.

Peretz D., Williamson R.A., Matsunaga Y., Serban H., Pinilla C., Bastidas R.B., Rozenshteyn R., James T.L., Houghten R.A., Cohen F.E., Prusiner S.B., and Burton D.R. 1997. A conformational transition at the N-terminus of the prion protein features in formation of the scrapie isoform. *J. Mol. Biol.* **273:** 614–622.

Peretz D., Williamson R.A., Kaneko K., Vergara J., Leclerc E., Schmitt-Ulms G., Mehlhorn I.R., Legname G., Wormald M.R., Rudd P.M., Dwek R.A., Burton D.R., and Prusiner S.B. 2001b. Antibodies inhibit prion propagation and clear cell cultures of prion infectivity. *Nature* **412:** 739–743.

Perrier V., Wallace A.C., Kaneko K., Safar J., Prusiner S.B., and Cohen F.E. 2000. Mimicking dominant negative inhibition of prion replication through structure-based drug design. *Proc. Natl. Acad. Sci.* **97:** 6073–6078.

Perrier V., Kaneko K., Safar J., Vergara J., Tremblay P., DeArmond S.J., Cohen F.E., Prusiner S.B., and Wallace A.C. 2002. Dominant-negative inhibition of prion replication in transgenic mice. *Proc. Natl. Acad. Sci.* **99:** 13079–13084.

Peters J., Miller J.M., Jenny A.L., Peterson T.I., and Carmichael K.P. 2000. Immunohistochemical diagnosis of chronic wasting disease in preclinically affected elk from a captive herd. *J. Vet. Diagn. Invest.* **12:** 579–582.

Petersen R.B., Tabaton M., Berg L., Schrank B., Torack R.M., Leal S., Julien J., Vital C., Deleplanque B., Pendlebury W.W., Drachman D., Smith T.W., Martin J.J., Oda M., Montagna P., Ott J., Autilio-Gambetti L., Lugaresi E., and Gambetti P. 1992. Analysis of the prion protein gene in thalamic dementia. *Neurology* **42:** 1859–1863.

Phillips N.A., Bridgeman J., and Ferguson-Smith M. 2000. Findings and conclusions. In *The BSE inquiry.* Stationery Office, London.

Poulter M., Baker H.F., Frith C.D., Leach M., Lofthouse R., Ridley R.M., Shah T., Owen F., Collinge J., Brown G., Hardy J., Mullan M.J., Harding A.E., Bennett C., Doshi R., and Crow T.J. 1992. Inherited prion disease with 144 base pair gene insertion. 1. Genealogical and molecular studies. *Brain* **115:** 675–685.

Priola S.A., Raines A., and Caughey W.S. 2000. Porphyrin and phthalocyanine antiscrapie compounds. *Science* **287:** 1503–1506.

Proske D., Gilch S., Wopfner F., Schatzl H.M., Winnacker E.L., and Famulok M. 2002. Prion-protein-specific aptamer reduces PrP^{Sc} formation. *Chembiochem* **3:** 717–725.

Prusiner S.B. 1978. An approach to the isolation of biological particles using sedimentation analysis. *J. Biol. Chem.* **253:** 916–921.

———. 1982. Novel proteinaceous infectious particles cause scrapie. *Science* **216:** 136–144.

———. 1989. Scrapie prions. *Annu. Rev. Microbiol.* **43:** 345–374.

———. 1991. Molecular biology of prion diseases. *Science* **252:** 1515–1522.

———. 1997. Prion diseases and the BSE crisis. *Science* **278:** 245–251.

———. 1998a. Prions (Les Prix Nobel Lecture). In *Les Prix Nobel* (ed. T. Frängsmyr), pp. 268–323. Almqvist and Wiksell, Stockholm, Sweden.

———. 1998b. Prions. *Proc. Natl. Acad. Sci.* **95:** 13363–13383.

———. 2001. Shattuck Lecture: Neurodegenerative diseases and prions. *N. Engl. J. Med.* **344:** 1516–1526.

Prusiner S.B., Scott M.R., DeArmond S.J., and Cohen F.E. 1998. Prion protein biology. *Cell* **93:** 337–348.

Prusiner S.B., Groth D.F., Bolton D.C., Kent S.B., and Hood L.E. 1984. Purification and structural studies of a major scrapie prion protein. *Cell* **38:** 127–134.

Prusiner S.B., Bolton D.C., Groth D.F., Bowman K.A., Cochran S.P., and McKinley M.P. 1982a. Further purification and characterization of scrapie prions. *Biochemistry* **21:** 6942–6950.

Prusiner S.B., Cochran S.P., Groth D.F., Downey D.E., Bowman K.A., and Martinez H.M. 1982b. Measurement of the scrapie agent using an incubation time interval assay. *Ann. Neurol.* **11:** 353–358.

Prusiner S.B., Groth D.F., Cochran S.P., Masiarz F.R., McKinley M.P., and Martinez H.M. 1980. Molecular properties, partial purification, and assay by incubation period measurements of the hamster scrapie agent. *Biochemistry* **21:** 4883–4891.

Prusiner S.B., Garfin D.E., Baringer J.R., Cochran S.P., Hadlow W.J., Race R.E., and Eklund C.M. 1978a. Evidence for multiple molecular forms of the scrapie agent. In *Persistent viruses* (ed. J. Stevens et al.), pp. 591–613. Academic Press, New York.

Prusiner S.B., Hadlow W.J., Garfin D.E., Cochran S.P., Baringer J.R., Race R.E., and Eklund C.M. 1978b. Partial purification and evidence for multiple molecular forms of the scrapie agent. *Biochemistry* **17:** 4993–4997.

Prusiner S.B., McKinley M.P., Bowman K.A., Bolton D.C., Bendheim P.E., Groth D.F., and Glenner G.G. 1983. Scrapie prions aggregate to form amyloid-like birefringent rods. *Cell* **35:** 349–358.

Prusiner S.B., Groth D., Serban A., Koehler R., Foster D., Torchia M., Burton D., Yang S.-L., and DeArmond S.J. 1993a. Ablation of the prion protein (PrP) gene in mice prevents scrapie and facilitates production of anti-PrP antibodies. *Proc. Natl. Acad. Sci.* **90:** 10608–10612.

Prusiner S.B., Fuzi M., Scott M., Serban D., Serban H., Taraboulos A., Gabriel J.-M., Wells G., Wilesmith J., Bradley R., DeArmond S.J., and Kristensson K. 1993b. Immunologic and molecular biological studies of prion proteins in bovine spongiform encephalopathy. *J. Infect. Dis.* **167:** 602–613.

Prusiner S.B., Scott M., Foster D., Pan K.-M., Groth D., Mirenda C., Torchia M., Yang S.-L., Serban D., Carlson G.A., Hoppe P.C., Westaway D., and DeArmond S.J. 1990. Transgenetic studies implicate interactions between homologous PrP isoforms in scrapie prion replication. *Cell* **63:** 673–686.

Prusiner S.B., Baron H., Carlson G., Cohen F.E., DeArmond S.J., Gabizon R., Gambetti P., Hope J., Kitamoto T., Kretzschmar H.A., Laplanche J.-L., Tateishi J., Telling G., and Will R. 2000. Prions. In *Virus taxonomy: Classification and nomenclature of viruses* (ed. M.H.V. van Regenmortel et al.), pp. 1032–1039. Academic Press, San Diego.

Public Health Service (PHS). 1997. Interagency Coordinating Committee report on human growth hormone and Creutzfeldt-Jakob disease. *U.S. Public Health Serv. Rep.* **14:** 1–11.

Raymond G.J., Hope J., Kocisko D.A., Priola S.A., Raymond L.D., Bossers A., Ironside J., Will R.G., Chen S.G., Petersen R.B., Gambetti P., Rubenstein R., Smits M.A., Lansbury P.T., Jr., and Caughey B. 1997. Molecular assessment of the potential transmissibilities of BSE and scrapie to humans. *Nature* **388:** 285–288.

Requena J.R., Groth D., Legname G., Stadtman E.R., Prusiner S.B., and Levine R.L. 2001. Copper-catalyzed oxidation of the recombinant SHa(29–231) prion protein. *Proc. Natl. Acad. Sci.* **98:** 7170–7175.

Ridley R.M. and Baker H.F. 1996. To what extent is strain variation evidence for an independent genome in the agent of the transmissible spongiform encephalopathies? *Neurodegeneration* **5:** 219–231.

Rieger R., Edenhofer F., Lasmézas C.I., and Weiss S. 1997. The human 37-kDa laminin receptor precursor interacts with the prion protein in eukaryotic cells. *Nat. Med.* **3:** 1383–1388.

Riek R., Hornemann S., Wider G., Glockshuber R., and Wüthrich K. 1997. NMR characterization of the full-length recombinant murine prion protein, *m*PrP(23–231). *FEBS Lett.* **413:** 282–288.

Riek R., Hornemann S., Wider G., Billeter M., Glockshuber R., and Wüthrich K. 1996. NMR structure of the mouse prion protein domain PrP(121–231). *Nature* **382:** 180–182.

Riek R., Wider G., Billeter M., Hornemann S., Glockshuber R., and Wüthrich K. 1998. Prion protein NMR structure and familial human spongiform encephalopathies. *Proc. Natl. Acad. Sci.* **95:** 11667–11672.

Riesner D., Kellings K., Post K., Wille H., Serban H., Groth D., Baldwin M.A., and Prusiner S.B. 1996. Disruption of prion rods generates 10-nm spherical particles having high α-helical content and lacking scrapie infectivity. *J. Virol.* **70:** 1714–1722.

Robakis N.K., Devine-Gage E.A., Kascsak R.J., Brown W.T., Krawczun C., and Silverman W.P. 1986. Localization of a human gene homologous to the PrP gene on the p arm of chromosome 20 and detection of PrP-related antigens in normal human brain. *Biochem. Biophys. Res. Commun.* **140:** 758–765.

Robinson M.M., Hadlow W.J., Knowles D.P., Huff T.P., Lacy P.A., Marsh R.F., and Gorham J.R. 1995. Experimental infection of cattle with the agents of transmissible mink encephalopathy and scrapie. *J. Comp. Path.* **113:** 241–251.

Rogers M., Yehiely F., Scott M., and Prusiner S.B. 1993. Conversion of truncated and elongated prion proteins into the scrapie isoform in cultured cells. *Proc. Natl. Acad. Sci.* **90:** 3182–3186.

Rogers M., Serban D., Gyuris T., Scott M., Torchia T., and Prusiner S.B. 1991. Epitope mapping of the Syrian hamster prion protein utilizing chimeric and mutant genes in a vaccinia virus expression system. *J. Immunol.* **147:** 3568–3574.

Rohwer R.G. 1984. Scrapie infectious agent is virus-like in size and susceptibility to inactivation. *Nature* **308:** 658–662.

Rosenthal N.P., Keesey J., Crandall B., and Brown W.J. 1976. Familial neurological disease associated with spongiform encephalopathy. *Arch. Neurol.* **33:** 252–259.

Ryou C., Legname G., Peretz D., Craig J.C., Baldwin M.A., and Prusiner S.B. 2003. Differential inhibition of prion propagation by enantiomers of quinacrine. *Lab. Invest.* **83:** 837–843.

Saeki K., Matsumoto Y., Hirota Y., Matsumoto Y., and Onodera T. 1996. Three-exon structure of the gene encoding the rat prion protein and its expression in tissues. *Virus Genes* **12:** 15–20.

Safar J., Roller P.P., Gajdusek D.C., and Gibbs C.J., Jr. 1993a. Conformational transitions, dissociation, and unfolding of scrapie amyloid (prion) protein. *J. Biol. Chem.* **268:** 20276–20284.

———. 1993b. Thermal-stability and conformational transitions of scrapie amyloid (prion) protein correlate with infectivity. *Protein Sci.* **2:** 2206–2216.

Safar J., Wille H., Itri V., Groth D., Serban H., Torchia M., Cohen F.E., and Prusiner S.B. 1998. Eight prion strains have PrP^Sc molecules with different conformations. *Nat. Med.* **4:** 1157–1165.

Safar J.G., Scott M., Monaghan J., Deering C., Didorenko S., Vergara J., Ball H., Legname G., Leclerc E., Solforosi L., Serban H., Groth D., Burton D.R., Prusiner S.B., and Williamson R.A. 2002. Measuring prions causing bovine spongiform encephalopathy or chronic wasting disease by immunoassays and transgenic mice. *Nat. Biotechnol.* **20:** 1147–1150.

Sakaguchi S., Katamine S., Shigematsu K., Nakatani A., Moriuchi R., Nishida N., Kurokawa K., Nakaoke R., Sato H., Jishage K., Kuno J., Noda T., and Miyamoto T. 1995. Accumulation of proteinase K-resistant prion protein (PrP) is restricted by the expression level of normal PrP in mice inoculated with a mouse-adapted strain of the Creutzfeldt-Jakob disease agent. *J. Virol.* **69:** 7586–7592.

Sakaguchi S., Katamine S., Nishida N., Moriuchi R., Shigematsu K., Sugimoto T., Nakatani A., Kataoka Y., Houtani T., Shirabe S., Okada H., Hasegawa S., Miyamoto T., and Noda T. 1996. Loss of cerebellar Purkinje cells in aged mice homozygous for a disrupted PrP gene. *Nature* **380:** 528–531.

Schatzl H.M., Da Costa M., Taylor L., Cohen F.E., and Prusiner S.B. 1995. Prion protein gene variation among primates. *J. Mol. Biol.* **245:** 362–374.

Schmitt-Ulms G., Legname G., Baldwin M.A., Ball H.L., Bradon N., Bosque P.J., Crossin K.L., Edelman G.M., DeArmond S.J., Cohen F.E., and Prusiner S.B. 2001. Binding of neural cell adhesion molecules (N-CAMs) to the cellular prion protein. *J. Mol. Biol.* **314:** 1209–1225.

Schuler B., Rachel R., and Seckler R. 1999. Formation of fibrous aggregates from a nonnative intermediate: The isolated P22 tailspike beta-helix domain. *J. Biol. Chem.* **274:** 18589–18596.

Scott M.R., Köhler R., Foster D., and Prusiner S.B. 1992. Chimeric prion protein expression in cultured cells and transgenic mice. *Protein Sci.* **1:** 986–997.

Scott M., Groth D., Foster D., Torchia M., Yang S.-L., DeArmond S.J., and Prusiner S.B. 1993. Propagation of prions with artificial properties in transgenic mice expressing chimeric PrP genes. *Cell* **73:** 979–988.

Scott M.R., Groth D., Tatzelt J., Torchia M., Tremblay P., DeArmond S.J., and Prusiner S.B. 1997a. Propagation of prion strains through specific conformers of the prion protein. *J. Virol.* **71:** 9032–9044.

Scott M.R., Will R., Ironside J., Nguyen H.-O.B., Tremblay P., DeArmond S.J., and Prusiner S.B. 1999. Compelling transgenetic evidence for transmission of bovine spongiform encephalopathy prions to humans. *Proc. Natl. Acad. Sci.* **96:** 15137–15142.

Scott M., Foster D., Mirenda C., Serban D., Coufal F., Wälchli M., Torchia M., Groth D., Carlson G., DeArmond S.J., Westaway D., and Prusiner S.B. 1989. Transgenic mice expressing hamster prion protein produce species-specific scrapie infectivity and amyloid plaques. *Cell* **59:** 847–857.

Scott M.R., Safar J., Telling G., Nguyen O., Groth D., Torchia M., Koehler R., Tremblay P., Walther D., Cohen F.E., DeArmond S.J., and Prusiner S.B. 1997b. Identification of a prion protein epitope modulating transmission of bovine spongiform encephalopathy prions to transgenic mice. *Proc. Natl. Acad. Sci.* **94:** 14279–14284.

Seckler R. 1998. Folding and function of repetitive structure in the homotrimeric phage P22 tailspike protein. *J. Struct. Biol.* **122:** 216–222.

Serban D., Taraboulos A., DeArmond S.J., and Prusiner S.B. 1990. Rapid detection of Creutzfeldt-Jakob disease and scrapie prion proteins. *Neurology* **40:** 110–117.

Sethi S., Lipford G., Wagner H., and Kretzschmar H. 2002. Postexposure prophylaxis against prion disease with a stimulator of innate immunity. *Lancet* **360:** 229–230.

Shibuya S., Higuchi J., Shin R.-W., Tateishi J., and Kitamoto T. 1998. Protective prion protein polymorphisms against sporadic Creutzfeldt-Jakob disease. *Lancet* **351:** 419.

Siakotos A.N., Gajdusek D.C., Gibbs C.J., Jr., Traub R.D., and Bucana C. 1976. Partial purification of the scrapie agent from mouse brain by pressure disruption and zonal centrifugation in sucrose-sodium chloride gradients. *Virology* **70:** 230–237.

Sigurdson C.J., Williams E.S., Miller M.W., Spraker T.R., O'Rourke K.I., and Hoover E.A. 1999. Oral transmission and early lymphoid tropism of chronic wasting disease PrP[res] in mule deer fawns (*Odocoileus hemionus*). *J. Gen. Virol.* **80:** 2757–2764.

Sigurdsson B. 1954. Rida, a chronic encephalitis of sheep with general remarks on infections which develop slowly and some of their special characteristics. *Br. Vet. J.* **110:** 341–354.

Singh A.C., Helms C., and Sherman F. 1979. Mutation of the non-Mendelian suppressor [PSI] in yeast by hypertonic media. *Proc. Natl. Acad. Sci.* **76:** 1952–1956.

Somerville R.A., Chong A., Mulqueen O.U., Birkett C.R., Wood S.C.E.R., and Hope J. 1997. Biochemical typing of scrapie strains. *Nature* **386:** 564.

Sparkes R.S., Simon M., Cohn V.H., Fournier R.E.K., Lem J., Klisak I., Heinzmann C., Blatt C., Lucero M., Mohandas T., DeArmond S.J., Westaway D., Prusiner S.B., and Weiner L.P. 1986. Assignment of the human and mouse prion protein genes to homologous chromosomes. *Proc. Natl. Acad. Sci.* **83:** 7358–7362.

Sparrer H.E., Santoso A., Szoka F.C., Jr., and Weissman J.S. 2000. Evidence for the prion hypothesis: Induction of the yeast [*PSI*+] factor by in vitro-converted Sup35 protein. *Science* **289:** 595–599.

Spencer M.D., Knight R.S., and Will R.G. 2002. First hundred cases of variant Creutzfeldt-Jakob disease: Retrospective case note review of early psychiatric and neurological features. *BMJ* **324:** 1479–1482.

Spudich S., Mastrianni J.A., Wrensch M., Gabizon R., Meiner Z., Kahana I., Rosenmann H., Kahana E., and Prusiner S.B. 1995. Complete penetrance of Creutzfeldt-Jakob disease in Libyan Jews carrying the E200K mutation in the prion protein gene. *Mol. Med.* **1:** 607–613.

Stahl N., Baldwin M.A., Burlingame A.L., and Prusiner S.B. 1990. Identification of glycoinositol phospholipid linked and truncated forms of the scrapie prion protein. *Biochemistry* **29:** 8879–8884.

Stahl N., Borchelt D.R., Hsiao K., and Prusiner S.B. 1987. Scrapie prion protein contains a phosphatidylinositol glycolipid. *Cell* **51:** 229–240.

Stahl N., Baldwin M.A., Teplow D.B., Hood L., Gibson B.W., Burlingame A.L., and Prusiner S.B. 1993. Structural analysis of the scrapie prion protein using mass spectrometry and amino acid sequencing. *Biochemistry* **32:** 1991–2002.

Stekel D.J., Nowak M.A., and Southwood T.R.E. 1996. Prediction of future BSE spread. *Nature* **381:** 119.

Stender A. 1930. Weitere Beiträge zum Kapitel "Spastische Pseudosklerose Jakobs." *Z. Gesamte Neurol. Psychiatrie* **128:** 528–543.

Stewart R.S. and Harris D.A. 2001. Most pathogenic mutations do not alter the membrane topology of the prion protein. *J. Biol. Chem.* **276:** 2212–2220.

Stöckel J., Safar J., Wallace A.C., Cohen F.E., and Prusiner S.B. 1998. Prion protein selectively binds copper (II) ions. *Biochemistry* **37:** 7185–7193.

Sulkowski E. 1985. Purification of proteins by IMAC. *Trends Biotechnol.* **3:** 1–7.

Supattapone S., Nguyen H.-O.B., Cohen F.E., Prusiner S.B., and Scott M.R. 1999a. Elimination of prions by branched polyamines and implications for therapeutics. *Proc. Natl. Acad. Sci.* **96:** 14529–14534.

Supattapone S., Wille H., Uyechi L., Safar J., Tremblay P., Szoka F.C., Cohen F.E., Prusiner S.B., and Scott M.R. 2001. Branched polyamines cure prion-infected neuroblastoma cells. *J. Virol.* **75:** 3453–3461.

Supattapone S., Bosque P., Muramoto T., Wille H., Aagaard C., Peretz D., Nguyen H.-O.B., Heinrich C., Torchia M., Safar J., Cohen F.E., DeArmond S.J., Prusiner S.B., and Scott M. 1999b. Prion protein of 106 residues creates an artificial transmission barrier for prion replication in transgenic mice. *Cell* **96:** 869–878.

Taraboulos A., Jendroska K., Serban D., Yang S.-L., DeArmond S.J., and Prusiner S.B. 1992. Regional mapping of prion proteins in brains. *Proc. Natl. Acad. Sci.* **89:** 7620–7624.

Taraboulos A., Scott M., Semenov A., Avrahami D., Laszlo L., and Prusiner S.B. 1995. Cholesterol depletion and modification of COOH-terminal targeting sequence of the prion protein inhibits formation of the scrapie isoform. *J. Cell Biol.* **129:** 121–132.

Tateishi J. and Kitamoto T. 1993. Developments in diagnosis for prion diseases. *Br. Med. Bull.* **49:** 971–979.

Tateishi J., Kitamoto T., Hoque M.Z., and Furukawa H. 1996. Experimental transmission of Creutzfeldt-Jakob disease and related diseases to rodents. *Neurology* **46:** 532–537.

Tatzelt J., Maeda N., Pekny M., Yang S.-L., Betsholtz C., Eliasson C., Cayetano J., Camerino A.P., DeArmond S.J., and Prusiner S.B. 1996. Scrapie in mice deficient in apolipoprotein E or glial fibrillary acidic protein. *Neurology* **47:** 449–453.

Taylor K.C. 1991. The control of bovine spongiform encephalopathy in Great Britain. *Vet. Rec.* **129:** 522–526.

Telling G.C., Haga T., Torchia M., Tremblay P., DeArmond S.J., and Prusiner S.B. 1996a. Interactions between wild-type and mutant prion proteins modulate neurodegeneration in transgenic mice. *Genes Dev.* **10:** 1736–1750.

Telling G.C., Scott M., Mastrianni J., Gabizon R., Torchia M., Cohen F.E., DeArmond S.J., and Prusiner S.B. 1995. Prion propagation in mice expressing human and chimeric PrP transgenes implicates the interaction of cellular PrP with another protein. *Cell* **83:** 79–90.

Telling G.C., Parchi P., DeArmond S.J., Cortelli P., Montagna P., Gabizon R., Mastrianni J., Lugaresi E., Gambetti P., and Prusiner S.B. 1996b. Evidence for the conformation of the pathologic isoform of the prion protein enciphering and propagating prion diversity. *Science* **274:** 2079–2082.

Telling G.C., Scott M., Hsiao K.K., Foster D., Yang S.-L., Torchia M., Sidle K.C.L., Collinge J., DeArmond S.J., and Prusiner S.B. 1994. Transmission of Creutzfeldt-Jakob disease from humans to transgenic mice expressing chimeric human–mouse prion protein. *Proc. Natl. Acad. Sci.* **91:** 9936–9940.

Ter-Avanesyan M.D., Dagkesamanskaya A.R., Kushnirov V.V., and Smirnov V.N. 1994. The *SUP35* omnipotent suppressor gene is involved in the maintenance of the non-Mendelian determinant [*psi*⁺] in the yeast *Saccharomyces cerevisiae. Genetics* **137:** 671–676.

Tobler I., Gaus S.E., Deboer T., Achermann P., Fischer M., Rülicke T., Moser M., Oesch B.,

McBride P.A., and Manson J.C. 1996. Altered circadian activity rhythms and sleep in mice devoid of prion protein. *Nature* **380:** 639–642.

Tremblay P., Ball H.L., Kaneko K., Groth D., Hegde R.S., Cohen F.E., DeArmond S.J., Pruisner S.B., and Safar J.G. 2003. Mutant PrP^{Sc} conformers induced by a synthetic peptide and various prion strains. *J. Virol.* (in press).

Tuite M.F. and Lindquist S.L. 1996. Maintenance and inheritance of yeast prions. *Trends Genet.* **12:** 467–471.

Tuite M.F., Cox B.S., and McLaughlin C.S. 1987. A ribosome-associated inhibitor of *in vitro* nonsense suppression in [psi]-strains of yeast. *FEBS Lett.* **225:** 205–208.

Tuite M.F., Mundy C.R., and Cox B.S. 1981. Agents that cause a high frequency of genetic change from [*psi*$^+$] to [*psi*$^-$] in *Saccharomyces cerevisiae*. *Genetics* **98:** 691–711.

Umland T.C., Taylor K.L., Rhee S., Wickner R.B., and Davies D.R. 2001. The crystal structure of the nitrogen regulation fragment of the yeast prion protein Ure2p. *Proc. Natl. Acad. Sci.* **98:** 1459–1464.

Vey M., Pilkuhn S., Wille H., Nixon R., DeArmond S.J., Smart E.J., Anderson R.G., Taraboulos A., and Prusiner S.B. 1996. Subcellular colocalization of the cellular and scrapie prion proteins in caveolae-like membranous domains. *Proc. Natl. Acad. Sci.* **93:** 14945–14949.

Viles J.H., Cohen F.E., Prusiner S.B., Goodin D.B., Wright P.E., and Dyson H.J. 1999. Copper binding to the prion protein: Structural implications of four identical cooperative binding sites. *Proc. Natl. Acad. Sci.* **96:** 2042–2047.

Vnencak-Jones C.L. and Phillips J.A. 1992. Identification of heterogeneous PrP gene deletions in controls by detection of allele-specific heteroduplexes (DASH). *Am. J. Hum. Genet.* **50:** 871–872.

Waggoner D.J., Drisaldi B., Bartnikas T.B., Casareno R.L., Prohaska J.R., Gitlin J.D., and Harris D.A. 2000. Brain copper content and cuproenzyme activity do not vary with prion protein expression level. *J. Biol. Chem.* **275:** 7455–7458.

Weissmann C. 1991a. Spongiform encephalopathies—The prion's progress. *Nature* **349:** 569–571.

———. 1991b. A "unified theory" of prion propagation. *Nature* **352:** 679–683.

Wells G.A.H. and Wilesmith J.W. 1995. The neuropathology and epidemiology of bovine spongiform encephalopathy. *Brain Pathol.* **5:** 91–103.

Wells G.A.H., Scott A.C., Johnson C.T., Gunning R.F., Hancock R.D., Jeffrey M., Dawson M., and Bradley R. 1987. A novel progressive spongiform encephalopathy in cattle. *Vet. Rec.* **121:** 419–420.

Westaway D., Cooper C., Turner S., Da Costa M., Carlson G.A., and Prusiner S.B. 1994a. Structure and polymorphism of the mouse prion protein gene. *Proc. Natl. Acad. Sci.* **91:** 6418–6422.

Westaway D., Goodman P.A., Mirenda C.A., McKinley M.P., Carlson G.A., and Prusiner S.B. 1987. Distinct prion proteins in short and long scrapie incubation period mice. *Cell* **51:** 651–662.

Westaway D., Zuliani V., Cooper C.M., Da Costa M., Neuman S., Jenny A.L., Detwiler L., and Prusiner S.B. 1994b. Homozygosity for prion protein alleles encoding glutamine-171 renders sheep susceptible to natural scrapie. *Genes Dev.* **8:** 959–969.

Westaway D., DeArmond S.J., Cayetano-Canlas J., Groth D., Foster D., Yang S.-L., Torchia M., Carlson G.A., and Prusiner S.B. 1994c. Degeneration of skeletal muscle, peripheral nerves, and the central nervous system in transgenic mice overexpressing wild-type prion proteins. *Cell* **76:** 117–129.

Westaway D., Mirenda C.A., Foster D., Zebarjadian Y., Scott M., Torchia M., Yang S.-L., Serban H., DeArmond S.J., Ebeling C., Prusiner S.B., and Carlson G.A. 1991. Paradoxical shortening of scrapie incubation times by expression of prion protein transgenes derived from long incubation period mice. *Neuron* **7:** 59–68.

Whittington M.A., Sidle K.C.L., Gowland I., Meads J., Hill A.F., Palmer M.S., Jefferys J.G.R., and Collinge J. 1995. Rescue of neurophysiological phenotype seen in PrP null mice by transgene encoding human prion protein. *Nat. Genet.* **9:** 197–201.

Wickner R.B. 1994. [URE3] as an altered URE2 protein: Evidence for a prion analog in *Saccharomyces cerevisiae*. *Science* **264:** 566–569.

Wickner R.B., Masison D.C., and Edskes H.K. 1995. [PSI] and [URE3] as yeast prions. *Yeast* **11:** 1671–1685.

Wilesmith J.W. 1991. The epidemiology of bovine spongiform encephalopathy. *Semin. Virol.* **2:** 239–245.

Wilesmith J.W., Ryan J.B.M., and Atkinson M.J. 1991. Bovine spongiform encephalopathy—Epidemiologic studies on the origin. *Vet. Rec.* **128:** 199–203.

Will R.G., Ironside J.W., Zeidler M., Cousens S.N., Estibeiro K., Alperovitch A., Poser S., Pocchiari M., Hofman A., and Smith P.G. 1996. A new variant of Creutzfeldt-Jakob disease in the UK. *Lancet* **347:** 921–925.

Wille H., Zhang G.-F., Baldwin M.A., Cohen F.E., and Prusiner S.B. 1996. Separation of scrapie prion infectivity from PrP amyloid polymers. *J. Mol. Biol.* **259:** 608–621.

Wille H., Michelitsch M.D., Guénebaut V., Supattapone S., Serban A., Cohen F.E., Agard D.A., and Prusiner S.B. 2002. Structural studies of the scrapie prion protein by electron crystallography. *Proc. Natl. Acad. Sci.* **99:** 3563–3568.

Williams E.S. and Miller M.W. 2002. Chronic wasting disease in deer and elk in North America. *Rev. Sci. Tech.* **21:** 305–316.

Williams E.S. and Young S. 1980. Chronic wasting disease of captive mule deer: A spongiform encephalopathy. *J. Wildl. Dis.* **16:** 89–98.

Williamson R.A., Peretz D., Smorodinsky N., Bastidas R., Serban H., Mehlhorn I., DeArmond S.J., Prusiner S.B., and Burton D.R. 1996. Circumventing tolerance to generate autologous monoclonal antibodies to the prion protein. *Proc. Natl. Acad. Sci.* **93:** 7279–7282.

Wyatt J.M., Pearson G.R., Smerdon T.N., Gruffydd-Jones T.J., Wells G.A.H., and Wilesmith J.W. 1991. Naturally occurring scrapie-like spongiform encephalopathy in five domestic cats. *Vet. Rec.* **129:** 233–236.

Xi Y.G., Ingrosso L., Ladogana A., Masullo C., and Pocchiari M. 1992. Amphotericin B treatment dissociates *in vivo* replication of the scrapie agent from PrP accumulation. *Nature* **356:** 598–601.

Yehiely F., Bamborough P., Costa M.D., Perry B.J., Thinakaran G., Cohen F.E., Carlson G.A., and Prusiner S.B. 1997. Identification of candidate proteins binding to prion protein. *Neurobiol. Dis.* **3:** 339–355.

Zahn R., Guntert P., von Schroetter C., and Wüthrich K. 2003. NMR structure of a variant human prion protein with two disulfide bridges. *J. Mol. Biol.* **326:** 225–234.

Zahn R., Liu A., Lührs T., Riek R., von Schroetter C., López García F., Billeter M., Calzolai L., Wider G., and Wüthrich K. 2000. NMR solution structure of the human prion protein. *Proc. Natl. Acad. Sci.* **97:** 145–150.

Zhang H., Kaneko K., Nguyen J.T., Livshits T.L., Baldwin M.A., Cohen F.E., James T.L., and Prusiner S.B. 1995. Conformational transitions in peptides containing two putative α-helices of the prion protein. *J. Mol. Biol.* **250:** 514–526.

Zhang Y., Swietnicki W., Zagorski M.G., Surewicz W.K., and Sönnichsen F.D. 2000. Solution structure of the E200K variant of human prion protein. Implications for the mechanism of pathogenesis in familial prion diseases. *J. Biol. Chem.* **275:** 33650–33654.

Zlotnik I. 1962. The pathology of scrapie: A comparative study of lesions in the brain of sheep and goats. *Acta Neuropathol.* (suppl.) **1:** 61–70.

Zulianello L., Kaneko K., Scott M., Erpel S., Han D., Cohen F.E., and Prusiner S.B. 2000. Dominant-negative inhibition of prion formation diminished by deletion mutagenesis of the prion protein. *J. Virol.* **74:** 4351–4360.

2

Development of the Prion Concept

Stanley B. Prusiner

Institute for Neurodegenerative Diseases
Departments of Neurology and of Biochemistry and Biophysics
University of California, San Francisco, California 94143

THE PRION CONCEPT WAS DEVELOPED IN THE AFTERMATH of many unsuccessful attempts to decipher the nature of the scrapie agent. In some
respects, the early development of the prion concept mirrors the story of
the discovery that DNA is the genetic material of life (Avery et al. 1944;
Stanley 1970; McCarty 1985). Prior to the acceptance that genes are composed of DNA (Hershey and Chase 1952; Watson and Crick 1953), some
scientists asserted that DNA preparations must be contaminated with
protein, which is the true genetic material (Mirsky and Pollister 1946).
For more than half a century, many biologists thought that genes were
composed of protein and that proteins reproduced as replicas of themselves (Stanley 1935; Haurowitz 1950). The prejudices of these scientists
were similar in some ways to those of investigators who vigorously
opposed the prion concept. However, the scientists who attacked the
hypothesis that genes are composed of DNA had no likely alternative;
they had only a set of feelings derived from poorly substantiated data sets
that genes are made of protein. In contrast, those who attacked the
hypothesis that the prion is composed only of protein had more than 30
years of cumulative evidence showing that genetic information in all
organisms on our planet is encoded in DNA. Studies of viruses and eventually viroids extended this concept and showed that genes could also be
composed of RNA in these small infectious pathogens (Fraenkel-Conrat
and Williams 1955; Gierer and Schramm 1956; Diener 1979).

It is on this background that investigators working on scrapie began
to unravel the curious and often puzzling properties of this infectious
pathogen. The resistance of the scrapie agent to inactivation by formalin

and heat treatments (Gordon 1946), which were commonly used to produce viral vaccines, was an important clue that the scrapie agent might be different from viruses. The resistance of the scrapie agent to procedures that inactivate viruses seemed to be dismissed as an interesting observation at a time when the structural features of viruses were beginning to be understood. The significance of these findings seemed to have been dulled by the appreciation that some relatively stable viruses could survive harsh treatments, such as formalin and heat. It was far easier to argue that the scrapie agent was an odd virus than to entertain the possibility that it was not a virus. Two decades passed before reports of the extreme resistance of the scrapie agent to irradiation again trumpeted the puzzling nature of this infectious pathogen (Alper et al. 1966, 1967).

WHAT MIGHT HAVE BEEN: AN ALTERNATIVE PATH OF DISCOVERY

By 1930, the high incidence of inherited Creutzfeldt-Jakob desease (CJD) in some families was known (Fig. 1) (Meggendorfer 1930; Stender 1930). Almost 60 years passed before the significance of this finding was appreciated (Roos et al. 1973; Masters et al. 1981; Hsiao et al. 1989; Prusiner 1989). CJD remained a curious, rare neurodegenerative disease of unknown etiology throughout this period of three-score years (Kirschbaum 1968). Only with transmission of disease to apes by the inoculation of brain extracts

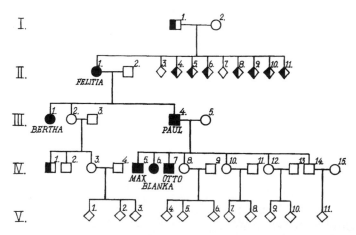

Figure 1. Pedigree of the Backer family with familial Creutzfeldt-Jakob disease (CJD) (Kirschbaum 1968). Sixty years after the first report that CJD could be a familial disease, the D178N mutation of the PrP gene was shown to be the cause (Kretzschmar et al. 1995). (Reprinted, with permission, from Kirschbaum 1968 [copyright Elsevier Science].)

prepared from patients who died of CJD did the story begin to unravel (Gibbs et al. 1968).

Once CJD was shown to be a transmissible disease, relatively little attention was paid to the familial form of the disease, because most cases were not found in families (Roos et al. 1973). It is interesting to speculate how the course of scientific investigation might have proceeded if transmission studies had not been performed until after the molecular genetic lesion was identified. If the prion protein (PrP) gene had been identified in families with prion disease by positional cloning or through the purification and sequencing of PrP in amyloid plaques before brain extracts were shown to be transmissible, the prion concept, which readily explains how a single disease can have a genetic or infectious etiology, might have been greeted with much less skepticism (Prusiner 1995).

Within the scenario of finding mutations in the PrP gene prior to learning that prion diseases are transmissible, it seems likely that investigators would have focused their efforts on explaining how a mutant gene product might stimulate modification of the wild-type (wt) protein after inoculation into a susceptible host (Prusiner 1998). The modified wt protein would, in turn, stimulate production of more of its modified self. Less likely would have been the postulate that a mutant protein unrelated to immune defenses would render the host more susceptible to an infectious pathogen with a foreign genome, such as a virus, bacterium, or fungus.

AN OVERVIEW: UNRAVELING THE ENIGMA OF THE PRION

The transmission of the scrapie agent from sheep to mice (Chandler 1961) made a series of radiobiological studies possible (Alper et al. 1966, 1967). With reports of the extreme resistance of the scrapie agent to inactivation by UV and ionizing radiation came a flurry of hypotheses to explain these curious observations. In some cases, these postulates ignored the lessons learned from studies of DNA, while others tried to accommodate them in rather obtuse but sometimes clever ways.

As the number of hypotheses about the molecular nature of the scrapie agent began to exceed the number of laboratories working on the problem (Table 1), the need for new experimental approaches became evident. Many of the available data on the properties of the scrapie agent were gathered using brain homogenates prepared from mice with clinical signs of scrapie. These mice were inoculated intracerebrally 4–5 months earlier with the scrapie agent, which originated in sheep but had been passaged multiple times in mice (Chandler 1963; Eklund et al. 1963). Once an experiment was completed with these homogenates, an additional 12

Table 1. Hypothetical structures proposed for the scrapie agent

1. Sarcosporidia-like parasite
2. "Filterable" virus
3. Small DNA virus
4. Replicating protein
5. Replicating abnormal polysaccharide with membrane
6. DNA subvirus controlled by a transmissible linkage substance
7. Provirus consisting of recessive genes generating RNA particles
8. Naked nucleic acid similar to plant viroids
9. Unconventional virus
10. Aggregated conventional virus with unusual properties
11. Replicating polysaccharide
12. Nucleoprotein complex
13. Nucleic acid surrounded by a polysaccharide coat
14. Spiroplasma-like organism
15. Multicomponent system with one component quite small
16. Membrane-bound DNA
17. Virino (viroid-like DNA complexed with host proteins)
18. Filamentous animal virus (scrapie-associated fibril [SAF])
19. Aluminum-silicate amyloid complex
20. Computer virus
21. Amyloid-inducing virus
22. Complex of apo- and co-prions (unified theory)
23. Nemavirus (SAF surrounded by DNA)
24. Retrovirus

months was required for endpoint titrations in mice. Typically, determining the titer of a single sample required 60 mice. These slow, tedious, and expensive experiments discouraged systematic investigation.

Purification of the Infectious Scrapie Agent

In 1972, I became interested in the biochemical nature of the scrapie agent while caring for a patient who was dying of CJD (Prusiner 1998). To determine the composition of the scrapie agent, purification seemed critical. Without fractions highly enriched for scrapie infectivity, it seemed unlikely that an accurate picture of the molecular architecture of this enigmatic pathogen would emerge. The purification of any unknown substance requires a quantitative assay for its activity, and the speed at which purification proceeds is directly proportional to the rapidity of the assay. In the case of the scrapie agent, the slow and expensive bioassays described above were an enormous impediment to purification.

Sedimentation Studies

Although many studies on the physicochemical nature of the scrapie agent using the mouse endpoint titration system had been published (Hunter 1972), the slow and expensive bioassays discouraged methodical investigations of the fundamental characteristics of the infectious scrapie particle. Perhaps not surprisingly, a careful study of the sedimentation behavior of the infectious scrapie particle was not pursued until 15 years after introduction of the mouse bioassay in 1961 (Chandler 1961). Because differential centrifugation is frequently used in the initial stages of the purification of many macromolecules, some knowledge of the sedimentation properties of the scrapie agent under defined conditions seemed mandatory (Prusiner 1978). To perform such studies, Swiss CD-1 mice were inoculated intracerebrally with the Chandler isolate of scrapie prions and the mice were sacrificed 30 and 140 days later, when the titers in their spleens and brains, respectively, were at maximum levels (Prusiner et al. 1977, 1978a). Spleen and brain tissues were homogenized, extracted with detergent, and centrifuged at increasing velocities and for progressively longer periods of time. The disappearance of scrapie infectivity was measured in supernatant fractions by endpoint titration, which required one year to score, as described above. Thus, a single set of experiments on the sedimentation properties of the scrapie agent required ~16 months to obtain results.

Prolonged Experimental Time Frames

To perform in triplicate a set of experiments like those described above requires four years. To shorten this duration and also ensure reproducibility, repetitions of experiments could be performed in parallel. No matter what strategy was employed, this long time frame severely retarded progress. Although the temptation to minimize the repetition of experiments was great because of prolonged durations, a balance was required due to the seriousness of drawing false interpretations from restricted data sets.

In a purification scheme, success requires that each step provides progressive enrichment of biological activity coupled with acceptable losses in activity. The greater the enrichment factor, the greater an acceptable loss in activity can be afforded, up to a certain point. Thus, the need for reliable quantification in developing a purification scheme is clear because misinterpretations of data can be very costly. This is especially true when several years might be required to reassess a faulty data set.

Impact of the Incubation-time Assay

From the foregoing discussion, it should be apparent why a more rapid and economical bioassay of the scrapie agent transformed investigations. This bioassay, or incubation-time assay as it is often called, had a particularly profound impact on the development of effective purification schemes for enriching fractions for scrapie infectivity (Prusiner et al. 1980b, 1982b). Instead of 60 mice that were typically used for endpoint titrations, four Syrian hamsters were employed. Instead of scoring the titer of the sample at one year after inoculation, only 70 days were required for samples with high titers. That fractions with high titers yielded shorter incubation times than those with low titers was fortuitous because the identification of fractions with high titers was most critical in developing an effective purification protocol. The incubation-time assay provided a much more rapid and economical means of quantifying fractions enriched for infectivity and those that were not. Such studies rather rapidly led to the formulation of a protocol for separating scrapie infectivity from most proteins and nucleic acids.

Probing the Molecular Structure of the Scrapie Agent

With a ~100-fold purification of infectivity relative to protein, >98% of the proteins and polynucleotides were eliminated, which permitted more reliable probing of the constituents of these enriched fractions than was previously possible using crude preparations. Data from several studies suggested that scrapie infectivity might depend on protein (Hunter et al. 1969; Millson et al. 1976; Cho 1980, 1983). As reproducible data began to accumulate, indicating that scrapie infectivity could be reduced by procedures that hydrolyze or modify proteins but was resistant to procedures that alter nucleic acids, a circumscribed family of hypotheses about the molecular architecture of the scrapie agent began to emerge (Prusiner 1982). These data established for the first time that a particular macromolecule, a protein, is required for infectivity. The experimental findings also extended the earlier observations on the resistance of scrapie infectivity to UV irradiation at 250 nm in two respects: (1) Four procedures were used to probe for a nucleic acid based on physical principles that are independent of UV radiation damage and (2) demonstration of a protein requirement provided a reference macromolecule. No longer could the scrapie agent be considered phlogiston or linoleum!

Once the requirement for a protein was established, it was possible to revisit the long list of hypothetical structures that had been proposed for the scrapie agent (Table 1) and to eliminate carbohydrates, lipids, and nucleic acids as the infective elements within a scrapie agent (Prusiner 1982). No longer could structures such as a viroid-like nucleic acid, a replicating polysaccharide, or a small polynucleotide surrounded by a carbohydrate be entertained as reasonable candidates to explain the enigmatic properties of the scrapie agent (Prusiner 1982). The family of hypotheses that remained was still large and required continued consideration of all possibilities.The prion concept, which posits that an infectious protein comprises the scrapie agent, evolved from this family of hypotheses.

In prion research as well as many other areas of scientific investigation, a single hypothesis is all too often championed at the expense of a reasoned approach, which involves continuing to entertain a series of complex arguments until one or more can be discarded on the basis of experimental data (Chamberlin 1890). Accordingly, many hypothetical structures were discarded as experimental data on the molecular properties of the scrapie agent accumulated.

Radiobiology of the Scrapie Agent

The experimental transmission of scrapie from sheep to mice (Chandler 1961) gave investigators a more convenient laboratory model, which yielded considerable information on the nature of the unusual infectious pathogen that causes scrapie (Alper et al. 1966, 1967, 1978; Gibbons and Hunter 1967; Pattison and Jones 1967; Millson et al. 1971).

Resistance to UV Radiation

The extreme resistance of the scrapie agent to both ionizing and UV irradiation suggested the infectious pathogen was quite different from all known viruses. The D_{37} value for irradiation at 254 nm was 42,000 J/m^2, which argued that the target was unlikely to be a nucleic acid (Bellinger-Kawahara et al. 1987a). Irradiation at different wavelengths of UV light showed that scrapie infectivity was equally resistant to inactivation at 250 nm and 280 nm (Alper et al. 1967). Because proteins in general and aldolase in particular (Setlow and Doyle 1957) are more sensitive to UV irradiation at 280 nm than at 250 nm, Tikvah Alper and her colleagues concluded that the scrapie agent was unlikely to contain protein (Alper et al. 1967). The sensitivity of proteins to inactivation at 280 nm is usually attributed to the destruction of amino acids with aromatic side chains.

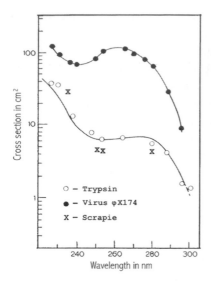

Figure 2. Spectrum for the inactivation of mouse scrapie prions ("x" symbols, multiply the ordinate by 10^{-19}) is superimposed on the inactivation spectrum for trypsin with 4% cysteine residues (*open circles*, multiply the ordinate by 10^{-19}). Both are shown with the spectrum for the ϕX174 bacteriophage (*filled circles*, multiply the ordinate by 10^{-15}). Data from Latarjet et al. (1970) for scrapie prions, from Setlow and Doyle (1957) for trypsin, and from Setlow and Boyce (1960) for ϕX174. (Reprinted, with permission, from Setlow 2002 [copyright Elsevier Science].)

Later, the scrapie agent was found to be six times more sensitive to inactivation at 237 nm than at either 250 nm or 280 nm (Fig. 2) (Latarjet et al. 1970). This finding was interpreted as further evidence that the scrapie agent contains neither a protein nor nucleic acid. Although the data were not sufficiently precise to eliminate protein as a candidate, a polysaccharide or polynucleotide composed of numerous modified nucleosides seemed more likely.

In October of 1996, during a visit to the Brookhaven National Laboratory, I was stunned when Richard Setlow showed me the inactivation spectrum of trypsin and pointed out its similarities to the inactivation spectrum of the scrapie agent. The UV inactivation spectrum for the scrapie agent is readily superimposed on that of trypsin (Setlow 2002). In contrast, the spectrum for the small single-stranded DNA bacteriophage ϕX174 is quite different from the spectra of the scrapie agent and trypsin.

Ironically, the UV inactivation spectrum of trypsin was published with the data on aldolase in the same paper (Setlow and Doyle 1957). Neither Alper nor Raymond Latarjet, both of whom are outstanding radiobiologists, had appreciated the differences between the spectra of aldolase and trypsin (Latarjet et al. 1970), which seem to be attributable to the cysteine content of the two proteins: Aldolase contains 1% cysteine and trypsin contains 4% cysteine (Setlow and Doyle 1957). If Alper and Latarjet had interpreted their data as Setlow had done, perhaps answering the question of what causes scrapie and CJD might not have been so challenging.

Target Size

Inactivation by ionizing radiation gave a target size of ~150 kD (Alper et al. 1966), which was later revised to 55 kD (Fig. 3) (Bellinger-Kawahara et al. 1988). These data argued that the scrapie agent is as small as a viroid and prompted speculation that a viroid might be the cause of scrapie (Diener 1972). Later, when purified preparations of prions became available, their properties were compared with those of viroids. The properties of prions and viroids were found to be antithetical—consistent with the notion that viroids are composed of polynucleotides and prions are proteins (Table 2) (Diener et al. 1982). Because inactivation by ionizing radiation of viruses had given spurious results due to repair of double-strand-

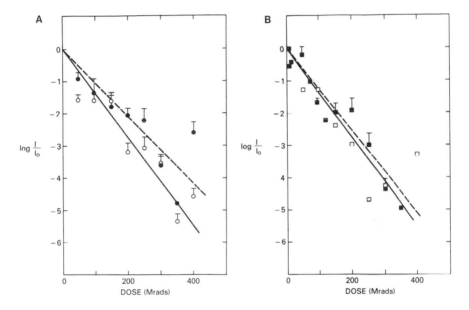

Figure 3. Resistance of scrapie prions to inactivation by ionizing radiation. (*A*) Microsomes (*filled circles*) isolated from scrapie-infected hamster brains were compared to detergent-extracted microsomes (*open circles*). Microsomes were isolated by a series of differential centrifugations as previously described. Sodium dodecyl sarcosinate (2%) was added to preparations and incubated for 90 minutes at room temperature prior to freezing. (*B*) Amyloid rods (*filled squares*) containing PrP 27-30 purified from scrapie-infected hamster brains were dissociated into liposomes (*open squares*). The rods and liposomes were prepared as described previously (Prusiner et al. 1983; Gabizon et al. 1987). All samples were frozen in ethanol dry-ice baths prior to storage at –70°C. The samples were irradiated with 13 MV electrons at –135°C. Controls receiving no irradiation were subjected to the same protocol. (Reprinted, with permission, from Bellinger-Kawahara et al. 1988 [copyright Elsevier Science].)

Table 2. Stabilities of prions and viroids after chemical and enzymatic treatment

Chemical treatment	Concentration	PSTV[a]	Scrapie agent
Et$_2$PC	10–20 mM	+	+
NH$_4$OH	0.1–0.5 M	+	–
Psoralen (AMT)	10–500 µg/ml	–	–
Phenol	saturated	–	+
SDS	1–10%	+	+
Zn^{++}	2 mM	–	–
Urea	3–8 mM	(–)	+
Alkali	pH 10	–	+
KSCN	1 M		+
RNase A	0.1–100 µg/ml	+	–
DNase	100 µg/ml	–	–
Proteinase K	100 µg/ml	–	+
Trypsin	100 µg/ml	–	+

"+" indicates sensitivity.
[a]Potato spindle tuber viroid.

ed genomes, the size of the putative scrapie virus was thought by a few investigators to be substantially larger than 150 kD (Rohwer 1984a, 1986). Once a protein was thought to be the most likely target of ionizing radiation, such arguments faded (Bellinger-Kawahara et al. 1988).

The target size of prions was determined to be independent of their physical state: Purified prions, either polymerized into amyloid or dispersed into liposomes, gave the same inactivation profile as prions found in either crude brain homogenates or microsomal fractions (Bellinger-Kawahara et al. 1988). The target size of 55 kD suggested that infectious prions are composed of either a dimer or trimer of PrP^{Sc}, the disease-causing isoform of the prion protein. Electron crystallographic studies of purified prions suggest that a trimer of amino-terminally truncated PrP^{Sc} (denoted PrP 27-30) molecules form the unit cell of the 2D crystals (Wille et al. 2002; Chapters 1 and 5). Because ionizing radiation inactivation of glycoproteins depends only on the size of the polypeptide chain and not on the covalently linked oligosaccharides (Kempner et al. 1986), an apparent molecular weight (M_r) of 55 kD would readily accommodate three polypeptide chains of PrP 27-30, each of which is ~17 kD (Oesch et al. 1985).

Hypotheses about the Nature of the Scrapie Agent

A fascinating array of structural hypotheses (Table 1) was offered to explain the unusual features of scrapie and its infectious agent. Among the earliest hypotheses was the notion that scrapie was a disease of muscle caused by the parasite *Sarcosporidia* (M'Gowan 1914; M'Fadyean 1918). After the successful experimental transmission of scrapie, the hypothesis that scrapie is caused by a "filterable" virus became popular (Cuillé and Chelle 1939; Wilson et al. 1950). Other suggestions were a provirus consisting of recessive genes generating RNA particles (Parry 1962, 1969); a small DNA virus (Kimberlin and Hunter 1967); a DNA subvirus controlled by a transmissible linkage substance (Adams and Field 1968; Adams 1970); membrane-bound DNA (Marsh et al. 1978); a virino (viroid-like DNA complexed with host proteins) (Dickinson and Outram 1979); and an aggregated, conventional virus with unusual properties (Rohwer and Gajdusek 1980). Small spherical particles ~10 nm in diameter were found in fractions said to be enriched for infectivity; these particles were thought to represent the smallest possible virus but were later shown to be ferritin (Cho and Greig 1975; Cho et al. 1977).

Other hypothetical structures proposed were a replicating protein (Griffith 1967; Pattison and Jones 1967; Lewin 1972, 1981); a replicating polysaccharide (Field 1967); an abnormal, replicating polysaccharide with membranes (Gibbons and Hunter 1967; Hunter et al. 1968); a nucleic acid surrounded by a polysaccharide coat (Adams and Caspary 1967; Narang 1974; Siakotos et al. 1979); a nucleoprotein complex (Latarjet et al. 1970); a naked nucleic acid similar to plant viroids (Diener 1972); a multicomponent system with one component quite small (Hunter et al. 1973; Somerville et al. 1976); and a spiroplasma-like organism (Bastian 1979; Humphery-Smith et al. 1992). As already noted, subsequent investigations showed the viroid suggestion to be incorrect (Diener et al. 1982), and many other studies argued for a cellular protein, as proposed earlier (Griffith 1967), that adopts an alternative conformation (Pan et al. 1993; Stahl et al. 1993; Telling et al. 1996a).

Unconventional Viruses

The term "unconventional virus" was proposed, but no structural details were ever given as to how these unconventional virions differ from conventional viral particles (Gajdusek 1977). Some investigators suggested that the term unconventional virus obscured the ignorance that continued to shroud the molecular nature of the infectious pathogens causing scrapie and CJD. The thoughts of Ian Pattison were particularly astute on this subject: "The fourth decade of my association with scrapie ended in 1978, with the causal agent still obscure, and virologists as adamant as ever that theirs was the only worthwhile point of view. To explain findings that did not fit with a virus hypothesis, they had rechristened the causal agent an 'unconventional virus.' Use of this ingenious cover-up made 'virus' meaningless—for is not a cottage an unconventional castle?" (Pattison 1988).

BIOASSAYS FOR PRIONS

Although experimental transmission of scrapie to mice allowed many more samples to be analyzed than was previously possible with sheep or goats, the one- to two-year intervals between designing experiments and obtaining results discouraged sequential studies, in which the results of one set of experiments were used as a foundation for the next. The measurement of scrapie infectivity by titration required 60 animals to evaluate a single sample and one year to establish the endpoint, as described above (Chandler 1961), which precluded big experiments and the pursuit of parallel studies. Such problems seem to encourage premature publication of experimental data, which sometimes forced investigators to defend their flawed interpretations.

From Mice to Hamsters and Back to Mice

The identification of an inoculum that produced scrapie in golden Syrian hamsters in ~70 days after intracerebral inoculation was the basis for an important advance (Marsh and Kimberlin 1975; Kimberlin and Walker 1977; Chapters 3 and 4). In earlier studies, Syrian hamsters had been inoculated with prions, but serial passage with short incubation times was not reported (Zlotnik 1963). Using the Syrian hamster, an incubation-time assay was developed that accelerated research by nearly a factor of 100 (Prusiner et al. 1980b, 1982b). Development of the incubation-time bioassay reduced the time required from 360 days to 70 days to measure prions in samples with high titers. Equally important, 4 animals could be used in place of the 60 that were required for endpoint titrations, making a large number of parallel experiments possible.

Notably, earlier attempts to develop more economical bioassays that relate titer to incubation times in mice were unsuccessful (Eklund et al. 1963; Hunter and Millson 1964). Some investigators used incubation times to characterize different "strains" of the scrapie agent, whereas others determined the kinetics of replication in rodents (Dickinson and Meikle 1969; Dickinson et al. 1969; Kimberlin and Walker 1978b, 1979), but they refrained from using incubation times to establish quantitative bioassays despite the successful application of such an approach for the measurement of picornaviruses and other animal viruses three decades earlier (Gard 1940).

Transgenic (Tg) mice expressing high levels of mouse (Mo) PrP or a foreign PrP on a null background ($Prnp^{0/0}$) are now used for most bioassays of prions. Incubation times as short as 50 days are possible for mice expressing high levels of either mouse or Syrian hamster (SHa) PrP (Prusiner et al. 1990; Carlson et al. 1994). Other Tg mice have been used to measure infectivity in bovine and human specimens (Telling et al. 1995; Scott et al. 1999; Buschmann et al. 2000; Korth et al. 2003).

PURIFICATION OF SCRAPIE INFECTIVITY

For over two decades, many investigators attempted to purify the scrapie agent but with relatively little success (Hunter et al. 1963, 1969, 1971; Hunter and Millson 1964, 1967; Kimberlin et al. 1971; Millson et al. 1971, 1976; Marsh et al. 1974, 1978, 1980; Siakotos et al. 1976; Gibbs and Gajdusek 1978; Millson and Manning 1979). The slow, cumbersome, and tedious bioassays in sheep and later in mice greatly limited the number of samples that could be analyzed. Little progress was made with sheep and goats because of very limited numbers of samples and incubation times that exceeded 18 months (Hunter 1972; Pattison 1988).

Hydrophobic Interactions

Although endpoint titrations in mice revealed some properties of the scrapie agent, development of an effective purification protocol was difficult because the interval between execution of the experiment and the availability of the results was nearly a year (Siakotos et al. 1976; Prusiner et al. 1984b; Prusiner 1988). The resistance of scrapie infectivity to nondenaturing detergents, nucleases, proteases, and glycosidases was determined by endpoint titrations in mice (Hunter and Millson 1964, 1967; Hunter et al. 1969; Millson et al. 1976).

Attempts to purify infectivity were complicated by the apparent size and charge heterogeneity of the scrapie agent, which was interpreted to be a consequence of hydrophobic interactions (Prusiner et al. 1978a,c). The scrapie agent presented a biochemical nightmare: Infectivity was spread from one end of a sucrose gradient to the other and from the void volume of chromatographic columns to fractions eluting at 5–10 times the included volume. Studies on the sedimentation properties of scrapie in mouse spleens and brains suggested that hydrophobic interactions were responsible for the nonideal physical behavior of the scrapie particle

(Prusiner et al. 1977, 1978b; Prusiner 1978). Such results demanded new approaches and better assays. Only the development of improved bioassays allowed purification of the infectious pathogen that causes scrapie and CJD (Prusiner et al. 1980b,1982b).

Fractions Enriched for Scrapie Infectivity

The discovery that the scrapie agent passaged in rats could be transmitted to Syrian hamsters in ~70 days not only led to the development of the incubation-time assay, but also provided a superior source of infectivity (Marsh and Kimberlin 1975). The level of scrapie infectivity in the brains of hamsters was found to be about 10-fold higher than in mice. Coupled with the development of an improved bioassay in hamsters, which allowed more rapid quantification of infectivity (Prusiner et al. 1980b, 1982b), it became possible to develop protocols for the significant enrichment of infectivity using a series of detergent extractions, limited digestions with proteases and nucleases, and differential centrifugation (Prusiner et al. 1980b), followed first by agarose gel electrophoresis (Prusiner et al. 1980a) and then by sucrose gradient centrifugation (Prusiner et al. 1982a, 1983). Using these methods, a protease-resistant polypeptide, later designated PrP 27-30, was identified in subcellular fractions of SHa brain enriched for infectivity; PrP 27-30 was absent in controls (Bolton et al. 1982; Prusiner et al. 1982a; McKinley et al. 1983a). Radioiodination of partially purified fractions revealed a protein unique to preparations from scrapie-infected brains (Bolton et al. 1982; Prusiner et al. 1982a). The existence of this protein was rapidly confirmed (Diringer et al. 1983).

THE PRION CONCEPT

For many years, the prion diseases were thought to be caused by slow-acting viruses, as described above. These diseases were often referred to as slow virus diseases, transmissible spongiform encephalopathies (TSEs), or unconventional viral diseases (Sigurdsson 1954; Gajdusek 1977, 1985). Considerable effort was expended searching for the "scrapie virus," yet none was found: Neither a virus-like particle nor a genome composed of RNA or DNA could be identified in preparations enriched for prion infectivity. Along with most investigators in this field, I believed that a small, sturdy virus causing scrapie and CJD would eventually be identified. As the search for the mythical scrapie virus progressed, the results of experimental studies constantly demanded more and more farfetched explanations to reconcile the data with the reasons the virus could not be found.

Distinguishing the Scrapie Agent from Viruses

Once an effective protocol was developed for preparation of partially purified fractions of scrapie agent from hamster brain, it became possible to demonstrate that procedures modifying or hydrolyzing proteins diminish scrapie infectivity (Table 2) (Prusiner et al. 1981; Prusiner 1982). Infectivity diminished as a function of the concentration of protease and duration of digestion. At the same time, tests completed in search of a scrapie-specific nucleic acid were unable to demonstrate any dependence of infectivity on a polynucleotide (Prusiner 1982), in agreement with earlier studies that reported the extreme resistance of infectivity to UV irradiation (Alper et al. 1967; Latarjet et al. 1970).

Choosing the Name "Prion"

On the basis of these findings, it seemed unlikely that the infectious pathogen of scrapie was either a virus or a viroid. Furthermore, other studies demonstrated that a polypeptide is required for propagation of the pathogen (Prusiner et al. 1981). For this reason, the term "prion" was introduced in order to distinguish the proteinaceous infectious particles that cause scrapie, CJD, and kuru from both viroids and viruses (Prusiner 1982); prions are "proteinaceous infectious particles that resist inactivation by procedures that modify nucleic acids." Hypotheses for the structure of the prion particle included (1) proteins surrounding a nucleic acid encoding them (a virus); (2) proteins associated with a small polynucleotide; and (3) proteins devoid of nucleic acid (Prusiner 1982). Mechanisms postulated for the replication of infectious prion particles ranged from those used by viruses, to the synthesis of polypeptides in the absence of a nucleic acid template, to the activation of transcription of cellular genes, and to posttranslational modifications of cellular proteins. Subsequent discoveries were used to narrow the hypotheses that could explain both the structure of the prion and the mechanism of replication.

A Family of Hypotheses

The prion hypothesis was not intended to champion a single entity as the structural explanation for the scrapie agent, but rather to embody a family of hypotheses (Chamberlin 1890), in which a protein was required for infectivity (Prusiner 1982). Almost immediately after I introduced the term "prion," some investigators redefined prion to signify an infectious particle composed exclusively of protein (Kimberlin 1982). The term "virino" was resuscitated to designate a particle composed of protein and a noncoding small nucleic acid similar to a viroid. Later, a few investigators adopted the term "protein-only hypothesis" to describe one possibility that had been postulated for the prion: an infectious particle composed exclusively of protein (Bolton and Bendheim 1988; Weissmann 1991).

Considerable evidence has accumulated over the past three decades in support of the prion concept and, in particular, the hypothesis that prions are composed entirely of protein (Prusiner 1991, 1997, 1998). Not only is the prion particle without precedent, but so are its mechanism of replication and mode of pathogenesis. A prudent, working definition of a prion is a "proteinaceous infectious particle that lacks nucleic acid." Perhaps this is an overly cautious and conservative definition, because we now understand much about the mechanism of prion diversity, and a second molecule is not required to explain prion strains (Prusiner 1997, 1998). Moreover, fungal prions, particularly those from yeast, have greatly strengthened the postulate that prions are infectious proteins (Chapter 7).

Attempts to Falsify the Prion Hypothesis

All attempts to falsify the prion hypothesis over the past three decades have failed. Experiments to separate scrapie infectivity from the protein and more specifically from PrP^{Sc} have been unsuccessful (McKinley et al. 1983a). No preparations of prions containing fewer than one PrP^{Sc} molecule per infectious dose that causes illness in 50% of inoculated animals (ID_{50}) have been reported (Akowitz et al. 1994). Prion replication in $Prnp^{0/0}$ mice has not been found, as described below. Some investigators have attempted to identify a scrapie-specific nucleic acid, but none has been found (Kellings et al. 1992; Manuelidis et al. 1995; Chapter 16).

Convergence of Data Supporting the Prion Concept

A remarkable convergence of experimental data accumulated over the last three decades convincingly argues that mammalian prions are composed of PrP^{Sc} molecules and, unlike all other infectious pathogens, are devoid of nucleic acid (Table 3). The copurification of scrapie infectivity and PrP^{Sc} was the first evidence that prions contained this macromolecule

Table 3. Arguments for prions being composed largely, if not entirely, of PrP^{Sc} molecules and devoid of nucleic acid

---

1. PrP^{Sc} and scrapie infectivity copurify using biochemical and immunologic procedures.
2. The unusual properties of PrP^{Sc} mimic those of prions. Many different procedures that modify or hydrolyze PrP^{Sc} inactivate prions.
3. Levels of PrP^{Sc} are directly proportional to prion titers. Nondenatured PrP^{Sc} has not been separated from scrapie infectivity.
4. No evidence for either a virus-like particle or a nucleic acid genome exists.
5. Accumulation of PrP^{Sc} is invariably associated with the pathology of prion diseases, including PrP amyloid plaques that are pathognomonic.
6. PrP gene mutations are linked to inherited prion disease and cause formation of PrP^{Sc}.
7. Overexpression of PrP^{C} increases the rate of PrP^{Sc} formation, which shortens the incubation time. Ablation of the PrP gene eliminates the substrate necessary for PrP^{Sc} formation and prevents both prion replication and prion disease.
8. Species variations in the PrP sequence are responsible, at least in part, for the species barrier that is found when prions are passaged from one host species to another.
9. PrP^{Sc} preferentially binds to PrP^{C} with an identical sequence, resulting in formation of nascent PrP^{Sc} and prion infectivity.
10. Chimeric and partially deleted PrP genes change susceptibility to prions from different species and support production of artificial prions with novel properties that are not found in nature.
11. Prion diversity is enciphered within the conformation of PrP^{Sc}. Strains can be generated by passage through hosts with different PrP genes. Prion strains are maintained by PrP^{C}–PrP^{Sc} interactions.
12. Human prions from fCJD(E200K) and FFI patients impart different properties to chimeric MHu2M PrP in transgenic mice, which provides a mechanism for strain propagation.

(Prusiner et al. 1982a, 1983), and implicit in this finding is that the molecular properties of prions and PrP^{Sc} are very similar, if not identical (Bolton et al. 1984). For example, both PrP^{Sc} and prion infectivity exhibited similar degradation kinetics by prolonged proteolysis (McKinley et al. 1983a). Various studies were performed attempting to separate prion infectivity from PrP^{Sc}, but no conditions were identified under which this could be accomplished.

The specificity of PrP^{Sc} for the prion diseases of humans and animals was a critical finding because it distinguished PrP^{Sc} from a variety of macromolecules that are expressed in response to CNS injury. A particularly important line of evidence implicating PrP^{Sc} as the causative agent was the discovery that PrP 27-30, the protease-resistant core of PrP^{Sc}, polymerizes into amyloid in vitro (Prusiner et al. 1983; McKinley et al. 1991) and that amyloid plaques in the brains of animals and humans with prion disease contain fragments of PrP, as shown by immunostaining and protein sequencing (Bendheim et al. 1984; DeArmond et al. 1985; Tagliavini et al. 1991, 1994).

The finding that purified PrP amyloid rods could disperse into liposomes without a loss of infectivity demonstrated that polymers of PrP 27-30 were not required for infectivity (Gabizon et al. 1987), in agreement with the small radiation target size of the prion. This liposomal dispersion permitted the immunoaffinity purification of PrP^{Sc} with a concomitant enrichment of prion infectivity by a procedure that was independent of the biochemical scheme initially used to purify infectivity (Gabizon et al. 1988a). Anti-PrP antibodies were also shown to neutralize infectivity when PrP^{Sc} was dispersed into liposomes.

Once the amino-terminal sequence of PrP 27-30 was determined (Prusiner et al. 1984a), recovery of cognate cDNAs encoding PrP was possible (Chesebro et al. 1985; Oesch et al. 1985). With PrP cDNAs, genetic linkage between PrP gene polymorphisms and the length of the disease incubation time was established (Carlson et al. 1986). Subsequently, mutations in the human PrP gene resulting in nonconservative amino acid substitutions were shown to be genetically linked to inherited prion diseases such as Gerstmann-Sträussler-Scheinker (GSS) disease and familial (f) CJD (Hsiao et al. 1989).

In other studies that provided strength to the prion concept, mutant PrP transgenes expressed in mice were shown to cause spontaneous neurodegeneration that is indistinguishable from experimental scrapie in mice (Hsiao et al. 1990). In studies on the transmission of prions from one species to another, species-specific differences in PrP sequences were found to be pivotal (Scott et al. 1989). Additionally, the length of the incubation time was found to be inversely proportional to the level of PrP^C expression (Prusiner et al. 1990). By abolishing PrP expression through genetic ablation, $Prnp^{0/0}$ mice were shown to be resistant to experimental prion disease (Büeler et al. 1993; Prusiner et al. 1993b). When $Prnp^{0/0}$ mice inoculated with prions were sacrificed at various intervals after inoculation, bioassay of their brain homogenates revealed no evidence of prion replication (Prusiner et al. 1993b; Sailer et al. 1994).

THE SEARCH FOR A PRION-SPECIFIC NUCLEIC ACID

For more than a decade after the introduction of the prion hypothesis, some investigators argued for the existence of a nucleic acid buried within the interior of a protein coat to explain the phenomenon of prion strains (Bruce and Dickinson 1987; Bruce et al. 1992). The search for a prion-specific nucleic acid has been intense, thorough, and comprehensive, yet unrewarding. The challenge to find a prion-specific polynucleotide was initiated by investigators who found that scrapie agent infectivity is highly resistant to UV and ionizing irradiation (Alper et al. 1966, 1967, 1978). Their results prompted speculation that the scrapie pathogen might be devoid of nucleic acid—a postulate initially dismissed by most scientists. Although some investigators argued that the interpretation of these data was flawed (Rohwer 1984a,b, 1986, 1991), they and others have failed to demonstrate the putative nucleic acid for the scrapie prion (Manuelidis and Fritch 1996; Chesebro 1998).

Selective Inactivation Studies

On the basis of the resistance of the scrapie agent to both UV and ionizing irradiation, the possibility was raised that the scrapie agent might contain a small polynucleotide similar in size and properties to viroids of plants (Diener 1972). Subsequently, evidence for a putative DNA-like viroid was published (Malone et al. 1978, 1979; Marsh et al. 1978), but the findings could not be confirmed (Prusiner et al. 1980a) and the properties of the scrapie agent were found to be incompatible with those of viroids (Diener et al. 1982). Besides UV irradiation, reagents that specifically modify or damage nucleic acids, such as nucleases, psoralens, hydroxylamine, and Zn^{++} ions, were found not to alter scrapie infectivity in homogenates (Table 2) (Prusiner 1982), microsomal fractions (Prusiner 1982), purified prion rod preparations, or detergent–lipid–protein complexes (McKinley et al. 1983b; Bellinger-Kawahara et al. 1987a,b, 1988; Gabizon et al. 1988b; Neary et al. 1991).

Physical and Molecular Cloning Studies

Attempts to find a scrapie-specific polynucleotide using physical techniques, such as SDS-PAGE, were as unsuccessful as molecular cloning approaches (Oesch et al. 1988). Subtractive hybridization studies identified several cellular genes with increased expression in the brains of prion-infected animals, but no unique sequence could be identified (Weitgrefe et al. 1985; Diedrich et al. 1987; Duguid et al. 1988). Extensively purified fractions were analyzed for a prion-specific nucleic acid by use of a specially developed technique, designated return refocusing gel electrophoresis, but none was found (Meyer et al. 1991). These studies argue that if such a molecule exists, it is composed of 80 or fewer nucleotides; only oligonucleotides of fewer than 50 bases were found in a concentration of one molecule per ID_{50} unit in prion preparations highly enriched for infectivity (Kellings et al. 1992, 1994). These small nucleic acids were of variable length and thought to be degradation by-products generated during purification of prions. Larger nucleic acids were excluded as components essential for infectivity (Riesner et al. 1992). Data from UV inactivation studies of scrapie prions argue that if prions contain a scrapie-specific nucleic acid, then a single-stranded molecule cannot exceed 6 bases in length and a double-stranded molecule can contain up to 40 bp (Bellinger-Kawahara et al. 1987a). This larger estimate for a double-stranded polynucleotide accounted for the possibility of repair of UV-damaged molecules after injection into rodents for bioassay.

The search for a component other than PrP[Sc] within the prion particle has been focused largely on a nucleic acid, because some properties of prions are similar to those of viruses, and a polynucleotide would most readily explain different isolates or "strains" (Kimberlin and Walker 1978a; Dickinson and Fraser 1979; Bruce and Dickinson 1987; Dickinson and Outram 1988). Specific scrapie isolates characterized by distinct incubation times retain this property when repeatedly passaged in mice or hamsters (Kimberlin and Walker 1978a; Dickinson and Fraser 1979; Bruce and Dickinson 1987; Dickinson and Outram 1988). Because evidence from many studies argues that prion diversity is enciphered within the conformation of PrP[Sc], the biological argument for a scrapie-specific polynucleotide now seems moot (Bessen and Marsh 1994; Telling et al. 1996b; Peretz et al. 2002; Chapter 9).

THE SEARCH FOR THE ELUSIVE SCRAPIE "VIRUS"

Despite these studies, some investigators chose to continue championing the idea that scrapie is caused by a "virus" (Kimberlin 1990; Chesebro 1992, 1998; Manuelidis and Fritch 1996). A few argued that the scrapie virus is similar to a retrovirus (Manuelidis and Manuelidis 1989; Sklaviadis et al. 1989, 1990, 1992, 1993; Akowitz et al. 1990; Murdoch et al. 1990), while others contended that it induces amyloid deposition in brain (Braig and Diringer 1985; Diringer 1991, 1992). Others argued that scrapie is caused by a larger pathogen similar to spiroplasma bacterium (Bastian 1979, 1993), and still others contended that elongated protein polymers covered by DNA are the etiologic agents in scrapie (Narang et al. 1987a,b, 1988; Narang 1992a,b). DNA molecules like the D-loop DNA of mitochondria were also suggested as the cause of scrapie (Aiken et al. 1989, 1990).

Table 4. Arguments to support the contention that the scrapie agent is a virus

1. Virus is tightly bound to PrP^{Sc} or possesses a coat protein that shares antigenic sites and physical properties with PrP^{Sc}, explaining copurification.
2. Structural properties of the virus are the same as PrP^{Sc}; procedures that modify PrP^{Sc} inactivate the virus.
3. Viral genome is protected from procedures modifying nucleic acid by PrP^{Sc}.
4. No nucleic acid >50 nucleotides has been found in purified preparations: It has unusual properties, might be small, and thus difficult to detect.
5. Virus is hidden by PrP^{Sc} and thus not found.
6. PrP amyloid plaques and spongiform degeneration result from accumulation of PrP^{Sc} induced by the virus.
7. Virus uses PrP^{C} as a receptor; virus has higher affinity for mutant than wt receptor.
8. At least two different ubiquitous viruses must exist.
9. Species specificity and artificial prions with a new host sequence demand that either PrP^{C} or PrP^{Sc} is tightly bound to the virus.
10. PrP^{Sc} is a cofactor for the virus, which is necessary for infectivity.

Many of the foregoing experimental results indicate that if the scrapie virus exists, then PrP^{Sc} must be an integral component of it (Table 4). Such a scenario would demand that PrP^{C} must function as a receptor for this elusive virus; furthermore, pathologic mutations in PrP^{C} that cause inherited prion diseases enhance the binding of the virus. Similarly, knockout of the PrP gene in mice prevents the scrapie virus from binding to its receptor PrP^{C}, thus explaining the resistance of these mice to experimental prion disease.

Copurification of PrP^{Sc} and Prion Infectivity

The copurification of PrP^{Sc} (or PrP 27-30) and scrapie infectivity demands that the physicochemical properties as well as antigenicity of these two entities be similar (Table 3) (Gabizon et al. 1988a). The results of a wide array of inactivation experiments demonstrated the similarities in the properties of PrP 27-30 and scrapie infectivity (McKinley et al. 1983a; Prusiner et al. 1983, 1993a; Bolton et al. 1984; Riesner et al. 1996). That PrP^{C} is required for prion replication was surmised from the large body of evidence showing that PrP^{Sc} is a major component of the infectious prion particle (Prusiner 1991, 1992), which was confirmed by experiments with $Prnp^{0/0}$ mice (Büeler et al. 1992). These mice were found to be resistant to prion disease and to not replicate prions. In two initial studies, no evidence of prion disease could be found many months after inoculation of $Prnp^{0/0}$ mice with RML prions (Büeler et al. 1993; Prusiner et al. 1993b). In one study, no evidence for prion replication was found in $Prnp^{0/0}$ mice (Prusiner et al. 1993b), but in the other, $Prnp^{0/0}$ mice sacrificed 20 weeks after inoculation were found to have $10^{3.6}$ ID_{50} units/ml of prions in a brain homogenate by bioassay (Büeler et al. 1993). This result could not be confirmed by the authors in a subsequent study (Sailer et al. 1994), but was ascribed to contamination or residual inoculum, or as evidence for the replication of prion infectivity in the absence of PrP, the last of which has been cited as evidence for the mythical "scrapie virus" (Chesebro and Caughey 1993; Manuelidis et al. 1995; Lasmézas et al. 1997). Two additional reports on the resistance of $Prnp^{0/0}$ mice to prion infection have been published (Manson et al. 1994; Sakaguchi et al. 1995).

A Mythical Virus Containing PrP[Sc]

For viruses to be a plausible cause of scrapie and CJD, they must mimic the properties of PrP[Sc]; thus, the virus must either encode PrP or acquire it from the cell during assembly. Such a virus would have a coat protein that is composed of PrP[C] or a protein that is highly similar to PrP. Alternatively, the virus might bind PrP[Sc]. In either case, the PrP-like coat protein or the PrP[Sc]–virus complex must display properties that are indistinguishable from PrP[Sc] alone (Table 4). With each new species that the virus invades, it must acquire a new PrP[Sc] sequence as well as be capable of incorporating artificial PrP molecules, such as chimeric mouse–hamster and mouse–human PrP (Scott et al. 1993; Telling et al. 1994).

That the infectious scrapie pathogen does not carry a PrP gene was established by hybridization studies. Fewer than 0.002 nucleic acid molecules encoding PrP per ID_{50} unit were found in purified preparations of SHa prions (Oesch et al. 1985). To circumvent these arguments, it was hypothesized that PrP is encoded by two very different genes: one that encodes a cellular protein and the other that contains the elusive scrapie virus. The genetic code used by the viral PrP gene differs so greatly from that used by the cellular PrP gene that the cellular PrP cDNA probe fails to detect the viral PrP gene in highly purified preparations. If this was the case, such a viral genome containing a PrP gene would also be expected to encode some specialized proteins required for replication as well as some unique tRNAs needed for translation of the viral PrP. Both UV and ionizing radiation inactivation studies, as well as physical studies that eliminated the possibility of a large nucleic acid hiding with purified preparations of prions, argued against this viral PrP genome (Bellinger-Kawahara et al. 1987a,b, 1988; Kellings et al. 1992). Additionally, no protein other than PrP^{Sc} has been found in purified preparations of prions.

Viruses and Familial Prion Diseases

In response to the discovery that PrP gene mutations cause familial prion diseases (Hsiao et al. 1989), it was conjectured that PrP^C is a receptor for a ubiquitous virus that binds more tightly to mutant than to wt PrP^C (Kimberlin 1990; Chesebro 1998). A similar hypothesis was proposed to explain why the length of the incubation time was found to be inversely proportional to the level of PrP expression in Tg mice and why $Prnp^{0/0}$ mice are resistant to prions (Chesebro and Caughey 1993): Higher levels of PrP expression resulted in the more rapid spread of the putative virus and shorter incubation times, and $Prnp^{0/0}$ mice lack the receptor required to spread the virus.

Studies on the transmission of mutant prions from fatal familial insomnia (FFI) and fCJD(E200K) patients to Tg(MHu2M) mice, which resulted in the formation of two different PrP^{Sc} molecules (Telling et al. 1996b), forced a corollary to the ubiquitous virus postulate. To accommodate this result, at least two different viruses must reside worldwide, each of which binds to a different mutant HuPrP^C and induces a different MHu2M PrP^{Sc} conformer when transmitted to Tg mice. Alternatively, one ubiquitous virus could acquire different mutant PrP^{Sc} molecules corresponding to FFI or fCJD(E200K) and then induce different MHu2M PrP^{Sc} conformers upon transmission to Tg mice.

PRION STRAINS

Unable to find evidence for a nucleic acid, I suggested that strains represent different forms of PrP^{Sc} (Prusiner 1991). The first experimental support for this conjecture came from work on two strains of prions isolated from mink and serially passaged in Syrian hamsters (Bessen and Marsh 1994). A particularly informative set of experiments emerged from studies of two human familial prion diseases. In those studies, distinct strains of prions generated de novo in the brains of patients carrying different PrP gene mutations were propagated in mice expressing a chimeric mouse–human (MHu2M) PrP transgene. Extracts from the brains of FFI and fCJD(E200K) patients transmitted disease to mice expressing MHu2M PrP ~200 days after inoculation. The FFI inoculum induced formation of a PrP^{Sc} fragment of 19 kD as measured by SDS-PAGE after limited proteolysis and removal of asparagine-linked carbohydrates, whereas the fCJD(E200K) inoculum produced a 21-kD PrP^{Sc} fragment (Telling et al. 1996b). On second passage, Tg(MHu2M) mice inoculated with FFI prions showed an incubation time of ~130 days and a 19-kD PrP^{Sc} fragment whereas those inoculated with fCJD(E200K) prions exhibited an incubation time of ~170 days and a 21-kD PrP^{Sc} fragment (Prusiner 1997). The experimental data demonstrated that MHu2MPrP^{Sc} can exist in two different conformations, based on the sizes of the protease-resistant fragments, within an invariant amino acid sequence. The results of these studies argue that PrP^{Sc} in the inoculum acts as a template for the conversion of PrP^{C} in the host into nascent PrP^{Sc}. Imparting the size of the protease-resistant fragment of PrP^{Sc} through conformational templating provides a mechanism for both the generation and propagation of prion strains.

Similar results were obtained with different strains of prions propagated in Syrian hamsters. Mice expressing an artificial transgene encoding chimeric mouse–hamster PrP, designated MH2M, were inoculated with two SHa prion strains, Sc237 and DY (Peretz et al. 2002). On first passage in Tg(MHM2)$Prnp^{0/0}$ mice, SHa(Sc237) prions exhibited prolonged incubation times, followed by a profound shortening on second passage, typical of a species barrier (Table 5). Moreover, PrP^{Sc} of the MH2M(Sc237) strain possess different structural properties from the original SHa(Sc237) strain, as demonstrated by relative conformational stability measurements (Table 6). Conversely, transmission of SHa(DY) prions to Tg(MHM2)$Prnp^{0/0}$ mice showed no species barrier, and the MH2M(DY) strain retained the conformational and phenotypic properties of SHa(DY) prions. These results extend the findings described above for the FFI and fCJD(E200K) strains of prions and contend that a change in PrP^{Sc} conformation is intimately associated with the emergence of a new prion strain.

Studies with prions from cattle with BSE and humans with variant (v) CJD make the notion of a set of ubiquitous viruses even more untenable. Both BSE and vCJD prions transmitted to Tg mice expressing bovine (Bo) PrP but not to Tg mice expressing either HuPrP or chimeric MHu2MPrP (Hill et al. 1997; Scott et al. 1999). Moreover, sCJD, iCJD, and fCJD(E200K) prions failed to transmit disease to Tg(BoPrP) mice after more than 400 days. To interpret these findings in terms of a ubiquitous virus, one must argue for two different viruses, both of which can bind to wt HuPrP^{C}, but one virus induces vCJD and the other causes other forms of CJD. The former virus must be distinct from the one that causes FFI. Thus, the virus hypothesis requires at least three different ubiquitous viruses to explain these prion strains, based on the molecular characteristics of PrP^{Sc} and the host range in Tg mice.

Table 5. Incubation times for Sc237 and DY prions passaged in Syrian hamsters, Tg(MH2M) mice, or Tg(SHa) mice

Inoculum[a]	Host	n/n_0 [b]	Days ± S.E.M.[c]
SHa(Sc237)	SHa	66/68	75 ± 0.3
SHa(Sc237)	Tg(MoPrP)4053	1/15	536
SHa(Sc237)	Tg(MH2M)229/$Prnp^{0/0}$	41/41	83 ± 1.6
MH2M(Sc237)	Tg(MH2M)229/$Prnp^{0/0}$	40/40	50 ± 0.1
MH2M(Sc237)	SHa	29/30	113 ± 1.8
MH2M(MH2M(Sc237))	SHa	17/20	133 ± 2.2
SHa(DY)	SHa	41/41	165 ± 1.0
SHa(DY)	Tg(MH2M)229/$Prnp^{0/0}$	28/29	70 ± 1.0
MH2M(DY)	Tg(MH2M)229/$Prnp^{0/0}$	28/28	76 ± 1.0
MH2M(DY)	SHa	26/26	176 ± 2.0
MH2M(MH2M(DY))	SHa	20/20	153 ± 2.0
SHa(Sc237)	Tg(SHa)3922/$Prnp^{0/0}$	17/17	44 ± 1.1
TgSHa(Sc237)	Tg(SHa)3922/$Prnp^{0/0}$	20/20	53 ± 0.5
TgSHa(Sc237)	SHa	20/20	80 ± 0.5

[a]Inoculum was a 1% (w/v) crude brain homogenate of prion-infected animals.

[b]n, number of animals developing clinical signs of prion disease; n_0, total number of animals inoculated. Animals dying atypically following inoculation were excluded (Prusiner and McKinley 1987).

[c]Mean incubation period in days ± standard error of the mean (S.E.M.).

The "Virus" Is a Prion

In summary, no single hypothesis arguing for a virus could explain the findings summarized in Table 3; instead, various ad hoc hypotheses, many of which could be refuted by new studies, were constructed to accommodate the available data (Table 4). Despite so much evidence to the con-

Table 6. Properties of PrP^{Sc} of Sc237 and DY prions passaged in Syrian hamsters and transgenic mice

Prion[a]	Host	M_r [b]	[GdnHCl]$_{1/2}$ ± S.D.[c]
SHa(Sc237)	SHa	21	1.8 ± 0.16 (6)
TgSHa(Sc237)	Tg(SHa)	21	1.7 ± 0.10 (5)
MH2M(Sc237)	Tg(MH2M)	21	1.1 ± 0.09 (5)
MH2M(MH2M(Sc237))	Tg(MH2M)	21	0.96 ± 0.14 (12)
SHa(DY)	SHa	19	0.98 ± 0.13 (6)
MH2M(DY)	Tg(MH2M)	19	1.1 ± 0.06 (8)
MH2M(MH2M(DY))	Tg(MH2M)	19	1.0 ± 0.09 (9)

[a]Nomenclature of prion strains.

[b]Apparent molecular weight (M_r) of the deglycosylated protease-resistant fragment of PrP^{Sc}.

[c]ELISA denaturation transitions were performed using monoclonal antibody fragment (Fab) HuM-D18 (0.25 µg/ml), and [GdnHCl]$_{1/2}$ values were interpolated with a sigmoid algorithm using a nonlinear least-square fit. The results are the mean, with the standard deviation (S.D.) calculated from the number of denaturation curves given in parentheses.

trary, some investigators persist in believing that prion diseases are caused by viruses (Kimberlin 1990; Rohwer 1991; Xi et al. 1992; Chesebro and Caughey 1993; Akowitz et al. 1994; Özel and Diringer 1994; Manuelidis et al. 1995; Narang 1996; Purdey 1996a,b; Caughey and Chesebro 1997; Lasmézas et al. 1997). Yet, we are left with only one logical conclusion from this debate: The 50-year quest for the scrapie virus (Gordon 1946) has failed because a scrapie virus does not exist!

MINIPRIONS

Prior to the availability of the NMR solution structures for recombinant PrPs (Riek et al. 1996; Donne et al. 1997; James et al. 1997), PrP was sub-divided based on a four-helix bundle model of the protein (Huang et al. 1994). Each region was systematically deleted and the mutant constructs were expressed in scrapie-infected neuroblastoma (ScN2a) cells and Tg mice (Muramoto et al. 1996, 1997). Deletion of any of the four putative helical regions prevented PrP^{Sc} formation, whereas deletion of the amino-terminal region containing residues 23–89 did not affect the yield of PrP^{Sc} in cultured cells. In addition to the 67 residues at the amino terminus, 36 residues from positions 141 to 176, including helix A and the S2 β-strand, could be deleted without altering PrP^{Sc} formation. The resulting PrP molecule of 106 amino acids was designated PrP106, or miniprion (James et al. 1997).

Transgene-specified Susceptibility

Tg(MHM2,Δ23-89,Δ141-176)$Prnp^{0/0}$ mice that express PrP106 developed neurological dysfunction ~300 days after inoculation with RML prions previously passaged in CD-1 Swiss mice (Supattapone et al. 1999, 2001). The resulting prions containing PrP^{Sc}106 produced CNS disease in ~66 days upon subsequent passage in Tg(MHM2,Δ23-89,Δ141-176)$Prnp^{0/0}$ mice (Table 7). Besides widespread spongiform degeneration and PrP deposits, hippocampal regions CA-1, CA-2, and CA-3 showed severe pyramidal cell loss in Tg(MHM2,Δ23-89,Δ141-176)$Prnp^{0/0}$ mice inoculated with prions containing PrP^{Sc}106. No similar neuropathologic lesions were seen in previous studies with Tg mice. Tg(MoPrP-A) mice overexpressing MoPrP were resistant to RML106 miniprions but highly susceptible to RML prions. These mice remained well for more than 450 days after inoculation with miniprions but developed disease in ~50 days when inoculated with RML prions containing full-length MoPrP^{Sc}.

Table 7. Susceptibility and resistance of transgenic mice to artificial mi

Host	Incubation time [Days ± S.E.M.] (		RM	
	RML106 miniprions			
Tg(PrP106)$Prnp^{0/0}$ mice	66 ± 3	(40/40)	300 ± 22	(8/8)
Tg(MoPrP-A) mice	>450	(0/10)	50 ± 2	(16/16)

Data from references (Carlson et al. 1994; Supattapone et al. 1999).

[a]n, number of ill mice; n_0, number of inoculated mice.

Miniprions and Mythical Viruses

The unique incubation times and neuropathology in Tg mice caused by miniprions are difficult to reconcile with the notion that prion disease is caused by an as-yet-unidentified virus. When the mutant or wt PrP^C sequence in the host matched the PrP^{Sc} sequence in the inoculum, the mice were highly susceptible (Table 7). However, when there was a mismatch between PrP^C and PrP^{Sc}, the mice were resistant to prions. This principle of PrP interactions based on sequence similarity, which underlies the species barrier, is recapitulated in studies of PrP106 in which the amino acid sequence was drastically changed by deleting nearly 50% of the residues. Indeed, the unique properties of the miniprions provide another persuasive argument that prions are infectious proteins.

PrP^{Sc} IS SPECIFIC FOR PRION DISEASE

With the discovery of PrP 27-30 and production of antiserum (Bendheim et al. 1984), brain tissues from humans and animals with putative prion diseases were examined for the presence of this protein. PrP 27-30 was found in each case of prion disease and was absent in other neurodegenerative disorders, such as Alzheimer's disease, Parkinson's disease, and amyotrophic lateral sclerosis (ALS) (Bockman et al. 1985, 1987; Manuelidis et al. 1985; Brown et al. 1986). An important feature of PrP is the specificity of PrP^{Sc} for prion disease, which is consistent with the postulated role of PrP^{Sc} in both the transmission and pathogenesis of these illnesses (Prusiner 1987).

The accumulation of PrP^{Sc} contrasts markedly with that of glial fibrillary acid protein (GFAP) in prion disease. In scrapie, GFAP mRNA and protein levels rise as the disease progresses (Manuelidis et al. 1987), but this is neither specific nor necessary for either the transmission or pathogenesis of disease (Gomi et al. 1995; Tatzelt et al. 1996b).

A protein designated 14-3-3 seems to increase in the cerebrospinal fluid of many but not all patients dying of CJD (M.D. Geschwind et al., in prep.). Elevations in 14-3-3 protein are not specific for CJD (Harrington et al. 1986; Hsich et al. 1996); moreover, increased 14-3-3 levels have not been found in vCJD patients (Chapter 13). Increased levels of 14-3-3 in the CSF have been found in patients with herpes simplex encephalitis, multi-infarct dementia, and stroke. 14-3-3 is a stress protein, levels of which increase nonspecifically in CNS injury.

With the exception of PrP^{Sc}, no macromolecule specific for prion disease has been found in tissues of patients dying of these encephalopathies. In searches for a prion-specific nucleic acid, cDNAs complementary to mRNAs encoding other proteins with increased expression in prion disease have been identified (Duguid et al. 1988, 1989; Diedrich et al. 1993), yet none of the proteins has been found to be specific for prion disease.

Prions Contain PrP^{Sc}

An abundance of evidence argues that PrP^{Sc} is an essential component of the infectious prion particle (Table 3). Attempts to find a second component of the prion particle have been unsuccessful; indeed, many lines of investigation have converged to contend that prions are composed solely of PrP^{Sc} molecules. Although some investigators proposed that PrP^{Sc} is merely a pathologic product of scrapie infection and that PrP^{Sc} coincidentally purifies with the "scrapie virus" (Braig and Diringer 1985; Aiken et al. 1989, 1990; Manuelidis and Manuelidis 1989; Sklaviadis et al. 1989, 1990; Akowitz et al. 1990; Murdoch et al. 1990), such views are not supported by the data. No infective fractions containing <1 PrP^{Sc} molecule per ID$_{50}$ unit have been found, which argues that PrP^{Sc} is required for infectivity. Some investigators report that PrP^{Sc} accumulation in hamsters occurs after the synthesis of many infectious units (Czub et al. 1986, 1988), but these results have been refuted (Jendroska et al. 1991). In another study, the kinetics of PrP^{Sc} and the production of infectivity in mice inoculated with mouse-passaged CJD prions were similar in brain, but thought to be different in the salivary gland (Sakaguchi et al. 1993). The discrepancies between PrP^{Sc} and infectivity levels in the above studies appear to be due to comparisons of infectivity in crude homogenates with PrP^{Sc} concentrations in purified fractions. Other investigators claim to have dissociated infectivity from PrP 27-30 in brains of Syrian hamsters treated with amphotericin B and inoculated with the 263K isolate; however, no dissociation was seen with either hamsters inoculated with 139H prions or mice inoculated with 139A prions (Xi et al. 1992). A subsequent study refuted the dissociation of PrP^{Sc} and infectivity in Syrian hamsters inoculated with 263K prions (McKenzie et al. 1994).

The covalent structure of PrP^{Sc} remains uncertain because purified fractions contain ~10^5 PrP 27-30 molecules per ID$_{50}$ unit (Bolton et al. 1982; Prusiner et al. 1982a; McKinley et al. 1983a). If <1% of the PrP^{Sc} molecules contains an amino acid substitution or posttranslational modification that confers infectivity, our methods would not detect such a change (Stahl et al. 1993).

PrP AMYLOID AND ULTRASTRUCTURE

In preparations highly enriched for infectivity and containing only PrP 27-30 by silver staining of gels after SDS-PAGE, numerous rod-shaped particles were seen by electron microscopy after negative staining (Fig. 4) (Prusiner et al. 1983). Each of the rods was slightly different, in contrast to viruses, which exhibit extremely uniform structures (Williams 1954). These irregular rods, composed largely if not entirely of PrP 27-30, were indistinguishable morphologically from many other purified amyloids (Cohen et al. 1982). Studies of the prion rods with Congo red dye demon-

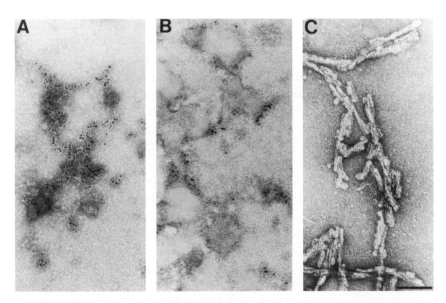

Figure 4. Electron micrographs of negatively stained and immunogold-labeled prion proteins: (A) PrP^C and (B) PrP^{Sc}. Neither PrP^C nor PrP^{Sc} forms recognizable, ordered polymers. (C) Prion rods composed of PrP 27-30 were negatively stained. The prion rods are indistinguishable from many purified amyloids. Bar, 100 nm. (Reprinted, with permission, from Prusiner 1998 [copyright The Nobel Foundation].)

strated that the rods also fulfilled the tinctorial criteria for amyloid (Prusiner et al. 1983), and immunostaining later showed that PrP is a major component of amyloid plaques in some animals and humans with prion disease (DeArmond et al. 1985; Kitamoto et al. 1986; Roberts et al. 1986). Subsequently, it was recognized that prion rods are not required for scrapie infectivity (Gabizon et al. 1987). Furthermore, the rods were shown to be an artifact of purification, during which limited proteolysis of PrP^{Sc} generated PrP 27-30, which polymerizes spontaneously in the presence of detergent (Fig. 4) (McKinley et al. 1991).

The idea that scrapie prions were composed of an amyloidogenic protein was truly heretical when it was introduced (Prusiner et al. 1983). Because the prevailing view at the time was that scrapie is caused by an atypical virus, the idea that the scrapie virus was composed of amyloid was unthinkable. Some investigators argued that amyloid proteins are mammalian polypeptides and not viral proteins! Ironically, they were correct because scrapie is not caused by a virus.

Scrapie-associated Fibrils

In crude extracts prepared from brains of rodents with scrapie and of humans with CJD, fibrillar structures composed of two or four helically wound subfilaments were found (Merz et al. 1981, 1983a). The crossing of these subfilaments occurred at specific intervals, and the distinctive ultrastructure of these fibers, designated scrapie-associated fibrils (SAF), permitted them to be distinguished from both intermediate filaments and amyloids (Merz et al. 1983b). The regular substructure of SAF prompted some investigators to propose that these particles might be the first example of a filamentous animal virus and that this virus causes scrapie and CJD (Merz et al. 1984).

Some investigators argued that SAF are synonymous with prion rods (Diringer et al. 1983; Merz et al. 1987; Somerville et al. 1989; Kimberlin 1990) although morphologic and tinctorial features of these fibrils clearly differentiate them from amyloid and, as such, from prion rods (Merz et al. 1981, 1983b). After the argument for a filamentous animal virus causing scrapie faded, it was hypothesized, in order to explain the accumulation of PrP^{Sc} in prion diseases, that a virus induces the formation of PrP amyloid (Diringer 1991). The term SAF was inappropriately used as a synonym for prion rods because it prompted the conclusion that SAF are composed of PrP (Diener 1987). Whether SAF contain PrP or are composed of one or more non-PrP proteins has not been studied; additionally, it is unknown whether SAF contain a polynucleotide.

Ultrastructural Studies in Search of a Virus

For many years, investigators searched for virus-like particles in brain sections from scrapie-infected sheep and rodents, as well as from humans who died of CJD. Despite the small target size of the infectious pathogen based on inactivation by ionizing radiation (Alper et al. 1966), there were many candidate structures reported (Bouteille et al. 1965; Vernon et al. 1970; Lamar et al. 1974; Jeffrey et al. 1992; Kato et al. 1992). Among these were tubulo-vesicular structures within postsynaptic evaginations, which seemed to be composed of arrays of spherical particles (David-Ferreira et al. 1968; Baringer and Prusiner 1978; Baringer et al. 1981; Liberski et al. 1990, 1992). These particles were relatively infrequent and could not be found in the brains of Syrian hamsters with clinical signs of scrapie (Baringer et al. 1983). Other structures, such as filamentous particles composed of protein internally and of DNA externally, have also been reported (Narang et al. 1987a,b, 1988; Narang 1992a,b,c, 1996), but these findings have not been confirmed (Bountiff et al. 1996).

AND MORE HYPOTHESES

After the prion concept was introduced and data from many lines of investigation were found to support it, additional hypotheses on the nature of the scrapie agent were proposed, perhaps in part because the properties of the prion were so unprecedented. The prion concept provided a challenge to investigators, the result of which was intensified research on the scrapie pathogen. To explain the phenomenon of prion strains, the virino was resuscitated, first in the form of a particle containing a foreign, viroid-like nucleic acid surrounded by PrP^{Sc} (Kimberlin 1982, 1990; Dickinson and Outram 1988) and later in the form of a small, cellular nucleic acid called a coprion that modifies the properties of PrP^{Sc}, labeled the apoprion (Weissmann 1991). No evidence for any polynucleotide that modifies the biological properties of PrP^{Sc} has been forthcoming. In another hypothesis, scrapie, CJD, and kuru were believed to be caused by an infectious amyloid; PrP was therefore renamed the scrapie amyloid protein and its infectious form was believed to be complexed with aluminum silicate (Gajdusek 1988). These infectious amyloids were likened to computer viruses on the premise that the term virus can be applied to any infectious entity (Gajdusek 1988), but use of the word virus in this context was as unhelpful as the term unconventional virus noted above. In another proposal, the amyloid-inducing virus is responsible for the conversion of PrP^{C} into PrP^{Sc}. Evidence for the existence of this virus was attributed to ~10-nm spherical particles in purified preparations of

PrP 27-30. Recent studies show that partially denatured PrP 27-30 can form such spheres that are devoid of infectivity (Riesner et al. 1996). Some investigators speculated that scrapie and CJD are caused by a retrovirus, for which they presented no credible evidence.

Prescient Speculation

Pattison suggested that the scrapie agent might be a basic protein based on the results of his attempts to purify the infectious pathogen and on findings of Alper and her colleagues (Alper et al. 1966, 1967; Pattison and Jones 1967). Pattison's speculation was enlarged by Griffith, who offered three possible mechanisms to explain how a protein might mimic an infectious pathogen (Griffith 1967). In this first scenario, a gene "G" might be silent normally, but its expression could be induced by its own product, protein "S." If protein S was the scrapie agent, it would induce the expression of gene G. Spontaneous disease can occur if gene G is not fully repressed, and a mutation in the gene may produce scrapie-resistant sheep. Griffith did not attempt to explain many different scrapie strains, but presumably, he would have hypothesized silent genes $G_1...G_n$ encoding proteins $S_1...S_n$, respectively. Furthermore, the evolutionary conservation of silent genes, which have no function but cause disease when expressed, was not addressed. In the second proposal, a protein can exist in two stable conformations, a and a′, in which a′ is the normal conformation. In this postulate, the dimer a_2 is the scrapie agent, which acts as a template for converting the conformation of a′ into a, which then dimerizes into a_2. Spontaneous disease could be explained in terms of the equilibrium between a and a_2. An alternate strain might be due to different isozymic subunits of the protein denoted b′ that can form mixed polymers, such as a_2b. Third, an antigen (A) may induce antibody A′, and A and A′ are identical, but the apparent absence of an immunologic component in the disease process makes this a less exciting proposal.

Some aspects of Griffith's first and second proposals were remarkably prescient. In agreement with his first proposal, we now know that changes in the amino acid sequence of PrP can render humans or animals resistant to prion disease whereas other mutations can produce inherited forms of the disease, which are fully penetrant. Several features of prions are in line with his second proposal, in which he suggested a conformational change in subunits of a multimeric protein could explain the infectious process. PrP^{Sc} and PrP^C represent a and a′, respectively, in his nomenclature, and the dimer a_2 is the prion. In further agreement with Griffith's second proposal, data support the notion that PrP^{Sc} acts as a template to direct the formation of nascent PrP^{Sc} (Prusiner et al. 1990; Telling et al. 1996b). What is especially interesting about Griffith's second proposal is that he seemed unencumbered by the proposition emerging at the time that the primary structure of a protein dictates a single tertiary structure under physiologic conditions (Anfinsen 1973). That prions violate this unitary relationship between primary and tertiary structure was genuinely unexpected (Pan et al. 1993). Even more surprising has been the finding that PrP^{Sc} can adopt more than one conformation, a phenomenon that explains the existence of prion strains (Prusiner 1991; Bessen and Marsh 1994; Telling et al. 1996b; Peretz et al. 2002).

Despite the remarkable insight of Griffith's proposals, several details and discrepancies have been uncovered by experimental data. The PrP gene, in which mutations can produce inherited prion disease, is not normally silent, but rather constitutively expressed in adults. Griffith posited in his second proposal that dimer a$_2$ is the disease agent; it is likely that the infectious prion particle is a trimer, not a dimer, based on ionizing radiation target data and electron crystallography findings. The ionizing radiation target size of 55 kD for the infectious prion particle accommodates a trimer of PrP 27-30 polypeptides (Bellinger-Kawahara et al. 1988). Similarly, the unit cell of 2D crystals is likely to comprise a trimer of PrP 27-30 (Wille et al. 2002; Chapter 5). In both his first and second proposals, Griffith attempted to explain strains by alternative amino acid sequences, which is clearly incorrect. Presumably, he would have invoked a similar explanation for his third proposal.

Were Griffith's proposals of value in deciphering the prion problem? The answer is probably no, but his truly prescient speculations serve to enrich the history of the field. In contrast, Alper's radiation inactivation data placed some important constraints on the physical features of the infectious pathogen of scrapie (Alper et al. 1966, 1967; Latarjet et al. 1970), but she and her colleagues failed to appreciate the similarity of the inactivation spectra for scrapie prions and trypsin, as described above (Fig. 2).

FUNGAL PRIONS

Although prions were originally defined in the context of an infectious pathogen (Prusiner 1982), it is now becoming widely accepted that prions are elements that impart and propagate conformational variability. Such a mechanism must surely not be restricted to a single class of transmissible pathogens. Indeed, it is likely that the original definition will need to be extended to encompass other situations in which a similar mechanism of information transfer occurs.

[URE3] and [PSI] in yeast as well as [Het-s*] in another fungus are the most well described fungal prion determinants (Wickner 1994; Chernoff et al. 1995; Coustou et al. 1997; Glover et al. 1997; King et al. 1997; Paushkin et al. 1997; Maddelein et al. 2002; Chapter 7). Studies of candidate prion proteins in yeast may prove to be particularly helpful in the dissection of some of the events that feature in PrP^{Sc} formation. Interestingly, different strains of yeast prions have been identified (Derkatch et al. 1996). Conversion to the [PSI$^+$] prion state in yeast requires the molecular chaperone Hsp104; however, no homolog of Hsp104 has been found in mammals (Chernoff et al. 1995; Patino et al. 1996). The amino-terminal prion domains of Ure2p and Sup35 that are responsible for the [URE3] and [PSI$^+$] phenotypes in yeast have been identified. In contrast to PrP, which is a glycosylphosphatidyl inositol (GPI)-anchored membrane protein, both Ure2p and Sup35p are cytosolic proteins (Wickner 1997). When the prion domains of these yeast proteins were expressed in E. coli, the proteins were found to polymerize into fibrils with properties similar to those of PrP 27-30 and other amyloids (Glover et al. 1997; King et al. 1997; Paushkin et al. 1997).

Whether prions explain some other examples of acquired inheritance in lower organisms is unclear (Sonneborn 1948; Landman 1991). For example, studies on the inheritance of positional order and cellular handedness on the surface of small organisms have demonstrated the epigenetic nature of these phenomena, but the mechanism remains unclear (Beisson and Sonneborn 1965; Frankel 1990).

"PROOF" OF THE PRION CONCEPT: DE NOVO FORMATION OF PRIONS

The de novo formation of prions from noninfectious materials has often been suggested as proof of the prion concept. Several approaches have been taken to develop experimental data. First, multiple attempts have been made to demonstrate prion infectivity in preparations of recombinant PrP. No fractions of purified recombinant PrP produced in E. coli or mammalian cells have exhibited prion infectivity when bioassayed in mice or hamsters (Baskakov et al. 2001, 2002).

Second, PrP molecules with pathologic mutations responsible for inherited prion diseases have been extensively studied. Different forms of mutant PrP^C are transformed into mutant PrP^{Sc} in humans and Tg mice. The E200K mutation in humans produces fCJD in 100% of carriers if they live long enough; in other words, this dominantly inherited disease is fully penetrant (Chapman et al. 1994; Spudich et al. 1995). Similar to humans with GSS, Tg mice expressing high levels of PrP with the analogous muta-tion (P101L) develop CNS dysfunction and accumulate a modified form of PrP^{Sc} (Hsiao et al. 1990). Prions in the brains of these Tg mice can be transmitted to another Tg line, designated Tg196, that expresses low lev-els of PrP(P101L) (Hsiao et al. 1994; Telling et al. 1996a). A few Tg196 mice develop spontaneous neurodegeneration at >500 days of age, but most remain healthy. In the brains of Tg mice expressing high levels of PrP(P101L) and of inoculated Tg196 mice, a modified form of PrP^{Sc} was found based on measurements of resistance to proteolytic digestion at 4°C and on measurements of abnormally folded PrP by the conforma-tion-dependent immunoassay (Tremblay et al. 2003).

Third, a PrP peptide composed of residues 89–143 has been shown to induce disease in 100% of inoculated Tg196 mice (Kaneko et al. 2000). This activity of the PrP peptide requires the P101L substitution and must be folded into a β-rich conformation. Serial transmission of disease in Tg196 mice initiated by the PrP(89–143,P101L) peptide argues that the 55-mer peptide is an artificial prion (Tremblay et al. 2003).

Fourth, experiments with the yeast prion [PSI$^+$] also argue for the de novo synthesis of prions (Sparrer et al. 2000). The prion domain of Sup35p and a portion of the adjacent M domain were expressed in *E. coli* and, after purification, aggregated into amyloid. The amyloid was then incorporated into liposomes, which were used to transform yeast into the [*PSI$^+$*] state. Like the PrP peptide studies described above, these studies argue for de novo synthesis of prions.

Taken together, these studies argue persuasively for proof of the prion concept and contend that prions are infectious proteins. Some investiga-tors would like to see studies in which recombinant wt PrP is converted into PrP^{Sc} in vitro and then inoculated into mice (Aguzzi and Weissmann 1997; Chesebro 1998) to produce prion disease. As the structure of PrP^{Sc} continues to be elucidated and the auxiliary proteins involved in the for-mation of PrP^{Sc} are identified, forming PrP^{Sc} in vitro should be possible.

LOOKING TO THE FUTURE

Although the study of prions has taken several unexpected directions over the past three decades, a rather novel and fascinating story of prion biol-ogy is emerging. Investigations of prions have elucidated a new principle of disease in humans and animals. Learning the details of the structures of PrP and deciphering the mechanism of PrP^C transformation into PrP^{Sc} will be important, but the fundamental principles of prion biology have become reasonably clear. Some investigators have preferred to view the composition of the infectious prion particle as unresolved (Aguzzi and Weissmann 1997; Chesebro 1998), but such a perspective denies an ever-enlarging body of data, none of which refutes the prion concept (Prusiner 1998). Moreover, the discovery of prion-like phenomena mediated by proteins unrelated to PrP in yeast and other fungi serves not only to strengthen the prion concept, but also to widen it (Wickner 1997).

Hallmarks of Prion Biology and Diseases

The hallmark of all prion diseases—whether sporadically occurring, dominantly inherited, or acquired by infection—is the accumulation of altered PrPs (Prusiner 1991, 1997, 1998). The conversion of PrP^C into PrP^{Sc} involves a profound conformational change whereby the α-helical content diminishes and the amount of β-sheet increases (Pan et al. 1993). These findings provide a reasonable mechanism to explain the conundrum presented by the three different manifestations of prion disease.

Understanding how PrP^C unfolds and refolds into PrP^{Sc} will be of paramount importance in transferring advances in the prion diseases to studies of other degenerative illnesses. The mechanism by which PrP^{Sc} is formed must involve a templating process whereby existing PrP^{Sc} directs the refolding of PrP^C into a nascent PrP^{Sc} with the same conformation.

Whether PrP^{Sc} is an aberrantly folded protein that is absent in normal, healthy cells or PrP^{Sc} is an alternatively folded protein present at low levels in some normal cells remains to be established. Studies with recombinant antibody fragments (Fabs) used to cure ScN2a cells in culture showed that these cells are capable of removing PrP^{Sc} when the substrate PrP^C is bound to the Fabs (Peretz et al. 2001). One hypothesis arising from these studies is that PrP^{Sc} is normally made at very low levels and is readily cleared by cells. These postulated low levels of PrP^{Sc} are below the threshold measurable by bioassays and, hence, extracts of normal brain do not cause disease when injected into normal recipients. When prion replication increases beyond the point where formation exceeds clearance, PrP^{Sc} accumulates and eventually reaches a threshold at which CNS dysfunction becomes evident.

If PrP^{Sc} is an alternatively folded protein that has a function in the metabolism of the cell, it is important to ask how many different PrP^{Sc} molecules might exist in any particular cell. From studies of prion strains, it is evident that PrP^{Sc} can adopt different conformations, each of which enciphers a distinct disease phenotype (Bessen and Marsh 1994; Telling et al. 1996b; Peretz et al. 2002). Might different conformations of PrP^{Sc} have different metabolic functions?

Multiple Conformers

The discovery that proteins may have multiple biologically active conformations may prove to be no less important than the existence of prions for diseases. How many different tertiary structures can PrP^{Sc} adopt? This query not only addresses the limits of prion diversity, but also applies to proteins as they normally function within the cell or act to affect homeostasis in multicellular organisms. The expanding list of chaperones that assist in the folding and unfolding of proteins promises much new knowledge about this process. For example, it is now clear that proproteases can carry their own chaperone activity, in which the *pro* portion of the protein functions as a chaperone in *cis* to guide the folding of the proteolytically active portion before it is cleaved (Shinde et al. 1997). Such a mechanism may feature in the maturation of polypeptide hormones. Interestingly, mutation of the chaperone portion of prosubtilisin resulted in the folding of a subtilisin protease with different properties from the one folded by the wt chaperone. Such chaperones have also been shown to work in *trans* (Shinde et al. 1997). In addition to transient metabolic regulation within the cell and hormonal regulation of multicellular organisms, alternative conformations of proteins might control, at least in part, the polymerization of proteins into multimeric structures, such as intermediate filaments. Such regulation of multimeric protein assemblies might occur in proteins that either form the polymers or function to facilitate the polymerization process. Additionally, alternative tertiary structures of proteins may regulate, at least in part, apoptosis during development and throughout adult life.

Applications to Other Degenerative Disorders

Not only will knowledge of PrP^{Sc} formation help in the rational design of drugs that interrupt the pathogenesis of prion diseases, but it may also open new approaches to deciphering the causes of, and to developing effective therapies for, the more common neurodegenerative diseases, including Alzheimer's disease, Parkinson's disease, and ALS (Prusiner 2001). Indeed, the expanding list of prion diseases and their novel modes of transmission and pathogenesis, as well as the unprecedented mechanisms of prion propagation and information transfer, indicate that much more attention to these fatal disorders of protein conformation is urgently needed.

Prions may have even wider implications than those noted for the common neurodegenerative diseases. If we think of prion diseases as disorders of protein conformation and do not require the diseases to be transmissible, what we have learned from the study of prions may reach far beyond neurodegeneration (Prusiner 2001). For example, the thyroid-stimulating hormone receptor (TSHR) has been shown to exist in two conformations that differ from the denatured state. Thyroid-stimulating hormone (TSH) in its active conformation binds to TSHR and can be inhibited by autoantibodies from the serum of patients with Graves' disease (Chazenbalk et al. 2001). The active conformation was found to be stabilized by chemical chaperones, which my colleagues and I previously used to stabilize PrP^C in ScN2a cells and prevent the conversion of PrP^C into PrP^{Sc} (Tatzelt et al. 1996a). Chemical chaperones (Welch and Brown 1996) were found to stabilize TSHR in the active state, which permitted binding of the Graves' autoantibodies. Whether Graves' disease is provoked by the shifting of the TSHR from one conformation to another remains to be determined. It is tempting to speculate that similar mechanisms might be operative in other autoimmune diseases, including multiple sclerosis, juvenile diabetes, lupus erythematosis, scleroderma, and rheumatoid arthritis (Prusiner 2001).

Shifting the Debate

The debate about prions and the diseases that they cause has shifted to such issues as how many biological processes are controlled by changes in protein conformation (Prusiner 1998). Although the extreme radiation-resistance of scrapie infectivity suggested that the pathogen causing this disease and related illnesses would be different from viruses, viroids, and bacteria (Alper et al. 1966, 1967), few thought that alternative protein conformations might even remotely feature in the pathogenesis of prion diseases (Griffith 1967). Indeed, an entirely new principle of disease was revealed in which an alternative conformation in a protein is propagated. The discovery of prions and their eventual acceptance by the community of scholars represent a triumph of the scientific process over prejudice. The future of this new and emerging area of biology should prove to be even more interesting and productive as a multitude of unpredicted discoveries emerge.

REFERENCES

Adams D.H. 1970. The nature of the scrapie agent: A review of recent progress. *Pathol. Biol.* **18:** 559–577.
Adams D.H. and Caspary E.A. 1967. Nature of the scrapie virus. *Br. Med. J.* **3:** 173.
Adams D.H. and Field E.J. 1968. The infective process in scrapie. *Lancet* **2:** 714–716.
Aguzzi A. and Weissmann C. 1997. Prion research: The next frontiers. *Nature* **389:** 795–798.

Aiken J.M., Williamson J.L., and Marsh R.F. 1989. Evidence of mitochondrial involvement in scrapie infection. *J. Virol.* **63:** 1686-1694.

Aiken J.M., Williamson J.L., Borchardt L.M., and Marsh R.F. 1990. Presence of mitochondrial D-loop DNA in scrapie-infected brain preparations enriched for the prion protein. *J. Virol.* **64:** 3265–3268.

Akowitz A., Sklaviadis T., and Manuelidis L. 1994. Endogenous viral complexes with long RNA cosediment with the agent of Creutzfeldt-Jakob disease. *Nucleic Acids Res.* **22:** 1101–1107.

Akowitz A., Sklaviadis T., Manuelidis E.E., and Manuelidis L. 1990. Nuclease-resistant polyadenylated RNAs of significant size are detected by PCR in highly purified Creutzfeldt-Jakob disease preparations. *Microb. Pathog.* **9:** 33–45.

Alper T., Haig D.A., and Clarke M.C. 1966. The exceptionally small size of the scrapie agent. *Biochem. Biophys. Res. Commun.* **22:** 278–284.

———. 1978. The scrapie agent: Evidence against its dependence for replication on intrinsic nucleic acid. *J. Gen. Virol.* **41:** 503–516.

Alper T., Cramp W.A., Haig D.A., and Clarke M.C. 1967. Does the agent of scrapie replicate without nucleic acid? *Nature* **214:** 764–766.

Anfinsen C.B. 1973. Principles that govern the folding of protein chains. *Science* **181:** 223–230.

Avery O.T., MacLeod C.M., and McCarty M. 1944. Studies on the chemical nature of the substance inducing transformation of pneumococcal types. Induction of transformation by a deoxyribonucleic acid fraction isolated from pneumococcus type III. *J. Exp. Med.* **79:** 137 157.

Baringer J.R. and Prusiner S.B. 1978. Experimental scrapie in mice: Ultrastructural observations. *Ann. Neurol.* **4:** 205–211.

Baringer J.R., Bowman K.A., and Prusiner S.B. 1983. Replication of the scrapie agent in hamster brain precedes neuronal vacuolation. *J. Neuropathol. Exp. Neurol.* **42:** 539–547.

Baringer J.R., Prusiner S.B. ,and Wong J. 1981. Scrapie-associated particles in post-synaptic processes. Further ultrastructural studies. *J. Neuropathol. Exp. Neurol.* **40:** 281–288.

Baskakov I.V., Legname G., Prusiner S.B., and Cohen F.E. 2001. Folding of prion protein to its native α-helical conformation is under kinetic control. *J. Biol. Chem.* **276:** 19687–19690.

Baskakov I.V., Legname G., Baldwin M.A., Prusiner S.B., and Cohen F.E. 2002. Pathway complexity of prion protein assembly into amyloid. *J. Biol. Chem.* **277:** 21140–21148.

Bastian F.O. 1979. Spiroplasma-like inclusions in Creutzfeldt-Jakob disease. *Arch. Pathol. Lab. Med.* **103:** 665–669.

———. 1993. Bovine spongiform encephalopathy: Relationship to human disease and nature of the agent. *ASM News* **59:** 235–240.

Beisson J. and Sonneborn T.M. 1965. Cytoplasmid inheritance of the organization of the cell cortex of *Paramecium aurelia*. *Proc. Natl. Acad. Sci.* **53:** 275–282.

Bellinger-Kawahara C., Cleaver J.E., Diener T.O., and Prusiner S.B. 1987a. Purified scrapie prions resist inactivation by UV irradiation. *J. Virol.* **61:** 159–166.

Bellinger-Kawahara C.G., Kempner E., Groth D.F., Gabizon R., and Prusiner S.B. 1988. Scrapie prion liposomes and rods exhibit target sizes of 55,000 Da. *Virology* **164:** 537–541.

Bellinger-Kawahara C., Diener T.O., McKinley M.P., Groth D.F., Smith D.R., and Prusiner S.B. 1987b. Purified scrapie prions resist inactivation by procedures that hydrolyze, modify, or shear nucleic acids. *Virology* **160:** 271–274.

Bendheim P.E., Barry R.A., DeArmond S.J., Stites D.P., and Prusiner S.B. 1984. Antibodies to a scrapie prion protein. *Nature* **310:** 418–421.

Bessen R.A. and Marsh R.F. 1994. Distinct PrP properties suggest the molecular basis of strain variation in transmissible mink encephalopathy. *J. Virol.* **68:** 7859–7868.

Bockman J.M., Prusiner S.B., Tateishi J., and Kingsbury D.T. 1987. Immunoblotting of Creutzfeldt-Jakob disease prion proteins: Host species-specific epitopes. *Ann. Neurol.* **21:** 589–595.

Bockman J.M., Kingsbury D.T., McKinley M.P., Bendheim P.E., and Prusiner S.B. 1985. Creutzfeldt-Jakob disease prion proteins in human brains. *N. Engl. J. Med.* **312:** 73–78.

Bolton D.C. and Bendheim P.E. 1988. A modified host protein model of scrapie. *Ciba Found. Symp.* **135:** 164–181.

Bolton D.C., McKinley M.P., and Prusiner S.B. 1982. Identification of a protein that purifies with the scrapie prion. *Science* **218:** 1309–1311.

———. 1984. Molecular characteristics of the major scrapie prion protein. *Biochemistry* **23:** 5898–5906.

Bountiff L., Levantis P., and Oxford J. 1996. Electrophoretic analysis of nucleic acids isolated from scrapie-infected hamster brain. *J. Gen. Virol.* **77:** 2371–2378.

Bouteille M., Fontaine C., Vedrenne C.L., and Delarue J. 1965. Sur un cas d'encephalite sub-aigue a inclusions. Étude anatomoclinique et ultrastructurale. *Rev. Neurol.* **118:** 454–458.

Braig H. and Diringer H. 1985. Scrapie: Concept of a virus-induced amyloidosis of the brain. *EMBO J.* **4:** 2309–2312.

Brown P., Coker-Vann M., Pomeroy K., Franko M., Asher D.M., Gibbs C.J., Jr., and Gajdusek D.C. 1986. Diagnosis of Creutzfeldt-Jakob disease by Western blot identification of marker protein in human brain tissue. *N. Engl. J. Med.* **314:** 547–551.

Bruce M.E. and Dickinson A.G. 1987. Biological evidence that the scrapie agent has an independent genome. *J. Gen. Virol.* **68:** 79–89.

Bruce M.E., Fraser H., McBride P.A., Scott J.R., and Dickinson A.G. 1992. The basis of strain variation in scrapie. In *Prion diseases in human and animals* (ed. S.B. Prusiner et al.), pp. 497–508. Ellis Horwood, London, United Kingdom.

Büeler H., Aguzzi A., Sailer A., Greiner R.-A., Autenried P., Aguet M., and Weissmann C. 1993. Mice devoid of PrP are resistant to scrapie. *Cell* **73:** 1339–1347.

Büeler H., Fisher M., Lang Y., Bluethmann H., Lipp H.-P., DeArmond S.J., Prusiner S.B., Aguet M., and Weissmann C. 1992. Normal development and behaviour of mice lacking the neuronal cell-surface PrP protein. *Nature* **356:** 577–582.

Buschmann A., Pfaff E., Reifenberg K., Müller H.M., and Groschup M.H. 2000. Detection of cattle-derived BSE prions using transgenic mice overexpressing bovine PrP^C. *Arch Virol.* (suppl.) **16:** 75–86.

Carlson G.A., Kingsbury D.T., Goodman P.A., Coleman S., Marshall S.T., DeArmond S., Westaway D., and Prusiner S.B. 1986. Linkage of prion protein and scrapie incubation time genes. *Cell* **46:** 503–511.

Carlson G.A., Ebeling C., Yang S.-L., Telling G., Torchia M., Groth D., Westaway D., DeArmond S.J., and Prusiner S.B. 1994. Prion isolate specified allotypic interactions between the cellular and scrapie prion proteins in congenic and transgenic mice. *Proc. Natl. Acad. Sci.* **91:** 5690–5694.

Caughey B. and Chesebro B. 1997. Prion protein and the transmissible spongiform encephalopathies. *Trends Cell Biol.* **7:** 56–62.

Chamberlin T.C. 1890. The method of multiple working hypotheses. *Science* (old series) **15:** 92–97.

Chandler R.L. 1961. Encephalopathy in mice produced by inoculation with scrapie brain material. *Lancet* **1:** 1378-1379.

———. 1963. Experimental scrapie in the mouse. *Res. Vet. Sci.* **4:** 276-285.

Chapman J., Ben-Israel J., Goldhammer Y., and Korczyn A.D. 1994. The risk of developing Creutzfeldt-Jakob disease in subjects with the *PRNP* gene codon 200 point mutation. *Neurology* **44:** 1683–1686.

Chazenbalk G.D., McLachlan S.M., Pichurin P., Yan X.M., and Rapoport B. 2001. A prion-like shift between two conformational forms of a recombinant thyrotropin receptor A-subunit module: Purification and stabilization using chemical chaperones of the form reactive with Graves' autoantibodies. *J. Clin. Endocrinol. Metab.* **86:** 1287–1293.

Chernoff Y.O., Lindquist S.L., Ono B., Inge-Vechtomov S.G., and Liebman S.W. 1995. Role of the chaperone protein Hsp104 in propagation of the yeast prion-like factor [psi^+]. *Science* **268:** 880–884.

Chesebro B. 1992. PrP and the scrapie agent. *Nature* **356:** 560.

———. 1998. BSE and prions: Uncertainties about the agent. *Science* **279:** 42–43.

Chesebro B. and Caughey B. 1993. Scrapie agent replication without the prion protein? *Curr. Biol.* **3:** 696–698.

Chesebro B., Race R., Wehrly K., Nishio J., Bloom M., Lechner D., Bergstrom S., Robbins K., Mayer L., Keith J.M., Garon C., and Haase A. 1985. Identification of scrapie prion protein-specific mRNA in scrapie-infected and uninfected brain. *Nature* **315:** 331–333.

Cho H.J. 1980. Requirement of a protein component for scrapie infectivity. *Intervirology* **14:** 213–216.

———. 1983. Inactivation of the scrapie agent by pronase. *Can. J. Comp. Med.* **47:** 494–496.

Cho H.J. and Greig A.S. 1975. Isolation of 14-nm virus-like particles from mouse brain infected with scrapie agent. *Nature* **257:** 685–686.

Cho H.J., Grieg A.S., Corp C.R., Kimberlin R.H., Chandler R.L., and Millson G.C. 1977. Virus-like particles from both control and scrapie-affected mouse brain. *Nature* **267:** 459–460.

Cohen A.S., Shirahama T., and Skinner M. 1982. Electron microscopy of amyloid. In *Electron microscopy of proteins* (ed. J.R. Harris), vol. 3, pp. 165–206. Academic Press, New York.

Coustou V., Deleu C., Saupe S., and Begueret J. 1997. The protein product of the *het-s* heterokaryon incompatibility gene of the fungus *Podospora anserina* behaves as a prion analog. *Proc. Natl. Acad. Sci.* **94:** 9773–9778.

Cuillé J. and Chelle P.L. 1939. Experimental transmission of trembling to the goat. *C.R. Seances Acad. Sci.* **208:** 1058–1060.

Czub M., Braig H.R., and Diringer H. 1986. Pathogenesis of scrapie: Study of the temporal development of clinical symptoms of infectivity titres and scrapie-associated fibrils in brains of hamsters infected intraperitoneally. *J. Gen. Virol.* **67:** 2005-2009.

———. 1988. Replication of the scrapie agent in hamsters infected intracerebrally confirms the pathogenesis of an amyloid-inducing virosis. *J. Gen. Virol.* **69:** 1753–1756.

David-Ferreira J.F., David-Ferreira K.L., Gibbs C.J., Jr., and Morris J.A. 1968. Scrapie in mice: Ultrastructural observations in the cerebral cortex. *Proc. Soc. Exp. Biol. Med.* **127:** 313–320.

DeArmond S.J., McKinley M.P., Barry R.A., Braunfeld M.B., McColloch J.R., and Prusiner S.B. 1985. Identification of prion amyloid filaments in scrapie-infected brain. *Cell* **41:** 221–235.

Derkatch I.L., Chernoff Y.O., Kushnirov V.V., Inge-Vechtomov S.G., and Liebman S.W.

1996. Genesis and variability of [*PSI*] prion factors in *Saccharomyces cerevisiae*. *Genetics* **144:** 1375–1386.

Dickinson A.G. and Fraser H. 1979. An assessment of the genetics of scrapie in sheep and mice. In *Slow transmissible diseases of the nervous system* (ed. S.B. Prusiner and W.J. Hadlow), vol. 1, pp. 367–386. Academic Press, New York.

Dickinson A.G. and Meikle V.M. 1969. A comparison of some biological characteristics of the mouse-passaged scrapie agents, 22A and ME7. *Genet. Res.* **13:** 213–225.

Dickinson A.G. and Outram G.W. 1979. The scrapie replication-site hypothesis and its implications for pathogenesis. In *Slow transmissible diseases of the nervous system* (ed. S.B. Prusiner and W.J. Hadlow), vol. 2, pp. 13–31. Academic Press, New York.

———. 1988. Genetic aspects of unconventional virus infections: The basis of the virino hypothesis. *Ciba Found. Symp.* **135:** 63–83.

Dickinson A.G., Meikle V.M., and Fraser H. 1969. Genetical control of the concentration of ME7 scrapie agent in the brain of mice. *J. Comp. Pathol.* **79:** 15–22.

Diedrich J.F., Carp R.I., and Haase A.T. 1993. Increased expression of heat shock protein, transferrin, and β_2-microglobulin in astrocytes during scrapie. *Microb. Pathog.* **15:** 1–6.

Diedrich J., Weitgrefe S., Zupancic M., Staskus K., Retzel E., Haase A.T., and Race R. 1987. The molecular pathogenesis of astrogliosis in scrapie and Alzheimer's disease. *Microb. Pathog.* **2:** 435–442.

Diener T.O. 1972. Is the scrapie agent a viroid? *Nature* **235:** 218–219.

———. 1979. *Viroids and viroid diseases.* John Wiley, New York.

———. 1987. PrP and the nature of the scrapie agent. *Cell* **49:** 719–721.

Diener T.O., McKinley M.P., and Prusiner S.B. 1982. Viroids and prions. *Proc. Natl. Acad. Sci.* **79:** 5220–5224.

Diringer H. 1991. Transmissible spongiform encephalopathies (TSE) virus-induced amyloidoses of the central nervous system (CNS). *Eur. J. Epidemiol.* **7:** 562–566.

———. 1992. Hidden amyloidoses. *Exp. Clin. Immunogenet.* **9:** 212–229.

Diringer H., Gelderblom H., Hilmert H., Ozel M., Edelbluth C., and Kimberlin R.H. 1983. Scrapie infectivity, fibrils and low molecular weight protein. *Nature* **306:** 476–478.

Donne D.G., Viles J.H., Groth D., Mehlhorn I., James T.L., Cohen F.E., Prusiner S.B., Wright P.E., and Dyson H.J. 1997. Structure of the recombinant full-length hamster prion protein PrP(29-231): The N terminus is highly flexible. *Proc. Natl. Acad. Sci.* **94:** 13452–13457.

Duguid J.R., Rohwer R.G., and Seed B. 1988. Isolation of cDNAs of scrapie-modulated RNAs by subtractive hybridization of a cDNA library. *Proc. Natl. Acad. Sci.* **85:** 5738–5742.

Duguid J.R., Bohmont C.W., Liu N., and Tourtellotte W.W. 1989. Changes in brain gene expression shared by scrapie and Alzheimer disease. *Proc. Natl. Acad. Sci.* **86:** 7260–7264.

Eklund C.M., Hadlow W.J., and Kennedy R.C. 1963. Some properties of the scrapie agent and its behavior in mice. *Proc. Soc. Exp. Biol. Med.* **112:** 974–979.

Field E.J. 1967. The significance of astroglial hypertrophy in scrapie, kuru, multiple sclerosis and old age together with a note on the possible nature of the scrapie agent. *Dtsch. Z. Nervenheilkd.* **192:** 265–274.

Fraenkel-Conrat H. and Williams R.C. 1955. Reconstitution of active tobacco virus from the inactive protein and nucleic acid components. *Proc. Natl. Acad. Sci.* **41:** 690–698.

Frankel J. 1990. Positional order and cellular handedness. *J. Cell Sci.* **97:** 205–211.

Gabizon R., McKinley M.P., and Prusiner S.B. 1987. Purified prion proteins and scrapie infectivity copartition into liposomes. *Proc. Natl. Acad. Sci.* **84:** 4017–4021.

Gabizon R., McKinley M.P., Groth D., and Prusiner S.B. 1988a. Immunoaffinity purification and neutralization of scrapie prion infectivity. *Proc. Natl. Acad. Sci.* **85:** 6617–6621.

Gabizon R., McKinley M.P., Groth D.F., Kenaga L., and Prusiner S.B. 1988b. Properties of scrapie prion protein liposomes. *J. Biol. Chem.* **263:** 4950–4955.

Gajdusek D.C. 1977. Unconventional viruses and the origin and disappearance of kuru. *Science* **197:** 943–960.

———. 1985. Subacute spongiform virus encephalopathies caused by unconventional viruses. In *Subviral pathogens of plants and animals: Viroids and prions* (ed. K. Maramorosch and J.J. McKelvey, Jr.), pp. 483–544. Academic Press, Orlando, Florida.

———. 1988. Transmissible and non-transmissible amyloidoses: Autocatalytic post-translational conversion of host precursor proteins to β-pleated sheet configurations. *J. Neuroimmunol.* **20:** 95–110.

Gard S. 1940. Encephalomyelitis of mice. II. A method for the measurement of virus activity. *J. Exp. Med.* **72:** 69-77.

Gibbons R.A. and Hunter G.D. 1967. Nature of the scrapie agent. *Nature* **215:** 1041–1043.

Gibbs C.J., Jr. and Gajdusek D.C. 1978. Atypical viruses as the cause of sporadic, epidemic, and familial chronic diseases in man: Slow viruses and human diseases. *Perspect. Virol.* **10:** 161–198.

Gibbs C.J., Jr., Gajdusek D.C., Asher D.M., Alpers M.P., Beck E., Daniel P.M., and Matthews W.B. 1968. Creutzfeldt-Jakob disease (spongiform encephalopathy): Transmission to the chimpanzee. *Science* **161:** 388–389.

Gierer A. and Schramm G. 1956. Infectivity of ribonucleic acid from tobacco mosaic virus. *Nature* **177:** 702 703.

Glover J.R., Kowal A.S., Schirmer E.C., Patino M.M., Liu J.-J., and Lindquist S. 1997. Self-seeded fibers formed by Sup35, the protein determinant of [*PSI*⁺], a heritable prion-like factor of *S. cerevisiae*. *Cell* **89:** 811–819.

Gomi H., Yokoyama T., Fujimoto K., Ikeda T., Katoh A., Itoh T., and Itohara S. 1995. Mice devoid of the glial fibrillary acidic protein develop normally and are susceptible to scrapie prions. *Neuron* **14:** 29–41.

Gordon W.S. 1946. Advances in veterinary research. *Vet. Res.* **58:** 516–520.

Griffith J.S. 1967. Self-replication and scrapie. *Nature* **215:** 1043–1044.

Harrington M.G., Merril C.R., Asher D.M., and Gajdusek D.C. 1986. Abnormal proteins in the cerebrospinal fluid of patients with Creutzfeldt-Jakob disease. *N. Engl. J. Med.* **315:** 279–283.

Haurowitz F. 1950. Protein synthesis. In *Chemistry and biology of proteins*, pp. 326–358. Academic Press, New York.

Hershey A.D. and Chase M. 1952. Independent functions of viral protein and nucleic acid in growth of bacteriophage. *J. Gen. Physiol.* **36:** 39–56.

Hill A.F., Desbruslais M., Joiner S., Sidle K.C.L., Gowland I., Collinge J., Doey L.J., and Lantos P. 1997. The same prion strain causes vCJD and BSE. *Nature* **389:** 448–450.

Hsiao K.K., Scott M., Foster D., Groth D.F., DeArmond S.J., and Prusiner S.B. 1990. Spontaneous neurodegeneration in transgenic mice with mutant prion protein. *Science* **250:** 1587–1590.

Hsiao K., Baker H.F., Crow T.J., Poulter M., Owen F., Terwilliger J.D., Westaway D., Ott J., and Prusiner S.B. 1989. Linkage of a prion protein missense variant to Gerstmann-Sträussler syndrome. *Nature* **338:** 342–345.

Hsiao K.K., Groth D., Scott M., Yang S.-L., Serban H., Rapp D., Foster D., Torchia M., DeArmond S.J. and Prusiner S.B. 1994. Serial transmission in rodents of neurodegeneration from transgenic mice expressing mutant prion protein. *Proc. Natl. Acad. Sci.* **91:** 9126–9130.

Hsich G., Kenney K., Gibbs C.J., Lee K.H., and Harrington M.G. 1996. The 14-3-3 brain protein in cerebrospinal fluid as a marker for transmissible spongiform encephalopathies. *N. Engl. J. Med.* **335:** 924–930.

Huang Z., Gabriel J.-M., Baldwin M.A., Fletterick R.J., Prusiner S.B., and Cohen F.E. 1994. Proposed three-dimensional structure for the cellular prion protein. *Proc. Natl. Acad. Sci.* **91:** 7139–7143.

Humphery-Smith I., Chastel C., and Le Goff F. 1992. Spirosplasmas and spongiform encephalopathies. *Med. J. Aust.* **156:** 142.

Hunter G.D. 1972. Scrapie: A prototype slow infection. *J. Infect. Dis.* **125:** 427–440.

Hunter G.D. and Millson G.C. 1964. Studies on the heat stability and chromatographic behavior of the scrapie agent. *J. Gen. Microbiol.* **37:** 251–258.

———. 1967. Attempts to release the scrapie agent from tissue debris. *J. Comp. Pathol.* **77:** 301–307.

Hunter G.D., Kimberlin R.H. and Gibbons R.A. 1968. Scrapie: A modified membrane hypothesis. *J. Theor. Biol.* **20:** 355–357.

Hunter G.D., Millson G.C., and Chandler R.L. 1963. Observations on the comparative infectivity of cellular fractions derived from homogenates of mouse-scrapie brain. *Res. Vet. Sci.* **4:** 543–549.

Hunter G.D., Gibbons R.A., Kimberlin R.H., and Millson G.C. 1969. Further studies of the infectivity and stability of extracts and homogenates derived from scrapie affected mouse brains. *J. Comp. Pathol.* **79:** 101–108.

Hunter G.D., Kimberlin R.H., Collis S., and Millson G.C. 1973. Viral and non-viral properties of the scrapie agent. *Ann. Clin. Res.* **5:** 262–267.

Hunter G.D., Kimberlin R.H., Millson G.C., and Gibbons R.A. 1971. An experimental examination of the scrapie agent in cell membrane mixtures. I. Stability and physicochemical properties of the scrapie agent. *J. Comp. Pathol.* **81:** 23–32.

James T.L., Liu H., Ulyanov N.B., Farr-Jones S., Zhang H., Donne D.G., Kaneko K., Groth D., Mehlhorn I., Prusiner S.B., and Cohen F.E. 1997. Solution structure of a 142-residue recombinant prion protein corresponding to the infectious fragment of the scrapie isoform. *Proc. Natl. Acad. Sci.* **94:** 10086–10091.

Jeffrey M., Scott J.R., Williams A., and Fraser H. 1992. Ultrastructural features of spongiform encephalopathy transmitted to mice from three species of bovidae. *Acta Neuropathol.* **84:** 559–569.

Jendroska K., Heinzel F.P., Torchia M., Stowring L., Kretzschmar H.A., Kon A., Stern A., Prusiner S.B., and DeArmond S.J. 1991. Proteinase-resistant prion protein accumulation in Syrian hamster brain correlates with regional pathology and scrapie infectivity. *Neurology* **41:** 1482–1490.

Kaneko K., Ball H.L., Wille H., Zhang H., Groth D., Torchia M., Tremblay P., Safar J., Prusiner S.B., DeArmond S.J., Baldwin M.A., and Cohen F.E. 2000. A synthetic peptide initiates Gerstmann-Sträussler-Scheinker (GSS) disease in transgenic mice. *J. Mol. Biol.* **295:** 997–1007.

Kato S., Hirano A., Umahara T., Llena J.F., Herz F., and Ohama E. 1992. Ultrastructural and immunohistochemical studies on ballooned cortical neurons in Creutzfeldt-Jakob disease: Expression of αB-crystallin, ubiquitin and stress-response protein 27. *Acta Neuropathol.* **84:** 443–448.

Kellings K., Prusiner S.B., and Riesner D. 1994. Nucleic acids in prion preparations:

Unspecific background or essential component? *Philos. Trans. R. Soc. Lond. B* **343:** 425–430.

Kellings K., Meyer N., Mirenda C., Prusiner S.B., and Riesner D. 1992. Further analysis of nucleic acids in purified scrapie prion preparations by improved return refocussing gel electrophoresis (RRGE). *J. Gen. Virol.* **73:** 1025–1029.

Kempner E.S., Miller J.H., and McCreery M.J. 1986. Radiation target analysis of glycoproteins. *Anal. Biochem.* **156:** 140–146.

Kimberlin R.H. 1982. Scrapie agent: Prions or virinos? *Nature* **297:** 107–108.

———. 1990. Scrapie and possible relationships with viroids. *Semin. Virol.* **1:** 153–162.

Kimberlin R.H. and Hunter G.D. 1967. DNA synthesis in scrapie-affected mouse brain. *J. Gen. Virol.* **1:** 115–124.

Kimberlin R. and Walker C. 1977. Characteristics of a short incubation model of scrapie in the golden hamster. *J. Gen. Virol.* **34:** 295–304.

———. 1978a. Evidence that the transmission of one source of scrapie agent to hamsters involves separation of agent strains from a mixture. *J. Gen. Virol.* **39:** 487–496.

———. 1978b. Pathogenesis of mouse scrapie: Effect of route of inoculation on infectivity titres and dose-response curves. *J. Comp. Pathol.* **88:** 39–47.

———. 1979. Pathogenesis of mouse scrapie: Dynamics of agent replication in spleen, spinal cord and brain after infection by different routes. *J. Comp. Pathol.* **89:** 551–562.

Kimberlin R.H., Millson G.C., and Hunter G.D. 1971. An experimental examination of the scrapie agent in cell membrane mixtures. III. Studies of the operational size. *J. Comp. Pathol.* **81:** 383–391.

King C.-Y., Tittman P., Gross H., Gebert R., Aebi M., and Wüthrich K. 1997. Prion-inducing domain 2-114 of yeast Sup35 protein transforms *in vitro* into amyloid-like filaments. *Proc. Natl. Acad. Sci.* **94:** 6618–6622.

Kirschbaum W.R. 1968. *Jakob-Creutzfeldt disease.* Elsevier, Amsterdam, The Netherlands.

Kitamoto T., Tateishi J., Tashima I., Takeshita I., Barry R.A., DeArmond S.J., and Prusiner S.B. 1986. Amyloid plaques in Creutzfeldt-Jakob disease stain with prion protein antibodies. *Ann. Neurol.* **20:** 204–208.

Korth C., Kaneko K., Groth D., Heye N., Telling G., Mastrianni J., Parchi P., Gambetti P., Will R., Ironside J., Heinrich C., Tremblay P., DeArmond S.J., and Prusiner S.B. 2003. Abbreviated incubation times for human prions in mice expressing a chimeric mouse–human prion protein transgene. *Proc. Natl. Acad. Sci.* **100:** 4784–4789.

Kretzschmar H.A., Neumann M., and Stavrou D. 1995. Codon 178 mutation of the human prion protein gene in a German family (Backer family): Sequencing data from 72 year-old celloidin-embedded brain tissue. *Acta Neuropathol.* **89:** 96–98.

Lamar C.H., Gustafson D.P., Krasovich M., and Hinsman E.J. 1974. Ultrastructural studies of spleens, brains, and brain cell cultures of mice with scrapie. *Vet. Pathol.* **11:** 13–19.

Landman O.E. 1991. The inheritance of acquired characteristics. *Annu. Rev. Genet.* **25:** 1–20.

Lasmézas C.I., Deslys J.-P., Robain O., Jaegly A., Beringue V., Peyrin J.-M., Fournier J.-G., Hauw J.-J., Rossier J., and Dormont D. 1997. Transmission of the BSE agent to mice in the absence of detectable abnormal prion protein. *Science* **275:** 402–405.

Latarjet R., Muel B., Haig D.A., Clarke M.C., and Alper T. 1970. Inactivation of the scrapie agent by near monochromatic ultraviolet light. *Nature* **227:** 1341–1343.

Lewin P. 1972. Scrapie: An infective peptide? *Lancet* **1:** 748.

———. 1981. Infectious peptides in slow virus infections: A hypothesis. *Can. Med. Assoc. J.* **124:** 1436–1437.

Liberski P.P., Yanagihara R., Gibbs C.J., Jr., and Gajdusek D.C. 1990. Appearance of tubulovesicular structures in experimental Creutzfeldt-Jakob disease and scrapie precedes the onset of clinical disease. *Acta Neuropathol.* **79:** 349–354.

Liberski P.P., Budka H., Sluga E., Barcikowska M., and Kwiecinski H. 1992. Tubulovesicular structures in Creutzfeldt-Jakob disease. *Acta Neuropathol.* **84:** 238–243.

Maddelein M.L., Dos Reis S., Duvezin-Caubet S., Coulary-Salin B., and Saupe S.J. 2002. Amyloid aggregates of the HET-s prion protein are infectious. *Proc. Natl. Acad. Sci.* **99:** 7402–7407.

Malone T.G., Marsh R.F., Hanson R.P., and Semancik J.S. 1978. Membrane-free scrapie activity. *J. Virol.* **25:** 933–935.

———. 1979. Evidence for the low molecular weight nature of the scrapie agent. *Nature* **278:** 575–576.

Manson J.C., Clarke A.R., McBride P.A., McConnell I., and Hope J. 1994. PrP gene dosage determines the timing but not the final intensity or distribution of lesions in scrapie pathology. *Neurodegeneration* **3:** 331–340.

Manuelidis L. and Fritch W. 1996. Infectivity and host responses in Creutzfeldt-Jakob disease. *Virology* **216:** 46–59.

Manuelidis L. and Manuelidis E.E. 1989. Creutzfeldt-Jakob disease and dementias. *Microb. Pathog.* **7:** 157–164.

Manuelidis L., Valley S., and Manuelidis E.E. 1985. Specific proteins associated with Creutzfeldt-Jakob disease and scrapie share antigenic and carbohydrate determinants. *Proc. Natl. Acad. Sci.* **82:** 4263–4267.

Manuelidis L., Sklaviadis T., Akowitz A., and Fritch W. 1995. Viral particles are required for infection in neurodegenerative Creutzfeldt-Jakob disease. *Proc. Natl. Acad. Sci.* **92:** 5124–5128.

Manuelidis L., Tesin D.M., Sklaviadis T., and Manuelidis E.E. 1987. Astrocyte gene expression in Creutzfeldt-Jakob disease. *Proc. Natl. Acad. Sci.* **84:** 5937–5941.

Marsh R.F. and Kimberlin R.H. 1975. Comparison of scrapie and transmissible mink encephalopathy in hamsters. II. Clinical signs, pathology and pathogenesis. *J. Infect. Dis.* **131:** 104–110.

Marsh R.F., Malone T.G., Semancik J.S., and Hanson R.P. 1980. Studies on the physicochemical nature of the scrapie agent. In *Search for the cause of multiple sclerosis and other chronic diseases of the central nervous system* (ed. A. Boese), pp. 314–320. Verlag Chemie, Weinheim, Germany.

Marsh R.F., Malone T.G., Semancik J.S., Lancaster W.D., and Hanson R.P. 1978. Evidence for an essential DNA component in the scrapie agent. *Nature* **275:** 146–147.

Marsh R.F., Semancik J.S., Medappa K.C., Hanson R.P., and Rueckert R.R. 1974. Scrapie and transmissible mink encephalopathy: Search for infectious nucleic acid. *J. Virol.* **13:** 993-996.

Masters C.L., Gajdusek D.C., and Gibbs C.J., Jr. 1981. Creutzfeldt-Jakob disease virus isolations from the Gerstmann-Sträussler syndrome. *Brain* **104:** 559–588.

McCarty M. 1985. *The transforming principle: Discovering that genes are made of DNA.* W.W. Norton, New York.

McKenzie D., Kaczkowski J., Marsh R., and Aiken J. 1994. Amphotericin B delays both scrapie agent replication and PrP-res accumulation early in infection. *J. Virol.* **68:** 7534–7536.

McKinley M.P., Bolton D.C., and Prusiner S.B. 1983a. A protease-resistant protein is a structural component of the scrapie prion. *Cell* **35:** 57–62.

McKinley M.P., Masiarz F.R., Isaacs S.T., Hearst J.E., and Prusiner S.B. 1983b. Resistance of the scrapie agent to inactivation by psoralens. *Photochem. Photobiol.* **37:** 539-545.

McKinley M.P., Meyer R.K., Kenaga L., Rahbar F., Cotter R., Serban A., and Prusiner S.B. 1991. Scrapie prion rod formation *in vitro* requires both detergent extraction and limited proteolysis. *J. Virol.* **65:** 1340–1351.

Meggendorfer F. 1930. Klinische und genealogische Beobachtungen bei einem Fall von spastischer Pseudosklerose Jakobs. *Z. Gesamte Neurol. Psychiatrie* **128:** 337-341.

Merz P.A., Somerville R.A., Wisniewski H.M., and Iqbal K. 1981. Abnormal fibrils from scrapie-infected brain. *Acta Neuropathol.* **54:** 63–74.

Merz P.A., Kascsak R.J., Rubenstein R., Carp R.I., and Wisniewski H.M. 1987. Antisera to scrapie-associated fibril protein and prion protein decorate scrapie-associated fibrils. *J. Virol.* **61:** 42–49.

Merz P.A., Somerville R.A., Wisniewski H.M., Manuelidis L., and Manuelidis E.E. 1983a. Scrapie-associated fibrils in Creutzfeldt-Jakob disease. *Nature* **306:** 474–476.

Merz P.A., Wisniewski H.M., Somerville R.A., Bobin S.A., Masters C.L., and Iqbal K. 1983b. Ultrastructural morphology of amyloid fibrils from neuritic and amyloid plaques. *Acta Neuropathol.* **60:** 113–124.

Merz P.A., Rohwer R.G., Kascsak R., Wisniewski H.M., Somerville R.A., Gibbs C.J., Jr., and Gajdusek D.C. 1984. Infection-specific particle from the unconventional slow virus diseases. *Science* **225:** 437–440.

Meyer N., Rosenbaum V., Schmidt B., Gilles K., Mirenda C., Groth D., Prusiner S.B., and Riesner D. 1991. Search for a putative scrapie genome in purified prion fractions reveals a paucity of nucleic acids. *J. Gen. Virol.* **72:** 37–49.

M'Fadyean J. 1918. Scrapie. *J. Comp. Pathol.* **31:** 102–131.

M'Gowan J.P. 1914. *Investigation into the disease of sheep called "scrapie".* William Blackwood and Sons, Edinburgh, Scotland.

Millson G.C. and Manning E.J. 1979. The effect of selected detergents on scrapie infectivity. In *Slow transmissible diseases of the nervous system* (ed. S.B. Prusiner and W.J. Hadlow), vol. 2, pp. 409–424. Academic Press, New York.

Millson G.C., Hunter G.D., and Kimberlin R.H. 1971. An experimental examination of the scrapie agent in cell membrane mixtures. II. The association of scrapie infectivity with membrane fractions. *J. Comp. Pathol.* **81:** 255–265.

———. 1976. The physico-chemical nature of the scrapie agent. In *Slow virus diseases of animals and man* (ed. R.H. Kimberlin), pp. 243–266. American Elsevier, New York.

Mirsky A.E. and Pollister A.W. 1946. Chromosin, a desoxyribose nucleoprotein complex of the cell nucleus. *J. Gen. Physiol.* **30:** 134–135.

Muramoto T., Scott M., Cohen F.E., and Prusiner S.B. 1996. Recombinant scrapie-like prion protein of 106 amino acids is soluble. *Proc. Natl. Acad. Sci.* **93:** 15457–15462.

Muramoto T., DeArmond S.J., Scott M., Telling G.C., Cohen F.E., and Prusiner S.B. 1997. Heritable disorder resembling neuronal storage disease in mice expressing prion protein with deletion of an α-helix. *Nat. Med.* **3:** 750–755.

Murdoch G.H., Sklaviadis T., Manuelidis E.E., and Manuelidis L. 1990. Potential retroviral RNAs in Creutzfeldt-Jakob disease. *J. Virol.* **64:** 1477–1486.

Narang H.K. 1974. Ruthenium red and lanthanum nitrate a possible tracer and negative stain for scrapie "particles"? *Acta Neuropathol.* **29:** 37–43.

———. 1992a. Relationship of protease-resistant protein, scrapie-associated fibrils and tubulofilamentous particles to the agent of spongiform encephalopathies. *Res. Virol.* **143:** 381–386.

———. 1992b. Scrapie-associated tubulofilamentous particles in human Creutzfeldt-Jakob disease. *Res. Virol.* **143:** 387–395.

——. 1992c. Scrapie-associated tubulofilamentous particles in scrapie hamsters. *Intervirology* **34:** 105–111.

——. 1996. The nature of the scrapie agent: The virus theory. *Proc. Soc. Exp. Biol. Med.* **212:** 208–224.

Narang H.K., Asher D.M. and Gajdusek D.C. 1987a. Tubulofilaments in negatively stained scrapie-infected brains: Relationship to scrapie-associated fibrils. *Proc. Natl. Acad. Sci.* **84:** 7730–7734.

——. 1988. Evidence that DNA is present in abnormal tubulofilamentous structures found in scrapie. *Proc. Natl. Acad. Sci.* **85:** 3575–3579.

Narang H.K., Asher D.M., Pomeroy K.L., and Gajdusek D.C. 1987b. Abnormal tubulovesicular particles in brains of hamsters with scrapie. *Proc. Soc. Exp. Biol. Med.* **184:** 504–509.

Neary K., Caughey B., Ernst D., Race R.E., and Chesebro B. 1991. Protease sensitivity and nuclease resistance of the scrapie agent propagated in vitro in neuroblastoma cells. *J. Virol.* **65:** 1031–1034.

Oesch B., Groth D.F., Prusiner S.B., and Weissmann C. 1988. Search for a scrapie-specific nucleic acid: A progress report. *Ciba Found. Symp.* **135:** 209–223.

Oesch B., Westaway D., Wälchli M., McKinley M.P., Kent S.B.H., Aebersold R., Barry R.A., Tempst P., Teplow D.B., Hood L.E., Prusiner S.B., and Weissmann C. 1985. A cellular gene encodes scrapie PrP 27-30 protein. *Cell* **40:** 735–746.

Özel M. and Diringer H. 1994. Small virus-like structure in fraction from scrapie hamster brain. *Lancet* **343:** 894–895.

Pan K.-M., Baldwin M., Nguyen J., Gasset M., Serban A., Groth D., Mehlhorn I., Huang Z., Fletterick R.J., Cohen F.E., and Prusiner S.B. 1993. Conversion of α-helices into β-sheets features in the formation of the scrapie prion proteins. *Proc. Natl. Acad. Sci.* **90:** 10962–10966.

Parry H.B. 1962. Scrapie: A transmissible and hereditary disease of sheep. *Heredity* **17:** 75–105.

——. 1969. Scrapie—Natural and experimental. In *Virus diseases and the nervous system* (ed. C.W.M. Whitty et al.), pp. 99–105. Blackwell, Oxford, United Kingdom.

Patino M.M., Liu J.-J., Glover J.R., and Lindquist S. 1996. Support for the prion hypothesis for inheritance of a phenotypic trait in yeast. *Science* **273:** 622–626.

Pattison I.H. 1988. Fifty years with scrapie: A personal reminiscence. *Vet. Rec.* **123:** 661–666.

Pattison I.H. and Jones K.M. 1967. The possible nature of the transmissible agent of scrapie. *Vet. Rec.* **80:** 1–8.

Paushkin S.V., Kushnirov V.V., Smirnov V.N., and Ter-Avanesyan M.D. 1997. In vitro propagation of the prion-like state of yeast Sup35 protein. *Science* **277:** 381–383.

Peretz D., Williamson R.A., Legname G., Matsunaga Y., Vergara J., Burton D., DeArmond S.J., Prusiner S.B., and Scott M.R. 2002. A change in the conformation of prions accompanies the emergence of a new prion strain. *Neuron* **34:** 921–932.

Peretz D., Williamson R.A., Kaneko K., Vergara J., Leclerc E., Schmitt-Ulms G., Mehlhorn I.R., Legname G., Wormald M.R., Rudd P.M., Dwek R.A., Burton D.R., and Prusiner S.B. 2001. Antibodies inhibit prion propagation and clear cell cultures of prion infectivity. *Nature* **412:** 739–743.

Prusiner S.B. 1978. An approach to the isolation of biological particles using sedimentation analysis. *J. Biol. Chem.* **253:** 916–921.

——. 1982. Novel proteinaceous infectious particles cause scrapie. *Science* **216:** 136–144.

——. 1987. Prions and neurodegenerative diseases. *N. Engl. J. Med.* **317:** 1571–1581.

———. 1988. Molecular structure, biology and genetics of prions. *Adv. Virus Res.* **35:** 83–136.

———. 1989. Scrapie prions. *Annu. Rev. Microbiol.* **43:** 345–374.

———. 1991. Molecular biology of prion diseases. *Science* **252:** 1515–1522.

———. 1992. Chemistry and biology of prions. *Biochemistry* **31:** 12278–12288.

———. 1995. The prion diseases. *Sci. Am.* **272:** 48–51, 54–57.

———. 1997. Prion diseases and the BSE crisis. *Science* **278:** 245–251.

———. 1998. Prions (Les Prix Nobel Lecture). In *Les Prix Nobel* (ed. T. Frängsmyr), pp. 268–323. Almqvist and Wiksell, Stockholm, Sweden.

———. 2001. Shattuck Lecture—Neurodegenerative diseases and prions. *N. Engl. J. Med.* **344:** 1516–1526.

Prusiner S.B. and McKinley M.P., Eds. 1987. *Prions—Novel infectious pathogens causing scrapie and Creutzfeldt-Jakob disease.* Academic Press, San Diego, California.

Prusiner S.B., Hadlow W.J., Eklund C.M., and Race R.E. 1977. Sedimentation properties of the scrapie agent. *Proc. Natl. Acad. Sci.* **74:** 4656–4660.

Prusiner S.B., Groth D.F., Bolton D.C., Kent S.B., and Hood L.E. 1984a. Purification and structural studies of a major scrapie prion protein. *Cell* **38:** 127–134.

Prusiner S.B., Groth D., Serban A., Stahl N., and Gabizon R. 1993a. Attempts to restore scrapie prion infectivity after exposure to protein denaturants. *Proc. Natl. Acad. Sci.* **90:** 2793–2797.

Prusiner S.B., Hadlow W.J., Eklund C.M., Race R.E., and Cochran S.P. 1978a. Sedimentation characteristics of the scrapie agent from murine spleen and brain. *Biochemistry* **17:** 4987–4992.

Prusiner S.B., Bolton D.C., Groth D.F., Bowman K.A., Cochran S.P. and McKinley M.P. 1982a. Further purification and characterization of scrapie prions. *Biochemistry* **21:** 6942–6950.

Prusiner S.B., Cochran S.P., Groth D.F., Downey D.E., Bowman K.A., and Martinez H.M. 1982b. Measurement of the scrapie agent using an incubation time interval assay. *Ann. Neurol.* **11:** 353–358.

Prusiner S.B., Groth D.F., Bildstein C., Masiarz F.R., McKinley M.P., and Cochran S.P. 1980a. Electrophoretic properties of the scrapie agent in agarose gels. *Proc. Natl. Acad. Sci.* **77:** 2984–2988.

Prusiner S.B., Groth D.F., Cochran S.P., Masiarz F.R., McKinley M.P., and Martinez H.M. 1980b. Molecular properties, partial purification, and assay by incubation period measurements of the hamster scrapie agent. *Biochemistry* **21:** 4883–4891.

Prusiner S.B., Garfin D.E., Baringer J.R., Cochran S.P., Hadlow W.J., Race R.E., and Eklund C.M. 1978b. Evidence for multiple molecular forms of the scrapie agent. In *Persistent viruses* (ed. J. Stevens et al.), pp. 591–613. Academic Press, New York.

Prusiner S.B., Hadlow W.J., Garfin D.E., Cochran S.P., Baringer J.R., Race R.E., and Eklund C.M. 1978c. Partial purification and evidence for multiple molecular forms of the scrapie agent. *Biochemistry* **17:** 4993–4997.

Prusiner S.B., McKinley M.P., Bowman K.A., Bolton D.C., Bendheim P.E., Groth D.F. and Glenner G.G. 1983. Scrapie prions aggregate to form amyloid-like birefringent rods. *Cell* **35:** 349-358.

Prusiner S.B., McKinley M.P., Groth D.F., Bowman K.A., Mock N.I., Cochran S.P., and Masiarz F.R. 1981. Scrapie agent contains a hydrophobic protein. *Proc. Natl. Acad. Sci.* **78:** 6675–6679.

Prusiner S.B., Groth D., Serban A., Koehler R., Foster D., Torchia M., Burton D., Yang S.-L., and DeArmond S.J. 1993b. Ablation of the prion protein (PrP) gene in mice prevents scrapie and facilitates production of anti-PrP antibodies. *Proc. Natl. Acad. Sci.* **90:** 10608–10612.

Prusiner S.B., McKinley M.P., Bolton D.C., Bowman K.A., Groth D.F., Cochran S.P., Hennessey E.M., Braunfeld M.B., Baringer J.R., and Chatigny M.A. 1984b. Prions: Methods for assay, purification and characterization. *Methods Virol.* **8:** 293–345.

Prusiner S.B., Scott M., Foster D., Pan K.-M., Groth D., Mirenda C., Torchia M., Yang S.-L., Serban D., Carlson G.A., Hoppe P.C., Westaway D. and DeArmond S.J. 1990. Transgenetic studies implicate interactions between homologous PrP isoforms in scrapie prion replication. *Cell* **63:** 673–686.

Purdey M. 1996a. The UK epidemic of BSE: Slow virus or chronic pesticide-initiated modification of the prion protein? 1. Mechanisms for a chemically induced pathogenesis/ transmissibility. *Med. Hypotheses* **46:** 429–443.

———. 1996b. The UK epidemic of BSE: Slow virus or chronic pesticide-initiated modification of the prion protein? 2. An epidemiological perspective. *Med. Hypotheses* **46:** 445–454.

Riek R., Hornemann S., Wider G., Billeter M., Glockshuber R., and Wüthrich K. 1996. NMR structure of the mouse prion protein domain PrP(121-231). *Nature* **382:** 180–182.

Riesner D., Kellings K., Meyer N., Mirenda C., and Prusiner S.B. 1992. Nucleic acids and scrapie prions. In *Prion diseases of humans and animals* (ed. S.B. Prusiner et al.), pp. 341–358. Ellis Horwood, London, United Kingdom.

Riesner D., Kellings K., Post K., Wille H., Serban H., Groth D., Baldwin M.A., and Prusiner S.B. 1996. Disruption of prion rods generates 10-nm spherical particles having high α-helical content and lacking scrapie infectivity. *J. Virol.* **70:** 1714–1722.

Roberts G.W., Lofthouse R., Brown R., Crow T.J., Barry R.A., and Prusiner S.B. 1986. Prion-protein immunoreactivity in human transmissible dementias. *N. Engl. J. Med.* **315:** 1231–1233.

Rohwer R.G. 1984a. Scrapie infectious agent is virus-like in size and susceptibility to inactivation. *Nature* **308:** 658–662.

———. 1984b. Virus-like sensitivity of the scrapie agent to heat inactivation. *Science* **223:** 600–602.

———. 1986. Estimation of scrapie nucleic acid molecular weight from standard curves for virus sensitivity to ionizing radiation. *Nature* **320:** 381.

———. 1991. The scrapie agent: "A virus by any other name". *Curr. Top. Microbiol. Immunol.* **172:** 195–232.

Rohwer R.G. and Gajdusek D.C. 1980. Scrapie—Virus or viroid: The case for a virus. In *Search for the cause of multiple sclerosis and other chronic diseases of the central nervous system* (ed. A. Boese), pp. 333–355. Verlag Chemie, Weinheim, Germany.

Roos R., Gajdusek D.C., and Gibbs C.J., Jr. 1973. The clinical characteristics of transmissible Creutzfeldt-Jakob disease. *Brain* **96:** 1–20.

Sailer A., Büeler H., Fischer M., Aguzzi A., and Weissmann C. 1994. No propagation of prions in mice devoid of PrP. *Cell* **77:** 967–968.

Sakaguchi S., Katamine S., Yamanouchi K., Kishikawa M., Moriuchi R., Yasukawa N., Doi T., and Miyamoto T. 1993. Kinetics of infectivity are dissociated from PrP accumulation in salivary glands of Creutzfeldt-Jakob disease agent-inoculated mice. *J. Gen. Virol.* **74:** 2117–2123.

Sakaguchi S., Katamine S., Shigematsu K., Nakatani A., Moriuchi R., Nishida N., Kurokawa K., Nakaoke R., Sato H., Jishage K., Kuno J., Noda T., and Miyamoto T. 1995. Accumulation of proteinase K-resistant prion protein (PrP) is restricted by the expression level of normal PrP in mice inoculated with a mouse-adapted strain of the Creutzfeldt-Jakob disease agent. *J. Virol.* **69:** 7586–7592.

Scott M., Groth D., Foster D., Torchia M., Yang S.-L., DeArmond S.J., and Prusiner S.B. 1993. Propagation of prions with artificial properties in transgenic mice expressing chimeric PrP genes. *Cell* **73:** 979–988.

Scott M.R., Will R., Ironside J., Nguyen H.-O.B., Tremblay P., DeArmond S.J., and Prusiner S.B. 1999. Compelling transgenetic evidence for transmission of bovine spongiform encephalopathy prions to humans. *Proc. Natl. Acad. Sci.* **96:** 15137–15142.

Scott M., Foster D., Mirenda C., Serban D., Coufal F., Wälchli M., Torchia M., Groth D., Carlson G., DeArmond S.J., Westaway D., and Prusiner S.B. 1989. Transgenic mice expressing hamster prion protein produce species-specific scrapie infectivity and amyloid plaques. *Cell* **59:** 847–857.

Setlow R.B. 2002. Shedding light on proteins, nucleic acids, cells, humans and fish. *Mutat. Res.* **511:** 1–14.

Setlow R. and Boyce R. 1960. The ultraviolet light inactivation of phi-X174 bacteriophage at different wave lengths and pH's. *Biophys. J.* **1:** 29–41.

Setlow R. and Doyle B. 1957. The action of monochromatic ultraviolet light on proteins. *Biochim. Biophys. Acta* **24:** 27–41.

Shinde U.P., Liu J.J., and Inouye M. 1997. Protein memory through altered folding mediated by intramolecular chaperones. *Nature* **389:** 520–522.

Siakotos A.N., Raveed D., and Longa G. 1979. The discovery of a particle unique to brain and spleen subcellular fractions from scrapie infected mice. *J. Gen. Virol.* **43:** 417–422.

Siakotos A.N., Gajdusek D.C., Gibbs C.J., Jr., Traub R.D., and Bucana C. 1976. Partial purification of the scrapie agent from mouse brain by pressure disruption and zonal centrifugation in sucrose-sodium chloride gradients. *Virology* **70:** 230–237.

Sigurdsson B. 1954. Rida, a chronic encephalitis of sheep with general remarks on infections which develop slowly and some of their special characteristics. *Br. Vet. J.* **110:** 341–354.

Sklaviadis T., Dreyer R., and Manuelidis L. 1992. Analysis of Creutzfeldt-Jakob disease infectious fractions by gel permeation chromatography and sedimentation field flow fractionation. *Virus Res.* **26:** 241–254.

Sklaviadis T.K., Manuelidis L., and Manuelidis E.E. 1989. Physical properties of the Creutzfeldt-Jakob disease agent. *J. Virol.* **63:** 1212–1222.

Sklaviadis T., Akowitz A., Manuelidis E.E., and Manuelidis L. 1990. Nuclease treatment results in high specific purification of Creutzfeldt-Jakob disease infectivity with a density characteristic of nucleic acid-protein complexes. *Arch. Virol.* **112:** 215–229.

———. 1993. Nucleic acid binding proteins in highly purified Creutzfeldt-Jakob disease preparations. *Proc. Natl. Acad. Sci.* **90:** 5713–5717.

Somerville R.A., Millson G.C., and Hunter G.D. 1976. Changes in a protein-nucleic acid complex from synaptic plasma membrane of scrapie-infected mouse brain. *Biochem. Soc. Trans.* **4:** 1112–1114.

Somerville R.A., Ritchie L.A., and Gibson P.H. 1989. Structural and biochemical evidence that scrapie-associated fibrils assemble *in vivo*. *J. Gen. Virol.* **70:** 25–35.

Sonneborn T.M. 1948. The determination of hereditary antigenic differences in genically identical *Paramecium* cells. *Proc. Natl. Acad. Sci.* **34:** 413–418.

Sparrer H.E., Santoso A., Szoka F.C., Jr., and Weissman J.S. 2000. Evidence for the prion hypothesis: Induction of the yeast [*PSI*⁺] factor by in vitro-converted Sup35 protein. *Science* **289:** 595–599.

Spudich S., Mastrianni J.A., Wrensch M., Gabizon R., Meiner Z., Kahana I., Rosenmann H., Kahana E., and Prusiner S.B. 1995. Complete penetrance of Creutzfeldt-Jakob disease in Libyan Jews carrying the E200K mutation in the prion protein gene. *Mol. Med.* **1:** 607–613.

Stahl N., Baldwin M.A., Teplow D.B., Hood L., Gibson B.W., Burlingame A.L., and Prusiner S.B. 1993. Structural analysis of the scrapie prion protein using mass spectrometry and amino acid sequencing. *Biochemistry* **32:** 1991–2002.

Stanley W.M. 1935. Isolation of a crystalline protein possessing the properties of tobacco mosaic virus. *Science* **81:** 644–645.

——. 1970. The "undiscovered" discovery. *Arch. Environ. Health* **21:** 256–262.

Stender A. 1930. Weitere Beiträge zum Kapitel "Spastische Pseudosklerose Jakobs". *Z. Gesamte Neurol. Psychiatrie* **128:** 528–543.

Supattapone S., Muramoto T., Legname G., Mehlhorn I., Cohen F.E., DeArmond S.J., Prusiner S.B., and Scott M.R. 2001. Identification of two prion protein regions that modify scrapie incubation time. *J. Virol.* **75:** 1408–1413.

Supattapone S., Bosque P., Muramoto T., Wille H., Aagaard C., Peretz D., Nguyen H.-O.B., Heinrich C., Torchia M., Safar J., Cohen F.E., DeArmond S.J., Prusiner S.B., and Scott M. 1999. Prion protein of 106 residues creates an artificial transmission barrier for prion replication in transgenic mice. *Cell* **96:** 869–878.

Tagliavini F., Prelli F., Ghiso J., Bugiani O., Serban D., Prusiner S.B., Farlow M.R., Ghetti B., and Frangione B. 1991. Amyloid protein of Gerstmann-Sträussler-Scheinker disease (Indiana kindred) is an 11 kd fragment of prion protein with an N-terminal glycine at codon 58. *EMBO J.* **10:** 513–519.

Tagliavini F., Prelli F., Porro M., Rossi G., Giaccone G., Farlow M.R., Dlouhy S.R., Ghetti B., Bugiani O., and Frangione B. 1994. Amyloid fibrils in Gerstmann-Sträussler-Scheinker disease (Indiana and Swedish kindreds) express only PrP peptides encoded by the mutant allele. *Cell* **79:** 695–703.

Tatzelt J., Prusiner S.B., and Welch W.J. 1996a. Chemical chaperones interfere with the formation of scrapie prion protein. *EMBO J.* **15:** 6363–6373.

Tatzelt J., Maeda N., Pekny M., Yang S.-L., Betsholtz C., Eliasson C., Cayetano J., Camerino A.P., DeArmond S.J., and Prusiner S.B. 1996b. Scrapie in mice deficient in apolipoprotein E or glial fibrillary acidic protein. *Neurology* **47:** 449–453.

Telling G.C., Haga T., Torchia M., Tremblay P., DeArmond S.J., and Prusiner S.B. 1996a. Interactions between wild-type and mutant prion proteins modulate neurodegeneration in transgenic mice. *Genes Dev.* **10:** 1736–1750.

Telling G.C., Scott M., Mastrianni J., Gabizon R., Torchia M., Cohen F.E., DeArmond S.J., and Prusiner S.B. 1995. Prion propagation in mice expressing human and chimeric PrP transgenes implicates the interaction of cellular PrP with another protein. *Cell* **83:** 79–90.

Telling G.C., Parchi P., DeArmond S.J., Cortelli P., Montagna P., Gabizon R., Mastrianni J., Lugaresi E., Gambetti P., and Prusiner S.B. 1996b. Evidence for the conformation of the pathologic isoform of the prion protein enciphering and propagating prion diversity. *Science* **274:** 2079–2082.

Telling G.C., Scott M., Hsiao K.K., Foster D., Yang S.-L., Torchia M., Sidle K.C.L., Collinge J., DeArmond S.J., and Prusiner S.B. 1994. Transmission of Creutzfeldt-Jakob disease from humans to transgenic mice expressing chimeric human-mouse prion protein. *Proc. Natl. Acad. Sci.* **91:** 9936–9940.

Tremblay P., Ball, H.L., Kaneko K., Groth D., Hegde R.S., Cohen F.E., DeArmond S.J.,

Prusiner S.B., and Safar J.G. 2003. Mutant PrP^{Sc} conformers induced by a synthetic peptide and various prion strains. *J. Virol.* (in press).

Vernon M.L., Horta-Barbosa L., Fuccillo D.A., Sever J.L., Baringer J.R., and Birnbaum G. 1970. Virus-like particles and nucleoprotein-type filaments in brain tissue from two patients with Creutzfeldt-Jakob disease. *Lancet* 1: 964–966.

Watson J.D. and Crick F.H.C. 1953. Genetical implication of the structure of deoxyribose nucleic acid. *Nature* 171: 964–967.

Weissmann C. 1991. A "unified theory" of prion propagation. *Nature* 352: 679–683.

Weitgrefe S., Zupancic M., Haase A., Chesebro B., Race R., Frey W., II, Rustan T., and Friedman R.L. 1985. Cloning of a gene whose expression is increased in scrapie and in senile plaques. *Science* 230: 1177–1181.

Welch W.J. and Brown C.R. 1996. Influence of molecular and chemical chaperones on protein folding. *Cell Stress Chaperones* 1: 109–115.

Wickner R.B. 1994. [URE3] as an altered URE2 protein: Evidence for a prion analog in *Saccharomyces cerevisiae*. *Science* 264: 566–569.

———. 1997. A new prion controls fungal cell fusion incompatibility (commentary). *Proc. Natl. Acad. Sci.* 94: 10012–10014.

Wille H., Michelitsch M.D., Guénebaut V., Supattapone S., Serban A., Cohen F.E., Agard D.A., and Prusiner S.B. 2002. Structural studies of the scrapie prion protein by electron crystallography. *Proc. Natl. Acad. Sci.* 99: 3563–3568.

Williams R.C. 1954. Electron microscopy of viruses. *Adv. Virus Res.* 2: 183–239.

Wilson D.R., Anderson R.D., and Smith W. 1950. Studies in scrapie. *J. Comp. Pathol.* 60: 267–282.

Xi Y.G., Ingrosso L., Ladogana A., Masullo C., and Pocchiari M. 1992. Amphotericin B treatment dissociates *in vivo* replication of the scrapie agent from PrP accumulation. *Nature* 356: 598–601.

Zlotnik I. 1963. Experimental transmission of scrapie to golden hamsters. *Lancet* 2: 1072.

3

Bioassays of Prions

Stanley B. Prusiner,[1,2,4] Jiri Safar,[1,2] and
Stephen J. DeArmond[1,3]
[1]Institute for Neurodegenerative Diseases
Departments of [2]Neurology, [3]Pathology, and
[4]Biochemistry and Biophysics
University of California, San Francisco, California 94143

P RIOR TO THE EXPERIMENTAL TRANSMISSION OF SCRAPIE to goats inoculated with tissue extracts prepared from sheep with scrapie (Cuillé and Chelle 1939), no experimental studies of the disease were possible. Despite the cumbersome bioassays in sheep and goats, some limited information was obtained. For example, the susceptibility of 24 different breeds of sheep was measured after subcutaneous inoculation with brain extract prepared from a sheep with scrapie (Gordon 1946). The resistance of the scrapie agent to inactivation by formalin and heat was also shown using bioassays performed in sheep (Pattison and Millson 1960). The first evidence for strains of prions accumulated with goats that presented with two different clinical syndromes (Pattison and Millson 1961b). One set of prion-infected goats was described as "drowsy" due to the lethargy manifest during the clinical phase of scrapie, and the other was called "hyper" because these animals were highly irritable and easily aroused.

One of the most distinctive and remarkable features of slow infections, such as those caused by prions, is their prolonged incubation periods, during which the host is free of recognizable clinical dysfunction. The onset of clinical symptoms marks the end of the incubation period and the beginning of a relatively short, progressive course of illness that ends in death. Although prolonged incubation periods are a fascinating phenomenon, they have been the biggest impediment to prion research. Because animals remain healthy throughout the incubation period, investigators must wait until signs of clinical illness appear before assigning a

Table 1. Endpoint titrations in goats

Inoculum[a]	Log dilution	n/n_0[b]	Incubation period (months)
Scratching	1	2/2	11, 12
	2	2/2	9, 22
	3	2/2	11, 12
	4	0/2	—
	5	1/3	14
	6	0/3	—
	7	1/3	15
	8	1/3	11
Drowsy	1	2/2	8, 8
	2	2/2	10, 11
	3	2/2	9, 11
	4	1/2	22
	5	0/3	—
	6	1/3	11
	7	0/3	—
	8	0/3	—

Data from Pattison (1966).

[a]Inocula prepared from pools of three goat brains. The goats had either the scratching or drowsy forms of scrapie.

[b]n, number of animals with scrapie; n_0, number of animals tested.

positive score. In early studies of scrapie using sheep, incubation periods of 1 to 3 years were required.

Goats seem to be better hosts for bioassays for prions than are sheep. Although the incubation periods were not substantially shorter, goats develop scrapie more consistently than sheep. In one study of endpoint titration of scrapie prions (Pattison 1966), an entire herd of goats was required to quantify the concentration of prions in a single sample (Table 1).

TRANSMISSION OF SCRAPIE TO RODENTS

Studies with rodents ushered in a new era of prion disease research. Laboratory studies of scrapie began with transmission in mice, then Syrian hamsters, and have now expanded to include mice expressing novel chimeric transgenes. At each phase, results have expanded our knowledge of prions and the diseases caused by them.

Mice

An important milestone in the history of prion research was the experimental transmission of scrapie from sheep to mice ~18 months after intracerebral inoculation of brain extracts (Chandler 1961). On second passage, the incubation periods shortened to 4–5 months and remained constant on subsequent passages. The demonstration that scrapie could be transmitted to a small laboratory rodent made possible many new experimental studies that were previously impractical in sheep or goats.

Syrian Hamsters

The identification of an inoculum that produced scrapie in the golden Syrian hamster in ~70 days after intracerebral inoculation represented another important advance in prion disease research (Marsh and Kimberlin 1975; Kimberlin and Walker 1977). In earlier studies, Syrian hamsters had been inoculated with prions, but serial passage with short incubation times was not reported (Zlotnik 1963). The relatively short incubation period in hamsters and the high prion titers in their brains made them the preferred animal for biochemical research on the nature of the scrapie agent.

Incubation-time Bioassay

The development of an incubation-time bioassay in Syrian hamsters reduced the time required to measure prions in samples with high titers by a factor of nearly 6; only 70 days were required instead of the 360 days previously needed. Equally important, 4 animals could be used in place of the 60 mice that were required for endpoint titrations, which made a large number of parallel experiments possible. However, there were disadvantages to using hamsters compared to mice: (1) the small number of inbred hamster strains, (2) the higher purchase price and more expensive care of the hamster, and (3) the lack of procedures for transfer and ablation of genes in the hamster.

Transgenic Mice

With the production of transgenic (Tg) mice that overexpress mouse (Mo) or Syrian hamster (SHa) PrP genes, an animal model for prion disease with shorter incubation times than Syrian hamster was found (Carlson et al. 1994). The expression of foreign and mutant PrP transgenes in mice has created a wealth of knowledge about prions that was previously unattainable. Such studies have elucidated the molecular mechanism of prion formation, begun to define the biochemical and genetic basis of the "species barrier," demonstrated an inverse relationship between the level of PrP^C expression and the incubation time, established the de novo synthesis of prion infectivity from mutant PrP, and revealed the molecular basis of prion strains. Additionally, in PrP-deficient ($Prnp^{0/0}$) mice, neither prion disease nor prion replication has been found.

Inbred and congenic strains of mice in conjunction with molecular genetics showed that the sequence of the PrP gene in mice can control the length of the incubation time (Carlson et al. 1994; Moore et al. 1998). For many years, two types of mice were studied by the Edinburgh group: Most mice had short incubation times when inoculated with the Me7 prion strain, whereas VM mice had long incubation times (Dickinson and Fraser 1977). Subsequently, it was shown that VM mice, like Iln/J mice from which they were derived, have different amino acids at positions 108 and 189 compared to mice with short incubation times, such as C57BL mice (Westaway et al. 1987; Carlson et al. 1988).

CULTURED CELLS

To date, no cell culture system has been devised for the bioassay of prions. No readily detectable, specific morphologic change in cultured cells infected with prions has been observed. Many reports of chronically infected cells with low levels of prions have been published (Clarke 1979); these low titers have prevented further studies using the cells as a source of prions. In contrast, mouse neuroblastoma cells (N2a) were found to propagate prions derived from sheep with scrapie and humans with Creutzfeldt-Jakob disease (CJD) if they were first passaged through mice (Kingsbury et al. 1984; Race et al. 1987; Butler et al. 1988). More recently, a mouse hypothalamic neuronal cell line (GT1) immortalized by expression of TAg has been found to support scrapie infection in culture (Schatzl et al. 1997). Rat pheochromocytoma cells (PC12) have been reported to propagate mouse prions (Rubenstein et al. 1984, 1991, 1992), but the biology of this system remains unclear.

SPECIES AND TRANSMISSION BARRIERS

In experimental studies of prion disease with mice and hamsters, the shortest incubation times are achieved with intracerebral inoculation of prions with the same amino acid sequence as the recipient animal. Prions adopt the amino acid sequence of the last animal in which they were passaged. Under these conditions, all of the animals develop prion disease

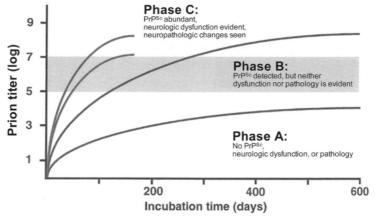

Figure 1. Kinetics of experimental prion infection in rodents. Phases A and B are preclinical whereas phase C is characterized by clinical signs of neurologic dysfunction. In phase A, only prion infectivity is detectable and the titer is generally $<10^5$ ID_{50} U/ml of a 10% (w/v) homogenate. In phase B, shown by the shaded area, the titer is typically 10^5 to 10^7 ID_{50} U/ml, which is sufficiently high for PrP^{Sc} detection by immunoblotting; some neuropathologic changes can be seen in this phase (Baringer et al. 1983; Jendroska et al. 1991). In phase C, the titer is $>10^7$ ID_{50} U/ml, PrP^{Sc} is readily detected, clinical signs of neurologic dysfunction are obvious, and neuropathologic changes are seen. The two curves on the left depict transmission of prions between homologous animals, and the two curves on the right depict heterologous transmission, i.e., crossing the species barrier, for which there is greater variation in transmission of disease.

within a narrow interval for a particular dose (Fig. 1). Conversely, when the PrP sequence of the donor prion differs from that of the recipient animal, the incubation time is prolonged. Although some of the factors controlling the length of the incubation time are well defined, the underlying mechanism is not understood (Dickinson et al. 1968; Carlson et al. 1986, 1989; Westaway et al. 1987; Telling et al. 1994, 1995; Tateishi et al. 1996; Moore et al. 1998; Scott et al. 1999).

When prions from one species are passaged into another, the PrP sequence of the donor differs from that of the recipient, and generally the incubation time is prolonged. This phenomenon is often referred to as the prion "species barrier" (Pattison 1965). The species barrier concept has been extended by creating artificial prions in Tg mice; in this case, the phenomenon of prolonged incubation time on first passage is referred to as the prion "transmission barrier" because the species of the host is unchanged (Supattapone et al. 1999). The strain of the prion can dramatically influence the size of the species or transmission barrier. For example, the prolongation of the incubation time can be longer for some strains and shorter for others (Scott et al. 1997a, 1999; Peretz et al. 2002).

The phenomenon of strain-dependent transmission barriers is presumably due to the differences between the tertiary and quaternary structures of prion strains, for which there is increasing evidence (Bessen and Marsh 1994; Telling et al. 1996b; Scott et al. 1997a, 1999; Peretz et al. 2002; Chapters 1 and 2). In other words, the interaction of PrP^{Sc} with PrP^{C} or an intermediate complex of PrP^{C} bound to a conversion cofactor, designated PrP^{*}, is governed largely by the conformation of PrP^{Sc}. PrP^{C} of the recipient animal and PrP^{Sc} of the donor prion may differ in sequence, but the conformation of PrP^{Sc} may largely or entirely negate these differences in primary structure (Scott et al. 1999). Conversely, PrP^{C} of the recipient and PrP^{Sc} of the donor may share the same sequence, but the conformations of PrP^{Sc} for different strains may be sufficiently different to produce differences in incubation times.

Kinetics of Prion Replication and Incubation Times

The length of the incubation time can be modified by (1) the dose of prions, (2) the route of inoculation, (3) the level of PrP^{C} expression, (4) the species of the prion (i.e., the sequence of PrP^{Sc}), and (5) the strain of prion. The dose of prions is inversely related to the incubation time: Higher doses shorten the time required to reach the threshold level for PrP^{Sc} at which signs of illness appear (Prusiner et al. 1982b). The intracerebral route of inoculation is substantially more efficient than other routes (Prusiner et al. 1985). Similar to prion dose, higher levels of PrP^{C} enable more rapid accumulation of PrP^{Sc} and, hence, result in shorter incubation times (Prusiner et al. 1990). When the amino acid sequence of PrP^{Sc} in the inoculum is the same as that of PrP^{C} expressed in the recipient, incubation times are generally abbreviated (Scott et al. 1989, 1997a). At present, the mechanism by which prion strains modify incubation time is not well understood, but interactions between PrP^{Sc} and PrP^{C} undoubtedly play a major role (DeArmond et al. 1997; Scott et al. 1997a).

ENDPOINT TITRATIONS

Although slow, tedious, and expensive, endpoint titrations of prions represent an important advance in prion disease research. An endpoint titration for prions is performed by serially diluting a sample with phosphate-buffered saline (PBS) at 10-fold increments (Fig. 2). Each of 10 dilutions is typically inoculated intracerebrally in the left parietal lobe of four to six animals. Animals are examined at regular intervals for clinical signs of disease, which may include bradykinesia, plasticity of the tail, waddling gait,

CONCENTRATION

10^1 10^2 10^3 10^4 10^5 10^6 10^7 10^8 10^9 10^{10}

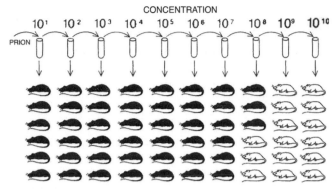

Figure 2. Endpoint titration of prions in mice. Ten serial dilutions of a prion sample are prepared, with each more dilute by a factor of 10. Each dilution is then inoculated into six mice. The titration is terminated when all or most of the mice destined to become ill show signs of CNS dysfunction. Upon neuropathologic examination, vacuolation of neurons, widespread astrocytic gliosis, and PrP^{Sc} accumulation are found. Reliable data from an endpoint titration require that no mice are ill at the highest dilution; otherwise, a valid endpoint cannot be calculated.

and a coarse, ruffled appearance of the coat. Histologic examination of the brain and spinal cord should be performed occasionally to confirm the clinical diagnosis. Typical endpoint titrations require 60 animals and one year to determine the titer.

The titration is terminated when all or most of the animals destined to become ill have shown signs of CNS dysfunction, and neuropathological examination reveals vacuolation of neurons, PrP^{Sc} accumulation, and widespread astrocytic gliosis. Reliable data from an endpoint titration require that no animals are ill at the highest dilution; otherwise, a valid endpoint cannot be calculated. Because the highest dilutions at which prion disease develops are the only observations of interest, all animals destined to become ill at this dilution must do so before the titration can be scored.

The concentration of prions in the original sample can be calculated from the score at the highest dilutions at which there is a positive result (Dougherty 1964). Titers are calculated according to the method of Spearman and Kärber (Dougherty 1964).

Despite the large number of mice and the extremely long time interval, much has been learned about the physicochemical properties of prions and the pathogenesis of disease (Hunter 1972; Kimberlin 1976; Dickinson and Fraser 1977; Hadlow et al. 1979). Not only are resources great with respect to animals and time, but accurate measurements are problematic because experiments require pipetting ten serial 10-fold dilutions. Unfortunately, the year-long intervals between designing experi ments and obtaining results discouraged sequential studies. Furthermore, the large number of mice needed to quantify a single sample prevented large experiments in which many studies were performed in parallel.

Endpoint Titrations in Mice

For endpoint titrations using mice, a 30-μl dose is intracerebrally inoculated using a 26-gauge needle inserted into the cranium to a depth of ~2 mm. Generally, the mice die within 4–6 weeks after the onset of disease. At the time of onset, the back of the sick mouse is painted with a 1% (w/v) solution of picric acid (Mallinckrodt, St. Louis).

Rocky Mountain Laboratory (RML) prions in CD-1 ($Prnp^{a/a}$) mice cause disease in ~130 days when the intracerebrally injected inoculum consists of a 10% brain homogenate from an RML-infected CD-1 mouse that was diluted 10-fold. At the highest positive dilution, CD-1 mice develop CNS dysfunction after ~250 days. An additional 100 days was allowed to ensure that any other mice developing clinical signs were included in the final titer determination. The standard errors of our titrations generally vary between 0.2 and 0.4 log units: 2 s.e. ≈ 0.4 to 0.8 log units (~95% confidence limits).

Ovine and Caprine Prions

Measuring scrapie prions in sheep or goat specimens by endpoint titrations in non-Tg mice is considerably more difficult than measuring mouse-adapted prions. The adaptation of the scrapie agent from sheep or goats into mice greatly extends the incubation period. When homogenates of sheep or goat brain are diluted 10-fold and Swiss mice are inoculated, incubation periods of 10–12 months are observed. To score endpoint titrations, experiments must run for 18–24 months (Hadlow et al. 1974, 1979). Typically, 10–12 mice are used at each dilution since intercurrent illnesses begin to produce significant mortality at ~18 months of age. Most confusing are chromophobe adenomas of the pituitary, which cause neurologic signs that can be misinterpreted as scrapie. Under these circumstances, histopathologic confirmation is necessary (W.J. Hadlow, pers. comm.). Because many mice die of illnesses other than scrapie during such prolonged experiments, increasing the number of mice inoculated is necessary in order to have at least 6 mice scored at each dilution for reliably determining the endpoint.

The use of Tg mice expressing ovine PrP genes promises to shorten greatly the transmission times and to give accurate titers. Although more studies are needed for the wide variety of strains in sheep scrapie samples, the initial studies with Tg(OvPrP)$Prnp^{0/0}$ mice look encouraging (Crozet et al. 2001; Vilotte et al. 2001).

Endpoint Titrations in Hamsters

Because of the increased cost, limited numbers of endpoint titrations have been performed in hamsters. Five to six months are required before an endpoint titration may be scored in hamsters. The principles of the methodology are the same as with endpoint titrations in mice, with the following exceptions: (1) a 50-μl dose is inoculated intracerebrally, (2) four to six hamsters are generally used for each dilution, (3) the clinical signs of scrapie are slightly different as discussed below, and (4) the abdomen of the sick hamster is painted with picric acid.

INCUBATION-TIME ASSAY

Faced with cumbersome, slow, and expensive endpoint titrations, we questioned whether a new approach might be devised that would decrease the number of animals and the time required for assaying each sample. Initially, we searched for a scrapie-specific alteration in the immune system of mice 20–30 days after inoculation, when the prion titer in the spleen reaches a maximum (Garfin et al. 1978a,b). Unable to find a scrapie-specific surrogate marker of sufficient magnitude to correlate with the prion titer, we turned to earlier observations in which the incubation period, or the time interval from inoculation to onset of illness, increased as the prion dose decreased (Eklund et al. 1963; Hunter et al. 1963; Dickinson et al. 1969). We asked whether it would be possible to use the length of the incubation period to measure the prion titer in a sample reliably.

Although the length of the incubation period was carefully recorded in early studies on scrapie in sheep and goats, its relationship to the volume of the injected dose was unclear (Table 1) (Pattison and Millson 1961a). In contrast, a clear relationship between the length of the incubation period and the dilution of the inoculum was evident in studies by Eklund and colleagues, who used mice infected with the scrapie agent (Eklund et al. 1963). Regular increases in the time intervals from inoculation to death with increasing dilutions of the inoculum prompted these investigators to suggest that this relationship might be used as an assay.

Encouraged by reports describing the measurement of viral titers using the time intervals from inoculation to either onset of symptoms or death (Bryan 1957; Luria and Darnell 1967), we determined the precision and predictability of this relationship for scrapie prions. Our studies clearly demonstrated that measurements of time intervals could be used to predict, with precision equal to or greater than that obtainable by end-point titration, the titer of scrapie prions in a sample (Prusiner et al. 1980b, 1982b).

It is of interest that prior to the advent of cell-culture assays for viruses, incubation-period assays were used successfully to measure encephalomyelitis viruses (Gard 1940), rabbit papillomavirus (Bryan and Beard 1939), avian erythromyeloblastic leukosis (Eckert et al. 1954), Rous sarcoma virus (Bryan 1956), and several viruses of the psittacosis-lymphogranuloma venereum group (Golob 1948; Gogolak 1953; Crocker 1954).

The incubation-time assay substantially reduced the number of test animals, the time required for bioassay, and potential pipetting errors of endpoint titration (Prusiner et al. 1980b, 1982b). With hamsters, studies on the structure of the scrapie agent were dramatically accelerated by the development of a bioassay based on measurements of incubation times. This bioassay made it possible to quantify the concentration of prions in a sample with four animals. If the prion titer in the sample was greater than 10^8 ID_{50} units (the median dose required for infectivity) per ml, only ~70 days was required for a positive response.

Nomenclature

In discussions of slow infectious diseases, the incubation period is the time from inoculation to the onset of illness. The term "incubation-time assay" was adopted to describe the bioassay procedure. The assay uses two measurements: (1) the time from inoculation to the onset of illness (y) and (2) the time from inoculation to death (z). The difference between z and y is the duration of clinical illness.

The dilution (D) of the sample to be assayed refers to the fractional concentration of prions. Conventional serial 10-fold dilutions were made and the dilution is expressed as $1/10^d$; thus, the $\log_{10}$ D equals $-d$. We recognize that the dilution process itself has a positive exponent. It is noteworthy that the ID_{50} and LD_{50} (median lethal dose) values for a given sample are identical because scrapie is uniformly fatal. The ID_{50} is expressed in units (U).

Preparation of Inocula

Rodent brains are generally homogenized in 320 mM sucrose to yield a 10% (w/v) homogenate. The homogenate is clarified by centrifugation at 1000g for 10 minutes at 4°C, and the supernatant fraction is diluted at least 3-fold to prevent toxicity and generally 10-fold so that the final inoculum is a 1% homogenate. The inocula are diluted with PBS. Other additives, including 0.5 U/ml of penicillin, 0.5 µg/ml of streptomycin, 2.5 µg/ml of amphotericin, and 50 µg/ml of recrystallized fraction V bovine serum albumin (Pentex-Miles Laboratories, Elkhart, Indiana) have been used, but have not been found to affect the outcome of our studies.

For intracerebral inoculations, a 50-µl dose is injected into female weanling hamsters and a 30-µl dose into weanling mice. For LVG/Lak Syrian hamsters (Charles River Laboratories) inoculated with Sc237 prions, a titer of $10^{9.5}$ ID_{50} U/g of brain tissue, as determined by endpoint titration using the method of Spearman and Kärber (Dougherty 1964), is used. For Swiss CD-1 mice (Charles River Laboratories) inoculated with RML prions, the titer is $10^{8.5}$ ID_{50} U/g of brain tissue.

Inoculation of Rodents

The shortest incubation periods are observed after intracerebral inoculation. Peripheral routes of inoculation, such as intraperitoneal, subcutaneous, and intravenous, result in longer incubation periods for a given dose (Prusiner et al. 1985; Kimberlin and Walker 1986). Studies on the intracerebral route of inoculation with bacteriophage or India ink have shown that 90% or more of the inoculum is absorbed systemically (Schlesinger 1949; Cairns 1950); less than 10% of the inoculum remains in the brain. Presumably, the extreme vascularity of the brain and the pressure during inoculation are responsible for this dispersal.

Weanling female hamsters or weanling mice of either sex are inoculated intracerebrally with 50 µl or 30 µl, respectively, using a 26-gauge needle inserted in the left parietal lobe. The needle is inserted into the cranium to an approximate depth of 3 mm for hamsters or 2 mm for mice. Immunohistochemical studies show that amyloid filaments of PrP accumulate along the needle tract where the inoculum was injected (Bendheim et al. 1984). These amyloid deposits are synthesized de novo because immunoreactive prion polymers were not detected until 60 days after inoculation (DeArmond et al. 1985).

In attempts to increase the deposition of inoculated prions in hamster brains, subarachnoid injections were compared to intraparenchymal parietal lobe injections. No difference in incubation periods was observed for the two different intracerebral routes of inoculation (Table 2). Additional studies using agarose plugs to hold the inoculum in the brain were performed, but neither the subarachnoid nor intraparenchymal injections showed any change in the incubation period.

Table 2. Comparison of intracerebral and subarachnoid inoculations

Route	Inoculum vol. (μl)	n	Onset of illness (mean days)	Death (mean days)	Agarose[a] (type/% conc.)	Temperature (°C)
i.c.	50	12	64	81	—	4
s.a.	50	12	63	84	—	4
s.a.	100	12	70	86	—	4
i.c.	50	12	69	89	—	45
s.a.	50	8	73	94	—	45
s.a.	100	8	75	89	—	45
i.c.	50	6	71	87	HGT 0.5%	45
s.a.	50	5	71	90	HGT 0.5%	45
i.c.	50	4	70	88	LGT 0.5%	45
s.a.	50	6	78	92	LGT 0.5%	45
i.c.	50	5	70	90	LGT 0.75%	45
s.a.	50	3	74	89	LGT 0.75%	45

Intracerebral (i.c.) or subarachnoid (s.a.) inoculations were performed with 27-gauge needles.
[a]Agarose solidifying at high temperature (HGT) or low temperature (LGT).

Oral Transmission of Prions

Studies on the oral transmission of prions have gained importance with the bovine spongiform encephalopathy (BSE) epidemic and the transmission of bovine prions to humans. Investigations of oral transmission of prions were performed using Syrian hamsters because they are avid cannibals, and this natural tendency can be exploited to examine the oral transmissibility of scrapie prions (Prusiner et al. 1985). Weanling hamsters were inoculated intracerebrally with ~10^7 ID_{50} units of scrapie prions and were allowed to cohabitate between 30 and 80 days later with the uninoculated hamsters. The hamsters inoculated with prions developed a progressive neurological disorder characterized by ataxia, difficulty righting from a supine position, generalized tremor, and head-bobbing, and were killed by their uninoculated cage mates. In all cases, the heads of the dead hamsters were consumed by their healthy cage mates. Frequently, the genital and abdominal organs were also eaten.

All of the uninoculated cannibal hamsters developed scrapie 110–135 days after eating their debilitated cage mates. Histopathological sections of the cerebral hemispheres of cannibal hamsters demonstrated mild but widespread vacuolation in the cerebral cortex and adjacent hippocampus. Vacuoles appeared to be localized to the neuropil and were occasionally seen within neuronal perikarya. The most conspicuous vacuolation was seen in the hippocampus adjacent to the layer of pyramidal cells and in the subiculum. This distribution of vacuolar changes was similar to that seen after intracerebral injection of scrapie prions. Adjacent to sections taken for histopathology, brain samples were taken from three different cannibal hamsters for bioassay. The prion titers were 9.7 ± 0.10, 9.5 ± 0.60, and 9.6 ± 0.13 ID_{50} U/g of tissue.

Dose-dependence of Oral Transmission

To determine whether a dose-dependent relationship for oral transmission of scrapie prions could be found, scrapie-infected hamsters were sacrificed by cervical dislocation and then placed in the cages with healthy animals. A dose-dependent relationship was not demonstrated when the number of experimental animals in each group was increased (Table 3). No significant differences were found among the incubation times of the experimental groups, which ranged in their cannibal-to-victim ratios from 16 to 1. Furthermore, differences between male and female cannibals with respect to incubation times were not discerned. For three experimental groups in which the ratios of cannibals to victims were 1, 2, and 16, more than 80% of the cannibals developed clinical signs of scrapie at 120–140 days. Likewise, more than 80% of the cannibals died after

Table 3. Oral transmission of prions to hamsters

| | | | | Incubation time intervals | |
	Sex	Ratio (cannibals/victims)	n	onset of illness (days $\pm$ s.e.m.)	death (days $\pm$ s.e.m.)
Scrapie	F	16[a]	80	128 ± 4.2	145 ± 6.6
	F	8	36	127 ± 6.4	147 ± 7.0
	F	4	20	126 ± 6.7	146 ± 6.7
	F	2	80	121 ± 6.6	139 + 3.9
	F	1	45	124 ± 10.2	144 ± 10.2
	M	16	58	132 ± 6.7	153 ± 5.7
	M	2	85	126 ± 6.7	152 ± 5.8
Normal	F	2[b]	42	>300	—
	F	2[c]	41	>300	—
	F	—	35	>300	—
	M	2[b]	28	>300	—

[a]Victims were inoculated with 10^7 ID_{50} units of scrapie prions, then sacrificed 70 days later when they displayed clinical signs of scrapie.
[b]Victims were uninoculated, normal animals sacrificed by cervical dislocation prior to cannibalism.
[c]Victims were inoculated with normal brain extract (0.05 ml), then sacrificed 70 days later.

130–150 days. In similar experiments with uninoculated, healthy victims as well as victims inoculated with normal brain extracts, all cannibalistic hamsters failed to develop scrapie.

Repeated oral consumption of scrapie-infected animals was also examined in an attempt to shorten the incubation times. In control experiments, healthy animals repeatedly cannibalized dead normal animals prior to eating scrapie-infected hamsters for the first time. A slight but probably insignificant reduction in the incubation times was observed for animals repeatedly fed scrapie-infected victims compared with those consuming scrapie-infected animals only once. Repeatedly eating uninfected controls prior to consuming scrapie-infected animals once resulted in incubation times comparable to those found with multiple prion-infected feedings. More than 80% of the cannibalistic hamsters developed clinical signs of scrapie after 105–125 days, and death occurred at 130–150 days. When hamsters cannibalized healthy, uninoculated cage mates, they never developed scrapie.

Factors Influencing Oral Transmission

Studies on the oral transmission of scrapie prions to mice showed a much higher rate of transmission after abrasion of the oral mucosa compared to unscarred controls (Carp 1982). To investigate the possibility that abrasion of the oral mucosa caused by the consumption of tissues such as the calvarium might have influenced oral transmission, healthy hamsters were fed pieces of infected brain. This feeding protocol slightly accelerated the onset of disease and death compared with cannibalism of infected animals (Table 4). Again, a dose-dependent relationship was not apparent,

Table 4. Transmission of prions by oral consumption of scrapie-infected brain tissue

| | Brain tissue consumed (grams) | n | Incubation time intervals | |
			onset of illness (days ± s.e.m.)	death (days ± s.e.m.)
Scrapie	1/6[a]	40	119 ± 17.5	140 ± 8.8
		36	114 ± 3.3	138 ± 6.0
	1[a]	38	113 ± 0.5	133 ± 4.8
		40	118 ± 4.5	138 ± 6.5
Normal	1/6[b]	18/40	>370	—
	1	13/40	>370	—

[a]Brain tissue was from scrapie-infected hamsters that were intracerebrally inoculated with 10^7 ID$_{50}$ units of prions and sacrificed 70 days later when they displayed clinical signs of scrapie.

[b]Brain tissue was from healthy hamsters that were sacrificed.

Table 5. Comparative efficiencies of prion transmission to Syrian hamsters by peripheral routes

Peripheral route of transmission	Dose (log ID$_{50}$ units)	Incubation time intervals		Equivalent intracerebral dose (log ID$_{50}$ units)	Comparative efficiency
		onset of illness (mean days)	death (mean days)		
Oral ingestion	9.5	120	147	0.4	$10^{-9.1}$
Intraperitoneal	8.3	92	112	2.9	$10^{-5.4}$

but the 6-fold range over which this was tested was probably too small to be measurable. More than 80% of the animals that developed clinical signs of scrapie at 110–120 days died at 130–150 days. When brains removed from healthy, uninoculated animals were used, the fed hamsters did not develop scrapie.

Comparison of Oral Transmission and Other Routes of Infection

Oral ingestion via cannibalism is 10^9 times, and intraperitoneal injection is 10^5 times, less efficient than intracerebral inoculation (Table 5). These findings, coupled with differences in the human PrP^{Sc} amino acid sequence in the inoculum (Goldfarb et al. 1990, 1994) and the PrP^C sequence of the recipient apes and monkeys (Schatzl et al. 1995), may explain the difficulties in transmitting kuru prions orally to nonhuman primates (Gibbs et al. 1980).

Care of Rodents

Long-term holding of rodents is mandatory for most studies of experimental scrapie and Creutzfeldt-Jakob disease (CJD). First passage of either scrapie prions from sheep to mice (Table 1) (W.J. Hadlow, unpub.) or CJD prions from humans to mice (Tateishi et al. 1983) may take 1.5 years or more. For studies of scrapie in sheep and goats, typically 10–12 mice were inoculated at each dilution in order to be able to score endpoint titrations at the end of 1.5–2 years (Hadlow et al. 1974, 1982). Approximately 50% of the mice will die of an illness other than scrapie prior to developing signs of neurologic dysfunction.

In our experience, holding hamsters is easily accomplished for 100–200 days. Long-term experiments extending beyond 400 days present difficulties. By the end of one year, ~40% of the uninoculated hamsters will have died. The attrition begins to be significant at ~6 months and progresses slowly (Fig. 2). The animals generally die of traumatic injuries because hamsters readily fight even when housed together since weaning. Male and female hamsters held in separate cages showed no difference in survival, although the males seemed to be more aggressive.

The hamsters are housed in polyurethane cages measuring 20 cm x 42 cm x 20 cm with a tight-fitting wire cover that holds the food. Eight animals are housed per cage, and groups of four are distinguished by the notching of both ears. Water is supplied by an automatic watering system and Purina rat chow is provided ad libitum. A generous supply of pine wood shavings is placed on the bottom of each cage and is changed two times per week. All personnel entering the room in which the hamsters are housed are required to wear masks, gowns, rubber boots, gloves, head covers, and eye protection.

Clinical Signs of Experimental Prion Disease

The most reliable clinical signs of prion disease in rodents consist of (1) rigidity of the tail, (2) limb and truncal ataxia, (3) forelimb flexion instead of extension when suspended by the tail, and (4) difficulty righting from a supine position. The diagnosis of probable prion disease requires progression of at least two of these signs over a period of a few days, although slow progression over 10–20 days is more reliable. In mice, the first three signs are routinely checked in assessing for disease. In some Tg mice, the clinical phase of disease is very rapid and lasts for ≤1 day; in such cases, histopathologic examination is necessary in a percentage of animals within the experimental group. In hamsters, the last three signs listed above are routinely checked in clinical assessment. Spongiform degeneration and reactive astrocytic gliosis are required to diagnose prion disease. Nonobligatory neuropathologic features include PrP amyloid plaques and protease-resistant PrP^{Sc} detected by histoblotting.

In addition to the four clinical signs listed above, generalized tremors and head bobbing are frequently seen. The bobbing movements of the head are progressive and may result from visual difficulties due to degenerative changes in the retina (Hogan et al. 1981). Between 1% and 10% of rodents show generalized convulsions at an early stage of illness. With further deterioration, the ataxia becomes so pronounced that balance is maintained with considerable difficulty. Kyphotic posture, bradykinesia, and weight loss appear 7–15 days after the onset of illness. Over the next week, the rodents lose their ability to maintain an erect posture; they lie quietly on their sides and exhibit frantic movements of the extremities when disturbed. Death follows in 3–5 days.

CALIBRATION CURVES

The development of a bioassay for prions, in which incubation periods are used to predict titers, requires the construction of calibration curves. For each combination of prion strain and host, a calibration curve relating prion dose to the incubation time must be created.

The construction of reliable calibration curves probably requires samples with a wide range of titers. The titers are determined by endpoint titrations, and the time intervals from inoculation to onset of illness and to death are measured for each dilution. In developing calibration curves for Syrian hamsters, each of 10 dilutions is inoculated intracerebrally into four animals. After 55 days, the animals are examined two times weekly for clinical signs of disease. From the number of animals positive at each dilution, the titer was calculated using the method of Spearman and Kärber (Dougherty 1964). The injected dose is then calculated by multiplying the titer by the dilution.

Curves relating the injected dose to the time intervals from inoculation to onset of clinical illness as well as from inoculation to time of death are shown in Figures 3 and 4. The titers of the samples used to construct

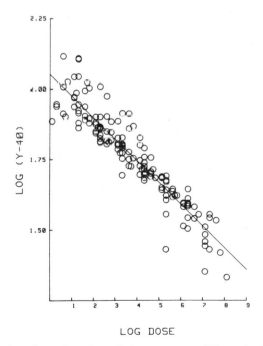

Figure 3. Incubation times from inoculation to onset of illness in Syrian hamsters as a function of the dose of Sc237 prions. (Reprinted, with permission, from Prusiner et al. 1982b [copyright John Wiley & Sons].)

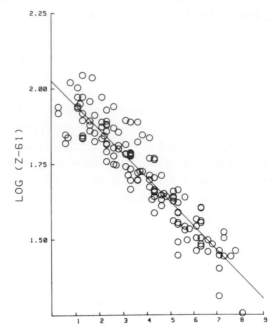

Figure 4. Incubation times from inoculation to death in Syrian hamsters as a function of the dose of Sc237 prions. (Reprinted, with permission, from Prusiner et al. 1982b [copyright John Wiley & Sons].)

these curves varied over a range from 10^3 to $10^{8.5}$ ID_{50} U/ml, as determined by endpoint titration. The interval from inoculation to onset of illness minus a time factor of 40 is a linear function of the inoculated dose (Fig. 3). The time factor was determined by maximizing the linear relationship between time interval and dose. With a factor of 40, the regression coefficient for the line is 0.87.

Whereas the onset of illness requires clinical judgment with respect to the diagnosis of disease, death is a completely objective measurement. The time interval from inoculation to death minus a time factor of 61 is a linear function of the injected dose (Fig. 4). As with the analysis of data for onset of illness, the time factor was determined by maximizing the linear relationship between this time interval and the dose. With a factor of 61, the regression coefficient for the line is 0.86.

Dose-response Relationships: Derivation of Equations

From the linear relationships described, equations relating titer, dilution, and time intervals can be expressed as:

$$\text{Log } T_y = 26.66 - (12.99) \log(y - 40) - \log D$$
$$\text{Log } T_z = 25.33 - (12.47) \log(z - 61) - \log D$$

in which T is the titer, expressed in ID_{50} U/ml; D is the dilution, defined as the fractional concentration of the diluted sample; y is the mean interval from inoculation to onset of clinical illness, in days; and z is the mean interval from inoculation to death, in days. The most precise estimate of titer is obtained by calculating a weighted average of T_y and T_z.

Similar linear relationships were obtained when the reciprocals of the time intervals were plotted as a function of the logarithm of the dose. Regression coefficients of 0.87 and 0.88 were obtained for lines relating $1/y$ and $1/z$, respectively, to the logarithms of the dose. Equations describing these functions gave similar results to those obtained with the two equations above.

A decade after developing the incubation-time bioassay for prions in Syrian hamsters, we reevaluated the system with a series of endpoint titrations in which the animals were scored only for the onset of neurologic dysfunction. After the hamsters showed a clear progression of neurologic signs, the animals were euthanized. In these studies, we were unable to find a substantial difference between the newly obtained titration data and those gathered in the original investigation. We found that the simple relationship shown below, which was derived from a plot of the incubation time as a function of the dose, is highly reliable. We have used the following equations for bioassays in Syrian hamsters over the past decade:

For incubation times <104 days: Log T = 17 + [Log D] – (0.138*Y)
For incubation times >104 days: Log T = 8.9 + [Log D] – (0.059*Y)

in which T is the titer, expressed in ID_{50} U/ml; D is the dilution, defined as the fractional concentration of the diluted sample; and Y is the mean interval from inoculation to onset of clinical illness, in days.

After developing the bioassay for Syrian hamster prions, we asked whether the same approach could be used for mouse prions, despite earlier studies that concluded such an approach would not be useful with mice. From endpoint titrations, the reciprocal of the incubation time was plotted against the dose of prions (Fig. 5). From those data, the equation for Swiss CD-1 mice was derived:

Log T = 1.52 + [Log D] + ((185 – Y)12.66)

in which T is the titer, expressed in ID_{50} U/ml; D is the dilution, defined as the fractional concentration of the diluted sample; and Y is the mean interval from inoculation to onset of clinical illness, in days.

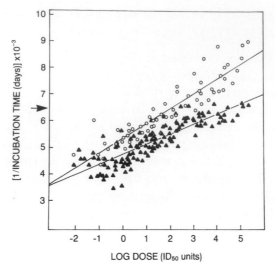

Figure 5. Calibration curve for the Chandler isolate of mouse scrapie prions in Swiss mice plotted as the reciprocal of the incubation time versus the average dose given to each animal. The curve was obtained from endpoint titration of 29 independent preparations (18 spleen samples, 10 sucrose gradient fractions, and 1 brain homogenate) (Butler et al. 1988). The experiments were terminated 226 days after inoculation; the ID_{50} units were calculated by dividing the dose as given above by the volume of the inoculum given to each mouse (30 μl). Circles represent incubation time, triangles represent the time of death. (Reprinted, with permission, from Butler et al. 1988 [copyright American Society for Microbiology].)

Validation

The validity of the time-interval measurements is supported by many lines of evidence. The titers of the scrapie prions in hamsters obtained by the time-interval method agree with those found by endpoint titration within $\pm 10^{0.3}$ ID_{50} U/ml. The precision of the incubation-time assays was estimated by calculating the 95% confidence intervals for each of the titer measurements at 10^{-1} dilution shown in Table 6. These confidence intervals were computed for two different degrees of freedom ($n - 1$, in which n is the number of observations). It can be argued that the number of observations for four hamsters is 8 because the times of both onset of illness and death were determined. Conversely, it can be argued that the onset of illness and death are closely linked and that each animal was inoculated only once; thus, only four observations are independent and can be used to compute the degrees of freedom. If seven degrees of freedom are chosen, the 95% confidence interval ranges from $10^{0.5}$ ID_{50} U/ml

Table 6. Comparison of titers determined by endpoint titration and incubation-time assays

| Endpoint titration ($\log_{10}$ titer) | Incubation-time assay | | | |
| | 10^{-1} dilution | | 10^{-3} dilution | |
	($\log_{10}$ titer)	(ρ value)	($\log_{10}$ titer)	(ρ value)
8.3 ± 0.35	8.3 ± 0.10	1.00	8.7 ± 0.16	0.27
8.3 ± 0.35	7.7 ± 0.012	0.08	8.2 ± 0.47	0.84
8.1 ± 0.25	8.2 ± 0.25	0.76	8.0 ± 0.28	0.76
8.3 ± 0.35	8.1 ± 0.33	0.69	7.6 ± 0.13	0.04
8.1 ± 0.25	8.2 ± 0.17	0.48	7.8 ± 0.11	0.23
8.0 ± 0.44	8.2 ± 0.16	0.62	8.5 ± 0.18	0.24

Titer is expressed as ID_{50} U/ml ± s.e.m. Probability (ρ) values represent the proportion of sample means farther from the mean titer from endpoint titration than the mean titer actually obtained from the incubation-time assay. Four Syrian hamsters were inoculated intracerebrally with Sc237 prions for each dilution of the endpoint titration.

to $10^{1.6}$ ID_{50} U/ml, with an average interval of $10^{0.9}$ ID_{50} U/ml. If three degrees of freedom are chosen, the 95% confidence interval ranges from $10^{0.6}$ to $10^{2.1}$ ID_{50} U/ml with an average interval of $10^{1.2}$ ID_{50} U/ml. The analysis agrees well with our experience, which dictates that samples must differ by more than 10^2 ID_{50} U/ml for the difference to be significant. On occasion, we have identified differences of $\sim 10^1$ ID_{50} U/ml that appear to be significant because they proved to be reproducible at least five times.

Scrapie prion titers were measured by the incubation-time assay and endpoint titration for a series of samples subjected to heat inactivation (Table 6). No significant difference between the two sets of data could be discerned. However, the endpoint titration assay required 40 hamsters for each of the eight points, for a total of 320 animals, whereas the time-interval assay gave the same data using 4 hamsters for each dilution point, or a total of 32 animals. For nearly 200 days after inoculation, no information about the titers of the agent was available using the endpoint titration method. In contrast, at ~75 days after inoculation, the stability of the agent up to 90°C was apparent using the incubation-time assay. From this comparison, the advantages of the incubation-time assay are obvious.

Measurements of the time intervals from inoculation to onset of illness are so reproducible that calibration curves developed 1.5 years apart are virtually superimposable (Prusiner et al. 1980b, 1982b). Differences between the molecular properties of prions derived from hamsters and mice have not been detected, using the time-interval method with the former and endpoint titration with the latter (Prusiner et al. 1980b). Sedimentation profiles, detergent stability studies, and gel electrophoresis experiments yield the same data whether assays are performed by the incubation-period method or by endpoint titration (Prusiner et al. 1978a,b, 1980a,b,c).

Some investigators have questioned the validity of the incubation-time assay (Lax et al. 1983; Somerville and Carp 1983). Criticisms of the incubation-time assay have involved situations in which prions are disaggregated by detergent treatment. Such difficulties with obtaining accurate measurements are not due to the assay but emanate from the amphipathic character of scrapie prions.

Advantages

The advantages of the incubation-time assay over the endpoint titration assay are numerous. First, the number of animals required to determine the titer in a particular sample is decreased by a factor of 10 to 15. Second, the imprecision that may arise by pipetting serial dilutions of samples is obviated by the incubation-time method. Third, determination of the titer in samples with high titers is considerably faster using the incubation-time method. As illustrated here, an inverse relationship exists between the length of the incubation period and the prion titer in the inoculum. To establish the endpoint of a titration for scrapie prions, it is necessary to wait until all animals become sick at the highest positive dilution. Fourth, the incubation-time assay always gives a quantitative estimate of the concentration of prions in a given sample, unlike the endpoint-titration assay. At times, attempts to economize with endpoint titrations will result in a series of dilutions that do not span the endpoint of the sample. In such cases, considerable time will have elapsed and a second titration will be necessary.

The incubation-time assay for prions depends on a database to maintain, calculate, and plot values. Semiweekly updating of titers has greatly facilitated our research efforts. By 70 days after inoculation, samples with high titers have begun to cause disease in animals. During the next 14 to 28 days, samples with substantially lower titers begin to induce disease. Calculations of titers and plots of experimental determinations are facilitated by the database.

The development of an incubation-time assay for measuring prions has had a profound effect on the progress of studies on the molecular biology of these infectious pathogens. Without the incubation-time assay, the purification of scrapie prions would not have advanced far enough to enable the identification of the amino-terminally truncated prion protein, PrP 27-30 (Prusiner et al. 1982a). The discovery of PrP 27-30 is the cornerstone upon which considerable progress in understanding the molecular structure of prions is based.

The economies of both time and resources afforded by the incubation-time assay are highly significant. We estimate that our research has been accelerated more than 100-fold by the incubation-time assay. It is doubtful that the purification and characterization methods described below would have been developed if the endpoint-titration method had been used to assay samples.

TRANSGENIC MICE

Tg mice overexpressing PrP^C from various species have proved to be of considerable value in the development of more rapid incubation-time assays. For instance, $Tg(SHaPrP^{+/+})7/Prnp^{0/0}$ mice, which are homozygous for the transgene array and overexpress $SHaPrP^C$ ~8-fold compared to Syrian hamsters, display incubation times of ~45 days when inoculated with ~10^7 ID_{50} units of Sc237 prions previously passaged in Syrian hamsters (Scott et al. 1997a). Similarly, Tg(MoPrP-A) mice overexpressing $MoPrP^C$-A exhibit incubation times of ~45 days upon inoculation with ~10^6 ID_{50} units of RML prions previously passaged in Swiss CD-1 mice (Carlson et al. 1994; Telling et al. 1997).

With such abbreviated incubation times, $Tg(SHaPrP^{+/+})7/Prnp^{0/0}$ and Tg(MoPrP-A)4053 mice were used for a series of endpoint titrations. The incubation times were plotted as a function of the dose of prions (Fig. 6) (Supattapone et al. 2001). At low doses, the incubation times were greatly prolonged, which makes the proof of sterility using bioassays difficult.

Although Tg(MoPrP-A)4053 mice offer a substantial advantage over non-Tg mice, the advantage of $Tg(SHaPrP^{+/+})7/Prnp^{0/0}$ mice compared to Syrian hamsters is less clear. Because Tg mice are extremely expensive to produce, the cost of maintaining a supply of such mice available for bioassays on demand is considerable. Thus, the availability of Syrian hamsters still makes them quite useful for many bioassays. When the incubation times in Tg mice decrease to fewer than 20 days for bioassays, the advantage will be so great that it seems very likely that such animals will become the system of choice for most basic studies.

Production of Transgenic Mice

Tg mice expressing a foreign or mutant PrP gene are generally produced using eggs from donor $Prnp^{0/0}$ mice (Büeler et al. 1992). These $Prnp^{0/0}$ mice were produced from an embryonic stem cell line derived from a

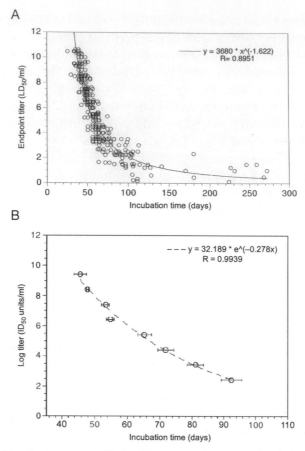

Figure 6. Calibration curves for Syrian hamster and mouse prions. (*A*) The correlation between incubation time and injected prion dose was obtained from 12 independent endpoint titration experiments (two hamster brain homogenates and two sucrose gradient fractions in triplicates) on 532 Tg(SHaPrP)7/*Prnp*[0/0] mice. The experiments were terminated 300 days after the inoculation, and the endpoint titer was calculated by the method of Spearman and Kärber (Dougherty 1964). The curve is the best fit of the data by nonlinear least-squares regression analysis. Circles represent incubation time of each transgenic mouse. (*B*) The calibration of Tg(MoPrP-A)4053 mice (Telling et al. 1996a) with RML prions was performed as described previously (Prusiner et al. 1982b). The brain homogenate used for calibration was prepared from a large pool of Swiss CD-1 mice inoculated intracerebrally with RML prions (Butler et al. 1988). Each data point is an average ± s.e.m. obtained from three endpoint titrations. Less than 100% of mice developed scrapie when the infectivity of the inoculum was <10^2 ID$_{50}$ U/ml. The data correlating the endpoint titer to the time intervals from inoculation to onset of clinical illness were best-fitted using the least-squares method (Supattapone et al. 2001). (Reprinted, with permission, from Supattapone et al. 2001 [copyright American Society for Microbiology].)

129Sv line and backcrossed to C57BL/6 before intercrossing to obtain homozygous $Prnp^{0/0}$ mice. In our experience, this genetic background (129Sv and C56BL/6) was suboptimal for the production of Tg mice; therefore, we crossed the $Prnp^{0/0}$ mice onto the FVB background. FVB mice produce large eggs that can be readily microinjected (Hogan et al. 1994). Typically, 20 superovulated females are placed with males and sacrificed early the next morning for collection of 0.5-day-old embryos. From these, we obtain 10–15 fertilized females, which generally provide 150–300 zygotes, of which 80–90% are amenable to microinjection. Embryos are cultured using standard M2/M16 media (Hogan et al. 1994). Microinjection is performed using an inverted Leica stereomicroscope and Nomarski illumination. Successfully microinjected embryos are selected after a few hours of culture and implanted back into recipient CD-1 pseudopregnant females (30 embryos/female).

Preparation of Transgenes and Vectors

The vector commonly used is cosmid-derived and contains 40 kb from the SHaPrP locus (CosSHa.tet) (Scott et al. 1992). The PrP open reading frame (ORF) was replaced by a convenient tetracycline cassette flanked by unique SalI sites, which facilitates the insertion of transgenes in this otherwise cumbersome vector. This vector offers the advantage of reliable copy-dependent transgene expression levels (Scott et al. 1989; Prusiner et al. 1990).

Transgenes are excised from plasmid sequences, separated by agarose gel electrophoresis, purified using β-agarase, extracted using phenol/chloroform and ether, precipitated, dissolved in TE (10 mM Tris, 0.1 mM EDTA; pH 7.5), and dialysed for 30 minutes against TE using a Millipore membrane filter (0.025 μm). The DNA is diluted to 1.5–2.5 ng/ml using TE and freed from residual particulate contaminants using a spin column (Spin-X, Costar).

With this protocol, we created 457 founder Tg mice over a two-year period. Using transgenes derived from the CosSHa.tet vector, the live birth rate was 10.7% and the percentage of transgenics (relative to the total number of eggs microinjected and transplanted) was 1.9%. Using plasmid-derived transgenes, the live birth rate was 12.9% and the transgenic rate was not significantly different at 2.3%.

Breeding of Tg Mice

All information pertaining to the genealogy, inoculation, and health status of animals is maintained on a custom-written animal database (based on Microsoft SQL Server and Macromedia ColdFusion). Entry of animal health status is facilitated by the use of bar codes (LabelRight, Qbar) and handheld bar code readers (Telzon). All the experimental information is uploaded to a central database and is accessible to all investigators.

Because we currently maintain 150–250 strains of Tg mice, production of these animals for experimental studies can easily become a logistical problem. To facilitate the breeding of animal strains, experimental requests are entered in the central animal database, which generates a list of Tg strains requested. Tg mice are produced and bred with $Prnp^{0/0}$ animals that have been backcrossed for 10 generations with FVB mice. This offers a homogeneous background that is suitable for both the production of new transgenic animals by microinjection and the maintenance of transgenic mouse lines by breeding.

Screening of Tg Mice

Some Tg lines are kept in the homozygous state and do not require constant screening for the transgene. Unfortunately, problems such as infertility, circling behavior, or embryonic death due to the transgene are not uncommon in homozygous lines. Therefore, most lines are maintained as heterozygotes, and the Tg status of every newborn mouse must be determined. Potential Tg mice are weaned at 21 days of age, and they each receive an individual identification number tattooed directly on the tail at 28 days. A small piece of tail is then cut off and frozen for screening purposes. To handle large numbers of Tg mice, some of the screening procedures, such as DNA extraction and slot blotting, are automated. A robotic workstation is used to decrease the labor costs, to eliminate human error, and to strengthen the consistency of the results. An automated pipetting station, Biomek 2000 (Beckman Instruments), has been adapted for these purposes. Potential contaminations are avoided by using disposable materials. A slot-blot apparatus has been added to the workstation to facilitate screening. On a weekly basis, over 1000 samples can be typed by two trained operators.

Genomic DNA is automatically extracted from the tail tissue with a phenol/chloroform-free method (Laird et al. 1991). The tails are lysed and digested with proteinase K; the DNA is directly precipitated in isopropanol, washed with 70% ethanol, dried, and dissolved in Tris-EDTA buffer. Determination of the Tg status is performed by polymerase chain reaction (PCR) with multiple primer pairs and/or slot blots of the genomic DNA followed by hybridization with ^{32}P-labeled probes. Although a variety of screening probes are routinely used, most Tg lines were established using the CosSHa.tet cosmid vector (Scott et al. 1992) and are screened using a probe that specifically recognizes the 3′ untranslated region of the SHaPrP ORF. Use of a vector-specific probe rather than an ORF-specific probe facilitates large-scale typing of the Tg mice. However, it also poses a potential disadvantage because mixing of animals between different Tg lines may remain undetected.

In addition to the hybridization procedure and PCR reactions, most of the screening is performed with a robotic workstation, although human intervention is still needed for the loading of the work surface and the mixing and spinning steps. The system is reasonably flexible and can be adapted to the constantly changing screening requirements that are mandated by new scientific questions.

TRANSMISSION OF HUMAN AND BOVINE PRIONS TO TRANSGENIC MICE

For many years, non-Tg mice were used in studies of human prions derived from the brains of patients with CJD. In such studies, the incubation times usually exceeded 500 days and a minority of the animals developed prion disease (Tateishi et al. 1996). Recently, lines of Tg mice expressing either chimeric mouse–human, human, or bovine PrP have been developed, which has facilitated the transmission and study of human prions.

Bioassay of Human Prions

Introduction of a chimeric mouse–human PrP transgene, designated MHu2M, created mice highly susceptible to human prions, in contrast to Tg mice expressing both HuPrP and MoPrP, which are resistant to human prions (Telling et al. 1994). MHu2M is similar to a transgene created earlier with SHaPrP and MoPrP sequences (Scott et al. 1992, 1993). Later, Tg(HuPrP) mice were rendered susceptible to human prions when they were crossed onto a $Prnp^{0/0}$ background (Telling et al. 1995). Incubation times for Tg(MHu2M) mice were similar whether or not MoPrP was co-expressed. Typically, Tg(MHu2M) mice and Tg(HuPrP)$Prnp^{0/0}$ mice exhibit signs of neurologic dysfunction ~200 days after inoculation with sporadic (s) CJD prions.

Transmission of vCJD Prions

vCJD prions inoculated into Tg(MHu2M)5378/$Prnp^{0/0}$ mice resulted in prolonged incubation times between 300 and 725 days with only 25% of the mice becoming ill by 500 days (Fig. 7a). On second passage of MHu2M(vCJD) prions into Tg(MHu2M)5378/$Prnp^{0/0}$ mice, the incubation times shortened substantially and showed a biphasic pattern (Fig. 7a), indicating a transmission barrier for at least two different strains of vCJD prions. Inocula prepared from different mouse brain homogenates, all of which received the same human vCJD inoculum on first passage, produced incubation times ranging between 156 days and 308 days on second passage. The two segments of the biphasic curve, indicating at least two strains of prions, gave mean incubation times of ~160 days and ~290 days.

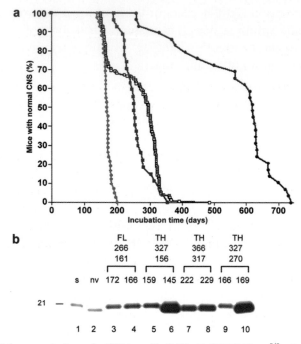

Figure 7. Serial transmission of vCJD into Tg(MHu2M)5378/$Prnp^{0/0}$ mice. (*a*) Survival curve of Tg(MHu2M)5378/$Prnp^{0/0}$ mice inoculated with vCJD. The black line with the filled circles represents the first passage of vCJD into Tg(MHu2M)5378/$Prnp^{0/0}$ mice, the black line with the open squares depicts the second passage, the blue line shows the third passage of the long-incubating substrain, and the red line illustrates the third passage of the short-incubating substrain. (*b*) Immunoblots of deglycosylated rPrP^{Sc} in selected brains from the third passage of vCJD into Tg(MHu2M)5378/$Prnp^{0/0}$ mice. The three sets of numbers above the immunoblot show, from top to bottom, incubation times for first, second, and third passage. For the third passage, deglycosylated rPrP^{Sc} from brains of two different mice is shown. Deglycosylated rPrP^{Sc} from human patients with sCJD(MM1) (lane *1*) and vCJD (lane *2*) are depicted. Tg mice were classified as short incubation time on second passage (lanes *3–6*; *red line* in panel *a*) and as long incubation time on second passage (lanes *7–10*; *blue line* in panel *a*). Deglycosylated rPrP^{Sc} has the electrophoretic mobility of either 21 kD (lanes *3–5* and *9*) or 19 kD (lanes *6–8* and *10*). The original vCJD prions used to inoculate the mice were either from the frontal lobe (FL) (lanes *3–4*) or from the thalamus (TH) (lanes *5–10*) of the human patient. (Reprinted, with permission, from Korth et al. 2003 [copyright National Academy of Sciences].)

On third passage in Tg(MHu2M)5378/$Prnp^{0/0}$ mice, the incubation time of the shorter strain (~160 days) remained constant. In contrast, the longer strain (~290 days) showed a further shortening of incubation time of up to 60 days. On the basis of these observations, we conclude that in Tg(MHu2M)5378/$Prnp^{0/0}$ mice, at least two vCJD prion strains could be identified. Of note, all inocula derived from the frontal lobe of one vCJD patient (RU 96/02) resulted in short (~160 days) incubation times on third passages, whereas inocula from thalamic tissue of another vCJD patient (RU 96/07) resulted in both long (250–320 days) and short incubation times upon third passage. Because only a total of eight second-passage brains derived from two vCJD patients were passaged a third time, specific strain characteristics cannot be assigned to a particular brain region based on the length of the incubation time.

In our initial study of vCJD passaged in Tg(MHu2M)5378/$Prnp^{0/0}$ mice, we found that on first and second passage, a deglycosylated, protease-resistant (r) PrP^{Sc} fragment of 19 kD was produced, which has been found in all vCJD cases examined to date. In third-passaged mouse brain homogenates, we were surprised to find deglycosylated rPrP^{Sc} fragments of 19 kD in some brains and of 21 kD in others (Fig. 7b). No correlation between the incubation time and the fragment size was found. In fact, different Tg(MHu2M)5378/$Prnp^{0/0}$ mice inoculated with the same homogenate from second passage had similar incubation times on third passage but different sizes of deglycosylated rPrP^{Sc} fragments (Fig. 7b, compare lane 5 with 6, and lane 9 with 10).

When extracts from the brains of patients who died of vCJD were inoculated in either Tg(HuPrP,V129)$Prnp^{0/0}$ or Tg(HuPrP,M129)$Prnp^{0/0}$ mice, only a few mice developed signs of neurologic dysfunction after more than 600 days (Hill et al. 1997; Asante et al. 2002). When BSE prions from cattle were inoculated into these lines of Tg(HuPrP)$Prnp^{0/0}$ mice, a minority of the mice developed signs of neurologic dysfunction after >300 days. When the brains of Tg(HuPrP)$Prnp^{0/0}$ mice were examined, some had a glycoform pattern typical of BSE and vCJD whereas others had a pattern similar to that of sCJD. This finding was interpreted as BSE prions having the ability to cause both vCJD and sCJD in humans, but the incidence of sCJD in European countries has not changed over the past six years and therefore this interpretation seems less plausible (Chapter 13).

Transmission of vCJD and BSE Prions to Mice with BoPrP Transgenes

Although non-Tg mice seem moderately susceptible to BSE prions, a considerable change in incubation period from first to second passage is always observed, shortening from >300 days to ~150 days (Lasmézas et al. 1996, 1997; Bruce et al. 1997). This change, indicative of a species barrier (Pattison 1965), provides clear evidence that the properties, and especially the pathogenicity, of the prion changed during passage. To demonstrate that expression of BoPrP abolishes the species barrier for passage of BSE prions from cattle to mice, Tg(BoPrP)$Prnp^{0/0}$ mice were inoculated with brain homogenate from Tg(BoPrP)$Prnp^{0/0}$ mice that became ill following inoculation with BSE prions. As anticipated, virtually identical incubation periods of ~250 days in Tg(BoPrP) mice were obtained upon serial transmission of BSE prions (Table 7). A control brain, containing BSE prions that had been transmitted directly to normal $Prnp^{a/a}$ (FVB) mice, failed to produce disease after >340 days (Table 7). These results show that the properties of BSE prions passaged in $Prnp^a$ mice, the type most frequently used (Lasmézas et al. 1996; Bruce et al. 1997), can vary widely depending on the passage history. Furthermore, the expression of BoPrP in mice eliminates the species barrier for BSE prions.

To test the hypothesis that vCJD is caused by the transmission of BSE prions to humans, Tg(BoPrP)$Prnp^{0/0}$ mice were inoculated with vCJD prions. The reintroduction of BSE prions from humans, i.e., vCJD, into Tg(BoPrP)$Prnp^{0/0}$ mice might restore the original strain properties of the BSE prions. If this transmission is negotiated successfully, the newly formed vCJD prions, which would be composed of BoPrP^{Sc}, would be indistinguishable from native BSE prions.

Each of three cases of vCJD produced an incubation period of 250–270 days on first passage in Tg(BoPrP)$Prnp^{0/0}$ mice (Table 7), which is comparable to those obtained with BSE prions from cattle as well as serially passaged BSE prions from Tg(BoPrP)$Prnp^{0/0}$ mice (Table 7) (Scott et al. 1999). The short incubation period came as a surprise because transmission was between species encoding PrP with different sequences. It is important to note that almost 100% of animals inoculated developed disease (Table 7) and that they exhibited incubation periods with a standard error of the mean (s.e.m.) similar to those observed with BSE or serially passaged BSE prions (Table 7). Furthermore, clinical signs observed in affected animals were indistinguishable from those in Tg(BoPrP)$Prnp^{0/0}$ mice inoculated with serially passaged BSE prions.

Transmission of Scrapie Prions to Mice with BoPrP Transgenes

A change in the processing of offal implemented in the late 1970s is believed to have allowed scrapie prions from sheep to survive rendering and to be passed into cattle (Wilesmith 1991; Kimberlin 1996). Because

Table 7. Susceptibility of transgenic mice to BSE, vCJD, and scrapie prions

Original source	Inoculum	Host	Incubation time (days ± s.e.m)	n/n_0[a]
BSE	PG31/90	Tg(BoPrP)	234 ± 8.3	10/10
BSE	PG31/90→Tg(BoPrP)E4125	Tg(BoPrP)	217 ± 6.1	15/15
BSE	GJ248/85	Tg(BoPrP)	265 ± 22.6	8/8
BSE	GJ248/85→Tg(BoPrP)E4125	Tg(BoPrP)	257 ± 6.6	15/15
BSE	SE1809/11	Tg(BoPrP)	254 ± 6.7	9/9
BSE	97/1612	Tg(BoPrP)	220 ± 7.8	9/9
BSE	97/1997	Tg(BoPrP)	227 ± 5.4	5/6
vCJD	RU96/02	Tg(BoPrP)	271 ± 6.1	5/6
vCJD	RU96/07	Tg(BoPrP)	274 ± 5.8	10/10
vCJD	RU96/110	Tg(BoPrP)	247 ± 4.3	8/8
Scrapie	Suffolk sheep Ewe 15	Tg(BoPrP)	234 ± 55.4	3/7
Scrapie	Suffolk sheep Ram 139	Tg(BoPrP)	215 ± 7.3	8/8
Scrapie	Suffolk sheep Ewe 340	Tg(BoPrP)	206 ± 5.5	8/8
Scrapie	Suffolk sheep Oreo	Tg(BoPrP)	203 ± 5.4	6/10
Scrapie	Suffolk sheep T97 1437	Tg(BoPrP)	245 ± 15.9	4/5
BSE	PG31/90FVB	Tg(MoPrP-A)FVB	>340	0/10

Mice used in this study and primary transmissions of the PG31/90 and GJ248/85 BSE isolates to Tg(BoPrP) and FVB mice were described previously (Scott et al. 1997b). Brain homogenates (10% [w/v] in sterile PBS without Ca or Mg) were prepared by repeated extrusion through syringe needles of successively smaller size, from 18- to 22-gauge. Inoculations of 30-μl doses were carried out with a 27-gauge, disposable hypodermic needle inserted into the right parietal lobe. Other BSE isolates were obtained from John Wilesmith at Central Veterinary Laboratory, U.K. vCJD cases were obtained through the U.K. National CJD Surveillance Unit. Sheep scrapie isolates (Westaway et al. 1994) have been described elsewhere except the Oreo and T97-1437 isolates, which were a generous gift from Dr. Pat Blanchard, UC Davis. The health status of the mice was monitored daily and the neurologic status was assessed semiweekly, as described previously (Carlson et al. 1988; Scott et al. 1993).

[a]n, number of ill mice; n_0, number of inoculated mice.

studies of natural scrapie in the U.S. showed that ~85% of the afflicted sheep are of the Suffolk breed (Hourrigan et al. 1979; Westaway et al. 1992), Tg(BoPrP)$Prnp^{0/0}$ mice were inoculated with scrapie prions from diseased Suffolk sheep. All five cases of natural Suffolk sheep scrapie (three from a herd in Illinois and two identified from California) transmitted disease to Tg(BoPrP)$Prnp^{0/0}$ mice (Table 7). Each of the five isolates transmitted with a relatively short incubation period of ~210 days. In two cases, 100% of the inoculated animals developed disease, and in two other cases, >60% of inoculated mice fell ill. Only one case (Ewe #15) transmitted relatively inefficiently (Table 7). Susceptibility in Suffolk sheep is governed primarily by the PrP codon-171 polymorphism; only sheep homozygous for glutamine at codon 171 are susceptible to scrapie prions (Westaway et al. 1994; Hunter et al. 1997a,b). BoPrP contains glutamine at codon 178, the position analogous to residue 171 in sheep PrP. Because all of the scrapie inocula were homozygous for glutamine at codon 171, the efficient transmission of sheep scrapie to Tg(BoPrP)$Prnp^{0/0}$ mice is consistent with the disposition of this polymorphic residue in controlling susceptibility to scrapie.

Conformation-dependent Immunoassay and Bioassays in Transgenic Mice

In addition to bioassays, Tg mice can be used to calibrate immunoassays measuring PrP^{Sc}. To measure BoPrP^{Sc}, we adapted the conformation-dependent immunoassay (CDI), which simultaneously measures specific antibody binding to denatured (D) and native (N) forms of PrP (Safar et al. 1998). We employed high-affinity recombinant antibody fragments (recFab) reacting with residues 95–105 of BoPrP for detection and another recFab that recognizes residues 132–156 for capture in the CDI (Safar et al. 2002). Using this configuration, the CDI was capable of measuring

---

Figure 8. Similar sensitivity of the CDI and bioassays in Tg(BoPrP)$Prnp^{0/0}$ mice for BSE prions. (A) Dynamic range and analytical sensitivity of the manual direct CDI in detecting PrP^{Sc} in BSE- and CWD-infected brains. Brain homogenates from either pooled BSE-infected Tg(BoPrP)$Prnp^{0/0}$ mice or CWD-infected deer were serially diluted into homologous normal brain homogenate and tested by the CDI. The (D-N) value measured in counts per minute (cpm) is directly proportional to the concentration of PrP^{Sc} (Safar et al. 1998). Data points and bars represent average ± standard deviation (s.d.) obtained from three or four independent measurements. Manual CDI cutoff values were calculated by (mean + 3 s.d.) using samples from uninoculated Tg(BoPrP) mice (*solid horizontal line*) and normal deer (*broken horizontal line*). (B) Direct and sandwich CDI protocols for the detection of BoPrP^{Sc} were compared using pooled BSE-infected brain stem homogenates, serially diluted into normal brain homogenate prepared from the same brain stem area. The automated CDI was used for these studies. Cutoff values for the direct aCDI (*solid horizontal line*) and sandwich aCDI (*broken horizontal line*) were calculated by (mean + 3 s.d.) and determined from 782 tests on 432 normal bovine brain stems. (C) Inverse exponential relationship between titers of BSE prions and incubation times in Tg(BoPrP$^{+/+}$)4092/$Prnp^{0/0}$ mice. The data points are the average ± standard error of the mean (s.e.m.) calculated from three independent endpoint titrations. (D) Direct relationship between BoPrP^{Sc} detected by CDI and BSE prions measured in Tg(BoPrP)$Prnp^{0/0}$ mice. The (D-N) value is directly proportional to the concentration of PrP^{Sc} (Safar et al. 1998). The percentage of ill mice at each BSE sample dilution used in the bioassay was calculated from three independent endpoint titrations. (Reprinted, with permission, from Safar et al. 2002 [copyright Nature Publishing Group].)

PrP^{Sc} in bovine brain stems with a sensitivity similar to that of endpoint titrations in Tg(BoPrP)$Prnp^{0/0}$ mice. To evaluate the sensitivity of the CDI in detecting BSE and chronic wasting disease (CWD) prions, pooled brain homogenates from BSE-infected Tg(BoPrP)$Prnp^{0/0}$ mice or CWD-infected deer were serially diluted into homogenates of normal mice or normal deer brain, respectively. The CDI was able to discriminate between PrP^{Sc} from BSE-infected cattle and Tg(BoPrP)$Prnp^{0/0}$ mice as well as from CWD-infected deer and elk (Safar et al. 2002). The sensitivity of the CDI in detecting CWD prions was equal to or greater than that for the detection of BSE prions in Tg(BoPrP)$Prnp^{0/0}$ mice (Fig. 8A). These results demonstrate a robust quantitative response over a dynamic range of several orders of magnitude. An extensive discussion of different immunoassays for PrP^{Sc} can be found in Chapter 16.

To determine the sensitivity of the automated CDI (aCDI) in detecting BSE prions, brain stem homogenates from BSE-infected cattle (Veterinary Laboratory Agency [VLA], U.K.) were serially diluted into

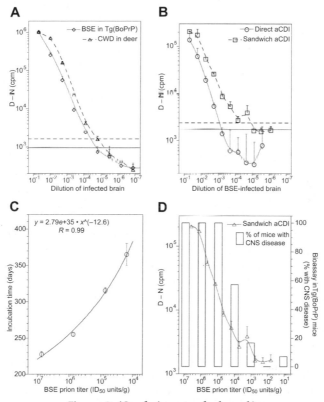

Figure 8. (*See facing page for legend.*)

homogenates of normal bovine brain. We found a robust quantitative response and sensitivity limit at $10^{-3.7}$ dilution of BSE-infected brain (Fig. 8B) (Safar et al. 2002). Introduction of the sandwich protocol using Fab D18 to capture BoPrP increased the sensitivity of the aCDI up to 20-fold (Fig. 8B).

Using a homogenate prepared from the medulla of a Hereford bull with BSE (case PG31/90), three separate titration series were performed in parallel in Tg(BoPrP$^{+/+}$)4092/$Prnp^{0/0}$ mice, giving an average endpoint titer of $10^{6.9}$ ID$_{50}$ U/g of brain tissue (Fig. 8C). This finding compares with $10^{3.1}$ ID$_{50}$ U/g of BSE-infected brain tissue titrated intracerebrally or intraperitoneally in RIII mice (Fraser et al. 1992; Bruce et al. 1994; Moynagh et al. 1999; Deslys et al. 2001), and with 10^{6} ID$_{50}$ U/g reported for endpoint titration in cattle (Wells et al. 1998). These data indicate that Tg(BoPrP$^{+/+}$)4092/$Prnp^{0/0}$ mice are ~10 times more sensitive than cattle and >1000 times more sensitive than RIII mice to infection with BSE prions. The possibility that the BSE isolate from PG31/90 contains an unusually high titer of prions seems highly remote, because previous comparisons of PG31/90 with other BSE brain samples from the VLA gave similar or shorter incubation periods in Tg(BoPrP)4125/$Prnp^{0/0}$ mice (Scott et al. 1999).

The sensitivity of the aCDI applied to brain tissues of BSE-infected Tg(BoPrP)$Prnp^{0/0}$ mice is similar to that of bioassay in cattle and Tg(BoPrP)$Prnp^{0/0}$ mice (Fig. 8A). Some variability from CDI and endpoint titration experiments observed at high dilutions (from 10^{-4} to 10^{-6}) of prion-infected brain homogenates may be attributed to the stochastic distribution of prions in the small aliquots used (bioassay, 30 μl/mouse; CDI, 100 μl/well). One ID$_{50}$ unit of BSE prions, which was calculated in Tg(BoPrP)$Prnp^{0/0}$ mice, is equal to ~10^{-4} dilution of a 10% brain homogenate. This 10^{-4} dilution in Tg mice correlates well with the $10^{-4.6}$ dilution that returns a positive result in 50% of aCDI tests (Fig. 8D).

Because the distribution of PrP^{Sc} in the CNS is quite heterogeneous (Prusiner et al. 1993), we employed the high sensitivity of the CDI to measure PrP^{Sc} in small sections of three BSE-infected brain stems. In each case, two adjacent, ~5-mm-thick cross-sections were cut at the level of the obex (Safar et al. 2002). Transverse slices were cut from the midline to the periphery, yielding an average of 11 samples per brain stem (Fig. 9). Greatest (D–N) differences were seen in samples taken from the midline of the brain stem, but these values progressively decreased in samples collected more laterally (Fig. 9). Significantly, we found up to an 8-fold variation in (D–N) values within one brain stem, indicating the importance of consistent sampling for accurate diagnostic performance.

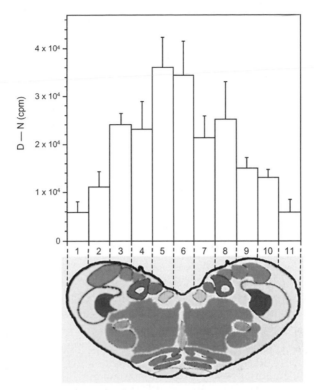

Figure 9. Transverse distribution of PrP^{Sc} in bovine brain stems at the level of the obex as determined by CDI analysis correlates with known BSE pathology. The averaged distribution of (D-N) values measured by CDI using six slices from three brain stems taken from BSE-infected cattle is shown in the upper panel. The numbered position of each slice corresponds to the anatomical location and structures in the lower panel drawing. Values (average ± s.e.m., $n = 6$) are directly proportional to the concentration of PrP^{Sc} (Safar et al. 1998). The data in the lower panel diagram were compiled from histoblots (Prusiner et al. 1993) and standard pathological techniques (Wells and Wilesmith 1995); the light-to-dark shading indicates, from mild to intense, respectively, the severity of vacuolation and degree of PrP^{Sc} staining. (Reprinted, with permission, from Safar et al. 2002 [copyright Nature Publishing Group].)

In addition to bovine prions, the aCDI also demonstrated a sensitivity in detecting sCJD prions similar to bioassay in Tg(MHu2M)*Prnp*$^{0/0}$ mice (Fig. 10). Brain homogenates from three sCJD cases were each diluted into normal human brain and measured by the aCDI. For bioassay, independent endpoint titrations were performed using brain homogenates from the same sCJD cases. Prion titers calculated from endpoint titrations and measured by the aCDI were similar.

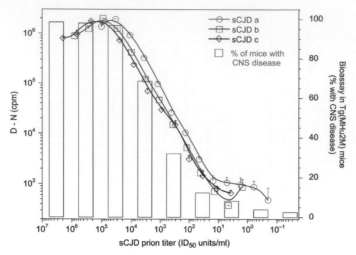

Figure 10. Similar sensitivity of the CDI and bioassays in Tg(MHu2M)*Prnp*[0/0] mice for sCJD prions. For bioassay, the cumulated percentage of ill mice at each sCJD sample dilution (right *y*-axis) was calculated from independent endpoint titrations of brain homogenates from three individual sCJD cases. CDI sensitivity for PrP[Sc] was assessed in the homogenates prepared from the same sCJD brains and diluted into normal human brain homogenate. The presence of PrP[Sc] was evaluated by the difference between denatured and native aliquots of each sample (D-N) (on the left *y*-axis), which is directly proportional to the concentration of PrP[Sc] (Safar et al. 1998, 2002). The Eu-labeled 3F4 mAb (Kascsak et al. 1987) was used as the detection antibody; Mar1 mAb, obtained from Aventis-Behring (Bellon et al. 2003), was used as the capture antibody. The dose of infectious human prions as determined by endpoint titration in Tg(MHu2M)*Prnp*[0/0] mice is plotted on the *x*-axis.

Public Health Implications of Prion Bioassays

Bioassays of prions in Tg mice are not only important for research studies, but they also have important practical implications for public health. All bioassays should detect one ID_{50} unit, but prion titers can be greatly altered by the host in which measurements are made. For example, BSE infectivity measured in cattle showed that earlier studies using an RIII mouse bioassay (Bruce et al. 1994; Moynagh et al. 1999) underestimated BSE prion titers by a factor of ~1000 as described above (Wells et al. 1998). Tg(BoPrP)*Prnp*[0/0] mice are ~10 times more sensitive to infection with BSE prions than cattle. This finding indicates that previous attempts to quantify BSE and scrapie prions in milk or nonneural tissues, such as muscle, may have underestimated infectious titers by up to 10^4, raising the possibility that prions could be present in these products in sufficient quantities to pose some risk to humans (Bosque et al. 2002).

Although the brain is known to contain the highest levels of PrP^{Sc} in all prion diseases, the inaccessibility of this tissue will continue to hinder epidemiologic investigations of these diseases. Using the CDI, we have detected protease-sensitive PrP^{Sc} consistently in the blood of rodents infected with prions (Prusiner and Safar 2000). Using western blots and ELISAs, we have detected rPrP^{Sc} in the hindlimbs of mice (Bosque et al. 2002). Whether blood or muscle will be a suitable matrix for the development of an antemortem test for prions in humans and livestock remains to be determined.

REFERENCES

Asante E.A., Linehan J.M., Desbruslais M., Joiner S., Gowland I., Wood A.L., Welch J., Hill A.F., Lloyd S.E., Wadsworth J.D., and Collinge J. 2002. BSE prions propagate as either variant CJD-like or sporadic CJD-like prion strains in transgenic mice expressing human prion protein. *EMBO J.* **21:** 6358–6366.

Baringer J.R., Bowman K.A., and Prusiner S.B. 1983. Replication of the scrapie agent in hamster brain precedes neuronal vacuolation. *J. Neuropathol. Exp. Neurol.* **42:** 539–547.

Bellon A., Seyfert-Brandt W., Lang W., Baron H., Groner A., and Vey M. 2003. Improved conformation-dependent immunoassay: Suitability for human prion detection with enhanced sensitivity. *J. Gen. Virol.* **84:** 1921–1925.

Bendheim P.E., Barry R.A., DeArmond S.J., Stites D.P., and Prusiner S.B. 1984. Antibodies to a scrapie prion protein. *Nature* **310:** 418–421.

Bessen R.A. and Marsh R.F. 1994. Distinct PrP properties suggest the molecular basis of strain variation in transmissible mink encephalopathy. *J. Virol.* **68:** 7859–7868.

Bosque P.J., Ryou C., Telling G., Peretz D., Legname G., DeArmond S.J., and Prusiner S.B. 2002. Prions in skeletal muscle. *Proc. Natl. Acad. Sci.* **99:** 3812–3817.

Bruce M., Chree A., McConnell I., Foster J., Pearson G., and Fraser H. 1994. Transmission of bovine spongiform encephalopathy and scrapie to mice: Strain variation and the species barrier. *Philos. Trans. R. Soc. Lond. B* **343:** 405–411.

Bruce M.E., Will R.G., Ironside J.W., McConnell I., Drummond D., Suttie A., McCardle L., Chree A., Hope J., Birkett C., Cousens S., Fraser H., and Bostock C.J. 1997. Transmissions to mice indicate that 'new variant' CJD is caused by the BSE agent. *Nature* **389:** 498–501.

Bryan W.R. 1956. Biological studies on the Rous sarcoma virus. IV. Interpretation of tumor-response data involving one inoculation site per chicken. *J. Natl. Cancer Inst.* **16:** 843–863.

———. 1957. Interpretation of host response in quantitative studies on animal viruses. *Ann. N.Y. Acad. Sci.* **69:** 698–728.

Bryan W.R. and Beard J.W. 1939. Estimation of purified papilloma virus protein by infectivity measurements. *J. Infect. Dis.* **65:** 306–321.

Büeler H., Fisher M., Lang Y., Bluethmann H., Lipp H.-P., DeArmond S.J., Prusiner S.B., Aguet M., and Weissmann C. 1992. Normal development and behaviour of mice lacking the neuronal cell-surface PrP protein. *Nature* **356:** 577–582.

Butler D.A., Scott M.R.D., Bockman J.M., Borchelt D.R., Taraboulos A., Hsiao K.K., Kingsbury D.T., and Prusiner S.B. 1988. Scrapie-infected murine neuroblastoma cells produce protease-resistant prion proteins. *J. Virol.* **62:** 1558–1564.

Cairns H.J.F. 1950. Intracerebral inoculation of mice: Fate of the inoculum. *Nature* **166:** 910.

Carlson G.A., Westaway D., DeArmond S.J., Peterson-Torchia M., and Prusiner S.B. 1989. Primary structure of prion protein may modify scrapie isolate properties. *Proc. Natl. Acad. Sci.* **86:** 7475–7479.

Carlson G.A., Goodman P.A., Lovett M., Taylor B.A., Marshall S.T., Peterson-Torchia M., Westaway D., and Prusiner S.B. 1988. Genetics and polymorphism of the mouse prion gene complex: Control of scrapie incubation time. *Mol. Cell. Biol.* **8:** 5528–5540.

Carlson G.A., Kingsbury D.T., Goodman P.A., Coleman S., Marshall S.T., DeArmond S., Westaway D., and Prusiner S.B. 1986. Linkage of prion protein and scrapie incubation time genes. *Cell* **46:** 503–511.

Carlson G.A., Ebeling C., Yang S.-L., Telling G., Torchia M., Groth D., Westaway D., DeArmond S.J., and Prusiner S.B. 1994. Prion isolate specified allotypic interactions between the cellular and scrapie prion proteins in congenic and transgenic mice. *Proc. Natl. Acad. Sci.* **91:** 5690–5694.

Carp R.I. 1982. Transmission of scrapie by oral route: Effect of gingival scarification. *Lancet* **1:** 170–171.

Chandler R.L. 1961. Encephalopathy in mice produced by inoculation with scrapie brain material. *Lancet* **1:** 1378–1379.

Clarke M.C. 1979. Infection of cell cultures with scrapie agent. In *Slow transmissible diseases of the nervous system* (ed. S.B. Prusiner and W.J. Hadlow), vol. 2, pp. 225–234. Academic Press, New York.

Crocker T.T. 1954. The number of elementary bodies per 50% lethal dose of meningo-pneumonitis virus as determined by electron microscopic counting. *J. Immunol.* **73:** 1–7.

Crozet C., Flamant F., Bencsik A., Aubert D., Samarut J., and Baron T. 2001. Efficient transmission of two different sheet scrapie isolates in transgenic mice expressing the ovine PrP gene. *J. Virol.* **75:** 5328–5334.

Cuillé J. and Chelle P.L. 1939. Experimental transmission of trembling to the goat. *C.R. Seances Acad. Sci.* **208:** 1058–1060.

DeArmond S.J., McKinley M.P., Barry R.A., Braunfeld M.B., McColloch J.R., and Prusiner S.B. 1985. Identification of prion amyloid filaments in scrapie-infected brain. *Cell* **41:** 221–235.

DeArmond S.J., Sánchez H., Yehiely F., Qiu Y., Ninchak-Casey A., Daggett V., Camerino A.P., Cayetano J., Rogers M., Groth D., Torchia M., Tremblay P., Scott M.R., Cohen F.E., and Prusiner S.B. 1997. Selective neuronal targeting in prion disease. *Neuron* **19:** 1337–1348.

Deslys J.P., Comoy E., Hawkins S., Simon S., Schimmel H., Wells G., Grassi J., and Moynagh J. 2001. Screening slaughtered cattle for BSE. *Nature* **409:** 476–478.

Dickinson A.G. and Fraser H.G. 1977. Scrapie: Pathogenesis in inbred mice: An assessment of host control and response involving many strains of agent. In *Slow virus infections of the central nervous system* (ed. V. ter Meulen and M. Katz), pp. 3–14. Springer Verlag, New York.

Dickinson A.G., Meikle V.M.H., and Fraser H. 1968. Identification of a gene which controls the incubation period of some strains of scrapie agent in mice. *J. Comp. Pathol.* **78:** 293–299.

———. 1969. Genetical control of the concentration of ME7 scrapie agent in the brain of mice. *J. Comp. Pathol.* **79:** 15–22.

Dougherty R. 1964. Animal virus titration techniques. In *Techniques in experimental virology* (ed. R.J.C. Harris), pp. 169–224. Academic Press, New York.

Eckert E.A., Beard D., and Beard J.W. 1954. Dose-response relations in experimental transmission of avian erythromyeloblastic leukosis. III. Titration of the virus. *J. Natl. Cancer Inst.* **14:** 1055–1066.

Eklund C.M., Hadlow W.J., and Kennedy R.C. 1963. Some properties of the scrapie agent and its behavior in mice. *Proc. Soc. Exp. Biol. Med.* **112:** 974–979.

Fraser H., Bruce M.E., Chree A., McConnell I., and Wells G.A.H. 1992. Transmission of bovine spongiform encephalopathy and scrapie to mice. *J. Gen. Virol.* **73:** 1891–1897.

Gard S. 1940. Encephalomyelitis of mice. II. A method for the measurement of virus activity. *J. Exp. Med.* **72:** 69–77.

Garfin D.E., Stites D.P., Zitnik L.A., and Prusiner S.B. 1978a. Suppression of polyclonal B cell activation in scrapie-infected C3H/HeJ mice. *J. Immunol.* **120:** 1986–1990.

Garfin D.E., Stites D.P., Perlman J.D., Cochran S.P., and Prusiner S.B. 1978b. Mitogen stimulation of splenocytes from mice infected with scrapie agent. *J. Infect. Dis.* **138:** 396–400.

Gibbs C.J., Jr., Amyx H.L., Bacote A., Masters C.L., and Gajdusek D.C. 1980. Oral transmission of kuru, Creutzfeldt-Jakob disease and scrapie to nonhuman primates. *J. Infect. Dis.* **142:** 205–208.

Gogolak F.M. 1953. A quantitative study of the infectivity of murine pneumonitis virus in mice infected in a cloud chamber of improved design. *J. Infect. Dis.* **92:** 240–253.

Goldfarb L.G., Brown P., Cervenakova L., and Gajdusek D.C. 1994. Genetic analysis of Creutzfeldt-Jakob disease and related disorders. *Philos. Trans. R. Soc. Lond. B* **343:** 379–384.

Goldfarb L.G., Brown P., Goldgaber D., Asher D.M., Rubenstein R., Brown W.T., Piccardo P., Kascsak R.J., Boellaard J.W., and Gajdusek D.C. 1990. Creutzfeldt-Jakob disease and kuru patients lack a mutation consistently found in the Gerstmann-Sträussler-Scheinker syndrome. *Exp. Neurol.* **108:** 247–250.

Golob O.J. 1948. A single-dilution method for the estimation of LD_{50} titers of the psittacosis-LGV group of viruses in chick embryos. *J. Immunol.* **59:** 71–82.

Gordon W.S. 1946. Advances in veterinary research. *Vet. Res.* **58:** 516–520.

Hadlow W.J., Kennedy R.C., and Race R.E. 1982. Natural infection of Suffolk sheep with scrapie virus. *J. Infect. Dis.* **146:** 657–664.

Hadlow W.J., Race R.E., Kennedy R.C., and Eklund C.M. 1979. Natural infection of sheep with scrapie virus. In *Slow transmissible diseases of the nervous system* (ed. S.B. Prusiner and W.J. Hadlow), vol. 2, pp. 3–12. Academic Press, New York.

Hadlow W.J., Eklund C.M., Kennedy R.C., Jackson T.A., Whitford H.W., and Boyle C.C. 1974. Course of experimental scrapie virus infection in the goat. *J. Infect. Dis.* **129:** 559–567.

Hill A.F., Desbruslais M., Joiner S., Sidle K.C.L., Gowland I., Collinge J., Doey L.J., and Lantos P. 1997. The same prion strain causes vCJD and BSE. *Nature* **389:** 448–450.

Hogan R.N., Baringer J.R., and Prusiner S.B. 1981. Progressive retinal degeneration in scrapie-infected hamsters: A light and electron microscopic analysis. *Lab. Invest.* **44:** 34–42.

Hogan B., Beddington R., Costantini F., and Lacy E. 1994. *Manipulating the mouse embryo: A laboratory manual.* Cold Spring Harbor Laboratory Press, New York.

Hourrigan J., Klingsporn A., Clark W.W., and de Camp M. 1979. Epidemiology of scrapie in the United States. In *Slow transmissible diseases of the nervous system* (ed. S.B. Prusiner and W.J. Hadlow), vol. 1, pp. 331–356. Academic Press, New York.

Hunter G.D. 1972. Scrapie: A prototype slow infection. *J. Infect. Dis.* **125:** 427–440.

Hunter G.D., Millson G.C., and Chandler R.L. 1963. Observations on the comparative infectivity of cellular fractions derived from homogenates of mouse-scrapie brain. *Res. Vet. Sci.* **4:** 543–549.

Hunter N., Moore L., Hosie B.D., Dingwall W.S., and Greig A. 1997a. Association between natural scrapie and PrP genotype in a flock of Suffolk sheep in Scotland. *Vet. Rec.* **140:** 59–63.

Hunter N., Cairns D., Foster J.D., Smith G., Goldmann W., and Donnelly K. 1997b. Is scrapie solely a genetic disease? *Nature* **386:** 137.

Jendroska K., Heinzel F.P., Torchia M., Stowring L., Kretzschmar H.A., Kon A., Stern A., Prusiner S.B., and DeArmond S.J. 1991. Proteinase-resistant prion protein accumulation in Syrian hamster brain correlates with regional pathology and scrapie infectivity. *Neurology* **41:** 1482–1490.

Kascsak R.J., Rubenstein R., Merz P.A., Tonna-DeMasi M., Fersko R., Carp R.I., Wisniewski H.M., and Diringer H. 1987. Mouse polyclonal and monoclonal antibody to scrapie-associated fibril proteins. *J. Virol.* **61:** 3688–3693.

Kimberlin R.H. 1976. Experimental scrapie in the mouse: A review of an important model disease. *Sci. Prog.* **63:** 461–481.

———. 1996. Speculations on the origin of BSE and the epidemiology of CJD. In *Bovine spongiform encephalopathy: The BSE dilemma* (ed. C.J. Gibbs, Jr.), pp. 155–175. Springer, New York.

Kimberlin R.H. and Walker C.A. 1977. Characteristics of a short incubation model of scrapie in the golden hamster. *J. Gen. Virol.* **34:** 295–304.

———. 1986. Pathogenesis of scrapie (strain 263K) in hamsters infected intracerebrally, intraperitoneally or intraocularly. *J. Gen. Virol.* **67:** 255–263.

Kingsbury D.T., Smeltzer D., and Bockman J. 1984. Purification and properties of the K. Fu. Isolate of the agent of Creutzfeldt-Jakob disease. In *Abstracts from the 6th International Congress of Virology,* Sendai, Japan, W47-6, p. 70.

Korth C., Kaneko K., Groth D., Heye N., Telling G., Mastrianni J., Parchi P., Gambetti P., Will R., Ironside J., Heinrich C., Tremblay P., DeArmond S.J., and Prusiner S.B. 2003. Abbreviated incubation times for human prions in mice expressing a chimeric mouse–human prion protein transgene. *Proc. Natl. Acad. Sci.* **100:** 4784–4789.

Laird P.W., Zijderveld A., Linders K., Rudnicki M.A., Jaenisch R., and Berns A. 1991. Simplified mammalian DNA isolation procedure. *Nucleic Acids Res.* **19:** 4293.

Lasmézas C.I., Deslys J.-P., Demaimay R., Adjou K.T., Lamoury F., Dormont D., Robain O., Ironside J., and Hauw J.-J. 1996. BSE transmission to macaques. *Nature* **381:** 743–744.

Lasmézas C.I., Deslys J.-P., Robain O., Jaegly A., Beringue V., Peyrin J.-M., Fournier J.-G., Hauw J.-J., Rossier J., and Dormont D. 1997. Transmission of the BSE agent to mice in the absence of detectable abnormal prion protein. *Science* **275:** 402–405.

Lax A.J., Millson G.C., and Manning E.J. 1983. Can scrapie titres be calculated accurately from incubation periods? *J. Gen. Virol.* **64:** 971–973.

Luria S.E. and Darnell J.E., Jr. 1967. *General virology.* John Wiley, New York.

Marsh R.F. and Kimberlin R.H. 1975. Comparison of scrapie and transmissible mink encephalopathy in hamsters. II. Clinical signs, pathology and pathogenesis. *J. Infect. Dis.* **131:** 104–110.

Moore R.C., Hope J., McBride P.A., McConnell I., Selfridge J., Melton D.W., and Manson J.C. 1998. Mice with gene targeted prion protein alterations show that *Prnp, Sinc* and *Prni* are congruent. *Nat. Genet.* **18:** 118–125.

Moynagh J., Schimmel H., and Kramer G.N. 1999. The evaluation of tests for the diagnosis of transmissible spongiform encephalopathy in bovines. *European Commission Report,* pp. 1–13.

Pattison I.H. 1965. Experiments with scrapie with special reference to the nature of the agent and the pathology of the disease. In *Slow, latent and temperate virus infections* (NINDB Monogr. 2) (ed. D.C. Gajdusek et al.), pp. 249–257. U.S. Government Printing Office, Washington, D.C.

———. 1966. The relative susceptibility of sheep, goats and mice to two types of the goat scrapie agent. *Res. Vet. Sci.* **7:** 207–212.

Pattison I.H. and Millson G.C. 1960. Further observations on the experimental production of scrapie in goats and sheep. *J. Comp. Pathol.* **70:** 182–193.

———. 1961a. Further experimental observations on scrapie. *J. Comp. Pathol.* **71:** 350–359.

———. 1961b. Scrapie produced experimentally in goats with special reference to the clinical syndrome. *J. Comp. Pathol.* **71:** 101–108.

Peretz D., Williamson R.A., Legname G., Matsunaga Y., Vergara J., Burton D., DeArmond S.J., Prusiner S.B., and Scott M.R. 2002. A change in the conformation of prions accompanies the emergence of a new prion strain. *Neuron* **34:** 921–932.

Prusiner S.B. and Safar J.G. 2000. Method of concentrating prion proteins in blood samples. U.S. Patent 6,166,187.

Prusiner S.B., Cochran S.P., and Alpers M.P. 1985. Transmission of scrapie in hamsters. *J. Infect. Dis.* **152:** 971–978.

Prusiner S.B., Groth D.F., Cochran S.P., McKinley M.P., and Masiarz F.R. 1980a. Gel electrophoresis and glass permeation chromatography of the hamster scrapie agent after enzymatic digestion and detergent extraction. *Biochemistry* **19:** 4892–4898.

Prusiner S.B., Hadlow W.J., Eklund C.M., Race R.E., and Cochran S.P. 1978a. Sedimentation characteristics of the scrapie agent from murine spleen and brain. *Biochemistry* **17:** 4987–4992.

Prusiner S.B., Bolton D.C., Groth D.F., Bowman K.A., Cochran S.P., and McKinley M.P. 1982a. Further purification and characterization of scrapie prions. *Biochemistry* **21:** 6942–6950.

Prusiner S.B., Cochran S.P., Groth D.F., Downey D.E., Bowman K.A., and Martinez H.M. 1982b. Measurement of the scrapie agent using an incubation time interval assay. *Ann. Neurol.* **11:** 353–358.

Prusiner S.B., Groth D.F., Cochran S.P., Masiarz F.R., McKinley M.P., and Martinez H.M. 1980b. Molecular properties, partial purification, and assay by incubation period measurements of the hamster scrapie agent. *Biochemistry* **21:** 4883–4891.

Prusiner S.B., Hadlow W.J., Garfin D.E., Cochran S.P., Baringer J.R., Race R.E., and Eklund C.M. 1978b. Partial purification and evidence for multiple molecular forms of the scrapie agent. *Biochemistry* **17:** 4993–4997.

Prusiner S.B., Garfin D.E., Cochran S.P., McKinley M.P., Groth D.F., Hadlow W.J., Race R.E., and Eklund C.M. 1980c. Experimental scrapie in the mouse: Electrophoretic and sedimentation properties of the partially purified agent. *J. Neurochem.* **35:** 574–582.

Prusiner S.B., Fuzi M., Scott M., Serban D., Serban H., Taraboulos A., Gabriel J.-M., Wells G., Wilesmith J., Bradley R., DeArmond S.J., and Kristensson K. 1993. Immunologic and molecular biological studies of prion proteins in bovine spongiform encephalopathy. *J. Infect. Dis.* **167:** 602–613.

Prusiner S.B., Scott M., Foster D., Pan K.-M., Groth D., Mirenda C., Torchia M., Yang S.-L., Serban D., Carlson G.A., Hoppe P.C., Westaway D., and DeArmond S.J. 1990.

Transgenetic studies implicate interactions between homologous PrP isoforms in scrapie prion replication. *Cell* **63:** 673–686.

Race R.E., Fadness L.H., and Chesebro B. 1987. Characterization of scrapie infection in mouse neuroblastoma cells. *J. Gen. Virol.* **68:** 1391–1399.

Rubenstein R., Carp R.I., and Callahan S.H. 1984. In vitro replication of scrapie agent in a neuronal model: Infection of PC12 cells. *J. Gen. Virol.* **65:** 2191–2198.

Rubenstein R., Deng H., Scalici C.L., and Papini M.C. 1991. Alterations in neurotransmitter-related enzyme activity in scrapie-infected PC12 cells. *J. Gen. Virol.* **72:** 1279–1285.

Rubenstein R., Deng H., Race R.E., Ju W., Scalici C.L., Papini M.C., Kascsak R., and Carp R.I. 1992. Demonstration of scrapie strain diversity in infected PC12 cells. *J. Gen. Virol.* **73:** 3027–3031.

Safar J., Wille H., Itri V., Groth D., Serban H., Torchia M., Cohen F.E., and Prusiner S.B. 1998. Eight prion strains have PrP^{Sc} molecules with different conformations. *Nat. Med.* **4:** 1157–1165.

Safar J.G., Scott M., Monaghan J., Deering C., Didorenko S., Vergara J., Ball H., Legname G., Leclerc E., Solforosi L., Serban H., Groth D., Burton D.R., Prusiner S.B., and Williamson R.A. 2002. Measuring prions causing bovine spongiform encephalopathy or chronic wasting disease by immunoassays and transgenic mice. *Nat. Biotechnol.* **20:** 1147–1150.

Schatzl H.M., Da Costa M., Taylor L., Cohen F.E., and Prusiner S.B. 1995. Prion protein gene variation among primates. *J. Mol. Biol.* **245:** 362–374.

Schatzl H.M., Laszlo L., Holtzman D.M., Tatzelt J., DeArmond S.J., Weiner R.I., Mobley W.C., and Prusiner S.B. 1997. A hypothalamic neuronal cell line persistently infected with scrapie prions exhibits apoptosis. *J. Virol.* **71:** 8821–8831.

Schlesinger R.W. 1949. The mechanism of active cerebral immunity to equine encephalomyelitis virus. I. Influence of the rate of viral multiplication. *J. Exp. Med.* **89:** 491–505.

Scott M.R., Köhler R., Foster D., and Prusiner S.B. 1992. Chimeric prion protein expression in cultured cells and transgenic mice. *Protein Sci.* **1:** 986–997.

Scott M., Groth D., Foster D., Torchia M., Yang S.-L., DeArmond S.J., and Prusiner S.B. 1993. Propagation of prions with artificial properties in transgenic mice expressing chimeric PrP genes. *Cell* **73:** 979–988.

Scott M.R., Groth D., Tatzelt J., Torchia M., Tremblay P., DeArmond S.J., and Prusiner S.B. 1997a. Propagation of prion strains through specific conformers of the prion protein. *J. Virol.* **71:** 9032–9044.

Scott M.R., Will R., Ironside J., Nguyen H.-O.B., Tremblay P., DeArmond S.J., and Prusiner S.B. 1999. Compelling transgenetic evidence for transmission of bovine spongiform encephalopathy prions to humans. *Proc. Natl. Acad. Sci.* **96:** 15137–15142.

Scott M., Foster D., Mirenda C., Serban D., Coufal F., Wälchli M., Torchia M., Groth D., Carlson G., DeArmond S.J., Westaway D., and Prusiner S.B. 1989. Transgenic mice expressing hamster prion protein produce species-specific scrapie infectivity and amyloid plaques. *Cell* **59:** 847–857.

Scott M.R., Safar J., Telling G., Nguyen O., Groth D., Torchia M., Koehler R., Tremblay P., Walther D., Cohen F.E., DeArmond S.J., and Prusiner S.B. 1997b. Identification of a prion protein epitope modulating transmission of bovine spongiform encephalopathy prions to transgenic mice. *Proc. Natl. Acad. Sci.* **94:** 14279–14284.

Somerville R.A. and Carp R.I. 1983. Altered scrapie infectivity estimates by titration and incubation period in the presence of detergents. *J. Gen. Virol.* **64:** 2045–2050.

Supattapone S., Wille H., Uyechi L., Safar J., Tremblay P., Szoka F.C., Cohen F.E., Prusiner S.B., and Scott M.R. 2001. Branched polyamines cure prion-infected neuroblastoma cells. *J. Virol.* **75:** 3453–3461.

Supattapone S., Bosque P., Muramoto T., Wille H., Aagaard C., Peretz D., Nguyen H.-O.B., Heinrich C., Torchia M., Safar J., Cohen F.E., DeArmond S.J., Prusiner S.B., and Scott M. 1999. Prion protein of 106 residues creates an artificial transmission barrier for prion replication in transgenic mice. *Cell* **96:** 869–878.

Tateishi J., Sato Y., and Ohta M. 1983. Creutzfeldt-Jakob disease in humans and laboratory animals. *Prog. Neuropathol.* **5:** 195–221.

Tateishi J., Kitamoto T., Hoque M.Z., and Furukawa H. 1996. Experimental transmission of Creutzfeldt-Jakob disease and related diseases to rodents. *Neurology* **46:** 532–537.

Telling G.C., Haga T., Torchia M., Tremblay P., DeArmond S.J., and Prusiner S.B. 1996a. Interactions between wild-type and mutant prion proteins modulate neurodegeneration in transgenic mice. *Genes Dev.* **10:** 1736–1750.

Telling G.C., Tremblay P., Torchia M., DeArmond S.J., Cohen F.E., and Prusiner S.B. 1997. N-terminally tagged prion protein supports prion propagation in transgenic mice. *Protein Sci.* **6:** 825–833.

Telling G.C., Scott M., Mastrianni J., Gabizon R., Torchia M., Cohen F.E., DeArmond S.J., and Prusiner S.B. 1995. Prion propagation in mice expressing human and chimeric PrP transgenes implicates the interaction of cellular PrP with another protein. *Cell* **83:** 79–90.

Telling G.C., Parchi P., DeArmond S.J., Cortelli P., Montagna P., Gabizon R., Mastrianni J., Lugaresi E., Gambetti P., and Prusiner S.B. 1996b. Evidence for the conformation of the pathologic isoform of the prion protein enciphering and propagating prion diversity. *Science* **274:** 2079–2082.

Telling G.C., Scott M., Hsiao K.K., Foster D., Yang S.-L., Torchia M., Sidle K.C.L., Collinge J., DeArmond S.J., and Prusiner S.B. 1994. Transmission of Creutzfeldt-Jakob disease from humans to transgenic mice expressing chimeric human-mouse prion protein. *Proc. Natl. Acad. Sci.* **91:** 9936–9940.

Vilotte J.L., Soulier S., Essalmani R., Stinnakre M.G., Vaiman D., Lepourry L., Da Silva J.C., Besnard N., Dawson M., Buschmann A., Groschup M., Petit S., Madelaine M.F., Rakatobe S., Le Dur A., Vilette D., and Laude H. 2001. Markedly increased susceptibility to natural sheep scrapie of transgenic mice expressing ovine PrP. *J. Virol.* **75:** 5977–5984.

Wells G.A.H. and Wilesmith J.W. 1995. The neuropathology and epidemiology of bovine spongiform encephalopathy. *Brain Pathol.* **5:** 91–103.

Wells G.A.H., Hawkins S.A.C., Green R.B., Austin A.R., Dexter I., Spencer Y.I., Chaplin M.J., Stack M.J., and Dawson M. 1998. Preliminary observations on the pathogenesis of experimental bovine spongiform encephalopathy (BSE): An update. *Vet. Rec.* **142:** 103–106.

Westaway D., Goodman P.A., Mirenda C.A., McKinley M.P., Carlson G.A., and Prusiner S.B. 1987. Distinct prion proteins in short and long scrapie incubation period mice. *Cell* **51:** 651–662.

Westaway D., Neuman S., Zuliani V., Mirenda C., Foster D., Detwiler L., Carlson G., and Prusiner S.B. 1992. Transgenic approaches to experimental and natural prion diseases. In *Prion diseases of humans and animals* (ed. S.B. Prusiner et al.), pp. 474–482. Ellis Horwood, London.

Westaway D., Zuliani V., Cooper C.M., Da Costa M., Neuman S., Jenny A.L., Detwiler L., and Prusiner S.B. 1994. Homozygosity for prion protein alleles encoding glutamine-171 renders sheep susceptible to natural scrapie. *Genes Dev.* **8:** 959–969.

Wilesmith J.W. 1991. The epidemiology of bovine spongiform encephalopathy. *Semin. Virol.* **2:** 239–245.

Zlotnik I. 1963. Experimental transmission of scrapie to golden hamsters. *Lancet* **2:** 1072.

4

Transmission and Replication of Prions

Stanley B. Prusiner,[1,2,3] Michael R. Scott,[1,2] and Stephen J. DeArmond[1,4]
[1]Institute for Neurodegenerative Diseases, Departments of [2]Neurology,
[3]Biochemistry and Biophysics, and [4]Pathology
University of California, San Francisco, California 94143

George Carlson
McLaughlin Research Institute
Great Falls, Montana 59405

INTRODUCTION

ONE OF THE MOST REMARKABLE FEATURES OF SLOW INFECTIONS is the clockwork precision with which the replication of prions occurs. Inoculation of numerous animals with the same dose of prions results in the onset of illness at the same time many months later. The molecular mechanisms controlling this extraordinarily precise process are unknown.

The term "infection" implies that a pathogen replicates during this process. When the infectious pathogen has achieved a high titer, a disease in the host generally appears. Prior to the recognition of the existence of prions, all infectious pathogens contained a nucleic acid genome that encoded their progeny. Copying of this genome by a polynucleotide polymerase provided a means of replicating the pathogen; in the case of prions, which lack nucleic acid, another mechanism clearly functions.

General Features of Prion Transmission and Replication

As we learn more about prions, some general features and rules of prion replication are beginning to emerge. Although many of the results described here can be found in other chapters, the consolidation of data on prion transmission and replication in this chapter is important.

Prion replication requires the prion protein (PrP); specifically, interaction between the central regions of its normal, cellular isoform (PrP^C)

Prion Biology and Diseases, 2nd Ed. ©2004 Cold Spring Harbor Laboratory Press 0-87969-693-1/04

ntly folded isoform (PrP^{Sc}) (Prusiner et al. 1990; Scott et al.
>rocess can replicate many different PrP^{Sc} conformations
rains breed true (Telling et al. 1996; Scott et al. 1999; Peretz
_____, ___mino acid substitutions modulate the rate of prion replica-
tion: In mouse (Mo) PrP, substitutions at residues 108 and 189 dramatically change the length of the incubation time in inbred mice (Westaway
et al. 1987; Moore et al. 1998). In the chimeric mouse–human (MHu2M)
transgene, substitution of mouse residues at positions 165 and 167 in the
"Hu2" insert results in a substantial reduction in incubation times in
transgenic (Tg) mice (Korth et al. 2003). Residues 167 and 218 are part
of a discontinuous epitope that was mapped by site-directed mutagenesis (Kaneko et al. 1997b). Substitution of a basic residue at 167 makes
PrP^C unavailable for conversion into PrP^{Sc}, resulting in a dominant negative (Perrier et al. 2002). When dominant-negative PrP is expressed with
wild-type (wt) PrP, the dominant-negative mutant inhibits the conversion of wt PrP^C into PrP^{Sc}. It is likely that dominant-negative PrP exerts
its effect by sequestering an auxiliary protein involved in prion replication, provisionally designated protein X. Presumably, protein X functions
as a chaperone and facilitates the refolding of PrP^C into PrP^{Sc}, analogous
to Hsp104 in the replication of the yeast prion [PSI] (Chernoff et al.
1995).

Prion Replication and Incubation Times

When the titer of prions reaches a critical threshold, animals develop signs
of neurologic dysfunction. The time interval from inoculation to clear
signs of CNS dysfunction is referred to as the incubation period.

The length of the incubation time can be modified by (1) the dose of
prions; (2) the route of inoculation; (3) the level of PrP^C expression; (4)
the species of the prion (i.e., the sequence of PrP^{Sc}); (5) polymorphisms
in the sequence of PrP^C (e.g., dominant-negative mutations); and (6) the
strain of prion. The incubation period is inversely related to the prion
dose (Prusiner et al. 1982b) (Chapter 3). The intracerebral route of inoculation is substantially more efficient than other routes (Prusiner et al.
1985). Higher levels of PrP^C expression result in more rapid accumulation
of PrP^{Sc} and, hence, shorter incubation times (Prusiner et al. 1990). When
the amino acid sequence of PrP^{Sc} in the inoculum is identical to that of
PrP^C in the recipient, incubation times are generally the most abbreviated
(Scott et al. 1989, 1997). Polymorphic residues in PrP can greatly modify
the length of the incubation time; $Prnp^{a/a}$ mice show different incubation
times from $Prnp^{b/b}$ mice for a particular prion strain (Dickinson et al.

1968; Carlson et al. 1986; Moore et al. 1998). Proteins encoded by $Prnp^a$ and $Prnp^b$ differ at positions 108 and 189 (Westaway et al. 1987). Dominant-negative mutations that inhibit prion formation result in prolonged incubation times (Perrier et al. 2002). At present, the mechanism by which strains modify the incubation time is not well understood, but interactions between PrP^{Sc} and PrP^C undoubtedly play a major role (DeArmond et al. 1997; Prusiner 1997; Scott et al. 1997, 1999; Peretz et al. 2002).

Transmission of Prions among Mammals

More than 180,000 cattle, primarily dairy cows, have died of bovine spongiform encephalopathy (BSE) over the past two decades (Anderson et al. 1996; Phillips et al. 2000) (Chapter 12). BSE is a massive common-source epidemic caused by prion-contaminated meat and bone meal (MBM) fed primarily to dairy cows (Wilesmith et al. 1991). MBM is prepared from the offal of sheep, cattle, pigs, and chickens as a high-protein nutritional supplement. In the late 1970s, the hydrocarbon-solvent extraction method used in the rendering of offal began to be abandoned, resulting in MBM with a much higher fat content (Wilesmith et al. 1991). Many believe that this change in the rendering process allowed scrapie prions from sheep to survive rendering and subsequently to be passed into cattle. An alternative hypothesis contends that the BSE epidemic began with a sporadic case of prion disease in a cow that was rendered into MBM (Phillips et al. 2000), which allowed BSE prions from a bovine source to survive and be repassaged to cattle. The mean incubation time for BSE is ~4–5 years. Most cattle therefore did not manifest disease because they were slaughtered between 2 and 3 years of age (Anderson et al. 1996; Stekel et al. 1996; Donnelly and Ferguson 2000).

The natural route by which scrapie prions are transmitted among sheep and goats is unknown. Some investigators have suggested that horizontal transmission occurs orally (Palsson 1979). Scrapie prions are hypothesized to originate from placentae from infected sheep and goats (Pattison et al. 1972). For many years, studies of experimental scrapie were performed exclusively with sheep and goats. The disease was first transmitted by intraocular inoculation (Cuillé and Chelle 1939) and later by intracerebral, oral, subcutaneous, intramuscular, and intravenous injections of brain extracts from sheep developing scrapie. Incubation periods of 1–3 years were common, and many of the inoculated animals often failed to develop disease (Dickinson and Stamp 1969; Hadlow et al. 1980, 1982). Different breeds of sheep exhibited markedly different susceptibilities to scrapie prions inoculated subcutaneously, suggesting that the genetic background might influence host permissiveness (Gordon 1966).

The transmission of chronic wasting disease (CWD) prions among deer and elk is equally perplexing (Spraker et al. 1997). The high incidence of this disease demands that horizontal transmission plays a large role in the spread of CWD prions. Up to 15% of free-ranging deer and elk in the western United States are estimated to carry CWD prions (Miller et al. 1998). Approximately 90% of culled deer in a captive population were found to have PrP^{Sc} in their brains, based on immunohistochemistry (Williams and Young 1980); this finding argues that prions in deer and possibly elk are highly communicable (Chapters 1 and 11).

Routes of Transmission

Oral transmission by ritualistic cannibalism of dead relatives was suggested to be the mechanism by which kuru was spread among New Guinea natives (Gajdusek 1977; Alpers 1979). However, the role of cannibalism in the etiology of kuru has been questioned because oral transmission of kuru to nonhuman primates has been demonstrated in only a few instances (Gibbs et al. 1980). Apes and monkeys regularly contracted a kuru-like illness many months or years after intracerebral or peripheral inoculation (Gajdusek 1977). The inefficient oral transmission of prions prompted hypotheses suggesting that the kuru prions might have entered only through abrasions of mucous membranes within the oropharynx or conjunctivae, as well as through open wounds on the hands (Gajdusek 1977).

It seems unlikely that Creutzfeldt-Jakob disease (CJD) is spread among humans by infection except in the cases of accidental inoculations; indeed, most CJD cases are sporadic and are likely to be the result of a somatic mutation, the spontaneous conversion of PrP^{C} to PrP^{Sc}, or reduced clearance of low levels of PrP^{Sc} that are normally present (Peretz et al. 2001b). Many attempts to identify an infectious source of prions to explain sporadic CJD have been unsuccessful (Brown et al. 1987; Harries-Jones et al. 1988; Cousens et al. 1990).

KINETICS OF PRION REPLICATION

Studies of prion replication in random-bred Swiss mice showed that the titers in systemic organs increased slowly after subcutaneous inoculation (Table 1) (Eklund et al. 1967). The prion titer accumulated most rapidly in the spleen, where it reached its maximum level approximately 1 month after inoculation, then remained constant. The titer in brain increased

Table 1. Kinetics of prion replication in mice

Weeks after inoculation	1	4	8	12	16	20	24	28	29[a]	32	36	42
Percent of surviving mice sick with scrapie								26	40	60	63	25
Percent of total mice dead of scrapie							6.8	4.5	6.9	25	61	73
Tissues examined[b]												
Spleen	4.5[c]	3.5	5.6	5.6	6.2	6.2	5.5	5.5	5.7	5.6	5.2	5.5
Peripheral lymph nodes		3.4	5.6	4.7	4.7	5.2	4.5	4.8	5.4	5.5	5.6	4.6
Thymus			4.2	4.8	5.5	5.0	5.4	4.5	0.8[d]	5.2	5.6	4.5
Submaxillary salivary gland		5.8	5.5	6.5	5.2	6.2	6.0	6.4	5.6	3.4	2.5	
Lung				3.5	3.4	3.2	3.2	2.4	2.5		2.2	3.8
Intestine				2.2	2.2	3.4	5.3	5.5	4.6	5.4	5.2	4.5
Spinal cord				1.4	5.6	4.5	6.6	6.5	7.4	7.4	6.7	6.6
Brain					4.4	3.2	5.7	6.3	6.7	6.5	7.2	7.4
Bone marrow (femur)							1.7	2.8	4.8	3.5	5.0	3.6
Uterus						+[e]	+		+			
Liver	NA	NA	NA	NA	NA	+	+	+	+	+	+	+
Kidney	NA	NA	NA	NA	NA	+	+					

Data from Eklund et al. (1967).

[a] From the 29th week on, only sick mice were examined.

[b] Blood clot, serum, and tests were also examined, but virus was never detected in them.

[c] Negative $\log_{10}$ of dilution of tissue suspension that contained 1 LD_{50} per 0.03 ml when inoculated intracerebrally into mice. Blank spaces indicate virus was not detected in any dilution. (NA) Not examined.

[d] Whether the thymus was removed is questionable.

[e] Virus was detected in 10^{-1} dilution only, and not all mice were affected.

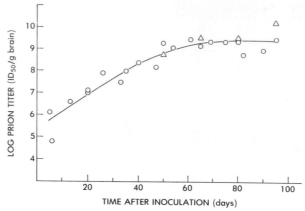

Figure 1. Kinetics of prion replication in brains of Syrian hamsters. Circles indicate titers determined by incubation-time assays; triangles depict titers determined by endpoint titrations. (Reprinted, with permission, from (Oesch et al. 1985 [copyright Cell Press].)

more slowly, but after approximately 3 months, exceeded titers in the spleen. By 4 months, the titer in brain reached its maximum. The kinetics of prion replication in murine spleen and brain after intracerebral inoculation are similar (Kimberlin 1976).

The kinetics of prion replication in brain can be determined by inoculating animals, sacrificing them in 1- to 2-week intervals, and then measuring titers by endpoint titration or the incubation-time assay (Fig. 1). In hamsters inoculated intracerebrally with a dose causing infection in half of inoculated animals (ID_{50}) of ~10^7 U/ml, the prion titer in brain increased 10^4-fold over 50 days and then plateaued. Histopathologic changes were first seen in brain ~50 days after inoculation. By 70 days, widespread vacuolation of the neuropil and extensive astrocytic proliferation were observed (Baringer et al. 1981). Between 75 and 85 days, most of the animals died.

Plots of survival curves for 200 hamsters inoculated intracerebrally with ~10^7 ID_{50} units show remarkable synchrony of the disease (Fig. 2A). All the animals were inoculated on day zero. Eighty percent of the hamsters developed clinical signs of disease between 65 and 69 days after inoculation; more than 70% of the animals died between 78 and 86 days after inoculation.

When the intracerebral inoculum was reduced to <10 ID_{50} units, the incubation period doubled (120–130 days; Fig. 2B). With the lower titer in the inoculum, the resulting titer in brain increases more slowly and the scatter of data points is much greater when compared to the inoculum of 10^7 ID_{50} U/ml (compare Fig. 2B with Fig. 1). The titer increases 10^8-fold

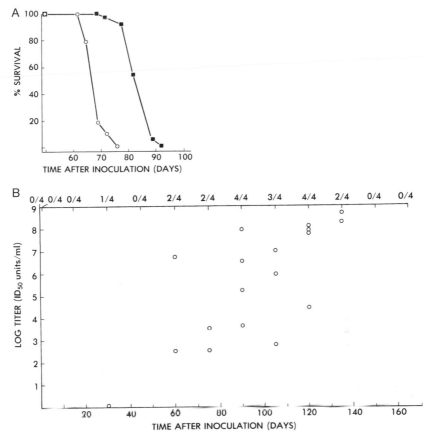

Figure 2. Survival curves for Syrian hamsters and the kinetics of prion replication. (*A*) Inoculated with ~10^7 ID$_{50}$ units of prions. Open circles indicate onset of clinical signs of disease; filled squares indicate death. (*B*) Inoculated with ~10 ID$_{50}$ units of prions. Four hamsters were sacrificed at times specified on the bottom *x*-axis and prion titers in brains were determined by bioassays. The number of hamsters with detectable levels of brain prions is denoted on the top *x*-axis. (Reprinted, with permission, from Prusiner 1987 [copyright Academic Press].)

over 130 days, clearly showing that prions can increase their number by a factor of ~100 million.

Prion Transmission between Species

A prolongation of the incubation time upon passage of prions from one species to another has been referred to as the "species barrier" (Pattison 1965). Once the first passage of prions in a new species occurs, second passage in the same species has frequently been called "adaptation" (Kimberlin and Walker 1977; Gibbs et al. 1979; Manuelidis and Manuelidis 1979a; Tateishi et al. 1979; Kingsbury et al. 1982). On third passage in the same host, the incubation time generally does not change, but occasionally a small decrement in the incubation time has been reported.

Scrapie prions passaged through mink retained their ability to infect goats but lost their ability to infect mice (W.J. Hadlow, unpubl.). Interestingly, prions causing mink encephalopathy have a similar host range (Marsh and Kimberlin 1975): They can cause disease in goats and Syrian hamsters but not in mice (Marsh and Kimberlin 1975; Marsh et al. 1991). The molecular changes that distinguish the scrapie prions propagated in mink from those found in goats and mice are undoubtedly due to the conformation adopted by PrP^{Sc} as they are passaged from one host to another. This scenario is similar to that described for the Me7 prion strain passaged either directly to Syrian hamsters (Kimberlin et al. 1987), or first through Tg mice expressing chimeric mouse–hamster PrP and subsequently to hamsters (Scott et al. 1997).

The incubation period for first passage of scrapie prions from sheep to Swiss mice is typically 40–48 weeks for a 10-fold dilution of 10% (w/v) brain homogenates (W.J. Hadlow, unpubl.). When brain homogenates prepared from goats with drowsy-type scrapie were inoculated into Swiss mice, 24–28 weeks elapsed before the mice showed signs of disease (Chandler and Fisher 1963). Passage of brain homogenates of diseased Swiss mice to other Swiss mice resulted in an incubation period of 16–20 weeks. In two subsequent passages to Swiss mice, the incubation period remained unchanged. Passage from brain homogenates of Swiss mice to white rats prolonged the incubation period to more than 64 weeks; however, subsequent passage to rats reduced the incubation time to 49 weeks (Pattison and Jones 1968). Passage of prions from rats back to Swiss mice resulted in an incubation period of 49 weeks. Subsequent passages to Swiss mice caused disease in 16–20 weeks (Table 1).

In studies on the transmission of CJD prions from humans to rodents, the species barrier is quite profound (Tateishi et al. 1996). Primary transmissions from humans to mice required between 300 and 800 days (Table 2). On second passage, with prions from mice passaged to mice, the incubation period was reduced to ~120 days; the same incubation period was observed on the third passage. Purification protocols and studies on the properties of CJD prions have been performed with mice. As with studies on scrapie prions, investigations on CJD prions have been dramatically accelerated by the use of incubation-time assays (Walker et al. 1983; Bendheim et al. 1985).

Table 2. Human prion transmission to nontransgenic mice

	Incubation times for transmission to mice					
	first passage		second passage		third passage	
Case number	(n)	(days ± s.d.)	(n)	(days ± s.d.)	(n)	(days ± s.d.)
1	6	318 ± 92	20	128 ± 14		
2	8	711 ± 115				
3	11	665 ± 122	18	111 ± 10	13	125 ± 43
4	7	447 ± 191	18	122 ± 9	3	120 ± 5
5	4	705 ± 53	2	215		
6	13	769 ± 71				

Data from Tateishi et al. (1996).

Transgenetics and Species Barriers

Prions synthesized de novo reflect the sequence of PrP encoded by the host and not that of the PrP^{Sc} molecules in the inoculum (Bockman et al. 1987). On subsequent passage in a homologous host, the incubation time shortens to a constant length observed for all subsequent passages, and transmission becomes a nonstochastic process. The species barrier is of practical importance in assessing the risk of humans acquiring prion disease after consumption of scrapie-infected lamb, BSE-infected beef, or CWD-infected venison.

To determine whether the differences in PrP gene sequence between two species is responsible for the prion transmission barrier, Tg mice expressing Syrian hamster (SHa) PrP were constructed (Scott et al. 1989; Prusiner et al. 1990). The PrP genes of Syrian hamsters and mice encode proteins that differ at 16 residues. Incubation times in four lines of Tg mice inoculated with mouse scrapie prions were prolonged compared to those observed for non-Tg, control mice (Fig. 3A). Tg mice inoculated with SHa prions showed abbreviated incubation times in a nonstochastic process (Fig. 3B) (Scott et al. 1989; Prusiner et al. 1990), with the length of the incubation time inversely proportional to the expression level of $SHaPrP^{C}$ in the brain (Fig. 3B,C) (Scott et al. 1989; Prusiner et al. 1990). $SHaPrP^{Sc}$ concentrations in the brains of clinically ill mice were similar in all four transgenic lines inoculated with hamster prions (Fig. 3D). Bioassays of brain extracts from clinically ill Tg mice inoculated with mouse prions revealed that mouse prions and not hamster prions were produced (Fig. 3E). Conversely, inoculation of Tg mice with hamster prions led to the synthesis of only hamster prions (Fig. 3F). Thus, in Tg mice, the de novo synthesis of prions is species-specific and reflects the genetics of the last host in which the prions were passaged.

Similarly, the neuropathology of Tg mice is determined by the genetic origin of the prion inoculum. Mouse prions injected into Tg mice produced neuropathologic changes characteristic of mice with scrapie. Moderate vacuolation in both the gray and white matter was found, whereas amyloid plaques were rarely detected (Fig. 3G). Inoculation of Tg mice with SHa prions produced intense vacuolation of the gray matter, sparing of the white matter, and numerous SHaPrP amyloid plaques, characteristic of Syrian hamsters with scrapie (Fig. 3H). These studies with Tg mice established that the PrP gene influences virtually all aspects of prion disease, including the species barrier, replication of prions, incubation times, synthesis of PrP^{Sc}, and neuropathologic changes.

VARIATIONS IN PATTERNS OF DISEASE

The length of the incubation time is one parameter used to characterize prion strains. Distinct strains characterized by particular incubation times have been studied in sheep, goats, mice, and hamsters. Dickinson and his colleagues developed a system for "strain typing" by which mice with

Figure 3. Transgenic mice expressing SHaPrP exhibit species-specific incubation times, prion synthesis, and neuropathology (Prusiner et al. 1990). (*A*) Incubation times in non-Tg mice, four lines of Tg(SHaPrP) mice, and Syrian hamsters inoculated intracerebrally with ~10^6 ID$_{50}$ units of Chandler mouse prions serially passaged in Swiss mice. The four lines of Tg mice have different copy numbers of the transgene: Tg69 and 71 mice have 2–4 copies of the transgene, Tg81 mice have 30–50 copies, and Tg7 mice have >60 copies. The incubation time is the number of days from inoculation to onset of neurologic dysfunction. (*B*) Incubation times in mice and hamsters inoculated with ~10^7 ID$_{50}$ units of Sc237 prions serially passaged in Syrian hamsters. (*C*) SHaPrP^C expressed in brain in Tg mice and hamsters. SHaPrP^C levels were quantified by ELISA. (*D*) SHaPrP^{Sc} in the brains of Tg mice and hamsters, determined by immunoassay. Animals were sacrificed after exhibiting clinical signs of disease. (*E*) Prion titers in brains of clinically ill non-Tg, Tg71, and Tg81 mice after inoculation with mouse prions, as determined by bioassays in mice (*left bar* of each set) and hamsters (*right bar* of each set). (*F*) Prion titers in brains of clinically ill Syrian hamsters, Tg71 mice, and Tg81 mice after inoculation with SHa prions as determined by bioassays in mice (*left bar* of each set) and hamsters (*right bar* of each set). (*G*) Neuropathology in non-Tg mice and Tg(SHaPrP) mice with clinical signs of disease after inoculation with Mo prions. Vacuolation in gray matter (*left bar* of each set) and white matter (*center bar* of each set); PrP amyloid plaques (*right bar* of each set). Vacuolation score: 0 = none, 1 = rare, 2 = modest, 3 = moderate, 4 = intense. PrP amyloid plaque frequency: 0 = none, 1 = rare, 2 = few, 3 = many, 4 = numerous. (*H*) Neuropathology in Syrian hamsters and Tg mice inoculated with SHa prions. Bars represent neuropathologic changes as in panel G. Vacuolation scores and amyloid plaque frequency as in panel G. (Reprinted, with permission, from Prusiner 1991 [copyright AAAS].)

genetically determined short and long incubation times were used in combination with the F_1 cross (Dickinson et al. 1968, 1984; Dickinson and Meikle 1971). When inoculated with either the Me7 or Chandler isolates, for example, C57BL mice exhibited short incubation times of ~150 days, whereas VM mice had prolonged incubation times of ~300 days. Because

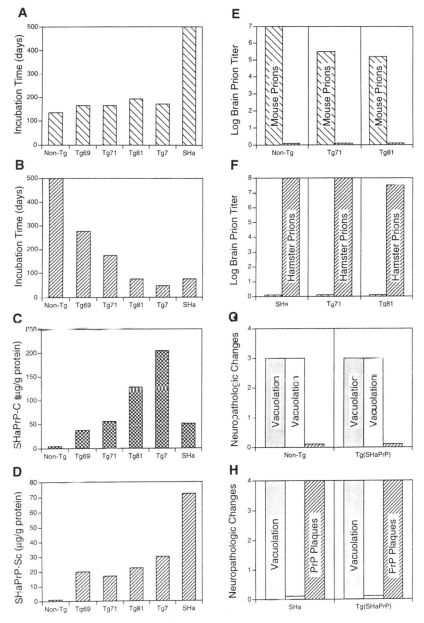

Figure 3. (*See facing page for legend.*)

of prolonged incubation times in F_1 mice, long incubation times were concluded to be a dominant trait. Prion strains were categorized into two groups based on their incubation times: (1) those causing disease more rapidly in "short"-incubation-time C57BL mice and (2) those causing disease more rapidly in "long"-incubation-time VM mice. The 22a and 87V prion strains are noteworthy because they can be passaged in VM mice while maintaining their distinct characteristics.

PrP Gene Dosage

Studies spanning more than a decade were required to unravel the mechanism responsible for the "dominance" of long incubation times. Not unexpectedly, long incubation times were found not to be dominant traits; instead, the apparent dominance of long incubation times was found to be due to a gene-dosage effect (Carlson et al. 1994).

The identification of a widely available mouse strain with long incubation times was a crucial step in elucidating the control mechanism of incubation time in mice. I/LnJ mice inoculated with RML prions were found to have incubation times exceeding 200 days (Kingsbury et al. 1983), a finding that was later confirmed by other workers (Carp et al. 1987). Once molecular clones of the PrP gene were available, a restriction fragment length polymorphism (RFLP) of the PrP gene was used to follow the segregation of MoPrP genes (*Prnp*) in the F_1 and F_2 progeny of either NZW or C57BL mice that harbor short incubation periods crossed with I/LnJ mice that demonstrate long incubation periods. This approach demonstrated genetic linkage between *Prnp* and a gene modulating incubation times, provisionally labeled *Prn-i* (Carlson et al. 1986). Other investigators confirmed the genetic linkage, and one group showed that the incubation-time gene *Sinc* is also linked to *Prnp* (Hunter et al. 1987; Race et al. 1990). Subsequently, *Prnp*, *Prn-i*, and *Sinc* genes were shown to be congruent (Ziegler 1993; Moore et al. 1998). The PrP gene sequences of NZW and I/LnJ with short and long incubation times, respectively, differ at positions 108 and 189 (Westaway et al. 1987). In NZW and C57BL mice, the PrP gene, designated *Prnp^a*, encodes PrP-A(L108,T189), and in I/LnJ mice, the PrP gene, designated *Prnp^b*, encodes PrP-B(F108,V189). The term "allotype" is used to describe these allelic variants of PrP.

Although the amino acid substitutions in PrP that distinguish *Prnp^a* from *Prnp^b* mice argued for the congruency of *Prnp* and *Prn-i*, experiments with *Prnp^a* mice expressing *Prnp^b* transgenes demonstrated a "paradoxical" shortening of incubation times (Westaway et al. 1991). We predicted that these Tg mice would exhibit prolonged incubation times after inoculation with RML prions based on ($Prnp^a$ × $Prnp^b$) F_1 mice, which exhibit long incubation times. We described these findings as a "paradoxical shortening," because we and other investigators had believed for many years that long incubation times are dominant traits (Dickinson et al. 1968; Carlson et al. 1986). From studies of congenic and transgenic mice expressing different numbers of the *a* and *b* alleles of *Prnp* (Table 3), we now realize that these findings were not paradoxical; indeed, they resulted from increased PrP gene dosage (Carlson et al. 1994). When RML prions were inoculated into congenic and transgenic mice, the increased number of copies of the *a* allele was found to be the major determinant in reducing the incubation time; however, increasing the number of copies of the *b* allele also reduced the incubation time, but not to the same extent as seen with the *a* allele (Table 3).

Table 3. MoPrP-A expression is a major determinant and MoPrP-B is a minor determinant of incubation times in mice inoculated with the RML prions

Mice	*Prnp* genotype	*Prnp* transgenes (copies)	Alleles *a* (copies)	Alleles *b* (copies)	Incubation time (days ± S.E.M.)	*n*
$Prnp^{0/0}$	0/0		0	0	>600	4
$Prnp^{+/0}$	a/0		1	0	426 ± 18	9[a]
B6.I-1	b/b		0	2	360 ± 16	7
B6.I-2	b/b		0	2	379 ± 8	10[b]
B6.I-3	b/b		0	2	404 ± 10	20
(B6 × B6.I-1)F1	a/b		1	1	268 ± 4	7
B6.I-1 × Tg(MoPrP-B$^{0/0}$)15	a/b		1	1	255 ± 7	11[c]
B6.I-1 × Tg(MoPrP-B$^{0/0}$)15	a/b		1	1	274 ± 3	9[d]
B6.I-1 × Tg(MoPrP-B$^{+/0}$)15	a/b	bbb/0	1	4	166 ± 2	11[c]
B6.I-1 × Tg(MoPrP-B$^{+/0}$)15	a/b	bbb/0	1	4	162 ± 3	8[d]
C57BL/6J (B6)	a/a		2	0	143 ± 4	8
B6.I-4	a/a		2	0	144 ± 5	8
non-Tg(MoPrP-B$^{0/0}$)15	a/a		2	0	130 ± 3	10
Tg(MoPrP-B$^{+/0}$)15	a/a	bbb/0	2	3	115 ± 2	18
Tg(MoPrP-B$^{+/+}$)15	a/a	bbb/bbb	2	6	111 ± 5	5
Tg(MoPrP-B$^{+/0}$)94	a/a	>30b	2	>30	75 ± 2	15[e]
Tg(MoPrP-A$^{+/0}$)B4053	a/a	>30a	>30	0	50 ± 2	16

[a]Data from Prusiner et al. (1993a).

[b]Data from Carlson et al. (1993).

[c]The homozygous Tg(MoPrP-B$^{+/+}$)15 mice were maintained as a distinct subline selected for transgene homozygosity two generations removed from the (B6 × LT/Sv)F$_2$ founder. Hemizygous Tg(MoPrP-B$^{+/0}$)15 mice were produced by crossing the Tg(MoPrP-B$^{+/+}$)15 line with B6 mice.

[d]Tg(MoPrP-B$^{+/0}$)15 mice were maintained by repeated backcrossing to B6 mice.

[e]Data from Westaway et al. (1991).

The discovery that incubation times are controlled by the relative dosage of $Prnp^a$ and $Prnp^b$ alleles was foreshadowed by studies of Tg(SHaPrP) mice in which the length of the incubation time after inoculation with SHa prions was inversely proportional to the transgene product, $SHaPrP^C$ (Prusiner et al. 1990). The PrP gene dose determines not only the length of the incubation time, but also the passage history of the inoculum, particularly in $Prnp^b$ mice (Table 3). The PrP^{Sc} allotype in the inoculum produced the shortest incubation times when it was the same as that of PrP^C in the host (Carlson et al. 1989).

To address the issue of whether gene products other than PrP might be responsible for these findings, we inoculated B6 and B6.I-4 mice carrying $Prnp^{a/a}$, as well as I/LnJ and B6.I-2 mice carrying $Prnp^{b/b}$ (Carlson et al. 1993, 1994), with RML prions passaged in the following homozygous $Prnp$ mice (Table 3): CD-1 and NZW/LacJ mice that produce PrP^{Sc}-A encoded by $Prnp^a$ and I/LnJ mice that produce PrP^{Sc}-B encoded by $Prnp^b$. The incubation times in the congenic mice reflected the PrP allotype rather than other factors acquired during prion passage. The effect of the allotype barrier was small when measured in $Prnp^{a/a}$ mice but was clearly demonstrable in $Prnp^{b/b}$ mice. B6.I-2 congenic mice inoculated with prions from I/LnJ mice had an incubation time of 237 ± 8 days compared to times of 360 ± 16 days and 404 ± 4 days for mice inoculated with prions passaged in CD-1 and NZW/LacJ mice, respectively. Thus, inoculation of prions from $Prnp^b$ mice shortened the incubation time by ~40% in $Prnp^b$ mice, compared to inoculation with prions passaged in $Prnp^a$ mice (Carlson et al. 1989).

Overdominance

The phenomenon of "overdominance" in which incubation times in F_1 hybrids are longer than those of either parent (Dickinson and Meikle 1969) contributed to the confusion surrounding control of incubation times. When the 22A scrapie isolate was inoculated into B6, B6.I-1, and F_1 from (B6 × B6.I-1) mice, overdominance was observed: the incubation times were 405 ± 2 days, 194 ± 10 days, and 508 ± 14 days for B6, B6.1, and (B6 × B6.I-1)F_1 mice, respectively (Table 4). Shorter incubation times were observed in homozygous and hemizygous Tg(MoPrP-B)15 mice. Hemizygous Tg(MoPrP-B$^{+/0}$)15 mice exhibited an incubation time of 395 ± 12 days, whereas the homozygous mice had an incubation time of 286 ± 15 days.

As with results from the RML isolate (Table 3), the findings with the 22A isolate can be explained on the basis of gene dosage; however, the relative effects of the a and b alleles differ in two respects. First, the b allele,

Table 4. Influence of MoPrP-B transgene expression on incubation times in mice inoculated with 22A scrapie prions

Mice	Prnp genotype	Prnp transgenes (copies)	Alleles a (copies)	b	Incubation time (days ± S.E.M.)	n
B6.I-1	b/b		0	2	194 ± 10	7
(B6 × B6.I-1)F1	a/b		1	1	508 ± 14	7
C57BL/6J (B6)	a/a		2	0	405 ± 2	8
non-Tg(MoPrP-B$^{0/0}$)15	a/a		2	0	378 ± 8	3[a]
Tg(MoPrP-B$^{+/0}$)15	a/a	bbb/0	2	3	318 ± 14	15[a]
Tg(MoPrP-B$^{+/0}$)15	a/a	bbb/0	2	3	395 ± 12	6[b]
Tg(MoPrP-B$^{+/+}$)15	a/a	bbb/bbb	2	6	266 ± 1	6[a]
Tg(MoPrP-B$^{+/+}$)15	a/a	bbb/bbb	2	6	286 ± 15	5[b]

[a]The homozygous Tg(MoPrP-B$^{+/+}$)15 mice were maintained as a distinct subline selected for transgene homozygosity two generations removed from the (B6 × LT/Sv)F$_2$ founder. Hemizygous Tg(MoPrP-B$^{+/0}$)15 mice were produced by crossing the Tg(MoPrP-B$^{+/+}$)15 line with B6 mice.

[b]Tg(MoPrP-B$^{+/0}$)15 mice were maintained by repeated backcrossing to B6 mice.

but not the *a* allele, is a major determinant of the incubation time with the 22A isolate. Second, increasing the number of copies of the *a* allele does not diminish the incubation time but prolongs it: The *a* allele inhibits prion disease caused by the 22A isolate (Table 4). With the 87V prion isolate, the inhibitory effect of the *Prnp^a* allele is even more pronounced; only a few *Prnp^a* and (*Prnp^a* × *Prnp^b*)F$_1$ mice developed disease >600 days postinoculation (Carlson et al. 1994).

The most interesting feature of the incubation time profile for 22A prions is the overdominance of *Prnp^a* in prolonging the incubation period. On the basis of overdominance, the replication site hypothesis was posited, postulating that dimers of the *Sinc* gene product feature in prion replication (Dickinson and Outram 1979). The interpretation that the target for PrP^{Sc} may be a PrP^C dimer or multimer is compatible with the results described here (Table 4). The assumptions under this model are that PrP^C-B dimers are more readily converted to PrP^{Sc} than are PrP^C-A dimers and that PrP^C-A:PrP^C-B heterodimers are even more resistant to conversion to PrP^{Sc} than PrP^C-A dimers. Increasing the ratio of PrP-B to PrP-A would lead to shorter incubation times by favoring the formation of PrP^C-B homodimers (Table 4). A similar mechanism may account for the relative paucity of individuals heterozygous for the Met/Val polymorphism at codon 129 of PrP in spontaneous and iatrogenic CJD (Palmer et al. 1991). Alternatively, the PrP^C–PrP^{Sc} interaction can be broken down into two distinct aspects: binding affinity and efficacy of conversion to PrP^{Sc}. If PrP-A has a higher affinity than PrP^C-B for 22A prions but inefficiently converts to PrP^{Sc}, the exceptionally long incubation time of *Prnp$^{a/b}$* heterozygotes might reflect a reduction in the supply of 22A prions available for interaction with PrP^C-B encoded by the single *Prnp^b* allele. Additionally, PrP^C-A may inhibit the interaction of 22A PrP^{Sc} with PrP^C-B, leading to prolongation of the incubation time. This interpretation is supported by prolonged incubation times in Tg(SHaPrP) mice inoculated with mouse prions, in which SHaPrP^C is thought to inhibit the binding of MoPrP^{Sc} to substrate MoPrP^C (Prusiner et al. 1990).

STRAINS OF PRIONS

Experimental transmission of prion diseases to laboratory animals has been extensively studied over the past three decades (Ridley and Baker 1996). The diversity of scrapie prions was first appreciated in goats inoculated with "hyper" and "drowsy" isolates (Pattison and Millson 1961b). Prion isolates or "strains" from goats with a drowsy syndrome transmitted a similar syndrome to inoculated recipients, whereas prions from goats with a "hyper" or ataxic syndrome transmitted an ataxic form of scrapie to recipient goats. Studies in mice also demonstrated the existence of prion strains, in which brain extracts producing a particular pattern of disease could be repeatedly passaged (Dickinson and Fraser 1979; Bruce and Dickinson 1987; Kimberlin et al. 1987; Dickinson and Outram 1988).

The existence of prion strains raises the question of how heritable biological information can be enciphered in any molecule other than nucleic acid (Dickinson et al. 1968; Kimberlin 1982, 1990; Dickinson and Outram 1988; Bruce et al. 1991; Weissmann 1991; Ridley and Baker 1996). These strains bred true through multiple passages in mice and thus suggested that the prions have a nucleic acid genome that encodes prion progeny (Dickinson et al. 1968; Bruce and Dickinson 1987). However, no evidence for a scrapie-specific nucleic acid encoding information that specifies the incubation time and the distribution of neuropathological lesions has emerged from considerable efforts using a variety of experimental approaches.

Prion strains were initially defined by incubation times and the distribution of neuronal vacuolation (Dickinson et al. 1968; Fraser and Dickinson 1973). Subsequently, patterns of PrP^{Sc} deposition were found to correlate with vacuolation profiles and were used as an additional characteristic to distinguish strains of prions (Bruce et al. 1989; Hecker et al. 1992; DeArmond et al. 1993). Moreover, mice expressing PrP transgenes demonstrated that the level of PrP^C expression is inversely related to the incubation time (Prusiner et al. 1990). Furthermore, the distribution of CNS vacuolation and accompanying astrocytic gliosis are a consequence of the pattern of PrP^{Sc} deposition, which can be altered by both PrP genes and non-PrP genes (Prusiner et al. 1990). Taken together, these observations began to build an argument for PrP^{Sc} as the information molecule, in which prion "strain"-specific information had to somehow be enciphered (Prusiner 1991; Cohen et al. 1994).

In retrospect, another important clue to the mechanism of prion strains arose from studies on passaging prions to a host of a different species. New strains of prions were isolated when inocula from mice were passaged into hamsters (Kimberlin and Walker 1978). With the isolation of PrP 27-30, the amino-terminally truncated fragment of PrP^{Sc}, and subsequently PrP^{Sc}, it became clear that each species encodes a different PrP. Tg mouse studies showed that the "species barrier" for transmission between mice and hamsters could be abrogated by expression of the hamster PrP gene (Scott et al. 1989). Subsequently, the passage of mouse prions into mice expressing chimeric mouse–hamster PrP transgenes, the product of which is designated MH2M, resulted in the isolation of new strains with novel features that become evident when passaged in hamsters (Scott et al. 1993, 1997; Peretz et al. 2002). For example, Me7 prions that had been passaged and cloned by limiting dilution in C57BL mice resulted in a strain designated Me7H with an incubation time of ~260 days and then inoculated into Syrian hamsters (Kimberlin et al. 1987). When Me7 prions from C57BL mice were passaged into Tg(MH2M) mice and then into Syrian hamsters, a new strain with an incubation time of ~80 days was isolated. By changing the PrP gene from non-Tg mice to a new host, different strains of prions were isolated, which argues that prion diversity resides in PrP (Scott et al. 1997).

PrP^{Sc} Conformation Enciphers Diversity

The tertiary structure of PrP^{Sc} was hypothesized to encipher strain-specific information (Prusiner 1991), support for which began to emerge with the isolation of the DY strain from mink with transmissible encephalopathy (Marsh et al. 1991; Bessen and Marsh 1992, 1994). Upon treatment with proteinase K (PK), PrP^{Sc} in DY prions showed both diminished resistance to digestion and a peculiar cleavage site. The DY strain presented a puzzling anomaly because other prion strains exhibiting similar incubation times did not show this altered susceptibility to PK digestion (Scott et al. 1997). For example, the 139H and Me7H strains passaged in hamsters exhibited prolonged incubation times similar to DY prions, but the protease resistance of PrP^{Sc} was similar to that of the Sc237 strain (McKinley et al. 1983; Hecker et al. 1992; Scott et al. 1997). Indeed, such comparisons argued for the existence of an auxiliary molecule, such as a scrapie-specific nucleic acid (Dickinson and Outram 1979, 1988; Kimberlin 1990; Weissmann 1991). Although the binding of radiolabeled PrP^C to PrP^{Sc} isolated from DY-infected hamster brain provided evidence for prion strain-specified binding, conversion of PrP^C to PrP^{Sc} was not demonstrable (Bessen et al. 1995). The binding of PrP^C with PrP^{Sc} in DY prions is of interest, but the inability to isolate the bound PrP^C and to measure its physical and biological properties severely limits the conclusions that can be legitimately drawn. Studies comparing the PrP^{Sc} structures of one strain of hamster prions and two strains of mouse prions show notable variations (Kascsak et al. 1985). However, the results are confusing because SHaPrP and MoPrP have different sequences and, thus, the interpretation was inconclusive (Merz et al. 1984a,b; Carp et al. 1985, 1994, 1997; Özel and Diringer 1994).

Transmission of two different inherited human prion diseases to mice expressing a chimeric mouse–human (MHu2M) PrP transgene provided persuasive evidence that strain-specific information is enciphered in the tertiary structure of PrP^{Sc} (Telling et al. 1996). In fatal familial insomnia (FFI), the protease-resistant (r) fragment of PrP^{Sc} after deglycosylation has an apparent molecular weight (M_r) of 19 kD, whereas in fCJD(E200K) and most sporadic prion diseases, the $rPrP^{Sc}$ fragment is 21 kD (see Chapter 1, Table 4) (Monari et al. 1994; Parchi et al. 1996). This difference in molecular mass was shown to be due to different sites of proteolytic cleavage at the amino termini of the two human PrP^{Sc} molecules, which reflects different tertiary structures (Monari et al. 1994). These distinct conformations are understandable because the PrP^{Sc} sequences of the inocula differ.

In Tg(MHu2M) mice, extracts from the brains of FFI patients transmitted disease ~200 days after inoculation and induced formation of the 19-kD $rPrP^{Sc}$, whereas fCJD(E200K) and sCJD inocula produced an incubation period of 170–190 days and 21-kD $rPrP^{Sc}$ (Telling et al. 1996). On second passage, Tg(MHu2M) mice inoculated with FFI prions showed an incubation time of ~130 days and a 19-kD $rPrP^{Sc}$, whereas those inoculated with fCJD(E200K) prions exhibited an incubation time of ~170 days and a 21-kD $rPrP^{Sc}$ (Prusiner 1997). The experimental data demonstrate that $MHu2MPrP^{Sc}$ can exist in two different conformations based on the sizes of the protease-resistant fragments within an invariant amino acid sequence. The results of these studies argue that PrP^{Sc} acts as a template for the conversion of PrP^C into nascent PrP^{Sc}. Imparting the size of the protease-resistant fragment of PrP^{Sc} through conformational templating provides a mechanism for both the generation and propagation of prion strains.

Interestingly, a 19-kD rPrP^{Sc} fragment after deglycosylation was found in a patient who developed sporadic fatal insomnia (sFI) (Mastrianni et al. 1997, 1999; Parchi et al. 1999a). At autopsy, spongiform degeneration, reactive astrocytic gliosis, and PrP^{Sc} deposition were confined to the thalamus (Gambetti and Parchi 1999). These findings argue that the clinicopathologic phenotype is determined by the conformation of PrP^{Sc}, in accord with the results of the transmission of prions from FFI patients to Tg mice (Telling et al. 1996; Mastrianni et al. 1999).

Studies demonstrating a correlation between the conformation of PrP^{Sc} and the strain-specified disease phenotype were accomplished using the DY and Sc237 prion strains passaged in Syrian hamsters and subsequently into Tg mice expressing chimeric mouse–hamster PrP, designated MHM2 (Peretz et al. 2002). On first passage in Tg(MHM2)$Prnp^{0/0}$ mice, the SHa(Sc237) prions exhibited prolonged incubation times, typical of a species barrier (Chapter 2). On subsequent passage in Tg(MHM2)$Prnp^{0/0}$ mice, the MH2M(Sc237) strain showed a profound shortening of incubation time. Moreover, the PrP^{Sc} of the MH2M(Sc237) strain possessed structural properties that differ from those of the SHa(Sc237) strain, as demonstrated by relative conformational stability measurements (Peretz et al. 2001a). Conversely, transmission of SHa(DY) prions to Tg(MHM2)$Prnp^{0/0}$ mice showed no species barrier, and the MH2M(DY) prions retained the conformational and phenotypic properties of SHa(DY). These results extend the findings described above for FFI and fCJD(E200K) prions and contend that a change in PrP^{Sc} conformation is intimately associated with the emergence of a new prion strain.

Selective Neuronal Targeting

Profiles of spongiform change (Fig. 4) have been used to characterize prion strains (Fraser and Dickinson 1968), but several studies argue that such profiles are not an intrinsic feature of strains (Carp et al. 1997; DeArmond et al. 1997). The mechanism by which prion strains modify the pattern of spongiform degeneration is perplexing because earlier investigations showed that PrP^{Sc} deposition precedes neuronal vacuolation and reactive astrocytic gliosis (Jendroska et al. 1991; Hecker et al. 1992). When FFI prions were inoculated into Tg(MHu2M) mice, PrP^{Sc} was confined largely to the thalamus (Fig. 5A), as is the case for FFI in humans (Medori et al. 1992; Telling et al. 1996). In contrast, fCJD(E200K) prions inoculated into Tg(MHu2M) mice produced widespread deposition of PrP^{Sc} throughout the cortical mantle and many of the deep structures of the CNS (Fig. 5B), as seen with fCJD(E200K) in humans.

To examine whether the diverse patterns of PrP^{Sc} deposition are influenced by Asn-linked glycosylation of PrP^C, we constructed Tg mice expressing PrP mutated at one or both of the Asn-linked glycosylation consensus sites (DeArmond et al. 1997). These mutations resulted in aberrant neuroanatomic topologies of PrP^C within the CNS, whereas patho-

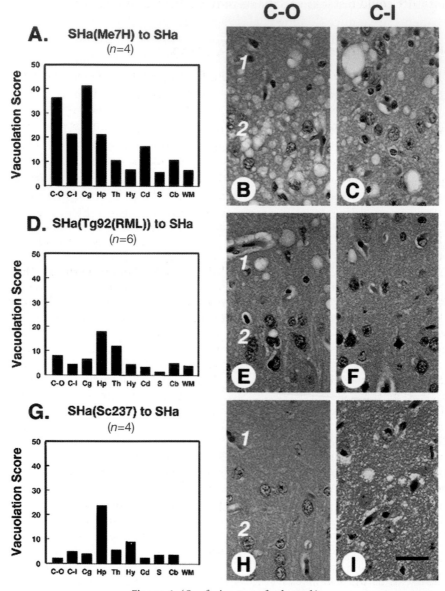

Figure 4. (*See facing page for legend.*)

logic point mutations adjacent to the consensus sites did not
tribution of PrP^C. Tg mice with mutation of the second PrP ₎
site exhibited prion incubation times of >500 days and unu₎
of PrP^{Sc} deposition. These findings raise the possibility that glycosylation
can modify the conformation of PrP and affect either the turnover of PrP^C
or the clearance of PrP^{Sc}. Regional differences in the rate of deposition or
clearance could result in specific patterns of PrP^{Sc} accumulation.

SPREAD OF PRIONS AMONG CELLS AND ORGANS

The mechanisms by which prions spread from one cell to another are not
known. It seems most likely that PrP^C on the surface of cells acts as a
receptor for PrP^{Sc}. Several studies suggest that PrP^C is converted into a
metastable state (PrP^*) that is capable of interacting with PrP^{Sc}. The
mechanism by which nascent PrP^{Sc} molecules are formed remains to be
established, as discussed in Chapters 1 and 2. Whether a PrP^*-like molecule has been generated in vitro by exposure of PrP^C to 3 M guanidine

Figure 4. Neuroanatomic distribution of spongiform degeneration in the gray and
white matter is specified by prion strains. Three prion strains, designated Me7H
(A–C), Tg92(RML) (D–F), and Sc237 (G–I), were passaged in Syrian hamsters (SHa)
and then repassaged by intrathalamic inoculation using 30 μl of brain extracts. The
Me7H strain was generated by passage of mouse Me7 prions into Syrian hamsters
(Kimberlin et al. 1989). The Tg92(RML) prion strain was generated by passage of
RML mouse prions through Tg(MH2M)92/$Prnp^{0/0}$ mice (Scott et al. 1997). (A, D, G)
Semiquantitative estimates of the intensity of vacuolation as a function of brain
region are displayed as histograms (Fraser and Dickinson 1968). The vacuolation
score is an estimate of the area of a hematoxylin and eosin (H&E)-stained brain section occupied by vacuoles (Carlson et al. 1994). (C-O) Outer half of the cerebral cortex; (C-I) inner half of the cerebral cortex; (Cg) cingulate gyrus; (Hp) hippocampus;
(Th) thalamus; (Hy) hypothalamus; (Cd) caudate nucleus; (S) septal nuclei; (Cb)
cerebellum; (WM) white matter; (n) number of animals examined. (B–C, E–F, H–I)
H&E-stained histological sections of layers 1 and 2 of the outer cerebral cortex (C-O)
and layer 5 of the inner cerebral cortex (C-I) are presented for each of the three
strains. With Me7H, intense vacuolation is present in all layers of the cerebral cortex,
including layers 1 and 2 of the outer cerebral cortex (panel B, labeled 1 and 2), and
layer 5 of the inner cortex (panel C). Although the vacuolation scores for Tg92(RML)
and Sc237 are similar (compare panels D and G), their distributions of vacuolation
differ. Layer 1 of C-O shows vacuolation from Tg92(RML) prions but not with Sc237
prions (compare panels E and H). Mild but definitive vacuolation occurred in layer 5
of the C-I with both Tg92(RML) and Sc237 strains (compare panels F and I). Bar in
I represents 25 μm and applies to all H&E-stained sections. Figure prepared by
Stephen J. DeArmond.

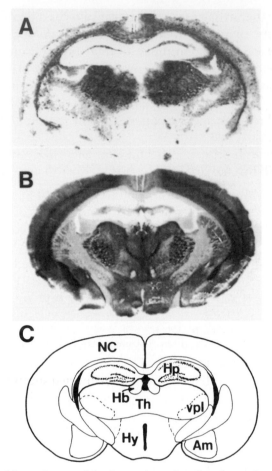

Figure 5. Histoblots of coronal brain sections showing the regional distribution of PrP^{Sc} deposition in Tg(MHu2M)*Prnp*$^{0/0}$ mice inoculated with prions from humans who died of (*A*) FFI and (*B*) fCJD(E200K). Cryostat sections were mounted on nitrocellulose and treated with proteinase K to eliminate PrP^{C} (Taraboulos et al. 1992). To enhance the antigenicity of PrP^{Sc}, the histoblots were exposed to 3 M guanidinium isothiocyanate before immunostaining using the α-PrP 3F4 monoclonal antibody (Kascsak et al. 1987). (*C*) Labeled diagram of a coronal section of the hippocampus/thalamus region. (Reprinted, with permission, from Prusiner et al. 1998 [copyright Cell Press].) (NC) Neocortex; (Hp) hippocampus; (vpl) ventral posterior lateral thalamic nucleic; (Am) amygdala; (Hy) hypothalamus; (Th) thalamus; (Hb) habernula.

hydrochloride (GdnHCl) and subsequent dilution to 0.3 M GdnHCl remains to be established (Kocisko et al. 1994, 1995; Kaneko et al. 1995, 1997a). It appears that PrP* formation can be prevented in cultured cells by compounds that stabilize protein conformation, such as glycerol and dimethylsulfoxide (DMSO) (Tatzelt et al. 1996a). The formation of PrP* in vivo probably occurs as a consequence of PrP^C binding to protein X; the PrP*–protein X complex is then capable of binding PrP^{Sc} (Telling et al. 1995; Kaneko et al. 1997b).

In animals with scrapie and humans with CJD, prions have been found in blood, but the levels are extremely low (Brown 1996, 2000; P. Brown et al. 1998, 1999). Studies with the buffy coat from rodents have shown the presence of CJD prions within white blood cells (Manuelidis et al. 1978; Tateishi 1985) (Chapter 13). Low levels of scrapie prions have been found in the blood of hamsters through the course of infection (Diringer 1984; Casaccia et al. 1989). Both scrapie and CJD prions have also been detected in cerebrospinal fluid, but, similar to blood, levels are extremely low (Pattison et al. 1959, 1964; Gajdusek et al. 1977).

In addition to experiments with mice, studies with sheep have produced important results on prion infectivity in blood. Transfusion of ~400 ml of whole blood from sheep inoculated with BSE or scrapie prions transmitted disease to ~20% of recipients (Hunter et al. 2002). Of four sheep transfused with blood from BSE-infected sheep, two developed disease at 538 and 610 days posttransfusion, and two other sheep began to show signs of neurologic dysfunction at the time the manuscript was prepared. Four of the sheep receiving blood from sheep with scrapie developed disease at 571 days, 614 days, 624 days, and 737 days posttransfusion.

Peripheral inoculation studies suggest that prions may also spread along neural tracts within both the central and peripheral nervous systems (Fraser 1982; Kimberlin et al. 1983; Fraser and Dickinson 1985; Beekes et al. 1996). Presumably, prions are spread by axonal retrograde transport systems (Borchelt et al. 1994). In cattle with BSE, histologic studies have argued that orally ingested prions spread from the mouth into the brain stem via retrograde transport up the trigeminal nerve (Wells et al. 1998). Presumably, abrasions in the mouths of cattle caused by sticks and stones that inadvertently enter the oral cavity with grass are the route of entry for prions from MBM. This hypothesis is supported by the constant presence of prions in the brain stem and wide variations in prion deposition in other parts of the CNS (Prusiner et al. 1993b).

Studies using mice with severe combined immunodeficiency (SCID) showed prolonged incubation times after intraperitoneal but not intracerebral inoculation with scrapie prions compared to control immuno-competent mice (O'Rourke et al. 1994; K.L. Brown et al. 1996; Fraser et al. 1996; Lasmézas et al. 1996). In studies of lymphoid cells, follicular den-dritic cells in the spleens of mice were found to accumulate PrP^{Sc} early after intraperitoneal inoculation with CJD prions (Kitamoto et al. 1991; McBride et al. 1992; Muramoto et al. 1993; K.L. Brown et al. 1999). In other investigations with immunocompromised mice, in which genes required for T- or B-cell function were ablated, B-lymphocytes were reported to play an important role in neuroinvasion (Klein et al. 1997), but this conclusion was later retracted by the authors (Klein et al. 1998; Frigg et al. 1999). Expression of anti-PrP antibodies in Tg mice prolonged the incubation times in mice after intraperitoneal but not intracerebral inoculation (Heppner et al. 2001).

In studies using neuroblastoma (N2a) and hippocampal (GT1) cells, prions have been found in media, which have been demonstrated to infect cultured cells that were previously uninfected (Schätzl et al. 1997). Studies with small pieces of steel wire either implanted in prion-infected brains or exposed to prion-infected brain homogenates have been used to transmit prion infectivity to N2a cells in culture (Zobeley et al. 1999; Weissmann et al. 2002).

ROLE OF PrP^{C} IN PRION INFECTION

The large body of evidence showing that PrP^{Sc} is a major component of the infectious prion particle led to the supposition that PrP^{C} expression is an absolute requirement for prion replication (Prusiner 1991, 1992), which was confirmed by experiments with $Prnp^{0/0}$ mice (Büeler et al. 1992).

PrP-deficient Mice

Two lines of $Prnp^{0/0}$ mice showed no untoward effects from ablation of the PrP gene (Büeler et al. 1992; Manson et al. 1994b). These apparently healthy mice were found to be resistant to prion disease and unable to replicate prions. In two initial studies, no evidence of prion disease could be found many months after inoculation of $Prnp^{0/0}$ mice with RML pri-ons (Büeler et al. 1993; Prusiner et al. 1993a) (Chapter 8). In one study, prion replication was reported at a low level in some $Prnp^{0/0}$ mice inocu-lated with RML prions (Büeler et al. 1993), but this was retracted later (Sailer et al. 1994). However, some investigators choose to argue that this artefact is an indication that PrP^{C} is not required for prion replication (Chesebro and Caughey 1993).

PrP^C Expression

In Syrian hamsters, the brain contains the highest levels of PrP poly(A)$^+$ RNA, whereas other organs express lower levels (Oesch et al. 1985). In the CNS, PrP mRNA is found primarily in neuronal cells (Kretzschmar et al. 1986). During prion infection, the highest prion titers in both hamsters and mice are found in the CNS prior to death (Table 1) (Eklund et al. 1967; Kimberlin and Walker 1977). These observations suggest that the accumulation of prions might be regulated by the level of PrP mRNA in cells; subsequently, studies of Tg(SHaPrP) mice showed that the length of the incubation time was inversely proportional to the level of PrP^C expression (Prusiner et al. 1990). Moreover, the level of PrP^{Sc} at the time of clinical illness was independent of the incubation time. These findings argue that the rate of PrP^{Sc} formation is proportional to the level of PrP^C expression and that signs of neurologic dysfunction appear when PrP^{Sc} levels reach a threshold (DeArmond et al. 1987).

Because the extracellular accumulation of PrP^{Sc} in the form of amyloid plaques was shown to be a nonobligatory feature of prion disease in humans and mice, it was postulated that intracellular PrP^{Sc} must be required for neuronal vacuolation to develop (Prusiner et al. 1990). This hypothesis is supported by the findings with prion-infected grafts of CNS tissue in $Prnp^{0/0}$ mice that develop vacuolar changes while the surrounding PrP-deficient neurons remain healthy (Brandner et al. 1996). Interestingly, primary cultures of neurons from $Prnp^{0/0}$ mice do not exhibit neurotoxicity when exposed to PrP peptides, whereas those from normal mice do, under these conditions (D.R. Brown et al. 1994, 1996). Whereas neurons undergo degeneration during prion infection, the surrounding astrocytes become hypertrophic and exhibit high levels of GFAP mRNA and protein (Mackenzie 1983). Genetic ablation of the GFAP gene in mice did not alter the incubation time, indicating that GFAP does not play a primary role in the neuronal degeneration found in prion disease (Gomi et al. 1995; Tatzelt et al. 1996b).

Consistent with the foregoing studies, results with newborn hamsters show no detectable brain PrP mRNA as measured by northern blot analysis and by immunoprecipitation of cell-free translation products. Detectable levels of PrP poly(A)$^+$ RNA appear ~1 day after birth and stay very low until 8 days after birth. PrP mRNA reaches a maximum level by 10 days of age (McKinley et al. 1987) and thereafter remains constant. These observations are of interest with respect to prion infection in newborn mice: Inoculation of newborn mice resulted in a delay of prion replication for almost 1 year (Hotchin and Buckley 1977).

Inducible Expression of PrP Transgenes

To control the expression of PrP^C in Tg mice, we used a tetracycline trans-activator (tTA) driven by the PrP gene control elements and a tTA-responsive operator linked to a PrP gene (Gossen and Bujard 1992). Adult Tg(tTA:PrP)3 mice showed no deleterious effects upon >90% repression of PrP^C expression by oral doxycycline (Tremblay et al. 1998). The same conclusion was reached using the Cre-loxP system (Sauer and Henderson 1988) in which bigenic $Prnp^{0/0}$ mice expressed $MoPrP^C$ from a transgene with a neurofilament heavy (NFH) subunit promoter until ~9 weeks of age, at which time the Cre recombinase excised the MoPrP gene expressed in neurons of the CNS (Mallucci et al. 2002). Like the Tg(tTA:PrP)3 mice on doxycycline with repressed PrP expression, the adult Tg(NFH-Cre-MloxP) mice suffered no deleterious effects from the acute absence of PrP^C expression in neurons during adulthood. These results argue that the lack of a disease phenotype in $Prnp^{0/0}$ mice is not due to adaptation to the absence of PrP^C during embryogenesis and development.

Tg(tTA:PrP)3 mice developed progressive ataxia at ~50 days after inoculation with prions unless maintained on doxycycline (Tremblay et al. 1998). Although Tg mice on doxycycline accumulated low levels of PrP^{Sc}, they showed no neurologic dysfunction, indicating that low levels of PrP^{Sc} can be tolerated. Use of the tTA system to control PrP expression allowed production of Tg mice with high levels of PrP^C that otherwise cause many embryonic and neonatal deaths.

When doxycycline was administered orally to adult Tg(tTA:PrP)3 mice, beginning 7 days prior to inoculation with RML prions, the mice remained free of any signs of CNS dysfunction for more than 380 days. When some of these mice were sacrificed at ~200 days after inoculation for neuropathologic examination, their brains showed no neurodegeneration. In contrast, mice from the same line not given doxycycline developed a progressive ataxia starting ~50 days after inoculation with prions (Table 5). Wild-type FVB mice developed signs of prion disease at ~120 days after inoculation, and $Prnp^{0/0}$ mice were not susceptible to prions, as reported previously (Prusiner et al. 1993a; Büeler et al. 1994; Manson et al. 1994a).

Brains of both untreated and doxycycline-treated Tg(tTA:PrP)3 mice were examined to assess the levels and distribution of PrP^C and PrP^{Sc}. In untreated mice, western blot and histoblot analysis showed high PrP^C and PrP^{Sc} levels in the brain (Tremblay et al. 1998). The brains from untreated Tg(tTA:PrP)3 mice exhibited extensive neuronal loss in the pyramidal cell layer and dentate gyrus of the hippocampus and focal loss of Purkinje cells and granule cells in the cerebellum. These changes were accompanied

Table 5. Inhibition of experimental prion disease in transgenic mice by doxycycline-mediated repression of *Prnp* gene expression

Recipient	Treatment	Inoculum	CNS dysfunction	Incubation time (mean days ± s.e.m)
FVB	—[a]	RML	11/11	122 ± 3
Tg(tTA:PrP)3	—[a]	RML	4/4[b]	51 ± 0
Tg(tTA:PrP)3	–Dox[c]	—	0/6	>100
Tg(tTA:PrP)3	+Dox[d]	RML	0/7	>350 (n = 5)[e]
Non-Tg(tTA:PrP)3[f]	—[a]	RML	0/10	>350 (n = 8)[e]
Tg(tetO-PrP/E6740)	—[a]	RML	0/8	>350 (n = 6)[e]
Prnp[0/0]	—[a]	—	0/10	>350

[a]Animals from these groups were kept at all times without doxycycline.

[b]Animals in this category first presented with ataxia at 51 days postinoculation and with additional signs of neurologic dysfunction at 69 days. Traditionally, the diagnosis of experimental scrapie in mice requires two signs of neurologic dysfunction as described previously (Carlson et al. 1988).

[c]Animals were born from parents maintained on doxycycline in the drinking water. Treatment was ceased at 3 weeks.

[d]Doxycycline was administered in the drinking water (2 mg/ml) 1 week prior to inoculation.

[e]Two animals from each of these groups were sacrificed for histopathology at 200 days postinoculation.

[f]Animals in this group did not harbor any transgene.

by moderate to severe astrocytic gliosis in all regions examined, including the neocortex, hippocampus, entorhinal cortex, thalamus, caudate nucleus, and substantia nigra, as well as in granule and molecular cell layers of the cerebellum. In treated Tg(tTA:PrP)3 mice, doxycycline repressed PrP^C to low levels similar to those observed in target Tg(tetO-PrP/E6740) mice. This residual PrP^C expression was nonetheless sufficient to permit conversion to low levels of PrP^{Sc}, which accumulated mostly within the forebrain and, in particular, in the corpus callosum and caudate nucleus. Histoblots showed PrP^{Sc} accumulation in the neocortex, hippocampus, corpus callosum, and white matter tracts of the cerebellum and pons.

Like the Tg(tTA:PrP)3 mice maintained on doxycycline, Tg(NFH-*Cre-MloxP*) mice, which did not express PrP after 9 weeks of age, were also resistant to prion disease when inoculated intracerebrally with RML prions. The Tg(NFH-*Cre-MloxP*) mice remained well for more than 400 days after inoculation, and no PrP^{Sc} accumulation was detected in the brain of the one mouse examined.

DOMINANT-NEGATIVE INHIBITION

Dominant-negative inhibition of PrP^{Sc} formation was demonstrated in Tg mice expressing either the analogous mutation that protects sheep

Table 6. Dominant-negative inhibition of prion formation in Tg(MoPrP,Q167R) mice

| Host | PrP expression level[a] | | | Incubation period | |
	mutant	wild type	Inoculum	(days ± S.E.M.)	n/n_0[b]
FVB	0	1x	RML	127 ± 1	50/50
Tg(MoPrP,Q167R)$Prnp^{0/0}$	1x	0	RML	>550	0/8
Tg(MoPrP,Q167R)$Prnp^{0/0}$	1x	0	RML	>557	0/4
Tg(MoPrP,Q167R)$Prnp^{0/0}$	1x	0	none	>557	0/10
FVB/$Prnp^{0/0}$	0	0	RML	>557	0/7
Tg(MoPrP,Q167R)$Prnp^{+/+}$	1x	1x	RML	447 ± 11	7/10[c]
Tg(MoPrP,Q167R)$Prnp^{+/+}$	1x	1x	none	>545	0/6

[a]Expression levels were determined by comparing serial dilutions of transgenic mouse brain homogenates to that of normal FVB mice (1x PrP level) by immunoblot.

[b](n) Number of sick mice; (n_0) number of inoculated mice.

[c]Seven animals died atypically of prolapsus at the mean incubation time indicated; three healthy animals were sacrificed at 300 days for western blot and histoblot analysis.

from scrapie or the mutation that protects humans from CJD. Tg mice that express MoPrP(Q167R) on the $Prnp^{0/0}$ background were constructed (Perrier et al. 2002); the Q167R mutation in MoPrP corresponds to Q171R in sheep PrP. In Tg(MoPrP,Q167R)$Prnp^{0/0}$ mice, mutant MoPrP(Q167R) is expressed at the same level as wt MoPrP in non-Tg FVB mice. None of the 12 Tg(MoPrP,Q167R)$Prnp^{0/0}$ mice inoculated with RML prions showed signs of disease after 550 days (Table 6); furthermore, no PrP^{Sc} was formed, showing that MoPrP^C(Q167R) is ineligible for conversion to PrP^{Sc} under these conditions. FVB control mice expressing wt PrP^C at levels similar to MoPrP^C(Q167R) in Tg mice developed signs of CNS dysfunction at 127 ± 1 days after inoculation. $Prnp^{0/0}$ mice inoculated with RML prions did not show signs of disease after >550 days (Table 6). Uninoculated Tg(MoPrP,Q167R)$Prnp^{0/0}$ mice showed no signs of spontaneous neurodegeneration.

Tg(MoPrP,Q167R)$Prnp^{+/+}$ mice were produced by crossing Tg(MoPrP,Q167R)$Prnp^{0/0}$ mice with wt FVB mice, resulting in similar expression levels of mutant MoPrP^C(Q167R) and wt MoPrP^C. None of six uninoculated Tg(MoPrP,Q167R)$Prnp^{+/+}$ mice showed signs of CNS dysfunction at more than 545 days (Table 6). Seven of 10 Tg(MoPrP,Q167R)$Prnp^{+/+}$ mice inoculated with RML prions died atypically with a mean incubation time of 447 ± 11 days (Table 6). We sacrificed the remaining three apparently healthy Tg(MoPrP,Q167R)$Prnp^{+/+}$ mice 300 days after RML inoculation to determine whether wt MoPrP^C supported prion replication. Brain homogenates subjected to limited PK digestion and immunoblot analysis revealed the presence of low amounts of protease-resistant PrP^{Sc}. The intensity of these bands corresponded to ~10% of the signal found in ill FVB mice inoculated with RML prions, arguing that the rate of PrP^{Sc} formation was greatly diminished in Tg(MoPrP,Q167R) *Prnp*$^{+/+}$ mice. Because we detected some wt PrP^{Sc} replication at 300 days, it seems likely that prion replication continued to occur over the subsequent 150 days in the Tg(MoPrP,Q167R)*Prnp*$^{+/+}$ mice that died atypically.

Tg mice expressing MoPrP with the Q218K mutation were also constructed; Q218K in MoPrP corresponds to E219K in human PrP. Like the Q167R mutation, the Q218K mutation exhibited dominant-negative inhibition of prion replication (Perrier et al. 2002).

It is noteworthy that dominant-negative inhibition of prion replication was demonstrated in cultured scrapie-infected neuroblastoma (ScN2a) cells when MoPrP(Δ114–121) carrying an eight-residue (AAAA-GAVV) deletion is expressed with wt MoPrP (Hölscher et al. 1998). MoPrP(Δ114–121) inhibited the conversion of wt MoPrP^C into PrP^{Sc}. Mutations in this region have been shown to produce spontaneous neurodegeneration through the accumulation of a carboxy-terminal transmembrane fragment of PrP, designated ^{Ctm}PrP (Hegde et al. 1998, 1999).

NATURAL AND EXPERIMENTAL SCRAPIE

Despite its recognition as a distinct disorder in 1738, scrapie in sheep remained enigmatic for more than two centuries (Parry 1983). Some veterinarians believed that scrapie was caused by parasites in muscle, whereas others believed that it was a dystrophic process (M'Gowan 1914).

Scrapie of sheep and goats was once thought to be unique among the prion diseases because it is communicable within flocks, but CWD has shown clear communicability within herds of deer and elk. Although the transmissibility of scrapie seems to be well established, the mechanism of its spread among sheep remains puzzling. The consumption of placenta has been implicated as one medium of transmission, which accounts for the horizontal spread of scrapie within flocks (Pattison and Millson 1961a; Pattison 1964; Pattison et al. 1972; Onodera et al. 1993). Whether or not this view is correct remains to be established. In Iceland, scrapied flocks of sheep were destroyed and the pastures were left vacant for several years. Reintroduction of sheep from flocks known to be free of scrapie to these pastures eventually led to a scrapie outbreak (Palsson 1979). The source of prions that infected the sheep is unknown.

Various hypotheses regarding the cause of natural scrapie in sheep have been posited over the last 40 years. Host genes were believed to be responsible for the development of scrapie in sheep, which led to the belief that it could be eradicated by proper breeding protocols (Parry 1962,

1983). Transmission by inoculation was therefore of importance primarily for laboratory studies, and communicable infection was of little consequence in nature. Other investigators viewed natural scrapie as an infectious disease and argued that host genetics only modulates susceptibility to an endemic infectious agent (Dickinson et al. 1965). The gene controlling incubation times for experimental scrapie in Cheviot sheep, denoted *Sip*, is believed to be linked to a PrP gene RFLP (Hunter et al. 1989); however, the null hypothesis of nonlinkage has yet to be tested, which is important in light of earlier studies which argue that susceptibility of sheep to scrapie is governed by a recessive gene (Parry 1962, 1983).

Polymorphisms of the PrP gene in sheep that produce amino acid substitutions at codons 136 and 171 have been studied with respect to the incidence of scrapie (Clousard et al. 1995). In Romanov and Ile-de-France breeds of sheep, an Ala/Val polymorphism in the PrP ORF was found at codon 136, which seems to correlate with scrapie (Laplanche et al. 1993b): Sheep expressing Val at codon 136 were susceptible to scrapie whereas those homozygous for Ala were resistant. Unexpectedly, only 1 of 74 scrapied autochthonous sheep among three breeds (Lacaune, Manech, and Presalpes) had a Val at codon 136 (Laplanche et al. 1993a).

In Suffolk sheep, a Gln/Arg polymorphism in the PrP ORF was found at codon 171 (Goldmann et al. 1990a,b). Studies of natural scrapie in the US have shown that ~85% of afflicted sheep are of the Suffolk breed. The incidence of scrapie was limited to Suffolk sheep homozygous for Gln at codon 171, although healthy controls of all genotypes have been found (Westaway et al. 1994). These results argue that susceptibility in Suffolk sheep is governed by the PrP codon-171 polymorphism, which has been supported by results from other studies of Suffolk sheep and other breeds (Hunter et al. 1997a,b, 1993; Goldmann et al. 1994; Belt et al. 1995; Clousard et al. 1995; Ikeda et al. 1995; Bossers et al. 1996; O'Rourke et al. 1997).

In sheep, Arg at PrP residue 171 has been suggested to act as a dominant negative by increasing the affinity of PrP^C for protein X, thus making protein X unavailable to assist in the PrP^C-to-PrP^{Sc} conversion (Kaneko et al. 1997b). The dominant-negative phenomenon is consistent with data showing that Glu/Arg heterozygous sheep rarely develop scrapie. The possibility of decreased binding of PrP^C(R171) to PrP^{Sc} in Glu/Arg heterozygotes seems unlikely (Bossers et al. 1997) because PrP^C(Q171) would be expected to be readily converted to PrP^{Sc} and cause disease.

In addition to polymorphisms at codons 136 and 171 in sheep PrP, additional polymorphisms have been found at positions 112, 127, 137, 138, 141, 143, 151, 154, 176, and 211 (Table 7) (Bossers et al. 2000). The reasons for and significance of these numerous polymorphisms in sheep are uncertain.

Table 7. PrP polymorphisms in sheep

Sheep PrP codon	Amino acid substitutions
112	Met, Thr
127	Gly, Ser
136	Ala, Val, Thr
137	Met, Thr
138	Ser, Asn
141	Leu, Phe
143	His, Arg
151	Arg, Cys
154	Arg, His
171	Gln, Arg, His, Lys
176	Arg, Lys
211	Arg, Gln

Data from Bossers et al. (2000).

One recent study reported that 3 of 19 sheep developed neurologic dysfunction at 1008, 1124, and 1127 days after intracerebral inoculation with BSE prions (Houston and Gravenor 2003). The sheep were homozygous for PrP genes encoding Ala (A) at positions 136 and 154 and Arg (R) at 171. Often referred to as AAR/AAR, such sheep are resistant to both natural and experimental scrapie, as described above. That these sheep are susceptible to BSE prions gives cause for European countries to reconsider breeding programs designed to create scrapie-resistant flocks populated with AAR/AAR sheep. A few AAR/AAR sheep from a very large population might produce the BSE strain of prions spontaneously that would then spread to other sheep in the flock. Whether the strain of prions found in such sporadic cases of sheep BSE is pathogenic to humans is unknown, but it seems likely, since sheep and bovine PrP differ at less than 3% of residues.

NATURAL AND EXPERIMENTAL BSE

BSE is thought to have been started by feeding cattle prion-infected MBM that was produced from the rendered offal of sheep, cattle, pigs, and chickens (Phillips et al. 2000). Whether the initial prions came from sheep with scrapie or a cow with a sporadic case of BSE is unknown. Epidemiological studies on an outbreak of transmissible mink encephalopathy (TME) near Stetsonville, Wisconsin, provide evidence for the sporadic case of BSE (Marsh et al. 1991). Evidence for the transmission of prions by ingestion of MBM is quite convincing because the number of cases of BSE began to decline steadily once a ban on MBM was imposed in 1988 after reaching a peak in 1992 (Phillips et al. 2000). This time lag can be explained by several experimental studies on cattle, which showed the development of clinical signs of BSE 3–4 years after oral feeding of BSE-infected brain tissue.

With the identification of asymptomatic cattle with prion disease using immunoassays to detect PrP^{Sc} in the brain stem, it may be possible to determine the incidence of sporadic BSE. Although all cases of BSE are currently attributed to the ingestion of prion-contaminated foodstuffs, it should be possible eventually to tease out those cases that are not due to exogenous prion infection. Whether or not some of the 17 cases of BSE in cattle born after July 1996 are cases of sporadic BSE is unknown (Wilesmith 2002). Eight of these BSE cases exhibited clinical signs with onset ranging between 46 months and 66 months of age, with a mean of 57 months. This is considerably older than the first 20 BSE cases in cattle born during the previous year, for which the age of onset ranged between 36 months and 49 months; the mean age was 43 months.

The pathogenesis of BSE in orally infected cattle is poorly understood. Some tissues have been studied using inbred, non-Tg RIII mice to assay the levels of prions in various tissues in orally infected cattle. In the first pathogenesis study performed in Britain, cattle were orally infected with BSE prions and sacrificed at 4- to 6-month intervals for bioassay. The low sensitivity of RIII mice to bovine prions (Chapter 3) makes the data difficult to interpret.

In addition to bioassays in RIII mice, bioassays of bovine prions were also performed in cattle (Wells 2002). In those studies, more than 60 samples harvested at various times after oral inoculation of cattle were bioassayed in groups of five cattle. Samples of the distal ileum taken at 6 months, 10 months, and 18 months after oral inoculation caused disease in all five cattle with incubation times of 27 months, 22 months, and 24 months, respectively. Samples of the distal ileum taken at 26 and 32 months after oral inoculation were negative for longer than 40 and 52 months, respectively (G. Wells, pers. comm.). Caudal medulla/spinal cord samples were harvested from cattle at 6 months, 10 months, 18 months, 22 months, 26 months, and 32 months after oral inoculation. Only the caudal medulla/spinal cord samples harvested at 32 months were found to be positive, to date. This sample caused disease in all five cattle with an incubation time of 23 months. The other caudal medulla/spinal cord samples have remained negative for longer than 40 months and some for longer than 52 months. These findings raise the possibility that BSE prion titers in the distal ileum during the period from 6 to 18 months after oral inoculation are as high as those found in the brain stem at 32 months. Endpoint titrations must be performed to define the relationship between

prion dose and incubation time for both distal ileum and br[ ]
ples and to interpret these findings. A single endpoint titr[ ]
infected brain stem from cattle has been performed in c[ ]
resulting titration indicates that studies in non-Tg mice underestimated
the titer of BSE prions in cattle brain stems by a factor of 500 to 1000
(Wells 2002). The nictitating membrane from one field case of BSE inoculated into five cattle has remained negative for more than 27 months
(G. Wells, pers. comm.). The nictitating membrane is of considerable
interest because it forms the basis of an antemortem test for prions in cattle as proposed for sheep prions (O'Rourke et al. 2000).

HUMAN PRION DISEASES

Less than 1% of human prion diseases seem to have an infectious etiology, whereas ~85% are sporadic cases and the remainder is dominantly
inherited through germ-line mutations of the PrP gene. Although transmission of prions to humans features in a small minority of cases, such
instances continue to receive considerable attention.

Kuru and Cannibalism

Kuru, a prion disease found among native Fore people in Papua New
Guinea, is believed to be transmitted by ritualistic cannibalism.
Cannibalism is one of the most fascinating and macabre of all human
activities. Cannibalism was a common but by no means universal practice
among groups of people inhabiting Papua New Guinea (Alpers 1979).
Among the Fore people and their neighbors in the Eastern Highlands,
endocannibalism, in which members of the group were eaten after death
by their relatives, was practiced.

The transmissibility of kuru and its disappearance in those born since
the cessation of cannibalism strongly implicate cannibalism in the spread
of this disorder (Alpers 1979). Cannibalism also explains the sex and age
distribution of kuru, because women and young children of both sexes,
who ate the internal organs of their relatives, were primarily afflicted
(Table 8). Only 2% of cases were found in adult males. Cannibalism also
explains the clustering of cases in space and time that has been noted in
multiple epidemiological studies of kuru (Alpers 1987).

Of several hundred orphans born since 1957 to mothers who later
died of kuru, none has yet developed the disease. Thus, the many children
with kuru seen in the 1950s were not infected prenatally, perinatally, or
neonatally by their mothers despite evidence for prions in the placenta

Table 8. Infectious prion diseases of humans

Diseases	No. Cases
Kuru (1957–1982)	
Adult females	1739
Adult males	248
Children and adolescents	597
Total	2584
Iatrogenic Creutzfeldt-Jakob disease	
Depth electrodes	2
Corneal transplants	2
Human pituitary growth hormone	>150
Human pituitary gonadotropin	5
Dura mater grafts	>120
Neurosurgical procedures	4
Total	>280

Data from Chapter 13 of this volume.

and colostrum of a pregnant woman who died of CJD (Tamai et al. 1992). Attempts to demonstrate consistent transmission of prion disease from mother to offspring in experimental animals have been unsuccessful (Morris et al. 1965; Pattison et al. 1972; Manuelidis and Manuelidis 1979b; Amyx et al. 1981; Taguchi et al. 1993).

Whereas patients currently afflicted with kuru exhibit greatly prolonged incubation periods, children with kuru, who were observed 30 years ago, provide some information on the minimum incubation period. The youngest patient with kuru was 4 years old at the onset of the disease and died at age 5, but it is not known at what age these young children were infected. Symptoms of CJD in humans developed 18 months after accidental intracerebral or intraoptic prion transmission (Duffy et al. 1974; Bernouilli et al. 1977). An incubation period of 18 months has been found in chimpanzees inoculated intracerebrally with kuru prions.

No individual born in the South Fore after 1959, when cannibalism ceased, has developed kuru (Alpers 1979, 1987; Klitzman et al. 1984). Kuru has progressively disappeared, first among children and thereafter among adolescents. The number of deaths in adult females has decreased steadily, and adult male deaths have remained almost invariant. Each year, the youngest new patients are older than those of the previous year.

The regular disappearance of kuru is inconsistent with the existence of a natural reservoir other than humans for the disease. Indeed, there is no evidence for animal or insect reservoirs. Thus, patients dying of kuru

Table 9. Oral transmission of prions

Peripheral route of inoculation	Inocula dose (log ID$_{50}$ units)	Incubation time intervals			
		onset of illness (mean days)	death (mean days)	equivalent intracerebral dose (log ID$_{50}$ units)	comparative efficiency of inoculation route
Cannibalism	9.5	120	147	0.4	$10^{-9.1}$
Intraperitoneal	8.3	92	112	2.9	$10^{-5.4}$

over the past decade seem to have incubation periods exceeding three decades (Prusiner et al. 1982a; Klitzman et al. 1984; Alpers 1987).

Numerous attempts to transmit kuru by feeding kuru-infected tissues to nonhuman primates have been unsuccessful, with a few exceptions (Gajdusek 1979; Gibbs et al. 1980). This contrasts with the rather uniform susceptibility of these animals to kuru after intracerebral or peripheral inoculation (Gajdusek 1977). In hamsters, intracerebral inoculation is 10^9 times more efficient than oral ingestion (Table 9), which may explain the results found in apes and monkeys (Prusiner et al. 1985).

Iatrogenic Creutzfeldt-Jakob Disease

Accidental transmission of CJD to humans appears to have occurred by corneal transplantation (Duffy et al. 1974), contaminated EEG electrode implantation (Bernouilli et al. 1977), and surgical procedures using contaminated instruments (Table 8) (Masters and Richardson 1978; Kondo and Kuroina 1982; Will and Matthews 1982; Davanipour et al. 1984). A cornea removed from a donor who unknowingly had CJD was transplanted to an apparently healthy recipient who developed CJD after a prolonged incubation period. Corneas of animals have significant levels of prions (Buyukmihci et al. 1980), making this scenario seem quite probable. Improperly decontaminated EEG electrodes, which caused CJD in two young patients with intractable epilepsy, were implanted in a chimpanzee and caused CJD 18 months later (Bernouilli et al. 1979; Gibbs et al. 1994).

Surgical procedures may have resulted in accidental transmission of prions to patients (Gajdusek 1977; Will and Matthews 1982; Brown et al. 1992), presumably because some instrument or apparatus in the operating theater became contaminated when a CJD patient underwent surgery. Although the epidemiology of these studies is highly suggestive, no proof of transmission exists.

Dura Mater Grafts

Since 1988, more than 120 cases of CJD occurring after implantation of dura mater grafts have been recorded (Otto 1987; Thadani et al. 1988; Masullo et al. 1989; Nisbet et al. 1989; Miyashita et al. 1991; Willison et al. 1991; Brown et al. 1992; Martínez-Lage et al. 1993; Clavel and Clavel 1996; Antoine et al. 1997; Defebvre et al. 1997; Shimizu et al. 1999). More than half of these cases have been reported from Japan (Centers for Disease Control 1997). All of the dura mater grafts are thought to have been acquired from a single manufacturer, whose preparative procedures seem to have been inadequate to inactivate human prions (Brown et al. 1992). One case of CJD occurred after repair of an eardrum perforation with a pericardium graft (Tange et al. 1989).

Thirty cases of CJD in physicians and health-care workers have been reported (Berger and David 1993); however, no occupational link has been established (Ridley and Baker 1993). Whether any of these cases represent infectious prion diseases contracted during care of patients with CJD or processing specimens from these patients remains uncertain.

Human Growth Hormone Therapy

The likelihood of transmission of CJD from contaminated human growth hormone (HGH) preparations derived from human pituitaries has been raised by the occurrence of fatal cerebellar disorders with dementia in more than 150 patients ranging in age from 10 to 41 years (Table 8) (Brown 1985; Buchanan et al. 1991; Fradkin et al. 1991; Brown et al. 1992; Billette de Villemeur et al. 1996; Public Health Service 1997). These patients received injections of HGH every 2–4 days for 4–12 years (Gibbs et al. 1985; Koch et al. 1985; Powell-Jackson et al. 1985; Titner et al. 1986; Croxson et al. 1988; Marzewski et al. 1988; New et al. 1988; Anderson et al. 1990; Billette de Villemeur et al. 1991; Macario et al. 1991; Ellis et al. 1992). Interestingly, most of the patients presented with cerebellar syndromes that progressed over periods varying from 24 to 72 weeks (Brown et al. 1992). Some patients became demented during the terminal phase of illness. In some respects, this clinical course resembles kuru more than ataxic CJD (Prusiner et al. 1982a). Assuming these patients developed CJD from injections of prion-contaminated HGH preparations, the possible incubation periods range from 4 to 30 years (Brown et al. 1992). The longest incubation periods are similar to those associated with recent cases of kuru (20–30 years) (Gajdusek et al. 1977; Prusiner et al. 1982a; Klitzman et al. 1984). Five cases of CJD have been reported in women receiving human pituitary gonadotropin (Cochius et al. 1990; Cochius et al. 1992; Healy and Evans 1993).

Because 10,000 human pituitaries were typically processed in a single HGH preparation, the possibility of hormone preparations contaminated with CJD prions is not remote (P. Brown et al. 1985, 1994; Brown 1988). Many patients received several common lots of HGH at various times during their prolonged therapies, but no single lot was administered to all patients. An aliquot of one lot of HGH has been reported to transmit CNS disease to a squirrel monkey after a prolonged incubation period (Gibbs et al. 1993). The number of lots of HGH contaminated with prions is unknown. Although CJD is a rare disease with an annual incidence of approximately one per million population (Masters and Richardson 1978), it is reasonable to assume that it is present with a proportional frequency among dead people. Because ~1% of the world population dies each year and most CJD patients die within 1 year of developing symptoms, we estimate that 1 per 10^4 dead people had CJD, which makes the processing of diseased pituitaries for HGH possible.

The concentration of CJD prions within infected human pituitaries is unknown; it is interesting that widespread degenerative changes have been observed in both the hypothalamus and pituitary of sheep with scrapie (Beck et al. 1964). Forebrains from scrapie-infected mice have been added to human pituitary suspensions to determine whether prions and HGH copurify (Jones et al. 1979). Bioassays in mice suggest that prions and HGH do not copurify with currently used protocols (Taylor et al. 1985). Although these results seem reassuring, especially for patients treated with HGH over much of the last decade, the relatively low titers of the murine scrapie prions used in these studies may not have provided an adequate test (Brown 1985). The extremely small size and charge heterogeneity exhibited by scrapie (Alper et al. 1966; Prusiner et al. 1978, 1980, 1983; Bolton et al. 1985) and presumably CJD prions (Bendheim et al. 1985; Bockman et al. 1985) may complicate procedures designed to separate pituitary hormones from these slow infectious pathogens. Even though additional investigations argue for the efficacy of inactivating prions in HGH fractions prepared from human pituitaries using 6 M urea (Pocchiari et al. 1991), it seems doubtful that such protocols will be used for purifying HGH because recombinant HGH is now available.

Molecular genetic studies have shown that most patients developing iatrogenic CJD after receiving pituitary-derived HGH are homozygous for either Met or Val at PrP codon 129 (Collinge et al. 1991; P. Brown et al. 1994; Deslys et al. 1994). Homozygosity at the codon 129 polymorphism has also been shown to predispose individuals to sporadic CJD (Palmer et al. 1991). Interestingly, Val homozygosity seems to be overrepresented in these HGH cases compared to the general population.

Replication of Human Prions in Transgenic Mice

Tg(MHu2M) mice are highly susceptible to human prions, in contrast to Tg(HuPrP) mice, which are resistant to human prions (Telling et al. 1994). Tg(HuPrP) mice were rendered susceptible to human prions when they were crossed onto a $Prnp^{0/0}$ background (Telling et al. 1995). Incubation times for Tg(MHu2M) mice were similar whether or not MoPrP was coexpressed. Typically, Tg(MHu2M) mice and Tg(HuPrP)$Prnp^{0/0}$ mice exhibit signs of neurologic dysfunction ~200 days after inoculation with sCJD prions (Table 10).

Mutants of the MHu2M Transgene Enable Abbreviated Incubation Times

Mutating some residues of the MHu2M transgene led to a shortening of the incubation time for human prions. Changing the two human residues to mouse at positions 165 and 167 at the carboxyl terminus of the "Hu2" insert yielded a transgene product with only seven Hu residues. The Tg(MHu2M,M165V,E167Q) mice were responsive to sCJD prions characterized by Met/Met at codon 129 and a 21-kD rPrP^{Sc} fragment (MM1), with incubation times between 106 and 114 days (Table 10). sCJD prions with Val/Val at codon 129 and a 19-kD rPrP^{Sc} fragment (VV2) did not transmit as rapidly as sCJD(MM1) prions, presumably reflecting the fact that the MHu2M,M165V,E167Q transgene encodes methionine at residue 129 (Table 10). The importance of homology between the host PrP^{C} and the prions in the inoculum at residue 129 also seems to be reflected by the differences in incubation times between Tg(HuPrP,M129)440/$Prnp^{0/0}$ and Tg(HuPrP,V129)152/$Prnp^{0/0}$ mice that received the same inocula (Table 10).

CONCLUDING REMARKS

Some investigators have difficulty embracing the many unprecedented principles of prion replication. Although several aspects of prion replication resemble viral replication superficially, the underlying principles are quite different. For example, in prion replication, the substrate is a host-encoded protein, PrP^{C}, which undergoes modification to form PrP^{Sc}, the only known component of the infectious prion particle. In contrast,

Table 10. Transmission of human prions to transgenic mice expressing MHu2M(M165V,E167Q), MHu2M, HuPrP(M129), or HuPrP(V129)

Transgene Tg Line	Incubation period (days) ± S.E.M. (n/n_0)[a]			
	MHu2M,M165V,E167Q Tg22372	MHu2M Tg5378	HuPrP,M129 Tg440	HuPrP,V129 Tg152
Inoculum				
MM1[b]				
(RG)	106 ± 2 (13/13)[c]	191 ± 3 (10/10)	165 ± 4 (7/7)	263 ± 2 (6/6)
(EC)	114 ± 2 (7/7)[d]	157 ± 3 (7/7)	254 ± 6 (9/9)	
(HS)	111 ± 2 (7/7)[d]	196 ± 4 (8/8)	163 + 2 (9/9)	
(Ho)		205 ± 7 (6/6)	155 ± 3 (8/8)	
(DG)	106 ± 2 (7/7)[d]			
(AM)	112 ± 2 (8/8)[e]			
MM2				
(A88-418)		>680 (0/10)	232 ± 5 (3/3)	368 ± 19 (9/9)
(094-3)		>650 (0/7)	>580 (0/8)	556 ± 63 (5/5)
(sFI-St)[f]	303 ± 20 (4/6)	221 ± 6 (4/4)	699 ± 30 (2/5)	
(vCJD-RU96/45)[g]	335 ± 23(7/7)	647 ± 35(2/7)		
(vCJD-RU96/02)[g]	380 ± 10(6/6)	563 ± 201(4/7)		
MV1				
(WP)	124 ± 3 (7/7)[d]	214 ± 3 (8/8)		
(Ghi)		215 ± 4 (5/5)		
(Ro)			176 ± 2 (9/9)	
MV2				
(093-25)		>640 (0/10)	350 ± 38 (3/6)	209 ± 3 (7/7)
(A94-311)		>640 (0/10)	419 ± 13 (9/9)	206 + 3 (6/6)
(096-48)		>640 (0/10)	307 ± 27 (7/9)	231 ± 4 (5/5)
(AMB)	>450 (0/10)			
VV2				
(RP)	>450 (1/10)	531 ± 46 (3/14)	248 ± 12 (3/7)	223 ± 7 (7/7)
(A90-332)		>500 (0/8)	448 ± 34 (3/7)	195 ± 3 (8/8)
(094-87)		433 (1/10)	378 ± 7 (3/7)	198 ± 5 (9/9)
(GF)	>450 (2/10)			

[a](n) Number of diseased animals; (n_0) number of inoculated animals.

[b]Strain typing as described in Parchi et al. (1999b). MM1 refers to the codon 129 polymorphic residues and the size of the deglycosylated, rPrP^{Sc} fragment. (M) methionine; (V) valine; (1) 21 kD; (2) 19 kD.

[c,d,e]Number of animals that died by a cause other than prion disease: [c]5, [d]3, [e]2.

[f]Inoculum from patient with sporadic FI.

[g]Inocula from patients with variant CJD.

viruses carry a DNA or RNA genome that is copied and directs the synthesis of most, if not all, of the viral proteins. The mature virus consists of a nucleic acid genome surrounded by a protein coat, whereas a prion is composed only of PrP^{Sc}. It seems likely that the smallest infectious prion particle is a dimer, based on ionizing radiation target-size studies (Bellinger-Kawahara et al. 1988).

When viruses pass from one species to another, they often replicate without any structural modification, whereas prions undergo a profound change. The prion adopts the PrP sequence encoded by the PrP gene of its current host. Differences in amino acid sequences can result in a restriction of transmission to some species while making the new prion permissive to others. Distinct strain properties for viruses are encoded in a genome, whereas strain-specific properties for prions are enciphered in the conformation of PrP^{Sc}.

As the body of data on prions continues to grow, changes in our understanding of how prions replicate and cause disease will undoubtedly emerge. But we hasten to add that many of the basic principles of prion biology are becoming well understood.

REFERENCES

Alper T., Haig D.A., and Clarke M.C. 1966. The exceptionally small size of the scrapie agent. *Biochem. Biophys. Res. Commun.* **22:** 278–284.

Alpers M.P. 1979. Epidemiology and ecology of kuru. In *Slow transmissible diseases of the nervous system* (ed. S.B. Prusiner and W.J. Hadlow), vol. 1, pp. 67–90. Academic Press, New York.

———. 1987. Epidemiology and clinical aspects of kuru. In *Prions: Novel infectious pathogens causing scrapie and Creutzfeldt-Jakob disease* (ed. S.B. Prusiner and M.P. McKinley), pp. 451–465. Academic Press, Orlando, Florida.

Amyx H.L., Gibbs C.J., Jr., Gajdusek D.C., and Greer W.E. 1981. Absence of vertical transmission of subacute spongiform viral encephalopathies in experimental primates. *Proc. Soc. Exp. Biol. Med.* **166:** 469–471.

Anderson J.R., Allen C.M.C., and Weller R.O. 1990. Creutzfeldt-Jakob disease following human pituitary-derived growth hormone administration. *Neuropathol. Appl. Neurobiol.* **16:** 543. (Abstr.)

Anderson R.M., Donnelly C.A., Ferguson N.M., Woolhouse M.E.J., Watt C.J., Udy H.J., MaWhinney S., Dunstan S.P., Southwood T.R.E., Wilesmith J.W., Ryan J.B.M., Hoinville L.J., Hillerton J.E., Austin A.R., and Wells G.A.H. 1996. Transmission dynamics and epidemiology of BSE in British cattle. *Nature* **382:** 779–788.

Antoine J.C., Michel D., Bertholon P., Mosnier J.F., Laplanche J.-L., Beaudry P., Hauw J.J., and Veyret C. 1997. Creutzfeldt-Jakob disease after extracranial dura mater embolization for a nasopharyngeal angiofibroma. *Neurology* **48:** 1451–1453.

Baringer J.R., Bowman K.A., and Prusiner S.B. 1981. Regional neuropathology and titers in hamster scrapie. *J. Neuropathol. Exp. Neurol.* **40:** 329.

Beck E., Daniel P.M., and Parry H.B. 1964. Degeneration of the cerebellar and hypothalamo-neurohypophysial systems in sheep with scrapie; and its relationship to human system degenerations. *Brain* **87:** 153–176.

Beekes M., Baldauf E., and Diringer H. 1996. Pathogenesis of scrapie in hamsters after oral and intraperitoneal infection. In *Transmissible subacute spongiform encephalopathies: Prion diseases* (ed. L. Court and B. Dodet), pp. 143–149. Elsevier, Paris.

Bellinger-Kawahara C.G., Kempner E., Groth D.F., Gabizon R., and Prusiner S.B. 1988. Scrapie prion liposomes and rods exhibit target sizes of 55,000 Da. *Virology* **164:** 537–541.

Belt P.B.G.M., Muileman I.H., Schreuder B.E.C., Ruijter J.B., Gielkens A.L.J., and Smits M.A. 1995. Identification of five allelic variants of the sheep PrP gene and their association with natural scrapie. *J. Gen. Virol.* **76:** 509–517.

Bendheim P.E., Bockman J.M., McKinley M.P., Kingsbury D.T., and Prusiner S.B. 1985. Scrapie and Creutzfeldt-Jakob disease prion proteins share physical properties and antigenic determinants. *Proc. Natl. Acad. Sci.* **82:** 997–1001.

Berger J.R. and David N.J. 1993. Creutzfeldt-Jakob disease in a physician: A review of the disorder in health care workers. *Neurology* **43:** 205–206.

Bernouilli C.C., Masters C.L., Gajdusek D.C., Gibbs C.J., Jr., and Harris J.O. 1979. Early clinical features of Creutzfeldt-Jakob disease (subacute spongiform encephalopathy). In *Slow transmissible diseases of the nervous system* (ed. S.B. Prusiner and W.J. Hadlow), vol. 1, pp. 229–251. Academic Press, New York.

Bernouilli C., Siegfried J., Baumgartner G., Regli F., Rabinowicz T., Gajdusek D.C., and Gibbs C.J., Jr. 1977. Danger of accidental person to person transmission of Creutzfeldt-Jakob disease by surgery. *Lancet* **1:** 478–479.

Bessen R.A. and Marsh R.F. 1992. Identification of two biologically distinct strains of transmissible mink encephalopathy in hamsters. *J. Gen. Virol.* **73:** 329–334.

———. 1994. Distinct PrP properties suggest the molecular basis of strain variation in transmissible mink encephalopathy. *J. Virol.* **68:** 7859–7868.

Bessen R.A., Kocisko D.A., Raymond G.J., Nandan S., Lansbury P.T., and Caughey B. 1995. Non-genetic propagation of strain-specific properties of scrapie prion protein. *Nature* **375:** 698–700.

Billette de Villemeur T., Beauvais P., Gourmelon M., and Richardet J.M. 1991. Creutzfeldt-Jakob disease in children treated with growth hormone. *Lancet* **337:** 864–865.

Billette de Villemeur T., Deslys J.-P., Pradel A., Soubrié C., Alpérovitch A., Tardieu M., Chaussain J.-L., Hauw J.-J., Dormont D., Ruberg M., and Agid Y. 1996. Creutzfeldt-Jakob disease from contaminated growth hormone extracts in France. *Neurology* **47:** 690–695.

Bockman J.M., Prusiner S.B., Tateishi J., and Kingsbury D.T. 1987. Immunoblotting of Creutzfeldt-Jakob disease prion proteins: Host species-specific epitopes. *Ann. Neurol.* **21:** 589–595.

Bockman J.M., Kingsbury D.T., McKinley M.P., Bendheim P.E., and Prusiner S.B. 1985. Creutzfeldt-Jakob disease prion proteins in human brains. *N. Engl. J. Med.* **312:** 73–78.

Bolton D.C., Meyer R.K., and Prusiner S.B. 1985. Scrapie PrP 27-30 is a sialoglycoprotein. *J. Virol.* **53:** 596–606.

Borchelt D.R., Koliatsis V.E., Guarnieri M., Pardo C.A., Sisodia S.S., and Price D.L. 1994. Rapid anterograde axonal transport of the cellular prion glycoprotein in the peripheral and central nervous systems. *J. Biol. Chem.* **269:** 14711–14714.

Bossers A., de Vries R., and Smits M.A. 2000. Susceptibility of sheep for scrapie as assessed by in vitro conversion of nine naturally occurring variants of PrP. *J. Virol.* **74:** 1407–1414.

Bossers A., Schreuder B.E., Muileman I.H., Belt P.B., and Smits M.A. 1996. PrP genotype contributes to determining survival times of sheep with natural scrapie. *J. Gen. Virol.* **77:** 2669–2673.

Bossers A., Belt P.B.G.M., Raymond G.J., Caughey B., de Vries R., and Smits M.A. 1997. Scrapie susceptibility-linked polymorphisms modulate the *in vitro* conversion of sheep prion protein to protease-resistant forms. *Proc. Natl. Acad. Sci.* **94:** 4931–4936.

Brandner S., Isenmann S., Raeber A., Fischer M., Sailer A., Kobayashi Y., Marino S., Weissmann C., and Aguzzi A. 1996. Normal host prion protein necessary for scrapie-induced neurotoxicity. *Nature* **379:** 339–343.

Brown D.R., Herms J., and Kretzschmar H.A. 1994. Mouse cortical cells lacking cellular PrP survive in culture with a neurotoxic PrP fragment. *Neuroreport* **5:** 2057–2060.

Brown D.R., Schmidt B., and Kretzschmar H.A. 1996. Role of microglia and host prion protein in neurotoxicity of a prion protein fragment. *Nature* **380:** 345–347.

Brown K.L., Stewart K., Bruce M.E., and Fraser H. 1996. Scrapie in immunodeficient mice. In *Transmissible subacute spongiform encephalopathies: Prion diseases* (ed. L. Court and B. Dodet), pp. 159–166. Elsevier, Paris.

Brown K.L., Stewart K., Ritchie D.L., Mabbott N.A., Williams A., Fraser H., Morrison W.I., and Bruce M.E. 1999. Scrapie replication in lymphoid tissues depends on prion protein-expressing follicular dendritic cells. *Nat. Med.* **5:** 1308–1312.

Brown P. 1985. Virus sterility for human growth hormone. *Lancet* **2:** 729–730.

———. 1988. The decline and fall of Creutzfeldt-Jakob disease associated with human growth hormone therapy. *Neurology* **38:** 1135–1137.

———. 1996. The risk of blood-borne Creutzfeldt-Jakob disease. In *Transmissible subacute spongiform encephalopathies: Prion diseases* (ed. L. Court and B. Dodet), pp. 447–450. Elsevier, Paris.

———. 2000. BSE and transmission through blood. *Lancet* **356:** 955–956.

Brown P., Preece M.A., and Will R.G. 1992. "Friendly fire" in medicine: Hormones, homografts, and Creutzfeldt-Jakob disease. *Lancet* **340:** 24–27.

Brown P., Gajdusek D.C., Gibbs C.J., Jr., and Asher D.M. 1985. Potential epidemic of Creutzfeldt-Jakob disease from human growth hormone therapy. *N. Engl. J. Med.* **313:** 728–731.

Brown P., Cathala F., Raubertas R.F., Gajdusek D.C., and Castaigne P. 1987. The epidemiology of Creutzfeldt-Jakob disease: Conclusion of a 15-year investigation in France and review of the world literature. *Neurology* **37:** 895–904.

Brown P., Cervenáková L., McShane L.M., Barber P., Rubenstein R., and Drohan W.N. 1999. Further studies of blood infectivity in an experimental model of transmissible spongiform encephalopathy, with an explanation of why blood components do not transmit Creutzfeldt-Jakob disease in humans. *Transfusion* **39:** 1169–1178.

Brown P., Rohwer R.G., Dunstan B.C., MacAuley C., Gajdusek D.C., and Drohan W.N. 1998. The distribution of infectivity in blood components and plasma derivatives in experimental models of transmissible spongiform encephalopathy. *Transfusion* **38:** 810–816.

Brown P., Cervenáková L., Goldfarb L.G., McCombie W.R., Rubenstein R., Will R.G., Pocchiari M., Martinez-Lage J.F., Scalici C., Masullo C., Graupera G., Ligan J., and Gajdusek D.C. 1994. Iatrogenic Creutzfeldt-Jakob disease: An example of the interplay between ancient genes and modern medicine. *Neurology* **44:** 291–293.

Bruce M.E. and Dickinson A.G. 1987. Biological evidence that the scrapie agent has an independent genome. *J. Gen. Virol.* **68:** 79–89.

Bruce M.E., McBride P.A., and Farquhar C.F. 1989. Precise targeting of the pathology of the sialoglycoprotein, PrP, and vacuolar degeneration in mouse scrapie. *Neurosci. Lett.* **102:** 1–6.

Bruce M.E., McConnell I., Fraser H., and Dickinson A.G. 1991. The disease characteristics of different strains of scrapie in *Sinc* congenic mouse lines: Implications for the nature of the agent and host control of pathogenesis. *J. Gen. Virol.* **72:** 595–603.

Buchanan C.R., Preece M.A., and Milner R.D.G. 1991. Mortality, neoplasia and Creutzfeldt-Jakob disease in patients treated with pituitary growth hormone in the United Kingdom. *Br. Med. J.* **302:** 824–828.

Büeler H., Raeber A., Sailer A., Fischer M., Aguzzi A., and Weissmann C. 1994. High prion and PrP^{Sc} levels but delayed onset of disease in scrapie-inoculated mice heterozygous for a disrupted PrP gene. *Mol. Med.* **1:** 19–30.

Büeler H., Aguzzi A., Sailer A., Greiner R.-A., Autenried P., Aguet M., and Weissmann C. 1993. Mice devoid of PrP are resistant to scrapie. *Cell* **73:** 1339–1347.

Büeler H., Fisher M., Lang Y., Bluethmann H., Lipp H.-P., DeArmond S.J., Prusiner S.B., Aguet M., and Weissmann C. 1992. Normal development and behaviour of mice lacking the neuronal cell-surface PrP protein. *Nature* **356:** 577–582.

Buyukmihci N., Rorvik M., and Marsh R.F. 1980. Replication of the scrapie agent in ocular neural tissues. *Proc. Natl. Acad. Sci.* **77:** 1169–1171.

Carlson G.A., Ebeling C., Torchia M., Westaway D., and Prusiner S.B. 1993. Delimiting the location of the scrapie prion incubation time gene on chromosome 2 of the mouse. *Genetics* **133:** 979–988.

Carlson G.A., Westaway D., DeArmond S.J., Peterson-Torchia M., and Prusiner S.B. 1989. Primary structure of prion protein may modify scrapie isolate properties. *Proc. Natl. Acad. Sci.* **86:** 7475–7479.

Carlson G.A., Goodman P.A., Lovett M., Taylor B.A., Marshall S.T., Peterson-Torchia M., Westaway D., and Prusiner S.B. 1988. Genetics and polymorphism of the mouse prion gene complex: Control of scrapie incubation time. *Mol. Cell. Biol.* **8:** 5528–5540.

Carlson G.A., Kingsbury D.T., Goodman P.A., Coleman S., Marshall S.T., DeArmond S., Westaway D., and Prusiner S.B. 1986. Linkage of prion protein and scrapie incubation time genes. *Cell* **46:** 503–511.

Carlson G.A., Ebeling C., Yang S.-L., Telling G., Torchia M., Groth D., Westaway D., DeArmond S.J., and Prusiner S.B. 1994. Prion isolate specified allotypic interactions between the cellular and scrapie prion proteins in congenic and transgenic mice. *Proc. Natl. Acad. Sci.* **91:** 5690–5694.

Carp R.I., Meeker H., and Sersen E. 1997. Scrapie strains retain their distinctive characteristics following passages of homogenates from different brain regions and spleen. *J. Gen. Virol.* **78:** 283–290.

Carp R.I., Kascsak R.J., Rubenstein R., and Merz P.A. 1994. The puzzle of PrP^{Sc} and infectivity—Do the pieces fit? *Trends Neurosci.* **17:** 148–149.

Carp R.I., Moretz R.C., Natelli M., and Dickinson A.G. 1987. Genetic control of scrapie: Incubation period and plaque formation in I mice. *J. Gen. Virol.* **68:** 401–407.

Carp R.I., Merz P.A., Moretz R.C., Somerville R.A., Callahan S.M., and Wisniewski H.M. 1985. Biological properties of scrapie: An unconventional slow virus. In *Subviral pathogens of plants and animals: Viroids and prions* (ed. K. Maramorosch and J.J. McKelvey, Jr.), pp. 425–463. Academic Press, Orlando, Florida.

Casaccia P., Ladogana A., Xi Y.G., and Pocchiari M. 1989. Levels of infectivity in the blood throughout the incubation period of hamsters peripherally injected with scrapie. *Arch. Virol.* **108:** 145–149.

Centers for Disease Control. 1997. Creutzfeldt-Jakob disease associated with cadaveric dura mater grafts, Japan, January 1979–May 1996. *Morb. Mortal. Wkly. Rep.* **46:** 1066–1069.

Chandler R.L. and Fisher J. 1963. Experimental transmission of scrapie to rats. *Lancet* **2:** 1165.

Chernoff Y.O., Lindquist S.L., Ono B., Inge-Vechtomov S.G., and Liebman S.W. 1995. Role of the chaperone protein Hsp104 in propagation of the yeast prion-like factor [*psi*⁺]. *Science* **268:** 880–884.

Chesebro B. and Caughey B. 1993. Scrapie agent replication without the prion protein? *Curr. Biol.* **3:** 696–698.

Clavel M. and Clavel P. 1996. Creutzfeldt-Jakob disease transmitted by dura mater graft. *Eur. Neurol.* **36:** 239–240.

Clousard C., Beaudry P., Elsen J.M., Milan D., Dussaucy M., Bounneau C., Schelcher F., Chatelain J., Launay J.-M., and Laplanche J.-L. 1995. Different allelic effects of the codons 136 and 171 of the prion protein gene in sheep with natural scrapie. *J. Gen. Virol.* **76:** 2097–2101.

Cochius J.I., Hyman N., and Esiri M.M. 1992. Creutzfeldt-Jakob disease in a recipient of human pituitary-derived gonadotrophin: A second case. *J. Neurol. Neurosurg. Psychiatry* **55:** 1094–1095.

Cochius J.I., Mack K., Burns R.J., Alderman C.P., and Blumbergs P.C. 1990. Creutzfeldt-Jakob disease in a recipient of human pituitary-derived gonadotrophin. *Aust. N. Z. J. Med.* **20:** 592–593.

Cohen F.E., Pan K.-M., Huang Z., Baldwin M., Fletterick R.J., and Prusiner S.B. 1994. Structural clues to prion replication. *Science* **264:** 530–531.

Collinge J., Palmer M.S., and Dryden A.J. 1991. Genetic predisposition to iatrogenic Creutzfeldt-Jakob disease. *Lancet* **337:** 1441–1442.

Cousens S.N., Harries-Jones R., Knight R., Will R.G., Smith P.G., and Matthews W.B. 1990. Geographical distribution of cases of Creutzfeldt-Jakob disease in England and Wales 1970–1984. *J. Neurol. Neurosurg. Psychiatry* **53:** 459–465.

Croxson M., Brown P., Synek B., Harrington M.G., Frith R., Clover G., Wilson J., and Gajdusek D.C. 1988. A new case of Creutzfeldt-Jakob disease associated with human growth hormone therapy in New Zealand. *Neurology* **38:** 1128–1130.

Cuillé J. and Chelle P.L. 1939. Experimental transmission of trembling to the goat. *C.R. Seances Acad. Sci.* **208:** 1058–1060.

Davanipour Z., Goodman L., Alter M., Sobel E., Asher D., and Gajdusek D.C. 1984. Possible modes of transmission of Creutzfeldt-Jakob disease. *N. Engl. J. Med.* **311:** 1582–1583.

DeArmond S.J., Mobley W.C., DeMott D.L., Barry R.A., Beckstead J.H., and Prusiner S.B. 1987. Changes in the localization of brain prion proteins during scrapie infection. *Neurology* **37:** 1271–1280.

DeArmond S.J., Yang S.-L., Lee A., Bowler R., Taraboulos A., Groth D., and Prusiner S.B. 1993. Three scrapie prion isolates exhibit different accumulation patterns of the prion protein scrapie isoform. *Proc. Natl. Acad. Sci.* **90:** 6449–6453.

DeArmond S.J., Sánchez H., Yehiely F., Qiu Y., Ninchak-Casey A., Daggett V., Camerino A.P., Cayetano J., Rogers M., Groth D., Torchia M., Tremblay P., Scott M.R., Cohen F.E., and Prusiner S.B. 1997. Selective neuronal targeting in prion disease. *Neuron* **19:** 1337–1348.

Defebvre L., Destee A., Caron J., Ruchoux M.M., Wurtz A., and Remy J. 1997. Creutzfeldt-Jakob disease after an embolization of intercostal arteries with cadaveric dura mater suggesting a systemic transmission of the prion agent. *Neurology* **48:** 1470–1471.

Deslys J.-P., Marcé D., and Dormont D. 1994. Similar genetic susceptibility in iatrogenic and sporadic Creutzfeldt-Jakob disease. *J. Gen. Virol.* **75:** 23–27.

Dickinson A.G. and Fraser H. 1979. An assessment of the genetics of scrapie in sheep and mice. In *Slow transmissible diseases of the nervous system* (ed. S.B. Prusiner and W.J. Hadlow), vol. 1, pp. 367–386. Academic Press, New York.

Dickinson A.G. and Meikle V.M. 1969. A comparison of some biological characteristics of the mouse-passaged scrapie agents, 22A and ME7. *Genet. Res.* **13:** 213–225.

———. 1971. Host-genotype and agent effects in scrapie incubation: Change in allelic interaction with different strains of agent. *Mol. Gen. Genet.* **112:** 73–79.

Dickinson A.G. and Outram G.W. 1979. The scrapie replication-site hypothesis and its implications for pathogenesis. In *Slow transmissible diseases of the nervous system* (ed. S.B. Prusiner and W.J. Hadlow), vol. 2, pp. 13–31. Academic Press, New York.

———. 1988. Genetic aspects of unconventional virus infections: The basis of the virino hypothesis. *Ciba Found. Symp.* **135:** 63–83.

Dickinson A.G. and Stamp J.T. 1969. Experimental scrapie in Cheviot and Suffolk sheep. *J. Comp. Pathol.* **79:** 23–26.

Dickinson A.G., Meikle V.M.H., and Fraser H. 1968. Identification of a gene which controls the incubation period of some strains of scrapie agent in mice. *J. Comp. Pathol.* **78:** 293–299.

Dickinson A.G., Bruce M.E., Outram G.W., and Kimberlin R.H. 1984. Scrapie strain differences: The implications of stability and mutation. In *Proceedings of Workshop on Slow Transmissible Diseases* (ed. J. Tateishi), pp. 105–118. Japanese Ministry of Health and Welfare, Tokyo.

Dickinson A.G., Young G.B., Stamp J.T., and Renwick C.C. 1965. An analysis of natural scrapie in Suffolk sheep. *Heredity* **20:** 485–503.

Diringer H. 1984. Sustained viremia in experimental hamster scrapie. Brief report. *Arch. Virol.* **82:** 105–109.

Donnelly C.A. and Ferguson N.M. 2000. *Statistical aspects of BSE and vCJD: Models for epidemics.* Chapman and Hall/CRC, Boca Raton, Florida.

Duffy P., Wolf J., Collins G., DeVoe A., Streeten B., and Cowen D. 1974. Possible person-to-person transmission of Creutzfeldt-Jakob disease. *N. Engl. J. Med.* **290:** 692–693.

Eklund C.M., Kennedy R.C., and Hadlow W.J. 1967. Pathogenesis of scrapie virus infection in the mouse. *J. Infect. Dis.* **117:** 15–22.

Ellis C.J., Katifi H., and Weller R.O. 1992. A further British case of growth hormone induced Creutzfeldt-Jakob disease. *J. Neurol. Neurosurg. Psychiatry* **55:** 1200–1202.

Fradkin J.E., Schonberger L.B., Mills J.L., Gunn W.J., Piper J.M., Wysowski D.K., Thomson R., Durako S., and Brown P. 1991. Creutzfeldt-Jakob disease in pituitary growth hormone recipients in the United States. *J. Am. Med. Assoc.* **265:** 880–884.

Fraser H. 1982. Neuronal spread of scrapie agent and targeting of lesions within the retinotectal pathway. *Nature* **295:** 149–150.

Fraser H. and Dickinson A.G. 1968. The sequential development of the brain lesions of scrapie in three strains of mice. *J. Comp. Pathol.* **78:** 301–311.

———. 1973. Scrapie in mice. Agent-strain differences in the distribution and intensity of grey matter vacuolation. *J. Comp. Pathol.* **83:** 29–40.

———. 1985. Targeting of scrapie lesions and spread of agent via the retino-tectal projection. *Brain Res.* **346:** 32–41.

Fraser H., Brown K.L., Stewart K., McConnell I., McBride P., and Williams A. 1996. Replication of scrapie in spleens of SCID mice follows reconstitution with wild-type mouse bone marrow. *J. Gen. Virol.* **77:** 1935–1940.

Frigg R., Klein M.A., Hegyi I., Zinkernagel R.M., and Aguzzi A. 1999. Scrapie pathogenesis in subclinically infected B-cell-deficient mice. *J. Virol.* **73:** 9584–9588.

Gajdusek D.C. 1977. Unconventional viruses and the origin and disappearance of kuru. *Science* **197:** 943–960.

———. 1979. Observations on the early history of kuru investigations. In *Slow transmissible diseases of the nervous system* (ed. S.B. Prusiner and W.J. Hadlow), vol. 1, pp. 7–36. Academic Press, New York.

Gajdusek D.C., Gibbs C.J., Jr., Asher D.M., Brown P., Diwan A., Hoffman P., Nemo G., Rohwer R., and White L. 1977. Precautions in medical care of, and in handling materials from, patients with transmissible virus dementia (Creutzfeldt-Jakob disease). *N. Engl. J. Med.* **297:** 1253–1258.

Gambetti P. and Parchi P. 1999. Insomnia in prion diseases: Sporadic and familial. *N. Engl. J. Med.* **340:** 1675–1677.

Gibbs C.J., Jr., Gajdusek D.C., and Amyx H. 1979. Strain variation in the viruses of Creutzfeldt-Jakob disease and kuru. In *Slow transmissible diseases of the nervous system* (ed. S.B. Prusiner and W.J. Hadlow), vol. 2, pp. 87–110. Academic Press, New York.

Gibbs C.J., Jr., Amyx H.L., Bacote A., Masters C.L., and Gajdusek D.C. 1980. Oral transmission of kuru, Creutzfeldt-Jakob disease and scrapie to nonhuman primates. *J. Infect. Dis.* **142:** 205–208.

Gibbs C.J., Jr., Asher D.M., Brown P.W., Fradkin J.E., and Gajdusek D.C. 1993. Creutzfeldt-Jakob disease infectivity of growth hormone derived from human pituitary glands. *N. Engl. J. Med.* **328:** 358–359.

Gibbs C.J., Jr., Asher D.M., Kobrine A., Amyx H.L., Sulima M.P., and Gajdusek D.C. 1994. Transmission of Creutzfeldt-Jakob disease to a chimpanzee by electrodes contaminated during neurosurgery. *J. Neurol. Neurosurg. Psychiatry* **57:** 757–758.

Gibbs C.J., Jr., Joy A., Heffner R., Franko M., Miyazaki M., Asher D.M., Parisi J.E., Brown P.W., and Gajdusek D.C. 1985. Clinical and pathological features and laboratory confirmation of Creutzfeldt-Jakob disease in a recipient of pituitary-derived human growth hormone. *N. Engl. J. Med.* **313:** 734–738.

Goldmann W., Hunter N., Manson J., and Hope J. 1990a. The PrP gene of the sheep, a natural host of scrapie. In *Abstracts from the Proceedings of the 8th International Congress of Virology,* Berlin, August 26–31, p. 284.

Goldmann W., Hunter N., Smith G., Foster J., and Hope J. 1994. PrP genotype and agent effects in scrapie: Change in allelic interaction with different isolates of agent in sheep, a natural host of scrapie. *J. Gen. Virol.* **75:** 989–995.

Goldmann W., Hunter N., Foster J.D., Salbaum J.M., Beyreuther K., and Hope J. 1990b. Two alleles of a neural protein gene linked to scrapie in sheep. *Proc. Natl. Acad. Sci.* **87:** 2476–2480.

Gomi H., Yokoyama T., Fujimoto K., Ikeda T., Katoh A., Itoh T., and Itohara S. 1995. Mice devoid of the glial fibrillary acidic protein develop normally and are susceptible to scrapie prions. *Neuron* **14:** 29–41.

Gordon W.S. 1966. Variation in susceptibility of sheep to scrapie and genetic implications (Report of Scrapie Seminar). *Agric. Res. Serv. Rep. 91-53,* pp. 53–67. U.S. Department of Agriculture, Washington, D.C.

Gossen M. and Bujard H. 1992. Tight control of gene expression in mammalian cells by tetracycline-responsive promoters. *Proc. Natl. Acad. Sci.* **89:** 5547–5551.

Hadlow W.J., Kennedy R.C., and Race R.E. 1982. Natural infection of Suffolk sheep with scrapie virus. *J. Infect. Dis.* **146:** 657–664.

Hadlow W.J., Kennedy R.C., Race R.E., and Eklund C.M. 1980. Virologic and neurohistologic findings in dairy goats affected with natural scrapie. *Vet. Pathol.* **17:** 187–199.

Harries-Jones R., Knight R., Will R.G., Cousens S., Smith P.G., and Matthews W.B. 1988.

Creutzfeldt-Jakob disease in England and Wales, 1980–1984: A case-control study of potential risk factors. *J. Neurol. Neurosurg. Psychiatry* **51:** 1113–1119.

Healy D.L. and Evans J. 1993. Creutzfeldt-Jakob disease after pituitary gonadotrophins. *Br. J. Med.* **307:** 517–518.

Hecker R., Taraboulos A., Scott M., Pan K.-M., Torchia M., Jendroska K., DeArmond S.J., and Prusiner S.B. 1992. Replication of distinct scrapie prion isolates is region specific in brains of transgenic mice and hamsters. *Genes Dev.* **6:** 1213–1228.

Hegde R.S., Tremblay P., Groth D., Prusiner S.B., and Lingappa V.R. 1999. Transmissible and genetic prion diseases share a common pathway of neurodegeneration. *Nature* **402:** 822–826.

Hegde R.S., Mastrianni J.A., Scott M.R., DeFea K.A., Tremblay P., Torchia M., DeArmond S.J., Prusiner S.B., and Lingappa V.R. 1998. A transmembrane form of the prion protein in neurodegenerative disease. *Science* **279:** 827–834.

Heppner F.L., Musahl C., Arrighi I., Klein M.A., Rülicke T., Oesch B., Zinkernagel R.M., Kalinke U., and Aguzzi A. 2001. Prevention of scrapie pathogenesis by transgenic expression of anti-prion protein antibodies. *Science* **294:** 178–182.

Hölscher C., Delius H., and Bürkle A. 1998. Overexpression of nonconvertible PrP^C Δ114-121 in scrapie-infected mouse neuroblastoma cells leads to *trans*-dominant inhibition of wild-type PrP^{Sc} accumulation. *J. Virol.* **72:** 1153–1159.

Hotchin J. and Buckley R. 1977. Latent form of scrapie virus: A new factor in slow virus disease. *Science* **196:** 668–671.

Houston E.F. and Gravenor M.D. 2003. Clinical signs in sheep experimentally infected with scrapie and BSE. *Vet. Rec.* **152:** 333–334.

Hunter N., Foster J.D., Dickinson A.G., and Hope J. 1989. Linkage of the gene for the scrapie-associated fibril protein (PrP) to the *Sip* gene in Cheviot sheep. *Vet. Rec.* **124:** 364–366.

Hunter N., Hope J., McConnell I., and Dickinson A.G. 1987. Linkage of the scrapie-associated fibril protein (PrP) gene and *Sinc* using congenic mice and restriction fragment length polymorphism analysis. *J. Gen. Virol.* **68:** 2711–2716.

Hunter N., Goldmann W., Benson G., Foster J.D., and Hope J. 1993. Swaledale sheep affected by natural scrapie differ significantly in PrP genotype frequencies from healthy sheep and those selected for reduced incidence of scrapie. *J. Gen. Virol.* **74:** 1025–1031.

Hunter N., Moore L., Hosie B.D., Dingwall W.S., and Greig A. 1997a. Association between natural scrapie and PrP genotype in a flock of Suffolk sheep in Scotland. *Vet. Rec.* **140:** 59–63.

Hunter N., Cairns D., Foster J.D., Smith G., Goldmann W., and Donnelly K. 1997b. Is scrapie solely a genetic disease? *Nature* **386:** 137.

Hunter N., Foster J., Chong A., McCutcheon S., Parnham D., Eaton S., MacKenzie C., and Houston F. 2002. Transmission of prion disease by blood transfusion. *J. Gen. Virol.* **83:** 2897–2905.

Ikeda T., Horiuchi M., Ishiguro N., Muramatsu Y., Kai-Uwe G.D., and Shinagawa M. 1995. Amino acid polymorphisms of PrP with reference to onset of scrapie in Suffolk and Corriedale sheep in Japan. *J. Gen. Virol.* **76:** 2577–2581.

Jendroska K., Heinzel F.P., Torchia M., Stowring L., Kretzschmar H.A., Kon A., Stern A., Prusiner S.B., and DeArmond S.J. 1991. Proteinase-resistant prion protein accumulation in Syrian hamster brain correlates with regional pathology and scrapie infectivity. *Neurology* **41:** 1482–1490.

Jones R.L., Benker G., Salacinski P.R., Lloyd T.J., and Lowry P.J. 1979. Large-scale preparation of highly purified pyrogen-free human growth hormone for clinical use. *Br. J. Endocrinol.* **82:** 77–86.

Kaneko K., Wille H., Mehlhorn I., Zhang H., Ball H., Cohen F.E., Baldwin M.A., and Prusiner S.B. 1997a. Molecular properties of complexes formed between the prion protein and synthetic peptides. *J. Mol. Biol.* **270:** 574–586.

Kaneko K., Zulianello L., Scott M., Cooper C.M., Wallace A.C., James T.L., Cohen F.E., and Prusiner S.B. 1997b. Evidence for protein X binding to a discontinuous epitope on the cellular prion protein during scrapie prion propagation. *Proc. Natl. Acad. Sci.* **94:** 10069–10074.

Kaneko K., Peretz D., Pan K.-M., Blochberger T., Wille H., Gabizon R., Griffith O.H., Cohen F.E., Baldwin M.A., and Prusiner S.B. 1995. Prion protein (PrP) synthetic peptides induce cellular PrP to acquire properties of the scrapie isoform. *Proc. Natl. Acad. Sci.* **32:** 11160–11164.

Kascsak R.J., Rubenstein R., Merz P.A., Carp R.I., Wisniewski H.M., and Diringer H. 1985. Biochemical differences among scrapie-associated fibrils support the biological diversity of scrapie agents. *J. Gen. Virol.* **66:** 1715–1722.

Kascsak R.J., Rubenstein R., Merz P.A., Tonna-DeMasi M., Fersko R., Carp R.I., Wisniewski H.M., and Diringer H. 1987. Mouse polyclonal and monoclonal antibody to scrapie-associated fibril proteins. *J. Virol.* **61:** 3688–3693.

Kimberlin R.H. 1976. Experimental scrapie in the mouse: A review of an important model disease. *Sci. Prog.* **63:** 461–481.

———. 1982. Reflections on the nature of the scrapie agent. *Trends Biochem. Sci.* **7:** 392–394.

———. 1990. Scrapie and possible relationships with viroids. *Semin. Virol.* **1:** 153–162.

Kimberlin R. and Walker C. 1977. Characteristics of a short incubation model of scrapie in the golden hamster. *J. Gen. Virol.* **34:** 295–304.

———. 1978. Evidence that the transmission of one source of scrapie agent to hamsters involves separation of agent strains from a mixture. *J. Gen. Virol.* **39:** 487–496.

Kimberlin R.H., Cole S., and Walker C.A. 1987. Temporary and permanent modifications to a single strain of mouse scrapie on transmission to rats and hamsters. *J. Gen. Virol.* **68:** 1875–1881.

Kimberlin R.H., Field H.J., and Walker C.A. 1983. Pathogenesis of mouse scrapie: Evidence for spread of infection from central to peripheral nervous system. *J. Gen. Virol.* **64:** 713–716.

Kimberlin R.H., Walker C.A., and Fraser H. 1989. The genomic identity of different strains of mouse scrapie is expressed in hamsters and preserved on reisolation in mice. *J. Gen. Virol.* **70:** 2017–2025.

Kingsbury D.T., Smeltzer D.A., Amyx H.L., Gibbs C.J., Jr., and Gajdusek D.C. 1982. Evidence for an unconventional virus in mouse-adapted Creutzfeldt-Jakob disease. *Infect. Immun.* **37:** 1050–1053.

Kingsbury D.T., Kasper K.C., Stites D.P., Watson J.D., Hogan R.N., and Prusiner S.B. 1983. Genetic control of scrapie and Creutzfeldt-Jakob disease in mice. *J. Immunol.* **131:** 491–496.

Kitamoto T., Muramoto T., Mohri S., Doh-Ura K., and Tateishi J. 1991. Abnormal isoform of prion protein accumulates in follicular dendritic cells in mice with Creutzfeldt-Jakob disease. *J. Virol.* **65:** 6292–6295.

Klein M.A., Frigg R., Raeber A.J., Flechsig E., Hegyi I., Zinkernagel R.M., Weissmann C., and Aguzzi A. 1998. PrP expression in B lymphocytes is not required for prion neuroinvasion. *Nat. Med.* **4:** 1429–1433.

Klein M.A., Frigg R., Flechsig E., Raeber A.J., Kalinke U., Bluethmann H., Bootz F., Suter

M., Zinkernagel R.M., and Aguzzi A. 1997. A crucial role for B cells in neuroinvasive scrapie. *Nature* **390:** 687–691.

Klitzman R.L., Alpers M.P., and Gajdusek D.C. 1984. The natural incubation period of kuru and the episodes of transmission in three clusters of patients. *Neuroepidemiology* **3:** 3–20.

Koch T.K., Berg B.O., DeArmond S.J., and Gravina R.F. 1985. Creutzfeldt-Jakob disease in a young adult with idiopathic hypopituitarism. Possible relation to the administration of cadaveric human growth hormone. *N. Engl. J. Med.* **313:** 731–733.

Kocisko D.A., Priola S.A., Raymond G.J., Chesebro B., Lansbury P.T., Jr., and Caughey B. 1995. Species specificity in the cell-free conversion of prion protein to protease-resistant forms: A model for the scrapie species barrier. *Proc. Natl. Acad. Sci.* **92:** 3923–3927.

Kocisko D.A., Come J.H., Priola S.A., Chesebro B., Raymond G.J., Lansbury P.T., Jr., and Caughey B. 1994. Cell-free formation of protease-resistant prion protein. *Nature* **370:** 471–474.

Kondo K. and Kuroina Y. 1982. A case control study of Creutzfeldt-Jakob disease: Association with physical injuries. *Ann. Neurol.* **11:** 377–381.

Korth C., Kaneko K., Groth D., Heye N., Telling G., Mastrianni J., Parchi P., Gambetti P., Will R., Ironside J., Heinrich C., Tremblay P., DeArmond S.J., and Prusiner S.B. 2003. Abbreviated incubation times for human prions in mice expressing a chimeric mouse–human prion protein transgene. *Proc. Natl. Acad. Sci.* **100:** 4784–4789.

Kretzschmar H.A., Prusiner S.B., Stowring L.E., and DeArmond S.J. 1986. Scrapie prion proteins are synthesized in neurons. *Am. J. Pathol.* **122:** 1–5.

Laplanche J.-L., Chatelain J., Beaudry P., Dussaucy M., Bounneau C., and Launay J.-M. 1993a. French autochthonous scrapied sheep without the 136Val PrP polymorphism. *Mamm. Genome* **4:** 463–464.

Laplanche J.-L., Chatelain J., Westaway D., Thomas S., Dussaucy M., Brugere-Picoux J., and Launay J. M. 1993b. PrP polymorphisms associated with natural scrapie discovered by denaturing gradient gel electrophoresis. *Genomics* **15:** 30–37.

Lasmézas C.I., Ceshron J.-Y., Deslys J.-P., Demaimay R., Adjou K., Lemaire C., Decavel J.-P., and Dormont D. 1996. Scrapie in severe combined immunodeficient mice. In *Transmissible subacute spongiform encephalopathies: Prion diseases* (ed. L. Court and B. Dodet), pp. 151–157. Elsevier, Paris.

Macario M.E., Vaisman M., Buescu A., Neto V.M., Araujo H.M.M., and Chagas C. 1991. Pituitary growth hormone and Creutzfeldt-Jakob disease. *Br. Med. J.* **302:** 1149.

Mackenzie A. 1983. Immunohistochemical demonstration of glial fibrillary acidic protein in scrapie. *J. Comp. Pathol.* **93:** 251–259.

Mallucci G.R., Ratte S., Asante E.A., Linehan J., Gowland I., Jefferys J.G., and Collinge J. 2002. Post-natal knockout of prion protein alters hippocampal CA1 properties, but does not result in neurodegeneration. *EMBO J.* **21:** 202–210.

Manson J.C., Clarke A.R., McBride P.A., McConnell I., and Hope J. 1994a. PrP gene dosage determines the timing but not the final intensity or distribution of lesions in scrapie pathology. *Neurodegeneration* **3:** 331–340.

Manson J.C., Clarke A.R., Hooper M.L., Aitchison L., McConnell I., and Hope J. 1994b. 129/Ola mice carrying a null mutation in PrP that abolishes mRNA production are developmentally normal. *Mol. Neurobiol.* **8:** 121–127.

Manuelidis E.E. and Manuelidis L. 1979a. Observations on Creutzfeldt-Jakob disease propagated in small rodents. In *Slow transmissible diseases of the nervous system* (ed. S.B. Prusiner and W.J. Hadlow), vol. 2, pp. 147–173. Academic Press, New York.

———. 1979b. Experiments on maternal transmission of Creutzfeldt-Jakob disease in guinea pigs. *Proc. Soc. Biol. Med.* **160:** 233–236.

Manuelidis E.E., Manuelidis L., Pincus I.H., and Collins W.F. 1978. Transmission, from man to hamster, of Creutzfeldt-Jakob disease with clinical recovery. *Lancet* **2:** 40–42.

Marsh R.F. and Kimberlin R.H. 1975. Comparison of scrapie and transmissible mink encephalopathy in hamsters. II. Clinical signs, pathology and pathogenesis. *J. Infect. Dis.* **131:** 104–110.

Marsh R.F., Bessen R.A., Lehmann S., and Hartsough G.R. 1991. Epidemiological and experimental studies on a new incident of transmissible mink encephalopathy. *J. Gen. Virol.* **72:** 589–594.

Martínez-Lage J.F., Sola J., Poza M., and Esteban J.A. 1993. Pediatric Creutzfeldt-Jakob disease: Probable transmission by a dural graft. *Child's Nerv. Syst.* **9:** 239–242.

Marzewski D.J., Towfighi J., Harrington M.G., Merril C.R., and Brown P. 1988. Creutzfeldt-Jakob disease following pituitary-derived human growth hormone therapy: A new American case. *Neurology* **38:** 1131–1133.

Masters C.L. and Richardson E.P., Jr. 1978. Subacute spongiform encephalopathy Creutzfeldt-Jakob disease—The nature and progression of spongiform change. *Brain* **101:** 333–344.

Mastrianni J., Nixon F., Layzer R., DeArmond S.J., and Prusiner S.B. 1997. Fatal sporadic insomnia: Fatal familial insomnia phenotype without a mutation of the prion protein gene. *Neurology* [suppl.] **48:** A296.

Mastrianni J.A., Nixon R., Layzer R., Telling G.C., Han D., DeArmond S.J., and Prusiner S.B. 1999. Prion protein conformation in a patient with sporadic fatal insomnia. *N. Engl. J. Med.* **340:** 1630–1638.

Masullo C., Pocchiari M., Macchi G., Alema G., Piazza G., and Panzera M.A. 1989. Transmission of Creutzfeldt-Jakob disease by dural cadaveric graft. *J. Neurosurg.* **71:** 954.

McBride P.A., Eikelenboom P., Kraal G., Fraser H., and Bruce M.E. 1992. PrP protein is associated with follicular dendritic cells of spleens and lymph nodes in uninfected and scrapie-infected mice. *J. Pathol.* **168:** 413–418.

McKinley M.P., Bolton D.C., and Prusiner S.B. 1983. A protease-resistant protein is a structural component of the scrapie prion. *Cell* **35:** 57–62.

McKinley M.P., Hay B., Lingappa V.R., Lieberburg I., and Prusiner S.B. 1987. Developmental expression of prion protein gene in brain. *Dev. Biol.* **121:** 105–110.

Medori R., Tritschler H.-J., LeBlanc A., Villare F., Manetto V., Chen H.Y., Xue R., Leal S., Montagna P., Cortelli P., Tinuper P., Avoni P., Mochi M., Baruzzi A., Hauw J.J., Ott J., Lugaresi E., Autilio-Gambetti L., and Gambetti P. 1992. Fatal familial insomnia, a prion disease with a mutation at codon 178 of the prion protein gene. *N. Engl. J. Med.* **326:** 444–449.

Merz P.A., Kascsak R., Rubenstein R., Carp R.I., and Wisniewski H.M. 1984a. Variations in SAF from different scrapie agents. In *Proceedings of Workshop on Slow Transmissible Diseases* (ed. J. Tateishi), pp. 137–145. Japanese Ministry of Health and Welfare, Tokyo.

Merz P.A., Rohwer R.G., Kascsak R., Wisniewski H.M., Somerville R.A., Gibbs C.J., Jr., and Gajdusek D.C. 1984b. Infection-specific particle from the unconventional slow virus diseases. *Science* **225:** 437–440.

M'Gowan J.P. 1914. *Investigation into the disease of sheep called "scrapie".* William Blackwood and Sons, Edinburgh, Scotland, United Kingdom.

Miller M.W., Wild M.A., and Williams E.S. 1998. Epidemiology of chronic wasting disease in captive Rocky Mountain elk. *J. Wildl. Dis.* **34:** 532–538.

Miyashita K., Inuzuka T., Kondo H., Saito Y., Fujita N., Matsubara N., Tanaka R.,

Hinokuma K., Ikuta F., and Miyatake T. 1991. Creutzfeldt-Jakob disease in a patient with a cadaveric dural graft. *Neurology* **41**: 940–941.

Monari L., Chen S.G., Brown P., Parchi P., Petersen R.B., Mikol J., Gray F., Cortelli P., Montagna P., Ghetti B., Goldfarb L.G., Gajdusek D.C., Lugaresi E., Gambetti P., and Autilio-Gambetti L. 1994. Fatal familial insomnia and familial Creutzfeldt-Jakob disease: Different prion proteins determined by a DNA polymorphism. *Proc. Natl. Acad. Sci.* **91**: 2839–2842.

Moore R.C., Hope J., McBride P.A., McConnell I., Selfridge J., Melton D.W., and Manson J.C. 1998. Mice with gene targetted prion protein alterations show that *Prn-p*, *Sinc* and *Prni* are congruent. *Nat. Genet.* **18**: 118–125.

Morris J.A., Gajdusek D.C., and Gibbs C.J., Jr. 1965. Spread of scrapie from inoculated to uninoculated mice. *Proc. Soc. Exp. Biol. Med.* **120**: 108–110.

Muramoto T., Kitamoto T., Hoque M.Z., Tateishi J., and Goto I. 1993. Species barrier prevents an abnormal isoform of prion protein from accumulating in follicular dendritic cells of mice with Creutzfeldt-Jakob disease. *J. Virol.* **67**: 6808–6810.

New M.I., Brown P., Temeck J.W., Owens C., Hedley-Whyte E.T., and Richardson E.P. 1988. Preclinical Creutzfeldt-Jakob disease discovered at autopsy in a human growth hormone recipient. *Neurology* **38**: 1133–1134.

Nisbet T.J., MacDonaldson I., and Bishara S.N. 1989. Creutzfeldt-Jakob disease in a second patient who received a cadaveric dura mater graft. *J. Am. Med. Assoc.* **261**: 1118.

Oesch B., Westaway D., Wälchli M., McKinley M.P., Kent S.B.H., Aebersold R., Barry R.A., Tempst P., Teplow D.B., Hood L.E., Prusiner S.B., and Weissmann C. 1985. A cellular gene encodes scrapie PrP 27-30 protein. *Cell* **40**: 735–746.

Onodera T., Ikeda T., Muramatsu Y., and Shinagawa M. 1993. Isolation of scrapie agent from the placenta of sheep with natural scrapie in Japan. *Microbiol. Immunol.* **37**: 311–316.

O'Rourke K.I., Huff T.P., Leathers C.W., Robinson M.M., and Gorham J.R. 1994. SCID mouse spleen does not support scrapie agent replication. *J. Gen. Virol.* **75**: 1511–1514.

O'Rourke K.I., Holyoak G.R., Clark W.W., Mickelson J.R., Wang S., Melco R.P., Besser T.E., and Foote W.C. 1997. PrP genotypes and experimental scrapie in orally inoculated Suffolk sheep in the United States. *J. Gen. Virol.* **78**: 975–978.

O'Rourke K.I., Baszler T.V., Besser T.E., Miller J.M., Cutlip R.C., Wells G.A., Ryder S.J., Parish S.M., Hamir A.N., Cockett N.E., Jenny A., and Knowles D.P. 2000. Preclinical diagnosis of scrapie by immunohistochemistry of third eyelid lymphoid tissue. *J. Clin. Microbiol.* **38**: 3254–3259.

Otto D. 1987. Jakob-Creutzfeldt disease associated with cadaveric dura. *J. Neurosurg.* **67**: 149–150.

Özel M. and Diringer H. 1994. Small virus-like structure in fraction from scrapie hamster brain. *Lancet* **343**: 894–895.

Palmer M.S., Dryden A.J., Hughes J.T., and Collinge J. 1991. Homozygous prion protein genotype predisposes to sporadic Creutzfeldt-Jakob disease. *Nature* **352**: 340–342.

Palsson P.A. 1979. Rida (scrapie) in Iceland and its epidemiology. In *Slow transmissible diseases of the nervous system* (ed. S.B. Prusiner and W.J. Hadlow), pp. 357–366. Academic Press, New York.

Parchi P., Capellari S., Chin S., Schwarz H.B., Schecter N.P., Butts J.D., Hudkins P., Burns D.K., Powers J.M., and Gambetti P. 1999a. A subtype of sporadic prion disease mimicking fatal familial insomnia. *Neurology* **52**: 1757–1763.

Parchi P., Castellani R., Capellari S., Ghetti B., Young K., Chen S.G., Farlow M., Dickson D.W., Sima A.A.F., Trojanowski J.Q., Petersen R.B., and Gambetti P. 1996. Molecular basis of phenotypic variability in sporadic Creutzfeldt-Jakob disease. *Ann. Neurol.* **39:** 767–778.

Parchi P., Giese A., Capellari S., Brown P., Schulz-Schaeffer W., Windl O., Zerr I., Budka H., Kopp N., Piccardo P., Poser S., Rojiani A., Streichemberger N., Julien J., Vital C., Ghetti B., Gambetti P., and Kretzschmar H. 1999b. Classification of sporadic Creutzfeldt-Jakob disease based on molecular and phenotypic analysis of 300 subjects. *Ann. Neurol.* **46:** 224–233.

Parry H.B. 1962. Scrapie: A transmissible and hereditary disease of sheep. *Heredity* **17:** 75–105.

———. 1983. *Scrapie disease in sheep.* Academic Press, New York.

Pattison I.H. 1964. The spread of scrapie by contact between affected and healthy sheep, goats or mice. *Vet. Rec.* **76:** 333–336.

———. 1965. Experiments with scrapie with special reference to the nature of the agent and the pathology of the disease. In *Slow, latent and temperate virus infections* (NINDB Monogr. 2) (ed D.C. Gajdusek et al.), pp. 249–257. U.S. Government Printing, Washington, D.C.

Pattison I.H. and Jones K.M. 1968. Modification of a strain of mouse-adapted scrapie by passage through rats. *Res. Vet. Sci.* **9:** 408–410.

Pattison I.H. and Millson G.C. 1961a. Experimental transmission of scrapie to goats and sheep by the oral route. *J. Comp. Pathol.* **71:** 171–176.

———. 1961b. Scrapie produced experimentally in goats with special reference to the clinical syndrome. *J. Comp. Pathol.* **71:** 101–108.

Pattison I.H., Gordon W.S., and Millson G.C. 1959. Experimental production of scrapie in goats. *J. Comp. Path. Ther.* **69:** 300–312.

Pattison I.H., Millson G.C., and Smith K. 1964. An examination of the action of whole blood, blood cells or serum on the goat scrapie agent. *Res. Vet. Sci.* **5:** 116–121.

Pattison I.H., Hoare M.N., Jebbett J.N., and Watson W.A. 1972. Spread of scrapie to sheep and goats by oral dosing with foetal membranes from scrapie-affected sheep. *Vet. Rec.* **90:** 465–468.

Peretz D., Scott M., Groth D., Williamson A., Burton D., Cohen F.E., and Prusiner S.B. 2001a. Strain-specified relative conformational stability of the scrapie prion protein. *Protein Sci.* **10:** 854–863.

Peretz D., Williamson R.A., Legname G., Matsunaga Y., Vergara J., Burton D., DeArmond S.J., Prusiner S.B., and Scott M.R. 2002. A change in the conformation of prions accompanies the emergence of a new prion strain. *Neuron* **34:** 921–932.

Peretz D., Williamson R.A., Kaneko K., Vergara J., Leclerc E., Schmitt-Ulms G., Mehlhorn I.R., Legname G., Wormald M.R., Rudd P.M., Dwek R.A., Burton D.R., and Prusiner S.B. 2001b. Antibodies inhibit prion propagation and clear cell cultures of prion infectivity. *Nature* **412:** 739–743.

Perrier V., Kaneko K., Safar J., Vergara J., Tremblay P., DeArmond S.J., Cohen F.E., Prusiner S.B., and Wallace A.C. 2002. Dominant-negative inhibition of prion replication in transgenic mice. *Proc. Natl. Acad. Sci.* **99:** 13079–13084.

Phillips N.A., Bridgeman J., and Ferguson-Smith M. 2000. Findings and conclusions. In *The BSE inquiry,* vol. 1. Stationery Office, London, United Kingdom.

Pocchiari M., Peano S., Conz A., Eshkol A., Maillard F., Brown P., Gibbs C.J., Jr., Xi Y.G., Tenham-Fisher E., and Macchi G. 1991. Combination ultrafiltration and 6 *M* urea treatment of human growth hormone effectively minimizes risk from potential Creutzfeldt-Jakob disease virus contamination. *Horm. Res.* **35:** 161–166.

Powell-Jackson J., Weller R.O., Kennedy P., Preece M.A., Whitcombe E.M., and Newsome-

Davis J. 1985. Creutzfeldt-Jakob disease after administration of human growth hormone. *Lancet* **2:** 244–246.

Prusiner S.B. 1987. The biology of prion transmission and replication. In *Prions: Novel infectious pathogens causing scrapie and Creutzfeldt-Jakob disease* (ed. S.B. Prusiner and M.P. McKinley), pp. 83–112. Academic Press, Orlando, Florida.

———.1991. Molecular biology of prion diseases. *Science* **252:** 1515–1522.

———. 1992. Chemistry and biology of prions. *Biochemistry* **31:** 12278–12288.

———. 1997. Prion diseases and the BSE crisis. *Science* **278:** 245–251.

———. 1998. Prions. *Proc. Natl. Acad. Sci.* **95:** 13363–13383.

Prusiner S.B., Cochran S.P., and Alpers M.P. 1985. Transmission of scrapie in hamsters. *J. Infect. Dis.* **152:** 971–978.

Prusiner S.B., Gajdusek D.C., and Alpers M.P. 1982a. Kuru with incubation periods exceeding two decades. *Ann. Neurol.* **12:** 1–9.

Prusiner S.B., Scott M.R., DeArmond S.J., and Cohen F.E. 1998. Prion protein biology. *Cell* **93:** 337–348.

Prusiner S.B., Groth D.F., Cochran S.P., McKinley M.P., and Masiarz F.R. 1980. Gel electrophoresis and glass permeation chromatography of the hamster scrapie agent after enzymatic digestion and detergent extraction. *Biochemistry* **19:** 4892–4898.

Prusiner S.B., Cochran S.P., Groth D.F., Downey D.E., Bowman K.A., and Martinez H.M. 1982b. Measurement of the scrapie agent using an incubation time interval assay. *Ann. Neurol.* **11:** 353–358.

Prusiner S.B., Hadlow W.J., Garfin D.E., Cochran S.P., Baringer J.R., Race R.E., and Eklund C.M. 1978. Partial purification and evidence for multiple molecular forms of the scrapie agent. *Biochemistry* **17:** 4993–4997.

Prusiner S.B., McKinley M.P., Bowman K.A., Bolton D.C., Bendheim P.E., Groth D.F., and Glenner G.G. 1983. Scrapie prions aggregate to form amyloid-like birefringent rods. *Cell* **35:** 349–358.

Prusiner S.B., Groth D., Serban A., Koehler R., Foster D., Torchia M., Burton D., Yang S.-L., and DeArmond S.J. 1993a. Ablation of the prion protein (PrP) gene in mice prevents scrapie and facilitates production of anti-PrP antibodies. *Proc. Natl. Acad. Sci.* **90:** 10608–10612.

Prusiner S.B., Fuzi M., Scott M., Serban D., Serban H., Taraboulos A., Gabriel J.-M., Wells G., Wilesmith J., Bradley R., DeArmond S.J., and Kristensson K. 1993b. Immunologic and molecular biological studies of prion proteins in bovine spongiform encephalopathy. *J. Infect. Dis.* **167:** 602–613.

Prusiner S.B., Scott M., Foster D., Pan K.-M., Groth D., Mirenda C., Torchia M., Yang S.-L., Serban D., Carlson G.A., Hoppe P.C., Westaway D., and DeArmond S.J. 1990. Transgenetic studies implicate interactions between homologous PrP isoforms in scrapie prion replication. *Cell* **63:** 673–686.

Public Health Service. 1997. Interagency Coordinating Committee report on human growth hormone and Creutzfeldt-Jakob disease. *U.S. Public Health Serv. Rep.* **14:** 1–11.

Race R.E., Graham K., Ernst D., Caughey B., and Chesebro B. 1990. Analysis of linkage between scrapie incubation period and the prion protein gene in mice. *J. Gen. Virol.* **71:** 493–497.

Ridley R.M. and Baker H.F. 1993. Occupational risk of Creutzfeldt-Jakob disease. *Lancet* **341:** 641–642.

———. 1996. To what extent is strain variation evidence for an independent genome in the agent of the transmissible spongiform encephalopathies? *Neurodegeneration* **5:** 219–231.

Sailer A., Büeler H., Fischer M., Aguzzi A., and Weissmann C. 1994. No propagation of prions in mice devoid of PrP. *Cell* **77:** 967–968.

Sauer B. and Henderson N. 1988. Site-specific DNA recombination in mammalian cells by the Cre recombinase of bacteriophage P1. *Proc. Natl. Acad. Sci.* **85:** 5166–5170.

Schätzl H.M., Laszlo L., Holtzman D.M., Tatzelt J., DeArmond S.J., Weiner R.I., Mobley W.C., and Prusiner S.B. 1997. A hypothalamic neuronal cell line persistently infected with scrapie prions exhibits apoptosis. *J. Virol.* **71:** 8821–8831.

Scott M., Groth D., Foster D., Torchia M., Yang S.-L., DeArmond S.J., and Prusiner S.B. 1993. Propagation of prions with artificial properties in transgenic mice expressing chimeric PrP genes. *Cell* **73:** 979–988.

Scott M.R., Groth D., Tatzelt J., Torchia M., Tremblay P., DeArmond S.J., and Prusiner S.B. 1997. Propagation of prion strains through specific conformers of the prion protein. *J. Virol.* **71:** 9032–9044.

Scott M.R., Will R., Ironside J., Nguyen H.-O.B., Tremblay P., DeArmond S.J., and Prusiner S.B. 1999. Compelling transgenetic evidence for transmission of bovine spongiform encephalopathy prions to humans. *Proc. Natl. Acad. Sci.* **96:** 15137–15142.

Scott M., Foster D., Mirenda C., Serban D., Coufal F., Wälchli M., Torchia M., Groth D., Carlson G., DeArmond S.J., Westaway D., and Prusiner S.B. 1989. Transgenic mice expressing hamster prion protein produce species-specific scrapie infectivity and amyloid plaques. *Cell* **59:** 847–857.

Shimizu S., Hoshi K., Muramoto T., Homma M., Ironside J.W., Kuzuhara S., Sato T., Yamamoto T., and Kitamoto T. 1999. Creutzfeldt-Jakob disease with florid-type plaques after cadaveric dura mater grafting. *Arch. Neurol.* **56:** 357–362.

Spraker T.R., Miller M.W., Williams E.S., Getzy D.M., Adrian W.J., Schoonveld G.G., Spowart R.A., O'Rourke K.I., Miller J.M., and Merz P.A. 1997. Spongiform encephalopathy in free-ranging mule deer (*Odocoileus hemionus*), white-tailed deer (*Odocoileus virginianus*), and Rocky Mountain elk (*Cervus elaphus nelsoni*) in north-central Colorado. *J. Wildl. Dis.* **33:** 1–6.

Stekel D.J., Nowak M.A., and Southwood T.R.E. 1996. Prediction of future BSE spread. *Nature* **381:** 119.

Taguchi F., Tamai Y., and Miura S. 1993. Experiments on maternal and paternal transmission of Creutzfeldt-Jakob disease in mice. *Arch. Virol.* **130:** 219–224.

Tamai Y., Kojima H., Kitajima R., Taguchi F., Ohtani Y., Kawaguchi T., Miura S., Sato M., and Ishihara Y. 1992. Demonstration of the transmissible agent in tissue from a pregnant woman with Creutzfeldt-Jakob disease. *N. Engl. J. Med.* **327:** 649.

Tange R.A., Troost D., and Limburg M. 1989. Progressive fatal dementia (Creutzfeldt-Jakob disease) in a patient who received homograft tissue for tympanic membrane closure. *Eur. Arch. Otorhinolaryngol.* **247:** 199–201.

Taraboulos A., Jendroska K., Serban D., Yang S.-L., DeArmond S.J., and Prusiner S.B. 1992. Regional mapping of prion proteins in brains. *Proc. Natl. Acad. Sci.* **89:** 7620–7624.

Tateishi J. 1985. Transmission of Creutzfeldt-Jakob disease from human blood and urine into mice. *Lancet* **2:** 1074.

Tateishi J., Kitamoto T., Hoque M.Z., and Furukawa H. 1996. Experimental transmission of Creutzfeldt-Jakob disease and related diseases to rodents. *Neurology* **46:** 532–537.

Tateishi J., Ohta M., Koga M., Sato Y., and Kuroiwa Y. 1979. Transmission of chronic spongiform encephalopathy with kuru plaques from humans to small rodents. *Ann. Neurol.* **5:** 581–584.

Tatzelt J., Prusiner S.B., and Welch W.J. 1996a. Chemical chaperones interfere with the formation of scrapie prion protein. *EMBO J.* **15:** 6363–6373.

Tatzelt J., Maeda N., Pekny M., Yang S.-L., Betsholtz C., Eliasson C., Cayetano J., Camerino A.P., DeArmond S.J., and Prusiner S.B. 1996b. Scrapie in mice deficient in apolipoprotein E or glial fibrillary acidic protein. *Neurology* **47:** 449–453.

Taylor D.M., Dickinson A.G., Fraser H., Robertson P.A., Salacinski P.R., and Lowry P.J. 1985. Preparation of growth hormone free from contamination with unconventional slow viruses. *Lancet* **2:** 260–262.

Telling G.C., Scott M., Mastrianni J., Gabizon R., Torchia M., Cohen F.E., DeArmond S.J., and Prusiner S.B. 1995. Prion propagation in mice expressing human and chimeric PrP transgenes implicates the interaction of cellular PrP with another protein. *Cell* **83:** 79–90.

Telling G.C., Scott M., Hsiao K.K., Foster D., Yang S.-L., Torchia M., Sidle K.C.L., Collinge J., DeArmond S.J., and Prusiner S.B. 1994. Transmission of Creutzfeldt-Jakob disease from humans to transgenic mice expressing chimeric human-mouse prion protein. *Proc. Natl. Acad. Sci.* **91:** 9936–9940.

Telling G.C., Parchi P., DeArmond S.J., Cortelli P., Montagna P., Gabizon R., Mastrianni J., Lugaresi E., Gambetti P., and Prusiner S.B. 1996. Evidence for the conformation of the pathologic isoform of the prion protein enciphering and propagating prion diversity. *Science* **274:** 2079–2082.

Thadani V., Penar P.L., Partington J., Kalb R., Janssen R., Schonberger L.B., Rabkin C.S., and Prichard J.W. 1988. Creutzfeldt-Jakob disease probably acquired from a cadaveric dura mater graft. Case report. *J. Neurosurg.* **69:** 766–769.

Titner R., Brown P., Hedley-Whyte E.T., Rappaport E.B., Piccardo C.P., and Gajdusek D.C. 1986. Neuropathologic verification of Creutzfeldt-Jakob disease in the exhumed American recipient of human pituitary growth hormone: Epidemiologic and pathogenetic implications. *Neurology* **36:** 932–936.

Tremblay P., Meiner Z., Galou M., Heinrich C., Petromilli C., Lisse T., Cayetano J., Torchia M., Mobley W., Bujard H., DeArmond S.J., and Prusiner S.B. 1998. Doxycyline control of prion protein transgene expression modulates prion disease in mice. *Proc. Natl. Acad. Sci.* **95:** 12580–12585.

Walker A.S., Inderlied C.B., and Kingsbury D.T. 1983. Conditions for the chemical and physical inactivation of the K. Fu. strain of the agent of Creutzfeldt-Jakob disease. *Am. J. Public Health* **73:** 661–665.

Weissmann C. 1991. A "unified theory" of prion propagation. *Nature* **352:** 679–683.

Weissmann C., Enari M., Klohn P.C., Rossi D., and Flechsig E. 2002. Transmission of prions. *Proc. Natl. Acad. Sci.* **99:** 16378–16383.

Wells G.A.H. 2002. European Commission report on TSE infectivity distribution in ruminant tissues (*State of Knowledge*, December 2001), pp. 10–37.

Wells G.A.H., Hawkins S.A.C., Green R.B., Austin A.R., Dexter I., Spencer Y.I., Chaplin M.J., Stack M.J., and Dawson M. 1998. Preliminary observations on the pathogenesis of experimental bovine spongiform encephalopathy (BSE): An update. *Vet. Rec.* **142:** 103–106.

Westaway D., Goodman P.A., Mirenda C.A., McKinley M.P., Carlson G.A., and Prusiner S.B. 1987. Distinct prion proteins in short and long scrapie incubation period mice. *Cell* **51:** 651–662.

Westaway D., Zuliani V., Cooper C.M., Da Costa M., Neuman S., Jenny A.L., Detwiler L., and Prusiner S.B. 1994. Homozygosity for prion protein alleles encoding glutamine-171 renders sheep susceptible to natural scrapie. *Genes Dev.* **8:** 959–969.

Westaway D., Mirenda C.A., Foster D., Zebarjadian Y., Scott M., Torchia M., Yang S.-L., Serban H., DeArmond S.J., Ebeling C., Prusiner S.B., and Carlson G.A. 1991. Paradoxical shortening of scrapie incubation times by expression of prion protein transgenes derived from long incubation period mice. *Neuron* **7:** 59–68.

Wilesmith J.W. 2002. Preliminary epidemiological analyses of the first 16 cases of BSE born after July 31, 1996, in Great Britain. *Vet Rec* **151:** 451–452.

Wilesmith J.W., Ryan J.B.M., and Atkinson M.J. 1991. Bovine spongiform encephalopathy: Epidemiologic studies on the origin. *Vet. Rec.* **128:** 199–203.

Will R.G. and Matthews W.B. 1982. Evidence for case-to-case transmission of Creutzfeldt-Jakob disease. *J. Neurol. Neurosurg. Psychiatry* **45:** 235–238.

Williams E.S. and Young S. 1980. Chronic wasting disease of captive mule deer: A spongiform encephalopathy. *J. Wildl. Dis.* **16:** 89–98.

Willison H.J., Gale A.N., and McLaughlin J.E. 1991. Creutzfeldt-Jakob disease following cadaveric dura mater graft. *J. Neurol. Neurosurg. Psychiatry* **54:** 940.

Ziegler D.R. 1993. In *Genetic maps: Locus maps of complex genomes*, 6th edition (ed. S.J. O'Brien), pp. 4.42–44.45. Cold Spring Harbor Laboratory Press, Cold Spring Harbor, New York.

Zobeley E., Flechsig E., Cozzio A., Enari M., and Weissmann C. 1999. Infectivity of scrapie prions bound to a stainless steel surface. *Mol. Med.* **5:** 240–243.

5

Structural Studies of Prion Proteins

Cédric Govaerts,[1] Holger Wille,[2,4]
Stanley B. Prusiner,[2,3,4] and Fred E. Cohen[1,3,4]
Departments of [1]Cellular and Molecular Pharmacology,
[2]Neurology, [3]Biochemistry and Biophysics and the
[4]Institute for Neurodegenerative Diseases,
University of California, San Francisco, California 94143

ALTHOUGH MANY ASPECTS OF PRION DISEASE BIOLOGY are unorthodox, perhaps the most fundamental paradox is posed by the coexistence of inherited, sporadic, and infectious forms of these diseases. Sensible molecular mechanisms for prion propagation must explain all three forms of prion diseases in a manner that is compatible with the formidable and ever-increasing array of experimental data derived from histopathologic, biochemical, biophysical, human genetic, and transgenetic studies. In this chapter, we explore the phenomenologic constraints on models of prion replication with a specific emphasis on biophysical studies of prion protein structures. We examine how an inherited disease can also present as a sporadic or infectious illness in the context of the structural data on PrPs that are currently available. Since the first version of this chapter was written, much more has been learned about prion biology. Many ideas have been reinforced and structural models have been revised. Exciting new advances toward the treatment of prion disease have resulted.

THEORY OF PRION DISEASES

The inherited prion diseases include Gerstmann-Sträussler-Scheinker disease (GSS), familial Creutzfeldt-Jakob disease (fCJD), and fatal familial insomnia (FFI). These patients present with characteristic clinical and neuropathologic findings as early as their third or fourth decade of life, and their family histories are compatible with an autosomal dominant pattern of inheritance (Chapter 14). Molecular genetic studies argue that

Prion Biology and Diseases, 2nd Ed. ©2004 Cold Spring Harbor Laboratory Press 0-87969-693-1/04

these diseases are caused by mutations in the prion protein (PrP) gene based on high LOD scores for 5 of the 30 known mutations (Hsiao et al. 1989; Dlouhy et al. 1992; Petersen et al. 1992; Poulter et al. 1992; Gabizon et al. 1993; Mastrianni et al. 1996; Windl et al. 1999; Finckh et al. 2000). As with many inherited disorders, the pathogenesis of the inherited prion disease is due to the aberrant behavior of the protein encoded by the mutant PrP gene. The altered physical properties of mutant PrP probably result from a change in the conformation of the mutant protein akin to an allosteric effect. In other pathologies resulting from protein misfolding events, the magnitude of this conformational change can be quite variable. For example, the conformational change in sickle cell hemoglobin is largely at the level of quaternary structure, whereas there is evidence that the conformational reorganization of transthyretin mutants associated with familial amyloid polyneuropathy (FAP) and the Alzheimer's βAPP fragment occurs at both the tertiary and quaternary structure levels (Lee et al. 1995; Colon et al. 1996; Kelly 1997). Unfortunately, these multimers are long-lived, exhibit pathologic properties, and have a histopathologic record of their existence. For instance, sickled hemoglobin corrupts the rheologic properties of red blood cells, leading to disruption of capillary blood flow, while pathologic aggregates of transthyretin deposit in the peripheral nerves. Considered in this context, it is not surprising that the conformation of the normal cellular isoform of the wild-type (wt) prion protein (PrP^C) is distinct from the disease-causing isoform of the mutant prion protein (PrP^{Sc}/fCJD, PrP^{Sc}/FFI, PrP^{Sc}/GSS) in both conformation and oligomerization states. However, the magnitude of the conformational rearrangement owing to a point mutation is unexpected.

To explain inherited prion diseases, one need only postulate that the protein can exist in two distinct folded conformations, one that prefers a monomeric state and a second that multimerizes, in which the wild type exhibits a dramatic preference for the monomeric state and the mutant preferentially adopts the multimeric state. The origin of this distinction could be kinetic or thermodynamic. Either the differential stability of the wt and mutant proteins in the monomeric and multimeric states is large, or a kinetic barrier that essentially precludes the conversion of the wt monomer is abrogated by the disease-causing mutations. These two scenarios are contrasted in Figure 1.

Results from a variety of site-directed mutagenesis studies of protein stability suggest that the impact of a single point mutation on the free energy of folding is unlikely to exceed 2–3 kcal (Matthews 1996). The work of Glockshuber and colleagues on the carboxy-terminal fragment of mouse PrP suggests that these mutations destabilize the helical conformation of PrP by 1–1.5 kcal (Liemann and Glockshuber 1999). Studies by Wüthrich

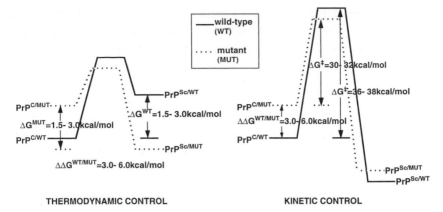

Figure 1. Illustration of the distinction between thermodynamic and kinetic models for the energetics of the conversion of PrP^C carrying the wild type (WT) and mutant (MUT) sequences into PrP^{Sc}. ΔG is the free-energy difference between the PrP^C and PrP^{Sc} states and $\Delta G^{\ddagger}$ is the activation energy barrier separating these two states. $\Delta\Delta G^{WT/MUT}$ is the difference between ΔG^{WT} and ΔG^{MUT}. The free-energy diagrams for the WT sequences are shown by solid lines and for the mutant sequences by broken lines.

and colleagues on the structure of mutants of human PrP indicate that mutations in the carboxyl terminus have little effect on the local structure (Calzolai et al. 2000). If the simple thermodynamic model was operative and the conversion of monomeric PrP^C into multimeric was controlled by the differences in the free energies of the ground state, one would expect that wt PrP^{Sc} production would be approximately 1% as efficient as mutant PrP^{Sc} formation. Given the resistance of the core of wt PrP^{Sc} to proteolytic digestion, PrP^{Sc} would accumulate in the wt setting due to the difficulties associated with the metabolic clearance of this molecule. This scenario is at odds with neuropathologic and epidemiologic data on the incidence of sporadic CJD (sCJD) which occurs in approximately one patient per million of population. Ionizing irradiation experiments have suggested that the minimally infectious PrP^{Sc} particle is a dimer or perhaps a trimer (Bellinger-Kawahara et al. 1988). If the energetics of oligomer formation is simply the sum of the monomeric terms, then wt PrP^{Sc} formation would be 0.01% as likely as mutant PrP^{Sc} production. This level of infectivity would be detectable using a variety of immunoassays and is in contrast to experimental results. For example, histoblots of brain slices taken from uninfected mice demonstrate no protease-resistant PrP (Taraboulos et al. 1992). Sensibly, the cooperativity component of oligomer formation is unlikely to be substantially different for the wt and mutant forms, because disease-causing mutations exist in several distinct regions of the sequence and are distributed throughout the core of the structure (Huang et al. 1994; Riek et al. 1996; James et al. 1997; Knaus et al. 2001). Thus, the cooperativity component of the free-energy difference between the cellular and pathogenic isoforms of the wt and mutant proteins ($\Delta\Delta G^{WT/MUT}$) is unlikely to exceed 2.0–3.0 kcal/mole. Under these extreme assumptions, wt PrP^{Sc} in normal cohorts should be 10^{-6} as common as mutant PrP^{Sc} in carriers from affected families. This level of wt PrP^{Sc} would be detected in bioassays of infectivity, a prediction that is at odds with a number of inoculation studies of PrP derived from natural and recombinant sources (Mehlhorn et al. 1996; Kaneko et al. 1997a).

In contrast, kinetic control over the conversion of PrP^{C} to PrP^{Sc} provides a simple explanation for the observed clinical and experimental results. PrP^{Sc} would need to be only marginally more stable than PrP^{C} ($\Delta G = -2.0$ to -3.0 kcal/mole). To explain the normal absence of PrP^{Sc}, the energetic barriers separating the two states would need to be quite large ($\Delta G^{\ddagger} = 36$ to 38 kcal/mole assuming a transition-state argument). In this setting, a small change in the activation barrier (e.g., $\Delta\Delta G^{\ddagger WT/MUT} = 2.0$ to 3.0 kcal) would result in ~100-fold slowing of the rate of conversion. Thus, the 30- to 40-year prodromal period for inherited prion disease in the mutant setting would become 3000–4000 years with the wt protein if this was the rate-limiting step in disease progression. Cooperative effects could further amplify the distinction between the normal cellular and disease-associated isoforms. For example, if the disease-associated isoform was a dimer, then interactions between the monomeric components could provide additional stability from the thermodynamic perspective. The disease-associated isoform could also affect the kinetic aspects of the conversion process by acting as a template that lowers the activation barrier ($\Delta G^{\ddagger}$) for the conformational change in a manner reminiscent of the way an enzyme's active site orchestrates the positioning of substrates to speed the rate of a reaction. We expect that a misfolded dimer or trimer could lower the activation barrier for the misfolding of subsequent monomers, accelerating the progression of the disease following its initiation. In this setting, the disease-associated isoform would also be disease-causing if it can traffic to the site of PrP production or accumulation, because its presence would dramatically enhance the likelihood of conversion of the normal cellular isoform.

In this context, the sporadic occurrence of CJD could arise for two reasons. First, a somatic cell mutation could give rise to a mutant PrP that would prefer the conformation of the disease-causing isoform (Prusiner 1989). Initially, propagation of the mutant PrP^{Sc}-like conformer would be limited to the cell in which the somatic mutation had occurred. Either within that cell or within surrounding cells, this mutant PrP^{Sc}-like conformer would have to be capable of triggering the conversion of wt PrP^{C} into PrP^{Sc}. If the particular somatic mutation produces a PrP^{Sc}-like conformer that cannot interact with wt PrP^{C}, then prion propagation will not occur and disease will not develop. This scenario of sporadic disease caused by a somatic mutation differs from the inherited prion diseases in which mutant PrPs with the characteristics of PrP^{Sc} accumulate. Because extracts from the brains of patients carrying the D178N, E200K, or V210I point mutation have transmitted to mice expressing the wt chimeric MHu2M transgene (Telling et al. 1996b; Mastrianni et al. 2001), it seems likely that any of these point mutations could initiate sporadic prion disease. In contrast, human prions carrying the P102L mutation could not be transmitted to Tg(MHu2M) mice, which argues that this mutation could not initiate sporadic prion disease. Although expression of the P102L mutation in either humans or mice causes neurodegeneration, detection of these prions was greatly facilitated when mice expressing PrPs carrying the same mutation were used as recipients of inocula from ill humans or Tg mice (Hsiao et al. 1994; Telling et al. 1995, 1996a). A second explanation for sporadic disease is a corollary of the kinetic ideas advanced in the discussion of inherited disease. The kinetic barrier separating the cellular and disease-causing conformers of PrP provides only a stochastic barrier that will be crossed given sufficient time. Although this event will be vanishingly rare in any individual human lifetime, following the logic of the ergodic theorem, the likelihood of a rare event will increase as the size of the population enlarges. In the case of sporadic CJD with an incidence of one per million people, a barrier height of 36–38 kcal/mole would be required following a transition-state argument. Once this misfolded conformer is formed, if it cannot be cleared by proteases, it would be available to act as a template to direct the rapid replication of the disease-causing conformer. If the infectious efficiency of an individual PrP^{Sc} oligomer was small, then this would act as a prefactor in the rate equation that would lower the expected $\Delta G^{\ddagger}$ for a single PrP^{Sc} formation event. Recent evidence suggests that newly formed PrP^{Sc} is protease-sensitive. Clearance provides another element to consider in the kinetic master equation for this system.

From the perspective of inherited and sporadic neurodegenerative diseases, Alzheimer's and the prion diseases could share similar pathogenic mechanisms. However, a point of departure arises from the transmissibility of the prion diseases, which has not been demonstrated for Alzheimer's disease (Goudsmit et al. 1980; Godec et al. 1994). This apparently fundamental difference has been recently challenged by work on mouse senile amyloidosis in which disease could have been induced by injection of amyloid fibrils (Xing et al. 2002). In prion diseases, it is clearly established that inoculation of tissues from sick animals causes disease in the recipient host. The infectious pathogen can be purified and treated with reagents that modify or hydrolyze polynucleotides without loss of infectivity. Although prions are remarkably resistant to proteolytic degradation, they can be inactivated by prolonged digestion with proteases or by exposure to high concentrations of salts known to denature proteins, such as guanidinium thiocyanate (GdnSCN) (Prusiner et al. 1981a,b, 1993b).

Although these initial biochemical results were viewed with great skepticism, it has become increasingly clear how this protein could replicate. The mechanism of infectious prion disease follows from the second explanation for sporadic disease with the minor modification that the initiation of the process is not truly stochastic at a molecular level but relates to facilitated industrial or ritualistic cannibalism (Gajdusek 1977; Wilesmith et al. 1991). The efficiency of the infection relates to the titer of the inoculum, the mode of entry, and the frequency of exposure; intracerebral inoculation of a large dose of prions most efficiently initiates prion replication (Prusiner et al. 1985). In contrast, a single ingestion of foodstuffs containing a small dose of prions is likely to be exceedingly inefficient. Recent results from Aguzzi and coworkers (Klein et al. 2001) suggest that lymphocytes play a key role in trafficking PrP^{Sc} from peripheral compartments to the central nervous system. The efficiency of this process is also governed by the strength of the interaction of PrP^{C} with PrP^{Sc}. When both isoforms contain the same sequence (homotypic interaction), experimental data confirm that conversion is most likely (Prusiner et al. 1990). When the sequences are different, especially in certain regions of the structure, conversion is less likely (Scott et al. 1993; Kocisko et al. 1995). The inefficiency of heterotypic conversion is commonly referred to as the "species barrier" and has been used to explain why humans have not contracted scrapie from sheep and why nontransgenic mice are largely resistant to human PrP^{Sc}/CJD inocula (Telling et al. 1995; Prusiner 1997). The recent cases of variant (v) CJD caused by bovine spongiform encephalopathy (BSE) prions represent a "violation" of the species barrier. In fact, this is entirely expected from the kinetic analysis of PrP^{Sc} replication. Ovine and bovine PrP differ by one octarepeat and 7 residues, with the bovine sequence more similar to human PrP in the region required for infectivity (residues 90–231). A 2–3 kcal change in $\Delta G^{\ddagger}$ could account for the incidence of vCJD, of about 100 cases for 50 million inhabitants exposed to BSE prions in the UK, versus none described from sheep scrapie.

A REPLICATION CYCLE FOR PrP^{Sc}

A simple replication cycle for PrP^{Sc} can be constructed. PrP^C exists in equilibrium with a second state, PrP*, which is best viewed as a transient intermediate that participates in PrP^{Sc} formation either through an encounter with PrP^{Sc} or with another PrP* molecule. Under normal circumstances, PrP^C dominates the conformational equilibrium. With infectious diseases, PrP^{Sc} is supplied exogenously. It can bind PrP* to create a heteromultimer that can be converted into a homomultimer of PrP^{Sc} (see Fig. 2A). Although many proteins are known to bind PrP (Edenhofer et al. 1996; Rieger et al. 1997; Schmitt-Ulms et al. 2001; Spielhaupter and Schatzl 2001), genetic evidence points to the existence of at least one auxiliary factor (protein X) in this conversion (Telling et al. 1995; Kaneko et al. 1997b). Protein X preferentially binds PrP^C and is liberated upon the conversion of PrP* to PrP^{Sc}. Protein X can then be recycled and join another heteromultimeric complex. The homomultimer can dissociate to form two replication-competent templates creating exponential growth of the PrP^{Sc} concentration. In inherited disease, the concentration of PrP* rises due to either the destabilizing effect of the mutation on PrP^C or the increased stability of a PrP^C or PrP* multimer. This increases the likelihood of the presence of a PrP*/PrP* complex that can form PrP^{Sc}/PrP^{Sc} (see Fig. 2B) and initiate the replication cycle. Sporadic disease requires merely a rare molecular event, formation of the PrP*/PrP* complex (Fig. 2B), or a somatic cell mutation that follows the mechanism for the initiation of inherited disease. Once formed, the replication cycle is primed for subsequent conversion.

From an analysis of the thermodynamics and kinetics of prion replication and the replication cycle for the inherited, sporadic, and infectious scenarios, several inferences can be made about the biophysical properties of the normal cellular and disease-causing PrP isoforms.

1. PrP^{Sc} replication requires the presence of the PrP gene in the host cell to direct PrP^C synthesis (Büeler et al. 1993; Prusiner et al. 1993b; Manson et al. 1994; Sakaguchi et al. 1995). However, although PrP^{Sc} replication requires a PrP gene in the host cell, this gene does not need to be carried by the infectious pathogen (Oesch et al. 1985).

2. PrP^{Sc} must be more stable than PrP^C, and a sensible origin for this distinction is an extensive network of intermolecular interactions between PrP monomers in a PrP^{Sc} multimer (Bellinger-Kawahara et al. 1988). Protease resistance could be a corollary of this increased stability and not necessarily the origin of the increased metabolic stability of PrP^{Sc} (Prusiner et al. 1982; McKinley et al. 1983).

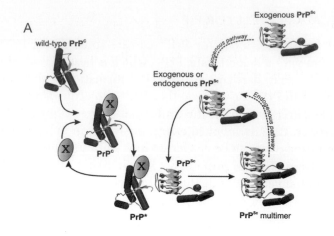

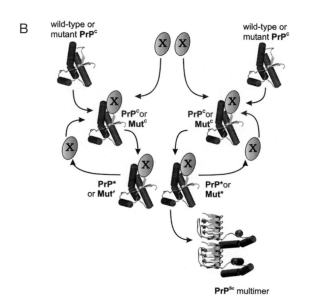

Figure 2. Initiation and replication of PrP^{Sc} synthesis. (*A*) Exogenous or endogenous PrP^{Sc} initiates PrP^{Sc} synthesis by binding to a PrP$^{C/*}$/X complex. Facilitated by protein X and directed by the PrP^{Sc} template, PrP* changes conformation and forms PrP^{Sc}. When PrP^{Sc} forms, the heteromultimeric complex dissociates, yielding recycled protein X and endogenous PrP^{Sc}. The newly generated PrP^{Sc} can then facilitate two replication cycles leading to an exponential rise in PrP^{Sc} formation. (*B*) Initiation of the replication process in spontaneous disease and in inherited disease. Wt PrP (PrP^{C}) or mutant PrP (Mut^{C}) can bind protein X and form the PrP*/X/PrP*/X (or Mut*/X/Mut*/X) encounter complex, which can then form PrP$^{Sc/Mut}$ in the absence of a PrP^{Sc} template. Once this event occurs, replication follows the pattern in panel *A*.

3. A large conformational distinction between PrP^C and PrP^{Sc} would create a substantial kinetic barrier rendering prion disease extremely uncommon in the wt setting. As a corollary, some region or regions of PrP must exhibit extreme conformational plasticity (Nguyen et al. 1995a; Zhang et al. 1995, 1997).

4. For PrP^{Sc} to provide a useful and efficient template to facilitate PrP^C conversion, the molecular interaction between PrP^{Sc} and PrP^C must be quite specific (Prusiner et al. 1990; Scott et al. 1993; Kocisko et al. 1995). Thus, differences in the sequences of PrP^C and PrP^{Sc} should disrupt or attenuate conversion, creating a species barrier.

5. When the PrP gene carries amino acid substitutions that destabilize the protein in the PrP^C isoform or stabilize PrP^* and lower the barrier to interconversion by stabilizing the transition state, the incidence of inherited disease should rise (Chapman et al. 1994; Spudich et al. 1995). Other amino acid substitutions that do not alter the stability of PrP^C, PrP^*, or the transition state significantly will provide sites for polymorphisms between or within species (for review, see Prusiner and Scott 1997).

6. If conformational conversion provides the rate-limiting step for prion replication, disease progression should depend on the concentration of PrP (Prusiner et al. 1993b; Büeler et al. 1994; Carlson et al. 1994; Manson et al. 1994). As this is most logically a first-order process, halving the concentration of PrP should double the time to disease and doubling the concentration of PrP should halve the time to disease. The standard deviation of measurements of time to disease (t) should be $t_{1/2}$.

7. Small molecules that stabilize the PrP^C isoform could act as therapeutic agents by decreasing the concentration of PrP^*, thereby slowing PrP^{Sc} replication (Cohen et al. 1994). Similarly, molecules that abrogate the PrP/protein X interaction or stabilize it sufficiently to prevent protein X recycling could have therapeutic potential (Kaneko et al. 1997b; Prusiner 1997; Perrier et al. 2000).

We return to these inferences in the context of the experimental results that follow.

PrP^C BINDS PrP^{Sc} DURING PRION FORMATION

A cardinal feature of prion transmission studies is the existence of a "species barrier" (Pattison 1965). That is, prions derived from a particular host species are more effective in transmitting disease upon inoculation into an animal of the same species than into one of a more evolutionarily distant species. This concept can be explicitly tested by creating mice expressing a Syrian hamster (SHa) PrP transgene (Scott et al. 1989). When inoculated with mouse PrP^{Sc} ($MoPrP^{Sc}$), this transgenic animal produces $MoPrP^{Sc}$. In contrast, when transgenic (Tg) mice expressing SHaPrP were inoculated with SHa prions containing $SHaPrP^{Sc}$, the Tg animals produced $SHaPrP^{Sc}$ (Prusiner et al. 1990). Thus, the transgene product $SHaPrP^{C}$ must recognize and bind $SHaPrP^{Sc}$ to template a sequence-specific conversion. A similar situation is obtained with the $MoPrP^{C}/MoPrP^{Sc}$ interaction. Thus, homotypic interactions between identical PrP sequences in the PrP^{C} and PrP^{Sc} isoforms are more favorable than the heterotypic alternatives (Prusiner 1991). This principle of prion replication was extended using chimeric SHa/Mo PrP transgenes in which only the central domain (residues 96–167) rendered the mice susceptible to SHa prions (Scott et al. 1992, 1993). Subsequent biochemical studies demonstrated that PrP^{C} binds PrP^{Sc} respecting sequence-specific preferences (Kocisko et al. 1994, 1995; Kaneko et al. 1995). These studies provide evidence in support of inference 4 concerning the specificity of the PrP^{C}/PrP^{Sc} heterodimer interface. Notably, although Tg(BoPrP) mice, bearing the bovine gene, are readily infected by BSE prions, they can also easily be infected by human vCJD prions (Scott et al. 1999) although these inocula transmit relatively inefficiently in Tg(HuPrP) mice (Hill et al. 1997). This suggests that, in addition to the sequence specificity (species barrier), a conformational specificity (strain barrier) exists, extending inference 4 to conformational properties of PrP^{Sc} and the ability of a specific strain to template preferentially a particular PrP sequence.

PROTEIN X IN PrP^{Sc} FORMATION

On the basis of the results with chimeric SHa/Mo PrP transgenes, mice expressing chimeric human–mouse PrP transgenes (MHu2M) were created. These Tg mice that coexpressed MHu2M and $MoPrP^{C}$ were susceptible to human (Hu) prions from the brains of patients who had died of inherited, sporadic, and infectious prion diseases, whereas mice coexpressing $HuPrP^{C}$ and $MoPrP^{C}$ were resistant to Hu prions (Telling et al. 1995). Only when Tg(HuPrP) mice were crossed onto the $Prnp^{0/0}$ background did they become susceptible to Hu prions. To achieve comparable incubation times, the expression of HuPrP needed to be at least 10-fold higher than that in Tg(MHu2M)$Prnp^{0/0}$ mice. These findings could be interpreted in terms of an auxiliary macromolecule (protein X) that facilitates the conversion of PrP^{Sc} formation, or a heterotypic interaction in which MoPrP blocks the formation of HuPrP^{Sc}. We consider the latter unlikely as it would need to be unidirectional; HuPrP does not block the formation of MoPrP^{Sc}, and homotypic interactions, HuPrP-HuPrP^{Sc} or MoPrP-PrP^{Sc}, are more stable than heterotypic interactions.

To explain the resistance of Tg mice expressing both MoPrP and HuPrP to Hu prions, MoPrP^{C} must bind to protein X with a higher avidity than does HuPrP^{C}. Because the conversion of MHu2M PrP^{C} into PrP^{Sc} was only weakly inhibited by MoPrP^{C}, it was surmised that either the amino- or carboxy-terminal domain of PrP, but not the central domain, binds to protein X (Telling et al. 1995). Since the amino terminus of PrP is not required for either transmission or propagation of prions (Rogers et al. 1993; Fischer et al. 1996; Muramoto et al. 1996), it was then postulated that the binding site for protein X lies at the carboxyl terminus (Telling et al. 1995). To test this hypothesis, a chimeric Hu/Mo gene was constructed in which the amino-terminal and central domains were composed of MoPrP and the carboxy-terminal region of HuPrP. This construct (M3Hu PrP) did not support PrP^{Sc} formation in scrapie-infected mouse neuroblastoma (ScN2a) cells, which suggests that its carboxyl terminus carrying the HuPrP sequence did not bind to Mo protein X. Subsequent studies of the PrP^{C}/protein X interaction in ScN2a cells have shown that MoPrP residues 167, 171, 214, and 218 are essential for protein X binding (Kaneko et al. 1997b). These results have been recently confirmed by work on transgenic animals involving PrP genes mutated at the expected protein X-binding sites (Perrier et al. 2002). Remarkably, not only were Tg(MoPrP,Q167R)$Prnp^{0/0}$ mice protected from infection, but the Tg(MoPrP,Q167R)$Prnp^{+/+}$ mice were partially protected from it, showing greatly slowed formation of PrP^{Sc} (Perrier et al. 2002). A similar effect was observed for Tg(MoPrP,Q218K)$Prnp^{0/0}$ and Tg(MoPrP,Q218K)$Prnp^{+/+}$ mice in which none of the transgenic animals showed signs of disease after 300 days. These results demonstrate a dominant-negative effect of the transgene, suggesting an increased affinity of the mutant PrP for protein X that would interfere with the "normal" conversion to PrP^{Sc}.

Efforts to identify protein X continue. Although many proteins have been shown to bind PrP, the key test for an auxiliary protein is whether a knockout mouse will be relatively resistant to PrP^{Sc} infection. For example, N-CAM has been shown to bind PrP. However, N-CAM–knockout mice develop infection with an incubation time indistinguishable from wt mice (Schmitt-Ulms et al. 2001).

EXPERIMENTAL AND COMPUTATIONAL STUDIES
OF PrP^C AND PrP^{Sc}

The failure to detect a polynucleotide associated with the infectious prion particle created a "replication" conundrum (Prusiner 1982, 1984, 1991). Purification of PrP 27-30 (Prusiner et al. 1982, 1983), the 27- to 30-kD, protease-resistant core of PrP^{Sc}, and subsequent microsequencing provided sufficient partial sequence information (Prusiner et al. 1984) to discover the endogenous PrP gene (Chesebro et al. 1985; Oesch et al. 1985). This gene resides on chromosome 20 in humans and on the syntenic chromosome 2 of mice (Sparkes et al. 1986). This is consistent with inference 1 on the requirements of prion replication. Genomic DNA sequencing revealed that the entire PrP coding region is contained in a single open reading frame (Basler et al. 1986). The sequence revealed many well-known features, including a signal sequence and two sites for N-linked glycosylation. In addition, several unusual features were noted, including an eight-residue repeating sequence (octarepeat) PHGGGWGQ that is unlike any known protein structural motif, and an alanine-rich region AGAAAAGA for residues 113 to 120. Until recently, no extended regions of the PrP sequence were recognizably analogous to non-PrP gene products (Bamborough et al. 1996). Subsequently, we uncovered a PrP homolog downstream of the PrP gene, called doppel (Dpl), which appears to have tissue distribution and a physiological role quite distinct from those of PrP (Moore et al. 1999). Although the sequence identity is low (~20%), NMR structures of PrP and Dpl show very similar overall topology (Mo et al. 2001).

Although PrP^C and PrP^{Sc} share a common sequence and pattern of posttranslational modification as judged by a variety of biochemical and mass spectroscopic studies, substantial differences at a structural level have been demonstrated between these two conformers (Table 1). Although the magnitude of these conformational distinctions was unexpected by most, it is clear that this feature is an essential aspect of prion biology (see inference 3).

Biochemical Characterization

PrP 27-30 was purified as the protease-resistant core of the major, and probably the only required, component of the infectious prion particle (Prusiner et al. 1982, 1983). PrP 27-30 aggregates into rod-shaped polymers that are insoluble in aqueous and organic solvents as well as non-ionic detergents. These purified prion rods exhibit the tinctorial and ultrastructural properties of amyloid (Prusiner et al. 1983). PrP 27-30 is

Table 1. Structural differences between PrP^C and PrP^{Sc}

Property	PrP^C	PrP^{Sc}	PrP^{Sc}106
Protease resistance	no	stable core containing residues 90–231	stable core containing residues 90–140, 177–231
Disulfide bridge	yes	yes	yes
Molecular mass after deglycosylation	16 kD (rPrP[90–231])	16 kD (PrP 27–30)	12 kD
Glycosylation	2 N-linked sugars	2 N-linked sugars	2 N-linked sugars
Glycoforms	mixture of un-, mono-, and diglycosylated forms	mixture of un-, mono-, and diglycosylated forms	only diglycosylated
Secondary structure	dominated by α-helices	rich in β-structure	rich in β-structure
Sedimentation properties	consistent with monomeric species	multimeric aggregated species (PrP 27-30)	multimeric aggregated species
Accessible epitopes[a]	109–112 138–155 225–231	225–231	225–231
Free energy of stabilization	$\Delta G = -6$–8 kcal/mole		
Secondary structure	at pH 4.5[b] α: 144–154; 179–193; 200–217 β: 128–131; 161–164 at pH 5.2[c] α: 144–157; 172–193; 200–227 β: 129–131; 161–163		

[a]Peretz et al. 1997.
[b]Riek et al. 1996.
[c]James et al. 1997.

stable at high temperatures for extended periods of time, consistent with the unusual stability of prion infectivity by comparison to globular proteins from mesophiles (see inference 2). Protein denaturants (e.g., 3 M GdnSCN) that modify the structure of PrP 27-30 also inactivate prion infectivity (Prusiner et al. 1993a). In contrast, prion infectivity resists inactivation by reagents that disrupt nucleic acid polymers including nucleases, psoralens, and UV irradiation (Alper et al. 1967; Latarjet et al. 1970; Diener et al. 1982; Prusiner 1982; Bellinger-Kawahara et al. 1987a,b). Microsequencing of the amino terminus of PrP 27-30 (Prusiner et al. 1984) led to the cloning and sequencing of the PrP gene (Chesebro et al. 1985; Oesch et al. 1985). The amino terminus of PrP 27-30 corresponds to approximately codon 90 in the full-length coding sequence (Basler et al. 1986; Locht et al. 1986).

Under physiologic conditions, cells synthesize PrP in the endoplasmic reticulum (ER) with the cleavage of a hydrophobic leader sequence (residues 1–22). Asparagine-linked carbohydrates that were attached to residues 181 and 197 in the ER are remodeled in the Golgi as PrP transits to the cell surface (Endo et al. 1989). A glycosylphosphatidylinositol (GPI) moiety is attached to residue 231 as 23 carboxy-terminal residues are removed in the ER (Stahl et al. 1990). The biogenesis of PrP is complicated by the presence of a stop transfer signal contained within residues 95–110 (Yost et al. 1990). Whereas the majority of PrP biosynthesis leads to a GPI-anchored form, the channel for PrP export into the ER may disassemble during biogenesis, resulting in a transmembrane form of PrP (De Fea et al. 1994; Hegde et al. 1998).

PrP secreted to the cell surface under normal conditions is known as PrP^C, a soluble, monomeric protein containing a single disulfide bridge between residues 179 and 214. This protein was initially purified from hamster brain in its glycosylated form (Turk et al. 1988; Pan et al. 1992, 1993) and subsequently produced by recombinant sources as either the truncated 142-residue molecule SHaPrP(90–231) corresponding to PrP 27-30, the nearly full-length molecule SHaPrP(29–231) (Mehlhorn et al. 1996; Donne et al. 1997; Riek et al. 1997), or the full-length MoPrP(23–231) (Hornemann et al. 1997). The recombinant protein can be oxidized and solubilized in guanidinium hydrochloride (GdnHCl) between pH 5 and pH 8. Dilution of the denaturant and incubation at room temperature for 2–12 hours yields a folded material that shares the spectral and immunologic properties of PrP^C purified from SHa brain (Mehlhorn et al. 1996; Peretz et al. 1997; Zhang et al. 1997). Thermal and GdnHCl denaturation of recPrP(90–231) indicates a three-state process with a well-defined intermediate (Zhang et al. 1997). The free-energy change between the folded and intermediate states was calculated to be

1.9 ± 0.4 kcal/mole. The second transition to the unfolded state was associated with a free-energy change of 6.5 ± 1.2 kcal/mole. PrP^C unfolds with a T_m of 50–55°C, substantially lower than PrP^{Sc}. Analytical ultracentrifugation studies reveal that recPrP(90–231) is monomeric with a monomer–dimer equilibrium association constant of 5.4 × 10^{-5} M^{-1}. recPrP(90–231) refolded under acidic conditions yields a molecule with less helicity as judged by circular dichroism (CD) spectroscopy. This form has a greater tendency to aggregate (Zhang et al. 1997).

In contrast, PrP^{Sc}, the full-length infectious conformer of PrP^C, has a tendency to form aggregates but not amyloid fibrils (McKinley et al. 1991). Attempts to identify a covalent distinction between PrP^C and PrP^{Sc} have been unsuccessful (Stahl et al. 1993). In both PrP^C and PrP^{Sc}, the two cysteines form a disulfide bridge that is needed for PrP^{Sc} formation (Turk et al. 1988; Muramoto et al. 1996). PrP^{Sc} can be denatured with GdnSCN, but refolding conditions that permit recovery of prion infectivity have not been identified (Prusiner et al. 1993a). In the absence of conditions for reversible unfolding of PrP^{Sc}, it has not been possible to establish the relative stabilities of PrP^{Sc}, PrP^C, and the unfolded state. However, the difference in T_m values between the molecules suggests that PrP^{Sc} is more stable than PrP^C. This is consistent with inference 2.

Unfolding–refolding experiments on recombinant MoPrP have shown that it can reach two folded states, one being α-rich, the other β-rich (Baskakov et al. 2001). Refolding to the α-isoform is a fast process that occurs under classical refolding conditions (dilution from 10 M urea, low salt), whereas direct refolding to the β-isoform requires high-salt conditions and is slower by several orders of magnitude. Following prolonged incubation of the α-form in high-salt buffer, one observes a slow conformational transition to the β-rich form, indicating that the latter is thermodynamically more stable. Notably, this transition is concentration-dependent and the β-form is probably not a global energy minimum, as it can further oligomerize into amyloid rods. These data indicate that the folding of PrP is under kinetic control strongly favoring, under normal physiologic conditions, the α-form (PrP^C-like) which would be separated from the β-form (PrP^{Sc}-like) by a large energy barrier.

Removal of a 36-residue stretch from PrP 27-30 resulted in the production of a "miniprion" called $PrP^{Sc}106$ (Muramoto et al. 1996). A Tg(PrP106)$Prnp^{0/0}$ mouse produced disease about 300 days after inoculation with full-length RML prions, but this incubation period was reduced to ~66 days using $PrP^{Sc}106$, on repeated passage, demonstrating that $PrP^{Sc}106$ is fully capable of prion infection but that the deletion introduced an artificial transmission barrier (akin to a species barrier) (Supattapone et al. 1999). Unstructured at low concentrations, recombinant unglycosylated PrP106 (rPrP106) undergoes a concentration-dependent conformational transition to a β-sheet-rich form (Baskakov et al. 2000). Following the conformational transition, recPrP106 possesses properties similar to those of PrP^{Sc}106, such as high β-sheet content, defined tertiary structure, resistance to limited digestion by proteinase K, and high thermodynamic stability, but it lacks infectivity.

CD and FTIR Spectroscopy

Studies of purified PrP 27-30 showed that it assembled into polymers with the properties of amyloid (Prusiner et al. 1983). Earlier investigations have shown that other amyloids have a high β-sheet content (Glenner et al. 1972; Glenner 1980). In the prion diseases, the amyloid deposits were subsequently shown to contain large amounts of the PrP gene product (Bendheim et al. 1984; DeArmond et al. 1985; Kitamoto et al. 1986; Roberts et al. 1988). Subsequently, Fourier transform infrared (FTIR) spectroscopy was used to measure the high β-sheet content of PrP 27-30 (see Fig. 3) (Caughey et al. 1991; Pan et al. 1993). FTIR and CD studies showed that PrP^{Sc} contains about 30% α-helical structure and about 40% β-sheet (Pan et al. 1993; Safar et al. 1993). These data are in marked contrast to the structural studies of PrP^{C} purified from normal brain (Pan et al. 1993; Pergami et al. 1996) and of recPrP(90–231) (Mehlhorn et al. 1996), which are soluble molecules with substantial α-helical structure (40%) and little β-structure (~3%), by CD and FTIR spectroscopy (illustrated in Fig. 3). Taken together, these studies confirm inference 3 concerning a large conformational distinction between PrP^{C} and PrP^{Sc}.

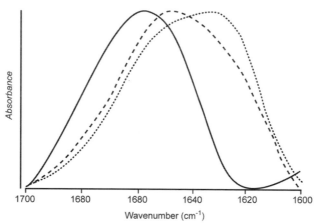

Figure 3. FTIR spectroscopy of prion proteins. Transmission FTIR spectra of the amide I′ band (1700–1600 cm^{-1}) of PrP^{C} (*solid line*), PrP^{Sc} (*dashed line*), and PrP 27-30 (*dotted line*) are shown. (Reprinted, with permission, from Pan et al. 1993 [copyright National Academy of Sciences].)

NMR Spectroscopy and X-ray Crystallography

Using solution phase nuclear magnetic resonance (NMR spectroscopy), several groups have been able to resolve the core structure of PrP^C. Originally, using an *Escherichia coli*-derived MoPrP(121–231) fragment, a three-dimensional structure was obtained by standard heteronuclear spectroscopic methods (Riek et al. 1996). At pH 4.5 and 20°C in the absence of buffer, three α-helical and two very short β-strand regions were identified with a disulfide bridge joining the two carboxy-terminal helices. The structure of SHaPrP(90–231), a recombinant protein derived from *E. coli* with a sequence corresponding to PrP 27-30, was also determined, showing a similar globular region with an unstructured amino-terminal part, between 90 and 124 (James et al. 1997). Interestingly, all attempts to resolve the structure of either the 90–231 region (James et al. 1997; Liu et al. 1999) or full-length PrP(23–231) (Donne et al. 1997, Riek et al. 1997) identified a similar structured region between residues 124 and 231, whereas the amino terminus is consistently extremely flexible. The NMR structure of human PrP demonstrated that this core structure is preserved throughout the species, although some local differences in the backbone conformations are present (Zahn et al. 2000). A typical representation of this globular region is shown in Figure 4. As expected from FTIR spectroscopy, PrP^C is mainly an α-helix-rich protein, with a core

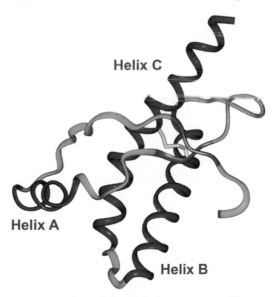

Figure 4. Ribbon representation of the globular structure of human PrP resolved by NMR (Zahn et al. 2000), comprising residues 125–228. The three α-helices are colored in red and the small β-sheet is in blue. The side chains of Cys-179 and Cys-214 forming the disulfide bridge joining helices B and C are shown in yellow.

made of three α-helices: A, B, and C. Helix A is short, spanning residues 144–154 (human sequence), helix B is a regular helix spanning 173–194, and helix C is longer (residues 200–228) but is slightly kinked around 210. Helices B and C are connected by a disulfide bridge between residues Cys-179 and Cys-214. The moderate β-signal observed in FTIR comes from a short antiparallel β-sheet involving residues 129–131 and 161–163.

Notable differences in the backbone conformations are observed when comparing structures from the different fragments (MoPrP[121–231] and SHaPrP[90–231]) or the full-length protein (SHaPrP[29–231]), as helices C and B originally appeared extended in the longer forms compared to the shorter MoPrP(121–231). Chemical shift differences were also observed between SHaPrP(29–231) and recPrP(90–231) at the level of helix B. More recently, comparisons of the NMR structures of HuPrP(23–230), HuPrP(90–230), and HuPrP(121–230) obtained under similar conditions also revealed a better defined structure for the longer fragments or full-length forms (Zahn et al. 2000). These data indicate transient contacts between the flexible amino terminus and the carboxy-terminal helices, suggesting a possible interaction between the octarepeat region and the globular part, at the level of the protein X–binding site (see below) and/or the carboxyl terminus.

Analysis of the NMR structure also reveals that disease-causing mutations are located throughout the globular region. Residues where point mutations lead to human diseases are highlighted in Figure 5A.

Information gained from mutagenesis and NMR studies performed in concert suggests that the side chains of MoPrP residues Gln-167, Gln-171, Val-214, and Gln-218, which correspond to SHaPrP residues Gln-168, Gln-172, Thr-215, and Gln-219, form the site at which protein X binds to PrP^C (James et al. 1997; Kaneko et al. 1997b). The importance of residues 167 and 218 has been recently demonstrated in transgenic studies in which Q167R and Q218K have dominant-negative effects on the wt protein, probably through sequestration of protein X (Perrier et al. 2002). Although sequentially distant, these four residues cluster on one face of the molecule. The glycosylation sites, Asn-181 and Asn-197, are evidently not very near this putative binding site. As mentioned above, a comparison of the structure of rPrP(90–231) with MoPrP(121–231) (Riek et al. 1996) suggests that omission of residues 90–120 destabilizes helix C, resulting in its truncation and consequent disordering of residues 167–176. Perhaps this explains why MoPrP(121–231) is ineligible for conversion into PrP^{Sc} and does not appear to be able to bind to protein X (Muramoto et al. 1996; Zulianello et al. 2000). As shown in Figure 5B, Thr-215 and Gln-219 lie in register one turn apart on helix C and interact with the loop containing residues 167

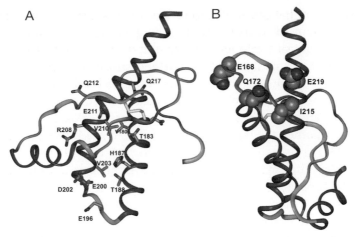

Figure 5. (*A*) NMR structure of human PrP (Zahn et al. 2000) showing the positions of the side chains corresponding to point mutations that lead to human diseases (Gly-131 is not shown). (*B*) Structure of human PrP highlighting the positions of Glu-168, Gln-172, Ile-215, and Glu-219, which correspond to MoPrP residues Gln-167, Gln-171, Val-214, and Gln-218, which have been implicated in the binding of protein X (Kaneko et al. 1997b). This view is rotated along a vertical axis compared to panel *A*.

and 171. This information also allows us to understand the structural basis of genetic resistance to scrapie. Residue 171 is a glutamine in most species. Studies of Suffolk sheep revealed codon 171 to be polymorphic, encoding either glutamine or arginine. All Suffolk sheep with scrapie were found to be Gln/Gln, which indicates that arginine conferred resistance (Westaway et al. 1994). These data suggest that the basic side chain of arginine acts to increase the affinity of PrP^C(R171) for protein X such that PrP^C is not readily released from the complex. Such a scenario in which R171 acts through a dominant-negative mechanism seems likely, since heterozygous Arg/Gln sheep are also resistant to scrapie. Susceptibility to scrapie in other breeds of sheep is also determined largely by the nature of residue 171 (Hunter et al. 1993, 1997a,b; Goldmann et al. 1994; Belt et al. 1995; Clousard et al. 1995; Ikeda et al. 1995; O'Rourke et al. 1997). Equally important is the observation that about 12% of the Japanese population encode lysine instead of glutamic acid at position 219 (Kitamoto and Tateishi 1994). No cases of CJD have been found in people with Lys-219 which, like arginine, is basic (Shibuya et al. 1998).

Recently, a crystal structure of a PrP^C dimer was solved at 2 Å resolution, showing a structured 119–226 region (Knaus et al. 2001). Although the local backbone conformation is similar to that of the (monomeric)

NMR structure, the oligomerization implies a reorganization of the tertiary structure as the disulfide bridge now links the two monomers, with helix C pointing out from each monomer into the other one. Although the authors claim that the covalent dimer is a minor species which is selectively selected by the crystallization process, other studies indicate that the disulfide bridge is exclusively intramolecular and, furthermore, not modified during an $\alpha \rightarrow \beta$ transition in a cell-free assay (Welker et al. 2002). It is unclear whether the covalently linked dimer is a biologically relevant structure.

Although the amino-terminal region appears unstructured in all the NMR structures and in the crystal structure, it is unlikely that the octarepeat region is truly flexible and disordered under biologically relevant conditions, given that full-length and not truncated PrP^C is normally purified from animal brain preparations and that the addition of a sixth octarepeat gives rise to an inherited prion disease in humans.

It is now established that PrP binds copper and is most probably involved in the regulation of copper (Brown et al. 1997; Pauly and Harris 1998). Recent studies clearly demonstrated that copper binding occurs in the amino-terminal region, more precisely in the four PHGGGWGQ octarepeats (Viles et al. 1999; Aronoff-Spencer et al. 2000; Burns et al. 2002). The HGGGW peptide was crystallized bound to copper, showing not only that it is sufficient for Cu^{++} binding, but the X-ray structure at 0.7 Å detailed the exact binding mode of the metal ion to the repeat (Burns et al. 2002). Although this binding mode, with four copper ions per PrP molecule, can occur at extracellular pH, lower pH (as in endosomes) would lower the affinity for the metal, changing the stoichiometry. Copper binding would favor stabilization of the octarepeat region. As a corollary, copper release in the endosome could destabilize PrP and facilitate its conversion to PrP^{Sc} in the presence of PrP^{Sc} multimers.

Antibody Studies

Early studies of distinctions in the antigenic surface of PrP^C and PrP^{Sc} were hampered by the limited antigenicity of PrP^{Sc} (Prusiner et al. 1993b; Williamson et al. 1996). Operationally, only two monoclonal antibodies (mAbs) have been developed that recognize PrP: 13A5 and 3F4 (Barry and Prusiner 1986; Kascsak et al. 1987). Although both bind to PrP^C and denatured PrP^{Sc}, neither binds to folded PrP^{Sc} effectively. We have surveyed a diverse set of Fab fragments of recombinant origin (Peretz et al. 1997). Seven Fabs were identified that bind to distinct linear PrP epitopes: D4, R10, D13, R72, R1, R2, and D2. The last three of these can reliably recognize active PrP^{Sc} as well as PrP^C. Epitope-mapping studies demonstrate that 3F4, D4, R10, and D13 bind to an epitope bounded by residues 94–112 (I) whereas 13A5 and R72 recognize a distinct epitope formed by residues 138–165 (II). The extreme carboxyl terminus of the molecule, residues 225–231, contains a binding site for R1, R2, and D2 (III). The large conformational rearrangement between PrP^C and PrP^{Sc} demonstrated spectroscopically is supported by a clearly differentiable antibody-binding pattern. From epitope-mapping studies, we can now localize the conformationally flexible region to include residues 90–112. This region could extend to residue 138. In contrast, the region between residues 225 and 231 is likely to adopt a similar structure in PrP^C and PrP^{Sc} based on the relatively uniform binding of recFabs R1, R2, and D2 to both isoforms. Whether this conformationally rigid zone extends to include residues 166–224 remains to be determined. This evaluation will require the identification of monoclonal antibodies or Fab fragments that recognize this region. Work by Oesch and colleagues has identified an IgM that appears to recognize PrP^{Sc} specifically (Korth et al. 1997). However, these results have been difficult to replicate by other teams, and various studies are still under way to identify antibodies that distinguish PrP^{Sc}.

Electron Microscopy and Electron Crystallography

Recent progress has been made in characterizing the structure of PrP^{Sc}. It is well known that PrP^{Sc} aggregates into insoluble material. Following cleavage by proteinase K, the protease-resistant core, PrP 27-30, forms rods, which bear all the tinctorial properties of amyloid fibers (Prusiner et al. 1983; McKinley et al. 1991) and resemble the amyloid fibrils found in vivo (DeArmond et al. 1985). In vitro, fibers form only after limited proteolysis and detergent extraction (McKinley et al. 1991; Pan et al. 1993). Typical prion rods are shown in Figure 6. Fiber diffraction studies of these fibers revealed the meridional reflection at 4.7 Å, characteristic of a cross-β structure (Nguyen et al. 1995b), indicating that PrP 27-30 shares the overall structural architecture of amyloids.

Although the high insolubility of PrP^{Sc} prevents its structural determination by X-ray crystallography or NMR spectroscopy, the discovery of two-dimensional crystals in preparations of PrP 27-30 led the way to its investigation by electron crystallography (Wille et al. 2002). Whereas most purifications of PrP 27-30 produce typical amyloid rods, some fractions, from preparations with high titers of infectivity, also contained 2D crystals. These crystals are specifically labeled by the R1, R2, and 3F4 anti-PrP mAbs, demonstrating that the crystals are indeed made of PrP

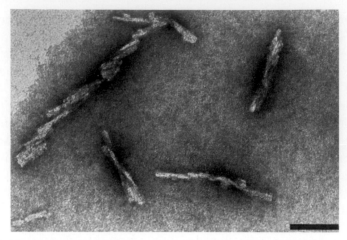

Figure 6. Electron micrographs of typical prion rods obtained from a proteinase-K-treated preparation of Syrian hamster PrP^{Sc} (PrP 27-30). The rods are stained with 2% uranyl acetate. Bar, 100 nm.

molecules. Decoration by the 3F4 mAb, a PrP^{C}-specific antibody, required denaturation by urea, suggesting that the crystal proteins are conformationally similar to PrP^{Sc}. Analysis of the crystals showed that, in several cases, prion rods appear to "emerge" from the crystals, suggesting that prion rods and 2D crystals are alternative polymerization products and that the crystals are infectious. Two-dimensional crystals isomorphous to the PrP 27-30 crystals were also found in preparations of PrP^{Sc}106, allowing direct comparison of the two types of crystals. Electron micrographs of these negatively stained crystals reveal a ~70-Å-wide subunit with an intense electron-dense center owing to bound uranyl ions. A P$_3$ symmetry is observed. Difference maps have been used to locate specific peptides and sugars in the unit cell. Signal observed in PrP 27-30 but absent in PrP^{Sc}106 is likely to arise from the region between residues 141 and 176, which is lacking in PrP106. Conversely, signal solely seen in PrP^{Sc}106 crystals is most probably due to the N-linked sugars, as it has been shown that PrP106 is always diglycosylated whereas PrP 27-30 is found either in di-, mono-, or nonglycosylated forms (Supattapone et al. 1999; Rudd et al. 1999). Subtraction of the PrP^{Sc}106 projection map from the PrP 27-30 projection map consistently showed signal in the center of the subunit, overlapping the uranyl ion staining, suggesting that the 141–176 region is located near the center of the multimeric structure. Difference maps of (PrP^{Sc}106)–(PrP 27-30) revealed that the sugars lie near the subunit boundary. This localization of the N-linked sugars was confirmed by Nanogold labeling of the oligosaccharides. Image processing shows a P_3 symmetry in the unit cells of both crystals. This suggests a trimeric or hexameric (dimer of trimers) structure for PrP^{Sc} and led us to revise the previous structural models developed on indirect structural measurements and extrapolations from biochemical data (Huang et al. 1996). Integration of the new structural requirements with the previously established biological and biochemical constraints were most compatible with a new structural arrangement for PrP^{Sc}, the β-helix (Wille et al. 2002).

Recently, new electron microscopic (EM) analysis of the crystals with a higher resolution microscope provided greater detail about the structure of PrP^{Sc}. We see evidence for a trimer (and not a hexamer) and better resolution in the localization of the sugars and of the peptide deletion (H. Wille et al., in prep.). In addition, the uranyl ions reveal a detailed tridentate staining in the center, in contrast to the large and ill-defined staining observed with older instrumentation. The development of bigger and better crystals by optimization of the crystallization conditions, coupled with high-performance cryo-electron microscopy, should lead to improvements in the structural characterization of PrP^{Sc}. Near-atomic detail has been observed with similar techniques in other systems (Kuhlbrandt et al. 1994; Mitsuoka et al. 1999).

Computational Studies

FTIR data demonstrate that PrP^{Sc} formation is associated with a dramatic increase in β structure compared to PrP^{C}, and also show that the α-helical content remains significant, with about 30–40 residues in α-helical conformations. It is important to note that antibodies directed against the carboxyl terminus of PrP^{C} also recognize PrP^{Sc}. Notably, it has been argued recently that the disulfide bridge linking residues 179 and 214 is indeed intramolecular, and that conversion to a β-rich form does not induce even temporary breakage of the bridge (Welker et al. 2002). It appears therefore likely that a portion of the carboxyl terminus of PrP, including a large part of helices B and C, is structurally preserved in the infectious isoform. This is energetically more favorable than a scenario in which this region should unfold while another part of the protein should refold into an α-helical conformation in order to account for the observed FTIR spectra. Therefore, we expect PrP^{Sc} to be composed of a β-rich region (approximately between residues 90 and 165) and an α-rich region comprising most of helices B and C. With this in mind, we have focused on modeling residues 90–165 in a β conformation.

The direct structural data provided by electron crystallography led us to revise our previously published models of PrP^{Sc}. Increasing evidence has suggested that amyloid β-sheet can follow a parallel architecture. The isomorphous form of PrP 27-30 and PrP^{Sc}106 crystals indicates that these molecules adopt similar structures and that residues 141 and 176 must be close together. The first analysis of the crystal data suggested that PrP^{Sc} could possibly include a β-helical fold (Wille et al. 2002). The β-helical folds are made of parallel β-sheets winding around a helical axis, forming repetitive patterns (Jenkins et al. 1998). They exist either as right-handed or left-handed structures and are found in bacterial and viral proteins. Interestingly, expression of the purely β-helical region of the P22 tailspike protein leads to the formation of fibrous aggregates (Schuler et al. 1999). As β-helices are intrinsically cross-β folds, they are being considered as a generic solution to the molecular arrangement of amyloid fibrils (Jenkins and Pickersgill 2001; Wetzel 2002). Although there may or may not be a generic amyloid tertiary structure, the β-helix could provide a relevant structural motif for a subset of amyloidogenic proteins.

Our group has now gone further in studying the β-helical folds as candidates for the structure of PrP^{Sc} (C. Govaerts et al., in prep.). A detailed survey of known β-helical structures shows that the sequence of PrP is compatible with the residue distribution observed in left-handed β-helices, whereas the right-handed fold appears less likely. Taking advantage of the observed "irregularities" of β-helices, like the bulged turns and protruding loops, it is possible to thread the sequence of PrP from residues 90 to 170 onto a left-handed β-helical fold. Using parts of known structures as guides, we constructed a β-helical model of the β-rich region of PrP^{Sc}. We then connected this part to the carboxy-terminal α-helical region, defined by the NMR studies, to produce a monomeric model of PrP 27-30 as shown in Figure 7.

The new EM data indicate a trimeric arrangement for PrP 27-30 in the 2D crystals. We note that some left-handed β-helical proteins have been crystallized in a trimeric form, in which the β-helices from each monomer converge to form the trimeric interface. Using this arrangement as a template, we have modeled the trimeric assembly by superimposing the atomic coordinates of the individual monomers onto those of the trimer of carbonic anhydrase. Remarkably, the resulting trimer of PrP^{Sc} molecules fits nicely into the images defined by the EM data without need for rescaling the model (Fig. 8). This superimposition satisfies the relevant distance constraints and positions the N-linked sugars in locations that agree with the locations defined by the subtraction maps. We conclude that the trimeric assembly of β-helical models appears to be a very promising candidate for the molecular arrangement of PrP^{Sc}.

A   B

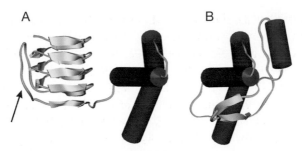

Figure 7. (*A*) Representation of the molecular model of the monomeric PrP^Sc^. The β-strands forming the β-helix are shown as yellow arrows, and the α-helices are represented as red cylinders. The β-helix is built based on templates extracted from known structures, and the carboxy-terminal α-helical region is taken from the NMR structure of SHaPrP (see text). The loop in the β-helix model (indicated by an arrow) was introduced to account for the charged residues in this region that are not compatible with a purely β-helical fold. This loop is specific to PrP 27-30 as it is part of the deleted region of PrP106. (*B*) Corresponding representation of PrP^C^. This view illustrates the proposed dramatic structural differences between the amino-terminal regions of the two conformations, while the carboxyl terminus is structurally conserved.

Figure 8. Superimposition of the trimeric model of PrP^Sc^ (PrP 27-30) onto an electron microscopy projection map obtained from the 2D crystals of PrP 27–30. The map shows the uranyl staining (*black*) with three visible dots in the center. The statistically significant differences between PrP 27 30 and PrP^Sc^106 (corresponding to the deletion of residues 141–176) are shown in maroon, and the signal seen in the PrP^Sc^106 map but not in PrP 27-30 (corresponding to the sugars) is shown in dark blue. The trimeric arrangement of the PrP^Sc^ monomers is based on the trimeric structure of carbonic

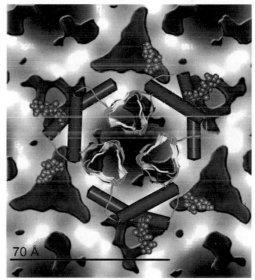

70 Å

anhydrase. The N-linked sugars are shown in blue space-filling representation. The size of the trimeric model matches the unit cell size of the crystals. The sugars of the model colocalize well with the putative positions measured from the subtraction map. The loop (specific to PrP 27-30) overlaps with the signal seen in the PrP 27-30 crystals but absent from the PrP^Sc^106 images. This trimeric model is therefore in good agreement with the structural constraints obtained from the 2D crystals and other biochemical sources.

PRION STRAINS

The existence of prion strains has been considered difficult to reconcile with the protein-only model and the evidence that prion diseases involve protein folding and distinct folded conformations (Bruce and Dickinson 1987; Dickinson and Outram 1988; Prusiner 1991; Ridley and Baker 1996). However, recent results on the creation of new prion strains and their subsequent biochemical characterization have shown how the strain phenomenon is entirely compatible with the protein-only conformational conversion model (Telling et al. 1996b). Prion strains were initially isolated on the basis of different clinical syndromes in goats with scrapie (Pattison and Millson 1961); subsequently, strains were isolated in rodents on the basis of different incubation times and neuropathologic profiles (Fraser and Dickinson 1973; Dickinson and Fraser 1977). New strains have been produced upon passage from one species to another (Kimberlin et al. 1987) or passage from non-Tg mice to mice expressing a foreign or artificial PrP transgene (Scott et al. 1997).

Studies of the drowsy (DY) and hyper (HY) prion strains isolated from mink by passage in Syrian hamsters showed that two strains produced PrP^{Sc} molecules with protease-resistant cores (PrP 27-30) of different molecular sizes as judged by gel electrophoresis (Bessen and Marsh 1994). Since the smaller size of the protease-resistant fragment of the DY strain did not correlate with a particular phenotype, the significance of these findings was unclear (Scott et al. 1997). In seemingly unrelated studies, brain extracts from patients who died of FFI exhibited a deglycosylated, protease-resistant core of PrP^{Sc}/FFI of 19 kD, whereas that found in extracts from patients with fCJD(E200K) and sCJD was 21 kD (Monari et al. 1994). The smaller size of the PrP^{Sc}/FFI fragment was attributed to the distinct sequence of the PrP^{Sc}/FFI protein. Studies on the transmission of prions from patients with FFI, fCJD(E200K), or sCJD to Tg(MHu2M)$Prnp^{0/0}$ mice have shown that different sequences were not required to maintain the structural distinctions as judged by the molecular sizes of the deglycosylated, protease-resistant core of PrP^{Sc} molecules (Telling et al. 1996b). The Tg(MHu2M)$Prnp^{0/0}$ mice inoculated with extracts from FFI patients developed signs of CNS dysfunction about 200 days later and exhibited a deglycosylated, protease-resistant core of PrP^{Sc} that was 19 kD in size, whereas mice inoculated with extracts from fCJD(E200K) patients developed signs at about 170 days and exhibited a PrP^{Sc} of 21 kD. On second passage in Tg(MHu2M)$Prnp^{0/0}$ mice, the animals receiving the mouse FFI extract developed neurologic disease after about 130 days and displayed a 19-kD PrP^{Sc} molecule, and those injected with the mouse fCJD extract exhibited disease after about 170 days and showed a 21-kD PrP^{Sc} molecule. The second passage of these distinct prion strains with identical PrP amino acid sequences demonstrates that the size of the protease-resistant core is a stable feature of the strain. Thus, the two prion strains represent distinct PrP^{Sc} structures that can be enciphered on a proteinaceous template that is stable in passaging experiments. Consistent with work on FFI, a sporadic form of fatal insomnia has been identified that behaves like the FFI strain (Mastrianni et al. 1999).

What is the structural basis of these alternative PrP^{Sc} conformers? Work on diphtheria toxin identified distinct crystal forms that displayed different tertiary and quaternary structures for a single polypeptide sequence (Bennett et al. 1995). To describe this observation, the notion of domain swapping was introduced, in which a region of one monomer displaces the corresponding region in another monomer to create an interlocking molecular handshake (Bennett et al. 1995). This phenomenon has now been observed in a variety of other protein structures with the swapped elements as small as an isolated α-helix or β-strand and as large as an entire folded domain. We suspect that a similar phenomenon is responsible for prion strains.

Imagine that the protease core of PrP^{Sc} has two subdomains joined by a linking region and that PrP^{Sc} exists as a dimer. Figure 9 shows a schematic of the fCJD-derived strain that shields a cleavage site near residue 105 but retains the cleavage site near residue 90. In the FFI-derived strain, the smaller subdomain of one monomer would swap with its partner in the dimer in a manner reminiscent of alternative ribonuclease A structures. In this way, the heterosubdomain interface (90–105:110–231) and the homosubdomain interface (90–105:90–105) are maintained, but the proteolytic remnant is quite different. Moreover, the templates provided by each protease-resistant core would present a distinct interface for the conversion of PrP^{C} to PrP^{Sc} during prion replication. Many lines of evidence suggest that the residues between 90 and 120 play a major role in the creation of the PrP^{C}/PrP^{Sc} interface.

In addition to a structure for PrP^{C} that is distinct from PrP^{Sc}, these results on prion strains argue that there are multiple, approximately isoenergetic, PrP^{Sc} conformers. This is an obvious point of departure from earlier work demonstrating that for most proteins, there was a single folded structure that was uniquely encoded in the sequence (Anfinsen 1973). In an effort to probe the limits of this conformational pluralism for PrP, we have explored this phenomenon on a thermodynamically well-understood model of protein structure, the H-P model on a square and cubic lattice (Chan and Dill 1996). In this model, two amino acid types are allowed, hydrophobic (H) and hydrophilic (P) residues, and the conformations of the chain are limited to those that can be exactly embedded upon the lattice. The energetics of this system are equally discrete, H-H

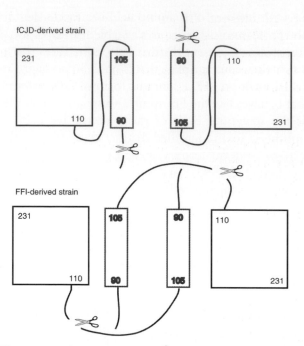

Figure 9. An illustration of two strains of PrP^{Sc} formed by domain swapping. The two schematic structures are approximately isoenergetic but are expected to have distinct proteolytic cleavage patterns.

interactions are worth one favorable energy unit, and all other interactions are negligible. Although this model lacks many features of real polypeptide chains, for relatively short chains, it is possible to enumerate all chain conformations and to identify all sequences with a unique lowest-energy monomeric structure.

For H-P sequences of length 27 on a cubic lattice, one could ask how many sequences that have a unique monomeric structure would prefer a different conformation upon forming a symmetric dimer. With this simplified system, we found that about 5% of all sequences are prion-like in that the lowest-energy structure for the monomer is unlike the most stable dimeric structures (Harrison et al. 1999). Taking this analogy one step further, it is possible to identify domain swaps between the amino- or carboxy-terminal regions of each monomeric member of the dimer that are isoenergetic with the new interleaved dimeric arrangement. This is reminiscent of the experimental work on strains. In this lattice limit, it is easy to see how the strains could act as templates to direct the formation of distinct "scrapie" molecules. It would also follow that proteolysis experiments could reveal different resistant cores. One could also ask how often a mutation to the 27-mer sequence would differentially affect the stability of the dimeric structures. Our preliminary calculations document that this is a relatively common phenomenon, holding true for 20–30% of the residues, especially when residues are mutated from Ps to Hs.

CONCLUSIONS

A wide array of biochemical and biophysical experimental results indicates that a dramatic change in protein conformation is a central feature of the prion diseases. Although the transmissible aspect of prion biology has led to a variety of quite disparate hypotheses about the molecular basis of prion diseases, these mechanistic issues are more easily understood when one begins with the tenet that these diseases are fundamentally autosomal dominant inherited diseases. From a historical perspective, the inherited prion diseases are likely to have been the first to arise, because the stochastic nature of sCJD would seem to require a substantial population of individuals over the age of 60, and cannibalism could lead only to infectious CJD if the victim had preexisting familial or sporadic CJD.

Many autosomal dominant inherited diseases act by distorting the structure of a protein. The experimental results on the prion diseases are entirely compatible with this notion. In general, the change in protein structure could be under kinetic or thermodynamic control. For the prion diseases, the evidence points toward kinetic control. A special feature of the prion diseases is that the disease-causing isoform, PrP^{Sc}, appears to be capable of acting as a template to lower the kinetic barriers that normally separate PrP^{C} from PrP^{Sc}. This provides a simple explanation for the inherited prion diseases that can be easily adapted to explain sporadic and infectious prion diseases.

Structural studies focusing on PrP^{C} have been quite successful, resolving the structure of its globular core with various techniques and providing insights into the role of the amino terminus. However, the function of the protein remains to be determined. Understanding the structure of PrP^{Sc} is more challenging, but recent progress suggests that it is a more realistic goal than ever before. Recent advances in the development of therapeutics for prion diseases (Perrier et al. 2000; Korth et al. 2001) highlight the need for a molecular insight into the conformation of PrP^{Sc}. We believe that additional structural information will facilitate antiprion drug design efforts and help in the generation of PrP^{Sc}-specific antibodies. Finally, the structural information obtained from studies on PrP^{Sc}, and the 2D crystals in particular, could prove to be relevant to other amyloid-forming diseases.

ACKNOWLEDGMENTS

We thank our colleagues for their help with the development of the ideas presented in this manuscript. In particular, we acknowledge A. Borovinsky, S. DeArmond, S. Doniach, D. Eisenberg, D. Engelman, R. Fletterick, P. Harrison, K. Kaneko, Z. Kanyo, S. Marqusee, B. May, M. Michelitsch, V. Perrier, F. Richards, M. Scott, T. Steitz, A. Wallace, D. Walther, A. Borovinsky, B. May, and L. Zulianello. This work was supported by grants from the National Institutes of Health and the Human Frontiers of Science Program. We gratefully acknowledge gifts from the Sherman Fairchild Foundation, the Keck Foundation, the G. Harold and Leila Y. Mathers Foundation, the Bernard Osher Foundation, the John D. French Foundation, and Centeon. C.G. is a fellow of the Belgian American Educational Foundation.

REFERENCES

Alper T., Cramp W.A., Haig D.A., and Clarke M.C. 1967. Does the agent of scrapie replicate without nucleic acid? *Nature* **214:** 764–766.

Anfinsen C.B. 1973. Principles that govern the folding of protein chains. *Science* **181:** 223–230.

Aronoff-Spencer E., Burns C.S., Avdievich N.I., Gerfen G.J., Peisach J., Antholine W.E., Ball H.L., Cohen F.E., Prusiner S.B., and Millhauser G.L. 2000. Identification of the Cu^{2+} binding sites in the N-terminal domain of the prion protein by EPR and CD spectroscopy. *Biochemistry* **39:** 13760–13771.

Bamborough P., Wille H., Telling G.C., Yehiely F., Prusiner S.B., and Cohen F.E. 1996. Prion protein structure and scrapie replication: Theoretical, spectroscopic and genetic investigations. *Cold Spring Harbor Symp. Quant. Biol.* **61:** 495–509.

Barry R.A. and Prusiner S.B. 1986. Monoclonal antibodies to the cellular and scrapie prion proteins. *J. Infect. Dis.* **154:** 518–521.

Baskakov I.V., Legname G., Prusiner S.B., and Cohen F.E. 2001. Folding of prion protein to its native alpha-helical conformation is under kinetic control. *J. Biol. Chem.* **276:** 19687–19690.

Baskakov I.V., Aagaard C., Mehlhorn I., Wille H., Groth D., Baldwin M.A., Prusiner S.B., and Cohen F.E. 2000. Self-assembly of recombinant prion protein of 106 residues. *Biochemistry* **39:** 2792–2804.

Basler K., Oesch B., Scott M., Westaway D., Wälchli M., Groth D.F., McKinley M.P., Prusiner S.B., and Weissmann C. 1986. Scrapie and cellular PrP isoforms are encoded by the same chromosomal gene. *Cell* **46:** 417–428.

Bellinger-Kawahara C., Cleaver J.E., Diener T.O., and Prusiner S.B. 1987a. Purified scrapie prions resist inactivation by UV irradiation. *J. Virol.* **61:** 159–166.

Bellinger-Kawahara C.G., Kempner E., Groth D.F., Gabizon R., and Prusiner S.B. 1988. Scrapie prion liposomes and rods exhibit target sizes of 55,000 Da. *Virology* **164:** 537–541.

Bellinger-Kawahara C., Diener T.O., McKinley M.P., Groth D.F., Smith D.R., and Prusiner S.B. 1987b. Purified scrapie prions resist inactivation by procedures that hydrolyze, modify, or shear nucleic acids. *Virology* **160:** 271–274.

Belt P.B., Muileman I.H., Schreuder B.E.C., Bos-de Ruijter J., Gielkens A.L.J., and Smits M.A. 1995. Identification of five allelic variants of the sheep PrP gene and their association with natural scrapie. *J. Gen. Virol.* **76:** 509–517.

Bendheim P.E., Barry R.A., DeArmond S.J., Stites D.P., and Prusiner S.B. 1984. Antibodies to a scrapie prion protein. *Nature* **310:** 418–421.

Bennett M.J., Schlunegger M.P., and Eisenberg D. 1995. 3D domain swapping: A mechanism for oligomer assembly. *Protein Sci.* **4:** 2455–2468.

Bessen R.A. and Marsh R.F. 1994. Distinct PrP properties suggest the molecular basis of strain variation in transmissible mink encephalopathy. *J. Virol.* **68:** 7859–7868.

Brown D.R., Qin K., Herms J.W., Madlung A., Manson J., Strome R., Fraser P.E., Kruck T., von Bohlen A., Schulz-Schaeffer W., Giese A., Westaway D., and Kretzschmar H. 1997. The cellular prion protein binds copper in vivo. *Nature* **390:** 684–687.

Bruce M.E. and Dickinson A.G. 1987. Biological evidence that the scrapie agent has an independent genome. *J. Gen. Virol.* **68:** 79–89.

Büeler H., Raeber A., Sailer A., Fischer M., Aguzzi A., and Weissmann C. 1994. High prion and PrP^{Sc} levels but delayed onset of disease in scrapie-inoculated mice heterozygous for a disrupted PrP gene. *Mol. Med.* **1:** 19–30.

Büeler H., Aguzzi A., Sailer A., Greiner R.-A., Autenried P., Aguet M., and Weissmann C. 1993. Mice devoid of PrP are resistant to scrapie. *Cell* **73:** 1339–1347.

Burns C.S., Aronoff-Spencer E., Dunham C.M., Lario P., Avdievich N.I., Antholine W.E., Olmstead M.M., Vrielink A., Gerfen G.J., Peisach J., Scott W.G., and Millhauser G.L. 2002. Molecular features of the copper binding sites in the octarepeat domain of the prion protein. *Biochemistry* **41:** 3991–4001.

Calzolai L., Lysek D.A., Guntert P., von Schroetter C., Riek R., Zahn R., and Wüthrich K. 2000. NMR structures of three single-residue variants of the human prion protein. *Proc. Natl. Acad. Sci.* **97:** 8340–8345.

Carlson G.A., Ebeling C., Yang S.-L., Telling G., Torchia M., Groth D., Westaway D., DeArmond S.J., and Prusiner S.B. 1994. Prion isolate specified allotypic interactions between the cellular and scrapie prion proteins in congenic and transgenic mice. *Proc. Natl. Acad. Sci.* **91:** 5690–5694.

Caughey B.W., Dong A., Bhat K.S., Ernst D., Hayes S.F., and Caughey W.S. 1991. Secondary structure analysis of the scrapie-associated protein PrP 27-30 in water by infrared spectroscopy. *Biochemistry* **30:** 7672–7680.

Chan H.S. and Dill K.A. 1996. Comparing folding codes for proteins and polymers. *Proteins* **24:** 335–344.

Chapman J., Ben-Israel J., Goldhammer Y., and Korczyn A.D. 1994. The risk of developing Creutzfeldt-Jakob disease in subjects with the *PRNP* gene codon 200 point mutation. *Neurology* **44:** 1683–1686.

Chesebro B., Race R., Wehrly K., Nishio J., Bloom M., Lechner D., Bergstrom S., Robbins K., Mayer L., Keith J.M., Garon C., and Haase A. 1985. Identification of scrapie prion protein-specific mRNA in scrapie-infected and uninfected brain. *Nature* **315:** 331–333.

Clousard C., Beaudry P., Elsen J.M., Milan D., Dussaucy M., Bounneau C., Schelcher F., Chatelain J., Launay J.M., and Laplanche J.L. 1995. Different allelic effects of the codons 136 and 171 of the prion protein gene in sheep with natural scrapie. *J. Gen. Virol.* **76:** 2097–2101.

Cohen F.E., Pan K.-M., Huang Z., Baldwin M., Fletterick R.J., and Prusiner S.B. 1994. Structural clues to prion replication. *Science* **264:** 530–531.

Colon W., Lai Z., McCutchen S.L., Miroy G.J., Strang C., and Kelly J.W. 1996. FAP mutations destabilize transthyretin facilitating conformational changes required for amyloid formation. *Ciba Found. Symp.* **199:** 228–238.

De Fea K.A., Nakahara D.H., Calayag M.C., Yost C.S., Mirels L.F., Prusiner S.B., and Lingappa V.R. 1994. Determinants of carboxyl-terminal domain translocation during prion protein biogenesis. *J. Biol. Chem.* **269:** 16810–16820.

DeArmond S.J., McKinley M.P., Barry R.A., Braunfeld M.B., McColloch J.R., and Prusiner S.B. 1985. Identification of prion amyloid filaments in scrapie-infected brain. *Cell* **41:** 221–235.

Dickinson A.G. and Fraser H.G. 1977. Scrapie: Pathogenesis in inbred mice: An assessment of host control and response involving many strains of agent. In *Slow virus infections of the central nervous system* (ed. V. ter Meulen and M. Katz), pp. 3–14. Springer Verlag, New York.

Dickinson A.G. and Outram G.W. 1988. Genetic aspects of unconventional virus infections: The basis of the virino hypothesis. *Ciba Found. Symp.* **135:** 63–83.

Diener T.O., McKinley M.P., and Prusiner S.B. 1982. Viroids and prions. *Proc. Natl. Acad. Sci.* **79:** 5220–5224.

Dlouhy S.R., Hsiao K., Farlow M.R., Foroud T., Conneally P.M., Johnson P., Prusiner S.B., Hodes M.E., and Ghetti B. 1992. Linkage of the Indiana kindred of Gerstmann-Sträussler-Scheinker disease to the prion protein gene. *Nat. Genet.* **1:** 64–67.

Donne D.G., Viles J.H., Groth D., Mehlhorn I., James T.L., Cohen F.E., Prusiner S.B., Wright P.E., and Dyson H.J. 1997. Structure of the recombinant full-length hamster prion protein PrP(29–231): The N terminus is highly flexible. *Proc. Natl. Acad. Sci.* **94:** 13452–13457.

Edenhofer F., Rieger R., Famulok M., Wendler W., Weiss S., and Winnacker E.L. 1996. Prion protein PrP^C interacts with molecular chaperones of the Hsp60 family. *J. Virol.* **70:** 4724–4728.

Endo T., Groth D., Prusiner S.B., and Kobata A. 1989. Diversity of oligosaccharide structures linked to asparagines of the scrapie prion protein. *Biochemistry* **28:** 8380–8388.

Finckh U., Muller-Thomsen T., Mann U., Eggers C., Marksteiner J., Meins W., Binetti G., Alberici A., Hock C., Nitsch R.M., and Gal A. 2000. High prevalence of pathogenic mutations in patients with early-onset dementia detected by sequence analyses of four different genes. *Am. J. Hum. Genet.* **66:** 110–117.

Fischer M., Rülicke T., Raeber A., Sailer A., Moser M., Oesch B., Brandner S., Aguzzi A., and Weissmann C. 1996. Prion protein (PrP) with amino-proximal deletions restoring susceptibility of PrP knockout mice to scrapie. *EMBO J.* **15:** 1255–1264.

Fraser H. and Dickinson A.G. 1973. Scrapie in mice. Agent-strain differences in the distribution and intensity of grey matter vacuolation. *J. Comp. Pathol.* **83:** 29–40.

Gabizon R., Rosenmann H., Meiner Z., Kahana I., Kahana E., Shugart Y., Ott J., and Prusiner S.B. 1993. Mutation and polymorphism of the prion protein gene in Libyan Jews with Creutzfeldt-Jakob disease. *Am. J. Hum. Genet.* **53:** 828–835.

Gajdusek D.C. 1977. Unconventional viruses and the origin and disappearance of kuru. *Science* **197:** 943–960.

Glenner G.G. 1980. Amyloid deposits and amyloidosis. *N. Engl. J. Med.* **302:** 1283–1292.

Glenner G.G., Eanes E.D., and Page D.L. 1972. The relation of the properties of Congo red-stained amyloid fibrils to the beta-conformation. *J. Histochem. Cytochem.* **20:** 821–826.

Godec M.S., Asher D.M., Kozachuk W.E., Masters C.L., Rubi J.U., Payne J.A., Rubi-Villa D.J., Wagner E.E., Rapoport S.I., and Schapiro M.B. 1994. Blood buffy coat from Alzheimer's disease patients and their relatives does not transmit spongiform encephalopathy to hamsters. *Neurology* **44:** 1111–1115.

Goldmann W., Hunter N., Smith G., Foster J., and Hope J. 1994. PrP genotype and agent effects in scrapie: Change in allelic interaction with different isolates of agent in sheep, a natural host of scrapie. *J. Gen. Virol.* **75:** 989–995.

Goudsmit J., Morrow C.H., Asher D.M., Yanagihara R.T., Masters C.L., Gibbs C.J., Jr., and Gajdusek D.C. 1980. Evidence for and against the transmissibility of Alzheimer's disease. *Neurology* **30:** 945–950.

Harrison P.M., Chan H.S., Prusiner S.B., and Cohen F.E. 1999. Thermodynamics of model prions and its implications for the problem of prion protein folding. *J. Mol. Biol.* **286:** 593–606.

Hegde R.S., Mastrianni J.A., Scott M.R., DeFea K.A., Tremblay P., Torchia M., DeArmond S.J., Prusiner S.B., and Lingappa V.R. 1998. A transmembrane form of the prion protein in neurodegenerative disease. *Science* **279:** 827–834.

Hill A.F., Desbruslais M., Joiner S., Sidle K.C., Gowland I., Collinge J., Doey L.J., and Lantos P. 1997. The same prion strain causes vCJD and BSE. *Nature* **389:** 448–450, 526.

Hornemann S., Korth C., Oesch B., Riek R., Wider G., Wüthrich K., and Glockshuber R. 1997. Recombinant full-length murine prion protein, mPrP(23–231): Purification and spectroscopic characterization. *FEBS Lett.* **413:** 277–281.

Hsiao K., Baker H.F., Crow T.J., Poulter M., Owen F., Terwilliger J.D., Westaway D., Ott J., and Prusiner S.B. 1989. Linkage of a prion protein missense variant to Gerstmann-Sträussler syndrome. *Nature* **338:** 342–345.

Hsiao K.K., Groth D., Scott M., Yang S.-L., Serban H., Rapp D., Foster D., Torchia M., DeArmond S.J., and Prusiner S.B. 1994. Serial transmission in rodents of neurodegeneration from transgenic mice expressing mutant prion protein. *Proc. Natl. Acad. Sci.* **91:** 9126–9130.

Huang Z., Prusiner S.B., and Cohen F.E. 1996. Scrapie prions: A three-dimensional model of an infectious fragment. *Fold. Des.* **1:** 13–19.

Huang Z., Gabriel J.-M., Baldwin M A., Fletterick R.J., Prusiner S.B., and Cohen F.E. 1994. Proposed three-dimensional structure for the cellular prion protein. *Proc. Natl. Acad. Sci.* **91:** 7139–7143.

Hunter N., Goldmann W., Benson G., Foster J.D., and Hope J. 1993. Swaledale sheep affected by natural scrapie differ significantly in PrP genotype frequencies from healthy sheep and those selected for reduced incidence of scrapie. *J. Gen. Virol.* **74:** 1025–1031.

Hunter N., Moore L., Hosie B.D., Dingwall W.S., and Greig A. 1997a. Association between natural scrapie and PrP genotype in a flock of Suffolk sheep in Scotland. *Vet. Rec.* **140:** 59–63.

Hunter N., Cairns D., Foster J.D., Smith G., Goldmann W., and Donnelly K. 1997b. Is scrapie solely a genetic disease? *Nature* **386:** 137.

Ikeda T., Horiuchi M., Ishiguro N., Muramatsu Y., Kai-Uwe G.D., and Shinagawa M. 1995. Amino acid polymorphisms of PrP with reference to onset of scrapie in Suffolk and Corriedale sheep in Japan. *J. Gen. Virol.* **76:** 2577–2581.

James T.L., Liu H., Ulyanov N.B., Farr-Jones S., Zhang H., Donne D.G., Kaneko K., Groth D., Mehlhorn I., Prusiner S.B., and Cohen F.E. 1997. Solution structure of a 142-residue recombinant prion protein corresponding to the infectious fragment of the scrapie isoform. *Proc. Natl. Acad. Sci.* **94:** 10086–10091.

Jenkins J. and Pickersgill R. 2001. The architecture of parallel beta-helices and related folds. *Prog. Biophys. Mol. Biol.* **77:** 111–175.

Jenkins J., Mayans O., and Pickersgill R. 1998. Structure and evolution of parallel beta-helix proteins. *J. Struct. Biol.* **122:** 236–246.

Kaneko K., Wille H., Mehlhorn I., Zhang H., Ball H., Cohen F.E., Baldwin M.A., and Prusiner S.B. 1997a. Molecular properties of complexes formed between the prion protein and synthetic peptides. *J. Mol. Biol.* **270:** 574–586.

Kaneko K., Zulianello L., Scott M., Cooper C.M., Wallace A.C., James T.L., Cohen F.E., and Prusiner S.B. 1997b. Evidence for protein X binding to a discontinuous epitope on the cellular prion protein during scrapie prion propagation. *Proc. Natl. Acad. Sci.* **94:** 10069–10074.

Kaneko K., Peretz D., Pan K.-M., Blochberger T., Wille H., Gabizon R., Griffith O.H., Cohen F.E., Baldwin M.A., and Prusiner S.B. 1995. Prion protein (PrP) synthetic peptides induce cellular PrP to acquire properties of the scrapie isoform. *Proc. Natl. Acad. Sci.* **32:** 11160–11164.

Kascsak R.J., Rubenstein R., Merz P.A., Tonna-DeMasi M., Fersko R., Carp R.I., Wisniewski H.M., and Diringer H. 1987. Mouse polyclonal and monoclonal antibody to scrapie-associated fibril proteins. *J. Virol.* **61:** 3688–3693.

Kelly J.W. 1997. Amyloid fibril formation and protein misassembly: A structural quest for insights into amyloid and prion diseases. *Structure* **5:** 595–600.

Kimberlin R.H., Cole S., and Walker C.A. 1987. Temporary and permanent modifications to a single strain of mouse scrapie on transmission to rats and hamsters. *J. Gen. Virol.* **68:** 1875–1881.

Kitamoto T. and Tateishi J. 1994. Human prion diseases with variant prion protein. *Philos. Trans. R. Soc. Lond. B Biol. Sci.* **343:** 391–398.

Kitamoto T., Tateishi J., Tashima I., Takeshita I., Barry R.A., DeArmond S.J., and Prusiner S.B. 1986. Amyloid plaques in Creutzfeldt-Jakob disease stain with prion protein antibodies. *Ann. Neurol.* **20:** 204–208.

Klein M.A., Kaeser P.S., Schwarz P., Weyd H., Xenarios I., Zinkernagel R.M., Carroll M.C., Verbeek J.S., Botto M., Walport M.J., Molina H., Kalinke U., Acha-Orbea H., and Aguzzi A. 2001. Complement facilitates early prion pathogenesis. *Nat. Med.* **7:** 488–492.

Knaus K.J., Morillas M., Swietnicki W., Malone M., Surewicz W.K., and Yee V.C. 2001. Crystal structure of the human prion protein reveals a mechanism for oligomerization. *Nat. Struct. Biol.* **8:** 770–774.

Kocisko D.A., Priola S.A., Raymond G.J., Chesebro B., Lansbury P.T., Jr., and Caughey B. 1995. Species specificity in the cell-free conversion of prion protein to protease-resistant forms: A model for the scrapie species barrier. *Proc. Natl. Acad. Sci.* **92:** 3923–3927.

Kocisko D.A., Come J.H., Priola S.A., Chesebro B., Raymond G.J., Lansbury P.T., Jr., and Caughey B. 1994. Cell-free formation of protease-resistant prion protein. *Nature* **370:** 471–474.

Korth C., May B.C., Cohen F.E., and Prusiner S.B. 2001. Acridine and phenothiazine derivatives as pharmacotherapeutics for prion disease. *Proc. Natl. Acad. Sci.* **98:** 9836–9841.

Korth C., Stierli B., Streit P., Moser M., Schaller O., Fischer R., Schulz-Schaeffer W., Kretzschmar H., Raeber A., Braun U., Ehrensperger F., Hornemann S., Glockshuber R., Riek R., Billeter M., Wüthrich K., and Oesch B. 1997. Prion (PrP^{Sc})-specific epitope defined by a monoclonal antibody. *Nature* **389:** 74–77.

Kühlbrandt W., Wang D.N., and Fujiyoshi Y. 1994. Atomic model of plant light-harvesting complex by electron crystallography. *Nature* **367:** 614–621.

Latarjet R., Muel B., Haig D.A., Clarke M.C., and Alper T. 1970. Inactivation of the scrapie agent by near monochromatic ultraviolet light. *Nature* **227:** 1341–1343.

Lee J.P., Stimson E.R., Ghilardi J.R., Mantyh P.W., Lu Y.A., Felix A.M., Llanos W., Behbin A., Cummings M., Van Criekinge M., Timms W., and Maggio J.E. 1995. ^{1}H NMR of Ab amyloid peptide congeners in water solution. Conformational changes correlate with plaque competence. *Biochemistry* **34:** 5191–5200.

Liemann S. and Glockshuber R. 1999. Influence of amino acid substitutions related to inherited human prion diseases on the thermodynamic stability of the cellular prion protein. *Biochemistry* **38:** 3258–3267.

Liu H., Farr-Jones S., Ulyanov N.B., Llinas M., Marqusee S., Groth D., Cohen F.E., Prusiner S.B., and James T.L. 1999. Solution structure of Syrian hamster prion protein rPrP(90–231). *Biochemistry* **38:** 5362–5377.

Locht C., Chesebro B., Race R., and Keith J.M. 1986. Molecular cloning and complete sequence of prion protein cDNA from mouse brain infected with the scrapie agent. *Proc. Natl. Acad. Sci.* **83:** 6372–6376.

Manson J.C., Clarke A.R., McBride P.A., McConnell I., and Hope J. 1994. PrP gene dosage determines the timing but not the final intensity or distribution of lesions in scrapie pathology. *Neurodegeneration* **3:** 331–340.

Mastrianni J.A., Iannicola C., Myers R.M., DeArmond S., and Prusiner S.B. 1996. Mutation of the prion protein gene at codon 208 in familial Creutzfeldt-Jakob disease. *Neurology* **47:** 1305–1312.

Mastrianni J.A., Capellari S., Telling G.C., Han D., Bosque P., Prusiner S.B., and DeArmond S.J. 2001. Inherited prion disease caused by the V210I mutation: Transmission to transgenic mice. *Neurology* **57:** 2198–2205.

Mastrianni J.A., Nixon R., Layzer R., Telling G.C., Han D., DeArmond S.J., and Prusiner S.B. 1999. Prion protein conformation in a patient with sporadic fatal insomnia. *N. Engl. J. Med.* **340:** 1630–1638.

Matthews B.W. 1996. Structural and genetic analysis of the folding and function of T4 lysozyme. *FASEB J.* **10:** 35–41.

McKinley M.P., Bolton D.C., and Prusiner S.B. 1983. A protease-resistant protein is a structural component of the scrapie prion. *Cell* **35:** 57–62.

McKinley M.P., Meyer R., Kenaga L., Rahbar F., Cotter R., Serban A., and Prusiner S.B. 1991. Scrapie prion rod formation in vitro requires both detergent extraction and limited proteolysis. *J. Virol.* **65:** 1440–1449.

Mehlhorn I., Groth D., Stöckel J., Moffat B., Reilly D., Yansura D., Willett W.S., Baldwin M., Fletterick R., Cohen F.E., Vandlen R., Henner D., and Prusiner S.B. 1996. High-level expression and characterization of a purified 142-residue polypeptide of the prion protein. *Biochemistry* **35:** 5528–5537.

Mitsuoka K., Hirai T., Murata K., Miyazawa A., Kidera A., Kimura Y., and Fujiyoshi Y. 1999. The structure of bacteriorhodopsin at 3.0 Å resolution based on electron crystallography: Implication of the charge distribution. *J. Mol. Biol.* **286:** 861–882.

Mo H., Moore R.C., Cohen F.E., Westaway D., Prusiner S.B., Wright P.E., and Dyson H.J. 2001. Two different neurodegenerative diseases caused by proteins with similar structures. *Proc. Natl. Acad. Sci.* **98:** 2352–2357.

Monari L., Chen S.G., Brown P., Parchi P., Petersen R.B., Mikol J., Gray F., Cortelli P., Montagna P., Ghetti B., Goldfarb L.G., Gajdusek D.C., Lugaresi E., Gambetti P., and Autilio-Gambetti L. 1994. Fatal familial insomnia and familial Creutzfeldt-Jakob disease: Different prion proteins determined by a DNA polymorphism. *Proc. Natl. Acad. Sci.* **91:** 2839–2842.

Moore R.C., Lee I.Y., Silverman G.L., Harrison P.M., Strome R., Heinrich C., Karunaratne A., Pasternak S.H., Chishti M.A., Liang Y., Mastrangelo P., Wang K., Smit A.F., Katamine S., Carlson G.A., Cohen F.E., Prusiner S.B., Melton D.W., Tremblay P., Hood L.E., and

Westaway D. 1999. Ataxia in prion protein (PrP)-deficient mice is associated with upregulation of the novel PrP-like protein doppel. *J. Mol. Biol.* **292:** 797–817.

Muramoto T., Scott M., Cohen F., and Prusiner S.B. 1996. Recombinant scrapie-like prion protein of 106 amino acids is soluble. *Proc. Natl. Acad. Sci.* **93:** 15457–15462.

Nguyen J., Baldwin M.A., Cohen F.E., and Prusiner S.B. 1995a. Prion protein peptides induce α-helix to β-sheet conformational transitions. *Biochemistry* **34:** 4186–4192.

Nguyen J.T., Inouye H., Baldwin M.A., Fletterick R.J., Cohen F.E., Prusiner S.B., and Kirschner D.A. 1995b. X-ray diffraction of scrapie prion rods and PrP peptides. *J. Mol. Biol.* **252:** 412–422.

O'Rourke K.I., Holyoak G.R., Clark W.W., Mickelson J.R., Wang S., Melco R.P., Besser T.E., and Foote W.C. 1997. PrP genotypes and experimental scrapie in orally inoculated Suffolk sheep in the United States. *J. Gen. Virol.* **78:** 975–978.

Oesch B., Westaway D., Wälchli M., McKinley M.P., Kent S.B.H., Aebersold R., Barry R.A., Tempst P., Teplow D.B., Hood L.E., Prusiner S.B., and Weissmann C. 1985. A cellular gene encodes scrapie PrP 27-30 protein. *Cell* **40:** 735–746.

Pan K.-M., Stahl N., and Prusiner S.B. 1992. Purification and properties of the cellular prion protein from Syrian hamster brain. *Protein Sci.* **1:** 1343–1352.

Pan K.-M., Baldwin M., Nguyen J., Gasset M., Serban A., Groth D., Mehlhorn I., Huang Z., Fletterick R.J., Cohen F.E., and Prusiner S.B. 1993. Conversion of α-helices into β-sheets features in the formation of the scrapie prion proteins. *Proc. Natl. Acad. Sci.* **90:** 10962–10966.

Pattison I.H. 1965. Experiments with scrapie with special reference to the nature of the agent and the pathology of the disease. In *Slow, latent and temperate virus infections* (NINDB Monogr. 2) (ed. D.C. Gajdusek et al.), pp. 249–257. U.S. Government Printing Office, Washington, D.C.

Pattison I.H. and Millson G.C. 1961. Scrapie produced experimentally in goats with special reference to the clinical syndrome. *J. Comp. Pathol.* **71:** 101–108.

Pauly P.C. and Harris D.A. 1998. Copper stimulates endocytosis of the prion protein. *J. Biol. Chem.* **273:** 33107–33110.

Peretz D., Williamson R.A., Matsunaga Y., Serban H., Pinilla C., Bastidas R., Rozenshteyn R., James T.L., Houghten R.A., Cohen F.E., Prusiner S.B., and Burton D.R. 1997. A conformational transition at the N terminus of the prion protein features in formation of the scrapie isoform. *J. Mol. Biol.* **273:** 614–622.

Pergami P., Jaffe H., and Safar J. 1996. Semipreparative chromatographic method to purify the normal cellular isoform of the prion protein in nondenatured form. *Anal. Biochem.* **236:** 63–73.

Perrier V., Wallace A.C., Kaneko K., Safar J., Prusiner S.B., and Cohen F.E. 2000. Mimicking dominant negative inhibition of prion replication through structure-based drug design. *Proc. Natl. Acad. Sci.* **97:** 6073–6078.

Perrier V., Kaneko K., Safar J., Vergara J., Tremblay P., DeArmond S.J., Cohen F.E., Prusiner S.B., and Wallace A.C. 2002. Dominant-negative inhibition of prion replication in transgenic mice. *Proc. Natl. Acad. Sci.* **99:** 13079–13084.

Petersen R.B., Tabaton M., Berg L., Schrank B., Torack R.M., Leal S., Julien J., Vital C., Deleplanque B., Pendlebury W.W., Drachman D., Smith T.W., Martin J.J., Oda M., Montagna P., Ott J., Autilio-Gambetti L., Lugaresi E., and Gambetti P. 1992. Analysis of the prion protein gene in thalamic dementia. *Neurology* **42:** 1859–1863.

Poulter M., Baker H.F., Frith C.D., Leach M., Lofthouse R., Ridley R.M., Shah T., Owen F., Collinge J., Brown G., Hardy J., Mullan M.J., Harding A.E., Bennett C., Doshi R., and

Crow T.J. 1992. Inherited prion disease with 144 base pair gene insertion. 1. Genealogical and molecular studies. *Brain* **115:** 675–685.

Prusiner S.B. 1982. Novel proteinaceous infectious particles cause scrapie. *Science* **216:** 136–144.

———. 1984. Prions. *Sci. Am.* **251:** 50–59.

———. 1989. Scrapie prions. *Annu. Rev. Microbiol.* **43:** 345–374.

———. 1991. Molecular biology of prion diseases. *Science* **252:** 1515–1522.

———. 1997. Prion diseases and the BSE crisis. *Science* **278:** 245–251.

Prusiner S.B. and Scott M.R. 1997. Genetics of prions. *Annu. Rev. Genet.* **31:** 139–175.

Prusiner S.B., Cochran S.P., and Alpers M.P. 1985. Transmission of scrapie in hamsters. *J. Infect. Dis.* **152:** 971–978.

Prusiner S.B., Groth D.F., Bolton D.C., Kent S.B., and Hood L.E. 1984. Purification and structural studies of a major scrapie prion protein. *Cell* **38:** 127–134.

Prusiner S.B., Groth D., Serban A., Stahl N., and Gabizon R. 1993a. Attempts to restore scrapie prion infectivity after exposure to protein denaturants. *Proc. Natl. Acad. Sci.* **90:** 2793–2797.

Prusiner S.B., Bolton D.C., Groth D.F., Bowman K.A., Cochran S.P., and McKinley M.P. 1982. Further purification and characterization of scrapie prions. *Biochemistry* **21:** 6942–6950.

Prusiner S.B., Groth D.F., McKinley M.P., Cochran S.P., Bowman K.A., and Kasper K.C. 1981a. Thiocyanate and hydroxyl ions inactivate the scrapie agent. *Proc. Natl. Acad. Sci.* **78:** 4606–4610.

Prusiner S.B., McKinley M.P., Bowman K.A., Bolton D.C., Bendheim P.E., Groth D.F., and Glenner G.G. 1983. Scrapie prions aggregate to form amyloid-like birefringent rods. *Cell* **35:** 349–358.

Prusiner S.B., McKinley M.P., Groth D.F., Bowman K.A., Mock N.I., Cochran S.P., and Masiarz F.R. 1981b. Scrapie agent contains a hydrophobic protein. *Proc. Natl. Acad. Sci.* **78:** 6675–6679.

Prusiner S.B., Groth D., Serban A., Koehler R., Foster D., Torchia M., Burton D., Yang S.-L., and DeArmond S.J. 1993b. Ablation of the prion protein (PrP) gene in mice prevents scrapie and facilitates production of anti-PrP antibodies. *Proc. Natl. Acad. Sci.* **90:** 10608–10612.

Prusiner S.B., Scott M., Foster D., Pan K.-M., Groth D., Mirenda C., Torchia M., Yang S.-L., Serban D., Carlson G.A., Hoppe P.C., Westaway D., and DeArmond S.J. 1990. Transgenetic studies implicate interactions between homologous PrP isoforms in scrapie prion replication. *Cell* **63:** 673–686.

Ridley R.M. and Baker H.F. 1996. To what extent is strain variation evidence for an independent genome in the agent of the transmissible spongiform encephalopathies? *Neurodegeneration* **5:** 219–231.

Rieger R., Edenhofer F., Lasmezas C.I., and Weiss S. 1997. The human 37-kDa laminin receptor precursor interacts with the prion protein in eukaryotic cells. *Nat. Med.* **3:** 1383–1388.

Riek R., Hornemann S., Wider G., Glockshuber R., and Wüthrich K. 1997. NMR characterization of the full-length recombinant murine prion protein, mPrP(23–231). *FEBS Lett.* **413:** 282–288.

Riek R., Hornemann S., Wider G., Billeter M., Glockshuber R., and Wüthrich K. 1996. NMR structure of the mouse prion protein domain PrP(121–231). *Nature* **382:** 180–182.

Roberts G.W., Lofthouse R., Allsop D., Landon M., Kidd M., Prusiner S.B., and Crow T.J. 1988. CNS amyloid proteins in neurodegenerative diseases. *Neurology* **38:** 1534–1540.

Rogers M., Yehiely F., Scott M., and Prusiner S.B. 1993. Conversion of truncated and elongated prion proteins into the scrapie isoform in cultured cells. *Proc. Natl. Acad. Sci.* **90:** 3182–3186.

Rudd P.M., Wormald M.R., Wing D.R., Prusiner S.B., and Dwek R.A. 2001. Prion glycoprotein: Structure, dynamics, and roles for the sugars. *Biochemistry* **40:** 3759–3766.

Safar J., Roller P.P., Gajdusek D.C., and Gibbs C.J., Jr. 1993. Conformational transitions, dissociation, and unfolding of scrapie amyloid (prion) protein. *J. Biol. Chem.* **268:** 20276–20284.

Sakaguchi S., Katamine S., Shigematsu K., Nakatani A., Moriuchi R., Nishida N., Kurokawa K., Nakaoke R., Sato H., Jishage K., Kuno J., Noda T., and Miyamoto T. 1995. Accumulation of proteinase K-resistant prion protein (PrP) is restricted by the expression level of normal PrP in mice inoculated with a mouse-adapted strain of the Creutzfeldt-Jakob disease agent. *J. Virol.* **69:** 7586–7592.

Schmitt-Ulms G., Legname G., Baldwin M.A., Ball H.L., Bradon N., Bosque P.J., Crossin K.L., Edelman G.M., DeArmond S.J., Cohen F.E., and Prusiner S.B. 2001. Binding of neural cell adhesion molecules (N-CAMs) to the cellular prion protein. *J. Mol. Biol.* **314:** 1209–1225.

Schuler B., Rachel R., and Seckler R. 1999. Formation of fibrous aggregates from a nonnative intermediate: The isolated P22 tailspike beta-helix domain. *J. Biol. Chem.* **274:** 18589–18596.

Scott M.R., Köhler R., Foster D., and Prusiner S.B. 1992. Chimeric prion protein expression in cultured cells and transgenic mice. *Protein Sci.* **1:** 986–997.

Scott M., Groth D., Foster D., Torchia M., Yang S.-L., DeArmond S.J., and Prusiner S.B. 1993. Propagation of prions with artificial properties in transgenic mice expressing chimeric PrP genes. *Cell* **73:** 979–988.

Scott M.R., Groth D., Tatzelt J., Torchia M., Tremblay P., DeArmond S.J., and Prusiner S.B. 1997. Propagation of prion strains through specific conformers of the prion protein. *J. Virol.* **71:** 9032–9044.

Scott M.R., Will R., Ironside J., Nguyen H.O., Tremblay P., DeArmond S.J., and Prusiner S.B. 1999. Compelling transgenetic evidence for transmission of bovine spongiform encephalopathy prions to humans. *Proc. Natl. Acad. Sci.* **96:** 15137–15142.

Scott M., Foster D., Mirenda C., Serban D., Coufal F., Wälchli M., Torchia M., Groth D., Carlson G., DeArmond S.J., Westaway D., and Prusiner S.B. 1989. Transgenic mice expressing hamster prion protein produce species-specific scrapie infectivity and amyloid plaques. *Cell* **59:** 847–857.

Shibuya S., Higuchi J., Shin R.-W., Tateishi J., and Kitamoto T. 1998. Protective prion protein polymorphisms against sporadic Creutzfeldt-Jakob disease. *Lancet* **351:** 419.

Sparkes R.S., Simon M., Cohn V.H., Fournier R.E.K., Lem J., Klisak I., Heinzmann C., Blatt C., Lucero M., Mohandas T., DeArmond S.J., Westaway D., Prusiner S.B., and Weiner L.P. 1986. Assignment of the human and mouse prion protein genes to homologous chromosomes. *Proc. Natl. Acad. Sci.* **83:** 7358–7362.

Spielhaupter C. and Schatzl H.M. 2001. PrP^C directly interacts with proteins involved in signaling pathways. *J. Biol. Chem.* **276:** 44604–44612.

Spudich S., Mastrianni J.A., Wrensch M., Gabizon R., Meiner Z., Kahana I., Rosenmann H., Kahana E., and Prusiner S.B. 1995. Complete penetrance of Creutzfeldt-Jakob disease in Libyan Jews carrying the E200K mutation in the prion protein gene. *Mol. Med.* **1:** 607–613.

Stahl N., Baldwin M.A., Burlingame A.L., and Prusiner S.B. 1990. Identification of glycoinositol phospholipid linked and truncated forms of the scrapie prion protein. *Biochemistry* **29:** 8879–8884.

Stahl N., Baldwin M.A., Teplow D.B., Hood L., Gibson B.W., Burlingame A.L., and Prusiner S.B. 1993. Structural analysis of the scrapie prion protein using mass spectrometry and amino acid sequencing. *Biochemistry* **32:** 1991–2002.

Supattapone S., Bosque P., Muramoto T., Wille H., Aagaard C., Peretz D., Nguyen H.-O.B., Heinrich C., Torchia M., Safar J., Cohen F.E., DeArmond S.J., Prusiner S.B., and Scott M. 1999. Prion protein of 106 residues creates an artifical transmission barrier for prion replication in transgenic mice. *Cell* **96:** 869–878.

Taraboulos A., Jendroska K., Serban D., Yang S.L., DeArmond S.J., and Prusiner S.B. 1992. Regional mapping of prion proteins in brain. *Proc. Natl. Acad. Sci.* **89:** 7620–7624.

Telling G.C., Haga T., Torchia M., Tremblay P., DeArmond S.J., and Prusiner S.B. 1996a. Interactions between wild-type and mutant prion proteins modulate neurodegeneration in transgenic mice. *Genes Dev.* **10:** 1736–1750.

Telling G.C., Scott M., Mastrianni J., Gabizon R., Torchia M., Cohen F.E., DeArmond S.J., and Prusiner S.B. 1995. Prion propagation in mice expressing human and chimeric PrP transgenes implicates the interaction of cellular PrP with another protein. *Cell* **83:** 79–90.

Telling G.C., Parchi P., DeArmond S.J., Cortelli P., Montagna P., Gabizon R., Mastrianni J., Lugaresi E., Gambetti P., and Prusiner S.B. 1996b. Evidence for the conformation of the pathologic isoform of the prion protein enciphering and propagating prion diversity. *Science* **274:** 2079–2082.

Turk E., Teplow D.B., Hood L.E., and Prusiner S.B. 1988. Purification and properties of the cellular and scrapie hamster prion proteins. *Eur. J. Biochem.* **176:** 21–30.

Viles J.H., Cohen F.E., Prusiner S.B., Goodin D.B., Wright P.E., and Dyson H.J. 1999. Copper binding to the prion protein: Structural implications of four identical cooperative binding sites. *Proc. Natl. Acad. Sci.* **96:** 2042–2047.

Welker E., Raymond L.D., Scheraga H.A., and Caughey B. 2002. Intramolecular versus intermolecular disulfide bonds in prion proteins. *J. Biol. Chem.* **277:** 33477–33481.

Westaway D., Zuliani V., Cooper C.M., Da Costa M., Neuman S., Jenny A.L., Detwiler L., and Prusiner S.B. 1994. Homozygosity for prion protein alleles encoding glutamine-171 renders sheep susceptible to natural scrapie. *Genes Dev.* **8:** 959–969.

Wetzel R. 2002. Ideas of order for amyloid fibril structure. *Structure* **10:** 1031.

Wilesmith J.W., Ryan J.B.M., and Atkinson M.J. 1991. Bovine spongiform encephalopathy-Epidemiologic studies on the origin. *Vet. Rec.* **128:** 199–203.

Wille H., Michelitsch M.D., Guénebaut V., Supattapone S., Serban A., Cohen F.E., Agard D.A., and Prusiner S.B. 2002. Structural studies of the scrapie prion protein by electron crystallography. *Proc. Natl. Acad. Sci.* **99:** 3563–3568.

Williamson R.A., Peretz D., Smorodinsky N., Bastidas R., Serban H., Mehlhorn I., DeArmond S.J., Prusiner S.B., and Burton D.R. 1996. Circumventing tolerance to generate autologous monoclonal antibodies to the prion protein. *Proc. Natl. Acad. Sci.* **93:** 7279–7282.

Windl O., Giese A., Schulz-Schaeffer W., Zerr I., Skworc K., Arendt S., Oberdieck C., Bodemer M., Poser S., and Kretzschmar H.A. 1999. Molecular genetics of human prion diseases in Germany. *Hum. Genet.* **105:** 244–252.

Xing Y., Nakamura A., Korenaga T., Guo Z., Yao J., Fu X., Matsushita T., Kogishi K., Hosokawa M., Kametani F., Mori M., and Higuchi K. 2002. Induction of protein conformational change in mouse senile amyloidosis. *J. Biol. Chem.* **277:** 33164–33169.

Yost C.S., Lopez C.D., Prusiner S.B., Myer R.M., and Lingappa V.R. 1990. Non-hydrophobic extracytoplasmic determinant of stop transfer in the prion protein. *Nature* **343:** 669–672.

Zahn R., Liu A., Lührs T., Riek R., von Schroetter C., Lopez G.F., Billeter M., Calzolai L., Wider G., and Wüthrich K. 2000. NMR solution structure of the human prion protein. *Proc. Natl. Acad. Sci.* **97:** 145–150.

Zhang H., Kaneko K., Nguyen J.T., Livshits T.L., Baldwin M.A., Cohen F.E., James T.L., and Prusiner S.B. 1995. Conformational transitions in peptides containing two putative α-helices of the prion protein. *J. Mol. Biol.* **250:** 514–526.

Zhang H., Stöckel J., Mehlhorn I., Groth D., Baldwin M.A., Prusiner S.B., James T.L., and Cohen F.E. 1997. Physical studies of conformational plasticity in a recombinant prion protein. *Biochemistry* **36:** 3543–3553.

Zulianello L., Kaneko K., Scott M., Erpel S., Han D., Cohen F.E., and Prusiner S.B. 2000. Dominant-negative inhibition of prion formation diminished by deletion mutagenesis of the prion protein. *J. Virol.* **74:** 4351–4360.

6

Doppel, a New PrP-like Mammalian Protein

David Westaway

Centre for Research in Neurodegenerative Diseases and
Department of Laboratory Medicine and Pathobiology
University of Toronto, Ontario, Canada

Leroy E. Hood

Institute for Systems Biology
Seattle, Washington 98105

Stanley B. Prusiner

Institute for Neurodegenerative Diseases and
Departments of Neurology and of Biochemistry and Biophysics
University of California
San Francisco, California 94143

INTRODUCTION

Discovery of the Prion Protein Gene, *Prnp*

Prior to 1985, the genetic origin of the infectious isoform of the prion protein (Prp) was unknown: Indeed, some had speculated that PrP would prove to be encoded by the genomic nucleic acid of hypothetical spongiform encephalopathy–causing "slow virus" (Rohwer 1984). Elucidation of the amino-terminal sequence of PrP 27-30 allowed the synthesis of degenerate oligonucleotide probes and the subsequent identification of a cDNA clone that encoded all but the first 11 amino acids of the mature prion protein (Prusiner et al. 1984; Cheseboro et al. 1985; Oesch et al. 1985). A PrP cDNA clone was then used as a hybridization probe to interrogate denatured preparations of purified infectious prions, genomic DNA, and total RNA preparations isolated from the brains of healthy and prion-infected hamsters. These analyses established that, whereas DNA or RNA molecules encoding prion proteins could not be detected in denatured prion preparations, PrP gene sequences were present in the genomic DNA of hamsters (Oesch et al. 1985). This chromosomal gene, cloned and mapped the following year (Basler et al. 1986), is transcribed in both healthy and infected hamsters to generate a 2.1-kb PrP mRNA. Subsequent work identified the translation product of this mRNA in healthy animals as the protease-sensitive, α-helical glycoprotein PrP^C, setting the stage in turn for the conformational hypotheses of prion replication elaborated elsewhere in this volume.

The Rise, Fall, and Rise of the Prion Gene Complex

Although discerning the exact function of PrP^C has proven challenging, it is abundantly clear that prion protein genes (designated *Prnp*) are ubiquitous in mammals and in a number of other vertebrates including chickens, turtles, and *Xenopus laevis* (Wopfner et al. 1999; Simonic et al. 2000; Strumbo et al. 2001). Analysis of genomic clones and unpublished hybridization analyses performed at low stringency failed to provide any evidence for additional prion protein genes. Thus, on one hand, from a practical point of view, *Prnp* has long been discerned as a single-copy gene. On the other hand, analysis of segregating crosses between mouse inbred strains (NZW and I/LnJ) with differing scrapie incubation times defined a linked prion incubation-time gene, *Prn-i*. Using flanking region polymorphisms, *Prn-i* was deduced to lie closely adjacent to the prion protein structural gene (Carlson et al. 1986). These data therefore suggested the potential existence of a prion gene complex (*Prn*), encoding proteins (other than PrP^C) also fulfilling a role in prion replication. To distinguish between these contrasting views, *Prnp* molecular clones were obtained from NZW and I/LnJ mice. Nucleotide sequencing defined the existence of an allelic dimorphism (L108F, T189V) in the PrP coding region distinguishing the mouse strains, and suggesting that *Prn-i* could be none other than *Prnp* itself (Westaway et al. 1987). This inference was eventually established by producing altered scrapie disease incubation times in "knockin" mice expressing the 108F, 189V *Prnp* "b" allele (*Prnp*^{b}) (Moore et al. 1998). By defining congruence between *Prnp* and *Prn-i*, these experiments appeared to dispel once and for all the notion of a *Prn* gene complex. In fact, a further twist in this long-running story has defined a bona fide and highly intriguing gene called *Prnd*, lying downstream of *Prnp*.

Prnd encodes a protein called doppel (Dpl). This new gene was discovered during the course of sequencing a genomic DNA cosmid clone isolated from the aforementioned I/LnJ inbred strain of mice (Westaway et al. 1991; Lee et al. 1998; Moore et al. 1999). This DNA sequencing project was undertaken with the express purpose of finding genes adjacent to

Prnp and was prompted at least in part because transgenic mice created with the I/LnJ-4 cosmid did not behave in accord with our original expectations for the dominant scrapie incubation time allele present in the I/LnJ strain, leading to speculation as to the existence of a second gene in this cosmid clone (Westaway et al. 1991). Although subsequent studies provided an explanation for the paradoxical scrapie incubation-time behavior of the I/LnJ-4 transgenic mice (sensitivity of prion replication to PrP^C expression levels outweighing effects of mismatched allelic type; Carlson et al. 1994) the I/LnJ-4 cosmid clone was also interesting because it extended 3′ further than the extant human, hamster, and sheep *Prnp* cosmid clones (Lee et al. 1998). Remarkably, the sequence of this 3′ flanking region yielded a sizeable "hit" in the form of an open reading frame (ORF) predicted to encode a protein related to PrP itself (Moore et al. 1999). Subsequent work established that this ORF is indeed expressed to make the cell-surface protein doppel (Silverman et al. 2000). The name "doppel" is derived from *downstream prion protein-like*.

PHYSIOLOGY AND BIOCHEMISTRY OF DOPPEL

Origin and Properties of the *Prnd* Gene

The Dpl coding sequence is conserved in different strains of mice, humans (Moore et al. 1999), and other mammals, including sheep and cattle (Tranulis et al. 2001). The mouse doppel gene *Prnd* lies ~16 kb downstream of the *Prnp* coding region (Fig. 1A), whereas the human Dpl gene *PRND* lies 27 kb downstream of *PRNP* (Moore et al. 1999). The bovine Dpl gene lies 16 kb downstream of *Prnp* (Comincini et al. 2001), with the analogous figure for the ovine gene reported as 20 or 52 kb (Comincini et al. 2001; Essalmani et al. 2002). Thus, the *Prn* gene complex comprises *Prnp* and *Prnd* and presumably arose by duplication of a "proto" prion protein gene. Insofar as essentially no nucleic acid homology remains between the *Prnp* and *Prnd* genes, and the respective proteins only exhibit 24% sequence identity, we may infer that the gene duplication from a proto-*Prn* gene was an ancient event. However, it is unclear whether the proto-*Prn* gene was more PrP-like, more Dpl-like, or in between, and whether the gene duplication predated the speciation of mammals. To help clarify the relationship between cellular prion proteins, domain maps of PrP and Dpl are presented in Figure 2B. PrP genes encoding proteins with a basic residue at the amino terminus, conserved central hydrophobic regions encoding the amino acids (M/V)AGAAAAGA, and S-S–linked, α-helical carboxy-terminal domains have been isolated from diverse species includ-

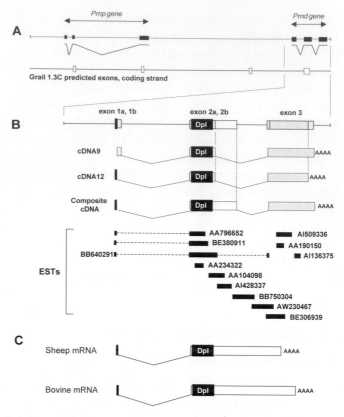

Figure 1. Structure of the mouse prion gene complex and *Prnd* mRNAs. (*A*) Structure of the *Prnp* and *Prnd* genes derived from analysis of the I/LnJ-4 cosmid. (*B*) Structure of mouse *Prnd* cDNAs and ESTs. Note some ESTs do not have a poly(A) tail or span a splice site and cannot be distinguished from genomic DNA contaminants. EST #BB750304 has an internal deletion corresponding to a $(TG)_{15}$ tract present in the wt sequence. (*C*) Putative two-exon structure of *Prnd* mRNAs from sheep and cattle.

ing turtles, birds, and the African clawed toad, *Xenopus laevis* (Gabriel et al. 1992; Simonic et al. 2000; Strumbo et al. 2001). None of these is predicted to encode proteins with two disulfide bonds, and they are thus more PrP-like (although the *X. laevis Prnp* gene lacks octarepeats, and in this sense is more Dpl-like). A cDNA isolated recently from the pufferfish *Fugu rubipes* encodes a protein with the first two of these features, but not the third (Suzuki et al. 2002). Since Dpl has only been cloned recently, the current inference of a progenitor gene more akin to PrP than Dpl is tentative, insofar as there has been longer to explore and retrieve PrP homologs from phylogenetically diverse species. Domain maps of PrP-like genes from amphibians and fish are presented in Figure 2A.

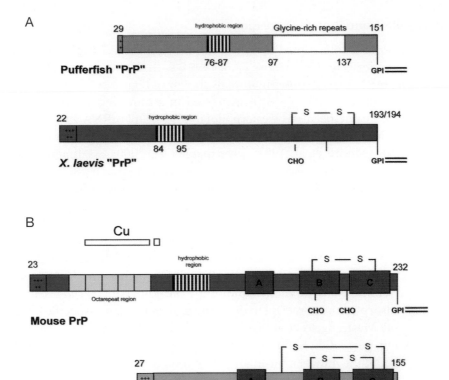

Figure 2. Structure of cellular prion proteins. (*A*) Structure of PrP-like proteins inferred from the sequence analysis of pufferfish and *X. laevis* genomic DNA and cDNA clones. α Helical regions in the *X. laevis* protein have not yet been determined by experimental analysis and are thus not shown. A hydrophobic region is denoted by a striped box and the + symbols indicate a cluster of basic residues. (*B*) Schematic of modern-day Dpl and PrP exemplified by analysis of mouse molecular clones and proteins. This panel shows the positions of the predicted features of mouse Dpl and its relationship to experimentally demonstrated features of mammalian PrP, including posttranslational modifications and secondary structure. Three α-helices found in PrP and Dpl are shown as boxes A, B, and C (Mo et al. 2001). The approximate locations of predicted disulfides and Asn-linked glycosylation sites (CHO) are also shown. The box with vertical stripes shows the transmembrane region containing the highly conserved sequence motif AGAAAAGA found in all PrPs. The numbering of both sequences is that of the mouse. The + symbols indicate a cluster of basic residues as in panel *A*. Positions of copper-binding domains (Cu) are indicated by open rectangles. In the case of PrP, these correspond to 4 histidine-containing octarepeats and a separate site involving histidine 95, whereas in the case of Dpl, a single binding site lies within a region defined by a Dpl(101–145) peptide.

Prnd mRNA and Protein Expression

In mice, *Prnd* is interrupted by introns and is transcribed into two major mRNA species (Fig. 1B) (Moore et al. 1999), with the structures of putative *Prnd* mRNAs from other mammals shown in Figure 1C. *Prnd* in mice (and most likely in sheep too) is expressed in a more restricted manner than *Prnp*. By northern and western blot analysis, Dpl mRNA and protein are found abundantly in adult testis, at lower levels in heart, but not in the adult CNS (Moore et al. 1999; Silverman et al. 2000; Essalmani et al. 2002). One study has raised the intriguing possibility that a pulse of CNS expression in neonatal mice corresponds to transcription within endothelial cells (Li et al. 2000a). Whether Dpl expression is induced under conditions of cellular stress remains unknown.

Recent studies have focused on the types of cells in the testis that express Dpl protein. We have used the antibody E6977 (Moore et al. 2001) directed against recombinant (rec) Dpl to perform immunohistochemistry of seminiferous tubules. Exploiting *Prnd*-knockout (*Prnd*$^{0/0}$) mice and preadsorption with recDpl to delineate and/or block nonspecific binding, we (P. Mastrangelo et al., in prep.) have localized Dpl expression to developing spermatids. Similar results have been obtained by Behrens et al. (2002). Using the mutagen busulfan to depopulate seminiferous tubules of germ cells, we infer that expression levels in Sertoli cells (support cells needed for spermatid maturation) are much lower than in developing spermatids. Seminiferous tubules in mice undergo waves of synchronous differentiation referred to as stages 1 to 12 (Russell et al. 1990), and, while studies to correlate Dpl staining to stages defined by periodic acid Schiff counterstaining are in progress, by defining adjacent Dpl-positive and Dpl-negative tubules, the studies of Behrens et al. and our studies, indicate stage-specific expression. Thus far, using the E6977 antibody, we have been unable to detect reproducible Dpl immunostaining in the epididimis, a site of sperm maturation (P. Mastrangelo and J. Coomaraswamy, unpubl.). This pattern of developmental expression may be different in other species. Laplanche and coworkers have reported Dpl expression in Sertoli cells of humans, using testis from a Sertoli-deficient syndrome as a point of reference (Peoc'h et al. 2002). Immunostaining on the tails of mature human sperm was also described, indicating another possible difference between humans and mice with regard to Dpl expression.

Biochemical Properties of Dpl

Sequence inspection reveals Dpl as an amino-terminally truncated version of PrP^C, lacking hexarepeat motifs, octarepeat motifs, and the conformationally plastic region lying in the center of the molecule (Fig. 2B). In contrast to this marked difference at the front end of the molecule, the carboxy-terminal domain is remarkably PrP-like, considering that the proteins have only ~24% sequence identity. Spectroscopic analysis of recombinant Dpl prepared by a similar oxidative refolding procedure yields an α-helical signature, virtually superimposable with that of recPrP (Silverman et al. 2000). The disulfide bond of mouse PrP linking residues 178 and 212 has a close cognate in Dpl (residues 109 and 143), although this is augmented by a second disulfide bond lying between residues 95 and 148. Broadly similar results have been obtained for human recombinant doppel (Lu et al. 2000). These low-resolution structural studies have now been augmented by NMR structures for mouse and human Dpl (Mo et al. 2001; Luhrs et al. 2003). These structures display three α-helices and two short β-strand motifs, even though many of the residues within analogous portions of the structure are not identical between PrP and Dpl. Nonetheless, there are differences that may be functionally significant. The angle of helix A compared to those of helices B and C is different, helix B has a marked interruption ("kink") and the loop between helices B and C is shorter and is "naked," lacking an N-glycosylation site (Fig. 3). Like PrP, Dpl is GPI-anchored, as demonstrated by treatment of neuroblastoma cells with PIPLC and also by use of Triton X-114 partitioning experiments (Silverman et al. 2000). More recently, using the E6977 antibody, we have demonstrated cell-surface staining of Dpl in neuroblastoma cell cultures and attenuation of this signal by PIPLC treatment (J. Coomaraswamy, in prep.). Unlike PrP, the exact site of covalent attachment of the GPI anchor has not been determined, with computer algorithms making slightly different predictions for human and mouse Dpl (Moore et al. 1999). An additional similarity between PrP^C and Dpl is selective copper binding (Fig. 2), and in the case of Dpl, the binding site is located within residues 101–145 (Qin et al. 2003).

Genetics and the Physiological Function of Dpl

What are the consequences of inactivation of the Dpl gene? In humans, besides a number of silent or missense polymorphisms (Fig. 4A), heterozygosity for a frameshift allele was detected in one patient in a survey of 29 Creutzfeldt-Jakob disease (CJD) patients and 111 normal subjects (Schröder et al. 2001). Whether this allele predisposes to familial (f) CJD or is asymptomatic in heterozygous form remains to be established, as no linkage analysis was presented. Inspection of the translated sequence of the frameshifted protein (actually misassigned in the original paper) prehuman PrP (90-230)   mouse Dpl (26–157)

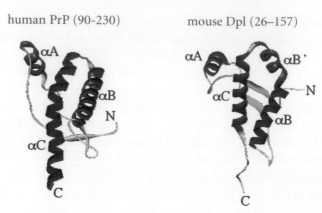

Figure 3. NMR structures of PrP and Dpl. Representations of the NMR structures derived from recombinant human PrP 90–230 (*left*) (Zahn et al. 2000) and mouse Dpl 26–157 (*right*) (Mo et al. 2001). α-Helices (αA, αB or B′, and αC) are shown in red, whereas short β-strand motifs are highlighted in turquoise. The assigned human structure corresponds to residues L125 to R228, and the assigned mouse Dpl structure corresponds to residues R51 to A157. This figure was compiled from the protein database (pdb) files 1QM0 and 1I17 using Web Lab™ Viewer light version 3.2. The protein structures are oriented with the carboxyl termini (site of GPI anchor addition and of membrane attachment) at the bottom. Despite a striking overall similarity, the Dpl structure differs from PrP in the division of helix αB into αB and B′, displacement of the β-strands by one or two residues, and an altered plane of the β-strands versus the axes of helices αB and αC. (Modified, with permission, from Westaway and Carlson 2002 [copyright Elsevier].)

dicts a protein that lacks the carboxy-terminal cysteine residues of both disulfide bonds (Figs. 4A and 5). We speculate that this protein would be highly unstable and that this frameshift mutation would therefore create a null allele. Further investigation of the frequency of this putative *PRND* null allele in the human population and the consequence of homozygosity will therefore be of interest.

With respect to *Prnd* null alleles in rodents, several laboratories have generated lines of *Prnd* $^{0/0}$ mice. In all cases, homozygous null animals derived from the mating of heterozygotes are viable, develop normally, and have normal neuroanatomy. However, male *Prnd* $^{0/0}$ mice (but not female *Prnd* $^{0/0}$ or male *Prnd* $^{0/+}$ mice) are sterile. This result is robust. In addition to studies from Aguzzi and coworkers (Behrens et al. 2002), sterility of male *Prnd* $^{0/0}$ Tg mice has been confirmed by Melton and coworkers at the University of Edinburgh and in mice created at the University of California by Moore and one of us (S.B.P.). In contrast to the phenotypes seem in *Prnp* $^{0/0}$ mice, which are subtle, disputed, or require exposure to an exogenous stimulus (however, for a different perspective,

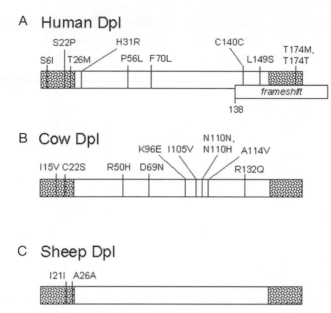

Figure 4. Polymorphisms in mammalian *PRND* genes. (*A*) The human *PRND* open reading frame (ORF) is indicated by a rectangle with amino- and carboxy-terminal signal peptides indicated by stippled boxes. Point-mutation polymorphisms in the human *PRND* gene are shown above the ORF, whereas a frameshift mutation is shown as an open rectangle below and to the right of the ORF (see also Fig. 5). (*B*) Multiple point-mutation polymorphisms detected in the bovine *Prnd* gene (representation as in panel *A*). (*C*) In contrast to *B*, missense polymorphisms are absent from the sheep *Prnd* gene alleles examined to date. Representations are as in panel *A*.

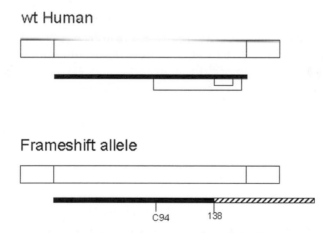

Figure 5. Protein structure of a putative frameshift allele of human *PRND* aligned with the analogous wild-type (wt) protein structure. Note that the frameshift mutation removes three of the four cysteine residues and is thus predicted to destroy both disulfide bonds and to leave one unpaired cysteine residue at position 94.

see Brown et al. 2002), this comprises the first overt phenotype ever to be assigned to a member of the *Prn* gene family. One possibility is that ancestral cellular prion proteins evolved first to perform a function in spermatogenesis, a role quite different from our typical perception of these proteins from their involvement in neurodegenerative diseases. Alternatively, subsequent to a duplication of a neurally expressed progenitor gene, the expression tropism of Dpl diverged such that it came to perform an essential function with regard to male reproductive ability.

Although a behavioral deficit could underlie the inability of male *Prnd*[0/0] mice to yield offspring, given that the cells expressing Dpl are spermatids, it is more plausible that infertility has a physiological basis and is a direct result of a deficit in male gametes. Studies are under way in several laboratories to pinpoint this lesion and thereby gain insight into Dpl's biochemical attributes. Deficit of Dpl may affect several aspects of spermatogenesis and sperm viability/activity. Malformed developing spermatids and malformed mature sperm with impaired ability to penetrate the oocyte zona pellucida have been reported by Behrens et al. (2002). Additional studies to focus on these deficits, the presence or absence of distinctions between different lines of *Prnd*[0/0] mice, and between mice and humans (e.g., the aforementioned presence or absence of Dpl expression in Sertoli cells) would appear to be in order. Another priority is to incorporate the pattern of PrP^C expression in the male reproductive tract into a larger picture. PrP^C expression has been reported at a later stage in gametogenesis (Shaked et al. 1999). Insofar as *Prnp*[0/0] males are not sterile, the role of PrP^C (being present at a later stage in gametogenesis) may be auxiliary, and it is therefore unclear whether a physical interaction between the two proteins is absolutely required for the production of healthy gametes. Nonetheless, it will be important to examine the phenotype of *Prnp*[0/0] + *Prnd*[0/0] "double" knockout mice and to determine whether expression of PrP^C at an earlier stage in gametogenesis, i.e., under the control of the *Prnd* promoter, can rescue the sterility of *Prnd*[0/0] mice.

THE PATHOBIOLOGY OF DOPPEL

Genetics and the Pathobiology of Dpl

A mooted Dpl null allele has been noted above (Figs. 4 and 5), but what of "activating" mutations in *Prnd*? In the case of the *Prnp*, point mutations cause familial prion diseases (Prusiner 1998), and overexpression causes a neuromyopathic syndrome (Westaway et al. 1994c). Currently, we do not know whether missense mutations in the human Dpl gene cause "doppelopathies." Given the expression tropism of Dpl, it is reasonable to assume that the phenotypes affect the male reproductive tract.

Like PrP^C, overexpression of wt Dpl is known to be pathogenic. Insights into this effect arose from close scrutiny of a puzzling feature of *Prnp*$^{0/0}$ mice (Sakaguchi et al. 1996; Moore et al. 1999, 2001; Rossi et al. 2001). No less than six independent lines have been constructed, but these have yielded not one but *two* phenotypes: either phenotypically normal in middle age or succumbing to a cerebellar ataxia marked by loss of Purkinje cell neurons (Table 1). The realization that the Dpl gene lies downstream of PrP, and can be affected by manipulations in the vicinity of the upstream gene, provides the solution to this curious divergence in outcome. In addition to "conventional" *Prnd* RNAs initiating from a proximal promoter, a second class of *Prnd* mRNAs contain exons deriving from the "intergene" region (i.e., between the *Prnp* polyadenylation site and the *Prnd* promoter) and initiate from the *Prnp* promoter (Westaway et al. 1994a; Moore et al. 1999; Li et al. 2000b). Although these chimeric *Prnp/Prnd* mRNAs generated by intergenic splicing are present at very low levels in the CNS of wt mice (Moore et al. 1999), their existence begged the question of the situation in *Prnp*$^{0/0}$ mice. Analysis of brain cDNAs derived from different lines of *Prnp*$^{0/0}$ mice revealed that intergenic splicing was strongly up-regulated in those lines exhibiting ataxia (see Fig. 6), yet remained at low levels in asymptomatic *Prnp*$^{0/0}$ mice. This finding of Dpl overexpression in adult neurons (considered as ectopic expression, given that Dpl is undetectable in these cells by standard blotting procedures) was confirmed by subsequent northern and western blot analyses (Moore et al. 1999, 2001; Silverman et al. 2000). A parsimonious explanation for the divergent behavior of *Prnp*$^{0/0}$ mice is that gene-targeting construct deletions that remove the *Prnp* exon-3 splice acceptor site (which normally competes effectively with the *Prnd* splice acceptor site) favor ectopic expression of *Prnd* from the *Prnp* promoter (Fig. 6, Table 1). The assumption of a causative role for Dpl in the cerebellar ataxia has been addressed in two ways. Rossi et al. performed studies of the ZrchII *Prnp*^{0} allele (differing from the ZrchI *Prnp*^{0} allele by a targeted deletion extending beyond the protein coding region) and a modified *Prnp* cosmid clone favoring expression of Dpl, leading to the conclusion that the level of Dpl expression in the CNS was proportional to the rapidity of onset of ataxia and cerebellar cell loss (Rossi et al. 2001). In another type of study, *Prnd* was excised from its usual chromosomal context and placed under the direct control of the (hamster) *Prnp* promoter, by using the cos.Tet expression vector (Scott et al. 1992). Subsequent to pronuclear microinjections, a variety of Dpl "founder" Tg mice was identified. Although some founders did not breed or died in adolescence, several stable Tg lines were established. Of note, four independent Tg lines exhibiting high levels of Dpl CNS expression exhibited an ataxic phenotype, often in early adult life. This was characterized by loss of cerebellar granule and Purkinje cells. Apoptosis was notable in the granule cell population, accompanied by loss of Purkinje cells, and these changes were accompanied by a florid activation of astrocytes (Moore et al. 2001). Morphological changes were also noted in hippocampal neurons. This neurodegenerative syndrome was not marked by myopathic changes, even though the hamster *Prnp* promoter is capable of driving expression in skeletal muscle (Westaway et al. 1994c). These data provide compelling evidence that the ataxia seen in Dpl-overexpressing *Prnp*$^{0/0}$ mice is a direct consequence of a toxic effect of this PrP-like protein. Furthermore, the observed pattern of neuropathology is distinct from that attributable to PrP^C overexpression or replication of rodent-adapted prions introduced by intracerebral inoculation. Extrapolating from the above, mutations in humans affecting the *PRNP* exon 2 splice acceptor or polyadenylation site could lead to overexpression of *PRND* and give rise to cerebellar ataxic syndromes.

Dpl, Conformational Change, and Prion Replication

If Dpl is a protein with PrP^C-like properties, can it be converted to an alternative, pathogenic conformer? Can Dpl participate in or modulate prion infections such as scrapie or CJD, or familial prion diseases such as fCJD, diseases all marked by accumulation of PrP^{Sc}? Although these possibilities

Table 1. PrP-ablated mice, ataxia, and Dpl overexpression

Prnp$^{0/0}$ Tg line	Ataxia in middle-age?	Deletion of *Prnp* exon 3 splice acceptor site?	Expression of Dpl in the CNS?	Reference
Wt mouse	no	no	no	n.a.
ZrchI	no	no	no	Büeler et al. (1992)
NPU	no	no	no	Manson et al. (1994)
Ngsk	yes	yes	yes	Sakaguchi et al. (1996); Moore et al. (1999); Li et al. (2000a)
Rcm0	yes	yes	yes	Moore et al. (1999)
ZrchII	yes	yes	yes	Weissmann and Aguzzi (1999); Rossi et al. (2001)
Riken	yes	yes	yes (S. Itohara, pers. comm.)	Yokoyama et al. (2001)

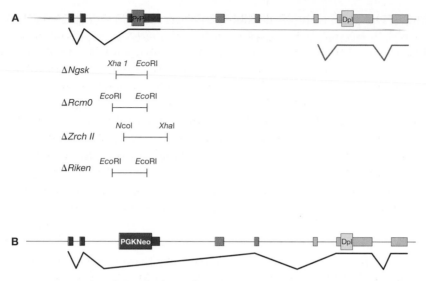

Figure 6. Intergenic splicing in the rodent *Prnd* gene complex. Panel *A* shows the wt mouse *Prn* gene complex with the protein coding exons ("PrP", "Dpl") as thick boxes and noncoding exons as thinner boxes (*red/pink* = PrP, *lemon/yellow* = Dpl, *orange* = intergenic). Spliced mRNAs are shown below the genes, with a dotted line extending beyond the *Prnp* mRNA indicating "leaky" transcriptional termination that may be germane to the genesis of intergenic splicing. Endpoints of the deletions (defined by restriction endonuclease sites) used to create four *Prnp⁰* alleles associated with cerebellar ataxia are shown below the panel. Panel *B* depicts the structure of a mutant *Prn* gene complex containing a *Prnp* knockout allele generated by interposition of a selectable marker ("PGKNeo", *purple*). Note that the *Prnp* coding exon (exon 3) splice acceptor is absent from the targeted allele. A chimeric mRNA resulting from intergenic splicing is depicted below the targeted allele. (Modified from Mastrangelo and Westaway 2001.)

have not been tested exhaustively, several observations argue against them. First, Dpl mRNAs are not detected in CNS neurons—a preferred site for prion replication—without recourse to sensitive PCR techniques to amplify cDNAs (RT-PCR) (Moore et al. 1999). Furthermore, the NMR structure of mouse Dpl (Mo et al. 2001) suggests three structural parameters that may preclude Dpl from undergoing a pathogenic conformational transformation. These are (1) the lack of the palindromic AGAAAAGA sequence present in the conformationally "plastic" region of PrP^C, (2) two disulfide bonds constraining the α-helical domain, whereas PrP^C has only one, and (3) a positively charged residue at the analog of position 171, which causes a dominant-negative effect upon prion replication in sheep, transgenic mice, and cultured cells (Westaway et al. 1994b; Kaneko et al. 1997; O'Rourke et al. 1997; Perrier et al. 2000, 2002). In the genetic realm, although nucleotide polymorphisms have now been delineated within the human Dpl gene, with one possible exception (Schröder et al. 2001), none has been suggested as being associated with altered susceptibility to CJD (Fig. 4) (Mead et al. 2000; Peoc'h et al. 2000). Transplantation experiments have been used to demonstrate that Dpl-deficient neural grafts can support replication of mouse-adapted prions (Behrens et al. 2001). Modest (nontoxic) overexpression of Dpl in the CNS of Tg mice does not affect incubation times for the Rocky Mountain Laboratory (RML) isolate of experimental scrapie (Moore et al. 2001), and incubation times after challenge with sheep scrapie prions do not differ between transgenic mice expressing PrP versus those expressing sheep Dpl as well as sheep PrP, at least when prions are administered by an intracerebral route (Vilotte et al. 2001; Essalmani et al. 2002). These findings all support the conclusion that Dpl may not undergo pathogenic conformational changes of the sort seen in PrP^C. However, outside of the context of prion infections, genetic and physical interactions between PrP^C and Dpl may nonetheless prove to be important.

INTERSECTIONS IN THE ACTIVITIES OF PrP^C AND DOPPEL

Genetic Interactions between Dpl and PrP^C in Neurons

Sequential expression of Dpl and PrP^C in the process of gametogenesis may prove interesting at the biochemical and cell biological level, but another intersection in the biology of these two proteins has arisen from a consideration of ataxic $Prnp^{0/0}$ mice. Ataxia in $Prnp^{0/0}$ lines such as the "Ngsk" (Nagasaki) line was originally attributed to a deficiency of PrP^C that, for reasons unknown, did not manifest itself in the ZrchI line of $Prnp^{0/0}$ mice. Accordingly, an attempt was made to "rescue" the ataxic phenotype by reintroduction of a wt PrP transgene (Nishida et al. 1999). The result of this experiment was abrogation of the ataxic phenotype, leading to the erroneous conclusion that ataxia was caused by a lack of PrP^C. With the benefit of hindsight, it is clear that the PrP transgene must somehow mitigate the cytotoxic effect of Dpl expressed in these Ngsk $Prnp^{0/0}$ mice (Moore et al. 1999). Since the notion of an interaction between the products of the Prnp and Prnd genes is intriguing, and plausible upon biochemical grounds given the similarities between the two proteins, two points bear elaboration.

First of all, the rescue phenomenon is robust. Besides the foregoing experiment, a protective effect of PrP transgenes has been noted in hippocampal cell lines derived from Riken $Prnp^{0/0}$ mice (Kuwahara et al. 1999). More recently, the phenomenon has been validated by crossing a wt hamster PrP transgene to ataxia-prone Tg(Dpl) mice (Moore et al. 2001). In some settings, Dpl's lethal effect on neurons is antagonized by wt PrP even when the PrP gene is present at only one copy per diploid genome (Rossi et al. 2001).

Second, levels of *Prnd* mRNA and Dpl protein are not affected by the introduction of PrP transgenes (Moore et al. 1999, 2001). Rather, these data are compatible with a posttranslational mechanism, presumably representing antagonistic biological activities of the two mature proteins. Some mechanisms are illustrated in Figure 7. One possibility is competitive binding for a common protein ligand or a common inorganic cofactor (Fig. 7A). Although the identity of this partner is unknown, it is not unreasonable to suggest that it might play a role in a signal transduction cascade. Another possibility is that by binding to Dpl, PrP^C might perturb a particular assembly state of Dpl (e.g., monomer, homodimer) required to initiate the same signaling cascade (Fig. 7B). In both of these scenarios, Dpl would play a positive role in activating a signal transduction pathway inappropriate for CNS neurons, and leading to apoptosis. Conversely, Dpl could have a "negative" effect on neuronal physiology by disrupting an endogenous pathway necessary for cell viability (Fig. 7C). Dpl could also be neurotoxic without requiring interactions with other proteins. For example, by virtue of multimerizing to form a membrane pore, PrP^C might intercalate into these assemblies and thereby disrupt their neurotoxicity (Behrens and Aguzzi 2002) (Fig. 7D). Another suggestion, prompted in part by the notion that PrP^C is protective and perhaps a metalloprotein (for review, see Lehmann 2002) or even a metalloenzyme (Brown et al. 1999), is that PrP^C and Dpl have antithetical properties with respect to the appearance of oxidative damage to macromolecules (Wong et al. 2001). However, reliance upon this hypothesis has to be tempered against the observation that oxidative damage to proteins present in Rcm0 $Prnp^{0/0}$ mice expressing Dpl (Wong et al. 2001) is not detected above non-Tg control levels in Tg10329 mice (Moore et al. 2001) expressing higher levels of Dpl (Qin et al. 2003).

Although many of these scenarios involve specificity in the interaction between PrP^C and Dpl, a rather different type of "interaction" also has to be considered. Extrapolating from the position that PrP^C possesses antiapoptotic properties (Bounhar et al. 2001), the specificity implied by experiments in which outcome can be determined by titrating the levels of PrP and Dpl (and implicit in models in which PrP and Dpl compete for ligands and cofactors) might be misleading. For example, Dpl might be proapoptotic by activation of an inappropriate pathway, with PrP blocking this effect via a general anti-apoptotic effect, but yet without any physical interactions or competition of a molecular nature between the two proteins. Again, this view of PrP^C as anti-apoptotic by virtue of interactions with the pro-apoptotic protein Bax has to be tempered against the fact that PrP is not normally a cytoplasmic or mitochondrial protein (Stahl et al.

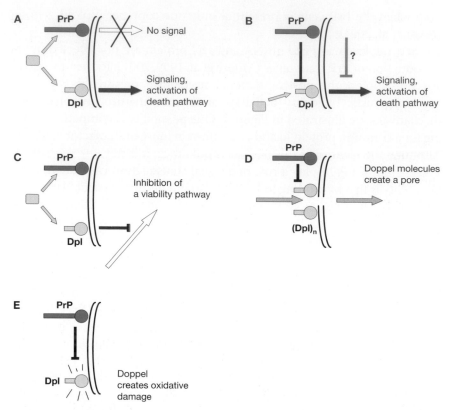

Figure 7. Models for neurotoxicity of doppel. Speculative models are represented in panels *A–E*. PrP is represented by a turquoise ball (carboxy-terminal α-helical domain) and stick (amino-terminal unstructured domain). Dpl is represented in a similar fashion but in a lighter color. Perpendicular symbols in panels *B–E* indicate inhibition. (*A*) PrP has no direct effect on a death pathway activated by Dpl but competes for a protein or inorganic coactivator of Dpl (rectangle with rounded corners). (*B*) PrP inhibits a death pathway by physical interaction with Dpl at the cell surface or interferes with an intracellular relay in the signaling pathway. (*C*) As in *A*, but instead of Dpl initiating a "new", inappropriate signaling cascade, it undercuts an endogenous pathway needed for cell viability. (*D*) Homomeric assemblies of Dpl ("$(Dpl)_n$") create pores allowing a toxic imbalance of metabolites (flux indicated by *yellow* arrows). PrP interferes with the formation of the homomeric protein complex resulting in assemblies without pores. (*E*) Dpl creates free radicals that cause oxidative damage (*dashes*) to proteins or membrane components, and PrP (perhaps by a scavenging effect involving copper chelation) antagonizes this effect and in consequence is neuroprotective.

1987; Vey et al. 1996). In any event, discerning between these possibilities will probably be useful as it could also clarify PrP^C's physiological activities. Additionally, as noted before, since PrP^C is a modulator of Dpl-induced CNS disease, "modifier" locus screens for Tg mice expressing Dpl might yield proteins that interact with PrP^C or modify Dpl-associated signal transduction pathways (Westaway and Carlson 2002).

"Shmerling Syndrome" and CNS Dpl Expression

The similarities between Shmerling syndrome and the effects of CNS expression of Dpl are sufficiently striking to require comment. Shmerling syndrome is an ataxia associated with an artificial interstitial deletion of mouse PrP (Shmerling et al. 1998). Shmerling et al. noted that whereas ZrchI $Prnp^{0/0}$ mice exhibit normal development, ZrchI mice expressing transgenes encoding PrP(Δ32–121) or PrP(Δ32–134) developed cerebellar ataxia. With regard to neuropathology, this was marked by loss of granule cells accompanied by prominent astrogliosis, as well as spongiosis in the cerebellar white matter tracts. Purkinje cells were unaffected. Proteinase K–resistant PrP was not described and attempts to transmit the disease to $tga20$ mice overexpressing wt PrP^C were negative. Pathology in these mice could be completely abrogated by reintroduction of full-length wt PrP^C.

Indeed, the many similarities in the genetics (PrP[Δ32–121] and PrP[Δ32–134] are amino-terminally reduced forms of PrP and thus Dpl-like; both syndromes are rescued by wt PrP expression) and pathology of Shmerling syndrome and the ectopic Dpl expression syndrome beg the question as to whether these disorders are really one and the same (Moore et al. 1999, Weissmann and Aguzzi 1999). The only notable difference is the lack of involvement of Purkinje cells in mice expressing PrP(Δ32–121) or PrP(Δ32–134). However, this discrepancy may have a trivial technical origin as the "half-genomic construct" vector used to express PrP(Δ32–121) or PrP(Δ32–134) may lack a Purkinje cell enhancer (Fischer et al. 1996). Also, in Tg(Dpl) mice made using the hamster PrP cosmid vector cos.Tet, the balance of pathology is slightly skewed from that seen in ataxic $Prnp^{0/0}$ mice, such that granule cell death is more prominent and Purkinje cell death less prominent (Moore et al. 2001). This again blurs a mooted dividing line between the two syndromes. One limitation to this story is that the PrP(Δ32–121) and PrP(Δ32–134) deletions are synthetic, having no exact equivalent in nature. Other internal deletions in PrP give rise to another distinct pathology (i.e., distinct from the pathologies of wt Dpl CNS expression, wt PrP overexpression, and prion infections), namely that of a lysosomal storage disease (Muramoto et al. 1997).

CONCLUSION AND PROSPECTS

Discovery of the Dpl protein has provided a new point of entry into the physiology of prion proteins. It also comprises a case study in the power and pitfalls of modern-day biology. On the one hand, it is remarkable that this gene lay undiscovered for so many years, located but a few kilobases downstream of the PrP gene. This situation was rectified by the application of directed large-scale sequencing, a testament (in our prion protein microcosm) to the power of chain-terminator sequencing reactions and high-throughput laser detection methods of sequencing reaction products. Subsequent to these studies, the complete sequence of chromosome 20 has confirmed these findings but failed to reveal any new members of the *Prn* gene family (Deloukas et al. 2001). On the other hand, Dpl exemplifies the difficulties still inherent in biological systems, specifically the area now designated as "functional genomics." At face value, the construction of four $Prnp^{0/0}$ lines all with the same phenotype, a phenotype "rescuable" by a wt PrP transgene, would appear to provide excellent evidence excluding the occurrence of ectopic mutations in the knockout ES cells and favoring the hypothesis that PrP is required for the viability of cerebellar cells. In fact, this logic is absolutely incorrect and all of these four targeted *Prnp* alleles produce a phenotype by activation of *Prnd* (Fig. 6). The toxicity of Dpl for CNS neurons is striking (given that it is modulated by PrP^C) and may provide insights into control of neuronal apoptosis. Since doppel-deficient mice have a phenotype in a developmental system (gametogenesis) that lends itself to experimental analysis, prospects for deciphering the function of the doppel protein are rosy and may prove a first in the realm of the *Prn* proteins. It will be interesting to see how the interactions of PrP^C and Dpl play out in these two experimental scenarios and whether conformational changes of a physiological nature are found to feature in the day-to-day life of the two cellular proteins.

ACKNOWLEDGMENTS

This work was supported by the Canadian Institutes of Health Research, the Bayer Blood Partnership Research Fund, the Canadian Red Cross, the Alzheimer Society of Ontario, and the National Institutes of Health.

REFERENCES

Basler K., Oesch B., Scott M., Westaway D., Wälchli M., Groth D.F., McKinley M.P., Prusiner S.B., and Weissmann C. 1986. Scrapie and cellular PrP isoforms are encoded by the same chromosomal gene. *Cell* **46:** 417–428.

Behrens A. and Aguzzi A. 2002. Small is not beautiful: Antagonizing functions for the prion protein PrP^C and its homologue Dpl. *Trends Neurosci.* **25:** 150–154.

Behrens A., Brandner S., Genoud N., and Aguzzi A. 2001. Normal neurogenesis and scrapie pathogenesis in neural grafts lacking the prion protein homologue Doppel. *EMBO Rep.* **2:** 347–352.

Behrens A., Genoud N., Naumann H., Rülicke T., Janett F., Heppner F.L., Ledermann B., and Aguzzi A. 2002. Absence of the prion protein homologue Doppel causes male sterility. *EMBO J.* **21**: 1–7.

Bounhar Y., Zhang Y., Goodyer C.G., and LeBlanc A. 2001. Prion protein protects human neurons against Bax-mediated apoptosis. *J. Biol. Chem.* **276**: 39145–39149.

Brown D.R., Nicholas R.S., and Canevari L. 2002. Lack of prion protein expression results in a neuronal phenotype sensitive to stress. *J. Neurosci. Res.* **67**: 211–224.

Brown D.R., Wong B.S., Hafiz F., Clive C., Haswell S.J., and Jones I.M. 1999. Normal prion protein has an activity like that of superoxide dismutase. *Biochem. J.* **344**: 1–5.

Büeler H., Fischer M., Lang Y., Bluethmann H., Lipp H.-P., DeArmond S.J., Prusiner S.B., Aguet M., and Weissmann C. 1992. Normal development and behaviour of mice lacking the neuronal cell-surface PrP protein. *Nature* **356**: 577–582.

Carlson G.A., Kingsbury D.T., Goodman P.A., Coleman S., Marshall S.T., DeArmond S.J., Westaway D., and Prusiner S.B. 1986. Linkage of prion protein and scrapie incubation time genes. *Cell* **46**: 503–511.

Carlson G.A., Ebeling C., Yang S.-L., Telling G., Torchia M., Groth D., Westaway D., DeArmond S.J., and Prusiner S.B. 1994. Prion isolate specified allotypic interactions between the cellular and scrapie prion proteins in congenic and transgenic mice. *Proc. Natl. Acad. Sci.* **91**: 5690–5694.

Chesebro B., Race R., Wehrly K., Nishio J., Bloom M., Lechner D., Bergstrom S., Robbins K., Mayer L., Keith J.M., Garon C., and Haase A. 1985. Identification of Scrapie prion protein–specific mRNA in scrapie-infected and uninfected brain. *Nature* **315**: 331–333.

Comincini S., Foti M.G., Tranulis M.A., Hills D., Di Guardo G., Vaccari G., Williams J.L., Harbitz I., and Ferretti L. 2001. Genomic organization, comparative analysis, and genetic polymorphisms of the bovine and ovine prion Doppel genes (*PRND*). *Mamm. Genome* **12**: 729–733.

Deloukas P., Matthews L.H., Ashurst J., Burton J., Gilbert J.G., Jones M., Stavrides G., Almeida J.P., Babbage A.K., Bagguley C.L., Bailey J., Barlow K.F., Bates K.N., Beard L.M., Beare D.M., Beasley O.P., Bird C.P., Blakey S.E., Bridgeman A.M., Brown A.J., Buck D., Burrill W., Butler A.P., Carder C., and Carter N.P., et al. 2001. The DNA sequence and comparative analysis of human chromosome 20. *Nature* **414**: 865–871.

Essalmani R., Taourit S., Besnard N., and Vilotte J.L. 2002. Sequence determination and expression of the ovine doppel-encoding gene in transgenic mice. *Gene* **285**: 287–290.

Fischer M., Rülicke T., Raeber A., Sailer A., Moser M., Oesch B., Brandner S., Aguzzi A., and Weissmann C. 1996. Prion protein (PrP) with amino-proximal deletions restoring susceptibility of PrP knockout mice to scrapie. *EMBO J.* **15**: 1255–1264.

Gabriel J.-M., Oesch B., Kretzschmar H., Scott M., and Prusiner S.B. 1992. Molecular cloning of a candidate chicken prion protein. *Proc. Natl. Acad. Sci.* **89**: 9097–9101.

Kaneko K., Zulianello L., Scott M., Cooper C.M., Wallace A.C., James T.L., Cohen F.E., and Prusiner S.B. 1997. Evidence for protein X binding to a discontinuous epitope on the cellular prion protein during scrapie prion propagation. *Proc. Natl. Acad. Sci.* **94**: 10069–10074.

Kuwahara C., Takeuchi A.M., Nishimura T., Haraguchi K., Kubosaki A., Matsumoto Y., Saeki K., Yokoyama T., Itohara S., and Onodera T. 1999. Prions prevent neuronal cell-line death. *Nature* **400**: 225–226.

Lee I., Westaway D., Smit A.F.A., Wang K., Seto J., Chen L., Acharya C., Ankener M., Baskin D., Cooper C., Yao H., Prusiner S.B., and Hood L. 1998. Complete genomic sequence and analysis of the prion protein gene region from three mammalian species. *Genome Res.* **8**: 1022–1037.

Lehmann S. 2002. Metal ions and prion diseases. *Curr. Opin. Chem. Biol.* **6**: 187–192.

Li A., Sakaguchi S., Atarashi R., Roy B.C., Nakaoke R., Arima K., Okimura N., Kopacek J., and Shigematsu K. 2000a. Identification of a novel gene encoding a PrP-like protein expressed as chimeric transcripts fused to PrP exon 1/2 in ataxic mouse line with a disrupted PrP gene. *Cell. Mol. Neurobiol.* **20:** 553–567.

Li A., Sakaguchi S., Shigematsu K., Atarashi R., Roy B.C., Nakaoke R., Arima K., Okimura N., Kopacek J., and Katamine S. 2000b. Physiological expression of the gene for PrP-like protein, PrPLP/Dpl, by brain endothelial cells and its ectopic expression in neurons of PrP-deficient mice ataxic due to Purkinje cell degeneration. *Am. J. Pathol.* **157:** 1447–1452.

Lu K., Wang W., Xie Z., Wong B.S., Li R., Petersen R.B., Sy M.S., and Chen S.G. 2000. Expression and structural characterization of the recombinant human doppel protein. *Biochemistry* **39:** 13575–13583.

Luhrs T., Riek R., Guntert P., and Wüthrich K. 2003. NMR structure of the human doppel protein. *J. Mol. Biol.* **326:** 1549–1557.

Manson J.C., Clarke A.R., Hooper M.L., Aitchison L., McConnel I., and Hope J. 1994. 129/Ola mice carrying a null mutation in PrP that abolishes mRNA production are developmentally normal. *Mol. Neurobiol.* **8:** 121–127.

Mastrangelo P. and Westaway D. 2001. The prion gene complex encoding PrP^C and doppel: Insights from mutational analysis. *Gene* **275:** 1–18.

Mead S., Beck J., Dickinson A., Fisher E.M., and Collinge J. 2000. Examination of the human prion protein-like gene doppel for genetic susceptibility to sporadic and variant Creutzfeldt-Jakob disease. *Neurosci. Lett.* **290:** 117–120.

Mo H., Moore R.C., Cohen F.E., Westaway D., Prusiner S.B., Wright P.E., and Dyson H.J. 2001. Two different neurodegenerative diseases caused by proteins with similar structures. *Proc. Natl. Acad. Sci.* **98:** 2352–2357.

Moore R.C., Hope J., McBride P.A., McConnell I., Selfridge J., Melton D.W., and Manson J.C. 1998. Mice with gene targetted prion protein alterations show that *Prnp, Sinc* and *Prni* are congruent. *Nat. Genet.* **18:** 118–125.

Moore R.C., Mastrangelo P., Bouzamondo E., Heinrich C., Legname G., Prusiner S.B., Hood L., Westaway D., DeArmond S.J., and Tremblay P. 2001. Doppel-induced cerebellar degeneration in transgenic mice. *Proc. Natl. Acad. Sci.* **98:** 15288–15293.

Moore R., Lee I., Silverman G.S., Harrison P., Strome R., Heinrich C., Karunaratne A., Pasternak S.H., Chishti M.A., Liang Y., Mastrangelo P., Wang K., Smit A.F.A., Katamine S., Carlson G.A., Cohen F.E., Prusiner S.B., Melton D.W., Tremblay P., Hood L.E., and Westaway D. 1999. Ataxia in prion protein (PrP) deficient mice is associated with upregulation of the novel PrP-like protein doppel. *J. Mol. Biol.* **293:** 797–817.

Muramoto T., DeArmond S.J., Scott M., Telling G.C., Cohen F.E., and Prusiner S.B. 1997. Heritable disorder resembling neuronal storage disease in mice expressing prion protein with deletion of an alpha-helix. *Nat. Med.* **3:** 750–755.

Nishida N., Tremblay P., Sugimoto T., Shigematsu K., Shirabe S., Petromilli C., Pilkuhn S., Nakaoke R., Atarashi R., Houtani T., Torchia M., Sakaguchi S., DeArmond S.J., Prusiner S.B., and Katamine S. 1999. Degeneration of Purkinje cells and demyelination in the spinal cord and peripheral nerves of mice lacking the prion protein gene (*Prnp*) is rescued by introduction of a transgene encoded for mouse *Prnp. Lab. Invest.* **79:** 689–697.

O'Rourke K.I., Holyoak G.R., Clark W.W., Mickelson J.R., Wang S., Melco R.P., Besser T.E., and Foote W.C. 1997. PrP genotypes and experimental scrapie in orally inoculated Suffolk sheep in the United States. *J. Gen. Virol.* **78:** 975–978.

Oesch B., Westaway D., Wälchli M., McKinley M.P., Kent S.B.H., Aebersold R., Barry R.A., Tempst P., Teplow D.B., Hood L.E., Prusiner S.B., and Weissmann C. 1985. A cellular gene encodes scrapie PrP 27-30 protein. *Cell* **40:** 735–746.

Peoc'h K., Guerin C., Brandel J.P., Launay J.M., and Laplanche J.L. 2000. First report of polymorphisms in the prion-like protein gene (*PRND*): Implications for human prion diseases. *Neurosci. Lett.* **286:** 144–148.

Peoc'h K., Serres C., Frobert Y., Martin C., Lehmann S., Chasseigneaux S., Sazdovitch V., Grassi J., Jouannet P., Launay J.M., and Laplanche J.L. 2002. The human "prion-like" protein Doppel is expressed in both Sertoli cells and spermatozoa. *J. Biol. Chem.* **277:** 43071–43078.

Perrier V., Wallace A.C., Kaneko K., Safar J., Prusiner S.B., and Cohen F.E. 2000. Mimicking dominant negative inhibition of prion replication through structure-based drug design. *Proc. Natl. Acad. Sci.* **97:** 6073–6078.

Perrier V., Kaneko K., Safar J., Vergara J., Tremblay P., DeArmond S.J., Cohen F.E., Prusiner S.B., and Wallace A.C. 2002. Dominant-negative inhibition of prion replication in transgenic mice. *Proc. Natl. Acad. Sci.* **99:** 13079–13084.

Prusiner S.B. 1998. Prions. *Proc. Natl. Acad. Sci.* **95:** 13363–13383.

Prusiner S.B., Groth D.F., Bolton D.C., Kent S.B., and Hood L.E. 1984. Purification and structural studies of a major scrapie prion protein. *Cell* **38:** 127–134.

Qin K., Coomaraswamy J., Mastrangelo P., Yang Y., Lugowski S., Petromilli C., Prusiner S.B., Fraser P.E., Goldberg J.M., Chakrabartty A., and Westaway D. 2003. The PrP-like protein Doppel binds copper. *J. Biol. Chem.* **278:** 8888–8896.

Rohwer R.G. 1984. Scrapie infectious agent is virus-like in size and susceptibility to inactivation. *Nature* **308:** 658–662.

Rossi D., Cozzio A., Flechsig E., Klein M.A., Rulicke T., Aguzzi A., and Weissmann C. 2001. Onset of ataxia and Purkinje cell loss in PrP null mice inversely correlated with Dpl level in brain. *EMBO J.* **20:** 694–702.

Russell L.D., Ettlin R.A., Hikim A.P., and Clegg E.D. 1990. *Histological and histopathological evaluation of the testis.* Cache River Press, Clearwater, Florida.

Sakaguchi S., Katamine S., Nishida N., Moriuchi R., Shigematsu K., Sugimoto T., Nakatani A., Kataoka Y., Houtani T., Shirabe S., Okada H., Hasegawa S., Miyamoto T., and Noda T. 1996. Loss of cerebellar Purkinje cells in aged mice homozygous for a disrupted PrP gene. *Nature* **380:** 528–531.

Schröder B., Franz B., Hempfling P., Selbert M., Jurgens T., Kretzschmar H.A., Bodemer M., Poser S., and Zerr I. 2001. Polymorphisms within the prion-like protein gene (*Prnd*) and their implications in human prion diseases, Alzheimer's disease and other neurological disorders. *Hum. Genet.* **109:** 319–325.

Scott M.R., Köhler R., Foster D., and Prusiner S.B. 1992. Chimeric prion protein expression in cultured cells and transgenic mice. *Protein Sci.* **1:** 986–997.

Shaked Y., Rosenmann H., Talmor G., and Gabizon R. 1999. A C-terminal-truncated PrP isoform is present in mature sperm. *J. Biol. Chem.* **274:** 32153–32158.

Shmerling D., Hegyi I., Fischer M., Blattler T., Brandner S., Gotz J., Rulicke, T., Flechsig E., Cozzio A., von Mering C., Hangartner C., Aguzzi A., and Weissmann C. 1998. Expression of amino-terminally truncated PrP in the mouse leading to ataxia and specific cerebellar lesions. *Cell* **93:** 203–214.

Silverman G.L., Qin K., Moore R.C., Yang Y., Mastrangelo P., Tremblay P., Prusiner S.B., Cohen F.E., and Westaway D. 2000. Doppel is an N-glycosylated, glycosylphosphatidylinositol-anchored protein. Expression in testis and ectopic production in the brains of *Prnp*[0/0] mice predisposed to Purkinje cell loss. *J. Biol. Chem.* **275:** 26834–26841.

Simonic T., Duga S., Strumbo B., Asselta R., Ceciliani F., and Ronchi S. 2000. cDNA cloning of turtle prion protein. *FEBS Lett.* **469:** 33–38.

Stahl N., Borchelt D.R., Hsiao K., and Prusiner S.B. 1987. Scrapie prion protein contains a phosphatidylinositol glycolipid. *Cell* **51:** 229–240.

Strumbo B., Ronchi S., Bolis L.C., and Simonic T. 2001. Molecular cloning of the cDNA coding for *Xenopus laevis* prion protein. *FEBS Lett.* **508:** 170–174.

Suzuki T., Kurokawa T., Hashimoto H., and Sugiyama M. 2002. cDNA sequence and tissue expression of Fugu rubripes prion protein-like: A candidate for the teleost orthologue of tetrapod PrPs. *Biochem. Biophys. Res. Commun.* **294:** 912–917.

Tranulis M.A., Espenes A., Comincini S., Skretting G., and Harbitz I. 2001. The PrP-like protein Doppel gene in sheep and cattle: cDNA sequence and expression. *Mamm. Genome* **12:** 376–379.

Vey M., Pilkuhn S., Wille H., Nixon R., DeArmond S.J., Smart E.J., Anderson R.G., Taraboulos A., and Prusiner S.B. 1996. Subcellular colocalization of the cellular and scrapie prion proteins in caveolae-like membranous domains. *Proc. Natl. Acad. Sci.* **93:** 14945–14949.

Vilotte J.L., Soulier S., Essalmani R., Stinnakre M.G., Vaiman D., Lepourry L., Da Silva J.C., Besnard N., Dawson M., Buschmann A., Groschup M., Petit S., Madelaine M.F., Rakatobe S., Le Dur A., Vilette D., and Laude H. 2001. Markedly increased susceptibility to natural sheep scrapie of transgenic mice expressing ovine PrP. *J. Virol.* **75:** 5977–5984.

Weissmann C. and Aguzzi A. 1999. Perspectives: Neurobiology. PrP's double causes trouble (erratum in *Science* [1999] **286:** 2086). *Science* **286:** 914–915.

Westaway D. and Carlson G.A. 2002. Mammalian prion proteins: Enigma, variation and vaccination. *Trends Biochem. Sci.* **27:** 301–307.

Westaway D., Cooper C., Turner S., Da Costa M., Carlson G.A., and Prusiner S.B. 1994a. Structure and polymorphism of the mouse prion protein gene. *Proc. Natl. Acad. Sci.* **91:** 6418–6422.

Westaway D., Goodman P.A., Mirenda C.A., McKinley M.P., Carlson G.A., and Prusiner S.B. 1987. Distinct prion proteins in short and long scrapie incubation period mice. *Cell* **51:** 651–662.

Westaway D., Zuliani V., Cooper C.M., Da Costa M., Neuman S., Jenny A.L., Detwiler L., and Prusiner S.B. 1994b. Homozygosity for prion protein alleles encoding glutamine-171 renders sheep susceptible to natural scrapie. *Genes Dev.* **8:** 959–969.

Westaway D., DeArmond S.J., Cayetano-Canlas J., Groth D., Foster D., Yang S.-L., Torchia M., Carlson G.A., and Prusiner S.B. 1994c. Degeneration of skeletal muscle, peripheral nerves, and the central nervous system in transgenic mice overexpressing wild-type prion proteins. *Cell* **76:** 117–129.

Westaway D., Mirenda C.A., Foster D., Zebarjadian Y., Scott M., Torchia M., Yang S.-L., Serban H., DeArmond S.J., Ebeling C., Prusiner S.B., and Carlson G.A. 1991. Paradoxical shortening of scrapie incubation times by expression of prion protein transgenes derived from long incubation period mice. *Neuron* **7:** 59–68.

Wong B.S., Liu T., Paisley D., Li R., Pan T., Chen S.G., Perry G., Petersen R.B., Smith M.A., Melton D.W., Gambetti P., Brown D.R., and Sy M.S. 2001. Induction of HO-1 and NOS in doppel-expressing mice devoid of PrP: Implications for doppel function. *Mol. Cell. Neurosci.* **17:** 768–775.

Wopfner F., Weidenhofer G., Schneider R., von Brunn A., Gilch S., Schwarz T.F., Werner T., and Schatzl H.M. 1999. Analysis of 27 mammalian and 9 avian PrPs reveals high conservation of flexible regions of the prion protein. *J. Mol. Biol.* **289:** 1163–1178.

Yokoyama T., Kimura K.M., Ushiki Y., Yamada S., Morooka A., Nakashiba T., Sassa T., and Itohara S. 2001. In vivo conversion of cellular prion protein to pathogenic isoforms, as monitored by conformation-specific antibodies. *J. Biol. Chem.* **276:** 11265–11271.

Zahn R., Liu A., Luhrs T., Riek R., von Schroetter C., Lopez Garcia F., Billeter M., Calzolai L., Wider G., and Wüthrich K. 2000. NMR solution structure of the human prion protein. *Proc. Natl. Acad. Sci.* **97:** 145–150.

7

Prions of Yeast and Filamentous Fungi: [URE3], [*PSI*⁺], [*PIN*⁺], and [Het-s]

Reed B. Wickner
Laboratory of Biochemistry & Genetics
National Institute of Diabetes, Digestive and Kidney Diseases
National Institutes of Health
Bethesda, Maryland 20892-0830

Susan W. Liebman
Laboratory of Molecular Biology
Department of Biological Sciences
University of Illinois at Chicago
Chicago, Illinois 60607

Sven J. Saupe
Laboratoire de Génétique Moléculaire des Champignons
Institut de Biochimie et de Génétique Cellulaires
UMR 5095 CNRS/Université de Bordeaux 2
33077 Bordeaux cedex, France

IN 1994, TWO NON-MENDELIAN GENETIC ELEMENTS of *Saccharomyces cerevisiae*, called [URE3] and [*PSI*⁺], were discovered to be prion (infectious protein) forms of the chromosomally encoded proteins, Ure2p and Sup35p, respectively (Wickner 1994). [Het-s], a non-Mendelian genetic element of the filamentous fungus *Podospora anserina*, is a prion of the HET-s protein (Coustou et al. 1997). Recently, [*PIN*⁺] was found to be a prion form of the yeast Rnq1p (Derkatch et al. 1997, 2001; Sondheimer and Lindquist 2000). Here we describe these phenomena, the evidence that they are prions, and their general implications for prion biology. The yeast system has made possible identification of many cellular factors affecting prion generation, propagation, and curing.

The concept of an infectious protein was first proposed to explain the unusual properties of the agent producing the transmissible spongiform encephalopathies (Griffith 1967). The term "prion" was coined to mean the scrapie agent, including the possibility that it may have no essential nucleic acid component, but was not restricted to this case (Prusiner 1982). We use the term prion here to mean "an infectious protein" (the protein-only model), regardless of the organism or protein, and making no assumptions about the mechanism involved.

One way in which a protein can be a prion is if it has undergone a change such that it no longer carries out its normal function but has acquired the ability to convert the normal form of the protein into the same form as itself, the prion form. By this definition, the prion change need not be one of conformation. A protein methylase might methylate another molecule of itself by mistake and, having done so, the methylated methylase may no longer be able to modify its normal target proteins but may have become quite efficient at acting on normal molecules of itself. This chain reaction would be self-propagating if the protein had a means to spread to other cells or individuals. Indeed, a prion of Prb1p based on such a mechanism has been described (Roberts and Wickner 2003) and is discussed below. We recognize that this usage of the term prion differs from its original definition (Prusiner 1982), and we hope that this does not cause unnecessary confusion.

The five prions (infectious proteins) discussed here were discovered by genetic analysis of the yeast, *S. cerevisiae*, and the filamentous fungus, *P. anserina*. Several other possible prions of *S. cerevisiae* have been described and, in these cases as well, the discovery was made by genetic means. However, with information about the properties of known prions becoming available, efforts are under way to use molecular cloning methods to screen for new prions, and [*PIN*$^+$] was independently discovered as [*RNQ*] by this method (Sondheimer and Lindquist 2000).

Even though infectious bacteriophage or viruses often integrate into the host genome or enter a plasmid state and are then inherited, it is customary to separate the concepts of infection and inheritance. The virus is being infectious when it passes from cell to cell as a virus particle and is being inherited when it is integrated into the host genome or becomes a plasmid. However, the study of yeast dsRNA viruses has blurred this distinction (for review, see Wickner 2001). These viruses are functionally and structurally similar to the cores of Reoviridae but have no extracellular route of infection. The same may be said of fungal viruses, all of which spread horizontally by mating or mitotic fusion of hyphae (cell processes). Thus, one would expect a prion of yeast or filamentous fungi to likewise spread by this means and to appear as a non-Mendelian genetic element. In the 1950s to 1970s, non-Mendelian genetic elements were considered of great interest because, in this pre-cloning era, they were thought to be potential sources of small nucleic acids that could be studied. This was particularly critical in yeast and filamentous fungi where no classic lytic viruses had (or have even now) been found. Among the non-Mendelian genetic elements described in this period were [*PSI*⁺] (Cox 1965), [URE3] (Lacroute 1971), and [Het-s] (Rizet 1952).

CHARACTERISTICS OF NON-MENDELIAN GENETIC ELEMENTS

A non-Mendelian genetic element is typically first recognized by its failure to segregate properly in meiosis. Mating a strain carrying the element with one lacking the element usually produces meiotic offspring, all of which have the element. This is called 4+:0 segregation in yeast, since there are four haploid meiotic progeny from a single diploid cell. In contrast, a single chromosomal gene difference between the parents leads to 2+:2– segregation.

A second characteristic of non-Mendelian genetic elements is their efficient transfer from cell to cell by cytoplasmic mixing (cytoduction or heterokaryon formation). In yeast, one uses a mutant that fails to undergo nuclear fusion after mating, the *kar1* mutant (Conde and Fink 1976). Cells with the element and the *kar1* mutation are mated for a few hours with cells of opposite mating type and lacking the element. The cells form mating pairs, but because of the *kar1* mutation, nuclei do not fuse in the heterokaryon formed. The nuclei separate into the daughter cells at the next cell division, but the cytoplasm of the two parent cells has mixed, so a cytoplasmic genetic element initially present in only one parent will be found in both daughter cells. Usually the donor parental strain has mitochondrial (mt) DNA (ρ^+) and so can grow on glycerol as a carbon source, but the recipient lacks mtDNA (ρ^0) and is glycerol⁻. Cells from the mating mixture that have the recipient nuclear genotype but have acquired mtDNA must have received cytoplasm from the donor. They will generally also have received other nonchromosomal genetic elements initially present in the donor. The same method is available in filamentous fungi, which naturally form heterokaryons when two strains grow together. This is discussed in some detail below for *Podospora*.

A third trait of non-Mendelian genetic elements is their curability by relatively nonmutagenic agents. These are specific to each element. For example, levels of ethidium bromide that are only mildly mutagenic to nuclear genes result in rapid and total elimination of the mitochondrial genome from all cells in the population (Goldring et al. 1970).

GENETIC PROPERTIES EXPECTED OF A PRION

As discussed above, an infectious protein should, like other infectious elements of yeast, appear as a non-Mendelian genetic element. Certain spe-

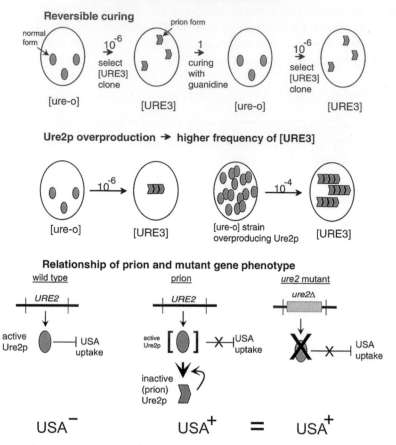

Figure 1. Genetic properties of a prion based on a self-inactivating altered protein form. The properties predicted for a prion are illustrated with the case of [URE3] (see Wickner 1994). (Reprinted, with permission, from Wickner 1997.)

cial properties of a non-Mendelian genetic element make it a candidate to be a prion (Fig. 1) (Wickner 1994). Each of these properties provides evidence against the non-Mendelian trait's being based on a nucleic acid replicon, such as a virus or plasmid.

Reversible Curability

If a prion can be cured from a strain, and the cured strain purified by multiple single-colony isolation, the cured strain should nonetheless be capable of giving rise to subclones that again carry the prion without its introduction from another cell. Whatever the change of the normal protein that constitutes the prion state, it can again occur in the cured cell (at some low frequency) and give rise to a clone of prion-containing cells. Curing of nucleic-acid based non-Mendelian genetic elements, such as the yeast viruses L-A and M or the mtDNA, is irreversible. Once gone, they will not return unless reintroduced from outside the cell.

Overproduction of the Normal Form Increases the Frequency of Prion Generation

If the cell has more of the normal form (a chromosomally encoded protein), then the stochastic process that gives rise to the prion form is expected to lead to the prion's arising with higher frequency. In contrast, overproducing one of the chromosomally encoded proteins on which a virus or plasmid depends would not give rise to the virus or plasmid in that cell.

Prion Phenotype Similar to Mutant Chromosomal Gene Phenotype

Deletion of the chromosomal gene encoding the protein capable of undergoing the prion change will result in loss of the function carried out by the normal form of the protein. The presence of the prion should likewise give rise to the same or a similar result, assuming that the prion form of the protein cannot carry out the normal function (Fig. 1). This phenotypic relation is opposite that customarily found between mutation of a chromosomal gene needed to propagate a non-Mendelian genetic element and the presence of the element (Table 1). In that case, the absence of the chromosomal gene results in loss of the nucleic acid and loss of whatever function it encodes. This is the opposite of the phenotype of the presence of the nucleic acid replicon. For example, chromosomal mutants that lose mtDNA cannot grow on glycerol, but the presence of the mtDNA results in the ability to grow on glycerol.

The chromosomal gene encoding the protein capable of undergoing the prion change will be found to be necessary for prion propagation. The prion propagates by converting the normal form into the prion form. Without the normal form, the prion form will be diluted out as the cells grow and will be lost. Thus, in looking for new prions, one should pay attention to chromosomal genes necessary for the propagation of the non-Mendelian genetic element that is a candidate prion (Wickner 1994).

Defective interfering mutants of viruses or plasmids can arise which complicate this picture (Table 1). For example, a deletion mutant of the mtDNA may interfere with the replication of the wild-type mitochondrial genome. In this case, the defective genome would be dominant. Mutation of a chromosomal gene needed to replicate the mutant mtDNA

Table 1. Relation of phenotypes of a prion and mutants in the gene for the protein

Non-Mendelian element	Presence of non-Mendelian element	Chromosomal replication mutant	Relation of phenotypes	Does replacing the chromosomal mutant gene restore the phenotype?
	phenotypes			
M dsRNA	killer$^+$	killer$^-$	opposite	no
mtDNA	glycerol$^+$	glycerol$^-$	opposite	no
mtDNA-DI	glycerol$^-$	glycerol$^-$	same	no
Theoretical prion	defective	defective	same	yes
[URE3]	USA$^+$	USA$^+$	same	yes
[PSI^+]	suppressor$^+$	suppressor$^+$	same	yes

See text for details. (Reprinted, with permission, from Wickner 1997.)

(e.g., Mip1p, the mitochondrial DNA polymerase; Lecrenier et al. 1997) would result in loss of either the wild-type mitochondrial genome or the mutant. Thus, the glycerol minus phenotype of the chromosomal mutant would be the same as that due to the presence of the defective interfering mtDNA mutant (Table 1) because each would result in loss of the normal mtDNA, and the phenotype relationship would mimic that of a prion. However, the distinction could still be made. If the chromosomal mutation was corrected by introducing the normal gene, the phenotype would remain defective because the mtDNA would not be replaced (Table 1). This contrasts with the prion case, in which a deletion of the chromosomal gene is complemented by introduction of the normal gene, with return of the normal phenotype, because there is no normal plasmid or virus to replace (Wickner 1994).

There are also prions whose activity is only apparent when the protein assumes the prion form. This is the case for the transmissible spongiform encephalopathies: Deletion of the gene for PrP has at most only a very subtle phenotype (Büeler et al. 1992), whereas presence of the prion causes a uniformly fatal disease. As discussed below in some detail, the [Het-s] prion results in heterokaryon incompatibility of *het-s* strains with *het-S* strains, but deletion of the *het-s* gene results in the opposite: a neutral phenotype with respect to heterokaryon formation (Coustou et al. 1997; Saupe 2000). The phenotype of the [PIN^+] prion, inducibility of the [PSI^+] prion by overproduction of Sup35p, is likewise a feature of the presence of the prion form of Rnq1p (Derkatch et al. 2001). In these cases, the relation of phenotypes is compatible with a viral or plasmid basis for the nonchromosomal gene and does not argue specifically for a prion basis.

Does satisfying the three genetic criteria constitute a proof of a prion? There are so far no cases of nonchromosomal genes that satisfy the three criteria and have proven not to be a prion. Of course, this does not imply that the three criteria constitute an absolute proof. However, when combined with biochemical evidence that the protein has changed form or structure, these genetic properties argue very strongly for the prion model.

[URE3], A PRION OF URE2P, A NITROGEN REGULATORY PROTEIN

In 1971, François Lacroute, studying uracil biosynthesis, described mutants of *S. cerevisiae*, called *ure* mutants, which could take up ureidosuccinate (USA) to allow growth of a mutant deficient in aspartate transcarbamylase (the first step in the uracil pathway, whose product is ureidosuccinate) (Lacroute 1971). This work uncovered the chromosomal *URE2* gene, a key player in regulation of nitrogen metabolism (Fig. 2) (Drillien et al. 1973), and the non-Mendelian genetic element, [URE3]. It was the early observation of Aigle and Lacroute (Aigle and Lacroute 1975) that propagation of [URE3] requires *URE2* which led eventually to the suggestion that [URE3] is a prion (Wickner 1994).

Wild-type yeast can take up USA from the medium if it is growing on a poor nitrogen source, such as proline, but the uptake system is repressed if there is a rich nitrogen source, such as asparagine, glutamine, glutamate, or ammonia. The control of uptake of USA, an intermediate in uracil biosynthesis, by the nitrogen source in the medium is an accident of the structural resemblance of USA and allantoate (Fig. 2). Allantoate is a poor, but usable, nitrogen source for yeast (Cooper 1982). Uptake of USA is carried out by the same *DAL5*-encoded permease that takes up allantoate, and thus USA uptake is subject to the same repression by good nitrogen sources as is that of allantoate (Turoscy and Cooper 1987). This repression occurs through the action of Ure2p, Gln3p, and other controlling proteins (Fig. 2) (for review, see Cooper 2002). Gln3p is a positive transcription regulator that recognizes the sequence GATAA upstream of many genes whose products are involved in catabolism of nitrogen sources. Ure2p is a cytoplasmic protein (Edskes et al. 1999a) that blocks the action of Gln3p and Gat1p/Nil1p by binding to them and keeping them in the cytoplasm (Beck and Hall 1999; Cardenas et al. 1999; Hardwick et al. 1999). Ure2p has sequence homology and marked structural similarity with glutathione-S-transferases (GST) (Coschigano and Magasanik 1991; Bousset et al. 2001; Umland et al. 2001), but GST activity of Ure2p has not been detected (Choi et al. 1998). Current evidence indicates that the nitrogen supply signal is transmitted to Ure2p by the Tor protein kinases (Beck and Hall 1999; Cardenas et al. 1999; Hardwick et al. 1999; Cooper 2002).

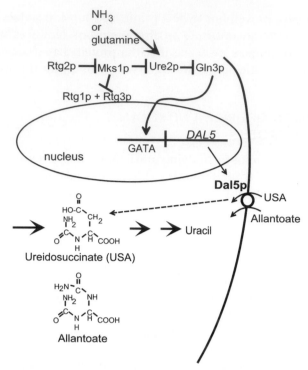

Figure 2. Ureidosuccinate, allantoate, Ure2p, and nitrogen regulation. Allantoate is a poor, but usable, nitrogen source for yeast. In the presence of a good nitrogen source, such as ammonia, glutamine, or asparagine, the proteins needed for allantoate utilization are transcriptionally turned off. These proteins include the allantoate transporter, Dal5p (Rai et al. 1987; Turoscy and Cooper 1987). Because allantoate chemically resembles ureidosuccinate, an intermediate in uracil biosynthesis, ureidosuccinate uptake (by the Dal5p transporter) is under control of the nitrogen catabolite repression system. Ure2p plays a key role in this system, sensing the nitrogen source and blocking positive transcription activation by Gln3p (Courchesne and Magasanik 1988; for review, see Cooper 2002).

Mks1p and Rtg2p Effects on Nitrogen Regulation

Overexpression of Mks1p allows cells to take up USA even in the presence of ammonia (Edskes et al. 1999b). This effect is countered by overexpression of Ure2p. Mks1p is negatively regulated by the RAS-cAMP system (Matsuura and Anraku 1993); moreover, *MKS1* is identical to *LYS80*, a negative regulator of lysine biosynthesis (Feller et al. 1997), and mutants have elevated pools of α-ketoglutarate and glutamate (Feller et al. 1997).

rtg2 mutants grow on USA in the presence of ammonia (Pierce et al. 2001). This effect depends on the presence of Mks1p and Gln3p and results in overexpression of *DAL5*, indicating that it acts through Ure2p (Pierce et al. 2001). Rtg2p is a cytoplasmic protein discovered through its involvement in signaling mitochondrial impairment to the nucleus (Liao and Butow 1993; Sekito et al. 2000). Deleting the mitochondrial genome triggers the "retrograde signaling pathway" and activates transcription of several tricarboxylic acid (TCA) cycle enzyme genes (Liao and Butow 1993; Liu and Butow 1999). Rtg2p transcription regulation requires Rtg1p and Rtg3p, and these regulatory effects are strongly inhibited by glutamate in the medium (Liao and Butow 1993; Jia et al. 1997; Liu and Butow 1999). Whereas *rtg2* mutants are USA⁺, *rtg1* and *rtg3* mutants are not, and glutamate in the medium does not make cells USA⁺, indicating that the effect of Rtg2p on nitrogen regulation is distinct from its action in the Rtg pathway (Pierce et al. 2001).

Rtg2p and Mks1p are found as a complex, and Mks1p is phosphorylated in response to the Rtg signals (Sekito et al. 2002). Mks1p is also involved in the Rtg signaling pathway, acting as a negative regulator functionally downstream of Rtg2, but upstream of Rtg1 and Rtg3 (Sekito et al. 2002), possibly as follows:

glutamate —| Rtg2p —| Mks1p —| Rtg1,3p → target genes → glutamate

For regulation of nitrogen catabolism, the following formal pathway has been suggested:

Rtg2 —| Mks1p —| Ure2p —| Gln3p → target genes

In the Rtg pathway, *rtg1* and *rtg3* mutations have the same effect as *rtg2*. However, unlike *rtg2* mutants, *rtg1* and *rtg3* mutants do not take up USA, nor does glutamate produce USA uptake. However, Mks1p overproduction or *rtg2* mutation does make cells USA⁺. We have thus suggested that the effect of Rtg2p on nitrogen regulation is distinct from its effect on the Rtg pathway (Fig. 2) (Pierce et al. 2001).

Genetic Properties of [URE3] Indicate It Is a Prion

Lacroute found mutations in two chromosomal genes (*ure1* and *ure2*) that made cells able to take up USA in the presence of ammonia and glutamate (Schoun and Lacroute 1969; Lacroute 1971), and one non-Mendelian "mutant," called [URE3-1] (Lacroute 1971). [URE3] was shown to segregate 4+:0 or irregularly in meiosis (Lacroute 1971), to be transferred by cytoduction (Aigle and Lacroute 1975), and to be cured by growth on low concentrations of guanidine (M. Aigle, cited in Cox et al. 1988), all properties of a nonchromosomal genetic element. [URE3] was shown to be unrelated to the mitochondrial genome (Lacroute 1971), the only well-documented nonchromosomal nucleic acid replicon in yeast at the time. Later studies showed it was not related to the yeast dsRNA viruses, the 2-micron DNA plasmid, or [PSI⁺] (Leibowitz and Wickner 1978).

Although *ure2* and [URE3] strains have the same phenotypes (Drillien et al. 1973), it was impossible to obtain *ure2* [URE3] strains (Aigle and Lacroute 1975). This key finding led eventually to the prion explanation for [URE3] (Wickner 1994).

[URE3] Is Reversibly Cured

[URE3] is efficiently cured by growth on rich medium containing 5 mM guanidine HCl (M. Aigle, cited in Cox et al. 1988; Wickner 1994). This in itself was certainly not evidence that [URE3] is a prion, since it is unlikely that 5 mM would be a high enough concentration to denature Ure2p and not other essential proteins in the cell. Moreover, similar low concentrations of guanidine have been shown to block replication of poliovirus by acting on the viral RNA-dependent RNA polymerase (Tamm and Eggers 1963). From a cured [URE3] strain, purified by single-colony isolation, [URE3] derivatives can again be isolated (Wickner 1994). These arise at a frequency of about 10^{-6} or 10^{-7}.

"Spontaneous" Creutzfeldt-Jakob disease (CJD) could be due to somatic mutation of the PrP gene, with the mutated cell killed early in the pathogenic process. This is nearly impossible to rule out in mammals. However, the reversible curability of [URE3] (Wickner 1994) (and [PSI⁺]; see below) shows that the analog of somatic mutation is not the basis for its arising. Spontaneous [URE3] clones arise from a wild-type strain with a frequency of about 10^{-6}. After curing [URE3] from such a clone, [URE3] again arises in the cured strain with a frequency that is the same or lower, not a higher frequency, showing that the original [URE3] did not arise from a particularly susceptible cell.

As discussed below, the discovery that guanidine curing acts by inhibition of the chaperone Hsp104 (Ferreira et al. 2001; Jung and Masison 2001; Jung et al. 2002) means that guanidine curing is now evidence for a prion.

Overexpression of Ure2p Increases the Frequency with Which [URE3] Arises

Introducing URE2 on a high-copy plasmid into a [ure-o] strain increased by 100-fold the frequency with which [URE3] arose (Wickner 1994). When URE2 was introduced under control of the galactose-inducible

GAL1 promoter, the increased frequency of [URE3] was seen only when cells had been grown on galactose. Thus, it was overexpression of the gene, not simply the high copy number of the gene, that induced [URE3] (Wickner 1994).

To show that it was Ure2p, and not the mRNA, whose overexpression induced [URE3], two frameshift mutations were introduced into the overexpressed URE2 gene. By removing one base in codon 44 and inserting a base in codon 80, most of the prion domain (see below) was completely changed in sequence. This resulted in only a minimal change in the mRNA sequence, no decrease in the mRNA amount, and an altered Ure2p sequence, which carried out nitrogen regulation normally but was reduced 10,000-fold (to undetectable levels) in its ability to induce [URE3] (Masison et al. 1997). This strongly indicates that it is the Ure2 protein and not the mRNA that is responsible for [URE3] induction.

Relation of Phenotypes of *ure2* and [URE3] Cells

The *ure2* mutants cannot propagate [URE3] (Aigle and Lacroute 1975; Wickner 1994), but *ure2* and [URE3] cause the same phenotypes (Drillien et al. 1973). This is the relationship expected for a prion and the gene encoding it (Wickner 1994).

Spontaneous Generation of [URE3]

The original [URE3] yeast strains were isolated as apparent mutants in a laboratory strain that did not have [URE3]. Could [URE3] be a mutant derivative of a nucleic acid replicon present in the parent that happens to depend on URE2 for its propagation and, like suppressive petites (deletion derivatives of mtDNA that prevent the normal functional mtDNA from replicating), has a phenotype dominant to the normal genome? In this case, a *ure2Δ* strain should lack this normal replicon (analogous to a ρ^0 strain) and so should be unable to give rise to [URE3] derivatives. A plasmid expressing Ure2p from a *GAL1* promoter was introduced into a *ure2Δ* strain. The resulting transformants did not grow on USA if the *GAL1* promoter was turned on by galactose, indicating that no normal Ure2p-dependent replicon is responsible for nitrogen regulation. Spontaneous USA$^+$ derivatives were isolated, and several were shown to carry [URE3] (Masison et al. 1997). Thus, it is Ure2p itself, and not some replicon dependent on Ure2p, which is responsible for nitrogen regulation and [URE3] appearance.

Protease Resistance of Ure2p in [URE3] Cells

The prion model for [URE3] predicts that Ure2p will be altered in [URE3] cells compared to [ure-o] yeast. The proteinase K resistance of PrP facilitated its discovery (Bolton et al. 1982). Although the alteration need not be one of conformation, sensitivity to proteases is a simple method to detect alterations when one has neither purified protein nor advance knowledge of the nature of the change.

Ure2p was more resistant to proteinase K digestion in extracts of [URE3] cells than in wild-type extracts (Fig. 3) (Masison and Wickner 1995). The 40-kD Ure2p was converted in less than 1 minute to fragments of 30 kD or 32 kD that persisted for 15 minutes. The most protease-resistant material, persisting for the duration of the experiment, comprised fragments of 7–10 kD detected by antibody specific for the amino-terminal prion domain (Fig. 3). The normal protein was digested in less than 1 minute to fragments too small to detect on the gel. The protease-resistant part of Ure2p in [URE3] strains included the amino-terminal part of Ure2p. Curing the [URE3] strain resulted in a return of the protease sensitivity to the wild-type pattern (Masison and Wickner 1995), and independent isolates of [URE3] showed similar patterns of protease resistance (Masison and Wickner 1995; Schlumpberger et al. 2001).

One alternate explanation of [URE3] is that it is a heritable state of a regulatory system, such as has been described for the lac operon, bacteriophage lambda, and *Drosophila* sex determination (see, e.g., Novick and Weiner 1957; for review, see Riggs and Porter 1996). This hypothesis is particularly important to consider because Ure2p is known to be a transcriptional regulator. In this model, the protease resistance of Ure2p would be explained as a change that happens as part of derepression of nitrogen metabolism, a normal alteration of the protein. However, it was found that growing normal cells on a poor nitrogen source such as proline did not make Ure2p detectably protease resistant, nor did the nitrogen source of pregrowth affect the frequency with which [URE3] arose, so this hypothesis is unlikely (Masison et al. 1997).

Prion Domain Versus Nitrogen Regulation Domain and Their Interactions

Overexpressed fragments of Ure2p were examined for their ability to complement a chromosomal *ure2Δ* mutation and for their ability to induce the appearance of the [URE3] non-Mendelian genetic element (Fig. 4) (Masison and Wickner 1995). The amino-terminal 65-residue fragment was sufficient to induce [URE3] at high frequency, about 100

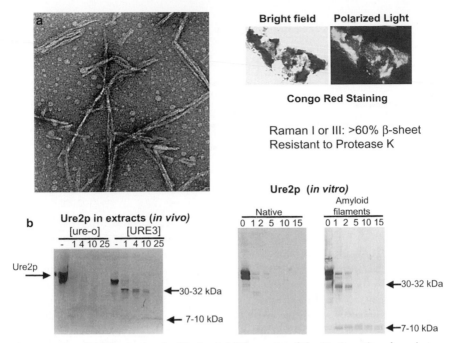

Figure 3. Amyloid formation by Ure2p. (*a*) Filaments of the Ure2p prion domain are protease-resistant, high in β-sheet, and show yellow-green birefringence on staining with Congo red, showing that they are amyloid. (*b*) Protease resistance of Ure2p in [URE3] cells is similar to that of amyloid of Ure2p formed in vitro, suggesting that Ure2p forms amyloid in [URE3] cells. The most resistant 7- to 10-kD fragment is the amino-terminal prion domain (Masison and Wickner 1995; Taylor et al. 1999).

times more efficiently than overexpression of the intact Ure2p. Deletion of the same amino-terminal 65 residues left a carboxy-terminal fragment that could completely, if modestly overproduced, complement the *ure2Δ* mutation but had no activity in inducing the appearance of [URE3] (Masison and Wickner 1995). Even a small deletion of the carboxyl terminus of Ure2p resulted in loss of complementing activity and increase of [URE3]-inducing activity, suggesting that in some way the active protein stabilizes the protein in the normal form (Fig. 4). The amino-terminal 65 residues was thus denoted the prion domain, since its overexpression was both necessary and sufficient for prion induction. The carboxy-terminal part of the molecule was named the nitrogen (N) regulation domain (Fig. 4). Further experiments showed that Ure2p$^{1-80}$ or Ure2p$^{1-89}$ was even more efficient at inducing [URE3] formation (Maddelein and Wickner 1999). Thus, the prion domain extends to about residue 89.

The roles of these domains have been further clarified by experiments modeled after the 1906 Chicago White Sox famous double-play combina-

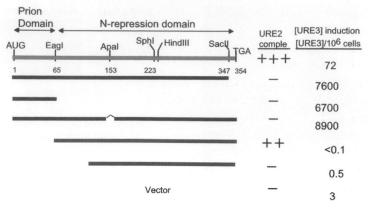

Figure 4. Domains of Ure2p. Parts of Ure2p were overexpressed from a GAL1 promoter and tested for induction of [URE3] appearance (in a wild-type host) or for complementation of a chromosomal deletion of URE2. The amino-terminal 65 amino acid residues were sufficient to induce [URE3], but could not complement ure2Δ, and so this part of Ure2p is called the prion domain. The carboxy-terminal domain lacking these first 65 residues could complement ure2Δ but could not induce [URE3] and is called the nitrogen regulatory domain. Deletion of the nitrogen regulatory domain increases the prion-inducing activity of the prion domain, suggesting that the native structure of Ure2p stabilizes the prion domain. (Reprinted, with permission, from Masison and Wickner 1995.)

tion, Tinker-to-Evers-to-Chance (Esteban and Wickner 1987; Masison et al. 1997). [URE3] was introduced from Tinker by cytoduction into a strain (Evers) deleted for its chromosomal copy of *URE2* but expressing the prion domain, the nitrogen regulation domain, or both (as separate molecules) from plasmids. The phenotypes of Evers cytoductants were noted, and then cytoplasm was transferred to Chance, a normal strain, to determine whether [URE3] was propagated in Evers.

It was found that the amino-terminal prion-inducing domain is sufficient to propagate [URE3], even in the complete absence of the carboxy-terminal N-regulation domain. The carboxy-terminal domain, lacking a covalently attached amino-terminal domain, could not respond to [URE3] when it was introduced. Thus, a strain in which the two domains are separately expressed shows repressed nitrogen metabolism (carboxy-terminal domain active). Nonetheless, [URE3] is propagated in this strain and is detected in strain Chance (Masison et al. 1997).

These results suggest that the prion domain and the N-regulation domain interact in the normal form of Ure2p. The carboxy-terminal part stabilizes the prion domain (reducing the frequency of prion generation), and the prion domain helps the nitrogen regulation function of the carboxy-terminal part. Using the yeast two-hybrid method, interaction between Ure2p$^{1-151}$ and Ure2p$^{153-354}$ was detected (Fernandez-Bellot et al. 1999), consistent with this idea. More detailed studies will be needed to fully define this interaction.

The glutathione-S-transferase homology of Ure2p (Coschigano and Magasanik 1991) is restricted to the carboxy-terminal domain. The prion domain of Ure2p is 40% asparagine and 20% threonine + serine, an unusual composition (Fig. 5). Indeed, as was first shown for Sup35p and [*PSI*$^+$] (see below), prion induction activity requires the runs of asparagine residues (Maddelein and Wickner 1999). The association of expanded polyglutamine domains in several dominant inherited diseases of humans suggested to M. Perutz that polyglutamine might mediate aggregation of the affected proteins (Perutz et al. 1994). This notion was supported by the finding that insertion of an artificial polyglutamine domain into a soluble monomeric protein promoted its self-association (Stott et al. 1995). It is now clear that the high asparagine and glutamine contents of the prion domains of Ure2p, Sup35p, and Rnq1 (see below and Fig. 5) are critical for their prion properties.

Ure2p Resembles Glutathione-S-Transferases

The crystal structure of the carboxy-terminal functional domain of Ure2p shows remarkable similarity to glutathione-S-transferases (GSTs) particularly those of *Escherichia coli* and *Proteus mirabilis* (Fig. 6) (Bousset et al. 2001; Umland et al. 2001), as had been suspected from sequence homology (Coschigano and Magasanik 1991). Like most GSTs, Ure2p is a dimer in solution (Perrett et al. 1999; Taylor et al. 1999) and in the crystal structure.

Despite the close resemblance of Ure2p to GSTs, no GST activity has been detected in Ure2p (Coschigano and Magasanik 1991; Choi et al. 1998). Although Ure2p has most of the consensus residues for Θ class GSTs (Rossjohn et al. 1996), Ala-122 of Ure2p should be serine, and mutation of the corresponding Ser-11 to alanine in *Arabidopsis thaliana* Θ class GST reduces activity to <0.5% (Rossjohn et al. 1998).

One dramatic difference between Ure2p and catalytically active GSTs is the presence in Ure2p of a loop with a short α-helix comprising residues 267–298 and called the "cap" (Bousset et al. 2001). The function of this region is not known.

It has not yet been possible to obtain crystals of the full-length Ure2p, probably because the amino-terminal prion domain is largely unstructured in the native molecule. Evidence for this view comes from denatu-

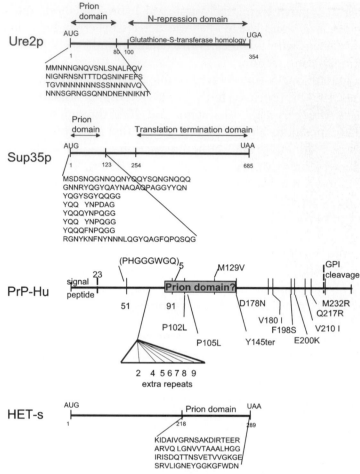

Figure 5. Comparison of Ure2p, Sup35p, HET-s, and PrP.

ration studies comparing the full-length protein and the carboxy-termi-nal domain which suggest that the prion domain is unstructured (Perrett et al. 1999). Furthermore, the prion domain is most sensitive to protease digestion (Bousset et al. 2002). In addition, comparison of circular dichroism (CD) spectra of full-length Ure2p and the carboxy-terminal domain suggests that the difference is due to an unstructured domain (Baxa et al. 2002).

Close Homologs of Ure2p

Although Ure2p has 10–20% identity with enzymatically active GSTs, it has very close homology with genes in other yeasts and fungi (Edskes and

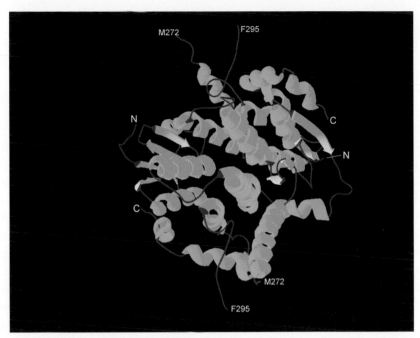

Figure 6. X-ray structure of the dimer of the globular part (carboxy-terminal functional region) of Ure2p.

Wickner 2002). The normal function of the *S. cerevisiae* Ure2p in nitrogen regulation is fully complemented by the Ure2p of *Candida albicans*, *Candida glabrata*, *Candida kefyr*, *Candida maltosa*, *Saccharomyces bayanus*, and *Saccharomyces paradoxus*, but not by homologs from *Candida lipolytica* or *Ashbya gossypii*. The non-*Saccharomyces* Ure2ps are 78–92% identical, whereas the *S. bayanus* and *S. paradoxus* Ure2ps are 99–100% identical to the *S. cerevisiae* Ure2p carboxy-terminal N-regulation domain. However, there is considerable divergence of their amino-terminal domains from that of Ure2p of *S. cerevisiae*. [URE3[Sc]] showed efficient transmission into *S. cerevisiae ure2Δ* cells only if expressing a Ure2p of a species within *Saccharomyces* (Edskes and Wickner 2002). The prion domain of Ure2p shows little variation within wild isolates of *S. cerevisiae*.

Interestingly, all of the Ure2 homologs that have an amino-terminal domain rich in asparagine and glutamine residues (like the *S. cerevisiae* Ure2p) also have the cap region between the equivalent of the *S. cerevisiae* residues 267 and 298 (Edskes and Wickner 2002). The Ure2 homologs of *Schizosaccharomyces pombe* and *Neurospora crassa* lack both amino-terminal extension and cap. The functional significance of this correlation remains to be seen.

[URE3] Is a Self-propagating Amyloid Form of Ure2p

The first biochemical evidence for the yeast prions (and [URE3] in particular) and the first hint of a connection with amyloid was the finding that Ure2p is partially protease-resistant in extracts of [URE3] strains (Fig. 3) (Masison and Wickner 1995). Expression of Ure2p–GFP fusion proteins at low levels showed that Ure2p is evenly distributed in the cytoplasm of wild-type strains, but formed aggregates in [URE3] strains (Edskes et al. 1999a). In addition, it is the prion domain that determines this aggregation. GFP fused to the carboxy-terminal domain does not display aggregation even in cells carrying the [URE3] prion. Of course, not all protein aggregation (or even all aggregation of Ure2p) is related to prion formation (Fernandez-Bellot et al. 2002). For example, unlike Ure2p in the [URE3] prion (Edskes et al. 1999a), a prion composed of only a Ure2p–GFP fusion protein does not make aggregates large enough to see by light microscopy (although smaller aggregates appear to be present) (Fernandez-Bellot et al. 2002). Furthermore, overproduction or misfolding of proteins in many systems is associated with aggregate formation that is not self-propagating. [URE3] involves aggregation, but aggregation of Ure2p is not necessarily [URE3]. [URE3] appears to be a specific kind of aggregate, namely amyloid (see below). However, the degree of aggregation might vary with culture conditions, strain properties, and other factors.

Chemically synthesized Ure2p$^{1-65}$, the prion domain peptide, forms amyloid filaments in vitro (Fig. 3a) (Taylor et al. 1999). These 45 Å diameter filaments show only β-sheet structure (>60%), are highly resistant to digestion with proteinase K, and show the birefringence on staining with Congo red typical of amyloid filaments (Fig. 3) (Taylor et al. 1999). Under the same conditions, the full-length native Ure2p, purified from yeast, is a stable dimer that does not form amyloid. However, addition of the Ure2p$^{1-65}$ prion domain peptide induces formation of "co-filaments," a 1:1 copolymer of full-length Ure2p and the peptide, which again shows all the properties of amyloid (Taylor et al. 1999). These co-filaments can seed the formation of amyloid by a large excess of added Ure2p. The amyloid of full-length Ure2p shows a pattern of protease-resistant fragments (Taylor et al. 1999) identical to that observed for Ure2p in extracts of [URE3] cells (Masison and Wickner 1995), with the most resistant part being the prion domain in each case (Fig. 3b).

It is the same amino-terminal part of Ure2p that induces prion formation in vivo which also induces amyloid formation in vitro. Neither peptides from other parts of Ure2p nor the Aβ peptide can induce amyloid formation by full-length Ure2p.

Thin-section electron microscopy of [URE3] cells reveals a network of filaments that can be specifically labeled by gold particles directed by antibody to the carboxy-terminal domain of Ure2p (Speransky et al. 2001). These filaments are only poorly detectable by gold particles labeled with antibody specific to the amino-terminal domain, suggesting that the prion domain is buried in the filament structure. Likewise, Ure2p filaments in extracts of [URE3] cells are detectable only with carboxy-terminus-specific antibody in blotting experiments (Speransky et al. 2001).

Structure of Ure2p Amyloid

The Ure2p prion domain plays a central role in prion formation in vivo and amyloid formation in vitro. Protease treatment of amyloid filaments produces thinner filaments composed of the prion domain (Taylor et al. 1999), and the most protease-resistant part of Ure2p in extracts of [URE3] cells is again the prion domain (Masison and Wickner 1995). These key findings indicate that in Ure2p amyloid (in vivo and in vitro), the connection between molecules of Ure2p is by interactions of the prion domain of one molecule with that of another molecule. The same conclusion is indicated by the ability of the prion domain to propagate [URE3] in the absence (or presence on separate molecules) of the carboxy-terminal domain (Masison et al. 1997). In addition, the prion domain is sufficient to direct amyloid formation by any of four carboxy-terminally attached enzymes, GST, green fluorescent protein (GFP), barnase, and carbonic anhydrase (Baxa et al. 2002) or amino-terminally attached GST (e.g., Fig. 7) (Schlumpberger et al. 2000).

The fact that Ure2p filaments in extracts of [URE3] cells are easily detected on blots by antibody to the carboxy-terminal domain of Ure2p, but not by antibody to the amino-terminal prion domain, suggests that the prion domain is buried in the amyloid filament structure (Speransky et al. 2001).

Non-amyloid Filaments of Ure2p

Ure2p (Taylor et al. 1999) and fusions of Ure2p with other proteins (Schlumpberger et al. 2000; Baxa et al. 2002) form filaments with all of the properties of amyloid. Moreover, these properties correspond closely to the properties of Ure2p in extracts of [URE3] strains (see preceding section). However, Ure2p is also reported to form (in vitro) a second type of filaments which are not amyloid in type (Thual et al. 1999; Bousset et al. 2002). These filaments are not resistant to proteinase K and do not show the increased β-sheet content typical of amyloid (Thual et al. 1999;

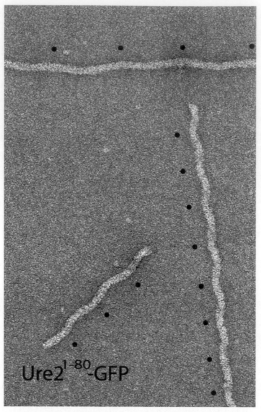

Figure 7. Amyloid filaments formed by the Ure2p prion domain fused to GFP can assume several morphologies. The helical repeat length in each case is consistent within each filament but varies between filaments, a phenomenon that may underlie strains of [URE3] (Baxa et al. 2002).

Bousset et al. 2002). Thus, these filaments are clearly distinct from those described by Taylor et al. (1999) and by Schlumpberger et al. (2000). The authors propose a model for [URE3] in which the filaments are formed by binding of the amino terminus of one monomer to the carboxyl terminus of the next monomer in the chain (Bousset et al. 2002). This model does not easily explain several key in vivo findings (the ability of the Ure2p prion domain to propagate [URE3] without the carboxy-terminal domain [Masison et al. 1997]; Ure2p is protease-resistant only in extracts of [URE3] cells [Masison and Wickner 1995]; the asymmetry between the role of prion domain and carboxy-terminal domain in generation and propagation of [URE3]; filaments of Ure2p in [URE3] cells are detectable by antibody to the carboxy-terminal domain of Ure2p, but not to antibody to the amino-terminal prion domain [Speransky et al. 2001]). Therefore, it is likely that the non-amyloid filaments described by Thual et al. and Bousset et al. are distinct in form from those found in [URE3] cells.

Mechanism of Inactivation of Ure2p by Amyloid Formation

[URE3] and [PSI^+] each produce a phenotype based on the inactivation of Ure2p and Sup35p, respectively. The conversion of PrP^C to PrP^{Sc} is believed to be accompanied by conversion of at least part of the α-helical structure of the native protein to β-sheet (Caughey et al. 1991; Pan et al. 1993). Alternatively, because both Ure2p and Sup35p have to interact with other macromolecules in their normal functions, it is possible that they are unable to do so simply because of steric blockage. To investigate this issue, fusions of the Ure2p prion domain were made with enzymes whose substrates are very small and therefore not subject to steric blockage: carbonic anhydrase, barnase, GFP, and GST (Baxa et al. 2002). The energy of denaturation of these enzymes is similar to that determined for Ure2p (Perrett et al. 1999), suggesting that if β-sheet formation by the prion domain could denature Ure2p, it could do so for these enzymes as well. The fusions with GST are of particular interest because the carboxyl terminus of Ure2p is very similar in sequence and especially in structure.

All of the fusion proteins formed amyloid filaments and all were active in both the soluble and amyloid forms. Fluorescence of Ure2–GFP was even slightly more efficient in the amyloid form than in the soluble form, and Ure2–GST was 80% as active in the amyloid form as in the soluble state (Baxa et al. 2002). The near full activity of the Ure2p$^{1-65}$–GST fusion protein amyloid argues strongly that the Ure2p carboxy-terminal domain is not converted to β-sheet form during amyloid formation, although this conclusion disagrees with the results of physical measurements of conformation of Ure2p amyloid (Taylor et al. 1999; Schlumpberger et al. 2000).

Both barnase and carbonic anhydrase are diffusion-limited enzymes. Incorporation of a diffusion-limited enzyme into a particle is expected to reduce the apparent activity because of the greater distance that the substrate must diffuse. This effect was estimated, and the results agreed well with the estimates (Baxa et al. 2002). That the apparently reduced activity of barnase was due to the diffusion limitation effect was confirmed by partial recovery of activity when filaments were made smaller by sonication, and by the expected temperature effects on diffusion (Baxa et al. 2002).

These results argue that conformational change is not the basis of the inactivation of Ure2p, and suggest that it is a steric effect, but the possibility was raised of a diffusion limitation effect on Ure2p, similar to that found for two enzymes in amyloid (Baxa et al. 2002).

Curing [URE3] by Fragments of Ure2p, Fusion Proteins, Meiosis, and Competition with Other Prions

Expression of fusions of Ure2p with GFP resulted in curing of [URE3] (Edskes et al. 1999a). Fusions that included the Ure2p prion domain were particularly active in curing, with efficient curing even when not overexpressed. Deletion studies showed that in fusions with GFP, residues 5–47 of the *S. cerevisiae* prion domain are necessary for curing the [URE3] prion (Edskes and Wickner 2002). Curing by the Ure2 carboxyl terminus fused to GFP required residues 116–333, most of the nitrogen regulation domain.

Overexpression of Ure2p–GFP fusion proteins resulted in an inactivation of the endogenous Ure2p even without the formation of stable [URE3] (Edskes and Wickner 2002). Residues 11–39 are necessary for this inactivating interaction. A nearly identical region is highly conserved among many of the yeasts examined in that study, despite the wide divergence of sequences found in other parts of the amino-terminal domains (Edskes and Wickner 2002). Whether this inactivating interaction is with the prion domain of the endogenous full-length Ure2p or with part of the carboxy-terminal domain remains to be seen.

The first curing method for [URE3] was its irregular segregation in meiosis (Aigle and Lacroute 1975). [URE3] can segregate 4[URE3]:0 in meiosis (Wickner 1994), like most other nonchromosomal genetic elements, but more often, it shows irregular segregation, or even 0:4[ure-o]. This may be a consequence of a small number of infectious seeds under meiotic growth conditions and the exclusion of much of the cytoplasm in formation of meiotic spores.

Recently, it has been found that [URE3] and [*PSI*⁺] destabilize and weaken each other (Schwimmer and Masison 2002). In the presence of [*PSI*⁺], the frequency of loss of [URE3] is higher than in its absence. Moreover, the growth-slowing effects of [URE3] are reduced in the presence of [*PSI*⁺] (Schwimmer and Masison 2002). [URE3] likewise weakens the suppression activity of [*PSI*⁺] and promotes its spontaneous loss (Schwimmer and Masison 2002). This may reflect competition of the prions for some prion-promoting cellular factor, or a cellular anti-prion response that is not substantially aroused by one prion, but is stimulated by two.

Strains of [URE3] with Different Strength and Stability

Strains of the scrapie agent, distinguished by incubation period and the distribution of lesions in the brain (Bruce and Fraser 1991), have been attributed to different conformers of PrP amyloid (Bessen and Marsh

1994). Similarly, strains of [PSI⁺] have been distinguished (Derkatch et al. 1996; see below).

Fusing the *DAL5* promoter to *ADE2* allows scoring of [URE3] by growth in the absence of adenine, or colony color with limiting levels of adenine, essentially the same assay used in studies of [PSI⁺] (Schlumpberger et al. 2001). Variation of color intensity gives a semi-quantitative measure of Ure2p activity, and loss of [URE3] can be assessed directly as change of colony color. It was critical that the *DAL5–ADE2* construct be integrated, as control of the plasmid-borne gene was too leaky to be useful.

Using this assay, independent isolates of [URE3] were examined for level of Ade2p expression and for mitotic stability of the [URE3] element (Schlumpberger et al. 2001). Type A isolates were pink or nearly white, and type B were red. This indicates that the Ure2p in type A isolates was more completely inactivated than in those of type B. The type B isolates were also more easily cured by expression of Ure2–GFP proteins than were type A [URE3] cells. However, type A cells, and the residual uncured cells of type B, both showed the characteristic punctate appearance of [URE3], indicative of aggregation of Ure2p. Both type A and type B showed partial protease resistance with similar digestion patterns, again indicating the formation of similar aggregates (Schlumpberger et al. 2001). The basis of the strain differences observed remains uncertain, but differences in the pattern of aggregation have been proposed as the explanation. The presence of unsolubilized Ure2p specifically in [URE3] strains has been reported (Speransky et al. 2001), and this is apparently specific for type A (Schlumpberger et al. 2001).

Interestingly, the [URE3] isolates induced by overexpression of the prion domain are all of type B, whereas those induced by overexpression of the full-length protein are of both type A and type B (Schlumpberger et al. 2001).

The physical basis for [URE3] strains is not known. However, Ure2p–GFP fusion proteins formed helical filaments whose repeat length was consistent within each filament, but varied more than twofold from one filament to another (Fig. 7) (Baxa et al. 2002). This suggests a self-propagating arrangement of molecules in the filaments that might be the basis of the [URE3] strains.

Chaperones Promote and Block [URE3] Propagation

Either deletion of *HSP104* or overproduction of its product results in the loss of [PSI⁺] (Chernoff et al. 1995; see below). [URE3] requires [PSI⁺] for its propagation, but overproduction of Hsp104 does not affect [URE3] stability (Moriyama et al. 2000). Overproduction of Ydj1p, a member of the Hsp40 family, also cures [URE3] (Moriyama et al. 2000). Remarkably, overproduction of the Hsp70 group chaperone Ssa1p cures [URE3], but the nearly identical Ssa2p does not cure [URE3] (Schwimmer and Masison 2002). Each similarly lowers Hsp104 levels modestly, suggesting that this is not the basis for the difference. Since Hsp104, Hsp70, and Hsp40 chaperones interact in their activities (Glover and Lindquist 1998), further work will be needed to dissect the mechanisms involved.

Mks1p and the Ras Signal Transduction Pathway Affect [URE3] Prion Generation

Because overexpression of Mks1p affects the activity of Ure2p in nitrogen regulation (Edskes et al. 1999b), its effects on the generation and propagation of [URE3] were tested. In an *mks1* deletion mutant, [URE3] arises very rarely, if ever (Edskes and Wickner 2000). Even overproduction of fragments of Ure2p that efficiently induce the de novo appearance of [URE3] at high frequency in a wild-type strain is ineffective in an *mks1* mutant. In the opposite direction, overproduction of Mks1p stimulates the appearance of [URE3] (Edskes and Wickner 2000). Neither of these effects operates by changing the level of Ure2p, nor are differences in the migration of Ure2p on SDS gels or isoelectric focusing gels seen in *mks1* mutants.

Since Mks1p was originally identified as a gene negatively regulated by the Ras–cAMP pathway (Matsuura and Anraku 1993), it was possible that this pathway might affect prion generation through its influence on Mks1p. Indeed, expression of Ras2^{val19}, a constitutively active ("oncogenic") allele, results in a marked reduction in the frequency of [URE3] generation (Edskes and Wickner 2000). The *mks1*Δ mutants have elevated levels of glutamate (Feller et al. 1997; Sekito et al. 2002), but elevated glutamate does not explain the effect of Mks1p on [URE3] generation. For example, providing glutamate as the sole source of nitrogen does not lower the frequency of [URE3] arising (Aigle 1979; H.K. Edskes, pers. comm.).

[PSI$^+$] IS A PRION FORM OF THE Sup35p TRANSLATION TERMINATION FACTOR

Discovery of [PSI$^+$]

In 1965, Cox described a non-Mendelian element which he named [PSI$^+$] that enhanced the efficiency of a Mendelian suppressor, SUQ5 (Cox

1965). In the presence of [*PSI*⁺] the weak *SUQ5* tRNA suppressor caused efficient readthrough of UAA stop codons (Liebman et al. 1975). [*PSI*⁺] also increased the efficiency of stronger UAA tRNA suppressors to the point where they were lethal (Cox 1971), and enhanced suppression of stop codons caused by non-tRNA suppressor (All-Robyn et al. 1990) or drugs (Palmer et al. 1979; A. Singh et al. 1979). In addition, [*PSI*⁺] acted as a nonsense suppressor by itself, without the presence of suppressor mutations or drugs (see Fig. 8) (Liebman and Sherman 1979). [*PSI*⁺] was dominant, was inherited by all meiotic progeny (Cox 1965), and was transmitted by cytoduction (Cox et al. 1988). Although this suggested that [*PSI*⁺] might be due to a nucleic acid located in the cytoplasm, no such nucleic acid was ever found. Indeed, several likely candidates, mitochondrial DNA, 2μ DNA, and the L-A and M double-stranded RNA viruses were eliminated (Young and Cox 1972; Leibowitz and Wickner 1978; Tuite et al. 1982; Cox 1993).

Mutations in the *SUP35* and *SUP45* loci cause a suppressor (Hawthorne and Mortimer 1968; Inge-Vechtomov and Andrianova 1970) or suppressor enhancer phenotype (Crouzet and Tuite 1987; Crouzet et al. 1988) similar to [*PSI*⁺]. However, these mutations differed from [*PSI*⁺] because they were recessive and segregated 2+:2– like Mendelian mutations. It is now clear that Sup35p (eRF3) and Sup45p (eRF1) together comprise the translational release factor that recognizes termination codons and promotes the release of polypeptide chains from the ribosome (Frolova et al. 1994, 1996; Stansfield et al. 1995; Zhouravleva et al. 1995). Because suppressor tRNA or noncognate tRNA competes with the release factors for termination codon recognition, readthrough of stop codons increases whenever the activity of the termination factor is compromised.

[*PSI*⁺] Meets the Genetic Criteria for a Prion

The similarity between the genetic properties of [*PSI*⁺] and [URE3] led to the hypothesis that [*PSI*⁺] is a prion form of Sup35p (Wickner 1994).

Reversible Curing of [PSI⁺]

Many treatments efficiently cure cells of [*PSI*⁺] without affecting Mendelian mutations. Exposure to high osmotic media caused efficient loss of [*PSI*⁺] (A.C. Singh et al. 1979), and the resulting [*psi*⁻] cells could give rise to [*PSI*⁺] revertants (Lund and Cox 1981). The most efficient

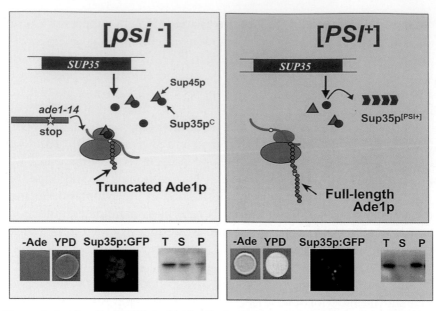

Figure 8. Scoring for [*PSI*⁺]. In [*psi*⁻] cells the Sup35p (*red circles*) and Sup45p (*green triangles*) translation termination factors complex together and efficiently promote termination at stop codons including, e.g., the premature stop codon in the *ade1-14* allele (*yellow star*). The unsuppressed *ade1-14* mutation prevents cells from growing in the absence of adenine (-Ade), and causes the cells to accumulate a red pigment when grown on complex medium (YPD). In [*PSI*⁺] cells much of the Sup35p is unavailable for termination because it is sequestered in the prion aggregate (*red arrowheads*). [*PSI*⁺] is scored by examining the level of suppression of nonsense mutations (see Cox 1993). [*PSI*⁺]-promoted readthrough of the *ade1-14* nonsense codon allows growth on -Ade and prevents the accumulation of the red pigment. A guanidine hydrochloride (GuHCl) curing test is used to distinguish [*PSI*⁺] from Mendelian mutations that cause a suppressor phenotype, because the suppressor phenotype associated with [*PSI*⁺] disappears following growth in the presence of 1–5 mM GuHCl. [*PSI*⁺] can also be scored even in the absence of suppressible markers by looking for aggregation of Sup35p. In one method, expression of Sup35p or its prion domain fused to GFP for 1–4 hours marks the presence of [*PSI*⁺] aggregates, which are detected under a fluorescent microscope as frequent dots within the cells compared with the diffuse fluorescence seen in [*psi*⁻] cells (Patino et al. 1996). In another method, the levels of Sup35p protein in lysate fractions (T, total lysate; S, supernatant; P, pellet) separated by centrifugation are compared by western blotting. Sup35p is enriched in supernatants of [*psi*⁻] vs. [*PSI*⁺] strains (Patino et al. 1996; Paushkin et al. 1996).

treatment to cure [*PSI*⁺] was growth in the presence of 5 mM guanidine HCl (Tuite et al. 1981). Although the resulting [*psi*⁻] cells were initially believed to be unable to revert to [*PSI*⁺] (Lund and Cox 1981), it is now clear that they can (Chernoff et al. 1993; Derkatch et al. 1997, 2000).

Transient Overproduction of Sup35 Induces [PSI⁺]

The presence of multicopy plasmids bearing the *SUP35* gene causes [*psi⁻*] cells to have the [*PSI⁺*] phenotype of nonsense suppression (Chernoff et al. 1993). Furthermore, when the plasmid is lost, a large fraction of the cells (as high as 20%) had become [*PSI⁺*] (Chernoff et al. 1993). This induction of [*PSI⁺*] was later shown to be caused by the overproduced Sup35p protein and not by an excess of the *SUP35* mRNA or DNA: A single-copy plasmid that overexpressed *SUP35* efficiently induced [*PSI⁺*], but the introduction of an early frameshift mutation that had no effect on the mRNA level prevented [*PSI⁺*] induction (Derkatch et al. 1996). Finally, insertion of purified Sup35p made in *E. coli* into [*psi⁻*] cells via a liposome technique also caused the de novo induction of [*PSI⁺*] (Sparrer et al. 2000).

The Relationship between Sup35p Mutations and [PSI⁺]

Because the *SUP35* termination factor activity is essential (Kushnirov et al. 1987, 1988; Wilson and Culbertson 1988), only *sup35* mutations that retain some activity are viable. As mentioned above, *sup35* mutants have the same nonsense suppressor phenotype as [*PSI⁺*], since both impair the activity of the Sup35p termination factor. Another crucial connection between [*PSI⁺*] and *SUP35* was the finding that a dominant mutation selected for causing the loss of [*PSI⁺*] was a missense mutation (called *PNM2*) in the *SUP35* gene (Young and Cox 1971; Doel et al. 1994). In addition, cells bearing deletions of small regions of the Sup35p protein near the *PNM2* missense mutation are unable to maintain [*PSI⁺*] (Ter-Avanesyan et al. 1994). The finding that loss-of-function mutations in the *SUP35* gene required for the maintenance of [*PSI⁺*] have the same phenotype as the presence of [*PSI⁺*] is the expected relationship if [*PSI⁺*] is a prion form of Sup35p, but not if *SUP35* is required for the maintenance of a cytoplasmic nucleic acid encoding [*PSI⁺*] (Wickner 1994).

Sup35p Prion Domain

The carboxy-terminal region of Sup35p (amino acids 254–685, Sup35C) is sufficient for viability and for translational release factor activity. This region is highly homologous to the translational elongation factor, EF-1α (Kushnirov et al. 1987, 1988; Wilson and Culbertson 1988). The amino-terminal nonessential 253 amino acids are composed of an N region (amino acids 1–123) which loosely corresponds to the prion domain (PrD), required for prion maintenance, and a charged middle domain, M

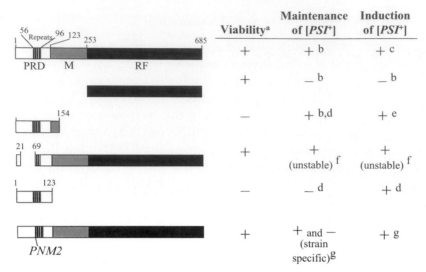

	Viability[a]	Maintenance of [*PSI*+]	Induction of [*PSI*+]
	+	+ [b]	+ [c]
	+	− [b]	− [b]
	−	+ [b,d]	+ [e]
	+	+ (unstable) [f]	+ (unstable) [f]
	−	− [d]	+ [d]
	+	+ and − (strain specific)[g]	+ [g]

Figure 9. Function of Sup35p domains. Shown are the Sup35p prion (PRD), middle (M), and release factor (RF) domains. Numbers indicate Sup35p amino acid residues. Vertical lines indicate the positions of the oligopeptide repeats in the prion domain. + or − respectively indicate the ability or inability of deletions of 1–253, 155–685, 21–69 (Kushnirov et al. 1990a; Ter-Avanesyan et al. 1993), 124–685 (M.E. Bradley and S.W. Liebman, in prep.), and the *PNM2* (G58D) mutation (Doel et al. 1994), to maintain cell viability or [*PSI*+] in the absence of the complete Sup35p protein, or to induce [*PSI*+] de novo when overexpressed. References: [a]Ter-Avanesyan et al. 1993; [b]Ter-Avanesyan et al. 1994; [c]Chernoff et al. 1993; [d]M.E. Bradley and S.W. Liebman, in prep.; [e]Derkatch et al. 1996; [f]Borchsenius et al. 2001; [g]Doel et al. 1994; Kochneva-Pervukhova et al. 1998b; Derkatch et al. 1999.

(aa 124–253) of unknown function (Fig. 9). As is true for other yeast prions, the prion domain of Sup35p is transferable (Wickner et al. 2000). When a fusion of the Sup35p prion domain and a mammalian glucocorticoid receptor protein was expressed in yeast, the chimeric protein behaved like a prion (Li and Lindquist 2000).

Although the smallest region sufficient to maintain [*PSI*+] initially seemed to be the first 114 amino acids of Sup35p (Ter-Avanesyan et al. 1994), it now appears that this was due to de novo induction and not maintenance of [*PSI*+] (Derkatch et al. 2000; M.E. Bradley and S.W. Liebman, in prep.). Currently the first 154 amino acids are the smallest region shown to be sufficient for [*PSI*+] maintenance (M.E. Bradley and S.W. Liebman, unpubl.). As described above for [URE3] (Masison and Wickner 1995), fragments containing the prion domain but lacking the C region are more efficient at inducing [*PSI*+] than is overproduction of the complete protein (Derkatch et al. 1996; Kochneva-Pervukhova et al. 1998a).

Two regions of Sup35p have been specifically implicated in prion activity. The first, amino acids 8–26, is very high in glutamine and asparagine residues. Propagation, in [*PSI*⁺] cells, of plasmids carrying mutations in this region either promotes [*PSI*⁺] curing (PNM for PSI-no-more), or causes a reduction in the [*PSI*⁺] phenotype of suppression (asu for antisuppression) (DePace et al. 1998). The specific sequence of the Gln/Asn-rich region does not appear to be critical, because a fusion of GFP with the first 253 amino acids of Sup35p, in which the entire Gln/Asn-rich region was replaced with polyGln, still aggregated into dots in [*PSI*⁺] but not [*psi*⁻] cells (DePace et al. 1998). The Ure2p prion domain is also Asn-rich. Such sequences are of special interest because Gln-rich sequences, found in triplet repeat diseases, have been shown to promote protein aggregation (Perutz et al. 1994).

The second Sup35p region critical for the [*PSI*⁺] prion, amino acids 40–97, contains five and a half imperfect oligopeptide repeats with the consensus sequence PQGGYQQ-YN (Fig. 5). A Gly to Asp mutation in the second repeat, *PNM2*, causes dominant curing of [*PSI*⁺] in some genetic backgrounds, although it still induces [*PSI*⁺] when overexpressed (Fig. 9) (Young and Cox 1971; Doel et al. 1994; Kochneva-Pervukhova et al. 1998b; Derkatch et al. 1999). Deletion of some of the oligopeptide repeats caused [*PSI*⁺] loss, whereas increasing the number of repeats stimulated the spontaneous appearance of [*PSI*⁺] (Liu and Lindquist 1999; Parham et al. 2001). A *SUP35* deletion covering the last several residues of the Gln/Asp-rich region and the first two and a half oligopeptide repeats (aa 22–69) still encodes a protein that can aggregate, but the aggregate does not propagate efficiently in vivo, probably because it is defective in the formation of prion seeds (Borchsenius et al. 2001). Similar oligopeptide repeats found in the mammalian PrP are not in the prion domain, but their mutation or duplication can cause inherited forms of prion diseases. Expansion of the repeat region causes disease, whereas reduction in the repeat number reduces susceptibility to the disease (Goldfarb et al. 1994; Prusiner and Scott 1997; Chiesa et al. 1998; Goldmann et al. 1998; Flechsig et al. 2000).

[*PSI*⁺] Causes Aggregation of Sup35p In Vivo

Sup35p in lysates from [*PSI*⁺] strains exists in a more rapidly sedimenting and more protease-K-resistant form than Sup35p in lysates from [*psi*⁻] strains (Patino et al. 1996; Paushkin et al. 1996). Furthermore, this rapid sedimentation is dependent only on the prion domain (Paushkin et al. 1996). In addition, Sup35p–GFP fusions rapidly aggregated into punctate dots in [*PSI*⁺] but not [*psi*⁻] living cells. As for sedimentation, the prion domain of Sup35p fused to GFP was sufficient to promote [*PSI*⁺]-specif-

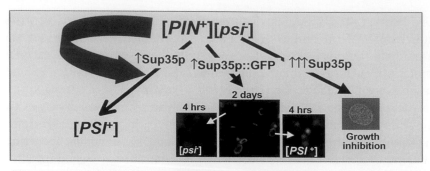

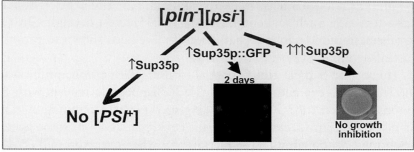

Figure 10. Properties of [*PIN*⁺]. The [*PIN*⁺] prion (also called [*RNQ*⁺]) provides an activity that promotes the de novo appearance of [*PSI*⁺]. In the presence of [*PIN*⁺] (*upper*) moderate overexpression of Sup35p leads to the induction of [*PSI*⁺] (Derkatch et al. 1997); long-term overexpression of Sup35p::GFP fusions in [*psi*⁻] cells leads to the appearance of ring, curve, and dot aggregates diagnostic for newly appearing [*PSI*⁺], cells with the aggregates give rise to [*PSI*⁺] progeny with characteristic dot (but not ring or curve) aggregates, cells without aggregates give rise to [*psi*⁻] progeny (Zhou et al. 2001); extreme overproduction of Sup35p causes growth inhibition in [*PIN*⁺] but not [*pin*⁻] cells (Derkatch et al. 1997).

ic aggregation (Fig. 8) (Patino et al. 1996). Upon longer overexpression of the Sup35p prion domain-GFP in [*psi*⁻] cells, large ring and line aggregates often appear, and cells carrying these aggregates have become [*PSI*⁺] (Fig. 10) (Zhou et al. 2001).

The composition of the [*PSI*⁺] aggregates is not known. Sup45p and Upf1p were found to be in the aggregate by one group (Paushkin et al. 1997b; Czaplinski et al. 1998), whereas a different group found that Sup45p sedimentation was not affected by the presence of [*PSI*⁺] (Patino et al. 1996).

In Vitro Propagation of Sup35p Aggregation

In early experiments, mixtures of [*psi*⁻] and [*PSI*⁺] extracts exhibited the low level of translational readthrough characteristic of [*psi*⁻] but not [*PSI*⁺] (Tuite et al. 1983, 1987). Since [*PSI*⁺] is dominant in vivo, this result stood in the face of classical genetics. Now, the result can be easily understood (Wickner et al. 1995), since according to the prion model the [*PSI*⁺] phenotype is caused by the absence of soluble Sup35p. Apparently in the mixture, the soluble Sup35p provided by [*psi*⁻] extract promoted translational termination before being recruited into [*PSI*⁺] aggregates.

More recently, soluble Sup35p found in lysates of [*psi*⁻] cells was shown to be efficiently recruited into aggregates when mixed with small amounts of [*PSI*⁺] extracts (Paushkin et al. 1997a). The [*PSI*⁺] cellular fraction associated with this activity was that containing aggregated Sup35p: Fractions containing only soluble Sup35p were inactive. Aggregates made by this procedure could again be used to seed aggregation of soluble Sup35p in [*psi*⁻] extracts.

Purified Sup35p, or Sup35p prion domain made in *E. coli*, self-assembles into ordered β-sheet-rich fibers that show a green birefringence with Congo red that is characteristic of amyloid. Fiber formation proceeds via a cooperative mechanism. There is a long lag, during which the first small aggregates or "seeds" presumably form. These then seed the rapid assembly of soluble Sup35p into fibers. The lag phase is significantly reduced by rotating the Sup35p solutions. The lag is also reduced by the addition of small amounts of in vitro made fibers or the addition of [*PSI*⁺], but not [*psi*⁻], extract (Glover et al. 1997; King et al. 1997; DePace et al. 1998; Serio et al. 2000; Uptain et al. 2001). Evidence suggests that at least some of the fibers grow in a bidirectional manner (Inoue et al. 2001; Scheibel et al. 2001; DePace and Weissman 2002).

Remarkably, the in vitro assembly properties of mutant Sup35p reflect their in vivo effects on [*PSI*⁺] maintenance, spontaneous appearance, or phenotype (DePace et al. 1998; Kochneva-Pervukhova et al. 1998b; Liu and Lindquist 1999). This suggests that the in-vitro-described amyloid fibers may correspond to the in vivo [*PSI*⁺] prion conformation.

Retention of the [*PSI*⁺] Prion Domain in Evolution

The carboxy-terminal region of Sup35p, which encodes the essential translational termination factor activity, has been highly conserved. Indeed, the *S. cerevisiae* and human Sup35C domains show 52% identity (Hoshino et al. 1989). The N and M domains, which are dispensable for termination factor activity, are less well conserved, but yeast homologs generally have the same unusual amino acid composition: the M domain is often charged; the N domain is often uncharged, very Gln/Asn-rich, and bears oligopeptide repeats (Kushnirov et al. 1990b; Chernoff et al. 1992; Santoso et al. 2000; Nakayashiki et al. 2001). The human Sup35N domain also has an unusual amino acid composition rich in glycine (Hoshino et al. 1989; Jean-Jean et al. 1996). A quantitative analysis of *SUP35* nucleotide polymorphisms in various *S. cerevisiae* strains compared with the sequence divergence in a related species, *S. paradoxus*, indicated that there is a weak purifying selection against mutations that alter the N and M portion of the protein. This suggests that the Sup35NM region has an adaptive function. (Jensen et al. 2001). Possibly N and M have some other important but nonessential role in the cell. Indeed, the N region is known to interact with a protein involved with the cytoskeleton, Sla1p, as well as with proteins that have a role in glucose metabolism, Reg1p and Eno2p (Bailleul et al. 1999). Alternatively, the adaptive function of NM may be prion formation (Eaglestone et al. 1999; True and Lindquist 2000). In addition, [*PSI*⁺] provides a benefit in heat tolerance in some backgrounds (Eaglestone et al. 1999).

Although aggregation of Sup35p indicative of the prion state was not observed among industrial or natural strains of yeast (Chernoff et al. 2000), the prion form of Sup35p was shown to be inducible in at least one other yeast species, *Kluyveromyces lactis* (Nakayashiki et al. 2001). When the prion domains of Sup35p from divergent yeast species, *Pichia methanolica*, *C. albicans*, and *K. lactis*, were fused with the *S. cerevisiae* Sup35C domain and expressed in *S. cerevisiae*, they behaved like [*PSI*⁺] prions (Chernoff et al. 2000; Kushnirov et al. 2000a; Santoso et al. 2000; Nakayashiki et al. 2001). Interestingly, within the *S. cerevisiae* host the prion form of a fusion with the NM domain from one yeast species generally failed to convert fusion proteins bearing NM domains from a different yeast species into the prion form, and efficient induction of the prion state by transient overproduction of the NM domain was species-specific, although some exceptions were noted. This species barrier was also demonstrated in vitro, since seeds of one species did not usually stimulate aggregation of another species' protein (Santoso et al. 2000). The conformation of the seeding aggregates, as well as their primary sequence, appears to be a factor in determining the species barrier, at least in vitro. This is because a purified chimeric fusion containing elements of both the *S. cerevisiae* and *C. albicans* Sup35p prion domain was shown to be able to aggregate in two distinct conformations, one of which preferentially seeded aggregation of *S. cerevisiae* Sup35p, whereas the other preferentially seeded *C. albicans* Sup35p aggregation (Chien and Weissman 2001).

The inhibition of Sup35p prion transmission across species lines has been compared to the "species barrier" described in transmission of the PrP prion state between different mammalian species (Chernoff et al. 2000; Kushnirov et al. 2000a; Santoso et al. 2000). Note, however, the yeast Sup35NM sequences are considerably more divergent than the highly conserved sequences of mammalian PrP.

[*PSI*⁺] Strains

One surprising aspect of PrP prions is the existence of different forms of the disease in inbred animals (Bruce and Fraser 1991). Evidence suggests that these so-called different prion "strains" result from infection with PrP aggregates in distinct conformations (Bessen and Marsh 1992, 1994; Kocisko et al. 1995; Safar et al. 1998). Thus, although the PrP protein has the same primary sequence, it appears to be able to form aggregates with distinct, heritable shapes associated with the different pathologies. Analogous findings of prion strain differences have now been reported for several yeast prions (Derkatch et al. 1996; Schlumpberger et al. 2001; Bradley et al. 2002). The first and best-studied such example is the different strains (sometimes called variants) of [*PSI*⁺] (Fig. 11).

Transient overproduction of Sup35p (or fragments of Sup35p containing the prion domain) induced the de novo appearance of [*PSI*⁺]

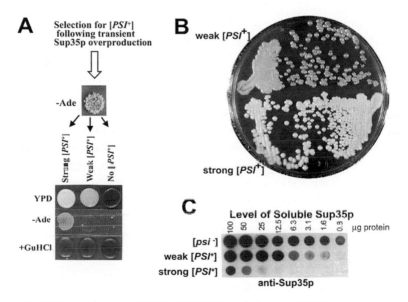

Figure 11. Different strains of [*PSI*⁺]. (*A*) Different strains of [*PSI*⁺] arise in the same genetic background following transient overexpression of Sup35p (Derkatch et al. 1996). Shown is yeast carrying the *ade1-14* nonsense mutation that is suppressed more efficiently by strong [*PSI*⁺] than by weak [*PSI*⁺], as shown by more efficient growth on adenineless media and a white vs. pink color. Both [*PSI*⁺]s are cured by growth in 5 mM guanidine hydrochloride (+GuHCl). (*B*) Colony purification of weak and strong [*PSI*⁺]. The more frequent appearance of red colonies and sectors from the weak [*PSI*⁺] shows that weak [*PSI*⁺] is less stable than strong [*PSI*⁺] (Derkatch et al. 1996). (*C*) Strong [*PSI*⁺] contains less soluble Sup35p than weak [*PSI*⁺], which has less soluble Sup35p than [*psi*⁻] (Zhou et al. 1999). (Adapted, with permission, from Derkatch et al. 1996; Zhou et al. 1999.)

derivatives with different characteristics (Derkatch et al. 1996). Weak [*PSI*⁺] causes a lower level of translational readthrough (scored as efficiency of nonsense suppression) than strong [*PSI*⁺]. This phenotypic difference is not caused by Mendelian mutations because cells cured of weak or strong [*PSI*⁺] factors can again give rise to the same array of [*PSI*⁺] (Derkatch et al. 1996). Furthermore, the [*PSI*⁺] strain-specific characteristics are faithfully cytoduced along with the [*PSI*⁺] factor (Kochneva-Pervukhova et al. 2001; Bradley et al. 2002). Interestingly, the relative frequency of weak versus strong [*PSI*⁺] induced varies depending on the Sup35p fragment overproduced (Kochneva-Pervukhova et al. 2001).

Quantitative western blotting experiments showed that there is more nonaggregated Sup35p in weak than in strong [*PSI*⁺] (Zhou et al. 1999; Uptain et al. 2001). Presumably, the nonaggregated Sup35p in the weak [*PSI*⁺] is available to facilitate translational termination resulting in the observed reduced readthrough. Another difference between the [*PSI*⁺] elements is that weak [*PSI*⁺] is less stable than strong [*PSI*⁺] and is more easily cured by growth in GuHCl or by overexpression of chaperone proteins (Hsp104, Ssb1p, and Ssa1p, together with Ydj1p) (Derkatch et al. 1996; Kushnirov et al. 2000b). Also, more cells contained punctate fluorescent aggregates when Sup35NM–GFP was expressed in strong [*PSI*⁺] than in weak [*PSI*⁺] (Derkatch et al. 1999).

[*PSI*⁺] strains have also been distinguished by their different responses to *SUP35* mutations. The *PNM2* allele of *SUP35* enhanced the readthrough caused by weak [*PSI*⁺], but reduced the readthrough caused by strong [*PSI*⁺] (Derkatch et al. 1999). Likewise, three [*PSI*⁺] strains were distinguished by their characteristic alterations in readthrough efficiency in response to 11 *SUP35* point mutations (King 2001). These [*PSI*⁺] strains were also distinguished by the different appearance of fluorescent aggregates induced by fusions of GFP with various Sup35p mutants and truncations.

Except for the occasional loss of [*PSI*⁺], diploids and all mitotic and meiotic progeny resulting from crosses between isogenic yeast bearing weak and strong [*PSI*⁺] elements were indistinguishable from the strong [*PSI*⁺] parent (Bradley et al. 2002). Moreover, the weak [*PSI*⁺] could not be recovered from such diploids even after stimulation of [*PSI*⁺] loss by short-term growth in medium containing GuHCl (M.E. Bradley and S.W. Liebman, unpubl.). Possibly only one stable [*PSI*⁺] strain can be maintained in a cell at a time. However, an unstable weak [*PSI*⁺] that continually throws off strong [*PSI*⁺] sectors has been described previously (Kochneva-Pervukhova et al. 2001).

It is generally believed that strains of [*PSI*⁺], like strains of PrP, result from distinct prion protein conformations. Several in vitro studies support this hypothesis. When purified Sup35NM forms spontaneous aggregates, individual fibers have either a wavy or straight morphology in electron micrographs. The absence of single fibers with both the wavy and straight characteristics suggests that distinct aggregate structures are propagated in vitro (Glover et al. 1997). Likewise, Sup35NM fibers have been shown to differ in overall growth rates and the degree of growth polarity from the fiber ends (Glover et al. 1997; DePace and Weissman 2002). To do this, fibers and the Sup35NM joining them were differentially labeled so that fiber growth could be examined by atomic force microscopy. Since these fiber differences were shown to be self-propagating, they mimic in vitro what is hypothesized to account for [*PSI*⁺] strains in vivo (DePace and Weissman 2002). No structural differences have yet been distinguished between the fiber types using atomic force microscopy (DePace and Weissman 2002). In addition, chimeric Sup35NM with the same primary structure can be seeded to form aggregates in vitro with distinct conformations, as demonstrated by their different proteolytic digestion patterns (Chien and Weissman 2001). Furthermore, fibers with the distinct digestion patterns can faithfully seed new fibers with the same distinct characteristics.

There is also evidence directly linking in vitro observations and the in vivo phenomenon of [*PSI*⁺] strains (Uptain et al. 2001). Protein extracts from strong [*PSI*⁺] cells converted purified Sup35NM into fibers much more efficiently than did protein extracts from weak [*PSI*⁺] cells. At least a 20-fold greater seeding efficiency was observed even when normalized to account for the excess aggregate present in the strong [*PSI*⁺] extracts. Although no structural differences were found between fibers seeded by weak or strong [*PSI*⁺] extracts, the lack of difference between these fibers could be because the [*PSI*⁺] strain difference is not faithfully reproduced in vitro. Indeed, the fibers formed in vitro did not retain the differential seeding efficiency when used as seed in a second round of seeding experiments.

Chaperones and [*PSI*⁺]

The first definitive connection between the propagation of any prion and a chaperone was the finding that an extra copy of *HSP104* antagonizes the suppression activity associated with [*PSI*⁺] and cures [*PSI*⁺], albeit inefficiently (Chernoff et al. 1995). Surprisingly, both transient high overproduction, and deletion or inactivation of Hsp104, caused efficient loss of [*PSI*⁺] (Fig. 12) (Chernoff et al. 1995). Moderate overexpression of Hsp104 was shown to solubilize some of the aggregated Sup35p in [*PSI*⁺] cultures (Patino et al. 1996; Paushkin et al. 1996), and measurements of CD spectra

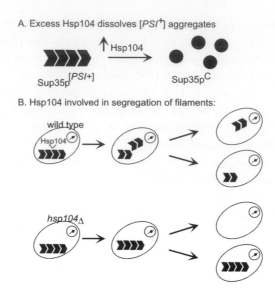

Figure 12. Proposed roles of Hsp104 chaperone in propagation of [*PSI*⁺]. (*A*) Excess Hsp104 may cure [*PSI*⁺] and other prions with relatively small aggregates by directly dissolving the aggregate. (*B*) Hsp104 may be required for propagation of [*PSI*⁺] because of its essential role in breaking the large aggregates into pieces or seeds that can be segregated into daughter cells. (Adapted, with permission, from Wickner et al. 1999.)

provide evidence for a transient interaction between purified Sup35p and Hsp104 (Schirmer and Lindquist 1997). Since Hsp104 promotes the disaggregation of aggregated proteins (Parsell et al. 1994), the finding that the level of Hsp104 is critical for the maintenance of [*PSI*⁺] inefficiently (Chernoff et al. 1995) strongly supports the prion model for [*PSI*⁺].

It seems likely that overexpressed Hsp104 cures cells of [*PSI*⁺] by disaggregating Sup35p prion complexes inefficiently (Chernoff et al. 1995). Low levels of Hsp104 may cause the loss of [*PSI*⁺] either because Hsp104 is required for soluble Sup35p to refold into the prion state inefficiently (Chernoff et al. 1995), and/or because Hsp104 is required to prevent prion aggregates from getting too big to be efficiently passed on to daughter cells (Paushkin et al. 1996). Recent evidence supports the latter hypothesis because inactivation of Hsp104 causes a rapid increase in [*PSI*⁺] prion aggregate size and the loss of [*PSI*⁺] in daughter cells (Wegrzyn et al. 2001). Also, unstable [*PSI*⁺] strains associated with larger than normal prion aggregates can be stabilized by a moderate increase in the level of Hsp104 (Borchsenius et al. 2001). Further evidence that Hsp104 is not required for formation of prion aggregates comes from the finding that overexpressed Sup35p–YFP aggregates even in cells lacking Hsp104, as long as a substitute for the [*PIN⁺*] prion required for induction of [*PSI⁺*] (see below) is present in the cell (Osherovich and Weissman 2001).

The yeast prion systems have been further exploited in an effort to discover the role other chaperones have in prion maintenance. Surprisingly, the effects of chaperones appear to differ for different prions and even for different strains of the same prion. Although deletion of Hsp104 has so far been found to cure the other known yeast prions, [URE3] (Moriyama et al. 2000) and [*PIN⁺*] (Derkatch et al. 1997; Sondheimer et al. 2001), Hsp104 overexpression only cures [*PSI⁺*]. Possibly [URE3] and [*PIN⁺*] prion aggregates are larger than [*PSI⁺*] aggregates, and are thus more resistant to disaggregation by elevated Hsp104 levels. Since Hsp104 works together with Hsp70 and Hsp40 to disaggregate aggregated proteins (Sanchez et al. 1993; Glover and Lindquist 1998), these chaperones have been extensively studied for their effects on [*PSI⁺*].

The Ssa–Hsp70 chaperones are encoded by a set of four conserved and functionally redundant genes (Boorstein et al. 1994). Ssa has been proposed to enhance prion aggregate formation by refolding Sup35p molecules that have been disaggregated by Hsp104 back into the prion form. This proposal arose from the finding that overproduction of Ssa1p enhanced the suppression efficiency of [*PSI⁺*] and protected [*PSI⁺*] from the curing effect of high levels of Hsp104 (Newnam et al. 1999). Although deletion of all four Ssa genes is lethal, and deletion of individual genes is compensated for by autoregulation of expression of the remaining Ssa loci, a dominant mutation in *SSA1* has been identified that interferes with [*PSI⁺*] propagation (Jung et al. 2000). Also consistent with this idea are two recent findings in which overproduced Ssa1p destabilizes [*PSI⁺*], presumably by making the [*PSI⁺*] prion aggregates so large that they cannot be segregated into daughter buds. First overproduction of Ssa1p enhanced destabilization of an already unstable, large aggregate, [*PSI¹*] (Borchsenius et al. 2001), and second overproduction of Ssa1p enhanced curing of [*PSI⁺*] caused by expression of a dominant-negative Hsp104 mutant (Wegrzyn et al. 2001).

The presence of Ssb–Hsp70 chaperones appears to interfere with prion formation. Indeed, overexpression of Ssb1p cures some strains of [*PSI⁺*] (Kushnirov et al. 2000b; Chacinska et al. 2001) and enhances the efficiency with which Hsp104 can cure [*PSI⁺*] (Chernoff et al. 1999). As expected, depletion of Ssb has the reverse effect, enhancing spontaneous and induced [*PSI⁺*] appearance, and stabilizing [*PSI⁺*] in the presence of Hsp104 overproduction (Chernoff et al. 1999). Since changes in Ssb levels are known to affect intracellular proteolysis (Ohba 1997), it has been suggested that Ssb may inhibit [*PSI*⁺] formation by targeting pre-prion Sup35p intermediates for degradation (Chernoff 2001). Alternatively, the Ssb chaperone may help fold Sup35p nascent protein into the soluble form.

Additional chaperone effects were uncovered by using artificial [*PSI*⁺] prions formed from a fusion of the *S. cerevisiae* Sup35p C-domain with the *Pichia methanolica* Sup35p NM-domain (Kushnirov et al. 2000b). Certain of these [*PSI*⁺] strains were cured by overexpression of the Hsp70 chaperones, Ssa1p or Ssb1p, and the Hsp40 chaperone, Ydj1p (Kushnirov et al. 2000b). Additional chaperones (Sis1p, Apj1p, and Sti1p) were uncovered in an overexpression screen for genes that can cure these [*PSI*⁺] (Kryndushkin et al. 2002).

Guanidine Cures Yeast Prions by Inhibition of Hsp104

[*PSI*⁺] was found to be curable by growth in media high in osmotic strength (A.C. Singh et al. 1979). When this curing was found to be reversible (Lund and Cox 1981), it was suspected that [*PSI*] was not actually cured. Growth in the presence of guanidine at concentrations of 1–5 mM was found to be an efficient method to cure [*PSI*⁺] (Tuite et al. 1981). This curing appeared to be irreversible, a result now explained by the discovery of the guanidine-curable [*PIN*⁺] prion of Rnq1p (Derkatch et al. 1997, 2000, 2001; Sondheimer and Lindquist 2000) (see below). In fact, guanidine curing was reversible (Derkatch et al. 1997), although not efficiently.

The mechanism of action of guanidine was suggested by the findings that the chaperone Hsp104 is necessary for [*PSI*⁺] propagation (Chernoff et al. 1995) and that guanidine could inhibit the ATPase activity of Hsp104 in vitro (Glover and Lindquist 1998). In fact, it was found that curing concentrations of guanidine inhibited the heat-shock protection function of Hsp104 (Ferreira et al. 2001; Jung and Masison 2001). The final proof came with the isolation of a mutant of *HSP104* with a single amino acid change (D184Y) that made cells completely incurable by guanidine (Jung et al. 2002). Other mutations in this residue, in the amino-terminal nucleotide-binding domain, led to loss of [*PSI*⁺], and there was no correlation of thermotolerance with effects on [PSI⁺] (Jung et al. 2002).

Guanidine inhibition was used to examine the role of Hsp104 in promoting the propagation of [*PSI*⁺]. Following addition of guanidine, cells only begin to lose [*PSI*⁺] after a delay of several generations (Eaglestone et al. 2000). A similar result was obtained by following loss of [*PSI*⁺] due to a dominant mutation of Sup35p (McCready et al. 1977). Assuming the delay was due to dilution out of seeds of [PSI^+], it was estimated that cells have about 60 seeds per cell. Guanidine treatment reduces the number of seeds per cell, indicating that Hsp104 acts by disaggregating the amyloid aggregates of Sup35p to produce seeds (Eaglestone et al. 2000), a mechanism first suggested by Paushkin et al. (1996).

Other Factors Influencing the Stability of [PSI^+]

When [PSI^+] is induced de novo by transient overexpression of Sup35p, many of the newly induced [PSI^+] elements are unstable and only become stabilized upon further propagation (I. Derkatch and S.W. Liebman, unpubl.). When [psi^-] cultures were plated nonselectively to score for [PSI^+] appearance following Sup35p overexpression, half of the colonies that were no longer the red color indicative of [psi^-] contained red sectors indicative of an unstable [PSI^+]. The other half of the non-red colonies were white or pink, indicative of stable strong or weak [PSI^+], respectively. Upon colony purification, the highly sectoring colonies continue to sector while throwing off some stable [PSI^+] (white or pink) and [psi^-] (red) progeny. Similar observations have been noted for the [URE3] prion (R. Wickner, pers. comm.), the [PIN^+] element (Derkatch et al. 2000), and artificial prions containing fusions of Sup35p prion domains from different yeast (Chernoff et al. 2000; Kushnirov et al. 2000a; Li and Lindquist 2000; Santoso et al. 2000). Apparently, instability upon initial appearance is a common characteristic of prions.

As mentioned above, different "strains" of [PSI^+] are associated with different levels of mitotic stability. Weak [PSI^+] is lost in about 0.1–1% of mitotically dividing cells, while strong [PSI^+] is lost in <0.03% of cells (Derkatch et al. 1996). In addition, meiosis destabilizes [PSI^+], causing a 10x increase in the loss of weak [PSI^+] (Bradley et al. 2002). Perhaps this loss in meiosis is caused by a change in chaperone levels and/or by the reduced level of cytoplasm (and therefore prion seeds) inherited by meiotic spores relative to mitotic daughter cells.

Although the most efficient treatment for curing [PSI^+] is growth in the presence of GuHCl, a variety of other agents (e.g., 1.8 M ethylene glycol, 1 M glutamate, 1.8 M KCl, 2.5% dimethylsulfoxide) also promote [PSI^+] loss (A.C. Singh et al. 1979; Tuite et al. 1981). Interestingly, when yeast strains contain another prion, [PIN^+], in addition to [PSI^+], loss of either prion, promoted by GuHCl (Derkatch et al. 2000) or inactivation of Hsp104 (Wegrzyn et al. 2001), is preferentially associated with loss of the other prion.

The carboxy-terminal region of the cytoskeletal assembly protein Sla1p (Sla1C) was retrieved from a two-hybrid screen for proteins that interact with the prion domain of Sup35p (Sup35N) (Bailleul et al. 1999). Like Sup35N, Sla1C is rich in glutamine, proline, and glycine. Deletion of *HSP104* interrupts this two-hybrid interaction. Possibly the dependence of the Sla1C–Sup35N two-hybrid interaction on Hsp104 is similar to that proposed for maintenance of [*PSI⁺*], i.e., inability to break large aggregates into heritable seeds. If this is true, deletion of *HSP104* may cause large Sla1C/Sla1C or Sup35N/Sup35N aggregates to form, preventing Sla1C/Sup35N binding, or cause the Sla1C/Sup35N aggregates that are formed to be too large to activate the reporter. Alternatively, the interaction between Sla1C and Sup35N may be dependent on the presence of a prion (e.g., [*PIN⁺*]; see below) cured by the absence of Hsp104. Deletion of *SLA1* reduces the frequency of de novo induction of [*PSI⁺*] and makes [*PSI⁺*] less stable in the presence of curing by DMSO or overexpression of Hsp104 without affecting curing by GuHCl.

The finding that the cytoskeletal assembly protein Sla1p interacts with Sup35p suggested a possible connection between cytoskeletal networks (Bailleul et al. 1999). Indeed, weak [*PSI⁺*] was efficiently cured by prolonged incubation in the presence of the anticytoskeletal drug, latrunculin A, as long as protein synthesis was not inhibited (Bailleul-Winslett et al. 2000). The curing effect of this drug did not seem to be mediated by altering chaperone levels.

Factors Influencing the De Novo Appearance of [*PSI⁺*]

Prolonged incubation at 4°C increases the frequency of de novo appearance of [*PSI⁺*] (unpublished observations cited in Derkatch et al. 2000; Chernoff 2001). Likewise, growth at 20–25°C (instead of the normal 30°C incubation temperature) (Chernoff et al. 1999, 2000), or growth in stationary phase at 30°C (Zhou et al. 2001), increased the frequency of [*PSI⁺*] appearance induced by overproduction of Sup35p.

The efficiency with which overproduced Sup35p (eRF3) induces the de novo appearance of [*PSI⁺*] is reduced by simultaneous overproduction of another component of the translational termination factor complex, Sup45p (eRF1) (Derkatch et al. 1998). Although overexpression of Sup45p inhibits induction of [*PSI⁺*], it does not destabilize [*PSI⁺*] once it is established. This suggests that Sup45p inhibits [*PSI⁺*] seed formation (Derkatch et al. 1998). Since Sup45p is known to bind directly to Sup35p (Paushkin et al. 1997b; Ebihara and Nakamura 1999), it seems likely that excess Sup45p sequesters Sup35p into the normal functional heterodimer, thereby reducing the level of free Sup35p. Since free Sup35p may be more likely to undergo the conformational flips and/or self-interactions leading to prion formation, this could explain how excess Sup45p inhibits [*PSI*+] formation.

Heterologous Prions Influence De Novo Appearance of Each Other

The first indication that different prions can influence each other was the discovery of a prion-like element, named [*PIN*+] (for [*PSI*+] inducibility). [*PIN*+] was identified because it enabled [*PSI*+] to be induced de novo by overproduction of Sup35p (see Fig. 10) (Derkatch et al. 1997). We now know that [*PIN*+] also enhances the de novo appearance of [URE3] following overexpression of the *URE2* prion domain (Bradley et al. 2002). As expected for a prion, [*PIN*+] was inherited in a non-Mendelian manner, could reappear after curing, and was cured by growth in GuHCl or by deletion of *HSP104*. Like [URE3] (Moriyama et al. 2000), but unlike [*PSI*+] (Chernoff et al. 1995), [*PIN*+] was not cured by overexpression of Hsp104 (Derkatch et al. 1997, 2000; Sondheimer et al. 2001).

[*PIN*+] has now been identified as the prion form of Rnq1p (Derkatch et al. 2001). Although no function for the soluble form of Rnq1p has been determined, Rnq1p was identified as a candidate prion on the basis of its sequence homology with the Sup35p prion domain and because aggregation of Rnq1p was inherited in a cytoplasmic manner (Sondheimer and Lindquist 2000). The Hsp40 chaperone, Sis1p, interacts with the prion form of Rnq1p, and the G/F domain of Sis1p is required for [*PIN*+] (Sondheimer et al. 2001). Furthermore, deletion of a different region of Sis1p causes a heritable alteration in the aggregation pattern of Rnq1 that persists in the presence of wild-type Sis1p (Sondheimer et al. 2001). Different strains of [*PIN*+] have also been identified that promote de novo [*PSI*+] formation with characteristic frequencies and which are associated with distinct levels of soluble Rnq1p (Bradley et al. 2002). In crosses between different [*PIN*+] strains, the strain associated with the higher level of soluble Rnq1p is always lost, presumably because soluble Rnq1p is sequestered by the faster-growing aggregates associated with the strain with less soluble Rnq1p.

Other prions, [URE3] (Derkatch et al. 2001) and [*NU*+] (an artificial prion composed of a Gln-rich domain from the New1p protein fused to the Sup35C reporter) (Osherovich and Weissman 2001), also exhibit the Pin+ phenotype of enhancing the de novo appearance of [*PSI*+] upon overexpression of Sup35p. In addition, [*PSI*+] allows the more efficient appearance of Rnq1p::GFP aggregates following incubation at 4°C

(Derkatch et al. 2001). In addition, overexpression of nine candidate prion genes, all carrying domains rich in glutamine, can substitute for [*PIN*⁺] (Derkatch et al. 2001). These findings suggest the possibility that heterologous prions or prion-like aggregates may promote the de novo appearance of another prion by an inefficient cross-seeding mechanism. Another possibility is that the heterologous aggregates titrate out an inhibitor of de novo prion formation.

In addition to the positive interactions between prions, negative interactions have also been reported. [*PSI*⁺] inhibits the appearance of [URE3] in the presence (Schwimmer and Masison 2002), or absence (Bradley et al. 2002), of [*PIN*⁺]. Also, [*PSI*⁺] enhances the loss of [URE3], and in some genetic backgrounds [URE3] inhibits the appearance of [*PSI*⁺] in the presence of [*PIN*⁺] (Schwimmer and Masison 2002). It has been suggested that the elevation of Hsp104 caused by [URE3] may enhance curing of [*PSI*⁺], whereas the elevation of Ssa1p by [*PSI*⁺] (Jung et al. 2000) may enhance curing of [URE3] (Schwimmer and Masison 2002). Factors affecting the frequency with which several prions arise de novo are summarized in Figure 13.

[HET-S] OF *PODOSPORA ANSERINA*, A PRION IN A FILAMENTOUS FUNGUS

The [Het-s] prion of *P. anserina* is involved in a cell death phenomenon that is very common in filamentous fungi and is known as heterokaryon incompatibility. In this section, we first describe some of the basic features of filamentous fungi that are relevant for the issue of prion propagation. We then recapitulate early genetic and physiological investigations on [Het-s] propagation and summarize recent molecular data. Finally, some hypotheses about the biological significance of [Het-s] are discussed.

What Is *het-s/het-S* Heterokaryon Incompatibility?

Filamentous fungi such as *Podospora* are evolutionarily related to yeast (both are ascomycete fungi) but display several specific biological characteristics. These fungi grow as a network of filaments (the mycelium) composed of multinucleated "cells," called articles. Articles are not individualized cells. The crosswalls (septa) between articles are perforated so that there exists a cytoplasmic continuity throughout the mycelium. Moreover, these filaments are able to fuse with one another. This syncytial cellular organization has obvious advantages, as it allows nutrient circulation

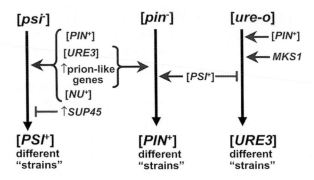

Figure 13. Summary of factors influencing the de novo appearance of prions. Overexpression of Sup35p can cause the de novo appearance of different variants (or strains) of the [*PSI*⁺] prion (Derkatch et al. 1996), presumably because the excess in Sup35p increases the chance that a [*PSI*⁺] seed will form (Wickner 1994). This de novo [*PSI*⁺] induction requires an additional factor (Derkatch et al. 1997, 2000), which can be supplied by the presence of the [*PIN*⁺], [URE3], or [*NU*⁺] prion, or by overexpression of any of a number of prion-like genes with Gln/Asn-rich regions (Derkatch et al. 2001; Osherovich and Weissman 2001). Furthermore, overexpression of the Sup35p-binding protein, Sup45p, inhibits the de novo appearance of [*PSI*⁺] (Derkatch et al. 1998). The efficiency of the de novo appearance of difference strains of the [*PIN*⁺] (Bradley et al. 2002) or [URE3] (Schlumpberger et al. 2001) prions is also affected by the presence of other prions as indicated (Derkatch et al. 2001; Bradley et al. 2002). Mks1p stimulates the de novo appearance of [URE3], and Ras inhibits Mks1p in this action (Edskes and Wickner 2000).

throughout the mycelium and construction of a three-dimensional interconnected network. This organization also has two major drawbacks. First, damage to a single article can compromise the whole mycelium. This peril is kept at bay by specialized organelles termed Woronin bodies. Upon wounding of a fungal article, the septal pores are sealed immediately by Woronin bodies, which correspond to modified peroxysomes filled with a crystallized protein termed HEX-1 (Jedd and Chua 2000). Thus, damage is limited to a single article and cytoplasmic bleeding is avoided. Second, their syncytial organization and their ability to fuse render filamentous fungi extremely susceptible to viruses and other cytoplasmic replicons. Such infectious elements are frequent in fungal populations and are readily transmitted from one strain to another through hyphal anastomoses (cell fusions) (McCabe et al. 1999). Filamentous fungi have developed genetic systems that limit such cytoplasmic exchanges to genetically related individuals. These systems are termed heterokaryon incompatibility systems. Each fungal species possesses a set of polymorphic genes termed *het* loci for heterokaryon incompatibility (for review, see Glass et al. 2000; Saupe 2000). Cytoplasmic exchange can only occur between strains of identical *het* genotype. If a fusion event involves two individuals that differ in *het* genotype, the fusion cell is formed but rapidly compartmentalized and destroyed by a lytic process. For some of the het systems, this cell death reaction involves induction of autophagy, a cellular self-digestion process taking place in the vacuole (Pinan-Lucarré et al. 2003). Since a single *het* genotype difference is sufficient to cause incompatibility, and because there are a high number of *het* loci (around 10 per species), the number of potential compatibility groups is enormous. In practice, two wild isolates of a given fungal species are most generally incompatible. Incompatibility systems can thus be considered as self versus non-self discrimination systems offering protection against infectious replicons. *het-s* is one of nine *het* loci in *P. anserina*. This heterokaryon incompatibility system displays two alternate alleles termed *het-s* and *het-S*. When *het-s* and *het-S* cells fuse, a cell death reaction occurs (Rizet 1952; Beisson-Schecroun 1962). This incompatibility reaction can easily be detected at the macroscopic level by the formation of an abnormal contact line termed "barrage" when the strains are confronted on solid medium (Fig. 14). *het-s* is a nonessential gene encoding a protein 289 amino acids in length that displays no homology with known proteins

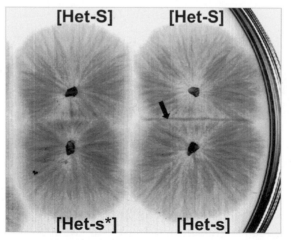

Figure 14. Phenotypic difference between [Het-s*] (prion-free) and [Het-s] (prion-infected) *het-s* strains. A [Het-s*] and a [Het-s] strain are confronted to a *het-S* strain on solid medium. In the [Het-s*]/*het-S* confrontation, the contact line is normal. In the [Het-s]/*het-S* confrontation, the incompatibility reaction leads to formation of an abnormal contact line termed "barrage" (*black arrow*).

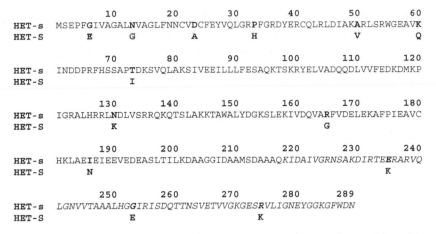

```
                10        20        30        40        50        60
HET-s   MSEPFGIVAGALNVAGLFNNCVDCFEYVQLGRPFGRDYERCQLRLDIAKARLSRWGEAVK
HET-S        E         G         A         H              V         Q

                70        80        90       100       110       120
HET-s   INDDPRFHSSAPTDKSVQLAKSIVEEILLLFESAQKTSKRYELVADQQDLVVFEDKDMKP
HET-S                I

               130       140       150       160       170       180
HET-s   IGRALHRRLNDLVSRRQKQTSLAKKTAWALYDGKSLEKIVDQVARFVDELEKAFPIEAVC
HET-S            K                                      G

               190       200       210       220       230       240
HET-s   HKLAEIEIEEVEDEASLTILKDAAGGIDAAMSDAAAQKIDAIVGRNSAKDIRTEERARVQ
HET-S        N                                                   K

               250       260       270       280       289
HET-s   LGNVVTAAALHGGIRISDQTTNSVETVVGKGESRVLIGNEYGGKGFWDN
HET-S              E              K
```

Figure 15. Sequences of the HET-s and HET-S proteins. The 13 polymorphic residues are in boldface, the region corresponding to the HET-s prion-forming domain is italicized.

(Turcq et al. 1990, 1991). HET-s and the antagonistic HET-S protein differ by 13 amino acids (Fig. 15). Coexpression of these two very similar proteins in the same cytoplasm leads to growth inhibition and cell death.

The HET-s Protein Is a Prion

The [Het-s] Character Is Infectious

In 1952, George Rizet showed that the [Het-s] genetic character (defined as the ability to produce a barrage reaction to a *het-S* strain) displays non-Mendelian traits (Rizet 1952). Although vegetatively incompatible, *het-s* and *het-S* are sexually compatible. In a cross between a *het-s* and a *het-S* strain, one recovers the expected 50% of *het-S* progeny, but a variable fraction of *het-s* progeny displays a novel phenotype termed [Het-s]. [Het-s] strains are neutral in incompatibility since they are compatible with [Het-S] strains (Fig. 14). *Podospora* develops male and female reproductive structures, and the female gamete contributes most if not all of the cytoplasm to the zygote. The [Het-s] and [Het-s] characters display maternal inheritance. In a [Het-s*] X [Het-s] cross, all progeny will display the phenotype of the maternal parent. Moreover, [Het-s] can be transmitted vegetatively to a [Het-s*] strain. Upon fusion with [Het-s], [Het-s*] strains turn into [Het-s] strains. This infectious propagation of the [Het-s] character was extensively analyzed by Janine Beisson-Schecroun (1962). She could show that the [Het-s] element spreads throughout the mycelium at an elevated rate. This rate increases with growth temperature but is always about 10 times faster than the radial growth rate of the fungus. In optimal conditions, it can reach 70 mm per day. [Het-s] propagation rate in the mycelium is maximal in the radial direction (the direction of filament growth) and an order of magnitude lower in the lateral direction (perpendicular to the direction of filament growth). Importantly, [Het-s] also arises spontaneously in [Het-s*] strains. After 3 days of growth on solid medium, about 1–2% of all [Het-s*] cultures spontaneously acquire the [Het-s] character. Ultimately, all [Het-s*] cultures become [Het-s]. [Het-s] can be reversibly cured. As stated above, the [Het-s] element can be lost during meiosis but also by regeneration of vegetative mycelial fragments or structures displaying very little cytoplasm (Beisson-Schecroun 1962; Belcour 1976). Léon Belcour had shown that when protoplasts are generated from a [Het-s] strain and plated out, a fraction of about 10% of the regenerating mycelia have lost the [Het-s] character. This is presumably due to sampling during protoplast generation of a cytoplasmic fraction lacking the [Het-s] element. These [Het-s*] regenerants can then spontaneously reacquire [Het-s] at the same low frequency. The thorough pioneering physiological and genetic investigation on [Het-s] performed by George Rizet and Jeanine Beisson in the 1950s and 1960s indicated that [Het-s] is a cytoplasmic infectious element specific to strains of the het-s genotype, which can emerge spontaneously and be reversibly cured. This infectious element kills the cell when the het-S gene product is present.

Genetic Evidence That [Het-s] Is a Prion

Janine Beisson-Schecroun wrote in 1962 "the transformation induced by the [Het-s]/[Het-s*] protoplasmic contact can be explained by multiplication of particles introduced by the [Het-s] strain. The [Het-s] to [Het-s*] transformation would then correspond to the loss of those particles. But the fact that the spontaneous [Het-s*] to [Het-s] reversion is still possible implies that all the information necessary for the reappearance of the particle is conserved, when the particle is lost" (Beisson-Schecroun 1962).

It was necessary to await molecular cloning of het-s and the demonstration that prions exist in microorganisms to understand that this "particle" is the prion form of the HET-s protein. It was shown that propagation of the [Het-s] character is strictly dependent on the het-s gene and that overexpression of het-s strongly increases the spontaneous emergence of [Het-s] (Coustou et al. 1997). As stated above, strains can be reversibly cured of [Het-s]. [Het-s] thus satisfied Wickner's genetic features of a microbial prion (Wickner 1994), with one notable exception; namely, the fact that inactivation of the het-s gene does not confer the phenotype of

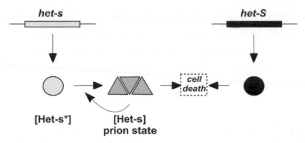

Figure 16. Schematic representation of the prion model for [Het-s]. The *het-s* allele leads to expression of the [Het-s*] state (neutral in incompatibility), then spontaneous emergence of the [Het-s] can occur. The [Het-s] state corresponds to a self-propagating aggregated state of the HET-s protein. Interaction of the HET-s protein in its prion form with the HET-S protein (which differs from HET-s by 13 residues) leads to a cell death reaction.

the presence of the prion element. This point is discussed below in this section.

The prion model for [Het-s] can be formulated as follows: In [Het-s*] strains, the HET-s protein is in its non-prion form. HET-s can then spontaneously switch to the infectious prion [Het-s] state. Transition to the prion state is detected by acquisition of incompatibility versus [Het-S] strains (Fig. 16). The prion model for [Het-s] readily explains the peculiar properties of the [Het-s] element, in particular infectivity, spontaneous appearance, and reversible curing.

HET-s Aggregates In Vivo and Forms Amyloid Fibrils In Vitro

Transition to the prion state involves aggregation of the HET-s protein. A HET-s–GFP fusion protein forms aggregates specifically in a [Het-s] strain but has a diffuse localization in a [Het-s*] strain (Coustou-Linares et al. 2001). However, this HET-s aggregate formation specific to the prion state is only evident in a strain overexpressing HET-s. It appears that in vivo upon transition to the prion state, the HET-s protein becomes prone to degradation in the vacuole. Depending on the expression level, massive aggregation or degradation is favored. In addition, the in vivo aggregation process is specific to HET-s, the HET-S form is soluble in vivo.

Purification of recombinant HET-s protein expressed in *E. coli* was performed and revealed that in vitro this protein undergoes a transition from a soluble form to an aggregated state. HET-s aggregates show Congo red binding and birefringence and appear as typical amyloids by electron microscopy (Dos Reis et al. 2002). Circular dichroism and infrared spectroscopy analyses revealed that the transition from the soluble to the aggregated state is accompanied by a decrease in α-helical content and an increase in antiparallel β-sheet secondary structure. These secondary structure analyses also indicated that in its soluble form HET-s displays a high content of random coil structure, suggesting that the protein might display an unstructured region.

Direct Validation of the Prion Hypothesis for [Het-s]

In vivo and in vitro aggregation of HET-s suggested that the molecular basis of prion propagation is similar for yeast prions and [Het-s]. Direct evidence that [Het-s] is a prion came from recent experiments in which recombinant HET-s protein was reintroduced into prion-free [Het-s*] *Podospora* strains by biolistics. Biolistic techniques based on bombardment of biological material with accelerated metal particles are often used for DNA-mediated transformation in plant systems. [Het-s*] mycelium was bombarded with tungsten particles together with recombinant HET-s protein. It was found that aggregates of recombinant HET-s protein induced emergence of the [Het-s] prion at an elevated frequency (Maddelein et al. 2002). This constitutes direct support for the protein-only hypothesis in the case of [Het-s] and shows that in this system infectivity can be generated de novo in vitro from recombinant sources. Only amyloid aggregates of HET-s were found to be infectious. Amorphous aggregates or soluble protein lacked infectivity in this assay. It also appears, as had been shown in the case of [*PSI*⁺] (Uptain et al. 2001), that aggregates formed in vivo are able to convert recombinant HET-s protein to amyloids, suggesting that the aggregates formed in vivo have characteristics of amyloids. As proposed for the yeast prions, the molecular basis of [Het-s] prion propagation thus also appears to be based on a self-perpetuating aggregation process into amyloid or amyloid-like aggregates.

Some Specific Features of the [Het-s] System

Although the molecular basis of prion propagation is apparently the same for [Het-s] and yeast prions, this system displays a number of specific features when compared to the yeast prion models.

[Het-s] Is Transmitted Horizontally

One feature of [Het-s] is directly connected to the differences in cellular organization between yeast and filamentous fungi. Because somatic cells in filamentous fungi spontaneously fuse, [Het-s] is truly infectious. It is propagated horizontally between strains and not only from mother to daughter cell in mitosis. Because filamentous fungi grow as a syncytium, emergence appears very frequent. After a few subcultures, a [HET-s*] strain will systematically switch to [HET-s]. This is easily understood if one remembers that emergence of a single HET-s particle anywhere in the mycelium will infect the whole mycelium. So, this does not necessarily mean that at the molecular level emergence of the individual HET-s particle is any more frequent than appearance of [URE3] or [PSI+]. In fact, in early studies this frequency had been estimated to be 10^{-7} per nucleus (Beisson-Schecroun 1962), which is comparable to what is known for [URE3] and [PSI+]. Similarly, in contrast to what was found for [PSI+] and [URE3], chemical curing experiments using GuHCl have been unsuccessful so far (V. Coustou and M.-L. Maddelein, unpubl.). A [Het-s] strain does not lose the prion when grown on GuHCl. This observation might reflect differences in propagation mechanisms for [Het-s] and yeast prion, but alternatively, this could be due to the fact that only complete curing of [Het-s] can be detected. Any residual [Het-s] particles will immediately reinfect the whole mycelium. Inactivation and overexpression of the P. anserina Hsp104 ortholog are under way and will allow further comparison of [Het-s] with the yeast prion models.

The Prion-forming Domain of HET-s Is Not N/Q-rich

When HET-s fibrils are submitted to proteinase K digestion, a resistant fragment of 7 kD is detected. These proteinase-K-digested fibrils display infectivity in the biolistic assay (Maddelein et al. 2002). This resistant fragment could correspond to the "amyloid core" of the HET-s aggregate, thus suggesting that only a subdomain of HET-s is involved in fibril formation and prion propagation. The 7-kD fragment resistant to proteinase K digestion in the amyloid fibrils corresponds to the region encompassing the 72 carboxy-terminal amino acids of HET-s (Balguerie et al. 2003). The corresponding peptide can form amyloid fibrils in vitro (Fig. 17) and cause aggregation of GFP in vivo. A carboxy-terminal deletion of this region in HET-s prevents protein aggregation both in vivo and in vitro and eliminates infectivity. Limited proteolysis experiments and nuclear magnetic resonance (NMR) analyses indicate that this region is poorly structured in the soluble form of the HET-s protein (Balguerie et al. 2003). The similarities between the yeast prions and [Het-s] are striking. The Ure2p and Sup35 yeast proteins both display a modular organization with a structured carboxy-terminal functional domain and an amino-terminal presumably unstructured prion-forming domain. Apparently the

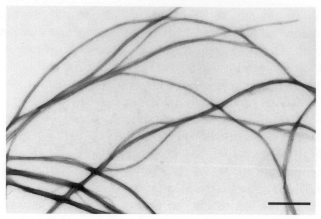

Figure 17. Fibrils of the HET-s prion-forming domain. Electron micrograph of fibers formed in vitro by the carboxy-terminal domain of HET-s (amino acid position 218–289). Bar, 500 nm.

same applies for HET-s, although in this case the prion-forming domain is carboxy-terminal. The prion-forming domains of Ure2p and Sup35 share a strong bias in amino acid composition, because both are N- and Q-rich, whereas the proposed prion-forming domain of HET-s lacks N- and Q-rich stretches.

HET-S, a Non-Prion Natural Variant

One essential feature of the [Het-s] model is the existence of the natural HET-S variant devoid of the prion properties. The natural HET-S is also antagonistic to [Het-s] in the incompatibility function. To determine which of the 13 amino acid polymorphisms are responsible for the functional difference between HET-s and HET-S, the properties of chimeric *het-s/het-S* alleles have been determined. It was thus shown that this functional difference is due to the amino acid differences located in the amino-terminal region (Deleu et al. 1993; Balguerie et al. 2003). Several single amino acid substitutions can switch a HET-S protein to the HET-s specificity. For instance, mutation of histidine 33 in HET-S to a proline—the residue found at that position in HET-s—confers the [Het-s] specificity (Deleu et al. 1993). In other words, this single amino acid substitution turns the HET-S non-prion form into a protein capable of switching to the prion state. Several other amino acid substitutions at that position achieve the same result (Coustou et al. 1999). Conversely, to turn a HET-s protein to the non-prion form state, two substitutions, P33H and D23A, are required. A HET-s P33H D23A mutant has lost the ability to propagate [Het-s]. In other words, although it contains an intact carboxy-terminal prion-forming domain, this mutant protein has lost its prion properties in vivo. These results suggest that there exists an interaction between the amino- and the carboxy-terminal regions of HET-s, and that the amino-terminal domain of HET-s can affect the prion properties conferred by the carboxy-terminal prion-forming domain responsible for amyloid formation (Balguerie et al. 2003).

HET-S is not only devoid of the prion properties, it also has the ability to render the [Het-s] prion toxic to the cell. [Het-s] strains containing HET-s aggregates grow normally but in the presence of HET-S apparently become harmful. How is this toxicity achieved? In vivo coalescence of HET-s-GFP into large foci is inhibited in the presence of HET-S (Coustou-Linares et al. 2001). Since HET-S is very similar to HET-s, HET-S might interact with the HET-s polymerization nucleus and thus prevent further aggregation of HET-s. This could lead to accumulation of toxic oligomeric HET-s aggregation intermediates. This hypothesis is attractive because it might connect cell death in the *het-s/het-S* system to cell death in amyloid diseases, but this awaits direct experimental support. At any rate, the mechanism of [Het-s] cell death is apparently connected with protein aggregation. The HET-s/HET-S interaction is only lethal when HET-s is in its prion form, and deletion of the putative amyloid domain of HET-s eliminates the incompatibility function (Balguerie et al. 2003).

[Het-s] as a "Prion with a Function"

One of the most puzzling questions raised by the discovery of prions in fungi is their biological significance. In yeast, transition to the prion form leads to partial or total loss of function of the Ure2p and Sup35 proteins. As noted previously, the emergence of the [Het-s] prion is not associated with a loss of function of the protein, but with acquisition of reactivity in incompatibility. The active form in incompatibility is the prion form. Therefore, if one admits that the function of *het-s* is self versus non-self discrimination, then [Het-s] is a prion with a biological function. In other words, the [Het-s] prion would be beneficial to the organism that supports its propagation. In most other systems, amyloid formation is understood as a cellular mishap with dramatic consequences. In the [Het-s] system, the ability to form toxic amyloids would have been recruited to perform a purposeful cell death reaction associated with self versus non-self discrimination. In this hypothesis, the equilibrated distribution between *het-s* and *het-S* alleles found in populations is easily explained, since this would be the optimal distribution for a self versus non-self discrimination locus. It is quite puzzling to envision amyloid formation as an integrated biological process. It should, however, be noted that this is not unprecedented. A class of very abundant fungal cell wall proteins termed hydrophobins is found in the hydrophobic aerial structures developed by filamentous fungi (Wosten and de Vocht 2000). These hydrophobins self-assemble into amphipathic films. It was found that the structure of the assembled state strongly resembles amyloid fibrils. This is clearly documented, for instance, for the EAS hydrophobin from *Neurospora crassa*, which undergoes a transition from an unstructured state in solution to a functional amyloid-like aggregated state (Mackay et al. 2001). Thus, in this remarkable example, the amyloid-like state of the protein clearly corresponds to the functional state of the protein. The same is true for bacterial cell-surface filaments like the Curli fibers of *E. coli*, which were shown to be amyloids (Chapman et al. 2002).

The [Het-s]-related cell death is presented as purposeful, but it should be noted that the biological meaning of heterokaryon incompatibility is far from being clearly understood (Saupe 2000). There is evidence that *het* genes can protect from viral infections (McCabe et al. 1999), and molecular evolution studies have shown that—like other self versus non-self discrimination loci—at least certain *het* loci are under selective pressure to maintain polymorphism (Wu et al. 1998). However, it is possible that despite its wide occurrence, heterokaryon incompatibility does not have a biological function. It may just exist with no purpose. Many of the known *het* genes have a cellular function; similarly, HET-s might well have a cellular function and—like Ure2p and Sup35p—lose it upon transition to the prion state. Unlike the inactivation of *URE2* or *SUP35*, inactivation of *het-s* by gene replacement does not lead to any particular phenotype except for the loss of the incompatibility function. Very much like mammalian prions, [Het-s] is not detected by the loss of function of HET-s, but by the toxic properties of the prion form. It is possible that the "true" function of the *het-s* gene remains to be established. Therefore, even though the fact that [Het-s] is the reactive form of the HET-s protein suggests that it is a "prion with a function," a better understanding of the biological significance of heterokaryon incompatibility is needed to firmly establish this conclusion.

When speculating about the biological significance of [Het-s], one should also consider a remarkable property [Het-s] displays during the sexual cycle. At 26°C, a *het-s* × *het-S* cross yields four spored asci with two *het-s* and two *het-S* spores. However, Jean Bernet observed that in a *het-s* × *het-S* cross performed at low temperature (18°C), in up to 50% of the

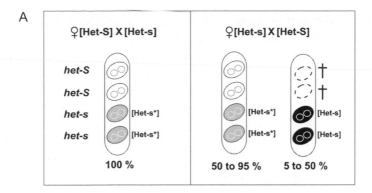

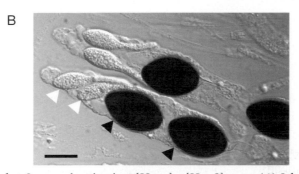

Figure 18. *het-S* spore abortion in a [Het-s] × [Het-S] cross. (*A*) Schematic representation of the *het-S* spore abortion phenomenon. In [Het-S] female × [Het-s] male cross, normal asci containing two *het-s* and two *het-S* spores are obtained. The het-s spores display the [Het-s*] phenotype, i.e., the [Het-s] prion has been lost during meiosis. In the reciprocal cross ([Het-s] female × |Het-S] male), in a variable proportion of asci, the two *het-S* spores abort. In asci in which *het-S* spore abortion occurs, the *het-s* spores always display the [Het-s] phenotype (prion infected), whereas in normal asci, the *het-s* spores are prion-free ([Het-s*]). (*B*) Micrograph of two asci from a [Het-s] female × [Het-S] male cross in which the *het-S* spores have aborted (*white arrowhead*) while the *het-s* spores have matured normally (*black arrowhead*). (Adapted from Bernet 1965.)

asci the two *het-S* spores abort (Fig. 18) (Bernet 1965; Dalstra et al. 2003). This only occurs when *het-s* is the maternal parent—contributing the cytoplasm to the zygote—and is in the [Het-s] prion infected state. In the asci in which *het-S* spore abortion occurs, the *het-s* spores contain the [Het-s] prion, whereas they display the [Het-s*] phenotype in asci in which the *het-S* spores survive. [Het-s] is capable of selectively killing the *het-S* spores within the maturing ascus. As a result, such a cross will yield more *het-s* than *het-S* progeny. In other words, presence of the [Het-s] prion skews the Mendelian segregation in favor of the *het-s* allele. This feature makes *het-s* a meiotic drive element (Dalstra et al. 2003). Meiotic drive alleles or haplotypes have been described as ultra-selfish DNA that are able to imbalance Mendelian segregation at the expense of the alternate allele (Lyttle 1993; Hurst and Werren 2001). A meiotic drive element can invade and even become fixed in a population, although it confers no fitness advantage. This property could well explain how the *het-s* allele has been maintained in populations even if [Het-s] is detrimental. It might be that the prion-forming *het-s* allele is on its way to fixation or that the fitness loss associated with inactivation of HET-s prevents loss of the "normal" (non-prion) *het-S* allele.

Many More Prions in Filamentous Fungi?

Because of their syncytial organization and ability to mix their cytoplasmic content, filamentous fungi should be exceptionally susceptible to prions. Moreover, these biological properties should make prions easier to detect in these organisms. A single spontaneously emerging propagon will infect the whole mycelium and can thus potentially be detected without the need for selection. In fact, contagious cytoplasmic elements are very frequent in filamentous fungi (Silar and Daboussi 1999). Very often they appear spontaneously as mycelial sectors with an altered color and/or morphology. The morphological alteration can be transmitted by contact to an unmodified strain. The molecular nature of the infectious determinant is not known for any of these elements. Still, at least some of them, like the "secteur" and the "anneau" infectious elements of the ascomycetes *Nectria haematococca* and the *Cripple Growth* element in *P. anserina*, can be reversibly cured by cytoplasmic sampling, making it hard to envision that they are nucleic-acid-based replicons. It is tempting to hypothesize that at least some of these spontaneously emerging infectious elements correspond to additional fungal prions.

A PRION BASED ON COVALENT AUTOACTIVATION OF AN ENZYME

Recently, a new type of prion (infectious protein), unrelated to amyloid formation, has been described (Roberts and Wickner 2003). In the absence of protease A (*PEP4*), the vacuolar protease B, encoded by *PRB1*, is necessary and sufficient for the proteolytic cleavages that activate its own proprotein (Nebes and Jones 1991; Zubenko and Jones 1982; for review, see Jones 2002). Cells that initially lack the active protease B cannot generally acquire activity, and give rise to progeny essentially all of whom lack this activity. Cells that have active protease B can activate newly made precursor, and so their offspring also have active protease B. Moreover, transmission of active protease B from one cell to another lacking it results in conversion of the recipient cell to one stably carrying the active form (Roberts and Wickner 2003). Since it is just the Prb1 protein that transmits this heritable information from one individual to another, active protease B is behaving as an infectious protein, i.e., a prion.

This system is primarily of interest as a "proof of principle" that a *trans* self-activating protein can behave as a prion. There are numerous protein kinases, protein transacetylases, protein methylases, protein glycosidases, and other protein-modifying enzymes that, given transmissibility and a requirement for self-modification, might become prions.

PROSPECTS FOR YEAST AND FUNGAL PRIONS

The discovery of prions in *Saccharomyces* and *Podospora* has provided dramatic new proof that the "protein-only" model is valid, and has facilitated rapid progress in our understanding of this interesting area. The ease with which the yeast and fungal systems may be manipulated contrasts with the long incubation times and expense of the mouse and tissue culture systems.

Because [URE3], [*PSI*⁺], [Het-s], and [*PIN*⁺] are all nonchromosomal genes, these results show that genes can be proteins. This complements the remarkable fact that nucleic acids can be enzymes (Kruger et al. 1982; Guerrier-Takada et al. 1983).

The yeast and fungal prions are self-propagating protein conformations. Self-propagating cellular structures were first described by Sonneborn as the cortical inheritance of *Paramecium* (Beisson and Sonneborn 1965). Altered patterns of ciliation on the surface of *Paramecium* cells due to accidents in mating or surgical alteration were stably propagated in the progeny. The parallel with the self-propagating protein structures of prions is obvious. One wonders whether there might not be other self-propagating cellular structures, such as organelles or sub-organellar parts. These might appear as mutants whose defect could not be repaired by simply replacing the defective gene.

Are prions and other amyloids of functional value to the individual? As discussed above, the [Het-s] prion can be viewed as carrying out a normal cellular function or as a pathogenic state (Saupe 2000). If indeed the [Het-s] prion is carrying out the normal function of this protein, it would explain why wild strains generally carry the prion form. Neither [*PSI*⁺]

nor [URE3] has yet been described in wild isolates (Chernoff et al. 2000), suggesting that they may provide very little (if any) evolutionary advantage. Dehydration of some fish eggs is prevented by an egg envelope layer composed of an amyloid material (Podrabsky et al. 2001), and *E. coli* use cell-surface amyloid filaments to colonize surfaces (Chapman et al. 2002). These are clearly adaptive functions using amyloid. However, no human amyloid is known to be adaptive.

How widespread will prions prove to be? A number of nonchromosomal traits of fungi have been described, and these may include some new prions (Silar and Daboussi 1999). The [Kil-d] nonchromosomal genetic element affecting the expression of the killer toxin of *S. cerevisiae* (Wickner 1976) has been suggested to be due to a prion (Talloczy et al. 1998, 2000), but the evidence is as yet incomplete. The non-Mendelian genetic element [*ISP*⁺] has an antisuppressor effect and is not associated with aggregation of Sup35p (Volkov et al. 2002). [*ISP*⁺] is cured by guanidine, but does not depend on *HSP104*. The authors suggest it may be a prion, but of a protein other than Sup35p. Knowing that the prion domains of Ure2p and Sup35p are rich in asparagine and glutamine residues, new prion domains were sought (and found) with the same properties (Santoso et al. 2000; Sondheimer and Lindquist 2000). Although some of these potential prion domains may be unable in the environment of the full-length protein to become a prion, one of them, Rnq1p (Sondheimer and Lindquist 2000), proved to be a prion identical to [*PIN*⁺] (Derkatch et al. 2001). This approach will likely turn up more prions. Since neither PrP nor HET-s has runs of asparagine or glutamine residues, there are likely to be other prions without this feature.

REFERENCES

Aigle M. 1979. "Contribution a l'etude de l'heredite non-chromosomique de *Saccharomyces cerevisiae* facteur [URE3] et plasmides hybrides." Ph.D. thesis. L'Universite Louis Pasteur de Strasbourg, Strasbourg, France.

Aigle M. and Lacroute F. 1975. Genetical aspects of [URE3], a non-Mendelian, cytoplasmically inherited mutation in yeast. *Mol. Gen. Genet.* **136:** 327–335.

All-Robyn J.A., Kelley-Geraghty D., Griffin E., Brown N., and Liebman S.W. 1990. Isolation of omnipotent suppressors in an [*eta*⁺] yeast strain. *Genetics* **124:** 505–514.

Bailleul P.A., Newnam G.P., Steenbergen J.N., and Chernoff Y.O. 1999. Genetic study of interactions between the cytoskeletal assembly protein Sla1 and prion-forming domain of the release factor Sup35 (eRF3) in *Saccharomyces cerevisiae*. *Genetics* **153:** 81–94.

Bailleul-Winslett P.A., Newnam G.P., Wegrzyn R.D., and Chernoff Y.O. 2000. An antiprion effect of the anticytoskeletal drug latrunculin A in yeast. *Gene Expr.* **9:** 145–156.

Balguerie A., Dos Reis S., Ritter C., Chaignepain S., Coulary-Salin B., Forge V., Bathany K., Lascu I., Schmitter J.M., Riek R., and Saupe S.J. 2003. Domain organization and structure-function relation of the HET-s prion protein of *Podospora anserina*. *EMBO J.* **22:** 2071–2081.

Baxa U., Speransky V., Steven A.C., and Wickner R.B. 2002. Mechanism of inactivation on prion conversion of the *Saccharomyces cerevisiae* Ure2 protein. *Proc. Natl. Acad. Sci.* **99:** 5253–5260.

Beck T. and Hall M.N. 1999. The TOR signalling pathway controls nuclear localization of nutrient-regulated transcription factors. *Nature* **402:** 689–692.

Beisson J. and Sonneborn T.M. 1965. Cytoplasmic inheritance of the organization of the cell cortex in *Paramecium aurelia*. *Proc. Natl. Acad. Sci.* **53:** 275–282.

Beisson-Schecroun J. 1962. Incompatibilte cellulaire et interactions nucleo-cytoplasmiques dans les phenomenes de barrage chez *Podospora anserina*. *Ann. Genet.* **4:** 3–50.

Belcour L. 1976. Loss of a cytoplasmic determinant through formation of protoplasts in *Podospora*. *Neurospora Newsl.* **23:** 26–27.

Bernet J. 1965. Mode d'action des gènes de barrage et relation entre l'incompatibilité cellulaire et l'incompatibilité sexuelle chez le *Podospora anserina*. *Ann. Sci. Natl. Bot.* **6:** 611–768.

Bessen R.A. and Marsh R.F. 1992. Identification of two biologically distinct strains of transmissible mink encephalopathy in hamsters. *J. Gen. Virol.* **73:** 329–334.

———. 1994. Distinct PrP properties suggest the molecular basis of strain variation in transmissible mink encephalopathy. *J. Virol.* **68:** 7859–7868.

Bolton D.C., McKinley M.P., and Prusiner S.B. 1982. Identification of a protein that purifies with the scrapie prion. *Science* **218:** 1309–1311.

Boorstein W.R., Ziegelhoffer T., and Craig E.A. 1994. Molecular evolution of the HSP70 multigene family. *J. Mol. Evol.* **38:** 1–17.

Borchsenius A.S., Wegrzyn R.D., Newnam G.P., Inge-Vechtomov S.G., and Chernoff Y.O. 2001. Yeast prion protein derivative defective in aggregate shearing and production of new 'seeds'. *EMBO J.* **20:** 6683–6691.

Bousset L., Thomson N.H., Radford S.E., and Melki R. 2002. The yeast prion Ure2p retains its native α-helical conformation upon assembly into protein fibrils *in vitro*. *EMBO J.* **21;** 2903–2911.

Bousset L., Beirhali H., Janin J., Melki R., and Morera S. 2001. Structure of the globular region of the prion protein Ure2 from the yeast *Saccharomyces cerevisiae*. *Structure* **9:** 39–46.

Bradley M.E., Edskes H.K., Hong J.Y., Wickner R.B., and Liebman S.W. 2002. Interactions among prions and prion "strains" in yeast. *Proc. Natl. Acad. Sci.* (suppl. 4) **99:** 16392–16399.

Bruce M.E. and Fraser H. 1991. Scrapie strain variation and its implications. *Curr. Top. Microbiol. Immunol.* **172:** 125–138.

Büeler H., Fischer M., Lang Y., Bluethmann H., Lipp H.P., DeArmond S.J., Prusiner S.B., Aguet M., and Weissmann C. 1992. Normal development and behavior of mice lacking the neuronal cell-surface PrP protein. *Nature* **356:** 577–582.

Cardenas M.E., Cutler N.S., Lorenz M.C., Di Como C.J., and Heitman J. 1999. The TOR signaling cascade regulates gene expression in response to nutrients. *Genes Dev.* **13:** 3271–3279.

Caughey B.W., Dong A., Bhat K.S., Ernst D., Hayes S.F., and Caughey W.S. 1991. Secondary structure analysis of the scrapie-associated protein PrP 27-30 in water by infrared spectroscopy. *Biochemistry* **30:** 7672–7680.

Chacinska A., Szczesniak B., Kochneva-Pervukhova N.V., Kushnirov V.V., Ter-Avanesyan M.D., and Boguta M. 2001. Ssb1 chaperone is a [PSI⁺] prion-curing factor. *Curr. Genet.* **39:** 62–67.

Chapman M.R., Robinson L.S., Pinkner J.S., Roth R., Heuser J., Hammar M., Normark S., and Hultgren S.J. 2002. Role of *Escherichia coli* curli operons in directing amyloid fiber formation. *Science* **295:** 851–855.

Chernoff Y.O. 2001. Mutation processes at the protein level: Is Lamarck back? *Mutat. Res.* **488:** 39–64.

Chernoff Y.O., Derkach I.L., and Inge-Vechtomov S.G. 1993. Multicopy SUP35 gene induces de-novo appearance of psi-like factors in the yeast *Saccharomyces cerevisiae.* *Curr. Genet.* **24:** 268–270.

Chernoff Y.O., Lindquist S.L., Ono B.-I., Inge-Vechtomov S.G., and Liebman S.W. 1995. Role of the chaperone protein Hsp104 in propagation of the yeast prion-like factor [psi$^+$]. *Science* **268:** 880–884.

Chernoff Y.O., Newnam G.P., Kumar J., Allen K., and Zink A.D. 1999. Evidence for a protein mutator in yeast: Role of the Hsp70-related chaperone Ssb in formation, stability and toxicity of the [PSI$^+$] prion. *Mol. Cell. Biol.* **19:** 8103–8112.

Chernoff Y.O., Galkin A.P., Lewitin E., Chernova T.A., Newnam G.P., and Belenkly S.M. 2000. Evolutionary conservation of prion-forming abilities of the yeast Sup35 protein. *Mol. Microbiol.* **35:** 865–876.

Chernoff Y.O., Ptyushkina M.V., Samsonova M.G., Sizonencko G.I., Pavlov Y.I., Ter-Avanesyan M.D., and Inge-Vectomov S.G. 1992. Conservative system for dosage-dependent modulation of translational fidelity in eukaryotes. *Biochimie* **74:** 455–461.

Chien P. and Weissman J.S. 2001. Conformational diversity in a yeast prion dictates its seeding specificity. *Nature* **410:** 223–227.

Chiesa R., Piccardo P., Ghetti B., and Harris D.A. 1998. Neurological illness in transgenic mice expressing a prion protein with an insertional mutation. *Neuron* **21:** 1339–1351.

Choi J.H., Lou W., and Vancura A. 1998. A novel membrane-bound glutathione S-transferase functions in the stationary phase of the yeast *Saccharomyces cerevisiae. J. Biol. Chem.* **273:** 29915–29922.

Conde J. and Fink G.R. 1976. A mutant of *Saccharomyces cerevisiae* defective for nuclear fusion. *Proc. Natl. Acad. Sci.* **73:** 3651–3655.

Cooper T.G. 1982. Nitrogen metabolism in *Saccharomyces cerevisiae*. In *The molecular biology of the yeast* Saccharomyces: *Metabolism and gene expression* (ed. J.N. Strathern et al.), pp. 39–99. Cold Spring Harbor Laboratory, Cold Spring Harbor, New York.

———. 2002. Transmitting the signal of excess nitrogen in *Saccharomyces cerevisiae* from the Tor proteins to the GATA factors: Connecting the dots. *FEMS Microbiol. Rev.* **26:** 223–238.

Coschigano P.W. and Magasanik B. 1991. The URE2 gene product of *Saccharomyces cerevisiae* plays an important role in the cellular response to the nitrogen source and has homology to glutathione S-transferases. *Mol. Cell. Biol.* **11:** 822–832.

Courchesne W.E. and Magasanik B. 1988. Regulation of nitrogen assimilation in *Saccharomyces cerevisiae:* Roles of the URE2 and GLN3 genes. *J. Bacteriol.* **170:** 708–713.

Coustou V., Deleu C., Saupe S., and Begueret J. 1997. The protein product of the *het-s* heterokaryon incompatibility gene of the fungus *Podospora anserina* behaves as a prion analog. *Proc. Natl. Acad. Sci.* **94:** 9773–9778.

———. 1999. Mutational analysis of the [Het-s] prion analog of *Podospora anserina:* A short N-terminal peptide allows prion propagation. *Genetics* **153:** 1629–1640.

Coustou-Linares V., Maddelein M.L., Begueret J., and Saupe S.J. 2001. In vivo aggregation of the HET-s prion protein of the fungus *Podospora anserina. Mol. Microbiol.* **42:** 1325–1335.

Cox B.S. 1965. PSI, a cytoplasmic suppressor of super-suppressor in yeast. *Heredity* **20:** 505–521.

———. 1971. A recessive lethal super-suppressor mutation in yeast and other PSI phenomena. *Heredity* **26:** 211–232.

———. 1993. Psi phenomena in yeast. In *The early days of yeast genetics* (ed. M.N. Hall and P. Linder), pp. 219–239. Cold Spring Harbor Laboratory Press, Cold Spring Harbor, New York.

Cox B.S., Tuite M.F., and McLaughlin C.S. 1988. The Psi factor of yeast: A problem in inheritance. *Yeast* **4:** 159–179.

Crouzet M. and Tuite M.F. 1987. Genetic control of translational fidelity in yeast: Molecular cloning and analysis of the allosuppressor gene *SAL3*. *Mol. Gen. Genet.* **210:** 581–583.

Crouzet M., Izgu F., Grant C.M., and Tuite M.F. 1988. The allosuppressor gene SAL4 encodes a protein important for maintaining translational fidelity in *Saccharomyces cerevisiae*. *Curr. Genet.* **14:** 537–543.

Czaplinski K., Ruiz-Echevarria M.J., Paushkin S.V., Han X., Weng Y., Perlick H.A., Dietz H.C., Ter-Avanesyan M.D., and Peltz S.W. 1998. The surveillance complex interacts with the translation release factors to enhance termination and degrade aberrant mRNAs. *Genes Dev.* **12:** 1665–1677.

Dalstra H.J.P., Swart K., Debets A.J.M, Saupe S.J., and Hoekstra R.F. 2003. Sexual transmission of the [Het-s] prion leads to meiotic drive in *Podospora anserina*. *Proc. Natl. Acad. Sci.* **100:** 6616–6621.

Deleu C., Clave C., and Begueret J. 1993. A single amino acid difference is sufficient to elicit vegetative incompatibility in the fungus *Podospora anserina*. *Genetics* **135:** 45–52.

DePace A.H. and Weissman J.S. 2002. Origins and kinetic consequences of diversity in Sup35 yeast prion fibers. *Nat. Struct. Biol.* **9:** 389–396.

DePace A.H., Santoso A., Hillner P., and Weissman J.S. 1998. A critical role for amino-terminal glutamine/asparagine repeats in the formation and propagation of a yeast prion. *Cell* **93:** 1241–1252.

Derkatch I.L., Bradley M.E., and Liebman S.W. 1998. Overexpression of the SUP45 gene encoding a Sup35p-binding protein inhibits the induction of the de novo appearance of the [PSI⁺] prion. *Proc. Natl. Acad. Sci.* **95:** 2400–2405.

Derkatch I.L., Bradley M.E., Hong J.Y., and Liebman S.W. 2001. Prions affect the appearance of other prions: The story of [PIN]. *Cell* **106:** 171–182.

Derkatch J.L., Bradley M.E., Zhou P., and Liebman S.W. 1999. The *PNM2* mutation in the prion protein domain of SUP35 has distinct effects on different variants of the [PSI⁺] prion in yeast. *Curr. Genet.* **35:** 59–67.

Derkatch I.L., Bradley M.E., Zhou P., Chernoff Y.O., and Liebman S.W. 1997. Genetic and environmental factors affecting the *de novo* appearance of the [PSI⁺] prion in *Saccharomyces cerevisiae*. *Genetics* **147:** 507–519.

Derkatch I.L., Chernoff Y.O., Kushnirov V.V., Inge-Vechtomov S.G., and Liebman S.W. 1996. Genesis and variability of [PSI] prion factors in *Saccharomyces cerevisiae*. *Genetics* **144:** 1375–1386.

Derkatch I.L., Bradley M.E., Masse S.V., Zadorsky S.P., Polozkov G.V., Inge-Vechtomov S.G., and Liebman S.W. 2000. Dependence and independence of [PSI(+)] and [PIN(+)]: A two-prion system in yeast? *EMBO J.* **19:** 1942–1952.

Doel S.M., McCready S.J., Nierras C.R., and Cox B.S. 1994. The dominant *PNM2⁻* mutation which eliminates the [PSI] factor of *Saccharomyces cerevisiae* is the result of a missense mutation in the *SUP35* gene. *Genetics* **137:** 659–670.

Dos Reis S., Coulary-Salin B., Forge V., Lascu I., Begueret J., and Saupe S.J. 2002. The HET-s prion protein of the filamentous fungus *Podospora anserina* aggregates in vitro into amyloid-like fibrils. *J. Biol. Chem.* **277:** 5703–5706.

Drillien R., Aigle M., and Lacroute F. 1973. Yeast mutants pleiotropically impaired in the regulation of the two glutamate dehydrogenases. *Biochem. Biophys. Res. Commun.* **53:** 367–372.

Eaglestone S.S., Cox B.S., and Tuite M.F. 1999. Translation termination efficiency can be regulated in *Saccharomyces cerevisiae* by environmental stress through a prion-mediated mechanism. *EMBO J.* **18:** 1974–1981.

Eaglestone S.S., Ruddock L.W., Cox B.S., and Tuite M.F. 2000. Guanidine hydrochloride blocks a critical step in the propagation of the prion-like determinant [*PSI*+] of *Saccharomyces cerevisiae*. *Proc. Natl. Acad. Sci.* **97:** 240–244.

Ebihara K. and Nakamura Y. 1999. C-terminal interaction of translational release factors eRF1 and eRF3 of fission yeast: G-domain uncoupled binding and the role of conserved amino acids. *RNA* **5:** 739–750.

Edskes H.K. and Wickner R.B. 2000. A protein required for prion generation: [URE3] induction requires the Ras-regulated Mks1 protein. *Proc. Natl. Acad. Sci.* **97:** 6625–6629.

———. 2002. Conservation of a portion of the *S. cerevisiae* Ure2p prion domain that interacts with the full-length protein. *Proc. Natl. Acad. Sci.* (suppl. 4) **99:** 16384–16391.

Edskes H.K., Gray V.T., and Wickner R.B. 1999a. The [URE3] prion is an aggregated form of Ure2p that can be cured by overexpression of Ure2p fragments. *Proc. Natl. Acad. Sci.* **96:** 1498–1503.

Edskes H.K., Hanover J.A., and Wickner R.B. 1999b. Mks1p is a regulator of nitrogen catabolism upstream of Ure2p in *Saccharomyces cerevisiae*. *Genetics* **153:** 585–594.

Esteban R. and Wickner R.B. 1987. A new non-Mendelian genetic element of yeast that increases cytopathology produced by M1 double-stranded RNA in ski strains. *Genetics* **117:** 399–408.

Feller A., Ramos F., Peirard A., and Dubois E. 1997. Lys80p of *Saccharomyces cerevisiae*, previously proposed as a specific repressor of *LYS* genes, is a pleiotropic regulatory factor identical to Mks1p. *Yeast* **13:** 1337–1346.

Fernandez-Bellot E., Guillemet E., Baudin-Baillieu A., Gaumer S., Komar A.A., and Cullin C. 1999. Characterization of the interaction domains of Ure2p, a prion-like protein of yeast. *Biochem. J.* **338:** 403–407.

Fernandez-Bellot E., Guillemet E., Ness F., Baudin-Baillieu A., Ripaud L., Tuite M., and Cullin C. 2002. The [URE3] phenotype: Evidence for a soluble prion in yeast. *EMBO Rep.* **3:** 76–81.

Ferreira P.C., Ness F., Edwards S.R., Cox B.S., and Tuite M.F. 2001. The elimination of the yeast [PSI+] prion by guanidine hydrochloride is the result of Hsp104 inactivation. *Mol. Microbiol.* **40:** 1357–1369.

Flechsig E., Shmerling D., Hegyi I., Raeber A.J., Fischer M., Cozzio A., von Mering C., Aguzzi A., and Weissmann C. 2000. Prion protein devoid of the octapeptide repeat region restores susceptibility to scrapie in PrP knockout mice. *Neuron* **27:** 399–408.

Frolova L., Le Goff X., Zhouravleva G., Davydova E., Philippe M., and Kisselev L. 1996. Eukaryotic polypeptide chain release factor eRF3 is an eRF1- and ribosome-dependent guanosine triphosphatase. *RNA* **2:** 334–341.

Frolova L., LeGoff X., Rasmussen H.H., Cheperegin S., Drugeon G., Kress M., Arman I., Haenni A.-L., Celis J.E., Philippe M., Justesen J., and Kisselev L. 1994. A highly conserved eukaryotic protein family possessing properties of polypeptide chain release factor. *Nature* **372:** 701–703.

Glass N.L., Jacobson D.J., and Shiu P.K. 2000. The genetics of hyphal fusion and vegetative incompatibility in filamentous ascomycete fungi. *Annu. Rev. Genet.* **34:** 165–186.

Glover J.R. and Lindquist S. 1998. Hsp104, Hsp70, and Hsp40: A novel chaperone system that rescues previously aggregated proteins. *Cell* **94:** 73–82.

Glover J.R., Kowal A.S., Shirmer E.C., Patino M.M., Liu J.-J., and Lindquist S. 1997. Self-seeded fibers formed by Sup35, the protein determinant of [*PSI*⁺], a heritable prion-like factor of *S. cerevisiae*. *Cell* **89:** 811–819.

Goldfarb L.G., Brown P., Cervenakova L., and Gajdusek D.C. 1994. Genetic analysis of Creutzfeldt-Jacob disease and related disorders. *Philos. Trans. R. Soc. Lond. B Biol. Sci.* **343:** 379–384.

Goldmann W., Chong A., Foster J., Hope J., and Hunter N. 1998. The shortest known prion protein gene allele occurs in goats, has only three octapeptide repeats and is non-pathogenic. *J. Gen. Virol.* **79:** 3173–3176.

Goldring E.S., Grossman L.I., Krupnick D., Cryer D.R., and Marmur J. 1970. The petite mutation in yeast: Loss of mitochondrial DNA during induction of petites with ethidium bromide. *J. Mol. Biol.* **52:** 323–335.

Griffith J.S. 1967. Self-replication and scrapie. *Nature* **215:** 1043–1044.

Guerrier-Takada C., Gardiner K., Marsh T., Pace N., and Altman S. 1983. The RNA moiety of ribonuclease P is the catalytic subunit of the enzyme. *Cell* **35:** 849–857.

Hardwick J.S., Kuruvilla F.G., Tong J.K., Shamji A.F., and Schreiber S.L. 1999. Rapamycin-modulated transcription defines the subset of nutrient-sensitive signaling pathways directly controlled by the tor proteins. *Proc. Natl. Acad. Sci.* **96:** 14866–14870.

Hawthorne D.C. and Mortimer R.K. 1968. Genetic mapping of nonsense suppressors in yeast. *Genetics* **60:** 735–742.

Hoshino S., Miyazawa H., Enomoto T., Hanoka F., Kikuchi Y., Kikuchi A., and Ui M. 1989. A human homologue of the yeast GST1 gene codes for a GTP-binding protein and is expressed in a proliferation-dependent manner in mammalian cells. *EMBO J.* **8:** 3807–3814.

Hurst G.D. and Werren J.H. 2001. The role of selfish genetic elements in eukaryotic evolution. *Nat. Rev. Genet.* **2:** 597–606.

Inge-Vechtomov S.G. and Andrianova V.M. 1970. Recessive super-suppressors in yeast. *Genetika* **6:** 103–115.

Inoue Y., Kishimoto A., Hirao J., Yoshida M., and Taguchi H. 2001. Strong growth polarity of yeast prion fiber revealed by single fiber imaging. *J. Biol. Chem.* **276:** 35227–35230.

Jean-Jean O., Le Goff X., and Philippe M. 1996. Is there a human [psl]? *C.R. Acad. Sci. Ser. III. Life Sci.* **319:** 487–492.

Jedd G. and Chua N.H. 2000. A new self-assembled peroxisomal vesicle required for efficient resealing of the plasma membrane. *Nat. Cell Biol.* **2:** 226–231.

Jensen M.A., True H.L., Chernoff Y.O., and Lindquist S. 2001. Molecular population genetics and evolution of a prion-like protein in *Saccharomyces cerevisiae*. *Genetics* **159:** 527–535.

Jia Y., Rothermel B., Thornton J., and Butow R.A. 1997. A basic helix-loop-helix zipper transcription complex functions in a signaling pathway from mitochondria to the nucleus. *Mol. Cell. Biol.* **17:** 1110–1117.

Jones E.W. 2002. Vacuolar proteases and proteolytic artifacts in *Saccharomyces cerevisiae*. *Methods Enzymol.* **351:** 127–150.

Jung G. and Masison D.C. 2001. Guanidine hydrochloride inhibits Hsp104 activity *in vivo*: A possible explanation for its effect in curing yeast prions. *Curr. Microbiol.* **43:** 7–10.

Jung G., Jones G., and Masison D.C. 2002. Amino acid residue 184 of yeast Hsp104 chaperone is critical for prion-curing by guanidine, prion propagation, and thermotolerance. *Proc. Natl. Acad. Sci.* **99:** 9936–9941.

Jung G., Jones G., Wegrzyn R.D., and Masison D.C. 2000. A role for cytosolic Hsp70 in yeast [PSI⁺] prion propagation and [PSI⁺] as a cellular stress. *Genetics* **156:** 559–570.

King C.-Y. 2001. Supporting the structural basis of prion strains: Induction and identification of [*PSI*] variants. *J. Mol. Biol.* **307:** 1247–1260.

King C.-Y., Tittmann P., Gross H., Gebert R., Aebi M., and Wüthrich K. 1997. Prion-inducing domain 2-114 of yeast Sup35 protein transforms in vitro into amyloid-like filaments. *Proc. Natl. Acad. Sci.* **94:** 6618–6622.

Kochneva-Pervukhova N.V., Poznyakovski A.I., Smirnov V.N., and Ter-Avanesyan M.D. 1998a. C-terminal truncation of the Sup35 protein increases the frequency of de novo generation of a prion-based [*PSI*⁺] determinant in *Saccharomyces cerevisiae. Curr. Genet.* **34:** 146–151.

Kochneva-Pervukhova N.V., Chechenova M.B., Valouev I.A., Kushnirov V.V., Smirnov V.N., and Ter-Avanesyan M.D. 2001. [Psi(+)] prion generation in yeast: Characterization of the 'strain' difference. *Yeast* **18:** 489–497.

Kochneva-Pervukhova N.V., Paushkin S.V., Kushnirov V.V., Cox B.S., Tuite M.F., and Ter-Avanesyan M.D. 1998b. Mechanism of inhibition of Psi+ prion determinant propagation by a mutation of the N-terminus of the yeast Sup35 protein. *EMBO J.* **17:** 5805–5810.

Kocisko D.A., Priola S.A., Raymond G.J., Chesebro B., Landsbury P.T., and Caughey B. 1995. Species specificity in the cell-free conversion of prion protein to protease-resistant forms: A model for the scrapie species barrier. *Proc. Natl. Acad. Sci.* **92:** 3923–3927.

Kruger K., Grabowski P.J., Zaug A.J., Sands J., Gottschling D.E., and Cech T.R. 1982. Self-splicing RNA: Autoexcision and autocyclization of the ribosomal RNA intervening sequence of *Tetrahymena. Cell* **31:** 147–157.

Kryndushkin D.S., Smirnov V.N., Ter-Avanesyan M.D., and Kushnirov V.V. 2002. Increased expression of hsp40 chaperones, transcriptional factors, and ribosomal protein rpp0 can cure yeast prions. *J. Biol. Chem.* **277:** 23702–23708.

Kushnirov V.V., Kochneva-Pervukhova N.V., Cechenova M.B., Frolova N.S., and Ter-Avanesyan M.D. 2000a. Prion properties of the Sup35 protein of yeast *Pichia methanolica. EMBO J.* **19:** 324–331.

Kushnirov V.V., Kryndushkin D.S., Boguta M., Smirnov V.N., and Ter-Avanesyan M.D. 2000b. Chaperones that cure yeast artificial [PSI⁺] and their prion-specific effects. *Curr. Biol.* **10:** 1443–1446.

Kushnirov V.V., Ter-Avanesyan M.D., Surguchov A.P., Smirnov V.N., and Inge-Vechtomov S.G. 1987. Localization of possible functional domains in the *SUP2* gene product of the yeast *Saccharomyces cerevisiae. FEBS Lett.* **215:** 257–260.

Kushnirov V.V., Ter-Avanesyan M.D., Dagkesamanskaya A.R., Chernoff Y.O., Inge-Vechtomov S.G., and Smirnov V.N. 1990a. Deletion analysis of *SUP2* gene of yeast *Saccharomyces cerevisiae. Mol. Biol.* **24:** 1037–1041.

Kushnirov V.V., Ter-Avanesyan M.D., Telckov M.V., Surguchov A.P., Smirnov V.N., and Inge-Vechtomov S.G. 1988. Nucleotide sequence of the *SUP2(SUP35)* gene of *Saccharomyces cerevisiae. Gene* **66:** 45–54.

Kushnirov V.V., Ter-Avanesyan M.D., Didichenko S.A., Smirnov V.N., Chernoff Y.O., Derkach I.L., Novikova O.N., Inge-Vechtomov S.G., Neistat M.A., and Tolstorukov I.I. 1990b. Divergence and conservation of *SUP2 (SUP35)* gene of yeasts *Pichia pinus* and *Saccharomyces cerevisiae. Yeast* **6:** 461–472.

Lacroute F. 1971. Non-Mendelian mutation allowing ureidosuccinic acid uptake in yeast. *J. Bacteriol.* **106:** 519–522.

Lecrenier N., Van Der Bruggen P., and Foury F. 1997. Mitochondrial DNA polymerases from yeast to man: A new family of polymerases. *Gene* **185:** 147–152.

Leibowitz M.J. and Wickner R.B. 1978. Pet18: A chromosomal gene required for cell growth and for the maintenance of mitochondrial DNA and the killer plasmid of yeast. *Mol. Gen. Genet.* **165:** 115–121.

Li L. and Lindquist S. 2000. Creating a protein-based element of inheritance. *Science* **287:** 661–664.

Liao X. and Butow R.A. 1993. *RTG1* and *RTG2:* Two yeast genes required for a novel path of communication from mitochondria to the nucleus. *Cell* **72:** 61–71.

Liebman S.W. and Sherman F. 1979. Extrachromosomal [PSI⁺] determinant suppresses nonsense mutations in yeast. *J. Bacteriol.* **139:** 1068–1071.

Liebman S.W., Stewart J.W., and Sherman F. 1975. Serine substitutions caused by an ochre suppressor in yeast. *J. Mol. Biol.* **94:** 595–610.

Liu J.J. and Lindquist S. 1999. Oligopeptide-repeat expansions modulate 'protein-only' inheritance in yeast. *Nature* **400:** 573–576.

Liu Z. and Butow R.A. 1999. A transcriptional switch in the expression of yeast tricarboxyic acid cycle genes in response to a reduction or loss of respiratory function. *Mol. Cell. Biol.* **19:** 6720–6728.

Lund P.M. and Cox B.S. 1981. Reversion analysis of [psi⁻] mutations in *Saccharomyces cerevisiae. Genet. Res.* **37:** 173–182.

Lyttle T.W. 1993. Cheaters sometimes prosper: Distortion of Mendelian segregation by meiotic drive. *Trends Genet.* **9:** 205–210.

Mackay J.P., Matthews J.M., Winefield R.D., Mackay L.G., Haverkamp R.G., and Templeton M.D. 2001. The hydrophobin EAS is largely unstructured in solution and functions by forming amyloid-like structures. *Structure* **9:** 83–91.

Maddelein M.-L. and Wickner R.B. 1999. Two prion-inducing regions of Ure2p are nonoverlapping. *Mol. Cell. Biol.* **19:** 4516–4524.

Maddelein M.-L., Dos Reis S., Duvezin-Caubet S., Coulary-Salin B., and Saupe S.J. 2002. Amyloid aggregates of the HET-s prion protein are infectious. *Proc. Natl. Acad. Sci.* **99:** 7402–7407.

Masison D.C. and Wickner R.B. 1995. Prion-inducing domain of yeast Ure2p and protease resistance of Ure2p in prion-containing cells. *Science* **270:** 93–95.

Masison D.C., Maddelein M. L., and Wickner R.B. 1997. The prion model for [URE3] of yeast. Spontaneous generation and requirements for propagation. *Proc. Natl. Acad. Sci.* **94:** 12503–12508.

Matsuura A. and Anraku Y. 1993. Characterization of the *MKS1* gene, a new negative regulator of the ras-cyclic AMP pathway in *Saccharomyces cerevisiae. Mol. Gen. Genet.* **238:** 6–16.

McCabe P.M., Pfeiffer P., and Van Alfen N.K. 1999. The influence of dsRNA viruses on the biology of plant pathogenic fungi. *Trends Microbiol.* **7:** 377–381.

McCready S.J., Cox B.S., and McLaughlin C.S. 1977. The extrachromosomal control of nonsense suppression in yeast: An analysis of the elimination of [psi⁺] in the presence of a nuclear gene PNM. *Mol. Gen. Genet.* **150:** 265–270.

Moriyama H., Edskes H.K., and Wickner R.B. 2000. [URE3] prion propagation in *Saccharomyces cerevisiae:* Requirement for chaperone Hsp104 and curing by overexpressed chaperone Ydj1p. *Mol. Cell. Biol.* **20:** 8916–8922.

Nakayashiki T., Ebihara K., Bannai H., and Nakamura Y. 2001. Yeast [PSI⁺] "prions" that are crosstransmissible and susceptible beyond a species barrier through a quasi-prion state. *Mol. Cell* **7:** 1121–1130.

Nebes V.L. and Jones E.W. 1991. Activation of the proteinase B precursor of the yeast *Saccharomyces cerevisiae* by autocatalysis and by an internal sequence. *J. Biol. Chem.* **266:** 22851–22857.

Newnam G.P., Wegrzyn R.D., Lindquist S.L., and Chernoff Y.O. 1999. Antagonistic inter-actions between yeast chaperones Hsp104 and Hsp70 in prion curing. *Mol. Cell. Biol.* **19:** 1325–1333.

Novick A. and Weiner M. 1957. Enzyme induction as an all-or-none phenomenon. *Proc. Natl. Acad. Sci.* **43:** 553–566.

Ohba M. 1997. Modulation of intracellular protein degradation by SSB1-SIS1 chaperon system in yeast *S. cerevisiae. FEBS Lett.* **409:** 307–311.

Osherovich L.Z. and Weissman J.S. 2001. Multiple Gln/Asn-rich prion domains confer susceptibility to induction of the yeast [*PSI⁺*] prion. *Cell* **106:** 183–194.

Palmer E., Wilhelm J.M., and Sherman F. 1979. Variation of phenotypic suppression due to the psi+ and psi- extrachromosomal determinants in yeast. *J. Mol. Biol.* **128:** 107–110.

Pan K.-M., Baldwin M., Nguyen J., Gasset M., Serban A., Groth D., Mehlhorn I., Huang Z., Fletterick R.J., Cohen F.E., and Prusiner S.B. 1993. Conversion of α-helices into β-sheets features in the formation of the scrapie prion proteins. *Proc. Natl. Acad. Sci.* **90:** 10962–10966.

Parham S.N., Resende C.G., and Tuite M.F. 2001. Oligopeptide repeats in the yeast protein Sup35p stabilize intermolecular prion interactions. *EMBO J.* **20:** 2111–2119.

Parsell D.A., Kowal A.S., Singer M.A., and Lindquist S. 1994. Protein disaggregation medi-ated by heat-shock protein Hsp104. *Nature* **372:** 475–478.

Patino M.M., Liu J.-J., Glover J.R., and Lindquist S. 1996. Support for the prion hypothe-sis for inheritance of a phenotypic trait in yeast. *Science* **273:** 622–626.

Paushkin S.V., Kushnirov V.V., Smirnov V.N., and Ter-Avanesyan M.D. 1996. Propagation of the yeast prion-like [*psi⁺*] determinant is mediated by oligomerization of the *SUP35*-encoded polypeptide chain release factor. *EMBO J.* **15:** 3127–3134.

———. 1997a. *In vitro* propagation of the prion-like state of yeast Sup35 protein. *Science* **277:** 381–383.

———. 1997b. Interaction between yeast Sup45p (eRF1) and Sup35p (eRF3) polypeptide chain release factors: Implications for prion-dependent regulation. *Mol. Cell. Biol.* **17:** 2798–2805.

Perrett S., Freeman S.J., Butler P.J.G., and Fersht A.R. 1999. Equilibrium folding properties of the yeast prion protein determinant Ure2. *J. Mol. Biol.* **290:** 331–345.

Perutz M.F., Johnson T., Suzuki M., and Finch J.T. 1994. Glutamine repeats as polar zip-pers: Their possible role in inherited neurodegenerative diseases. *Proc. Natl. Acad. Sci.* **91:** 5355–5358.

Pierce M.M., Maddelein M.L., Roberts B.T., and Wickner R.B. 2001. A novel Rtg2p activi-ty regulates nitrogen catabolism in yeast. *Proc. Natl. Acad. Sci.* **98:** 13213–13218.

Pinan-Lucarré B., Paoletti M., Dementhon K., Coulary-Salin B., and Clavé C. 2003. Autophagy is induced during cell death by incompatibility and is essential for differen-tiation in the filamentous fungus *Podospora anserina. Mol. Microbiol.* **47:** 321–333.

Podrabsky J.E., Carpenter J.F., and Hand S.C. 2001. Survival of water stress in annual fish embryos: Dehydration avoidance and egg amyloid fibers. *Am. J. Physiol. Regul. Integr. Comp. Physiol.* **280:** R123–R131.

Prusiner S.B. 1982. Novel proteinaceous infectious particles cause scrapie. *Science* **216:** 136–144.

Prusiner S.B. and Scott M.R. 1997. Genetics of prions. *Annu. Rev. Genet.* **31:** 139-175.

Rai R., Genbauffe F., Lea H.Z., and Cooper T.G. 1987. Transcriptional regulation of the *DAL5* gene in *Saccharomyces cerevisiae. J. Bacteriol.* **169:** 3521–3524.

Riggs A.D. and Porter T.N. 1996. Overview of epigenetic mechanisms. In *Epigenetic mech-*

anisms of gene regulation (ed. V.E.A. Russo et al.), pp. 29–45. Cold Spring Harbor Laboratory Press, Cold Spring Harbor, New York.

Rizet G. 1952. Les phenomenes de barrage chez *Podospora anserina:* Analyse genetique des barrages entre les souches s et S. *Rev. Cytol. Biol. Veg.* **13:** 51–92.

Roberts B.T. and Wickner R.B. 2003. Heritable activity: A prion that propagates by covalent autoactivation. *Genes Dev.* **17:** 2083–2087.

Rossjohn J., Board P.G., Parker M.W., and Wilce M.C. 1996. A structurally derived consensus pattern for theta class gluthione transferases. *Protein Eng.* **9:** 327–332.

Rossjohn J., McKinstry W.J., Oakley A.J., Verger D., Flanagan J., Chelvanayagam G., Tan K.L., Board P.G., and Parker M.W. 1998. Human theta glass glutathione transferase: The crystal structure reveals a sulfate-binding pocket within a buried active site. *Structure* **6:** 309–322.

Safar J., Wille H., Itri V., Groth D., Serban H., Torchia M., Cohen F.E., and Prusiner S.B. 1998. Eight prion strains have PrP(Sc) molecules with different conformations (comments). *Nat. Med.* **4:** 1157–1165.

Sanchez Y., Parsell D.A., Taulien J., Vogel J.L., Craig E.A., and Lindquist S. 1993. Genetic evidence for a functional relationship between Hsp104 and Hsp70. *J. Bacteriol.* **175:** 6484–6491.

Santoso A., Chien P., Osherovich L.Z., and Weissman J.S. 2000. Molecular basis of a yeast prion species barrier. *Cell* **100:** 277–288.

Saupe S.J. 2000. Molecular genetics of heterokaryon incompatibility in filamentous ascomycetes. *Microbiol. Mol. Biol. Rev.* **64:** 489–502.

Scheibel T., Kowal A.S., Bloom J.D., and Lindquist S.L. 2001. Bidirectional amyloid fiber growth for a yeast prion determinant. *Curr. Biol.* **11:** 366–369.

Schirmer E.C. and Lindquist S. 1997. Interactions of the chaperone Hsp104 with yeast Sup35 and mammalian PrP. *Proc. Natl. Acad. Sci.* **94:** 13932–13937.

Schlumpberger M., Prusiner S.B., and Herskowitz I. 2001. Induction of distinct [URE3] yeast prion strains. *Mol. Cell. Biol.* **21:** 7035–7046.

Schlumpberger M., Wille H., Baldwin M.A., Butler D.A., Herskowitz I., and Prusiner S.B. 2000. The prion domain of yeast Ure2p induces autocatalytic formation of amyloid fibers by a recombinant fusion protein. *Protein Sci.* **9:** 440–451.

Schoun J. and Lacroute F. 1969. Etude physiologique d'une mutation permettant l'incoporation d'acide ureidosuccinique chez la levure. *C.R. Acad. Sci.* **269:** 1412–1414.

Schwimmer C. and Masison D.C. 2002. Antagonistic interactions between yeast [PSI+] and [URE3] prions and curing of [URE3] by Hsp70 protein chaperone Ssa1p but not by Ssa2p. *Mol. Cell. Biol.* **22:** 3590–3598.

Sekito T., Thornton J., and Butow R.A. 2000. Mitochondria-to-nuclear signaling is regulated by the subcellular localization of the transcription factors Rtg1p and Rtg3p. *Mol. Biol. Cell* **11:** 2103–2115.

Sekito T., Zhengchang L., Thornton J., and Butow R.A. 2002. RTG-dependent mitochondria-to-nucleus signaling is regulated by *MKS1* and is linked to formation of yeast prion [URE3]. *Mol. Biol. Cell* **13:** 795–804.

Serio T.R., Cashikar A.G., Kowal A.S., Sawicki G.J., Moslehi J.J., Serpell L., Arnsdorf M.F., and Lindquist S.L. 2000. Nucleated conformational conversion and the replication of conformational information by a prion determinant. *Science* **289:** 1317–1321.

Silar P. and Daboussi M.J. 1999. Non-conventional infectious elements in filamentous fungi. *Trends Genet.* **15:** 141–145.

Singh A., Ursic D., and Davies J. 1979. Phenotypic suppression and misreading *Saccharomyces cerevisiae. Nature* **277:** 146–148.

Singh A.C., Helms C., and Sherman F. 1979. Mutation of the non-Mendelian suppressor [PSI] in yeast by hypertonic media. *Proc. Natl. Acad. Sci.* **76**: 1952–1956.

Sondheimer N. and Lindquist S. 2000. Rnq1: An epigenetic modifier of protein function in yeast. *Mol. Cell* **5**: 163–172.

Sondheimer N., Lopez N., Craig E.A., and Lindquist S. 2001. The role of Sis1 in the maintenance of the [*RNQ*⁺] prion. *EMBO J.* **20**: 2435–2442.

Sparrer H.E., Santoso A., Szoka F.C., and Weissman J.S. 2000. Evidence for the prion hypothesis: Induction of the yeast [*PSI*⁺] factor by in vitro-converted Sup35 protein. *Science* **289**: 595–599.

Speransky V., Taylor K.L., Edskes H.K., Wickner R.B., and Steven A. 2001. Prion filament networks in [URE3] cells of *Saccharomyces cerevisiae. J. Cell Biol.* **153**: 1327–1335.

Stansfield I., Jones K.M., Kushnirov V.V., Dagkesamanskaya A.R., Poznyakovski A.I., Paushkin S.V., Nierras C.R., Cox B.S., Ter-Avanesyan M.D., and Tuite M.F. 1995. The products of the *SUP45* (eRF1) and *SUP35* genes interact to mediate translation termination in *Saccharomyces cerevisiae. EMBO J.* **14**: 4365–4373.

Stott K., Blackburn J.M., Butler P.J.G., and Perutz M. 1995. Incorporation of glutamine repeats makes protein oligomerize: Implications for neurodegenerative diseases. *Proc. Natl. Acad. Sci.* **92**: 6509–6513.

Talloczy Z., Menon S., Neigeborn L., and Leibowitz M.J. 1998. The [KIL-d] cytoplasmid genetic element of yeast results in epigenetic regulation of viral M double-stranded RNA gene expression. *Genetics* **150**: 21–30.

Talloczy Z., Mazar R., Georgopoulos D.E., Ramos F., and Leibowitz M.J. 2000. The [KIL-d] element specifically regulates viral gene expression in yeast. *Genetics* **155**: 601–609.

Tamm I. and Eggers H.J. 1963. Specific inhibition of replication of animal viruses. *Science* **142**: 24–33.

Taylor K.L., Cheng N., Williams R.W., Steven A.C., and Wickner R.B. 1999. Prion domain initiation of amyloid formation *in vitro* from native Ure2p. *Science* **283**: 1339–1343.

Ter-Avanesyan A., Dagkesamanskaya A.R., Kushnirov V.V., and Smirnov V.N. 1994. The *SUP35* omnipotent suppressor gene is involved in the maintenance of the non-Mendelian determinant [*psi*⁺] in the yeast *Saccharomyces cerevisiae. Genetics* **137**: 671–676.

Ter-Avanesyan M.D., Kushnirov V.V., Dagkesamanskaya A.R., Didichenko S.A., Chernoff Y.O., Inge-Vechtomov S.G., and Smirnov V.N. 1993. Deletion analysis of the *SUP35* gene of the yeast *Saccharomyces cerevisiae* reveals two non-overlapping functional regions in the encoded protein. *Mol. Microbiol.* **7**: 683–692.

Thual C., Komar A.A., Bousset L., Fernandez-Bellot E., Cullin C., and Melki R. 1999. Structural characterization of *Saccharomyces cerevisiae* prion-like protein Ure2. *J. Biol. Chem.* **274**: 13666–13674.

True H.L. and Lindquist S.L. 2000. A yeast prion provides a mechanism for genetic variation and phenotypic diversity. *Nature* **407**: 477–483.

Tuite M.F., Cox B.S., and McLaughlin C.S. 1983. *In vitro* nonsense suppression in [*psi*⁺] and [*psi*⁻] cell-free lysates of *Saccharomyces cerevisiae. Proc. Natl. Acad. Sci.* **80**: 2824–2828.

———. 1987. A ribosome-associated inhibitor of in vitro nonsense suppression in [psi⁻] strains of yeast. *FEBS Lett.* **225**: 205–208.

Tuite M.F., Mundy C.R., and Cox B.S. 1981. Agents that cause a high frequency of genetic change from [*psi*⁺] to [*psi*⁻] in *Saccharomyces cerevisiae. Genetics* **98**: 691–711.

Tuite M.F., Lund P.M., Futcher A.B., Dobson M.J., Cox B.S., and McLaughlin C.S. 1982. Relationship of the [*psi*] factor with other plasmids of *Saccharomyces cerevisiae. Plasmid* **8**: 103–111.

Turcq B., Denayrolles M., and Begueret J. 1990. Isolation of two alleles incompatibility genes *s* and *S* of the fungus *Podospora anserina*. *Curr. Genet.* **17:** 297–303.

Turcq B., Deleu C., Denayrolles M., and Begueret J. 1991. Two allelic genes responsible for vegetative incompatibility in the fungus *Podospora anserina* are not essential for cell viability. *Mol. Gen. Genet.* **288:** 265–269.

Turoscy V. and Cooper T.G. 1987. Ureidosuccinate is transported by the allantoate transport system in *Saccharomyces cerevisiae*. *J. Bacteriol.* **169:** 2598–2600.

Umland T.C., Taylor K.L., Rhee S., Wickner R.B., and Davies D.R. 2001. The crystal structure of the nitrogen catabolite regulatory fragment of the yeast prion protein Ure2p. *Proc. Natl. Acad. Sci.* **98:** 1459–1464.

Uptain S.M., Sawicki G.J., Caughey B., and Lindquist S. 2001. Strains of [*PSI*⁺] are distinguished by their efficiencies of prion-mediated conformational conversion. *EMBO J.* **20:** 6236–6245.

Volkov K.V., Aksenova Y.A., Soom M.J., Sipov K.V., Svitin A.V., Kurischko C., Shkundina I.S., Ter-Avanesyan M.D., Inge-Vechtomov S.G., and Mironova L.N. 2002. Novel non-Mendelian determinant involved in the control of translation accuracy in *Saccharomyces cerevisiae*. *Genetics* **160:** 25–36.

Wegrzyn R.D., Bapat K., Newnam G.P., Zink A.D., and Chernoff Y.O. 2001. Mechanism of prion loss after Hsp104 inactivation in yeast. *Mol. Cell. Biol.* **21:** 4656–4669.

Wickner R.B. 1976. Mutants of the killer plasmid of *Saccharomyces cerevisiae* dependent on chromosomal diploidy for expression and maintenance. *Genetics* **82:** 273–285.

———. 1994. Evidence for a prion analog in *S. cerevisiae:* The [URE3] non-Mendelian genetic element as an altered *URE2* protein. *Science* **264:** 566–569.

———. 1997. *Prion diseases of mammals and yeast: Molecular mechanisms and genetic features.* R.G. Landes, Austin, Texas.

———. 2001. Viruses of yeasts, fungi and parasitic microorganisms. In *Field's virology* (ed. D.M. Knipe and P.M. Howley), pp. 629–658. Lippincott, Williams & Wilkins, Philadelphia, Pennsylvania.

Wickner R.B., Masison D.C., and Edkes H.K. 1995. [PSI] and [URE3] as yeast prions. *Yeast* **11:** 1671–1685.

Wickner R.B., Taylor K.L., Edkes H.K., and Maddelein M.L. 2000. Prions: Portable prion domains. *Curr. Biol.* **10:** R335–R337.

Wickner R.B., Taylor K.L., Edkes H K , Maddelein M.-L., Moriyama H., and Roberts B.T. 1999. Prions in *Saccharomyces* and *Podospora:* Protein-based inheritance. *Microbiol, Mol. Biol. Rev.* **63:** 844–861.

Wilson P.G. and Culbertson M.R. 1988. *SUF12* suppressor protein of yeast: A fusion protein related to the EF-1 family of elongation factors. *J. Mol. Biol.* **199:** 559–573.

Wosten H.A. and de Vocht M.L. 2000. Hydrophobins, the fungal coat unravelled. *Biochim. Biophys. Acta* **1469:** 79–86.

Wu J., Saupe S.J., and Glass N.L. 1998. Evidence for balancing selection operating at the *het-c* heterokaryon incompatibility locus in a group of filamentous fungi. *Proc. Natl. Acad. Sci.* **95:** 12398–12403.

Young C.S.H. and Cox B.S. 1971. Extrachromosomal elements in a super-suppression system of yeast. 1. A nuclear gene controlling the inheritance of the extrachromosomal elements. *Heredity* **26:** 413–422.

———. 1972. Extrachromosomal elements in a super-suppression system of yeast. II. Relations with other extrachromosomal elements. *Heredity* **28:** 189–199.

Zhou P., Derkatch I.L., and Liebman S.W. 2001. The relationship between visible intracellular aggregates that appear after overexpression of Sup35 and the yeast prion-like elements [*PSI*⁺] and [*PIN*⁺]. *Mol. Microbiol.* **39:** 37–46.

Zhou P., Derkatch I.L., Uptain S.M., Patino M.M., Lindquist S., and Liebman S.W. 1999. The yeast non-Mendelian factor [*ETA*⁺] is a variant of [*PSI*⁺], a prion-like form of release factor eRF3. *EMBO J.* **18:** 1182–1191.

Zhouravleva G., Frolova L., LeGoff X., LeGuellec R., Inge-Vectomov S., Kisselev L., and Philippe M. 1995. Termination of translation in eukaryotes is governed by two interacting polypeptide chain release factors, eRF1 and eRF3. *EMBO J.* **14:** 4065–4072.

Zubenko G. and Jones E.W. 1982. Genetic properties of mutations at the *PEP4* locus *Saccharomyces cerevisiae*. *Genetics* **102:** 679–690.

8

Knockouts, Knockins, Transgenics, and Transplants in Prion Research

E. Flechsig
Institut für Virologie und Immunbiologie
D-97078 Würzburg, Germany

J.C. Manson and R. Barron
Institute for Animal Health
AFRC and MRC Neuropathogenesis Unit
Edinburgh EH9 3JF, United Kingdom

A. Aguzzi
Institut für Neuropathologie
CH-8091 Zürich, Switzerland

C. Weissmann
MRC Prion Unit, Department of Neurodegenerative Disease
Institute of Neurology
London WC1N 3BG, United Kingdom

THE *"PROTEIN ONLY"* HYPOTHESIS PROPOSES THAT the prion is a conformational isoform of the normal host protein PrP^C and that the abnormal conformer, when introduced into the organism, causes the conversion of PrP^C into a likeness of itself. PrP^C is encoded by a single-copy gene (Basler et al. 1986) located on chromosome 2 in mouse (*Prnp*) and on chromosome 20 (*PRNP*) in humans. A PrP gene has been found in all vertebrates examined (Wopfner et al. 1999), including birds (Wopfner et al. 1999), amphibians (Strumbo et al. 2001), and reptiles (Simonic et al. 2000); moreover, a PrP-like gene was identified in fugu fish (Suzuki et al. 2002).

Maturation of the primary translation product results in removal of an amino-terminal signal sequence of 22 amino acids, replacement of 23

Prion Biology and Diseases, 2nd Ed. ©2004 Cold Spring Harbor Laboratory Press 0-87969-693-1/04

amino acids at the carboxyl terminus by a glycosylphosphatidylinositol (GPI) residue, and glycosylation at two asparagine residues. Mature PrP^C comprises a highly flexible amino-terminal half, made up of octapeptide repeats, and a globular domain consisting of three α-helices, one short antiparallel β-sheet, and a single disulfide bond (Riek et al. 1996, 1997; Donne et al. 1997). Cu^{++} binds both to the octapeptide region and to histidines 96 and 111 of the PrP flexible tail (Sulkowski 1992; Hornshaw et al. 1995; Brown et al. 1997b; Jackson et al. 2001). PrP^C is anchored to the outer surface of the cell membrane, in cholesterol-rich microdomains or caveolae (Vey et al. 1996; Naslavsky et al. 1997), and undergoes endocytosis and recycling (Shyng et al. 1993, 1995; Harris 1999). In cell culture, endocytosis of PrP^C is stimulated by copper ions (Pauly and Harris 1998); the stimulation is abrogated if the copper-binding octarepeats of PrP^C are deleted (Lee et al. 2001; Perera and Hooper 2001).

In mouse embryonic development, the PrP gene is expressed by midgestation in the developing central and peripheral nervous tissue. It continues to be expressed throughout development, and mRNA can be detected in a number of nonneuronal tissues both within the embryo and in the extraembryonic tissue (Manson et al. 1992). PrP mRNA is expressed at high levels in neurons of the adult brain and has also been detected in astrocytes and oligodendrocytes (Moser et al. 1995). Lower levels of PrP mRNA can be detected in other tissues such as heart, lung, and spleen (Oesch et al. 1985). In the mouse and hamster, PrP^C has been detected predominantly in brain tissue (Bendheim et al. 1992), in particular in some (but not all) neurons (Ford et al. 2002a); in the sympathetic, parasympathetic, enteric and peripheral nervous systems (Ford et al. 2002a); as well as in Schwann cells (Follet et al. 2002). In neurons, PrP^C is localized at synapses (Sales et al. 1998; Herms et al. 1999; Moya et al. 2000) but also in the soma, dendrites (Laine et al. 2001), and elongating axons (Sales et al. 2002). PrP^C is also present at quite high levels in heart and skeletal muscle (Bendheim et al. 1992), follicular dendritic cells (FDCs) (McBride et al. 1992), nonfollicular dendritic cells (Ford et al. 2002a), and some lymphocytes (Cashman et al. 1990; Mabbott et al. 1997; Ford et al. 2002a), but it is barely detectable in liver (Bendheim et al. 1992). The abundance of PrP mRNA in a cell is not necessarily reflected in the level of PrP^C (Ford et al. 2002b).

GENERATION AND PROPERTIES OF MICE DEVOID OF PrP

Compelling linkage between the infectious scrapie agent, or prion, and PrP was established by biochemical and genetic data (Prusiner 1982, 1989;

Gabizon et al. 1988; Hsiao et al. 1989; Weissmann 1989), leading to the prediction that animals devoid of PrP should be resistant to experimental scrapie and fail to propagate infectivity. This prediction was indeed borne out, adding substantial support to the "protein only" hypothesis. In addition, the availability of PrP knockout mice provided an approach to carry out reverse genetics on PrP, with regard both to prion disease and to its physiological role.

Generation of PrP Knockout Mice

The *Prnp* gene of the mouse consists of three exons, with the entire reading frame contained in the third exon (Fig. 1b). Several lines of mice devoid of PrP^C have been generated by homologous recombination in embryonic stem cells, using either of two strategies. The conservative strategy involves deletions or replacements within the open reading frame (ORF) only. Thus, Büeler et al. (1992) replaced PrP codons 4–187 (of altogether 254 codons), about 80% of the sequence encoding the mature protein (residues 23–231), by an expression cassette for neomycin phosphotransferase (*neo*). Mice homozygous for the disrupted gene (designated in this chapter as *Prnp*$^{0/0}$[Zürich I]) express truncated PrP mRNA but no detectable PrP fragment and develop normally. Manson et al. (1994b) prepared a second line of PrP knockout mice (Npu, here designated as *Prnp*$^{-/-}$[Edinburgh]) in which the PrP gene was disrupted by the insertion of a *neo* cassette into a *Kpn*I site following codon 93 of the PrP ORF. Mice homozygous for the knockout allele also developed normally, and no PrP mRNA or PrP-related protein was detected in the brain. Two lines of "conditional PrP knockout" mice have been prepared on the basis of the conservative strategy. A PrP-expressing transgene under the control of a doxycycline-repressible promoter was introduced into *Prnp*$^{0/0}$[Zürich I] mice; however, repression was incomplete and resulted in basal PrP^C levels about 5–15% of wild type. No histopathological changes appeared after repression, and no clinical disease after prion inoculation was noted, but prions accumulated nonetheless, albeit at a low level (Tremblay et al. 1998). Mallucci et al. (2002) generated *Prnp*$^{0/0}$ mice that were transgenic for both a *loxP*-flanked cassette containing the PrP ORF under control of a *Prnp* promoter element and Cre recombinase under the control of the neurofilament heavy chain (NFH) promoter. Expression of the recombinase resulted in the ablation of PrP in neurons at 9 weeks of age without entailing histopathological changes, but expression of PrP was still detectable in nonneuronal cells. No clinical disease was observed for up to 400 days after prion inoculation.

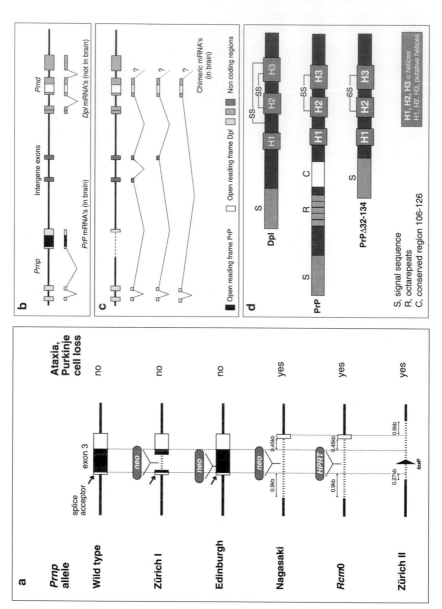

Figure 1. (*See facing page for legend.*)

PrP knockout by the radical strategy involved deletion not only of the reading frame, but also of its flanking regions. In the $Prnp^{-/-}$[Nagasaki] line, a 2.1-kb genomic DNA segment comprising 0.9 kb of intron 2, 10 bp of 5′ noncoding region, the entire PrP ORF, and 0.45 kb of the 3′ noncoding region, was replaced with a *neo* cassette (Sakaguchi et al. 1996). As described below, these mice developed normally but exhibited severe ataxia and Purkinje cell loss in later life (Sakaguchi et al. 1996). A similar phenotype was found in the RcmO (Moore et al. 1999; Silverman et al. 2000) PrP knockout line that resembles both the Nagasaki line with respect to the extensive PrP gene replacement and the $Prnp^{-/-}$[Zürich II] line, where 0.26 kb of intron 2, 10 bp of the 5′ flanking sequence, the entire ORF, the 3′ noncoding region of exon 3, and 0.6 kb of 3′ adjacent sequence were replaced by a 34-bp *loxP* sequence (Fig. 1a) (Rossi et al. 2001).

The Phenotype of PrP Knockout Mice

The hope that the generation of PrP null mice would cast light on the physiological role of PrP^C has not been fully realized. The offspring of mating pairs of mice hemizygous for the [Zürich I] PrP allele yielded about one-quarter homozygous knockout offspring, showing that embryonic development of these mice was not impaired. $Prnp^{0/0}$[Zürich I] and $Prnp^{-/-}$[Edinburgh] mice developed and reproduced normally (Büeler et al. 1992; Manson et al. 1994b), but old mice showed demyelination in the peripheral nervous system (Nishida et al. 1999). Behavioral studies revealed no significant differences from wild-type mice in the Morris water maze, the Y maze discrimination test, and the two-way avoidance test (Büeler et al. 1992; Lipp et al. 1998). However, mice devoid of PrP

Figure 1. Knockout strategies and their consequences. (*a*) Various strategies used to target *Prnp* by homologous recombination. The black boxes represent PrP ORFs; white boxes, noncoding *Prnp* regions; gray boxes, inserted sequences; dotted line, deleted regions; *neo*, neomycin phosphotransferase; HPRT, hypoxanthine phosphoribosyltransferase; *loxP* (*black arrowhead*), a 34-bp recombination site from phage P1. (*b*) Coding and noncoding exons of *Prnp*, *Prnd*, and intergenic exons of unknown function. (*c*) Exon skipping leads to expression of Doppel under the direction of the *Prnp* promoter. Deletion of the splice receptor site upstream of the third *Prnp* exon entails the formation of several chimeric mRNAs comprising the first two exons of *Prnp*, which are spliced directly or indirectly to the Dpl-encoding exon (Moore et al. 1999). (*d*) Comparison of domains of Dpl with full-length PrP and with PrP lacking residues 32–134. (*a*, Reprinted, with permission, from Rossi et al. 2001 [copyright Oxford University Press]; *b–d*, reprinted, with permission, from Weissmann and Aguzzi 1999 [copyright AAAS].)

were reported to have alterations in both circadian activity rhythms and sleep that were abolished by the introduction of PrP transgenes (Tobler et al. 1996, 1997). A change in slow-wave activity (SWA) prior to sleep–wake transitions in PrP knockout mice following sleep deprivation was attributed to diminished recovery from stress (Huber et al. 1999, 2002). Furthermore, $Prnp^{0/0}$[Zürich I] mice were found to be more susceptible than wild-type mice to acute seizures induced by drugs such as kainic acid (Walz et al. 1999).

Electrophysiological studies on $Prnp^{0/0}$[Zürich I] mice showed that GABA-A receptor-mediated fast inhibition was weakened, long-term potentiation (LTP) was impaired (Collinge et al. 1994; Manson et al. 1995; Whittington et al. 1995; Herms and Kretzschmar 2001), and Ca^{++}-activated K^+ currents were disrupted in hippocampal cells (Colling et al. 1996) and cerebellar Purkinje cells (Herms et al. 2001). Introduction of high copy numbers of the human (Whittington et al. 1995) or murine (Herms and Kretzschmar 2001) PrP gene into PrP null mice restored the LTP response to that seen in the wild-type controls. Slow after-hyperpolarization in hippocampal CA1 cells was effectively abolished by the conditional knockout of PrP at 9 weeks, suggesting that this is due to loss of a differentiated neuronal function rather than to a developmental deficit (Mallucci et al. 2002).

Biochemical changes reported for Zürich-I-type knockout mice include a change in the localization of neuronal nitric oxide synthase (nNOS) and a decrease in its activity (Keshet et al. 1999). Reduced mitochondrial numbers, unusual mitochondrial morphology, and elevated levels of mitochondrial Mn-dependent superoxide dismutase (SOD) have been found in some tissues (Miele et al. 2002). Furthermore, levels of nuclear factor NF-κB and MnSOD are increased, while Cu/ZnSOD activity and p53 are decreased (Brown et al. 2002). Protein and lipid oxidation is increased in brain, skeletal muscles, heart, and liver of PrP knockout mice, in association with a reduced catalase activity (Klamt et al. 2001). The impairment of enzymatic activity required for antioxidant defense suggests that the absence of PrP^C alters oxidative stress homeostasis.

In contrast to the moderate phenotypic alterations in Zürich I and Edinburgh knockout mice, the $Prnp^{-/-}$[Nagasaki], Rcm0, and $Prnp^{-/-}$ [Zürich II] lines exhibit severe ataxia and Purkinje cell loss in later life (Sakaguchi et al. 1996; Moore et al. 1999; Silverman et al. 2000; Rossi et al. 2001), as well as demyelination of peripheral nerves (Nishida et al. 1999). Because the Nagasaki phenotype was abolished by introduction of a PrP transgene (Nishida et al. 1999), it was attributed to the absence of PrP^C. However, the fact that the cerebellar symptoms were not observed in the

Zürich I and Edinburgh knockout lines argued that they were caused not by the absence of PrP, but rather by the partial deletion of sequences flanking the PrP open reading frame (Weissmann 1996). The discovery of *Prnd* and the ectopic expression of its product, Doppel (Dpl), in the brains of all three ataxic PrP knockout lines, Nagasaki, Zürich II, and Rcm0, led to the resolution of this problem (Moore et al. 1999).

Dpl is an N-glycosylated, GPI-anchored protein (Lu et al. 2000; Silverman et al. 2000) normally expressed in a variety of tissues but not in postnatal brain (Moore et al. 1999). Dpl and PrP show about 25% sequence similarity (Moore et al. 1999) and share similar globular domains with only minor differences in secondary structure (Lu et al. 2000; Silverman et al. 2000; Mo et al. 2001). However, Dpl lacks a counterpart to the flexible amino-terminal segment of PrP (Fig. 1d). Its physiological function is still elusive, but the absence of Dpl caused male sterility in mice (Behrens et al. 2002). Dpl is unlikely to be involved in prion pathogeneses (Moore et al. 2001; Tuzi et al. 2002).

Dpl is encoded by *Prnd*, located 16 kb downstream of *Prnp* (Moore et al. 1999). In wild-type mice, Dpl mRNA is very weakly expressed, under control of its own promoter. As a consequence of the radical PrP gene deletion strategy that gives rise to ataxic knockout lines, the acceptor splice site of the third exon is lost, causing exon skipping and formation of chimeric transcripts with the first two noncoding exons of the *Prnp* locus linked to the Dpl-encoding *Prnd* exon. This places Dpl expression under the control of the *Prnp* promoter (Fig. 1b) (Moore et al. 1999; Li et al. 2000). Time to appearance of disease is inversely correlated with the expression level of Dpl in brain (Rossi et al. 2001), and the phenotype is rescued by coexpression of PrP^C (Nishida et al. 1999; Rossi et al. 2001). *Prnd*-encoding mRNA was expressed undiminished in brain of Nagasaki-type mice "cured" by the introduction of a PrP-expressing transgene (Rossi et al. 2001). Thus, ectopic expression of Dpl in the absence of PrP^C, rather than absence of PrP^C itself, causes Purkinje cell loss in Nagasaki-type PrP knockout mice. This interpretation is supported by the finding that transgenic $Prnp^{0/0}$ mice expressing Dpl ubiquitously in brain developed severe ataxia associated with loss of both granule and Purkinje cells (Moore et al. 2001). As in the case of Nagasaki-type mice, age of onset was inversely related to the level of Dpl expression, and introduction of a hamster-PrP-encoding transgene resulted either in complete abrogation of the cerebellar syndrome in mice expressing moderate Dpl levels or partial abrogation in animals expressing high levels of Dpl (Moore et al. 2001).

How does Dpl cause damage in the brain? It has been proposed that because overexpression of Dpl entrains an increase of heme oxygenase 1 and both neuronal and inducible synthase (NOS), it would cause oxidative stress deleterious to sensitive neurons; this effect would be counteracted by the antioxidant properties of PrP^C (Wong et al. 2001). Another explanation is discussed below, in connection with the pathogenic properties of an amino-proximally truncated PrP that resembles Dpl in some respects (Fig. 4c) (Moore et al. 1999; Silverman et al. 2000).

The Phenotype of Dpl Knockout Mice

Inactivation of the *Prnd* gene in brain grafts or in whole mice did not yield any discernible neurological phenotype, and prion replication was unimpaired (Behrens et al. 2001, 2002). However, no progeny resulted from intercrosses of *Prnd*$^{-/-}$ mice. Female *Prnd*$^{-/-}$ mice, when crossed to *Prnd*$^{-/-}$ or *Prnd*$^{+/-}$ males, yielded litter sizes similar to those of wild type. In contrast, male *Prnd*$^{-/-}$ mice were infertile. Their sexual activity was similar to that of controls, as shown by a normal number of copulation plugs. However, the number of spermatozoa in the cauda epididymis of *Prnd*$^{-/-}$ males was reduced, and motility of mutant sperm was decreased. Therefore, sterility of Dpl-mutant males is not due to behavioral abnormalities, but may be due to a spermatogenesis defect. Indeed, *Prnd*$^{-/-}$ sperm heads were severely malformed and lacked a discernible well-developed acrosome (Behrens et al. 2002). Because the acrosome is essential for sperm–egg interaction, this defect could explain the sterility of Dpl-deficient males.

In vitro fertilization (IVF) experiments confirmed that spermatozoa isolated from *Prnd*$^{-/-}$ males were unable to fertilize wild-type oocytes. Spermatozoa of *Prnd*$^{-/-}$ males never penetrated the zona pellucida. However, if the zona pellucida was partially dissected and IVF was performed with sperm suspension from *Prnd*$^{-/-}$ males, fertility was partially rescued. These data indicate that *Prnd*$^{-/-}$ spermatozoa are capable of oocyte fertilization, albeit at a lower frequency than controls, but that they cannot overcome the barrier imposed by the zona pellucida.

A significant amount of PrP^C is expressed in mature spermatozoa. PrP^C found in testes was truncated in the vicinity of residue 200 (Shaked et al. 1999). A protective role for PrP^C against copper toxicity has been proposed: Sperm cells originating from *Prnp*-ablated mice were more susceptible to high copper concentrations than wild-type sperm. However, male *Prnp*$^{0/0}$ knockout mice are not sterile and produce normal litter sizes. PrP expressed in testes is clearly not capable of compensating for the absence of Dpl, suggesting nonredundant functions for the two proteins. However, it will be very interesting to explore whether males lacking both

PrP^C and Dpl might display a more severe defect than $Prnd^{-/-}$ single mutant mice.

At present, the molecular mechanism of Dpl-regulated acrosome development is unclear. Dpl may be present on the acrosomic vesicles through its GPI anchor and may participate in acrosome morphogenesis. Alternatively, Dpl may regulate acrosome function in a more indirect way. Oligosaccharides have been implicated in sperm binding and signaling for the acrosome reaction, but the composition and structure of the essential carbohydrate moieties remain controversial. Because Dpl is a highly glycosylated protein located at the outside of the plasma membrane (Moore et al. 1999; Silverman et al. 2000), it may be directly involved in sperm–egg interaction.

In contrast to the mice generated in Zürich, Dpl-deficient homozygous males in Edinburgh appear to be able to fertilize eggs. In a series of in vitro fertilizations, the Melton laboratory has reported that progression to the early cleavage divisions occurred, but was soon followed by death of the embryos at the preimplantation stage. In addition, there were no obvious malformations of sperms (communicated by Derek Paisley and David Melton at the International TSE Conference in Edinburgh, September 2002). Whether this discrepancy is related to different genetic backgrounds of the mice utilized, whether it is due to slightly different targeting strategies, or, finally, whether it might reflect yet another surprising phenomenon in the genetics of prion-related genes, is at the time of writing wholly unclear.

Possible Physiological Functions of PrP^C

Several physiological roles for PrP^C have been proposed; in particular, cell adhesion, signaling, neuroprotection, and metabolic functions related to its copper-binding properties. Because PrP binds Cu^{++} tightly (Hornshaw et al. 1995; Brown et al. 1997b; Jackson et al. 2001), and Cu^{++} stimulates endocytosis of PrP^C, a role in copper ion transport (Pauly and Harris 1998; Brown 1999), in particular in the re-uptake at the synaptic cleft, has been proposed (Herms and Kretzschmar 2001). Another suggestion is that PrP^C protects cells against Cu^{++} by scavenging it (Brown et al. 1997a; Herms et al. 1999) or against oxidative stress by enhancing Cu,Zn superoxide dismutase (SOD) (Brown and Besinger 1998) and/or glutathione reductase activity (White et al. 1999). Moreover, it has been reported that recombinant PrP possesses Cu-dependent SOD activity, albeit at low levels (D.R. Brown et al. 1999, 2000). The claim that the level of SOD activity in brain tissue significantly varied with expression level of PrP^C is a subject of controversy (Waggoner et al. 2000; Wong et al. 2000a; Brown et al.

2002). Despite the missing link between the physiological function of PrP^C and its copper-binding properties, there is considerable evidence that oxidative stress homeostasis is altered in the absence of PrP^C (Wong et al. 2000b; Brown et al. 2002).

It has also been proposed that Cu^{++} may serve to impart a defined tertiary structure to PrP^C (Stöckel et al. 1998) and/or to target it to caveolae-like domains (Pauly and Harris 1998; Perera and Hooper 2001) and thereby support some other function, in particular signaling, as shown for other GPI-anchored proteins (Horejsi et al. 1999). The proposal that PrP^C is a signal transduction protein is supported by studies on differentiated neuronal cells, in which antibody-mediated cross-linking of PrP^C led to caveolin-dependent activation of the tyrosine kinase Fyn (Mouillet-Richard et al. 2000). PrP interacts with the 37-kD/67-kD laminin receptor precursor (Rieger et al. 1997; Gauczynski et al. 2001; Hundt et al. 2001) and heparan sulfate (Warner et al. 2002), supporting a possible role in cell adhesion and/or signaling.

The findings that neuronal cultures of Nagasaki-type PrP knockout mice, which presumably express Dpl, undergo apoptosis more readily than wild-type cultures (Kuwahara et al. 1999) and that PrP^C protects against Bax-mediated apoptosis (Bounhar et al. 2001), suggest an anti-apoptotic role for PrP^C. Moreover, PrP^C reportedly mediates rescue of retinal neurons from apoptosis (Chiarini et al. 2002); stress-inducible protein 1 (STI1) was proposed to be the 67-kD cell-surface ligand for PrP (Martins et al. 1997) that triggers this neuroprotection (Zanata et al. 2002).

Almost a dozen other proteins have been found to interact with PrP, using various screening strategies. However, the physiological significance, if any, of interactions with proteins such as APLP1 or a tyrosine phosphatase (non-receptor type) (Yehiely et al. 1997), N-CAM (Schmitt-Ulms et al. 2001), Bcl-2 (Kurschner and Morgan 1995, 1996), synapsin Ib or Grb2 (Spielhaupter and Schatzl 2001), and the laminin γ-1 chain (Graner et al. 2000) needs to be established.

The profusion of interactions and functions attributed to PrP^C underlines the current state of perplexity.

PrP Knockout Mice Are Resistant to Scrapie

Prnp$^{0/0}$[Zürich] mice, *Prnp*$^{+/+}$ littermates (both with a genetic background derived from 129/Sv × C57BL/6J animals), as well as Swiss CD1 control mice, were inoculated intracerebrally (i.c.) with about 10^7 LD$_{50}$ units of the Chandler isolate of mouse-adapted prions (RML) (Chandler 1961). CD-1 mice showed typical neurological symptoms at 140 ± 6 days and died at 153 ± 7 days. $Prnp^{+/+}$ mice with the C57BL/6–129/Sv background developed clinical symptoms at 158 ± 11 days and died at 171 ± 11 days. In stark contrast, 23 of 25 $Prnp^{0/0}$ mice remained alive and free of clinical symptoms for at least 2 years. Two mice died of intercurrent disease other than scrapie. None of the prion-inoculated $Prnp^{0/0}$ mice showed scrapie-specific pathology as late as 57 weeks after inoculation; they were indistinguishable from controls inoculated with normal brain homogenate, while Prnp+/+ mice showed typical neuropathological changes such as astrogliosis and vacuolation mainly in the cortex, thalamus, and hippocampus 23–25 weeks after inoculation (Büeler et al. 1993). Similar results were reported for the $Prnp^{0/0}$[Zürich] mice bred in San Francisco and inoculated with the Chandler isolate (Prusiner et al. 1993), for the $Prnp^{-/-}$[Edinburgh] mice inoculated with the ME7 mouse strain (Manson et al. 1994a), and for the $Prnp^{-/-}$ [Nagasaki] mice challenged with the mouse-adapted Fukuoka-1 strain of CJD prions (Sakaguchi et al. 1995). All three lines of PrP null mice showed a complete lack of scrapie typical neuropathology following intracerebral inoculation with prions.

An additional prediction of the "protein only" hypothesis is that PrP knockout mice should be unable to propagate prions. Indeed, $Prnp^{0/0}$ [Zürich] mice challenged with the Chandler isolate (heated at 80°C for 20 minutes to inactivate any adventitious infectious agent) were unable to propagate prions in brain and spleen, whereas prion levels in brain and spleen of $Prnp^{+/+}$ mice increased to about 8.6 and 6.9 log LD_{50} units/ml, respectively, by 20 weeks postinfection (Büeler et al. 1993). In a further experiment with non-heated Chandler isolate, a pooled sample (from 4 brains) taken at 20 weeks gave a titer of 3.2 log LD_{50} units/ml, whereas all other samples, taken at 8, 12, 25, 33, and 48 weeks, were negative (Büeler et al. 1993). The experiment was repeated; analyses for infectivity were performed after 16, 18, 20, and 22 weeks, on 4 mice for each time point. Only the 20-week sample gave rise to scrapie disease in 1 of 8 indicator mice; all other samples in 56 indicator mice were negative (Sailer et al. 1994). Prusiner et al. (1993) performed a similar experiment and found no infectivity in the brains of 8 PrP null mice tested between 120 and 315 days after inoculation (d.p.i.), but borderline infectivity was detected in 4 out of 5 mice examined between 5 and 60 d.p.i. Inoculation experiments performed by Sakaguchi et al. (1995) on $Prnp^{-/-}$ [Nagasaki] mice showed that low levels of infectivity were found even as late as 29 weeks, in all but 1 of 7 samples (each sample was a pool of 5 mouse brains for each time point), and that this infectivity was diminished or abolished if the samples were first heat-treated. Infectivity has been recovered on occasion from Edinburgh PrP null mice up to 600 days after inoculation (J. Manson, unpubl.).

According to these reports, there is no evidence for net generation of scrapie prions in PrP null mice; the occasional low-level infectivity detected in the brains of such mice after intracerebral inoculation may be due to residual inoculum or, less likely, to contamination. It has, however, been speculated that in the absence of PrP, a temperature-sensitive form or otherwise labile form of scrapie agent might be generated (Chesebro and Caughey 1993; Sakaguchi et al. 1995).

Properties of Mice Hemizygous for the $Prnp^0$ Allele

Mice carrying a single $Prnp$ allele ($Prnp^{0/+}$ mice) showed no abnormal phenotype as regards behavior and development (Büeler et al. 1992; Manson et al. 1994b; Sakaguchi et al. 1996). Surprisingly, they showed prolonged incubation times of about 290 days to appearance of disease, as compared to 170 days in the case of $Prnp^{+/+}$ mice. The $Prnp^{0/+}$ mice harbored high levels of infectivity and PrP^{Sc} by 140 days after inoculation, as did wild-type controls, but survived thereafter for at least another 140 days without showing severe clinical symptoms (Büeler et al. 1994). Mice hemizygous for the "Zürich" ablated PrP allele but bred in San Francisco (Prusiner et al. 1993) developed disease 400–465 d.p.i. The reason for the difference between the results with the San Francisco and the Zürich mice, both of which were inoculated with the Chandler RML strain, is not clear; the criteria for diagnosing clinical disease may have been different and/or the genetic background of the mice may play a role. Prolonged incubation times of 259 ± 27 days in $Prnp^{0/+}$[Nagasaki] mice as compared to 138 ± 13 days in $Prnp^{+/+}$ mice were also reported for the mouse-adapted Fukuoka-1 strain of CJD prions (Sakaguchi et al. 1995). $Prnp^{0/+}$[Edinburgh] mice challenged either with the ME7 mouse strain or the mouse-adapted BSE strains 301V and 301C also developed clinical symptoms with a delay of 80–150 days, as compared with wild-type mice (Manson et al. 1994a). There appeared, however, to be no difference in the severity of the clinical signs and the pathology in the brain at the terminal stages of disease. PrP gene dosage seems, therefore, to affect the timing of disease but not the final pathology.

In summary, whereas in wild-type animals and $Prnp^{0/+}$[Edinburgh] mice inoculated with the ME7 strain the increase in prion titer and PrP^{Sc} levels is followed within weeks by scrapie symptoms and death, $Prnp^{0/+}$[Zürich] mice remained free of symptoms for many months despite similar levels of scrapie infectivity and PrP^{Sc} (Fig. 2). These findings suggest that clinical symptoms are not necessarily correlated with the overall accumulation of PrP^{Sc} in brain; it has been suggested that the pathological processes must extend to a so-called "clinical target area"

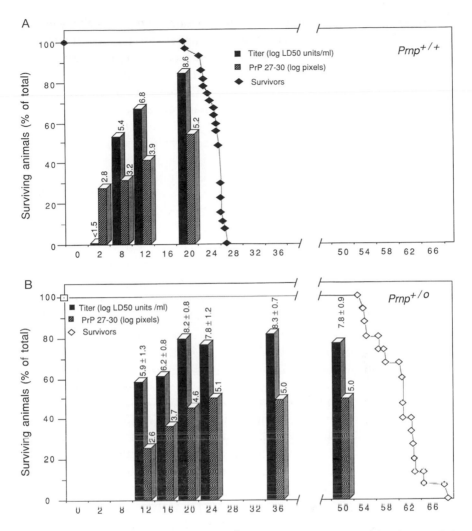

Figure 2. Survival, prion titers, and PrP^Sc in brains of scrapie-inoculated mice. (*A*) Wild-type and (*B*) *Prnp^{0/+}* mice at various times after inoculation with mouse prions. Although both types of mice have similar levels of prions and PrP^Sc at 20 weeks after infection, the wild-type mice succumb within the following 8 weeks, while the hemizygous knockout mice survive for another 30 weeks or more without clinical symptoms. (Reprinted, with permission, from Büeler et al. 1994 [copyright Johns Hopkins University Press].)

before disease and death ensue (Kimberlin et al. 1987). There are in fact several reports in which prion disease leads to death without substantial accumulation or even detectable levels of PrP^Sc in the brain (Collinge et al. 1995; Lasmézas et al. 1997; Manuelidis et al. 1997; Manson et al. 1999;

Flechsig et al. 2000) and where discrete processes in the postulated target area, which might be in the brain stem or upper spinal cord (Flechsig et al. 2000), have not been routinely searched for. It is of practical interest that under certain conditions clinically healthy mice can harbor high levels of infectivity for long periods of time (Raeber et al. 1997; Frigg et al. 1999; Hill et al. 2000; Race et al. 2001; Thackray et al. 2002), because it suggests that cattle or humans in apparent good health also could contain infectious agent in their central nervous and lymphoreticular systems and perhaps fail to show clinical symptoms. This is a point to consider when animal or human tissues are used for preparing pharmaceuticals or grafts, or when blood is transfused.

TRANSGENESIS AND GENE REPLACEMENT

Modifications of the genome may be achieved by transgenesis; i.e., insertion of cloned genes, or by in situ gene replacement using homologous recombination in embryonic stem cells. Gene insertion is usually based on nuclear injection into a one-cell embryo of a DNA segment containing the gene, which may be modified in a variety of ways. In particular, modifications may be introduced into the coding region, or the natural promoter may be replaced in order to target expression to other tissues. Following injection, the DNA segment containing the cloned gene usually concatenates prior to integration, resulting in random insertion of multiple gene copies into one or a few sites of the genome; therefore, each resulting transgenic mouse line is unique. Depending on the site of insertion, expression of the transgene can be silenced or modulated by regulatory elements in the neighboring regions. Perhaps these effects can be mitigated by flanking the transgene with so-called locus control regions (LCRs) (Li et al. 2002). Moreover, by using mice carrying *loxP* sequences in a characterized, favorable locus of their genome, it is possible to direct insertion of one or a few gene copies into that site by transient expression of Cre recombinase (Araki et al. 1997; Soukharev et al. 1999).

In situ gene replacement is based on homologous recombination rather than random integration and allows modification of a resident gene in a variety of fashions. It has the advantage that the normal copy number is preserved and that the modified gene, because it remains in its physiological environment, will not be influenced by alien enhancers or by position-dependent silencing. However, protein expression levels depend not only on the properties of the promoter, but also on the stability of the mRNA and the protein, both of which may be profoundly affected by sequence changes, so that the cellular levels of a protein encoded by the normal gene and by its modified counterpart need not be the same.

A double replacement gene-targeting strategy was used to introduce point mutations into the *Prnp* locus of a 129/Ola embryonic stem (ES) cell line lacking hypoxanthine phosphoribosyl transferase (HPRT). In the first step, the entire PrP ORF was replaced with an HPRT minigene by homologous recombination, and successfully targeted clones were isolated by selection in a HAT minus and gancyclovir-containing medium that is lethal for cells lacking HPRT. In the second step, the HPRT minigene was replaced with a mutated PrP coding sequence, again by homologous recombination. Cells devoid of HPRT, selected by virtue of their resistance to 6-thioguanine, contained the altered *Prnp* allele that was indistinguishable from the wild-type allele, except for the desired mutation (Manson et al. 1999, 2000; Moore et al. 1995, 1998). Selected ES cell clones were injected into 3.5-day-old blastocysts from C57BL/6 mice, and chimeric offspring derived from both 129 and C57BL/6 cells, identified by the presence of two coat colors, were bred with 129/Ola mice to detect germ-line transmission of the transgene and establish the line.

A different "knockin" technology was used by Kitamoto et al. to replace the murine PrP ORF by a mutated human counterpart. In the first step, a murine *Prnp* DNA segment containing a cassette with the human PrP ORF and a neomycin-resistance gene flanked by *loxP* sites was introduced into the *Prnp* locus of ES cells by homologous recombination. The neomycin-resistance gene allowed the selection of positive clones from which mice carrying the cassette were generated. In a second step, a Cre expression construct was injected into fertilized oocytes carrying the modified allele, which led to elimination of the neomycin-resistance gene, leaving a single *loxP* site in the 3′ noncoding region of the human gene (Kitamoto et al. 1996). This type of approach can be simplified by eliminating the resistance gene in the modified ES cells prior to generating mice (Luo et al. 2001). Elimination of selection genes is recommended because it has been reported that in at least one instance a neomycin-resistance cassette caused embryonal lethality (Fiering et al. 1995) and that the full-length (albeit not a truncated; Cohen et al. 1998) herpes simplex virus thymidine kinase (HSV-TK) gene may lead to male sterility (Braun et al. 1990).

Transgene Vectors for PrP Expression

When a certain phenotype is generated by gene ablation—in this case resistance to scrapie—it is important to show that this is indeed the consequence of the targeted genetic intervention and not of some unintended or subsidiary event, such as obliteration of an enhancer governing another gene or disruption of an unidentified reading frame. The most effective, albeit not infallible (Weissmann and Aguzzi 1999; and see below), strategy to link a phenotype to the ablation of a specific protein is to abrogate the effect of the deletion by introducing into the knockout animal a cDNA encoding the protein in question. Although this may be a problem when a gene gives rise to various protein species as a consequence of transcriptional or splicing variants, it is not an issue in the case of PrP because it is encoded in a single exon.

It is desirable to reproduce the expression pattern of the wild-type gene as closely as possible but, as is the case for most mammalian genes, the location of all elements required for regulation of tissue-specific expression is not known. A long Syrian-hamster-derived cosmid clone, comprising *Prnd* as well as *Prnp*, was used to express hamster PrP in mice (Scott et al. 1989); from it was derived the 43-kb "cos.Tet" vector (Scott et al. 1992) that encompasses the two exons and one intron of hamster *Prnp* and ~24 and 6 kb of the 5′ and 3′ flanking sequence, respectively, but lacks the Dpl ORF. The 40-kb mouse cosmid I/InJ, derived from the *Prnp^b* allele of an I/InJ mouse, contains 6 kb of upstream sequence, the three exons and two introns of mouse *Prnp,* and ~18 kb of 3′ downstream sequence that include *Prnd* (Westaway et al. 1991; Moore et al. 1999). Because such large cosmid vectors are laborious to work with, the "half-genomic PrP" expression vector (Fischer et al. 1996) was constructed from the cosmid I/InJ by deleting the large intron 2, replacing exon 3 by its counterpart from the *Prnp^a* locus of NMRI and the 3′ flanking sequence by 2.2 kb of its counterpart from the NZW/Lac strain of mice. Thus, the half-genomic PrP vector contains about 6 kb of *Prnp^b* promoter sequence controlling expression of PrP encoded by the *Prnp^a* locus (Fischer et al. 1996).

Efficient PrP expression was achieved with both the I/InJ cosmid and the half-genomic *Prnp* constructs (Fischer et al. 1996), but not with a vector lacking both introns derived from the latter. At least one intron seems essential for efficient expression of a PrP-encoding transgene under control of its own promoter, as in the case of a hamster PrP cDNA vector (Scott et al. 1989) and various other transgenes (Brinster et al. 1988). Whereas initially the expression pattern elicited by the half-genomic construct appeared the same as that in I/InJ-cosmid-transgenic mice, it later emerged that cerebellar Purkinje cells expressed neither PrP nor PrP mRNA at detectable levels (Fischer et al. 1996; Rossi et al. 2001). Thus, the half-genomic construct may lack a Purkinje-cell-specific enhancer, which could be located in the large intron or in the distal part of the 3′ noncoding region, both of which are absent in the half-genomic construct. The lack of Purkinje-cell-specific PrP expression has also been found in mice expressing mutated PrP (Chiesa et al. 1998; Ma et al. 2002) under control of MoPrP vector.*Xho*I (Borchelt et al. 1996), a construct similar to the half-genomic vector described here. A construct consisting of a 214-bp human 5′ upstream *PRNP* promoter fragment followed by 125 bp of the first PrP exon, a generic intron, and a *lacZ* ORF, expressed β-galactosidase in Purkinje cells, but also at high levels in kidney and spleen (Asante et al. 2002); the latter is not the case for PrP in wild-type mice, suggesting that the expression pattern was largely ubiquitous. Gene expression may be controlled by an interplay of negative and positive elements, and removal of both may result in ubiquitous expression.

Whereas overexpression of the cosmid transgenes such as HaPrP and PrP-B leads to a profound necrotizing myopathy involving skeletal muscle, demyelinating polyneuropathy, and focal vacuolation of the CNS as the mice age (Westaway et al. 1994), no impairments were observed with mice transgenic for the half-genomic construct, even though PrP-A expression was equally high (up to seven times the wild-type PrP level) (Fischer et al. 1996). Because a similar phenotype occurred in mice harboring hamster PrP cosmid transgenes lacking *Prnd* (Westaway et al. 1994), and because mice expressing Dpl under control of the cos.Tet do not exhibit myopathic changes (Moore et al. 2001), the myopathy is not due to expression of Dpl. In view of the fact that expression of PrP-A at levels similar to those of Ha PrP or PrP-B is not pathogenic (Carlson et al. 1994; Fischer et al. 1996), the neuromyopathy could be due to overexpression of PrP-B or some unknown feature of the cosmid.

Restoration of Susceptibility of *Prnp*[0/0] Mice to Prions by PrP Transgenes

Two lines of *Prnp*[0/0] mice transgenic for the half-genomic *Prnp* construct, expressing PrP-A at about 3–4 and 6–7 times the level of wild-type mice, were challenged with mouse prions. As shown in Figure 3A, they succumbed to scrapie even more rapidly than wild-type CD-1 mice, confirming that the incubation times are inversely related to PrP expression levels (Scott et al. 1989). Because of its short incubation time, the mouse line *tga20* was rendered homozygous and used for a rapid scrapie infectivity assay (Brandner et al. 1996b; Fischer et al. 1996).

These experiments sustain the conclusion that the scrapie-resistant phenotype of the Zürich PrP knockout mice is indeed due to ablation of PrP. However, it is worth emphasizing the pitfalls that may beset the interpretation of knockout experiments. In the case of the Nagasaki-type PrP

Table 1. Expression of wild-type and mutant PrP in mice

PrP transgenes	Copy no.	Phenotype in PrP null mice (in the presence of PrP, if determined)	Refs.	Susceptibility of transgenic PrP null mice to prions (strain)	Refs.
Wild-type PrP^C					
MoPrP-A[a]	multiple	none	(1,2)	yes (Mo prions)	(1,2)
MoPrP-B[b] in *Prnp^b* cosmid	multiple	ataxia (yes)[c]	(1,3)	yes (Mo prions)	(1,3)
MoPrP-B in genomic *Prnp^a* locus	two[d]	none	(20)	yes (Mo adapted Bo prions)	(20)
HaPrP	multiple	ataxia (yes)[c]	(3)	yes (Ha prions)	(4,5)
BoPrP	multiple	none	(37)	yes (Bo, Hu, Ov prions)	(37,29,30)
OvPrP (VRQ) or (ARQ)[e]	multiple	none	(31,34)	yes (Ov, Bo prions)	(31,34,35)
HuPrP (VV129)[f]	multiple	none	(14)	yes (Hu, Bo prions)	(14,19,36,38)
PrP^C hybrids					
MoPrP (3F4 tag)[g]	multiple	none	(4,5)	yes (Mo prions)[g]	(4,5)
MH2M[h]	multiple	none	(4,5)	yes (Ha prions)	(4,5)
MHu2M[h] (MM129)[f]	multiple	none	(14)	yes (Hu prions)	(14,33)
MBo2M[h]	multiple	none	(37)	no (Bo prions)	(37)
PrP^C deletions or insertions					
MoPrP-A, Δ1-22, Δ231-254[i]	multiple	cerebellar disorder[j] (yes)[k]	(28)	n/a	
MoPrP-A, 145stop	multiple	none	(1)	n/a	
MoPrP-A, Δ32-80	multiple	none	(6)	yes (Mo prions)	(6)
MoPrP-A, Δ32-93	multiple	none	(8)	yes (Mo prions)	(22)
MoPrP-A, Δ32-106	multiple	none	(8)	no (Mo prions)	(26)
MoPrP-A, Δ32-121	multiple	cerebellar disorder (no)[c]	(8)	n/a	
MoPrP-A, Δ32-134	multiple	cerebellar disorder (no)[c]	(8)	(graft) no (Mo prions)	(27)
MoPrP-A, Δ23-88	multiple	none	(25)	yes (Mo prions)[g]	(25)
MoPrP (3F4-tag) 145stop	multiple	none	(6)	no (Mo prions)	(6)
MoPrP (3F4-tag) Δ23-88	multiple	none	(6)	no (Mo prions)[g]	(21)
MoPrP (3F4-tag) Δ23-88, Δ141-176	multiple	none	(6)	yes (Mo prions)[g]	(21)

MoPrP (3F4-tag) Δ23–88, Δ141–221	multiple	neurodegeneration (yes)[c]	(7)	n/a	
MoPrP (3F4-tag) Δ23–88, Δ177–200	multiple	storage disease[k]	(6)	n/a	
MoPrP (3F4-tag) Δ23–88, Δ201–217	multiple	storage disease[k]	(6)	n/a	
MoPrP (3F4-tag) 9 repeats inserted	multiple	cerebellar disorder (yes)[c]	(9,10)	n/a	
HaPrP, Δ104–113	multiple	none	(17)	yes (Ha prions)	(24)
PrP[C] point mutations					
MoPrP-A, P101L	two[d]	none	(11)	yes (Mo, Ha, Ov, Hu prions)	(11,23)
MoPrP-A, P101L in Prnp[b] locus	multiple	ataxia (delayed)[l]	(12,13)	yes (Mo prions)	(13,18)
MoPrP-A, P101L in Ha cosmid	multiple	ataxia (delayed)[l]	(2)	yes (Mo prions)	(16,18)
MoPrP-A, Q167R[m]	multiple	none	(32)	no (Mo prions)	(32)
MH2M, AV3 (A113V, A115V, A118V)	multiple	neurodegeneration[c]	(15,17)	n/a	
HaPrP, A117V	multiple	neurodegeneration[c]	(24)	yes (Ha prions)	(24)
HaPrP, K110I, H111I	multiple	neurodegeneration[c]	(17)	yes (Ha prions)	(24)
HaPrP, T183A[n]	multiple	none	(18)	no (Ha prions)	(18)
HaPrP, T199A[n]	multiple	none	(18)	yes (Ha prions)	(18)
MHu2M[h] E199K (MM129)[f]	multiple	none	(19)	yes (Ha prions)	(18)

(n/a) Not applicable. PrP sequences and prion strains are abbreviated: Mo (mouse), Ha (hamster), Hu (human), Bo (bovine), or (Ov) sheep. Residues are numbered based on the wild-type PrP sequence of the given species.

[a] MoPrP-A (short incubation time allele L108 and T189); [b] MoPrP-B (long incubation time allele L108F and T189V); [c] disease is not transmissible to mice expressing wild-type PrP; [d] targeted mutation; [e] polymorphisms at positions 136, 154, and 171; [f] polymorphisms at position 129; [g] MoPrP (3F4-tag). The 3F4 tag (L108M, V111M) alters the efficiency of the PrP molecule to support prion replication. After inoculation, mice expressing MoPrP (3F4-tag) developed disease after 120 days, while mice expressing MoPrP-A get sick after 50 d.p.i. Deletion of amino acids 23–88 in MoPrP (3F4 tag) confers resistance to mice (21), while in the MoPrP-A still sustains prion replication, albeit with longer incubation times (25); [h] mouse PrP containing the PrP sequence of the indicated species between amino acids 97 and 167; [i] deletions yielding PrP 23–230, a cytosolic form of PrP; [j] phenotype in wild-type mice; [k] transmissibility not determined; [l] disease is transmissible to mice expressing PrP with the same mutation, but not to mice expressing wild-type PrP; [m] residue corresponds to sheep position 171; [n] mutation deletes one glycosylation site.

(1) Fischer et al. 1996; (2) Telling et al. 1996a; (3) Westaway et al. 1994; (4) Scott et al. 1989; (5) Scott et al. 1993; (6) Muramoto et al. 1997; (7) Supattapone et al. 2001b; (8) Shmerling et al. 1998; (9) Chiesa et al. 1998; (10) Chiesa et al. 2000; (11) Manson et al. 1999; (12) Hsiao et al. 1990; (13) Telling et al. 1995; (15) Prusiner and Scott 1997; (16) Kaneko et al. 2000; (17) Hedge et al. 1998b; (18) DeArmond et al. 1997; (19) Telling et al. 1994; (20) Moore et al. 1998; (21) Supattapone et al. 1999; (22) Flechsig et al. 2000; (23) Barron et al. 2001; (24) Hegde et al. 1999; (25) Supattapone et al. 2001a; (26) E. Flechsig et al., unpubl.; (27) I. Hegyi et al., unpubl.; (28) Ma et al. 2002; (29) Buschmann et al. 2000; (30) Scott et al. 1999; (31) Vilotte et al. 2001; (32) Perrier et al. 2002; (33) Mastrianni et al. 2001; (34) Crozet et al. 2001a (36); (35) Crozet et al. 2001b; (37) Scott et al. 1997; (38) Collinge et al. 1996.

knockout mice described above, extensive deletions of the PrP-encoding exon gave rise to an ataxic phenotype that was reversed by introduction of an intact PrP-expressing transgene, the classical experiment correlating a phenotype with ablation of a gene. Nonetheless, in that case, the conclusion was misleading, because the ataxic phenotype did not result from the deletion of the PrP ORF but from the incidental up-regulation of a deleterious gene product whose pathogenicity was offset by wild-type PrP^C.

EFFECT OF PrP MUTATIONS ON HEALTH AND PRION DISEASE IN THE MOUSE

The ability to reconstitute susceptibility to scrapie in mice devoid of PrP or to replace the wild-type gene by modified counterparts paved the way to the analysis of various consequences of modified PrP expression. Table 1 gives an overview of the mutant PrP transgenes introduced into mice and their effects.

Delineation of PrP Regions Dispensable for Susceptibility to Prion Disease

Treatment of scrapie prion preparations with protease cleaves off about 60 amino-terminal residues of PrP^{Sc}, up to residue 88 (Hope et al. 1988), but does not abrogate infectivity of the sample (McKinley et al. 1983). To ascertain whether amino-terminally truncated PrP^C could serve as substrate for the conversion to PrP^{Sc} and sustain susceptibility to scrapie in the mouse, transgenic $Prnp^{0/0}$[Zürich I] mice expressing PrP molecules that retained the signal sequence but had a deletion between residues 32 and 80 (Δ32-80), 93 (Δ32-93), and 106 (Δ32-106) of the mature sequence were challenged with scrapie prions (Fig. 3C). PrP knockout mice overexpressing PrP with deletions to positions 80 (Fischer et al. 1996) and 93 (Shmerling et al. 1998) yielded normal mice that after intracerebral inoculation with scrapie prions developed disease, propagated prions, and exhibited protease-resistant, truncated PrP^{Sc}, albeit with longer incubation times and with a lower level of both infectivity and PrP^{Sc} in the case of the deletion to residue 93 (Flechsig et al. 2000). PrP with a deletion extending to position 106 was unable to restore susceptibility to scrapie prions (E. Flechsig et al., unpubl.). Brain grafts expressing PrPΔ32-134 in PrP knockout mice failed to support propagation of RML prions (I. Hegyi et al., unpubl.; as described below, mice expressing this truncated PrP succumb to spontaneous neurological disease and can therefore not be assayed for susceptibility to prions).

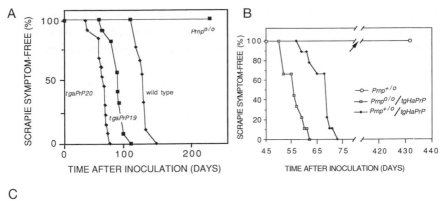

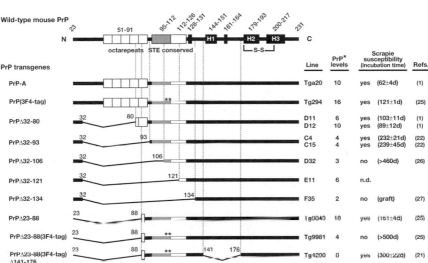

Figure 3. Susceptibility to scrapie of PrP knockout mice carrying various transgenes encoding full-length or truncated PrP. (*A*) *Prnp*[0/0] mice were rendered transgenic for *Prnp* genes. *tga19/+* mice had 3–4 times the normal PrP[C] level; *tga20/+* mice had 6–7 times the normal level. (*B*) *Prnp*[0/0] and *Prnp*[0/+] mice with hamster PrP transgenes were inoculated with the Sc237 isolate of hamster prions. (*Arrow*) One animal died spontaneously without scrapie symptoms and one was killed because of a tumor. Mice with a mouse PrP allele are completely resistant to hamster prions, whereas the presence of hamster PrP in a PrP knockout mouse renders them very susceptible. The additional presence of a mouse PrP allele reduces susceptibility to hamster prions, as evidenced by increased incubation times. (*C*) Susceptibility to scrapie prions of *Prnp*[0/0] mice expressing various *Prnp* transgenes. (*) Relative to wild-type; (**) 3F4 epitope tag; (Refs) references given in Table 1. (*A*, Reprinted, with permission, from Fischer et al. 1996 [copyright Oxford University Press]; *B*, modified, with permission, from Büeler et al. 1993 [copyright Elsevier].)

In apparent contradiction to our findings that mice expressing murine PrPΔ32-93 are susceptible to RML (Flechsig et al. 2000), it was reported that mice expressing PrPΔ23-88 (tagged with an epitope for the monoclonal antibody 3F4) were resistant to RML prions (Supattapone et al. 1999). However, it was subsequently found that mice expressing PrP-A (Δ23-88) under control of the hamster cos.Tet vector were in fact susceptible to RML prions and developed disease after prolonged incubation time (Supattapone et al. 2001a), similar to mice expressing PrPΔ32-93. It is now clear that the 3F4 tag in PrP-A (L108M and V111M) extended incubation times, possibly because L108 is associated with the short incubation time allele of PrP-A in mice.

It is interesting that the histopathological changes typical of scrapie in PrPΔ32-93 PrP$^{0/0}$ mice inoculated with RML were limited to the spinal cord (Flechsig et al. 2000) whereas scrapie-ill mice expressing PrP-A(Δ23-88) displayed vacuolation and gliosis in the hippocampus. The different histopathology in the latter could be due to (1) the absence of residues 89–93, (2) the presence of the amino-terminal motif ($_{23}$KKRPKP$_{29}$) (Zulianello et al. 2000), (3) a difference between the hamster cos.Tet and the half-genomic vector, or (4) a difference in expression levels. Nevertheless, both studies demonstrated that the prolonged incubation times result from reduced conversion efficiency of the truncated PrP, because the prolonged incubation times are retained with prions passaged through mice expressing the same truncated PrP (Flechsig et al. 2000; Supattapone et al. 2001a). The same conclusion was drawn from cell-free conversion experiments (Lawson et al. 2001). Thus, at least 60 residues of the amino-proximal region of mature PrP^{C} are expendable; they include not only the segment that is cleaved off PrP^{Sc} by proteinase K, but also the entire octarepeat region. This is remarkable because amplification of the number of octarepeats is associated with familial Creutzfeldt-Jakob disease (CJD) and Gerstmann-Sträussler Scheinker disease (GSS) (Goldfarb et al. 1992a; Collinge 2001) and appears to affect the type of cerebellar amyloid deposits (Vital et al. 1998). As shown by nuclear magnetic resonance (NMR), the first 98 residues of mature murine PrP^{C} form a flexible random coil (Donne et al. 1997; Riek et al. 1997), and it was proposed that this segment would be converted to β-sheets in PrP^{Sc} (Riek et al. 1997). If so, our results show that at most only 38 of these residues, starting at position 93, are essential for this conversion. Interestingly, a peptide comprising residues 106–126, which are contained in the flexible region, can form β-sheet structures (Tagliavini et al. 1993; Nguyen et al. 1995) and is toxic to primary neuronal cultures (Forloni et al. 1993).

In addition to a partial deletion of the flexible tail (Δ23-88), the first α-helix (Δ141-176) could also be removed from 3F4-tagged PrP without abrogating its capacity to restore prion susceptibility to $Prnp^{0/0}$ mice. The resulting "PrP106" contains only 106 amino acids, compared to the 208 residues in full-length PrP (Supattapone et al. 1999).

PrP Mutations Affecting the "Species Barrier"

In many cases, prions derived from one species fail to elicit disease when inoculated into a different species, or do so only inefficiently and/or after a long incubation time. This phenomenon is attributed to a "species barrier" (Pattison 1965). However, at least in the case of mice inoculated intracerebrally with hamster prions, PrP^{Sc} and infectivity may accumulate in the brain despite the absence of clinical symptoms (Hill et al. 2000; Race et al. 2001) so that the barrier is not absolute. In many cases, introduction into mice of PrP transgenes containing all or part of the PrP sequence of the prion donor can overcome the species barrier. Thus, $Prnp^{+/+}$ mice transgenic for hamster PrP transgenes are susceptible to both mouse and hamster prions (Scott et al. 1989; Prusiner et al. 1990) whereas $Prnp^{0/0}$ mice containing Syrian hamster PrP transgenes are susceptible to hamster, but not mouse-derived prions; interestingly, the presence of a mouse PrP allele diminished the susceptibility to hamster prions (Fig. 3B) (Büeler et al. 1993). Mouse–hamster hybrid PrP (MH2M), containing 5 hamster-specific residues between positions 96 and 167, conferred susceptibility to hamster-derived prions (Table 1). Similarly, $Prnp^{0/0}$ mice expressing human transgenes were susceptible to human CJD prions, whereas $Prnp^{+/+}$ mice with the same transgene cluster were as resistant to human prions as wild-type mice (Telling et al. 1994, 1995). MHu2M, a chimeric PrP that contains 9 human PrP-specific residues between positions 96 and 167 in a murine framework, was almost as effective in rendering mice susceptible to human prions on a $Prnp^{+/+}$ background as it was on a $Prnp^{0/0}$ background. This showed that the critical region for "human species specificity" lay in the region 96–168 of human PrP and, along with other data, led to the suggestion that a species-specific "factor X" that bound to residues 168, 172, 215, and 219 near the carboxyl terminus participated in prion formation (Telling et al. 1995; Kaneko et al. 1997b). $Prnp^{0/0}$ mice expressing MHu2M have been used as recipients for a variety of CJD prions (Mastrianni et al. 2001). Surprisingly, transgenic mice overexpressing human PrP are less susceptible to human variant CJD (vCJD) than wild-type mice (Hill et al. 1997b). $Prnp^{0/0}$ mice transgenic for bovine PrP genes are susceptible to bovine (Scott et al. 1999; Buschmann et al. 2000) as well as to vCJD and sheep scrapie prions, whereas mice containing ovine $Prnp^{VRQ}$ or $Prnp^{ARQ}$ transgenes showed vastly decreased incubation time for sheep scrapie prions as compared to wild-type mice (Crozet et al. 2001b; Vilotte et al. 2001).

Inasmuch as a decrease in incubation time is viewed as a lowering of the species barrier, it is noteworthy that this can be achieved by mutations in murine PrP that do not increase the similarity to the PrP of the prion donor. Mice homozygous for the targeted PrP mutation P101L ($Prnp^{a[P101L,P101L]}$) had shorter incubation times with both sheep (SSBP1) and hamster (263K) prions, but extended incubation times with murine ME7 and human vCJD prions, as compared with wild-type mice. Thus, the P101L mutation can alter incubation times across three species barriers in a strain-dependent manner (Barron et al. 2001). Interestingly, this mutation is located in the flexible region and not in the domain postulated to bind the conjectural protein X.

Polymorphisms in the PrP gene have been shown to alter incubation time and susceptibility to prion disease in mice (Moore et al. 1998), sheep (Goldmann et al. 1994), and man (Palmer et al. 1991; Goldfarb et al. 1992b). These polymorphisms may control the rate of replication of the infectious agent (Dickinson and Outram 1979), but the mechanism by which this is achieved is not known.

PrP Mutations Linked to Incubation Time of Prion Strains

As described above, concordance between the sequence of the PrP gene of the prion donor and the recipient, as well as the expression level of PrP^C, has a major influence on susceptibility and incubation time. Moreover, two different prion strains, even when derived from the same inbred mouse line and thus associated with same PrP^{Sc} sequence, may exhibit different incubation times and patterns of histopathology in the same mouse strain. This shows that features other than PrP sequence concordance influence incubation time. It has previously been suggested that an essential nucleic acid ("virino" hypothesis; Dickinson and Outram 1988) or a nonessential nucleic acid ("co-prion" hypothesis; Weissmann 1991b) is the crucial feature; currently the prevalent idea is that different conformations of the PrP^{Sc} molecule, imposed on the same sequence (or even on a different one), specify strain properties (Bessen and Marsh 1992; Telling et al. 1996b; Caughey et al. 1998; Safar et al. 1998).

The *Sinc* gene has been shown to be the major gene controlling survival time of mice exposed to scrapie strains (Dickinson et al. 1968; Carlson et al. 1986). Animals homozygous for the *Sinc s7* allele have short incubation times when infected with the Chandler isolate of mouse prions, and those with the *Sinc p7* allele have long incubation times. With a different agent, the mouse-adapted bovine spongiform encephalopathy (BSE) strain 301V, incubation times are long for the *Sinc s7* allele and short for the *Sinc p7* allele (Bruce et al. 1994). The *Sinc s7* allele is genetically linked to the *Prnp^a* allele characterized by the residues Leu-108 and Thr-189, whereas the *Sinc p7* allele is linked to the *Prnp^b* allele with Phe-108 and Val-189 (Westaway et al. 1987). The close linkage of the *Sinc* and PrP genes suggested that they might be congruent (Hunter et al. 1987, 1992; Westaway et al. 1987); however, analysis using transgenes of the two alleles was inconclusive because the effects of variable gene copy numbers and PrP^C levels were superimposed on those of the alleles (Carlson et al. 1994).

Using the double-replacement gene-targeting technique, PrP codons 108 and 189 in murine embryonic stem cells from a *Sinc s7* (*Prnp^a* allele) mouse were altered to Phe-108 and Val-189, which are characteristic for the *Prnp^b* allele. ES clones with the mutant PrP gene were used to produce chimeras that were bred with 129/Ola mice to generate an inbred line of mice carrying a PrP gene with Phe-108 Val-189 (*Prnp^{a[108F189V]}*). Mice homozygous for the *Prnp^a* allele or the mutated allele were inoculated with the 301V agent. The incubation time in the mutant mice carrying the (*Prnp^{a[108F189V]}*) allele was 133 days, about 110 days shorter than that of their wild-type littermates carrying the *Prnp^a* allele (Moore et al. 1998). As mentioned above, mice homozygous for the targeted PrP mutation P101L had extended incubation times with murine ME7 as compared with wild type (Barron et al. 2001).

The finding that point mutations in the PrP gene in mice with otherwise identical genetic background caused a dramatic change in incubation time supports the view that *Sinc* and *Prnp* are indeed congruent (Moore et al. 1998). However, it is also clear that loci other than *Prnp* contribute importantly to incubation time (Stephenson et al. 2000; Lloyd et al. 2001; Manolakou et al. 2001).

PrP Mutations Leading to Dominant-negative Inhibition of Prion Formation

Certain configurations of polymorphic residues near the carboxyl terminus of PrP in sheep (positions 136, 154, and 171) afford protection against prion disease (Hunter et al. 1997). For example, Cheviot sheep with two 171Q alleles (171Q/Q) are susceptible to scrapie, whereas the configuration 171Q/R and 171R/R confers protection (Hunter 1996); thus, the 171R allele may be considered as dominant negative. In cultured murine N2a cells persistently infected with scrapie prions, a single hamster-specific V138M change in transgenically expressed murine PrP prevents its conversion into the scrapie form and interferes with the formation of PrP^Sc from the endogenous, wild-type PrP^C (Priola and Chesebro 1995).

Similarly, basic amino acid substituents of the carboxy-proximal residues 167 (which corresponds to sheep position 171), 171, 214, or 218 of mouse PrP^C act as dominant-negative inhibitors of PrP^{Sc} formation (Kaneko et al. 1997b); amino acid replacements at multiple sites were less effective than single-residue substitutions. Interestingly, an intact amino-terminal region is required to sustain the dominant-negative effect (Zulianello et al. 2000). Transgenic mice expressing murine PrP^{Q167R} at about wild-type level on a $Prnp^{0/0}$ background remained healthy and failed to accumulate PrP^{Sc} after inoculation with mouse scrapie prions. Mice with the same transgene on a $Prnp^{+/+}$ background had an extended life span after inoculation, but developed atypical neurological symptoms at about 450 days (wild-type controls had an incubation time of 127 days) and showed accumulation of PrP^{Sc} and neuropathological changes typical for scrapie no later than 300 days. Thus, PrP^{Q167R} is not converted to the scrapie form by RML prions and affords partial protection to mice containing wild-type PrP; it was suggested that this is due to sequestration of the conjectural protein X (Perrier et al. 2002). This protective effect should not be confused with the partial protection against sCJD afforded by the M/V heterozygosity at position 129 of human PrP, because neither M/M nor V/V is protective (Palmer and Collinge 1993; Windl et al. 1996), and therefore this is not due to dominance of an allele. Because all cases of vCJD occur in humans with the 129M/M polymorphism (Collinge et al. 1996), it is possible that with this prion strain 129V exerts a dominant-negative effect. The human polymorphism 219K (which corresponds to murine 218K) appears to be dominant negative because all of 85 Japanese sCJD cases were 219E/E and none 219E/K, although 12% of the population are heterozygous at this position (Shibuya et al. 1998).

PATHOLOGICAL MOUSE PHENOTYPES ELICITED BY PrP MUTATIONS

Three types of PrP mutations have given rise to pathological phenotypes when overexpressed in Zürich-I-type mice: (1) deletion of the amino-proximal part comprising the flexible tail, (2) internal deletions of the α-helices, and (3) point mutations in the highly conserved region.

Amino-proximal Deletions of the Flexible PrP Tail Causing a Cerebellar Syndrome

Transgenic mice expressing murine PrP with various amino-proximal deletions on a $Prnp^{0/0}$[Zürich I] background were generated and analyzed for phenotypic abnormalities (Table 1). The "half-genomic" PrP vector used to express the modified PrPs mimics the endogenous expression pattern, except for Purkinje cells, where neither PrP nor PrP mRNA could be detected (Fischer et al. 1996; Rossi et al. 2001). To ensure correct processing and localization in the cell, the signal peptide and the first 9 amino acids were retained in the set of amino-proximal truncated PrP genes described below. Mice expressing PrP with deletions from codons 32 to 80, 93, or 106 (the latter two lacking the five octarepeats) were healthy and showed normal behavior. Unexpectedly, mice expressing PrP with deletions of the flexible tail extending to amino acid 121 or 134, comprising a region that is conserved among various species, developed severe ataxia and apoptosis of the cerebellar granule cell layer as early as 1–3 months of age (Fig. 4A) (Shmerling et al. 1998). Neurons in the cortex and elsewhere expressed truncated PrP at similar levels as in granule cells but did not undergo cell death, arguing against a nonspecific toxic effect. Strikingly, the pathological phenotype was completely abolished by the introduction of one copy of a wild-type *Prnp* allele, although there was an unchanged level of truncated PrP, exceeding that of the wild-type counterpart. When the truncated PrP was specifically targeted to Purkinje cells of $Prnp^{0/0}$[Zürich I] mice using the L7 promoter, ataxia and Purkinje cell degeneration developed, while the cerebellar granule layer remained unaffected (Fig. 4B) (Flechsig et al. 2003). Surprisingly, the phenotype resembled that observed in the three lines of Nagasaki-type PrP knockout mice with up-regulation of Dpl in brain, namely ataxia and Purkinje cell degeneration, which was also abrogated by PrP (Sakaguchi et al. 1996; Nishida et al. 1999; Rossi et al. 2001). Because the overall structure of Dpl is remarkably similar to that of the globular domain of PrP lacking the flexible amino terminus, the mechanism of pathogenesis might be the same in both ataxic syndromes (Moore et al. 1999; Weissmann and Aguzzi 1999). The results have been explained by a model in which truncated PrP^C acts as dominant-negative inhibitor of a conjectured functional homolog of PrP^C, with both competing for the same putative PrP^C ligand (Fig. 4C) (Shmerling et al. 1998). In the case of Nagasaki-type PrP knockouts, it has been proposed that Dpl promotes oxidative damage that is counteracted by PrP^C (Wong et al. 2001). Whether the observed increases in heme oxygenase 1 and NOS systems are the cause or the consequence of primary neuronal damage, how such damage might come about, and whether it also occurs in the case of truncated PrP^C are questions that need to be addressed.

Internal Deletions of the α-Helices Causing a Storage Disease

Mice expressing PrP with the octarepeats (PrPΔ23-88) deleted or even, in addition, lacking α-helix-1 and β-sheet-2 (Δ141-176, "PrP106"), residues

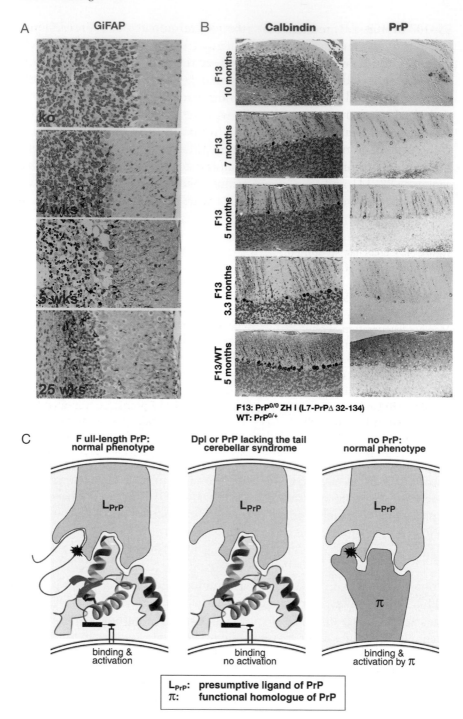

F13: PrP$^{0/0}$ ZH I (L7-PrPΔ 32-134)
WT: PrP$^{0/+}$

L$_{PrP}$:   presumptive ligand of PrP
π: functional homologue of PrP

Figure 4. (*See facing page for legend.*)

95–107 or 108–121, remained healthy (Muramoto et al. 1997). However, deletion of either α-helix-2 (Δ177-202) or α-helix-3 (Δ201-217) strongly affected correct PrP folding (Muramoto et al. 1997) (Table 1). Transgenic mice carrying either carboxy-terminal α-helix deletion spontaneously developed a lethal illness resembling neuronal storage disease between 90 and 220 days, characterized by proliferation of the endoplasmic reticulum and enlarged neurons accumulating mutant PrP within cytoplasmic inclusions (Muramoto et al. 1997). $Prnp^{0/0}$[Zürich I] mice expressing low levels of a 61-amino-acid PrP peptide (PrP61) containing the mouse PrP sequence 88–141 and the GPI-attachment site (residues 221–231) developed a progressive neurological disease within a few months of age and

Figure 4. Cerebellar sections from $Prnp^{0/0}$ mice expressing truncated PrPΔ32-134 under the control of different promoters. Development of the cerebellum proceeds normally until early postnatal life, leading to formation of molecular layer, Purkinje cell layer, and granule cell layer. (A) Under control of the half-genomic PrP promoter, truncated PrP is expressed ubiquitously in brain except for the Purkinje cells. Massive degeneration of cerebellar granule cell layer is apparent beginning at 5 weeks. Note strong gliosis affecting also the molecular layer as shown by immunostaining for glial fibrillary acidic protein (GFAP). At the end stage of disease, mice suffer from a profound cerebellar syndrome and the thickness of the granule cell layer is considerably reduced, but Purkinje cells appear intact throughout this process (Shmerling et al. 1998). (B) Under control of the Purkinje-cell-specific L7 promoter, truncated PrP^C is expressed in Purkinje cells, but nowhere else in the brain. Three-week-old mice show normal cerebellar structure with an intact Purkinje cell layer, as revealed by immunostaining for calbindin. After onset of the cerebellar syndrome, 14-week-old mice present with Purkinje cell loss and moderate gliosis. At the end stage of the disease, there is an almost complete Purkinje cell loss and shrinkage of the molecular layer; a single, remaining Purkinje cell is shown positive for PrP. A slight decrease of granule cells may be secondary to the extensive Purkinje cell loss. Age-matched $Prnp^{0/0}$ and $Prnp^{+/+}$ mice show an intact Purkinje cell layer. The syndrome is abrogated in the presence of full-length PrP^C (Flechsig et al. 2003). Nagasaki-type PrP knockout mice exhibit similar Purkinje cell loss and ataxia, caused by ectopic expression of Dpl. The phenotype caused by Dpl is also abrogated by full-length PrP^C. (C) Model to explain the pathogenic effect of truncated PrP^C or Dpl. PrP^C, consisting of a globular part and a flexible tail, interacts with a presumed ligand, L_{PrP}, via these two domains, thereby eliciting a signal (*black starburst*). The same signal is elicited by binding of L_{PrP} with π, a conjectural protein with the functional properties of PrP^C. This explains why Zürich-I-type PrP knockout mice show no obvious phenotype. However, truncated PrP lacking the flexible tail or Dpl can interact with L_{PrP} without eliciting a signal and competes efficiently with π, thus acting as a dominant inhibitor of the latter. If PrP^C is coexpressed with the truncated PrP^C or Dpl, it displaces the latter and restores the signal. (A,C, Reprinted, with permission, from Weissmann et al. 2001 [copyright Cambridge University Press].)

died a few days later. Neuropathological examination revealed accumulation of protease-resistant PrP61 within neuronal dendrites and cell bodies of hippocampal neurons, apparently causing apoptosis. PrP61-mediated neurotoxicity was not prevented by coexpression of full-length PrP, and the disease was not transmissible to transgenic $Prnp^{0/0}$ mice expressing PrP106 (Supattapone et al. 2001b).

Point Mutations Leading to Neurodegeneration with Astrocytic Gliosis

In an attempt to generate spontaneous prion disease in transgenic mice, Scott et al. substituted alanine by valine at positions 113, 115, and 118 of PrP ("AV3") to promote de novo β-sheet formation in the flexible tail (Prusiner and Scott 1997). Founders indeed developed a fatal neurological disorder (Hegde et al. 1998b). Several $PrP^{0/0}$ founders carrying the transgene Tg(MH2M-AV3) developed neurological signs and died within 2–4 weeks after birth. Founders expressing about 2- to 4-fold more PrP than normal hamsters succumbed to disease within 2 months after birth. However, no disease transmission was achieved and no protease-resistant PrP was detected (Prusiner and Scott 1997).

Three topological forms of PrP have been identified, using a cell-free translation system containing microsomal membranes: secreted, GPI-anchored prion protein (^{Sec}PrP), and two transmembrane forms, with either the carboxy-terminal (^{Ctm}PrP) or the amino-terminal region (^{Ntm}PrP) in the lumen (Hay et al. 1987a,b; Lopez et al. 1990). Both forms appear to span the membrane with the same hydrophobic stretch comprising residues 110 to 135. The triple mutation (AV3) in this region increased the formation of ^{Ctm}PrP. A correlation between elevated ^{Ctm}PrP levels in the cell-free system and the spontaneous development of clinical and neuropathological signs in transgenic mice expressing ^{Ctm}PrP-favoring mutations was proposed (Hegde et al. 1998a,b). The same phenotype arose in transgenic mice expressing HaPrP with substitutions at either position 110 (K→I), 111 (H→I) (Hegde et al. 1998b), or 108 (N→I) (Hegde et al. 1999). PrP lacking this region (PrPΔ104-113,"PrPΔSTE") or PrP^{G123P} failed to yield ^{Ctm}PrP in vitro, and neither PrP mutant caused neuropathology in transgenic $Prnp^{0/0}$[Zürich I] mice (Hegde et al. 1998b). It was suggested that accumulation of ^{Ctm}PrP in a post-ER compartment may represent a step common to transmissible and inherited prion disease (Hegde et al. 1999). The pathway involving ^{Ctm}PrP may not be the only mechanism of neurodegeneration in PrP-linked diseases, but it was implicated in the case of the A117V mutation responsible for GSS, because ^{Ctm}PrP, but not PrP^{Sc} was found in the brain of human patients (Hegde et al. 1999). $Prnp^{0/0}$ mice expressing the human PrP A117V muta tion (HaPrP^{A117V}) also accumulated ^{Ctm}PrP and developed nontransmissible neurodegenerative disease (Hegde et al. 1999).

As shown in cultured cells, mature PrP^C is subject to retrograde transport to the cytosol and degradation by proteasomes. Cytosolic PrP was strongly increased after inhibition of this pathway. Accumulation of even small amounts of cytosolic PrP was strongly neurotoxic in cultured cells (Ma and Lindquist 2001). PrP^{D178N}, which is associated with inherited forms of prion disease, accumulates in the cytoplasm even without inhibition of proteasomes. C57BL/6 mice transgenic for murine PrP23-230, which facilitates accumulation of cytosolic PrP, developed normally but acquired severe ataxia, with cerebellar degeneration and gliosis. This suggests a new pathway for seemingly diverse PrP neurodegenerative disorders (Ma et al. 2002).

Transgenic Mouse Models of Inherited Prion Diseases

More than 20 mutations of the PrP gene have been identified in families suffering from inherited human prion diseases. It has been speculated that mutations in the PrP gene give rise to an unstable PrP^C protein that can spontaneously convert into the abnormal conformer, PrP^{Sc}. Sporadic forms of the disease would be explained by a rarely occurring spontaneous transition of PrP^C into PrP^{Sc} or by somatic mutations in the PrP gene. Studies on thermodynamic stability of the carboxy-terminal domain of PrP with those mutations do not support the concept of mutation-induced destabilization of PrP^C generally (Swietnicki et al. 1998; Liemann and Glockshuber 1999). Studies with mice transgenic for mutant PrPs associated with inheritable prion diseases in humans have failed to produce transmissible disease, with the possible exception of PrP^{P101L}.

P101L

The mouse $Prnp^a$ ORF carrying the mutation 101 (Pro→Leu) (corresponding to the P102L mutation causing familial GSS; Hsiao et al. 1989) was placed in a cosmid derived from either a $Prnp^b$ allele (Hsiao et al. 1990) or hamster $Prnp$ (Telling et al. 1996a). Mice overexpressing mouse PrP^{P101L} at least 8-fold, on a $Prnp^{+/+}$ background, spontaneously developed neurodegeneration between 150 and 300 days of age, as evidenced by vacuolation, astrocytic gliosis, and PrP amyloid plaques, but other than in the human disease, little if any protease-resistant PrP could be detected. When introduced into $Prnp^{0/0}$[Zürich I] mice, the transgene cluster caused highly synchronous onset of disease at 145 days of age (Telling et al. 1996a). The disease could not be transmitted to wild-type mice, but only to transgenic mice that expressed the same mutation at low levels (2-fold) and developed neurodegeneration only late in life if at all (Hsiao et al. 1994; Telling et al. 1996a; Kaneko et al. 2000). It was suggested that the point mutation constituted a barrier to transmission to wild-type mice. However, human GSS disease with P102L can be transmitted to both monkeys and wild-type mice (Brown et al. 1994; Tateishi et al. 1996).

Mice carrying the P101L mutation introduced by a knockin procedure (Manson et al. 1999) did not come down with spontaneous disease, arguably because they did not overexpress the mutated PrP. However, when they were infected with human GSS prions, the incubation time was shortened as compared to wild-type mice, and the resulting disease could be transmitted to both the knockin and wild-type mice. To reconcile these results with those from the Prusiner group, one has to assume that the spontaneous disease of the transgenic PrP^{P101L} mice is caused by a different prion strain than that resulting from infection of Manson's mice with GSS. It has been argued that the P101L mutation may be an important susceptibility factor rather than a direct cause of GSS (Chesebro 1999).

T182A

Mice expressing hamster PrP encoding the inherited CJD mutation at codon 183 (Thr→Ala) on a PrP null background do not develop any pathological signs and are even resistant to hamster prions (DeArmond et al. 1997). In a study of thermodynamic stability of recombinant PrP (121-231) with point mutations, murine PrP^{T182A} showed the strongest reduction in stability of the eight variants analyzed, but the general conclusion was that destabilization of PrP^C is neither a general mechanism underlying the formation of PrP^{Sc} nor the basis of disease phenotypes in inherited human TSEs (Liemann and Glockshuber 1999). The T182A mutation destroys the first glycosylation site and has a profound effect on the anatomical distribution of the mutated PrP in brain (DeArmond et al. 1997). Aberrant trafficking results in its accumulation in the body of the cell and complete absence in its dendritic trees, which may prevent it from moving to caveolae-like domains on the cell surface where PrP^{Sc} formation is thought to occur (Vey et al. 1996; Kaneko et al. 1997a). Studies in cell culture confirmed that PrP^{T182A} fails to reach the cell surface after synthesis (Lehmann and Harris 1997), but this is ascribed to altered protein folding rather than to absence of glycosylation (Capellari et al. 2000).

Y144Stop

Independent attempts to generate transgenic mice expressing PrP with a nonsense mutation corresponding to the human *Y145Stop* mutation associated with GSS (Kitamoto et al. 1993; Ghetti et al. 1996) in mice have failed; neither protein expression nor disease was found in various mouse lines (Fischer et al. 1996; Muramoto et al. 1997). This mutation deletes the site for the GPI anchor necessary for cell-surface attachment, and the very short protein is rapidly degraded by the proteosomal pathway (Zanusso et al. 1999).

E199K

Transgenic mice expressing chimeric PrP (MHu2M) with a mutation equivalent to the human codon 200 (Glu→Lys) mutation linked to inherited CJD (Hsiao et al. 1991), on a wild-type background, did not develop any signs of scrapie-like disease (Telling et al. 1996a) although PrP^{E199K} is properly transported to the cell surface and located in rafts (Rosenmann et al. 2001).

Octarepeat Insertions

Expression of a mouse PrP version of a 9-octapeptide insertion (resulting in a total of 14 repeats) associated with "prion dementia" in humans (Owen et al. 1992; Krasemann et al. 1995) produced a slowly progressive cerebellar disorder in transgenic mice. Mice homozygous for the transgene cluster developed ataxia at 2 months and died at 4 months of age. They show apoptosis of cerebellar granule cells, accumulation of detergent-insoluble PrP that is resistant to low levels of protease but not to the high levels of protease characteristic for PrP^{Sc} (Chiesa et al. 1998, 2000), and progressive, primary myopathy (Chiesa et al. 2001). Disease progression was not affected by coexpression of wild-type PrP (Chiesa et al. 2000). Because the disease was not transmissible, it qualifies as a "proteinopathy" rather than a "prion disease."

Transmissible Versus Nontransmissible PrP-linked Disease

Many mutant forms of murine PrP are pathogenic, mostly or perhaps only when overexpressed. It is not surprising that neurons accumulating abnormal forms of protein may suffer damage; the only unusual case is that of the amino-terminal deletions Δ32-121 and Δ32-134, where coexpression of the truncated molecule with a relatively small amount of normal PrP abrogates the disease phenotype (Shmerling et al. 1998). With the possible exception of the neurodegeneration associated with highly overexpressed PrP^{P101L} (Telling et al. 1996a), none of the PrP-linked diseases in mice was claimed to be transmissible. It is therefore appropriate to distinguish between prion diseases, which are transmissible, and "nontransmissible conformational diseases," or "nontransmissible proteinopathies," which are not (Tateishi et al. 1996).

In the case of human prion diseases, transmission to experimental animals has been achieved for kuru (Gajdusek et al. 1966; Tateishi et al. 1979), sCJD (Gibbs et al. 1968), GSS (Tateishi et al. 1984), fatal familial insomnia (FFI) (Collinge et al. 1995; Tateishi et al. 1995; Mastrianni et al. 2001), fCJD$^{[V210I]}$ (Mastrianni et al. 2001), fCJD$^{[E200K]}$ (Mastrianni et al. 2001), and vCJD (Bruce et al. 1997; Hill et al. 1997b). The experimental transmission to mice of hereditary cases with P102L, P105L, A117V, Y145stop, and octarepeat insertions was "difficult," except for one-third of the cases with P102L (Tateishi et al. 1996). Doubtless, in any particular instance, lack of transmission may be due to inadequate recipients, but in view of the results with mice, it is possible that at least some of the familial human spongiform encephalopathies may be truly nontransmissible conformational diseases and, therefore, in view of the definition of prions as transmissible agents, not prion diseases.

ECTOPIC EXPRESSION OF PrP TRANSGENES

Several questions arising in connection with the propagation of prions have been addressed by generating mice that express PrP in only one type of tissue or organ, rather than almost ubiquitously, as is the case in wild-type mice. In some cases this can be achieved by transplantation, as described below; in other cases the PrP ORF has been linked to tissue- or organ-specific promoters and introduced into PrP knockout mice.

PrP-expressing Astrocytes in a PrP Knockout Mouse Propagate Prions and Cause Scrapie-like Symptoms

PrP is expressed in neurons and in astrocytes (Moser et al. 1995), but it is not evident in which cell type infectious agent is generated. Expression of hamster PrP under direction of the astrocyte-specific glial fibrillary acid protein (GFAP) promoter in PrP knockout mice rendered the animals susceptible to hamster prions and led to clinical disease and prion propagation. Because no expression of PrP could be detected in neurons (within the sensitivity of the immunohistochemical analysis), it would seem that astrocytes are competent for prion replication. Interestingly, the neuropathology exhibited by these transgenic mice is quite similar to that found in scrapie-infected wild-type mice (Raeber et al. 1997). $Prnp^{+/+}$ mice expressing the same transgene cluster did not develop clinical disease after inoculation with hamster prions for at least 750 days, despite a prion titer only 10- to 100-fold lower than in transgenic knockout mice that developed disease after about 230 d.p.i. (Raeber et al. 1997).

Source of Prions in the Lymphoreticular System: PrP Expression in B or T Lymphocytes Does Not Suffice for Prion Propagation

In most mouse scrapie models, infectivity appears in the spleen within days after intracerebral inoculation and rises slowly or remains constant throughout the lifetime of the mouse, whereas infectivity in the brain starts rising a few weeks after inoculation and reaches titers two orders of magnitude higher than in the spleen (Eklund et al. 1967; Kimberlin and Walker 1979; Büeler et al. 1993). In the spleen, PrP is present on the surface of T and B cells, albeit at very low levels (Cashman et al. 1990; Liu et al. 2001); after infection, PrP^{Sc} is found mainly in follicular dendritic cells (FDCs) (Kitamoto et al. 1991; McBride et al. 1992). Fractionation of scrapie-infected spleen showed that infectivity was associated with B and T lymphocytes and with the FDC-containing stromal fraction, but not with neutrophils. Surprisingly, no infectivity was detected in circulating lymphocytes (Raeber et al. 1996). This raised the question as to whether B and T cells were able to propagate prions or whether they acquired them from another source.

Transgenic mice expressing PrP under the control of an Eμ enhancer/IRF-1 promoter (Yamada et al. 1991) overexpress PrP in the spleen, including T and B lymphocytes. Infectivity appeared in the spleen 2 weeks after intraperitoneal (i.p.) inoculation and persisted at a level similar to that in wild-type mice, but was not detectable in the brain at 6 months postinoculation (p.i.) (Raeber et al. 1999b). Thus, prions in the spleen did not originate in the CNS. Transgenic mice overexpressing PrP under the control of the T-cell-specific lck promoter (Chaffin et al. 1990) showed high levels of PrP in T cells, both in thymus and in spleen. No PrP was detected in brain. Inoculation led to no symptoms, and no infectivity was detected in spleen, thymus, or brain up to one year after inoculation. Because T cells in these mice expressed at least 10-fold more PrP than those in mice expressing PrP under the direction of the Eμ enhancer/IRF-1 promoter, it would seem that T cells alone are unable to generate infectivity, but are able to pick it up from other sources (Raeber et al. 1999b). Transgenic mice overexpressing PrP under control of the B-cell-specific CD19 promoter had 10- to 20-fold higher PrP levels on B cells compared to wild-type mice (Montrasio et al. 2001), yet no pathology or prion propagation was observed after i.p. inoculation with scrapie prions.

From these experiments, it can be concluded that the presence of PrP on the cell surface does not suffice to support prion replication; maybe location within a particular plasma membrane region is required (Taraboulos et al. 1995) and/or other components are necessary, such as the postulated protein X (Telling et al. 1995) or a receptor. Moreover, it follows that prions associated with splenic T and B lymphocytes must stem from another source, almost certainly FDCs (Mabbott et al. 2000; Montrasio et al. 2000). Perhaps splenic lymphocytes, some of which interdigitate with FDCs, carry away prion-associated membrane fragments when mechanically separated from the latter, whereas circulating lymphocytes disengage physiologically without doing so.

Expression of PrP in Neuronal Cells Suffices for Transmission of Prions from Periphery to the Brain

Expression of hamster PrP (HaPrP) under the control of the NSE promoter, which is considered to be neuron-specific, rendered transgenic mice susceptible to hamster prions (Race et al. 1995). To determine whether spread of prions from the periphery to the CNS is dependent on hematogenous or lymphoreticular tissue (LRS), PrP knockout mice transgenic for NSE-promoter-controlled HaPrP, which express HaPrP in brain and nerve but not in spleen, lymph nodes, or bone marrow, were exposed to 263K hamster prions orally or by i.p. inoculation. In both cases, scrapie disease ensued with the same incubation time as in mice expressing HaPrP under the control of the PrP promoter. Thus, at least after peripheral exposure to high doses of 263K hamster prions, the LRS was not required for prion amplification, nor were blood cells required for transport (Race et al. 2000).

Prion Formation in PrP-overexpressing Muscle of PrP Knockout Mice

Wild-type mice inoculated intracerebrally (i.c.) with Me7 murine prions show distinct levels of infectivity and PrP^{Sc} in hindleg, but not in other muscles (Bosque et al. 2002). To determine whether prions were formed in muscle or imported from elsewhere, PrP knockout mice carrying the murine PrP ORF under the control of the β-actin promoter were inoculated intramuscularly with RML prions; high prion levels were found in the contralateral as well as the ipsilateral hindleg muscles, but not in any other organs. Similar results, albeit at a lower level, were obtained in knockout mice expressing hamster PrP under the control of the muscle creatine kinase promoter, after inoculation with hamster prions (Bosque et al. 2002).

Prion Formation in PrP-overexpressing Liver of PrP Knockout Mice

Earlier attempts to overexpress PrP in liver of $Prnp^{0/0}$ mice using the albumin promoter resulted in hepatic PrP levels less than 1/100th those in wild-type brain. Intraperitoneal inoculation of the transgenic mice did not result in clinical disease, and no infectivity was detected in the liver, brain, or spleen (Raeber et al. 1999b).

$Prnp^{0/0}$ mice transgenic for PrP under the control of the transthyretin promoter/enhancer expressed PrP in liver and to a lesser degree in brain. PrP levels in the liver were similar to those found in wild-type brain. After i.c. or i.p. inoculation, prion titers in the liver were about 10^3 ID_{50} units/g. Although hepatic prion titers were very low compared to those in brain, it was concluded that expression of high levels of PrP^C in liver enhances prion formation (Bosque et al. 2002).

TRANSPLANTATION AS A TOOL IN PRION RESEARCH

Among the many unresolved questions about prion diseases are those concerning the molecular mechanisms of pathogenesis in the CNS and the mechanism of prion spread in the infected organism. A useful approach to address some of these problems is based on the transplantation of tissue expressing PrP into $Prnp^{0/0}$ mice and vice versa; in particular, of neuroectodermal or hematopoietic tissue.

The neurografting procedure is quite straightforward; neuroectoderm is derived from embryos at defined stages of gestation, ideally E12.5–13.5, and injected into the caudoputamen or lateral ventricles of recipient mice using a stereotaxic frame (Aguzzi et al. 1991; Isenmann et al. 1996). Thereby, tissue from a donor mouse that may die young from a lethal disease can be kept alive in a healthy recipient. Another important technique is transplantation of hematopoietic tissue, either bone marrow or fetal liver, into lethally irradiated mice.

PrP^{Sc}-producing Neurografts Do Not Cause Neuropathological Changes in Surrounding *Prnp*$^{0/0}$ Brain

An interesting question regards the molecular mechanism underlying neuropathological changes, in particular cell death, resulting from prion disease. Depletion of PrP^{C} is an unlikely cause in view of the finding that abrogation of PrP does not cause scrapie-like neuropathological changes (Büeler et al. 1992), even when elicited postnatally (Mallucci et al. 2002). More likely, toxicity of PrP^{Sc} or some PrP^{C}-dependent process is responsible.

To address the question of neurotoxicity, brain tissue of *Prnp*$^{0/0}$ mice was exposed to a continuous source of PrP^{Sc}. For this purpose, telencephalic tissue from transgenic mice overexpressing PrP (Fischer et al. 1996) was transplanted into the forebrain of *Prnp*$^{0/0}$ mice, and the "pseudochimeric" brains were inoculated with scrapie prions. All grafted and scrapie-inoculated mice remained free of scrapie symptoms for at least 70 weeks; this exceeded by at least sevenfold the survival time of scrapie-infected donor mice. Therefore, the presence of a continuous source of PrP^{Sc} and of scrapie prions does not exert any clinically detectable adverse effects on a mouse devoid of PrP^{C}. On the other hand, the grafts developed characteristic histopathological features of scrapie after inoculation (Fig. 5). The course of the disease in the graft was very similar to that observed in the brain of scrapie-inoculated wild-type mice. Importantly, grafts had extensive contact with the recipient brain, and prions could navigate between the two compartments, as shown by the fact that inoculation of wild-type animals engrafted with PrP-expressing neuroectodermal tissue resulted in scrapie pathology in both graft and host tissue. Nonetheless, histopathological changes never extended into host tissue, even at the latest stages (>450 days), although PrP^{Sc} was detected in both grafts and recipient brain, and immunohistochemistry revealed PrP deposits in the hippocampus, and occasionally in the parietal cortex, of all animals (Brandner et al. 1996b). Thus, prions moved from the grafts to some regions of the PrP-deficient host brain without causing pathological changes or clinical disease. The distribution of PrP^{Sc} in the white-matter tracts of the host brain suggests diffusion within the extracellular space (Jeffrey et al. 1994) rather than axonal transport (S. Brandner and A. Aguzzi, unpubl.).

These findings suggest that the expression of PrP^{C} by an infected cell, rather than the extracellular deposition of PrP^{Sc}, is the critical prerequisite for the development of scrapie pathology. Perhaps PrP^{Sc} is inherently nontoxic, and PrP^{Sc} plaques found in spongiform encephalopathies are an epiphenomenon rather than a cause of neuronal damage. One may therefore propose that availability of PrP^{C} for some intracellular process elicit-

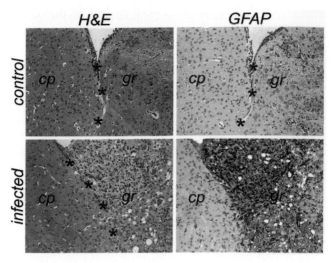

Figure 5. Typical histological appearance of a PrP^C-expressing graft (*left* side of each panel) at the interface to the PrP knockout host brain (*right* side). Knockout mice were grafted with PrP-expressing neuroectodermal tissue and inoculated intracerebrally with mouse scrapie prions where indicated. Animals were sacrificed 232 days after inoculation. Note the spongiform microcystic changes (*left* panel: hematoxylineosin) and the brisk astrocytic reaction evidenced by the immunocytochemical stain for glial fibrillary acidic protein (*right* panel: GFAP). The host tissue is not affected. For further details see Brandner et al. (1996b).

ed by the infectious agent, perhaps the formation of a toxic form of PrP (PrP*; Weissmann 1991a) differing from the bulk of protease-resistant, aggregated material designated PrP^{Sc} is responsible for spongiosis, gliosis, and neuronal death. This would be in agreement with the fact that in several instances, and especially in FFI, spongiform pathology is detectable even though very little PrP^{Sc} is present (Collinge et al. 1995; Aguzzi and Weissmann 1996).

Spread of Prions in the Central Nervous System Requires PrP^C-expressing Tissue

Intracerebral (i.c.) inoculation of tissue homogenate into suitable recipients is the most effective method for transmission of spongiform encephalopathies and may even facilitate circumvention of the species barrier. However, prion diseases can also be initiated from the eye by conjunctival instillation (Scott et al. 1993), corneal grafts (Duffy et al. 1974), and intraocular (i.o.) injection (Fraser 1982). The latter method has proved particularly useful to study neural spread of the agent, since the retina is a part of the CNS and i.o. injection does not produce direct physical trauma to the brain. Diachronic spongiform changes along the retinal pathway following i.o. infection argue in favor of axonal spread (Fraser 1982).

As discussed above, expression of PrP^C is required for prion replication and also for neurodegenerative changes to occur. This raised the question as to whether spread of prions within the CNS is also dependent on PrP^C expression in neural pathways. Again, intracerebrally placed grafts served as an indicator for infectivity in an otherwise scrapie-resistant host. After inoculation of prions into the eyes of engrafted $Prnp^{0/0}$ mice, no signs of spongiform encephalopathy or deposition of PrP^{Sc} were detected in the grafts. In control experiments, unilateral i.o. inoculation led to progressive appearance of scrapie pathology along the optic nerve and optic tract, which then extended into the rest of the brain as described earlier (Fraser 1982; Kimberlin and Walker 1986). Thus, infectivity administered to the eye of PrP-deficient hosts does not induce scrapie in a PrP-expressing brain graft (Brandner et al. 1996a).

Since $Prnp^{0/0}$ mice are not exposed to PrP during maturation of their immune system, engraftment with PrP^C-producing tissue might lead to an immune response to PrP and possibly to neutralization of infectivity. Indeed, analysis of sera from engrafted $Prnp^{0/0}$ mice revealed significant anti-PrP antibody titers. Because even mock-inoculated and uninoculated engrafted $Prnp^{0/0}$ mice showed an immune response to PrP, whereas i.c. inoculated non-engrafted $Prnp^{0/0}$ mice did not (Büeler et al. 1993), PrP^C presented by the intracerebral graft (rather than the inoculum or graft-borne PrP^{Sc}) was clearly the offending antigen. To test whether grafts would develop scrapie if infectivity were administered before they elicited a potentially neutralizing immune response, mice were inoculated 24 hours after grafting. Again, no disease was detected in the graft of two mice inoculated intraocularly.

To definitively rule out the possibility that prion transport was disabled by a neutralizing immune response, the experiments were repeated in mice tolerant to PrP. Transgenic $Prnp^{0/0}$ mice overexpressing PrP in T lymphocytes (tg33) are resistant to scrapie and do not contain scrapie infectivity in brain or spleen after inoculation with scrapie prions (Raeber et al. 1999b). When engrafted with PrP-overexpressing $tga20$ neuroectoderm, these mice did not develop antibodies to PrP after i.c. or i.o. inoculation. As before, i.o. inoculation with prions did not provoke scrapie in the graft, supporting the conclusion that lack of PrP^C, rather than an immune response to PrP, prevented spread (Brandner et al. 1996a).

The prion itself is surprisingly sessile, and spread might proceed along a PrP^C-paved chain of cells. Perhaps prions require PrP^C for propagation across synapses; PrP^C is present in the synaptic region (Fournier et al. 1995) and, as discussed above, certain synaptic properties are altered in $Prnp^{0/0}$ mice (Collinge et al. 1994; Whittington et al. 1995). Perhaps transport of prions within (or on the surface of) neuronal processes is PrP^C-dependent. Within the framework of the protein-only hypothesis, these findings may be accommodated by a "domino-stone" model in which spreading of scrapie prions in the CNS occurs per continuitatem through conversion of PrP^C by adjacent PrP^{Sc} (Aguzzi 1997).

Spread of Prions from Extracerebral Sites to the CNS

As discussed above, PrP^C seems indispensable for prion spread within the CNS. But prions need not be delivered directly to the CNS. Intraspinal, intraperitoneal, intramuscular, intravenous (i.v.), and subcutaneous injections (Kimberlin and Walker 1979, 1986; Kimberlin et al. 1987; Buchanan et al. 1991) or scarification (Taylor et al. 1996) are effective, albeit less efficient. Peroral infection has been demonstrated in many animal species (Prusiner et al. 1985; Foster et al. 1993; Anderson et al. 1996; Bradley 1996; Ridley and Baker 1996; Bons et al. 1999; Maignien et al. 1999; Jeffrey et al. 2001).

In mice and sheep (Kimberlin and Walker 1988), and probably also in man in the case of vCJD (Hill et al. 1997a), prions accumulate in the spleen before doing so in the brain, even when infectivity is administered i.c. Prions can accumulate in all components of the LRS, including lymph nodes and intestinal Peyer's patches, where they are found almost immediately following oral infection (Kimberlin and Walker 1989). In SCID (severe combined immune deficiency) mice, i.p. infection does not lead to replication of prions in spleen or brain; however, transfer of wild-type spleen cells to SCID mice restores susceptibility of the CNS to i.p. inoculation (Lasmézas et al. 1996). This suggests that components of the immune system are required for efficient transfer of prions from the site of peripheral infection to the CNS. Perhaps prions injected i.p. or ingested are first brought to lymphatic organs, specifically to germinal centers, by mobile immune cells: dendritic cells are being discussed for this role (Aucouturier et al. 2001; Huang et al. 2002). Opsonization by complement system components is likely to be relevant, since mice genetically engineered to lack complement factors (Klein et al. 2001), or mice depleted of the C3 complement component by administration of cobra venom (Mabbott et al. 2001), exhibit a remarkable resistance to peripheral prion inoculation. It is unlikely that lymphocytes transport prions to the CNS (Raeber et al. 1999a). Rather, it would seem that in spleen, B-cell-dependent, mature follicular dendritic cells (FDCs) allow prions to multiply to a relatively high titer (K.L. Brown et al. 1999; Mabbott et al. 2000; Montrasio et al. 2000), facilitating transfer to the peripheral nervous system. However, prion amplification in lymph nodes can occur in cells other than FDCs (Oldstone et al. 2002; Prinz et al. 2002). Following invasion of peripheral nerve endings in the LRS, the CNS is reached and further spread occurs trans-synaptically and along fiber tracts. Prions are found first in the CNS segments to which the sites of peripheral inoculation project (Kimberlin and Walker 1980; Beekes et al. 1996).

Because PrP^C is crucial for prion spread within the CNS (Brandner et al. 1996a), it may be required also for spread of prions from peripheral sites to CNS. Indeed, PrP-expressing neurografts in $Prnp^{0/0}$ mice did not develop scrapie histopathology after i.p. or i.v. inoculation with scrapie prions. After reconstitution of the hematopoietic system of $Prnp^{0/0}$ mice with PrP-expressing cells, prion accumulation in the spleen was as in wild-type animals. Surprisingly, however, i.p. or i.v. inoculation failed to produce scrapie pathology in the neurografts of 27 out of 28 animals, in contrast to i.c. inoculation. Thus, transfer of infectivity from spleen to the CNS is crucially dependent on the expression of PrP in a tissue compartment that cannot be reconstituted by bone marrow transfer (Blättler et al. 1997), likely the peripheral nervous system (Kimberlin and Walker 1980; Beekes et al. 1996). In particular, the autonomic nervous system has been held responsible for transport from lymphoid organs to the CNS (Clarke and Kimberlin 1984; Cole and Kimberlin 1985; Beekes et al. 1998; McBride and Beekes 1999), in a PrP^C-dependent fashion (Blättler et al. 1997; Glatzel and Aguzzi 2000; Race et al. 2000). Chemical sympathectomy and immunosympathectomy significantly delayed the development of scrapie, and mice with sympathetic hyperinnervation of immune organs had a shortened incubation time and a 100-fold increased prion titer in spleen (Glatzel et al. 2001).

It is unclear how prions are actually transported by peripheral nerves. Axonal or nonaxonal transport mechanisms may be involved, and non-neuronal cells (such as Schwann cells) may play a role. The "domino" mechanism implies that incoming PrP^{Sc} converts resident PrP^C on the axolemmal surface, thereby propagating the infection along the axon. Although speculative, this model accommodates the findings that the velocity of neural prion spread is much slower than expected from fast axonal transport (Kimberlin et al. 1983) and that PrP^{Sc} is deposited peri-axonally (Hainfellner and Budka 1999; Glatzel and Aguzzi 2000).

Transgenic Mice Expressing PrP Antibodies Are Protected against Peripheral Prion Inoculation

The central role of PrP in the genesis of transmissible spongiform encephalopathies suggested several potential strategies for treatment or prophylaxis of these diseases. Ablation of the PrP gene was suggested early on as a possible approach to generate prion-resistant farm animals (Büeler et al. 1993), although the implementation of this strategy would only be practical for particular purposes, such as production of recombinant proteins from transgenic sheep or of serum or collagen for pharmaceutical purposes in cattle. Therapeutic strategies involve identification of drugs that would lower PrP^C levels by specifically interfering with PrP mRNA transcription, translation, or processing of PrP^C, by preventing its correct trafficking, by promoting its endocytosis and degradation, interfering with the conversion of PrP^C to its pathogenic form or accelerating the degradation of PrP^{Sc}. Many substances have proved effective at lowering levels of PrP^{Sc} and/or of prion infectivity in cell culture, and some have been effective in extending incubation time in mice or hamsters, especially when administered early after infection. None has, so far, proved particularly effective for actual therapy of sick animals—let alone human patients (Brown 2002).

Antibodies against PrP could provide a promising approach to prophylaxis and therapy. In vitro preincubation with anti-PrP antisera was reported to reduce the prion titer of infectious hamster brain homogenates by up to 2 log units (Gabizon et al. 1988), and an anti-PrP antibody inhibited formation of PrP^{Sc} in a cell-free system (Horiuchi and Caughey 1999). Two recent papers reported that antibodies (Enari et al. 2001) and Fab fragments raised against certain domains of PrP (Peretz et al. 2001) abrogate PrP^{Sc} and infectivity in cultured cells. However, it is difficult to induce humoral immune responses against PrP^C and PrP^{Sc} in wild-type mice because of tolerance of the mammalian immune system to PrP^C. Accordingly, ablation of the Prnp gene (Büeler et al. 1992) renders mice highly susceptible to immunization with PrP (Prusiner et al. 1993; Brandner et al. 1996a), and many monoclonal antibodies to the prion protein have been generated in $Prnp^{0/0}$ mice. However, $Prnp^{0/0}$ mice are unsuitable for testing vaccination regimens since they do not support prion pathogenesis (Büeler et al. 1993).

Heppner et al. (2001) therefore prepared transgenic mice expressing the variable V(D)J region of the heavy chain of the anti-PrP monoclonal antibody 6H4 (Korth et al. 1997). At the age of 150 days, transgenic $Prnp^{0/0}$, $Prnp^{0/+}$, and $Prnp^{+/+}$ mice expressed similar substantial titers of anti-PrP antibodies, indicating that deletion of autoreactive B cells does not prevent anti-PrP immunity. The buildup of anti-PrP^C titers, however, was more sluggish in the presence of endogenous PrP^C, most likely indicating that some clonal deletion was actually occurring.

The total anti-PrP^C titer results from pairing of one transgenic μ heavy chain with a large repertoire of endogenous κ and λ chains; some pairings may lead to reactive moieties, and others may be anergic. Maybe the B-cell clones with the highest affinity to PrP^C are eliminated by the immune tolerization mechanism, and only clones with medium affinity are retained. This would explain the delay in titer buildup in the presence of PrP^C and would be in agreement with the affinity measurements that indicate that the total molar avidity of 6H4μ serum is ~100-fold lower than that of the original 6H4 antibody from which the transgene was derived.

Expression of the 6H4μ heavy chain conferred protection from scrapie upon i.p. inoculation of the prion agent (Heppner et al. 2001). This delivers proof of principle that a protective humoral response against prions can be mounted by the mammalian immune system and suggests that B cells are not intrinsically tolerant to PrP^C. If the latter is generally true, lack of immunity to prions may be due to T-helper tolerance and could perhaps be overcome by presenting PrP^C to the immune system along with highly active adjuvants (Sigurdsson et al. 2002) or in a highly ordered and repetitive form. These findings, therefore, encourage a reassessment of the possible value of active and passive immunization, and perhaps of reprogramming B-cell repertoires by μ-chain transfer, in prophylaxis or in therapy of prion diseases.

ACKNOWLEDGMENTS

The Zürich/London prion groups are indebted to M. Aguet, A. Behrens, T. Blättler, H. Blüthmann (Hoffmann LaRoche, Basel), H.R. Büeler, S. Brandner, A. Cozzio, M. Fischer, M. Glatzel, I. Hegyi, M.A. Klein, Y. Kobayashi, H.-P. Lipp (Anatomisches Institut, Universität Zürich), F. Montrasio, D. Rossi, T. Rülicke (Biologisches Zentrallabor, Universitätsspital Zürich), M. Prinz, A. Raeber, A. Sailer, S. Sakaguchi, D. Shmerling, and P. Schwarz for their valuable contributions. The work carried out in Edinburgh is also the work of H. Baybutt, L. Aitchison, P. McBride, I. McConnell, J. Hope (Institute for Animal Health, Neuropathogenesis Unit, Edinburgh), A. Clarke (Department of Pathology, Edinburgh University), A. Johnston, N. MacLeod (Department of Physiology, Edinburgh University), D. Melton, R. Moore, N. Redhead (Institute for Cell and Molecular Biology, Edinburgh University), and I. Tobler (Institute for Pharmacology, University of Zurich).

REFERENCES

Aguzzi A. 1997. Neuro-immune connection in spread of prions in the body? *Lancet* **349:** 742–743.

Aguzzi A. and Weissmann C. 1996. Sleepless in Bologna: Transmission of fatal familial insomnia. *Trends Microbiol.* **4:** 129–131.

Aguzzi A., Kleihues P., Heckl K., and Wiestler O.D. 1991. Cell type-specific tumor induction in neural transplants by retrovirus-mediated oncogene transfer. *Oncogene* **6:** 113–118.

Anderson R.M., Donnelly C.A., Ferguson N.M., Woolhouse M.E., Watt C.J., Udy H.J., MaWhinney S., Dunstan S.P., Southwood T.R., Wilesmith J.W., Ryan J.B., Hoinville L.J., Hillerton J.E., Austin A.R., and Wells G.A. 1996. Transmission dynamics and epidemiology of BSE in British cattle (comments). *Nature* **382:** 779–788.

Araki K., Araki M., and Yamamura K. 1997. Targeted integration of DNA using mutant lox sites in embryonic stem cells. *Nucleic Acids Res.* **25:** 868–872.

Asante E.A., Gowland I., Linehan J.M., Mahal S.P., and Collinge J. 2002. Expression pattern of a mini human PrP gene promoter in transgenic mice. *Neurobiol. Dis.* **10:** 1–7.

Aucouturier P., Geissmann F., Damotte D., Saborio G.P., Meeker H.C., Kascsak R., Carp R.I., and Wisniewski T. 2001. Infected splenic dendritic cells are sufficient for prion transmission to the CNS in mouse scrapie. *J. Clin. Invest.* **108:** 703–708.

Barron R.M., Thomson V., Jamieson E., Melton D.W., Ironside J., Will R., and Manson J.C. 2001. Changing a single amino acid in the N-terminus of murine PrP alters TSE incubation time across three species barriers. *EMBO J.* **20:** 5070–5078.

Basler K., Oesch B., Scott M., Westaway D., Wälchli M., Groth D.F., McKinley M.P., Prusiner S.B., and Weissmann C. 1986. Scrapie and cellular PrP isoforms are encoded by the same chromosomal gene, *Cell* **46:** 417–428.

Beekes M., Baldauf E., and Diringer H. 1996. Sequential appearance and accumulation of pathognomonic markers in the central nervous system of hamsters orally infected with scrapie. *J. Gen. Virol.* **77:** 1925–1934.

Beekes M., McBride P.A., and Baldauf E. 1998. Cerebral targeting indicates vagal spread of infection in hamsters fed with scrapie. *J. Gen. Virol.* **79:** 601–607.

Behrens A., Brandner S., Genoud N., and Aguzzi A. 2001. Normal neurogenesis and scrapie pathogenesis in neural grafts lacking the prion protein homologue Doppel. *EMBO Rep.* **2:** 347–352.

Behrens A., Genoud N., Naumann H., Rulicke T., Janett F., Heppner F.L., Ledermann B., and Aguzzi A. 2002. Absence of the prion protein homologue Doppel causes male sterility. *EMBO J.* **21:** 3652–3658.

Bendheim P.E., Brown H.R., Rudelli R.D., Scala L.J., Goller N.L., Wen G.Y., Kascsak R.J., Cashman N.R., and Bolton D.C. 1992. Nearly ubiquitous tissue distribution of the scrapie agent precursor protein. *Neurology* **42:** 149–156.

Bessen R.A. and Marsh R.F. 1992. Biochemical and physical properties of the prion protein from two strains of the transmissible mink encephalopathy agent. *J. Virol.* **66:** 2096–2101.

Blättler T., Brandner S., Raeber A.J., Klein M.A., Voigtlander T., Weissmann C., and Aguzzi A. 1997. PrP-expressing tissue required for transfer of scrapie infectivity from spleen to brain. *Nature* **389:** 69–73.

Bons N., Mestre-Frances N., Belli P., Cathala F., Gajdusek D.C., and Brown P. 1999. Natural and experimental oral infection of nonhuman primates by bovine spongiform encephalopathy agents. *Proc. Natl. Acad. Sci.* **96:** 4046–4051.

Borchelt D.R., Davis J., Fischer M., Lee M.K., Slunt H.H., Ratovitsky T., Regard J., Copeland N.G., Jenkins N.A., Sisodia S.S., and Price D.L. 1996. A vector for expressing foreign genes in the brains and hearts of transgenic mice. *Genet. Anal.* **13:** 159–163.

Bosque P.J., Ryou C., Telling G., Peretz D., Legname G., DeArmond S.J., and Prusiner S.B. 2002. Prions in skeletal muscle. *Proc. Natl. Acad. Sci.* **99:** 3812–3817.

Bounhar Y., Zhang Y., Goodyer C.G., and LeBlanc A. 2001. Prion protein protects human neurons against Bax-mediated apoptosis. *J. Biol. Chem.* **276:** 39145–39149.

Bradley R. 1996. Bovine spongiform encephalopathy-distribution and update on some transmission and decontamination studies. In *Bovine spongiform encephalopathy: The BSE dilemma* (ed. C.J. Gibbs, Jr.), pp. 11. Springer-Verlag, New York.

Brandner S., Raeber A., Sailer A., Blättler T., Fischer M., Weissmann C., and Aguzzi A. 1996a. Normal host prion protein (PrP^C) is required for scrapie spread within the central nervous system. *Proc. Natl. Acad. Sci.* **93:** 13148–13151.

Brandner S., Isenmann S., Raeber A., Fischer M., Sailer A., Kobayashi Y., Marino S., Weissmann C., and Aguzzi A. 1996b. Normal host prion protein necessary for scrapie-induced neurotoxicity. *Nature* **379:** 339–343.

Braun R.E., Lo D., Pinkert C.A., Widera G., Flavell R.A., Palmiter R.D., and Brinster R.L. 1990. Infertility in male transgenic mice: Disruption of sperm development by HSV-tk expression in postmeiotic germ cells. *Biol. Reprod.* **43:** 684–693.

Brinster A.L., Allen J.M., Behringer R.R., Gelinas R.E., and Palmiter R.D. 1988. Introns increase transcriptional efficiency in transgenic mice. *Proc. Natl. Acad. Sci.* **85:** 836–840.

Brown D.R. 1999. Prion protein expression aids cellular uptake and veratridine-induced release of copper. *J. Neurosci. Res.* **58:** 717–725.

Brown D.R. and Besinger A. 1998. Prion protein expression and superoxide dismutase activity. *Biochem. J.* **334:** 423–429.

Brown D.R., Nicholas R.S., and Canevari L. 2002. Lack of prion protein expression results in a neuronal phenotype sensitive to stress. *J. Neurosci. Res.* **67:** 211–224.

Brown D.R., Schmidt B., and Kretzschmar H.A. 1997a. Effects of oxidative stress on prion protein expression in PC12 cells. *Int. J. Dev. Neurosci.* **15:** 961–972.

Brown D.R., Wong B.S., Hafiz F., Clive C., Haswell S.J., and Jones I.M. 1999. Normal prion protein has an activity like that of superoxide dismutase. *Biochem. J.* **344:** 1–5.

Brown D.R., Iordanova I.K., Wong B.S., Venien-Bryan C., Hafiz F., Glasssmith L.L., Sy M.S., Gambetti P., Jones I.M., Clive C., and Haswell S.J. 2000. Functional and structural differences between the prion protein from two alleles prnp(a) and prnp(b) of mouse. *Eur. Biophys. J.* **267:** 2452–2459.

Brown D.R., Qin K., Herms J.W., Madlung A., Manson J., Strome R., Fraser P.E., Kruck T., von Bohlen A., Schulz-Schaeffer W., Giese A., Westaway D., and Kretzschmar H. 1997b. The cellular prion protein binds copper in vivo. *Nature* **390:** 684–687.

Brown K.L., Stewart K., Ritchie D.L., Mabbott N.A., Williams A., Fraser H., Morrison W.I., and Bruce M.E. 1999. Scrapie replication in lymphoid tissues depends on prion protein-expressing follicular dendritic cells. *Nat. Med.* **5:** 1308–1312.

Brown P. 2002. Drug therapy in human and experimental transmissible spongiform encephalopathy. *Neurology* **58:** 1720–1725.

Brown P., Gibbs C.J., Jr., Rodgers J.P., Asher D.M., Sulima M.P., Bacote A., Goldfarb L.G., and Gajdusek D.C. 1994. Human spongiform encephalopathy: The National Institutes of Health series of 300 cases of experimentally transmitted disease. *Ann. Neurol.* **35:** 513–529.

Bruce M., Chree A., McConnell I., Foster J., Pearson G., and Fraser H. 1994. Transmission of bovine spongiform encephalopathy and scrapie to mice: Strain variation and the species barrier. *Philos. Trans. R. Soc. Lond. B Biol. Sci.* **343:** 405–411.

Bruce M.E., Will R.G., Ironside J.W., McConnell I., Drummond D., Suttie A., McCardle L., Chree A., Hope J., Birkett C., Cousens S., Fraser H., and Bostock C.J. 1997. Transmissions to mice indicate that 'new variant' CJD is caused by the BSE agent. *Nature* **389:** 498–501.

Buchanan C.R., Preece M.A., and Milner R.D. 1991. Mortality, neoplasia, and Creutzfeldt-Jakob disease in patients treated with human pituitary growth hormone in the United Kingdom. *Br. Med. J.* **302:** 824–828.

Büeler H., Raeber A., Sailer A., Fischer M., Aguzzi A., and Weissmann C. 1994. High prion and PrP^{Sc} levels but delayed onset of disease in scrapie-inoculated mice heterozygous for a disrupted PrP gene. *Mol. Med.* **1:** 19–30.

Büeler H., Aguzzi A., Sailer A., Greiner R.A., Autenried P., Aguet M., and Weissmann C. 1993. Mice devoid of PrP are resistant to scrapie. *Cell* **73:** 1339–1347.

Büeler H., Fischer M., Lang Y., Bluethmann H., Lipp H.-P., DeArmond S.J., Prusiner S.B., Aguet M., and Weissmann C. 1992. Normal development and behaviour of mice lacking the neuronal cell-surface PrP protein. *Nature* **356:** 577–582.

Buschmann A., Pfaff E., Reifenberg K., Muller H.M., and Groschup M.H. 2000. Detection of cattle-derived BSE prions using transgenic mice overexpressing bovine PrP(C). *Arch. Virol. Suppl.* (16): 75–86.

Capellari S., Zaidi S.I., Long A.C., Kwon E.E., and Petersen R.B. 2000. The Thr183Ala mutation, not the loss of the first glycosylation site, alters the physical properties of the prion protein. *J. Alzheimer's Dis.* **2:** 27–35.

Carlson G.A., Kingsbury D.T., Goodman P.A., Coleman S., Marshall S.T., DeArmond S., Westaway D., and Prusiner S.B. 1986. Linkage of prion protein and scrapie incubation time genes. *Cell* **46:** 503–511.

Carlson G.A., Ebeling C., Yang S.L., Telling G., Torchia M., Groth D., Westaway D., DeArmond S.J., and Prusiner S.B. 1994. Prion isolate specified allotypic interactions between the cellular and scrapie prion proteins in congenic and transgenic mice. *Proc. Natl. Acad. Sci.* **91:** 5690–5694.

Cashman N.R., Loertscher R., Nalbantoglu J., Shaw I., Kascsak R.J., Bolton D.C., and Bendheim P.E. 1990. Cellular isoform of the scrapie agent protein participates in lymphocyte activation. *Cell* **61:** 185–192.

Caughey B., Raymond G.J., and Demon R.A. 1998. Strain-dependent differences in beta-sheet conformations of abnormal prion protein. *J. Biol. Chem.* **273:** 32230–32235.

Chaffin K.E., Beals C.R., Wilkie T.M., Forbush K.A., Simon M.I., and Perlmutter R.M. 1990. Dissection of thymocyte signaling pathways by in vivo expression of pertussis toxin ADP-ribosyltransferase. *EMBO J.* **9:** 3821–3829.

Chandler R.L. 1961. Encephalopathy in mice produced by inoculation with scrapie brain material. *Lancet* **1:** 1378–1379.

Chesebro B. 1999. Prion protein and the transmissible spongiform encephalopathy diseases. *Neuron* **24:** 503–506.

Chesebro B. and Caughey B. 1993. Scrapie agent replication without the prion protein? *Curr. Biol.* **3:** 696–698.

Chiarini L.B., Freitas A.R., Zanata S.M., Brentani R.R., Martins V.R., and Linden R. 2002. Cellular prion protein transduces neuroprotective signals. *EMBO J.* **21:** 3317–3326.

Chiesa R., Piccardo P., Ghetti B., and Harris D.A. 1998. Neurological illness in transgenic mice expressing a prion protein with an insertional mutation. *Neuron* **21:** 1339–1351.

Chiesa R., Drisaldi B., Quaglio E., Migheli A., Piccardo P., Ghetti B., and Harris D.A. 2000. Accumulation of protease-resistant prion protein (PrP) and apoptosis of cerebellar granule cells in transgenic mice expressing a PrP insertional mutation. *Proc. Natl. Acad. Sci.* **97:** 5574–5579.

Chiesa R., Pestronk A., Schmidt R.E., Tourtellotte W.G., Ghetti B., Piccardo P., and Harris D.A. 2001. Primary myopathy and accumulation of PrPSc-like molecules in peripheral tissues of transgenic mice expressing a prion protein insertional mutation. *Neurobiol. Dis.* **8:** 279–288.

Clarke M.C. and Kimberlin R.H. 1984. Pathogenesis of mouse scrapie: Distribution of agent in the pulp and stroma of infected spleens. *Vet. Microbiol.* **9:** 215–225.

Cohen J.L., Boyer O., Salomon B., Onclerco R., Depetris D., Lejeune L., Dubus-Bonnet V., Bruel S., Charlotte F., Mattei M.G., and Klatzmann D. 1998. Fertile homozygous transgenic mice expressing a functional truncated herpes simplex thymidine kinase delta TK gene. *Transgenic Res.* **7:** 321–330.

Cole S. and Kimberlin R.H. 1985. Pathogenesis of mouse scrapie: Dynamics of vacuolation in brain and spinal cord after intraperitoneal infection. *Neuropathol. Appl. Neurobiol.* **11:** 213–227.

Colling S.B., Collinge J., and Jefferys J.G. 1996. Hippocampal slices from prion protein null mice: Disrupted Ca(2+)-activated K+ currents. *Neurosci. Lett.* **209:** 49–52.

Collinge J. 2001. Prion diseases of humans and animals: Their causes and molecular basis. *Annu. Rev. Neurosci.* **24:** 519–550.

Collinge J., Beck J., Campbell T., Estibeiro K., and Will R.G. 1996. Prion protein gene analysis in new variant cases of Creutzfeldt-Jakob disease (letter). *Lancet* **348:** 56.

Collinge J., Palmer M.S., Sidle K.C., Gowland I., Medori R., Ironside J., and Lantos P. 1995. Transmission of fatal familial insomnia to laboratory animals. *Lancet* **346:** 569–570.

Collinge J., Whittington M.A., Sidle K.C.L., Smith C.J., Palmer M.S., Clarke A.R., and Jefferys J.G.R. 1994. Prion protein is necessary for normal synaptic function. *Nature* **370:** 295–297.

Crozet C., Bencsik A., Flamant F., Lezmi S., Samarut J., and Baron T. 2001a. Florid plaques in ovine PrP transgenic mice infected with an experimental ovine BSE. *EMBO Rep.* **2:** 952–956.

Crozet C., Flamant F., Bencsik A., Aubert D., Samarut J., and Baron T. 2001b. Efficient transmission of two different sheep scrapie isolates in transgenic mice expressing the ovine PrP gene. *J. Virol.* **75:** 5328–5334.

DeArmond S.J., Sanchez H., Yehiely F., Qiu Y., Ninchak-Casey A., Daggett V., Camerino A.P., Cayetano J., Rogers M., Groth D., Torchia M., Tremblay P., Scott M.R., Cohen F.E., and Prusiner S.B. 1997. Selective neuronal targeting in prion disease. *Neuron* **19:** 1337–1348.

Dickinson A.G. and Outram G.W. 1979. The scrapie replication-site hypothesis and its implications for pathogenesis. In *Slow transmissible diseases of the nervous system* (ed. S.B. Prusiner et al.), pp. 13–31. Academic Press, New York.

———. 1988. Genetic aspects of unconventional virus infections: The basis of the virino hypothesis. *Ciba Found. Symp.* **135:** 63–83.

Dickinson A.G., Meikle V.M., and Fraser H. 1968. Identification of a gene which controls the incubation period of some strains of scrapie agent in mice. *J. Comp. Pathol.* **78:** 293–299.

Donne D.G., Viles J.H., Groth D., Mehlhorn I., James T.L., Cohen F.E., Prusiner S.B., Wright P.E., and Dyson H.J. 1997. Structure of the recombinant full-length hamster prion protein PrP(29- 231): The N terminus is highly flexible. *Proc. Natl. Acad. Sci.* **94:** 13452–13457.

Duffy P., Wolf J., Collins G., DeVoe A.G., Streeten B., and Cowen D. 1974. Letter: Possible person-to-person transmission of Creutzfeldt-Jakob disease. *N. Engl. J. Med.* **290:** 692–693.

Eklund C.M., Kennedy R.C., and Hadlow W.J. 1967. Pathogenesis of scrapie virus infection in the mouse. *J. Infect. Dis.* **117:** 15–22.

Enari M., Flechsig E., and Weissmann C. 2001. Scrapie prion protein accumulation by scrapie-infected neuroblastoma cells abrogated by exposure to a prion protein antibody. *Proc. Natl. Acad. Sci.* **98:** 9295–9299.

Fiering S., Epner E., Robinson K., Zhuang Y., Telling A., Hu M., Martin D.I., Enver T., Ley T.J., and Groudine M. 1995. Targeted deletion of 5′HS2 of the murine beta-globin LCR reveals that it is not essential for proper regulation of the beta-globin locus. *Genes Dev.* **9:** 2203–2213.

Fischer, M., Rülicke T., Raeber A., Sailer A., Moser M., Oesch B., Brandner S., Aguzzi A., and Weissmann C. 1996. Prion protein (PrP) with amino-proximal deletions restoring susceptibility of PrP knockout mice to scrapie. *EMBO J.* **15:** 1255–1264.

Flechsig E., Shmerling D., Hegyi I., Raeber A.J., Fischer M., Cozzio A., von Mering C., Aguzzi A., and Weissmann C. 2000. Prion protein devoid of the octapeptide repeat region restores susceptibility to scrapie in PrP knockout mice. *Neuron* **27:** 399–408.

Flechsig E., Hegyi L., Leimeroth R., Zuniga A., Rossi D., Cozzio A., Schwarz P., Rülicke T., Götz J., Aguzzi A., and Weissmann C. 2003. Expression of truncated PrP targeted to Purkinje cells of PrP knockout mice causes Purkinje cell death and ataxia. *EMBO J.* **22:** 3095–3101.

Follet J., Lemaire-Vieille C., Blanquet-Grossard F., Podevin-Dimster V., Lehmann S., Chauvin J.P., Decavel J.P., Varea R., Grassi J., Fontes M., and Cesbron J.Y. 2002. PrP expression and replication by Schwann cells: Implications in prion spreading. *J. Virol.* **76:** 2434–2439.

Ford M.J., Burton L.J., Morris R.J., and Hall S.M. 2002a. Selective expression of prion protein in peripheral tissues of the adult mouse. *Neuroscience* **113:** 177–192.

Ford M.J., Burton L.J., Li H., Graham C.H., Frobert Y., Grassi J., Hall S.M., and Morris R.J. 2002b. A marked disparity between the expression of prion protein and its message by neurones of the CNS. *Neuroscience* **111:** 533–551.

Forloni G., Angeretti N., Chiesa R., Monzani E., Salmona M., Bugiani O., and Tagliavini F. 1993. Neurotoxicity of a prion protein fragment. *Nature* **362:** 543–546.

Foster J.D., Hope J., and Fraser H. 1993. Transmission of bovine spongiform encephalopathy to sheep and goats. *Vet. Rec.* **133:** 339–341.

Fournier J.G., Escaig-Haye F., Billette de Villemeur T., and Robain O. 1995. Ultrastructural localization of cellular prion protein (PrPc) in synaptic boutons of normal hamster hippocampus. *C.R. Acad. Sci. III* **318:** 339–344.

Fraser H. 1982. Neuronal spread of scrapie agent and targeting of lesions within the retinotectal pathway. *Nature* **295:** 149–150.

Frigg R., Klein M.A., Hegyi I., Zinkernagel R.M., and Aguzzi A. 1999. Scrapie pathogenesis in subclinically infected B-cell-deficient mice. *J. Virol.* **73:** 9584–9588.

Gabizon R., McKinley M.P., Groth D., and Prusiner S.B. 1988. Immunoaffinity purification and neutralization of scrapie prion infectivity. *Proc. Natl. Acad. Sci.* **85:** 6617–6621.

Gajdusek D.C., Gibbs C.J., and Alpers M. 1966. Experimental transmission of a Kuru-like syndrome to chimpanzees. *Nature* **209:** 794–796.

Gauczynski S., Peyrin J.M., Haik S., Leucht C., Hundt C., Rieger R., Krasemann S., Deslys J.P., Dormont D., Lasmézas C.I., and Weiss S. 2001. The 37-kDa/67-kDa laminin receptor acts as the cell-surface receptor for the cellular prion protein. *EMBO J.* **20:** 5863–5875.

Ghetti B., Piccardo P., Spillantini M.G., Ichimiya Y., Porro M., Perini F., Kitamoto T., Tateishi J., Seiler C., Frangione B., Bugiani O., Giaccone G., Prelli F., Goedert M.,

Dlouhy S.R., and Tagliavini F. 1996. Vascular variant of prion protein cerebral amyloid-osis with tau -positive neurofibrillary tangles: The phenotype of the stop codon 145 mutation in PRNP. *Proc. Natl. Acad. Sci.* **93**: 744–748.

Gibbs C.J., Gajdusek D.C., Asher D.M., Alpers M.P., Beck E., Daniel P.M., and Matthews W.B. 1968. Creutzfeldt-Jakob disease (spongiform encephalopathy); Transmission to the chimpanzee. *Science* **161**: 388–389.

Glatzel M. and Aguzzi A. 2000. PrP(C) expression in the peripheral nervous system is a determinant of prion neuroinvasion. *J. Gen. Virol.* **81**: 2813–2821.

Glatzel M., Heppner F.L., Albers K.M., and Aguzzi A. 2001. Sympathetic innervation of lymphoreticular organs is rate limiting for prion neuroinvasion. *Neuron* **31**: 25–34.

Goldfarb L.G., Brown P., and Gajdusek D.C. 1992a. The molecular genetics of human transmissible spongiform encephalopathy. In *Prion diseases of humans and animals* (ed. S.B. Prusiner et al.), pp.139–153. Ellis Horwood, London.

Goldfarb L.G., Petersen R.B., Tabaton M., Brown P., LeBlanc A.C., Montagna P., Cortelli P., Julien J., Vital C., and Pendelbury W.W., et al. 1992b. Fatal familial insomnia and famil-ial Creutzfeldt-Jakob disease: Disease phenotype determined by a DNA polymorphism. *Science* **258**: 806–808.

Goldmann W., Hunter N., Smith G., Foster J., and Hope J. 1994. PrP genotype and agent effects in scrapie: Change in allelic interaction with different isolates of agent in sheep, a natural host of scrapie. *J. Gen. Virol.* **75**: 989–995.

Graner E., Mercadante A.F., Zanata S.M., Forlenza O.V., Cabral A.L., Veiga S.S., Juliano M.A., Roesler R., Walz R., Minetti A., Izquierdo I., Martins V.R., and Brentani R.R. 2000. Cellular prion protein binds laminin and mediates neuritogenesis. *Brain Res. Mol. Brain Res.* **76**: 85–92.

Hainfellner J.A. and Budka H. 1999. Disease associated prion protein may deposit in the peripheral nervous system in human transmissible spongiform encephalopathies. *Acta Neuropathol.* **98**: 458–460.

Harris D.A. 1999. Cell biological studies of the prion protein. *Curr. Issues Mol. Biol.* **1**: 65–75.

Hay B., Prusiner S.B., and Lingappa V.R. 1987a. Evidence for a secretory form of the cellu-lar prion protein. *Biochemistry* **26**: 8110–8115.

Hay B., Barry R.A., Lieberburg I., Prusiner S.B., and Lingappa V.R. 1987b. Biogenesis and transmembrane orientation of the cellular isoform of the scrapie prion protein. *Mol. Cell. Biol.* **7**: 914–920.

Hegde R.S., Voigt S., and Lingappa V.R. 1998a. Regulation of protein topology by trans-acting factors at the endoplasmic reticulum. *Mol. Cell* **2**: 85–91.

Hegde R.S., Tremblay P., Groth D., DeArmond S.J., Prusiner S.B., and Lingappa V.R. 1999. Transmissible and genetic prion diseases share a common pathway of neurodegenera-tion. *Nature* **402**: 822–826.

Hegde R.S., Mastrianni J.A., Scott M.R., DeFea K.A., Tremblay P., Torchia M., DeArmond S.J., Prusiner S.B., and Lingappa V.R. 1998b. A transmembrane form of the prion pro-tein in neurodegenerative disease. *Science* **279**: 827–834.

Heppner F.L., Musahl C., Arrighi I., Klein M.A., Rülicke T., Oesch B., Zinkernagel R.M., Kalinke U., and Aguzzi A. 2001. Prevention of scrapie pathogenesis by transgenic expression of anti-prion protein antibodies. *Science* **294**: 178–182.

Herms J.W. and Kretzschmar H. 2001. Die Funktion des zellularen Prion-Proteins PrP^C als kupferbindendes Protein an der Synapse. In *Prionen und Prionenkrankheiten* (ed. B. Hornlimann et al.), pp. 74–80.

Herms J.W., Tings T., Dunker S., and Kretzschmar H.A. 2001. Prion protein affects Ca2+-activated K+ currents in cerebellar Purkinje cells. *Neurobiol. Dis.* **8**: 324–330.

Herms J.W., Tings T., Gall S., Madlung A., Giese A., Siebert H., Schurmann P., Windl O., Brose N., and Kretzschmar H. 1999. Evidence of presynaptic location and function of the prion protein. *J. Neurosci.* **19:** 8866–8875.

Hill A.F., Zeidler M., Ironside J., and Collinge J. 1997a. Diagnosis of new variant Creutzfeldt-Jakob disease by tonsil biopsy. *Lancet* **349:** 99–100.

Hill A.F., Joiner S., Linehan J., Desbruslais M., Lantos P.L., and Collinge J. 2000. Species-barrier-independent prion replication in apparently resistant species. *Proc. Natl. Acad. Sci.* **97:** 10248–10253.

Hill A.F., Desbruslais M., Joiner S., Sidle K.C.L., Gowland I., Collinge J., Doey L.J., and Lantos P. 1997b. The same prion strain causes vCJD and BSE. *Nature* **389:** 448–450.

Hope J., Multhaup G., Reekie L.J., Kimberlin R.H., and Beyreuther K. 1988. Molecular pathology of scrapie-associated fibril protein (PrP) in mouse brain affected by the ME7 strain of scrapie. *Eur. J. Biochem.* **172:** 271–277.

Horejsi V., Drbal K., Cebecauer M., Cerny J., Brdicka T., Angelisova P., and Stockinger H. 1999. GPI-microdomains: A role in signalling via immunoreceptors. *Immunol. Today* **20:** 356–361.

Horiuchi M. and Caughey B. 1999. Specific binding of normal prion protein to the scrapie form via a localized domain initiates its conversion to the protease-resistant state. *EMBO J.* **18:** 3193–3203.

Hornshaw M.P., McDermott J.R., Candy J.M., and Lakey J.H. 1995. Copper binding to the N-terminal tandem repeat region of mammalian and avian prion protein: Structural studies using synthetic peptides. *Biochem. Biophys. Res. Commun.* **214:** 993–999.

Hsiao K.K., Scott M., Foster D., Groth D.F., DeArmond S.J., and Prusiner S.B. 1990. Spontaneous neurodegeneration in transgenic mice with mutant prion protein. *Science* **250:** 1587–1590.

Hsiao K., Baker H.F., Crow T.J., Poulter M., Owen F., Terwilliger J.D., Westaway D., Ott J., and Prusiner S.B. 1989. Linkage of a prion protein missense variant to Gerstmann-Sträussler syndrome. *Nature* **338:** 342–345.

Hsiao K.K., Groth D., Scott M., Yang S.L., Serban H., Rapp D., Foster D., Torchia M., DeArmond S.J., and Prusiner S.B. 1994. Serial transmission in rodents of neurodegeneration from transgenic mice expressing mutant prion protein. *Proc. Natl. Acad. Sci.* **91:** 9126–9130.

Hsiao K., Meiner Z., Kahana E., Cass C., Kahana I., Avrahami D., Scarlato G., Abramsky O., Prusiner S.B., and Gabizon R. 1991. Mutation of the prion protein in Libyan Jews with Creutzfeldt-Jakob disease. *N. Engl. J. Med.* **324:** 1091–1097.

Huang F.P., Farquhar C.F., Mabbott N.A., Bruce M.E., and MacPherson G.G. 2002. Migrating intestinal dendritic cells transport PrP(Sc) from the gut. *J. Gen. Virol.* **83:** 267–271.

Huber R., Deboer T., and Tobler I. 1999. Prion protein: A role in sleep regulation? *J. Sleep Res.* **8:** 30–36.

———. 2002. Sleep deprivation in prion protein deficient mice and control mice: Genotype dependent regional rebound. *Neuroreport* **13:** 1–4.

Hundt C., Peyrin J.M., Haik S., Gauczynski S., Leucht C., Rieger R., Riley M.L., Deslys J.P., Dormont D., Lasmézas C.I., and Weiss S. 2001. Identification of interaction domains of the prion protein with its 37-kDa/67-kDa laminin receptor. *EMBO J.* **20:** 5876–5886.

Hunter N. 1996. Genotyping and susceptibility of sheep to scrapie. In *Prion diseases* (ed. H.F. Baker et al.), pp. 211–221. Humana Press Totowa, New Jersey.

Hunter N., Hope J., McConnell I., and Dickinson A.G. 1987. Linkage of the scrapie-associated fibril protein (PrP) gene and Sinc using congenic mice and restriction fragment length polymorphism analysis. *J. Gen. Virol.* **68:** 2711–2716.

Hunter N., Moore L., Hosie B.D., Dingwall W.S., and Greig A. 1997. Association between natural scrapie and PrP genotype in a flock of Suffolk sheep in Scotland. *Vet. Rec.* **140:** 59–63.

Hunter N., Dann J.C., Bennett A.D., Somerville R.A., McConnell I., and Hope J. 1992. Are Sinc and the PrP gene congruent? Evidence from PrP gene analysis in Sinc congenic mice. *J. Gen. Virol.* **73:** 2751–2755.

Isenmann S., Brandner S., and Aguzzi A. 1996. Neuroectodermal grafting: A new tool for the study of neurodegenerative diseases. *Histol. Histopathol.* **11:** 1063–1073.

Jackson G.S., Murray I., Hosszu L.L., Gibbs N., Waltho J.P., Clarke A.R., and Collinge J. 2001. Location and properties of metal-binding sites on the human prion protein. *Proc. Natl. Acad. Sci.* **98:** 8531–8535.

Jeffrey M., Goodsir C.M., Bruce M.E., McBride P.A., Fowler N., and Scott J.R. 1994. Murine scrapie-infected neurons in vivo release excess prion protein into the extracellular space. *Neurosci. Lett.* **174:** 39–42.

Jeffrey M., Ryder S., Martin S., Hawkins S.A., Terry L., Berthelin-Baker C., and Bellworthy S.J. 2001. Oral inoculation of sheep with the agent of bovine spongiform encephalopathy (BSE). 1. Onset and distribution of disease-specific PrP accumulation in brain and viscera. *J. Comp. Pathol.* **124:** 280–289.

Kaneko K., Vey M., Scott M., Pilkuhn S., Cohen F.E., and Prusiner S.B. 1997a. COOH-terminal sequence of the cellular prion protein directs subcellular trafficking and controls conversion into the scrapie isoform. *Proc. Natl. Acad. Sci.* **94:** 2333–2338.

Kaneko K., Zulianello L., Scott M., Cooper C.M., Wallace A.C., James T.L., Cohen F.E., and Prusiner S.B. 1997b. Evidence for protein X binding to a discontinuous epitope on the cellular prion protein during scrapie prion propagation. *Proc. Natl. Acad. Sci.* **94:** 10069–10074.

Kaneko K., Ball H.L., Wille H., Zhang H., Groth D., Torchia M., Tremblay P., Safar J., Prusiner S.B., DeArmond S.J., Baldwin M.A., and Cohen F.E. 2000. A synthetic peptide initiates Gerstmann-Sträussler-Scheinker (GSS) disease in transgenic mice. *J. Mol. Biol.* **295:** 997–1007.

Keshet G.I., Ovadia H., Taraboulos A., and Gabizon R. 1999. Scrapie-infected mice and PrP knockout mice share abnormal localization and activity of neuronal nitric oxide synthase. *J. Neurochem.* **72:** 1224–1231.

Kimberlin R.H. and Walker C.A. 1979. Pathogenesis of mouse scrapie: Dynamics of agent replication in spleen, spinal cord and brain after infection by different routes. *J. Comp. Pathol.* **89:** 551–562.

———. 1980. Pathogenesis of mouse scrapie: Evidence for neural spread of infection to the CNS. *J. Gen. Virol.* **51:** 183–187.

———. 1986. Pathogenesis of scrapie (strain 263K) in hamsters infected intracerebrally, intraperitoneally or intraocularly. *J. Gen. Virol.* **67:** 255–263.

———. 1988. Pathogenesis of experimental scrapie. In *Novel infectious agents and the central nervous system* (ed. G. Bock and J. Marsh), pp. 37–62. Wiley & Sons, Chichester, United Kingdom.

———. 1989. Pathogenesis of scrapie in mice after intragastric infection. *Virus Res.* **12:** 213–220.

Kimberlin R.H., Cole S., and Walker C.A. 1987. Pathogenesis of scrapie is faster when infection is intraspinal instead of intracerebral. *Microb. Pathog.* **2:** 405–415.

Kimberlin R.H., Hall S.M., and Walker C.A. 1983. Pathogeneses of mouse scrapie: Evidence for direct neural spread of infection to the CNS after injection of sciatic nerve. *J. Neurol. Sci.* **61:** 315–325.

Kitamoto T., Iizuka R., and Tateishi J. 1993. An amber mutation of prion protein in

Gerstmann-Sträussler syndrome with mutant PrP plaques. *Biochem. Biophys. Res. Commun.* **192:** 525–531.

Kitamoto T., Muramoto T., Mohri S., Dohura K., and Tateishi J. 1991. Abnormal isoform of prion protein accumulates in follicular dendritic cells in mice with Creutzfeldt-Jakob disease. *J. Virol.* **65:** 6292–6295.

Kitamoto T., Nakamura K., Nakao K., Shibuya S., Shin R.W., Gondo Y., Katsuki M., and Tateishi J. 1996. Humanized prion protein knock-in by Cre-induced site-specific recombination in the mouse. *Biochem. Biophys. Res. Commun.* **222:** 742–747.

Klamt F., Dal-Pizzol F., Conte da Frota M.J., Walz R., Andrades M.E., da Silva E.G., Brentani R.R., Izquierdo I., and Fonseca Moreira J.C. 2001. Imbalance of antioxidant defense in mice lacking cellular prion protein. *Free Radic. Biol. Med.* **30:** 1137–1144.

Klein M.A., Kaeser P.S., Schwarz P., Weyd H., Xenarios I., Zinkernagel R.M., Carroll M.C., Verbeek J.S., Botto M., Walport M.J., Molina H., Kalinke U., Acha-Orbea H., and Aguzzi A. 2001. Complement facilitates early prion pathogenesis. *Nat. Med.* **7:** 488–492.

Korth C., Stierli B., Streit P., Moser M., Schaller O., Fischer R., Schulz-Schaeffer W., Kretzschmar H., Raeber A., Braun U., Ehrensperger F., Hornemann S., Glockshuber R., Riek R., Billeter M., Wüthrich K., and Oesch B. 1997. Prion (PrPSc)-specific epitope defined by a monoclonal antibody. *Nature* **390:** 74–77.

Krasemann S., Zerr I., Weber T., Poser S., Kretzschmar H., Hunsmann G., and Bodemer W. 1995. Prion disease associated with a novel nine octapeptide repeat insertion in the PRNP gene. *Brain Res. Mol. Brain Res.* **34:** 173–176.

Kurschner C. and Morgan J.I. 1995. The cellular prion protein (PrP) selectively binds to Bcl-2 in the yeast two-hybrid system. *Brain Res. Mol. Brain Res.* **30:** 165–168.

———. 1996. Analysis of interaction sites in homo- and heteromeric complexes containing Bcl-2 family members and the cellular prion protein. *Brain Res. Mol. Brain Res.* **37:** 249–258.

Kuwahara C., Takeuchi A.M., Nishimura T., Haraguchi K., Kubosaki A., Matsumoto Y., Saeki K., Yokoyama T., Itohara S., and Onodera T. 1999. Prions prevent neuronal cell-line death. *Nature* **400:** 225–226.

Laine J., Marc M.F., Sy M.S., and Axelrad H. 2001. Cellular and subcellular morphological localization of normal prion protein in rodent cerebellum. *Eur. J. Neurosci.* **14:** 47–56.

Lasmézas C.I., Cesbron J.Y., Deslys J.P., Demaimay R., Adjou K.T., Rioux R., Lemaire C., Locht C., and Dormont D. 1996. Immune system-dependent and -independent replication of the scrapie agent. *J. Virol.* **70:** 1292–1295.

Lasmézas C.I., Deslys J.P., Robain O., Jaegly A., Beringue V., Peyrin J.M., Fournier J.G., Hauw J.J., Rossier J., and Dormont D. 1997. Transmission of the BSE agent to mice in the absence of detectable abnormal prion protein. *Science* **275:** 402–405.

Lawson V.A., Priola S.A., Wehrly K., and Chesebro B. 2001. N-terminal truncation of prion protein affects both formation and conformation of abnormal protease-resistant prion protein generated in vitro. *J. Biol. Chem.* **276:** 35265–35271.

Lee K.S., Magalhaes A.C., Zanata S.M., Brentani R.R., Martins V.R., and Prado M.A. 2001. Internalization of mammalian fluorescent cellular prion protein and N-terminal deletion mutants in living cells. *J. Neurochem.* **79:** 79–87.

Lehmann S. and Harris D.A. 1997. Blockade of glycosylation promotes acquisition of scrapie-like properties by the prion protein in cultured cells. *J. Biol. Chem.* **272:** 21479–21487.

Li A., Sakaguchi S., Atarashi R., Roy B.C., Nakaoke R., Arima K., Okimura N., Kopacek J., and Shigematsu K. 2000. Identification of a novel gene encoding a PrP-like protein expressed as chimeric transcripts fused to PrP exon 1/2 in ataxic mouse line with a disrupted PrP gene. *Cell. Mol. Neurobiol.* **20:** 553–567.

Li Q., Zhang M., Han H., Rohde A., and Stamatoyannopoulos G. 2002. Evidence that DNase I hypersensitive site 5 of the human beta-globin locus control region functions as a chromosomal insulator in transgenic mice. *Nucleic Acids Res.* **30:** 2484–2491.

Liemann S. and Glockshuber R. 1999. Influence of amino acid substitutions related to inherited human prion diseases on the thermodynamic stability of the cellular prion protein. *Biochemistry* **38:** 3258–3267.

Lipp H.P., Stagliar-Bozicevic M., Fischer M., and Wolfer D.P. 1998. A 2-year longitudinal study of swimming navigation in mice devoid of the prion protein: No evidence for neurological anomalies or spatial learning impairments. *Behav. Brain Res.* **95:** 47–54.

Liu T., Li R., Wong B.-S., Liu D., Pan T., Petersen R.B., Gambetti P., and Sy M.-S. 2001. Normal cellular Prion protein is preferentially expressed on subpopulations of murine hemopoietic cells. *J. Immunol.* **166:** 3733–3742.

Lloyd S.E., Onwuazor O.N., Beck J.A., Mallinson G., Farrall M., Targonski P., Collinge J., and Fisher E.M. 2001. Identification of multiple quantitative trait loci linked to prion disease incubation period in mice. *Proc. Natl. Acad. Sci.* **98:** 6279–6283.

Lopez C.D., Yost C.S., Prusiner S.B., Myers R.M., and Lingappa V.R. 1990. Unusual topogenic sequence directs prion protein biogenesis. *Science* **248:** 226–229.

Lu K., Wang W., Xie Z., Wong B.S., Li R., Petersen R.B., Sy M.S., and Chen S.G. 2000. Expression and structural characterization of the recombinant human doppel protein. *Biochemistry* **39:** 13575–13583.

Luo J.L., Yang Q., Tong W.M., Hergenhahn M., Wang Z.Q., and Hollstein M. 2001. Knock-in mice with a chimeric human/murine p53 gene develop normally and show wild-type p53 responses to DNA damaging agents: A new biomedical research tool. *Oncogene* **20:** 320–328.

Ma J. and Lindquist S. 2001. Wild-type PrP and a mutant associated with prion disease are subject to retrograde transport and proteasome degradation. *Proc. Natl. Acad. Sci.* **98:** 14955–14960.

Ma J., Wollmann R., and Lindquist S. 2002. Neurotoxicity and neurodegeneration when PrP accumulates in the cytosol. *Science* **298:** 1781–1785.

Mabbott N.A., Brown K.L., Manson J., and Bruce M.E. 1997. T-lymphocyte activation and the cellular form of the prion protein. *Immunology* **92:** 161–165.

Mabbott N.A., Mackay F., Minns F., and Bruce M.E. 2000. Temporary inactivation of follicular dendritic cells delays neuroinvasion of scrapie. *Nat. Med.* **6:** 719–720.

Mabbott N.A., Bruce M.E., Botto M., Walport M.J., and Pepys M.B. 2001. Temporary depletion of complement component C3 or genetic deficiency of C1q significantly delays onset of scrapie. *Nat. Med.* **7:** 485–487.

Maignien T., Lasmézas C.I., Beringue V., Dormont D., and Deslys J.P. 1999. Pathogenesis of the oral route of infection of mice with scrapie and bovine spongiform encephalopathy agents. *J. Gen. Virol.* **80:** 3035–3042.

Mallucci G.R., Ratte S., Asante E.A., Linehan J., Gowland I., Jefferys J.G., and Collinge J. 2002. Post-natal knockout of prion protein alters hippocampal CA1 properties, but does not result in neurodegeneration. *EMBO J.* **21:** 202–210.

Manolakou K., Beaton J., McConnell I., Farquar C., Manson J., Hastie N.D., Bruce M., and Jackson I.J. 2001. Genetic and environmental factors modify bovine spongiform encephalopathy incubation period in mice. *Proc. Natl. Acad. Sci.* **98:** 7402–7407.

Manson J.C., Clarke A.R., McBride P.A., McConnell I., and Hope J. 1994a. PrP gene dosage determines the timing but not the final intensity or distribution of lesions in scrapie pathology. *Neurodegeneration* **3:** 331–340.

Manson J.C., Clarke A.R., Hooper M.L., Aitchison L., McConnell I., and Hope J. 1994b. 129/Ola mice carrying a null mutation in PrP that abolishes mRNA production are developmentally normal. *Mol. Neurobiol.* **8:** 121–127.

Manson J.C., Hope J., Clarke A.R., Johnston A., Black C., and MacLeod N. 1995. PrP dosage and long term potentiation. *Neurodegeneration* **4:** 113–114.

Manson J.C., West J.D., Thomson V., McBride P., Kaufman M.H., and Hope J. 1992. The prion protein gene a role in mouse embryogenesis? *Development* **115:** 117–122.

Manson J.C., Barron R., Jamieson E., Baybutt H., Tuzi N., McConnell I., Melton D., Hope J., and Bostock C. 2000. A single amino acid alteration in murine PrP dramatically alters TSE incubation time. *Arch. Virol. Suppl.* (16): 95–102.

Manson J.C., Jamieson E., Baybutt H., Tuzi N.L., Barron R., McConnell I., Somerville R., Ironside J., Will R., Sy M.S., Melton D.W., Hope J., and Bostock C. 1999. A single amino acid alteration (101L) introduced into murine PrP dramatically alters incubation time of transmissible spongiform encephalopathy. *EMBO J.* **18:** 6855–6864.

Manuelidis L., Fritch W., and Xi Y.G. 1997. Evolution of a strain of CJD that induces BSE-like plaques. *Science* **277:** 94–98.

Martins V.R., Graner E., Garcia-Abreu J., de Souza S.J., Mercadante A.F., Veiga S.S., Zanata S.M., Neto V.M., and Brentani R.R. 1997. Complementary hydropathy identifies a cellular prion protein receptor. *Nat. Med.* **3:** 1376–1382.

Mastrianni J.A., Capellari S., Telling G.C., Han D., Bosque P., Prusiner S.B., and DeArmond S.J. 2001. Inherited prion disease caused by the V210I mutation: Transmission to transgenic mice. *Neurology* **57:** 2198–2205.

McBride P.A. and Beekes M. 1999. Pathological PrP is abundant in sympathetic and sensory ganglia of hamsters fed with scrapie. *Neurosci. Lett.* **265:** 135–138.

McBride P.A., Eikelenboom P., Kraal G., Fraser H., and Bruce M.E. 1992. PrP protein is associated with follicular dendritic cells of spleens and lymph nodes in uninfected and scrapie-infected mice. *J. Pathol.* **168:** 413–418.

McKinley M.P., Bolton D.C., and Prusiner S.B. 1983. A protease-resistant protein is a structural component of the scrapie prion. *Cell* **35:** 57–62.

Miele G., Jeffrey M., Turnbull D., Manson J., and Clinton M. 2002. Ablation of cellular prion protein expression affects mitochondrial numbers and morphology. *Biochem. Biophys. Res. Commun.* **291:** 372–377.

Mo H., Moore R.C., Cohen F.E., Westaway D., Prusiner S.B., Wright P.E., and Dyson H.J. 2001. Two different neurodegenerative diseases caused by proteins with similar structures. *Proc. Natl. Acad. Sci.* **98:** 2352–2357.

Montrasio F., Frigg R., Glatzel M., Klein M.A., Mackay F., Aguzzi A., and Weissmann C. 2000. Impaired prion replication in spleens of mice lacking functional follicular dendritic cells. *Science* **288:** 1257–1259.

Montrasio F., Cozzio A., Flechsig E., Rossi D., Klein M.A., Rülicke T., Raeber A.J., Vosshenrich C.A., Proft J., Aguzzi A., and Weissmann C. 2001. B lymphocyte-restricted expression of prion protein does not enable prion replication in prion protein knockout mice. *Proc. Natl. Acad. Sci.* **98:** 4034–4037.

Moore R.C., Redhead N.J., Selfridge J., Hope J., Manson J.C., and Melton D.W. 1995. Double replacement gene targeting for the production of a series of mouse strains with different prion protein alterations. *Nat. Biotechnol.* **13:** 999–1004.

Moore R.C., Hope J., McBride P.A., McConnell I., Selfridge J., Melton D.W., and Manson J.C. 1998. Mice with gene targetted prion protein alterations show that Prnp, Sinc and Prni are congruent. *Nat. Genet.* **18:** 118–125.

Moore R.C., Mastrangelo P., Bouzamondo E., Heinrich C., Legname G., Prusiner S.B.,

Hood L., Westaway D., DeArmond S.J., and Tremblay P. 2001. Doppel-induced cerebellar degeneration in transgenic mice. *Proc. Natl. Acad. Sci.* **98:** 15288–15293.

Moore R.C., Lee I.Y., Silverman G.L., Harrison P.M., Strome R., Heinrich C., Karunaratne A., Pasternak S.H., Chishti M.A., Liang Y., Mastrangelo P., Wang K., Smit A.F., Katamine S., Carlson G.A., Cohen F.E., Prusiner S.B., Melton D.W., Tremblay P., Hood L.E., and Westaway D. 1999. Ataxia in prion protein (PrP)-deficient mice is associated with upregulation of the novel PrP-like protein doppel. *J. Mol. Biol.* **292:** 797–817.

Moser M., Colello R.J., Pott U., and Oesch B. 1995. Developmental expression of the prion protein gene in glial cells. *Neuron* **14:** 509–517.

Mouillet-Richard S., Ermonval M., Chebassier C., Laplanche J.L., Lehmann S., Launay J.M., and Kellermann O. 2000. Signal transduction through prion protein. *Science* **289:** 1925–1928.

Moya K.L., Sales N., Hassig R., Creminon C., Grassi J., and Di Giamberardino L. 2000. Immunolocalization of the cellular prion protein in normal brain. *Microsc. Res. Tech.* **50:** 58–65.

Muramoto T., DeArmond S.J., Scott M., Telling G.C., Cohen F.E., and Prusiner S.B. 1997. Heritable disorder resembling neuronal storage disease in mice expressing prion protein with deletion of an alpha-helix. *Nat. Med.* **3:** 750–755.

Naslavsky N., Stein R., Yanai A., Friedlander G., and Taraboulos A. 1997. Characterization of detergent-insoluble complexes containing the cellular prion protein and its scrapie isoform. *J. Biol. Chem.* **272:** 6324–6331.

Nguyen J., Baldwin M.A., Cohen F.E., and Prusiner S.B. 1995. Prion protein peptides induce alpha-helix to beta-sheet conformational transitions. *Biochemistry* **34:** 4186–4192.

Nishida N., Tremblay P., Sugimoto T., Shigematsu K., Shirabe S., Petromilli C., Erpel S.P., Nakaoke R., Atarashi R., Houtani T., Torchia M., Sakaguchi S., DeArmond S.J., Prusiner S.B., and Katamine S. 1999. A mouse prion protein transgene rescues mice deficient for the prion protein gene from Purkinje cell degeneration and demyelination. *Lab. Invest.* **79:** 689–967.

Oesch B., Westaway D., Wälchli M., McKinley M.P., Kent S.B., Aebersold R., Barry R.A., Tempst P., Teplow D.B., Hood L.E., Prusiner S.B., and Weissmann C. 1985. A cellular gene encodes scrapie PrP 27-30 protein. *Cell* **40:** 735–746.

Oldstone M.B., Race R., Thomas D., Lewicki H., Homann D., Smelt S., Holz A., Koni P., Lo D., Chesebro B., and Flavell R. 2002. Lymphotoxin-alpha- and lymphotoxin-beta-deficient mice differ in susceptibility to scrapie: Evidence against dendritic cell involvement in neuroinvasion. *J. Virol.* **76:** 4357–4363.

Owen F., Poulter M., Collinge J., Leach M., Lofthouse R., Crow T.J., and Harding A.E. 1992. A dementing illness associated with a novel insertion in the prion protein gene. *Brain Res. Mol. Brain Res.* **13:** 155–157.

Palmer M.S. and Collinge J. 1993. Mutations and polymorphisms in the prion protein gene. *Hum. Mutat.* **2:** 168–173.

Palmer M.S., Dryden A.J., Hughes J.T., and Collinge J. 1991. Homozygous prion protein genotype predisposes to sporadic Creutzfeldt-Jakob disease. *Nature* **352:** 340–342.

Pattison I.H. 1965. Experiments with scrapie with special reference to the nature of the agent and the pathology of the disease. In *Slow, latent and temperate virus infections* (NINDB Monogr. 2) (ed. D.C. Gajdusek et al.), pp. 249–257. U.S. Government Printing Office, Washington, D.C.

Pauly P.C. and Harris D.A. 1998. Copper stimulates endocytosis of the prion protein *J. Biol. Chem.* **273:** 33107–33110.

Perera W.S. and Hooper N.M. 2001. Ablation of the metal ion-induced endocytosis of the prion protein by disease-associated mutation of the octarepeat region. *Curr. Biol.* **11:** 519–523.

Peretz D., Williamson R.A., Kaneko K., Vergara J., Leclerc E., Schmitt-Ulms G., Mehlhorn I.R., Legname G., Wormald M.R., Rudd P.M., Dwek R.A., Burton D.R., and Prusiner S.B. 2001. Antibodies inhibit prion propagation and clear cell cultures of prion infectivity. *Nature* **412:** 739–743.

Perrier V., Kaneko K., Safar J., Vergara J., Tremblay P., DeArmond S.J., Cohen F.E., Prusiner S.B., and Wallace A.C. 2002. Dominant-negative inhibition of prion replication in transgenic mice. *Proc. Natl. Acad. Sci.* **99:** 13079–13084.

Prinz M., Montrasio F., Klein M.A., Schwarz P., Priller J., Odermatt B., Pfeffer K., and Aguzzi A. 2002. Lymph nodal prion replication and neuroinvasion in mice devoid of follicular dendritic cells. *Proc. Natl. Acad. Sci.* **99:** 919–924.

Priola S.A. and Chesebro B. 1995. A single hamster PrP amino acid blocks conversion to protease-resistant PrP in scrapie-infected mouse neuroblastoma cells. *J. Virol.* **69:** 7754–7758.

Prusiner S.B. 1982. Novel proteinaceous infectious particles cause scrapie. *Science* **216:** 136–144.

———. 1989. Scrapie prions. *Annu. Rev. Microbiol.* **43:** 345–374.

Prusiner S.B. and Scott M.R. 1997. Genetics of prions. *Annu. Rev. Genet.* **31:** 139–175.

Prusiner S.B., Cochran S.P., and Alpers M.P. 1985. Transmission of scrapie in hamsters. *J. Infect. Dis.* **152:** 971–978.

Prusiner S.B., Groth D., Serban A., Koehler R., Foster D., Torchia M., Burton D., Yang S.L., and DeArmond S.J. 1993. Ablation of the prion protein (PrP) gene in mice prevents scrapie and facilitates production of anti-PrP antibodies. *Proc. Natl. Acad. Sci.* **90:** 10608–10612.

Prusiner S.B., Scott M., Foster D., Pan K.M., Groth D., Mirenda C., Torchia M., Yang S.L., Serban D., and Carlson G.A., et al. 1990. Transgenetic studies implicate interactions between homologous PrP isoforms in scrapie prion replication. *Cell* **63:** 673–686.

Race R., Oldstone M., and Chesebro B. 2000. Entry versus blockade of brain infection following oral or intraperitoneal scrapie administration: Role of prion protein expression in peripheral nerves and spleen. *J. Virol.* **74:** 828–833.

Race R., Raines A., Raymond G.J., Caughey B., and Chesebro B. 2001. Long-term subclinical carrier state precedes scrapie replication and adaptation in a resistant species: Analogies to bovine spongiform encephalopathy and variant Creutzfeldt-Jakob disease in humans. *J. Virol.* **75:** 10106–10112.

Race R.E., Priola S.A., Bessen R.A., Ernst D., Dockter J., Rall G.F., Mucke L., Chesebro B., and Oldstone M.B. 1995. Neuron-specific expression of a hamster prion protein minigene in transgenic mice induces susceptibility to hamster scrapie agent. *Neuron* **15:** 1183–1191.

Raeber A.J., Klein M.A., Frigg R., Flechsig E., Aguzzi A., and Weissmann C. 1999a. PrP-dependent association of prions with splenic but not circulating lymphocytes of scrapie-infected mice, *EMBO J.* **18:** 2702–2706.

Raeber A., Sailer A., Fischer M., Rülicke T., Brandner S., Aguzzi A., and Weissmann C. 1996. Prion propagation in PrP null mice with ectopic PrP expression. *Experientia* **52:** A44.

Raeber A.J., Sailer A., Hegyi I., Klein M.A., Rülicke T., Fischer M., Brandner S., Aguzzi A., and Weissmann C. 1999b. Ectopic expression of prion protein (PrP) in T lymphocytes or hepatocytes of PrP knockout mice is insufficient to sustain prion replication. *Proc. Natl. Acad. Sci.* **96:** 3987–3992.

Raeber A.J., Race R.E., Brandner S., Priola S.A., Sailer A., Bessen R.A., Mucke L., Manson J., Aguzzi A., Oldstone M.B., Weissmann C., and Chesebro B. 1997. Astrocyte-specific expression of hamster prion protein (PrP) renders PrP knockout mice susceptible to hamster scrapie. *EMBO J.* **16:** 6057–6065.

Ridley R.M. and Baker H.F. 1996. Oral transmission of BSE to primates. *Lancet* **348:** 1174.

Rieger R., Edenhofer F., Lasmézas C.I., and Weiss S. 1997. The human 37-kDa laminin receptor precursor interacts with the prion protein in eukaryotic cells. *Nat. Med.* **3:** 1383–1388.

Riek R., Hornemann S., Wider G., Glockshuber R., and Wüthrich K. 1997. NMR characterization of the full-length recombinant murine prion protein, *m*PrP(23-231). *FEBS Lett.* **413:** 282-288.

Riek R., Hornemann S., Wider G., Billeter M., Glockshuber R., and Wüthrich K. 1996. NMR structure of the mouse prion protein domain PrP(121-321). *Nature* **382:** 180–182.

Rosenmann H., Talmor G., Halimi M., Yanai A., Gabizon R., and Meiner Z. 2001. Prion protein with an E200K mutation displays properties similar to those of the cellular isoform PrP(C). *J. Neurochem.* **76:** 1654–1662.

Rossi D., Cozzio A., Flechsig E., Klein M.A., Rülicke T., Aguzzi A., and Weissmann C. 2001. Onset of ataxia and Purkinje cell loss in PrP null mice inversely correlated with Dpl level in brain. *EMBO J.* **20:** 694–702.

Safar J., Wille H., Itri V., Groth D., Serban H., Torchia M., Cohen F.E., and Prusiner S.B. 1998. Eight prion strains have PrP(Sc) molecules with different conformations. *Nat. Med.* **4:** 1157–1165.

Sailer A., Büeler H., Fischer M., Aguzzi A., and Weissmann C. 1994. No propagation of prions in mice devoid of PrP. *Cell* **77:** 967–968.

Sakaguchi S., Katamine S., Shigematsu K., Nakatani A., Moriuchi R., Nishida N., Kurokawa K., Nakaoke R., Sato H., Jishage K., Kuno J., Noda T., and Miyamoto T. 1995. Accumulation of proteinase K-resistant prion protein (PrP) is restricted by the expression level of normal PrP in mice inoculated with a mouse-adapted strain of the Creutzfeldt-Jakob disease agent. *J. Virol.* **69:** 7586–7592.

Sakaguchi S., Katamine S., Nishida N., Moriuchi R., Shigematsu K., Sugimoto T., Nakatani A., Kataoka Y., Houtani T., Shirabe S., Okada H., Hasegawa S., Miyamoto T., and Noda T. 1996. Loss of cerebellar Purkinje cells in aged mice homozygous for a disrupted PrP gene. *Nature* **380:** 528–531.

Sales N., Rodolfo K., Hassig R., Faucheux B., Di Giamberardino L., and Moya K.L. 1998. Cellular prion protein localization in rodent and primate brain. *Eur. J. Neurosci.* **10:** 2464–2471.

Sales N., Hassig R., Rodolfo K., Di Giamberardino L., Traiffort E., Ruat M., Fretier P., and Moya K.L. 2002. Developmental expression of the cellular prion protein in elongating axons. *Eur. J. Neurosci.* **15:** 1163–1177.

Schmitt-Ulms G., Legname G., Baldwin M.A., Ball H.L., Bradon N., Bosque P.J., Crossin K.L., Edelman G.M., DeArmond S.J., Cohen F.E., and Prusiner S.B. 2001. Binding of neural cell adhesion molecules (N-CAMs) to the cellular prion protein. *J. Mol. Biol.* **314:** 1209–1225.

Scott J.R., Foster J.D., and Fraser H. 1993. Conjunctival instillation of scrapie in mice can produce disease. *Vet. Microbiol.* **34:** 305–309.

Scott M.R., Kohler R., Foster D., and Prusiner S.B. 1992. Chimeric prion protein expression in cultured cells and transgenic mice. *Protein Sci.* **1:** 986–997.

Scott M.R., Will R., Ironside J., Nguyen H.O., Tremblay P., DeArmond S.J., and Prusiner S.B. 1999. Compelling transgenetic evidence for transmission of bovine spongiform encephalopathy prions to humans. *Proc. Natl. Acad. Sci.* **96:** 15137–15142.

Scott M., Foster D., Mirenda C., Serban D., Coufal F., Wälchli M., Torchia M., Groth D., Carlson G., DeArmond S.J., Westaway D., and Prusiner S.B. 1989. Transgenic mice expressing hamster prion protein produce species-specific scrapie infectivity and amyloid plaques. *Cell* **59:** 847–857.

Scott M.R., Safar J., Telling G., Nguyen O., Groth D., Torchia M., Koehler R., Tremblay P., Walther D., Cohen F.E., DeArmond S.J., and Prusiner S.B. 1997. Identification of a prion protein epitope modulating transmission of bovine spongiform encephalopathy prions to transgenic mice. *Proc. Natl. Acad. Sci.* **94:** 14279–14284.

Shaked Y., Rosenmann H., Talmor G., and Gabizon R. 1999. A C-terminal-truncated PrP isoform is present in mature sperm. *J. Biol. Chem.* **274:** 32153–32158.

Shibuya S., Higuchi J., Shin R.W., Tateishi J., and Kitamoto T. 1998. Protective prion protein polymorphisms against sporadic Creutzfeldt-Jakob disease. *Lancet* **351:** 419.

Shmerling D., Hegyi I., Fischer M., Blättler T., Brandner S., Götz J., Rülicke T., Flechsig E., Cozzio A., von Mering C., Hangartner C., Aguzzi A., and Weissmann C. 1998. Expression of amino-terminally truncated PrP in the mouse leading to ataxia and specific cerebellar lesions. *Cell* **93:** 203–214.

Shyng S.L., Huber M.T., and Harris D.A. 1993. A prion protein cycles between the cell surface and an endocytic compartment in cultured neuroblastoma cells. *J. Biol. Chem.* **268:** 15922–15928.

Shyng S.L., Lehmann S., Moulder K.L., and Harris D.A. 1995. Sulfated glycans stimulate endocytosis of the cellular isoform of the prion protein, PrPC, in cultured cells. *J. Biol. Chem.* **270:** 30221–30229.

Sigurdsson E.M., Brown D.R., Daniels M., Kascsak R.J., Kascsak R., Carp R., Meeker H.C., Frangione B., and Wisniewski T. 2002. Immunization delays the onset of prion disease in mice. *Am. J. Pathol.* **161:** 13–17.

Silverman G.L., Qin K., Moore R.C., Yang Y., Mastrangelo P., Tremblay P., Prusiner S.B., Cohen F.E., and Westaway D. 2000. Doppel is an N-glycosylated, glycosylphosphatidyl-inositol-anchored protein. Expression in testis and ectopic production in the brains of *Prnp0/0* mice predisposed to Purkinje cell loss. *J. Biol. Chem.* **275:** 26834–26841.

Simonic T., Duga S., Strumbo B., Asselta R., Ceciliani F., and Ronchi S. 2000. cDNA cloning of turtle prion protein. *FEBS Lett.* **469:** 33–38.

Soukharev S., Miller J.L., and Sauer B. 1999. Segmental genomic replacement in embryonic stem cells by double lox targeting. *Nucleic Acids Res.* **27:** e21.

Spielhaupter C. and Schatzl H.M. 2001. PrPC directly interacts with proteins involved in signaling pathways. *J. Biol. Chem.* **276:** 44604–44612.

Stephenson D.A., Chiotti K., Ebeling C., Groth D., DeArmond S.J., Prusiner S.B., and Carlson G.A. 2000. Quantitative trait loci affecting prion incubation time in mice. *Genomics* **69:** 47–53.

Stöckel J., Safar J., Wallace A.C., Cohen F.E., and Prusiner S.B. 1998. Prion protein selectively binds copper(II) ions. *Biochemistry* **37:** 7185–7193.

Strumbo B., Ronchi S., Bolis L.C., and Simonic T. 2001. Molecular cloning of the cDNA coding for *Xenopus laevis* prion protein. *FEBS Lett.* **508:** 170–174.

Sulkowski E. 1992. Spontaneous conversion of PrPC to PrPSc. *FEBS Lett.* **307:** 129–130.

Supattapone S., Muramoto T., Legname G., Mehlhorn I., Cohen F.E., DeArmond S.J., Prusiner S.B., and Scott M.R. 2001a. Identification of two prion protein regions that modify scrapie incubation time. *J. Virol.* **75:** 1408–1413.

Supattapone S., Bouzamondo E., Ball H.L., Wille H., Nguyen H.O., Cohen F.E., DeArmond S.J., Prusiner S.B., and Scott M. 2001b. A protease-resistant 61-residue prion peptide causes neurodegeneration in transgenic mice. *Mol. Cell. Biol.* **21:** 2608–2616.

Supattapone S., Bosque P., Muramoto T., Wille H., Aagaard C., Peretz D., Nguyen H.O., Heinrich C., Torchia M., Safar J., Cohen F.E., DeArmond S.J., Prusiner S.B., and Scott M. 1999. Prion protein of 106 residues creates an artifical transmission barrier for prion replication in transgenic mice. *Cell* **96:** 869–878.

Suzuki T., Kurokawa T., Hashimoto H., and Sugiyama M. 2002. cDNA sequence and tissue expression of Fugu rubripes prion protein-like: A candidate for the teleost orthologue of tetrapod PrPs. *Biochem. Biophys. Res. Commun.* **294:** 912–917.

Swietnicki W., Petersen R.B., Gambetti P., and Surewicz W.K. 1998. Familial mutations and the thermodynamic stability of the recombinant human prion protein. *J. Biol. Chem.* **273:** 31048–31052.

Tagliavini F., Prelli F., Verga L., Giaccone G., Sarma R., Gorevic P., Ghetti B., Passerini F., Ghibaudi E., Forloni G., Salmona M., Bugiani O., and Frangione B. 1993. Synthetic peptides homologous to prion protein residues 106-147 form amyloid-like fibrils in vitro. *Proc. Natl. Acad. Sci.* **90:** 9678–9682.

Taraboulos A., Scott M., Semenov A., Avrahami D., Laszlo L., Prusiner S.B., and Avraham D. 1995. Cholesterol depletion and modification of COOH-terminal targeting sequence of the prion protein inhibit formation of the scrapie isoform (erratum in *J. Cell Biol.* [1995] **130:** 501). *J. Cell Biol.* **129:** 121–132.

Tateishi J., Kitamoto T., Hoque M.Z., and Furukawa H. 1996. Experimental transmission of Creutzfeldt-Jakob disease and related diseases to rodents. *Neurology* **46:** 532–537.

Tateishi J., Ohta M., Koga M., and Sato Y. 1979. Transmission of chronic spongiform encephalopathy with Kuru plaques from humans to small rodents. *Ann. Neurol.* **5:** 581–584.

Tateishi J., Sato Y., Nagara H., and Boellaard J.W. 1984. Experimental transmission of human subacute spongiform encephalopathy to small rodents. IV. Positive transmission from a typical case of Gerstmann-Sträussler-Scheinker's disease. *Acta Neuropathol.* **64:** 85–88.

Tateishi J., Brown P., Kitamoto T., Hoque Z.M., Roos R., Wollman R., Cervenakova L., and Gajdusek D.C. 1995. First experimental transmission of fatal familial insomnia. *Nature* **376:** 434–435.

Taylor D.M., McConnell I., and Fraser H. 1996. Scrapie infection can be established readily through skin scarification in immunocompetent but not immunodeficient mice. *J. Gen. Virol.* **77:** 1595–1599.

Telling G.C., Haga T., Torchia M., Tremblay P., DeArmond S.J., and Prusiner S.B. 1996a. Interactions between wild-type and mutant prion proteins modulate neurodegeneration in transgenic mice. *Genes Dev.* **10:** 1736–1750.

Telling G.C., Scott M., Mastrianni J., Gabizon R., Torchia M., Cohen F.E., DeArmond S.J., and Prusiner S.B. 1995. Prion propagation in mice expressing human and chimeric PrP transgenes implicates the interaction of cellular PrP with another protein. *Cell* **83:** 79–90.

Telling G.C., Parchi P., DeArmond S.J., Cortelli P., Montagna P., Gabizon R., Mastrianni J., Lugaresi E., Gambetti P., and Prusiner S.B. 1996b. Evidence for the conformation of the pathologic isoform of the prion protein enciphering and propagating prion diversity. *Science* **274:** 2079–2082.

Telling G.C., Scott M., Hsiao K.K., Foster D., Yang S.L., Torchia M., Sidle K.C., Collinge J., DeArmond S.J., and Prusiner S.B. 1994. Transmission of Creutzfeldt-Jakob disease from humans to transgenic mice expressing chimeric human-mouse prion protein. *Proc. Natl. Acad. Sci.* **91:** 9936–9940.

Thackray A.M., Klein M.A., Aguzzi A., and Bujdoso R. 2002. Chronic subclinical prion disease induced by low-dose inoculum. *J. Virol.* **76:** 2510–2517.

Tobler I., Deboer T., and Fischer M. 1997. Sleep and sleep regulation in normal and prion protein-deficient mice. *J. Neurosci.* **17:** 1869–1879.

Tobler I., Gaus S.E., Deboer T., Achermann P., Fischer M., Rülicke T., Moser M., Oesch B.,

McBride P.A., and Manson J.C. 1996. Altered circadian activity rhythms and sleep in mice devoid of prion protein. *Nature* **380:** 639–642.

Tremblay P., Meiner Z., Galou M., Heinrich C., Petromilli C., Lisse T., Cayetano J., Torchia M., Mobley W., Bujard H., DeArmond S.J., and Prusiner S.B. 1998. Doxycycline control of prion protein transgene expression modulates prion disease in mice. *Proc. Natl. Acad. Sci.* **95:** 12580–12585.

Tuzi N.L., Gall E., Melton D., and Manson J.C. 2002. Expression of doppel in the CNS of mice does not modulate transmissible spongiform encephalopathy disease. *J. Gen. Virol.* **83:** 705–711.

Vey M., Pilkuhn S., Wille H., Nixon R., DeArmond S.J., Smart E.J., Anderson R.G., Taraboulos A., and Prusiner S.B. 1996. Subcellular colocalization of the cellular and scrapie prion proteins in caveolae-like membranous domains. *Proc. Natl. Acad. Sci.* **93:** 14945–14949.

Vilotte J.L., Soulier S., Essalmani R., Stinnakre M.G., Vaiman D., Lepourry L., Da Silva J.C., Besnard N., Dawson M., Buschmann A., Groschup M., Petit S., Madelaine M.F., Rakatobe S., Le Dur A., Vilette D., and Laude H. 2001. Markedly increased susceptibility to natural sheep scrapie of transgenic mice expressing ovine prp. *J. Virol.* **75:** 5977–5984.

Vital C., Gray F., Vital A., Parchi P., Capellari S., Petersen R.B., Ferrer X., Jarnier D., Julien J., and Gambetti P. 1998. Prion encephalopathy with insertion of octapeptide repeats: The number of repeats determines the type of cerebellar deposits. *Neuropathol. Appl. Neurobiol.* **24:** 125–130.

Waggoner D.J., Drisaldi B., Bartnikas T.B., Casareno R.L., Prohaska J.R., Gitlin J.D., and Harris D.A. 2000. Brain copper content and cuproenzyme activity do not vary with prion protein expression level. *J. Biol. Chem.* **275:** 7455–7458.

Walz R., Amaral O.B., Rockenbach I.C., Roesler R., Izquierdo I., Cavalheiro E.A., Martins V.R., and Brentani R.R. 1999. Increased sensitivity to seizures in mice lacking cellular prion protein. *Epilepsia* **40:** 1679–1682.

Warner R.G., Hundt C., Weiss S., and Turnbull J.E. 2002. Identification of the heparan sulfate binding sites in the cellular prion protein. *J. Biol. Chem.* **277:** 18421–18430.

Weissmann C. 1989. Sheep disease in human clothing. *Nature* **338:** 298–299.

———. 1991a. Spongiform encephalopathies. The prion's progress. *Nature* **349:** 569–571.

———. 1991b. A "unified theory" of prion propagation. *Nature* **352:** 679–683.

———. 1996. PrP effects clarified. *Curr. Biol.* **6:** 1359.

Weissmann C. and Aguzzi A. 1999. Perspectives: Neurobiology. PrP's double causes trouble. *Science* **286:** 914–915.

Weissmann C., Shmerling D., Rossi D., Cozzio A., Hegyi I., Fischer M., Leimeroth R., and Flechsig E. 2001. Structure-function analysis of prion protein. *Symp. Soc. Gen. Microbiol.* **60:** 179–193.

Westaway D., Goodman P.A., Mirenda C.A., McKinley M.P., Carlson G.A., and Prusiner S.B. 1987. Distinct prion proteins in short and long scrapie incubation period mice. *Cell* **51:** 651–662.

Westaway D., DeArmond S.J., Cayetano C.J., Groth D., Foster D., Yang S.L., Torchia M., Carlson G.A., and Prusiner S.B. 1994. Degeneration of skeletal muscle, peripheral nerves, and the central nervous system in transgenic mice overexpressing wild-type prion proteins. *Cell* **76:** 117–129.

Westaway D., Mirenda C.A., Foster D., Zebarjadian Y., Scott M., Torchia M., Yang S.L., Serban H., DeArmond S.J., and Ebeling C., et al. 1991. Paradoxical shortening of scrapie incubation times by expression of prion protein transgenes derived from long incubation period mice. *Neuron* **7:** 59–68.

White A.R., Bush A.I., Beyreuther K., Masters C.L., and Cappai R. 1999. Exacerbation of copper toxicity in primary neuronal cultures depleted of cellular glutathione. *J. Neurochem.* **72**: 2092–2098.

Whittington M.A., Sidle K.C., Gowland I., Meads J., Hill A.F., Palmer M.S., Jefferys J.G., and Collinge J. 1995. Rescue of neurophysiological phenotype seen in PrP null mice by transgene encoding human prion protein. *Nat. Genet.* **9**: 197–201.

Windl O., Dempster M., Estibeiro J.P., Lathe R., de Silva R., Esmonde T., Will R., Springbett A., Campbell T.A., Sidle K.C., Palmer M.S., and Collinge J. 1996. Genetic basis of Creutzfeldt-Jakob disease in the United Kingdom: A systematic analysis of predisposing mutations and allelic variation in the PRNP gene. *Hum. Genet.* **98**: 259–264.

Wong, B.S., Pan T., Liu T., Li R., Gambetti P., and Sy M.S. 2000a. Differential contribution of superoxide dismutase activity by prion protein in vivo. *Biochem. Biophys. Res. Commun.* **273**: 136–139.

Wong B.S., Pan T., Liu T., Li R., Petersen R.B., Jones I.M., Gambetti P., Brown D.R., and Sy M.S. 2000b. Prion disease: A loss of antioxidant function? *Biochem. Biophys. Res. Commun.* **275**: 249–252.

Wong B.S., Liu T., Paisley D., Li R., Pan T., Chen S.G., Perry G., Petersen R.B., Smith M.A., Melton D.W., Gambetti P., Brown D.R., and Sy M.S. 2001. Induction of HO-1 and NOS in doppel-expressing mice devoid of PrP: Implications for doppel function. *Mol. Cell. Neurosci.* **17**: 768–775.

Wopfner F., Weidenhofer G., Schneider R., von Brunn A., Gilch S., Schwarz T.F., Werner T., and Schatzl H.M. 1999. Analysis of 27 mammalian and 9 avian PrPs reveals high conservation of flexible regions of the prion protein. *J. Mol. Biol.* **289**: 1163–1178.

Yamada G., Ogawa M., Akagi K., Miyamoto H., Nakano N., Itoh S., Miyazaki J., Nishikawa S., Yamamura K., and Taniguchi T. 1991. Specific depletion of the B-cell population induced by aberrant expression of human interferon regulatory factor 1 gene in transgenic mice. *Proc. Natl. Acad. Sci.* **88**: 532–536.

Yehiely F., Bamborough P., Da Costa M., Perry B.J., Thinakaran G., Cohen F.E., Carlson G.A., and Prusiner S.B. 1997. Identification of candidate proteins binding to prion protein (erratum in *Neurobiol. Dis.* [2002] **10**: 67–68). *Neurobiol. Dis.* **3**: 339–355.

Zanata S.M., Lopes M.H., Mercadante A.F., Hajj G.N., Chiarini L.B., Nomizo R., Freitas A.R., Cabral A.L., Lee K.S., Juliano M.A., de Oliveira E., Jachieru S.G., Burlingame A., Huang L., Linden R., Brentani R.R., and Martins V.R. 2002. Stress-inducible protein 1 is a cell surface ligand for cellular prion that triggers neuroprotection. *EMBO J.* **21**: 3307–3316.

Zanusso G., Petersen R.B., Jin T., Jing Y., Kanoush R., Ferrari S., Gambetti P., and Singh N. 1999. Proteasomal degradation and N-terminal protease resistance of the codon 145 mutant prion protein. *J. Biol. Chem.* **274**: 23396–23404.

Zulianello L., Kaneko K., Scott M., Erpel S., Han D., Cohen F.E., and Prusiner S.B. 2000. Dominant-negative inhibition of prion formation diminished by deletion mutagenesis of the prion protein. *J. Virol.* **74**: 4351–4360.

9

Transgenetic Investigations of the Species Barrier and Prion Strains

Michael Scott and David Peretz
Institute for Neurodegenerative Diseases
Department of Neurology, University of California
San Francisco, California 94143

Rosalind M. Ridley and Harry F. Baker
School of Clinical Veterinary Medicine
University of Cambridge
Cambridge CB3 0ES, United Kingdom

Stephen J. DeArmond
Institute for Neurodegenerative Diseases
Department of Pathology, University of California
San Francisco, California 94143

Stanley B. Prusiner
Institute for Neurodegenerative Diseases
Department of Neurology and Biochemistry and Biophysics
University of California
San Francisco, California 94143

THE HISTORY OF PRION STRAINS BEGINS IN 1936 with the first reported claim by Cuillé and Chelle (1939) of the successful transmission of scrapie by intraocular injection of healthy sheep with spinal cord from an afflicted sheep. This milestone event provided the first evidence that the disease was transmissible and led inevitably to a search for the infectious agent responsible. At first, the logical conclusion was that it must be a virus. Because of the extended incubation times, these diseases were often referred to as either slow virus diseases or unconventional viral diseases (Sigurdsson 1954; Gajdusek 1977, 1985). As detailed in other chapters in this volume, considerable effort was expended searching for the "scrapie virus"; yet nothing was found either with respect to a virus-like particle or a genome composed of RNA or DNA. Over the ensuing period of more

Prion Biology and Diseases, 2nd Ed. ©2004 Cold Spring Harbor Laboratory Press 0-87969-693-1/04

than 60 years, it has become clear that scrapie is only one of a growing number of diseases known to be caused by prions and that prions comprise a unique class of pathogens that multiply by an entirely novel mechanism (Prusiner 1982a). In this chapter, we focus on the mechanisms that determine susceptibility to prions, that support prion diversity, and that govern the emergence of distinct prion strains during serial transmission.

EVIDENCE AGAINST A CONVENTIONAL GENOME

Aficionados of the virus hypothesis were challenged by a large body of evidence showing that the agent had properties unlike any known virus. The infectious agent showed a remarkable resistance to treatments, such as UV irradiation or treatment with nucleases (Alper et al. 1966; Hunter 1972; Prusiner 1982b), that would normally inactivate the nucleic acid genome of a virus composed of either RNA or DNA. Following the discovery of the prion protein, PrP (Bolton et al. 1982; Prusiner et al. 1982a), came the formulation of the prion hypothesis (Prusiner 1982a). The original definition of prions was "small *proteinaceous infectious* particles that resist inactivation by procedures which modify nucleic acids." A more contemporary working definition of a prion might be "a proteinaceous infectious particle that lacks an essential nucleic acid." However, one of us (M.S.) prefers not to think about prions as infectious particles at all, since transmission of the disease is accomplished not through the action of a foreign agent, but rather through the action of an abnormal conformation of a protein, which is itself a normal component of the host.

THE PRION STRAIN PROBLEM

For many years, it has been appreciated that the etiological agent of the transmissible spongiform encephalopathies, or prion diseases, is unusual: There was no detectable immune response, fever, or other sign of infection, and the resistance to treatments that inactivate viruses was also well established. In recognition of the unusual nature of the agent responsible, a number of hypotheses were advanced, including the "unconventional virus" alluded to earlier (Gajdusek 1977) and the slightly more radical "virino" (Kimberlin 1982; Dickinson and Outram 1988), a particle composed partly of proteins encoded by the host. However, even the virino hypothesis stipulated that a nucleic acid, albeit small, must be involved. Later, it was conceded that the "genome" might not actually comprise a nucleic acid, but that a separate "informational component" distinct from PrP must still be present in the infectious particle (Bruce and Dickinson 1987; Bruce et al. 1992; Bruce 1993).

Why was it so difficult to lay to rest the notion of a nucleic acid genome, or at least an "independent self-determining informational component?" One would think that the biophysical evidence against an essential nucleic acid would have been sufficiently compelling. The concept of a "protein-only" nature for the agent was not hard to grasp and had even been entertained previously (Griffith 1967). Indeed, in the era preceding the elucidation of the structure of DNA, it was widely believed that genes must be made of protein, since only proteins seemed capable of forming structures of sufficient complexity (Watson 1968). If there is a single obstacle to the idea that prions contain no genome, let alone one composed of nucleic acid, it has to be the existence of distinct strains of the agent. Since the sequence of the PrP gene is contributed by the host, it is easy to visualize how changes in prion properties could occur during serial passage from one host to another, in response to changes in the sequence of PrP, and some aspects of strain behavior can be attributed to this mechanism (Westaway et al. 1987). However, the existence of prion strains with properties that remain distinct in hosts that share a PrP gene of identical sequence has been championed as evidence for an independently replicating informational molecule or genome (Bruce and Dickinson 1987). To accommodate multiple strains in the context of the protein-only model, PrP must be able to sustain separate informational states within the same amino acid sequence faithfully. In the absence of another molecule, this can be accomplished only by a covalent modification or through the adoption of multiple conformations (Prusiner 1991). Opponents of the prion hypothesis have used the argument that the protein-only model could never support the high degree of variation observed. However, as we argue, this level of diversity may be largely illusory, as the number of truly distinct prion strains seems entirely consistent with experimental data showing that prion strain characteristics are enciphered in multiple conformations of PrP^{Sc} (Bessen and Marsh 1994; Telling et al. 1996; Safar et al. 1998; Peretz et al. 2001, 2002).

PRION STRAIN CHARACTERISTICS

Prion strains with distinct biological properties have long been recognized since the identification of two strains of scrapie in goats, described originally as "scratching" and "drowsy" (Pattison and Millson 1961b). In this case, the difference in clinical presentation of the disease is apparent from the terminology. However, behavioral differences are only one of several criteria that can be adopted to classify strains. Others include differences in the histopathology observed in animals inoculated with different strains (Fraser and Dickinson 1973) and variations in the pattern of deposition of PrP^{Sc} (Hecker et al. 1992; DeArmond et al. 1993). Perhaps the most reliable and important characteristic for classifying prion strains has been that of incubation period (Dickinson et al. 1968).

The incubation period can be defined as the time interval between inoculation and appearance of clinical signs or death. There are several remarkable properties of the incubation period that should be noted. First, the length of the incubation period is related inversely to the titer of the inoculum (Pattison and Smith 1963; Prusiner et al. 1981). Second, the relationship between incubation period and titer can differ between distinct prion strains (Hecker et al. 1992), and the incubation period observed using a standardized inoculum (usually 1% brain homogenate) is a reliable marker for distinguishing some prion strains. Third, the incubation period in animals of one species was always found to be shorter if the donor animal was of the same species as the host animal, a phenomenon termed the "species barrier" (Pattison 1965). Typically, in the first passage from animals of one species to another, the longer incubation period was accompanied by atypical clinical signs and unusual histopathology. Once the initial passage had been accomplished, the incubation period shortened and became fixed, as did other characteristics, such as histopathology (Pattison 1965). The concept of the species barrier is of great importance in understanding both the mechanisms that underlie prion propagation and the etiology of prion diseases, and is discussed at great length below.

STRAINS OF PRIONS DERIVED FROM NATURAL SCRAPIE

Parry argued that host genes were responsible for the development of scrapie in sheep. He was convinced that natural scrapie is a genetic disease that could be eradicated by proper breeding protocols (Parry 1962, 1983). He considered its transmission by inoculation of importance primarily for laboratory studies and considered communicable infection of little consequence in nature. Other investigators viewed natural scrapie as an infectious disease and argued that host genetics only modulates susceptibility to an endemic infectious agent (Dickinson et al. 1965).

In sheep, polymorphisms that produce amino acid substitutions at codons 136, 154, and 171 of the PrP gene in sheep have been studied with respect to the occurrence of scrapie in sheep (Goldmann et al. 1990a,b; Laplanche et al. 1993; Clousard et al. 1995). Studies of natural scrapie in the US have shown that about 85% of the afflicted sheep are of the Suffolk breed. Only those Suffolk sheep homozygous for glutamine (Q) at codon

171 were found with scrapie, although healthy controls with QQ, QR, and RR genotypes were also found (Westaway et al. 1994; Hunter et al. 1997a,b). These results argue that susceptibility in Suffolk sheep is governed by the PrP codon-171 polymorphism. In Cheviot sheep, as in Suffolks, the PrP codon-171 polymorphism has a profound influence on susceptibility to scrapie, and codon 136 seems to play a less pronounced role (Goldmann et al. 1991a; Hunter et al. 1991).

DEPENDENCE OF SCRAPIE PRION STRAIN PROPERTIES ON PASSAGE HISTORY

Before we review the published history of many established scrapie strains, we first attempt to clarify the interpretation of these results, in light of our current understanding of the mechanism of prion propagation. As previously stated, much of the work on scrapie strains reported in the literature was performed with the apparent intent being to prove that the agent possessed an independent genome. Accordingly, the nomenclature used, as well as the sheer enormity of the number of passages, can become bewildering. For example, the convention has been to give new isolates alphanumeric names, such as 22A or 139A, which in large part reflect an identity that the investigators believed best explains the relationships between strains. We would argue that this can be misleading. For example, we should not assume that 22A and 22C are any more closely related in structure than 22A is to 139A. In some cases, the opposite is true; for example, the following strains, 87A, 31A, 51C, 125A, and 138A, are proposed to represent "re-isolation of the same strain" (Bruce and Dickinson 1979). Other examples abound and have been extensively reviewed previously (Ridley and Baker 1996).

Since an extensive review of the history of mouse prion isolates has already been presented (Ridley and Baker 1996), we focus on specific examples that seem to be particularly significant. Strikingly, many of the well-characterized prion strains appear to be derived from the same original pool of brains of scrapied sheep, termed SSBP/1. This was a pooled brain homogenate derived from one Cheviot and two Cheviot x Border Leicester crossbred sheep (Dickinson 1976). Passaging of SSBP/1 in sheep and then to mice resulted in the so-called "22 family" of strains (Dickinson and Fraser 1979). Transmission of SSBP/1 to goats at Compton led to isolation of the scratching and drowsy strains.

The drowsy strain in the goat line is of particular importance. Passaging of the goat drowsy source to mice led to isolation of the

"Chandler" mouse scrapie isolate (Chandler 1961), from which the strain we term RML was obtained and, through a series of passages performed over a number of years, to isolation of several other well-characterized mouse (Mo) prion strains, including 139A, which seems to be similar to RML. In addition, passaging of the drowsy goat isolate to $Prnp^{a/a}$ mice led to the 79A strain, and passage of the drowsy isolate to $Prnp^{b/b}$ animals resulted in the 87V strain (Bruce et al. 1976; Bruce and Dickinson 1985).

In contrast to the experience with the drowsy inoculum, passaging of the goat scratching isolate to mice led to re-isolation of members of the 22 strain family. To explain this, Dickinson and colleagues proposed that in reality, the drowsy goat source had become contaminated with a case of natural scrapie in one of the original goats, since he apparently believed that the passage in mice was simply isolating strains that previously exist-ed in the original SSBP/1 inoculum (Dickinson 1976). An alternative explanation is possible, and seems more likely to us on the basis of our current knowledge. The action of passaging prions from one species to another can cause the generation of new strains because of differences in the sequences of the host and donor PrP (Scott et al. 1997a). If a general picture from all of this emerges, it is that prion strains are frequently cre-ated during passage, and that in many instances in which distinct prion strains have been identified, the two strains being compared differ with respect to their passage history. In other instances, the concept of agent "mutation" was used to explain the frequent emergence of new strains (Bruce and Dickinson 1987; Kimberlin et al. 1987). Once again, we believe these results can be better explained by changes enforced upon the prion by a change in the sequence of PrP encoded by the host. It is intriguing to note that mutation alone does not seem sufficient to explain all of the vari-ation observed, prompting speculation that some of the data could best be explained if a host-encoded protein formed a functional part of the infec-tious agent (Kimberlin et al. 1987).

Because all of the above-mentioned strains may be considered to orig-inate from a single source, strains that can be traced to a completely differ-ent natural source are of particular biological interest. Passage of the spleen of a scrapied Suffolk sheep into Moredun ($Prnp^{a/a}$) mice and subsequent cloning by limiting dilution led eventually to isolation of the Me7 strain (Zlotnik and Rennie 1965; Dickinson and Meikle 1969; Dickinson et al. 1969). This same strain was reportedly obtained from several independent isolations and appears identical to 7D and 58A (Ridley and Baker 1996). It is important to emphasize that 87A, which was derived from the Cheviot source, was found to produce a strain indistinguishable from Me7 even fol-lowing passage at high dilution ("cloning") (Ridley and Baker 1996). This introduces a point we shall revisit later: Identical strains can be obtained from entirely different primary sources, consistent with the view that prion diversity is fundamentally limited in scope.

The epidemic of bovine spongiform encephalopathy (BSE) among cattle in Britain has provided a new source of mouse prion strains; transmission of BSE prions to mice yielded two new isolates, 301V and 301C (Bruce et al. 1994). As discussed below, 301V and 301C are probably identical, differing only with respect to the sequence of PrP supplied by the host.

STRAINS OF PRIONS IN RODENTS: THE *Sinc* GENE

Studies of scrapie in inbred mice demonstrated the existence of a genetic locus, *Sinc* (for *s*crapie *inc*ubation), which profoundly influenced the scrapie incubation period of mouse prion strains (Dickinson et al. 1968; Dickinson and Meikle 1969). Two alleles of the *Sinc* gene, termed s7 and p7, were found to produce short and prolonged incubation periods, respectively, upon inoculation with the Chandler scrapie isolate. In a later study using different strains of mice, a major determinant of scrapie incubation period, termed *Prni*, was found to be either congruent with or closely linked to the structural gene for PrP, designated *Prnp* (Carlson et al. 1986; Westaway et al. 1987). Subsequent molecular genetic studies showed that the *Sinc* locus was probably synonymous with the scrapie incubation period determinant *Prni* (Carlson et al. 1986; Hunter et al. 1987). Mice with different alleles of the *Sinc/Prni* gene encoded PrPs that differed at two amino acid residues; these *Prnp* alleles were designated *Prnp^a* and *Prnp^b* (Westaway et al. 1987). The *Prnp^a* allele corresponds to the s7 allele of the *Sinc* locus and encodes leucine at codon 108 and threonine at codon 189 of the MoPrP open reading frame (ORF), whereas the *Prnp^b* allele corresponds to p7 and encodes phenylalanine and valine in MoPrP at these locations, respectively. It is now widely accepted that the scrapie incubation-time genes (*Sinc* and *Prni*) are congruent with the PrP gene; hence the use of the *Sinc/Prni/Prnp* terminology is redundant. Indeed, it is often more convenient to refer to the short and long allelic products simply as PrP-A and PrP-B, respectively.

PARADOXICAL RESULTS FROM TRANSGENIC STUDIES

Although the amino acid substitutions in PrP that distinguish *Prnp^a* from *Prnp^b* mice argued for the congruency of *Prnp* and *Prni*, experiments with *Prnp^a* mice expressing *Prnp^b* transgenes demonstrated a paradoxical shortening of incubation times (Westaway et al. 1991). We had predicted that these Tg mice would exhibit a prolongation of the incubation time after inoculation with RML prions on the basis of our previous studies with ($Prnp^a$ × $Prnp^b$) F_1 mice, which exhibit long incubation times. We described those findings as paradoxical shortening because we and other investigators had believed for many years that long incubation times are dominant traits (Dickinson et al. 1968; Carlson et al. 1986). From studies of congenic and transgenic mice expressing different numbers of the *a* and *b* alleles of *Prnp*, we learned that these findings were not paradoxical; indeed, they result from increased PrP gene dosage (Carlson et al. 1994). As discussed in more detail below, increased gene dosage was found to be inversely related to the length of the incubation period in studies using transgenic mice expressing hamster PrP (Prusiner et al. 1990). When the RML strain was inoculated into congenic and transgenic mice, increasing the number of copies of the *a* allele was found to be the major determinant in reducing the incubation time; however, increasing the number of copies of the *b* allele also reduced the incubation time, but not to the same extent as seen with the *a* allele.

PRION STRAIN VARIATION AND INCUBATION PERIOD

The relationship between prion titer and incubation period is of great significance in tracing prion strain properties during passage, particularly when comparing inocula prepared by different methods. For example, inefficient routes of inoculation, such as oral dosing, can prolong the incubation period, and titers of prions in tissues other than brain may be much lower, leading to prolongation of incubation period. It is therefore crucial to be always aware that a variation in incubation period could be caused by a change in titer, a change in the strain of prion, or changes in both.

To clarify the relationship between incubation period and titer, two distinct Syrian hamster prion strains, Sc237 and 139H, were compared (Hecker et al. 1992). Sc237 and 139H exhibit incubation periods of about 75 and 165 days, respectively, in hamsters. Since previous studies had disclosed that a single LD_{50} of the Sc237 strain causes disease at about 140 days after inoculation (Marsh and Kimberlin 1975; Kimberlin and Walker 1977; Prusiner et al. 1982b), the possibility that the relatively long incubation period of 139H could arise due to a defect in 139H that prevented accumulation of high titers of infectious prions was investigated by endpoint titration (Fig. 1) (Hecker et al. 1992). Significantly, no difference was found in the titer of 10% (w/v) brain homogenates prepared from hamsters showing clinical signs of scrapie after inoculation with either strain.

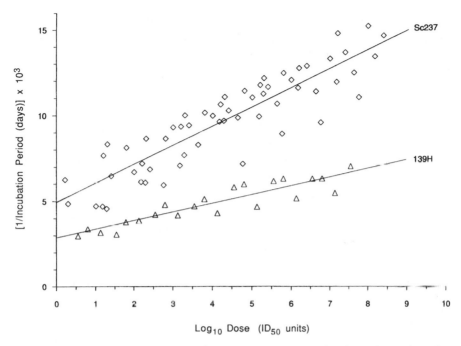

Figure 1. Scrapie incubation times plotted as a function of the dose of inoculum for two distinct prion isolates. The $\log_{10}$ dose, which is calculated from the titer × dilution, is on the x axis; the reciprocal of incubation time is on the y axis. The curves were fitted by linear regression analysis. Sc237 regression coefficient = 0.88, slope = 1.11, and y intercept = 4.88; 139H regression coefficient = 0.91, slope = 0.51, and y intercept = 2.82. (Reprinted, with permission, from Hecker et al. 1992.)

The relationship between incubation period and titer was different, however; a comparison of any two doses of 139H gave a larger change in incubation period than between corresponding doses of Sc237 (Fig. 1) (Hecker et al. 1992). Thus, it seems that distinct incubation periods can arise because of differences in the rate of accumulation of prion strains. It is important to stress that PrP gene dosage also affects the kinetics of prion propagation, as discussed below.

PRION PROTEIN INTERACTIONS AND THE SPECIES BARRIER

The molecular mechanisms that underlie the species barrier are of great importance with respect to both the mechanism of prion propagation and our understanding of the etiology of prion diseases. The latter is of particular significance in assessing the risk to the human population from the

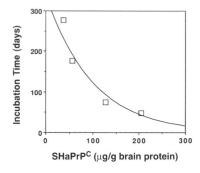

Figure 2. Plot of scrapie incubation times as a function of the SHaPrP^C steady-state levels in the four Tg(SHaPrP) lines. An exponential curve was fitted to the data points. (Reprinted, with permission, from Prusiner et al. 1990 [copyright Cell Press].)

ongoing epidemic of BSE in Britain and is discussed in depth below. Much has been learned about the mechanisms that underlie the species barrier and prion susceptibility through the use of transgenic mice expressing foreign PrP genes. When transgenic mice were constructed expressing Syrian hamster (SHa) PrP, they were found to be highly susceptible to prions that had been passaged in hamsters, in contrast to control nontransgenic mice (Scott et al. 1989; Prusiner et al. 1990). The incubation period was found to be inversely proportional to the level of expression of SHaPrP (Fig. 2) (Prusiner et al. 1990). Significantly, there was no difference in incubation period whether the prions were derived from hamsters or by prior transmission in Tg(SHaPrP) mice, establishing that no detectable change in the prion could be attributed to other host factors. Prions passaged through Tg(SHaPrP) mice also appeared unaltered when transmitted back to hamsters.

From these studies, we can conclude that the PrP gene governs the species barrier, and expression of a PrP gene from one species can be both necessary and sufficient to confer susceptibility to prions derived from the species of animal from which the PrP transgene was derived (Scott et al. 1989; Prusiner et al. 1990). Exceptions to this general rule exist, however. On the basis of observations with Tg mice discussed in the next section, it was proposed that PrP^C may need to bind specifically to a host-encoded cellular factor termed protein X as a prerequisite for conversion to PrP^{Sc} (Telling et al. 1995). If PrP and protein X are derived from distantly related species, the affinity of interaction is reduced (Telling et al. 1995). It appears that variations in the sequence of PrP that affect the binding site to protein X can prevent PrP from being efficiently converted (Kaneko et al. 1997). These observations have importance to the design of efficient transgenic mouse models for bioassays of prions from humans and ruminant sources and underscore an unfortunate weakness in the terminology coined by Pattison. Classically, the species barrier refers to the specific situation in which prions undergo a change during passage from one species to another. It is clear from the aforementioned transgenic mouse studies, as well as from extensive DNA sequence analysis of the PrP genes of numerous species (Gabriel et al. 1992; Schätzl et al. 1995), that this change is attributable to a corresponding difference in the sequence of PrP of the donor and host animals. In other cases, however, the host species may be simply incapable of supporting prion propagation because it expresses a PrP molecule that acts in a dominant-negative fashion, preventing prion propagation by sequestering protein X (Kaneko et al. 1997). Certain breeds of sheep, generated through selective breeding to be resistant to scrapie, appear to derive this resistance due to a polymorphism (Q171R) affecting the binding site to protein X (Kaneko et al. 1997). A similar mechanism may explain the absence of Creutzfeldt-Jakob disease (CJD) in Japanese individuals who carry the E219K polymorphism (Kaneko et al. 1997). It is important to stress that this type of resistance to foreign prions does not arise through the same mechanism as that observed during most interspecies transmissions. However, it may be of great practical significance to the production of farm animals resistant to prions, as well as to the development of novel therapies for patients suffering from prion disease or bearing a prion mutation causing familial prion disease.

It has become popular to equate the use of the term species barrier with the relative lack of susceptibility of some host species to prions derived from another species. A more complex picture is emerging, however. As discussed below, the species barrier originally described by Pattison may sometimes reflect a modulation from one strain into another, enforced by a change in the primary structure of PrP (Scott et al. 1997a). In other cases, it may arise due to differences in the sequence of PrP that prevent interaction of PrP^C encoded by the host and PrP^{Sc} in the inoculum (Prusiner et al. 1990; Scott et al. 1993).

CHIMERIC PrP TRANSGENES AND PROTEIN X

Attempts to abrogate the prion species barrier between humans and mice by using an approach similar to that described for the abrogation of the species barrier between Syrian hamsters and mice were initially unsuccessful. Mice expressing human (Hu)PrP transgenes did not develop signs of CNS dysfunction more rapidly or frequently when inoculated with human prions than non-Tg controls (Telling et al. 1994). Like the Tg(SHaPrP) mice that preceded them, these mice were created prior to the availability of mouse strains that lacked endogenous MoPrP (Büeler et al. 1992). However, in studies with Tg(SHaPrP) mice, although SHaPrP^C

Table 1. Incubation times for chimeric prions passaged in Tg(MH2M PrP)92 mice or Syrian hamsters

		Scrapie incubation times			
		illness		death	
Inoculum	Host	(n/n_0)	(days ± s.e.)	(n/n_0)[a]	(days ± s.e.)
SHa(Sc237)	Tg(MH2M PrP)92	34/34	134 ± 3.6	26/26	142 ± 4.4
SHa(Sc237)→MH2M	Tg(MH2M PrP)92	22/22	73 ± 0.7	12/12	85 ± 1.2
SHa(Sc237)→MH2M→ MH2M	Tg(MH2M PrP)92	10/10	64 ± 1.9	8/8	76 ± 2.9
SHa(Sc237)	Syrian hamsters	48/48	77 ± 1.1[b]	48/48	89 ± 1.7
SHa(Sc237)→MH2M	Syrian hamsters	23/23	116 ± 1.9	10/10	136 ± 3.5
SHa(Sc237)→MH2M→ MH2M	Syrian hamsters	6/6	161 ± 3.8	4/4	187 ± 2.4
SHa(Sc237)→MH2M→ SHa	Syrian hamsters	8/8	77 ± 0.6	4/4	85 ± 1.2

Reprinted, with permission, from Scott et al. (1993) (copyright Cell Press).

[a]n, Number of ill animals; n_0, number of inoculated animals. The reduced number of animals in the death column reflects sacrifice of some animals for immunoblotting and neuropathology.

[b]Scott et al. (1989).

appeared to compete with MoPrP^C for binding to SHaPrP^{Sc}, the effect of this competition was a slight prolongation of incubation period (Prusiner et al. 1990). Viewed in this context, the complete resistance of Tg(HuPrP) mice to Hu prions was initially puzzling.

The successful breaking of the species barrier between humans and mice has its origins in a set of studies with Tg mice expressing chimeric PrP genes derived from SHa and Mo PrP genes (Scott et al. 1992). One SHa/Mo PrP gene, designated MH2M PrP, contains five amino acid substitutions encoded by SHaPrP, whereas another construct designated MHM2 PrP has two Sha substitutions. Tg(MH2M PrP) mice were susceptible to both SHa and Mo prions, whereas three lines expressing MHM2 PrP were resistant to SHa prions (Table 1) (Scott et al. 1993). The brains of Tg(MH2M PrP) mice with prion disease contained chimeric PrP^{Sc} and prions with an artificial host range favoring propagation in mice that express the corresponding chimeric PrP but were also transmissible, albeit at reduced efficiency, to non-Tg mice and hamsters. These findings provided additional genetic evidence for homophilic interactions between PrP^{Sc} in the inoculum and PrP^C synthesized by the host (Scott et al. 1993).

With the recognition that Tg(HuPrP) mice were not suitable recipients for the transmission of Hu prions, we constructed Tg(MHu2M) mice analogous to the Tg(MH2M) mice described above. HuPrP differs from MoPrP at 28 of 254 positions (Kretzschmar et al. 1986), whereas chimeric

MHu2M PrP differs from MoPrP at 9 residues. The mice expressing the MHu2M transgene are susceptible to human prions and exhibit abbreviated incubation times of about 200 days (Telling et al. 1994). In these initial studies, the chimeric MHu2M transgene encoded a methionine at codon 129, and all of the patients from whom inocula were derived were homozygous for methionine at this residue.

From Tg(SHaPrP) mouse studies, prion propagation is thought to involve the formation of a complex between PrP^{Sc} and the homotypic substrate PrP^{C} (Prusiner et al. 1990). Propagation of prions may require the participation of other proteins, such as chaperones, which might be involved in catalyzing the conformational changes that feature in the formation of PrP^{Sc} (Pan et al. 1993). Notably, efficient transmission of CJD prions to Tg(HuPrP)/$Prnp^{0/0}$ mice was obtained when the endogenous MoPrP gene was inactivated, suggesting that $MoPrP^{C}$ competes with $HuPrP^{C}$ for binding to a cellular component (Telling et al. 1995). In contrast, the sensitivity of Tg(MHu2M) mice to CJD prions was not affected by the expression of $MoPrP^{C}$. One explanation for the difference in susceptibilities of Tg(MHu2M) and Tg(HuPrP) mice to Hu prions may be that mouse chaperones catalyzing the refolding of PrP^{C} into PrP^{Sc} can readily interact with the $MHu2MPrP^{C}$/$HuPrP^{CJD}$ complex but not with $HuPrP^{C}$/$HuPrP^{CJD}$.

To accomplish the conversion of PrP^{C} into PrP^{Sc}, participation of one or more molecular chaperones may be required. The identification of protein X is an important avenue of research, since isolation of this protein or complex of proteins would presumably facilitate studies of PrP^{Sc} formation. Scrapie infected cells in culture display marked differences in the induction of heat-shock proteins (Tatzelt et al. 1995, 1996), and Hsp70 mRNA has been reported to increase during prion disease in mice (Kenward et al. 1994). By two-hybrid analysis in yeast, PrP has been shown to interact with Bcl-2 and Hsp60 (Edenhofer et al. 1996; Kurschner and Morgan 1996). Weiss and coworkers have used a similar approach to show that PrP binds the laminin receptor protein (Rieger et al. 1997). Although these studies are suggestive, no direct identification of a molecular chaperone involved in prion formation in mammalian cells has been accomplished. Many other proteins that bind to PrP have been identified but none has been shown to participate in prion formation (Oesch et al. 1990; Martins et al. 1997; Yehiely et al. 1997; Graner et al. 2000; Keshet et al. 2000; Mouillet-Richard et al. 2000; Gauczynski et al. 2001; Hundt et al. 2001; Schmitt-Ulms et al. 2001). Whether or not identification of protein X will require isolation of a ternary complex composed of PrP^{C}, PrP^{Sc}, and protein X remains to be determined.

SELECTIVE NEURONAL TARGETING OF PRION STRAINS?

In addition to incubation times, neuropathologic profiles of spongiform change have been used to characterize prion strains (Fraser and Dickinson 1968). Although this is still an essential parameter in categorizing prion strains, the development of a new procedure for in situ detection of PrP^{Sc}, designated histoblotting (Taraboulos et al. 1992), made it possible to localize and quantify PrP^{Sc} as well. This technique makes it possible to determine whether or not distinct strains produce different, reproducible patterns of PrP^{Sc} accumulation; the patterns of PrP^{Sc} accumulation were found to be different for each prion strain if the genotype of the host was held constant (Hecker et al. 1992; DeArmond et al. 1993). This finding was in accord with earlier studies showing that the distribution of spongiform degeneration is strain-specific (Fraser and Dickinson 1973), since PrP^{Sc} accumulation precedes vacuolation and reactive gliosis (DeArmond et al. 1987; Jendroska et al. 1991; Hecker et al. 1992).

Although studies with both mice and Syrian hamsters established that each strain has a specific signature as defined by a specific pattern of PrP^{Sc} accumulation in the brain (Hecker et al. 1992; DeArmond et al. 1993; Carlson et al. 1994), any such comparison must be done on an isogenic background (Scott et al. 1993; Hsiao et al. 1994). When a single strain is inoculated into mice expressing different PrP genes, variations in the patterns of PrP^{Sc} accumulation were found to be at least equal to those seen between two strains (DeArmond et al. 1997). On the basis of the initial studies, which were performed in animals of a single genotype, we suggested that PrP^{Sc} synthesis occurs in specific populations of cells for a given distinct prion isolate (Prusiner 1989; Hecker et al. 1992). More recent data using transgenic mice supplied another striking example of the localized accumulation of prion strains. When fatal familial insomnia (FFI) prions were inoculated into Tg(MHu2M) mice, PrP^{Sc} was confined largely to the thalamus, as is the case for FFI in humans (Medori et al. 1992b; Telling et al. 1996). In contrast, fCJD(E200K) prions inoculated into Tg(MHu2M) mice produced widespread deposition of PrP^{Sc} throughout the cortical mantle and many of the deep structures of the CNS, as is seen in fCJD(E200K) of humans (Telling et al. 1996). Thus, the distinct pattern of PrP^{Sc} accumulation exhibited by each strain can be faithfully reproduced in another species.

Although these data appear to support the hypothesis of targeted propagation of distinct prion isolates, other studies with PrP transgenes argue that such profiles do not depend entirely on the strain (Carp et al. 1997; DeArmond et al. 1997). To examine whether the diverse patterns of PrP^{Sc} deposition are influenced by asparagine-linked glycosylation of PrP^{C}, we constructed Tg mice expressing PrPs mutated at one or both of the asparagine-linked glycosylation consensus sites (DeArmond et al. 1997). These mutations resulted in aberrant neuroanatomic topologies of PrP^C within the CNS, whereas pathologic point mutations adjacent to the consensus sites did not alter the distribution of PrP^C. Tg mice with mutation of the second PrP glycosylation site exhibited prion incubation times of more than 500 days and unusual patterns of PrP^{Sc} deposition. These findings raise the possibility that glycosylation can modify the conformation of PrP and affect either the turnover of PrP^C or the clearance of PrP^{Sc}. Regional differences in the rate of deposition or clearance would result in specific patterns of PrP^{Sc} accumulation. Because a single prion strain produced many different patterns when inoculated into mice expressing various PrP transgenes, we concluded that the pattern of PrP^{Sc} deposition is a manifestation of the particular strain but not an essential feature for successful propagation (DeArmond et al. 1997). The results of studies of three prion strains prepared from three brain regions and spleens of inbred mice support this contention (Carp et al. 1997).

ISOLATION OF NEW HAMSTER PRION STRAINS USING MICE AND TRANSGENIC MICE

The successful passaging of scrapie to Syrian hamsters led to the discovery and purification of the prion protein (Bolton et al. 1982; Prusiner et al. 1982a). Hamsters are unrivaled as a source for purified prions. In addition to Syrian hamsters, prions have been passaged to two other species of hamsters, Armenian and Chinese hamsters. These three distinct species carry different PrP alleles and exhibit distinctive characteristic incubation periods when inoculated with prions passaged in either of the other two species (Lowenstein et al. 1990). Initial transmission of rat-adapted 139A strain to Syrian hamsters reportedly yielded two distinct hamster strains, termed 302K and 431K (Kimberlin and Walker 1978), leading the authors to suggest that the original inoculum must have been a mixture of the two strains. Following serial transmission in hamsters for several passages, however, a new strain, termed 263K, emerged that was no longer transmissible to mice (Kimberlin and Walker 1978). The 263K strain was derived from the same source as our standard hamster strain Sc237 (Marsh and Kimberlin 1975; Kimberlin and Walker 1977, 1978; Kimberlin et al. 1987). Passaging of the 139A mouse strain (indistinguishable from RML) to hamsters led to isolation of 139H, and serial transmission of Me7 to Syrian hamsters led to the stable isolate Me7-H (Kimberlin et al. 1987).

Although a few mouse-passaged scrapie prion strains had been successfully transmitted directly to hamsters, as described above, a new approach was developed for increasing the number of distinct prion isolates available in the Syrian hamster genetic background. Transgenic mice expressing the MH2M gene are susceptible to Mo prion strains (Scott et al. 1993). Following a single passage in these Tg(MH2M) mice, the chimeric prions are able to infect hamsters efficiently (Scott et al. 1993). In the course of these studies, evidence was obtained showing that strain variation could be created by altering the sequence of PrP encoded by the host animal, even using prion strains that had been previously cloned by limiting dilution in mice (Dickinson et al. 1969). When the "cloned" Me7 strain was transmitted to hamsters, a stable strain with an extremely long incubation period, termed Me7-H, was obtained (Kimberlin et al. 1987). It is important to note that Me7-H could also be transmitted back to mice, and that the resultant mouse isolate was restored to a state indistinguishable from the Me7 strain (Kimberlin et al. 1989). However, a single passage of Me7 through Tg(MH2M) mice followed by transmission to hamsters led to formation of a new strain of hamster-passaged Me7, which we termed SHa(Me7). In contrast to Me7-H, SHa(Me7) produced short incubation periods and a noticeably distinct pattern of neuropathology (Scott et al. 1997a). In each passage, 100% of the inoculated animals exhibited clinical signs with minimal spread in the time of first onset of symptoms, making it extremely unlikely that selection for a rare variant occurred. Thus, it seems that the change in the strain was caused entirely by a change in the primary structure of PrP. In the same study, strains once thought to be distinct, which had originally been isolated from different breeds of scrapied sheep, were shown to have indistinguishable properties. Thus, prion strain properties depend on passage history with respect to the sequence of PrP, not on a separate, distinct informational component (Scott et al. 1997a), and both the generation and propagation of prion strains seem to depend on the structure of PrP^{Sc} and PrP^{C}.

HAMSTER PRION STRAINS DERIVED FROM TRANSMISSIBLE MINK ENCEPHALOPATHY AND SCRAPIE EXHIBIT BIOCHEMICAL DIFFERENCES

A remarkable series of discoveries followed the successful transmission of transmissible mink encephalopathy (TME) to hamsters. An epidemic of TME in farmed mink in North America occurred simultaneously at five farms sharing the same feed source, strongly suggesting a food-borne route of infection (Marsh 1992). Interestingly, although BSE prions have been successfully transmitted to mink, the resulting disease does not resemble TME (Robinson et al. 1994). Attempts to transmit TME prions to mice were unsuccessful (Ridley and Baker 1996); however, the disease could be transmitted to Chinese hamsters (Kimberlin et al. 1986) and subsequently to Syrian hamsters (Bessen and Marsh 1992b). Two distinct strains were obtained, termed "hyper" (HY) and "drowsy" (DY), creating a remarkable parallel to the initial passage of sheep scrapie to goats years earlier, which had produced two distinct strains, termed "scratching" and "drowsy" (Pattison and Millson 1961a). The drowsy syndrome in hamsters bears a striking resemblance to that of its namesake in goats (Bessen and Marsh 1992a,b).

In contrast to the HY isolate, the DY strain was found to differ significantly from other known hamster prion strains in its biochemical and physical properties. Marked differences were identified by sedimentation analysis, protease sensitivity, and the migration pattern of PrP^{Sc} proteolytic fragments on SDS gels (Bessen and Marsh 1992b). PrP^{Sc} produced by the DY prions showed diminished resistance to proteinase K digestion and truncation of the amino terminus compared to HY and many other strains (Fig. 3) (Bessen and Marsh 1994). Because both HY and DY were prepared by passaging in outbred Syrian hamsters, the markedly different properties of the two isolates could not be attributed to differences in the sequence of PrP. For the first time, there was evidence that different strains might represent different conformers of PrP^{Sc} (Prusiner 1991; Bessen and Marsh 1992b). An alternative explanation, that the distinct properties of HY and DY were caused by differences in posttranslational modification, was also considered (Bessen and Marsh 1992a). As far as we are aware, the remarkable properties of DY remain unique among hamster prion strains; when several distinct hamster prion isolates were studied in an attempt to identify a similar change in the size of the partially protease-resistant fragment, all except the DY isolate produced a fragment of the same size as the HY strain (Scott et al. 1997a).

The unusual properties of DY were also valuable in attempts to replicate prions in cell-free systems. The altered sensitivity to protease displayed by the DY strain in vivo was mimicked in vitro when partially denatured, radiolabeled PrP^{C} was mixed with an excess of PrP^{Sc} (Bessen et al. 1995). Although, to our knowledge, it has not yet been possible to demonstrate the propagation of infectious prions using this system, the faithful reproduction of this evidence of a conformational change is suggestive of a highly specific interaction.

Recently, a new method for comparing prion strains by assessment of their relative conformational stability was developed (Peretz et al. 2001). The HY and DY strains were two among eight Syrian hamster prion isolates compared using this assay (Peretz et al. 2001), with the surprising conclusion that the eight strains were subdivided into four distinct groups

A.

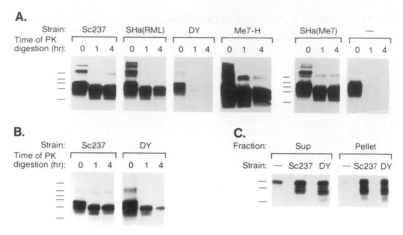

B.

C.

Figure 3. Comparison of protease sensitivity and protease-resistant PrP of different prion strains. (*A*) Brain homogenates (1 mg ml^{-1} total protein) of Syrian hamsters infected with several SHa prion strains were treated with 100 µg ml^{-1} proteinase K for 0, 1, or 4 hours, and analyzed by western blotting, using the α-PrP 3F4 mAb, which recognizes SHaPrP (Kascsak et al. 1987). The inocula used in this experiment were Sc237, SHa(RML), DY, Me7-H, SHa(Me7) prions, and uninoculated control. (*B*) Brain microsomal membrane fractions (1 mg ml^{-1} total protein) of Syrian hamsters infected with Sc237 or DY were treated with 100 µg ml^{-1} proteinase K for 0, 1, or 4 hours, and analyzed by western blotting, using the α-PrP 3F4 mAb. (*C*) Brains from normal or scrapie-ill hamsters infected with either the Sc237 or DY inoculum were solubilized in cold buffer A (1% Triton X-100 and 1% sodium deoxycholate in phosphate-buffered saline) incubated on ice and then centrifuged at 15,000g for 20 minutes at 4°C. The supernatant was removed, and then an equal percentage of the supernatant (Sup) and the pellet (Pellet) fractions were analyzed by western blotting using the 3F4 anti-PrP antibody. Molecular-size markers are indicated on the left of the panels and, in descending order, represent 84, 53, 35, 29, and 21 kD (panels *A,B*) or 35, 29, and 21 kD (panel *C*). (Reprinted, with permission, from Scott et al. 1997a [copyright American Society for Microbiology].)

based on their conformational characteristics. Remarkably, the groupings established by assessment of conformational stability (Peretz et al. 2001) correlated perfectly with similarities of prion strain properties established in a previous study (Scott et al. 1997a); prion strains with similar conformational characterisitics also exhibited similar strain properties. Strains once thought to be distinct, which had originally been isolated from different breeds of scrapied sheep or derived from mink with TME, were shown to have virtually indistinguishable biochemical, clinical, and neuropathological properties (Peretz et al. 2001). Together, these observations further illustrate how, for any given sequence of PrP, prion strain diversity may be a very limited resource.

DE NOVO GENERATION OF HUMAN PRION
STRAINS FROM MUTANT PrP GENES

In humans, a mutation of the PrP gene at codon 178 that results in the substitution of asparagine for aspartic acid causes FFI, provided the polymorphic codon 129 encodes methionine (Goldfarb et al. 1992; Medori et al. 1992a). In this disease, adults generally over age 50 present with a progressive sleep disorder and die within about a year of onset (Lugaresi et al. 1986). In their brains, deposition of PrP^{Sc} is confined largely within the anteroventral and the dorsal medial nuclei of the thalamus. In contrast, the same D178N mutation with a valine encoded at position 129 produces fCJD, in which the patients present with dementia, and widespread deposition of PrP^{Sc} is found postmortem (Goldfarb et al. 1991). The first family to be recognized with CJD was found to carry the D178N mutation (Meggendorfer 1930; Kretzschmar et al. 1995).

When the PrP^{Sc} molecules produced in the two prion diseases with the D178N mutations were examined by western immunoblotting after limited digestion with proteinase K and deglycosylation with PNGase F, the PrP^{Sc}(D178N,M129) of FFI exhibited an apparent molecular weight (M_r) of 19 kD, and PrP^{Sc}(D178N,V129) of fCJD had an M_r of 21 kD (Monari et al. 1994). This difference in molecular size was shown to be due to different sites of proteolytic cleavage at the amino termini of the two PrP^{Sc} molecules and interpreted as reflecting different tertiary structures (Monari et al. 1994). These distinct conformations were not entirely unexpected, since the amino acid sequences of the PrPs differ. Extracts from the brains of FFI patients transmitted disease to mice expressing a chimeric MHu2M PrP gene about 200 days after inoculation and induced formation of the 19-kD PrP^{Sc}; fCJD(E200K) and sCJD produced the 21-kD PrP^{Sc} in these mice (see Chapter 1, Table 4) (Telling et al. 1996). These experimental data demonstrated that $MHu2MPrP^{Sc}$ can exist in two different conformations based on the sizes of the protease-resistant fragments; yet the amino acid sequence of $MHu2MPrP^{Sc}$ was invariant. These findings, together with those of the DY strain discussed earlier, argue that propagation of prion strains results in the faithful copying of specific conformations of PrP from a "template" molecule of PrP^{Sc} into a "substrate" molecule that we have termed PrP^*, which may itself derive from PrP^C following interaction with protein X (Prusiner 1998; Prusiner et al. 1998).

Interestingly, the 19-kD protease-resistant fragment of PrP^{Sc} formed after deglycosylation has also been found in a patient who died after developing a clinical disease similar to FFI. Since both PrP alleles encoded the wild-type sequence and a methionine at position 129, we labeled this case sporadic fatal insomnia (sFI). At autopsy, the spongiform degeneration, reactive astrogliosis, and PrP^{Sc} deposition were confined to the thalamus (Mastrianni et al. 1997). These findings argue that the clinico-pathologic phenotype is determined by the conformation of PrP^{Sc}, in accord with the results of the transmission of human prions from patients with FFI to Tg mice (Telling et al. 1996).

CONSERVATION OF BSE "PHENOTYPE" DURING SERIAL PASSAGE

Beginning in 1986, BSE or "mad cow" disease appeared in Great Britain (Wells et al. 1987). It has been proposed that BSE represents a massive common-source epidemic that was caused by prion-contaminated meat and bone meal (MBM) fed primarily to dairy cows (Wilesmith et al. 1991). The MBM was prepared from the offal of sheep, cattle, pigs, and chickens as a high-protein nutritional supplement. In the late 1970s, the solvent extraction method used in the rendering of offal began to be abandoned, resulting in an MBM with a much higher fat content (Wilesmith et al. 1992). It is possible that some scrapie prion strains originating in sheep were able to survive rendering and to pass into cattle. Alternatively, bovine prions that had caused clinical CNS dysfunction at such a low level as not to be recognized or prions derived from the rare case of spontaneous prion disease survived the rendering process and were passed back to cattle through the MBM. Recent statistics indicate that the epidemic is now disappearing as a result of the 1988 food ban.

Protease-resistant PrP has been found in the brains of cattle with BSE (Hope et al. 1988; Prusiner et al. 1993). Brain extracts from ill cattle have transmitted disease to mice, with an incubation period of about 400 days (Fraser et al. 1992). On subsequent passages in mice, the incubation period shortens considerably and eventually stabilizes at a value that depends on the *Sinc/Prnp* genotype of the host mice (Bruce et al. 1994). Although the initial report concluded that two distinct strains, termed 301V and 301C, could be discriminated in several separate transmissions from BSE-afflicted cows, it now appears more likely that 301V and 301C are fundamentally the same strain, differing only with respect to the *Sinc/Prnp* identity of PrP encoded by the donor and recipient mice used in these transmissions (Bruce et al. 1994; Ridley and Baker 1996).

It is interesting to note that, in contrast to sheep, different breeds of cattle have no specific PrP polymorphisms other than a variation in the number of octarepeats: Most cattle, like humans, have five octarepeats, but some have six (Goldmann et al. 1991b; Prusiner et al. 1993). Humans with seven octarepeats develop fCJD (Goldfarb et al. 1993), but the presence of six octarepeats does not seem to be overrepresented in BSE cases (Goldmann et al. 1991b; Prusiner et al. 1993; Hunter et al. 1994). The number of octarepeats does not appear to affect the specificity of transmission of prions (Fischer et al. 1996), in contrast to polymorphic changes within the core of the protein spanning residues 90–231, which we have explained above. Since the majority of evidence for prion strain diversity can be traced to differences in the sequence of PrP in successive hosts for transmission, it is possible that the lack of PrP polymorphisms in cattle (Heaton et al. 2003) may to some degree account for the remarkable consistency of BSE strain properties observed.

TRANSGENIC MOUSE MODELS OF BSE

Mice inoculated intracerebrally with BSE prions require more than a year to develop disease (Taylor 1991; Fraser et al. 1992; Lasmézas et al. 1997). Other attempts at assaying BSE prions have used animals from various species. Brain extracts from BSE cattle cause disease in cattle, sheep, mice, pigs, and mink after intracerebral inoculation (Fraser et al. 1988; Dawson et al. 1990a,b; Bruce et al. 1993). All of these alternative bioassay systems suffer from severe limitations to their usefulness. Apart from the cost involved, the long incubation periods and low efficiency of transmission of prions, heightened in most cases by a species barrier caused by lack of PrP sequence identity, have conspired to severely impede progress in performing routine measurements of titers of BSE prions. Since no alternative bioassay offered any significant advantage over the use of normal mice, these became the species of choice for the majority of BSE bioassays, until recently. Of available mouse strains, RIII mice give the shortest incubation periods, with mean incubation periods reportedly ranging from 302 to 335 days for transmission of frozen brain samples (Bruce et al. 1994). Even RIII mice suffer from limitations, however, which restrict their usefulness in bioassays of bovine prions. BSE prions that have been serially passaged in mice exhibit dramatically shortened incubation periods, in some cases, of less than 140 days (Bruce et al. 1994). Although some investigators appear to disagree that this is problematic (Bruce et al. 1994), it seems clear to us from these data that BSE prions passaged in mice and those passaged in cattle are not equivalent; instead there is clear evidence of a species barrier (Pattison 1965) to the transmission of BSE prions from cattle to mice.

The evidence for a significant species barrier to the transmission of BSE prions to mice has important ramifications that make mice inappropriate for use in bioassays aimed at detecting low doses of BSE prions. As we have argued elsewhere (Scott et al. 1997a) and in this chapter, a species barrier can be indicative of a change in the strain of prion caused by differences in the sequences of PrP of the donor and recipient animals. Some investigators argue that this is not the case with BSE; the characteristic BSE strain "signature" is reportedly unchanged by serial transmission from numerous species to mice (Bruce et al. 1994). Even if we assume this to be true, only one other explanation for the enormous difference in incubation period seems possible: Since the incubation period and the titer of inoculum are inversely related (Pattison and Smith 1963; Prusiner et al. 1981), the extended incubation period during primary transmission must represent an effective reduction in titer, caused presumably by the inefficient interaction of BoPrP^{Sc} in the inoculated prions with MoPrP^C of the host (Prusiner et al. 1990). In support of this reasoning, endpoint titration of BSE prions in cattle gave a much higher titer of prions in bovine brain than those determined by mouse bioassays (Wells et al. 1998). To overcome this limitation, we created transgenic mice expressing BoPrP transgenes (Scott et al. 1997b).

Three lines of Tg(BoPrP)$Prnp^{0/0}$ mice expressing BoPrP were inoculated with a 10% homogenate derived from the medulla of a Hereford bull (case PG31/90) clinically ill with BSE (Scott et al. 1997b). One line, Tg(BoPrP)E4125/$Prnp^{0/0}$, which showed the highest level of BoPrP expression, was found to be highly susceptible to BSE prions, with 100% of animals exhibiting clinical signs within 250 days after inoculation. Another line, Tg(BoPrP)E4092/$Prnp^{0/0}$, had an intermediate level of BoPrP expression and exhibited a longer incubation period with the same inoculum; these mice had a mean incubation period of about 320 days (Table 2). Similar incubation periods were obtained when the lines were inoculated with other BSE isolates. Like most cattle with BSE, vacuolation and astrocytic gliosis were confined in the brain stems of these Tg mice (Fig. 4) (Scott et al. 1997b). Many previous studies with transgenic mice over the last decade have clearly established that expression of a foreign PrP transgene abrogates the species barrier (Scott et al. 1989, 1993, 1997b; Prusiner et al. 1990; Hsiao et al. 1994; Telling et al. 1994, 1995); hence it was entirely reasonable to expect that these Tg(BoPrP) mice would make possible, for the first time, an accurate determination of BSE prion titers in brain and other tissues.

In the interests of public safety, prion bioassays must be capable of detecting a single infectious unit. BSE infectivity measured in cattle showed that earlier efforts to quantify BSE prion titers using an RIII mouse bioassay may have underestimated prion titers by a factor of ~1000 (Wells et al. 1998).

After we were successful in breeding the Tg(BoPrP)E4092/$Prnp^{0/0}$ line to homozygosity for the BoPrP transgene, studies with these mice were greatly facilitated. The new Tg(BoPrP$^{+/+}$)E4092/$Prnp^{0/0}$ mice show a similar level of expression of BoPrP to the Tg(BoPrP)E4125/$Prnp^{0/0}$ mice used previously (Scott et al. 1999) and give similar incubation periods when inoculated with BSE prions (Safar et al. 2002). The Tg(BoPrP$^{+/+}$)E4092/$Prnp^{0/0}$ mice appear to be ~10 times more sensitive to infection with BSE prions than cattle (Safar et al. 2002). This finding raises concerns that previous attempts to quantify BSE and scrapie prions in milk or nonneural tissues, such as muscle, may have underestimated infectious titers by as much as 10^3 or 10^4, raising the possibility that prions could be present in these products in sufficient quantities to pose risk to humans (Middleton and Barlow 1993). Recently, the detection of infectious prions in hind limb tissue of mice infected with scrapie (Bosque et al. 2002) has highlighted the need for a comprehensive screen for infectivity in tissues from BSE-infected cattle using Tg(BoPrP$^{+/+}$)E4092/

Table 2. Susceptibility and resistance of transgenic mice to BSE prions

Inoculum	Recipient	Transgene expression	Incubation time (days ± s.e.)	n/n_0
A. Mice deficient of Mo PrP (Prnp$^{0/0}$)				
BSE(PG31/90)	Tg(BoPrP)4125	8–16x	234 ± 8	10/10
BSE(PG31/90)	Tg(BoPrP)4092	4–8x	319 ± 15	8/8
BSE(GJ248/85)	Tg(BoPrP)4125	8–16x	281 ± 19	10/10
BSE(GJ248/85)	Tg(BoPrP)4092	4–8x	343 ± 18	8/8
BSE(GJ248/85)	Tg(MBo2M)14586	8–16x	>600	0/15
BSE(PG31/90)	Tg(MBo2M)14586	8–16x	>600	0/13
BSE(574C)	Tg(MBo2M)14586	8–16x	>600	0/13
B. Mice expressing MoPrP-A				
BSE(GJ248/85)	FVB	0	628 ± 47	2/3
BSE(PG31/90)	FVB	0	448 ± 29	2/2
BSE(574C)	FVB	0	525 ± 34	4/4
BSE(PG31/90)	Tg(MoPrP-A)4053	8–16x	>310	4/10
BSE(GJ248/85)	Tg(MoPrP-A)4053	8–16x	322 ± 6	8/9
BSE(PG31/90)	Tg(BoPrP)333	0[a]	426 ± 11	8/8
BSE(PG31/90)	Tg(BoPrP)833	0[a]	395 ± 22	9/9
BSE→Tg(BoPrP)333	CD-1	0	163 ± 5	8/8
BSE→Tg(BoPrP)333	CD-1	0	148 ± 0	8/8

Reprinted, with permission, from Scott et al. (1997b) (copyright National Academy of Sciences).
n, number of ill mice; n_0, number of inoculated mice.
[a]No BoPrP was detected by western immunoblotting.

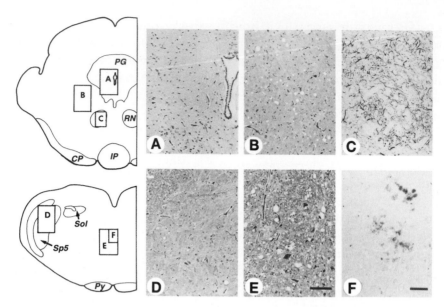

Figure 4. Neuropathology of Tg(BoPrP)$Prnp^{0/0}$ mice inoculated with BSE prions. (*A*) No pathologic changes were found in the periaqueductal gray of the midbrain. (*B*) Mild to moderate vacuolar degeneration was found in the reticular formation of the midbrain tegmentum. (*C*) Reactive astrocytic gliosis colocalized with sites of vacuolar degeneration: Astrogliosis in the red nucleus is shown here. (*D*) Little or no vacuolar degeneration was found in the tract or the nucleus of the spinal tract of the trigeminal nerve in the medulla. (*E*) Moderate to severe vacuolar degeneration occurred in the medial tegmentum of the medullary reticular formation. (*F*) Small PrP-immunoposi-tive primitive plaque-like deposits colocalized with sites of the most severe vacuolar degeneration. Diagrams show locations of photomicrographs. Hematoxylin and eosin stain used in *A,B,D,E*. Glial fibrillary acidic protein immunohistochemistry used in *C*. PrP immunohistochemistry used in *F*. Bar in *E*, 100 μm; also applies to *A*, *B*, and *D*. Bar in *F*, 50 μm; also applies to *C*. Italicized letters identify selected brain stem struc-tures: (*CP*) cerebral peduncle; (*IP*) interpeduncular nucleus; (*PG*) periaqueductal gray; (*Py*) pyramidal tract; (*RN*) red nucleus; (*Sol*) nucleus and tractus solitarius; (*Sp5*) nucleus of the spinal tract of the trigeminal nerve. (Reprinted, with permission, from Scott et al. 1997b [copyright National Academy of Sciences].)

$Prnp^{0/0}$ mice. It is often assumed that the maximal exposure of humans to BSE prions must have occurred prior to the specified bovine offals ban of November 1989 that prohibited CNS and lymphoid tissues from cattle older than 6 months of age to be used in food destined for human con-sumption. This legislation was based on studies in sheep showing that the highest titers of scrapie prions are found in these tissues (Hadlow et al. 1982). In those scrapie studies, sheep tissues were inoculated into non-Tg Swiss mice, which are slightly more susceptible to sheep prions than bovine prions. Because the bioassay for bovine prions in ordinary mice is so insensitive (Taylor 1991), the levels of prions in bovine muscle remain unknown. If the distribution of bovine prions proves to be different from that presumed for sheep, then assumptions about the efficacy of the offal ban may need to be reassessed.

THE LINK BETWEEN BSE AND vCJD

The emergence of variant (v)CJD in teenagers and young adults in the UK and France raised the possibility that transmission of BSE prions to humans may have occurred (Chazot et al. 1996; Will et al. 1996). The average age of these individuals was 27 years, much younger than for any other group of people who have died of CJD except for those who received pituitary-derived human growth hormone. Why such cases should be confined to young people is unclear. Whether the young CNS is more vulnerable to invasion by bovine prions or the dietary habits of these young individuals exposed them to a greater dose of bovine prions is unknown. Not only does age set these teenagers and young adults apart from other individuals who died of prion disease, but so does the neuropathology. The deposition of PrP^{Sc} in the brains of these patients is extreme, and numerous multinucleated PrP amyloid plaques surrounded by intense spongiform degeneration have been observed. These neuropathologic changes seem to be unlike any observed in other forms of prion disease.

It has been estimated that about 750,000 cattle infected with BSE were slaughtered for human consumption in Great Britain, during a period spanning 1980 to 1996 (Anderson et al. 1996; Ferguson et al. 1997). It now seems clear that during this period, the disease passed to a section of the human population through the consumption of contaminated beef products, leading to the emergence of vCJD. Evidence for this link now seems indisputable. Epidemiological studies (Will et al. 1996, 1999) and prion strain "typing" experiments of scrapie and BSE prions passaged to inbred mice (Collinge et al. 1996; Bruce et al. 1997; Hill et al. 1997) provided evidence that vCJD is caused by the transmission of BSE prions to humans. Transmissions of BSE prions to primates were also performed; BSE could be transmitted to marmosets and macaques after intracerebral inoculation (Baker et al. 1993; Lasmézas et al. 1996). In addition, a young adult monkey born and reared in Britain died of a neurodegenerative disorder thought to be a prion disease in a Montpelier zoo (Bons et al. 1996, 1999).

Transgenic mice expressing foreign PrP (Scott et al. 1989, 1993; Prusiner et al. 1990; Telling et al. 1995) and PrP knockout (Büeler et al. 1993) mice have conclusively shown that the PrP gene is the primary determinant controlling susceptibility to foreign prions. Transgenic mice expressing HuPrP with the V129 polymorphism have been reported to be somewhat susceptible to human vCJD prions, with incubation periods of ~220 days (Hill et al. 1997), but in the same report, these same mice were not found to be susceptible to BSE prions. To test the hypothesis that vCJD is caused by BSE prions, we inoculated Tg(BoPrP) mice with BSE and vCJD prions. We reasoned that the reintroduction of BSE prions from humans (i.e., vCJD) into Tg(BoPrP)$Prnp^{0/0}$ mice might restore the original strain properties of BSE prions. If this transmission was performed without disruption of the original strain properties, the newly formed vCJD prions, which would now be composed of BoPrP^{Sc}, would be indistinguishable from native BSE prions. Surprisingly, we found that the Tg(BoPrP)$Prnp^{0/0}$ mice were not only susceptible to BSE and vCJD prions, but also highly susceptible to natural scrapie prions of sheep. The incubation times of ~210 days for sheep scrapie prions were shorter than ~240 days for BSE prions and ~270 days for vCJD prions (Scott et al. 1999). Of note, the five cases of natural Suffolk sheep scrapie were all from the US: three from a flock in Illinois and two from California. It will be of great interest to transmit scrapie isolates obtained from sheep in the UK into these mice to determine whether any of these transmissions lead to reisolation of the "BSE strain."

A comparison of vCJD and BSE transmissions to Tg(BoPrP)$Prnp^{0/0}$ mice provided compelling evidence for a link between vCJD and BSE. The neuropathology in Tg(BoPrP)$Prnp^{0/0}$ mice inoculated with BSE prions was indistinguishable from that found in Tg(BoPrP)$Prnp^{0/0}$ mice inoculated with vCJD prions (Scott et al. 1999). In contrast, the neuropathology of Tg(BoPrP)$Prnp^{0/0}$ mice inoculated with sheep scrapie prions was quite different. The PrP^{Sc} isoforms in Tg(BoPrP)$Prnp^{0/0}$ mice inoculated with vCJD or BSE brain extracts were indistinguishable and differed dramatically from those seen in Tg(BoPrP)$Prnp^{0/0}$ mice injected with natural scrapie prions (Scott et al. 1999). Each of three cases of vCJD produced an incubation period of ~270 days on first passage, which shortened to ~225 days on second passage, in contrast to cattle BSE and sheep scrapie prions for which the length of the incubation time remained unchanged on second passage.

Although a link between vCJD and BSE has certainly been forged by the preceding studies, it is still impossible to evaluate the potential risk to humans exposed to BSE-tainted beef. Although an ominous increase in the number of reported cases of vCJD in the last quarter of 1998 was reported (Will et al. 1999), this did not herald the dramatic increase in the incidence of vCJD that was feared; the number of reported cases of vCJD does not seem to be increasing exponentially. It is also important to note that epidemiological studies over nearly three decades have failed to establish convincing evidence for transmission of sheep prions to humans (Cousens et al. 1990; Malmgren et al. 1979). Also of note is the high incidence of CJD among Libyan Jews that was initially attributed to the consumption of lightly cooked sheep brain (Kahana et al. 1974); however, subsequent studies showed that this geographical cluster of CJD is due to the E200K mutation (Goldfarb et al. 1990; Hsiao et al. 1991). At present, therefore, vCJD remains the only known example of a human prion disease caused by consumption of infected animal products.

VARIATIONS IN GLYCOSYLATION OF PrP^{Sc} ASSOCIATED WITH PARTICULAR STRAINS

Recent studies of PrP^{Sc} from brains of patients who died of vCJD show a pattern of PrP^{Sc} glycoforms different from those found for sporadic or iatrogenic CJD: It closely resembled that found in the brains of mice and cats infected with BSE (Collinge et al. 1996, 1997). We have found a similar pattern of glycoforms in cattle with BSE, and this pattern appears to be reproduced upon transmission to Tg(BoPrP) mice, providing further support for the notion that these mice represent a faithful biological model for the bovine prion disease (Scott et al. 1997b). The significance of this pattern is questionable, however, in trying to relate BSE to vCJD (Collinge et al. 1997; Parchi et al. 1997; Somerville et al. 1997). From our own studies, prion strains are often unstable when transmissions are effected between animals with distinct PrP sequences (Scott et al. 1997a). Although the glycoform ratios appear similar when BSE or vCJD prions passaged into normal mice are compared with extracts of Hu vCJD patients, the larger glycoform from extracts derived from the murine background appears to migrate more rapidly than that from human extracts (Collinge et al. 1997), suggesting that BSE and vCJD PrP^{Sc} preparations from mice are underglycosylated compared to their corresponding protease-resistant fragments in Hu vCJD brain. In addition, since PrP^{Sc} is formed after the protein is glycosylated (Borchelt et al. 1990; Caughey and Raymond 1991), and enzymatic deglycosylation of PrP^{Sc} requires denaturation (Endo et al. 1989; Haraguchi et al. 1989), either some form of selective conversion of PrP^{C} into PrP^{Sc} (DeArmond et al. 1997) or variations in the rate of clearance of different PrP^{Sc} glycoforms (Prusiner et al. 1998) would have to account for the glycoform ratios observed.

The glycoform ratio is a useful surrogate marker for strain typing, despite the fact that many strains cannot be discriminated on this basis alone (Somerville et al. 1997). However, in some cases, two-dimensional gel electrophoresis may be used to identify variations in glycoforms that are not apparent by conventional western blotting (Parchi et al. 2000; Pan et al. 2001). If different PrP^{Sc} glycoform ratios do arise, as we suspect, due to variations in either the rate of synthesis or the rate of clearance of the various PrP^{Sc} glycoforms, then it seems reasonable to suggest this is itself dependent on the strain of prion and hence the conformation of PrP^{Sc}. Apparently, the BSE strain will preferentially form fully glycosylated PrP^{Sc}, producing the "type 4" pattern (Collinge et al. 1996). By optimizing the sequence of PrP to facilitate formation of the higher-molecular-mass glycoforms, we may be able to accelerate the rate of conversion and consequently reduce the incubation period dramatically. PrP mutations that selectively increase the conversion of diglycosylated PrP^{C} may assist in the creation of Tg mice with more rapid incubation times when inoculated with BSE and/or vCJD prions.

PrP STRUCTURAL CONSTRAINTS MODULATING PRION TRANSMISSION

Because of our success in transmitting Hu CJD prions to Tg mice expressing MHu2M (Telling et al. 1994, 1995), we also created mice expressing an analogous Bo/Mo PrP transgene, which we termed MBo2M (Scott et al. 1997b). Unexpectedly, when Tg(MBo2M) mice expressing high levels of chimeric PrP were inoculated with BSE prions, they did not develop signs of disease, even after more than 600 days (Table 2) (Scott et al. 1997b). This was surprising, since the MBo2M ORF comprised only BoPrP and MoPrP-A sequences, and both mice and cattle have been shown to be susceptible to BSE prions. A comparison of the MoPrP-A, MBo2M PrP, and MHu2M Prp translated sequences showed that Hu residue substitutions in MHu2M extended from 97 to 168, whereas Bo substitutions in MBo2M included additional differences between 168 and 186. This finding raised the possibility that residues 184 and 186, which are not homologous in MoPrP and BoPrP and lie at the end of the chimeric region, might account for this difference in susceptibility. Alternatively, residue 203, which is valine in MoPrP and HuPrP and isoleucine in BoPrP, might be responsible for this difference in susceptibility. Since residue 203 is valine in MBo2M, and isoleucine in BoPrP, this difference might prevent efficient formation of MBo2M PrP^{Sc} (Scott et al. 1997b).

When viewed in a spatial context, the potential impact of these differences is more easily appreciated (Fig. 5) (Scott et al. 1997b). Residues 184, 186, 203, and 205 were identified within the three-dimensional structure of SHaPrP(90–231) derived by solution NMR (Fig. 5) (James et al. 1997). These residues were seen to cluster on one side of the PrP^{C} structure and

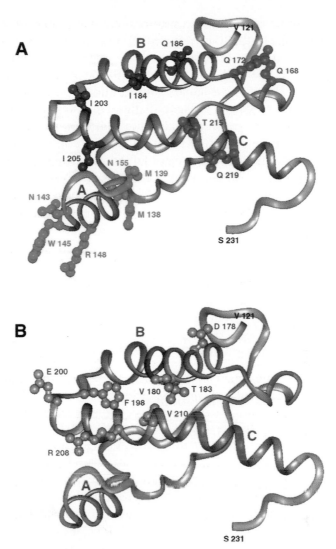

Figure 5. PrP residues governing the transmission of prions. (*A*) NMR structure of recombinant SHaPrP region 121–231 (James et al. 1997) shown with the putative epitope formed by residues 184, 186, 203, and 205 highlighted in red. Residue numbers correspond to SHaPrP. Additional residues (138, 139, 143, 145, 148, and 155) that might participate in controlling the transmission of prions across species are depicted in green. Residues 168, 172, 215, and 219 that form the epitope for the binding of protein X are shown in blue. (*B*) Within and adjacent to the epitope formed by residues 184, 186, 203, and 205 shown in red are known pathologic mutations at residues 178, 180, 183, 198, 200, 208, and 210 colored in magenta. The three helices (A, B, and C) are highlighted in brown. Illustration was generated with Biosym/Insight II. (*A*, Reprinted, with permission, from Scott et al. 1997b [copyright National Academy of Sciences.])

are spatially distinct from the discontinuous epitope comprising residues 168, 172, 215, and 219 that binds protein X (Fig. 5) (Kaneko et al. 1997). We next compared these residues to others known to be polymorphic from sequence comparisons of the PrP ORFs of many other species (Bamborough et al. 1996), which are thus candidates for contributing to the species barrier. As shown in Figure 5, these residues appear to cluster and enlarge the epitope formed by residues 184, 186, 203, and 205. It is also interesting to note that seven point mutations (at positions 178, 180, 183, 198, 200, 208, 210) known to cause inherited prion diseases, including those that have been shown to create a transmissible encephalopathy, map to this region of structure as well (Fig. 5) (Scott et al. 1997b).

The identification of a species-specific epitope that modulates the conversion of PrP^C into PrP^{Sc} has important implications for the future design of PrP transgenes. It remains to be determined whether optimal Bo/Mo chimeric transgenes may be created by substituting Mo or Bo residues at positions 184, 186, 203, and/or 205. Similarly, it is unknown whether improved Hu/Mo chimeric transgenes can be constructed by simultaneously mutating Hu residues at these same positions to Mo residues (Fig. 5). Mutagenesis at any or all of these positions may overcome the resistance of Tg(MBo2M) mice to Bo prions.

PRION STRAINS AND SPECIES BARRIERS

Sequence similarity between the PrP^C of the host and PrP^{Sc} of the inoculum is clearly the major determinant in controlling susceptibility to prion disease (Scott et al. 1989, 1993; Prusiner et al. 1990), hence it is generally assumed that humans are in some measure protected from exposure to prions of other species because of differences in the sequences of PrP. However, it is important to note that differences in PrP sequence alone cannot be the only determinant of susceptibility. Detection of a species barrier between hamsters and Tg mice expressing chimeric SHa/Mo PrP depends on the strain used for the inoculum (Peretz et al. 2002), and changing a single amino acid near the amino terminus of MoPrP^{Sc} was sufficient to render Tg mice expressing this modified protein susceptible to human prions (Barron et al. 2001). It must also be emphasized that the presence of clinical disease following interspecies transmission is not an absolute measure for transmission; evidence of subclinical propagation of prions in rodents raises concerns that significant quantities of prions pathogenic to humans could be accumulated in another species in the human food chain without any evidence of disease in the "carrier" species (Hill et al. 2000; Asante et al. 2002).

It is well established that a given species can tolerate multiple distinct prion strains, demonstrating that an informational variable independent of the amino acid sequence of PrP^{Sc} exists. The SHa prion strains Sc237 and DY, each representing a distinct conformer of PrP^{Sc} but sharing the same amino acid sequence (Peretz et al. 2001), were transmitted to Tg mice expressing MH2M (Scott et al. 1993). Following transmission to Tg(MH2M) mice, the incubation period to onset of disease and neuropathological characteristics of the original and Tg(MH2M)-passaged strains were noted, and these properties were correlated with changes in the conformation of PrP^{Sc} for each strain using a conformational stability assay (Peretz et al. 2001). In this assay, the relative conformational stability is expressed as a $[GdnHCl]_{1/2}$ value, which corresponds to the midpoint of the transition between the native and the denatured state following exposure of PrP^{Sc} to increasing concentrations of guanidium hydrochloride. SHa(Sc237) and MH2M(Sc237) produced different incubation periods in Syrian hamsters (75 and 113 days, respectively) and displayed distinct $[GdnHCl]_{1/2}$ values (1.8 and 1.1 M, respectively) (Table 3 and Fig.

Table 3. Incubation times and PrP^{Sc} properties for Sc237 and DY prions passaged in Syrian hamsters (SHa) or Tg(MH2M) mice

Inoculum[a]	Recipient animal	Incubation time[b] (days ± S.E.M.)	M_r[c]	Conformational stability[d] ($[GdnHCl]_{1/2}$ ± S.D.)
SHa(Sc237)	SHa	75 ± 0.3	21	1.8 ± 0.16
SHa(Sc237)	Tg(MH2M PrP)229/$Prnp^{0/0}$	83 ± 1.6	21	1.1 ± 0.09
MH2M(Sc237)	Tg(MH2M PrP)229/$Prnp^{0/0}$	50 ± 0.1	21	0.96 ± 0.14
MH2M(Sc237)	SHa	113 ± 1.8	n.d.	n.d.
MH2M(MH2M(Sc237))	SHa	133 ± 2.2	n.d.	n.d.
SHa(DY)	SHa	165 ± 1.0	19	0.98 ± 0.13
SHa(DY)	Tg(MH2M PrP)229/$Prnp^{0/0}$	70 ± 1.0	19	1.1 ± 0.06
MH2M(DY)	Tg(MH2M PrP)229/$Prnp^{0/0}$	76 ± 1.0	19	1.0 ± 0.09
MH2M(DY)	SHa	176 ± 2.0	n.d.	n.d.
MH2M(MH2M(DY)	SHa	153 ± 2.0	n.d.	n.d.

Reprinted, with permission, from Peretz et al. (2002) (copyright Cell Press).

[a]Nomenclature of prion strains. Inoculum was a 1% (w:v) crude brain homogenate of prion-infected animals.

[b]Mean incubation period in days ± standard error of the mean (S.E.M.), and n.d. (not determined).

[c]Apparent molecular weight (M_r) of the deglycosylated protease-resistant fragment of PrP^{Sc}, and n.d.

[d]ELISA denaturation transitions were performed using monoclonal antibody fragment (Fab) HuM-D18 (0.25 µg/ml), and $[GdnHCl]_{1/2}$ values were interpolated with a sigmoid algorithm using a nonlinear least-square fit. The results are the mean, with the standard deviation (S.D.) calculated from more than 4 denaturation curves.

6C,D). Furthermore, the patterns of deposition of PrP^{Sc} observed in the brain were strikingly different, with small amyloid plaques over the hippocampus in Tg(SHaPrP) mice and hamsters infected with SHa(Sc237) but much larger, GSS-type plaques in Tg(MH2M) mice infected with MH2M(Sc237) (Fig. 6A,B) (Peretz et al. 2002). In contrast, the other strain, SHa(DY), when passaged to the same line of Tg(MH2M) mice, resulted in neither a species barrier nor any apparent change in strain properties. MH2M(DY) PrP^{Sc} had a relative conformational stability similar to that of SHa(DY) PrP^{Sc} and produced a similar disease phenotype in Tg(MH2M) mice (Table 3 and Fig. 7) (Peretz et al. 2002).

In these studies, a single Tg(MH2M) mouse host "species" exhibited markedly different levels of susceptibility to two prion strains sharing the same SHa amino acid sequence (Peretz et al. 2002). Limited host sequence homology, as seen in this model system, has a variable effect on interspecies prion transmission, and therefore, the degree of resistance must depend on both the conformation of PrP^{Sc} and the sequence of PrP. These findings also suggest that inefficient transmission of prions between species with distinct PrP genes leads the emergence of new prion strains with conformations that are different from the original, and that this conformational change coincides with new disease phenotypes in infected hosts (Peretz et al. 2002). Hence, in transmissions that exhibited a species barrier, the resultant prion strain displayed biochemical and neuropathological properties that differed from those of the original inoculum. However, transmissions between hosts with different PrP sequences could also be accomplished without evidence of a species barrier, and in these cases, the original prion strain characteristics were retained (Peretz et al. 2002).

A transmission that produces a species barrier could arise if the sequence of the recipient host PrP is energetically unfavorable for folding into the appropriate conformation for that particular strain (Scott et al. 1997b; Collinge 1999). In this circumstance, we envisage two likely mechanisms by which a new prion strain could emerge (Peretz et al. 2002). First, a change in the conformation of PrP^{Sc} might arise following the binding of donor PrP^{Sc} to nonhomologous PrP^{C}, leading to the de novo formation of an energetically more favorable conformer of PrP^{Sc} than the parental PrP^{Sc}. Second, if most prion isolates were actually mixtures of more than one PrP^{Sc} conformer, the resulting strain could be selected from a heterogeneous population of strains already existing within the original inoculum, and a small subpopulation of PrP^{Sc} in the prion would be preferentially expanded in the new species. We presume that the process is inefficient in either scenario and thereby leads to the prolonged incubation period seen upon first passage. Some measure of support exists for the latter hypothesis: Bartz and colleagues have shown that a biologically "cloned" hamster TME isolate could produce either the HY or DY prion strain following serial propagation, depending on degree of dilution of the inoculum (Bartz et al. 2000). Further evidence that each

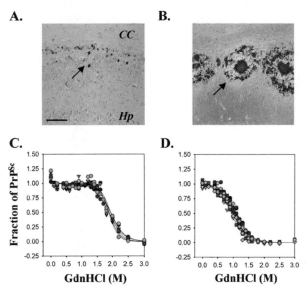

Figure 6. PrP^{Sc} deposition phenotypes and denaturation transitions of Sc237 PrP^{Sc} in Syrian hamsters and in Tg(MH2M) mice. (*A,B*) Transmission of SHa(Sc237) prions to Tg(MH2M) mice significantly altered its PrP^{Sc} deposition phenotype. (*A*) Syrian hamsters inoculated with SHa(Sc237) prions. (*B*) Tg(MH2M)229 mice inoculated with MH2M(Sc237) obtained from a single passage of SHa(Sc237) prions in Tg(MH2M)229. Immunological detection of PrP^{Sc} extracellular deposits (indicated by arrow) in the brain was conducted by the hydrolytic autoclaving method using the 3F4 mAb as described (Muramoto et al. 1992). The brains were immersion fixed in 10% buffered formalin upon dissection from the animal. Bar in *A*, 50 μm; also applies to *B*. Italicized letters identify selected cerebrum structures: (*Hp*) gray matter of the hippocampus; (*CC*) white matter of the corpus callosum. (*C,D*) Denaturation transitions of Sc237 prion strains. P2 fractions were prepared from brains of (*C*) Syrian hamsters (*n* = 10) inoculated with SHa(Sc237) prions; and (*D*) Tg(MH2M) mice (*n* = 5) inoculated with SHa(Sc237) prions and treated with increasing concentrations of GdnHCl. Following 1 hour of incubation, the samples were diluted, treated with proteinase K for 1 hour at 37°C, and precipitated with methanol/chloroform. ELISA wells were coated with 50 μl of either 1, 2.5, 5, or 10 μg/ml of denatured proteins (Peretz et al. 2001). PrP^{Sc} was detected with Fab HuM-D18 (0.25 μg/ml), and absorbency was read after 60 minutes of color development. The apparent fraction of PrP^{Sc} was calculated for each ELISA optical density (OD) mean value, and the sigmoidal patterns of PrP were plotted with a four-parameter algorithm using a nonlinear least-square fit and correlation coefficient greater than 0.97. (*C,D*, Reprinted, with permission, from Peretz et al. 2002 [copyright Cell Press].)

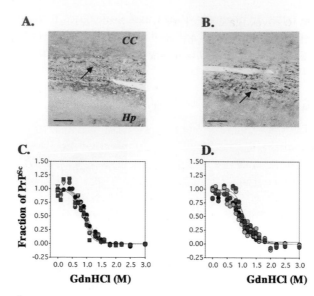

Figure 7. PrP^{Sc} deposition phenotypes and denaturation transitions of DY PrP^{Sc} in Syrian hamsters and in Tg(MH2M) mice. (*A,B*) Transmission of SHa(DY) prions to Tg(MH2M) mice did not alter its PrP^{Sc} deposition phenotype. (*A*) Syrian hamsters inoculated with SHa(DY) prions. (*B*) Tg(MH2M)229 mice inoculated with MH2M(DY) obtained from a single passage of SHa(DY) prions in Tg(MH2M)229. Immunological detection of PrP^{Sc} (indicated by arrow) in the brain was conducted by the hydrolytic autoclaving method using the 3F4 mAb as described (Muramoto et al. 1992). The brains were snap-frozen upon dissection from the animal and later thawed at room temperature and fixed with 10% buffered formalin. Bar in *A*, 50 μm, and bar in *B*, 25 μm (the magnification in this micrograph was increased to better visualize the perivascular PrP deposits). Italicized letters identify selected cerebrum structures: (*Hp*) gray matter of the hippocampus; (*CC*) white matter of the corpus callosum. (*C, D*) Denaturation transitions of DY prion strains. P2 fractions prepared from brains of (*C*) Syrian hamsters (*n* = 10) inoculated with SHa(DY) prions; (*D*) Tg(MH2M) mice (*n* = 4) inoculated with SHa(DY) prions. Proteins were denatured, PK-digested, and detected in ELISA as described in Fig. 6. The apparent fraction of DY PrP^{Sc} was calculated for each ELISA OD mean value, and sigmoidal curves were plotted as described in Fig. 6, except that absorbency was read after 90 minutes of color development. (Reprinted, with permission, from Peretz et al. 2002 [copyright Cell Press].)

individual prion strain itself comprises multiple distinct conformers of PrP^{Sc} might exist in the work of Safar and colleagues (Safar et al. 1998). When several strains were compared using a conformation-dependent immunoassay (CDI) following limited GdnHCl denaturation and protease digestion, each strain produced a profile (Safar et al. 1998) that could be interpreted as representing multiple discrete conformers. It will be of great interest to investigate whether most prion strains are composed of multiple distinct conformers, and whether these conformers exist in a stoichiometric arrangement unique to each strain.

ESTIMATING THE LIMITS OF PRION DIVERSITY

As we stated at the outset, the only significant obstacle to understanding prion strains within the context of a protein-only model for prion replication has been that of the existence of prion diversity. It now seems clear that prion strains encipher their properties in PrP conformers (Telling et al. 1996; Safar et al. 1998; Peretz et al. 2001). When eight distinct hamster prion isolates were analyzed using a CDI (Safar et al. 1998), the ratio of denatured/native PrP as a function of PrP^{Sc} concentration before and after limited digestion with proteinase K (Safar et al. 1998) could be used to discriminate each of the eight prion isolates. Notably, only the DY strain could be distinguished from the other seven isolates by western blotting after limited proteolysis (Scott et al. 1997a). However, a later comparison of these same eight distinct hamster prion isolates revealed that the eight strains could be segregated into four discrete groups based on both conformational stability measurements and comparisons of prion strain properties (Peretz et al. 2001). The apparent discrepancy between these two studies can be easily reconciled if, as proposed above, each isolate corresponded to a mixture of more than one basic conformer or "substrain." We might expect that the CDI as employed by Safar and colleagues (Safar et al. 1998) would be extremely sensitive to fluctuations in the relative proportions of minor conformers in different isolates sharing a common dominant PrP^{Sc} conformer. In contrast, the conformational stability assay measures the midpoint of the conformational transition denaturation of the predominant conformer (Peretz et al. 2001, 2002).

In this review, and in others (Ridley and Baker 1996), we have argued that, in reality, the number of possible strains may be much lower than has been claimed. If prion strains within a single PrP genetic background are found to appear coincident in their properties, despite having been derived from completely different primary sources, geographically isolated from one another, prion diversity must be restricted. The frequent "reisolation" of the Me7 strain illustrates this point (Ridley and Baker 1996). In other studies, the apparent identity of Sc237 and SHa(Me7) when passaged between identical hosts (Scott et al. 1997a) is made all the more striking in view of the fact that Me7 appears to be derived from an entirely different primary source, a Suffolk sheep spleen, whereas the ancestry of both Sc237 and RML can be traced to the Compton "drowsy goat" source, which in turn was derived by transmission from scrapied Cheviot sheep (Pattison et al. 1959). In contrast, we were able to obtain two completely different SHa strains from the same, "cloned" primary source, Me7, by including one additional passage in Tg(MH2M) mice (Scott et al. 1997a). Thus, the passage history, rather than the original source, determines strain characteristics in any particular host.

We have also described how differences in prion strain characteristics can be directly traced back to changes in PrP sequence during serial transmission, suggesting that strain characteristics are maintained and propagated through protein–protein interactions between PrP isoforms. This is entirely consistent with a model in which prion strain diversity is contained entirely within PrP. The spectrum of possible PrP conformations must be constrained by the sequence of PrP, and it seems likely that only a limited number of distinct conformations for PrP^{Sc} of any particular sequence are possible. In turn, each prion strain may well be composed of several of these individual conformers of PrP^{Sc}. Admittedly, more detailed analysis of the molecular architecture of distinct prion strains will be required before the exact mechanism of strain diversity is resolved. Despite this, the picture that is rapidly emerging from these studies illustrates a mechanism that seems easily capable of supporting a variety of distinct prion strains encoded by a single PrP polypeptide sequence.

THE SPECIES AND STRAIN OF PRIONS REFLECT THE RELATIONSHIP BETWEEN SEQUENCE AND CONFORMATION

The recent advances described above in our understanding of the role of the primary and tertiary structures of PrP in the transmission of disease have given new insights into the pathogenesis of the prion diseases. It appears that the amino acid sequence of PrP encodes the species of the prion (Scott et al. 1989; Telling et al. 1995) and the prion derives its PrP^{Sc} sequence from the last mammal in which it was passaged (Bockman et al. 1987). Whereas the primary structure of PrP is likely to be the most important, or even the sole, determinant of the tertiary structure of PrP^{C}, existing PrP^{Sc} seems to function as a template in determining the tertiary structure of nascent PrP^{Sc} molecules as they are formed from PrP^{C} (Prusiner 1991; Cohen et al. 1994). In turn, prion diversity is enciphered in the conformation of PrP^{Sc}, and prion strains may represent different conformers of PrP^{Sc} (Bessen and Marsh 1992a; Telling et al. 1996). Furthermore, we have argued that the species barrier observed during transmission of prions between hosts with distinct PrP genes may, at least in some instances, reflect a change in the strain of prion (Scott et al.

1997a). This can easily explain how multiple different isolates are identified when prions cross species barriers—they are either generated or selected from a preexisting mixture. Presumably, new strains are created because the original strain conformer is energetically unfavorable when folded from the new PrP sequence (Scott et al. 1997a).

Although the foregoing scenario seems to be unprecedented in biology, considerable experimental data now support these concepts. Two prion-like elements, [URE3] and [PSI], have been described in yeast, and another in another fungus denoted [Het-s*] (Wickner 1994; Chernoff et al. 1995; Coustou et al. 1997; Glover et al. 1997; King et al. 1997; Paushkin et al. 1997). Conversion to the prion-like [PSI+] state in yeast can be modified by the participation of the molecular chaperone Hsp104 (Chernoff et al. 1995; Patino et al. 1996). It may prove to be highly significant that different strains of yeast prions have already been identified (Derkatch et al. 1996). There are also remarkable similarities between the behavior of chimeric yeast prion strains during in vitro propagation (Chien and Weissmann 2001) and the transmission of scrapie and TME prion strains to Tg mice expressing chimeric PrP genes (Scott et al. 1993; Peretz et al. 2002). Other parallels exist in higher organisms. Refolding studies of hen and human lysozyme have clearly shown that the folding pathways of homologous proteins can differ markedly as a result of amino acid replacements (Hooke et al. 1994).

REFERENCES

Alper T., Haig D.A., and Clarke M.C. 1966. The exceptionally small size of the scrapie agent. Biochem. Biophys. Res. Commun. 22: 278–284.

Anderson R.M., Donnelly C.A., Ferguson N.M., Woolhouse M.E.J., Watt C.J., Udy H.J., MaWhinney S., Dunstan S.P., Southwood T.R.E., Wilesmith J.W., Ryan J.B.M., Hoinville L.J., Hillerton J.E., Austin A.R., and Wells G.A.H. 1996. Transmission dynamics and epidemiology of BSE in British cattle. Nature 382: 779–788.

Asante E.A., Linehan J.M., Desbruslais M., Joiner S., Gowland I., Wood A.L., Welch J., Hill A.F., Lloyd S.E., Wadsworth J.D., and Collinge J. 2002. BSE prions propagate as either variant CJD-like or sporadic CJD-like prion strains in transgenic mice expressing human prion protein. EMBO J. 21: 6358–6366.

Baker H.F., Ridley R.M., and Wells G.A.H. 1993. Experimental transmission of BSE and scrapie to the common marmoset. Vet. Rec. 132: 403–406.

Bamborough P., Wille H., Telling G.C., Yehiely F., Prusiner S.B., and Cohen F.E. 1996. Prion protein structure and scrapie replication: Theoretical, spectroscopic and genetic investigations. Cold Spring Harbor Symp. Quant. Biol. 61: 495–509.

Barron R.M., Thomson V., Jamieson E., Melton D.W., Ironside J., Will R., and Manson J.C. 2001. Changing a single amino acid in the N-terminus of murine PrP alters TSE incubation time across three species barriers. EMBO J. 20: 5070–5078.

Bartz J.C., Bessen R.A., McKenzie D., Marsh R.F. and Aiken J.M. 2000. Adaptation and selection of prion protein strain conformations following interspecies transmission of transmissible mink encephalopathy. *J. Virol.* **74:** 5542–5547.

Bessen R.A. and Marsh R.F. 1992a. Biochemical and physical properties of the prion protein from two strains of the transmissible mink encephalopathy agent. *J. Virol.* **66:** 2096–2101.

———. 1992b. Identification of two biologically distinct strains of transmissible mink encephalopathy in hamsters. *J. Gen. Virol.* **73:** 329–334.

———. 1994. Distinct PrP properties suggest the molecular basis of strain variation in transmissible mink encephalopathy. *J. Virol.* **68:** 7859–7868.

Bessen R.A., Kocisko D.A., Raymond G.J., Nandan S., Lansbury P.T., and Caughey B. 1995. Non-genetic propagation of strain-specific properties of scrapie prion protein. *Nature* **375:** 698–700.

Bockman J.M., Prusiner S.B., Tateishi J., and Kingsbury D.T. 1987. Immunoblotting of Creutzfeldt-Jakob disease prion proteins: Host species-specific epitopes. *Ann. Neurol.* **21:** 589–595.

Bolton D.C., McKinley M.P., and Prusiner S.B. 1982. Identification of a protein that purifies with the scrapie prion. *Science* **218:** 1309–1311.

Bons N., Mestre-Francés N., Charnay Y., and Tagliavini F. 1996. Spontaneous spongiform encephalopathy in a young adult rhesus monkey. *Lancet* **348:** 55.

Bons N., Mestre-Francés N., Belli P., Cathala F., Gajdusek D.C., and Brown P. 1999. Natural and experimental oral infection of nonhuman primates by bovine spongiform encephalopathy agents. *Proc. Natl. Acad. Sci.* **96:** 4046–4051.

Borchelt D.R., Scott M., Taraboulos A., Stahl N., and Prusiner S.B. 1990. Scrapie and cellular prion proteins differ in their kinetics of synthesis and topology in cultured cells. *J. Cell Biol.* **110:** 743–752.

Bosque P.J., Ryou C., Telling G., Peretz D., Legname G., DeArmond S.J., and Prusiner S.B. 2002. Prions in skeletal muscle. *Proc. Natl. Acad. Sci.* **99:** 3812–3817.

Bruce M.E. 1993. Scrapie strain variation and mutation. *Br. Med. Bull.* **49:** 822–838.

Bruce M.E. and Dickinson A.G. 1979. Biological stability of different classes of scrapie agent. In *Slow transmissible diseases of the nervous system* (ed. S.B. Prusiner and W.J. Hadlow), vol. 2, pp. 71–86. Academic Press, New York.

———. 1985. Genetic control of amyloid plaque production and incubation period in scrapie-infected mice. *J. Neuropathol. Exp. Neurol.* **44:** 285–294.

———. 1987. Biological evidence that the scrapie agent has an independent genome. *J. Gen. Virol.* **68:** 79–89.

Bruce M.E., Dickinson A.G., and Fraser H. 1976. Cerebral amyloidosis in scrapie in the mouse: Effect of agent strain and mouse genotype. *Neuropathol. Appl. Neurobiol.* **2:** 471–478.

Bruce M., Chree A., McConnell I., Foster J., and Fraser H. 1993. Transmissions of BSE, scrapie and related diseases to mice. In *Abstracts from the Proceedings of the 9th International Congress of Virology*, Glasgow, Scotland, p. 93.

Bruce M.E., Fraser H., McBride P.A., Scott J.R., and Dickinson A.G. 1992. The basis of strain variation in scrapie. In *Prion diseases in human and animals* (ed. S.B. Prusiner et al.), pp. 497–508. Ellis Horwood, London.

Bruce M.E., Chree A., McConnell I., Foster J., Pearson G., and Fraser H. 1994. Transmission of bovine spongiform encephalopathy and scrapie to mice: Strain variation and the species barrier. *Philos. Trans. R. Soc. Lond. B Biol. Sci.* **343:** 405–411.

Bruce M.E., Will R.G., Ironside J.W., McConnell I., Drummond D., Suttie A., McCardle L., Chree A., Hope J., Birkett C., Cousens S., Fraser H., and Bostock C.J. 1997.

Transmissions to mice indicate that "new variant" CJD is caused by the BSE agent. *Nature* **389:** 498–501.

Büeler H., Aguzzi A., Sailer A., Greiner R.-A., Autenried P., Aguet M., and Weissmann C. 1993. Mice devoid of PrP are resistant to scrapie. *Cell* **73:** 1339–1347.

Büeler H., Fischer M., Lang Y., Bluethmann H., Lipp H.-P., DeArmond S.J., Prusiner S.B., Aguet M., and Weissmann C. 1992. Normal development and behaviour of mice lacking the neuronal cell-surface PrP protein. *Nature* **356:** 577–582.

Carlson G.A., Kingsbury D.T., Goodman P.A., Coleman S., Marshall S.T., DeArmond S., Westaway D., and Prusiner S.B. 1986. Linkage of prion protein and scrapie incubation time genes. *Cell* **46:** 503–511.

Carlson G.A., Ebeling C., Yang S.-L., Telling G., Torchia M., Groth D., Westaway D., DeArmond S.J., and Prusiner S.B. 1994. Prion isolate specified allotypic interactions between the cellular and scrapie prion proteins in congenic and transgenic mice. *Proc. Natl. Acad. Sci.* **91:** 5690–5694.

Carp R.I., Meeker H., and Sersen E. 1997. Scrapie strains retain their distinctive characteristics following passages of homogenates from different brain regions and spleen. *J. Gen. Virol.* **78:** 283–290.

Caughey B. and Raymond G.J. 1991. The scrapie-associated form of PrP is made from a cell surface precursor that is both protease- and phospholipase-sensitive. *J. Biol. Chem.* **266:** 18217–18223.

Chandler R.L. 1961. Encephalopathy in mice produced by inoculation with scrapie brain material. *Lancet* **I:** 1378–1379.

Chazot G., Broussolle E., Lapras C.I., Blättler T., Aguzzi A., and Kopp N. 1996. New variant of Creutzfeldt-Jakob disease in a 26-year-old French man. *Lancet* **347:** 1181.

Chernoff Y.O., Lindquist S.L., Ono B., Inge-Vechtomov S.G., and Liebman S.W. 1995. Role of the chaperone protein Hsp104 in propagation of the yeast prion-like factor [*psi*+]. *Science* **268:** 880–884.

Chien P. and Weissman J.S. 2001. Conformational diversity in a yeast prion dictates its seeding specificity. *Nature* **410:** 223–227.

Clousard C., Beaudry P., Elsen J.M., Milan D., Dussaucy M., Bounneau C., Schelcher F., Chatelain J., Launay J.M., and Laplanche J.L. 1995. Different allelic effects of the codons 136 and 171 of the prion protein gene in sheep with natural scrapie. *J. Gen. Virol.* **76:** 2097–2101.

Cohen F.E., Pan K.-M., Huang Z., Baldwin M., Fletterick R.J., and Prusiner S.B. 1994. Structural clues to prion replication. *Science* **264:** 530–531.

Collinge J. 1999. Variant Creutzfeldt-Jakob disease. *Lancet* **354:** 317–323.

Collinge J., Hill A.F., Sidle K.C.L., and Ironside J. 1997. Biochemical typing of scrapie strains (reply). *Nature* **386:** 564.

Collinge J., Sidle K.C.L., Meads J., Ironside J., and Hill A.F. 1996. Molecular analysis of prion strain variation and the aetiology of "new variant" CJD. *Nature* **383:** 685–690.

Cousens S.N., Harries-Jones R., Knight R., Will R.G., Smith P.G., and Matthews W.B. 1990. Geographical distribution of cases of Creutzfeldt-Jakob disease in England and Wales 1970–1984. *J. Neurol. Neurosurg. Psychiatry* **53:** 459–465.

Coustou V., Deleu C., Saupe S., and Begueret J. 1997. The protein product of the *het-s* heterokaryon incompatibility gene of the fungus *Podospora anserina* behaves as a prion analog. *Proc. Natl. Acad. Sci.* **94:** 9773–9778.

Cuillé J. and Chelle P.L. 1939. Experimental transmission of trembling to the goat. *C.R. Seances Acad. Sci.* **208:** 1058–1060.

Dawson M., Wells G.A.H., and Parker B.N.J. 1990a. Preliminary evidence of the experimental transmissibility of bovine spongiform encephalopathy to cattle. *Vet. Rec.* **126:** 112–113.

Dawson M., Wells G.A.H., Parker B.N.J., and Scott A.C. 1990b. Primary parenteral transmission of bovine spongiform encephalopathy to the pig. *Vet. Rec.* **127:** 338.

DeArmond S.J., Mobley W.C., DeMott D.L., Barry R.A., Beckstead J.H., and Prusiner S.B. 1987. Changes in the localization of brain prion proteins during scrapie infection. *Neurology* **37:** 1271–1280.

DeArmond S.J., Yang S.-L., Lee A., Bowler R., Taraboulos A., Groth D., and Prusiner S.B. 1993. Three scrapie prion isolates exhibit different accumulation patterns of the prion protein scrapie isoform. *Proc. Natl. Acad. Sci.* **90:** 6449–6453.

DeArmond S.J., Sánchez H., Yehiely F., Qiu Y., Ninchak-Casey A., Daggett V., Camerino A.P., Cayetano J., Rogers M., Groth D., Torchia M., Tremblay P., Scott M.R., Cohen F.E., and Prusiner S.B. 1997. Selective neuronal targeting in prion disease. *Neuron* **19:** 1337–1348.

Derkatch I.L., Chernoff Y.O., Kushnirov V.V., Inge-Vechtomov S.G., and Liebman S.W. 1996. Genesis and variability of [*PSI*] prion factors in *Saccharomyces cerevisiae*. *Genetics* **144:** 1375–1386.

Dickinson A.G. 1976. Scrapie in sheep and goats. In *Slow virus diseases of animals and man* (ed. R.H. Kimberlin), pp. 209–241. Elsevier/North-Holland, Amsterdam.

Dickinson A.G. and Fraser H. 1979. An assessment of the genetics of scrapie in sheep and mice. In *Slow transmissible diseases of the nervous system* (ed. S.B. Prusiner and W.J. Hadlow), vol. 1, pp. 367–386. Academic Press, New York.

Dickinson A.G. and Meikle V.M. 1969. A comparison of some biological characteristics of the mouse-passaged scrapie agents, 22A and ME7. *Genet. Res.* **13:** 213–225.

Dickinson A.G. and Outram G.W. 1988. Genetic aspects of unconventional virus infections: The basis of the virino hypothesis. *Ciba Found. Symp.* **135:** 63–83.

Dickinson A.G., Meikle V.M.H., and Fraser H. 1968. Identification of a gene which controls the incubation period of some strains of scrapie agent in mice. *J. Comp. Pathol.* **78:** 293–299.

———. 1969. Genetical control of the concentration of ME7 scrapie agent in the brain of mice. *J. Comp. Pathol.* **79:** 15–22.

Dickinson A.G., Young G.B., Stamp J.T., and Renwick C.C. 1965. An analysis of natural scrapie in Suffolk sheep. *Heredity* **20:** 485–503.

Edenhofer F., Rieger R., Famulok M., Wendler W., Weiss S., and Winnacker E.-L. 1996. Prion protein PrP^C interacts with molecular chaperones of the Hsp60 family. *J. Virol.* **70:** 4724–4728.

Endo T., Groth D., Prusiner S.B., and Kobata A. 1989. Diversity of oligosaccharide structures linked to asparagines of the scrapie prion protein. *Biochemistry* **28:** 8380–8388.

Ferguson N.M., Donnelly C.A., Woolhouse M.E.J. and Anderson R.M. 1997. The epidemiology of BSE in cattle herds in Great Britain. II. Model construction and analysis of transmission dynamics. *Phil. Trans. R. Soc. Lond. B Biol. Sci.* **352:** 803–838.

Fischer M., Rülicke T., Raeber A., Sailer A., Moser M., Oesch B., Brandner S., Aguzzi A., and Weissmann C. 1996. Prion protein (PrP) with amino-proximal deletions restoring susceptibility of PrP knockout mice to scrapie. *EMBO J.* **15:** 1255–1264.

Fraser H. and Dickinson A.G. 1968. The sequential development of the brain lesions of scrapie in three strains of mice. *J. Comp. Pathol.* **78:** 301–311.

———. 1973. Scrapie in mice. Agent-strain differences in the distribution and intensity of grey matter vacuolation. *J. Comp. Pathol.* **83:** 29–40.

Fraser H., McConnell I., Wells G.A.H., and Dawson M. 1988. Transmission of bovine spongiform encephalopathy to mice. *Vet. Rec.* **123:** 472.

Fraser H., Bruce M.E., Chree A., McConnell I., and Wells G.A.H. 1992. Transmission of bovine spongiform encephalopathy and scrapie to mice. *J. Gen. Virol.* **73:** 1891–1897.

Gabriel J.-M., Oesch B., Kretzschmar H., Scott M., and Prusiner S.B. 1992. Molecular cloning of a candidate chicken prion protein. *Proc. Natl. Acad. Sci.* **89:** 9097–9101.

Gajdusek D.C. 1977. Unconventional viruses and the origin and disappearance of kuru. *Science* **197:** 943–960.

———. 1985. Subacute spongiform virus encephalopathies caused by unconventional viruses. In *Subviral pathogens of plants and animals: Viroids and prions* (ed. K. Maramorosch and J.J. McKelvey, Jr.), pp. 483–544. Academic Press, Orlando, Florida.

Gauczynski S., Peyrin J.M., Haik S., Leucht C., Hundt C., Rieger R., Krasemann S., Deslys J.P., Dormont D., Lasmézas C., and Weiss S. 2001. The 37-kDa/67-kDa laminin receptor acts as the cell-surface receptor for the cellular prion protein. *EMBO J.* **20:** 5863–5875.

Glover J.R., Kowal A.S., Schirmer E.C., Patino M.M., Liu J.-J., and Lindquist S. 1997. Self-seeded fibers formed by Sup35, the protein determinant of [*PSI*⁺], a heritable prion-like factor of *S. cerevisiae. Cell* **89:** 811–819.

Goldfarb L., Korczyn A., Brown P., Chapman J., and Gajdusek D.C. 1990. Mutation in codon 200 of scrapie amyloid precursor gene linked to Creutzfeldt-Jakob disease in Sephardic Jews of Libyan and non-Libyan origin. *Lancet* **336:** 637–638.

Goldfarb L.G., Brown P., Little B.W., Cervenáková L., Kenney K., Gibbs C.J., Jr., and Gajdusek D.C. 1993. A new (two-repeat) octapeptide coding insert mutation in Creutzfeldt-Jakob disease. *Neurology* **43:** 2392–2394.

Goldfarb L.G., Haltia M., Brown P., Nieto A., Kovanen J., McCombie W.R., Trapp S., and Gajdusek D.C. 1991. New mutation in scrapie amyloid precursor gene (at codon 178) in Finnish Creutzfeldt-Jakob kindred. *Lancet* **337:** 425.

Goldfarb L.G., Petersen R.B., Tabaton M., Brown P., LeBlanc A.C., Montagna P., Cortelli P., Julien J, Vital C., Pendelbury W.W., Haltia M., Wills P.R., Hauw J.J., McKeever P.E., Monari L., Schrank B., Swergold G.D, Autilio-Gambetti L., Gajdusek D.C., Lugaresi E., and Gambetti P. 1992. Fatal familial insomnia and familial Creutzfeldt-Jakob disease: Disease phenotype determined by a DNA polymorphism. *Science* **258:** 806–808.

Goldmann W., Hunter N., Manson J., and Hope J. 1990a. The PrP gene of the sheep, a natural host of scrapie. In *Abstracts from the Proceedings of the 8th International Congress of Virology,* Berlin, p. 284.

Goldmann W., Hunter N., Benson G., Foster J.D., and Hope J. 1991a. Different scrapie-associated fibril proteins (PrP) are encoded by lines of sheep selected for different alleles of the *Sip* gene. *J. Gen. Virol.* **72:** 2411–2417.

Goldmann W., Hunter N., Martin T., Dawson M., and Hope J. 1991b. Different forms of the bovine PrP gene have five or six copies of a short, G-C-rich element within the protein-coding exon. *J. Gen. Virol.* **72:** 201–204.

Goldmann W., Hunter N., Foster J.D., Salbaum J.M., Beyreuther K., and Hope J. 1990b. Two alleles of a neural protein gene linked to scrapie in sheep. *Proc. Natl. Acad. Sci.* **87:** 2476–2480.

Graner E., Mercadante A.F., Zanata S.M., Forlenza O.V., Cabral A.L.B., Veiga S.S., Juliano M.A., Roesler R., Walz R., Mineti A., Izquierdo I., Martins V.R., and Brentani R.R. 2000. Cellular prion protein binds laminin and mediates neuritogenesis. *Mol. Brain Res.* **76:** 85–92.

Griffith J.S. 1967. Self-replication and scrapie. *Nature* **215:** 1043–1044.

Hadlow W.J., Kennedy R.C., and Race R.E. 1982. Natural infection of Suffolk sheep with scrapie virus. *J. Infect. Dis.* **146:** 657–664.

Haraguchi T., Fisher S., Olofsson S., Endo T., Groth D., Tarantino A., Borchelt D.R., Teplow D., Hood L., Burlingame A., Lycke E., Kobata A., and Prusiner S.B. 1989. Asparagine-linked glycosylation of the scrapie and cellular prion proteins. *Arch. Biochem. Biophys.* **274:** 1–13.

Heaton M.P., Leymaster K.A., Freking B.A., Hawk D.A., Keele J.W., Snelling W.M., Fox J.M., Chitko-McKown C.G., and Laegreid W.W. 2003. Prion gene sequence variation within diverse groups of US sheep, beef cattle, and deer. *Mamm. Genome* (in press).

Hecker R., Taraboulos A., Scott M., Pan K.-M., Torchia M., Jendroska K., DeArmond S.J., and Prusiner S.B. 1992. Replication of distinct prion isolates is region specific in brains of transgenic mice and hamsters. *Genes Dev.* **6:** 1213–1228.

Hill A.F., Joiner S., Linehan J., Desbruslais M., Lantos P.L., and Collinge J. 2000. Species-barrier-independent prion replication in apparently resistant species. *Proc. Natl. Acad. Sci.* **97:** 10248–10253.

Hill A.F., Desbruslais M., Joiner S., Sidle K.C.L., Gowland I., Collinge J., Doey L.J., and Lantos P. 1997. The same prion strain causes vCJD and BSE. *Nature* **389:** 448–450.

Hooke S.D., Radford S.E., and Dobson C.M. 1994. The refolding of human lysozyme: A comparison with the structurally homologous hen lysozyme. *Biochemistry* **33:** 5867–5876.

Hope J., Reekie L.J.D., Hunter N., Multhaup G., Beyreuther K., White H., Scott A.C., Stack M.J., Dawson M., and Wells G.A.H. 1988. Fibrils from brains of cows with new cattle disease contain scrapie-associated protein. *Nature* **336:** 390–392.

Hsiao K.K., Groth D., Scott M., Yang S.-L., Serban H., Rapp D., Foster D., Torchia M., DeArmond S.J., and Prusiner S.B. 1994. Serial transmission in rodents of neurodegeneration from transgenic mice expressing mutant prion protein. *Proc. Natl. Acad. Sci.* **91:** 9126–9130.

Hsiao K., Meiner Z., Kahana E., Cass C., Kahana I., Avrahami D., Scarlato G., Abramsky O., Prusiner S.B., and Gabizon R. 1991. Mutation of the prion protein in Libyan Jews with Creutzfeldt-Jakob disease. *N. Engl. J. Med.* **324:** 1091–1097.

Hundt C., Peyrin J.M., Haik S., Gauczynski S., Leucht C., Rieger R., Riley M.L., Deslys J.P., Dormont D., Lasmézas C.I., and Weiss S. 2001. Identification of interaction domains of the prion protein with its 37-kDa/67-kDa laminin receptor. *EMBO J.* **20:** 5876–5886.

Hunter G.D. 1972. Scrapie: A prototype slow infection. *J. Infect. Dis.* **125:** 427–440.

Hunter N., Foster J.D., Benson G., and Hope J. 1991. Restriction fragment length polymorphisms of the scrapie-associated fibril protein (PrP) gene and their association with susceptiblity to natural scrapie in British sheep. *J. Gen. Virol.* **72:** 1287–1292.

Hunter N., Goldmann W., Smith G., and Hope J. 1994. Frequencies of PrP gene variants in healthy cattle and cattle with BSE in Scotland. *Vet. Rec.* **135:** 400–403.

Hunter N., Hope J., McConnell I., and Dickinson A.G. 1987. Linkage of the scrapie-associated fibril protein (PrP) gene and *Sinc* using congenic mice and restriction fragment length polymorphism analysis. *J. Gen. Virol.* **68:** 2711–2716.

Hunter N., Moore L., Hosie B.D., Dingwall W.S., and Greig A. 1997a. Association between natural scrapie and PrP genotype in a flock of Suffolk sheep in Scotland. *Vet. Rec.* **140:** 59–63.

Hunter N., Cairns D., Foster J.D., Smith G., GoldmannW., and Donnelly K. 1997b. Is scrapie solely a genetic disease? *Nature* **386:** 137.

James T.L., Liu H., Ulyanov N.B., Farr-Jones S., Zhang H., Donne D.G., Kaneko K., Groth D., Mehlhorn I., Prusiner S.B., and Cohen F.E. 1997. Solution structure of a 142-residue recombinant prion protein corresponding to the infectious fragment of the scrapie iso-form. *Proc. Natl. Acad. Sci.* **94:** 10086–10091.

Jendroska K., Heinzel F.P., Torchia M., Stowring L., Kretzschmar H.A., Kon A., Stern A., Prusiner S.B., and DeArmond S.J. 1991. Proteinase-resistant prion protein accumulation in Syrian hamster brain correlates with regional pathology and scrapie infectivity. *Neurology* **41:** 1482–1490.

Kahana E., Milton A., Braham J., and Sofer D. 1974. Creutzfeldt-Jakob disease: Focus among Libyan Jews in Israel. *Science* **183:** 90–91.

Kaneko K., Zulianello L., Scott M., Cooper C.M., Wallace A.C., James T.L., Cohen F.E., and Prusiner S.B. 1997. Evidence for protein X binding to a discontinuous epitope on the cellular prion protein during scrapie prion propagation. *Proc. Natl. Acad. Sci.* **94:** 10069–10074.

Kascsak R.J., Rubenstein R., Merz P.A., Tonna-DeMasi M., Fersko R., Carp R.I., Wisniewski H.M., and Diringer H. 1987. Mouse polyclonal and monoclonal antibody to scrapie-associated fibril proteins. *J. Virol.* **61:** 3688–3693.

Kenward N., Hope J., Landon M., and Mayer R.J. 1994. Expression of polyubiquitin and heat-shock protein 70 genes increases in the later stages of disease progression in scrapie-infected mouse brain. *J. Neurochem.* **62:** 1870–1877.

Keshet G.I., Bar-Peled O., Yaffe D., Nudel U., and Gabizon R. 2000. The cellular prion protein colocalizes with the dystroglycan complex in the brain. *J. Neurochem.* **75:** 1889–1897.

Kimberlin R.H. 1982. Scrapie agent: Prions or virinos? *Nature* **297:** 107–108.

Kimberlin R. and Walker C. 1977. Characteristics of a short incubation model of scrapie in the golden hamster. *J. Gen. Virol.* **34:** 295–304.

———. 1978. Evidence that the transmission of one source of scrapie agent to hamsters involves separation of agent strains from a mixture. *J. Gen. Virol.* **39:** 487–496.

Kimberlin R.H., Cole S., and Walker C.A. 1986. Transmissible mink encephalopathy (TME) in Chinese hamsters: Identification of two strains of TME and comparisons with scrapie. *Neuropathol. Appl. Neurobiol.* **12:** 197 206.

———. 1987. Temporary and permanent modifications to a single strain of mouse scrapie on transmission to rats and hamsters. *J. Gen. Virol.* **68:** 1875–1881.

Kimberlin R.H., Walker C.A., and Fraser H 1989 The genomic identity of different strains of mouse scrapie is expressed and preserved on reisolation in mice. *J. Gen. Virol.* **70:** 2017–2025.

King C.-Y., Tittman P., Gross H., Gebert R., Aebi M., and Wüthrich K. 1997. Prion-inducing domain 2-114 of yeast Sup35 protein transforms *in vitro* into amyloid-like filaments. *Proc. Natl. Acad. Sci.* **94:** 6618–6622.

Kretzschmar H.A., Neumann M., and Stavrou D. 1995. Codon 178 mutation of the human prion protein gene in a German family (Backer family): Sequencing data from 72 year-old celloidin-embedded brain tissue. *Acta Neuropathol.* **89:** 96–98.

Kretzschmar H.A., Stowring L.E., Westaway D., Stubblebine W.H., Prusiner S.B., and DeArmond S.J. 1986. Molecular cloning of a human prion protein cDNA. *DNA* **5:** 315–324.

Kurschner C. and Morgan J.I. 1995. The cellular prion protein (PrP) selectively binds to Bcl-2 in the yeast two-hybrid system. *Mol. Brain Res.* **30:** 165–168.

———. 1996. Analysis of interaction sites in homo- and heteromeric complexes containing Bcl-2 family members and the cellular prion protein. *Mol. Brain Res.* **37:** 249–258.

Laplanche J.-L., Chatelain J., Beaudry P., Dussaucy M., Bounnea C., and Launay J.-M. 1993. French autochthonous scrapied sheep without the 136Val PrP polymorphism. *Mamm. Genome* **4:** 463–464.

Lasmézas C.I., Deslys J.-P., Demaimay R., Adjou K.T., Lamoury F., Dormont D., Robain O., Ironside J., and Hauw J.-J. 1996. BSE transmission to macaques. *Nature* **381:** 743–744.

Lasmézas C.I., Deslys J.-P., Robain O., Jaegly A., Beringue V., Peyrin J.-M., Fournier J.-G., Hauw J.-J., Rossier J., and Dormont D. 1997. Transmission of the BSE agent to mice in the absence of detectable abnormal prion protein. *Science* **275:** 402–405.

Lowenstein D.H., Butler D.A., Westaway D., McKinley M.P., DeArmond S.J., and Prusiner S.B. 1990. Three hamster species with different scrapie incubation times and neuropathological features encode distinct prion proteins. *Mol. Cell. Biol.* **10:** 1153–1163.

Lugaresi E., Medori R., Montagna P., Baruzzi A., Cortelli P., Lugaresi A., Tinuper P., Zucconi M., and Gambetti P. 1986. Fatal familial insomnia and dysautonomia with selective degeneration of thalamic nuclei. *N. Engl. J. Med.* **315:** 997–1003.

Malmgren R., Kurland L., Mokri B., and Kurtzke J. 1979. The epidemiology of Creutzfeldt-Jakob disease. In *Slow transmissible diseases of the nervous system* (ed. S.B. Prusiner and W.J. Hadlow), vol. 1, pp. 93–112. Academic Press, New York.

Marsh R.F. 1992. Transmissible mink encephalopathy. In *Prion diseases of humans and animals* (ed. S.B. Prusiner et al.), pp. 300–307. Ellis Horwood, London.

Marsh R.F. and Kimberlin R.H. 1975. Comparison of scrapie and transmissible mink encephalopathy in hamsters. II. Clinical signs, pathology and pathogenesis. *J. Infect. Dis.* **131:** 104–110.

Martins V.R., Graner E., Garcia-Abreu J., de Souza S.J., Mercadante A.F., Veiga S.S., Zanata S.M., Neto V.M., and Brentani R.R. 1997. Complementary hydropathy identifies a cellular prion protein receptor. *Nat. Med.* **3:** 1376–1382.

Mastrianni J., Nixon F., Layzer R., DeArmond S.J., and Prusiner S.B. 1997. Fatal sporadic insomnia: Fatal familial insomnia phenotype without a mutation of the prion protein gene. *Neurology* (suppl.) **48:** A296.

Medori R., Montagna P., Tritschler H.J., LeBlanc A., Cortelli P., Tinuper P., Lugaresi E., and Gambetti P. 1992a. Fatal familial insomnia: A second kindred with mutation of prion protein gene at codon 178. *Neurology* **42:** 669–670.

Medori R., Tritschler H.-J., LeBlanc A., Villare F., Manetto V., Chen H.Y., Xue R., Leal S., Montagna P., Cortelli P., Tinuper P., Avoni P., Mochi M., Baruzzi A., Hauw J.J., Ott J., Lugaresi E., Autilio-Gambetti L., and Gambetti P. 1992b. Fatal familial insomnia, a prion disease with a mutation at codon 178 of the prion protein gene. *N. Engl. J. Med.* **326:** 444–449.

Meggendorfer F. 1930. Klinische und genealogische Beobachtungen bei einem Fall von spastischer Pseudosklerose Jakobs. *Z. Gesamte Neurol. Psychiatr.* **128:** 337–341.

Middleton D.J. and Barlow R.M. 1993. Failure to transmit bovine spongiform encephalopathy to mice by feeding them with extraneural tissues of affected cattle. *Vet. Rec.* **132:** 545–547.

Monari L., Chen S.G., Brown P., Parchi P., Petersen R.B., Mikol J., Gray F., Cortelli P., Montagna P., Ghetti B., Goldfarb L.G., Gajdusek D.C., Lugaresi E., Gambetti P., and Autilio-Gambetti L. 1994. Fatal familial insomnia and familial Creutzfeldt-Jakob disease: Different prion proteins determined by a DNA polymorphism. *Proc. Natl. Acad. Sci.* **91:** 2839–2842.

Mouillet-Richard S., Ermonual M., Chebassier C., Laplanche J.L., Lehmann S., Launay J.M., and Kellermann O. 2000. Signal transduction through prion protein. *Science* **289:** 1925–1928.

Muramoto T., Kitamoto T., Tateishi J., and Goto I. 1992. The sequential development of abnormal prion protein accumulation in mice with Creutzfeldt-Jakob disease. *Am. J. Pathol.* **140:** 1411–1420.

Oesch B., Teplow D.B., Stahl N., Serban D., Hood L.E., and Prusiner S.B. 1990. Identification of cellular proteins binding to the scrapie prion protein. *Biochemistry* **29:** 5848–5855.

Pan T., Colucci M., Wong B.-S., Li R., Liu T., Petersen R.B., Chen S., Gambetti P., and Sy M.-S. 2001. Novel differences between two human prion strains revealed by two-dimensional gel electrophoresis. *J. Biol. Chem.* **276:** 37284–37288.

Pan K.-M., Baldwin M., Nguyen J., Gasset M., Serban A., Groth D., Mehlhorn I., Huang Z., Fletterick R.J., Cohen F.E., and Prusiner S.B. 1993. Conversion of α-helices into β-sheets features in the formation of the scrapie prion proteins. *Proc. Natl. Acad. Sci.* **90:** 10962–10966.

Parchi P., Capellari S., Chen S.G., Petersen R.B., Gambetti P., Kopp P., Brown P., Kitamoto T., Tateishi J., Giese A., and Kretzschmar H. 1997. Typing prion isoforms (letter). *Nature* **386:** 232–233.

Parchi P., Zou W., Wang W., Brown P., Capellari S., Ghetti B., Kopp N., Schulz-Schaeffer W.J., Kretzschmar H.A., Head M.W., Ironside J.W., Gambetti P., and Chen S.G. 2000. Genetic influence on the structural variations of the abnormal prion protein. *Proc. Natl. Acad. Sci.* **97:** 10168–10172.

Parry H.B. 1962. Scrapie: A transmissible and hereditary disease of sheep. *Heredity* **17:** 75–105.

———. 1983. *Scrapie disease in sheep.* Academic Press, New York.

Patino M.M., Liu J.-J., Glover J.R., and Lindquist S. 1996. Support for the prion hypothesis for inheritance of a phenotypic trait in yeast. *Science* **273:** 622–626.

Pattison I.H. 1965. Experiments with scrapie with special reference to the nature of the agent and the pathology of the disease. In *Slow, latent and temperate virus infections* (NINDB Monogr. 2) (ed. D.C. Gajdusek et al.), pp. 249–257. U.S. Government Printing Office, Washington, D.C.

Pattison I.H. and Millson G.C. 1961a. Experimental transmission of scrapie to goats and sheep by the oral route. *J. Comp. Pathol.* **71:** 171–176.

———. 1961b. Scrapie produced experimentally in goats with special reference to the clinical syndrome. *J. Comp. Pathol.* **71:** 101–108.

Pattison I.H. and Smith K. 1963. Histological observations on experimental scrapie in the mouse. *Res. Vet. Sci.* **4:** 269–275.

Pattison I.H., Gordon W.S., and Millson G.C. 1959. Experimental production of scrapie in goats. *J. Comp. Pathol. Ther.* **69:** 300–312.

Paushkin S.V., Kushnirov V.V., Smirnov V.N., and Ter-Avanesyan M.D. 1997. In vitro propagation of the prion-like state of yeast Sup35 protein. *Science* **277:** 381–383.

Peretz D., Scott M., Groth D., Williamson A., Burton D., Cohen F.E., and Prusiner S.B. 2001. Strain-specified relative conformational stability of the scrapie prion protein. *Protein Sci.* **10:** 854–863.

Peretz D., Williamson R.A., Legname G., Matsunaga Y., Vergara J., Burton D.R., DeArmond S.J., Prusiner S.B., and Scott M.R. 2002. A change in the conformation of prions accompanies the emergence of a new prion strain. *Neuron* **34:** 921–932.

Prusiner S.B. 1982a. Novel proteinaceous infectious particles cause scrapie. *Science* **216:** 136–144.

———. 1982b. On prions causing dementia—Molecular studies of the scrapie agent. *Curr. Neurol.* **4:** 201–224.

———. 1989. Scrapie prions. *Annu. Rev. Microbiol.* **43:** 345–374.

———. 1991. Molecular biology of prion diseases. *Science* **252:** 1515–1522.

———. 1997. Prion diseases and the BSE crisis. *Science* **278:** 245–251.

———. 1998. Prions. *Proc. Natl. Acad. Sci.* **95:** 13363–13383.

Prusiner S.B., Cochran S.P., Downey D.E., and Groth D.F. 1981. Determination of scrapie agent titer from incubation period measurements in hamsters. In *Hamster immune responses in infectious and oncologic diseases* (ed. J.W. Streilein et al.), pp. 385–399. Plenum Press, New York.

Prusiner S.B., Scott M.R., DeArmond S.J., and Cohen F.E. 1998. Prion protein biology. *Cell* **93:** 337–348.

Prusiner S.B., Bolton D.C., Groth D.F., Bowman K.A., Cochran S.P., and McKinley M.P. 1982a. Further purification and characterization of scrapie prions. *Biochemistry* **21:** 6942–6950.

Prusiner S.B., Cochran S.P., Groth D.F., Downey D.E., Bowman K.A., and Martinez H.M. 1982b. Measurement of the scrapie agent using an incubation time interval assay. *Ann. Neurol.* **11:** 353–358.

Prusiner S.B., Fuzi M., Scott M., Serban D., Serban H., Taraboulos A., Gabriel J.-M., Wells G., Wilesmith J., Bradley R., DeArmond S.J., and Kristensson K. 1993. Immunologic and molecular biological studies of prion proteins in bovine spongiform encephalopathy. *J. Infect. Dis.* **167:** 602–613.

Prusiner S.B., Scott M., Foster D., Pan K.-M., Groth D., Mirenda C., Torchia M., Yang S.-L., Serban D., Carlson G.A., Hoppe P.C., Westaway D., and DeArmond S.J. 1990. Transgenetic studies implicate interactions between homologous PrP isoforms in scrapie prion replication. *Cell* **63:** 673–686.

Ridley R.M. and Baker H.F. 1996. To what extent is strain variation evidence for an independent genome in the agent of the transmissible spongiform encephalopathies? *Neurodegeneration* **5:** 219–231.

Rieger R., Edenhofer F., Lazmézas C., and Weiss S. 1997. The human 37-kDa laminin receptor precursor interacts with the prion protein in eukaryotic cells. *Nat. Med.* **3:** 1383–1388.

Robinson M.M., Hadlow W.J., Huff T.P., Wells G.A.H., Dawson M., Marsh R.F., and Gorham J.R. 1994. Experimental infection of mink with bovine spongiform encephalopathy. *J. Gen. Virol.* **75:** 2151–2155.

Safar J., Wille H., Itri V., Groth D., Serban H., Torchia M., Cohen F.E., and Prusiner S.B. 1998. Eight prion strains have PrP^{Sc} molecules with different conformations. *Nat. Med.* **4:** 1157–1165.

Safar J.G., Scott M., Monaghan J., Deering C., Didorenko S., Vergara J., Ball H., Legname G., Leclerc E., Solforosi L., Serban H., Groth D., Burton D.R., Prusiner S.B., and Williamson R.A. 2002. Measuring prions causing bovine spongiform encephalopathy or chronic wasting disease by immunoassays and transgenic mice. *Nat. Biotechnol.* **20:** 1147–1150.

Schätzl H.M., Da Costa M., Taylor L., Cohen F.E., and Prusiner S.B. 1995. Prion protein gene variation among primates. *J. Mol. Biol.* **245:** 362–374.

Schmitt-Ulms G., Legname G., Baldwin M.A., Ball H.L., Bradon N., Bosque P.J., Crossin K.L., Edelman G.M., DeArmond S.J., Cohen F.E., and Prusiner S.B. 2001. Binding of neural cell adhesion molecules (N-CAMs) to the cellular prion protein. *J. Mol. Biol.* **314:** 1209–1225.

Scott M.R., Köhler R., Foster D., and Prusiner S.B. 1992. Chimeric prion protein expression in cultured cells and transgenic mice. *Protein Sci.* **1:** 986–997.

Scott M., Groth D., Foster D., Torchia M., Yang S.-L., DeArmond S.J., and Prusiner S.B. 1993. Propagation of prions with artificial properties in transgenic mice expressing chimeric PrP genes. *Cell* **73:** 979–988.

Scott M.R., Groth D., Tatzelt J., Torchia M., Tremblay P., DeArmond S.J., and Prusiner S.B. 1997a. Propagation of prion strains through specific conformers of the prion protein. *J. Virol.* **71:** 9032–9044.

Scott M.R., Will R., Ironside J., Nguyen H.-O.B., Tremblay P., DeArmond S.J., and Prusiner S.B. 1999. Compelling transgenetic evidence for transmission of bovine spongiform encephalopathy prions to humans. *Proc. Natl. Acad. Sci.* **96:** 15137–15142.

Scott M., Foster D., Mirenda C., Serban D., Coufal F., Wälchli M., Torchia M., Groth D., Carlson G., DeArmond S.J., Westaway D., and Prusiner S.B. 1989. Transgenic mice expressing hamster prion protein produce species-specific scrapie infectivity and amyloid plaques. *Cell* **59:** 847–857.

Scott M.R., Safar J., Telling G., Nguyen O., Groth D., Torchia M., Koehler R., Tremblay P., Walther D., Cohen F.E., DeArmond S.J., and Prusiner S.B. 1997b. Identification of a prion protein epitope modulating transmission of bovine spongiform encephalopathy prions to transgenic mice. *Proc. Natl. Acad. Sci.* **94:** 14279–14284.

Sigurdsson B. 1954. Rida, a chronic encephalitis of sheep with general remarks on infections which develop slowly and some of their special characteristics. *Br. Vet. J.* **110:** 341–354.

Somerville R.A., Chong A., Mulqueen O.U., Birkett C.R., Wood S.C., and Hope J. 1997. Biochemical typing of scrapie strains. *Nature* **386:** 564.

Stender A. 1930. Weitere Beiträge zum Kapitel "Spastische Pseudosklerose Jakobs." *Z. Gesamte Neurol. Psychiatr.* **128:** 528–543.

Taraboulos A., Jendroska K., Serban D., Yang S.-L., DeArmond S.J., and Prusiner S.B. 1992. Regional mapping of prion proteins in brains. *Proc. Natl. Acad. Sci.* **89:** 7620–7624.

Tatzelt J., Prusiner S.B., and Welch W.J. 1996. Chemical chaperones interfere with the formation of scrapie prion protein. *EMBO J.* **15:** 6363–6373.

Tatzelt J., Zuo J., Voellmy R., Scott M., Hartl U., Prusiner S.B., and Welch W.J. 1995. Scrapie prions selectively modify the stress response in neuroblastoma cells. *Proc. Natl. Acad. Sci.* **92:** 2944–2948.

Taylor K.C. 1991. The control of bovine spongiform encephalopathy in Great Britain. *Vet. Rec.* **129:** 522–526.

Telling G.C., Scott M., Mastrianni J., Gabizon R., Torchia M., Cohen F.E., DeArmond S.J., and Prusiner S.B. 1995. Prion propagation in mice expressing human and chimeric PrP transgenes implicates the interaction of cellular PrP with another protein. *Cell* **83:** 79–90.

Telling G.C., Parchi P., DeArmond S.J., Cortelli P., Montagna P., Gabizon R., Mastrianni J., Lugaresi E., Gambetti P., and Prusiner S.B. 1996. Evidence for the conformation of the pathologic isoform of the prion protein enciphering and propagating prion diversity. *Science* **274:** 2079–2082.

Telling G.C., Scott M., Hsiao K.K., Foster D., Yang S.-L., Torchia M., Sidle K.C.L., Collinge J., DeArmond S.J., and Prusiner S.B. 1994. Transmission of Creutzfeldt-Jakob disease from humans to transgenic mice expressing chimeric human-mouse prion protein. *Proc. Natl. Acad. Sci.* **91:** 9936–9940.

Watson J.D. 1968. *The double helix: A personal account of the discovery of the structure of DNA.* Atheneum, New York.

Weiss R. 1997. Study of "mad cow" disease challenges prevailing theory. *International Herald Tribune* (January 18–19), p. 2.

Wells G.A.H., Scott A.C., Johnson C.T., Gunning R.F., Hancock R.D., Jeffrey M., Dawson M., and Bradley R. 1987. A novel progressive spongiform encephalopathy in cattle. *Vet. Rec.* **121:** 419–420.

Wells G.A.H., Hawkins S.A.C., Green R.B., Austin A.R., Dexter I., Spencer Y.I., Chaplin M.J., Stack M.J., and Dawson M. 1998. Preliminary observations on the pathogenesis of experimental bovine spongiform encephalopathy (BSE): An update. *Vet. Rec.* **142:** 103–106.

Westaway D., Goodman P.A., Mirenda C.A., McKinley M.P., Carlson G.A., and Prusiner S.B. 1987. Distinct prion proteins in short and long scrapie incubation period mice. *Cell* **51:** 651–662.

Westaway D., Zuliani V., Cooper C.M., Da Costa M., Neuman S., Jenny A.L., Detwiler L., and Prusiner S.B. 1994. Homozygosity for prion protein alleles encoding glutamine-171 renders sheep susceptible to natural scrapie. *Genes Dev.* **8:** 959–969.

Westaway D., Mirenda C.A., Foster D., Zebarjadian Y., Scott M., Torchia M., Yang S.-L., Serban H., DeArmond S.J., Ebeling C., Prusiner S.B., and Carlson G.A. 1991. Paradoxical shortening of scrapie incubation times by expression of prion protein transgenes derived from long incubation period mice. *Neuron* **7:** 59–68.

Wickner R.B. 1994. [URE3] as an altered URE2 protein: Evidence for a prion analog in *Saccharomyces cerevisiae*. *Science* **264:** 566–569.

Wilesmith J.W., Ryan J.B.M., and Atkinson M.J. 1991. Bovine spongiform encephalopathy—Epidemiologic studies on the origin. *Vet. Rec.* **128:** 199–203.

Wilesmith J.W., Ryan J.B.M., and Hueston W.D. 1992. Bovine spongiform encephalopathy: Case-control studies of calf feeding practices and meat and bonemeal inclusion in proprietary concentrates. *Res. Vet. Sci.* **52:** 323–331.

Will R.G., Cousens S.N., Farrington C.P., Smith P.G., Knight R.S.G., and Ironside J.W. 1999. Deaths from variant Creutzfeldt-Jakob disease. *Lancet* **353:** 979.

Will R.G., Ironside J.W., Zeidler M., Cousens S.N., Estibeiro K., Alperovitch A., Poser S., Pocchiari M., Hofman A., and Smith P.G. 1996. A new variant of Creutzfeldt-Jakob disease in the UK. *Lancet* **347:** 921–925.

Yehiely F., Bamborough P., Costa M.D., Perry B.J., Thinakaran G., Cohen F.E., Carlson G.A., and Prusiner S.B. 1997. Identification of candidate proteins binding to prion protein. *Neurobiol. Dis.* **3:** 339–355.

Zlotnik I. and Rennie J.C. 1965. Experimental transmission of mouse passaged scrapie to goats, sheep, rats and hamsters. *J. Comp. Pathol.* **75:** 147–157.

10

Cell Biology of Prions

David A. Harris

Department of Cell Biology and Physiology
Washington University School of Medicine
St. Louis, Missouri 63110

Peter J. Peters

The Netherlands Cancer Institute
Amsterdam 1066, The Netherlands

Albert Taraboulos

Hebrew University-Hadassah Medical School
Jerusalem 91120, Israel

Vishwanath Lingappa,[1,2] Stephen J. DeArmond,[1,3] and Stanley B. Prusiner[1,4]

[1]Institute for Neurodegenerative Diseases and
Departments of [2]Physiology, [2]Medicine, [3]Pathology,
[4]Neurology, and [4]Biochemistry and Biophysics,
University of California, San Francisco, California 94143

THE CELLULAR PRION PROTEIN (PrP^C) is a cell-surface glycoprotein anchored by a glycosylphosphatidylinositol (GPI) moiety (Stahl et al. 1987). PrP^C is expressed throughout the brain, particularly in neurons (Kretzschmar et al. 1986; Moser et al. 1995), and to a lesser extent in extraneural tissues (Bendheim et al. 1992; Ford et al. 2002). In the prion diseases, PrP^C is converted to an abnormal, conformationally altered isoform (PrP^{Sc}), which subsequently accumulates in the brain and results in extensive neurodegeneration with an inevitably fatal outcome (Prusiner 1996). Therefore, localizing PrP^C in the brain is an important step in understanding the biology of the normal protein and in mapping changes in models of experimental prion diseases.

EXPRESSION AND LOCALIZATION OF PrP^C IN BRAIN

The precise localization of PrP^C remains enigmatic, due to conflicting data obtained using different techniques. Immunohistochemical studies have described a somatic expression of PrP^C in neurons with no or only a minor signal in the neuropil (DeArmond et al. 1987; Piccardo et al. 1990; Safar et al. 1990; Bendheim et al. 1992; Verghese-Nikolakaki et al. 1999; Ford et al. 2002). Contradictory findings probably reflect the peculiarities inherently associated with pre-embedding techniques. Many immuno-electron microscopic procedures may result in a destruction of cellular membranes, possibly leading to an artificial redistribution of GPI-anchored proteins within the membrane.

These uncertainties with regard to the precise subcellular localization of PrP^C therefore encouraged two of us (P.J.P. and S.B.P.) to perform a quantitative study of the ultrastructural localization of PrP^C in the mouse brain. We used a sensitive and high-resolution detection method combining immunofluoresence and Immunogold labeling of 500- and 60-nm cryosections at light microscopy (LM) and electron microscopy (EM) levels, respectively. The method uses glutaraldehyde for both optimal fixation and to prevent the migration of GPI-anchored proteins, and circumvents the need for alcohol dehydration. We focused on the localization of PrP^C in the hippocampus and collected quantitative data of PrP^C distribution at the ultrastructural level (Mironov et al. 2003).

Brain tissue from four different mouse lines with the FVB background: (1) wild-type (wt) mice; (2) PrP-ablated ($Prnp^{0/0}$) mice (Büeler et al. 1993); (3) transgenic (Tg) 4053 mice overexpressing mouse PrP^C (Telling et al. 1996); and (4) Tg3045 mice overexpressing hamster PrP^C (Telling et al. 1996), were examined. For immunolabeling, PrP-specific recombinant antibody fragments (Fabs) D13, D18, R1, R2, E123, and E149 were used (Peretz et al. 1997, 2001; Williamson et al. 1998; Leclerc et al. 2001). Cryosections were prepared as described previously (Peters 2001), then incubated with protein A–gold (10 nm) for 20 minutes. Cellular and subcellular profiles were identified and defined according to previously described criteria (Peters et al. 1991).

Quantitative evaluation of hippocampal labeling was carried out on ultrathin cryosections of wt FVB mouse brain fixed with 2% para-formaldehyde–0.2% glutaraldehyde in order to avoid overexpression artifacts of the transgenes. The relative distributions of labeled PrP^C were determined by counting gold particles over plasma and intracellular membranes of selected hippocampal cells. Membrane (gold/μm) and area (gold/μm^2) labeling densities were estimated on micrographs with a final magnification of 32,000 by using point and intersection counting with a line and point lattice (10-mm distance) overlay, as described by Griffiths (1993). We analyzed cells taken from a vertical strip running through the CA1 area, from the stratum oriens to the hilus of the dentate gyrus. This included cells from the pyramidal cell layer, dentate granule cells, and cells from the hilus. The gold particles were counted in the following subcellular structures: endoplasmic reticulum (ER; including the nuclear envelope), Golgi complex, endosomes and lysosomes, tubules and vesicles without definite coat, clathrin-coated vesicles and pits, plasma membrane, mitochondria, and nucleus. Immunogold labeling on mitochondria was treated as background labeling, as PrP^C has never been observed on these organelles by biochemical or morphological methods by us or by other investigators.

In addition, the distribution of gold particles in the neuropil of the strata radiatum, oriens, and moleculare of the dentate gyrus, on membrane profiles in dendrites and axons, and on membranes of synaptic and perisynaptic profiles was analyzed. Each class was further subdivided into the plasma membrane, internal transport vesicles, spines (only for dendrites), endosomes (only for dendrites), mitochondria, and myelin sheaths (only for axons). Membranes of synaptic complexes were classified into synaptic vesicles, synaptic specialization (that included the two closely opposed membranes in the synapse), and the pre- and postsynaptic membranes (which include the membranes outside the synaptic specialization region). For our cryosections, we used the same standard criteria for subcellular structures in brain cells as in epon sections (Peters et al. 1991). Unidentified membrane compartments were not taken into account, as they did not show substantial labeling and represented only ~5% of all cellular membranes.

By EM, PrP^C labeling in the neuropil was predominantly found on the plasma membrane of dendrites, including spines, as well as dendritic transport vesicles, endosomes, axolemma, axonal transport vesicles, and myelin sheaths. In addition, the membranes of synaptic specializations, including pre- and postsynaptic membranes, and of synaptic vesicles labeled positively for PrP^C (Fig. 1). $Prnp^{0/0}$ mice displayed no immunopositive profiles for PrP^C. Being aware that PrP^C has been previously described as enriched in synapses (Sales et al. 1998; Fournier et al. 2000; Haeberle et al. 2000; Moya et al. 2000), we checked the accessibility of other synaptic proteins for immunolabeling on ultrathin cryosections. Sections were co-labeled for the synaptic vesicle–specific protein synaptobrevin (Vamp2) and for PrP^C. As expected, Vamp2 was enriched in synaptic vesicles, whereas PrP^C was mainly seen on the plasma membrane.

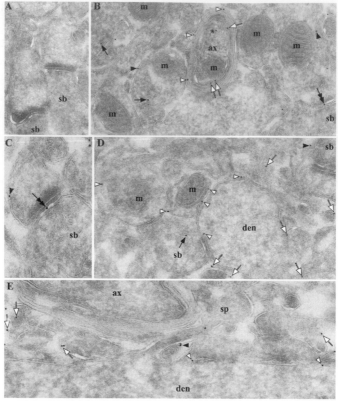

Figure 1. Ultrastructural localization of PrP^C in the neuropil in hippocampal CA1 area. Labeling was carried out using Fab D18 and protein A–gold (10 nm). (*A*) Absence of PrP^C labeling in a *Prnp*$^{0/0}$ mouse. (*B*) Stratum radiatum of a wt FVB mouse. Gold particles are localized on the axolemma (*white arrows*), myelin sheaths (*white arrowheads*), putative axon terminal membranes (*black arrowheads*), synaptic vesicles (*black arrows*), axonal transport vesicle (*near asterisk*), and synaptic specialization (*double-headed arrow*). Gold particles without arrows belong to processes that are difficult to identify as being either dendritic or axonal. (*C*) Stratum radiatum from a wt FVB mouse. As in panel *B*, gold particles can be found at the synaptic specialization (*double-headed arrow*) and on the membranes of the postsynaptic profile (*arrowhead*). (*D*) Stratum oriens of a wt FVB mouse. Gold particles are localized on the dendritic shaft (*white arrowheads*), small processes (*white arrows*), and synaptic vesicles (*black arrow*). (*E*) Stratum radiatum of a wt FVB mouse, longitudinal section of a dendrite. Gold particles are localized on the dendritic shaft (*white arrowheads*), membrane of a spine (*black arrowhead*), and small processes (*white arrows*). The particle inside the spine probably sits on the spinal apparatus, which is not clearly identifiable due to the tangential orientation. Relative distribution of the Immunogold labels is provided in Table 1; panels *B* and *D* do not reflect quantitative information. Abbreviations: (ax) myelinated axon; (den) dendrite; (m) mitochondria; (sb) synaptic bouton; (sp) spine.

In both neuronal and glial cells, PrP was detected on the ER, Golgi complex, endosomes, uncoated transport vesicles, and the plasma membrane, which are important parts of the biosynthetic and endocytic pathways (Fig. 2). No labeling was detected in the coated rims of Golgi cisternae, clathrin-coated pits of the plasma membrane, or clathrin-coated vesicles. Although the distribution of PrP^C in glial cells (confirmed by glial fibrillary acid protein [GFAP] labeling) was not quantified, EM analysis suggests that glia do not express PrP^C at levels comparable to neurons because surrounding neuropil structures often had more gold particles than glial cells and their processes.

A small population (2%) of cell bodies was intensely labeled by both the Immunogold and immunofluorescence procedures (Fig. 3). These cells with a high PrP^C content within the cell bodies are referred to as cytosolic PrP (CPrP) cells and were found with similar frequency in the brains of three different mouse lines (wt FVB, Tg4053, Tg3405) and were absent in $Prnp^{0/0}$ mice. CPrP cells were not seen in the cerebellum of any of the murine lines analyzed. Electron micrographs from ultrathin sections, which were cut directly after semithin sections, revealed the subcellular location of PrP in CPrP cells identified by light microscopy. Strikingly, the majority of the Immunogold particles was not associated with surrounding membranous structures but was located in the cytosol; therefore, we designate this form of PrP as "cytPrP."

Morphologically, CPrP cells appear to have abundant, dense cytoplasms; long cisternae of the ER; well-developed Golgi complexes, endosomal and lysosomal structures; well-structured mitochondria; and often, a nucleus with an irregular shape. The cells show an irregular distribution of cytPrP· shifted to the periphery of the cell and mostly excluded from the pericentrosomal region. The cytosolic labeling can be classified as specific, as it exceeded more than 100 times the background labeling in pyramidal neurons. cytPrP was detected with Fabs recognizing the central region (D18 and D13), amino-terminal region (Est123 and Est149), and carboxy-terminal region (R1 and R2) of PrP^C. D18 conjugated with UltraSmall gold gave the same pattern of labeling. Furthermore, SAF32 and 8H4 monoclonal antibodies were equally able to detect cytPrP (data not shown). This argues that full-length PrP molecules present in the cytosol probably bound to some factor or aggregated into multimeric complexes that prevent diffusion into the nucleus. Theoretically, a proteasome could cut PrP molecules into fragments, which are recognizable by all applied antibodies. However, this scenario seems much less likely because either fragments of degraded proteins are destroyed very quickly by various peptidases present in the cytosol or the peptides should be

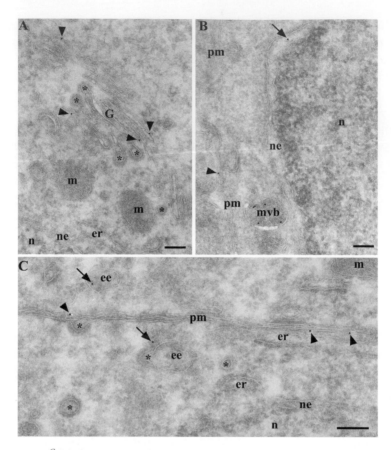

Figure 2. PrP^C labeling using Fab D18 and protein A–gold (10 nm) in neuronal cell bodies in the hippocampus of a wt FVB mouse. (*A*) Pyramidal neuron. Gold particles are found in Golgi complex (*black arrowheads*) and absent from coated rims (asterisks) of Golgi cisternae. (*B*) Granule neuron from the dentate gyrus. Gold particles are present in the nuclear envelope (*arrow*) that is a part of the ER, on late endosomes/multivesicular bodies, and on the plasma membrane (*arrowhead*). (*C*) Pyramidal neuron. Gold particles are found on the plasma membrane (*arrowheads*) and early endosomes (*arrows*) but not in clathrin-coated pits or clathrin-coated vesicles (*asterisks*). Abbreviations: (ee) early endosome; (er) endoplasmic reticulum; (G) Golgi complex; (m) mitochondria; (mvb) late endosomal multivesicular body; (n) nucleus; (ne) nuclear envelope; (pm) plasma membrane; (asterisks) lumen of coated vesicles, pits and rims. Bars, 200 nm.

detectable in the nucleus, where they are unavailable for degradation (Reits et al. 2003). The soma and dendritic processes of cells with cytPrP receive synaptic input from other neurons. Occasionally, axonal terminals with unusually high labeling for PrP^C were also found in the neuropil. We assume that these axonal terminals are derived from CPrP cells.

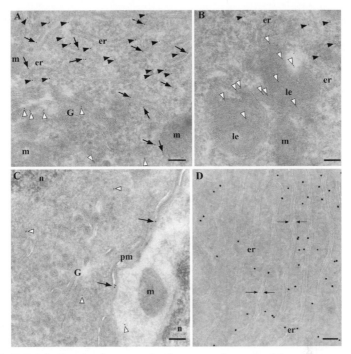

Figure 3. PrP labeling in the cytosol of neurons from the hippocampus of a FVB mouse. Labeling was accomplished using Fab D18 and protein A–gold (10 nm). (*A*) Gold particles label the cytosol (*black arrowheads*), the ER lumen (*black arrows*), and the Golgi cisternae and vesicles (*white arrowheads*). (*B*) Gold particles are found in the cytosol (*black arrowheads*) and late endosomes (*white arrowheads*). (*C*) Although cells positive for cytPrP have a denser cytosol than surrounding neurons, this is not a distinctive feature because there are cells with a dense cytosol displaying only PrP^C labeling on the plasma membrane (*black arrows*) and intracellular organelles (*white arrowheads*). (*D*) Direct immunolabeling with Fab D18 demonstrates the same cytPrP abundance as seen with indirect labeling methods (compare to *A* and *B*). The space indicated by the opposing arrows shows the lumen of the ER. Abbreviations: (er) endoplasmic reticulum; (G) Golgi complex; (le) late endosome; (m) mitochondria; (n) nucleus; (pm) plasma membrane. Bars, 200 nm.

It is possible that PrP accumulation in the cytosol may reflect stress or damage to the cells, but morphological examination of these neurons did not reveal organelle swelling (mitochondria, ER, Golgi apparatus), disaggregation of polyribosomes, or cell and nuclear membrane breaks that are clearly indicative of neuronal necrosis. Furthermore, no apoptotic signs, such as chromatin clumping, condensation of cytoplasmic content, or accumulation of autophagic lysosomes, were observed. Mitochondria displayed organized structures with well-preserved cristae, inner and outer membranes. Although morphological analysis remains the "gold standard" for assessment and quantification of apoptosis (Hall 1999), we nevertheless checked CPrP cells on semithin sections, using an apoptosis detection kit (based on TUNEL methodology). None of the CPrP cells appeared to be apoptotic.

QUANTIFICATION OF PrP^C IN THE HIPPOCAMPUS

The use of gold particles and ultrastructural preservation produced by ultracryomicrotomy provided the opportunity to quantify the distribution of PrP^C in the hippocampus. All segments of dendritic membrane—dendritic shaft, spines, transport vesicles, and endocytic structures—showed approximately the same density of gold particles per unit of membrane (Table 1). Interestingly, similar PrP^C concentrations were found on the membrane of pre- and postsynaptic profiles and on the membranes within the synapse. No preferential labeling was observed within the synaptic specialization. Synaptic vesicles were labeled just above the back-

Table 1. Quantification of PrP^C labeling density on the membranes of and within dendritic, axonal, and synaptic profiles from the neuropil of the CA1 area and dentate gyrus

Membranous profiles	Neuropil area		
	stratum oriens ($n = 17$)	stratum radiatum ($n = 20$)	stratum moleculare ($n = 17$)
Dendrites			
Dendritic shaft	1.54 ± 0.04	1.63 ± 0.06	0.62 ± 0.04
Spines	1.58 ± 0.1	1.52 ± 0.09	0.58 ± 0.1
Transport vesicles/tubules	1.08 ± 0.06	1.12 ± 0.09	0.48 ± 0.1
Endosomes/lysosomes	1.46 ± 0.2	1.38 ± 0.08	0.58 ± 0.05
Axons			
Axolemma	0.94 ± 0.09	0.84 ± 0.07	0.55 ± 0.06
Transport vesicles/tubules	0.44 ± 0.04	0.37 ± 0.03	0.35 ± 0.03
Myelin sheaths	0.38 ± 0.05	0.35 ± 0.03	0.36 ± 0.04
Synaptic complexes			
Presynaptic bouton	1.6 ± 0.08	1.65 ± 0.02	0.82 ± 0.11
Synaptic specialization	1.51 ± 0.15	1.48 ± 0.2	0.79 ± 0.12
Postsynaptic bouton	1.64 ± 0.1	1.63 ± 0.12	0.71 ± 0.1
Synaptic vesicles	0.12 ± 0.012	0.14 ± 0.013	0.096 ± 0.015
Mitochondria	0.056 ± 0.014	0.073 ± 0.016	0.04 ± 0.008

Values represent the number of gold particles per 1 μm of membrane (gold/μm). Mitochondria were used to assess background labeling. The results are presented as mean ± S.E.M.

ground level determined for mitochondria, suggesting an exclusion of PrP^C from this structure.

On the basis of these observations, it seems that (1) PrP^C generally follows the standard biosynthetic trafficking pathway in brain neurons with prominent presence in endosomes and the plasma membrane of dendrites and axons; (2) PrP^C has ubiquitous distribution on the neuronal plasma membrane and cellular processes without preferential accumulation at synaptic specializations; (3) PrP^C is found with the same frequency on presynaptic as well as postsynaptic membranes and within the synapse; (4) PrP^C is almost excluded from synaptic vesicle membrane; and (5) PrP is expressed in the cytosol in a small population of neurons in the hippocampus, thalamus, and somatosensory neocortex but not in the cerebellum.

The almost ubiquitous distribution of PrP^C on the neuronal plasma membrane and cellular processes without a preferential accumulation at synaptic specializations suggests the absence of active retention mechanisms, allowing unhampered diffusion of PrP^C along cellular membranes. This diffusion could play a major role in PrP^{Sc} propagation, as it was shown that a defective fast axonal transport did not interfere with prion neuroinvasion (Kunzi et al. 2002). It is entirely plausible that PrP^{Sc}, which retains the GPI anchor, could physically contact PrP^C on adjacent cells or even be itself physically translocated to the membranes from neighboring cells (Liu et al. 2002) at sites of very close apposition to cellular membranes, including at the synapse. Additionally, the finding of significant PrP^C labeling in myelin sheaths points to the possible involvement of oligodendrocytes in the propagation of prion diseases.

Studies by three of us (P.J.P., S.J.DeA., and S.B.P.) revealed the existence of neurons containing cytPrP. These cells showed a very different morphology from glial cells and are negative for GFAP (an astroglial marker), CNPase, and S100 (oligodendrocytic markers). We observed synapses on cell bodies and processes as well as axonal terminals with a high PrP^C content, which probably belong to CPrP cells. On the basis of our LM and EM observations, we assumed CPrP cells to be of neuronal nature. However, we are still focusing our efforts on identifying the neuronal subtype. Although it seems unusual for a cell to have a protein in two such distinct locations, the membranes of the trafficking pathway and the cytosol, it is entirely plausible that a protein such as PrP^C could have roles in more than one compartment of a cell (Hegde and Lingappa 1999). As described here and in many publications, PrP^C has a rather complex array of cellular topologies (Zhang and Ling 1995; Hegde et al. 1998; Hölscher et al. 2001; Kim et al. 2001). Therefore, it is possible that the synthesis of the different topological forms of PrP^C varies among different cell types according to the expression of different cytoplasmic components of the translocation machinery (Hegde and Lingappa 1999).

It was recently shown that transfected cytPrP appears to be toxic both in cell culture and in transgenic animals, in a cell type–dependent manner. Only cerebellar cells appeared to be affected in mice that expressed PrP^C without an ER translocation signal (Ma et al. 2002). Accumulation of the cytPrP in "susceptible" neurons has been suggested to be responsible for some of the variants of prion diseases in which cytPrP aggregates have been postulated to kill cells and produce prions. However, our studies show that cytPrP is present in normal rodent brains in a population of neurons that appear healthy and show no degeneration. Thus, we can infer that cytPrP is not toxic in some neurons, but might be toxic when overexpressed in specific cell populations. More work is needed to elucidate the physiologic and possibly pathologic roles of cytPrP.

STRUCTURE AND BIOSYNTHESIS OF PrP^C

The mammalian PrP gene encodes a protein of ~250 amino acids that contains several distinct domains, including an amino-terminal signal peptide, a series of five peptide repeats, a central hydrophobic segment that is highly conserved, and a carboxy-terminal hydrophobic region that is a signal for addition of a GPI anchor (Fig. 4). In the human protein, the peptide repeats comprise one copy of the nonapeptide PQGGGGWGQ and four copies of the octapeptide PHGGGWGQ.

Most PrP^C molecules are normally found on the cell surface, where they are attached by their carboxy-terminal GPI anchor. The biosynthetic pathway followed by PrP^C is similar to that of other membrane and secreted proteins. PrP^C is synthesized in the rough ER and transits the Golgi on its way to the cell surface. During its biosynthesis, PrP^C is subject to cleav-

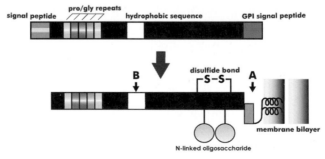

Figure 4. Structure and posttranslational processing of PrP.

age of the amino-terminal signal peptide, addition of N-linked oligosaccharide chains at two sites, formation of a single disulfide bond, and attachment of the GPI anchor (Fig. 4) (Stahl et al. 1987; Turk et al. 1988; Haraguchi et al. 1989). The N-linked oligosaccharide chains added initially in the ER are of the high-mannose type and are sensitive to digestion by endoglycosidase H. These are subsequently modified in the Golgi to yield complex-type chains that contain sialic acid and are resistant to endoglycosidase H (Caughey et al. 1989). The N-linked glycans of PrP^C comprise over 50 distinct bi-, tri-, and tetra-antennary structures (Rudd et al. 1999; Stimson et al. 1999). The GPI anchor is added in the ER after cleavage of the carboxy-terminal hydrophobic segment. It has a core structure common to other glycolipid-anchored proteins, but is modified by the addition of mannose, ethanolamine, and sialic acid residues (Stahl et al. 1992). PrP^C can be released from the cell surface by treatment with the bacterial enzyme phosphatidylinositol-specific phospholipase C (PIPLC), which specifically cleaves the GPI anchor.

CELLULAR TRAFFICKING OF PrP^C

PrP^C Undergoes Endocytosis

Several lines of evidence demonstrate that PrP^C does not remain on the cell surface after its delivery there, but rather constitutively cycles between the plasma membrane and an endocytic compartment (Fig. 5). First, it has been observed that PrP^C undergoes a physiological cleavage near residue 110, within a highly conserved segment of hydrophobic amino acids (site B in Fig. 4) (Caughey et al. 1989; Harris et al. 1993; Chen et al. 1995; Taraboulos et al. 1995). Evidence shows that this cleavage is carried out by members of the disintegrin family of metalloproteases (Vincent et al. 2001). One of us (D.A.H.) showed proteolytic cleavage of PrP^C to be significantly inhibited by lysosomotropic amines, leupeptin and brefeldin A, consistent with this processing step occurring in an endocytic compartment (Shyng et al. 1993). Second, PrP molecules on the cell surface were found to undergo internalization and recycling after labeling with membrane-impermeant iodination or biotinylation reagents (Shyng et al. 1993). Kinetic analyses of the data demonstrate that PrP^C molecules cycle through the cell with a transit time of ~60 minutes, and during each passage, 1–5% of the PrP^C molecules undergo cleavage near residue 110. Internalization has also been demonstrated by tracking surface PrP^C molecules that have been labeled by binding of fluorescently labeled antibodies (Shyng et al. 1993).

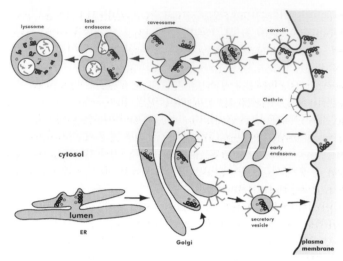

Figure 5. PrP traverses the secretory pathway en route to the cell surface. Some PrP molecules are endocytosed via a caveolar pathway. Evidence exists for internalization of PrP via clathrin-coated pits, and recycling of molecules from early endosomes back to the plasma membrane.

Involvement of Clathrin in Endocytosis of PrP^C

According to studies by one of us (D.A.H.), clathrin-coated pits and vesicles appear to be the morphological structures responsible for endocytic uptake of PrP^C. This conclusion is based on Immunogold localization of PrP^C in these organelles by electron microscopy (Shyng et al. 1994; Madore et al. 1999; Lainé et al. 2001); inhibition of PrP^C internalization by incubation of cells in hypertonic sucrose, which disrupts clathrin lattices (Shyng et al. 1994); detection of PrP^C in purified preparations of coated vesicles from brain (Shyng et al. 1994); and transfection with a dynamin I mutant (K44A; Magalhães et al. 2002).

The involvement of clathrin-coated pits in the endocytosis of PrP^C is surprising (and in contrast to findings of other coauthors of this chapter), because GPI-anchored proteins like PrP^C lack a cytoplasmic domain that could interact directly with adapter proteins and clathrin. Dynamin has recently been found to be required for caveolae-mediated uptake.

PrP^C Is Associated with Membrane Rafts and Caveolae

Studies of its intracellular trafficking are critical to understanding the cellular role of the protein, as well as to resolving the transformation of PrP^C into PrP^{Sc} and how this process eventually leads to neurodegeneration.

During refolding, regions of the primarily α-helical PrP^C are converted to β-rich structures (Pan et al. 1993). Whether these β-rich structures are composed of β-helices remains to be established (Wille et al. 2002). Studies of PrP^{Sc}-infected neuroblastoma (ScN2a) cells have shown that this conversion is a late event in the life cycle of PrP^C and occurs after PrP^C has passed through the Golgi complex (Borchelt et al. 1990; Caughey and Raymond 1991). After conversion, PrP^{Sc} accumulates in late endocytic and lysosomal compartments of the infected cell (Taraboulos et al. 1990; McKinley et al. 1991; Arnold et al. 1995). The mechanism involved in the conversion of PrP^C to PrP^{Sc} and the exact subcellular site of PrP^{Sc} formation have not yet been determined.

After posttranslational GPI anchoring, disulfide bonding, and glycosylation in the ER, PrP^C transits the Golgi complex and is incorporated into post-Golgi, cholesterol-rich, caveolae-like domains (CLDs) (Fig. 6) (Vey et al. 1996). Caveolae, specialized membrane microdomains that contain members of the caveolin protein family, are rich in cholesterol and glycosphingolipids. They often appear as flask-shaped invaginations on the plasma membrane or form more complex interconnected chains of several caveolae with a single neck connecting the chains to the outer surface of the cell (Parton 2003). Caveolae, which can be considered a special class of raft, mediate a number of key physiological processes, such as signal transduction and transcytosis (Anderson 1998; Shaul and Anderson 1998; Kurzchalia and Parton 1999; Brown 2001). CLDs, also called lipid rafts (Simons and Ikonen 1997), are biochemically defined in terms of their Triton X-100 detergent insolubility at 4°C. Because Triton insolubility is also considered the standard hallmark of cytoskeletal association of membrane proteins (Scherfeld et al. 1998), one cannot assume that CLDs are equivalent to purified caveolae.

CLDs appear to be crucial to the PrP^C-to-PrP^{Sc} conversion process (Taraboulos et al. 1995; Kaneko et al. 1997a). Enrichment of PrP^{Sc} in CLDs (Vey et al. 1996; Naslavsky et al. 1997) under conditions in which PrP^C is converted to nascent PrP^{Sc} further suggests that these membrane domains are the subcellular sites at which endogenous PrP^C interacts with PrP^{Sc}. Moreover, PrP^C with modifications in the cytoplasmic tail that cause it to be redirected to clathrin-coated pits is not converted to PrP^{Sc} (Taraboulos et al. 1995; Kaneko et al. 1997a). Significantly, neurons in culture (Mouillet-Richard et al. 2000), astrocytes, and basal ganglion cells contain caveolae, raising the possibility that PrP^C resides in these structures. In addition, many different caveolae-containing cell types outside the nervous system that express PrP^C may be required for the transmission of prion diseases.

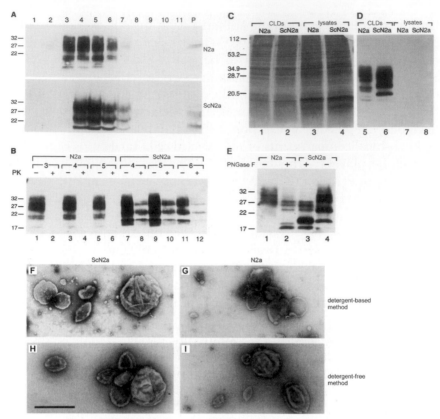

Figure 6. (*A–D*) PrP^C and PrP^{Sc} are concentrated in CLDs isolated from neuroblastoma cells. (*A*) Distribution of PrP after lysis of N2a and ScN2a cells in cold Triton X-100 and separation of CLDs by flotation into sucrose gradients. Aliquots of gradient fractions were immunoblotted with the α-PrP polyclonal RO73 rabbit antiserum (Serban et al. 1990). (Lanes *1–7*) Fractions of the sucrose gradients; (lanes *8–11*) lysate fractions; (lane *P*), pellets. (*B*) Detection of PrP^{Sc} in gradient fractions from panel *A* before (–) and after (+) treatment with proteinase K (PK). (*C,D*) Silver-stained gel (*C*) and immunoblot (*D*) show the concentration of PrP and other proteins in CLDs. (*E*) Detection of PrP^C degradation products in CLDs before (–) and after (+) treatment with PNGase F (Tarentino et al. 1985). (*F–I*) Ultrastructure of isolated CLDs, by the cold Triton X-100 detergent method (*F, G*) and by the detergent-free procedure (*H, I*). Samples were prepared from ScN2a (*F, H*) and N2a (*G, I*) cells. Bar, 0.5 μM and applies to panels *F–I*. (Reprinted, with permission, from Vey et al. 1996 [copyright National Academy of Sciences].)

To evaluate the trafficking of PrP^C in Chinese hamster ovary (CHO) cells stably transfected with Syrian hamster (SHa) PrP^C (Scott et al. 1992), a study using cryoimmunogold electron microscopy was initiated by some of us (P.J.P. and S.B.P.). PrP^C was found to be highly enriched in the

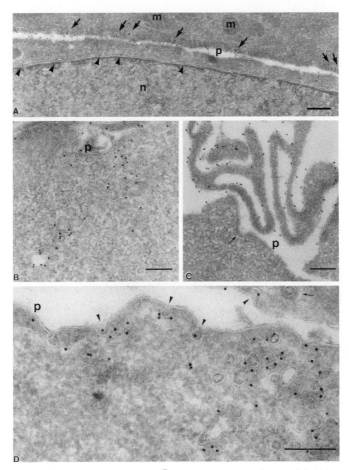

Figure 7. Immunogold labeling of PrP^C and caveolin-1 on ultrathin cryosections of CHO/30C3 cells. (A–C) Immunogold labeling using the R1 recombinant antibody fragment (Fab) shows the steady-state distribution of PrP^C. (A) Labeling was observed on the ER (*thick arrowheads*), plasma membrane, caveolae-like structures (*arrows*), caveolae like structures that appeared as flask-shaped invaginations on the plasma membrane, and interconnecting chains of caveolae-like structures deeper into the cytoplasm (B) and microvilli at the leading edge of the cell (C). (C) Clathrin-coated pits (*thin arrow*) did not label. (D) Immunogold labeling using anti-caveolin-1–gold shows caveolin-1 in caveolae-like structures and interconnecting chains of caveolae-like structures deeper into the cytoplasm. Abbreviations: (m) mitochondria; (n) nucleus; (p) plasma membrane.

caveolae and caveolae-containing endocytic structures of these cells (Fig. 7) (Peters et al. 2003). To our surprise, we also found that protein A–gold bound directly to PrP^C (Fig. 8). PrP^C gained access to a nonclassical endocytic pathway by internalization through caveolin-coated caveolae.

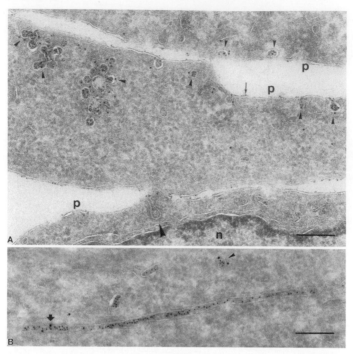

Figure 8. (*A*) Endocytosed protein A–gold is found in caveolae of SHaPrP^C-expressing CHO/30C3 cells. Cells were incubated at 37°C for 10 minutes with protein A–gold (5 nm) and chased for 50 minutes prior to fixation. Ultrathin cryosections were labeled with anti-caveolin-1 (10-nm gold). Protein A–gold (5 nm) detecting PrP^C labeling was observed on the plasma membrane and was enriched in caveolae (*arrowheads*) and caveolin-1-positive (10-nm gold) structures. Clathrin-coated pit (*large arrowhead*) did not label for either PrP^C or caveolin-1. (*B*) Tubule and vesicular structures (*small arrowhead*) that contained endocytosed protein A–gold (5 nm) labeled for caveolin-1 (10 nm, *large arrow*). Abbreviations: (n) nucleus; (p) plasma membrane. Bars, 200 nm.

Uptake of caveolae was slow, and degradation of PrP^C after 1-hour incubation with protein A–gold was undetectable by western blot analysis. This caveolae-dependent endocytic pathway was PIPLC- and filipin-dependent and was not observed with several other GPI-anchored proteins, as demonstrated by CD59 and aerolysin, a monovalent probe for the glycan core of several GPI-anchored proteins. This pathway is therefore likely to determine the subcellular location at which PrP^C and PrP^{Sc} interact to lead to the generation of nascent PrP^{Sc}.

The foregoing data raise the question of whether other GPI-anchored proteins follow the same route as PrP^C in CHO cells. To address this issue, two of us (P.J.P and S.B.P.) used a CHO cell line stably expressing human GPI-anchored CD59 (Fig. 9A–C) (Wheeler et al. 2002). We found a uni-

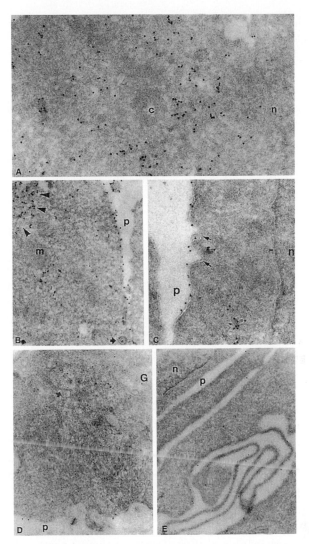

Figure 9. Human GPI-anchored CD59 protein under steady state does not reside in caveolae of CHO cells. Uniform distribution of anti-CD59 on the ER, Golgi complex, TGN (*A*), and the plasma membrane (*B, C*), but with no enrichment in caveolae (*arrows, C*). (*B*) Clathrin-coated pits at the plasma membrane (*arrow*) labeled for anti-CD59. The bacterial toxin aerolysin, a monovalent probe for several GPI-anchored proteins (but not PrP) was used on ultrathin cryosections of PrP^C-expressing CHO cells as described previously (Peters et al. 2003). (*D*) Labeling was seen on the plasma membrane, Golgi complex, and late endocytic multivesicular bodies. (*E*) Greater intensity in labeling was seen on the plasma membrane lamellae, but no labeling in caveolae was detected. Abbreviations: (c) centriole; (G) Golgi complex; (n) nucleus; (m) mitochondria; (p) plasma membrane.

form distribution of anti-CD59 on the ER, Golgi complex, trans-Golgi network (TGN), and the plasma membrane, with a higher labeling density on the lamellae of the leading edge (not shown) but without any enrichment in caveolae. Furthermore, clathrin-coated pits at the plasma membrane were occasionally labeled (~5% of 200 coated structures) and enrichment was seen at steady state in late endocytic multivesicular bodies.

To track the fate of a collection of GPI-anchored proteins, the bacterial toxin aerolysin, a monovalent probe for the glycan core of several GPI-anchored proteins, was applied (Fivaz et al. 2002). Remarkably, this probe does not bind GPI-anchored PrP^C (G. van der Goot, pers. comm.). Ultrathin cryosections of CHO cells expressing PrP^C were incubated with biotinylated aerolysin followed by antibiotin–gold. Moderate but specific labeling was seen on the plasma membrane, Golgi complex, and late endocytic multivesicular bodies, as described previously (Fivaz et al. 2002). A distinct higher-labeling intensity on plasma membrane lamellae, but no detectable label in caveolae, was found. A clathrin- and caveolin-independent pathway for internalization of CD59 or aerolysin could not be determined morphologically. These data confirm the differential sorting and fate of endocytosed GPI-anchored proteins described previously (Fivaz et al. 2002), including a different port of entry of PrP^C relative to other GPI-anchored proteins of the same cell type (Peters et al. 2003).

Earlier reports describing GPI-anchored proteins in caveolae had been criticized because antibody labeling was performed on living cells or with non-crosslinking fixation reagents, such as paraformaldehyde, which allow lateral diffusion after fixation. Under these conditions, clustering of the antigen by antibodies used in immunodetection could have artificially concentrated the GPI-anchored proteins in caveolae (Mayor et al. 1994). In our studies, a mixture of glutaraldehyde and paraformaldehyde was used for fixation and ultrathin cryosectioning prior to incubation with recombinant anti-PrP Fabs. Because Fabs are monovalent molecules, the artificial mAb-mediated clustering of antigen on the cell surface is not possible. Moreover, the PrP antigen was chemically crosslinked and immobilized in an ultrathin section on a carbon-coated formvar grid. On the basis of the foregoing considerations, we conclude that the labeling pattern represents the distribution of PrP^C before fixation.

In addition to the clustering of PrP^C in caveolae, we found gold labeling for PrP^C over undifferentiated areas of the plasma membrane with a higher density at the lamellae of leading edges that was independent of caveolin-1 (Fig. 10). Lamellae of leading edges have been proposed to be rich in cholesterol- and glycosphingolipid-enriched rafts. Two of us (P.J.P. and S.B.P.) did not find PrP^C to be enriched in clathrin-coated pits in brain (Mironov et al. 2003) or CHO cells (Peters et al. 2003), in contrast to the findings of one of us (D.A.H.) who used the N2a cell line transfected with chicken (Chk) PrP^C (Shyng et al. 1994). ChkPrP^C was reported to be enriched in CLDs (Gorodinsky and Harris 1995), but no caveolae were observed in those cells, suggesting that detergent-resistant complexes might not be equivalent to caveolae. Results from studies by two of us (P.J.P. and S.B.P.) also argue that detergent-resistant complexes are not equivalent to caveolae in all cell types, and that in neuronal cells, caveolin is not essential for the integrity of these complexes (Peters et al. 2003).

The localization of PrP^C in caveolae generates important questions about the physiological function of PrP^C and raises the possibility that caveolae are the site of pathological events that ultimately lead to neurodegeneration in prion disease. Caveolin-1 has been found in both serotonergic and noradrenergic neurons in culture as well as in cultured astrocytes, glial cells (Silva et al. 1999), and Schwann cells (Mikol et al. 1999). Caveolae have also been identified in astrocytes and endothelial cells in the brain and other tissues, such as intestine and muscle, implicated in prion propagation and transmission (A. Mironov and P.J. Peters, unpubl.). Significantly, interest in caveolae localization also has arisen from studies on trafficking and processing of amyloid precursor protein in Alzheimer's disease (Bouillot et al. 1996) and on the colocalization of huntingtin with caveolin in Huntington's disease (Peters et al. 2002). Thus, caveolae-type structures and rafts might be important subcellular structures in the pathogenesis of several neurodegenerative diseases.

The active internalization of PrP^C via caveolae raises some important questions in the study of prion biology. For example, does the reported signaling activity of PrP^C occur at the plasma membrane or after caveolae-mediated endocytosis (Mouillet-Richard et al. 2000)? Similarly, is endocytosis via caveolae a necessary step in the conversion of PrP^C into PrP^{Sc}, perhaps because of a change in milieu or in the relative concentration of PrP conformers or other auxiliary molecules involved in prion propagation (Telling et al. 1995; Kaneko et al. 1997c)? Is the role of PrP^{Sc} in promoting neuronal cell dysfunction performed during or after endocytosis? Is the caveolin-knockout mouse resistant to oral prion infection? Future studies of this internalization pathway may lead to the identification of inhibitors of the caveolae-mediated endocytosis of PrP^C, and such molecules may be useful as reagents for the treatment of prion diseases and other neurodegenerative disorders.

PrP^C AND CELLULAR TRAFFICKING OF COPPER IONS

The endocytic recycling pathway followed by PrP^C (Fig. 5) suggests a role for the protein in cellular uptake or efflux of an extracellular ligand, with

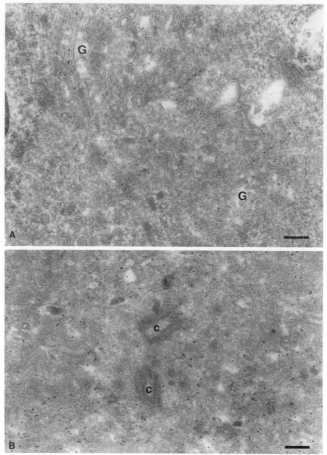

Figure 10. Immunogold labeling on ultrathin cryosections of CHO/30C3 cells demonstrates the steady-state distribution of PrP^C using the R1 recombinant Fab. Labeling was observed on the (*A*) Golgi cisternae, (*B*) TGN area around pericentriolar region, (*C*) lamellae at the leading edge of the cell, (*D*) caveolae-containing structures that appeared as flask-shaped invaginations on the plasma membrane, (*E*) interconnecting chains of endocytic caveolae-containing structures deeper into the cytoplasm, and *(F)* multivesicular-body-type (late endocytic) structures. (*D*) Clathrin-coated pits did not label for PrP^C (*arrows*). Abbreviations: (c) centriole; (G) Golgi complex; (la) lamellae; (m) mitochondria; (n) nucleus; (p) plasma membrane. (*Figure continues on facing page.*)

copper ions perhaps being one such ligand. PrP^C binds copper with low micromolar affinity, and in a pH-sensitive manner, via the amino-terminal histidine-containing octapeptide repeats (Brown et al. 1997b; Stöckel et al. 1998; Jackson et al. 2001; Kramer et al. 2001; Burns et al. 2002; Qin et al. 2002). Whether Cu^{++} binds to the carboxy-terminal portion of PrP^C

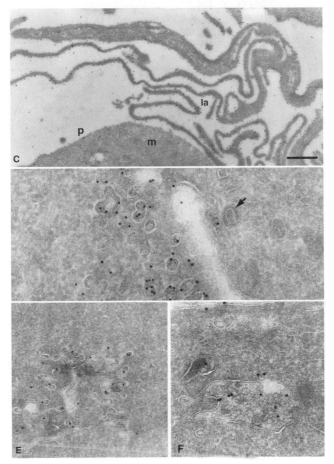

Figure 10. (*See facing page for legend.*)

seems unlikely but has been suggested using amino-terminally truncated fragments of the protein (Jackson et al. 2001; Burns et al. 2003). Although several other lines of evidence have emerged in the past few years suggesting a connection between PrP^C and copper metabolism (for review, see Brown and Harris 2002; Lehmann 2002), the high concentration of Cu^{++} required to demonstrate binding to PrP raises questions about the physiological role of PrP in copper metabolism.

One of us (D.A.H.) found that incubation of transfected neuroblastoma cells with Cu^{++} concentrations greater than 100 μM for 30–90 minutes caused a marked reduction in the total amount of PrP^C on the cell surface, as determined by biotinylation or by immunofluorescence staining (Pauly and Harris 1998; Brown and Harris 2003). This effect was temperature-dependent (it did not occur at 4°C) and was rapidly reversible (within minutes). It was also seen with Zn^{++}, but not with Co^{++}, Mn^{++}, Cd^{++}, Ni^{++}, or Fe^{++}. Neither copper nor zinc had any effect on the distribution of the transferrin receptor, suggesting that the metals were not causing a generalized stimulation of endocytosis. Copper-induced internalization of PrP^C depended on the presence of the histidine-containing repeats (Pauly and Harris 1998; Lee et al. 2001; Perera and Hooper 2001), suggesting that the effect may be due to binding of the metal to PrP^C rather than to some other cellular protein that indirectly modulates endocytosis. Interestingly, an insertionally mutated form of PrP containing 14 octapeptide repeats that is associated with familial Creutzfeldt-Jakob disease (CJD) is refractory to copper-induced endocytosis (Perera and Hooper 2001), implying that the normal complement of repeats is necessary either for optimal copper binding or for whatever structural change the metal induces to trigger endocytosis.

Several pieces of evidence indicate that the primary effect of copper is to stimulate the endocytosis of PrP^C, with relatively little change in the rate of recycling. To measure endocytosis and recycling, biochemical methods were used to follow the internalization of surface PrP^C molecules that had been labeled by iodination or biotinylation (Pauly and Harris 1998). Alternatively, immunocytochemical techniques were applied to visualize the metal-induced redistribution of surface PrP^C molecules that had been prelabeled with anti-PrP antibodies (Brown and Harris 2003). The antibody-tagged protein was found to translocate from the cell surface to punctate intracellular compartments in the presence of copper (Fig. 11). The internalized PrP^C partially colocalized with both fluorescent transferrin and fluorescent wheat germ agglutinin, but not with LysoTracker (a fluorescent lysosomotropic amine), implying that the protein was being delivered to early endosomes and the Golgi, but not to lysosomes.

Two other groups have also investigated the intracellular fate of PrP^C that is endocytosed in response to copper. One study, using PrP–GFP, suggests that conventional, Rab5-positive early endosomes are involved in the initial stage of internalization, with the protein ultimately being delivered to a perinuclear compartment whose identity was not determined (Lee et al. 2001; Magalhães et al. 2002). In contrast, the second study, which monitored the movement of antibody-tagged PrP^C, concluded that a caveolin-dependent pathway is responsible for copper-induced endocytosis (Marella et al. 2002). One of us (D.A.H.) has hypothesized that copper binding may enhance the affinity of PrP^C for a receptor that targets it to clathrin-coated pits (Pauly and Harris 1998). Further work will be required to clarify the endocytic pathway followed by PrP^C after copper treatment.

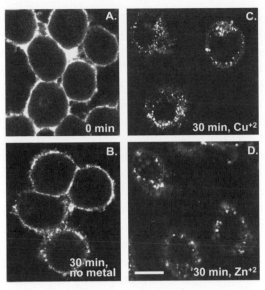

Figure 11. Antibody-bound PrP^C is internalized in response to copper and zinc. N2a cells expressing chicken PrP were first labeled with anti-PrP antibody at 4°C, and were then stained immediately (*A*), or were warmed to 37°C for 30 minutes in the absence of metal (*B*) or in the presence of 250 μM $CuSO_4$ (*C*) or $ZnSO_4$ (*D*) prior to staining. To reveal antibody-bound PrP, cells were fixed, permeabilized, and reacted with secondary antibody. Scale bar, 30 μm and applies to all panels. (Reprinted, with permission, from Brown and Harris 2003 [copyright International Society for Neurochemistry].)

Considering the available data, one can envision at least two possible models for the role of PrP^C in the cellular trafficking of copper ions. In an uptake model, PrP^C on the plasma membrane binds Cu^{++} via the peptide repeats, and then delivers the metal by endocytosis to an acidic, endosomal compartment. Copper ions then dissociate from PrP^C by virtue of the low endosomal pH and, after reduction to Cu^+, are transported into the cytoplasm by a transmembrane transporter. PrP^C subsequently returns to the cell surface to bind additional copper, and the cycle is repeated. This proposed function for PrP^C is analogous to that of the transferrin receptor in uptake of iron, with the exception that metal ions bind directly to the receptor in the case of PrP^C rather than to a protein carrier comparable to transferrin. In a second model, PrP^C serves as a receptor that facilitates cellular efflux of copper via the secretory pathway. PrP^C is first delivered via endosomal vesicles to the TGN or other post-Golgi compartments, where it then serves to bind copper ions that have been pumped into the secretory pathway via the ATP-dependent, Menkes or Wilson transporters. In addition to acting as a carrier for copper ions during their transit to the cell surface, PrP^C could also play a role in specifically transferring the metal from the transporters to secreted cuproproteins such as ceruloplasmin by physically interacting with these molecules. The immunocytochemical localization of copper-internalized PrP^C in both endosomes and the Golgi is consistent with either an uptake or efflux model.

It remains to be proven whether either of these models, or other ones, will turn out to be correct. It was originally reported that the brains of $Prnp^{0/0}$ mice have a greatly reduced content of ionic copper, and decreased enzymatic activity of Cu-Zn superoxide dismutase (SOD) compared to wt mice (Brown et al. 1997a,b, 2002; Brown and Besinger 1998; Herms et al. 1999). However, one of us (D.A.H.) has not observed either of these abnormalities in $Prnp^{0/0}$ mice (Waggoner et al. 2000). Thus, these results indicate that PrP^C is not likely to be the primary carrier for net uptake or efflux of brain copper, and it is not likely to be involved in the specialized pathway responsible for copper delivery to SOD. This conclusion is also consistent with the finding that cells expressing different amounts of PrP^C do not show differences in net uptake of ^{64}Cu (Rachidi et al. 2003; C. Pauly and D.A. Harris, unpubl.). Since several pieces of evidence indicate that neurons from $Prnp^{0/0}$ mice are more susceptible to oxidative stress (Brown 2001), it is possible that PrP^C-mediated copper uptake plays a role in delivery of the metal to other enzymes capable of protecting cells from oxidative damage. It has been reported that recombinant PrP displays SOD activity when it is refolded in the presence of copper (Brown et al. 1999, 2000). However, the high concentrations of metal used in these experiments call into question the physiological relevance of the effect. In addition, copper binds much more weakly to PrP^C than to known cuproenzymes like Cu-Zn SOD, and even small organic molecules like amino acids can bind copper and exhibit weak dismutase activity. Clearly, additional work is needed to define the role, if any, played by PrP^C in copper transport, metabolism, storage, or enzymatic catalysis.

CELL BIOLOGY OF MUTANT PrP

In humans, familial prion diseases are linked to dominantly inherited mutations in the PrP gene on chromosome 20 (Young et al. 1999). Point mutations in the carboxy-terminal half of the protein are associated with CJD, Gerstmann-Sträussler-Scheinker syndrome (GSS), or fatal familial insomnia (FFI), whereas octapeptide insertions in the amino-terminal half cause a mixed CJD–GSS phenotype. The presence of the mutation is presumed to favor spontaneous conversion of PrP^C to the PrP^{Sc} state without contact with exogenous prions. To understand how mutant PrP molecules cause neurological dysfunction in familial prion diseases, analysis of the biochemical properties, metabolism, and cellular localization of these proteins has been carried out in cultured cells.

Biochemical Properties of Mutant PrPs

One of us (D.A.H.) has constructed stably transfected lines of CHO, baby hamster kidney (BHK), and PC12 cells that express murine homologs of mutant PrPs believed to cause familial prion diseases of humans (Lehmann and Harris 1995, 1996a,b, 1997; Daude et al. 1997; Lehmann et al. 1997; Chiesa et al. 2000; Ivanova et al. 2001). Each PrP harboring a pathogenic mutation displays three biochemical properties reminiscent of PrP^{Sc} (Fig. 12). These properties include (1) detergent-insolubility, manifested by sedimentation at 265,000g from Triton/deoxycholate lysates; (2) protease-resistance, evidenced by production of an amino-terminally truncated core of 27–30 kD after treatment with proteinase K (PK); and (3) PIPLC-resistance, defined by the failure of PIPLC to release the protein from the cell surface or to render it hydrophilic in biochemical assays. This last property reflects structural alterations that make the GPI anchor physically inaccessible to phospholipase, since SDS denaturation of the protein restores PIPLC susceptibility (Narwa and Harris 1999). Neither wt PrP nor PrP carrying a M128V mutation analogous to a nonpathogenic polymorphism of human PrP displays any of the three properties. Differences have been noted between the mutant PrPs in their degree of detergent insolubility and other biochemical properties. Interestingly, molecules with the most prominent PrP^{Sc} properties are those carrying mutations that have been found to decrease maximally the thermodynamic stability of the molecule (Liemann and Glockshuber 1999). Several other laboratories have also found that mutant PrPs synthesized in cultured cell types of both neuronal origin (N2a, M17 neuroblastoma cells) and nonneuronal origin (3T3, BHK cells) display PrP^{Sc}-like biochemical properties (Singh et al. 1997; Priola and Chesebro 1998; Zanusso et al. 1999; Capellari et al. 2000a,b; Gauczynski et al. 2002; Lorenz et al. 2002). It thus seems likely that cell-culture systems are modeling important features of the PrP^{C}→PrP^{Sc} conversion process that occurs in vivo.

This contention has been borne out by studies of mice (Chiesa et al. 1998, 2000), designated Tg(PG14) mice, that express a MoPrP molecule harboring a nine-octapeptide insertion associated with a familial prion disease with a mixed CJD–GSS phenotype. Tg(PG14) mice spontaneously develop a progressive and ultimately fatal neurological illness characterized by ataxia, massive apoptosis of cerebellar granule neurons, and abnormal deposition of PrP in several brain regions. PG14 PrP progressively accumulates in the brains of the animals, eventually reaching levels

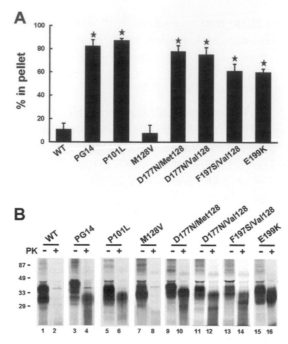

Figure 12. PrP carrying pathogenic mutations is detergent-insoluble and protease-resistant when expressed in cultured cells. (*A*) CHO cells expressing wild-type (wt) and mutant mouse PrPs were labeled with [^{35}S]methionine for 20 minutes and then chased for 3 hours. Detergent lysates of the cells were centrifuged first at 16,000*g* for 5 minutes, and then at 265,000*g* for 40 minutes. PrP in the supernatants and pellets from the second centrifugation was immunoprecipitated and analyzed by SDS-PAGE. PrP-specific bands were quantified using a Phosphor-Imager, and the percentage of PrP in the pellet was calculated. Each bar represents the mean ± S.D. from three experiments. Values that are significantly different from wt PrP by t-test (*p*<0.001) are indicated by a star. PrPs carrying disease-related mutations sediment (are detergent-insoluble), whereas wt and M128V PrPs remain largely in the supernatant. Mutant PrPs analyzed here and the analogous human phenotypes are: PG14 (9-octapeptide insertion), mixed CJD–GSS phenotype; P101L, GSS; M128V, normal; D177N/Met128, FFI; D177N/Val128, CJD; F197S/Val128, GSS; E199K, CJD. (*B*) CHO cells expressing each PrP construct were labeled for 3 hours with [^{35}S]methionine, and chased for 4 hours. Proteins in cell lysates were either undigested (–) or were digested at 37°C for 10 minutes with 3.3 μg/ml of proteinase K (+) prior to recovery of PrP by immunoprecipitation. Five times as many cell equivalents were loaded in the + lanes as in the – lanes. Molecular masses of migrated fragments are shown in kD. PrPs carrying pathogenic mutations yield a protease-resistant fragment of 27–30 kD, while wt and M128V PrPs are completely degraded. (Reprinted, with permission, from Harris and Lehmann 1997 [copyright John Wiley and Sons, Chichester, United Kingdom].)

that are 10 times those seen in wt mice. Importantly, the mutant protein displays the same three biochemical abnormalities as the mutant PrPs expressed in cultured cells (PK resistance, detergent insolubility, and PIPLC resistance). Thus, it is clear that mutant PrPs with these properties are pathogenic in an in vivo setting.

It is important to point out, however, that mutant PrPs synthesized in cultured cells and in Tg(PG14) mice differ in at least two key respects from authentic PrP^{Sc}: The mutant PrPs display a considerably lower degree of protease-resistance, and they are not infectious in transmission experiments. In a recent study by one of us (D.A.H.) (Chiesa et al. 2003), we have begun to clarify the molecular relationship between PrP^{Sc} and the weakly PK-resistant form of mutant PrP that accumulates spontaneously in Tg(PG14) mice (which we designate PG14^{spon}). As part of our analysis, we inoculated Tg(PG14) mice with RML prions, which resulted in accumulation of an alternate form of mutant PrP (designated PG14^{RML}) that has the properties of authentic PrP^{Sc}: It is infectious in transmission experiments and highly protease-resistant. We found that, like PrP^{Sc}, both PG14^{spon} and PG14^{RML} display conformationally masked epitopes in the central and octapeptide repeat regions, based on their behavior in a conformation-dependent immunoassay. However, the two forms differ profoundly in their oligomeric state: PG14^{RML} aggregates are much larger (up to 120 S, compared to 20 S for PG14^{spon}) and are more resistant to dissociation by urea.

These results have been interpreted to mean that aggregated, β-rich forms of PrP are neurotoxic, but that propagation of infectivity requires a larger, more tightly packed polymeric structure. Thus, a distinction exists between the neurotoxic and infectious properties of PrP. This conclusion is consistent with the observation that some familial prion diseases lack infectious PrP^{Sc}, and with a number of other experimental situations in which there is a dissociation between the development of pathology and the amount of PrP^{Sc} (for review, see Chiesa and Harris 2001). These examples also suggest that alternative forms of PrP, distinct from both PrP^{C} and PrP^{Sc}, may be the ultimate causes of neurodegeneration in many prion diseases. PG14^{spon}, and the analogous mutant PrPs synthesized in cultured cells, may be candidates for such neurotoxic intermediates.

Time Course of Changes in Mutant PrP Properties

Using pulse-chase labeling of cultured cells, one of us (D.A.H.) has identified three intermediate biochemical steps in the conversion of mutant PrPs to a PrP^{Sc}-like state (Daude et al. 1997). The earliest biochemical change in mutant PrP, one that is observable within minutes of pulse-labeling cells, is the acquisition of PIPLC resistance. As mentioned above, PIPLC resistance reflects conformational alterations at the carboxyl terminus of the protein (Narwa and Harris 1999). The timing of this step suggests that it takes place in the ER, a conclusion consistent with the observation that treatment with brefeldin A or incubation at 18°C, manipulations that block transit of proteins beyond the Golgi, did not affect acquisition of PIPLC resistance. The second step in the pathway is acquisition of detergent insolubility, which is not maximal until 1 hour of chase (Lehmann and Harris 1996b), arguing that it occurs after the acquisition of PIPLC resistance. Detergent insolubility reflects aggregation of PrP molecules. By sucrose gradient fractionation, oligomers ranging in size from 4 S (monomeric) to ~20 S (~30 PrP molecules) have been detected. The third step is acquisition of protease resistance, which is not maximal until several hours after labeling (Lehmann and Harris 1996b). Because detergent-insolubility and protease-resistance do not reach their highest levels until later times of chase, and because they are reduced by brefeldin A and 18°C incubation, it is likely that these properties continue to develop after arrival of the protein at the cell surface, either on the plasma membrane itself or in endocytic compartments. However, these results do not rule out the possibility that small aggregates of protease-resistant PrP begin forming in earlier compartments of the secretory pathway.

Taken together, the results of these kinetic experiments suggest that mutant PrP molecules begin to misfold early in the secretory pathway, perhaps even while they are being synthesized in the ER. The misfolded proteins then begin to aggregate, with the oligomerization process occurring over an extended period of time as the molecules reach the cell surface and beyond. Protease resistance may develop gradually as the aggregates reach a size or tightness of subunit packing that is sufficient to render the protein inaccessible to protease (except in the region near residue 90).

Subcellular Localization of Mutant PrP

Although extensive work on the biochemical properties of mutant PrPs in cultured cells has been done, less attention has been paid to the precise localization of these proteins at a subcellular level. To investigate this subject further, a detailed study of the localization of mutant PrPs in cultured cells was undertaken using light and electron microscopic techniques (Ivanova et al. 2001). To visualize PrP on the cell surface, BHK cells expressing wt or PG14 PrP were stained with anti-PrP antibody without permeabilization (Fig. 13, Surface). Despite equivalent expression levels

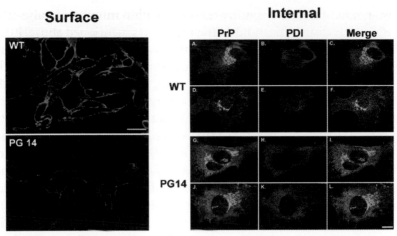

Figure 13. Mutant PrP is present at low levels on the cell surface and is concentrated in the ER. *Surface:* BHK cells expressing wt or PG14 PrP were stained with the 3F4 mAb prior to fixation. Bar represents 50 μm and applies to both panels. *Internal:* Cells expressing (*A–F*) wt PrP or (*G–L*) PG14 PrP were fixed, permeabilized, and stained with mouse anti-PrP mAb 3F4 (*A, D, G, J*) and a rabbit anti-PDI antibody (*B, E, H, K*). In the merged images (*C, F, I, L*), there is no yellow color for wt PrP, demonstrating the absence of the protein from the ER. In contrast, PG14 PrP extensively colocalizes with PDI, producing a yellow color throughout the cytoplasm, except in a region near the nucleus, which appears green due to the presence of the protein in the Golgi. Bar, 25 μm, applies to all panels. (Adapted from Ivanova et al. 2001.)

by western blotting, virtually all cells synthesizing PG14 PrP showed much weaker surface staining than those expressing wt PrP. To visualize intracellular PrP, fixed cells were permeabilized with Triton X-100 prior to application of primary and secondary antibodies (Fig. 13, Internal). In cells expressing wt PrP, staining was restricted to the perinuclear Golgi, a localization that is observed for other plasma membrane proteins, which probably reflects the relatively slow transit of secretory proteins through this compartment on their way to the cell surface. In contrast, many cells expressing PG14 PrP showed a much more widespread pattern of staining that largely colocalized with the ER marker protein disulfide isomerase. A similar ER localization has been observed for four different, disease-causing mutations (PG14, P101L, D177N/M128, and F197S/M128V) in CHO as well as BHK cells, and at both low and high expression levels. Ultrastructural studies confirm these immunofluorescence results. Immunogold labeling of the plasma membrane was abundant for wt PrP, but was virtually absent for PG14 PrP. Inside the cell, labeling of wt PrP was seen over the Golgi stacks, whereas labeling of PG14 PrP was observed over tubular elements of the ER as well as over the Golgi complex.

One of us (D.A.H.) has utilized PrP–EGFP fusion proteins to monitor the subcellular distribution and trafficking of wt and mutant PrPs (Ivanova et al. 2001). Insertion of a GFP reporter at either of two positions (adjacent to the signal sequence cleavage site or the GPI addition site) does not affect a number of key attributes of these molecules, including their glycosylation, GPI anchoring, subcellular distribution, detergent insolubility, and protease resistance. These results lay the foundation for the use of PrP–EGFP chimeras in a wide range of experiments on the distribution, trafficking, and molecular transformations of PrP in living cells and in animals. Importantly, PG14 PrP–EGFP shows reduced expression on the cell surface and concentration in the ER, analogous to what is seen with the nonfluorescent protein. It is noteworthy that to our knowledge, there are no reports of a PrP–EGFP fusion protein that has been converted into PrP^{Sc}–EGFP. Additionally, there are no bioassay data to argue that any PrP–EGFP fusion protein is infectious. Certainly, development of PrP fusion proteins that support prion replication is a worthwhile landmark goal.

The foregoing studies argue that several different pathogenic mutations share the property of impairing delivery of PrP molecules to the plasma membrane, which results in their accumulation within the ER. Similar findings have been obtained by several other laboratories (Petersen et al. 1996; Singh et al. 1997; Capellari et al. 2000b; Jin et al. 2000; Negro et al. 2001).

Delayed Biosynthetic Maturation of Mutant PrP

To complement cellular localization studies, one of us (D.A.H.) has carried out an analysis of the maturation and turnover of PrP molecules with disease-causing mutations using pulse–chase metabolic labeling (Drisaldi et al. 2003). The oligosaccharide chains of mutant PrPs became resistant to endoglycosidase H more slowly than those of wt PrP. wt PrP was initially synthesized as a 33-kD, endo H–sensitive species that matured into a 38-kD, endo H–resistant form by 20 minutes of chase. In contrast, PG14 and D177N PrPs were not fully converted into endo H–resistant forms until 60 minutes and 40 minutes of chase, respectively. Slower maturation of PG14 PrP was also seen in cerebellar granule neurons cultured from the brains of Tg(PG14) mice.

These results indicate that the mutant proteins are delayed in their transit through the ER or early Golgi compartments. Mutant PrPs do not seem to be irreversibly retained in the ER, since a substantial percentage of the initially labeled protein (50–70%) eventually becomes endo H–resistant, an efficiency that is similar to that for wt PrP. The delayed maturation of mutant PrP molecules correlates with their early acquisition of PIPLC resistance and likely reflects abnormal folding of the polypeptide chains in the ER. Slow transit through the early secretory pathway is also consistent with the altered distribution of mutant PrPs observed by immunofluorescence microscopy, since at steady state, there will be fewer molecules on the cell surface and more in the ER. The combined results of localization and biosynthetic studies therefore suggest that mutant PrP molecules are delayed in their export from the ER. This delay results in a steady-state distribution in which the proteins are concentrated in the ER and are expressed at lower levels on the cell surface.

ROLE OF THE PROTEASOME IN METABOLISM OF PrP

Protein Folding and Quality Control

The potential dangers presented by misfolded PrP^C have spurred an interest in its folding during biosynthesis. In the ER, an efficient quality-control mechanism detects misfolded or unassembled nascent proteins and prevents their export to the secretory pathway. These defective proteins are usually retrotranslocated into the cytosol for proteasomal degradation, a process known as ER-associated degradation (ERAD) (for review, see Ellgaard et al. 1999). Several PrP^C mutants are either retained in the ER or are processed through ERAD. The disulfide bond appears to be especially important for the proper folding of PrP^C. Replacing one or both of its cysteines results in the retention of PrP in large distensions of the ER (Yanai et al. 1999). Like many misfolded PrP molecules, these disulfide mutants form tight aggregates that are insoluble in Sarkosyl and possess a core resembling PrP 27–30 that resists stringent proteolysis. This dependence of native PrP folding on an intact disulfide bridge may help explain the tendency of PrP to aggregate when it is directed to the cytosol by means of removing the amino-terminal signal peptide (since the cytosol is vastly more reducing than the ER lumen) or in cells treated with dithiothreitol (DTT) (Ma and Lindquist 1999).

Proteasomal Degradation of Mutant PrP

Given the fact that PrP molecules carrying disease-causing mutations are delayed in their exit from the ER (see above), it was of interest to know whether these mutant proteins are subject to ERAD. One of us (D.A.H.) tested the effect of proteasome inhibitors on the turnover of wt, PG14, and D177N PrP in CHO cells (Drisaldi et al. 2003). Neither PSI 1 (Z-Ile-Glu(OtBu)-Ala-Leu-CHO) nor lactacystin had any significant effect on either the half-lives or maturation kinetics of the PrPs. The half-life of each of these proteins was 3–5 hours in the presence or absence of inhibitor. In addition, the inhibitors did not alter the maximal percentage of initially labeled molecules that matured to an endo H–resistant form. Proteasome inhibitors also had no effect on the maturation or turnover of wt and PG14 PrP expressed by cultured cerebellar granule neurons from Tg mice. It was concluded from these studies that, although mutant PrP molecules are delayed in their exit from the ER, they are not substrates for proteasomal degradation via a retrotranslocation pathway. These results led to the hypothesis that accumulation of mutant PrP molecules in the lumen of the ER could play a pathogenic role in some inherited prion disorders. Numerous other inherited human diseases are attributable to defects in export of a mutant protein from the ER (Aridor and Hannan 2002). In a subset of these disorders, the retained protein accumulates in the ER without being degraded. In these cases, the disease phenotype is due to a toxic effect of the accumulated protein, which stimulates one or more ER stress-response pathways.

Two PrP mutants associated with GSS (Y145stop and Q217R) and one associated with a familial early-onset dementia (Q160stop) have been shown to be subject to proteasomal degradation in the cytosol. Y145stop and Q160stop stay largely cytosolic and are rapidly degraded by proteasomes (Zanusso et al. 1999; Lorenz et al. 2002). In contrast to the complete failure of the two stop mutants to exit the ER by vesicular traffic, most Q217R molecules are successfully exported to the secretory pathway. A minority of Q217R molecules, however, is retained in the ER, where it interacts with the ER chaperone BiP for an unusually long time, and is then degraded by the proteasomal pathway (Jin et al. 2000). Prolonged interaction with BiP is a hallmark of many proteins that cannot attain native folding. The more extensive shunting of the two stop mutants to the proteasomal pathway suggests that carboxy-terminally truncated molecules interact with the ER quality-control machinery in a way that is different from full-length molecules.

Proteasomal Degradation of wt PrP

Whether or not wt PrP^C is subject to ERAD has also been studied. When cells were treated with proteasome inhibitors, one of us (A.T.) found that PrP accumulates in the cytosol where it forms aggregates of a 26-kD species with a 19-kD protease-resistant core (Ma and Lindquist 2001; Yedidia et al. 2001), a molecular weight that is consistent with unglycosylated PrP. A small portion of these molecules is mono- or polyubiquitinated (Yedidia et al. 2001). One interpretation of these results is that a minority of PrP^C molecules is constitutively subject to ERAD. In this process, they would be translocated from the ER lumen into the cytosol through the translocon, deglycosylated by a cytosolic N-glycosidase, ubiquitinated, and directed to the proteasome for degradation. When proteasomes are inhibited, PrP accumulates throughout the cytosol.

However, this interpretation has been recently challenged by one of us (D.A.H.) (Drisaldi et al. 2003). Using a signal peptide antiserum, it was shown that the 26-kD PrP species that accumulates in cells during treatment with proteasome inhibitors carries an intact amino-terminal signal peptide (Fig. 14). This species is also unprocessed at its carboxyl terminus, containing an intact GPI addition sequence and no GPI anchor. These characteristics suggest that the inhibitor-induced form of PrP has never entered the ER lumen where signal peptidase and GPI transamidase reside. Thus, a small fraction of PrP chains fail to be translocated into the ER lumen during their synthesis. These chains, which remain closely associated with the cytoplasmic face of the ER membrane, are rapidly degraded by proteasome. This phenomenon of abortive translocation is not unique to PrP and is likely to reflect saturation of one or more components of the translocation machinery at the elevated expression levels typ-

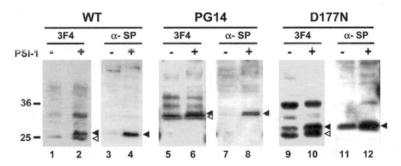

Figure 14. An unglycosylated, signal peptide–bearing form of PrP accumulates after treatment of CHO cells with a proteasome inhibitor. Transiently transfected CHO cells expressing wt, PG14, or D177N PrPs were treated for 8 hours with either ethanol vehicle (– lanes) or PSI 1 (Z-Ile-Glu[OtBu]-Ala-Leu-CHO; 20 μM) (+ lanes). Cells were then lysed, and PrP analyzed by western blotting using either the 3F4 antibody or an anti–signal peptide antibody (α-SP). The white and black arrowheads indicate the positions, respectively, of processed (signal peptide–cleaved) and unprocessed (signal peptide–bearing) forms of unglycosylated PrP. These two species are not completely resolved for PG14 PrP, because of the higher M_r of this protein. The slightly faster migration of all bands in lane 9 compared to those in lane 10 is an artifact of gel smiling. (Reprinted, with permission, from Drisaldi et al. 2003 [copyright American Society for Biochemistry and Molecular Biology].)

ical of transfected cells. Untranslocated PrP does not accumulate in cultured cerebellar granule cells treated with proteasome inhibitors, implying that this species is unlikely to be an obligate by-product of PrP biosynthesis in neurons. Moreover, untranslocated, signal peptide–bearing forms of other proteins have been found to accumulate in transfected cells treated with proteasome inhibitors.

A complicating factor that potentially affects studies using proteasome inhibitors is that these drugs have been found to artifactually increase transcription of transfected genes expressed from strong viral promoters (Drisaldi et al. 2003). This phenomenon provides an alternative explanation for the purported "self-perpetuating" properties of cytoplasmic PrP that accumulates in the presence of proteasome inhibitors (Ma et al. 2002). In the latter study, it is possible that initial treatment with inhibitor caused a sustained increase in the synthesis of PrP that continued even after the inhibitor was removed. Thus, increased PrP synthesis, rather than a self-propagating conformational change, probably accounts for the continued accumulation of PrP after transient treatment with proteasome inhibitors. The detergent insolubility and protease resistance of cytosolic PrP probably reflect protein aggregation induced by the presence of the hydrophobic signal peptide, rather than a conformational change related to acquisition of a PrP^{Sc}-like state.

Cyclophilins and PrP

PrP^C also accumulates in the cytosol of cells treated with the cyclophilin inhibitor, cyclosporin A (CsA). Cyclophilins are *cis–trans* proline isomerases that are found in most cellular compartments, assist the folding of many proteins, and are also involved in other cellular processes. Notably, CsA is used as an immunosuppressant for transplantation patients. Upon treatment with CsA, ~10% of N2a or CHO cells overexpressing PrP developed a juxtanuclear, PrP-containing spot (Cohen and Taraboulos 2003), which resembled aggresomes. Aggresomes are cytosolic inclusion bodies in which some misfolded proteins accumulate when proteasomes are inhibited (Johnston et al. 1998). The CsA-induced PrP bodies resemble aggresomes in that (1) their formation depends on intact microtubules, (2) they are encased within a network of collapsed vimentin, and (3) they form at the microtubule organizing center (MTOC), where they colocalize with γ-tubulin. However, the CsA-induced PrP bodies differ from the classic aggresomes in that they form even while proteasomes are apparently still functioning. The CsA-induced PrP is probably severely misfolded because it is largely insoluble in

Sarkosyl and contains a 19-kD protease-resistant core. The accumulation of PrP in aggresomes is not restricted to CsA-treated cells. A variety of disease-linked PrP mutants also accumulated in aggresomes, but this accumulation was dependent on the presence of proteasome inhibitors (Cohen and Taraboulos 2003; Mishra et al. 2003).

Because the cytosolic forms of PrP described above were obtained only when cells were treated with inhibitors, it is unlikely that they occur in large amounts in normal brain or other organs. However, it is not inconceivable that cytosolic PrP aggregates could slowly accumulate in aging neurons, when degradation and/or quality-control mechanisms progressively weaken. Whether PrP within such aggregates can acquire the conformation of PrP^{Sc} over time remains to be determined. Conceivably, such a mechanism could contribute to the etiology of the age-related, spontaneous prion diseases. This idea is supported by the finding of large amounts of PrP in the cytosol of ~2% of neurons in the brains of normal mice (Mironov et al. 2003).

Membrane Topology of PrP

PrP^{C} is made at the ER in at least three different folded conformers (Fig. 15). These conformers of PrP are identical in sequence but differ in their transmembrane topology and folding, allowing them to be distinguished by probing either (1) their orientation across the membrane with proteases or (2) their conformation in mild nondenaturing detergent solution (Hegde et al. 1998). Observed in cell-free translation systems, the three conformers are (1) ^{Sec}PrP that is fully translocated and glycolipid-anchored to the membrane; (2) ^{Ctm}PrP that spans the membrane with its carboxy-terminal domain in the ER lumen and elevated levels appear to produce neurodegeneration (Hegde et al. 1998); and (3) ^{Ntm}PrP that spans the membrane with its amino terminus in the ER lumen. A fourth topological variant, cytosolic PrP, represents unprocessed molecules that have been synthesized in the cytoplasm and have never been translocated into the ER lumen (Drisaldi et al. 2003; Mironov et al. 2003).

The significance of ^{Ctm}PrP has been clarified by a series of studies in Tg mice. Mutations in the mature coding region that favored ^{Ctm}PrP and ^{Sec}PrP in cell-free translation systems supplemented with microsomal membranes were identified and introduced as transgenes on the $Prnp^{0/0}$ background (Hegde et al. 1998). Only those mutations that favored production of ^{Ctm}PrP in the cell-free translation system were found to cause neurodegeneration in mice, and ^{Ctm}PrP was observed upon western blotting of protease-digested brain homogenates (Hegde et al. 1998).

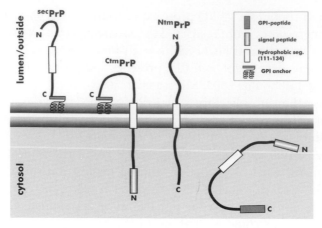

Figure 15. Four topological forms of PrP.

In a subsequent study, Tg mice were inoculated with prions and a striking inverse correlation was observed between the level of PrP^{Sc} accumulated at the onset of clinical illness and what was termed the Ctm index (Hegde and Lingappa 1999). The Ctm index was defined as the propensity of a mutation to favor (or disfavor) production of ^{Ctm}PrP compared to wt PrP x the level of transgene expression. This finding suggested that PrP^{Sc} triggered ^{Ctm}PrP production and that ^{Ctm}PrP initiated a final common pathway culminating in neurodegeneration in both spontaneous and infectious prion disease (Hegde and Lingappa 1999). Consistent with this hypothesis, the amount of ^{Ctm}PrP was found to increase during the course of prion infection of mice carrying a SHaPrP transgene that served as a reporter of ^{Ctm}PrP levels (Fig. 16).

It should be noted that ^{Ctm}PrP is *not* likely to be a misfolded protein because it is found to exit the ER, as evidenced by maturation of its carbohydrate, suggesting that it passes muster by quality-control machinery. Others have found ^{Ctm}PrP on the cell surface, consistent with this hypothesis (Mishra et al. 2002). It should also be noted that the pathophysiological mechanism supported by these studies predicts that some but not all forms of spontaneous prion disease will be due to ^{Ctm}PrP overproduction. In other cases, neurodegeneration could be due to events downstream of ^{Ctm}PrP in a multistep cascade.

Regulation of PrP Topology

Transmembrane and secretory forms of PrP were first noted in the mid-1980s in cell-free translation extracts (Hay et al. 1987a,b). Subsequent

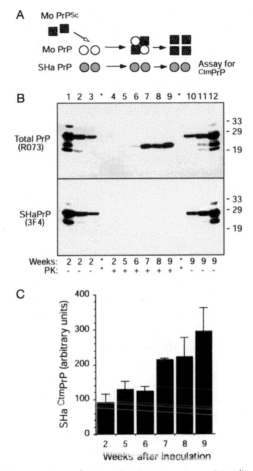

Figure 16. CtmPrP generation during the time course of PrPSc accumulation. (*A*) Diagram of experimental design. Transgenic mice expressing both SHaPrP (*shaded circles*) and MoPrP (*open circles*) were inoculated with RML mouse prions (*crosshatched squares*). Over time, host MoPrP is converted to MoPrPSc and accumulates. During this time course, the SHaPrPC is not converted to SHaPrPSc owing to the species barrier and may therefore be assayed for CtmPrP. (*B*) Relative amounts of total PrPSc and SHaPrPSc in mice at various times (in weeks) after inoculation with RML prions. Homogenate was digested using harsh PK conditions, treated with PNGase, and analyzed by SDS-PAGE and immunoblotting with either the RO73 polyclonal antibody (to detect total PrP) or the 3F4 monoclonal antibody (to selectively probe for SHaPrP). An equivalent amount of homogenate is analyzed in each lane except lanes *2* and *11* (which contain one-fourth as much) and lanes *3* and *10* (which contain one-tenth as much). (*C*) Relative amounts of Syrian hamster CtmPrP (detected selectively using the 3F4 monoclonal antibody) at various times after inoculation with RML prions. Each bar represents the average ± s.e.m. of three determinations. (Reprinted, with permission, from Hegde et al. 1999 [copyright Nature Publishing Company].)

work revealed that a complex regulation at the ER membrane directed initially homogeneous PrP chains to alternate pathways of biogenesis and folding, with the ^{Ctm}PrP form being the default in the absence of accessory factors to redirect the chain to the secretory topology (Lopez et al. 1990; Yost et al. 1990; Hegde et al. 1998). Some of the machinery involved in the translocational regulation of PrP has been identified recently (Fons et al. 2003).

Recent studies have demonstrated that the signal sequence and the transmembrane domains are major determinants of PrP topology. These two determinants act in mechanistically distinct ways (Kim et al. 2001; Kim and Hegde 2002). The signal sequence serves a dual function, first targeting the nascent polypeptide chain to the translocon channel in the ER membrane via binding to the signal-recognition particle (SRP), and subsequently gating the translocon to allow passage of the amino terminus into the ER lumen. In contrast, the transmembrane domain acts primarily to trigger integration of the polypeptide into the lipid bilayer. The combined action of both domains operating during the translocation process serves to regulate the proportions of the three topological variants of PrP.

One of us (D.A.H.) has carried out a mutational analysis of topological determinants in PrP that further extends this model (Stewart and Harris 2003). The substitution of charged residues in the hydrophobic core of the signal peptide increases synthesis of ^{Ctm}PrP and also reduces the efficiency of translocation into microsomes. Combining these mutations with substitutions in the transmembrane domain causes the protein to be synthesized exclusively with the ^{Ctm}PrP topology. Reducing the spacing between the signal peptide and the transmembrane domain by deletion of octapeptide repeats also increases ^{Ctm}PrP, consistent with an interaction between these two topological determinants during translocation. In contrast, topology is not altered by mutations that prevent signal peptide cleavage (G20W, C22Y) or by deletion of the carboxy-terminal signal for GPI anchor addition. Removal of the signal peptide completely blocks translocation, indicating that the central hydrophobic domain cannot directly mediate membrane integration.

Cell Biology of ^{Ctm}PrP

To investigate further whether or not ^{Ctm}PrP plays an important role in the pathogenesis of prion diseases, it is necessary to characterize the cell biological properties of this form, because very little is known about its localization, metabolism, or mode of synthesis and processing in cells. Part of the difficulty in addressing these issues has been that it was not possible to produce ^{Ctm}PrP in the absence of ^{Ntm}PrP and ^{Sec}PrP. To overcome this limitation, mutations have been identified that cause PrP to be synthesized exclusively with the ^{Ctm}PrP topology. Using these and other mutant forms, some of us have attained considerable insight into the cell biology of ^{Ctm}PrP, as well as into general mechanisms that control the topology of membrane proteins during their synthesis in the ER. From these results, it has been possible to develop hypotheses about the mechanisms by which ^{Ctm}PrP may cause neurodegeneration.

The discovery of a novel structural feature—that ^{Ctm}PrP has an uncleaved, amino-terminal signal peptide—served as a starting point for studies by one of us (D.A.H.) (Stewart et al. 2001). The existence of this feature was suspected because of the observation that after in vitro translation of PrP mRNA in the presence of canine pancreatic microsomes, the ^{Ctm}PrP band migrated on SDS-PAGE gels with a molecular size ~2 kD larger than mature, glycosylated ^{Sec}PrP. To confirm the presence of the uncleaved signal peptide, one of us (D.A.H.) used a version of PrP in which a FLAG epitope tag was inserted at the signal peptide cleavage site, between residues 22 and 23. The ^{Ctm}PrP band failed to react with antibody M1, which recognizes the FLAG epitope only if it displays a free amino terminus. In contrast, the protein reacted with a second antibody (M2) that recognizes the FLAG epitope regardless of sequence context. Recently, ^{Ctm}PrP has been shown to react with an antibody raised against the signal peptide, providing clear evidence that ^{Ctm}PrP retains a signal peptide (Stewart and Harris 2003). The presence of an uncleaved, amino-terminal signal peptide makes ^{Ctm}PrP unusual among other type II transmembrane proteins, most of which have internal signal-anchor sequences. However, this feature can be rationalized by the fact that the amino terminus of the polypeptide chain does not enter the ER lumen, where signal peptidase is located.

The proportion of ^{Ctm}PrP generated during the biosynthesis of PrP is normally quite low, but previous work demonstrated that the percentage could be increased by introduction of mutations into the central hydrophobic region of the molecule, including A116V (the mouse homolog of a GSS-linked human mutation) and 3AV (an artificial mutation in which Val residues are substituted for Ala residues at positions 112, 114, and 117) (Hegde et al. 1998, 1999). However, these mutations increase the proportion of ^{Ctm}PrP to, at most, 30% of the total PrP chains. Given the conclusion that ^{Ctm}PrP contains an uncleaved, amino-terminal signal peptide, one of us (D.A.H.) reasoned that mutations in the signal peptide itself might also influence the amount of ^{Ctm}PrP. Substitution of a

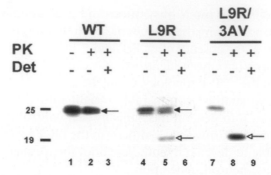

Figure 17. The L9R mutation in the signal peptide increases the amount of ^{Ctm}PrP, and molecules bearing an L9R/3AV mutation are synthesized exclusively as ^{Ctm}PrP. BHK cells were transiently transfected with plasmids encoding wt or mutant PrPs. Postnuclear supernatants prepared from cells 24 hours after transfection were incubated with (lanes *2, 3, 5, 6, 8, 9*) or without (lanes *1, 4, 7*) PK in the presence (lanes *3, 6, 9*) or absence (lanes *1, 2, 4, 5, 7, 8*) of Triton X-100 (Det). Proteins were then solubilized in SDS, deglycosylated with PNGase F, and subjected to western blotting with the 3F4 mAb. The protease-protected forms of ^{Sec}PrP and ^{Ctm}PrP are indicated by the filled and unfilled arrows, respectively. (Adapted from Stewart et al. 2001.)

charged residue for a hydrophobic residue within the signal sequence (L9R) markedly increased the proportion of ^{Ctm}PrP to ~50% (Fig. 17). Combining the L9R mutation with the 3AV mutation in the transmembrane domain to create L9R/3AV resulted in a protein that was synthesized exclusively as ^{Ctm}PrP, in both in vitro translation reactions and transfected cells (Fig. 17) (Stewart et al. 2001).

The availability of L9R/3AV PrP provided one of us (D.A.H.) with the ability to analyze the properties of ^{Ctm}PrP in a cellular context in the absence of the other two topological variants (Stewart et al. 2001). By labeling cells expressing L9R/3AV PrP with [^{3}H]palmitate, ^{Ctm}PrP was demonstrated to contain a GPI anchor, implying that ^{Ctm}PrP has an unusual, dual mode of membrane attachment, including both a membrane-spanning domain and a carboxy-terminal GPI anchor. L9R/3AV PrP (and hence ^{Ctm}PrP) was absent from the cell surface and was completely retained in the ER when expressed in transfected cells. This conclusion is supported by immunofluorescence localization studies as well as by the observation that the mutant protein remains completely endo H–sensitive. Surprisingly, ^{Ctm}PrP is not subject to retrotranslocation from the ER and degradation by the proteasome. The metabolic half-life of L9R/3AV is similar to that of wt PrP and is not altered by treatment with proteasome inhibitors (R.S. Stewart and D.A. Harris, in prep.).

Most Pathogenic Mutations Do Not Alter
the Membrane Topology of PrP

Mutations associated with familial prion diseases are found throughout the length of the PrP sequence. Although mutations in or around the central, hydrophobic region were known to increase the amount of ^{Ctm}PrP, the effect of mutations outside this area had not been examined. Therefore, in vitro translations of PrP mRNA encoding disease-causing mutations that lie both amino- and carboxy-terminal to the central, hydrophobic segment were completed (Stewart and Harris 2001). In these experiments, rabbit reticulocyte lysate supplemented with microsomes from either canine pancreas or murine thymoma cells was employed. Microsomes from the latter source are much more efficient at attachment of the GPI anchor. As positive controls, PrP containing several different mutations in the central region (A116V, 3AV, and K109I/H110I) that have been shown to increase the amount of transmembrane PrP were analyzed. None of the mutations outside the central, hydrophobic domain resulted in an increased proportion of ^{Ctm}PrP over wt levels. As expected, the A116V, 3AV, and K109I/H110I mutations resulted in significantly increased levels of ^{Ctm}PrP. Taken together, these observations argue that ^{Ctm}PrP does not play an obligate role in all forms of inherited prion disease. However, ^{Ctm}PrP could be involved in a subset of inherited cases due to mutations in the transmembrane domain (e.g., A117V, P105L).

Expression of ^{Ctm}PrP-favoring Mutations in Tg Mice

The role of the different topological forms of PrP was explored using Tg mice carrying PrP mutations that alter the relative ratios of these forms. Expression of ^{Ctm}PrP produced neurodegenerative changes in mice similar to some inherited human prion diseases (Hegde et al. 1998). Brains from these mice contained ^{Ctm}PrP, but not PrP^{Sc}, as defined by the lack of protease resistance and transmissibility. Furthermore, in one heritable prion disease of humans denoted GSS(A117V), brain tissue contained ^{Ctm}PrP but not PrP^{Sc}. Thus, considerable evidence argues that aberrant regulation of PrP biogenesis and topology is involved in the pathogenesis of at least some prion diseases (Hegde et al. 1998).

Synthesis of two different transmembrane forms of PrP is dependent on discrete sequences within the PrP coding region (Hay et al. 1987a; Yost et al. 1990; DeFea et al. 1994). Two adjacent domains within PrP, the hydrophobic, potentially membrane-spanning stretch from amino acids A113 to S135 (termed TM1) and the preceding hydrophilic domain (termed STE, and presently narrowed to residues L104 to M112), appear

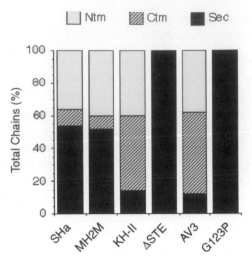

Figure 18. Quantitative representation of the relative amounts of ^{sec}PrP (*black bars*), ^{Ctm}PrP (*hatched bars*), and ^{Ntm}PrP (*gray bars*) for six different PrP constructs: SHaPrP, MH2MPrP, PrP(K109I/H110I), STE, AV3, and PrP(G123P).

to act in concert to generate both transmembrane forms of PrP (Fig. 4). Mutations, deletions, or insertions within these domains can alter the relative amounts of each topological form of PrP that is synthesized at the ER (Lopez et al. 1990; Yost et al. 1990). Given these complex and unusual features of PrP biogenesis, it seems plausible that these different forms of PrP may have dramatic consequences for the physiology of an organism.

To explore the hypothesis that misregulation of PrP biogenesis might cause neurodegeneration, four mutations within the STE–TM1 region were identified (Fig. 18) that greatly alter the ratio of the topological forms when assayed by cell-free translation. Two of these mutations (K109I/H110I and ΔSTE) were engineered into SHaPrP whereas the other two (AV3 and G123P) were introduced in MH2MPrP, a mouse–hamster chimera in which residues 94–188 are from SHaPrP (Scott et al. 1993). The species variation between SHaPrP and MH2MPrP (differing at eight residues) had little effect on topology. However, a comparison of mutant SHaPrP(K109I/H110I) with wt SHaPrP showed a dramatic increase (from ~10% to ~50%) in the relative amount of ^{Ctm}PrP synthesized and a concomitant decrease in ^{Sec}PrP (Fig. 18). The amount of ^{Ntm}PrP remained essentially unchanged. Similar results were obtained when MH2MPrP(AV3) was compared to MH2MPrP. In contrast, both SHaPrP(ΔSTE) and MH2MPrP(G123P) were synthesized exclusively in the ^{Sec}PrP form. These results, quantified and summarized (Fig. 18), provided the basis for examination of the effects of aberrant ^{Ctm}PrP synthesis in vivo. To accomplish this, PrP transgenes encoding each of these mutations were expressed in mice that lacked the PrP gene (FVB/$Prnp^{0/0}$). These mice were then observed for clinical signs and symptoms and examined for histopathology, and the PrP molecules in their brains were analyzed biochemically for transmembrane topology.

^{Ctm}PrP and the Development of Neurodegeneration

Mice expressing the SHaPrP(K109I/H110I) transgene (designated Tg[SHaPrP,K109I/H110I]H1198 mice) developed signs of neurodegeneration. All 29 Tg mice spanning three generations developed clinical signs of neurologic dysfunction, including ataxia and paresis. In the F_2 generation of mice harboring the transgene ($n = 24$), the average age of disease onset was 58 ± 11 days, with the earliest development of symptoms at 41 days. In contrast, none of the non-Tg littermates exhibited any signs of illness. Neither the FVB/$Prnp^{0/0}$ mice nor FVB/$Prnp^{0/0}$ mice expressing the wt SHaPrP transgene designated Tg(SHaPrP)3922 developed any signs of neurologic dysfunction. Quantification of PrP expression levels demonstrated that Tg(SHaPrP,K109I/H110I)H1198 mice expressed PrP at approximately half the level of Tg(SHaPrP)3922/$Prnp^{0/0}$ mice, indicating that disease was not a consequence of massive overexpression, which has been shown to cause neuromyopathy at the age of ~1 year (Westaway et al. 1994).

Tg mice expressing MH2MPrP(AV3) at high levels also showed neurological signs of illness and death within 8 weeks (Hegde et al. 1998). Biochemical analyses of mutant PrP in selected ill founder animals revealed the expression level to be equal to or lower than that of Tg(SHaPrP) mice, again militating against the explanation of simple overexpression. Thus, similar to the K109I/H110I mutation, expression of the AV3 mutation caused neurodegeneration in mice.

Tg mice expressing SHaPrP(ΔSTE) and MH2MPrP(G123P) were also constructed, and a high-expressing line of each—designated Tg(SHaPrP,ΔSTE)1788 and Tg(MH2MPrP,G123P)13638, respectively— was selected for further study. In contrast to mice carrying the K109I/H110I or AV3 mutation, neither Tg(SHaPrP,ΔSTE) nor Tg(MH2MPrP,G123P) mice showed any signs of illness. Furthermore, even at ages significantly beyond the life span of the Tg(SHaPrP,K109I/H110I)H mice, histological analysis of Tg(SHaPrP,ΔSTE) and Tg(MH2MPrP,G123P) mice revealed no abnormal neuropathologic changes. Taken together with the above findings, these results are suggestive of the notion that favored synthesis in the ^{Ctm}PrP form, as judged by the cell-free translocation assay, is indicative of the pathogenicity of the PrP mutation in vivo.

The topology of PrP in the brains of these Tg mice was examined using microsomal membranes prepared from brain. These intact vesicles were then subjected to protease digestion, and the accessibility of PrP to protease was assessed by immunoblotting. Generation of a proteolytic fragment encompassing the carboxyl terminus of PrP would indicate that those molecules were in the ^{Ctm}PrP orientation, whereas full protection from protease would indicate that the PrP was in the ^{Sec}PrP orientation. Analysis of the proteolytic fragments was simplified by the removal, just prior to SDS-PAGE, of the highly heterogeneous carbohydrate trees with the enzyme PNGase F. This allowed a search for the 18-kD carboxy-terminal fragment characteristic of the ^{Ctm}PrP form without the complications of differential electrophoretic migration of variably glycosylated PrP molecules. To ensure that vesicle integrity was maintained during the proteolytic reaction, the accessibility of GRP94, an ER luminal protein (Welch et al. 1983), to protease was also evaluated.

Only the Tg(SHaPrP,K109I/H110I)H and Tg(MH2MPrP,AV3) samples contained PrP molecules spanning the membrane, which produced an 18-kD fragment after protease digestion in the absence of detergent (Hegde et al. 1998). This fragment was determined to be a carboxy-terminal fragment of PrP, based on its detection by the 13A5 monoclonal antibody (mAb) and its observed slower migration on SDS-PAGE as a heterogeneous set of bands if polysaccharides were not removed from PrP prior to analysis. Thus, brains from Tg(SHaPrP,K109I/H110I)H and Tg(MH2MPrP,AV3) mice contained ^{Ctm}PrP (comprising 20–30% of the PrP in the microsomes). Similar results were obtained from multiple Tg(SHaPrP,K109I/H110I)H and Tg(MH2MPrP,AV3) animals, but in no instance was any ^{Ctm}PrP detected in Tg(SHaPrP), Tg(SHaPrP,ΔSTE), or Tg(MH2MPrP,G123P) mice. Taken together, these data indicate that the presence of ^{Ctm}PrP in the brains of Tg mice correlates well with observations made using the cell-free translocation system, although the absolute amount of ^{Ctm}PrP in brain was consistently less than that observed in cell-free assays.

Level of ^{Ctm}PrP Expression Modulates Disease

The observation that the percentage of PrP molecules found in the ^{Ctm}PrP topology in vivo was consistently lower than that found in vitro raises the possibility that cells normally have mechanisms to prevent the accumulation of this potentially pathogenic form. Thus, the basis for the modest ^{Ctm}PrP accumulation in brain in the K109I/H110I or AV3 mutant may be a combination of overexpression and the severe skew toward ^{Ctm}PrP synthesis, which together exceed the capacity of the cell to eliminate ^{Ctm}PrP (e.g., by rapid degradation at the ER). As a result, ^{Ctm}PrP would accumulate, exit the ER, and trigger disease. If this was the case, one would predict that lower levels of expression of a ^{Ctm}PrP-favoring mutant should fall below such a threshold and, thus, produce only ^{Sec}PrP. Such mice would be predicted not to get sick despite the mutation in the PrP gene, due to the absence of ^{Ctm}PrP.

This idea was explored using Tg(SHaPrP,K109I/H110I)L12485 mice expressing mutant SHaPrP(K109I/H110I) at a low level, approximately one-fourth to one-half the level of PrP found in normal Syrian hamsters and approximately fivefold lower than Tg(SHaPrP,K109I/H110I)H mice. Upon biochemical examination of the brains of Tg(SHaPrP,K109I/H110I)L mice, no ^{Ctm}PrP was detected, and all PrP appeared in the ^{Sec}PrP form. Thus, by decreasing the level of transgene expression by a factor of ~5, the percentage of PrP in the ^{Ctm}PrP form was reduced from ~30% to undetectable levels, even upon overexposure of the blots, under conditions in which ^{Sec}PrP was readily detectable. Corresponding to this lack of ^{Ctm}PrP generation, observation of Tg(SHaPrP,K109I/H110I)L mice has revealed no signs of illness at ages greater than 300 days, in sharp contrast to Tg(SHaPrP,K109I/H110I)H mice, which showed both ^{Ctm}PrP and signs of disease at ~60 days of age. These data support the hypothesis that generation of ^{Ctm}PrP leads to neurodegeneration in mice, with the role of the mutation being limited to one of favored synthesis of ^{Ctm}PrP.

Spontaneous Disease without PrP^{Sc} Accumulation

Whether an increase in ^{Ctm}PrP production is the basis of disease pathogenesis for the K109I/H110I and AV3 mutants remains to be determined. Whether the spontaneous disease caused by the K109I/H110I and AV3 mutations is transmissible remains to be determined. The brains of these ill Tg mice were analyzed for the presence of protease-resistant PrP^{Sc}, but none was detected in either the Tg(SHaPrP,K109I/H110I)H or Tg(MH2MPrP,AV3) mice, even after overloading the gels. Immunohistochemistry of brain sections from Tg(SHaPrP,K109I/H110I)H and Tg(MH2MPrP,AV3) mice following hydrolytic autoclaving failed to detect PrP^{Sc}. The lack of evidence for PrP^{Sc} lends support to the notion that ^{Ctm}PrP accumulation is responsible for the neurodegeneration observed in these Tg mice.

^{Ctm}PrP Accumulation in GSS(A117V)

Given that the distribution of newly synthesized PrP between transmembrane and secretory topologic forms is readily manipulated by mutations, we asked whether an inherited human prion disease might be caused by elevated levels of a TM form of PrP. The A117V mutation causing GSS (Doh-ura et al. 1989; Hsiao et al. 1991; Mastrianni et al. 1995) was a likely candidate for several reasons. First, this mutation lies in the hydrophobic TM1 domain that has been shown to be crucial to the biogenesis of transmembrane forms of PrP (Yost et al. 1990). Second, the pathologic findings in these cases of GSS(A117V) (Doh-ura et al. 1989; Hsiao et al. 1991; Mastrianni et al. 1995) appear to share some features with mice that become ill due to [Ctm]PrP overexpression. Third, the biochemical examination of brain tissue from these cases of GSS has revealed little or no protease-resistant PrP (Tateishi et al. 1990; Hegde et al. 1998). These observations raise the possibility that a mechanism other than PrP[Sc] accumulation is involved in the pathogenesis of GSS(A117V).

To explore the mechanism by which the A117V mutation causes disease, the biogenesis in a cell-free system of this mutant PrP was compared with its wt counterpart (both of which contain Val at polymorphic codon 129). Translocation reactions demonstrated that the A117V mutation significantly favored the synthesis of both [Ctm]PrP and [Ntm]PrP, with a concordant decrease of the [Sec]PrP form (Hegde et al. 1998).

Assay of PrP topology in human brain is problematic because fresh tissue suitable for subcellular fractionation is not readily available. To exploit differences in the native conformations that would be expected of the different topologic forms of PrP, limited proteolysis was employed. Although harsh digestion at elevated levels of PK (at >500 µg/ml) or elevated temperature (37°C) digested all topologic forms of PrP (but not PrP[Sc]), a subset of PrP was only partially digested under milder conditions (250 µg/ml PK and 0°C). The fragment generated under these conditions comigrated on SDS-PAGE with the protected carboxy-terminal domain generated in the cell-free translation-topology assay.

Digestion of each of the topological mutants of PrP under the same conditions revealed that the generation of the protease-resistant fragment correlated well with the amount of [Ctm]PrP form. Mutations (A117V, K109I/H110I, and AV3) that increase the relative amount of [Ctm]PrP resulted in increased generation of the PK-resistant fragment, whereas mutations (ΔSTE and G123P) that abolished synthesis of [Ctm]PrP did not yield a PK-resistant fragment. As expected, all of the topological forms were completely digested by harsher treatment under which PrP[Sc] is amino-terminally truncated to form PrP 27–30.

To confirm that the procedure described above could distinguish [Ctm]PrP from both [Sec]PrP and PrP[Sc] in frozen brain tissue samples, assays were performed on samples from Tg mice in which the distribution of PrP among the topologic forms had previously been established. Brain tissues from Tg(SHaPrP), Tg(SHaPrP,K109I/H110I)L, and Tg(SHaPrP,

K109I/H110I)H mice were analyzed by the conformational assay, and a protease-protected fragment was detected only in the Tg(SHaPrP, K109I/H110I)H sample. Thus, the presence of this carboxy-terminal fragment, generated under mild but not harsh digestion conditions, appeared to be diagnostic of the presence of ^{Ctm}PrP.

With the ability to distinguish ^{Ctm}PrP from PrP^C, ^{Sec}PrP, and PrP^{Sc} in frozen brain tissue, samples from GSS(A117V) brains were analyzed for ^{Ctm}PrP. Although similar levels of PrP were found in both normal and GSS human brain, the GSS brain contained increased levels of ^{Ctm}PrP. Neither control nor GSS(A117V) brain contained detectable protease-resistant PrP^{Sc} under conditions in which PrP^{Sc} was readily found in brain tissue from a sporadic CJD patient. These results were confirmed by analysis of multiple samples of tissue from the same patient, and also with brain tissue from a second patient carrying the A117V mutation. The lack of accumulation of protease-resistant PrP^{Sc} was also confirmed for the second patient. Thus, consistent with observations in vitro, the A117V mutation resulted in increased generation of ^{Ctm}PrP in vivo, suggesting that this is the basis of at least some of the neuropathological changes seen in cases of GSS(A117V).

Measurement of Transmembrane and Cytosolic PrP during Prion Infection

To test more directly the hypothesis that ^{Ctm}PrP and cytosolic PrP are toxic intermediates in prion infection, one of us (D.A.H.) measured the amounts of these two species in ScN2a cells and in rodent brain (Stewart and Harris 2003). An anti–signal peptide antibody was used to detect ^{Ctm}PrP and cytosolic PrP by virtue of the uncleaved signal peptide carried by both of these forms. No difference was discerned between infected and uninfected cells in the amounts of signal-peptide-reactive PrP, even when PrP expression was increased by transfection of plasmids encoding wt PrP or PrP carrying ^{Ctm}PrP-favoring mutations (A116V, 3AV, and L9R/3AV) (Fig. 19A).

To assay ^{Ctm}PrP and cytosolic PrP in mouse brain, SDS-PAGE was used to detect the small (~2 kD) difference in size between PrP molecules with and without a signal peptide (Fig. 19B). Proteins were enzymatically deglycosylated prior to SDS-PAGE to eliminate size differences due to differential glycosylation. When the gels were run long enough, it was possible to reliably detect the difference in migration between PrP molecules containing a cleaved signal peptide (25 kD) and an uncleaved signal peptide (27 kD), as demonstrated by visualization of two bands in brain samples from Tg(L9R/3AV) mice. These mice synthesize ^{Ctm}PrP as well as signal–peptide cleaved forms (R.S. Stewart and D.A. Harris, in prep.). Importantly, the 27-kD band was not detected in brain samples from

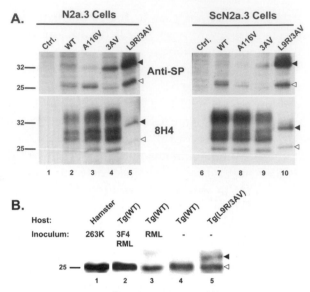

Figure 19. (*A*) The amounts of ^{Ctm}PrP and untranslocated PrP are not altered by scrapie infection of N2a cells. Uninfected N2a.3 cells (lanes *1–5*) and scrapie-infected N2a.3 cells (lanes *6–10*) were untransfected (lanes *1* and *6*), or were transiently transfected to express wt (lanes *2* and *7*), A116V (lanes *3* and *8*), 3AV (lanes *4* and *9*), or L9R/3AV (lanes *5* and *10*) PrP. Cells were labeled for 6 hours with [^{35}S]methionine and then lysed. PrP was immunoprecipitated from cell lysates using either an anti-signal peptide antibody (anti-SP) (*upper* panels) or the 8H4 antibody (*lower* panels), and analyzed by SDS-PAGE and autoradiography. The filled and open arrowheads indicate the positions of ^{Ctm}PrP and untranslocated PrP, respectively. (*B*) The amount of signal peptide–bearing PrP is not altered in scrapie-infected brain. Brain homogenates were prepared from a Syrian hamster infected with 263K prions (lane *1*), a Tg(WT) mouse infected with 3F4-tagged RML prions (lane *2*), a Tg(WT) mouse infected with RML prions (lane *3*), an uninfected Tg(WT) mouse (lane *4*), and an uninfected Tg(L9R/3AV) mouse (lane *5*). Infected animals were terminally ill at the time of sacrifice. Homogenates were treated with PNGase F and subjected to western blotting with the 3F4 antibody. The filled and open arrowheads indicate the positions of signal peptide–bearing PrP and signal peptide–cleaved PrP, respectively. (Adapted from Stewart and Harris 2003.)

uninfected mice, or from mice and hamsters infected with several different scrapie prions. Because samples were deglycosylated prior to SDS-PAGE, this analysis would not distinguish ^{Ctm}PrP from untranslocated (cytosolic) PrP. These data have led one of us (D.A.H.) to conclude that scrapie infection of mice and hamsters does not increase the amount of either ^{Ctm}PrP or cytosolic PrP in the brain to detectable levels. Thus, ^{Ctm}PrP and cytosolic PrP are unlikely to be obligate intermediates in infectious forms of prion diseases.

LOCALIZATION OF PrP^{Sc}

Neuroanatomical Level

Pathologic changes in prion diseases seem to be confined to the nervous system. Such changes include vacuolation of neurons and hypertrophy of astrocytes, as well as the extracellular accumulation of fragments of PrP^{Sc} that polymerize to form amyloid fibrils. The vacuolation of neurons and hypertrophy of reactive astrocytes are generally obligatory features of prion disease, whereas PrP amyloid deposits are a variable feature (Prusiner et al. 1990). When PrP deposits in the form of amyloid plaques are present, the diagnosis of prion disease cannot be in question.

The cell biology of PrP^{Sc} accumulation that leads to neuronal vacuolation remains unclear. Various studies argue that the intracellular accumulation of PrP^{Sc} induces vacuolation of neurons (DeArmond et al. 1987). As vacuolated neurons die, these vacuoles coalesce to form the spongiform change that is seen in prion disease, with rare exception. However, it is important to note that spongiform degeneration is frequently confined to very limited regions of the brain, as in FFI of humans, scrapie of sheep, and bovine spongiform encephalopathy of cattle (Zlotnik and Stamp 1961; Lugaresi et al. 1986; Wells et al. 1989). The reactive astrocytic gliosis that accompanies neuronal degeneration is most easily assessed by GFAP immunostaining. In the prion diseases, increased GFAP mRNA levels are found in addition to elevated levels of the protein and intermediate filaments composed of GFAP. It is notable that GFAP-deficient (GFAP$^{0/0}$) mice were as susceptible to inoculated prions as their nonablated controls (Gomi et al. 1995; Tatzelt et al. 1996). A much more extensive description of the processes underlying the pathology of the prion diseases can be found in Chapter 15.

When cultured neurons isolated from mouse brains were exposed to a PrP peptide corresponding to residues 100–126, the cells died (Forloni et al. 1993). When cultured neurons isolated from brains of $Prnp^{0/0}$ mice were exposed to the same PrP peptide, the cells were unharmed (Brown et al. 1994). Consistent with these results are studies with brain grafts infected with prions in which the graft was transplanted into the brains of $Prnp^{0/0}$ mice (Brandner et al. 1996). Although the cells of the graft produced large amounts of PrP^{Sc}, the surrounding cells remained healthy. These studies argue that PrP^{Sc} is cytotoxic only for cells that express PrP^{C} and that extracellular PrP^{Sc} does not cause cellular dysfunction (Prusiner et al. 1990).

Subcellular Level

Several studies suggest that >50% of PrP^{Sc} in ScN2a cells may remain attached to the plasma membrane (S.J. DeArmond, unpubl.). Whether a similar localization will be found in the brains of mammals with prion diseases remains to be determined. It is unknown whether the accumulation of PrP^{Sc} exerts its effect on neuronal function when it is newly formed in CLDs or after it is transported into the interior of the cell. The mechanisms by which prions spread through the CNS are also unknown. Several studies argue that prions transit along axons by both retrograde and anterograde transport (Kimberlin and Walker 1979; Fraser 1982; DeArmond et al. 1987; DeArmond and Prusiner 1997).

KINETICS OF PrP^{Sc} PRODUCTION

In an uninfected cell, PrP^{C} with the wt sequence exists in equilibrium in its monomeric α-helical, protease-sensitive state or bound to protein X (Fig. 2 in Chapter 5). The conformation of PrP^{C} that is bound to protein X is denoted PrP* (Cohen et al. 1994); this conformation is likely to be different from that determined under aqueous conditions for monomeric recombinant PrP. The PrP*–protein X complex will bind PrP^{Sc}, creating a replication-competent assembly. Order-of-addition experiments demonstrate that for PrP^{C}, protein X binding precedes productive PrP^{Sc} interactions (Kaneko et al. 1997c). A conformational change takes place wherein PrP, in a shape competent for binding to protein X and PrP^{Sc} (denoted PrP*$^{/Sc}$), represents the initial phase in the formation of infectious PrP^{Sc}. It is noteworthy that PrP* has also been used to denote a subgroup of PrP^{Sc} molecules that are infectious (Weissmann 1991); however, we have no evidence for such a subset.

Several lines of evidence argue that the smallest infectious prion particle is an oligomer of PrP^{Sc}, perhaps as small as a dimer (Prusiner et al. 1978; Bellinger-Kawahara et al. 1988). Upon purification, PrP^{Sc} tends to aggregate into insoluble multimers that can be dispersed into liposomes (Gabizon et al. 1988; McKinley et al. 1991).

In attempts to form PrP^{Sc} in vitro, PrP^{C} has been exposed to 3 M guanidine hydrochloride (GdnHCl) and then diluted tenfold prior to binding to PrP^{Sc} (Kocisko et al. 1994; Kaneko et al. 1997b). Exposure of PrP^{C} to GdnHCl presumably converts it into a PrP*-like molecule. Although the PrP*-like protein bound to PrP^{Sc} is protease-resistant and insoluble, it has not been re-isolated in order to assess whether or not it was converted into PrP^{Sc}. It is noteworthy that recombinant PrP can be refolded into forms that are rich in either α-helical or β-structure, but none has been found to possess prion infectivity as judged by bioassay (Zhang et al. 1997; Baskakov et al. 2000, 2001, 2002).

Whether attempts to produce PrP^{Sc} using normal homogenates from mouse or hamster brain combined with homogenates from prion-infected rodent brains are able to amplify prion infectivity is unknown. In one set of such studies, repeated sonication was thought to be crucial (Saborio et al. 2001), whereas in another protocol, sonication was stated to be not particularly useful (Lucassen et al. 2003).

SUBCELLULAR SITE OF PrP^{Sc} FORMATION

Because infectious prions seem to be composed entirely of PrP^{Sc} molecules (Prusiner 1991), the identification of the site of PrP^{Sc} synthesis and the definition of the molecular events involved in this process are critical. Earlier observations argued that PrP^{Sc} synthesis occurs within the endocytic pathway (Borchelt et al. 1992; Shyng et al. 1994), but more recent studies have refined those data and contend that PrP^{Sc} formation takes place in CLDs.

How exogenous prions initiate infection is unknown, but the presence of PrP^{C} on the cell surface suggests a mechanism whereby PrP^{Sc} binds to either PrP^{C} or PrP*, which is already bound to protein X. PrP* is thought to be a conformationally distinct, metastable variant of PrP^{C} that is competent for conversion into PrP^{Sc}. Whether PrP^{Sc} can bind to PrP^{C} on the cell surface or whether PrP^{C} must first be transformed into PrP* remains uncertain. Once PrP^{Sc} is formed in CLDs, it enters the endocytic pathway and eventually is deposited in lysosomes. The foregoing studies are all consistent with the hypothesis that PrP^{Sc} formation is confined to CLDs, and although this is a pathologic process, it occurs within a specific subcellular domain. The apparent restriction of PrP^{Sc} formation to CLDs would seem to argue that such a process is likely to involve auxiliary macromolecules found within this compartment. Such an auxiliary factor has been implicated in the conversion of PrP^{C} into PrP^{Sc} based on the results of transgenetic studies in which a chimeric human–mouse PrP molecule but not human (Hu) PrP rendered mice susceptible to Hu prions from patients who died of CJD (Telling et al. 1995). Additionally, the recognition site on PrP^{C} to which protein X binds has been mapped using a series of chimeric PrP constructs expressed in ScN2a cells (Kaneko et al. 1997c).

CONCLUSIONS

Although many questions concerning the cell biology of prion proteins remain to be answered, much progress has been made in understanding the processing, trafficking, and pathologic roles of these proteins. PrP^{C} can assume several different topologies, the most important of which seem to be ^{Sec}PrP and ^{Ctm}PrP. ^{Sec}PrP is the form of PrP^{C} that migrates to the cell surface, where it is thought to be concentrated in cholesterol-rich rafts, which can coalesce into caveolae. Defining the factors that govern the topology and trafficking of PrP^C is an important area of further investigation.

A wealth of data argues that rafts or caveolae are the subcellular sites where PrP^C is converted into PrP^{Sc}. Presently, none of the auxiliary proteins thought to be involved in this process have been identified, and the sequence of events, by which either PrP^{Sc} or ^{Ctm}PrP causes neurodegeneration, remains to be elucidated.

Defining mechanisms that govern the formation of ^{Ctm}PrP will be an important avenue of research. A variety of mutations in PrP have been shown to alter the levels of ^{Ctm}PrP as well as create a spectrum of neurodegenerative diseases. The finding of elevated levels of ^{Ctm}PrP in both GSS(A117V) patients and Tg(SHaPrP,K109I/H110I)H mice argues that this topologic form of PrP may be pathogenic.

How many different forms of prion disease will be found is of considerable interest. The prion diseases are the only disorders that present as sporadic, infectious, and genetic illnesses. Moreover, at least three different forms of heritable prion disease can now be defined by the nature of the PrP molecule that accumulates. First, mutant PrP^{Sc} accumulates in such disorders as fCJD(E200K) and FFI. Second, ^{Ctm}PrP has been found to be elevated in both GSS(A117V) in humans and Tg(SHaPrP,K109I/H110I)H mice. Third, protease-sensitive MoPrP^{Sc}(P101L) is found in Tg mice as well as in the brains of patients in GSS(P102L).

REFERENCES

Anderson R.G. 1998. The caveolae membrane system. *Annu. Rev. Biochem.* **67:** 199–225.

Aridor M. and Hannan L.A. 2002. Traffic jams II: An update of diseases of intracellular transport. *Traffic* **3:** 781–790.

Arnold J.E., Tipler C., Laszlo L., Hope J., Landon M., and Mayer R.J. 1995. The abnormal isoform of the prion protein accumulates in late-endosome-like organelles in scrapie-infected mouse brain. *J. Pathol.* **176:** 403–411.

Baskakov I.V., Legname G., Prusiner S.B., and Cohen F.E. 2001. Folding of prion protein to its native α-helical conformation is under kinetic control. *J. Biol. Chem.* **276:** 19687–19690.

Baskakov I.V., Legname G., Baldwin M.A., Prusiner S.B., and Cohen F.E. 2002. Pathway complexity of prion protein assembly into amyloid. *J. Biol. Chem.* **277:** 21140–21148.

Baskakov I.V., Aagaard C., Mehlhorn I., Wille H., Groth D., Baldwin M.A., Prusiner S.B., and Cohen F.E. 2000. Self-assembly of recombinant prion protein of 106 residues. *Biochemistry* **39:** 2792–2804.

Bellinger-Kawahara C.G., Kempner E., Groth D.F., Gabizon R., and Prusiner S.B. 1988. Scrapie prion liposomes and rods exhibit target sizes of 55,000 Da. *Virology* **164:** 537–541.

Bendheim P.E., Brown H.R., Rudelli R.D., Scala L.J., Goller N.L., Wen G.Y., Kascsak R.J., Cashman N.R., and Bolton D.C. 1992. Nearly ubiquitous tissue distribution of the scrapie agent precursor protein. *Neurology* **42:** 149–156.

Borchelt D.R., Taraboulos A., and Prusiner S.B. 1992. Evidence for synthesis of scrapie prion proteins in the endocytic pathway. *J. Biol. Chem.* **267:** 16188–16199.

Borchelt D.R., Scott M., Taraboulos A., Stahl N., and Prusiner S.B. 1990. Scrapie and cellular prion proteins differ in their kinetics of synthesis and topology in cultured cells. *J. Cell Biol.* **110:** 743–752.

Bouillot C., Prochiantz A., Rougon G., and Allinquant B. 1996. Axonal amyloid precursor protein expressed by neurons *in vitro* is present in a membrane fraction with caveolae-like properties. *J. Biol. Chem.* **271:** 7640–7644.

Brandner S., Isenmann S., Raeber A., Fischer M., Sailer A., Kobayashi Y., Marino S., Weissmann C., and Aguzzi A. 1996. Normal host prion protein necessary for scrapie-induced neurotoxicity. *Nature* **379:** 339–343.

Brown D.R. 2001. Prion and prejudice: Normal protein and the synapse. *Trends Neurosci.* **24:** 85–90.

Brown D.R. and Besinger A. 1998. Prion protein expression and superoxide dismutase activity. *Biochem. J.* **334:** 423–429.

Brown D.R., Herms J., and Kretzschmar H.A. 1994. Mouse cortical cells lacking cellular PrP survive in culture with a neurotoxic PrP fragment. *Neuroreport* **5:** 2057–2060.

Brown D.R., Nicholas R.S., and Canevari L. 2002. Lack of prion protein expression results in a neuronal phenotype sensitive to stress. *J. Neurosci. Res.* **67:** 211–224.

Brown D.R., Schulz-Schaeffer W.J., Schmidt B., and Kretzschmar H.A. 1997a. Prion protein-deficient cells show altered response to oxidative stress due to decreased SOD-1 activity. *Exp. Neurol.* **146:** 104–112.

Brown D.R., Wong B.-S., Hafiz F., Clive C., Haswell S.J., and Jones I.M. 1999. Normal prion protein has an activity like that of superoxide dismutase. *Biochem. J.* **344:** 1–5.

Brown D.R., Hafiz F., Glasssmith L.L., Wong B.S., Jones I.M., Clive C., and Haswell S.J. 2000. Consequences of manganese replacement of copper for prion protein function and proteinase resistance. *EMBO J.* **19:** 1180–1186.

Brown D.R., Qin K., Herms J.W., Madlung A., Manson J., Strome R., Fraser P.E., Kruck T., von Bohlen A., Schulz-Schaeffer W., Giese A., Westaway D., and Kretzschmar H. 1997b. The cellular prion protein binds copper *in vivo*. *Nature* **390:** 684–687.

Brown L.R. and Harris D.A. 2002. The prion protein and copper: What is the connection? In *Handbook of copper pharmacology and toxicology* (ed. E.J. Massaro), pp. 103–113. Humana Press, Totowa, New Jersey.

———. 2003. Copper and zinc cause delivery of the prion protein from the plasma membrane to a subset of early endosomes and the Golgi. *J. Neurochem.* **87:** 353–363.

Büeler H., Aguzzi A., Sailer A., Greiner R.-A., Autenried P., Aguet M., and Weissmann C. 1993. Mice devoid of PrP are resistant to scrapie. *Cell* **73:** 1339–1347.

Burns C.S., Aronoff-Spencer E., Legname G., Prusiner S.B., Antholine W.E., Gerfen G.J., Peisach J., and Millhauser G.L. 2003. Copper coordination in the full-length, recombinant prion protein. *Biochemistry* **42:** 6794–6803.

Burns C.S., Aronoff-Spencer E., Dunham C.M., Lario P., Avdievich N.I., Antholine W.E., Olmstead M.M., Vrielink A., Gerfen G.J., Peisach J., Scott W.G., and Millhauser G.L. 2002. Molecular features of the copper binding sites in the octarepeat domain of the prion protein. *Biochemistry* **41:** 3991–4001.

Capellari S., Zaidi S.I., Long A.C., Kwon E.E., and Petersen R.B. 2000a. The Thr183Ala mutation, not the loss of the first glycosylation site, alters the physical properties of the prion protein. *J. Alzheimers Dis.* **2:** 27–35.

Capellari S., Parchi P., Russo C.M., Sanford J., Sy M.S., Gambetti P., and Petersen R.B. 2000b. Effect of the E200K mutation on prion protein metabolism. Comparative study of a cell model and human brain. *Am. J. Pathol.* **157:** 613–622.

Caughey B. and Raymond G.J. 1991. The scrapie-associated form of PrP is made from a cell surface precursor that is both protease- and phospholipase-sensitive. *J. Biol. Chem.* **266:** 18217–18223.

Caughey B., Race R.E., Ernst D., Buchmeier M.J., and Chesebro B. 1989. Prion protein biosynthesis in scrapie-infected and uninfected neuroblastoma cells. *J. Virol.* **63:** 175–181.

Chen S.G., Teplow D.B., Parchi P., Teller J.K., Gambetti P., and Autilio-Gambetti L. 1995. Truncated forms of the human prion protein in normal brain and in prion diseases. *J. Biol. Chem.* **270:** 19173–19180.

Chiesa R. and Harris D.A. 2001. Prion diseases: What is the neurotoxic molecule? *Neurobiol. Dis.* **8:** 743–763.

Chiesa R., Piccardo P., Ghetti B., and Harris D.A. 1998. Neurological illness in transgenic mice expressing a prion protein with an insertional mutation. *Neuron* **21:** 1339–1351.

Chiesa R., Drisaldi B., Quaglio E., Migheli A., Piccardo P., Ghetti B., and Harris D.A. 2000. Accumulation of protease-resistant prion protein (PrP) and apoptosis of cerebellar granule cells in transgenic mice expressing a PrP insertional mutation. *Proc. Natl. Acad. Sci.* **97:** 5574–5579.

Chiesa R., Piccardo P., Quaglio E., Drisaldi B., Si-Hoe S.L., Takao M., Ghetti B., and Harris D.A. 2003. Molecular distinction between pathogenic and infectious properties of the prion protein. *J. Virol.* **77:** 7611–7622.

Cohen E. and Taraboulos A. 2003. Scrapie-like prion protein accumulates in aggresomes of cyclosporin A–treated cells. *EMBO J.* **22:** 404–417.

Cohen F.E., Pan K.-M., Huang Z., Baldwin M., Fletterick R.J., and Prusiner S.B. 1994. Structural clues to prion replication. *Science* **264:** 530–531.

Daude N., Lehmann S., and Harris D.A. 1997. Identification of intermediate steps in the conversion of a mutant prion protein to a scrapie-like form in cultured cells. *J. Biol. Chem.* **272:** 11604–11612.

DeArmond S.J. and Prusiner S.B. 1997. Molecular neuropathology of prion diseases. In *The molecular and genetic basis of neurological disease,* 2nd edition (ed. R.N. Rosenberg et al.), pp. 145–163. Butterworth Heinemann, Stoneham, Massachusetts.

DeArmond S.J., Mobley W.C., DeMott D.L., Barry R.A., Beckstead J.H., and Prusiner S.B. 1987. Changes in the localization of brain prion proteins during scrapie infection. *Neurology* **37:** 1271–1280.

DeFea K.A., Nakahara D.H., Calayag M.C., Yost C.S., Mirels L.F., Prusiner S.B., and Lingappa V.R. 1994. Determinants of carboxyl-terminal domain translocation during prion protein biogenesis. *J. Biol. Chem.* **269:** 16810–16820.

Doh-ura K., Tateishi J., Sasaki H., Kitamoto T., and Sakaki Y. 1989. Pro→Leu change at position 102 of prion protein is the most common but not the sole mutation related to Gerstmann-Sträussler syndrome. *Biochem. Biophys. Res. Commun.* **163:** 974–979.

Drisaldi B., Stewart R.S., Adles C., Stewart L.R., Quaglio E., Biasini E., Fioriti L., Chiesa R., and Harris D.A. 2003. Mutant PrP is delayed in its exit from the endoplasmic reticulum, but neither wild-type nor mutant PrP undergoes retrotranslocation prior to proteasomal degradation. *J. Biol. Chem.* **278:** 21732–21743.

Ellgaard L., Molinari M., and Helenius A. 1999. Setting the standards: Quality control in the secretory pathway. *Science* **286:** 1882–1888.

Fivaz M., Vilbois F., Thurnheer S., Pasquali C., Abrami L., Bickel P.E., Parton R.G., and van der Goot F.G. 2002. Differential sorting and fate of endocytosed GPI-anchored proteins. *EMBO J.* **21:** 3989–4000.

Fons R.D., Bogert B.A., and Hegde R.S. 2003. Substrate-specific function of the translocon-associated protein complex during translocation across the ER membrane. *J. Cell Biol.* **160:** 529–539.

Ford M.J., Burton L.J., Morris R.J., and Hall S.M. 2002. Selective expression of prion protein in peripheral tissues of the adult mouse. *Neuroscience* **113:** 177–192.

Forloni G., Angeretti N., Chiesa R., Monzani E., Salmona M., Bugiani O., and Tagliavini F. 1993. Neurotoxicity of a prion protein fragment. *Nature* **362:** 543–546.

Fournier J.G., Escaig-Haye F., and Grigoriev V. 2000. Ultrastructural localization of prion proteins: Physiological and pathological implications. *Microsc. Res. Tech.* **50:** 76–88.

Fraser H. 1982. Neuronal spread of scrapie agent and targeting of lesions within the retino-tectal pathway. *Nature* **295:** 149–150.

Gabizon R., McKinley M.P., Groth D., and Prusiner S.B. 1988. Immunoaffinity purification and neutralization of scrapie prion infectivity. *Proc. Natl. Acad. Sci.* **85:** 6617–6621.

Gauczynski S., Krasemann S., Bodemer W., and Weiss S. 2002. Recombinant human prion protein mutants huPrP D178N/M129 (FFI) and huPrP+9OR (fCJD) reveal proteinase K resistance. *J. Cell Sci.* **115:** 4025–4036.

Gomi H., Yokoyama T., Fujimoto K., Ikeda T., Katoh A., Itoh T., and Itohara S. 1995. Mice devoid of the glial fibrillary acidic protein develop normally and are susceptible to scrapie prions. *Neuron* **14:** 29–41.

Gorodinsky A. and Harris D.A. 1995. Glycolipid-anchored proteins in neuroblastoma cells form detergent-resistant complexes without caveolin. *J. Cell Biol.* **129:** 619–627.

Griffiths G. 1993. *Fine structure immunocytochemistry.* Springer-Verlag, New York.

Haeberle A.M., Ribaut-Barassin C., Bombarde G., Mariani J., Hunsmann G., Grassi J., and Bailly Y. 2000. Synaptic prion protein immuno-reactivity in the rodent cerebellum. *Microsc. Res. Tech.* **50:** 66–75.

Hall P.A. 1999. Assessing apoptosis: A critical survey. *Endocr. Relat. Cancer* **6:** 3–8.

Haraguchi T., Fisher S., Olofsson S., Endo T., Groth D., Tarantino A., Borchelt D.R., Teplow D., Hood L., Burlingame A., Lycke E., Kobata A., and Prusiner S.B. 1989. Asparagine-linked glycosylation of the scrapie and cellular prion proteins. *Arch. Biochem. Biophys.* **274:** 1–13.

Harris D.A. and Lehmann S. 1997. Mutant prion proteins acquire PrP^{Sc}-like properties in cultured cells: An experimental model of familial prion diseases. In *Alzheimer's disease: Biology, diagnosis and therapeutics* (ed. K. Iqbal et al.), pp. 631–643. John Wiley and Sons, Chichester, United Kingdom.

Harris D.A., Huber M.T., van Dijken P., Shyng S.-L., Chait B.T., and Wang R. 1993. Processing of a cellular prion protein: Identification of N- and C-terminal cleavage sites. *Biochemistry* **32:** 1009–1016.

Hay B., Prusiner S.B., and Lingappa V.R. 1987a. Evidence for a secretory form of the cellular prion protein. *Biochemistry* **26:** 8110–8115.

Hay B., Barry R.A., Lieberburg I., Prusiner S.B., and Lingappa V.R. 1987b. Biogenesis and transmembrane orientation of the cellular isoform of the scrapie prion protein. *Mol. Cell. Biol.* **7:** 914–920.

Hegde R.S. and Lingappa V.R. 1999. Regulation of protein biogenesis at the endoplasmic reticulum membrane. *Trends Cell Biol* **9:** 132–137.

Hegde R.S., Tremblay P., Groth D., Prusiner S.B., and Lingappa V.R. 1999. Transmissible and genetic prion diseases share a common pathway of neurodegeneration. *Nature* **402:** 822–826.

Hegde R.S., Mastrianni J.A., Scott M.R., DeFea K.A., Tremblay P., Torchia M., DeArmond S.J., Prusiner S.B., and Lingappa V.R. 1998. A transmembrane form of the prion protein in neurodegenerative disease. *Science* **279:** 827–834.

Herms J.W., Tings T., Gall S., Madlung A., Giese A., Siebert H., Schurmann P., Windl O., Brose N., and Kretzschmar H. 1999. Evidence of presynaptic location and function of the prion protein. *J. Neurosci.* **19:** 8866–8875.

Hölscher C., Bach U.C., and Dobberstein B. 2001. Prion protein contains a second endoplasmic reticulum targeting signal sequence located at its C terminus. *J. Biol. Chem.* **276:** 13388–13394.

Hsiao K.K., Cass C., Schellenberg G.D., Bird T., Devine-Gage E., Wisniewski H., and Prusiner S.B. 1991. A prion protein variant in a family with the telencephalic form of Gerstmann-Sträussler-Scheinker syndrome. *Neurology* **41:** 681–684.

Ivanova L., Barmada S., Kummer T., and Harris D.A. 2001. Mutant prion proteins are partially retained in the endoplasmic reticulum. *J. Biol. Chem.* **276:** 42409–42421.

Jackson G.S., Murray I., Hosszu L.L., Gibbs N., Waltho J.P., Clarke A.R., and Collinge J. 2001. Location and properties of metal-binding sites on the human prion protein. *Proc. Natl. Acad. Sci.* **98:** 8531–8535.

Jin T., Gu Y., Zanusso G., Sy M., Kumar A., Cohen M., Gambetti P., and Singh N. 2000. The chaperone protein BiP binds to a mutant prion protein and mediates its degradation by the proteasome. *J. Biol. Chem.* **275:** 38699–38704.

Johnston J.A., Ward C.L., and Kopito R.R. 1998. Aggresomes: A cellular response to misfolded proteins. *J. Cell Biol.* **143:** 1883–1898.

Kaneko K., Vey M., Scott M., Pilkuhn S., Cohen F.E., and Prusiner S.B. 1997a. COOH-terminal sequence of the cellular prion protein directs subcellular trafficking and controls conversion into the scrapie isoform. *Proc. Natl. Acad. Sci.* **94:** 2333–2338.

Kaneko K., Wille H., Mehlhorn I., Zhang H., Ball H., Cohen F.E., Baldwin M.A., and Prusiner S.B. 1997b. Molecular properties of complexes formed between the prion protein and synthetic peptides. *J. Mol. Biol.* **270:** 574–586.

Kaneko K., Zulianello L., Scott M., Cooper C.M., Wallace A.C., James T.L., Cohen F.E., and Prusiner S.B. 1997c. Evidence for protein X binding to a discontinuous epitope on the cellular prion protein during scrapie prion propagation. *Proc. Natl. Acad. Sci.* **94:** 10069–10074.

Kim S.J. and Hegde R.S. 2002. Cotranslational partitioning of nascent prion protein into multiple populations at the translocation channel. *Mol. Biol. Cell* **13:** 3775–3786.

Kim S.J., Rahbar R., and Hegde R.S. 2001. Combinatorial control of prion protein biogenesis by the signal sequence and transmembrane domain. *J. Biol. Chem.* **276:** 26132–26140.

Kimberlin R.H. and Walker C.A. 1979. Pathogenesis of mouse scrapie: Dynamics of agent replication in spleen, spinal cord and brain after infection by different routes. *J. Comp. Pathol.* **89:** 551–562.

Kocisko D.A., Come J.H., Priola S.A., Chesebro B., Raymond G.J., Lansbury P.T., Jr., and Caughey B. 1994. Cell-free formation of protease-resistant prion protein. *Nature* **370:** 471–474.

Kramer M.L., Kratzin H.D., Schmidt B., Romer A., Windl O., Liemann S., Hornemann S., and Kretzschmar H. 2001. Prion protein binds copper within the physiological concentration range. *J. Biol. Chem.* **276:** 16711–16719.

Kretzschmar H.A., Prusiner S.B., Stowring L.E., and DeArmond S.J. 1986. Scrapie prion proteins are synthesized in neurons. *Am. J. Pathol.* **122:** 1–5.

Kunzi V., Glatzel M., Nakano M.Y., Greber U.F., Van Leuven F., and Aguzzi A. 2002. Unhampered prion neuroinvasion despite impaired fast axonal transport in transgenic mice overexpressing four-repeat tau. *J. Neurosci.* **22:** 7471–7477.

Kurzchalia T.V. and Parton R.G. 1999. Membrane microdomains and caveolae. *Curr. Opin. Cell Biol.* **11:** 424–431.

Lainé J., Marc M.E., Sy M.S., and Axelrad H. 2001. Cellular and subcellular morphological localization of normal prion protein in rodent cerebellum. *Eur. J. Neurosci.* **14:** 47–56.

Leclerc E., Peretz D., Ball H., Sakurai H., Legname G., Serban A., Prusiner S.B., Burton D.R., and Williamson R.A. 2001. Immobilized prion protein undergoes spontaneous rearrangement to a conformation having features in common with the infectious form. *EMBO J.* **20:** 1547–1554.

Lee K.S., Magalhaes A.C., Zanata S.M., Brentani R.R., Martins V.R., and Prado M.A. 2001. Internalization of mammalian fluorescent cellular prion protein and N-terminal deletion mutants in living cells. *J. Neurochem.* **79:** 79–87.

Lehmann S. 2002. Metal ions and prion diseases. *Curr. Opin. Chem. Biol.* **6:** 187–192.

Lehmann S. and Harris D.A. 1995. A mutant prion protein displays aberrant membrane association when expressed in cultured cells. *J. Biol. Chem.* **270:** 24589–24597.

———. 1996a. Mutant and infectious prion proteins display common biochemical properties in cultured cells. *J. Biol. Chem.* **271:** 1633–1637.

———. 1996b. Two mutant prion proteins expressed in cultured cells acquire biochemical properties reminiscent of the scrapie isoform. *Proc. Natl. Acad. Sci.* **93:** 5610–5614.

———. 1997. Blockade of glycosylation promotes acquisition of scrapie-like properties by the prion protein in cultured cells. *J. Biol. Chem.* **272:** 21479–21487.

Lehmann S., Daude N., and Harris D.A. 1997. A wild-type prion protein does not acquire properties of the scrapie isoform when coexpressed with a mutant prion protein in cultured cells. *Brain Res. Mol. Brain Res.* **52:** 139–145.

Liemann S. and Glockshuber R. 1999. Influence of amino acid substitutions related to inherited human prion diseases on the thermodynamic stability of the cellular prion protein. *Biochemistry* **38:** 3258–3267.

Liu T., Li R., Pan T., Liu D., Petersen R.B., Wong B.S., Gambetti P., and Sy M.S. 2002. Intercellular transfer of the cellular prion protein. *J. Biol. Chem.* **277:** 47671–47678.

Lopez C.D., Yost C.S., Prusiner S.B., Myers R.M., and Lingappa V.R. 1990. Unusual topogenic sequence directs prion protein biogenesis. *Science* **248:** 226–229.

Lorenz H., Windl O., and Kretzschmar H.A. 2002. Cellular phenotyping of secretory and nuclear prion proteins associated with inherited prion diseases. *J. Biol. Chem.* **277:** 8508–8516.

Lucassen R., Nishina K., and Supattapone S. 2003. In vitro amplification of protease-resistant prion protein requires free sulfhydryl groups. *Biochemistry* **42:** 4127–4135.

Lugaresi E., Medori R., Montagna P., Baruzzi A., Cortelli P., Lugaresi A., Tinuper P., Zucconi M., and Gambetti P. 1986. Fatal familial insomnia and dysautonomia with selective degeneration of thalamic nuclei. *N. Engl. J. Med.* **315:** 997–1003.

Ma J. and Lindquist S. 1999. De novo generation of a PrP^{Sc}-like conformation in living cells. *Nat. Cell Biol.* **1:** 358–361.

———. 2001. Wild-type PrP and a mutant associated with prion disease are subject to retrograde transport and proteasome degradation. *Proc. Natl. Acad. Sci.* **98:** 14955–14960.

Ma J., Wollmann R., and Lindquist S. 2002. Neurotoxicity and neurodegeneration when PrP accumulates in the cytosol. *Science* **298:** 1781–1785.

Madore N., Smith K.L., Graham C.H., Jen A., Brady K., Hall S., and Morris R. 1999. Functionally different GPI proteins are organized in different domains on the neuronal surface. *EMBO J.* **18:** 6917–6926.

Magalhães A.C., Silva J.A., Lee K.S., Martins V.R., Prado V.F., Ferguson S.S., Gomez M.V., Brentani R.R., and Prado M.A. 2002. Endocytic intermediates involved with the intracellular trafficking of a fluorescent cellular prion protein. *J. Biol. Chem.* **277:** 33311–33318.

Marella M., Lehmann S., Grassi J., and Chabry J. 2002. Filipin prevents pathological prion protein accumulation by reducing endocytosis and inducing cellular PrP release. *J. Biol. Chem.* **277:** 25457–25464.

Mastrianni J.A., Curtis M.T., Oberholtzer J.C., Da Costa M.M., DeArmond S., Prusiner S.B., and Garbern J.Y. 1995. Prion disease (PrP-A117V) presenting with ataxia instead of dementia. *Neurology* **45:** 2042–2050.

Mayor S., Rothberg K.G., and Maxfield F.R. 1994. Sequestration of GPI-anchored proteins in caveolae triggered by cross-linking. *Science* **264:** 1948–1951.

McKinley M.P., Taraboulos A., Kenaga L., Serban D., Stieber A., DeArmond S.J., Prusiner S.B., and Gonatas N. 1991. Ultrastructural localization of scrapie prion proteins in cytoplasmic vesicles of infected cultured cells. *Lab. Invest.* **65:** 622–630.

Mikol D.D., Hong H.L., Cheng H.L., and Feldman E.L. 1999. Caveolin-1 expression in Schwann cells. *Glia* **27:** 39–52.

Mironov A., Jr., Latawiec D., Wille H., Bouzamondo-Bernstein E., Legname G., Williamson R.A., Burton D., DeArmond S.J., Prusiner S.B., and Peters P.J. 2003. Cytosolic prion protein in neurons. *J. Neurosci.* **23:** 7183–7193.

Mishra R.S., Bose S., Gu Y., Li R., and Singh N. 2003. Aggresome formation by mutant prion proteins: The unfolding role of proteasomes in familial prion disorders. *J. Alzheimers Dis.* **5:** 15–23.

Mishra R.S., Gu Y., Bose S., Verghese S., Kalepu S., and Singh N. 2002. Cell surface accumulation of a truncated transmembrane prion protein in Gerstmann-Sträussler-Scheinker disease P102L. *J. Biol. Chem.* **277:** 24554–24561.

Moser M., Colello R.J., Pott U., and Oesch B. 1995. Developmental expression of the prion protein gene in glial cells. *Neuron* **14:** 509–517.

Mouillet-Richard S., Ermonval M., Chebassier C., Laplanche J.L., Lehmann S., Launay J.M., and Kellermann O. 2000. Signal transduction through prion protein. *Science* **289:** 1925–1928.

Moya K.L., Sales N., Hassig R., Creminon C., Grassi J., and Di Giamberardino L. 2000. Immunolocalization of the cellular prion protein in normal brain. *Microsc. Res. Tech.* **50:** 58–65.

Narwa R. and Harris D.A. 1999. Prion proteins carrying pathogenic mutations are resistant to phospholipase cleavage of their glycolipid anchors. *Biochemistry* **38:** 8770–8777.

Naslavsky N., Stein R., Yanai A., Friedlander G., and Taraboulos A. 1997. Characterization of detergent-insoluble complexes containing the cellular prion protein and its scrapie isoform. *J. Biol. Chem.* **272:** 6324–6331.

Negro A., Ballarin C., Bertoli A., Massimino M.L., and Sorgato M.C. 2001. The metabolism and imaging in live cells of the bovine prion protein in its native form or carrying single amino acid substitutions. *Mol. Cell. Neurosci.* **17:** 521–538.

Pan K.-M., Baldwin M., Nguyen J., Gasset M., Serban A., Groth D., Mehlhorn I., Huang Z., Fletterick R.J., Cohen F.E., and Prusiner S.B. 1993. Conversion of α-helices into β-sheets features in the formation of the scrapie prion proteins. *Proc. Natl. Acad. Sci.* **90:** 10962–10966.

Parton R.G. 2003. Caveolae—From ultrastructure to molecular mechanisms. *Nat. Rev. Mol. Cell Biol.* **4:** 162–167.

Pauly P.C. and Harris D.A. 1998. Copper stimulates endocytosis of the prion protein. *J. Biol. Chem.* **273:** 33107–33110.

Perera W.S. and Hooper N.M. 2001. Ablation of the metal ion-induced endocytosis of the prion protein by disease-associated mutation of the octarepeat region. *Curr. Biol.* **11:** 519–523.

Peretz D., Williamson R.A., Matsunaga Y., Serban H., Pinilla C., Bastidas R.B., Rozenshteyn R., James T.L., Houghten R.A., Cohen F.E., Prusiner S.B., and Burton D.R. 1997. A conformational transition at the N-terminus of the prion protein features in formation of the scrapie isoform. *J. Mol. Biol.* **273:** 614–622.

Peretz D., Williamson R.A., Kaneko K., Vergara J., Leclerc E., Schmitt-Ulms G., Mehlhorn I.R., Legname G., Wormald M.R., Rudd P.M., Dwek R.A., Burton D.R., and Prusiner S.B. 2001. Antibodies inhibit prion propagation and clear cell cultures of prion infectivity. *Nature* **412:** 739–743.

Peters P.J. 2001. Cryo-immunogold electron microscopy. In *Current protocols in cell biology* (ed. J.S. Bonifacino), pp. 4.7.1–4.7.12. John Wiley, New York.

Peters P.J., Neefjes J.J., Oorschot V., Ploegh H.L., and Geuze H.J. 1991. Segregation of MHC class II molecules from MHC class I molecules in the Golgi complex for transport to lysosomal compartments. *Nature* **349:** 669–678.

Peters P.J., Ning K., Palacios F., Boshans R.L., Kazantsev A., Thompson L.M., Woodman B., Bates G.P., and D'Souza-Schorey C. 2002. Arfaptin 2 regulates the aggregation of mutant huntingtin protein. *Nat. Cell. Biol.* **4:** 240–245.

Peters P.J., Mironov A., Peretz D., van Donselaar E., Leclerc E., Erpel S., DeArmond S.J., Burton D.R., Williamson R.A., Vey M., and Prusiner S.B. 2003. Trafficking of prion proteins through a caveolae-mediated endosomal pathway. *J. Cell Biol.* **162:** 703–717.

Petersen R.B., Parchi P., Richardson S.L., Urig C.B., and Gambetti P. 1996. Effect of the D178N mutation and the codon 129 polymorphism on the metabolism of the prion protein. *J. Biol. Chem.* **271:** 122661–122668.

Piccardo P., Safar J., Ceroni M., Gajdusek D.C., and Gibbs C.J., Jr. 1990. Immunohistochemical localization of prion protein in spongiform encephalopathies and normal brain tissue. *Neurology* **40:** 518–522.

Priola S.A. and Chesebro B. 1998. Abnormal properties of prion protein with insertional mutations in different cell types. *J. Biol. Chem.* **273:** 11980–11985.

Prusiner S.B. 1991. Molecular biology of prion diseases. *Science* **252:** 1515–1522.

———. 1996. Molecular biology and pathogenesis of prion diseases. *Trends Biochem. Sci.* **252:** 482–487.

Prusiner S.B., Hadlow W.J., Garfin D.E., Cochran S.P., Baringer J.R., Race R.E., and Eklund C.M. 1978. Partial purification and evidence for multiple molecular forms of the scrapie agent. *Biochemistry* **17:** 4993–4997.

Prusiner S.B., Scott M., Foster D., Pan K.-M., Groth D., Mirenda C., Torchia M., Yang S.-L., Serban D., Carlson G.A., Hoppe P.C., Westaway D., and DeArmond S.J. 1990. Transgenetic studies implicate interactions between homologous PrP isoforms in scrapie prion replication. *Cell* **63:** 673–686.

Qin K., Yang Y., Mastrangelo P., and Westaway D. 2002. Mapping Cu(II) binding sites in prion proteins by diethyl pyrocarbonate modification and matrix-assisted laser desorption ionization-time of flight (MALDI-TOF) mass spectrometric footprinting. *J. Biol. Chem.* **277:** 1981–1990.

Rachidi W., Vilette D., Guiraud P., Arlotto M., Riondel J., Laude H., Lehmann S., and Favier A. 2003. Expression of prion protein increases cellular copper binding and antioxidant enzyme activities but not copper delivery. *J. Biol. Chem.* **278:** 9064–9072.

Reits E., Griekspoor A., Neijssen J., Groothuis T., Jalink K., van Veelen P., Janssen H., Calafat J., Drijfhout J.W., and Neefjes J. 2003. Peptide diffusion, protection, and degradation in nuclear and cytoplasmic compartments before antigen presentation by MHC class I. *Immunity* **18:** 97–108.

Rudd P.M., Endo T., Colominas C., Groth D., Wheeler S.F., Harvey D.J., Wormald M.R.,

Serban H., Prusiner S.B., Kobata A., and Dwek R.A. 1999. Glycosylation differences between the normal and pathogenic prion protein isoforms. *Proc. Natl. Acad. Sci.* **96:** 13044–13049.

Saborio G.P., Permanne B., and Soto C. 2001. Sensitive detection of pathological prion protein by cyclic amplification of protein misfolding. *Nature* **411:** 810–813.

Safar J., Ceroni M., Piccardo P., Liberski P.P., Miyazaki M., Gajdusek D.C., and Gibbs C.J., Jr. 1990. Subcellular distribution and physicochemical properties of scrapie-associated precursor protein and relationship with scrapie agent. *Neurology* **40:** 503–508.

Sales N., Rodolfo K., Hassig R., Faucheux B., Di Giamberardino L., and Moya K.L. 1998. Cellular prion protein localization in rodent and primate brain. *Eur. J. Neurosci.* **10:** 2464–2471.

Scherfeld D., Schneider G., Guttmann P., and Osborn M. 1998. Visualization of cytoskeletal elements in the transmission X-ray microscope. *J. Struct. Biol.* **123:** 72–82.

Scott M.R., Köhler R., Foster D., and Prusiner S.B. 1992. Chimeric prion protein expression in cultured cells and transgenic mice. *Protein Sci.* **1:** 986–997.

Scott M., Groth D., Foster D., Torchia M., Yang S.-L., DeArmond S.J., and Prusiner S.B. 1993. Propagation of prions with artificial properties in transgenic mice expressing chimeric PrP genes. *Cell* **73:** 979–988.

Serban D., Taraboulos A., DeArmond S.J., and Prusiner S.B. 1990. Rapid detection of Creutzfeldt-Jakob disease and scrapie prion proteins. *Neurology* **40:** 110–117.

Shaul P.W. and Anderson R.G. 1998. Role of plasmalemmal caveolae in signal transduction. *Am. J. Physiol.* **275:** L843–L851.

Shyng S.-L., Heuser J.E., and Harris D.A. 1994. A glycolipid-anchored prion protein is endocytosed via clathrin-coated pits. *J. Cell Biol.* **125:** 1239–1250.

Shyng S.-L., Huber M.T., and Harris D.A. 1993. A prion protein cycles between the cell surface and an endocytic compartment in cultured neuroblastoma cells. *J. Biol. Chem.* **21:** 15922–15928.

Silva W.I., Maldonado H.M., Lisanti M.P., Devellis J., Chompre G., Mayol N., Ortiz M., Velazquez G., Maldonado A., and Montalvo J. 1999. Identification of caveolae and caveolin in C6 glioma cells. *Int. J. Dev. Neurosci.* **17:** 705–714.

Simons K. and Ikonen E. 1997. Functional rafts in cell membranes. *Nature* **387:** 569–572.

Singh N., Zanusso G., Chen S.G., Fujioka H., Richardson S., Gambetti P., and Petersen R.B. 1997. Prion protein aggregation reverted by low temperature in transfected cells carrying a prion protein gene mutation. *J. Biol. Chem.* **272:** 28461–28470.

Stahl N., Borchelt D.R., Hsiao K., and Prusiner S.B. 1987. Scrapie prion protein contains a phosphatidylinositol glycolipid. *Cell* **51:** 229–240.

Stahl N., Baldwin M.A., Hecker R., Pan K.-M., Burlingame A.L., and Prusiner S.B. 1992. Glycosylinositol phospholipid anchors of the scrapie and cellular prion proteins contain sialic acid. *Biochemistry* **31:** 5043–5053.

Stewart R.S. and Harris D.A. 2001. Most pathogenic mutations do not alter the membrane topology of the prion protein. *J. Biol. Chem.* **276:** 2212–2220.

———. 2003. Mutational analysis of topological determinants in PrP, and measurement of transmembrane and cytosolic PrP during prion infection. *J. Biol. Chem.* (in press).

Stewart R.S., Drisaldi B., and Harris D.A. 2001. A transmembrane form of the prion protein contains an uncleaved signal peptide and is retained in the endoplasmic reticulum. *Mol. Biol. Cell* **12:** 881–889.

Stimson E., Hope J., Chong A., and Burlingame A.L. 1999. Site-specific characterization of the N-linked glycans of murine prion protein by high-performance liquid chromatography/electrospray mass spectrometry and exoglycosidase digestions. *Biochemistry* **38:** 4885–4895.

Stöckel J., Safar J., Wallace A.C., Cohen F.E., and Prusiner S.B. 1998. Prion protein selectively binds copper (II) ions. *Biochemistry* **37:** 7185–7193.

Taraboulos A., Serban D., and Prusiner S.B. 1990. Scrapie prion proteins accumulate in the cytoplasm of persistently infected cultured cells. *J. Cell Biol.* **110:** 2117–2132.

Taraboulos A., Scott M., Semenov A., Avrahami D., Laszlo L., and Prusiner S.B. 1995. Cholesterol depletion and modification of COOH-terminal targeting sequence of the prion protein inhibits formation of the scrapie isoform. *J. Cell Biol.* **129:** 121–132.

Tarentino A.L., Gomez C.M., and Plummer T.H. 1985. Deglycosylation of asparagine linked glycans by peptide: *N*-glycosidase F. *Biochemistry* **24:** 4665–4671.

Tateishi J., Kitamoto T., Doh-ura K., Sakaki Y., Steinmetz G., Tranchant C., Warter J.M., and Heldt N. 1990. Immunochemical, molecular genetic, and transmission studies on a case of Gerstmann-Sträussler-Scheinker syndrome. *Neurology* **40:** 1578–1581.

Tatzelt J., Maeda N., Pekny M., Yang S.-L., Betsholtz C., Eliasson C., Cayetano J., Camerino A.P., DeArmond S.J., and Prusiner S.B. 1996. Scrapie in mice deficient in apolipoprotein E or glial fibrillary acidic protein. *Neurology* **47:** 449–453.

Telling G.C., Haga T., Torchia M., Tremblay P., DeArmond S.J., and Prusiner S.B. 1996. Interactions between wild-type and mutant prion proteins modulate neurodegeneration in transgenic mice. *Genes Dev.* **10:** 1736–1750.

Telling G.C., Scott M., Mastrianni J., Gabizon R., Torchia M., Cohen F.E., DeArmond S.J., and Prusiner S.B. 1995. Prion propagation in mice expressing human and chimeric PrP transgenes implicates the interaction of cellular PrP with another protein. *Cell* **83:** 79–90.

Turk E., Teplow D.B., Hood L.E., and Prusiner S.B. 1988. Purification and properties of the cellular and scrapie hamster prion proteins. *Eur. J. Biochem.* **176:** 21–30.

Verghese-Nikolakaki S., Michaloudi H., Polymenidou M., Groschup M.H., Papadopoulos G.C., and Sklaviadis T. 1999. Expression of the prion protein in the rat forebrain: An immunohistochemical study. *Neurosci. Lett.* **272:** 9–12.

Vey M., Pilkuhn S., Wille H., Nixon R., DeArmond S.J., Smart E.J., Anderson R.G., Taraboulos A., and Prusiner S.B. 1996. Subcellular colocalization of the cellular and scrapie prion proteins in caveolae-like membranous domains. *Proc. Natl. Acad. Sci.* **93:** 14945–14949.

Vincent B., Paltel E., Saftig P., Frobert Y., Hartmann D., De Strooper B., Grassi J., Lopez-Perez E., and Checler F. 2001. The disintegrins ADAM10 and TACE contribute to the constitutive and phorbol ester-regulated normal cleavage of the cellular prion protein. *J. Biol. Chem.* **276:** 37743–37746.

Waggoner D.J., Drisaldi B., Bartnikas T.B., Casareno R.L., Prohaska J.R., Gitlin J.D., and Harris D.A. 2000. Brain copper content and cuproenzyme activity do not vary with prion protein expression level. *J. Biol. Chem.* **275:** 7455–7458.

Weissmann C. 1991. Spongiform encephalopathies: The prion's progress. *Nature* **349:** 569–571.

Welch W.J., Garrels J.I., Thomas G.P., Lin J.J., and Feramisco J.R. 1983. Biochemical characterization of the mammalian stress proteins and identification of two stress proteins as glucose- and Ca^{2+}-ionophore-regulated proteins. *J. Biol. Chem.* **258:** 7102–7111.

Wells G.A.H., Hancock R.D., Cooley W.A., Richards M.S., Higgins R.J., and David G.P. 1989. Bovine spongiform encephalopathy: Diagnostic significance of vacuolar changes in selected nuclei of the medulla oblongata. *Vet. Rec.* **125:** 521–524.

Westaway D., DeArmond S.J., Cayetano-Canlas J., Groth D., Foster D., Yang S.-L., Torchia M., Carlson G.A., and Prusiner S.B. 1994. Degeneration of skeletal muscle, peripheral nerves, and the central nervous system in transgenic mice overexpressing wild-type prion proteins. *Cell* **76:** 117–129.

Wheeler S.F., Rudd P.M., Davis S.J., Dwek R.A., and Harvey D.J. 2002. Comparison of the N-linked glycans from soluble and GPI-anchored CD59 expressed in CHO cells. *Glycobiology* **12:** 261–271.

Wille H., Michelitsch M.D., Guénebaut V., Supattapone S., Serban A., Cohen F.E., Agard D.A., and Prusiner S.B. 2002. Structural studies of the scrapie prion protein by electron crystallography. *Proc. Natl. Acad. Sci.* **99:** 3563–3568.

Williamson R.A., Peretz D., Pinilla C., Ball H., Bastidas R.B., Rozenshteyn R., Houghten R.A., Prusiner S.B., and Burton D.R. 1998. Mapping the prion protein using recombinant antibodies. *J. Virol.* **72:** 9413–9418.

Yanai A., Meiner Z., Gahali I., Gabizon R., and Taraboulos A. 1999. Subcellular trafficking abnormalities of a prion protein with a disrupted disulfide loop. *FEBS Lett.* **460:** 11–16.

Yedidia Y., Horonchik L., Tzaban S., Yanai A., and Taraboulos A. 2001. Proteasomes and ubiquitin are involved in the turnover of the wild-type prion protein. *EMBO J.* **20:** 5383–5391.

Yost C.S., Lopez C.D., Prusiner S.B., Myers R.M., and Lingappa V.R. 1990. Non-hydrophobic extracytoplasmic determinant of stop transfer in the prion protein. *Nature* **343:** 669–672.

Young K., Piccardo P., Dlouhy S.R., Bugiani O., Tagliavini F., and Ghetti B. 1999. The human genetic prion diseases. In *Prions: Molecular and cellular biology* (ed. D.A. Harris), pp. 139–175. Horizon Scientific Press, Wymondham, United Kingdom.

Zanusso G., Petersen R.B., Jin T., Jing Y., Kanoush R., Ferrari S., Gambetti P., and Singh N. 1999. Proteasomal degradation and N-terminal protease resistance of the codon 145 mutant prion protein. *J. Biol. Chem.* **274:** 22396–22404.

Zhang H., Stöckel J., Mehlhorn I., Groth D., Baldwin M.A., Prusiner S.B., James T.L., and Cohen F.E. 1997. Physical studies of conformational plasticity in a recombinant prion protein. *Biochemistry* **36:** 3543–3553.

Zhang J.T. and Ling V. 1995. Involvement of cytoplasmic factors regulating the membrane orientation of P-glycoprotein sequences. *Biochemistry* **34:** 9159–9165.

Zlotnik I. and Stamp J.L. 1961. Scrapie disease of sheep. *World Neurol.* **2:** 895–907.

11

Scrapie, Chronic Wasting Disease, and Transmissible Mink Encephalopathy

Stanley B. Prusiner

Institute for Neurodegenerative Diseases
Departments of Neurology and of Biochemistry and Biophysics
University of California, San Francisco
San Francisco, California 94143

Elizabeth Williams

State Veterinary Laboratory
Department of Veterinary Sciences
University of Wyoming
Laramie, Wyoming 82070

Jean-Louis Laplanche

Laboratoire de Biologie Cellulaire, CNRSEP1591
Faculté des Sciences Pharmaceutiques
et Biologiques (Paris V)
75006 Paris, France

Morikazu Shinagawa

Laboratory of Veterinary Public Health
Department of Veterinary Medicine
Obihiro University of Agriculture and Veterinary Medicine
Inada-cho W-2, Obihiro, Hokkaido 080-8555, Japan

SCRAPIE IN SHEEP AND GOATS, TRANSMISSIBLE MINK ENCEPHALOPATHY
(TME), and chronic wasting disease (CWD) of cervids are all caused by
prions. These diseases are of great interest both with respect to issues of
animal health as well as the safety of the human food supply. In contrast
to bovine spongiform encephalopathy (BSE), many attempts to link the

ingestion of scrapie prions to the development of Creutzfeldt-Jakob disease (CJD) have been unsuccessful (Chapter 13). The high frequency of CWD among cervids has raised the possibility of cervid prion transmission to cattle and sheep through grazing on shared grasslands. In addition, epidemiologic investigations continue to focus on the frequency of CJD in deer hunters.

SCRAPIE

Epidemiology

Sheep, goats, and moufflon (*Ovis musimon*) (Wood et al. 1992) are susceptible to natural scrapie, and the disease occurs primarily in sheep of breeding age. Scrapie is an enzootic fatal neurodegenerative disorder caused by prions. The disease is also called *tremblante* (trembling) in France, *traberkrankheit* in Germany (trotting disease), or *rida* in Iceland (ataxia or tremor). Scrapie was the first spongiform encephalopathy for which transmissibility was demonstrated (Cuillé and Chelle 1936). Epidemiological studies have been conducted on the potential risk of transmission of the scrapie agent to humans but have never supported a causal relationship (Chatelain et al. 1981).

Scrapie was initially reported in Europe in 1732 in England and in 1759 in Germany. The etiology of scrapie was already a matter of debate at the beginning of the 19th century. Some authors considered that the disease appeared spontaneously and became transmissible, while others supposed it was caused by an animalcule such as an amoeba (Girard 1830).

In the following decades, scrapie endemically affected flocks in several countries following the importation of affected sheep, but its exact prevalence and incidence remained obscure. In the Netherlands, the prevalence of flocks with scrapie cases was recently established between 3.8% and 8.4%, with an incidence in the sheep population of 1 case per 1000 ewes annually (Schreuder et al. 1993). The disease appears to be much more prevalent in the northern hemisphere than in the southern hemisphere. Whether this reflects the influence of climate on scrapie development or is an expression of the differences among breeding strategies is unknown. New Zealand and Australia are fortunate because they are among the few sheep-raising countries recognized by most as free from scrapie. This status results from extensive protective measures designed to prevent the introduction of scrapie from imported sheep (MacDiarmid 1996). The importation strategy is based on embryo transfer, bioassay of lymph nodes of donor sheep in young goats, and a 3-year quarantine period for the embryo-derived offspring. Other countries such as Denmark, Finland, Sweden, Israel, and Zimbabwe have a low scrapie risk. Scrapie was also eradicated from the Republic of South Africa before it was widely disseminated through the national flock (MacDiarmid 1996).

There is little information on scrapie in goats and moufflon. Most reported occurrences of natural scrapie in goats have been associated with the presence of scrapie in sheep (Brotherston et al. 1968). However, Hourrigan and colleagues (Hourrigan et al. 1979) observed that scrapie could be spread from goat to goat without sheep contact. In the UK, six cases of scrapie have been confirmed in two different flocks of moufflon, and it has not been possible to determine the origin of the disease in either case. No relationship has been made with sheep scrapie.

Despite progress made on many aspects of prion diseases, important points concerning scrapie remain obscure. How does scrapie begin in a flock and how does it spread horizontally as well as vertically among other flock members? It seems likely that scrapie begins in a flock with a sporadic case of scrapie in a susceptible sheep. As discussed below, among the most susceptible sheep are those with a VRQ genotype. One study showed that placenta and amniotic fluid harbors prions, suggesting that the placenta might contaminate pastures with prions. The bioassays used in studies of fetal membranes utilized sheep so that the absence of a species barrier did not diminish the sensitivity of the measurements, as described below (Pattison et al. 1972, 1974). Alternatively, the feces might be a source of contamination since prion replication in the lymphoid tissue of the gut after oral inoculation is well documented in sheep and goats (Hadlow et al. 1980, 1982). Whether vertical transfer of prions from mother to offspring occurs is unclear because conflicting results on prion transfer from prion-infected cattle and sheep have been reported (Foote et al. 1993; Wilesmith et al. 2000; Hunter et al. 2002).

Clinical Presentation of Scrapie

Typical clinical signs of scrapie in sheep begin with mildly impaired social behavior such as unusual restlessness and signs of nervousness, which are often noticeable only to an experienced shepherd (Zlotnik and Stamp 1961; Dickinson 1976; Parry 1983). Overt illness can last from 2 weeks to 6 months. In later stages of the clinical course, the general condition of the animal begins to deteriorate and is sometimes accompanied by a change in fleece color. Pruritis can result from the animal's attempting to relieve what seems to be an intense irritation by scratching against fence posts or by biting the affected area. Despite ataxia, affected sheep will walk considerable distances to indulge in bouts of scratching (Parry 1983). Specific parts of the skin are continually rubbed and suffer wool loss; for example, around the base of the tail and, occasionally, an entire side of the body. In the late stages of scrapie, sheep waste away, cannot walk very far, and become very agitated by any (even mild) stress. The appetite may appear normal, but the animals lose the ability to feed themselves. Scrapie does not alter reproductive ability until muscle wasting interferes with motor activity. Lambs can be born successfully to mothers in the clinical phase of the disease, and rams remain fertile and active even when showing scrapie symptoms (Parry 1983; J.D. Foster and C. MacKenzie, unpubl.).

Reported clinical descriptions of scrapie often vary considerably. In a group of scrapie-affected sheep from Shetland between 1985 and 1991 (Clark and Moar 1992), most animals showed pruritis and emaciation; others had pruritis, emaciation, and hyperaesthesia, while others showed all of these signs in addition to ataxia. These authors also described sheep apparently with scrapie that exhibited such a short clinical course that they were simply "found dead." In a report of clinical signs in some Japanese sheep (Onodera and Hayashi 1994), some animals (Suffolks and Corriedales) showed signs of pruritis, but others (Corriedales) died for no obvious reason; scrapie was diagnosed after histopathologic examination. These differences both in symptoms and clinical course may simply be due to breed characteristics, or they may indicate the presence of different strains of scrapie prions.

Affected goats are less likely to rub against fixed objects, but scratch vigorously with hind feet and horns (Detwiler 1992). In moufflon, clinical signs were indistinguishable from sheep scrapie (Wood et al. 1992).

Prior to the discovery of the disease-causing isoform of the prion protein (PrP^{Sc}), scrapie was often diagnosed in asymptomatic animals by histopathologic examination. At the terminal stage of disease, common neuropathologic lesions in the brain include neuronal degeneration with the formation of intracytoplasmic vacuoles and development of an intense, reactive astrocytic gliosis. Neither demyelination nor an inflammatory response is found. Vacuolation is not present in all areas of the brain; for example, one study of scrapie-affected sheep in Britain described seven patterns of vacuolation (Wood et al. 1997) in 10 brain areas: medulla, pons, cerebellum, substantia nigra, mesencephalon, hypothalamus, thalamus, septal area, corpus striatum, and neocortex. Some patterns of vacuolar damage were very similar to each other; however, others were markedly different. In another study of both naturally affected and experimentally challenged sheep, vacuolation occurred in areas such as the dorsal vagus nucleus, cerebellum, and thalamic nuclei (Foster et al. 1996a). Vacuolation was described as "seldom apparent" without detection of PrP^{Sc} in the vicinity; however, PrP^{Sc} was sometimes present in areas with no vacuolation. The presence of PrP^{Sc} can be detected in the preclinical phase and is therefore of greater potential interest for diagnosis than for vacuolation.

Distribution of Scrapie Infectivity in Sheep Tissues

Cuillé and Chelle (1936) first demonstrated scrapie infectivity by transmitting it to sheep using scrapie-affected sheep brain. Most subsequent studies on prion infectivity in the tissues of scrapie-infected sheep and goats used bioassays in non-transgenic (Tg) mice. These bioassays required the crossing of a species barrier, which resulted in greatly diminished sensitivity. For example, bioassays for prions from cattle with BSE conducted in non-Tg, inbred RIII mice are 1,000 times less sensitive than bioassays using cattle and 10,000 times less sensitive than in Tg(BoPrP)$Prnp^{0/0}$ mice (Fraser et al. 1988; Safar et al. 2002; Wells 2002). Using bioassays performed in Swiss mice, the distribution of infectivity in sheep and goat tissues was determined (Hadlow et al. 1979, 1980, 1982; Groschup et al. 1996).

Based on bioassays in non-Tg mice (Hadlow et al. 1980, 1982), the WHO (World Health Organization 1992) classified tissues into four categories based on relative levels of infectivity: (I) highly infective, (II) moderately infective, (III) minimally infective, and (IV) no detectable infectivity. Category I tissues include brain and spinal cord; category II tissues are the spleen, tonsils, lymph nodes, ileum, and proximal colon; category III includes the sciatic nerve, pituitary gland, adrenal glands, distal colon, nasal mucosa, cerebrospinal fluid, thymus, bone marrow, liver, lung, and pancreas; and category IV tissues are skeletal muscles, heart, mammary glands, colostrum, milk, blood clots, serum, feces, kidneys, thyroid gland, salivary glands, saliva, ovaries, uterus, testes, and seminal vesicles. Skin is classified as category IV (Stamp et al. 1959). Sometimes, the ovaries, uterus, placenta, and amniotic fluid are positive for scrapie infectivity (Hourrigan 1990; Onodera et al. 1993). Groschup et al. (1996) have recently shown that substantial amounts of infectivity could also be detectable in several peripheral nerves except nervus saphenicus. Recent studies of muscle in mice have shown that hind limbs contain ~10^5 ID$_{50}$ units/g (Bosque et al. 2002). Although murine hind limb muscle is ~100 times less infectious than brain, it should be classified as category I or II but certainly not category IV. Because bioassays in non-Tg mice are so insensitive, the WHO classification should be viewed with considerable skepticism, especially when formulating food safety regulations.

As noted above, fetal membranes, in which infectivity is demonstrable by sheep bioassay (Pattison et al. 1972, 1974), the placenta, and amniotic fluid from infected ewes are all thought to be important sources of natural infection.

Hadlow and colleagues (Hadlow et al. 1979, 1982) examined temporal distribution of scrapie infectivity in Suffolk sheep from scrapie-infected flocks. Using non-Tg mouse bioassays, they described no infectivity before 8 months of age. From 10 to 11 months of age, infectivity was detected at a low level in various lymph nodes, spleen, and tonsil, and in intestinal tissue. Subsequently, the infectivity in these tissues increased and eventually reached a moderate level. The appearance of the infectivity in the tissues of the central nervous system (CNS), including the spinal cord, occurs later than that in the lymphoid tissues but before the onset of clinical disease; i.e., a low level of infectivity was detected in the diencephalon and medulla oblongata in a 25-month-old clinically normal sheep. By the time animals showed clinical signs of CNS disease, the infectivity in the CNS was at its highest levels. Among the parts of the CNS with the highest levels of infectivity were the diencephalon, midbrain, medulla oblongata, and cerebellar cortex. Outside the CNS, infectivity appeared first in the tonsils, retropharyngeal lymph nodes, and intestines, possibly reflecting an oral route of infection. In many cases of natural scrapie, sheep seem to become infected with prions at or soon after birth.

PrP^{Sc} in Sheep

The discovery of PrP^{Sc} revolutionized studies of prions in sheep. PrP^{Sc} is detected in the tissues of sheep that are clinically infected and in preclinically infected sheep tissues by immunochemical methods and immunohistochemical staining using antibodies to prion or synthetic PrP-based peptides. Although these methods are less sensitive than the bioassay for scrapie prion infectivity detection, they have significant advantages over bioassay in time and expense. Relative concentrations of PrP^{Sc} in ovine tissues have been assayed by western blots and enzyme-linked immunosorbent assay (ELISA), whereas the distribution and localization of PrP^{Sc} in tissues are revealed by conventional immunohistochemistry and histoblots, in which cryostat sections are blotted onto nitrocellulose membranes (Taraboulos et al. 1992).

Sheep PrP^{Sc} was first detected in brain extracts by western blot (Takahashi et al. 1986; Rubenstein et al. 1987), and detection of PrP^{Sc} in CNS is frequently used for the diagnosis of scrapie (Farquhar et al. 1989; Ikegami et al. 1991; Mohri et al. 1992; Race et al. 1992; Skarphedinsson et al. 1994). Although data describing concentrations of PrP^{Sc} in different parts of sheep brain are not available, PrP^{Sc} seems to distribute unevenly, with the brain stem being the best tissue for reliable western-blot detection of PrP^{Sc}. Immunohistochemistry of PrP^{Sc} in CNS revealed that it is detected preferentially in the brain stem, especially at the obex, even when PrP^{Sc} detection in other parts of the brain is difficult. PrP^{Sc} deposits have been found in neuronal cells, gray matter neuropile, sometimes around blood vessels, and outside neuronal cells as plaques (McBride et al. 1988; Miller et al. 1993).

van Keulen and colleagues (van Keulen et al. 1996) used immunohistochemistry to examine the distribution of PrP^{Sc} in lymphoid tissues from scrapie-affected sheep and found that PrP^{Sc} is consistently detected in the spleen, retropharyngeal lymph node, mesenteric lymph node, and the palatine tonsil. PrP^{Sc} is deposited in a reticular pattern in the center of lymphoid follicles. As in mice (Kitamoto et al. 1991; McBride et al. 1992), follicular dendritic cells appear to be the target of PrP^{Sc} deposition. Macrophages associated with lymphoid follicles contain PrP^{Sc} in their cytoplasm.

In a preclinical-stage analysis of naturally infected sheep using western blots, PrP^{Sc} was detected in the spleen and in visceral and surface lymph nodes, but not in the CNS from euthanized, apparently healthy 5-month-old (Fig. 1, sheep A), 8-month-old, and 40-month-old Suffolk sheep (Ikegami et al. 1991). Lymphoid tissues can be collected by biopsy without killing sheep. Muramatsu and colleagues (Muramatsu et al. 1994) and Schreuder and colleagues (Schreuder et al. 1996) examined the lymphoid tissues from sheep that were born and raised on farms experiencing scrapie outbreaks. Muramatsu's group detected PrP^{Sc} by western blotting in a subiliac lymph node from a 12-month-old Suffolk-Corriedale mixed-breed sheep that developed scrapie 2 months after the biopsy (Fig. 1, sheep B), and Schreuder's group used immunohistochemistry to detect PrP^{Sc} in the palatine tonsils from six sheep genetically determined to be susceptible (see below) that were ~10 months of age, which was less than half the length of the expected incubation period. PrP^{Sc} was also detected in the lymph node from a Suffolk sheep experimentally infected intravenously at 14 months after infection (Ikegami et al. 1991).

Distribution of scrapie infectivity and PrP^{Sc} in affected sheep tissues is summarized in Table 1. Generally, less is known about the temporal distribution of PrP^{Sc} in sheep tissues than about scrapie infectivity. The

Brain   Spleen   L.N.   L.N.

M P A N P A N P A N P B N

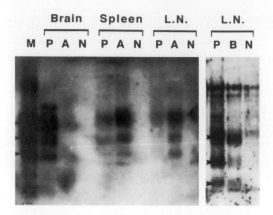

Figure 1. Detection of PrP^{Sc} in sheep spleen and lymph nodes at preclinical stage of scrapie. Apparently healthy sheep tissues were obtained by autopsy after euthanasia (sheep A, 5 months old) or by biopsy (sheep B, 12 months old). Sheep B showed scrapie signs 2 months after the biopsy. Brain, spleen, and mesenteric lymph nodes (L.N.) from sheep A and a subiliac lymph node from sheep B were examined for the presence of PrP^{Sc} by western blot. (Lane *A*) Apparently healthy sheep A; (lane *B*) apparently healthy sheep B; (lane *P*) scrapie-affected sheep, (lane *N*) healthy sheep. Lane *M* is a partially purified prion protein fraction prepared from scrapie sheep brain, as a positive control. PrP^{Sc} was detected as three broad bands between 19 kD and 30 kD, indicated by arrowheads.

appearance of PrP^{Sc} in lymphoid tissues precedes its appearance in the CNS, and the appearance of PrP^{Sc} seems to coincide with infectivity. Improvement of the biochemical methods for detecting PrP^{Sc}, especially in developing new antibodies (Somerville et al. 1997b) and methods of tissue processing for western blot and ELISA (Grathwohl et al. 1996, 1997), permits increased sensitivity of PrP^{Sc} detection. For some purposes, such as early diagnosis of scrapie through examination of peripheral lymphoid tissues and determination of the kinetics of prion accumulation, the infectivity bioassays using mice have been replaced by assays for PrP^{Sc}.

A few studies have reported the appearance of scrapie-associated fibrils (SAF) in detergent extracts of brain and spleen from natural and experimentally affected sheep. The extracts were negatively stained and viewed by electron microscopy (Gibson et al. 1987; Rubenstein et al. 1987; Stack et al. 1991). In many cases, the fibrils referred to as SAF did not meet the ultrastructural criteria published for SAF (Merz et al. 1981, 1983a). In fact, the fibrils proved to be indistinguishable from amyloid and the prion rods found in highly purified fractions of prions (Prusiner et al. 1983). Moreover, a comparative ultrastructural study on amyloid and SAF

Table 1. Distribution of scrapie prions in sheep tissues

Tissues	Infectivity in mice[a]	PrP^{Sc} WB	IH
Brain	+++	+	+
Spinal cord	+++	+	+
Nervus axillaris	++	ND	ND
Nervus ulnaris	++[b]	ND	ND
Nervus medianus	++[b]	ND	ND
Nervus ischiadicus	+ to ++	ND	ND
Nervus tibialis	++[b]	ND	ND
Nervus fibularis	++[b]	ND	ND
Nervus saphenus	− to +[b]	ND	ND
Spleen	++	+	+
Lymph nodes	++	+	+
Tonsil	++	ND	+
Ileum	++	ND	ND
Proximal colon	++	ND	ND
Cerebrospinal fluid	+	ND	ND
Pituitary gland	+	ND	ND
Thymus	+	ND	ND
Bone marrow	+	ND	ND
Nasal mucosa	+	ND	ND
Adrenal glands	+	ND	ND
Distal colon	+	ND	ND
Liver	+	ND	ND
Lung	+	ND	ND
Pancreas	+	ND	ND
Ovaries	+	ND	ND
Uterus	+	ND	ND
Skeletal muscle	−	ND	ND
Heart muscle	−	ND	ND
Mammary gland	−	ND	ND
Colostrum	−	ND	ND
Milk	−	ND	ND
Thyroid gland	−	ND	ND
Salivary gland	−	ND	ND
Saliva	−	ND	ND
Kidney	−	ND	ND
Testis	−	ND	ND
Blood clot	−	ND	ND
Serum	−	ND	ND
Feces	−	ND	ND
Skin	−	ND	ND
Placenta	+[c]	−	ND
Amniotic fluid	+[c]	ND	ND

[a]Infectivity titers (ID_{50} units/g) were determined in non-Tg mice and are indicated by + and − marks as follows: +++, >10^5; ++, 10^5 > infectivity > 10^2; +, <10^2; −, no detectable infectivity. Infectivity was tested after intracerebral inoculation or after [b]intraperitoneal inoculation.

[c]Infectivity titers not determined. In WB and IH rows, + or − indicates that PrP^{Sc} was detectable or undetectable by western blot (WB) and/or immunohistochemistry (IH). (ND) Not done. The data are from the following references: Scrapie infectivity assayed in mice by intracerebral inoculation (Hadlow et al. 1982; Hourrigan 1990; Onodera et al. 1993; Groschup et al. 1996), western blot detection (Ikegami et al. 1991; Race and Ernst 1992; Onodera et al. 1993), and immunohistochemical detection (McBride et al. 1988; van Keulen et al. 1996).

showed that the two entities could be readily differentiated (Merz et al. 1983b). Subsequently, fibril detection has been shown to be an insensitive and impractical method compared to immunoassays for PrP^{Sc} in the diagnosis of scrapie (Cooley et al. 1998).

Host Control of Scrapie in Sheep and Goats

Early experimental studies of scrapie in sheep left the scrapie world with a selection of contradictory findings. In a study published in 1966 (Gordon 1966), more than 1000 sheep of 24 different breeds were injected intracerebrally (i.c.) with a source of scrapie known as SSBP/1 (Dickinson 1976). The animals were observed for 2 years, during which time some of them developed scrapie with incubation periods ranging from 3.5 months to 23 months and some remained completely healthy. It was thought at the time that some breeds were more resistant than others, and the results apparently confirmed this idea: Herdwicks, Scottish Blackface, Suffolks, and Dorset Horn breeds had disease incidences of 78%, 18%, 12%, and zero, respectively. Different results were obtained when others repeated the work with separate groups of sheep; for example, Herdwicks were later found to be only 30% susceptible (Pattison 1966). There were also problems with the original Dorset Downs, which were supposed to be resistant and were kept alive after the experimental observation period ended at 24 months: Some of these Dorset Downs developed SSBP/1 scrapie at later dates (Dickinson 1976). There was a clear need to develop lines of sheep with more predictable responses to scrapie if any progress was to be made, and three such flocks were developed in Britain—Cheviots, Herdwicks, and Swaledales.

From a Cheviot sheep flock, two lines of sheep were selected: a positive line and a negative line, depending on the incubation period in the animals following injection with SSBP/1 (Dickinson 1976). The gene that controls experimental scrapie incubation period in these sheep was called *Sip* (for *s*crapie *i*ncubation *p*eriod) and had two alleles denoted sA and pA (Dickinson and Outram 1988). The negative line of Cheviots was described as *Sip*^{pApA} and these survived subcutaneous (s.c.) injection of SSBP/1. The positive line of sheep was *Sip*^{sAsA} or *Sip*^{sApA}, with *Sip*^{sA} being partially dominant (Foster and Hunter 1991). These sheep from the positive line developed scrapie in 150–400 days after s.c. inoculation with SSBP/1 prions.

Different sources of scrapie prion inocula were found to exhibit different transmission characteristics. For example, SSBP/1 prions (termed an A group isolate) have shorter incubation periods in *Sip*^{sA} carriers (positive line) than in the *Sip*^{pApA} (negative line) sheep; in contrast, CH1641 (a

C group scrapie isolate) and BSE prions have shorter incubation periods in some negative sheep lines than in positive lines (Foster and Dickinson 1988; Foster et al. 1993).

PrP Gene Polymorphisms in Sheep Modulate Susceptibility to Scrapie

The complex interactions between the prion strain and the host genotype described above for sheep were modeled extensively in mice (Dickinson et al. 1968). With the discovery of the prion protein (PrP) and, subsequently, the chromosomal gene that encodes this protein, studies were undertaken in mice with short and long incubation times. First, control of the length of the incubation time was demonstrated to be genetically linked to the PrP gene (Carlson et al. 1986). Next, a dimorphism was identified that distinguished the PrP-A gene found in mice with short incubation times from the PrP-B gene found in mice with long incubation times (Westaway et al. 1987). Subsequently, the incubation time was shown to be governed by the relative number of copies of the PrP-A and PrP-B genes in Tg mice, in which PrP-A was found to be converted into PrP^{Sc} more efficiently than PrP-B for most, but not all, prion strains (Westaway et al. 1991; Carlson et al. 1994).

In accord with the findings in mice, the *Sip* gene was shown to be the PrP gene in sheep. Studies of the PrP gene in Cheviot sheep revealed three polymorphisms at codons 136 (A/V), 154 (R/H), and 171 (Q/R) (Goldmann et al. 1991). Based on these polymorphisms, a study of natural scrapie in Suffolk sheep in the US was performed. It was found that Q171 correlated with susceptibility to scrapie and R171 with resistance. Sheep homozygous for R at 171 did not develop scrapie, a finding that was shown to be statistically significant (Westaway et al. 1994). Although similar results were reported for experimental scrapie in Cheviot sheep, the sample size was too small to demonstrate statistical significance (Goldmann et al. 1994). Subsequent studies by many investigators have confirmed the finding that RR171 sheep are resistant to scrapie and that QR171 sheep are relatively resistant.

Studies in cultured cells and Tg mice demonstrated that MoPrP(R167), which is equivalent to OvPrP(R171), was not converted into PrP^{Sc} when inoculated with RML prions (Kaneko et al. 1997; Perrier et al. 2002). The RML prions were derived from the brain of a scrapied sheep and repeated passage in PrP-A mice (Chandler 1961). Additionally, MoPrP(R167) was found to act as a dominant negative in that it inhibited the conversion of MoPrP(Q167) into PrP^{Sc} in both cultured cells and Tg mice (Kaneko et al. 1997; Zulianello et al. 2000; Perrier et al. 2002). As described below, RR171 sheep inoculated i.c. with some strains of prions are susceptible to prion disease and, thus, additional studies with other murine passaged strains of prions need to be performed, particularly in Tg(MoPrP,R167)$Prnp^{0/0}$ and Tg(MoPrP,R167)$Prnp^{+/+}$ mice. Studies with a murine isolate of BSE (Fraser et al. 1992) may prove to be particularly informative in view of the passage of BSE to sheep that is described below.

In Cheviot sheep, one PrP allele encoding valine at codon 136 is linked to Sip^{sA}, but there are so many Sip^{pA}-related PrP alleles encoding alanine at codon 136 but differing at other codons that it has become more precise to use the PrP genotype rather than Sip terminology. Therefore, in Cheviot sheep, valine at codon 136 (V136) on both PrP alleles (VV136 or Sip^{sAsA}) is linked to short incubation following s.c. inoculation with SSBP/1, whereas those with longer incubation periods are heterozygotes with one V136 allele and one encoding alanine (AV136 or Sip^{sApA}) (Goldmann et al. 1994). Alanine encoded on both alleles (AA136 or Sip^{pApA}) is linked with resistance to s.c. challenge with SSBP/1 prions. The linkage of SSBP/1-induced experimental scrapie in Cheviot sheep with the V136 allele was confirmed in the US (Maciulis et al. 1992), and thus it clearly is not restricted to a Scottish flock in which it was initially studied.

Incidence of prion disease in Cheviots following challenge with CH1641 or BSE prions does not seem to be governed by polymorphisms at codon 136. Instead, susceptibility correlated much better with the codon 171 genotype when sheep were challenged with CH1641 or BSE prions. QQ171 sheep had the shortest incubation times after i.c. inoculation, whereas QR171 sheep with one arginine allele had much longer incubation periods (Goldmann et al. 1994). The QQ171 sheep that developed disease following inoculation with CH1641 or BSE prions could be AA136 (or Sip^{pApA}); alternatively, they could be AV136 (or Sip^{sApA}) or VV136 (or Sip^{sAsA}). The complexities of the "Sip" genotyping system have led to its being supplanted by the use of $PRNP$ genotypes that are more informative. For example, the AQ/AQ designation refers to alanine homozygosity at codon 136 and glutamine homozygosity at codon 171.

It is noteworthy that codon 171 does influence the length of the incubation period in sheep inoculated with SSBP/1 prions (Table 2). AV136 sheep that are QR171 have longer incubation times than those that are QQ171. Although the major effect on incubation period depends on the codon 171 genotype, with QQ171 being the most susceptible genotype, sheep with VV136 or AV136 have longer incubation periods than AA136 sheep when inoculated with CH1641 prions (Goldmann et al. 1994).

In Suffolk sheep, codon 136 has limited polymorphism, and the V136 allele is rare (Ikeda et al. 1995; O'Rourke et al. 1996; Hunter et al. 1997a).

Table 2. Sheep PrP genotypes of Cheviots and incidence of experimental scrapie

Infection source	Route	*Sip* genotype	*PRNP* genotype	n	Incubation time (days ± s.d.)
SSBP/1	s.c.	sAsA	VQ/VQ	9	170 ± 16
		sApA	VQ/AQ	10	260 ± 15
		sApA	VQ/AR	15	364 ± 17
		pApA	AQ/AQ	>20	survive
		pApA	AQ/AR	>20	survive
		pApA	AR/AR	>20	survive

Data simplified and taken from Goldmann et al. (1994) and Foster et al. (2001c). PrP genotypes are given at codons 136 (A = alanine, V = valine) and 171 (Q = glutamine, R = arginine). (s.c.) Subcutaneous.

Suffolk sheep do, however, succumb to scrapie, and one study in Suffolk sheep in the US demonstrated that orally administered infectious prions would cause disease in a proportion of animals (O'Rourke et al. 1997). The inoculum was pooled brain and spleen from scrapie-affected Suffolk sheep. Inoculated animals succumbed to disease in 622 ± 240 days and were all AQ/AQ genotype. Not all AQ/AQ sheep succumbed, but the oral route is known to be less efficient than other routes of infection. No AQ/AR or AR/AR animals developed scrapie.

Besides positions 136 and 171, residue 154 has also been found to influence the susceptibility of sheep to prion disease. In addition, nine other polymorphic residues at positions 112, 127, 137, 138, 141, 143, 151, 176, and 211 (Fig. 2) have been identified (Laplanche et al. 1992; Belt et al. 1995; Clousard et al. 1995; Bossers et al. 1996; Hunter et al. 1996). All these polymorphisms seemed independent of one another. Some of them are rare and detected in only a few animals, making it difficult to conclude whether they influence the development of prion disease. For codons 136, 154, and 171, five alleles have been defined: ARQ, VRQ, ARH, AHQ, and ARR. The frequency and distribution of these allelic variants can differ widely from breed to breed and also between flocks of the same breed, depending on the breeding strategy of each country. For example, in Suffolk sheep in the US (Westaway et al. 1994; O'Rourke et al. 1996) and Lacaune sheep in France (Clousard et al. 1995), the VRQ allele is rarely found; the ARR allele has not yet been observed in Icelandic sheep (A. Palsdottir, pers. comm.).

PrP Gene Polymorphisms in Goats

Four variants of the caprine PrP gene have been found at codons 142, 143, and 240 (Obermaier et al. 1995; Goldmann et al. 1996). One caprine PrP

PrP codon	112	127	**136**	137	138	141	143	151	**154**	**171**	176	211
wild-type	**M**	**G**	**A**	**M**	**S**	**L**	**H**	**R**	**R**	**Q**	**N**	**R**
T112	T	G	A	M	S	L	H	R	R	Q	N	R
S127	M	S	A	M	S	L	H	R	R	Q	N	R
V136 (VRQ)	M	G	V	M	S	L	H	R	R	Q	N	R
T136	M	G	T	M	S	L	H	R	R	Q	N	R
T137	M	G	A	T	S	L	H	R	R	Q	N	R
N138	M	G	A	M	N	L	H	R	R	Q	N	R
F141	M	G	A	M	S	F	H	R	R	Q	N	R
R143	M	G	A	M	S	L	R	R	R	Q	N	R
C151	M	G	A	M	S	L	H	C	R	Q	N	R
H154 (AHQ)	M	G	A	M	S	L	H	R	H	Q	N	R
R171 (ARR)	M	G	A	M	S	L	H	R	R	R	N	R
H171 (ARH)	M	G	A	M	S	L	H	R	R	H	N	R
K171	M	G	A	M	S	L	H	R	R	K	N	R
K176	M	G	A	M	S	L	H	R	R	Q	K	R
Q211	M	G	A	M	S	L	H	R	R	Q	N	Q

Figure 2. Schematic map of the allelic variants in the sheep PrP gene coding sequence and predicted sheep PrPs resulting from their combination. Polymorphisms at codons 136, 154, and 171 influence survival times of sheep with experimental or natural scrapie. (A) alanine; (C) cysteine; (G) glycine; (H) histidine; (F) phenylalanine; (L) leucine; (M) methionine; (N) asparagine; (R) arginine; (S) serine; (T) threonine; (Q) glutamine; (V) valine.

variant is identical to the ARQ PrP in sheep. The dimorphism at codon 142 appears to be associated with different incubation periods in goats experimentally infected with isolates of BSE, sheep scrapie CH1641, or sheep-passaged ME7 scrapie prions (Table 3). All experiments showed that the change from isoleucine to methionine at codon 142 is associated with prolongation of the incubation period. Preliminary analysis indicated that none of four goats with cases of natural scrapie carried the M142 allele (Goldmann et al. 1996).

Table 3. Association of goat PrP codon 142 genotype with experimental prion disease

Infection source	Route	PrP genotype	n	Incubation time (days $\pm$ S.D.)
BSE	i.c.	II	7	551 $\pm$ 29
		IM	2	984, 985
CH1641	i.c.	II	4	396 $\pm$ 64
		IM	2	675, 894
spME7	i.c.	II	9	378 $\pm$ 45
		IM	2	640, 895

Adapted from Goldmann et al. (1996). (i.c.) Intracerebral; (spME7) sheep-passaged ME7 scrapie strain; (I) isoleucine; (M) methionine.

Molecular Genetics of Natural Scrapie in Sheep

Among the first breeds endemically affected with natural scrapie that were examined for PrP gene mutations were French Romanov and Ile-de-France (Laplanche et al. 1992, 1993b). In these two breeds, susceptibility to scrapie appeared highly associated with the V136 allele, which was also evidenced as shortening the survival time of Cheviot sheep subcutaneously inoculated with SSBP/1 (Goldmann et al. 1991). No homozygous Romanov or Ile-de-France VV136 sheep were found among their apparently healthy flockmates that were beyond the mean age at onset, nor in an Ile-de-France scrapie-free flock (Table 4), suggesting a complete disease penetrance of scrapie in VRQ/VRQ animals.

The strong association between VRQ and susceptibility to natural scrapie was then found in many other breeds in the UK (Bleu du Maine, Cheviot, Herdwick, Merino × Shetland, Scottish halfbred, Shetland, Swaledale) and the Netherlands (Flemish, Texel Swifter) (Hunter et al.

Table 4. PrP genotype frequencies (%) and natural scrapie

PrP genotype[a]	Breeds with polymorphic PrP codon 136				Breeds with limited polymorphic PrP codon 136			
	Ile-de-France		Swaledales		Suffolk		Lacaune	
	S (20)	H[b] (33)	S (24)	Hc (48)	S (31)	Hd (57)	S (58)	He (105)
VRQ/VRQ	35	0	37.5	4	0	0	0	0
ARQ/VRQ	50	3	62.5	15	0	0	14	2
ARQ/ARQ	0	0	0	35.5	100	56	76	9
VRQ/ARR	15	39.5	0	15	0	0	0	0
ARQ/ARH	0	0	0	0	0	0	10	1
ARQ/AHQ	0	0	0	8	0	0	0	2
VRQ/AHQ	0	0	0	4	0	0	0	0
ARQ/ARR	0	12	0	10.5	0	37	0	44.5
AHQ/ARR	0	0	0	2	0	0	0	2
AHQ/ARH	0	0	0	0	0	0	0	1
ARH/ARR	0	0	0	0	0	0	0	4
ARR/ARR	0	45.5	0	6	0	7	0	24.5
AHQ/AHQ	0	0	0	0	0	0	0	0

Data adapted from Laplanche et al. (1993b and unpubl.), Westaway et al. (1994), Clousard et al. (1995), and Hunter et al. (1997a).

[a]PrP genotypes are given at codon 136 (A=alanine, V=valine), 154 (R=arginine, H=histidine), and 171 (Q=glutamine, R=arginine, H=histidine). (S) Scrapie-affected sheep; (H) healthy sheep. Values given in parentheses indicate number of sheep genotyped.

[b]From a single scrapie-free flock.

[c]Healthy flockmates.

[d]From three scrapie-free flocks.

[e]All the rams used for artificial insemination.

1993, 1994, 1997a; Belt et al. 1995; Ikeda et al. 1995; Bossers et al. 1996). In all these case-control studies, scrapie-affected groups had a proportion of VRQ/VRQ or ARQ/VRQ sheep higher than the control groups.

This defined the VRQ allele as a susceptibility factor for natural scrapie, as it was also described in experimental scrapie following SSBP/1 inoculation (Goldmann et al. 1991). VRQ/VRQ and ARQ/VRQ genotypes also influenced the survival time of affected animals with a gene dose effect: 77% of the 35 natural scrapie cases in a closed Cheviot flock between 1986 and 1995 were VRQ/VRQ and died at the mean age of 907 days (range 497–1631), whereas 23% were ARQ/VRQ and died later, at 1462 days (range 1107–2250) (Hunter et al. 1996). A similar effect was also observed by Bossers et al. (1996) in a Flemish and Swifter sheep flock.

As noted above, the first report to demonstrate a statistically significant correlation between a PrP genotype and the occurrence of natural scrapie was based on studies of natural scrapie in Suffolk sheep in the US (Westaway et al. 1994). In those investigations, PrP alleles encoding Q171 were found to render sheep susceptible to natural scrapie. The probability that the PrP genotype is unrelated to the susceptibility to develop natural scrapie was 0.000004.

Deciphering the pathogenesis of natural scrapie is complicated because prion diseases can be genetic, infectious, or sporadic disorders. The recessive effect of the VRQ allele in V136 breeds is reminiscent of the hypothesis made by Parry (1960), who believed that scrapie resulted from a recessive mutation. The 100% penetrance of scrapie observed in sheep homozygous for V136 and the development of the disease in VRQ/VRQ and ARQ/VRQ sheep within defined times of birth led to the question of whether scrapie could be a genetic disease caused by V136. The identification of healthy Cheviot sheep carrying the VRQ allele in Australia, a country said to be free of scrapie, has been used as an argument against scrapie being a genetic disease caused by VV136 (Hunter et al. 1997b). Viewing VV136 as simply modulating susceptibility to a particular strain of exogenous prions like SSBP/1 ignores the multiple mechanisms of prion disease pathogenesis. It seems quite likely that scrapie is initiated in a flock by a VRQ/VRQ sheep that develops a sporadic case of scrapie. Whether scrapie prions then spread through the flock depends on the strain of prions produced in the sheep with sporadic scrapie.

In contrast to VRQ/VRQ sheep, ARQ/ARQ sheep are relatively resistant to scrapie (Laplanche et al. 1992; Hunter et al. 1993, 1994). ARQ/ARQ Cheviot sheep in a closed experimental flock did not spontaneously develop scrapie (Hunter et al. 1996). The ARQ genotype was not found among 83 scrapie-affected sheep belonging to six different British flocks and

Table 5. Sheep PrP genotypes of two Romanov flocks and incidence of scrapie

PrP genotype[a]	P flock[b]		NP flock[b]	
	exposed	affected (%)	exposed	affected (%)
VRQ/VRQ	8	8 (100)	66	48 (73)
ARQ/VRQ	15	12 (80)	252	128 (51)
ARQ/ARQ	13	10 (77)	171	67 (39)
ARQ/AHQ	2	0 (0)	89	1 (1)
VRQ/AHQ	5	1 (20)	49	0 (0)
ARQ/ARR	3	0 (0)	93	0 (0)
VRQ/ARR	2	0 (0)	76	0 (0)
AHQ/ARR	0	0 (0)	37	0 (0)
ARR/ARR	0	0 (0)	47	0 (0)
AHQ/AHQ	0	0 (0)	11	0 (0)

[a]PrP genotype at codons 136, 154, 171.

[b](P) Experimentally parasited flock; (NP) non-parasited flock (see text for details). The youngest age at scrapie onset was 352 days. For this reason, only sheep that lived at least 1 year in the flock were considered. Modified from Elsen et al. (1998) and J.M. Elsen (pers. comm.).

breeds (Hunter et al. 1997a), although it was reported in diseased Dutch Texel sheep in a percentage (9%, $n = 34$) not different from that observed in the healthy control group (11%, $n = 91$) (Belt et al. 1995).

A scrapie outbreak occurred following an experimental oral challenge with nematode parasites in a French Romanov subflock previously free from scrapie (Clousard et al. 1995; Laplanche et al. 1996). Not only all VRQ/VRQ ($n = 8$) and 80% of ARQ/VRQ ($n = 15$), but also 77% of ARQ/ARQ ($n = 13$) sheep developed scrapie at ages ranging from 704 to 1232 days (Table 5) (Elsen et al. 1998). The nearly complete penetrance of the disease in these three genotypes is not understood, but one suggestion involves lesions made by the parasite in the gastrointestinal tract that facilitated the entry of ingested prions. Counter to this view was the finding of scrapie in a flock not exposed to the parasites that was living in close contact with the exposed flock (Laplanche et al. 1996). In the flock not exposed to parasites, 891 sheep, including 244 scrapie-affected sheep born between 1986 and 1995 that lived at least 1 year in the flock after the outbreak, were genotyped. Sheep carrying the VRQ/VRQ genotype appeared the most susceptible, followed by those with ARQ/VRQ and ARQ/ARQ (Table 5) (Elsen et al. 1998). These observations suggested that sheep carrying the ARQ/ARQ genotype might develop scrapie only when exposed to high levels of scrapie prions.

As PrP genotyping increased, sheep breeds were identified in which the PrP genotype was invariant (Hunter et al. 1993; Laplanche et al.

1993a). Such studies indicated that the VRQ genotype is not the sole determinant of scrapie susceptibility. The frequency of the VRQ allele, when present, was particularly low in Lacaune, Manech, and Préalpes (Laplanche et al. 1993a; Clousard et al. 1995), Suffolk (Westaway et al. 1994; Ikeda et al. 1995; O'Rourke et al. 1996), and Soay (Hunter et al. 1997a) breeds. Scrapie-affected sheep from these breeds mainly carried the ARQ/ARQ genotype, which defines it as the major susceptibility factor to scrapie in the breeds with limited polymorphism at codon 136 (Table 4). It is difficult to determine how susceptible to scrapie are the few heterozygous or homozygous VRQ sheep. However, there are reports of scrapie in these rare sheep (Clousard et al. 1995; Ikeda et al. 1995; Hunter et al. 1997a), indicating that they are also susceptible. Unfortunately, it was not mentioned whether their survival time was longer than that of their ARQ/ARQ flockmates, as was observed when CH1641 prions were used for challenging Cheviots (Goldmann et al. 1994; Hunter et al. 1997a).

Since AV136 Ile-de-France or Romanov sheep living in three scrapied flocks survived well beyond the maximal age of clinical onset (Romanov >3.5 years; Ile-de-France >7 years), these findings suggested that heterozygosity at codon 154 or 171 was responsible for this resistance to scrapie (Laplanche et al. 1992, 1993b). Sheep with the AHQ or ARR alleles were found at low frequency, or not at all, in scrapied sheep, whereas they were significantly represented in the corresponding healthy population (Table 4) (Westaway et al. 1994; Belt et al. 1995; Clousard et al. 1995; Ikeda et al. 1995; Hunter et al. 1996, 1997a; Laplanche et al. 1996).

Analysis of twin pairs from distinct breeds, Lacaune (Clousard et al. 1995) and Cheviot (Hunter et al. 1996), living in their respective flocks, also indicated that healthy animals were genetically distinct from scrapie cases with respect to the PrP genotype. None of the scrapied sheep carried an ARR allele.

In Europe and the US, no case of natural scrapie has been reported in ARR/ARR sheep. Additionally, prolonged survival has been observed in ARR sheep that were inoculated with SSBP/1, CH1641 (Goldmann et al. 1991, 1994), or Suffolk-passaged scrapie prions (O'Rourke et al. 1997). None of five Romanov sheep carrying at least one ARR allele developed scrapie after an outbreak of scrapie following the experimental parasitic infestation described above (Table 5). This finding suggested that ARR sheep are resistant to scrapie even with high levels of exposure to prions. In the Romanov flock not exposed to parasites, no case of scrapie was found in the 217 sheep heterozygous for the ARR allele (Table 5) (Elsen et al. 1998). This finding is consistent with the dominant-negative inhibition of prion formation by PrP(R171).

There is only one report in Japan of a single scrapie-affected ARR/ARR sheep (Ikeda et al. 1995). This sheep was the second scrapie case examined in this country (1982) and died at 4 years and 3 months of age. PrP^{Sc} was detected in the brain, and the disease was successfully transmitted to mice. This observation could reflect the existence in Japan of scrapie strains different from those in western Europe or the US, but the sheep might also be different genetically. Hunter et al. (1996) have noted that in the Japanese Suffolk breed, PrP(R171) is linked to a particular series of restriction fragment length polymorphisms not found in British sheep (Muramatsu et al. 1992).

The AHQ allele is found at a low frequency in most sheep breeds, making it difficult to establish whether it confers resistance to scrapie. In the Romanov-affected flock, 2 of 186 ARQ/AHQ sheep and 1 of 5 VRQ/AHQ sheep developed scrapie (Table 5) (Elsen et al. 1998). These observations suggest that the AHQ allele may, like ARR, confer reduced susceptibility to scrapie; and like ARR, AHQ may act as a dominant negative and inhibit the conversion of the PrP(VRQ) and PrP(ARQ) into PrP^{Sc}.

Transmission of Scrapie Prions

From a multitude of studies, it seems most likely that natural scrapie spreads among the sheep of a flock by oral transmission. Suffolk sheep fed brain from scrapied sheep were found to have prions in the Peyer's patches of the ileum soon after feeding (Hadlow et al. 1982). In addition, PrP^{Sc} has been found in sheep tonsils at early preclinical stages examined (Schreuder et al. 1996). Additional evidence for oral transmission of prions to sheep comes from studies in which the placentas from scrapied Swaledale ewes were fed to Herdwick sheep (Pattison et al. 1972, 1974; Onodera et al. 1993).

Some studies suggest that the spread of scrapie is encouraged by close contact between animals (Brotherston et al. 1968). Whether this is due to abrasions of the skin is unclear. Unpublished work of D.R. Wilson in the 1950s suggested that it was possible to inoculate sheep successfully with scrapie using scarification (Dickinson 1976). Although the scarification route has not been explored experimentally in sheep, scratches of the skin have been shown to be effective routes to experimental infection of mice, with much the same efficiency as inoculation by the intraperitoneal route (Taylor et al. 1996).

It is part of the dogma of scrapie that the disease in sheep is maternally transmitted, with the placenta as the prime suspect that brings scrapie infectivity in utero to the developing lamb. Telling the difference between genuine maternal transmission and lateral transmission by close contact during the perinatal period is difficult, and it is not always possible to know when an animal has acquired its infection. However, in a natural scrapie outbreak in a Cheviot flock (Hunter et al. 1996), it seems likely that most animals were infected at about the same age. The ages at death from scrapie were constant, with VRQ/VRQ animals dying at 700–900 days of age and heterozygotes ARQ/VRQ dying at 1100–1200 days of age. This difference between homozygotes and heterozygotes, expected from the experimental data, is not always seen in other flocks (Hunter et al. 1994). However, the survival of three VRQ/VRQ animals to more than 1900 days in this flock, after stringently clean procedures were used around the time of their birth, suggests that the perinatal period is a dangerous time for a susceptible lamb.

Embryo transfer (ET) procedures have also been investigated in the hope of bypassing any infection cycle. Although work with the Neuropathogenesis Unit (NPU) Cheviot flock was compromised by natural scrapie in the flock (Foster et al. 1992, 1996b), it is clear that ET procedures coupled with "over the top" cleanliness at birth do not prevent scrapie from occurring in all lambs. These experiments generated the three surviving VRQ/VRQ animals discussed above but also generated 10 VRQ/VRQ sheep that did develop scrapie (9 at mean age 826 ± 24 days of age and 1 at 1267 days of age). This group of 10 animals was not protected by ET procedures even following International Embryo Transfer Society (IETS) protocols (Stringfellow and Seidel 1990). A similar experiment set up in the US (Foote et al. 1993) is difficult to interpret because of the lack of PrP genotype information. However, one of the groups involved Cheviot sheep inoculated with SSBP/1 scrapie, and it has been shown that these animals give incubation periods similar to those of Cheviots in a Scottish flock inoculated with SSBP/1 prions (Maciulis et al. 1992).

The incubation periods in scrapie-inoculated donors and recipients ranging from 6.1 months (183 days) to 14.9 months (447 days), plus survivors to 25 months (750 days) suggest that a mixture of PrP genotypes is present in the group. Of the positive control lambs (gestated and born naturally), 2 of 9 developed scrapie at 31 months (930 days) and 49 months (1470 days) of age, whereas none of the 22 ET-derived Cheviot lambs succumbed to scrapie, with ages at death ranging from 74.5 months (2235 days) to 96 months (2880 days). It is of course possible that there were no susceptible lambs among the ET group, but there is at least a suggestion that ET procedures have protected lambs from scrapie in this case, and this again implicates the possibility of maternal transmission as a route for transmission of scrapie prions.

Nematodes and hay mites have also been considered as scrapie vectors, but the data in support of such postulates are at best circumstantial (Fitzsimmons and Pattison 1968; Clousard et al. 1995; Wisniewski et al. 1996). In one report, it was suggested that sheep become infected by eating the mites along with hay. In that report, it was stated that proteinase-K-resistant PrP was found in mites taken from scrapie-affected Icelandic farms, and injection of the mites into mice produced a few positive transmissions.

SSBP/1 and CH1641 prions used for studies of experimental scrapie have not transmitted well to non-Tg mice. It will be of interest to transmit both of these isolates to Tg(OvPrP) and Tg(BoPrP) mice described below. SSBP/1 originated from a pool of brains from about eight scrapie-affected positive-line Cheviots, all of which encode V136 but differ at other codons. CH1641 originated from a natural scrapie case in a Cheviot positive line but has subsequently been passaged in a negative line of AA136 sheep to provide the inoculum currently in use. These sheep-passaged sources of scrapie prions tend to cause disease more readily in animals of similar PrP genotype to that of the sheep that were used to produce the infectious inoculum (W. Goldmann and N. Hunter, unpubl.).

For most of the studies with Cheviot sheep described above, these sheep belong to a closed flock that has been studied at the NPU for more than three decades. Two other selected flocks of sheep have not produced as much information as the Cheviots but have confirmed the results from the larger set of studies. The Herdwick lines, also positive and negative responders to SSBP/1, have similar PrP genetics to the Cheviots, in that V136 is associated with susceptibility (N. Hunter, unpubl.). The Swaledale flock has only a negative line, selected for scrapie resistance using two pools of Swaledale natural scrapie (SW73 and SW75) as inocula (Davies and Kimberlin 1985). SW73 and SW75 were clearly selecting sheep with the same PrP genotypes as does SSBP/1, as the prevailing genotypes in the resistant animals are AA136.

Transmission of BSE Prions to Sheep and Goats

Sheep have been exposed to BSE prions by oral and i.c. inoculation. In two independent studies, Cheviot and Romney sheep were given 5 g of BSE brain as a 10% homogenate. Seven of the 13 inoculated Cheviot sheep developed CNS dysfunction between 553 and 1073 days after inoculation. In all 7 sheep, PrP immunostaining was widespread throughout the CNS, including the spinal cord and in lymphoid tissues, including the tonsils and mesenteric lymph nodes (Foster et al. 2001a). ARQ/ARQ Romney

Table 6. Clinical onset of disease and accumulation of disease-specific PrP in the CNS and viscera of inoculated PrP$^{ARQ/ARQ}$ sheep

Sheep no.[a]	Time (months) between inoculation and		Presence of visceral PrP	Presence of CNS	
	necroscopy	clinical onset[b]		vacuolation	PrP
1	4	N	+/−	−	−
2	4	N	−	−	−
3	4	N	−	−	−
4	4	N	−	−	−
5	10	N	−	−	−
6	10	N	+/−	−	−
7	10	N	−	−	−
8	10	N	−	−	−
9	16	N	−	−	−
10	16	N	−	−	−
11	16	N	+/−	−	+/−
12	16	N	+	−	+/−
13	22	N	−	−	+/−
14	22	N	−	−	+/−
15	22	21	+	+	+
16	22	20	+	+	+
17	22	N	+	+	+
18	24	24	+	+	+
19	27	27	+	+	+
20	28	28	+	+	+
21[c]	. . .	. . .	. . .	. . .	. . .

Reprinted with permission from Jeffrey et al. (2001), copyright 2001.

[a]Control sheep nos. 22, 23, 24, and 25 were killed and examined 4, 10, 16, and 22 months, respectively, after the other 21 animals had been inoculated; the results were entirely negative.

[b]N, clinically normal.

[c]Sheep 21 was alive but clinically sick 36 months after inoculation.

sheep developed CNS dysfunction by 20–28 months after oral inoculation with BSE prions (Table 6), but ARQ/ARR or ARR/ARR Romney sheep remained well for more than 24 months (Jeffrey et al. 2001). As in the Cheviots, PrP immunostaining was found throughout the CNS in the ARQ/ARQ Romney sheep and was widespread in the lymphoid system (Table 7).

The incubation time in sheep inoculated i.c. with BSE prions was greatly prolonged by the presence of R171 in one of the two PrP alleles, consistent with OvPrP(R171) acting as a dominant negative (Table 8). When ARQ/ARQ Suffolk sheep were inoculated i.c. with BSE prions, they devel-

Table 7. Distribution of disease-specific PrP in lymph nodes, tonsil, spleen, and alimentary tract of inoculated PrP$^{ARQ/ARQ}$ sheep

Sheep no.[a]	Months between inoculation and necroscopy	lymph nodes[b]						tonsil	spleen	intestine	abomasun	forestomachs
		SM	RP	PS	MES	IC	MED					
1–8	4–10	–	–[c]	–	–	–	–	–	–	–	–	–
9	16	–	–	–	–	–	–	–	–	–	–	–
10	16	–	–	–	–	–	–	–	–	–	–	–
11	16	–	–	–	–	–	–	–	+	+	ND	–
12	16	+	+	+	–	+	+	+	+	+[d]	ND	+
13	22	–	–	–	–	–	–	–	–	–	–	–
14	22	–	–	–	–	–	–	–	–	+[e]	–	–
15	22	+	+	+	+	+	+	+	+	+	–	+
16	22	+	+	+	+	+	+	+	+	+[d]	+	+
17	22	+	+	+	+	+	+	+	+	+	+	+
18	25	+	+	+	+	+	+	+	+	+	+	+
19	28	+	+	+	+	+	+	+	+	+	+	+
20	29	+	+	+	+	+	+	+	+	+	+	+
21[f]	…	…	…	…	…	…	…	…	…	…	…	…

Reprinted with permission from Jeffrey et al. (2001a, copyright 2001.

[a] Control sheep nos 22, 23, 24, and 25 were killed and examined 4, 10, 16, and 22 months, respectively, after the other 21 animals had been inoculated; the results were entirely negative.

[b] (SM) Submandibular; (RP) retropharyngeal; (PS) prescapular; (MES) mesenteric; (IC) ileocaecal; (MED) mediastinal; (ND) not determined.

[c] RP lymph nodes positive in sheep nos. 1 and 6.

[d] Peyer's patches not represented in the same tissue block.

[e] Immunolabeling confined to a single enteric neuron.

[f] Sheep 21 was alive but clinically sick 36 months after inoculation.

Table 8. Sheep PrP genotypes of Cheviots and incidence of experimental prion disease

Infection source	Route	*Sip* genotype	*PRNP* genotype	n	Incubation time (days)
BSE	i.c.	sAsA	VQ/VQ	ND	—
		sApA	VQ/AQ	2	424, 880
		sApA	VQ/AR	2	1874, 1874
		pApA	AQ/AQ	2	440, 487
		pApA	AQ/AR	3	1886, 1923, 2353
		pApA	AR/AR	ND	—

Data simplified and taken from Goldmann et al. (1994) and Foster et al. (2001c). PrP genotypes are given at codon 136 (A = alanine, V = valine) and 171 (Q = glutamine, R = arginine). (i.c.) Intracerebral; (ND) not done.

oped disease in 556 ± 23 days (8 of 10 inoculated sheep), whereas ARQ/ARQ Cheviot sheep developed disease in 608 ± 38 days (5/5). VRQ/VRQ Cheviot sheep developed disease in 1071 ± 22 days (4/5) (Table 9).

In contrast to ARR/ARR sheep inoculated with SSBP/1 and CH1641 prions, BSE prions did cause CNS disease in these scrapie-resistant animals. One of 10 ARR/ARR Suffolk sheep developed prion disease at 1008 days after i.c. inoculation and 2 of 5 ARR/ARR Cheviot sheep developed prion disease at 1124 and 1127 days after i.c. inoculation.

Eight goats inoculated i.c. with BSE prions developed signs of CNS dysfunction between 506 and 985 days, with a mean incubation time of 665 ± 47 days (Foster et al. 2001b). In another study, 4 female goats were inoculated s.c. with BSE prions and exhibited incubation times between 760 and 1284 days. In a study of 10 female goats inoculated with BSE prions, none of the 37 progeny developed prion disease, and 21 of them lived for more than 2000 days (Foster et al. 1999).

Molecular Diagnosis of Scrapie and BSE in Sheep and Goats

Detection of PrP^{Sc} is the most specific, sensitive, and reliable means of making the diagnosis of scrapie in any animal. PrP^{Sc} is found only in prion diseases, and thus, its presence in sheep or goats is diagnostic of prion infection. The histopathology of prion diseases in livestock can be quite variable, with vacuolation of the neuropil and reactive astrocytic gliosis varying considerably with respect to intensity and location.

In natural scrapie, lesions in the medulla of sheep are always present late in the disease (Fraser 1976; Wood et al. 1997). Immunohistochemical studies of the medulla demonstrated that PrP immunostaining was found

Table 9. Sheep challenged intracerebrally with BSE brain homogenate

Breed	Genotype	Number challenged	Age (days) at inoculation	Number of clinical cases	Incubation period (mean days ± s.e.m.)	Survival period of remaining sheep (days postinfection)
Suffolk	ARQ/ARQ	10	745–1283	8	556 ± 23	1133
	ARQ/ARR	9	1110–1283	0	—	966–1133
	ARR/ARR	10	742–1107	1	1008	959–1136
Cheviot	VRQ/VRQ	5	194–355	4	1071 ± 22	945
	VRQ/ARQ	5	733–883	2	881, 1008	965–1115
	VRQ/ARR	5	695–902	0	—	945–1153
	ARQ/ARQ	5	695–902	5	608 ± 38	—
	ARQ/ARR	5	733–883	0	—	965–1114
	ARR/ARR	5	164–325	2	1124, 1127	945–1107
Poll Dorset	VRQ/VRQ	5	316–438	0	—	134
	VRQ/ARQ	5	418–448	0	—	132–134
	VRQ/ARR	4	294–325	0	—	945
	ARQ/ARQ	4	715–918	4	500 ± 4	—
	ARQ/ARR	6	646–918	0	—	959–1163
	ARR/ARR	4	646–1283	0	—	959–1231

Data provided by Nora Hunter.

in all 97 cases of natural scrapie in the dorsal motor nucleus of the vagus (DMV) (Ryder et al. 2001). Interestingly, these parasympathetic motor neurons of the vagus nerve innervate the gastrointestinal tract, a finding that supports the gut as the portal of entry for scrapie prions. Further support for this hypothesis comes from several studies in sheep and goats which show that oral exposure to prions renders the Peyer's patches of the ileum positive for prion infectivity and PrP^{Sc} deposition (Hadlow et al. 1980, 1982; Andreoletti et al. 2000; Heggebo et al. 2000).

In a study of goats inoculated with the SSBP/1 and CH1641 scrapie isolates and a BSE isolate, western blotting detected PrP^{Sc} in the brains of all samples examined (Foster et al. 2001a). In contrast, immunohistochemistry was negative for some samples from the brains of goats inoculated with the SSBP/1 and CH1641 isolates, especially where the vacuolation was particularly intense. Such findings are not unexpected, since severe neuronal loss accompanied by widespread vacuolation can result in diminished local levels of PrP^{Sc}; moreover, it has been shown that PrP^{Sc} can be transported from its site of formation to other areas of the CNS (Brandner et al. 1996).

Western blots have been used to analyze the glycoform profiles of the protease-resistant fragment of PrP^{Sc}, designated PrP 27-30. High levels of the diglycosylated PrP 27-30 have been used in conjunction with a 19-kD unglycosylated PrP 27-30 as a molecular signature for the BSE strain of prions (Collinge et al. 1996; Asante et al. 2002). Other investigators have argued that this signature lacks the specificity that it was originally thought to have, since a number of scrapie isolates from sheep, as well as some passaged in mice, exhibit similar western blot patterns (Somerville et al. 1997a; Baron et al. 1999). Studies of fatal insomnia (FI) have shown that the spontaneous generation of FI prions is independent of the PrP glycoform profile. Because the mutation causing familial FI is close to the first glycosylation site, the glycoform profile of PrP^{Sc} from familial FI is much different from that of sporadic FI (Gambetti and Parchi 1999). Yet familial and sporadic FI are caused by the same strain of prion. Both diseases present with the same clinical neurological symptoms and signs, and both exhibit neuropathologic changes consisting of vacuolation and gliosis restricted to the dorsal medial nuclei of the thalamus (Medori et al. 1992; Mastrianni et al. 1999; Parchi et al. 1999). In human brains from patients with familial or sporadic FI, unglycosylated as well as deglycosylated PrP 27-30 has an M_r of 19 kD (Parchi et al. 1999). Passage of either familial or sporadic FI prions into Tg(MHu2M) mice produces an incubation time of ~200 days and a 19-kD band after limited proteolysis and deglycosylation of PrP^{Sc} (Telling et al. 1996; Mastrianni et al. 1999). In the brains of both humans and Tg mice, PrP^{Sc} deposition is restricted to the thalamus.

Besides using inbred and Tg mice, strains have been studied using the conformation-dependent immunoassay (CDI) (Safar et al. 1998, 2002) and conformational stability profiles (Peretz et al. 2001, 2002). The CDI has been adapted to a robotic platform and used to measure BSE prions in a high-throughput system (Safar et al. 2002). Because the CDI is based on measuring a PrP epitope that is exposed in native PrP^C, buried in native PrP^{Sc}, and exposed in denatured PrP^{Sc}, it can be used to measure different prion strains by adjusting the concentration of guanidine used in the denaturation step. Whether the CDI will prove useful in differentiating scrapie from BSE prion strains in sheep and goats remains to be established.

The CDI may also offer the possibility for an antemortem test for prion disease in sheep and goats. The CDI has been used to detect prions in muscle of mice (Bosque et al. 2002). PrP^{Sc} is found primarily in the hind limbs of mice and is not readily detectable in other muscle groups. In the hind limb, the level of PrP^{Sc} is lower by about a factor of 100 compared to brain (C. Ryou et al., unpubl.). The molecular basis for this highly restricted distribution is unknown. Learning which muscle groups contain readily measurable levels of PrP^{Sc} in sheep and goats will be very important, as will be defining the kinetics of PrP^{Sc} formation with respect to the strain of prion, the PrP genotype, and the breed of animal.

It may be possible to develop an antemortem test for prions using blood as well as muscle. A protease-sensitive (s) form of PrP^{Sc} has been found in rodent blood (Prusiner and Safar 2000), as well as in human blood (J. Safar and S.B. Prusiner, unpubl.). In both rodent and human blood, the level of prion infectivity is thought to be ~1 ID$_{50}$ unit/ml (Brown et al. 1998; Brown 2000), and the level of sPrP^{Sc} is also quite low. Whether a blood test for sPrP^{Sc} in the blood of sheep and goats can be devised is unknown.

Eradication of Scrapie

The eradication of scrapie has typically been accomplished by the killing of all animals in a flock determined to have scrapie. However, studies of pastures inhabited by scrapied flocks have shown that the disease can reappear when the pasture is repopulated years later (Palsson 1979). Presumably, prions that contaminate the topsoil persist and remain infectious. In one study, prion infectivity was recovered from soil after more than a year (Brown and Gajdusek 1991).

The transmission of BSE to ARR/ARR sheep, as described above (Houston and Gravenor 2003; Houston et al. 2003) has major implications for the development of scrapie-resistant flocks. Large-scale breeding programs are already under way in some European countries and are being contemplated in the US. Such programs are already being undertaken in the Netherlands, the UK, and France.

The transmission of the BSE strain of prions to scrapie-resistant sheep forces a reconsideration of the wisdom of breeding large national flocks of ARR/ARR sheep. Such sheep might prove to be much more dangerous with respect to human food safety than, for instance, ARQ and VRQ animals.

Despite numerous attempts to show a relationship between CJD and sheep scrapie, no link between these diseases has been demonstrable. In contrast, it is now clear the BSE prions have been transmitted to more than 140 teenagers and young adults who have developed vCJD. Whether the BSE strain of prions must have bovine PrP to be transmissible to humans, or whether ovine PrP is sufficient, is unknown. Bovine and ovine PrP share 97% sequence identity, making it likely that ovine PrP^{Sc} folded into the BSE strain is also transmissible to humans.

Since Tg mice expressing HuPrP as well as chimeric Hu/Mo PrP are available (Telling et al. 1995; Hill et al. 1997; Asante et al. 2002; Korth et al. 2003), it seems mandatory that experiments be undertaken to determine whether Tg mice expressing HuPrP are equally susceptible to BSE prions from cattle and sheep. Certainly the results of such experiments need to be known before proceeding with any selective breeding programs. In addition to these Tg mouse experiments, perhaps a limited number of similar studies with monkeys and possibly some apes should be contemplated, in which BSE prions from cattle and sheep are inoculated. Moreover, BSE prions from sheep as well as from Tg(OvPrP,ARR)$Prnp^{0/0}$ mice should be carefully studied.

New Experimental Approaches to Scrapie

Transgenic mice that have been so useful in many areas of prion research promise also to be helpful in studies of sheep with scrapie. Tg mice expressing OvPrP have been constructed and inoculated with prions from sheep. In one study, Tg mice expressing OvPrP(ARQ) were inoculated with two different sheep scrapie isolates (Crozet et al. 2001b). With one isolate, the Tg(OvPrP,ARQ)$Prnp^{0/0}$ mice were said to develop signs of CNS dysfunction after only 3–6 weeks that persisted for another 200 days until the animals died. With the second isolate, the Tg(OvPrP,ARQ)$Prnp^{0/0}$ mice remained healthy for ~230 days and died within a few days after exhibiting neurologic dysfunction. In another study, nine lines of Tg(OvPrP,VRQ)$Prnp^{0/0}$ mice were constructed; they exhibited incubation times ranging from 73 to 399 days after inoculation with prions from two scrapied sheep, both of which were homozygous for PrP(VRQ) (Vilotte et al. 2001).

Tg(OvPrP,ARQ)$Prnp^{0/0}$ mice were inoculated with a brain extract from an ARQ/ARQ sheep inoculated with BSE prions (Crozet et al. 2001a). The Tg mice ($n = 20$) developed CNS dysfunction and died 300 ± 40 days after inoculation. In the brains of the Tg(OvPrP,ARQ)$Prnp^{0/0}$ mice, numerous florid plaques that are pathognomonic of vCJD were found. None of the other inocula derived from the brains of six scrapied sheep produced such plaques. Additionally, neither BSE nor vCJD inocula produced florid plaques in the brains of Tg(BoPrP)$Prnp^{0/0}$ mice but did produce CNS degeneration (Scott et al. 1999).

Like BSE or vCJD inocula, scrapie prions from sheep produced CNS degeneration in the Tg(BoPrP)$Prnp^{0/0}$ mice (Scott et al. 1999). Although both BSE and vCJD inocula induced profound deposition of PrP amyloid in the corpus callosum and pons, the scrapie prions did not. It is not unexpected that Tg(BoPrP)$Prnp^{0/0}$ mice are susceptible to sheep scrapie prions, since the bovine and ovine PrP sequences share ~97% identity. Surprisingly, the Tg(BoPrP)$Prnp^{0/0}$ mice developed disease in ~210 days when inoculated with brain extracts from scrapied Suffolk sheep, whereas ~240 days were required when BSE brain extracts were inoculated.

Tg(OvPrP) mice expressing different OvPrP polymorphisms are likely to prove useful in discerning the mechanisms of susceptibility and resistance conferred by changes in the amino acid sequence of OvPrP. Constructing Tg lines expressing combinations of these polymorphic residues might prove to be quite valuable.

Whether studies incubating radiolabeled OvPrP with OvPrP^{Sc} from the brains of scrapied sheep are informative remains to be determined. In one study, OvPrP^{Sc}(VRQ) protected radiolabeled OvPrP^{C}(VRQ) from proteolytic digestion approximately three times more efficiently than it protected OvPrP^{C}(ARQ). A PrP^{Sc} preparation prepared from the brain of an ARQ/VRQ sheep gave similar results. OvPrP^{Sc}(ARQ) protected radiolabeled OvPrP^{C}(VRQ) from proteolytic digestion with the same efficiency as it protected OvPrP^{C}(ARQ). All of the OvPrP^{Sc} preparations exhibited relatively little ability to protect radiolabeled OvPrP^{C}(ARR) from proteolytic digestion (Bossers et al. 1997). Similar studies with BoPrP^{Sc} showed no ability to protect OvPrP^{C}(ARR) from limited proteolysis (Bossers et al. 2000), but as noted above, BSE prions have transmitted to ARR sheep (Houston et al. 2003).

CHRONIC WASTING DISEASE AND TRANSMISSIBLE MINK ENCEPHALOPATHY

Chronic Wasting Disease

Host Range of CWD

CWD is a prion disease of captive and free-ranging mule deer (*Odocoileus hemionus*), white-tailed deer (*Odocoileus virginianus*), and Rocky Mountain elk (*Cervus elaphus nelsoni*) (Fig. 3) (Williams and Young 1992; Spraker et al. 1997). The clinical syndrome of wasting and eventual death was first recognized in the late 1960s by biologists working with captive mule deer in Colorado who observed the complex of clinical signs associated with the syndrome they designated chronic wasting disease. It was initially thought to be associated with stresses of captivity, nutritional deficiencies, or intoxications. CWD in mule deer was first recognized as a spongiform encephalopathy in 1977 (Williams and Young 1980). This was followed in the early 1980s by recognition of CWD in captive and free-ranging elk (Williams and Young 1982, 1992). By the mid-1980s, CWD was found in free-ranging mule deer in Colorado and Wyoming, and in 1990, the host range of CWD was expanded to include white-tailed deer. These species are the most abundant wild cervids (members of the deer family) in North America.

CWD has been transmitted by intracerebral inoculation into a variety of species. Of domestic livestock species, cattle (Hamir et al. 2001), goats (Williams and Young 1992), and domestic sheep are susceptible (A.N. Hamir et al., pers. comm.) albeit with relative inefficiency. Laboratory species found susceptible to CWD by intracerebral inoculation include domestic ferrets (*Mustela putorius*), domestic mink (*Mustela vison*), squirrel monkey (*Saimiri sciureus*) (R. Marsh, pers. comm.), mice (M. Bruce, pers. comm.), and hamsters following passage through ferrets (Bartz et al.

Figure 3. Wild cervids. (*A*) Mule deer (*Odocoileus hemionus*); (*B*) white-tailed deer (*Odocoileus virginianus*); (*C*) elk (*Cervus elaphus nelsoni*).

1998). Although scrapie and transmissible mink encephalopathy (TME) were transmitted by intracerebral inoculation into raccoons (*Procyon lotor*), transmission of CWD into raccoons did not occur (A.N. Hamir et al., unpubl.). CWD is readily transmitted by intracerebral or oral inoculation from deer to deer or elk to elk (Williams and Young 1992; Sigurdson et al. 1999; E.S. Williams et al., unpubl.). Additional studies of host range by the oral route, particularly testing those species found in habitats where CWD exists in free-ranging cervids and other farmed cervid species, are warranted.

Epidemiology of CWD

Modes of transmission of CWD are unclear. Epidemiologic evidence and modeling strongly suggest that CWD is contagious and can be horizontally transmitted (Miller et al. 1998, 2000; Gross and Miller 2001). It probably can be freely transmitted among the three known susceptible species, although this has not been definitively demonstrated.

Unlike the human TSEs, BSE, and TME, environmental contamination probably is important. Modeling suggests that residual contamination may be even more important in transmission than direct animal-to-animal contact (M. Miller and T. Hobbs, pers. comm.). The natural route of exposure to the CWD agent is probably oral, in keeping with what has been found with other animal TSEs. However, dermal, subcutaneous, or gingival exposure may also play a role. Recent demonstration of rapid involvement of the CNS following experimental inoculation of TME in the tongue of hamsters (Bartz et al. 2003) supports this possibility. Rough forage, contaminated with CWD agent, could result in scarification of the oral mucosa and inoculation of cervids.

Oral transmission of CWD to mule deer, white-tailed deer, and elk via affected brain inoculum is readily accomplished (Sigurdson et al. 1999; E.S. Williams et al., unpubl.), suggesting this as a natural route of exposure. The means of exit of the agent from live affected hosts remains to be determined. Because of the presence of PrP^{Sc} in lymphoid tissues lining the gastrointestinal tract and in tonsil (Miller and Williams 2002; Spraker et al. 2002a,b; Wild et al. 2002), fecal or salivary shedding of the agent is plausible.

Since 1983, and with much greater intensity since 1992, surveillance for CWD in hunter-harvested deer and elk has been conducted in the endemic areas of Colorado, Wyoming, and Nebraska to develop prevalence estimates and better understanding of distribution (Fig. 4). With recognition of CWD in both captive and free-ranging cervids, surveillance

Figure 4. Locations of CWD in North America, 1977–2003. Areas where CWD has been found in free-ranging deer and elk are shown in red. Gray-shaded states and provinces indicate jurisdictions where CWD has been found in captive and farmed cervids. Many of these captive cervid premises have been depopulated. In addition, a case of CWD was found in an elk in Korea imported from Canada.

among these populations has greatly increased across the continent. In the past, brains were examined by routine microscopy and, beginning in 1995, by immunohistochemistry (IHC) for PrP^{Sc}. During the 2002 hunting season, based on evidence of diffuse lymphoid involvement in the pathogenesis of CWD (Sigurdson et al. 1999; Miller and Williams 2002), surveillance for CWD in deer in many jurisdictions used IHC of retropharyngeal lymph nodes. In addition, a commercially available ELISA was used for testing thousands of cervids in Colorado (M.W. Miller, pers. comm.). Results of extensive testing indicated prevalences ranging from <1% to ~20% in some free-ranging deer management areas in Colorado and Wyoming. Continued surveillance for CWD is planned throughout North America in coming years.

Clinical Presentation of CWD

Clinical signs of CWD in captive cervids have been described previously (Williams and Young 1992). The majority of CWD-affected animals are prime-aged animals. Natural incubation period is not precisely known, but the youngest animal diagnosed with the disease was 16 months of age. Maximum incubation period has not been determined, but the old-

Figure 5. Female elk with CWD. Notice the lowered head and ears, depressed demeanor, and poor body condition.

est animal with CWD was >15 years of age. As suggested by its name, the most prominent clinical sign of the disease is loss of body condition that progresses to a state of emaciation if the disease is allowed to run its course (Fig. 5). Characteristic behavioral changes include alteration in behavior toward humans, changes in interaction with other members of the herd, stupor, walking repetitive patterns, and (rarely) hyperexcitability, especially when restrained. Polydipsia and consequent polyuria are frequently present. Hypersalivation and difficulty swallowing may occur. A smaller percentage of affected animals, especially elk, displays ataxia, incoordination, and/or fine head tremors. Pruritus has never been observed. Duration of clinical disease is usually protracted, but a few white-tailed deer have died acutely or after only a few days of recognizable illness. Pneumonia, often aspiration pneumonia, may be the actual cause of death in affected animals, even before they reach a stage of emaciation.

Distribution of CWD

Since publication of the first edition of this book, the known distribution of CWD has expanded considerably (Fig. 4). This geographic expansion may be attributed to increased surveillance as well as movement of CWD-affected animals into new areas. Two related CWD outbreaks are occurring concurrently—one in farmed cervids and one in free-ranging cervids (Williams and Miller 2002).

As the cervid farming industry grew rapidly during the last three decades, elk and deer have been moved in commerce. In the early 1990s, animal health concerns were raised because of an epidemic of bovine tuberculosis among farmed elk, and the industry became more regulated at state, provincial, and federal levels. This resulted in requirements in many jurisdictions for individual animal identification and, in some places, record keeping that allowed determination of animal movement throughout life. When CWD was found in farmed elk in Saskatchewan (Canada, 1996; Canadian Food Inspection Agency 2002), the extent of CWD in the industry was not known; surveillance since that time suggests that CWD was present prior to the bovine tuberculosis epidemic but was not recognized. This temporal assessment is based on finding CWD in a farmed elk herd in Alberta (Canada), a province that had been closed to importation of elk since 1990. Following recognition of CWD in farmed elk in Saskatchewan, CWD has been found, as of early 2003, in farmed elk from South Dakota, Montana, Nebraska, Oklahoma, Colorado, Kansas, Alberta, Minnesota, and Wisconsin. In addition, a case of CWD was diagnosed in an elk imported from Canada into Korea (Sohn et al. 2002). In most of these jurisdictions, affected herds have been depopulated.

CWD has been diagnosed in farmed white-tailed deer on several farms in Wisconsin and in Alberta. The farms in Wisconsin were epidemiologically linked, but it is not known how CWD entered the white-tailed deer industry. Until recently, the farmed deer industry has been less concerned about CWD than the elk industry, even though the species was known to be susceptible. It seems likely that CWD will be found in other farmed white-tailed deer and possibly mule deer or other farmed cervids as surveillance increases and regulations are implemented. Many of the regulations or proposed regulations cover only species known to be susceptible to CWD; however, it is not known whether some farmed exotic cervid species in North America, such as fallow deer (*Dama dama*), sika deer (*Cervus nippon*), or axis deer (*Axis axis*), are susceptible to CWD. Regulations for control and eradication of CWD in farmed cervids have been implemented by most state animal health regulatory agencies (often through state departments of agriculture or wildlife departments); a federal plan for CWD is nearing completion (US Department of Agriculture [L. Creekmore, pers. comm.]).

The situation in free-ranging deer and elk is quite different. CWD has been found in free-ranging elk in Colorado, Wyoming, and South Dakota. It has been found in free-ranging mule deer in Colorado, Wyoming, Nebraska, South Dakota, Utah, and New Mexico, and in free-ranging white-tailed deer in Colorado, Wyoming, Wisconsin (Joly et al. 2003),

South Dakota, Illinois, and Nebraska (Fig. 4). In populations of these animals, geographic expansion of CWD is determined by natural movement of animals along migration corridors or in natural areas of concentration such as on winter range. This has been demonstrated by recognition of CWD spreading east of the historic endemic area in southeastern Wyoming and northcentral and northeastern Colorado along the Platte River drainages into western Nebraska and westward across the continental divide to the western slope of the Rocky Mountains in Colorado and Wyoming. Migration patterns of elk and/or mule deer to the west probably explain finding CWD in mule deer in eastern Utah adjacent to CWD areas in Colorado in 2002. Explanation for the occurrence of CWD in localized areas such as the focus in New Mexico remains to be determined. Continued geographic expansion can be expected along these animal movement corridors unless specific techniques can be developed to manage or prevent movement of infected animals.

The association between occurrence of CWD in farmed and free-ranging cervids is not clear at this time. The geographic overlap between the CWD epidemic in farmed and free-ranging cervids suggests an association between the two, but this could be due to biased sampling because surveillance is intensified locally in both segments when CWD is recognized in either group. Biologically, the disease could move from farmed cervids to the wild or vice versa; this is most likely to have occurred via ingress or egress from farmed facilities. An ear-tagged escaped white-tailed deer with CWD was found outside the fence of a deer farm with CWD in Wisconsin (Wisconsin Department of Natural Resources 2003), raising suspicion of escape of the disease from the facility along with the deer.

Pathological and Biochemical Diagnosis of CWD

Morphologic pathology of CWD has been described previously (Guiroy et al. 1993, 1994; Williams and Young 1993; Hadlow 1996; Liberski et al. 2001; Spraker et al. 2002a,b). Spongiform lesions are most striking in the medulla oblongata, especially the parasympathetic vagal nucleus; thalamus and hypothalamus; and olfactory cortex. Prominent lesions in the medulla oblongata at the obex are found in all cases of CWD, and examination of this region is used for surveillance and diagnostic purposes. PrP plaques surrounded by spongiform change are prominent in some cases of CWD (Williams and Young 1993), particularly in white-tailed deer or white-tailed x mule deer hybrids (Fig. 6). These plaques morphologically resemble "florid plaques," which are a prominent feature of vCJD and BSE in macaques (*Macaca mulatta*) (Lasmézas et al. 1996; Will et al. 1996), but the

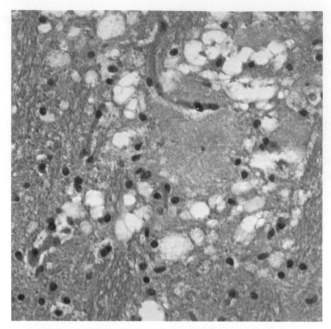

Figure 6. Amyloid plaque from the medulla oblongata of a mule deer with CWD. Note the marked spongiform change at the periphery of the plaque.

distribution of the plaques in CWD-affected cervid brain is considerably different, generally sparing the cerebellum, which is extensively affected in primates. In addition to the CNS, PrP^{Sc} may be found by IHC in myenteric plexes, vagosympathetic trunk, pars nervosa and intermedia of the pituitary, islets of Langerhans in the pancreas, and adrenal medulla (Sigurdson et al. 2001). SAFs are readily identified in brains and spleen of clinically affected deer and elk (Spraker et al. 1997). Immunohistochemistry (Guiroy et al. 1991a,b; Spraker et al. 1997, 2002a,b,c; O'Rourke et al. 1998) and immunoblotting (Race et al. 2002) for PrP^{Sc} in brains and lymphoid tissues of affected and preclinical deer and elk are used for diagnostic, surveillance, and research purposes.

Antemortem tests for scrapie based on detection of PrP^{Sc} in third eyelids of sheep (O'Rourke et al. 1998) led to investigation of the use of lymphoid tissues for a live animal test for CWD. Lymphoid tissue in the nictitating membrane of cervids is scant in comparison to domestic sheep and thus is not suitable for use in a live animal test; however, biopsy of tonsil is a useful antemortem diagnostic tool for deer (Wild et al. 2002; Wolfe et al. 2002).

The relationship of CWD prions with those of other transmissible spongiform encephalopathies is not yet clear. Strain typing by incubation period and lesion profiles in genetically characterized mice has shown CWD prions differ from BSE prions (Bruce et al. 1997; M.E. Bruce, unpubl.). The CWD agent also differs from all strains of scrapie tested in this system thus far (M. Bruce, pers. comm.). PrP glycoform patterns from mule deer, white-tailed deer, and elk could not be distinguished (Race et al. 2002); however, there was more variation in glycoform patterns among the deer than elk, possibly indicating some differences among the CWD strains. In all cervids, the diglycosylated bands gave the strongest signal. The conformation-dependent immunoassay distinguished PrP^{Sc} from elk and deer (Safar et al. 2002), suggesting strain differences between the species. The similarities in glycoform patterns from cervids with CWD and sheep with scrapie provide support for the hypothesis that CWD may have arisen from scrapie (Race et al. 2002). This origin of CWD also has been suggested on the basis of in vitro protection experiments (Raymond et al. 2000).

Genetics

The influence of genetics on CWD susceptibility and resistance is being studied, and only preliminary information is currently available. Sequencing of the PrP genes shows marked homology between mule deer, white-tailed deer, and elk (Cervenakova et al. 1997; R. Rohwer, pers. comm.) with an amino acid polymorphism recognized at codon 138 (serine or asparagine) of mule deer (Cervenakova et al. 1997; K. O'Rourke, pers. comm.). Sequences of the PrP gene of three captive elk from Germany (Schätzl et al. 1997) identified leucine at codon 132 (analogous to human codon 129); Cervenakova et al. (1997) reported methionine at this location. In a larger sample of captive and free-ranging elk, codon 132 was polymorphic (leucine or methionine) and CWD-positive elk had both the MM and LM genotypes (O'Rourke et al. 1999). It is not yet known whether LL homozygotes are susceptible to CWD; this is a rare genotype in free-ranging elk populations.

Transmissible Mink Encephalopathy

Epidemiology of TME

TME is a rare sporadic disease of ranched mink. It was first recognized in 1947 in Wisconsin and Minnesota; outbreaks occurring in the 1960s in Wisconsin were epidemiologically associated with consumption of feed of bovine or ovine origin (Hartsough and Burger 1965). Additional outbreaks have been identified in Idaho, Canada, Finland, East Germany, and Russia (Hadlow and Karstad 1968; Marsh 1976). The most recent occurrence of TME was in 1985 in Stetsonville, Wisconsin (Marsh et al. 1991). The origin of TME was hypothesized to be from ingestion of scrapie agent. Epidemiologic investigation failed to find evidence of exposure of the mink to sheep products. The Stetsonville incident was specially interesting because the rancher was a dead-stock feeder who used mostly dairy cows collected around his ranch. This led to the suggestion that TME may arise by exposure to an unrecognized case of BSE (Marsh et al. 1991).

Clinical Presentation and Pathology of TME

The clinical signs and features of TME were summarized by Marsh (1976). Only adult mink are affected by TME, and the minimum incubation period is 7 months, with a maximum incubation period of 12 months observed in two outbreaks (Marsh 1976). Mortality may be nearly 100% of adult animals. Clinical signs of TME are characterized by behavioral alterations that include confusion, loss of cleanliness, and aimless circling. Affected animals lose weight, may develop matted fur, experience hind-quarter ataxia, and carry the tail arched over the back. Somnolence and progressive debilitation occur until death after a clinical course of 6–8 weeks.

Lesions of TME in mink are typical of other spongiform encephalopathies although the distribution is somewhat different than in ruminants (Eckroade et al. 1979; Hadlow 1996). Spongiform degeneration involves the cerebral cortex, telencephalon, diencephalon, and mesencephalon with decreasing intensity in the caudal portions of the brain. No plaques are visible. Experimental studies of TME in hamsters demonstrated retrograde and transsynaptic transport and spread of PrP^{Sc} from the sciatic nerve to the lumbar spinal cord and then to the brain (Bartz et al. 2002).

Transmission Studies

Mink inoculated intramuscularly, or fed brain tissue from animals affected with the Wisconsin source of TME, developed TME-like disease in 5 and 7 months, respectively. TME has also been successfully transmitted to other species (ferret, hamster, nonhuman primates, sheep, goat, and cattle).

TME was not reproduced in mink via oral exposure to various strains of sheep scrapie, although mink were susceptible by intracerebral route to certain sheep scrapie sources (Hanson et al. 1971; Marsh and Hanson

1979), showing that different sources of the agent can vary in their pathogenicity for mink. Interestingly, scrapie agent from naturally infected Suffolk sheep that was passaged three times in mink became nonpathogenic for mice (Marsh and Hadlow 1992). Relatively rapid incubation (15–25 months) and development of severe spongiform encephalopathy were found in cattle inoculated intracerebrally with TME agent (Marsh et al. 1991; Robinson et al. 1995), suggesting lack of a significant species barrier. In addition, bovine-passaged TME was highly pathogenic for orally infected mink, with incubation of 7 months (Marsh et al. 1991).

The mink PrP gene was cloned in 1992 (Kretzschmar et al. 1992). At the amino acid level, the greatest interspecific similarities are observed between the mink, sheep (~97%), and bovine (~96%) PrP sequences (Bartz et al. 1994). No information is available about possible mink PrP polymorphisms and their role in TME susceptibility.

Molecular Characterization of Hamster-adapted TME Agent Strains

Comprehensive biological characterization of the TME agent has been conducted by Marsh and colleagues. Through series of experimental infections in hamsters, mink, and primates, multiple strains of TME agent were identified from single outbreaks (Marsh and Hanson 1979; Bessen and Marsh 1992). Two TME strains from the Stetsonville outbreak produced distinctly different clinical syndromes in Syrian hamsters and had distinct brain titers. PrP from hamsters infected with these strains, called "drowsy (DY)" and "hyper (HY)" strains, differed in sedimentation in N-laurosylsarcosine, sensitivity to and differences in protein sequencing after proteinase-K digestion, migration on polyacrylamide gels, immunoreactivity, and targeting in the brain, which suggested that the structure of PrP^{Sc} determines strain variation (Bessen and Marsh 1992, 1994; Bartz et al. 2002). Fractions enriched for PrP^{Sc} prepared from the brains of hamsters infected with the HY or DY strains, when mixed with radiolabeled PrP^{C}, gave patterns of PrP^{C} protection from proteolysis similar to HY and DY PrP^{Sc}, respectively (Bessen et al. 1995).

ACKNOWLEDGMENTS

The authors thank J. Chatelain, J.M. Launay (University of Paris V), J.M. Elsen (INRA, Toulouse), J. Foster, W. Goldmann (IAH/NPU), M. Miller (Colorado Department of Wildlife), T. Thorne (Wyoming Game and Fish Department), and M.B. Foster for their contributions and helpful discussions.

REFERENCES

Andreoletti O., Berthon P., Marc D., Sarradin P., Grosclaude J., van Keulen L., Schelcher F., Elsen J.M., and Lantier F. 2000. Early accumulation of PrP^{Sc} in gut-associated lymphoid and nervous tissues of susceptible sheep from a Romanov flock with natural scrapie. *J. Gen. Virol.* **81:** 3115–3126.

Asante E.A., Linehan J.M., Desbruslais M., Joiner S., Gowland I., Wood A.L., Welch J., Hill A.F., Lloyd S.E., Wadsworth J.D., and Collinge J. 2002. BSE prions propagate as either variant CJD-like or sporadic CJD-like prion strains in transgenic mice expressing human prion protein. *EMBO J.* **21:** 6358–6366.

Baron T.G.M., Madec J.Y., and Calavas D. 1999. Similar signature of the prion protein in natural sheep scrapie and bovine spongiform encephalopathy-linked diseases. *J. Clin. Microbiol.* **37:** 3701–3704.

Bartz J.C., Kincaid A.E., and Bessen R.A. 2002. Retrograde transport of transmissible mink encephalopathy within descending motor tracts. *J. Virol.* **76:** 5759–5768.

———. 2003. Rapid prion neuroinvasion following tongue infection. *J. Virol.* **77:** 583–591.

Bartz J.C., Marsh R.F., McKenzie D.I., and Aiken J.M. 1998. The host range of chronic wasting disease is altered on passage in ferrets. *Virology* **251:** 297–301.

Bartz J.C., McKenzie D.I., Bessen R.A., Marsh R.F., and Aiken J.M. 1994. Transmissible mink encephalopathy species barrier effect between ferret and mink: PrP gene and protein analysis. *J. Gen. Virol.* **75:** 2947–2953.

Belt P.B.G.M., Muileman I.H., Schreuder B.E.C., Ruijter J.B., Gielkens A.L.J., and Smits M.A. 1995. Identification of five allelic variants of the sheep PrP gene and their association with natural scrapie. *J. Gen. Virol.* **76:** 509–517.

Bessen R.A. and Marsh R.F. 1992. Biochemical and physical properties of the prion protein from two strains of the transmissible mink encephalopathy agent. *J. Virol.* **66:** 2096–2101.

———. 1994. Distinct PrP properties suggest the molecular basis of strain variation in transmissible mink encephalopathy. *J. Virol.* **68:** 7859–7868.

Bessen R.A., Kocisko D.A., Raymond G.J., Nandan S., Lansbury P.T., and Caughey B. 1995. Non-genetic propagation of strain-specific properties of scrapie prion protein. *Nature* **375:** 698–700.

Bosque P.J., Ryou C., Telling G., Peretz D., Legname G., DeArmond S.J., and Prusiner S.B. 2002. Prions in skeletal muscle. *Proc. Natl. Acad. Sci.* **99:** 3812–3817.

Bossers A., Schreuder B.E.C., Muileman I.H., Belt P.B.G.M., and Smits M.A. 1996. PrP genotype contributes to determining survival times of sheep with natural scrapie. *J. Gen. Virol.* **77:** 2669–2673.

Bossers A., Belt P.B.G.M., Raymond G.J., Caughey B., de Vries R., and Smits M.A. 1997. Scrapie susceptibility-linked polymorphisms modulate the *in vitro* conversion of sheep prion protein to protease-resistant forms. *Proc. Natl. Acad. Sci.* **94:** 4931–4936.

Bossers A., de Vries R., and Smits M.A. 2000. Susceptibility of sheep for scrapie as assessed by in vitro conversion of nine naturally occurring variants of PrP. *J. Virol.* **74:** 1407–1414.

Brandner S., Isenmann S., Raeber A., Fischer M., Sailer A., Kobayashi Y., Marino S., Weissmann C., and Aguzzi A. 1996. Normal host prion protein necessary for scrapie-induced neurotoxicity. *Nature* **379:** 339–343.

Brotherston J.G., Renwick C.C., Stamp J.T., Zlotnik I., and Pattison I.H. 1968. Spread of scrapie by contact to goats and sheep. *J. Comp. Pathol.* **78:** 9–17.

Brown P. 2000. BSE and transmission through blood. *Lancet* **356:** 955–956.

Brown P. and Gajdusek D.C. 1991. Survival of scrapie virus after 3 years' interment. *Lancet* **337:** 269–270.

Brown P., Rohwer R.G., Dunstan B.C., MacAuley C., Gajdusek D.C., and Drohan W.N. 1998. The distribution of infectivity in blood components and plasma derivatives in experimental models of transmissible spongiform encephalopathy. *Transfusion* **38:** 810–816.

Bruce M.E., Will R.G., Ironside J.W., McConnell I., Drummond D., Suttie A., McCardle L., Chree A., Hope J., Birkett C., Cousens S., Fraser H., and Bostock C.J. 1997. Transmissions to mice indicate that 'new variant' CJD is caused by the BSE agent. *Nature* **389:** 498–501.

Canadian Food Inspection Agency. 2002. *Chronic wasting disease (CWD) of deer and elk.* http://www.inspection.gc.ca/english/anima/heasan/disemala/cwdmdc/cwdmdce.shtm.

Carlson G.A., Kingsbury D.T., Goodman P.A., Coleman S., Marshall S.T., DeArmond S., Westaway D., and Prusiner S.B. 1986. Linkage of prion protein and scrapie incubation time genes. *Cell* **46:** 503–511.

Carlson G.A., Ebeling C., Yang S.-L., Telling G., Torchia M., Groth D., Westaway D., DeArmond S.J., and Prusiner S.B. 1994. Prion isolate specified allotypic interactions between the cellular and scrapie prion proteins in congenic and transgenic mice. *Proc. Natl. Acad. Sci.* **91:** 5690–5694.

Cervenakova L., Rohwer R., Williams S., Brown P., and Gajdusek D.C. 1997. High sequence homology of the PrP gene in mule deer and Rocky Mountain elk. *Lancet* **350:** 219–220.

Chandler R.L. 1961. Encephalopathy in mice produced by inoculation with scrapie brain material. *Lancet* **1:** 1378–1379.

Chatelain J., Cathala F., Brown P., Raharison S., Court L., and Gajdusek D.C. 1981. Epidemiologic comparisons between Creutzfeldt-Jakob disease and scrapie in France during the 12-year period 1968–1979. *J. Neurol. Sci.* **51:** 329–337.

Clark A.M. and Moar J.A. 1992. Scrapie: A clinical assessment. *Vet. Rec.* **130:** 377–378.

Clousard C., Beaudry P., Elsen J.M., Milan D., Dussaucy M., Bounneau C., Schelcher F., Chatelain J., Launay J.-M., and Laplanche J.-L. 1995. Different allelic effects of the codons 136 and 171 of the prion protein gene in sheep with natural scrapie. *J. Gen. Virol.* **76:** 2097–2101.

Collinge J., Sidle K.C.L., Meads J., Ironside J., and Hill A.F. 1996. Molecular analysis of prion strain variation and the aetiology of "new variant" CJD. *Nature* **383:** 685–690.

Cooley W.A., Clark J.K., and Stack M.I. 1998. Comparison of scrapie-associated fibril detection and Western immunoblotting for the diagnosis of natural ovine scrapie. *J. Comp. Pathol.* **118:** 41–49.

Crozet C., Bencsik A., Flamant F., Lezmi S., Samarut J., and Baron T. 2001a. Florid plaques in ovine PrP transgenic mice infected with an experimental ovine BSE. *EMBO Rep.* **2:** 952–956.

Crozet C., Flamant F., Bencsik A., Aubert D., Samarut J., and Baron T. 2001b. Efficient transmission of two different sheet scrapie isolates in transgenic mice expressing the ovine PrP gene. *J. Virol.* **75:** 5328–5334.

Cuillé J. and Chelle P.L. 1936. La maladie dite tremblante du mouton est-elle inoculable? *C.R. Seances Acad. Sci.* **203:** 1552–1554.

Davies D.C. and Kimberlin R.H. 1985. Selection of Swaledale sheep of reduced susceptibility to experimental scrapie. *Vet. Rec.* **116:** 211–214.

Detwiler L.A. 1992. Scrapie. *Rev. Sci. Tech.* **11:** 491–537.

Dickinson A.G. 1976. Scrapie in sheep and goats. In *Slow virus diseases of animals and man* (ed. R.H. Kimberlin), pp. 209–241. North-Holland, Amsterdam.

Dickinson A.G. and Outram G.W. 1988. Genetic aspects of unconventional virus infections: The basis of the virino hypothesis. *Ciba Found. Symp.* **135:** 63–83.

Dickinson A.G., Meikle V.M.H., and Fraser H. 1968. Identification of a gene which controls the incubation period of some strains of scrapie agent in mice. *J. Comp. Pathol.* **78:** 293–299.

Eckroade R.J., Zurhein G.M., and Hanson R.P. 1979. Experimental transmissible mink encephalopathy: Brain lesions and their sequential development. In *Slow transmissible diseases of the nervous system* (ed. S.B. Prusiner and W.J. Hadlow), vol. 1, pp. 409–449. Academic Press, New York.

Elsen J.M., Vu Tien Kang J., Schelcher F., Amigues Y., Eychenne F., Piovey J.P., Ducrocq V., Lantier F., and Laplanche J.-L. 1998. A scrapie epidemic in a closed flock of Romanov. *Proceedings of the 6th World Congress on Genetics Applied to Livestock Production.* **27:** 269–273.

Farquhar C.F., Somerville R.A., and Ritchie L.A. 1989. Post-mortem immunodiagnosis of scrapie and bovine spongiform encephalopathy. *J. Virol. Methods* **24:** 215–222.

Fitzsimmons W.M. and Pattison I.H. 1968. Unsuccessful attempts to transmit scrapie by nematode parasites. *Res. Vet. Sci.* **9:** 281–283.

Foote W.C., Clark W., Maciulis A., Call J.W., Hourrigan J., Evans R.C., Marshall M.R., and de Camp M. 1993. Prevention of scrapie transmission in sheep, using embryo transfer. *Am. J. Vet. Res.* **54:** 1863–1868.

Foster J.D. and Dickinson A.G. 1988. The unusual properties of CH1641, a sheep-passaged isolate of scrapie. *Vet. Rec.* **123:** 5–8.

Foster J.D. and Hunter N. 1991. Partial dominance of the sA allele of the *Sip* gene for controlling experimental scrapie. *Vet. Rec.* **128:** 548–549.

Foster J.D., Hope J., and Fraser H. 1993. Transmission of bovine spongiform encephalopathy to sheep and goats. *Vet. Rec.* **133:** 339–341.

Foster J.D., Wilson M., and Hunter N. 1996a. Immunolocalisation of the prion protein (PrP) in the brains of sheep with scrapie. *Vet. Rec.* **139:** 512–515.

Foster J.D., Parnham D.W., Hunter N., and Bruce M. 2001a. Distribution of the prion protein in sheep terminally affected with BSE following experimental oral transmission. *J. Gen. Virol.* **82:** 2319–2326.

Foster J., Goldmann W., Parnham D., Chong A., and Hunter N. 2001b. Partial dissociation of PrP^{Sc} deposition and vacuolation in the brains of scrapie and BSE experimentally affected goats. *J. Gen. Virol.* **82:** 267–273.

Foster J.D., Parnham D., Chong A., Goldmann W., and Hunter N. 2001c. Clinical signs, histopathology and genetics of experimental transmission of BSE and natural scrapie to sheep and goats. *Vet. Rec.* **148:** 165–171.

Foster J.D., McKelvey W.A.C., Mylne M.J.A., Williams A., Hunter N., Hope J., and Fraser H. 1992. Studies on maternal transmission of scrapie in sheep by embryo transfer. *Vet. Rec.* **130:** 341–343.

Foster J.D., Hunter N., Williams A., Mylne M.J., McKelvey W.A., Hope J., Fraser H., and Bostock C. 1996b. Observations on the transmission of scrapie in experiments using embryo transfer. *Vet. Rec.* **138:** 559–562.

Foster J., McKelvey W., Fraser H., Chong A., Ross A., Parnham D., Goldmann W., and Hunter N. 1999. Experimentally induced bovine spongiform encephalopathy did not transmit via goat embryos. *J. Gen. Virol.* **80:** 517–524.

Fraser H. 1976. The pathology of natural and experimental scrapie. In *Slow virus diseases of animals and man* (ed. R.H. Kimberlin), pp. 267–305. North-Holland, Amsterdam.

Fraser H., McConnell I., Wells G.A.H., and Dawson M. 1988. Transmission of bovine spongiform encephalopathy to mice. *Vet. Rec.* **123:** 472.

Fraser H., Bruce M.E., Chree A., McConnell I., and Wells G.A.H. 1992. Transmission of bovine spongiform encephalopathy and scrapie to mice. *J. Gen. Virol.* **73:** 1891–1897.

Gambetti P. and Parchi P. 1999. Insomnia in prion diseases: Sporadic and familial. *N. Engl. J. Med.* **340:** 1675–1677.

Gibson P.H., Somerville R.A., Fraser H., Foster J.D., and Kimberlin R.H. 1987. Scrapie associated fibrils in the diagnosis of scrapie in sheep. *Vet. Rec.* **120:** 125–127.

Girard J. 1830. Notice sur quelques maladies peu connues des bêtes à laine. *Rec. Med. Vet.* **VII:** 26–39.

Goldmann W., Hunter N., Benson G., Foster J.D., and Hope J. 1991. Different scrapie-associated fibril proteins (PrP) are encoded by lines of sheep selected for different alleles of the *Sip* gene. *J. Gen. Virol.* **72:** 2411–2417.

Goldmann W., Hunter N., Smith G., Foster J., and Hope J. 1994. PrP genotype and agent effects in scrapie: Change in allelic interaction with different isolates of agent in sheep, a natural host of scrapie. *J. Gen. Virol.* **75:** 989–995.

Goldmann W., Martin T., Foster J., Hughes S., Smith G., Hughes K., Dawson M., and Hunter N. 1996. Novel polymorphisms in the caprine PrP gene: A codon 142 mutation associated with scrapie incubation period. *J. Gen. Virol.* **77:** 2885–2891.

Gordon W.S. 1966. Variation in susceptibility of sheep to scrapie and genetic implications (Report of Scrapie Seminar). *Agric. Res. Serv. Rep.* #91-53, pp. 53–67. U.S. Department of Agriculture, Washington, D.C.

Grathwohl K.U., Horiuchi M., Ishiguro N., and Shinagawa M. 1996. Improvement of PrP^{Sc}-detection in mouse spleen early at the preclinical stage of scrapie with collagenase-completed tissue homogenization and Sarkosyl-NaCl extraction of PrP^{Sc}. *Arch. Virol.* **141:** 1863–1874.

———. 1997. Sensitive enzyme-linked immunosorbent assay for detection of PrP^{Sc} in crude tissue extracts from scrapie-affected mice. *J. Virol. Methods* **64:** 205–216.

Groschup M.H., Weiland F., Straub O.C., and Pfaff E. 1996. Detection of scrapie agent in the peripheral nervous system of a diseased sheep. *Neurobiol. Dis.* **3:** 191–195.

Gross J.E. and Miller M.W. 2001. Chronic wasting disease in mule deer: Disease dynamics and control. *J. Wildl. Manag.* **65:** 205–215.

Guiroy D.C., Liberski P.P., Williams E.S., and Gajdusek D.C. 1994. Electron microscopic findings in brain of Rocky Mountain elk with chronic wasting disease. *Folia Neuropathol.* **32:** 171–173.

Guiroy D.C., Williams E.S., Yanagihara R., and Gajdusek D.C. 1991a. Immunolocalization of scrapie amyloid (PrP27-30) in chronic wasting disease of Rocky Mountain elk and hybrids of captive mule deer and white-tailed deer. *Neurosci. Lett.* **126:** 195–198.

———. 1991b. Topographic distribution of scrapie amyloid-immunoreactive plaques in chronic wasting disease in captive mule deer (*Odocoileus hemionus hemionus*). *Acta Neuropathol.* **81:** 475–478.

Guiroy D.C., Williams E.S., Liberski P.P., Wakayama I., and Gajdusek D.C. 1993. Ultrastructural neuropathology of chronic wasting disease in captive mule deer. *Acta Neuropathol.* **85:** 437–444.

Hadlow W.J. 1996. Differing neurohistologic images of scrapie, transmissible mink encephalopathy, and chronic wasting disease of mule deer and elk. In *Bovine spongiform encephalopathy: The BSE dilemma* (ed. C.J. Gibbs, Jr.), pp. 122–137. Springer Verlag, New York.

Hadlow W.J. and Karstad L. 1968. Transmissible encephalopathy of mink in Ontario. *Can. Vet. J.* **9:** 193–196.

Hadlow W.J., Kennedy R.C., and Race R.E. 1982. Natural infection of Suffolk sheep with scrapie virus. *J. Infect. Dis.* **146:** 657–664.

Hadlow W.J., Kennedy R.C., Race R.E., and Eklund C.M. 1980. Virologic and neurohisto-logic findings in dairy goats affected with natural scrapie. *Vet. Pathol.* **17:** 187–199.

Hadlow W.J., Race R.E., Kennedy R.C., and Eklund C.M. 1979. Natural infection of sheep with scrapie virus. In *Slow transmissible diseases of the nervous system* (ed. S.B. Prusiner and W.J. Hadlow), vol. 2, pp. 3–12. Academic Press, New York.

Hamir A.N., Cutlip R.C., Miller J.M., Williams E.S., Stack M.J., Miller M.W., O'Rourke K.I., and Chaplin M.J. 2001. Preliminary findings on the experimental transmission of chronic wasting disease agent of mule deer to cattle. *J. Vet. Diagn. Invest.* **13:** 91–96.

Hanson R.P., Eckroade R.J., Marsh R.F., ZuRhein G.M., Kanitz C.L., and Gustafson D.P. 1971. Susceptibility of mink to sheep scrapie. *Science* **172:** 859–861.

Hartsough G.R. and Burger D. 1965. Encephalopathy of mink. I. Epizootiologic and clini-cal observations. *J. Infect. Dis.* **115:** 387–392.

Heggebo R., Press C.M., Gunnes G., Lie K.I., Tranulis M.A., Ulvund M., Groschup M.H., and Landsverk T. 2000. Distribution of prion protein in the ileal Peyer's patch of scrapie-free lambs and lambs naturally and experimentally exposed to the scrapie agent. *J. Gen. Virol.* **81:** 2327–2337.

Hill A.F., Desbruslais M., Joiner S., Sidle K.C.L., Gowland I., Collinge J., Doey L.J., and Lantos P. 1997. The same prion strain causes vCJD and BSE. *Nature* **389:** 448–450.

Hourrigan J.L. 1990. Experimentally induced bovine spongiform encephalopathy in cattle in Mission, Tex, and the control of scrapie. *J. Am. Vet. Med. Assoc.* **196:** 1678–1679.

Hourrigan J., Klingsporn A., Clark W.W., and de Camp M. 1979. Epidemiology of scrapie in the United States. In *Slow transmissible diseases of the nervous system* (ed. S.B. Prusiner and W.J. Hadlow), vol. 1, pp. 331–356. Academic Press, New York.

Houston E.F. and Gravenor M.B. 2003. Clinical signs in sheep experimentally infected with scrapie and BSE. *Vet. Rec.* **152:** 333–334.

Houston F., Goldmann W., Chong A., Jeffrey M., Gonzalez L., Foster J., Parnham D., and Hunter N. 2003. Prion diseases: BSE in sheep bred for resistance to infection. *Nature* **423:** 498.

Hunter N., Goldmann W., Smith G., and Hope J. 1994. The association of a codon 136 PrP gene variant with the occurrence of natural scrapie. *Arch. Virol.* **137:** 171–177.

Hunter N., Goldmann W., Benson G., Foster J.D., and Hope J. 1993. Swaledale sheep affect-ed by natural scrapie differ significantly in PrP genotype frequencies from healthy sheep and those selected for reduced incidence of scrapie. *J. Gen. Virol.* **74:** 1025–1031.

Hunter N., Goldmann W., Foster J.D., Cairns D., and Smith G. 1997a. Natural scrapie and PrP genotype: Case-control studies in British sheep. *Vet. Rec.* **141:** 137–140.

Hunter N., Cairns D., Foster J.D., Smith G., Goldmann W., and Donnelly K. 1997b. Is scrapie solely a genetic disease? *Nature* **386:** 137.

Hunter N., Foster J.D., Goldmann W., Stear M.J., Hope J., and Bostock C. 1996. Natural scrapie in a closed flock of Cheviot sheep occurs only in specific PrP genotypes. *Arch. Virol.* **141:** 809–824.

Hunter N., Foster J., Chong A., McCutcheon S., Parnham D., Eaton S., MacKenzie C., and Houston F. 2002. Transmission of prion disease by blood transfusion. *J. Gen. Virol.* **83:** 2897–2905.

Ikeda T., Horiuchi M., Ishiguro N., Muramatsu Y., Kai-Uwe G.D., and Shinagawa M. 1995. Amino acid polymorphisms of PrP with reference to onset of scrapie in Suffolk and Corriedale sheep in Japan. *J. Gen. Virol.* **76:** 2577–2581.

Ikegami Y., Ito M., Isomura H., Momotani E., Sasaki K., Muramatsu Y., Ishiguro N., and Shinagawa M. 1991. Pre-clinical and clinical diagnosis of scrapie by detection of PrP protein in tissues of sheep. *Vet. Rec.* **128:** 271–275.

Jeffrey M., Ryder S., Martin S., Hawkins S.A., Terry L., Berthelin-Baker C., and Bellworthy S.J. 2001. Oral inoculation of sheep with the agent of bovine spongiform encephalopathy (BSE). 1. Onset and distribution of disease-specific PrP accumulation in brain and viscera. *J. Comp. Pathol.* **124:** 280–289.

Joly D.O., Ribic C.A., Langenberg J.A., Beheler K., Batha C.A., Dhuey B.J., Rolley R.E., Bartelt G., Van Deelen T.R., and Samuel M.D. 2003. Chronic wasting disease in free-ranging Wisconsin white-tailed deer. *Emerg. Infect. Dis.* **9:** 599–601.

Kaneko K., Zulianello L., Scott M., Cooper C.M., Wallace A.C., James T.L., Cohen F.E., and Prusiner S.B. 1997. Evidence for protein X binding to a discontinuous epitope on the cellular prion protein during scrapie prion propagation. *Proc. Natl. Acad. Sci.* **94:** 10069–10074.

Kitamoto T., Muramoto T., Mohri S., Doh-Ura K., and Tateishi J. 1991. Abnormal isoform of prion protein accumulates in follicular dendritic cells in mice with Creutzfeldt-Jakob disease. *J. Virol.* **65:** 6292–6295.

Korth C., Kaneko K., Groth D., Heye N., Telling G., Mastrianni J., Parchi P., Gambetti P., Will R., Ironside J., Heinrich C., Tremblay P., DeArmond S.J., and Prusiner S.B. 2003. Abbreviated incubation times for human prions in mice expressing a chimeric mouse-human prion protein transgene. *Proc. Natl. Acad. Sci.* **100:** 4784–4789.

Kretzschmar H.A., Neumann M., Riethmüller G., and Prusiner S.B. 1992. Molecular cloning of a mink prion protein gene. *J. Gen. Virol.* **73:** 2757–2761.

Laplanche J.-L., Chatelain J., Beaudry P., Dussaucy M., Bounneau C., and Launay J.-M. 1993a. French autochthonous scrapied sheep without the 136Val PrP polymorphism. *Mamm. Genome* **4:** 463–464.

Laplanche J.-L., Chatelain J., Thomas S., Dussaucy M., Brugere-Picoux J., and Launay J.M. 1992. PrP gene allelic variants and natural scrapie in French Ile-de-France and Romanov sheep. In *Prion diseases of humans and animals* (ed. S.B. Prusiner et al.), pp. 329–337. Ellis Horwood, New York.

Laplanche J.-L., Chatelain J., Westaway D., Thomas S., Dussaucy M., Brugere Picoux J., and Launay J.-M. 1993b. PrP polymorphisms associated with natural scrapie discovered by denaturing gradient gel electrophoresis. *Genomics* **15:** 30–37.

Laplanche J.-L., Elsen J.M., Eychenne F., Schelcher F., Richard S., Amigues Y., and Launay J.M. 1996. Scrapie outbreak in sheep after oral exposure to infective gastrointestinal nematode larvae. In *Transmissible subacute spongiform encephalopathies. Prion diseases* (ed. L. Court and B. Dodet), pp. 42–46. Elsevier, Paris.

Lasmézas C.I., Deslys J.-P., Demaimay R., Adjou K.T., Lamoury F., Dormont D., Robain O., Ironside J., and Hauw J.-J. 1996. BSE transmission to macaques. *Nature* **381:** 743–744.

Liberski P.P., Guiroy D.C., Williams E.S., Walis A., and Budka H. 2001. Deposition patterns of disease-associated prion protein in captive mule deer brains with chronic wasting disease. *Acta Neuropathol.* **102:** 496–500.

MacDiarmid S.C. 1996. Scrapie: The risk of its introduction and effects on trade. *Aust. Vet. J.* **73:** 161–164.

Maciulis A., Hunter N., Wang S., Goldmann W., Hope J., and Foote W.C. 1992. Polymorphisms of a scrapie-associated fibril protein (PrP) gene and their association with susceptibility to experimentally induced scrapie in Cheviot sheep in the United States. *Am. J. Vet. Res.* **53:** 1957–1960.

Marsh R.F. 1976. The subacute spongiform encephalopathies. *Front. Biol.* **44:** 359–380.

Marsh R.F. and Hadlow W.J. 1992. Transmissible mink encephalopathy. *Rev. Sci. Tech.* **11:** 539–550.

Marsh R.F. and Hanson R.P. 1979. On the origin of transmissible mink encephalopathy. In

590   Prusiner et al.

Slow transmissible diseases of the nervous system (ed. S.B. Prusiner and W.J. Hadlow), vol. 1, pp. 451–460. Academic Press, New York.

Marsh R.F., Bessen R.A., Lehmann S., and Hartsough G.R. 1991. Epidemiological and experimental studies on a new incident of transmissible mink encephalopathy. *J. Gen. Virol.* **72:** 589–594.

Mastrianni J.A., Nixon R., Layzer R., Telling G.C., Han D., DeArmond S.J., and Prusiner S.B. 1999. Prion protein conformation in a patient with sporadic fatal insomnia. *N. Engl. J. Med.* **340:** 1630–1638.

McBride P.A., Bruce M.E., and Fraser H. 1988. Immunostaining of scrapie cerebral amyloid plaques with antisera raised to scrapie associated fibrils (SAF). *Neuropathol. Appl. Neurobiol.* **14:** 325–336.

McBride P.A., Eikelenboom P., Kraal G., Fraser H., and Bruce M.E. 1992. PrP protein is associated with follicular dendritic cells of spleens and lymph nodes in uninfected and scrapie-infected mice. *J. Pathol.* **168:** 413–418.

Medori R., Tritschler H.-J., LeBlanc A., Villare F., Manetto V., Chen H.Y., Xue R., Leal S., Montagna P., Cortelli P., Tinuper P., Avoni P., Mochi M., Baruzzi A., Hauw J.J., Ott J., Lugaresi E., Autilio-Gambetti L., and Gambetti P. 1992. Fatal familial insomnia, a prion disease with a mutation at codon 178 of the prion protein gene. *N. Engl. J. Med.* **326:** 444–449.

Merz P.A., Somerville R.A., Wisniewski H.M., and Iqbal K. 1981. Abnormal fibrils from scrapie-infected brain. *Acta Neuropathol.* **54:** 63–74.

Merz P.A., Carp R.I., Kascsak R., Somerville R.A., and Wisniewski H.M. 1983a. Scrapie-associated fibrils in spleens of preclinical and clinical scrapie-infected mice. *Am. Soc. Virol. Abstr.*, no. 246.

Merz P.A., Wisniewski H.M., Somerville R.A., Bobin S.A., Masters C.L., and Iqbal K. 1983b. Ultrastructural morphology of amyloid fibrils from neuritic and amyloid plaques. *Acta Neuropathol.* **60:** 113–124.

Miller J.M., Jenny A.L., Taylor W.D., Marsh R.F., Rubenstein R., and Race R.E. 1993. Immunohistochemical detection of prion protein in sheep with scrapie. *J. Vet. Diagn. Invest.* **5:** 309–316.

Miller M.W. and Williams E.S. 2002. Detection of PrP(CWD) in mule deer by immunohistochemistry of lymphoid tissues. *Vet. Rec.* **151:** 610–612.

Miller M.W., Wild M.A., and Williams E.S. 1998. Epidemiology of chronic wasting disease in captive Rocky Mountain elk. *J. Wildl. Dis.* **34:** 532–538.

Miller M.W., Williams E.S., McCarty C.W., Spraker T.R., Kreeger T.J., Larsen C.T., and Thorne E.T. 2000. Epizootiology of chronic wasting disease in free-ranging cervids in Colorado and Wyoming. *J. Wildl. Dis.* **36:** 676–690.

Mohri S., Farquhar C.F., Somerville R.A., Jeffrey M., Foster J., and Hope J. 1992. Immunodetection of a disease-specific PrP fraction in scrapie-affected sheep and BSE-affected cattle. *Vet. Rec.* **131:** 537–539.

Muramatsu Y., Onodera A., Horiuchi M., Ishiguro N., and Shinagawa M. 1994. Detection of PrP^{Sc} in sheep at the preclinical stage of scrapie and its significance for diagnosis of insidious infection. *Arch. Virol.* **134:** 427–432.

Muramatsu Y., Tanaka K., Horiuchi M., Ishiguro N., Shinagawa M., Matsui T., and Onodera T. 1992. A specific RFLP type associated with the occurrence of sheep scrapie in Japan. *Arch. Virol.* **127:** 1–9.

Obermaier G., Kretzschmar H.A., Hafner A., Heubeck D., and Dahme E. 1995. Spongiform central nervous system myelinopathy in African dwarf goats. *J. Comp. Pathol.* **113:** 357–372.

Onodera T. and Hayashi T. 1994. Diversity of clinical signs in natural scrapie cases occurring in Japan. *Jpn. Agric. Res. Q.* **28:** 59–61.

Onodera T., Ikeda T., Muramatsu Y., and Shinagawa M. 1993. Isolation of scrapie agent from the placenta of sheep with natural scrapie in Japan. *Microbiol. Immunol.* **37:** 311–316.

O'Rourke K.I., Melco R.P., and Mickelson J.R. 1996. Allelic frequencies of an ovine scrapie susceptibility gene. *Anim. Biotechnol.* **7:** 155–162.

O'Rourke K.I., Baszler T.V., Miller J.M., Spraker T.R., Sadler-Riggleman I., and Knowles D.P. 1998. Monoclonal antibody F89/160.1.5 defines a conserved epitope on the ruminant prion protein. *J. Clin. Microbiol.* **36:** 1750–1755.

O'Rourke K.I., Holyoak G.R., Clark W.W., Mickelson J.R., Wang S., Melco R.P., Besser T.E., and Foote W.C. 1997. PrP genotypes and experimental scrapie in orally inoculated Suffolk sheep in the United States. *J. Gen. Virol.* **78:** 975–978.

O'Rourke K.I., Besser T.E., Miller M.W., Cline T.F., Spraker T.R., Jenny A.L., Wild M.A., Zebarth G.L., and Williams E.S. 1999. PrP genotypes of captive and free-ranging Rocky Mountain elk (*Cervus elaphus nelsoni*) with chronic wasting disease. *J. Gen. Virol.* **80:** 2765–2769.

Palsson P.A. 1979. Rida (scrapie) in Iceland and its epidemiology. In *Slow transmissible diseases of the nervous system* (ed. S.B. Prusiner and W.J. Hadlow), vol. 1, pp. 357–366. Academic Press, New York.

Parchi P., Capellari S., Chin S., Schwarz H.B., Schecter N.P., Butts J.D., Hudkins P., Burns D.K., Powers J.M., and Gambetti P. 1999. A subtype of sporadic prion disease mimicking fatal familial insomnia. *Neurology* **52:** 1757–1763.

Parry H.B. 1960. Scrapie: A transmissible hereditary disease of sheep. *Nature* **185:** 441–443.

———. 1983. *Scrapie disease in sheep.* Academic Press, New York.

Pattison I.H. 1966. The relative susceptibility of sheep, goats and mice to two types of the goat scrapie agent. *Res. Vet. Sci.* **7:** 207–212.

Pattison I.H., Hoare M.N., Jebbett J.N., and Watson W.A. 1972. Spread of scrapie to sheep and goats by oral dosing with foetal membranes from scrapie-affected sheep. *Vet. Rec.* **90:** 465–468.

———. 1974. Further observations on the production of scrapie in sheep by oral dosing with foetal membranes from scrapie-infected sheep. *Br. Vet. J.* **130:** lxv–lxvii.

Peretz D., Scott M., Groth D., Williamson A., Burton D., Cohen F.E., and Prusiner S.B. 2001. Strain specified relative conformational stability of the scrapie prion protein. *Protein Sci.* **10:** 854–863.

Peretz D., Williamson R.A., Legname G., Matsunaga Y., Vergara J., Burton D., DeArmond S.J., Prusiner S.B., and Scott M.R. 2002. A change in the conformation of prions accompanies the emergence of a new prion strain. *Neuron* **34:** 921–932.

Perrier V., Kaneko K., Safar J., Vergara J., Tremblay P., DeArmond S.J., Cohen F.E., Prusiner S.B., and Wallace A.C. 2002. Dominant-negative inhibition of prion replication in transgenic mice. *Proc. Natl. Acad. Sci.* **99:** 13079–13084.

Prusiner S.B. and Safar J.G. 2000. Method of concentrating prion proteins in blood samples. U.S. Patent No. 6, 166, 187; 16 pp.

Prusiner S.B., McKinley M.P., Bowman K.A., Bolton D.C., Bendheim P.E., Groth D.F., and Glenner G.G. 1983. Scrapie prions aggregate to form amyloid-like birefringent rods. *Cell* **35:** 349–358.

Race R.E. and Ernst D. 1992. Detection of proteinase K-resistant prion protein and infectivity in mouse spleen by 2 weeks after scrapie agent inoculation. *J. Gen. Virol.* **73:** 3319–3323.

Race R., Ernst D., Jenny A., Taylor W., Sutton D., and Caughey B. 1992. Diagnostic implications of detection of proteinase K-resistant protein in spleen, lymph nodes, and brain of sheep. *Am. J. Vet. Res.* **53:** 883–889.

Race R.E., Raines A., Baron T.G., Miller M.W., Jenny A., and Williams E.S. 2002. Comparison of abnormal prion protein glycoform patterns from transmissible spongiform encephalopathy agent-infected deer, elk, sheep, and cattle. *J. Virol.* **76:** 12365–12368.

Raymond G.J., Bossers A., Raymond L.D., O'Rourke K.I., McHolland L.E., Bryant P.K., III, Miller M.W., Williams E.S., Smits M., and Caughey B. 2000. Evidence of a molecular barrier limiting susceptibility of humans, cattle and sheep to chronic wasting disease. *EMBO J.* **19:** 4425–4430.

Robinson M.M., Hadlow W.J., Knowles D.P., Huff T.P., Lacy P.A., Marsh R.F., and Gorham J.R. 1995. Experimental infection of cattle with the agents of transmissible mink encephalopathy and scrapie. *J. Comp. Pathol.* **113:** 241–251.

Rubenstein R., Merz P.A., Kascsak R.J., Carp R.L., Scalici C.L., Fama C.L., and Wisniewski H.M. 1987. Detection of scrapie-associated fibrils (SAF) and SAF proteins from scrapie-affected sheep. *J. Infect. Dis.* **156:** 36–42.

Ryder S.J., Spencer Y.I., Bellerby P.J., and March S.A. 2001. Immunohistochemical detection of PrP in the medulla oblongata of sheep: The spectrum of staining in normal and scrapie-affected sheep. *Vet. Rec.* **148:** 7–13.

Safar J., Wille H., Itri V., Groth D., Serban H., Torchia M., Cohen F.E., and Prusiner S.B. 1998. Eight prion strains have PrP[Sc] molecules with different conformations. *Nat. Med.* **4:** 1157–1165.

Safar J.G., Scott M., Monaghan J., Deering C., Didorenko S., Vergara J., Ball H., Legname G., Leclerc E., Solforosi L., Serban H., Groth D., Burton D.R., Prusiner S.B., and Williamson R.A. 2002. Measuring prions causing bovine spongiform encephalopathy or chronic wasting disease by immunoassays and transgenic mice. *Nat. Biotechnol.* **20:** 1147–1150.

Schätzl H.M., Wopfner F., Gilch S., von Brunn A., and Jager G. 1997. Is codon 129 of prion protein polymorphic in human beings but not in animals? *Lancet* **349:** 1603–1604.

Schreuder B.E.C., de Jong M.C.M., Vellema P.P., Bröker A.J.M., and Betcke H. 1993. Prevalence and incidence of scrapie in the Netherlands: A questionnaire survey. *Vet. Rec.* **133:** 211–214.

Schreuder B.E., van Keulen L.J., Vromans M.E., Langeveld J.P., and Smits M.A. 1996. Preclinical test for prion diseases. *Nature* **381:** 563.

Scott M.R., Will R., Ironside J., Nguyen H.-O.B., Tremblay P., DeArmond S.J., and Prusiner S.B. 1999. Compelling transgenetic evidence for transmission of bovine spongiform encephalopathy prions to humans. *Proc. Natl. Acad. Sci.* **96:** 15137–15142.

Sigurdson C.J., Spraker T.R., Miller M.W., Oesch B., and Hoover E.A. 2001. PrP(CWD) in the myenteric plexus, vagosympathetic trunk and endocrine glands of deer with chronic wasting disease. *J. Gen. Virol.* **82:** 2327–2334.

Sigurdson C.J., Williams E.S., Miller M.W., Spraker T.R., O'Rourke K.I., and Hoover E.A. 1999. Oral transmission and early lymphoid tropism of chronic wasting disease PrP[res] in mule deer fawns (*Odocoileus hemionus*). *J. Gen. Virol.* **80:** 2757–2764.

Skarphedinsson S., Johannsdottir R., Gudmundsson P., Sigurdarson S., and Georgsson G. 1994. PrP[Sc] in Icelandic sheep naturally infected with scrapie. *Ann. N.Y. Acad. Sci.* **724:** 304–309.

Sohn H.J., Kim J.H., Choi K.S., Nah J.J., Joo Y.S., Jean Y.H., Ahn S.W., Kim O.K., Kim D.Y., and Balachandran A. 2002. A case of chronic wasting disease in an elk imported to Korea from Canada. *J. Vet. Med. Sci.* **64:** 855–858.

Somerville R.A., Chong A., Mulqueen O.U., Birkett C.R., Wood S.C.E.R., and Hope J. 1997a. Biochemical typing of scrapie strains. *Nature* **386:** 564.

Somerville R.A., Birkett C.R., Farquhar C.F., Hunter N., Goldmann W., Dornan J., Grover D., Hennion R.M., Percy C., Foster J., and Jeffrey M. 1997b. Immunodetection of PrP^{Sc} in spleens of some scrapie-infected sheep but not BSE-infected cows. *J. Gen. Virol.* **78:** 2389–2396.

Spraker T.R., Zink R.R., Cummings B.A., Sigurdson C.J., Miller M.W., and O'Rourke K.I. 2002a. Distribution of protease-resistant prion protein and spongiform encephalopathy in free-ranging mule deer (*Odocoileus hemionus*) with chronic wasting disease. *Vet. Pathol.* **39:** 546–556.

Spraker T.R., Zink R.R., Cummings B.A., Wild M.A., Miller M.W., and O'Rourke K.I. 2002b. Comparison of histological lesions and immunohistochemical staining of proteinase-resistant prion protein in a naturally occurring spongiform encephalopathy of free-ranging mule deer (*Odocoileus hemionus*) with those of chronic wasting disease of captive mule deer. *Vet. Pathol.* **39:** 110–119.

Spraker T.R., O'Rourke K.I., Balachandran A., Zink R.R., Cummings B.A., Miller M.W., and Powers B.E. 2002c. Validation of monoclonal antibody F99/97.6.1 for immunohistochemical staining of brain and tonsil in mule deer (*Odocoileus hemionus*) with chronic wasting disease. *J. Vet. Diagn. Invest.* **14:** 3–7.

Spraker T.R., Miller M.W., Williams E.S., Getzy D.M., Adrian W.J., Schoonveld G.G., Spowart R.A., O'Rourke K.I., Miller J.M., and Merz P.A. 1997. Spongiform encephalopathy in free-ranging mule deer (*Odocoileus hemionus*), white-tailed deer (*Odocoileus virginianus*), and Rocky Mountain elk (*Cervus elaphus nelsoni*) in north-central Colorado. *J. Wildl. Dis.* **33:** 1–6.

Stack M.J., Scott A.C., Done S.H., and Dawson M. 1991. Natural scrapie: Detection of fibrils in extracts from the central nervous system of sheep. *Vet. Rec.* **128:** 539–540.

Stamp J.T., Brotherston J.G., Zlotnik I., Mackay J.M., and Smith W. 1959. Further studies on scrapie. *J. Comp. Pathol.* **69:** 268–280.

Stringfellow D.A. and Seidel S.M. 1990. *Manual of the International Embryo Transfer Society.* International Embryo Transfer Society, Champaign, Illinois.

Takahashi K., Shinagawa M., Doi S., Sasaki S., Goto H., and Sato G. 1986. Purification of scrapie agent from infected animal brains and raising of antibodies to the purified fraction. *Microbiol. Immunol.* **30:** 123–131.

Taraboulos A., Jendroska K., Serban D., Yang S.-L., DeArmond S.J., and Prusiner S.B. 1992. Regional mapping of prion proteins in brains. *Proc. Natl. Acad. Sci.* **89:** 7620–7624.

Taylor D.M., McConnell I., and Fraser H. 1996. Scrapie infection can be established readily through skin scarification in immunocompetent but not immunodeficient mice. *J. Gen. Virol.* **77:** 1595–1599.

Telling G.C., Scott M., Mastrianni J., Gabizon R., Torchia M., Cohen F.E., DeArmond S.J., and Prusiner S.B. 1995. Prion propagation in mice expressing human and chimeric transgenes implicates the interaction of cellular PrP with another protein. *Cell* **83:** 79–90.

Telling G.C., Parchi P., DeArmond S.J., Cortelli P., Montagna P., Gabizon R., Mastrianni J., Lugaresi E., Gambetti P., and Prusiner S.B. 1996. Evidence for the conformation of the pathologic isoform of the prion protein enciphering and propagating prion diversity. *Science* **274:** 2079–2082.

van Keulen L.J., Schreuder B.E., Meloen R.H., Mooij-Harkes G., Vromans M.E., and Langeveld J.P. 1996. Immunohistochemical detection of prion protein in lymphoid tissues of sheep with natural scrapie. *J. Clin. Microbiol.* **34:** 1228–1231.

Vilotte J.L., Soulier S., Essalmani R., Stinnakre M.G., Vaiman D., Lepourry L., Da Silva J.C., Besnard N., Dawson M., Buschmann A., Groschup M., Petit S., Madelaine M.F., Rakatobe S., Le Dur A., Vilette D., and Laude H. 2001. Markedly increased susceptibility to natural sheep scrapie of transgenic mice expressing ovine PrP. *J. Virol.* **75:** 5977–5984.

Wells G.A.H. 2002. European Commission report on TSE infectivity distribution in ruminant tissues (*State of Knowledge*, December 2001), pp. 10–37.

Westaway D., Goodman P.A., Mirenda C.A., McKinley M.P., Carlson G.A., and Prusiner S.B. 1987. Distinct prion proteins in short and long scrapie incubation period mice. *Cell* **51:** 651–662.

Westaway D., Zuliani V., Cooper C.M., Da Costa M., Neuman S., Jenny A.L., Detwiler L., and Prusiner S.B. 1994. Homozygosity for prion protein alleles encoding glutamine-171 renders sheep susceptible to natural scrapie. *Genes Dev.* **8:** 959–969.

Westaway D., Mirenda C.A., Foster D., Zebarjadian Y., Scott M., Torchia M., Yang S.-L., Serban H., DeArmond S.J., Ebeling C., Prusiner S.B., and Carlson G.A. 1991. Paradoxical shortening of scrapie incubation times by expression of prion protein transgenes derived from long incubation period mice. *Neuron* **7:** 59–68.

Wild M.A., Spraker T.R., Sigurdson C.J., O'Rourke K.I., and Miller M.W. 2002. Preclinical diagnosis of chronic wasting disease in captive mule deer (*Odocoileus hemionus*) and white-tailed deer (*Odocoileus virginianus*) using tonsillar biopsy. *J. Gen. Virol.* **83:** 2629–2634.

Wilesmith J.W., Ryan J.B., Stevenson M.A., Morris R.S., Pfeiffer D.U., Lin D., Jackson R., and Sanson R.L. 2000. Temporal aspects of the epidemic of bovine spongiform encephalopathy in Great Britain: Holding-associated risk factors for the disease. *Vet. Rec.* **147:** 319–325.

Will R.G., Ironside J.W., Zeidler M., Cousens S.N., Estibeiro K., Alperovitch A., Poser S., Pocchiari M., Hofman A., and Smith P.G. 1996. A new variant of Creutzfeldt-Jakob disease in the UK. *Lancet* **347:** 921–925.

Williams E.S. and Miller M.W. 2002. Chronic wasting disease in deer and elk in North America. *Rev. Sci. Tech.* **21:** 305–316.

Williams E.S. and Young S. 1980. Chronic wasting disease of captive mule deer: A spongiform encephalopathy. *J. Wildl. Dis.* **16:** 89–98.

———. 1982. Spongiform encephalopathy of Rocky Mountain elk. *J. Wildl. Dis.* **18:** 465–471.

———. 1992. Spongiform encephalopathies in *Cervidae*. *Rev. Sci. Tech. Off. Int. Epiz.* **11:** 551–567.

———. 1993. Neuropathology of chronic wasting disease of mule deer (*Odocoileus hemionus*) and elk (*Cervus elaphus nelsoni*). *Vet. Pathol.* **30:** 36–45.

Wisconsin Department of Natural Resources. 2003. Environmental impact statement on rules to eradicate chronic wasting disease from Wisconsin's free-ranging white-tailed deer herd, 176 pp.

Wisniewski H.M., Sigurdarson S., Rubenstein R., Kascsak R.J., and Carp R.I. 1996. Mites as vectors for scrapie. *Lancet* **347:** 1114.

Wolfe L.L., Conner M.M., Baker T.H., Dreitz V.J., Burnham K.P., Williams E.S., Hobbs N.T., and Miller M.W. 2002. Evaluation of antemortem sampling to estimate chronic wasting disease prevalence in free-ranging mule deer. *J. Wildl. Manag.* **66:** 564–573.

Wood J.L., Lund L.J., and Done S.H. 1992. The natural occurrence of scrapie in moufflon. *Vet. Rec.* **130:** 25–27.

Wood J.L., McGill I.S., Done S.H., and Bradley R. 1997. Neuropathology of scrapie: a study of the distribution patterns of brain lesions in 222 cases of natural scrapie in sheep, 1982–1991. *Vet. Rec.* **140:** 167–174.

World Health Organization. 1992. Public health issues related to animal and human spongiform encephalopathies: Memorandum from a WHO meeting. *Bull. W.H.O.* **70:** 183–190.

Zlotnik I. and Stamp J.L. 1961. Scrapie disease of sheep. *World Neurol.* **2:** 895–907.

Zulianello L., Kaneko K., Scott M., Erpel S., Han D., Cohen F.E., and Prusiner S.B. 2000. Dominant-negative inhibition of prion formation diminished by deletion mutagenesis of the prion protein. *J. Virol.* **74:** 4351–4360.

12

Bovine Spongiform Encephalopathy and Related Diseases

Gerald A. H. Wells

Veterinary Laboratories Agency
Department of the Environment, Fisheries and Rural Affairs
Surrey, United Kingdom

John W. Wilesmith

Epidemiology Department, Veterinary Laboratories Agency
Department of the Environment, Fisheries and Rural Affairs
Surrey, United Kingdom

IN NOVEMBER 1986, AT THE VETERINARY LABORATORIES AGENCY, Weybridge, United Kingdom (formerly the Central Veterinary Laboratory), the routine histopathological examination of the brains of three cows from dairy farms in southern England was the critical juncture in the recognition of a new neurologic disease of cattle, subsequently termed bovine spongiform encephalopathy (BSE). The neurohistological changes observed bore a striking resemblance to those of scrapie of sheep, the historic archetype of the transmissible spongiform encephalopathies (TSEs) (Wells et al. 1987). Important in the confirmation of this diagnostic discovery were not only the characteristic pathological changes in the brain, but also the individual case histories, which included details of the clinical signs of the disorder provided by veterinarians (Wells et al. 1987, 1992; Wells 1989). Clinically, there was the insidious onset of altered behavior (apprehension and sometimes, aggressive responses), ataxia (uncoordinated gait with falling), and dysesthesia or reflex hyperesthesia (abnormal responses to touch and sound). In retrospect, clinical cases of BSE were seen in England in 1985, but probably not before that year (Bradley 2001). The relentlessly progressive and prolonged course (1–6 months) of the signs made affected animals impossible to handle, necessitating their slaughter (Wilesmith et al. 1988). Through 1987, further cases were diag-

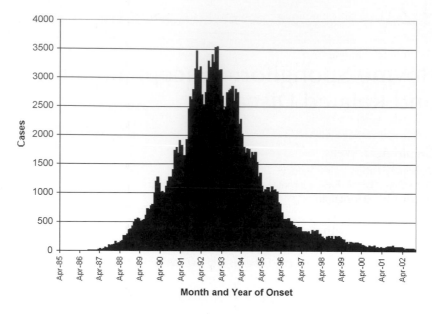

Figure 1. Distribution of confirmed cases of BSE by month of clinical onset, UK, April, 1985, to October 2002.

nosed in Great Britain and the number of cases continued to increase each month, indicating the onset of a major epidemic (Fig. 1). The new disease, BSE (Wells et al. 1987), was made statutorily notifiable in June, 1988 (Anonymous 1996a; Wilesmith 1996a).

The many aspects of BSE, including occurrence, epidemiology, clinical signs, pathology, diagnosis, transmissibility, prevention, and control, have been reviewed on a number of occasions previously (Kimberlin 1992; Dawson et al. 1994; Fraser and Foster 1994; Collee and Bradley 1997a,b; Wilesmith 1998; Heim and Kihm 1999; Wells et al. 2000; Bradley 2001). In this chapter, current information on the epidemiology, pathology, transmissibility, and pathogenesis of BSE is discussed.

EPIDEMIOLOGY OF BSE

The United Kingdom

Establishing the Source of Infection

The advent of a new prion disease in epidemic form led to a detailed investigation of its possible cause. An important clue was offered by map-

Table 1. Risk factors among 169 early cases of BSE, UK, 1986

Risk factor	Number of cases
MBM dietary supplement	169
Dictocaulus viviparus vaccine	82
Leptospira hardjo vaccine	8
Salmonella dublin vaccine	5
Clostridia species vaccine	18
Viral respiratory vaccines	5
Bacterial antisera	9
Hormones	49
Pyrethroid insecticides	6
Organophosphorus insecticides	121

Data from Wilesmith et al. (1991).

ping of the cases occurring within the first 18 months of the outbreak (Wilesmith 1991), which showed that they were widely distributed throughout much of Great Britain and began almost simultaneously in all regions. This is the classic pattern of a common source rather than a propagated epidemic and led to the search for a common source (Wilesmith et al. 1988). A series of some 200 case studies explored numerous possible sources (Table 1) and suggested that one common exposure was the use of a dietary protein supplement, meat and bone meal (MBM), that was regularly fed to dairy cattle from the first weeks of life. A comparison of dairy with beef suckler herds (cow–calf operations) (Fig. 2) showed a strikingly higher incidence of disease in dairy herds, consistent with the possible role of MBM, which comprises a much larger proportion of the

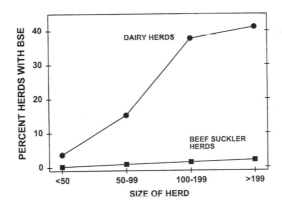

Figure 2. Frequency of BSE among cattle herds in the UK, by herd size, for dairy and beef cattle. Data taken from Wilesmith (1991).

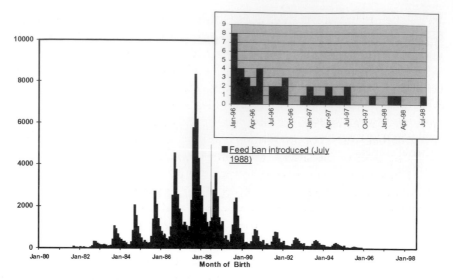

Figure 3. Distribution of confirmed cases of BSE by month of birth (as of 1 October, 2002).

diet of dairy animals than of beef animals. The latter often receive no concentrate ration. Another clue was provided (Wilesmith et al. 1991) by the age at onset of animals in the first wave of BSE cases; most of these cases were 3–5 years old. If it is assumed that exposure to a putative causal agent began shortly after birth, which is consistent with the initiation of MBM feeding, it may be inferred that exposure began in the early 1980s. A computer simulation, based on the early incidence of the epidemic, estimated 1980–1981 as the date of first exposures (Wilesmith et al. 1988), and this is borne out when cases are plotted by month and year of birth (Fig. 3).

What could explain the date of initial transmission of BSE, and how does this relate to the possible incrimination of MBM? MBM is manufactured in rendering plants, of which there were about 45 in Great Britain in the mid 1980s. These plants convert slaughterhouse refuse (offal) into two products: tallow (fat) and a defatted mixture of MBM. The process involves mixing, heating with steam, milling, and sometimes, extraction with hydrocarbon organic compounds that act as fat solvents while precipitating the protein. The organic solvent extraction process demanded a high energy input, and the increase in fuel prices during the 1970s made this process inefficient. This, together with a loss of the differential price between tallow and MBM, meant that extracting tallow from greaves (partially rendered offal), resulting in a fat content of about 1% in the MBM, was uneconomic. Other factors responsible for the decline in the use of the organic solvent extraction process were the increase in energy density provided by residual fat in MBM and the introduction of more stringent health and safety measures in industrial processes. The proportion of MBM processed with fat solvents fell from at least 70% in the mid 1970s to about 10% in the early 1980s, concurrent with the postulated first transmissions of BSE to cattle (Wilesmith et al. 1991). In the 1980s, there was also an increase in the use of MBM as a dietary supplement relative to other sources of protein, such as soya and fishmeal, because of the escalating costs of the latter (Taylor and Woodgate 1997).

At this point, it is necessary to note the physical properties of prions, the putative causal agents of the TSEs. Their properties are known mainly from studies of scrapie of sheep (Ernst and Race 1993; Taylor 1993; Taylor et al. 1994). Prions consist of a single protein, designated PrP (prion protein), which is attached to cellular lipid membranes by a glycosyl phosphatidylinositol (GPI) anchor. The infectivity of prions is notoriously resistant to heat (including steam under pressure) and treatment with some harsh denaturing agents, such as formaldehyde, but can be inactivated by lipid solvents (Ernst and Race 1993). It may be postulated that if petroleum distillates were used in preparing MBM, they would substantially reduce the infectivity of any prions in the raw tissues used by rendering plants, but that infectivity could survive heating in the absence of solvents (Safar et al. 1993; Taylor 1996).

Another critical fact is that there is a large sheep population in Great Britain (GB), so that ovine waste constituted a substantial part of the offal treated by rendering plants. Furthermore, in GB, scrapie is endemic in the ovine population. An abattoir survey of 2809 healthy sheep in GB in 1997–1998 (Simmons et al. 2000) did not reveal any cases of scrapie using standard pathological confirmatory tests, but application of a mathematical model indicated that the data did not rule out a prevalence of infection in the slaughter population of up to 11%. These considerations are consistent with the hypothesis that the scrapie agent had always been present in slaughterhouse offal but was inactivated during the production of MBM. With the change in rendering practices in the late 1970s, it is postulated that inactivation became less effective, leading to the contamination of some batches of MBM beginning about 1980 (Wilesmith et al. 1988, 1991; Taylor 1989, 1996).

Circumstantial evidence in support of this hypothesis is provided by studies of the geographic distribution of BSE in GB (Table 2). There was a marked gradient, with highest cumulative incidence in southern England decreasing toward the north, with the lowest rates in Scotland. MBM was produced in about 45 rendering plants in GB, and most of the

Table 2. Geographic differences in BSE in GB, 1986–1989, and relationship to the production process used by rendering plants

Region	Dairy herds with BSE (%)	Use of solvents to process MBM	MBM from greaves (%)
Southern England	12.6	no	0.2
Midlands	3.9	no	8.6
Northern England	2.8	no	25.5
Scotland	1.8	yes	39.0

Data from Wilesmith et al. (1991).

product was distributed locally. A survey revealed (Wilesmith et al. 1991) that in the late 1980s rendering practices differed in different regions, with the use of petroleum solvents and reprocessing of greaves inversely related to BSE incidence (Table 2).

Incubation Period

Determination of the incubation period of BSE is complicated by several problems. First, it is difficult to determine the exact age of infection of BSE for individual animals. The low rate of BSE even in herds with the highest incidence indicates that infection was a relatively infrequent event, from which it may be inferred that transmission did not necessarily occur at the age that cattle began to consume potentially contaminated MBM. Second, the high rate of slaughter of cattle, particularly between ages 1.5 and 6 years, means that no intact cohort could be followed throughout the incubation period. Although this is somewhat corrected by using age-specific incidence, it must be assumed that living cattle are representative of the total population. Thus, the age-specific incidence has limitations as a surrogate for incubation period. A statistical analysis (Anderson et al. 1996) took these confounding variables into account and produced the distribution of incubation times. The mean incubation period was estimated at about 5 years with a variance of about 1.6 years.

Control of the Epidemic

The recognition that BSE was a common-source epidemic, and the identification of contaminated MBM as the putative cause, led to a decision to ban ruminant offal as a raw material in the preparation of dietary supplements destined for feeding to cattle and other ruminant species. The ruminant feed ban, introduced in July, 1988 (extended to all livestock species and to all mammalian-derived meat and bone meal by August 1, 1996), represented a decisive administrative action to control the problem (Anonymous 1997a), as well as the ultimate test of the hypothesis that MBM was the source of the outbreak. This measure was supplemented in September, 1990, and by a ban on feeding of animals and birds with specified bovine offals (SBO) or any products derived there-from. The SBO ban was extended to include intestine and thymus in November, 1994. In GB, the epidemic peaked in January, 1993, and has undergone a dramatic decline so that incidence in October, 2002, was about 0.4% of that at the peak of the epidemic (Fig. 1). The waning of the epidemic has followed a path that was roughly predicted at the time of the feed ban (Nathanson et al. 1993), providing strong support for the MBM causal hypothesis.

A mathematical and statistical study of the BSE epidemic (Anderson et al. 1996) has used back-calculation to reconstruct the dynamics of the outbreak, including the probable numbers of infected animals. This analysis indicates that infections did not cease in July, 1988, immediately following the feed ban, but continued in substantial but declining numbers in animals born through 1991 (estimated new infections. >300,000 in 1988 and ~30,000 in 1991). These results reflect observations on the occurrence of BSE cases in birth cohorts (Fig. 3), which show that cases occurred in animals born in 1989 through 1991, but in sharply decreasing numbers (Anonymous 1996b). In addition, these analyses were used to predict the course of the epidemic (Anderson et al. 1996; Donnelly et al. 1997a), which projected a waning to minimal numbers by the year 2000. A routinely run age-period-cohort model, constructed in 1991, has been used to provide estimates of both the number of confirmed clinical cases and the number of suspect cases reported in each calendar and financial year (M.S. Richards and J.W. Wilesmith, unpubl.). The resulting estimates have been used to provide supporting evidence for budgetary bids to the UK Treasury to secure the necessary funding for indemnity and disposal costs. From an epidemiological perspective, this model also predicted a waning epidemic with relatively few cases in the first years of the new millennium.

The occurrence of new infections after 1988 can be ascribed in part to contaminated feedstuffs, manufactured before July, 1988, that were already in the feed supply chain. In addition, there is evidence after 1988 of accidental cross-contamination in feed mills of rations produced for consumption by monogastric animals (poultry and pigs) and ruminants (cattle and sheep), and of the possible feeding of cattle with rations produced for monogastric animals (Wilesmith 1996a,b; Hörnlimann et al.

1997). The risks of accidental contamination were greatest in the northern and eastern regions of England where pig and poultry populations are concentrated, and this has resulted in a remarkable change in the geographical variation in risk for animals born since the feed ban (Wilesmith 1996b).

Further studies of the temporal aspects of the epidemic in GB have examined holding-associated and individual animal-associated risk factors (Stevenson et al 2000a,b; Wilesmith et al. 2000). Analyses of the change over time (up to the end of June, 1997) of cattle holdings that had experienced at least one confirmed case of BSE showed different rates of onset of BSE in different regions and in holdings of different sizes and types, with the epidemic being propagated predominantly in the southern part of the country and the pattern of growth essentially the same in all regions. The findings confirmed the effectiveness of the 1988 and 1990 control measures in stopping expansion of the epidemic but also showed that the rate of progress of control was slowed by regions where control measures took longest to be effective. When the cumulative incidence of BSE in the British cattle population (July 1986–June 1987) and the individual animal-associated risk factors influencing the onset of clinical signs in confirmed cases were examined, a reduction of 67% in the monthly hazard of being a confirmed case of BSE was seen in cattle born in the first 12 months after July, 1988. Successive cohorts in 1989 through 1991 also experienced further reductions in this monthly hazard. An additional reduction of a further 46%, attributable to the effect of the Specified Bovine Offal ban of September, 1990, was experienced in the 1990 cohort; i.e., cattle born in the first 12 months after the introduction of the ban (Stevenson et al. 2000b). But for these two measures, the burden of BSE infectivity circulating in the feed and food chains would have produced exposures sufficient to have made control of BSE in Britain almost impossible, simply because the necessary veterinary resources would have not been able to visit all suspect cases. These would have amounted to some 6,000 animals (and farms) to be visited in April, 1994, and these animals would have had to be disposed of by incineration or other approved means at considerable cost.

There are salutary lessons to be learned from the experience of the control of BSE in the UK and other European Union member states. Although the starting point in the control measures must clearly be appropriate legislation, based on sound epidemiological premise, the long course of establishing the effectiveness of the legislation must not be underestimated. The effectiveness of legislation depends on many interrelated factors, including the practicality of implementation of the measures, the levels of compliance that can be attained, and detail with which monitoring of the measures can be achieved. The difficulties of sustaining the necessary long-term vigilance in the control strategy cannot be overemphasized. This is particularly so given the relatively small amounts of infective material required for effective exposure, as indicated both from epidemiological studies (Wilesmith 1991; Kimberlin and Wilesmith 1994) and results of an experimental attack rate study, discussed below, and the associated difficulty in removing contaminated material from industrial plants used for the production of feedstuffs.

Changes in Epidemiological Patterns

Several actions were taken early in the epidemic in GB to attempt to terminate exposure of cattle to the BSE agent; some of these have been discussed already. BSE was made a reportable disease, with compensation to farmers for animals destroyed because of the disease, a diagnostic service was established to screen all suspect cases by neurohistologic examination of the brain, a ruminant feed ban was introduced in July, 1988, and in September, 1990, a specified bovine offal (SBO) ban preventing the use of certain offals in any animal feeds. In parallel, measures were taken to minimize the theoretical risk of BSE transmission to humans (Anonymous 1996a, 1997a; Wilesmith 1996a). Apart from preventing clinical cases entering the human food chain, a ban on SBO was imposed initially in November, 1989, forbidding entry into the human food chain of certain bovine tissues (including brain and spleen) that might contain the BSE agent. This ban was widened in 1992 (to prohibit the use of head meat if the skull had been opened), also in 1994 (to prohibit human consumption of thymus and intestines), and again in 1995 (to include the whole bovine head as a specified banned offal) (Anonymous 1996a; Taylor and Woodgate 1997). With the announcement of the occurrence of a variant form of Creutzfeldt-Jakob disease (vCJD) in April, 1996 (Will et al. 1996), and the recognition of the partial nature of the measures to prevent further exposures in cattle, as discussed above, additional legislation in 1996 introduced prohibition of the supply of mammalian meat and bone meal, or any feedstuff in which it was a constituent, for the purpose of feeding to farmed animals and prohibition of the possession of meat and bone meal on premises where livestock feedstuffs were kept.

Prior to the effective time (August, 1996) of the start of this last piece of legislation, two distinct phases of the epidemic had occurred (Fig. 4). In the first phase, there were those cases born before the initial ruminant-derived protein ban on 18 July, 1988; the "born before the ban" (BBB)

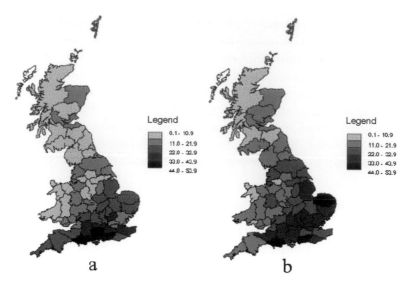

Figure 4. (*a*) Cumulative incidence (%) of dairy herds with BSE in home-bred cattle (November 1986 to December 1990) (the BBBs). (*b*) Cumulative incidence (%) of dairy herds with at least one case of BSE in home-bred cattle born after the feed ban (July, 1988) (from 1991 to December 1997) (the BABS).

cases. These have formed the major part of the epidemic, and their greatest incidence was in the southern part of England. Second, cases born after 18 July, 1988, and before 1 August, 1996, formed the so-called "born after the ban" (BAB) cases. Characterizing the latter was the shift in incidence to the eastern region of England and the major, and only significant, risk factor, of the ratio of cattle to pigs in the region, as discussed previously. Third, the occurrence of a small number of cases born after 1 August, 1996, with distinct epidemiological features (Wilesmith 2002) has introduced a third epidemiological group, those "born after the reinforced ban" (BARB) cases. These cases are not associated with the BAB risk factor; rather, their geographical distribution is associated with the distribution of dairy herds—suggestive of a random risk consistent with a widespread distribution of a low risk of exposure. What then are the possible sources of such continuing exposures? The absence of cases in the dams or siblings of these cases argues against maternal transmission. Horizontal transmission resulting from environmental exposure is also very unlikely, as there have been no previous cases in approximately one-third of BARB affected herds, and none has occurred in high-incidence herds. Is there a persistent source of feed-borne exposure? The current rigor of controls in GB, together with the change in geographical distribution compared to the BAB cases, reduces the likelihood of an indigenous source of mammalian MBM. Although imported mammalian MBM for pet food is a legal theoretical foodborne exposure source, it is apparently not used or imported by the pet food industry. Given that the handling of mammalian MBM was not illegal in continental Europe until January, 2001, the cross-contamination of imported feed ingredients during handling, storage, and transport, possibly at European ports, remains an intriguing possible explanation for this third epidemiological phase of the UK BSE epidemic. Although as yet, epidemiological analyses can focus only on BARB cases in the UK, similar analysis of the future occurrence of such cases in mainland Europe may shed further light on this tail of the BSE epidemic.

Possible Natural Transmission

The question of whether BSE could be transmitted horizontally from cow to cow or maternally from cow to calf has been considered from the beginning of the epidemic (Hoinville et al. 1995; Wilesmith 1996a). Because all animals in each age group in a herd would be exposed to the same dietary supplements, it could be difficult to distinguish horizontal from common-source transmission. The low frequency of BSE in affected herds (overall <5% annual incidence) suggests that horizontal transmission was not occurring at a high frequency, although it does not preclude transmission at a low rate.

It has long been suggested that scrapie in sheep is transmitted from ewe to lamb (Hoinville 1996), perhaps via the placenta from which the scrapie agent has been isolated, although recent critical reviews (Ridley and Baker 1995; Hoinville 1996) concluded that vertical transmission may be more myth than fact. The epidemiological study of this possible method of transmission began in 1987 when the collection of detailed epidemiological data and information of all cases of BSE was initiated (Wilesmith et al. 1988). This included obtaining information on the identity and survival of offspring of all cows in which BSE was confirmed. In GB in 1994, a case-control study (Hoinville et al. 1995) was conducted of cases of BSE in animals born after the ban on feeding ruminant-derived protein (July, 1988). This study indicated that maternal transmission—if it occurred—would be at a rate between 0 and 13% and would not be a major factor in the occurrence of cases of BSE in animals born after the feed ban. These data collected from the beginning of the epidemic facilitated the design and execution of a cohort study to investigate specifically the possibility of a maternal risk factor. The objective of this study was to determine whether the offspring of dams that developed clinical BSE were more likely to develop signs of BSE, clinically and/or histopathologically, than offspring of cows that survived to at least 6 years of age and did not develop clinical signs of BSE. For the required statistical power, and allowing for intercurrent mortality, 300 pairs of offspring were required. In the event, 301 pairs were purchased, each member of the pair having been born in the same calving season. The animals were allowed to live until 7 years of age unless death or the need to slaughter them intervened. The results indicated that there was a statistically significant risk difference between the two cohorts of 9.7% (95% confidence interval [CI]: 5.1–14.2%) and relative risk of 3.2 (95% CI: 1.8–5.9) (Wilesmith et al. 1997). This indicated that there was a maternally associated risk factor for the occurrence of BSE in this population of animals. It was not possible to determine conclusively at this time whether this factor was due to true maternal transmission or a genetic predisposition or a combination of these two effects (Donnelly et al. 1997b; Gore et al. 1997; Curnow et al. 1997). The result of the cohort study was in agreement with the results of analyses of the main epidemiological database of cases. These involved estimations and comparisons of the observed and expected number of cases in offspring using information on dam–calf pairs, and the results indicated a 10% rate in the data accumulated by 1997 (Donnelly et al. 1997b; M.S. Richards and J.W. Wilesmith, unpubl.).

Additional analyses of the cohort study and of the main epidemiology database indicated that offspring born near the onset of clinical signs in the dam were at most risk of developing BSE. As a result, modeling studies incorporated an enhanced risk of 10% for offspring born in the last 6 months of the dams' incubation period. This allowed predictions of the number of cases, in animals born after July, 1996, in GB, in 2001 due to the maternally associated factor when the feedborne risk was thought to have been eliminated. The upper estimate of the number of cases, allowing for the offspring and selective culls, was 55 (European Commission 2001). In the outcome, the actual number of cases was less than the predicted (Wilesmith 2002). This suggests that for the more recent-born cohorts, the maternal risk factor has reduced. In the original cohort study, there was some evidence for a declining maternal risk with successive birth cohorts (Wilesmith et al. 1997). Recent modeling studies also indicate that the cumulative maternal risk, throughout the epidemic in GB, is now reduced to 2%, but with a confidence interval including zero. The occurrence of true maternal transmission remains questionable. There is little or no evidence from affected countries outside the UK concerning "maternal transmission." One hypothesis that has emerged is that the maternal effect is only important when the risk from feedborne exposure is large, as occurred in the UK, and therefore there is a low-level genetic predisposition whose molecular basis is currently unknown. In terms of controlling BSE in low-incidence countries, the maternal effect, observed at the height of the epidemic in GB, can be ignored.

Origin of the Epidemic

As a result of the occurrence of vCJD (Will et al. 1996), the evidence of an association with BSE in cattle, and a change in the political administration in GB, a public inquiry was initiated. The main objective of the inquiry was to investigate and comment on how the BSE crisis was handled by the government and therefore identify the lessons to be learned. The process of this inquiry naturally included an examination of the scientific understanding of BSE in particular and the TSEs in general, which could have had a bearing on the control measures adopted and the advice given to the public. The inquiry report (Inquiry Report 2000) included a commentary on the origin of BSE. Perhaps surprisingly, the Inquiry Committee concluded that the origin of BSE was unassociated with sheep scrapie, but rather that it was due to the sporadic occurrence of BSE in the British cattle population. This conclusion prompted a further specific review (Horn 2001) of the origins of BSE, commissioned as part of the government's initial response to the Inquiry Report. The review considered all potential hypotheses, and although no firm conclusion was made, the report tended to favor the sheep scrapie origin hypothesis. This remains as the most supported hypothesis by the research community.

We have discussed above the circumstances relative to rendering processes that could have resulted in the effective exposure of cattle to a scrapie agent and the limited epidemiologic data in support of this hypothesis. The epidemiologic observations and theoretical reconstruction of the source of the BSE epidemic raise the question of whether BSE can be reproduced by the experimental transmission of sheep scrapie to cattle. A number of transmission experiments have been reported (Cutlip et al. 1996, 1997; Gibbs et al. 1996; Prusiner 1997) that have produced certain consistent results. Various USA sources of sheep and goat scrapie can be transmitted to cattle by intracerebral or multiple parenteral routes, but not by the oral route (Cutlip et al. 2001). The resulting pathologic picture differs from both BSE in cattle and scrapie in sheep and does not change on first passage of the agent from cattle to cattle by intracerebral inoculation (Cutlip et al. 1997).

Difference between the neuropathological changes produced in cattle by experimental transmission of sheep scrapie on primary transmission and first passage and those of BSE does not undermine the postulated sheep scrapie origin of the BSE epidemic. It is well recognized in rodent models of scrapie that TSE agents often change their properties upon serial passage in a new host species, but that the biological properties of the isolate then become stabilized, usually after the third or later passage. The BSE isolate is a TSE agent strain that has had several serial passages in cattle and has consequently become "adapted" to cattle. Sheep scrapie agent is believed to occur in several different strains, and of the transmissions to cattle so far reported, none has been characterized by biological strain-typing methods. Until such time, therefore, as scrapie transmissions to cattle have been extended to serial passages of multiple characterized sources through cattle, we will not know the answer. These considerations suggest that the experimental data do not exclude the possibility that a strain of sheep scrapie could produce the BSE phenotype upon serial passage in cattle. Studies in progress of the primary transmission of UK isolates of scrapie to cattle may provide further information, but are as yet incomplete.

BSE in Other Countries

BSE has occurred in a number of European countries, but the reported incidence has been much lower than in GB. It is likely that the higher incidence was due to the concatenation of several circumstances, some of which occurred only in GB. These include a high ratio of sheep to cattle, a high enzootic prevalence of scrapie in sheep, the intensive feeding of MBM to dairy cattle, and the changes in rendering practices. A comparison of BSE in GB and Switzerland illustrates these differences. The cumulative incidence of BSE in Switzerland is about 100-fold lower than in GB. There were three rendering plants in Switzerland that supplied MBM to the feedmill industry. A comparison was made of the relative risk of scrapie contamination of slaughterhouse waste produced in the two countries, based on the ratio of sheep to cattle and the relative prevalence of scrapie in sheep (Hörnlimann et al. 1996). If this relative ratio is normalized to 1 for GB, it is estimated at 0.002 for Switzerland, indicating that the likelihood of scrapie entering rendering plants was much lower in Switzerland. The BSE outbreak in Switzerland occurred about 3 years later than the outbreak in GB. Since the Swiss imported considerable amounts of MBM from other countries, it is likely that most of the BSE cases were due to imported rather than domestically manufactured protein supplement (Hörnlimann et al. 1994; Butler 1996). This inference is supported by studies of the transmission of BSE to mice, which indicate that isolates from Swiss cows produced lesions identical to those produced by UK isolates of BSE (Taylor and Woodgate 1997).

By the end of 2002, BSE had been identified in indigenous cattle in 21 countries outside the UK (O.I.E. 2003). The 21 countries are shown in Table 3 by the year of the first identification of BSE infection in indigenous cattle. In addition, BSE has been detected only in cattle exported from GB in Canada, the Falkland Islands, and the Sultanate of Oman. The evidence is that BSE has not been propagated in these countries. Epidemiological studies that might inform on the precise means of introduction of infection in the countries with indigenous cases have not been documented, and therefore comparisons, other than those above between GB and Switzerland, have not been pursued. The means of introduction of infection will undoubtedly have included the importation of infected animals, contaminated meat and bone meal, and bovine carcasses. Within the member states of the European Union there is evidence of secondary transmission of infected animals, at least, between countries outside the UK.

Table 3. Indigenous cases of BSE arranged by country and order of first occurrence (as of December 2002)

Country	1985–1989	1990–1994	1995–1999	2000	2001	2002	Total
United Kingdom	10,181	136,574	32,640	1,443	1,202	534	182,574
Rep. of Ireland	10	78	347	149	246	254	1,084
France		10	69	161	274	177	691
Portugal		12	362	149	110	49	682
Switzerland		118	215	33	42	12	420
Belgium			10	9	46	30	95
Netherlands			6	2	20	15	43
Liechtenstein			2	0	0	0	?
Luxembourg			1	0	0	1	2
Germany				7	125	83	215
Spain				2	82	109	193
Denmark				1	6	1	8
Italy					50	10	60
Slovakia					5	5	10
Japan					3	2	5
Czech Rep.					2	2	4
Slovenia					1	1	2
Austria					1	0	1
Finland					1	0	1
Greece					1	0	1
Poland						4	4
Israel						1	1

Numbers are still incomplete for the year 2002; regular updates of global BSE figures are available on the OIE website: www.oie.int

The availability of rapid screening tests (Moynagh and Schimmel 1999) has improved our knowledge of the worldwide occurrence of BSE. Of the 21 countries, 10 were first detected as having indigenous infected cattle in 2001, and later, when the screening tests began to be used extensively, particularly in Europe, for surveillance. As indicated above, there are considerable difficulties associated with preventing cattle from becoming infected. This is highlighted in the incomplete control of the disease in some countries in mainland Europe. As a result, the feeding of mammalian-derived MBM to any livestock was prohibited in January, 2001, throughout the European Union. It is likely that the disease will persist for some years to come in affected countries, and other countries may well be identified as infected as surveillance is intensified. It is unlikely that large epidemics as experienced in the UK will occur in the future.

TRANSMISSIBLE SPONGIFORM ENCEPHALOPATHY IN OTHER SPECIES

Over the course of the BSE epidemic, there have been a total of 43 cases (through December, 2002) of spongiform encephalopathy diagnosed in captive wild exotic animal species in which such diseases had not been previously diagnosed (Table 4) (Kirkwood and Cunningham 1994a; Taylor and Woodgate 1997; Anonymous 2003; G. Wells, unpubl.). Affected species are confined to two phylogenetic families: the bovidae and felidae. The bovid species in which cases have occurred were all fed MBM as a dietary supplement in a manner similar to domestic cattle. The initial cases of TSE in captive wild felids were routinely fed bovine carcass material, including spinal cord (Kirkwood and Cunningham 1994a,b). The more recent cases appear to have been fed high-risk bovine tissues, such as brain in the case of the lions (J.W. Wilesmith and G.A.H. Wells, unpubl.), but further investigations are needed on some of these cases.

By December 2002, a total of 93 cases of spongiform encephalopathy had been diagnosed in domestic cats (Table 5). The first observed case had an onset of clinical signs in November, 1989, and was born in 1983 (Wyatt et al. 1991; Anonymous 2003; J.W. Wilesmith and G.A.H. Wells, unpubl.). Of the 93 cases of feline spongiform encephalopathy (FSE) confirmed worldwide, 89 occurred in GB, including the Channel Islands. The last case observed in GB, in December, 2002, developed clinical signs in June, 2001, and was a cat born in 1995.

The results of an epidemiological study of FSE in domestic cats indicated that the observed cases were due to exposure to high-risk bovine tissues contained in commercially produced cat food or to bovine tissues,

Table 4. Transmissible spongiform encephalopathy in zoo animal species, 1986–2002

Species	Number of cases
Bovidae	
Ankole (*Bos taurus*)	2
Arabian oryx (*Oryx leucoryx*)	1
Eland (*Taurotragus oryx*)	6
Gemsbok (*Oryx gazella*)	1
Kudu (*Tragelaphus strepsiceros*)	6
Nyala (*Tragelaphus angasi*)	1
Scimitar-horned oryx (*Oryx dammah*)	1
Bison (*Bison bison*)	1
Felidae	
Cheetah (*Acinonyx jubatus*)	10[a]
Ocelot (*Felis pardalis*)	3
Puma (*Felis concolor*)	3
Tiger (*Panthera tigris*)	3
Lion (*Panthera leo*)	4
Asian golden cat (*Felis temmincki*)	1[b]

Cases occurred and were born in GB unless stated otherwise.

[a]The initial case of TSE in cheetah occurred in Australia, a further case has occurred in the Republic of Ireland, and three cases have occurred in France. All are reported to have been born in Britain.

[b]Reported from Australia, but born in Germany (Anonymous 2002).

After Kirkwood and Cunningham (1994a); Anonymous (2002); Baron et al. (1997) and DEFRA, UK, Monthly Summary Statistics: BSE and Scrapie (December, 2002).

including heads, intended for feeding to exotic species kept in zoological collections (J.W. Wilesmith, unpubl.). This study did not reveal evidence of a large, but unobserved epidemic of TSE in domestic cats, but there has been a degree of under-ascertainment of cases.

In 1989, prior to the diagnosis of the first case of FSE in 1990, the UK pet food industry instituted a voluntary ban on the inclusion of certain bovine offals as a precaution against the exposure of domestic pets, notably cats and dogs (Kimberlin 1992). By December, 2002, a total of 8 of the reported cases in domestic cats were in individuals born after September, 1990, when legislation banned the use of specified bovine offals (central nervous system, spleen, thymus, tonsils, and intestines) in animal feed, suggesting that although the epidemic was largely controlled by the exclusion of the specified bovine offals, contamination of pet food has continued to a lesser extent. TSE in domestic and wild felid species may have provided the first evidence of the contamination of beef carcasses and bovine tissue with the agent of BSE.

Table 5. Total number and temporal distribution of occurrence of cases of feline spongiform encephalopathy in domestic cats

Year	1989	1990	1991	1992	1993	1994	1995	1996	1997	1998	1999	2000	2001
Cases[a]	1	16	11	14	10	14	4	7(1)[b]	8(4)	1(1)	1	1(1)	1(1)

Total number of cases in GB and the Channel Islands: 89. Four additional cases have occurred in cats outside Britain: N Ireland (1), Norway (1), Liechtenstein (1), Switzerland (1).

[a]By date of onset of clinical signs.

[b]Number of cases born after September, 1990 (N.B. No cases recorded with known date of birth after August, 1996).

NEUROPATHOLOGY OF BSE

Qualitative Features

The pathologic changes in BSE are confined to the central nervous system. Qualitatively, they include spongiform change, neuronal vacuolation, an astrocytic reaction, and neuronal degeneration. Demonstrable amyloidosis, based on examination for apple green birefringence in Congo red-stained sections and a feature of some PrP aggregations in scrapie, is very infrequent and localized in BSE (Wells and Wilesmith 1995). Spongiform change is the most characteristic morphological feature in BSE and can be seen with routine hematoxylin and eosin (HE) staining in histological sections, as a vacuolation of the neuropil. By light microscopy this vacuolar change shows no obvious cellular association, but in ultrastructural studies (Liberski et al. 1992) membrane-bounded vacuoles are evident within neuronal processes. Readily visible intraneuronal vacuoles, associated with the perikarya, are frequent only in certain neuroanatomic locations. Using special stains, astrocytic hypertrophy is regularly seen, although usually it is not as marked as in sheep scrapie. Evidence of neuronal degeneration is seen occasionally, including necrotic neurons, basophilic shrunken neurons, neurophagia, and dystrophic neurites. Substantial neuronal loss has been demonstrated only by morphometric studies of the vestibular nuclei (Jeffrey et al. 1992; Jeffrey and Halliday 1994). Consistent with the light-microscopic appearances, ultrastructural changes are mainly seen in gray matter (Liberski et al. 1992) and resemble closely those of other TSEs. Additional to the vacuoles within neuritis there are hypertrophic astrocytes and dystrophic cells containing accumulations of neurofilaments, mitochondria, lamellar bodies, and other electron-dense profiles. Immunohistochemical examination for PrP^{Sc} demonstrates widespread immunostaining, which generally approximates to the distribution of vacuolar changes (Wells et al 1994a; Wells and Wilesmith 1995).

From the first preliminary identification of BSE as a new disease in cattle, its distinctive pathological presentation played a key role in establishing the syndrome as a nosological entity (Wells et al. 1987). Subsequent studies (Wells et al. 1991, 1994a; Wells and Wilesmith 1995; Hawkins et al. 1996; Simmons et al. 1996a,b) established that BSE produced a consistent neuropathological pattern of changes, a feature which was in contrast to the variable lesion patterns reported in scrapie of sheep. In the 1980s the application of methods of detection of the prion protein (disease-specific PrP) was not a routine approach to diagnosis of the TSE; histopathological examination of the brain was considered the standard. Ascertainment of suspected cases necessarily relied on recognition of the spectrum of clinical signs of BSE and although, over the course of the disease, these are in many respects distinctive, they can be confused with the signs of several other neurologic disorders of cattle. This is particularly the case when the clinical diagnosis is based on few, even single, examinations by a veterinarian. Consequently, throughout the epidemic a variable proportion of clinically suspect cases remained unconfirmed based on the neuropathologic examination. The stereotypical pattern of the brain lesions was important, however, since it had the potential to simplify the histopathologic diagnosis examination; the consistency of the brain pathology made it possible to restrict the examination to a single brain stem section (medulla) where neuropil vacuolation in the solitary tract nucleus or the nucleus of the spinal tract of the trigeminal nerve occurred in 99.6% of clinically suspect cases confirmed by a more extensive histopathological examination of the brain stem (Wells et al. 1989). This permitted the use of the histopathological examination as a routine confirmatory method, despite the very large number of suspect cases confirmed throughout the epidemic.

Quantitative Analysis

The remarkable consistency of the brain lesions among BSE cases prompted studies, using semi-quantitative methods previously applied only to mouse models of scrapie, to assess the severity of the vacuolar changes relative to their distribution. An initial study examined 100 confirmed cases of BSE slaughtered from 1986 to 1989 (Wells et al. 1991, 1992, 1994a; Wells and Wilesmith 1995). In HE-stained sections of the brain, vacuolar changes were scored in 88 neuroanatomical locations on a 0–4 scale of intensity, based on the method originally introduced for study of experimental scrapie in mice (Fraser and Dickinson 1968). The results were used to construct a lesion profile. Subsequently, the method was

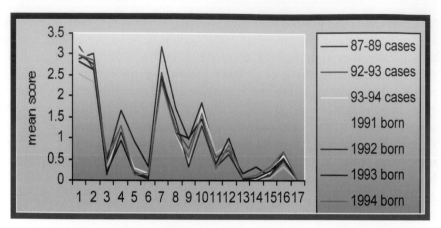

Figure 5. The unchanging lesion profile of BSE. From studies of the neuropathology of sequential longitudinal samples of cases from the UK epidemic. 1–17 on the horizontal axis represent a caudal-to-rostral series of neuroanatomic sites in the bovine brain (Simmons et al. 1996a,b; M.M. Simmons, pers. comm.).

modified to reduce the number of anatomic areas examined to 17 key gray-matter areas (Simmons et al. 1996a) and used to monitor the pattern of the vacuolar changes in the brains of a longitudinal series of case samples taken sequentially through the epidemic. These studies have shown no significant changes in the lesion profile of BSE over the course of the epidemic in Britain (Fig. 5). This provided crucial evidence of an apparent biological stability of the phenotype of disease in cattle with BSE and proved important in supporting the notion, considered further below, that the epidemic was sustained by a single, or major, strain of TSE agent. Areas with the most severe lesion scores are in midbrain (central gray matter), pons and medulla (nucleus of the spinal tract of the trigeminal nerve), diencephalon (hypothalamic nuclei), and hippocampus. The severity of changes in the brain invariably diminishes rostrally. The spinal cord was evaluated for vacuolar changes in only 10 confirmed cases (Wells et al. 1994a). Lesions were confined to the gray matter and were seen at cervical, thoracic, and lumbar regions; the most marked changes were always in the substantia gelatinosa of the dorsal gray horns.

BIOLOGICAL CHARACTERIZATION OF THE BSE AGENT

Studies in mice and sheep have shown that there are two major determinants of the lesion profile: the host genotype, principally with respect to the PrP gene, and the agent or prion strain (Fraser and Dickinson 1973).

The marked uniformity of BSE lesions, discussed above, implies that there is little variation in host genotype, at least as regards susceptibility to BSE and disease phenotype, and this is consistent with genetic studies that show little variation in the PrP gene among different breeds of cattle (Prusiner et al. 1993; Hunter et al. 1994). In addition, the stereotypic lesion pattern suggests that, regardless of origin, cattle have been infected with, or have selected, a single agent or prion strain (Fraser et al. 1992; Bruce et al. 1994). Consistent with this interpretation, when different BSE isolates are transmitted to a panel of different strains of mice, all isolates produce a constant incubation period and lesion profile in each of the mouse strains (Bruce et al. 1994, 1997; Hill et al. 1997). BSE in cattle and FSE in domestic cats each have their own characteristic pathological phenotype, which, together with evidence from "strain typing" of their agents in mice, has led to the conclusion that the epidemics of these diseases, and probably those also in the exotic species, have been due to a common exposure to a stable and apparently unique, major strain of a scrapie-like agent. These and other findings suggest a common epidemiological relationship between these "new" TSE, but do not confirm, or refute, a scrapie origin of BSE.

TRANSMISSIBILITY

BSE has been transmitted experimentally by parenteral routes from cattle to mice (Fraser et al. 1992), cattle (Dawson et al. 1990a), pigs (Dawson et al. 1990b; Ryder et al. 2000; Wells et al. 2003), sheep and goats (Foster et al. 1993; Fraser and Foster 1994), mink (*Mustela vison*) (Robinson et al. 1994), marmoset (*Callithrix jacchus*) (Baker et al. 1993), cynomolgus macaque (Lasmézas et al. 1996), and guinea pig (M. Dawson et al., unpubl.). There is also a considerable body of evidence that BSE has transmitted, by natural or accidental means, via foodstuffs to several other animal species and to man (Bruce et al. 1994, 1997; Kirkwood and Cunningham 1994b; Hill et al. 1997). It has also been transmitted by feeding BSE-affected brain to mice (Barlow and Middleton 1990) and mink (Robinson et al. 1994) and by the oral dosing of sheep and goats (Foster et al. 1993; Fraser and Foster 1994) and cattle (Wells et al. 1994b, 1996, 1998). Transmission of BSE to lemurs (*Microcebus murinus*) after oral dosing with infected brain has been reported (Bons et al. 1999). Thus, there is no doubt as to the susceptibility of several species of mammals to BSE infection after oral exposure. Some species, notably pigs (Wells et al. 2003), dogs, and probably also farmed deer, which almost certainly have been exposed in Britain to feed contaminated by the BSE agent, are, it appears, exceptions.

PATHOGENESIS

The initial expectation of close similarities between scrapie and BSE, with respect to pathogenesis and the distribution and relative titers of infectivity in tissues, was not borne out by the limited mouse bioassays of tissue from naturally occurring cases of BSE, in which infectivity was detected only in the CNS (Middleton and Barlow 1993; Fraser and Foster 1994).

The pathogenesis of BSE has been studied in a long series of experiments that began in 1991. Such studies, conducted in cattle, were facilitated by the evidence (Bruce et al. 1994; Simmons et al. 1996a,b) supporting the conclusion that the BSE epidemic was due to a single, stable, cattle-adapted strain of a scrapie-like agent. The previously reported initial study of the pathogenesis of BSE (Wells et al. 1994b, 1996, 1998, 1999; European Commission 2002) examined the spatial and temporal development of infectivity and pathological changes in cattle after oral exposure to a single 100-g dose of BSE-affected brain homogenate (titer: $10^{3.5}$ mouse (i.c./i.p.) LD_{50}/g) at 4 months of age. Groups of challenged and control calves were killed sequentially, at intervals of mainly 4 months, and a large range of tissues was collected for standard qualitative infectivity assays in wild-type mice (RIII and/or C57BL). The earliest onset of clinical signs in the cattle was 35 months after dosing. The qualitative bioassays of a large range of tissues (~45) have been completed, confirming a limited anatomical distribution of infectivity. For all tissues in which infectivity has not been detected, it can be stated that they contain less than $10^{1.4}$ mouse (i.c./i.p.) log10 LD_{50}/g, the calculated limit of detection. Infectivity in nonneural tissues was detected repeatedly only in the distal ileum (6–18 months and 36–40 months postexposure). Traces of infectivity were also shown in sternal bone marrow but only from cattle killed 38 months postexposure. In the CNS, the earliest presence of PrP^{Sc} (32 months postexposure) was coincident with the earliest detected infectivity and occurred prior to evidence of typical diagnostic histopathological changes in the brain in association with clinical disease, at 36, 38, and 40 months postexposure. Infectivity was also demonstrated in certain sensory ganglia of the peripheral nervous system (32–40 months postexposure).

The study also suggests a close temporal association between the detection of infection, abnormal PrP, and pathological changes in the CNS, all first apparent only at a late stage (about 90%) of the incubation period. Were this apparent relationship true and applicable to naturally occurring cases of BSE, then, for risk assessment purposes, mathematical modeling studies could provide more accurate estimates of the total load of infectivity in cattle carcasses relative to age. However, the pathogenesis study described above does not provide interpretable data on the relationship between the earliest detectable infectivity in CNS (or any other tissue) and incubation period. This is because the two observations, of clinical onset and tissue infectivity, cannot be compared directly since (given the sequential kill protocol of the study) the incubation period range of all animals in the study cannot be determined. A further study has examined the effect of oral inoculum dose on the attack rate and incubation period of BSE in cattle. The dose response data from this study suggest a mean incubation of almost 45 months (range 33–55 months) for an inoculum dose comparable with that received by cattle in the pathogenesis study. Because there are no direct experimental data relating infectivity of tissues to incubation period in BSE, there is no equation that might be applicable to calculate the initial time of detectability of tissue infectivity in relation to incubation of the natural disease. In certain experimental mouse models of scrapie, after peripheral routes of exposure, a constant relationship can be shown between the initial detection of infectivity in CNS and incubation, with infectivity detectable at ~50% of the incubation period (Kimberlin and Walker 1988, 1989). Data from naturally occurring sheep scrapie suggest a similar value for this relationship (European Commission 2000). It is not known whether such a constant relationship might be applicable to BSE of cattle, but for risk assessment purposes it would seem prudent to accept that infectivity may be first detectable in the CNS in natural BSE well in advance of clinical onset. This might be as little as 3 months before clinical signs, by conventional mouse bioassay, but theoretically at least, it could be 30 months, in an animal with an average estimated field case incubation of 60 months.

The oral attack rate experiment in cattle was also directed toward obtaining an estimate of the limiting dilution of infectivity of BSE for cattle, an essential, but hitherto missing, component of epidemiological modeling studies. Interim results suggest a cattle oral ID_{50} of 0.38 g of brain tissue containing $10^{3.5}$ mouse ID_{50}/g, with a wide 95% confidence interval. The study continues with examination of the attack rate of doses less than 1 g.

A rapid diagnostic immunoassay (Bio-Rad BSE test) to detect PrP^{Sc} in CNS, evaluated for the postmortem diagnosis of BSE using tissues taken at various times throughout the disease course from the pathogenesis study (Grassi et al. 2001), proved comparable in sensitivity to the wild-type mouse bioassay and to the immunohistochemical examination results of CNS tissue. A recently reported rapid immunoassay (CDI) has been shown to be capable of detecting PrP^{BSE} in the brain stems of cattle with a sensitivity similar to that of the infectivity levels determined by end-point titration in Tg(BoPrP) mice (Safar et al. 2002). This and possibly other emerging rapid immunoassays may offer prospects for extremely sensitive detection of disease-related PrP as a proxy for infectivity assays, and thereby enable further examinations of tissues from studies of the pathogenesis of BSE.

The localization of infectivity in the distal ileum of animals from the pathogenesis study was examined by the application of the immunohistochemical detection of PrP^{Sc} as a proxy for the agent (Terry et al. 2003). PrP^{Sc} was detected in macrophages within a small proportion of follicles of Peyer's patches (PP) throughout much of the disease course, in agreement with the mouse bioassay results for the time points in the study at which distal ileum was positive. No immunostaining was detected in the lymphoid tissue of distal ileum of naturally occurring clinical cases of BSE, nor was it detected in mesenteric lymph nodes or other levels of intestine of experimentally exposed calves killed 6 months postexposure.

Evidence from the study of the pathogenesis of BSE and from further experimental studies (Somerville et al. 1997) suggests that in BSE there may be a much more restricted involvement of the lymphoreticular system (LRS), compared to natural scrapie or experimental BSE in sheep.

A study was conducted to provide a measure of the underestimation of the titer of infectivity in tissues across a species barrier in mice, as the species barrier for BSE infection of wild-type mice could explain the negative results obtained in the bioassay of many tissues from both natural and experimental cases of BSE. The study also gave the opportunity to produce an approximate dose-incubation curve for infectivity of brain from BSE-affected cattle by the simultaneous titration of a primary inoculum of BSE-affected brain tissue in cattle and in mice. The underestimation of the infectivity titer of BSE tissue when titrated across a species barrier in wild-type mice was determined to be a factor of 500-fold (see European Commission 2002). That this relative degree of insensitivity of the mouse bioassay could explain the absence of widespread LRS infectivity in BSE is not supported by the results of assays by intracerebral inoculation of cattle with pooled lymph nodes (retropharyngeal, mesenteric, and popliteal) or pooled spleens from five terminal clinical cases of BSE. In this study, survival data suggest that, if present, the concentration of infectivity in these tissues is below the limit of detection of the cattle-to-cattle assay.

Additional studies have also assayed selected tissues from the original pathogenesis study by intracerebral inoculation of cattle. These studies remain in progress and so far have confirmed infectivity in tissues that were found to be positive by the mouse bioassay. Very recently they have also shown preliminary evidence of traces of infectivity in palatine tonsil of cattle killed 10 months after experimental oral exposure (European Commission 2002). End-point titrations of infectivity in tissues of BSE-infected cattle, other than for CNS, have not been done. Nevertheless, by applying mean incubation times to dose response curves of transmissions from BSE-affected brain to mice and cattle, levels of infectivity in cattle appear to be low in the few tissues that have assayed positive (European Commission 2002).

A further experimental oral exposure of cattle to the BSE agent, con-ducted in the UK, is providing tissue from sequentially killed recipients for the evaluation of rapid diagnostic tests and could also inform further on the pathogenesis of BSE in cattle. Investigations of the pathogenesis of experimental scrapie in mice and hamsters (Kimberlin and Walker 1980, 1989; Kimberlin et al. 1983; Beekes et al. 1996, 1998; Baldauf et al. 1997; Blattler et al. 1997; McBride and Beekes 1999; Beeks and McBride 2000; Glatzel and Aguzzi 2000; McBride et al. 2001), and natural scrapie of sheep (van Keulen et al. 2000), indicate that after oral intake the agents replicate in the lymphatic and/or neural tissue of the intestine and access the brain and spinal cord via the tracts of the autonomic nervous system. However, the only evidence to date of infectivity in the autonomic ner-vous system of cattle with BSE is that of inconsistently detectable PrP accumulation in the ganglion cells of the myenteric plexus (but not the submucosal plexus) of naturally and experimentally affected cattle (Terry et al. 2003). Therefore, although it cannot be excluded that parts of the autonomic nervous system may contain infectivity at some point in the course of BSE in cattle, it seems likely that infectivity levels can be expect-ed to be low or undetectable. Further examinations of elements of the autonomic nervous system for infectivity over the course of the disease are required either by the intracerebral inoculation of cattle, or by assays conducted in transgenic mice expressing bovine PrP (Safar et al. 2002), to address this issue.

On the basis of experience with field cases of BSE and studies of the distribution and relative severity of vacuolar changes, it has been specu-lated that the solitary tract nucleus and the spinal tract nucleus of the trigeminal nerve are possible primary sites of neural pathogenesis in BSE (Wells 1993). Subsequent study of early disease-specific PrP distribution in the brains of cattle infected with BSE agent by the oral route (Ryder et al. 1999) indicated PrP accumulation initially in these same medullary nuclei, suggesting possibly neuroinvasion via sensory pathways (solitary tract nucleus: general visceral afferent and special visceral afferent of cranial nerves VII, IX, X, and the spinal tract nucleus of V: general somatic afferent of cranial nerves V, VII, IX, X), in contrast to natural scrapie (Ryder et al. 2001) and experimental BSE in sheep (Ryder et al. 1999), in which preclinical accumulation of PrP was found solely in the dorsal nucleus of the vagus nerve (autonomic/parasympathetic, general visceral efferent). This might suggest some differences in pathogenetic events between BSE in cattle and scrapie and BSE in sheep with respect to entry of the agent into the CNS. The late occurrence and paucity of PrP immunostaining in neurons of the enteric nervous system in BSE could be consistent with the apparent less pivotal role of the vagus, suggested from studies in other TSE hosts. The recent finding of infectivity by cattle bioassay in palatine tonsil of cattle killed 10 months after oral exposure in the pathogenesis study (European Commission 2002) raises the possibility that such a pathway for infection could arise in the innervation of the soft palate/tonsil.

SUMMARY

Bovine spongiform encephalopathy (BSE) is a transmissible spongiform encephalopathy (TSE) or prion disease of cattle that was first recognized in GB in 1986, where it produced a common-source epidemic that peaked in January, 1993, and has subsided markedly since that time. By December, 2002, more than 180,000 cases had been confirmed, representing over 99% of the total projected incidence. The epidemic began simultaneously at many geographic locations and was traced to contamination of meat and bone meal (MBM), a dietary supplement prepared from rendering of slaughterhouse offal. It appears that the epidemic was initiated by the presence of a TSE agent that was first transmitted to cattle in the late 1970s to early 1980s, when most rendering plants ceased the use of organic solvents in the preparation of MBM. The epidemic was accelerated by the recycling of infected bovine tissues, prior to the recognition of BSE. To control the epidemic, a prohibition on the feeding of ruminant-derived protein to ruminants was introduced in GB in July, 1988, and accounts for the major decline of the epidemic after an interval of about 5 years, approximately equivalent to the average incubation period of BSE. BSE has now been confirmed in indigenous cattle in a total of 21 countries, in part due to the availability of rapid diagnostic screening tests. Additional countries may in the future be found to have infected cattle populations as the result of active surveillance, but it seems unlikely that major epidemics as occurred in the UK will develop elsewhere. That the epidemic has been mainly confined to the UK is because of a unique combination of risk factors which have included a high population ratio of sheep to cattle, a relatively high rate of endemic scrapie, the inclusion of MBM at high rates in feed for dairy cattle, and changes in the rendering process used to prepare MBM.

The initial pathogenesis of BSE in cattle appears similar to that of scrapie and some other animal TSEs after oral exposure, but questions remain concerning the biology of BSE in cattle. The apparent lack of any widespread tissue distribution of infectivity in cattle with BSE argues against a potential for shedding of agent, horizontal transmission, and carrier status. A low and restricted infectivity load in tissues of cattle infected with BSE will clearly also have resulted in a lower exposure of humans than might have been anticipated from the example of scrapie of sheep. But questions remain: Do the observed differences in agent distribution and apparent titer between species reflect true pathogenetic differences? What is the role of the autonomic nervous system, or other peripheral nervous system components, in BSE pathogenesis? With the application of increasingly sensitive detection methods for infectivity and disease-related PrP, how should future results implicating potential risks from bovine tissue be assessed?

ACKNOWLEDGMENTS

This chapter has been substantially revised from Nathanson et al. (1999). The authors thank Mrs. L.P. Cooper for her typographical skills in preparing the chapter, and Ms. J. Ryan for provision of data and preparation of epidemiological figures. The research work of the authors was funded by Defra, UK, formerly MAFF, UK.

REFERENCES

Anderson R.M., Donnelly C.A., Ferguson N.M., Woolhouse M.E.J., Watt C.J., Udy H.J., MaWhinney S., Dunstan S.P., Southwood T.R.E., Wilesmith J.W., Ryan J.B.M., Hoinville L.J., Hillerton J.E., Austin A.R., and Wells G.A. 1996. Transmission dynamics and epidemiology of BSE in British cattle. *Nature* **382:** 779–788.

Anonymous. 1996a. *Bovine spongiform encephalopathy in Great Britain: A progress report, May 1996,* pp. 1–51. Ministry of Agriculture, Fisheries and Food, London, United Kingdom.

———. 1996b. *BSE reports through May 1996,* pp. 1–3. Central Veterinary Laboratory, London, United Kingdom.

———. 1997a. *Bovine spongiform encephalopathy in Great Britain: A progress report, June 1997,* pp. 1–144. Ministry of Agriculture, Fisheries and Food, London, United Kingdom.

———. 2002. Imported zoo cat falls victim to rare disease (editorial). *Aust. Vet. J.* **80:** 445.

————. 2003. *Bovine spongiform encephalopathy in Great Britain: A progress report, June 2002*, pp. 1–181. Department for Environment, Food and Rural Affairs (DEFRA) Publications, London, United Kingdom.

Baker H.F., Ridley R.M., and Wells G.A.H. 1993. Experimental transmission of BSE and scrapie to the common marmoset. *Vet. Rec.* **132:** 403–406.

Baldauf E., Beekes M., and Diringer H. 1997. Evidence for an alternative direct route of access for the scrapie agent to the brain bypassing the spinal cord. *J. Gen. Virol.* **78:** 1187–1197.

Barlow R.M. and Middleton D. 1990. Dietary transmission of bovine spongiform encephalopathy to mice. *Vet. Rec.* **126:** 111–112.

Baron T., Belli P., Madec J.Y., Moutou F., Vitaud C., and Savey M. 1997. Spongiform encephalopathy in an imported cheetah in France. *Vet. Rec.* **141:** 270–271.

Beekes M. and McBride P.A. 2000. Early accumulation of pathological PrP in the enteric nervous system and gut-associated lymphoid tissue of hamsters orally infected with scrapie. *Neurosci. Lett.* **278:** 181–184.

Beekes M., Baldauf E., and Diringer H. 1996. Sequential appearance and accumulation of pathognomonic markers in the central nervous system of hamsters orally infected with scrapie. *J. Gen. Virol.* **77:** 1925–1934.

Beekes M., McBride P.A., and Baldauf E. 1998. Cerebral targeting indicates vagal spread of infection in hamsters fed with scrapie. *J. Gen. Virol.* **79:** 601–607.

Blattler T., Brandner S., Raeber A.J., Klein M.A., Voigtlander T., Weissmann C., and Aguzzi A. 1997. PrP-expressing tissue required for transfer of scrapie infectivity from spleen to brain. *Nature* **389:** 69–73.

Bons N., Mestre-Frances N., Belli P., Cathala F., Gajdusek C., and Brown P. 1999. Natural and experimental oral infection of nonhuman primates by bovine spongiform encephalopathy agents. *Proc. Natl. Acad. Sci.* **96:** 4046–4051.

Bradley R. 2001. Bovine spongiform encephalopathy and its relationship to the new variant form of Creutzfeldt-Jakob disease. An account of bovine spongiform encephalopathy, its cause, the clinical signs and epidemiology including the transmissibility of prion diseases with special reference to the relationship between bovine spongiform encephalopathy and the variant form of Creutzfeldt-Jakob disease. *Contrib. Microbiol.* **7:** 105–144.

Bruce M.E., Chree A., McConnell I., Foster J., Pearson G., and Fraser H. 1994. Transmission of bovine spongiform encephalopathy and scrapie to mice: Strain variation and the species barrier. *Philos. Trans. R. Soc. Lond. B Biol. Sci.* **343:** 405–411.

Bruce M.E., Will R.G., Ironside J.W., McConnell I., Drummond D., Suttie A., McCardle L., Chree A., Hope J., Birkett C., Cousens S., Fraser H., and Bostock C.J. 1997. Transmissions to mice indicate that 'new variant' CJD is caused by the BSE agent. *Nature* **389:** 498–501.

Butler D. 1996. Did UK "dump" contaminated feed after ban? *Nature* **381:** 544–545.

Collee J.G. and Bradley R. 1997a. BSE: A decade on: Part 1. *Lancet* **349:** 636–641.

————. 1997b. BSE: A decade on: Part 2. *Lancet* **349:** 715–721.

Curnow R.N., Hodge A., and Wilesmith J.W. 1997. Analysis of the Bovine Spongiform Encephalopathy Maternal Cohort Study: The discordant case-control pairs. *Appl. Stat.* **46:** 345–349.

Cutlip R.C., Miller J.M., and Lehmkuhl H.D. 1997. Second passage of a US scrapie agent in cattle. *J. Comp. Pathol.* **117:** 271–275.

Cutlip R.C., Miller J.C., Race R.E., Jenny A.L., Lehmkuhl H.D., and Robinson M.M. 1996. Experimental transmission of scrapie to cattle. In *Bovine spongiform encephalopathy: The BSE dilemma* (ed. C.J. Gibbs, Jr.), pp. 92–96. Springer-Verlag, New York.

Cutlip R.C., Miller J.M., Hamir A.N., Peters J., Robinson M.M., Jenny A.L., Lehmkuhl H.D., Taylor W.D., and Bisplinghoff F.D. 2001. Resistance of cattle to scrapie by the oral route. *Can. J. Vet. Res.* **65:** 131–132.

Dawson M., Wells G.A.H., and Parker B.N.J. 1990a. Preliminary evidence of the experimental transmissibility of bovine spongiform encephalopathy to cattle. *Vet. Rec.* **126:** 112–113.

Dawson M., Wells G.A.H., Parker B.N.J., and Scott A.C. 1990b. Preliminary parenteral transmission of bovine spongiform encephalopathy to the pig. *Vet. Rec.* **127:** 338–339.

Dawson M., Wells G.A.H., Parker B.N.J., Francis M.E., Scott A.C., Hawkins S.A.C., Martin T.C., Simmons M.M., and Austin A.R. 1994. Transmission studies of BSE in cattle, pigs and domestic fowl. In *Transmissible spongiform encephalopathies* (a consultation on BSE with the Scientific Veterinary Committee of the Commission of the European Communities held in Brussels, Belgium, September 14–15, 1993) (ed. R. Bradley and B. Marchant), pp. 161–167. Document VI/4131/94-EN. European Commission on Agriculture, Brussels, Belgium.

Donnelly C.A., Ghani A.C., Ferguson N.M., and Anderson R.M. 1997a. Recent trends in the BSE epidemic. *Nature* **389:** 903–904.

Donnelly C.A, Ghani A.C., Ferguson N.M., Wilesmith J.W., and Anderson R.M. 1997b. Analysis of the Bovine Spongiform Encephalopathy Maternal Cohort Study: Evidence for direct maternal transmission. *Appl. Stat.* **46:** 321-344.

European Commission. 2000. Specified risk materials of small ruminants (scientific opinion adopted by the Scientific Steering Committee at its meeting of 13–14, April 2000). http://europa.eu.int./comm/food/fs/sc/ssc/outcome_en.html

———. 2001. Opinion on the six BARB BSE cases in the UK since 1 August 1996 (Scientific opinion adopted by the Scientific Steering Committee at its meeting of 29–30, November 2001). http://europa.eu.int./comm/food/fs/sc/ssc/outcome_en.html

———. 2002. Update of the opinion on TSE infectivity distribution in ruminant tissues (initially adopted by the Scientific Steering Committee at its meeting of 10–11, January 2002 and amended at its meeting of 7–8, November 2002, following the submission of (1) a risk assessment by the German Federal Ministry of Consumer Protection, Food and Agriculture and (2) new scientific evidence regarding BSE infectivity distribution in tonsils). http://europa.eu.int./comm/food/fs/sc/ssc/outcome_en.html

Ernst D.R. and Race R.E. 1993. Comparative analysis of scrapie agent inactivation methods. *J. Virol. Methods* **41:** 193–201.

Foster J..D., Hope J., and Fraser H. 1993. Transmission of bovine spongiform encephalopathy in sheep and goats. *Vet. Rec.* **133:** 339–341.

Fraser H. and Dickinson A.G. 1968. The sequential development of the brain lesions of scrapie in three strains of mice. *J. Comp. Pathol.* **78:** 301–311.

———. 1973. Scrapie in mice. Agent-strain differences in the distribution and intensity of grey matter vacuolation. *J. Comp. Pathol.* **83:** 29-40.

Fraser H. and Foster J. 1994. Transmission to mice, sheep and goats and bioassay of bovine tissues. In *Transmissible spongiform encephalopathies* (a consultation on BSE with the Scientific Veterinary Committee of the Commission of the European Communities held in Brussels, September 14–15, 1993) (ed. R. Bradley and B. Marchant), pp. 145–159. Document VI/4131/94-EN. European Commission on Agriculture, Brussels, Belgium.

Fraser H., Bruce M.E., Chree A., McConnell I., and Wells G.A. 1992. Transmission of bovine spongiform encephalopathy and scrapie to mice. *J. Gen. Virol.* **73:** 1891–1897.

Gibbs C.J., Jr., Safar J., Sulima M.P., Bacote A., and san Martin R.A. 1996. Transmission of sheep and goat strains of scrapie from experimentally infected cattle to hamsters and mice. In *Bovine spongiform encephalopathy: The BSE dilemma* (ed. C.J. Gibbs, Jr.), pp. 84–96. Springer-Verlag, New York.

Glatzel M. and Aguzzi A. 2000. PrP(C) expression in the peripheral nervous system is a determinant of prion neuroinvasion. *J. Gen. Virol.* **81:** 2813–2821.

Gore S.M., Gilks W.R., and Wilesmith J.W. 1997. Bovine Spongiform Encephalopathy Maternal Cohort Study: Exploratory analysis. *Appl. Stat.* **46:** 305–320.

Grassi J., Comoy E., Simon S., Créminon C., Frobert Y., Trapmann S., Schimmel H., Hawkins S.A.C., Moynagh J., Deslys J.P., and Wells G.A.H. 2001. Rapid test for the preclinical postmortem diagnosis of BSE in central nervous system tissue. *Vet. Rec.* **149:** 577–582.

Hawkins S.A.C., Wells G.A., Simmons M.M., Blamire I.W.H., Meek S.C., and Harris P. 1996. The topographical distribution pattern of vacuolation in the central nervous system of cattle infected orally with bovine spongiform encephalopathy. *Cattle Pract.* **4:** 365–368.

Heim D. and Kihm U. 1999. Bovine spongiform encephalopathy in Switzerland: The past and present. *Rev. Sci. Tech. Off. Int. Epizoot.* **18:** 135–144.

Hill A.F., Desbrusliers M., Joiner S., Sidle K.C.L., Gowland I., Collinge J., Doey L.J., and Lantos P.L. 1997. The same prion strain causes vCJD and BSE. *Nature* **389:** 448–451.

Hoinville L.J. 1996. A review of the epidemiology of scrapie in sheep. *Rev. Sci. Tech. Off. Int. Epizoot.* **15:** 827–852.

Hoinville L.J., Wilesmith J.W., and Richards M.S. 1995. An investigation of risk factors for cases of bovine spongiform encephalopathy born after the introduction of the 'feed ban.' *Vet. Rec.* **136:** 312–318.

Horn G. 2001. *Review of the origin of BSE*, pp. 66. Department for Environment, Food and Rural Affairs (DEFRA) Publications, London, United Kingdom.

Hörnlimann B., Guidon D., and Griot C. 1994. Risk assessment on the import of BSE. *Dtsch. Tieraertztl. Wochenschr.* **101:** 295–298.

Hörnlimann B., Heim D., and Griot C. 1996. Evaluation of BSE risk factors among European countries. In *Bovine spongiform encephalopathy: The BSE dilemma* (ed. C.J. Gibbs), pp. 384–394. Springer-Verlag, New York.

Hörnlimann B., Audige L., Somaini B., Guidon D., and Griot C. 1997. Case study of BSE in animals born after the feed ban (BAB) in Switzerland. Epidémiologie et Santé Animale. In *Proceedings of the 8th International Symposium on Veterinary Epidemiology,* vol. 31–32.

Hunter N., Goldmann W., Smith G., and Hope J. 1994. Frequencies of PrP gene variants in healthy cattle and cattle with BSE in Scotland. *Vet. Rec.* **135:** 400–403.

Inquiry Report. 2000. Report, evidence and supporting papers of the inquiry into the emergence and identification of BSE and variant CJD and the action taken in response to it up to March 20, 1996. Stationery Office, Norwich, United Kingdom.

Jeffrey M. and Halliday W.G. 1994. Numbers of neurons in vacuolated and non-vacuolated neuroanatomical nuclei in bovine spongiform encephalopathy-affected brains. *J. Comp. Pathol.* **110:** 287–293.

Jeffrey M., Halliday W.G., and Goodsir C.M. 1992. A morphometric and immunohistochemical study of the vestibular nuclear complex in bovine spongiform encephalopathy. *Acta Neuropathol.* **84:** 651–657.

Kimberlin R.H. 1992. Bovine spongiform encephalopathy. *Rev. Sci. Tech. Off. Int. Epizoot.* **11:** 347–390 (English); 391–439 (French); and 441–489 (Spanish).

Kimberlin R.H. and Walker C.A. 1980. Pathogenesis of mouse scrapie: Evidence for neural spread of infection to the CNS. *J. Gen. Virol.* **51:** 183–187.

———. 1988. Pathogenesis of experimental scrapie. *Ciba Found. Symp.* **135:** 37–53.

———. 1989. Pathogenesis of scrapie in mice after intragastric infection. *Virus Res.* **12:** 213–220.

Kimberlin R.H. and Wilesmith J.W. 1994. Bovine spongiform encephalopathy. Epidemiology, low dose exposure and risks. *Ann. N.Y. Acad. Sci.* **724:** 210–220.

Kimberlin R.H., Field, H.J., and Walker C.A. 1983. Pathogenesis of mouse scrapie: Evidence for spread of infection from central to peripheral nervous system. *J. Gen. Virol.* **64:** 713–716.

Kirkwood J.K. and Cunningham A.A. 1994a. Spongiform encephalopathy in captive wild animals in Britain: Epidemiological observations. In *Transmissible spongiform encephalopathies* (a consultation on BSE with the Scientific Veterinary Committee of the Commission of the European Communities held in Brussels, Belgium, September 14–15, 1993) (ed. R. Bradley and B. Marchant), pp. 29–47. Document VI/4131/94-EN. European Commission on Agriculture, Brussels, Belgium.

———. 1994b. Epidemiological observations on spongiform encephalopathies in captive wild animals in the British Isles. *Vet. Rec.* **135:** 296–303.

Lasmézas C.I., Deslys J.-P., Demalmay R., Adjou K.T., Lamoury F., Dormont D., Robain O., Ironside J.W., and Hauw J.-J. 1996. BSE transmission to macaques. *Nature* **381:** 743–744.

Liberski P.P., Yanagihara R., Wells G.A., Gibbs C.J., Jr., and Gajdusek D.C. 1992. Comparative ultrastructural neuropathology of naturally occurring bovine spongiform encephalopathy and experimentally induced scrapie and Creutzfeldt-Jakob disease. *J. Comp. Pathol.* **106:** 361–381.

McBride P.A. and Beekes M. 1999. Pathological PrP is abundant in sympathetic and sensory ganglia of hamsters fed with scrapie. *Neurosci. Lett.* **265:** 135–138.

McBride P.A., Schulz-Schaeffer W.J., Donaldson M., Bruce M., Diringer H., Kretzschmar H.A., and Beekes M. 2001. Early spread of scrapie from the gastrointestinal tract to the central nervous system involves autonomic fibers of the splanchnic and vagus nerves. *J. Virol.* **75:** 9320-9327.

Middleton D.J. and Barlow R.M. 1993. Failure to transmit bovine spongiform encephalopathy to mice by feeding them with extraneural tissues of affected cattle. *Vet Rec.* **132:** 545–547.

Moynagh J. and Schimmel H. 1999. Tests for BSE evaluated. *Nature* **400:** 105.

Nathanson N., Wilesmith J.W., Wells G.A.H., and Griot C. 1999. Bovine spongiform encephalopathy and related diseases. In *Prion biology and diseases* (ed. S.B. Prusiner), pp. 431–463. Cold Spring Harbor Laboratory Press, Cold Spring Harbor, New York.

Nathanson N., McGann K.A., Wilesmith J.W., Desrosiers R.C., and Brookmeyer R. 1993. The evolution of virus diseases: Their emergence, epidemicity, and control. *Virus Res.* **29:** 3–20.

O.I.E. (Office International des Epizooties). 2003. http://www.oie.int/eng/info/en_esb.htm

Prusiner S.B. 1997. Prion diseases and the BSE crisis. *Science* **278:** 245–251.

Prusiner S.B., Fuzi M., Scott M., Serban D., Taraboulos A., Gabriel J.M., Wells G.A., Wilesmith J.W., and Bradley R. 1993. Immunologic and molecular biologic studies of prion proteins in bovine spongiform encephalopathy. *J. Infect. Dis.* **167:** 602–613.

Ridley R.M. and Baker H.F. 1995. The myth of maternal transmission of spongiform encephalopathy. *Br. Med. J.* **311:** 1071–1075.

Robinson M.M., Hadlow W.J., Huff T.P., Wells G.A.H., Dawson M., Marsh R.F., and Gorham, J.R. 1994. Experimental infection of mink with bovine spongiform encephalopathy. *J. Gen. Virol.* **75:** 2151–2155.

Ryder S., Bellworthy S., and Wells G.A.H. 1999. A comparison of early PrP distribution in the CNS in natural scrapie and experimental BSE in both cattle and sheep. In *Abstracts from the European Commission Symposium on Characterisation and Diagnosis of Prion Diseases in Animals and Man*, Tübingen, September 23–25, 1999, p.56.

Ryder S.J., Hawkins S.A.C., Dawson M., and Wells G.A.H. 2000. The neuropathology of experimental bovine spongiform encephalopathy in the pig. *J. Comp. Pathol.* **122:** 131–143.

Ryder S.J., Spencer Y.I., Bellerby P.J., and Marsh S.A. 2001. Immunohistochemical detection of PrP in the medulla oblongata of sheep: The spectrum of staining in normal and scrapie-affected sheep. *Vet. Rec.* **148:** 7–13.

Safar J., Roller P.P., Gajdusek D.C., and Gibbs C.J., Jr. 1993. Thermal stability and conformational transitions of scrapie amyloid (prion) protein correlate with infectivity. *Protein Sci.* **2:** 2206–2216.

Safar J. G., Scott M., Monaghan J., Deering C., Didorenko S., Vergara J., Ball H., Legname G., Leclerc E., Solforosi L., Serban H., Groth D., Burton D.R., Prusiner S.B., and Williamson R.A. 2002. Measuring prions causing bovine spongiform encephalopathy or chronic wasting disease by immunoassays and transgenic mice. *Nat. Biotechnol.* **20:** 1147–1150.

Simmons M.M., Harris P., Jeffrey M., Meek S.C., Blamire I.W.H., and Wells G.A. 1996a. BSE in Great Britain: Consistency of the neurohistological findings in two random annual samples of clinically suspect cases. *Vet. Rec.* **138:** 175–177.

Simmons M.M., Harris P., Meek S.C., Blamire I.W.H., Jeffrey M., and Wells G.A. 1996b. The BSE epidemic: Consistencies of neurohistological findings in annual samples of clinical suspects. *Cattle Pract.* **4:** 361–364.

Simmons M.M., Ryder S.J., Chaplin M.C., Spencer Y.I., Webb C.R., Hoinville L.J., Ryan J., Stack M.J., Wells G.A.H., and Wilesmith J.W. 2000. Scrapie surveillance in Great Britain: Results of an abattoir survey, 1997/98. *Vet. Rec.* **146:** 391–395.

Somerville R.A., Birkett C.R., Farquhar C.F., Hunter N., Goldmann W., Dornan J., Grover D., Hennion R.M., Percy C., Foster J., and Jeffrey M. 1997. Immunodetection of PrP^{Sc} in spleens of some scrapie-infected sheep but not BSE-infected cows. *J. Gen. Virol.* **78:** 2389–2396.

Stevenson M.A., Wilesmith J.W., Ryan J.B.M., Morris R.S., Lawson A.B., Pfeiffer D.U., and Lin D. 2000a. Descriptive spatial analysis of the epidemic of bovine spongiform encephalopathy in Great Britain to June 1997. *Vet. Rec.* **147:** 379–384.

Stevenson M.A., Wilesmith J.W., Ryan J.B.M., Morris R.S., Lockhart J.W., Lin D., and Jackson R. 2000b. Temporal aspects of the epidemic of bovine spongiform encephalopathy in Great Britain: Individual animal-associated risk factors for the disease. *Vet. Rec.* **147:** 349–354.

Taylor D.M. 1989. Scrapie agent decontamination: Implications for bovine spongiform encephalopathy. *Vet. Rec.* **124:** 291–292.

———. 1993. Inactivation of SE agents. *Br. Med. Bull.* **49:** 810–821.

———. 1996. Exposure to, and inactivation of, the unconventional agents that cause transmissible degenerative encephalopathies. In *Prion diseases* (ed. H.F. Baker and R.M. Ridley), pp. 105–118. Humana Press, Totowa, New Jersey.

Taylor D.M. and Woodgate S.L. 1997. Bovine spongiform encephalopathy: The causal role of ruminant-derived protein in cattle diets. *Rev. Sci. Tech. Off. Int. Epizoot.* **16:** 187–198.

Taylor D.M., Fraser H., McConnell I., Brown D.A., Brown K.L., Lamza K.A., and Smith G.R. 1994. Decontamination studies with the agents of bovine spongiform encephalopathy and scrapie. *Arch. Virol.* **139:** 313–326.

Terry L.A., Marsh S., Ryder S.J., Hawkins S.A.C., Wells G.A.H., and Spencer Y.I. 2003. Detection of disease-specific PrP in the distal ileum of cattle exposed orally to the agent of bovine spongiform encephalopathy. *Vet. Rec.* **152:** 387–392.

van Keulen L.J.M., Schreuder B.E.C., Vromans M.E.W., Langeveld J.P.M., and Smiths M.A. 2000. Pathogenesis of natural scrapie in sheep. *Arch. Virol. Suppl.* **16:** 57–71.

Wells G.A.H. 1989. Bovine spongiform encephalopathy. *Vet. Annu.* **29:** 59–63.

———. 1993. Pathology of non-human spongiform encephalopathies: Variations and their implications for pathogenesis. *Dev. Biol. Stand.* **80:** 61–69.

Wells G.A. and Wilesmith J.W. 1995. The neuropathology and epidemiology of bovine spongiform encephalopathy. *Brain Pathol.* **5:** 91–103.

Wells G.A.H., Bradley R., and Wilesmith J.W. 2000. Bovine spongiform encephalopathy. In *OIE manual of standards for diagnostic tests and vaccines,* chapt. 2.3.13, pp. 457–466. Office International des Epizooties, Paris, France.

Wells G.A., Wilesmith J.W., and McGill I.S. 1991. Bovine spongiform encephalopathy: A neuropathological perspective. *Brain Pathol.* **1:** 69–78.

Wells G.A., Hawkins S.A.C., Hadlow W.J., and Spencer Y.I. 1992. The discovery of bovine spongiform encephalopathy and observations on the vacuolar changes. In *Prion diseases of humans and animals* (ed. S.B. Prusiner et al.), pp. 256–274. Ellis-Horwood, Chichester, United Kingdom.

Wells G.A.H., Hancock R.D., Cooley W.A., Richards M.S., Higgins R.J., and David G.P. 1989. Bovine spongiform encephalopathy: Diagnostic significance of vacuolar changes in selected nuclei of the medulla oblongata. *Vet. Rec.* **125:** 521–524.

Wells G.A.H., Hawkins S.A.C., Green R.B., Spencer Y.I., Dexter I., and Dawson M. 1999. Limited detection of sternal bone marrow infectivity in the clinical phase of experimental bovine spongiform encephalopathy (BSE). *Vet. Rec.* **144:** 292–294.

Wells G.A., Hawkins S.A.C., Cunningham A.A., Blamire I.W.H., Wilesmith J.W., Sayers A.R., Harris P.B., and Marchant B. 1994a. Comparative pathology of the new spongiform encephalopathies. In *Transmissible spongiform encephalopathies* (a consultation on BSE with the Scientific Veterinary Committee of the Commission of the European Communities held in Brussels, Belgium, September 14–15, 1993) (ed. R. Bradley and B. Marchant), pp. 327–345. Document VI/4131/94-EN. European Commission on Agriculture, Brussels, Belgium.

Wells G.A., Dawson M., Hawkins S.A.C., Austin A.R., Green R.B., Dexter I., Horigan M.W., and Simmons M.M. 1996. Preliminary observations on the pathogenesis of experimental bovine spongiform encephalopathy. In *Bovine spongiform encephalopathy: The BSE dilemma* (ed. C.J. Gibbs, Jr.), pp. 28–44. Springer-Verlag, New York.

Wells G.A.H., Dawson M., Hawkins S.A.C., Green R.B., Francis M.E., Simmons M.M., Austin A.R., and Horigan M.W. 1994b. Infectivity in the ileum of cattle challenged orally with bovine spongiform encephalopathy. *Vet. Rec.* **135:** 40–41.

Wells G.A., Scott A.C., Johnson C.T., Gunning R.F., Hancock R.D., Jeffrey M., Dawson M., and Bradley R. 1987. A novel progressive spongiform encephalopathy in cattle. *Vet. Rec.* **121:** 419–420.

Wells G.A.H., Hawkins S.A.C., Austin A.R., Ryder S.J., Done S.H., Green R.B., Dexter I., Dawson M., and Kimberlin R.H. 2003. Studies of the transmissibility of the agent of bovine spongiform encephalopathy to pigs. *J. Gen. Virol.* **84:** 1021–1031.

Wells G.A.H., Hawkins S.A.C., Green R.B., Austin A.R., Dexter I., Spencer Y.I., Chaplin M.J., Stack M.J., and Dawson M. 1998. Preliminary observations on the pathogenesis of experimental bovine spongiform encephalopathy (BSE): An update. *Vet. Rec.* **142:** 103–106.

Wilesmith J.W. 1991. The epidemiology of bovine spongiform encephalopathy. *Semin. Virol.* **2:** 239–245.

———. 1996a. Bovine spongiform encephalopathy: Methods of analyzing the epidemic in the United Kingdom. In *Prion diseases* (eds. H.F. Baker and R.M. Ridley), pp. 155–173. Humana Press, Totowa, New Jersey.

———. 1996b. Recent observations on the epidemiology of bovine spongiform encephalopathy. In *Bovine spongiform encephalopathy: The BSE dilemma* (ed. C.J. Gibbs, Jr.), pp. 45–58. Springer-Verlag, New York.

———. 1998. Manual on bovine spongiform encephalopathy. *FAO Animal Health Manual*, no. 2, pp. 51. Food and Agriculture Organisation of the United Nations, Rome, Italy.

———. 2002. Preliminary epidemiological analyses of the first 16 cases of BSE born after July 31, 1996, in Great Britain. *Vet. Rec.* **151:** 451–452.

Wilesmith J.W., Ryan J.B., and Atkinson M.J. 1991. Bovine spongiform encephalopathy: Epidemiological studies on the origin. *Vet. Rec.* **128:** 199–203.

Wilesmith J.W., Wells G.A., Cranwell M.P., and Ryan J.B. 1988. Bovine spongiform encephalopathy: Epidemiological studies. *Vet. Rec.* **123:** 638–644.

Wilesmith J.W., Wells G.A., Ryan J.B.M., and Simmons M.M. 1997. A cohort study to examine maternally-associated risk factors for bovine spongiform encephalopathy. *Vet. Rec.* **141:** 239–243.

Wilesmith J.W., Ryan J.B.M., Stevenson M.A., Morris R.S., Pfeiffer D.U., Lin D., Jackson R., and Sanson R.L. 2000. Temporal aspects of the epidemic of bovine spongiform encephalopathy in Great Britain: Holding-associated risk factors for the disease. *Vet. Rec.* **147:** 319–325.

Will R.G., Ironside J.W., Zeidler M., Cousens S.N., Estibeiro K., Alperovitch A., Poser S., Pocchiari M., Hofman A., and Smith P.G. 1996. A new variant of Creutzfeldt-Jakob disease in the UK. *Lancet* **347:** 921–925.

Wyatt J.M., Pearson G.R., and Smerdon T.N. 1991. Naturally occurring scrapie-like spongiform encephalopathy in five domestic cats. *Vet. Rec.* **129:** 233–236.

13

Infectious and Sporadic Prion Diseases

Robert G. Will

National CJD Surveillance Unit
Western General Hospital
Edinburgh EH4 2XU, Scotland

Michael P. Alpers

Centre for International Health
Division of Health Sciences
Curtin University of Technology
Perth WA 6845, Australia

Dominique Dormont

Commisseriat A L'Energie Atomique
BP6 92265 Fontenay Aux Roses Cedex, France

Lawrence B. Schonberger

Division of Viral and Rickettsial Diseases (MS A39)
National Center for Infectious Diseases
Centers for Disease Control and Prevention
Atlanta, Georgia 30333

CREUTZFELDT-JAKOB DISEASE (CJD) WAS FIRST IDENTIFIED in 1921 by Jakob (Jakob 1921), who referred to a previous case described by Creutzfeldt in 1920 (Creutzfeldt 1920). These original cases were clinically heterogeneous, and review of pathologic material provided confirmation of the diagnosis of CJD in only two of the original five cases (Masters and Gajdusek 1982). Over subsequent decades, the nosology of CJD was confused and confusing. Wilson regarded CJD as a "dumping ground for several rare cases of presenile dementia" (Wilson 1940). Although an important monograph by Kirschbaum (1968) listed the clinical and pathologic features of all 150 cases identified before 1965, it included cases such as Creutzfeldt's original case, which would not now fulfill clinical or pathologic criteria for the diagnosis of CJD. In 1954 (Jones and Nevin 1954) and 1960 (Nevin et al. 1960), Nevin and Jones described the typical clinical course, the "characteristic" electroencephalogram (EEG), and neuropathologic changes, including spongiform change, which in combination are now recognized as the paradigm features of sporadic CJD.

In the late 1950s and early 1960s, with remarkable perseverance in difficult conditions, Gajdusek and colleagues investigated kuru, a fatal ataxic syndrome restricted to the Okapa area of the highlands of Papua New Guinea (Gajdusek and Zigas 1957). The similarity of the neuropathologic findings in scrapie and kuru was recognized in 1959 by Hadlow (Hadlow 1959) with the implication that kuru, like scrapie, might be transmissible in the laboratory. Successful laboratory transmission of kuru in 1966 (Gajdusek et al. 1966) was followed in 1968 (Gibbs et al. 1968) by the laboratory transmission of CJD following intracerebral inoculation of primates, leading to a transformation in the level of scientific interest in these diseases. CJD had been regarded as a nontransmissible degenerative condition, because there was no evidence of a host immune response either serologically or histologically. The confirmation that CJD was transmissible in the laboratory led to a range of important scientific questions, including whether and if so, to what extent, CJD might be naturally transmitted. Epidemiological research in a number of countries established baseline characteristics for CJD, such as incidence rates, but no environmental source of infection was identified (Brown et al. 1987). Paradoxically for a transmissible disease, about 5–15% of cases were found to be familial with a dominant pattern of inheritance (Brown et al. 1987). By the 1970s, convincing evidence implicated iatrogenic transmission of the disease in a small number of cases through transfer of a corneal transplant (Duffy et al. 1974) or CNS tissue from patient to patient in the course of medical practice (Will and Matthews 1982).

The identification of prion protein (PrP) accumulation as a specific feature of CJD and other prion diseases (McKinley et al. 1983) has led to a revolution in the understanding of the pathogenesis of these conditions. The discovery that hereditary forms of human prion diseases were linked to mutations of the PrP gene (Hsiao et al. 1989) provided critical support for a central role of PrP in disease causation and expression.

Although these advances stimulated much scientific debate, a major increase in public interest in prion diseases, particularly in the UK, only began after confirmation of bovine spongiform encephalopathy (BSE) as an emerging epizootic disease in UK cattle in 1987 (Wells et al. 1987). The leading hypothesis for the origin of BSE included the transmission of scrapie to cattle by ingestion of contaminated meat and bone meal (Wilesmith et al. 1988). The emergence of BSE raised important questions regarding the possibility of a risk to public health. Many scientists regarded this possibility as remote, in part because of the presumed link between BSE and scrapie (Southwood Committee 1989). Epidemiological studies had not identified scrapie in sheep as a risk factor for the development of CJD (Brown et al. 1987). Apart from its economic importance, scrapie was largely of scientific interest because many of the advances in the understanding of prion diseases were based on laboratory models of scrapie transmission in rodents. In 1996, however, a new type of CJD, designated variant CJD (vCJD), was reported in the UK (Will et al. 1996). The emergence of vCJD raised concerns that its temporal and geographic association with the emergence of BSE might reflect a causal relationship and indicate that BSE had different properties from scrapie in its pathogenicity to humans. The accumulating reports of vCJD and a series of scientific studies strengthening the link between vCJD and BSE (Bruce et al. 1997; Hill et al.1997b; Scott et al. 1999) have led to an exponential increase in public interest and concern about CJD and other prion diseases.

Systematic studies of CJD in many countries have established that the crude average annual incidence of CJD is about one case per million population and that there are three main forms of disease: (1) sporadic cases with no known environmental source, (2) familial cases, and (3) cases transmitted from a known or presumed environmental source, such as iatrogenic cases (Brown et al. 1992). Sporadic CJD accounts for about 85% of all cases (Will et al. 1998), and these cases are characterized by a relatively stereotyped and rapidly progressive clinical course, although rare variants with an extended duration of illness are well recognized (Brown et al. 1984). The cause of sporadic CJD is unknown. Cases appear to occur without predictable geographic or temporal clustering (Will et al. 1986). Case-control studies have failed to establish any common environmental risk factor for the development of disease (Wientjens et al. 1996).

Molecular biological analyses have demonstrated an increased risk of disease in relation to the codon-129 polymorphism of the PrP gene (Palmer et al. 1991). Familial CJD, unlike sporadic CJD, is associated with mutations of the PrP gene (Pocchiari 1994). Since the original mutation in human prion disease was found in two kindreds with Gerstmann-Sträussler-Scheinker syndrome (GSS), more than 35 point or insertional mutations have now been discovered, the majority in CJD families (Kovacs et al. 2002). Genetic forms of CJD account for about 5–15% of cases (Will et al. 1998).

Transmitted CJD from a known environmental source such as iatrogenic CJD is rare (Brown et al. 1992). Transmission has involved cross-contamination through neurosurgical instruments, corneal grafts, human dura mater grafts, and human pituitary-derived hormones. Iatrogenic forms of CJD associated with a CNS route of inoculation are clinically similar to sporadic CJD. In contrast, iatrogenic forms of CJD associated with a peripheral route of inoculation result in a predominantly ataxic syndrome. Kuru is clinically similar to these latter cases. Unlike them, however, kuru has been specifically related to direct human-to-human transmission through ingestion or peripheral inoculation of the kuru agent during ritual endocannibalism practiced by the Fore people of Papua New Guinea (Alpers 1968; Gajdusek 1990). vCJD is clinically similar to human growth hormone (hGH)-related CJD and kuru, consistent with a probable oral or peripheral route of exposure. Currently available data are insufficient to estimate the future number of cases of vCJD (Cousens et al. 1997).

There is no in vivo diagnostic test that is noninvasive and generally practical for determining the presence of infectivity in prion diseases, including CJD. A firm diagnosis rests on identifying the characteristic neuropathologic changes, usually postmortem. The accuracy of routine histologic diagnosis has been enhanced by techniques, including immunocytochemical staining for PrP and the identification of protease-resistant PrP from brain tissue by histoblot or western blot analysis. The transmissibility of individual cases or subtypes of CJD can be confirmed by laboratory experiments, but these investigations are time-consuming and expensive. The development of transgenic rodents with copies of human PrP genes may allow more sensitive bioassay, but this remains to be established. Diagnostic criteria for CJD were first proposed in 1979 (Masters et al. 1979) and have been adapted in the light of scientific developments (Budka et al. 1995), including the occurrence of vCJD (Will et al. 2000; World Health Organization 2002). These criteria are useful in epidemiologic studies for ensuring data consistency and comparability. In any clinically suspect case of CJD, a firm diagnosis still depends on neuropathologic examination.

SPORADIC PRION DISEASE

Sporadic CJD is typified clinically by progressive multifocal neurologic dysfunction, myoclonic involuntary movements, a terminal state of severe cognitive impairment, and death within a few months (Roos et al. 1973). Prodromal symptoms prior to the occurrence of frank neurologic dysfunction are common and include alteration of personality or behavior, insomnia, anorexia, and depression (Table 1). Because such symptoms may occur as part of other background illnesses, their presence may not be related to CJD. In about 5% of patients with sporadic CJD, the onset of neurologic dysfunction is abrupt, mimicking stroke (McNaughton and

Table 1. Prodromal symptoms in sporadic CJD ($n = 93$) and comparison with frequency of these symptoms in hospital controls ($n = 186$): England and Wales 1980–1984

	CJD cases (%)	Controls (%)
Depression	37	34
Tiredness	47	50
Sleep disturbance	28	58
Abnormal sweating	13	16
Abnormal appetite	22	19
Weight loss	46	24
Diarrhea	6	8
Headache	12	9

Will 1997). For example, patients may present with the acute development of hemiparesis or brain stem dysfunction. In these patients, the diagnosis may become apparent when their condition fails to stabilize or improve and progressive neurologic dysfunction, particularly myoclonus, develops.

The presenting symptoms and signs of sporadic CJD are varied, presumably reflecting the disparate areas of cerebral cortex that may be initially involved in the pathologic process. The most common initial symptoms and signs are listed in Table 2, with cognitive impairment and ataxia predominating. Rare presentations include initial cortical blindness (the Heidenhain variant) (Heidenhain 1929), which is usually associated with visual hallucinations, and a pure cerebellar syndrome (the Brownell-Oppenheimer variant) (Brownell and Oppenheimer 1965). The most striking feature of sporadic CJD is the relentless and multifocal progres-

Table 2. Clinical characteristics of sporadic CJD: Symptoms at presentation (percentages)

	England and Wales 1970–1979 ($n = 124$)	France 1968–1977 ($n = 124$)	US 1963–1993 ($n = 232$)	UK 1990–1994 ($n = 144$)
Dementia	21	29	48	19
Ataxia	19	29	33	39
Behavioral disturbance	18	30	29	15
Dizziness	11	8	13	N/A
Visual	9	17	19	10
Involuntary movements	5	1	4	2
Dysphasia	5	N/A	N/A	4
Sensory	4	2	6	N/A
Headache	3	10	11	N/A

Table 3. Clinical characteristics of sporadic CJD: Signs during course of illness (percentage)

	England and Wales 1970–1979 ($n = 124$)	France 1968–1977 ($n = 124$)	US 1963–1993 ($n = 232$)	UK[a] 1990–1994 ($n = 144$)
Dementia	100	100	100	100
Myoclonus	82	84	78	85
Pyramidal	79[b]	44	62	62
Dysphasia	62	N/A	N/A	58
Cerebellar	42	56	71	85
Akinetic mutism	39	N/A	N/A	75
Primitive reflexes	30	N/A	N/A	58
Cortical blindness	13	N/A	N/A	52
Extrapyramidal	3	60	56	34
Lower motor neuron	3	12	12	17
Seizures	9	9	19	13

[a]In this series, cases of CJD were examined in life, which may explain the higher incidence of primitive reflexes and cortical blindness.
[b]Rigidity alone classified as pyramidal.

sion of neurologic deficits: Patients who are initially ataxic may develop cognitive impairment, dysphasia, and myoclonus, whereas patients presenting initially with cognitive impairment may develop ataxia and cortical blindness. Although there is great variation in the symptoms in individual cases, the rapid evolution of deficits involving multiple cerebral areas is typical of CJD and relatively distinctive.

The signs in sporadic CJD reflect this variation in clinical symptomatology. Initial cognitive impairment or ataxia occurs most frequently (Table 3), but as the disease evolves, a range of neurologic signs are seen, most importantly myoclonus (Brown et al. 1986). In addition to the progressive cognitive impairment, which occurs in all cases of sporadic CJD, the great majority of cases develop ataxia and myoclonus, and a proportion develop dysphasia, cortical blindness, primitive reflexes, and paratonic rigidity of the limbs. Lower motor neuron signs are rare, although peripheral neuropathy has been described in some cases (Sadeh et al. 1990), and epilepsy occurs in only a small minority of cases despite the predominantly cortical neuropathology.

Historically, cases with a combination of dementia and motor neuron disease, the amyotrophic variant, were regarded as a form of CJD, but these cases are no longer classified as prion diseases because transmission studies have been almost uniformly unsuccessful (Salazar et al. 1983).

The differential diagnosis of CJD is wide (Table 4), and the clinical distinction from other forms of dementia rests on the occurrence of focal

Table 4. Differential diagnosis of CJD in systematic surveys

	England and Wales[a] 1970–1979 ($n = 42$)	England and Wales[b] 1980–1984 ($n = 95$)	UK[c] 1990–1994 ($n = 41$)
Alzheimer's disease (ATD)	20 (3)[d]	28 (5)[d]	17
ATD + multi infarct dementia (MID)	—	—	5
MID	3 (1)[d]	4	4
Motor neuron disease	5 (1)[d]	2 (1)[d]	2
Corticostriatonigral degeneration	3 (3)[d]	—	—
Idiopathic encephalopathy	2 (2)[d]	45 (5)[d]	1
Hydrocephalus	—	4 (3)[d]	—
Parkinson's disease	3 (2)[d]	1	—
Pick's disease	1 (1)[d]	1 (1)[d]	1
Others: ($n = 1$)	ATD and MS limbic encephalitis herpes simplex encephalitis familial spinocerebellar degeneration multiple cerebral abscess	stroke limbic encephalitis atypical demyelination carcinomatous meningitis glioblastoma post-anoxic encephalopathy cerebral metastases hepatic encephalopathy catatonic schizophrenia progressive supranuclear palsy	diffuse Lewy body diseases (DLBD) DLBD + ATD cerebellar degeneration multi-system atrophy cortico-basal degeneration viral encephalomyelitis metastatic carcinoma hypoxia epilepsy progressive supranuclear palsy spongiform myelinopathy

[a]Case certified as dying of CJD in which the diagnosis was reclassified after review of clinical data or pathology.

[b]Cases referred as suspect CJD in which the diagnosis was reclassified after review of clinical data or pathology.

[c]Cases referred as suspect CJD with neuropathologic confirmation of alternative diagnosis.

[d]Numbers of cases with neuropathologic data in parentheses.

neurologic signs early in the course of the illness, the rapid clinical evolution, and, in particular, the total illness duration. In systematic surveys the mean duration from first symptom to death in CJD is about 5 months (Will and Matthews 1984), which is relatively distinct from other more common forms of dementia. Furthermore, in CJD many patients decline rapidly to a state of akinetic mutism and then survive for some weeks or months. In these cases, the rapidity of disease evolution is quite different from more common forms of dementia and raises the possibility of other diagnoses including stroke, encephalitis, and brain tumor.

About 90% of CJD cases are recognized clinically, but there are cases with a prolonged clinical illness lasting for months or years, and in some of these cases, the insidious progression may mimic other conditions such as Alzheimer's disease (Brown et al. 1984). The clinical distinction from Alzheimer's disease may be particularly difficult if there is associated myoclonus, which is often not recognized to be a feature of Alzheimer's disease. A definitive diagnosis of CJD requires neuropathologic confirmation. In cases of CJD with long clinical duration the diagnosis may only be recognized postmortem.

FAMILIAL CJD

The features of familial CJD vary according to the underlying mutation associated with the particular case. Overall, there is an earlier age of onset in familial cases and a more prolonged duration of illness. A description of the clinical features of familial CJD can be found in Chapter 10.

EPIDEMIOLOGY

In 1968 it was established that CJD was experimentally transmissible (Gibbs et al. 1968), and this stimulated epidemiologic research aimed at identifying the mechanism of natural transmission, should this indeed occur. Epidemiologic surveys of CJD have varied in their scope, ranging from localized study to systematic national studies, and also in their methodologies for case ascertainment and criteria for case definition. Nonetheless, the conclusions of these studies are remarkably consistent, and baseline epidemiologic parameters for CJD have been established (Brown et al. 1979; Galvez et al. 1980; Kondo 1985; Will et al. 1986), which have been important in assessing hypotheses generated from basic scientific research and in the identification of any change due to novel risk factors for human prion disease.

The majority of studies indicate that males and females are affected with equal frequency, although some studies have shown a relative female excess of cases (Will et al. 1986). In the US, 1979–1994 (Holman et al.

1996), 52.9% of the reported CJD deaths were in women, but the age-adjusted CJD death rate in men was slightly higher than in women. The higher absolute number of deaths from CJD in women primarily reflected the greater proportion of women in the older age groups that were at highest risk for CJD. The annual incidence of CJD is often estimated to be about one case per million population. Annual incidence and mortality rates are similar in CJD because the illness is invariably fatal and its mean duration is relatively short. Annual mortality rates in some early studies were lower than the quoted figures, ranging from 0.09 (Matthews 1975) case per million in one study including only neuropathologically verified cases to about 0.5 case per million in more systematic and comprehensive surveys (Brown et al. 1979; Cousens et al. 1990). More recent systematic surveys of CJD have identified incidence or mortality rates (Table 5) with

Table 5. Systematic studies of CJD worldwide

Country	Period	Incidence: cases/ million	Country	Period	Incidence: cases/ million
Austria	1969–1985	0.18	Italy	1958–1971	0.05
	1986–1994	0.67		1993–2001	1.10
	1995–2001	1.15			
Australia	1970–1980	0.66	Japan	1975–1977	0.45
	1987–1996	1.07		1985–1996	0.58
	1997 2001	1.39			
Chile	1955–1972	0.10	Netherlands	1993–2001	1.00
	1973–1977	0.31			
	1978–1983	0.69	New Zealand	1980 1989	0.88
Czechoslovakia	1972–1986	0.66	Slovakia	1993–2001	1.17
France	1968–1977	0.34	Switzerland	1995–2001	1.50
	1978–1982	0.58	United	1964–1973	0.09
	1993–2001	1.52	Kingdom	1970–1979	0.31
Germany	1979–1990	0.31		1980–1984	0.47
	1993[a]–2001	1.14		1985–1989	0.46
				1993–2001	0.99
Israel	1963–1972	0.75			
	1963–1987	0.91	US	1973–1977	0.26
				1983–1990[b]	1.1
				1991–1998[b]	1.1

[a]Extrapolated from part-year data.
[b]Age-adjusted to the standard US projected 2000 population.

figures approaching or exceeding one case per million in many countries. The efficiency of case ascertainment may have improved with increased awareness of CJD among the neurologic community and, more recently, by other health practitioners and indeed the community at large. In Switzerland, annual incidence rates of between 2.7 and 3.9 per million have been reported for 2001 and the first quarter of 2002 (Glatzel et al. 2002). Whether these estimates reflect variation about a mean or some other cause is not yet established. In the US, based on reviews of national death certificate data, the recent annual CJD death rate has been 0.97 (1999–2000) (L.B. Schonberger, pers. comm.).

CJD almost certainly occurs worldwide, and countries in which the condition has not been reported are mainly in central Africa where diagnostic facilities are limited and in which the majority of the population may not live to the age at which CJD is likely to develop (Fig. 1). Since the age-adjusted CJD mortality rate for African-Americans in the US, however, is only 40% of that for whites, the age-adjusted incidence of CJD in many African countries may, in fact, be unusually low.

There are two populations, Libyan-born Israelis (Goldberg et al. 1979) and certain populations in restricted areas of Slovakia (Ferak et al. 1981), in which the incidence of CJD is 60–100 times greater than expected. The hypothesis that these clusters might be causally related to dietary exposure to the scrapie agent was not supported by a case-control study, and a high index of inbreeding was identified many years ago in the Slovak population (Ferak et al. 1981). The crucial study linking familial

■ Country Reporting CJD

Figure 1. Countries in which CJD has been reported.

forms of prion disease to a specific mutation of the PrP gene (Hsiao et al. 1989) led to genetic investigation of the high-incidence populations in Israel and Slovakia. It is now clear that the increased incidence in these populations is linked to a high prevalence of codon-200 mutations of the PrP gene (Goldfarb et al. 1990; Hsiao et al. 1991).

Systematic national surveys of CJD suggest that cases are randomly distributed within individual countries (Fig. 2) (Cousens et al. 1990), and cluster analysis in many surveys indicates an apparently random distribution of cases in space and time. An increased incidence has been found in urban populations in some studies (Brown et al. 1979; Galvez et al. 1980), but this may be due to improved ascertainment as a result of more available medical services in urban areas. In the US, 1979–1990, the proportion of CJD deaths in metropolitan counties of residence (75.4%) was similar to the proportion of the US population in such areas in 1980 and 1990 (76.2% and 77.5%, respectively) (Holman et al. 1995). With the potentially long incubation period, analysis of case distribution by residence of death might be misleading. One study that analyzed residence throughout life showed no convincing evidence of space-time clustering (Cousens et al. 1990) although a cluster of cases of sporadic CJD has recently been reported in Australia (Collins et al. 2002).

Sporadic CJD is largely a disease of late middle age with a peak incidence and mean age at death in the 65–79-year age group (Fig. 3). The

Figure 2. Map of the UK showing the distribution of cases of definite or probable sporadic CJD according to residence at death (1990–2002).

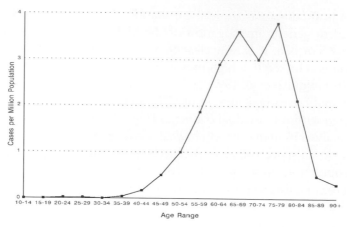

Figure 3. Age-specific mortality rates for sporadic CJD (UK, 1990–2002).

rarity of cases of sporadic CJD in patients aged less than 30 years has been confirmed in all systematic surveys, although there have been occasional case reports of CJD in teenagers (Monreal et al. 1981; Brown et al. 1985). A sharp drop in incidence over the age of 79 years has been consistent in all surveys, although some doubt persists about the efficacy of case identification in older patients. Recent studies in Europe have shown higher incidence rates in the elderly than in previous surveys (Will et al. 1998), but the age-specific incidence rate in those over the age of 79 years remains significantly lower than in the 65–79-year age group. The identification of clinically unrecognized cases of CJD in postmortem studies in the elderly demented is exceptional. Current evidence suggests that the progressive fall in incidence in older patients is accurate and not simply an artifact of less efficient case ascertainment.

Case-control studies have been carried out in a number of populations in order to look for risk factors for the development of CJD (Bobowick et al. 1973; Kondo and Kuroiwa 1982; Davanipour et al. 1985; Harries-Jones et al. 1988; van Duijn et al. 1998; Collins et al. 1999; Ward et al. 2002). The methodologies of these studies have varied; all have necessarily included only small numbers of cases and controls, and the findings have been contradictory (Table 6). A meta-analysis of case-control studies has found an increased risk of CJD in those individuals with a family history of the condition and also an increase in risk in those with a family history of other neurodegenerative diseases including Parkinson's disease (Wientjens et al. 1996). The latter finding is unexpected and may, in some studies, be related to the exclusion of control subjects with dementia. Such an exclusion could artificially lower the prevalence of a

Table 6. Significant risk factors for CJD in controlled studies

Author	Method	Risk factors
Bobowick et al. (1973)	38 "selected" cases; healthy controls	none
Kondo and Kuroiwa (1982)	population study: 60 cases; healthy controls	trauma in males
Kondo (1985)	88 autopsied cases; autopsied controls	organ resection
Davanipour et al. (1985)	26 cases; 40 controls	trauma or surgery to head or neck; other trauma surgery needing sutures tonometry
Davanipour et al. (1985)	as above	roast pork, ham, underdone meat, hot dogs
Davanipour et al. (1985)	as above	contact with fish, rabbits, squirrels
Harries-Jones et al. (1988)	92 cases; 184 controls	herpes zoster keeping cats contact with pets other than cats/dogs dementia in family
van Duijn et al. (1998)	405 cases; 405 controls	consumption of raw meat consumption of brain frequent exposure to leather products exposure to fertilizer consisting of hoof and horn
Collins et al. (1999)	241 cases; 784 controls	number of surgical procedures residence or employment on a farm or market garden
Ward et al. (2002)	326 cases; 326 controls	surgery, especially in females ear piercing psychiatric consultation

family history of dementia in control subjects in comparison to the CJD cases. Although fewer than 1000 CJD cases have been enrolled in published case-control studies to date and not all risk factors have been consistently examined, these studies are important. They provide no convincing evidence of an increased risk of CJD through dietary habits, occupation, animal contact, or a history of blood transfusion. Because of inherent statistical and other limitations of case-control studies, however, they cannot rule out possibly small increases in risk, for example, in relation to prior surgery (Collins et al. 1999; Ward et al. 2002).

The epidemiologic studies have not resulted in an explanation for the cause of sporadic CJD. The apparently random geographic distribution of cases and the absence of any common risk factor for disease development have led some experts to hypothesize that most sporadic cases do not have an environmental source of infection and that a spontaneous alteration in the structure of PrP is the mechanism of disease initiation, possibly facilitated by an occasional somatic mutation (Prusiner 1994). This theory fits much, but not all, of the epidemiologic evidence. For example, the decrease in incidence of CJD over the age of 79 years remains puzzling. It is also important to note that exclusion of risk factors can be difficult in a disease with a prolonged incubation period, particularly when evidence of risk factors is derived from surrogate witnesses.

INFECTIOUS PRION DISEASES

Iatrogenic CJD

The very fact that CJD has been demonstrated to be transmissible in laboratory studies raises the possibility that transmission from person to person might occur in the course of medical treatment. Iatrogenic transmission of CJD was first suggested in 1974 by a case of CJD in a recipient of a corneal transplant derived from a patient who also died of CJD (Duffy et al. 1974), although in retrospect it is likely that three of Nevin and Jones's original cases were transmitted by contaminated neurosurgical instruments in the early 1950s (Jones and Nevin 1954; Nevin et al. 1960). Transmission of CJD by depth electrodes in two cases was suggested by an article in 1977 (Bernoulli et al. 1977), and the tragedy of CJD in human growth hormone recipients was first recognized in 1985 (Koch et al. 1985). More recently, the iatrogenic transmission of CJD by human dura mater grafts has been identified (CDC 1987; Thadani et al. 1988). All these transmissions have involved cross-contamination with material in or adjacent to the brain where the expected titers of infectivity would be highest (Table 7). All have involved parenteral inoculation either by surgery or by intramuscular injection. Strong evidence exists that the risk, if any, of transmission of CJD by blood products is low; no such transmission has yet been documented. It is uncertain, however, whether this undetected risk reflects the absence of the agent of CJD in the blood of infected persons or the safety of a low-titer tissue.

Table 7. Total cases of iatrogenic CJD worldwide

Mode	Number of cases	Mean incubation period (years)	Clinical
Neurosurgery	4	1.6	visual/cerebellar/dementia
Depth electrodes	2	1.5	dementia
Corneal transplant	3	15.5[a]	dementia
Dura mater	136	6[b,c]	visual/cerebellar/dementia
Human growth hormone	165	12[b]	cerebellar
Human gonadotrophin	5	13	cerebellar

Data courtesy of Dr. P. Brown and Dr. L.B. Schonberger.
[a]Range 1.5–30 years.
[b]Estimated on incomplete data; mean is 21 years for 25 US hGH-related cases.
[c]Median is 8–9 years for initial 65 dura mater-associated cases in Japan.

Neurosurgical Transmission

The seminal publications by Nevin and Jones were crucial to the definition of clinical, investigative, and neuropathologic features of sporadic CJD. In retrospect, three of these cases may have been caused by iatrogenic transmission via contaminated neurosurgical instruments. Review of the cases in the original papers indicated that some of the cases were investigated in the same hospitals and that hospital admissions overlapped. A small number of patients with CJD underwent invasive procedures, including brain biopsy, and concurrent inpatients underwent neurosurgery for other conditions; e.g., removal of meningioma or cortical undercut. The readmission of three of these cases 18 months to 2 years later to the same hospitals with CJD provides strong circumstantial evidence of transmission through contaminated neurosurgical instruments. A similar case has been described in France (Foncin et al. 1980). The clinical features are indistinguishable from sporadic CJD in these putative cases of neurosurgical transmission by an intracerebral inoculation.

Corneal Grafts and Depth Electrodes

The first description of potential iatrogenic transmission of CJD was in 1974 (Duffy et al. 1974). A 55-year-old woman developed CJD 18 months after transplantation of a cadaveric corneal graft, which had been obtained from a 55-year-old man who had died of pathologically confirmed CJD. A second case of CJD in association with corneal grafting was published in 1994 (Uchiyama et al. 1994), but in this case the diagnosis in the corneal donor was not established. A third case was described in 1997 in which both donor and recipient died from pathologically confirmed CJD, although the latency between corneal transplant and the development of CJD was 30 years (Heckmann et al. 1997). These cases, particularly the case described in 1974, provide strong circumstantial evidence of transmission of CJD through cadaveric corneal grafts. Although epidemiologic data indicate that the risks to any individual patient from corneal graft are very low, the importance of screening donors is underscored by the documentation that this route of transmission is possible.

In 1977 two patients, aged 25 years and 19 years, developed CJD 2 years after stereotactic EEG recordings (Bernoulli et al. 1977). The instruments had previously been used in a 69-year-old patient with rapidly progressive dementia and myoclonus, who was later confirmed as having died of CJD. The electrodes had been "sterilized" with ethanol and formaldehyde vapor. The suspicion that CJD had been transmitted through inadequately sterilized depth electrodes, raised by the young age of the two secondary cases, was subsequently supported by transmission of CJD to a chimpanzee 18 months after intradural implantation of the suspect electrodes (Brown et al. 1992).

Dura Mater Grafts

The potential for the transmission of CJD from patient to patient by cadaveric dura mater grafts was first recognized after a case report in the US in 1987 (CDC 1987). There have now been at least 136 cases worldwide, including 88 cases identified in Japan (Brown et al. 2000; L.B. Schonberger, pers. comm.). Almost all of these cases, including at least 54 out of the initial 57 of the Japanese cases (Hoshi et al. 2000), involved the insertion of Lyodura brand grafts processed before May, 1987. Other known or possible exceptions to the use of such Lyodura grafts include one case in the UK in which the source was unknown (Will and Matthews 1982), one case in Italy in which locally produced dura was implanted (Masullo et al. 1989), and one case in the US in which a Tutoplast brand of dura mater graft was used (Hannah et al. 2001). The latter case led to a voluntary US recall of Tutoplast lots processed before April 1994. In 1987, the US Centers for Disease Control (CDC) published a description of differences between the processing of Lyodura and other similar products and suggested that Lyodura may be associated with a higher risk for transmitting CJD than other dura mater products used in the US (CDC 1987). Also in June, 1987, representatives of B. Braun Melsungen A.G. reported that as of May 1, 1987, their procedures for collecting and processing dura were revised to reduce the risk of CJD transmission (Janssen and Schonberger 1991). These revised procedures included conversion from batch to individual processing of dura mater and treatment of each dura mater graft with 1.0 N sodium hydroxide. The large number of Lyodura-associated cases of CJD has provided strong evidence for cross-contamination within the production process. The risk of transmission of CJD through dura mater grafts had been thought to be low because only a small number of recipients would receive contaminated material from any individual infected donor.

For the initially reported 57 dura mater-associated cases of CJD in Japan, the mean latency period from receipt of the graft to the onset of CJD was 8.2 years, range 1.3–16.1 years (Hoshi et al. 2000). For the 114 reported cases worldwide by the millenium, the interval from the implantation of the graft to the development of clinical disease ranged from 1.5 to 18 years, with a mean of about 6 years (Brown et al. 2000).

Clinically, the majority of patients with dura mater-related CJD present with symptoms and signs consistent with sporadic CJD. In some cases the presentation is less typical, and in a few patients a cerebellar syndrome has been described. The anatomical site of the graft may influence clinical presentation, and two cases in the UK presented with a cerebellar syndrome 5–10 years after receiving posterior fossa grafts following excision of cerebellar tumors.

Estimating the risk of developing CJD following dura mater grafts is hampered by the absence of precise information on the number of persons who received grafts each year and their subsequent survival. The major risk, however, appears to be in those individuals who received grafts between 1981 and 1987 (Fig. 4). It was roughly estimated in Japan that between 1983 and 1987, up to 100,000 patients may have received a Lyodura graft. By the year 2000, within a 14-year period after the receipt

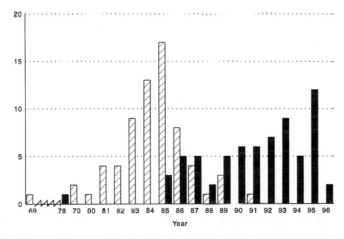

Figure 4. Dura mater cases worldwide (as of 1998) shown by year of operation and year of onset of symptoms for CJD.

of a dural graft in Japan, 1983–1987, 51 of these recipients had died of CJD. These cases indicated that within 14 years of receipt of a dural graft, the minimum risk was approximately 1 case of CJD per 2000 recipients.

In many countries, human dura mater grafts have been withdrawn from use, but some countries continue to use this material. In the US, it has been recommended that human dura mater grafts may still be used provided there is stringent donor screening, including neuropathologic evaluation of donors, and the processing of each dura separately to avoid possible cross-contamination. In addition, treatment of dura with 1 N sodium hydroxide for one hour, or some other comparable inactivating procedure, has been recommended. Because even these steps may not completely eliminate the potential for an infectious graft, some surgeons may choose the alternative use of autologous fascia lata grafts, fascia temporalis, or synthetic substitutes. Whether autologous fascia lata grafts, which involve more extensive surgery, or other forms of dural implant that may have other disadvantages carry significantly lower risks to neurosurgical patients than human dura mater grafts remains controversial.

Human Pituitary Hormones

The treatment of short stature in children, frequently endocrinologically determined, with human pituitary-derived growth hormone was initiated in the late 1950s, and the effectiveness of this therapy resulted in the subsequent treatment of large numbers of individuals. About 30,000 children had been treated with hGH worldwide by 1985. Small numbers of women were treated for infertility with human pituitary gonadotrophin over a similar period in a number of countries. In 1985, the occurrence of CJD in a hGH recipient (Koch et al. 1985) indicated the possibility of iatrogenic transmission. The remarkably young age of the patient, and the occurrence of a further case in the US (Gibbs et al. 1985) and a case in the UK (Powell-Jackson et al. 1985) in the same year provided strong circumstantial evidence of transmission of CJD via hGH. Since then, CJD has developed in over 150 hGH recipients in a number of countries including the US, UK, France, New Zealand, and the Netherlands (Table 8). As of December 2002, the reported overall proportion of CJD cases in the total recipient population of these five countries was about 1 in 75, consistent with laboratory decontamination studies (Pocchiari et al. 1991), but this proportion varies between countries for reasons that have not been fully established. In the US, the proportion of cases in relation to the total number of recipients was approximately 1 in 300; in the UK, 1 in 45; and in France, 1 in 20 (Brown 1996; L.B. Schonberger, pers. comm.). Evidence

Table 8. Numbers of deaths from HGH-related CJD by country

Country	Number of deaths
France	89[a]
UK	41
US	25[b]
New Zealand	5[a]
Netherlands	2
Brazil	1[a]
Australia	1[a]
Quatar	1[a]

[a]Data courtesy of Dr. P. Brown.
[b]Includes two likely HGH-related CJD cases under investigation.

is accumulating that differences in the methods of hormone preparation may be influencing these proportions. Through 2002 in the US, for example, no case of CJD had occurred among persons who began hGH treatment after 1977, the year when the method used to extract hormone from pituitary glands was changed. The change included use of a column purification step now known to significantly reduce but not eliminate experimentally introduced infectious material. Sufficient time has elapsed to have expected at least one case of CJD to have occurred among the post-1977 hGH recipients were their CJD risk identical to that of the pre-1977 hormone recipients. The estimated proportion of CJD cases among those receiving some of the hormone produced before 1977 was slightly below 1 case in 100 recipients. In addition to differences in hormone preparation, these differences in proportions may reflect differences in the periods of follow-up and the characteristics of the donor and recipient populations. In New Zealand, the proportion of cases to recipients was 1 in 37, but all 5 hGH-associated cases in this country appeared in 46 persons (11%) who received hormone from the US. This higher proportion of cases to recipients in New Zealand compared to the US is unexplained, but presumably reflects differences primarily in the final products used in the two countries rather than the susceptibilities of the recipients.

The methodology of growth hormone production has varied with time and between production centers, and it had been hoped that hGH produced using a column chromatography step might be of low risk on the basis of decontamination studies. This hope remains in the US because of the absence of CJD cases among recipients exposed only to US hormone purified with a column chromatographic method. However, cases of CJD have developed in recipients of hGH produced using at least one column chromatographic method (Billette de Villemeur et al. 1991).

Contamination of the hormone preparation occurred when pituitary glands derived from patients who died from or were incubating CJD were included in the production process. The necessity to pool many thousands of glands in order to produce hGH, and cross-contamination of multiple lots during production, presumably aggravated the risk. The powerful circumstantial evidence suggesting a causal link between hGH and CJD has been supported by transmission studies in which a squirrel monkey developed a spongiform encephalopathy following inoculation with 1 of 76 potentially contaminated lots of hGH in the US (Gibbs et al. 1993). The long incubation period in this animal experiment and the inability to transmit CJD from any of the other potentially contaminated lots of hGH support the conclusion that the infectivity levels in affected lots were extremely low. In the US, the mean incubation period calculated from the midpoint of therapy to the onset of CJD has been increasing slightly over time; for the initial 25 US cases, including two likely cases under investigation, this mean is 21 years.

Cases of hGH-related CJD continue to occur, but the current numbers of cases indicate that the large majority of hGH recipients are unlikely to develop the disease. In Europe, in contrast to the US, the incubation period ranges from 4.5 to 38 years with a mean of about 12 years. The reasons for the shorter incubation periods in Europe are unknown but suggest exposure to possibly higher levels of infectivity. In the absence of any test for the presence of infectivity, there is a population of many young individuals at risk of CJD, and even low dosage exposure may have resulted in transmission (Croes et al. 2002). hGH was withdrawn in most countries in 1985 and human pituitary gonadotrophin has also been withdrawn in many countries following the occurrence of CJD in four recipients in Australia (Cochius et al. 1990, 1992).

The clinical features of human pituitary hormone-related CJD are distinct from sporadic CJD. In the great majority of cases, the initial presentation involves a progressive cerebellar syndrome, and other features, including dementia, develop late, if at all. It is possible that the route of inoculation of the infectious agent may be an important determinant to clinical expression of disease. In kuru, presumed to be due to a peripheral route of infection, cerebellar signs predominate in the early stages as in human pituitary hormone recipients, whereas in iatrogenic CJD due to central inoculation, the clinical features are similar to classic CJD.

Possibility of Transmission through Other Routes, e.g., Blood

All cases of iatrogenic CJD have involved cross-contamination with potentially high titers of infectivity by a parenteral route. However, the possibility of transmission of CJD via other tissues must be considered. Some laboratory transmission studies of prion diseases have demonstrated the presence of infectivity in blood, although the validity of the positive studies in patients with CJD has been questioned (Brown 1995). In rodent models, the infectivity of blood is present early in the incubation period well before development of clinical disease (Diringer 1984; Casaccia et al. 1989). These studies of infectivity usually maximize the chances of successful transmission by the inoculation of concentrated blood samples, such as buffy coat, by the most efficient, intracerebral route. Infectivity has been reported in blood derived from CJD patients in four separate laboratories involving rodent inoculations. In contrast, no infectivity of blood was reported with CJD or kuru in tests using primates as the assay animal. The primate assays are more sensitive to human prions than rodent assays. Thus, the negative results using the primate assays, and a number of "puzzling" findings in the four positive experiments using rodents, indicate that the presence of CJD infectivity in human blood is not clearly established. In contrast to the difficulties in clearly demonstrating the infectivity of blood in CJD patients, the infectivity of blood in rodent models of CJD is well established.

The studies of the infectivity of blood in prion diseases suggest a theoretical possibility that blood or blood products might pose a risk of iatrogenic transmission of CJD. The existing reports of infectivity in the blood of CJD patients, whether valid or not, however, do not establish the magnitude or even the existence of the possible risk of iatrogenic transmission through blood or blood products. To help assess this risk, epidemiologic data are used.

Up until the end of 2002, no convincing instance of transmission of CJD by blood or blood products has been reported. A claim that there were four transfusion-related cases of CJD in Australia (Klein and Dumble 1993) has not been substantiated, and the report of a French case that potentially implicated transmission by albumin or liver transplant was most likely fortuitous, as suggested by the remarkably short potential incubation period (Creange et al. 1995). There have been no cases of CJD identified in hemophiliacs, and case-control studies do not suggest an increased risk of CJD through previous blood transfusion (Esmonde et al. 1993; Wientjens et al. 1996; van Duijn et al. 1998; Collins et al. 1999). This evidence cannot preclude the possibility that CJD may be occasionally transmitted through blood or blood products, but the balance of evidence indicates that if such transmission occurs at all, it is very rare. As a precaution, in many countries patients at higher risk of CJD, including those with a family history of CJD or those with a past history of hGH treatment, are excluded as blood donors.

The relatively reassuring evidence in relation to classic CJD may not be applicable to vCJD, as this condition is caused by a different infectious agent (Bruce et al. 1997) and recent studies have established the transmission of BSE infection by blood transfusion in sheep (Houston et al. 2000; Hunter et al. 2002). There is evidence that the level of PrP in peripheral tissues and infectivity is higher in vCJD than in classic CJD (Hill et al. 1997b; Bruce et al. 2001). The proportion of the population in the UK who are incubating vCJD and may be acting as blood donors is unknown. Blood products derived from patients who subsequently develop classic CJD have been withdrawn in some countries, and similar withdrawals of blood products derived from vCJD blood donors have been initiated in the UK. The degree to which such a preventive measure reduces the theoretical risk of transmission of CJD or vCJD largely depends on the proportion of the potentially infectious blood or blood products that are withdrawn before they are used. Since for the endemic form of CJD this proportion is low, CJD-related withdrawals can have only a very limited effect on reducing the theoretical risk of transmission of CJD by blood or blood products.

Infectivity titers in blood are relatively low in experimental studies, and procedures that reduce infectivity, for example, filtration, centrifugation, or column chromatography, may significantly reduce risk. In the UK other measures such as leuco-depletion of all blood donations have been introduced in response to the theoretical risks of vCJD. Blood and blood products are an important part of medical treatment and may be lifesaving. Efforts to reduce or avoid the theoretical risk of transmission of CJD or vCJD must be balanced against the public health consequences of such efforts, including the potential creation or aggravation of shortages of blood products for specific therapies.

Iatrogenic CJD: Precautions and Implications for Public Health

The occurrence of iatrogenic CJD is a medical tragedy. The first recommendations for precautions in tissue transfer in CJD were published in 1977 (Gajdusek et al. 1977), years before the identification of hGH- or dura mater-related CJD. The long latency between the initiation of specific therapies and the identification of a CJD-related risk indicates the need for caution and continued vigilance regarding the case-to-case transmission of human prion diseases.

Guidelines have been drawn up in some countries to minimize the risk of iatrogenic transmission of CJD and modern sterilization techniques, for example porous load autoclaving at 134°C for 18 minutes, have been demonstrated to significantly reduce infectivity. In some countries the destruction of all neurosurgical or ophthalmological instruments used on cases of CJD is obligatory. However, no cases of transmission of CJD via neurosurgical instruments have been documented since 1980.

hGH was withdrawn in many countries in 1985 and has been replaced by recombinant therapy. Dura mater grafts are no longer used in many countries, and donor screening will significantly reduce any risk in countries that continue to use this material. There has been no well-documented case of transmission of CJD through blood or blood products, and individuals at greater risk of developing CJD have been excluded as blood donors in many countries.

These measures significantly reduce the possibility of iatrogenic transmission of CJD but do not exclude all risk. Guidelines may not be followed and tissues may be derived from individuals incubating CJD with no clinical evidence of disease. It is important to continue to study CJD systematically in order to maintain confidence in the safety of blood products and to identify any novel and perhaps unexpected route of iatrogenic transmission.

Kuru

Kuru is a fatal CNS disease that is geographically restricted to the Okapa area of the highlands of Papua New Guinea (Gajdusek and Zigas 1957). At the peak of the kuru epidemic in the late 1950s and early 1960s, 200 people died each year of kuru, which caused over half the deaths in the most affected communities (Alpers and Gajdusek 1965; Alpers 1987). The total population within the kuru region (all communities with a history of kuru) was 36,000 at that time, although the population actually at risk of kuru was probably closer to 20,000. Over 2,700 cases of kuru have been documented since 1957 in a case file of well over 3,000 patients investigated. The identification of the mechanism of disease transmission followed years of meticulous research carried out in very difficult conditions by Dr. Carleton Gajdusek and colleagues (Glasse 1967), and the transmissibility of human prion disease was first confirmed in 1966 following the intracerebral inoculation of kuru brain tissue in primates (Gajdusek et al. 1966).

The epidemiologic pattern of kuru has been unusual (Alpers 1987). The disease predominantly affected women and their children of either sex in the early years of the epidemic. With time, the incidence of the disease has declined and the proportion of affected adult males and females has become more similar (Fig. 5), and no children born after 1959 have been affected. The cause of kuru was unknown, and theories included a

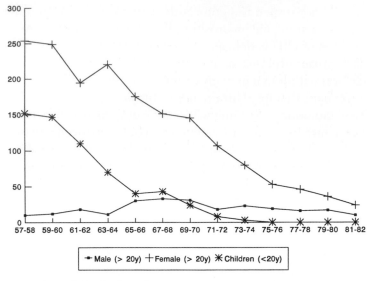

Figure 5. Number of deaths from kuru in 2-year periods between 1957 and 1982 in adults aged >20 years and children aged <20 years.

toxic, dietary exposure, infection with conventional viruses or bacteria, or dietary deficiency. All these possibilities were excluded by investigation including a detailed study of dietary factors and examination of tissues, including brain histology. Neuropathologically there was no evidence of an inflammatory response (Fowler and Robertson 1959), and the possibility of a genetic basis of kuru was raised by the unusual sex and age distribution of cases. The possibility of transmission through endocannibalism was raised by anthropological enquiry (Glasse and Lindenbaum 1992), and the decline in the epidemic following the cessation of endocannibalism in the late 1950s is consistent with this proposed mechanism of transmission.

Women and children consumed diseased relatives as a mark of respect, leading to a familial aggregation of cases. Virtually all tissues were eaten, including brain, viscera, and even powdered bone. In addition to the oral transmission of kuru, there is also the possibility of transcutaneous transmission through rubbing infected material on the skin. Males, and in particular elderly males, were involved at times in the mortuary rituals, but the excess of cases in females and children is consistent with the available descriptions of the rituals, since they, and not the men, ate the internal organs, in particular the brain. In recent years the decline in incidence of kuru has led to the linking of cases with mortuary feasts in which several members of a family, including those resident in different areas, have been affected by kuru (Klitzman et al. 1984). The incubation period ranges from 4 1/2 years to at least 40 years (based on the onset of disease 40 years after transmission ceased), and cases are still occurring, albeit at a very low rate. No precise figure for the mean incubation period is available, as affected individuals may have taken part in multiple funeral rituals and the critical exposure cannot be identified. The decline in incidence in the age group 15–20 years in the distribution of cases seen around 1960 suggests that the modal incubation period (at least in childhood transmission) is about 12 years. Over the past 5 years the incidence of kuru has been only 1 or 2 a year, and in 1998 and 2001 there were periods without a current living case, so the disease is dying out; however, epidemiological surveillance continues, in order to establish the full extent of the incubation period, which is relevant to other prion diseases.

Clinically, cases of kuru presented with a pure cerebellar syndrome (Alpers 1987). A prodromal phase characterized by headache and limb pains was universal. Initial signs of the cerebellar syndrome included astasia, midline trunkal ataxia, trunkal tremors, and titubation, followed by gait ataxia and dysarthria. Hypotonia was common in the early stages, and occasional patients developed marked flaccidity of limb musculature. In contrast to CJD, other neurologic signs such as upper motor neuron signs and myoclonus did not occur. In the later stages communication was often difficult because of severe dysarthria, but in the majority of patients, dementia was conspicuous by its absence. Even in the terminal moribund, akinetic, and mute state, most patients could make eye contact and attempt to carry out simple commands. The total illness duration in adults ranged from 6 to 36 months, mostly under 24 months.

In juveniles the clinical picture was similar to the adult form of disease. Brain stem signs were common and included titubation, nystagmus, strabismus, facial weakness, and sometimes ptosis. Other features included transient generalized hyper-reflexia and clonus. The clinical course in children was shorter than in adults, ranging from 3 to 15 months.

The scientific study of kuru has been crucial to current understanding of prion diseases. Following Hadlow's recognition of a similarity of the pathology in kuru and scrapie, transmission studies were initiated which proved that human prion diseases were caused by transmissible agents, despite no evidence of infection histologically or serologically. This remarkable discovery led to the successful laboratory transmission of CJD and initiated research into the epidemiology and pathogenesis of human prion disease and to further investigation of the nature of the infectious agent. From a clinical perspective, there is a remarkable consistency in the clinical picture in kuru, following exposure to an exogenous infectious agent, and the epidemiologic pattern of kuru has demonstrated that human diseases can be transmitted through a peripheral route, either orally or transdermally. It is also important to note that there was no evidence of case-to-case transmission of kuru through routes other than endocannibalism, and in particular, there was no evidence of vertical transmission of infectivity in kuru, despite the breastfeeding of infants by many hundreds of clinically affected mothers (Alpers 1968; Gajdusek 1990).

Variant CJD

Surveillance of CJD was instituted in the UK in 1990 in order to identify any change in this condition that might indicate transmission of BSE to the human population. In 1993, a collaborative system for the study of CJD, sharing common methodologies and criteria for case definition, was established in Europe and included national surveillance programs in France, Germany, Italy, the Netherlands, Slovakia, and the UK (Wientjens et al. 1994). One aim of this collaborative study was to identify any change in CJD that might correlate with presumed prior exposure to the BSE agent.

In 1995 and early 1996, a number of cases of CJD were identified in the UK with a clinicopathologic phenotype distinct from previous experience. By March, 1996, 10 cases of CJD had been identified with a young age at onset, mean 29 years in comparison to 66 years in sporadic CJD; a long duration of illness, mean 14 months in comparison to 4.5 months in sporadic CJD; and an unusual and remarkably uniform clinical presentation, relatively different from that previously seen in CJD (Will et al. 1996). The cases all developed early psychiatric symptoms including depression and withdrawal for months prior to the development of neurologic signs. The terminal stages were similar to classic CJD, but the EEG did not show the characteristic changes so often seen in these cases. Crucially, the neuropathologic appearances, including extensive florid plaque formation, were thought to be distinct from the neuropathology observed in CJD in the past.

As of January 6, 2003, there were 129 cases of this form of CJD, designated vCJD, identified in the UK, 6 cases in France, and single cases in the Republic of Ireland, Italy, US, and Canada. It is of note that there is consensus that cases of vCJD are classified according to the country of normal residence at the time of onset of symptoms, rather than the country of presumed exposure to infection with BSE. The cases of vCJD in Ireland, US, and Canada, but not those in France and Italy, had a history of extended residence in the UK where they are likely to have been

Table 9. Differences between classic CJD and vCJD

	Classic	New variant
Mean age at death (years)	66	28
Median illness duration (months)	4	13
EEG	typical	atypical
Pathology	kuru plaques (10%)	florid plaques (100%)

exposed to BSE infection. The occurrence of cases of vCJD in France may be related to human exposure to indigenous BSE infection in cattle, travel to the UK, or exports of BSE-infected material from the UK. The lower number of cases in France in relation to the UK is consistent with the lower level of human exposure to BSE infection in France (Alperovitch and Will 2002).

The distinction between individual clinical features in vCJD and classic CJD (Table 9) is not absolute, but as a group, the cases of vCJD exhibit a different phenotype (Zeidler et al. 1997b,c; Will et al. 2000). Occasionally, cases of classic CJD may share many of the features of vCJD, and the identification of vCJD cases still rests on neuropathologic verification. The hypothesis that the neuropathologic features are distinctive is supported by the failure to identify cases with similar pathologic features in other countries, even after review of archival material (Budka et al. 2002).

The causal link with BSE depends from an epidemiologic perspective on the timing of the occurrence of vCJD and the colocalization of vCJD with BSE, as a potential novel risk factor for human prion disease (Will 1998). It is of interest that the first cases of vCJD in Canada and the US had a history of potential exposure to BSE in the UK. The US case patient was born in the UK in 1979 where she lived until 1992 before moving to the US (CDC 2002). Thus, the interval between this patient's exposure to BSE and onset of illness was almost certainly between 9 and 21 years. Other evidence on incubation periods for human prion diseases comes from iatrogenic cases and, in particular, hGH-related CJD and kuru, which both involved a peripheral route of infection. The minimum incubation period in these types of diseases is 4.5 years, with an estimated mean in hGH-related CJD cases of from 8 years (first 41 cases in France) (Brown 1996) to 21 years (first 25 cases in the US) (L.B. Schonberger, pers. comm.). The transmission of BSE to the human population would necessarily involve crossing a species barrier, which usually results in a relative prolongation of incubation period compared to within-species transmission. Human exposure to the BSE agent probably started around 1983, probably peaked in 1989, and declined subsequently (Fig. 6). The first

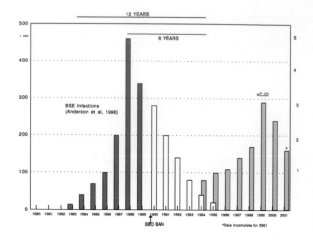

Figure 6. BSE "exposure" and the incidence of vCJD by year of onset.

cases of vCJD had disease onset in 1994, implying a minimum incubation period of CJD to the human population of ~6–12 years, consistent with data from hGH-related CJD and kuru. The slightly longer incubation period in vCJD may reflect a species barrier effect.

Analysis of risk factors has been carried out in vCJD. None of the cases has a history of neurosurgery, corneal graft, or human pituitary hormone therapy. Full sequencing of the PrP gene has excluded all known mutations of the PrP gene (Zeidler et al. 1997a), and preliminary analysis of the promoter region has also failed to identify any specific abnormality. On current evidence, these cases are not associated with mutations of the PrP gene. All tested cases in the UK and other countries (120/139) have been methionine homozygotes at codon 129 of the PrP gene, but evidence from the French cohort of hGH recipients indicates that variation at this locus may influence incubation period rather than overall susceptibility to exogenous infection (Deslys et al. 1996). It is possible that cases of vCJD expressing valine at codon 129 may occur in the future, and it is possible that the clinicopathologic phenotype may be different from the cases already identified.

vCJD occurs throughout the UK (Fig. 7), and cluster analysis suggests that these cases are randomly distributed, with the implication that any risk factor must have been widely disseminated (Cousens et al. 2001). A case-control study comparing risk factors in vCJD cases with age- and sex-matched hospital controls has identified no increased risk for vCJD in relation to past medical history, including operative procedures, occupation, or dietary factors, although all cases to date did eat beef or beef products in the 1980s. The negative findings of the dietary study do not exclude the possibility that vCJD is caused by oral exposure to the BSE agent. Information on dietary history is necessarily obtained from surro-

Figure 7. Map of the UK showing the distribution of cases of definite or probable vCJD according to residence at disease onset.

gate witnesses. In vCJD any causal exposure is likely to have taken place many years ago, probably in the 1980s, and information on past dietary exposure is known to be unreliable even with direct interview and, furthermore, there is evidence of recall bias in the study of dietary factors in sporadic CJD (National CJD Surveillance Unit 1995). Most importantly, there may have been dietary exposure to the BSE agent in a range of products that may have been contaminated intermittently. The retrospective identification of specific dietary risk factors in vCJD may be impossible.

The epidemiologic evidence implying a causal link between BSE and vCJD has been supported by other research. The neuropathologic features in macaque monkeys inoculated with BSE are similar to vCJD (Lasmézas et al. 1996), and the protein subtypes deposited in the brains of vCJD cases are similar to BSE and distinct from other forms of CJD (Collinge et al. 1996). Transmission studies in mice have demonstrated that the incubation periods in inbred strains of mice are almost identical in BSE and vCJD (Bruce et al. 1997) and that the anatomic distribution of neuropathology is also very similar in mice following transmission of either BSE or vCJD. Incubation periods and anatomic lesion profile in vCJD transmissions are distinct from sporadic CJD. These and similar studies in transgenetic mice provide strong evidence supporting the proposed causal link (Bruce et al. 1997; Hill et al. 1997b; Scott et al. 1999).

The mechanism of transmission of BSE to the human population remains speculative, although transmission through the oral route is the only reasonable current hypothesis. Information is available on the use of bovine material in the human food chain in the UK in the 1980s (Cooper and Bird 2002a,b), with particular reference to bovine CNS tissue, which on current data is the only tissue known to contain infectivity in BSE, apart from retina, dorsal root ganglia, and possibly bone marrow. Furthermore, the high titers of infectivity found in bovine CNS tissues in comparison to any other tissue suggest that consumption of CNS tissue is the most likely mechanism of cross-contamination to humans. Brain tissue is unlikely to have entered the human food chain in large quantities, but spinal cord tissue almost certainly did enter the human food chain in the UK through the production of mechanically recovered meat (Cooper and Bird 2002a). This material was used in a range of foodstuffs including burgers, sausages, meat pies, and pate, and there may have been widespread human consumption of this material in the 1980s and the early 1990s.

Current evidence suggests that vCJD is a novel condition, clinically and pathologically. The epidemiologic evidence is consistent with BSE as the causal agent, and the recent laboratory evidence provides strong support to the hypothesis of a causal link between BSE and vCJD. The number of future cases of vCJD that may occur is uncertain. Mathematical modeling indicates that the possible future number of cases could vary widely, with estimates ranging from a few hundred cases to tens of thousands of cases or more (Cousens et al. 1997; Ghani et al. 2000; d'Aignaux et al. 2001; Valleron et al. 2001), with predictions crucially dependent on variables including the incubation period. Cases of vCJD were occurring at an increasing rate in the UK (Fig. 8) (Andrews et al. 2000), but there has been a reduction in the annual number of cases identified in 2001 and 2002.

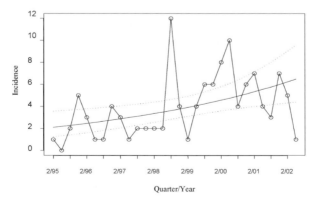

Figure 8. Observed (-o-) quarterly incidence of vCJD deaths. Fitted underlying trend (—) is given with its 95% confidence limits (···).

DIAGNOSIS AND INVESTIGATION OF HUMAN PRION DISEASES

The diagnosis of CJD is usually suspected because of the relatively stereo-typed and characteristic features, including multifocal neurologic dysfunction, involuntary myoclonic movements, and rapid progression. A small proportion of sporadic cases present atypically with relatively insidious onset and progression, and in familial cases the clinical features may depend on PrP gene analysis. Iatrogenic cases may be diagnosed on the basis of progressive neurologic dysfunction occurring in the context of a patient with a recognized risk factor for CJD. In vCJD the age of the patient remains an important clue to diagnosis, and progressive and unexplained neurologic deficits are not common in neurologic practice in patients under the age of 50 years. In all forms of CJD the diagnosis may be supported by investigations; a major research objective is to identify early diagnostic tests.

Routine hematological and biochemical indices are usually normal in all forms of CJD. Liver function tests are abnormal in about a third of cases of sporadic CJD, with either a raised bilirubin, raised liver enzymes, or both (Roos et al. 1971). These abnormalities may be an epiphenomenon reflecting side effects of medication; for example, chlorpromazine therapy or the poor general state of patients with CJD in the terminal stages of the illness.

EEG shows triphasic generalized periodic complexes (Fig. 9) in about two-thirds of all cases of sporadic CJD (Will and Matthews 1984), and the chances of finding this abnormality are enhanced if serial EEG recordings are obtained. The EEG appearances are not specific to CJD, occurring in a variety of conditions, including metabolic diseases such as hyponatremia or hepatic encephalopathy, and in toxic states; for example, lithium overdose or metrizamide encephalopathy (Will 1991). However, the clinical distinction between CJD and these other conditions is usually apparent, and in the appropriate clinical context, a "typical" EEG suggests that the diagnosis of CJD is highly likely. Preliminary criteria for the classification of the EEG changes in CJD have been proposed previously (Steinhoff et al. 1996).

Cerebrospinal fluid (CSF) examination is usually normal. An elevated CSF protein is found in about one-third of cases (Will and Matthews 1984), and rarely, evidence of localized IGG synthesis within the CNS has been observed. The CSF is acellular, and a pleocytic CSF is very much against the diagnosis of any form of CJD. Immunoassay for the 14-3-3 protein has been developed from the two-dimensional gel electrophoresis abnormalities originally described by Harrington (Blisard et al. 1990). The original publication describing the 14-3-3 immunoassay implied a

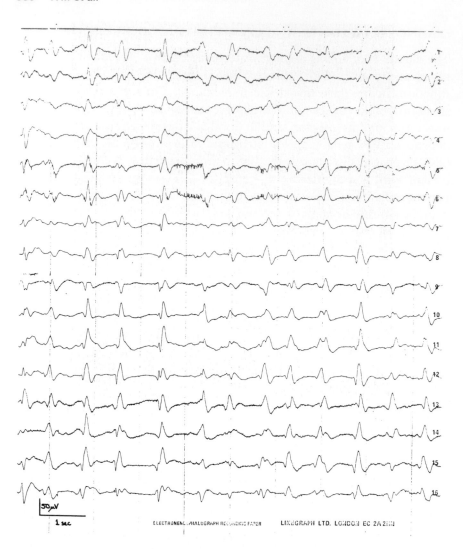

Figure 9. Typical electroencephalogram in sporadic CJD.

high diagnostic sensitivity and specificity of over 90% (Hsich et al. 1996), which has been supported in classic CJD by other studies. In the appropriate clinical context, i.e., in patients with rapidly progressive and unexplained dementia, the 14-3-3 immunoassay may be diagnostically helpful, but the test does not distinguish between subtypes of CJD, and preliminary analysis of 14-3-3 immunoassay in vCJD suggests a lower sensitivity and specificity than in classic CJD (Green 2002).

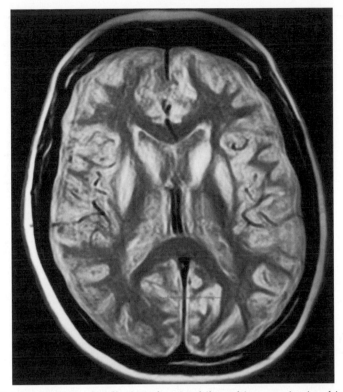

Figure 10. MRI scan in sporadic CJD showing bilateral increase in signal in the caudate and putamen.

Cerebral imaging is important in suspect cases of CJD in order to exclude other conditions such as brain tumor. Computed tomography (CT) scanning does not provide specific diagnostic information in any form of CJD, but recent findings suggest that magnetic resonance imaging (MRI) brain scanning may be more useful. The Gottingen group has reported a high incidence of high signal abnormalities on T2 imaging in the putamen and caudate regions in classic CJD (Fig. 10) (Finkenstaedt et al. 1996). In vCJD, a high proportion of cases exhibit high signal in the pulvinar region of the posterior thalamus (Fig. 11), which correlates with the neuropathologic findings in that the most marked gliotic changes are found in this region of the thalamus (Sellar et al. 1997; Collie et al. 2001). Brain biopsy has been used to diagnose CJD in life and may have a high diagnostic accuracy. However, a negative biopsy result does not exclude the diagnosis of CJD because of the possibility of sampling error, and the procedure does have risks to the patient; for example, brain hemorrhage

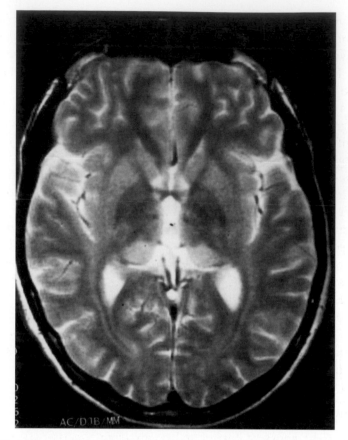

Figure 11. MRI scan in vCJD showing increased signal in the posterior thalamic regions.

or abscess formation. There are also theoretical risks to surgical staff and risk of onward transmission of infection to other patients via surgical instruments. The main role of brain biopsy is the exclusion of other potentially treatable conditions, such as cerebral vasculitis, and the procedure is mainly indicated in younger suspect cases.

Tonsillar tissue obtained postmortem from a case of vCJD has been shown to exhibit PrP immunostaining in germinal centers (Hill et al. 1997a), and limited information suggests that such staining is not found in classic CJD. Tonsillar biopsy may have to be performed using general anesthesia in suspect cases of vCJD, and there are attendant risks of hemorrhage; biopsy instruments should be destroyed after use. Whether or not this procedure should be included as part of the diagnostic workup of suspect vCJD cases remains controversial, as a positive biopsy is of no therapeutic value to the patient, although it may allow early diagnosis.

In systematic surveys of CJD, about 15% of cases are found to be familial. In these cases, about one-third have a positive family history of CJD, about one-third a family history of some other form of neurologic disease, and about one-third no significant family history of any form of neurologic disease (Laplanche et al. 1994). Utilizing a family history alone to identify hereditary forms of human prion disease is unreliable, and screening of the PrP gene in suspect cases is important in systematic surveys if hereditary forms of CJD are to be identified. Because of the implications for other family members, in most centers informed consent is necessary prior to PrP gene analysis.

A firm diagnosis of CJD or vCJD depends on neuropathologic examination postmortem, and the accurate identification of cases of CJD in population surveys is dependent on a high postmortem rate. The clinical diagnosis of sporadic CJD is relatively accurate, particularly if there is an atypical EEG, and hereditary forms of CJD may be confirmed by PrP gene analysis. Recognition of a relevant risk factor is essential to the diagnosis of iatrogenic CJD. A young age at onset, prominent early psychiatric features (Spencer et al. 2002), and the presence of posterior thalamic abnormalities on MRI may be important in diagnosis of vCJD.

SUMMARY AND CONCLUSIONS

Infectious and sporadic forms of human prion disease are rare and, although symptomatic treatment may be helpful in individual cases, these conditions are incurable and invariably fatal. There has been a remarkable increase in awareness and interest in these diseases, partly because the conditions are caused by unique causal agents, prions, and partly because of concerns about the potential public health implications of the transmission of an animal prion disease to the human population. Basic scientific research, particularly using transgenic models of prion disease, has led to insights into mechanisms of disease causation that may be relevant to other currently unexplained neurologic diseases. A major objective is to identify a marker for the presence of infectivity prior to the onset of clinical illness, and this may in turn lead to novel therapeutic approaches. The occurrence of vCJD is a stimulus to further research in view of the uncertainty about the future course of this disease and the possibility of large numbers of affected individuals. CJD in any form is a tragedy for the individuals and their relatives, and it is hoped that further scientific developments may lead to progress in therapeutic strategies (Korth et al. 2001).

REFERENCES

Alperovitch A. and Will R.G. 2002. Predicting the size of the vCJD epidemic in France. *C.R. Acad. Sci. III* **325:** 33–36.

Alpers M.P. 1968. Kuru: Implications of its transmissibility for the interpretation of its changing epidemiological pattern. In *The central nervous system, some experimental models of neurological diseases* (ed. O.T. Bailey and D.E. Smith), pp. 234–251. Williams and Wilkins, Baltimore, Maryland.

———. 1987. Epidemiology and clinical aspects of kuru. In *Prions—Novel infectious pathogens causing scrapie and Creutzfeldt-Jakob disease* (ed. S.B. Prusiner and M.P. McKinley), pp. 451–465. Academic Press, Orlando, Florida.

Alpers M.P. and Gajdusek D.C. 1965. Changing patterns of kuru: Epidemiological changes in the period of increasing contact of the Fore people with western civilization. *Am. J. Trop. Med. Hyg.* **14:** 852–879.

Andrews N.J., Farrington C.P., Cousens S.N., Smith P.G., Ward H., Knight R.S.G., Ironside J.W., and Will R.G. 2000. Incidence of variant Creutzfeldt-Jakob disease in the UK. *Lancet* **356:** 481–482.

Bernoulli C., Sigfried J., Baungarten G., Regli F., Rabinowitz T., Gajdusek D.C., and Gibbs C.J., Jr. 1977. Danger of accidental person to person transmission of Creutzfeldt-Jakob disease by surgery. *Lancet* **I:** 478–479.

Billette de Villemeur T., Beauvais P., Gourmelen M., and Richardet J.M. 1991. Creutzfeldt-Jakob disease in children treated with growth hormone (letter). *Lancet* **337:** 864–865.

Blisard K.S., Davis L.E., Harrington M.G., Lovell J.K., Kornfeld M., and Berger M.L. 1990. Pre-mortem diagnosis of Creutzfeldt-Jakob disease by detection of abnormal cerebrospinal fluid proteins. *J. Neurol. Sci.* **99:** 75–81.

Bobowick A.R., Brody J.A., Matthews M.R., Roos R., and Gajdusek D.C. 1973. Creutzfeldt-Jakob disease: A case-control study. *Am. J. Epidemiol.* **98:** 381–394.

Brown P. 1995. Can Creutzfeldt-Jakob disease be transmitted by transfusion? *Curr. Opin. Hematol.* **2:** 472–477.

———. 1996. Environmental causes of human spongiform encephalopathy. In *Prion diseases* (ed. H.F. Baker and R.M. Ridley), pp. 139–154, Humana Press, Totowa, New Jersey.

Brown P., Cathala F., and Gajdusek D.C. 1979. Creutzfeldt-Jakob disease in France. III. Epidemiological study of 170 patients dying during the decade 1968–1977. *Ann. Neurol.* **6:** 438–446.

Brown P., Preece M.A., and Will R.G. 1992. "Friendly fire" in medicine: Hormones, homografts, and Creutzfeldt-Jakob disease. *Lancet* **340:** 24–27.

Brown P., Cathala F., Castaigne P., and Gajdusek D.C. 1986. Creutzfeldt-Jakob disease: Clinical analysis of a consecutive series of 230 neuropathologically verified cases. *Ann. Neurol.* **20:** 597–602.

Brown P., Cathala F., Raubertas R.F., Gajdusek D.C., and Castaigne P. 1987. The epidemiology of Creutzfeldt-Jakob disease: Conclusion of a 15-year investigation in France and review of the world literature. *Neurology* **37:** 895–904.

Brown P., Rodgers-Johnson P., Cathala F., Gibbs C.J., Jr., and Gajdusek D.C. 1984. Creutzfeldt-Jakob disease of long duration: Clinicopathological characteristics, transmissibility, and differential diagnosis. *Ann. Neurol.* **16:** 295–304.

Brown P., Cathala F., Labauge R., Pages M., Alary J.C., and Baron H. 1985. Epidemiologic implications of Creutzfeldt-Jakob disease in a 19 year old girl. *Eur. J. Epidemiol.* **1:** 42–47.

Brown P., Preece M., Brandel J.-P., Sato T., McShane L., Zerr I., Fletcher A., Will R.G., Pocchiari M., Cashman N.R., D'Aignaux J.H., Cervenakova L., Fradkin J., Schonberger L.B., and Collins S.J. 2000. Iatrogenic Creutzfeldt-Jakob disease at the millenium. *Neurology* **55:** 1075–1081.

Brownell B. and Oppenheimer D.R. 1965. An ataxic form of subacute presenile polioencephalopathy (Creutzfeldt-Jakob disease). *J. Neurol. Neurosurg. Psychiatry* **28:** 350–361.

Bruce M.E., McConnell I., Will R.G., and Ironside J.W. 2001. Detection of variant Creutzfeldt-Jakob disease infectivity in extraneural tissues. *Lancet* **358:** 208–209.

Bruce M.E., Will R.G., Ironside J.W., McConnell I., Drummond D., Suttie A., McCardle L., Chree A., Hope J., Birkett C., Cousens S., Fraser H., and Bostock C.J. 1997. Transmissions to mice indicate that "new variant" CJD is caused by the BSE agent. *Nature* **389:** 498–501.

Budka H., Dormont D., Kretzschmar H., Pocchiari M., and van Duijn C. 2002. BSE and variant Creutzfeldt-Jakob disease: Never say never. *Acta Neuropathol* **103:** 627–628.

Budka H., Aguzzi A., Brown P., Brucher J.-M., Bugiani O., Collinge J., Diringer H., Gullotta F., Haltia M., Hauw J.-J., Ironside J.W., Kretzschmar H.A., Lantos P.L., Masullo C., Pocchiari M., Schlote W., Tateishi J., and Will R.G. 1995. Tissue handling in suspected Creutzfeldt-Jakob disease (CJD) and other human spongiform encephalopathies (prion diseases). *Brain Pathol.* **5:** 319–322.

Casaccia P., Ladogana L., Xi Y.G., and Pocchiari M. 1989. Levels of infectivity in the blood throughout the incubation period of hamsters peripherally injected with scrapie. *Arch. Virol.* **108:** 145–149.

CDC. 1987. Update: Creutzfeldt-Jakob disease in a patient receiving a cadaveric dura mater graft. *Morb. Mortal. Wkly. Rep.* **36:** 324–325.

———. 1997. Creutzfeldt-Jakob disease associated with cadaveric dura mater grafts in Japan 1979–1996. *Morb. Mortal. Wkly. Rep.* **46:** 1066–1069.

———. 2002. Probable variant Creutzfeldt-Jakob disease in a U.S. resident, Florida, 2002. *Morb. Mortal. Wkly. Rep.* **51:** 927 929.

Cochius J.I., Hyman N., and Esiri M.M. 1992. Creutzfeldt-Jakob disease in a recipient of human pituitary-derived gonadotrophin: A second case. *J. Neurol. Neurosurg. Psychiatry* **55:** 1094–1095.

Cochius J.I., Burns R.J., Blumbergs P.C., Mack K., and Alderman C.P. 1990. Creutzfeldt-Jakob disease in a recipient of human pituitary derived gonadotropin *Aust. N.Z. J. Med.* **20:** 592–593.

Collie D.A., Sellar R.J., Zeidler M., Colchester A.C.F., Knight R., and Will R.G. 2001. MRI of Creutzfeldt-Jakob disease: Imaging features and recommended MRI protocol. *Clin. Radiol.* **56:** 726–739.

Collinge J., Sidle K.C.L., Meads J., Ironside J., and Hill A.F. 1996. Molecular analysis of prion strain variation and the aetiology of "new variant" CJD. *Nature* **383:** 685–690.

Collins S., Law M.G., Fletcher A., Boyd A., Kaldor J., and Masters C.L. 1999. Surgical treatment and risk of sporadic Creutzfeldt-Jakob disease: A case-control study. *Lancet* **353:** 693–697.

Collins S., Boyd A., Fletcher A., Kaldor J., Hill A., Farish S., McLean C., Ansari Z., Smith M., and Masters C.L. 2002. Creutzfeldt-Jakob disease cluster in an Australian rural city. *Ann. Neurol.* **52:** 115–118.

Cooper J.D. and Bird S.M. 2002a. UK dietary exposure to BSE in beef mechanically recovered meat: By birth-cohort and gender. *J. Cancer Epidemiol. Prev.* **7:** 59–70.

———. 2002b. UK dietary exposure to BSE in head meat: By birth-cohort and gender. *J. Cancer Epidemiol. Prev.* **7:** 71–83.

Cousens S.N., Vynnycky E., Zeidler M., Will R.G., and Smith P.G. 1997. Predicting the CJD epidemic in humans. *Nature* **385:** 197–198.

Cousens S.N., Harries-Jones R., Knight R., Will R.G., Smith P.G., and Matthews W.B. 1990. Geographical distribution of cases of Creutzfeldt-Jakob disease in England and Wales 1970–1984. *J. Neurol. Neurosurg. Psychiatry* **53:** 459–465.

Cousens S.N., Smith P.G., Ward H., Everington D., Knight R.S.G., Zeidler M., Stewart G., Smith-Bathgate E.A.B., Macleod M.A., Mackenzie J., and Will R.G. 2001. Geographical distribution of variant CJD in Great Britain, 1994–2000. *Lancet* **357:** 1002–1007.

Creange A., Gray F., Cesaro P., Adle-Biassette H., Duvoux C., Cherqui D., Bell J., Parchi P., Gambetti P., and Degos J.-D. 1995. Creutzfeldt-Jakob disease after liver transplantation. *Ann. Neurol.* **38:** 269–272.

Creutzfeldt H.G. 1920. Über eine eigenartige herdförmige Erkrankung des Zentralnervensystems. *Z. Gesamte Neurol. Psychiatr.* **57:** 1–18.

Croes E.A., Roks G., Jansen J.H., Nijssen P.C.G., and van Duijn C.M. 2002. Creutzfeldt-Jakob disease 38 years after diagnostic use of human growth hormone. *J. Neurol. Neurosurg. Psychiatry* **72:** 792–793.

d'Aignaux J.N.H., Cousens S.N., and Smith P.G. 2001. Predictibility of the UK variant CJD epidemic. *Science* **294:** 1729–1731.

Davanipour Z., Alter M., Sobel E., Asher D.M., and Gajdusek D.C. 1985. A case-control study of Creutzfeldt-Jakob disease: Dietary risk factors. *Am. J. Epidemiol.* **122:** 443–451.

Deslys J-P., Lasmézas C.I., Jaegly A., Marce D., Lamoury F., and Dormont D. 1996. Growth hormone related Creutzfeldt-Jakob disease: Genetic susceptibility and molecular pathogenesis. In *Transmissible subacute spongiform encephalopathies: Prion diseases* (ed. L. Court and B. Dodet), pp. 465–470. Elsevier, Paris.

Diringer H. 1984. Sustained viraemia in experimental hamster scrapie. *Arch. Virol.* **82:** 105–109.

Duffy P., Wolf J., Collins G., DeVoe A.G., Streeten B., and Cowen D. 1974. Possible person-to-person transmission of Creutzfeldt-Jakob disease. *N. Engl. J. Med.* **290:** 692–693.

Esmonde T.F.G., Will R.G., Slattery J.M., Knight R., Harries-Jones R., De Silva R., and Matthews W.B. 1993. Creutzfeldt-Jakob disease and blood transfusion. *Lancet* **341:** 205–207.

Ferak V., Kroupova Z., and Mayer V. 1981. Are population-genetic mechanisms responsible for clustering of Creutzfeldt-Jakob disease. *Br. Med. J.* **282:** 521–522.

Finkenstaedt M., Szudra A., Zerr I., Poser S., Hise J.H., Stoebner J.M., and Weber T. 1996. MR imaging of Creutzfeldt-Jakob disease. *Radiology* **199:** 793–798.

Foncin J.F., Gaches J., Cathala F., El Sherif E., and Le Beau J. 1980. Transmission iatrogene interhumaine possible de maladie de Creutzfeldt-Jakob avec atteinte des grains de cervelet. *Rev. Neurol.* **136:** 280.

Fowler M. and Robertson E.G. 1959. Observations on kuru. III. Pathological features in five cases. *Australas. Ann. Med.* **8:** 16–26.

Gajdusek D.C. 1990. Subacute spongiform encephalopathies: Transmissible cerebral amyloidoses caused by unconventional viruses. In *Field's virology* (ed. B.N. Fields and D.M. Knipe), pp. 2289–2324. Raven Press, New York.

Gajdusek D.C. and Zigas V. 1957. Degenerative disease of the central nervous system in New Guinea: The endemic occurrence of "kuru" in the native population. *N. Engl. J. Med.* **257:** 974–978.

Gajdusek D.C., Gibbs C.J., Jr., and Alpers M. 1966. Experimental transmission of a kuru-like syndrome in chimpanzees. *Nature* **209:** 794–796.

Gajdusek D.C., Gibbs C.J., Jr., Asher D.M., Brown P., Diwan A., Hoffman P., Nemo G., Rohwer R., and White L. 1977. Precautions in medical care of, and in handling materials from, patients with transmissible virus dementia (Creutzfeldt-Jakob disease). *N. Engl. J. Med.* **297:** 1253–1258.

Galvez S., Masters C., and Gajdusek D.C. 1980. Descriptive epidemiology of Creutzfeldt-Jakob disease in Chile. *Arch. Neurol.* **37:** 11–14.

Ghani A.C., Ferguson N.M., Donnelly C.A., and Anderson R.M. 2000. Predicted vCJD mortality in Great Britain. *Nature* **406:** 583–584.

Gibbs C.J., Jr., Asher D.M., Brown P., Fradkin J.E., and Gajdusek D.C. 1993. Creutzfeldt-Jakob disease infectivity of growth hormone derived from human pituitary glands. *N. Engl. J. Med.* **328:** 358–359.

Gibbs C.J., Jr., Gajdusek D.C., Asher D.M., Alpers M.P., Beck E., Daniel P.M., and Matthews W.B. 1968. Creutzfeldt-Jakob disease (spongiform encephalopathy): Transmission to the chimpanzee. *Science* **161:** 388–389.

Gibbs C.J., Jr., Joy A., Heffner R., Franko M., Miyazaki M., Asher D.M., Parisi J.E., Brown P.W., and Gajdusek D.C. 1985. Clinical and pathological features and laboratory confirmation of Creutzfeldt-Jakob disease in a recipient of pituitary-derived human growth hormone. *N. Engl. J. Med.* **313:** 734–738.

Glasse R.M. 1967. Cannibalism in the kuru region of New Guinea. *Trans. N.Y. Acad. Sci.* **29:** 748–754.

Glasse R. and Lindenbaum S. 1992. Field work in the South Fore: The process of ethnographic inquiry. In *Prion diseases of humans and animals* (ed. S.B. Prusiner et al.), pp. 77–91. Ellis Horwood, London, United Kingdom.

Glatzel M., Rogivue C., Ghani A., Streffer R., Amsier L., and Aguzzi A. 2002. Incidence of Creutzfeldt-Jakob disease in Switzerland. *Lancet* **360:** 139–141.

Goldberg H., Alter M., and Kahana E. 1979. The Libyan Jewish focus of Creutzfeldt-Jakob disease: A search for the mode of natural transmission. In *Slow transmissible diseases of the nervous system* (ed. S.B. Prusiner and W.J. Hadlow), pp. 195–211. Academic Press, New York.

Goldfarb L.G., Mitrova E., Brown P., Toh B.H., and Gajdusek D.C. 1990. Mutation in codon 200 of scrapie amyloid protein gene in two clusters of Creutzfeldt-Jakob disease in Slovakia (letter). *Lancet* **336:** 514–515.

Green A.J.E. 2002. Use of 14-3-3 in the diagnosis of Creutzfeldt-Jakob disease. *Biochem. Soc. Trans.* **30:** 382–386.

Hadlow W.J. 1959. Scrapie and kuru. *Lancet* **II:** 289–290.

Hannah E.L., Belay E.D., Gambetti P., Krause G., Parchi P., Capellari S., Hoffman R.E., and Schonberger L.B. 2001. Creutzfeldt-Jakob disease after receipt of a previously unimplicated brand of dura mater graft. *Neurology* **56:** 1080–1083.

Harries-Jones R., Knight R., Will R.G., Cousens S., Smith P.G., and Matthews W.B. 1988. Creutzfeldt-Jakob disease in England and Wales, 1980–1984: A case-control study of potential risk factors. *J. Neurol. Neurosurg. Psychiatry.* **51:** 1113–1119.

Heckmann J.G., Lang C.J.G., Petruch F., Druschky A., Erb C., Brown P., and Nuendorfer B. 1997. Transmission of Creutzfeldt-Jakob disease via a corneal transplant. *J. Neurol. Neurosurg. Psychiatry* **63:** 388–390.

Heidenhain A. 1929. Klinische und Anatomische Untersuchungen uber eine eigenartige Erkrankung des Zentralnervensystems in Praesenium. *Z. Gesamte Neurol. Psychiatr.* **118:** 49–114.

Hill A.F., Zeidler M., Ironside J., and Collinge J. 1997a. Diagnosis of new variant Creutzfeldt-Jakob disease by tonsil biopsy. *Lancet* **349:** 99–100.

Hill A.F., Desbruslais M., Joiner S., Sidle K.A.C., Gowland I., and Collinge J. 1997b. The same prion strain causes vCJD and BSE. *Nature* **389:** 448–450.

Holman R.C., Khan A.S., Belay E.D., and Schonberger L.B. 1996. Creutzfeldt-Jakob disease in the United States, 1979–1994: Using national mortality data to assess the possible occurrence of variant cases. *Emerg. Infect. Dis.* **2:** 333–337.

Holman R.C., Khan A.S., Kent J., Strine T.W., and Schonberger L.B. 1995. Epidemiology of Creutzfeldt-Jakob disease in the United States, 1979–1990: Analysis of national mortality data. *Neuroepidemiology* **14:** 174–181.

Hoshi K., Yoshino H., Urata J., Nakamura Y., Yanagawa H., and Sato T. 2000. Creutzfeldt-Jakob disease associated with cadaveric dura mater grafts in Japan. *Neurology* **55:** 718–721.

Houston F., Foster J.D., Chong A., Hunter N., and Bostock C.J. 2000. Transmission of BSE by blood transfusion in sheep. *Lancet* **356:** 999–1000.

Hsiao K., Baker H.F., Crow T.J., Poulter M., Owen F., Terwilliger J.D., Westaway D., Ott J., and Prusiner S.B. 1989. Linkage of a prion protein missense variant to Gerstmann-Sträussler syndrome. *Nature* **338:** 342–345.

Hsiao K., Meiner Z., Kahana E., Cass C., Kahana I., Avrahami D., Scarlatto G., Abramsky O., Prusiner S.B., and Gabizon R. 1991. Mutation of the prion protein in Libyan Jews with Creutzfeldt-Jakob disease. *N. Engl. J. Med.* **324:** 1091–1097.

Hsich G., Kenney K., Gibbs C.J., Jr., Lee K.H., and Harrington M.G. 1996. The 14-3-3 brain protein in cerebrospinal fluid as a marker for transmissible spongiform encephalopathies. *N. Engl. J. Med.* **335:** 924–930.

Hunter N., Foster J., Chong A., McCutcheon S., Parnham D., Eaton S., MacKenzie C., and Houston F. 2002. Transmission of prion diseases by blood transfusion. *J. Gen. Virol.* **83:** 2897–2905.

Jakob A. 1921. Über eine der multiplen sklerose klinisch nahenstehende erkrankung des zentralnervensystems (spastiche pseudosklerose) mit bemerkswertem anatomishen befunde. *Med. Klin.* **13:** 372–376.

Janssen R.S. and Schonberger L.B. 1991. Discussion—Creutzfeldt-Jakob disease from allogeneic dura: A review of risks and safety. *J. Oral Maxillofac. Surg.* **49:** 274–275.

Jones D.P. and Nevin S. 1954. Rapidly progressive cerebral degeneration (subacute vascular encephalopathy) with mental disorder, focal disturbance, and myoclonic epilepsy. *J. Neurol. Neurosurg. Psychiatry* **17:** 148–159.

Kirschbaum W.R. 1968. *Jakob-Creutzfeldt disease*. Elsevier, Amsterdam.

Klein R. and Dumble L.J. 1993. Transmission of Creutzfeldt-Jakob disease by blood transfusion. *Lancet* **341:** 768.

Klitzman R.L., Alpers M.P., and Gajdusek D.C. 1984. The natural incubation period of kuru and the episodes of transmission in three clusters of patients. *Neuroepidemiology* **3:** 3–20.

Koch T.K., Berg B.O., DeArmond S.J., and Gravina R.F. 1985. Creutzfeldt-Jakob disease in a young adult with idiopathic hypopituitarism. *N. Engl. J. Med.* **313:** 731–733.

Kondo K. 1985. Epidemiology of Creutzfeldt-Jakob disease in Japan. In *Creutzfeldt-Jakob disease* (ed. T. Mizutani and H. Shiraki), pp. 17–30. Elsevier, Amsterdam and Nishimura, Niigate, Japan.

Kondo K. and Kuroiwa Y. 1982. A case-control study of Creutzfeldt-Jakob disease: Association with physical injuries. *Ann. Neurol.* **11:** 377–381.

Korth C., May B.C.H., Kohen F.E., and Prusiner S.B. 2001. Acridine and phenothiazine derivatives as pharmacotherapeutics for prion disease. *Proc. Natl. Acad. Sci.* **98:** 9836–9841.

Kovacs G.G., Trabattoni G., Hainfellner J.A., Ironside J.W., Knight R.S.G., and Budka H. 2002. Mutations of the prion protein gene: Phenotypic spectrum. *J. Neurol.* **249:** 1567–1582.

Laplanche J.-L., Delasnerie-Laupretre N., Brandel J.P., Chatelain J., Beaudry P., Alperovitch A., and Launay J.-M. 1994. Molecular genetics of prion diseases in France. *Neurology* **44:** 2347–2351.

Lasmézas C.I., Deslys J.-P., Demaimay R., Adjou K.T., Lamoury F., Dormont D., Robain I., Ironside J.W., and Hauw J.-J. 1996. BSE transmission to macaques. *Nature* **381:** 743–744.

Masters C.L. and Gajdusek D.C. 1982. The spectrum of Creutzfeldt-Jakob disease and virus induced subacute spongiform encephalopathies. In *Recent advances in neuropathology* (ed. W. Thomas Smith and J.B. Cavanagh), vol. 2, pp. 139–163. Churchill Livingstone, Edinburgh, Scotland.

Masters C.L., Harris J.O., Gajdusek D.C., Gibbs C.J., Jr., Bernoulli C., and Asher D.M. 1979. Creutzfeldt-Jakob disease: Patterns of worldwide occurrence and the significance of familial and sporadic clustering. *Ann. Neurol.* **5:** 177–188.

Masullo C., Pocchiari M., Macchi G., Alema G., Piazza G., and Panzera M.A. 1989. Transmission of Creutzfeldt-Jakob disease by dural cadaveric graft (letter). *J. Neurosurg.* **71:** 954–955.

Matthews W.B. 1975. Epidemiology of Creutzfeldt-Jakob disease in England and Wales. *J. Neurol. Neurosurg. Psychiatry* **38:** 210–213.

McKinley M.P., Bolton D.C., and Prusiner S.B. 1983. A protease-resistant protein is a structural component of the scrapie prion. *Cell* **35:** 57–62.

McNaughton H.K. and Will R.G. 1997. Creutzfeldt-Jakob disease presenting acutely as stroke: An analysis of 30 cases. *Neurol. Infect. Epidemiol.* **2:** 19–24.

Monreal J., Collins G.H., Masters C.L., Fisher C.M., Kim R.C., Gibbs C.J., Jr., and Gajdusek D.C. 1981. Creutzfeldt-Jakob disease in an adolescent. *J. Neurol. Sci.* **52:** 341–350.

National CJD Surveillance Unit. 1995. *Creutzfeldt-Jakob disease surveillance in the United Kingdom.* National CJD Surveillance Unit and London School of Hygiene and Tropical Medicine: Fourth Annual Report.

Nevin S., McMenemey W.H., Behrman S., and Jones D.P. 1960. Subacute spongiform encephalopathy—A subacute form of encephalopathy attributable to vascular function (spongiform cerebral atrophy). *Brain* **83:** 519–563.

Palmer M.S., Dryden A.J., Hughes J.T., and Collinge J. 1991. Homozygous prion protein genotype predisposes to sporadic Creutzfeldt-Jakob disease. *Nature* **352:** 340–341.

Pocchiari M. 1994. Prions and related neurological diseases. *Mol. Asp. Med.* **15:** 195–291.

Pocchiari M., Peano S., Conz A., Eshkol A., Maillard F., Brown P., Gibbs C.J., You G.X., Tenhamfisher E., and Macchi G. 1991. Combination ultrafiltration and 6-M-urea treatment of human growth hormone effectively minimizes risk from potential Creutzfeldt-Jakob disease virus contamination. *Horm. Res.* **35:** 161–166.

Powell-Jackson J., Weller R.O., Kennedy P., Preece M.A., Whitcombe E.M., and Newsom-Davis J. 1985. Creutzfeldt-Jakob disease after administration of human growth hormone. *Lancet* **II:** 244–246.

Prusiner S.B. 1994. Prion diseases of humans and animals. *J. R. Coll. Physicians Lond.* (suppl.) **28:** 1–30.

Roos R., Gajdusek D.C., and Gibbs C.J., Jr. 1971. Liver disease in CJD (subacute spongiform encephalopathy). *Neurology* **21:** 397–398.

———. 1973. The clinical characteristics of transmissible Creutzfeldt-Jakob disease. *Brain* **96:** 1–20.

Sadeh M., Chagnac Y., and Goldhammer Y. 1990. Creutzfeldt-Jakob disease associated with peripheral neuropathy. *Isr. J. Med. Sci.* **26:** 220–222.

Salazar A.M., Masters C.L., Gajdusek D.C., and Gibbs C.J., Jr. 1983. Syndromes of amyotrophic lateral sclerosis and dementia: Relation to transmissible Creutzfeldt-Jakob disease. *Ann. Neurol.* **14:** 17–26.

Sellar R.J., Will R.G., and Zeidler M. 1997. MR imaging of new variant Creutzfeldt-Jakob disease: The pulvinar sign. *Neuroradiology* **39:** S53.

Scott M.R., Will R.G., Ironside J., Nguyen H-O.B., Tremblay P., DeArmond S.J., and Prusiner S.B. 1999. Compelling transgenetic evidence for transmission of bovine spongiform encephalopathy prions to humans. *Proc. Natl. Acad. Sci.* **96:** 15137–15142.

Southwood Committee. 1989. *Report of the Working Party on bovine spongiform encephalopathy.* Department of Health and Ministry of Agriculture, Fisheries and Food. (ISBN 185197 405 9).

Spencer M.D., Knight R.S.G., and Will R.G. 2002. First hundred cases of variant Creutzfeldt-Jakob disease: Retrospective case note review of early psychiatric and neurological features. *Br. Med. J.* **324:** 1479–1482.

Steinhoff B.J., Racker S., Herrendorf G., Poser S., Grosche S., Zerr I., Kretzschmar H., and Weber T. 1996. Accuracy and reliability of periodic sharp wave complexes in Creutzfeldt-Jakob disease. *Arch. Neurol.* **53:** 162–165.

Thadani V., Penar P.L., Partington J., Kalb R., Janssen R., Schonberger L.B., Rabkin C.S., and Prichard J.W. 1988. Creutzfeldt-Jakob disease probably acquired from a cadaveric dura mater graft. *J. Neurosurg.* **69:** 766–769.

Uchiyama K., Ishida C., Yago S., Kurumaya H., and Kitamoto T. 1994. An autopsy case of Creutzfeldt-Jakob disease associated with corneal transplantation. *Dementia* **8:** 466–473.

Valleron A.J., Boelle P.Y., Will R., and Cesbron J.Y. 2001. Estimation of epidemic size and incubation time based on age characteristics of vCJD in the United Kingdom. *Science* **294:** 1726–1728.

van Duijn C.M., Delasnerie-Laupretre N., Masullo C., Zerr I., De Silva R., Wientjens D.P.W.M., Brandel J.-P., Weber T., Bonavita V., Zeidler M., Alperovitch A., Poser S., Granieri E., Hofman A., and Will R.G. 1998. Case-control study of risk factors of Creutzfeldt-Jakob disease in Europe during 1993–1995. *Lancet* **351:** 1081–1085.

Ward H.J.T., Everington D., Croes E.A., Alperovtich A., Delasnerie-Lauptretre N., Zerr I., Poser S., and van Duijn C.M. 2002. Sporadic Creutzfeldt-Jakob disease and surgery. *Neurology* **59:** 543–548.

Wells G.A.H., Scott A.C., Johnson C.T., Gunning R.F., Hancock R.D., Jeffrey M., Dawson M., and Bradley R. 1987. A novel progressive spongiform encephalopathy in cattle. *Vet. Rec.* **121:** 419–420.

Wientjens D.P.W.M., Will R.G., and Hofman A. 1994. Creutzfeldt-Jakob disease: A collaborative study in Europe. *J. Neurol. Neurosurg. Psychiatry* **57:** 1285–1299.

Wientjens D.P.W.M., Davanipour Z., Hofman A., Kondo K., Matthews W.B., Will R.G., and van Duijn C.M. 1996. Risk factors for Creutzfeldt-Jakob disease: A reanalysis of case-control studies. *Neurology* **46:** 1287–1291.

Wilesmith J.W., Wells G.A.H., Cranwell M.P., and Ryan J.B. 1988. Bovine spongiform encephalopathy: Epidemiological studies. *Vet. Rec.* **123:** 638–644.

Will R.G. 1991. Epidemiology of Creutzfeldt-Jakob disease in the United Kingdom. *Eur. J. Epidemiol.* **7:** 460–465.

———. 1998. New variant Creutzfeldt-Jakob disease. In *Safety of biological products prepared from mammalian cell culture* (ed. F. Brown et al.), pp. 79–84. Karger, Basel.

Will R.G. and Matthews W.B. 1982. Evidence for case-to-case transmission of Creutzfeldt-Jakob disease. *J. Neurol. Neurosurg. Psychiatry* **45:** 235–238.

———. 1984. A retrospective study of Creutzfeldt-Jakob disease in England and Wales 1970–1979. I. Clinical features. *J. Neurol. Neurosurg. Psychiatry* **47:** 134–140.

Will R.G., Matthews W.B., Smith P.G., and Hudson C. 1986. A retrospective study of Creutzfeldt-Jakob disease in England and Wales 1970–1979. II. Epidemiology. *J. Neurol. Neurosurg. Psychiatry* **49:** 749–755.

Will R.G., Ironside J.W., Zeidler M., Cousens S.N., Estibeiro K., Alperovitch A., Poser S., Pocchiari M., Hofman A., and Smith P.G. 1996. A new variant of Creutzfeldt-Jakob disease in the UK. *Lancet* **347:** 921–925.

Will R.G., Zeidler M., Stewart G.E., Macleod M.A., Ironside J.W., Cousens S.N., Mackenzie J., Estibeiro K., Green A.J.E., and Knight R.S.G. 2000. Diagnosis of new variant Creutzfeldt-Jakob disease. *Ann. Neurol.* **47:** 575–582.

Will R.G., Alperovitch A., Poser S., Pocchiari M., Hofman A., Mitrova E., De Silva R., D'Alessandro M., Delasnerie-Laupretre N., Zerr I., and van Duijn C. 1998. Descriptive epidemiology of Creutzfeldt-Jakob disease in six European countries, 1993–1995. *Ann. Neurol.* **43:** 763–767.

Wilson S.A.K. 1940. Syndrome of Jakob: Cortico-striato-spinal degeneration. In *Neurology* (ed. A.N. Bruce), pp. 907–910. Edward Arnold, London, United Kingdom.

World Health Organization. 2002. *The revision of the surveillance case definition for variant Creutzfeldt-Jakob disease (vCJD)*. Report of a WHO Consultation, Edinburgh, United Kingdom, May 17, 2001.

Zeidler M., Stewart G., Cousens S.N., Estibeiro K., and Will R.G. 1997a. Codon 129 genotype and new variant CJD. *Lancet* **350:** 668.

Zeidler M., Stewart G.E., Barraclough C.R., Bateman D.E., Bates D., Burn D.J., Colchester A.C., Durward W., Fletcher N.A., Hawkins S.A., Mackenzie J.M., and Will R.G. 1997b. New variant Creutzfeldt-Jakob disease: Neurological features and diagnostic tests. *Lancet* **350:** 903–907.

Zeidler M., Johnstone E.C., Bamber R.W.K., Dickens C.M., Fisher C.J., Francis A.F., Goldbeck R., Higgo R., Johnson-Sabine E.C., Lodge G.J., McGarry P., Mitchell S., Tarlo I., Turner M., Ryley P., and Will R.G. 1997c. New variant Creutzfeldt-Jakob disease: Psychiatric features. *Lancet* **350:** 908–910.

14

Inherited Prion Diseases

Qingzhong Kong,[1] Witold K. Surewicz,[2] Robert B. Petersen,[1]
Wenquan Zou,[1] Shu G. Chen,[1] and Pierluigi Gambetti[1]
[1]Division of Neuropathology, Institute of Pathology
[2]Department of Physiology and Biophysics
Case Western Reserve University
Cleveland, Ohio 44106

Piero Parchi and Sabina Capellari
Department of Neurological Sciences
University of Bologna
Bologna 40123, Italy

Lev Goldfarb
National Institutes of Health
National Institute of Neurological Disorders and Stroke
Bethesda, Maryland 20892

Pasquale Montagna and Elio Lugaresi
Instituto di Clinica Neurologica
Università di Bologna
Bologna 40123, Italy

Pedro Piccardo and Bernardino Ghetti
Department of Pathology and Laboratory Medicine
Indiana University School of Medicine
Indianapolis, Indiana 46202

HISTORICAL OUTLINE

THE FIRST REPORT OF AN INHERITED PRION DISEASE can be traced to an affected member of the "H" family carrying a neurologic disorder through multiple generations. It was presented at a meeting of the Viennese Neurological and Psychiatric Association in 1912 (Dimitz

1913). Two decades passed before Gerstmann in 1928 and Gerstmann, Sträussler, and Scheinker in 1936 reported clinical presentations and neuropathologic findings for several affected members of the "H" family. These reports established the disease that is currently referred to as Gerstmann-Sträussler-Scheinker disease or GSS (Gerstmann 1928; Gerstmann et al. 1936).

The subject reported by Creutzfeldt in 1920 and 1921, who had a positive family history (Creutzfeldt 1920, 1921), is unlikely to have been affected by the condition now called Creutzfeldt-Jakob disease (CJD). The first authentic familial case of CJD was recorded in 1924 by Kirschbaum (1924). However, it was Meggendorfer in 1930 who showed that the subject described by Kirschbaum was a member of a large kindred that became known as the "Backer" family, proven in subsequent publications to be affected by an inherited form of CJD (Meggendorfer 1930; Stender 1930; Jakob et al. 1950).

In 1973, Gajdusek, Gibbs, and their colleagues first demonstrated the transmissibility of inherited prion diseases with a CJD-like phenotype to nonhuman primates (Roos et al. 1973). This finding followed earlier studies that reported the transmissibility of the sporadic form of CJD and kuru, a prion disease of the Fore tribe of New Guinea propagated through endocannibalism (Gajdusek et al. 1966; Gibbs et al. 1968). Subsequently, the experimental transmission of at least one subtype of GSS was also achieved (Masters et al. 1981a).

The experimental transmission of familial CJD and GSS to animals by an infectious mechanism was a defining moment in the history of medicine because it presented the first example of a disease that was at the same time heritable and infectious. At that time, the transmissibility of familial CJD and GSS was interpreted within the framework of a viral illness. It was postulated that the affected subjects carried a genetically determined susceptibility to an infection that was acquired in early infancy or childhood (Masters et al. 1981b). Subsequent work, largely carried out by Prusiner and his colleagues, has provided a novel molecular basis for diseases that are inherited and infectious. A protein unique to the scrapie-infected hamster brain was discovered in fractions enriched for scrapie infectivity (Bolton et al. 1982; Prusiner 1982). Sequencing of the protein, designated prion protein (PrP), allowed the subsequent cloning of the PrP coding sequence (Chesebro et al. 1985; Oesch et al. 1985). This work paved the way for the cloning and sequencing of the human PrP gene (*PRNP*) and for the discovery of *PRNP* pathogenic mutations in inherited forms of prion diseases (Hsiao et al. 1989; Owen et al. 1989). Analysis of genetic linkage, which was initially carried out in a subtype of

GSS associated with a *PRNP* mutation at codon 102, and subsequently in other *PRNP* mutations, indicated that the mutations are likely to be the cause of GSS and other inherited prion diseases such as inherited CJD and fatal familial insomnia (FFI) (Hsiao et al. 1989; Goldfarb et al. 1990; Dlouhy et al. 1992; Medori et al 1992; Poulter et al. 1992; Gabizon et al. 1993).

The discovery of *PRNP* mutations in familial prion diseases provided an easy explanation for the coexistence of genetic and infectious forms within the framework of the prion hypothesis (Prusiner 1989). This new framework received additional support from a wealth of other studies, but especially from the production of transgenic mice that developed spontaneously, or were susceptible to develop, a prion disease when overexpressing mutant PrP (Hsiao et al. 1990; Hegde et al. 1999; Barron et al. 2001; Supattapone et al. 2001). However, some investigators interpret the *PRNP* mutations as affecting a viral receptor in a way that increases the affected individual's susceptibility to infection by a highly ubiquitous virus (Chesebro 1997).

GENOTYPES AND PHENOTYPES

The genetic and physical map of the *PRNP* region on the short arm of chromosome 20, and diagrams of *PRNP* and normal or cellular PrP (PrP^C), are shown in Figure 1. Currently, 55 pathogenic mutations and 16 polymorphisms have been identified in the *PRNP* gene (Fig. 2 and Table 1). They include (1) 24 missense point mutations; (2) 27 insertion mutations consisting of 1, 2, and 4–9 repeats of 24 base pairs (bp), which are either distinct or in distinct order, and are located between codons 51 and 91; (3) 2 deletion mutations including distinct 48 bp deletions between codons 51 and 91; and (4) 2 nonsense mutations, which result in the premature termination of synthesis and yield truncated PrP. Twelve polymorphisms located at codons 68, 117, 124, 161, 173, 177, 188, 202, 208, 212, 228, and 230 do not result in amino acid substitutions (Prusiner 1997; Collinge 2001; Parchi et al. 2003; http://www.mad-cow.org/prion_point_mutations.html). Four polymorphisms that do alter the amino acid sequence are (1) the methionine/valine (M/V) polymorphism at codon 129; (2) the asparagine/serine (N/S) polymorphism at codon 171 (Samaia et al. 1997); (3) the glutamic acid/lysine (E/K) polymorphism at codon 219; and (4) the deletion of one 24-bp repeat. The M/V polymorphism at codon 129 is common. The M/M and V/V homozygous and the heterozygous subjects account for 43%, 8%, and 49%, respectively, in a normal Caucasian population (Zimmermann et al. 1999). These values change to 92%, 0%, and

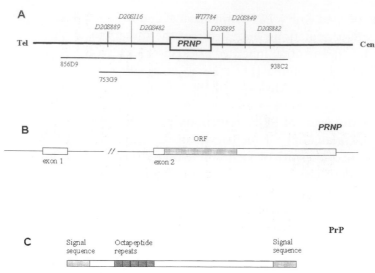

Figure 1. Genetic and physical maps of the *PRNP* region on chromosome 20p12-pter. (*A*) D20-S889, D20S116, D20S482, D20S895, D20S849, and D20S882 are genetic markers flanking the *PRNP* gene; 856D9, 753G8, and 938C2 are human YAC clones; W17784 is an expressed sequence tag located within the PRNP coding sequence in an overlapping region of YAC clones 753G8 and 938C2. (*B*) Schematic representation of the *PRNP* gene consisting of exons 1 and 2, with the open reading frame (ORF) located entirely within exon 2. (*C*) The prion protein includes two signal sequences and a five-octapeptide repeat.

8% in a Japanese population and to 97%, 0%, and 3% in a Chinese population (Doh-ura et al. 1991; Tsai et al. 2001). The 219 E/K polymorphism has a 6% prevalence of the lysine allele in the Japanese population (Kitamoto and Tateishi 1994; Barbanti et al. 1996), but this allele was not observed in an Italian population of 204 subjects (Petraroli and Pocchiari 1996). The deletion of a single repeat is found with the same 1–2.5% prevalence in the general population and in affected patients, as expected for a polymorphism (Fig. 2) (Laplanche et al. 1990; Puckett et al. 1991; Vnencak-Jones and Phillips 1992; Palmer et al. 1993; Cervenáková et al. 1996). The 129 M/V and the 219 E/K, as well as, possibly, the deletion of one 24-bp repeat, are of particular interest because in at least some inherited prion diseases these polymorphisms have been shown to modify basic aspects of the disease phenotype when they are located on the mutant allele, whereas on the normal allele the 129 polymorphism influences age at onset and duration of the disease. In FFI, the 129 polymorphism on the normal allele has also been shown to affect other aspects of the disease phenotype (see below). Because at least two polymorphisms modify basic aspects of the disease phenotype, it is appropriate to identify each *PRNP*

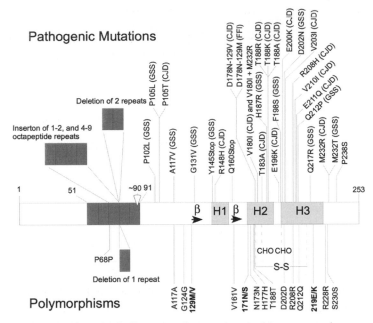

Figure 2. Diagram of PrP including mutations, polymorphisms, secondary structures, and posttranslational modifications. (β) β-sheet (*arrows*); (H) α-helix (*lightly shaded boxes*); (S-S) disulfide bond; (CHO) N-glycans; octapeptide repeats (dark-shaded). Pathogenic mutations are indicated above the diagram, with the type of prion diseases (GSS, CJD, FFI) indicated where classification is available. Polymorphisms are indicated below, where bold face indicates amino acid change. Mutations involving the octapeptide repeats have variable phenotypes. The I138M and G142S mutations are not included because they seem unrelated to prion disease (http://www.mad-cow.org/prion_point_mutations.html).

genotype associated with inherited prion diseases not only by the mutation, but also by codons 129 and 219 (or other modifying polymorphic codons) present on the mutant allele; i.e., the haplotype. Accordingly, three point mutations (at codons 102, 178, and 200), one four-repeat insertion mutation (R122322234), and one five-repeat insertion mutation (R1223222a234) are each associated with two haplotypes determined by the M/V polymorphism at codon 129. The point mutation located at codon 102 is associated with three haplotypes, as determined by the polymorphisms at codons 129 and 219 (Table 1) (Furukawa et al. 1995). Affected subjects carrying either the M or the V 129 codon on the normal allele have been observed in virtually all inherited prion diseases. Including base-pair insertion and deletion mutations, 61 disease-associated *PRNP* haplotypes have been reported, which are associated with distinct phenotypes.

Table 1. Genotype and phenotype of inherited prion diseases

Genotype	Onset (yrs)	Duration	• Clinical and •• Pathological features
Creutzfeldt-Jakob disease phenotype			
P105T	30–42	NA	• NA
			•• NA
R148H, 129M	63	18 months	• like sCJDMV2
			•• like sCJDMV2
D178N, 129V	26–56	9–51 months	• dementia, ataxia, myoclonus, extrapyramidal and pyramidal signs
			•• spongiosis, neuronal loss, and astrogliosis in the cerebral cortex (most severe), striatum, and thalamus (least severe), while the cerebellum is spared
V180I, 129M	66–85	1–2 years	• similar to typical sCJD but with a slower progression
			•• like typical sCJD
T183A, 129M	45	4 years	• personality changes followed by dementia and parkinsonism
			•• atrophy with spongiform degeneration in the cerebral cortex and, to a lesser extent, in the basal ganglia
T188A, 129M	82	4 months	• like sCJDMM1
			•• immunohistochemistry for PrP negative, no immunoblot
T188K	59		• dysphasia, rapidly progressive dementia, and negative family history
			•• NA
T188R			• NA
			•• NA
E196K, 129M	63–77	~1 year	• rapidly progressive dementia, ataxia, no PSW in EEG recording
			•• NA
E200K-129M	35–66	2–41	• similar to typical sCJD; atypical signs such as supranuclear palsy and peripheral

			Clinical phenotype	Neuropathology
V203I, 129M	69	~1 month	• neuropathy in some cases	•• like typical sCJD
H208R, 129M	60	7 months	• sudden confusion hallucinations, abnormal motor functions, myoclonus, EEG showed PSW, negative family history	•• NA
V210I, 129M	49–70	3–5 months	• like typical sCJD	•• like typical sCJD
E211Q, 129M	42–81	3–32 months	• like typical sCJD • like typical sCJD, EEG showed PSW	•• like typical sCJD •• NA
M232R, 129M	55–70	4–24 months	• like typical sCJD	•• like typical sCJD
Fatal familial insomnia phenotype				
D178N, 129M	20–71	6–23 months	• reduction of total sleep time, enacted dreams, sympathetic hyperactivity, myoclonus, ataxia; late dementia, pyramidal and extrapyramidal signs in the cases with a relatively long duration (>1 year)	•• preferential thalamic and olivary atrophy; spongiform changes in the cerebral cortex in the subjects with a duration of symptoms longer than 1 year
Gerstmann-Sträussler-Scheinker disease and PrP cerebral amyloid angiopathy phenotypes				
P102L, 129M	30–62	1–10 years	• slowly progressive cerebellar syndrome with late dementia, extrapyramidal and pyramidal signs; some cases (shorter duration) overlap with CJD	•• PrP amyloid deposits in the cerebellum and, to a lesser extent, in the cerebrum; variable degree of spongiosis, neuronal loss, and astrogliosis; no NFT

(Continued on following pages)

Table 1. (*continued*)

Genotype	Onset (yrs)	Duration	• Clinical and •• Pathological features
P102L, 129M, 219K	31–34	4 years	• differ from the above P102L form for the less prominent cerebellar signs •• few PrP plaques in the cerebral and cerebellar cortices; no spongiosis
P102L, 129V	33	12 years	• seizures, numbness, gait difficulties, dysarthria, long tract signs; no dementia •• widespread PrP plaques with no spongiosis
P105L, 129V	40–50	6–12 years	• spastic paraparesis progressing to quadriparesis; late dementia; no myoclonus and only mild cerebellar signs •• PrP amyloid deposits, neuronal loss, and gliosis in the cerebral cortex and, to a lesser extent, in the striatum and thalamus; no spongiform changes and NFT
A117V, 129V	20–64	1–11 years	• dementia, parkinsonism, pyramidal signs; occasional cerebellar signs •• widespread PrP amyloid deposits in the cerebrum and, more rarely, in the cerebellum associated with variable degree of spongiform changes, neuronal loss, and astrogliosis; no NFT
G131V, 129M	42	9 years	• dementia and cerebellar ataxia, uncontrollable aggressive behavior •• PrP deposits in the neocortex, caudate nucleus, putamen and thalamus; PrP amyloid plaques in the cerebellum
Y145Stop, 129M	38	21 years	• slowly progressive dementia •• PrP amyloid deposits in the cerebral and cerebellar cortices associated with NFT in the neocortex, hippocampus, and subcortical nuclei; PrP amyloid angiopathy; no spongiosis
H187R, 129V	33–50	8–19 years	• early progressive cognitive impairment, cerebellar ataxia, myoclonus, seizures, pyramidal and extrapyramidal signs •• PrP deposition in the neocortex, hippocampus, PrP amyloid plaques in the cerebellum, atrophy and astrogliosis in the subcortical white matter

F198S, 129V	34–71	3–11 years	• Like 102 GSS subtype, but with a more chronic course (no overlap with CJD) •• like 102 GSS subtype but with more extensive PrP amyloid deposits, NFT in the cerebral cortex and subcortical nuclei, and inconspicuous spongiosis
D202N, 129V	73	6 years	• dementia and cerebellar signs •• PrP amyloid deposits in the cerebrum and cerebellum, neurofibrillary tangles in the cerebral cortex
Q212P, 129M	60	8 years	• progressive ataxia, dysarthria with no dementia •• PrP plaque-like deposits neocortex and cerebellum; few PrP amyloid deposits; no NFT
Q217R, 129V	62–66	5–5 years	• slowly progressive dementia, cerebellar and extrapyramidal signs •• like 198 GSS subtype but with the most severe lesions in the cerebral cortex, thalamus, and amygdala
M232T	50	6 years	• cerebellar signs, spastic paraparesis, dementia •• PrP amyloid plaques and diffuse PrP deposits in the neocortex, basal ganglia and cerebellum
Other			
Q160Stop, 129M	32–48	>6 years	• progressive dementia •• reduced brain weight, diffuse cortical atrophy, extensive enlargement of ventricles
Heterogeneous phenotype: The deletional mutations			
del 24 bp, 129M	64, 86	18, 23 months	• memory impairment, dizziness, tremors and minor ataxia, myoclonus •• fine spongiform degeneration and gliosis, like sCJDMM1, "synaptic" PrP plaques in cerebral and cerebellar cortices

(Continued on following pages)

Table 1. (*continued*)

Genotype	Onset (yrs)	Duration	• Clinical and •• Pathological features
Heterogeneous phenotype: The insertional mutations			
ins 24 bp, 129M	58–73	4–10 months	• like typical sCJD
			•• spongiosis, gliosis, and neuronal loss throughout the cerebral gray matter associated with "synaptic" PrP immunoreactivity; plaque-like PrP deposits in basal ganglia, thalamus, hypothalamus, and cerebellum
ins 48 bp, 129M	58	3 months	• like typical sCJD
			•• like typical sCJD
ins 48 bp, 129V	52–64	4 months–7 years	• dementia, behavior problem, ataxia
			•• atrophy of the neocortex
ins 96 bp, 129M	56–65	2 months–7 years	• like typical sCJD
			•• like typical sCJD
ins 96 bp, 129V	82	4 months	• examined at terminal stage showed akinetic mutism, diffuse myoclonus, and pyramidal signs
			•• NA
ins 120 bp, 129M	26–61	4 months–19 years	• progressive dementia, myoclonus, cerebellar and extrapyramidal signs
			•• spongiosis, gliosis, and neuronal loss (no information on topography, severity and presence of PrP deposits); CJD phenotype

ins 120 bp, 129V	46	4 years	•	NA
			••	NA
ins 144 bp, 129M	22–53	3 months–18 years	•	similar to 120-bp insertion subtype
			••	most cases show a CJD phenotype with spongiosis, gliosis, and neuronal loss; one case had kuru-like plaques in the cerebellum; some cases show only mild aspecific gliosis and neuronal loss; PrP patches in the cerebellum
ins 168 bp, 129M	23–35?	7–16 years	•	similar to 120-bp insertion subtype
			••	mild gliosis and neuronal loss, and no spongiosis in one case, CJD phenotype in another
ins 192 bp, 129M	21–34	12 months–7 years	•	similar to 120-bp insertion subtype
			••	spongiosis, gliosis, and neuronal loss, PrP multicentric amyloid plaques with widespread distribution; GSS-like phenotype
ins 192 bp, 129V	21–54	3 months–13 years	•	similar to 120-bp insertion subtype
			••	spongiosis, gliosis, and neuronal loss, PrP multicentric amyloid plaques with widespread distribution; GSS-like phenotype
ins 216 bp, 129M	32–55	2.5–>4 years	•	similar to 120-bp insertion subtype
			••	PrP amyloid plaques in the cerebellum, cerebral cortex, and striatum; no obvious neuronal loss, gliosis, or spongiosis; GSS-like phenotype

(NA) Not available; (PrP) prion protein; (NFT) neurofibrillary tangles.

CHARACTERISTICS OF THE PROTEASE-RESISTANT PRION PROTEIN

The central event in prion diseases is the conversion of PrP^C (and its truncated forms) into the pathogenic and infectious isoform, commonly identified as scrapie PrP (PrP^{Sc}) (Prusiner et al. 1998). Actually, this conversion
leads to the formation of several PrP^{Sc} isoforms that can be distinguished
on the basis of physicochemical features such as the conformation, glycosylation, and size of the fragment resistant to treatment with proteases
(Safar et al 1998; Parchi et al. 2000; Pan et al. 2001; Peretz et al. 2002). In
inherited prion diseases, evidence indicates that the formation of the various PrP^{Sc} isoforms is influenced by the *PRNP* genotype, especially the
haplotype resulting from the pathogenic mutation and the genotype at
codon 129. The most practical, although simplistic, approach to characterizing the PrP^{Sc} isoforms is the identification of the PrP^{Sc} fragments
resistant to proteases associated with the individual prion diseases. When
the homogenate from the brain of prion disease-affected subjects is treated with a protease, such as proteinase K (PK), the amino-terminal region
is digested while the carboxy-terminal portion remains intact (Parchi et
al. 2000). Thus, the PK-resistant fragments of PrP^{Sc} migrate faster on gel
than the full-length PrP^{Sc} (Figs. 3, 4). This applies to each of the three
major bands into which PrP^{Sc} migrates on gel following treatment with
PK, which include (1) mostly a highly glycosylated form thought to contain two complex glycan chains, (2) an intermediate form mostly with
only one complex chain, and (3) a form that is unglycosylated (Prusiner
1982; Chen et al. 1995; Petersen et al. 1996). The size of the PrP^{Sc} fragments and the ratio of the PrP^{Sc} forms that are differently glycosylated,
also referred to as glycoforms, are features that help in characterizing the
various inherited prion diseases. The first observation that the PrP^{Sc} iso-

PK RESISTANT PrP IN CJD[178] AND FFI

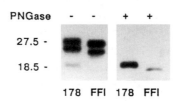

Figure 3. Immunoblot of PrP^{Sc} following treatment with proteinase K. The three
bands in the left lane labeled 178 correspond to the diglycosylated, monoglycosylated, and the unglycosylated forms, respectively, which all comigrate as the unglycosylated form following PNGase treatment. Note the underrepresentation of the unglycosylated form (two left lanes) in both conditions but more pronounced in FFI.

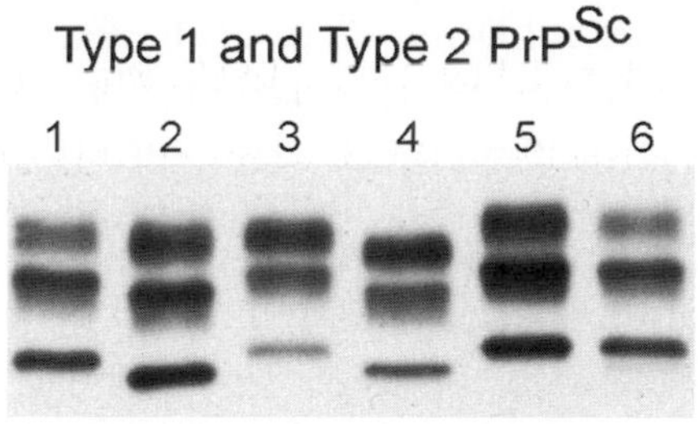

Type 1 and Type 2 PrP^{Sc}

Figure 4. Immunoblot of PrP^{Sc} in sporadic CJD and fCJD with PrP^{Sc} type 1 or 2. (Lane *1*) sCJD 129MM, PrP^{Sc} type 1; (lane *2*) sCJD 129VV, PrP^{Sc} type 2; (lane *3*) fCJD E200K-129M, PrP^{Sc} type 1; (lane *4*) fCJD E200K-129V, PrP^{Sc} type 2; (lane *5*) fCJD 144bp-Insert-129M, PrP^{Sc} type 1; (lane *6*) fCJD V210I-129M, PrP^{Sc} type 1. All samples were PK-treated. (Reprinted, with permission, from Parchi et al. 2003 [copyright Neuropath Press].)

lated from different human prion diseases could be distinguished on the basis of the size of the fragment generated by the PK digestion and the ratio of the glycoforms was made in 1992 and was further validated subsequently (Figs. 3, 4) (Medori et al. 1992; Monari et al. 1994; Parchi et al. 1999c, 2000, 2003). It has been shown that following PK treatment PrP^{Sc} is most frequently digested either at residue 82 or 97 (Parchi et al. 2000). However, a variety of minor fragments cleaved at different sites between residues 74 and 103 also are generated (Parchi et al. 2000). Because of the distinct cleavage sites, two major PK-resistant PrP^{Sc} core fragments have been identified in human prion diseases that, following deglycosylation to eliminate the heterogeneity generated by the presence of the glycoforms, migrate on gel at 20–21 kD and 18–19 kD, respectively. These two PrP^{Sc} fragments are respectively identified as PrP^{Sc} type 1 and type 2 (Parchi et al. 1996c). PrP^{Sc} type 1 and type 2 may be further subdivided according to the pattern determined by the ratio of the three glycoforms. For example, the PrP^{Sc} associated with the insertion mutations is characterized by the predominance of the intermediate glycoform, whereas in the CJDs linked to the D178N and E200K mutations, and in fatal familial insomnia (FFI), the glycoform ratio is consistently dominated by the highly glycosylated form, whereas the unglycosylated form is markedly underrepresented (Monari et al. 1994; Parchi et al. 1996a). Therefore, it seems appropriate to characterize the inherited prion diseases not only according to the haplotype, but also according to the type of PrP^{Sc} as determined by the size of the PrP^{Sc} fragment and the ratio of the three glycoforms. The issue of the PK-resistant PrP^{Sc} species is further compounded by the presence of at least two other sets of fragments in many human prion diseases. These fragments are commonly identified according to the mobility on gel following PK treatment as 7/8-kD and 12/13-kD fragments. However, in PK-untreated PrP^{Sc} preparations, these fragments likely represent protease-resistant domains located in different regions of PrP^{Sc} (Zou et al. 2003). Since these fragments are present in different amounts in different inherited prion diseases, they emphasize the heterogeneity of PrP^{Sc} and also demonstrate how this heterogeneity is influenced by the haplotype associated with the inherited prion diseases. Furthermore, PrP^{Sc} isoforms that are associated with distinct disease phenotypes, but appear to differ only by the type of their glycans, have been recently reported, suggesting that PrP^{Sc}-linked glycans may act as disease determinants (Pan et al. 2001).

The precise role of the PrP^{Sc} type in the phenotypic expression of the inherited prion diseases remains to be determined. Since the different sizes of the PrP^{Sc} fragments following treatment with PK have been shown to result from cleavage at different sites of the full-length PrP^{Sc}, it is likely that the full-length PrP^{Sc} associated with different diseases has different conformations that expose distinct cleavage sites (Parchi et al. 2000; Zou et al. 2003). The different ratio of the glycoforms may be due to a variety of causes, such as different topographic origin of the PrP^{Sc} within the brain in different prion diseases; preferential conversion of PrP^{C} to PrP^{Sc} of one glycoform over the others; or underrepresentation of one of the glycoforms due to the altered metabolism caused by the mutation, the underrepresented form becoming thus less available for conversion to PrP^{Sc} (also see below). Transmission of inherited and sporadic prion diseases to receptive animals has underlined the importance of the PrP^{Sc} typing (Telling et al. 1996). Transgenic mice expressing a chimeric wild-type human/mouse PrP^{C} develop a prion disease after intracerebral inoculation with human brain homogenates carrying PrP^{Sc} of either type 1 or type 2 (Telling et al. 1996). The size of the PrP^{Sc} fragment expressed by the inoculated mice corresponds precisely to the PrP^{Sc} type present in the inoculum. Since the recipient mice did not carry any of the donor's *PRNP* mutations and were isogenic, this remarkable finding indicates that the distinct conformations associated with PrP^{Sc} type 1 and type 2 can be reproduced independently of the genetic information. However, the ratio of the glycoforms present in the recipient's PrP^{Sc} was different from that of the donor's PrP^{Sc}, indicating that conformation and glycoform ratio are independently determined (Telling et al. 1996). The information necessary for the specification of the cleavage site determining the size of the PrP^{Sc} fragment is likely to be contained in the conformation of the donor's PrP^{Sc}, whereas the glycosylation is probably controlled by the host according to the cell population involved and glycoform convertibility, as mentioned above.

A major determinant of the PrP^{Sc} type in sporadic CJD has been shown to be the genotype at codon 129, since over 95% of subjects that carry PrP^{Sc} type 1 are homozygous for methionine at codon 129. Conversely, PrP^{Sc} type 2 is present in over 85% of the subjects that carry either one or two 129 valine alleles. Therefore, in sporadic CJD, the presence of methionine at position 129 of PrP^{C} favors its conversion to PrP^{Sc} type 1, whereas the formation of PrP^{Sc} type 2 is favored in the presence of valine in the same position (Gambetti et al. 2001). In inherited prion diseases, the pathogenic mutation is likely to be the major determinant of the PrP^{Sc} type, because there seems to be no tight correlation between PrP^{Sc} type and the methionine or valine codon 129 coupled with the mutation. However, codon 129 appears to be capable of modifying the PrP^{Sc} type determined by the mutation, since generally the PrP^{Sc} type is different when the same mutation is coupled with the methionine or valine 129 codon (see below).

REFLECTIONS ON THE PATHOGENESIS OF INHERITED PRION DISEASES: THE CELL MODELS

Subjects with inherited prion diseases become symptomatic in the adult or at an advanced age. Therefore, the disease remains clinically silent for many years despite the presence of the mutant PrP (PrP^{M}). Presently, it is not known whether the extended presence of the abnormal PrP^{M} is sufficient to initiate the clinical disease and the conversion occurs subsequently, or whether PrP^{Sc} is required for the expression of the disease phenotype. The cell models have provided considerable information and insight on the effect that *PRNP* mutations have on the metabolism and physicochemical properties of PrP^{M}, including its convertibility to PrP^{Sc} (Table 2).

Different cell models have been used to study the characteristics of the mutant PrP. They include neuroblastoma cells transfected with the pathogenic *PRNP* mutations D178N, E200K, Q217R, F198S, T183A, Y145Stop, and P102L (Petersen et al. 1996; Singh et al. 1997; Zanusso et al. 1999; Capellari et al 2000a,b; Rosenmann et al. 2001; Mishra et al. 2002; S.I. Zaidi et al., unpubl.); Chinese hamster ovary cells carrying *PRNP* mutations corresponding to the human D178N, E200K, T183A, P102L, 6- and 9-octapeptide insertion mutations (Lehmann and Harris 1995, 1996a,b, 1997); mouse N2A neuroblastoma cells carrying murine *PRNP* with mutation corresponding to the human T188R, T188K, and Q160Stop (Lorenz et al. 2002), or carrying hamster PrP with 5-, 7-, 9-, or 11-octapeptide insertion mutations (Priola and Chesebro 1998); and human fibroblasts from patients with the E200K *PRNP* mutation (Meiner et al. 1992; Gabizon et al. 1996).

Table 2. Changes in PrP^M metabolism observed in cell models of inherited prion disease

Haplotype	System	Common	Specific
E200K, 129M			
Meiner et al. (1992) Gabizon et al. (1996)	fibroblast from patients	PrP^M releasable by PI-PLC and PK resistant in monocytes	abnormal glycosylation
Lehmann and Harris (1996a)	transfected CHO (mouse gene)	PrP^{Sc}-like; underrepresentation of U form at cell surface	NR
Capellari et al. (2000b)	transfected M17 (human gene)		abnormal glycans at position 197
E200K, 129V			
Capellari et al. (2000b)	transfected M17 (human gene)	PrP^{Sc}-like	abnormal glycans at position 197; increased carboxy-terminal fragments
D178N, 129M			
Lehmann and Harris (1996a)	transfected CHO (mouse gene)	PrP^{Sc}-like	NR
Petersen et al. (1996)	transfected neuroblastoma (human gene)	underrepresentation of U form at cell surface	preferential degradation in ER
Ma and Lindquist (2001)	transfected NT2 and COS-1 (mouse gene)	underrepresentation at cell surface; insolubility	abnormal accumulation in the cytoplasm
D178N, 129V			
Petersen et al. (1996)	transfected M17 (human gene)	PrP^{Sc}-like; underrepresentation of U form at cell surface	preferential degradation in a post-ER compartment
T188R, 129M			
Lorenz et al. (2002)	transfected N (mouse gene)	PrP^{Sc}-like	
T188K, 129M			
Lorenz et al. (2002)	transfected N (mouse gene)	PrP^{Sc}-like	
T183A, 129M			
Rogers et al. (1990) Lehmann and Harris (1997) Capellari et al. (2000a)	transfected CHO (mouse gene) transfected M17 (human gene)	PrP^{Sc}-like	intracellular retention in a pre-Golgi compartment
P102L, 129M			
Lehmann and Harris (1996a) Mishra et al. (2002)	transfected CHO (mouse gene) transfected M17 (human gene)	PrP^{Sc}-like	aberrant recycling of PrP^M, decrease of 18 kD, increase of Ctm20 kD, carboxy-terminal fragments at the cell surface

A117, 129V Hegde et al. (1998b) Stewart and Harris (2001)	transfected CHO (mouse gene)	PrP^{Sc}-like	enhanced production of Ctm20 kD a transmembrane carboxy-terminal fragment
Insertion (+9), 129M Lehmann and Harris (1995, 1996a)	transfected CHO (mouse gene)	PrP^{Sc}-like; abnormal membrane insertion	NR
Insertion (+5, 7, 9, 11), 129M Priola and Chesebro (1998)	transfected NIH3T3 (hamster gene) transfected N2a (hamster gene)	PrP^{Sc}-like	the amount of PrP^{Sc}-like increase with increasing number of extra repeats
F198S, 129V S.I. Zaidi et al. (unpubl.)	transfected M17 (human gene)	PrP^{Sc}-like; underrepresentation of U form at cell surface	hyperglycosylation
Q217R, 129V Singh et al. (1997)	transfected M17 (human gene)	PrP^{Sc}-like; underrepresentation of U form at cell surface	formation of an additional fragment with GPI anchor; presence of two aberrant anchor carrying fragments
Y145Stop, 129M Zanusso et al. (1999)	transfected M17 (human gene)		high instability of PrP^{M}, with preservation of signal peptide; nuclear degradation through the proteasomal pathway
Y160Stop, 129M Lorenz et al. (2002)	transfected N2a (mouse gene)		high instability of PrP^{M}, with preservation of signal peptide; nuclear degradation through the proteasomal pathway

(NR) Not reported.

The studies of Petersen, Singh, and coworkers have shown shared metabolic changes as well as changes specific to each mutation. There are four major common changes: (1) Mutant PrP is unstable and may undergo degradation or aggregation. In particular, the unglycosylated form of mutant PrP is less stable than the mono- and diglycosylated forms. Furthermore, when glycosylation is abolished with tunicamycin, the entire expressed mutant PrP exhibits the same instability as the original unglycosylated form. This finding indicates that mutations destabilize PrP^M and that glycosylation partially corrects this effect. (2) Since the unglycosylated form of PrP^M is less stable, presumably indicating that PrP^M is difficult to fold properly, it is partially retained in intracellular compartments. As a result, less PrP is present at the surface of mutant cells, and the ratio of the three glycoforms is altered due to preferential underrepresentation of the unglycosylated form. (3) Overall, more PrP^M is recovered in the detergent-insoluble (P2) fraction, showing that PrP^M has the tendency to aggregate. (4) A small fraction of PrP^M displays a mild resistance to digestion with PK (3.3 µg/ml for 5–10 minutes), an order of magnitude lower than that of the PrP^{Sc} present in the corresponding human disease (50 µg/ml for 30 minutes). The PK-digested PrP^{Sc} fragments classically seen in the corresponding disease, i.e., type 2 in FFI or type 1 in the E200K, 129M haplotype, are not evident in the cell system after protease digestion. The PrP^M invariably results in a 20-kD truncated fragment intermediate between type 1 and 2 PrP^{Sc}. Most of the PrP^M alterations are corrected by maintaining the mutant cells at 24°C, indicating that they are secondary to the misfolding of the mutant PrP, which is known to be partially corrected at lower temperatures (Denning et al. 1992).

In addition, transfected neuroblastoma cells display PrP^M changes more specifically associated with the individual mutations. In some mutations, abnormal PrP^M fragments are formed in significant amounts. For example, a relatively stable 32-kD PrP^M fragment lacking the glycosyl phosphatidylinositol (GPI) anchor, as well as two unstable anchor-carrying fragments, has been found in cells transfected with the Q217R mutant construct and not with other constructs (Singh et al. 1997). Additionally, the Y145Stop mutation is associated with the corresponding truncated PrP^M fragment, which is rapidly degraded by the proteasomal system, unstable even at 15°C, and, under treatment with proteasomal inhibitors, seems to accumulate in the ER as well as in the nucleus (Zanusso et al. 1999). In the F198S mutation, the unglycosylated PrP^M is underrepresented not because it is degraded as in other mutations, but apparently because PrP^M is over-glycosylated (S.I. Zaidi et al., unpubl.). The T183A mutant is totally retained in intracellular locations probably in the ER

(Rogers et al. 1990; Lehmann and Harris 1997; Capellari et al. 2000a). The E200K mutation is characterized by retarded and abnormal maturation of the N-glycans, resulting in altered gel mobility of the diglycosylated or mature PrP^M form (Capellari et al. 2000b). However, Rosenmann et al. (2001) reported that PrP^{E200K} displayed no difference from PrP^C in neuroblastoma cells. In the P102L mutation, the surface expression of the 18-kD carboxy-terminal fragment is decreased fivefold, probably as a result of aberrant recycling of PrP^M from the cell surface, while there is a four-fold increase of a 20-kD carboxy-terminal fragment (Mishra et al. 2002).

Finally, in transfected cells carrying the D178N, 129M and the D178N, 129V haplotypes, corresponding to FFI and familial CJD linked to the D178N mutation, respectively, the instability of the PrP^M unglycosylated form is especially severe and results in an altered glycoform ratio at the cell surface. These changes are more conspicuous in the FFI than in the CJD^{D178N} cell model. Moreover, the unglycosylated form of PrP^M is also preferentially underrepresented in the membrane fraction isolated from FFI brains, but not from brains of control subjects (Petersen et al. 1996). This finding suggests that the underrepresentation of the PrP^{Sc} unglycosylated form in FFI and CJD^{D178N} is due to the relative unavailability of this form for conversion to PrP^{Sc} resulting from the mutation, rather than the preferential conversion of the highly glycosylated forms to PrP^S (Petersen et al. 1996; Parchi et al. 1998a). Ma and Lindquist (2001) have confirmed that the mutant PrP^{D178N} is less efficiently trafficked to the surface. Moreover, they found that the mutant PrP^{D178N} protein is subject to retrograde transport to the cytoplasm where it becomes pathogenic and neurotoxic (Ma and Lindquist 2002; Ma et al. 2002).

Using a mouse cell model transfected with a green fluorescent PRNP construct carrying mutations homologous to the human T188R and T188K, Lorenz et al. (2002) found that these two proteins are processed and localized in the cell like PrP^C. However, they possess PrP^{Sc}-like properties, i.e., enhanced PK resistance (5 µg/ml, 30 minutes at 37°C) and detergent insolubility. The truncated protein homologous to the Q160Stop mutant is described as a partially insoluble, short-lived (10 minutes) protein that is rapidly degraded by the proteasomal system and, when proteasomal activity is inhibited, has a direct influx to the nucleus (Lorenz et al. 2002).

Overall, comparable results have been obtained by Harris, Lehmann, and coworkers (Lehmann and Harris 1995, 1996a,b, 1997; Daude et al. 1997). These authors observed a number of characteristics shared by all these mutations, which include the following: (1) All PrP^M forms are in part hydrophobic, detergent-insoluble, and mildly protease-resistant. (2)

PrP^M acquires these characteristics through distinct and successive steps. A finding that is at variance with that of other studies is that the GPI anchor of the mutant PrP can be cleaved by PI-PLC, but PrP^M remains attached to the cell membrane, consistent with an abnormal PrP^M-membrane association (Lehmann and Harris 1995). (3) Delivery of PrP^M to the surface is impaired, causing PrP^M accumulation in the ER (Ivanova et al. 2001).

Recently, an alternative topological variant of PrP called ^{Ctm}PrP has been proposed as a key intermediate in the infectious and inherited forms of prion disease. Whereas most molecules of PrP are anchored to the cell membrane by a carboxy-terminal GPI anchor, two alternative forms, ^{Ctm}PrP and ^{Ntm}PrP, span the membrane once via a conserved, hydrophobic segment encompassing residues 111–134, with the carboxyl or amino terminus on the exofacial surface (Hegde et al. 1998a,b). ^{Ctm}PrP could be implicated in the pathogenesis of prion diseases for the following reasons: (1) Transgenic mice expressing PrP molecules carrying mutations in or near the transmembrane domain that favor the formation of ^{Ctm}PrP developed a spontaneous neurological disease with scrapie-like features, but without detectable PrP^{Sc} (Hegde et al. 1998b, 1999). (2) When these mice were inoculated with scrapie prions, the amount of PrP^{Sc} produced was inversely related to the amount of ^{Ctm}PrP (Hegde et al. 1999). (3) After scrapie inoculation of mice that carried wild-type hamster PrP, ^{Ctm}PrP was found to accumulate during the disease (Hegde et al. 1999). In different cell models it has been shown that the A117V (Hegde et al. 1998b; Stewart and Harris 2001) and P102L mutations (Mishra et al. 2002) would increase the relative proportion of ^{Ctm}PrP. Other mutants (D178N, F198S, E200K, insertion of 9 extra repeats), on the contrary, do not change the proportion of ^{Ctm}PrP (Stewart and Harris 2001).

Despite some differences, the cell models demonstrate that *PRNP* mutations cause profound and early changes in the metabolism and physicochemical characteristics of PrP^M. These changes appear to result from an altered conformation that PrP^M acquires cotranslationally: The misfolded protein is recognized by the cell as abnormal and subject to retention in the ER and destruction by the proteasomal system. A possible age-dependent impairment of the proteasomal function would cause PrP to accumulate in the cytoplasm, where it becomes pathogenic (Yedidia et al. 2001; Ma and Lindquist 2002; Ma et al. 2002). The altered conformation and/or the altered location/environmental condition confer characteristics of increased hydrophobicity, aggregability, and resistance to proteases reminiscent of those of the PrP^{Sc}. It remains to be established whether the "PrP^{Sc}-like" PrP^M is an irrelevant early by-product of the mutation, pathogenic itself, or the direct precursor of the pathogenic PrP^{Sc} that in due time would form the "real" pathogenic PrP^{Sc}.

BIOPHYSICAL BACKGROUND OF INHERITED PRION DISEASES

Although the link between hereditary prion diseases and specific mutations in *PRNP* provides strong support for the protein-only model, the molecular mechanism by which pathogenic mutations facilitate the PrP^{C}–PrP^{Sc} conversion remains unclear. Since the conversion process in hereditary spongiform encephalopathies appears to occur spontaneously (i.e., does not require exogenous PrP^{Sc}), understanding how individual pathogenic mutations affect the conformation, thermodynamic stability, and folding pathway of the prion protein could provide fundamentally important clues regarding the molecular basis of the disease.

It has been postulated that point mutations associated with inherited prion diseases promote the spontaneous PrP^{C}–PrP^{Sc} conversion by decreasing the stability of the native PrP^{C} isoform (Cohen et al. 1994; Huang et al. 1994). This is a very logical and attractive hypothesis. However, recent experiments in a cell-free environment have revealed that familial mutations do not exert a uniform effect on the thermodynamic stability of the prion protein, and there is no correlation between mutation-dependent PrP^{C} stability and the phenotype of the disease (Swietnicki et al. 1998; Liemann and Glockshuber 1999). Although some disease-related mutations (e.g., F198S, D178N, T183A) are indeed destabilizing, many others were found to have a negligible effect on the global thermodynamic stability of PrP (Fig. 5). The latter group includes, among others, P102L, E200K, V210I, and V180I. The experimental results on the stability of prion protein variants are in full agreement with the predictive analysis based on nuclear magnetic resonance (NMR) structure of the recombinant murine PrP (Riek et al. 1998). Overall, available experimental data suggest that although some familial spongiform encephalopathies could be rationalized by the mechanism based on mutation-induced decrease in global stability of PrP^{C}, this simple thermodynamic concept clearly does not apply to all inherited prion diseases.

Studies with classical amyloid-forming proteins such as transthyretin, immunoglobulin light chain, or lysozyme variants indicate that the pathway of amyloid formation by these proteins likely involves partially folded intermediates (Booth et al. 1997; Kelly et al. 1997; Khurana et al. 2001; Canet et al. 2002). In the presence of disease-associated mutations, these intermediates are believed to become more populated, resulting in an increased amyloidogenicity of the mutant proteins. Folding intermediates

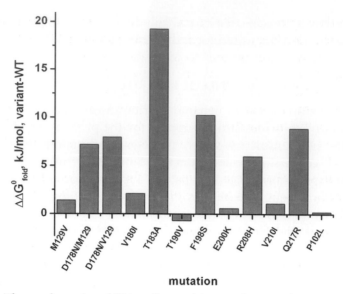

Figure 5. Thermodynamic stabilities of recombinant prion protein mutants relative to the wild-type protein. $\Delta\Delta G^0{}_{fold}$ represents the difference in the stability of the mutant protein relative to the wild-type PrP. Positive values indicate that the variant is less stable than the wild-type protein. Data for all mutant proteins except P102L variants are from Liemann and Glockshuber (1999). Those for the P102L are from Swietnicki et al. (1998).

have also been postulated to play a key role in the PrP^C–PrP^{Sc} conversion (Cohen et al. 1994). However, unlike many other amyloidogenic proteins, partially folded intermediates for prion protein have proven to be difficult to detect and characterize. This led to the hypothesis that the mechanism of prion protein conversion is fundamentally different from that of classical amyloid-forming proteins and that it is the fully unfolded state of PrP that is a direct monomeric precursor of PrP^{Sc} (Hosszu et al. 1999; Wildegger et al. 1999). Recent NMR (Kuwata et al. 2002; Nicholson et al. 2002) and kinetic stopped-flow (Apetri and Surewicz 2002) data have, however, disputed this model, providing experimental evidence for a monomeric folding intermediate for the prion protein. The finding of an intermediate in prion protein folding is of potentially crucial importance for understanding the mechanism of inherited prion diseases. Mutations associated with these diseases could facilitate the PrP^C–PrP^{Sc} conversion reaction not by increasing the population of an unfolded protein (as implied by the simple two-state thermodynamic model), but rather by increasing the stability and/or affecting the conformation of the partially folded intermediate species.

A potentially crucial role of partially folded intermediates in prion protein conversion is consistent with the observation that the transition of the recombinant prion protein to an oligomeric β-sheet-rich (scrapie-like) form is facilitated in the presence of low and medium concentrations of urea; i.e., under conditions that could increase the population of partially folded intermediates. In contrast, conditions favoring the native state (no denaturant) or those favoring a fully unfolded state (high concentration of urea) are not conducive to the conversion reaction (Morillas et al. 2001).

A recent study has demonstrated that the conversion of the recombinant human prion protein to a scrapie-like form is strongly promoted by the F198S mutation, a substitution linked to GSS with extensive PrP amyloidosis (Vanik and Surewicz 2002). Importantly, the conversion of the mutant protein was found to occur spontaneously and under physiologically relevant buffer conditions. The oligomeric, β-sheet-rich structures formed in vitro by the F198S PrP variant were thioflavine-T-positive and morphologically appeared similar to those found in brain tissue of GSS patients. The increased conversion propensity of the F198S variant appears to correlate with a mutation-induced destabilization of the native state and an increased stability of a folding intermediate (A.C. Apetri and W.K. Surewicz, unpubl.). This correlation supports the notion that partial destabilization of the native PrP^C state is important for prion protein conversion to a β-sheet-rich oligomeric form. Whereas for the wild-type PrP, formation of a conversion-conducive intermediate in vitro appears to require partially denaturing buffer conditions, a similar state for the F198S variant may be populated under native conditions because of the destabilizing effect of the Phe-198-Ser substitution.

A dramatic increase in the propensity for conformational conversion to a scrapie-like form has also been observed for the recombinant PrP with D178N mutation. Importantly, the kinetics of this conversion was found to depend strongly on the Met/Val polymorphism at position 129 (D.L. Vanik and W.K. Surewicz, unpubl.). The latter observation provides a direct link between biophysical properties of prion protein variants and polymorphism-dependent phenotypic variability of familial prion diseases (D178N mutation with the Met-129 polymorphism correlates with FFI, whereas the same mutation with the Val-129 polymorphism is associated with familial CJD). However, the structural basis of these polymorphism-dependent effects is not fully understood. Based on the analysis of NMR structural data for the wild-type murine PrP, Riek et al. (1998) noted that Asp-178 forms a salt bridge with Arg-164 and a hydrogen bond with Tyr-128, a residue sequentially adjacent to position 129. These authors have also proposed that the hydrogen bond network involving residues 128 and 178 could be affected by the Asp-178-Asn substitution, providing a structural rationale for the polymorphism-dependent phenotypic variability of prion diseases with the D178N mutations. However, this attractive hypothesis is yet to be verified by direct structural studies of the human prion protein with amino acid substitutions at positions 178 and 129.

A full understanding of the molecular basis of inherited prion diseases requires structural data for individual prion protein mutants. To date, high-resolution structural data are available only for E200K PrP, a variant associated with familial CJD. Apart from minor differences in the flexible loop regions, the three-dimensional structure of the E200K variant (Zhang et al. 2000) appears to be nearly identical to that of the wild-type protein (Zahn et al. 2000). The only major consequence of the Glu-200-Lys substitution appears to be the redistribution of surface charges, resulting in a dramatically altered electrostatic potential in the mutant protein (Fig. 6). The above changes in surface properties could profoundly affect the interactions of the prion protein with other molecules present in the cellular environment. On the basis of the structural data, it has been postulated that abnormalities in cellular interactions of PrP^C should be

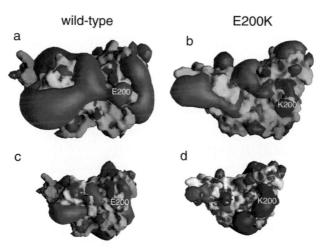

Figure 6. The electrostatic potential for the residue-200-containing surface in the folded domain of the wild-type human prion protein (*left*) and the E200K variant (*right*). The potentials in the two upper panels (*a* and *b*) were calculated with His side chains unprotonated (neutral pH), whereas those in the bottom panels (*c* and *d*) were calculated with His side chains protonated. Blue and red colors represent positive and negative potential, respectively. (Reprinted, with permission, from Zhang et al. 2000 [Copyright American Society for Biochemistry and Molecular Biology].)

considered as a potentially key determinant in the spontaneous PrP^C-PrP^{Sc} conversion in the E200K form of CJD (Zhang et al. 2000).

Overall, available biophysical data suggest that there may be no common pattern by which individual mutations linked to familial prion diseases affect the structure, stability, and folding pathway of PrP. Whereas some of the mutations appear to act by destabilizing the native PrP^C conformer and/or increasing the population of a folding intermediate, it is also possible that formation of a conversion-competent state could be promoted by mutation-dependent abnormalities in the interactions of PrP^C with membranes, chaperones, or other components of the cellular environment.

CLASSIFICATION

Inherited prion diseases may be associated with a phenotype (1) that has the clinical and histopathological features of CJD, FFI, or GSS; (2) that blends the features of both CJD and GSS; or (3) that lacks distinctive histopathological features. Furthermore, within each of these phenotypic groups, inherited prion diseases are classified according to the haplotype as determined by the pathogenic mutation and the codon 129 coupled with the mutated codon (Table 1).

INHERITED PRION DISEASES WITH CJD PHENOTYPE

CJD with the E200K Mutation Coupled to the 129M Codon (CJD^{E200K,129M})

Epidemiology

The epidemiology of the CJD^{E200K,129M} is of particular interest. The largest cluster of CJD^{E200K,129M} occurs among Jews of Libyan and Tunisian origin (Kahana et al. 1974). In this community, the incidence of CJD is 100 times higher than the worldwide incidence (50 per million as opposed to 0.5 per million), the highest incidence of CJD in the world. Although early speculation attributed the high incidence to diet or other environmental factors, a series of epidemiologic studies done in Israel and other countries revealed the unusually high incidence of familial CJD in this community (Neugut et al. 1979), which indicated a genetic origin (Cathala et al. 1985; Radhakrishnan and Mousa 1988; Nisipeanu et al. 1990; Hsiao et al. 1991b; Zilber et al. 1991; Gabizon et al. 1992).

After the E200K mutation was discovered, new data were obtained regarding the penetrance and other genetic details of the disease, the biochemistry of the mutant and normal protein in these patients, and the transmission rate of prions from these patients to experimental animals. These discoveries strengthened the causative relationship between the E200K mutation and familial CJD, and began to suggest the modalities by which this pathogenic *PRNP* mutation causes a prion disease.

Genetics

DNA analysis of patients with familial CJD revealed the presence of the E200K mutation linked to the disease with a logarithm of odds (LOD) score exceeding 4.85, where a LOD score above 3 is considered significant evidence for linkage (Lathrop et al. 1984; Goldfarb et al. 1990; Chapman and Korczyn 1991; Hsiao et al. 1991b; Korczyn et al. 1991). Five patients homozygous for the E200K mutation (due to consanguinity) have been identified (Simon et al. 2000).

In addition to the Jews of Libyan and Tunisian origins, the E200K mutation has been identified in other large clusters of patients with familial CJD in Slovakia, Chile, and Italy (Goldfarb et al. 1990; Brown et al. 1992a; Lee at al. 1999), raising the possibility of a common founder for all these clusters. Genetic analyses indicate that Libyan, Tunisian, Italian, Chilean, and Spanish families share a major haplotype, suggesting that the E200K mutation originated from a single mutational event in Spain and was dispersed by the expulsion of the Sephardic Jews to other parts of the world in 1492 (Goldfarb et al. 1991c; Lee et al. 1999; Colombo 2000). Slovakian and Polish families carry a distinct E200K haplotype, and the haplotypes of families from Germany, Sicily, Austria, and Japan are still different (Lee et al. 1999). Therefore, the present geographic distribution of the E200K mutation is likely to reflect a founder effect as well as independent mutational events (Lee et al. 1999).

In early genetic studies, presymptomatic mutation carriers above the mean age of onset of the familial CJD[E200K,129M] were identified (Goldfarb et al. 1991c; Gabizon et al. 1993), giving the false impression that the penetrance of the disease is low and raising the possibility that other factors play a role in the pathogenesis of the disease. However, recent studies using life-table analysis have shown that the penetrance is age-dependent: 0.45 by age 60, 0.89 at age 80, and 0.96 above age 80 years (Chapman et al. 1994; Spudich et al. 1995). Such results argue that the E200K mutation is sufficient to cause familial CJD. The correlation between the M/V polymorphism at codon 129 and age of onset or disease duration in CJD[E200K,129M] is controversial. Although lack of correlation was reported by Gabizon et al. (1993) in a study of 19 Libyan Jewish patients, in 95

Slovak affected subjects carrying the E200K,129M haplotype, Mitrova and Belay (2002) reported a shorter disease duration in M/M homozygous than in M/V heterozygous patients, as well as different PrP immunohisto-chemical staining patterns between these two populations. A Japanese family was reported to have the rare E200K,219K haplotype, but the influence of the 219K polymorphism could not be ascertained (Seno et al. 2000).

Clinical Manifestation

$CJD^{E200K,129M}$ resembles the typical form of the sporadic CJD (sCJD), also identified as sCJDMM1 (Parchi et al. 1996c, 1999c). Slightly younger mean age at onset and shorter disease duration have been reported in Libyan than in non-Libyan patients (Kahana et al. 1991; Simon et al. 2000). However, overall, the mean age at onset in patients heterozygous for the mutation is 58 years (range 33–84) and the mean duration 6 months (Kahana et al. 1991; Simon et al. 2000). Furthermore, mean age at onset was younger (58 years) and the duration longer (16 months) in five Libyan Jewish patients who are homozygous for the E200K mutation as compared to heterozygous patients (Simon et al. 2000). According to another study based on Slovak patients, the disease duration is shorter in patients who are 129M/M homozygous (3.7 ± 2.0 months) than in MV heterozygous patients (7.8 ± 7.3 months) (Mitrova and Belay 2002).

As in sCJDMM1, presenting signs include cognitive impairment and psychiatric changes (80–83% of patients), cerebellar signs (43–55%), visual signs (19%), and myoclonic jerks (12%) (Brown et al. 1991a; Kahana et al. 1991). Although in most of the patients the course of the disease is insidious, 10% of patients present with an acute, sudden onset. During the course of the disease, all patients develop dementia as well as other cognitive and psychiatric disturbances, 73% have myoclonus, 79% cerebellar signs, 40% seizures, and 24% sensory and cranial nerve involvement (Brown et al. 1991a). A puzzling clinical feature of the $CJD^{E200K,129M}$ is the involvement of the peripheral nervous system, which is rare in the course of sCJD (Lope et al. 1977; Schoene et al. 1981; Guiroy et al. 1989). The peripheral neuropathy is both motor and sensory and is often accompanied by protein elevation in the cerebrospinal fluid (Sadeh et al. 1990; Neufeld et al. 1992; Antonine et al. 1996; Worrall et al. 2000).

Electrophysiological and neuroimaging findings are also similar to those of sCJD. The typical electroencephalogram (EEG) activity with periodic spike and wave (PSW) complexes is found in ~75% of $CJD^{E200K,129M}$ patients, a prevalence comparable to that of sCJDMM1

(Brown et al. 1986, 1991a; Parchi et al. 1999b). Slowing of the background is found in all patients. In some of the CJD^{E200K,129M} patients, the pattern seen in routine EEG is asymmetric and not strictly generalized (Neufeld et al. 1992). Computed tomography (CT) scans have shown brain atrophy in one-third of the patients (Chapman et al. 1993). Single-photon emission computed tomography (SPECT) of one patient's brain that was normal in a brain CT scan showed bilateral perfusion defects (Cohen et al. 1989). All these nonspecific diagnostic findings are similar to those seen in sCJD (Galvez and Cartier 1984). The test for the 14-3-3 protein in the cerebrospinal fluid (CSF) has been reported positive in almost all the cases of CJD^{E200K} (Rosenmann et al. 1997; Oberndorfer et al. 2002). Insomnia, which in one case was indistinguishable by polysomnography from that associated with FFI, has been reported in two cases carrying the E200K,129M haplotype that also showed prominent histopathological involvement of the thalamus (Chapman et al. 1996; Taratuto et al. 2002). Therefore, insomnia may be a prominent and early sign in CJD^{E200K,129M} when the thalamus is severely involved.

Histopathology

The histological changes associated with CJD^{E200K,129M} are very similar to those of the typical sCJDMM1 (Parchi et al. 1996c, 1999c) and are invariably characterized by spongiosis, astrogliosis, and neuronal loss (Fig. 7).

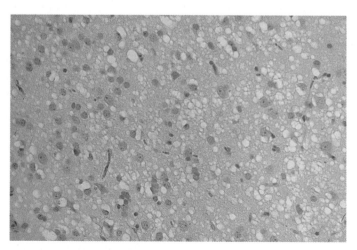

Figure 7. Typical histopathology of CJD^{E200K,129M}. The cerebral cortex shows spongiform degeneration indistinguishable from that of the typical sporadic CJD (Group 1 of Parchi et al. 1996c). (Hematoxylin and eosin.)

The severity of the astrogliosis and of the neuronal loss appears to be a function of the disease duration. These lesions are severe and widely distributed in the cerebral cortex. They are also present with decreasing severity in the striatum, diencephalon, and cerebellum. Immunostaining is consistently positive throughout the brain, with the punctate or "synaptic" pattern and a severity that appears to be directly related to that of the histological lesions (K. Young et al., unpubl.). In addition, punctate PrP immunostaining is also present in the substantia gelatinosa of the spinal cord. A difference in PrP immunostaining pattern in the cerebellum has been reported between $CJD^{E200K,129M}$ subjects that are M/M or M/V at codon 129 (Mitrova and Belay 2002). Whereas the homozygous subjects show a synaptic pattern of immunostaining, in the heterozygous patients the pattern includes granules and plaque-like structures (Mitrova and Belay 2002). The peripheral neuropathy is both axonal and demyelinating (Chapman et al. 1993); the latter is characterized by segmental demyelination (Sadeh et al. 1990; Neufeld et al. 1992).

Characteristics and Allelic Origin of PrP^{Sc}

The availability of patients homozygous for the E200K mutation has provided the opportunity to examine whether the E200K PrP^{M} is formed as a soluble, nonpathogenic protein and is converted into mutant PrP^{Sc} during the long incubation period, or whether it is initially formed as mutant PrP^{Sc}. The indirect evidence for conversion during the incubation period includes the similarity in age of onset of $CJD^{E200K,129M}$ and sCJD. A more direct finding is the absence of PrP^{Sc} in the brain biopsy of a presymptomatic homozygous carrier of the E200K mutation (R. Gabizon, unpubl.).

Studies have also been carried out to determine whether both the E200K PrP^{M} and the PrP^{C} expressed by the normal allele are converted into PrP^{Sc} and participate in the pathologic process, or if only the PrP^{M} is converted. PrP^{M} was found to be converted to PrP^{Sc}, whereas PrP^{C} did not acquire protease-resistance but became insoluble in detergents (Table 3). Since insolubility is also a characteristic of PrP^{Sc} (Meyer et al. 1986), these findings raise the possibility that both (E200K) PrP^{M} and PrP^{C} participate in the pathogenesis of $CJD^{E200K,129M}$.

The biochemical properties of the E200K PrP^{M} have been examined in brain samples, lymphocytes, and cultured fibroblasts of $CJD^{E200K,129M}$ patients, and in brain samples of transgenic mice carrying the E200K mutation. In brains of subjects with $CJD^{E200K,129M}$, the PrP^{Sc} has the gel migration pattern of the PrP^{Sc} type 1. $CJD^{E200K,129M}$ migrates at 21 kD on gel and shows an underrepresentation of the unglycosylated form

Table 3. Allelic origin of abnormal PrP in inherited prion diseases

Haplotype	Phenotype	PrP^{insol}	PrP^{Sc}	References
A117V, 129V	GSS	mutant	mutant	Tagliavini et al. (2001)
D178N, 129V	CJD	mutant	mutant	Chen et al. (1997)
D178N, 129M	FFI	mutant	mutant	Chen et al. (1997)
P102L, 129M	GSS	mutant	mutant	Parchi et al. (1998b)
F198S, 129V[a]	GSS	mutant	mutant	Tagliavini et al. (1994)
Q217R, 129V[a]	GSS	mutant	mutant	Tagliavini et al. (1994)
E 200K, 129M	CJD	wild-type and mutant	mutant	Gabizon et al. (1996)
V210I, 129M	CJD	wild-type and mutant	wild-type and mutant	Silvestrini et al. (1997)
Insertion, 129M	CJD	wild-type and mutant	wild-type and mutant	Chen et al. (1997) Rossi et al. (2000); Pietrini et al. (2003)

[a]Only the amyloid subfraction was studied.

(Monari et al. 1994; Parchi et al. 1996a). This glycoform ratio is similar to that observed in FFI (see below) and in the new variant CJD (Monari et al 1994; Collinge et al. 1996). It was also shown that the highly glycosylated form of E200K PrP^M migrates faster on SDS-PAGE than does the corresponding PrP^C form (Gabizon et al. 1996), a difference that disappears following deglycosylation. This implies a difference in the glycosylation pattern between the E200K PrP^M and PrP^C, which is likely due to the proximity of the mutation to the glycosylation site at residue N197. This conclusion is also supported by the data obtained with the E200K cell transfectants (see above). Solution structure of human PrP^{E200K} revealed only minor differences in flexible regions, and the most significant change was surface electrostatic potential (Zhang et al. 2000), suggesting that perturbation of PrP interaction with other proteins or the cell membrane may be the cause of spontaneous PrP^C–PrP^{Sc} conversion in CJD^{E200K} patients.

The amount of PrP, but not of PrP mRNA, was found to be higher in lymphocytes and fibroblasts from subjects with CJD^{E200K,129M} than in controls (Meiner et al. 1992), suggesting that PrP accumulates in CJD^{E200K,129M} patients, and that the accumulation is not the result of accelerated synthesis, but rather of a slower degradation process. This hypothesis is further supported by the slower degradation rate of the E200K PrP^M in fibroblasts from a CJD^{E200K,129M} homozygous patient compared to controls (Gabizon et al. 1996).

The PrP that accumulates in lymphocytes and fibroblasts of CJD^{E200K,129M} patients has been found to have the biochemical properties of PrP^C (Meiner et al. 1992), in agreement with a recent report indicating that PrP^{E200K} in all tissues of transgenic mice is very similar to PrP^C (Rosenmann et al. 2001). These findings are in disagreement with a study in which E200K PrP^M expressed in transfected Chinese hamster ovary (CHO) cells was found to be partially protease-resistant, insoluble, and not released from the plasma membrane by phosphatidylinositol-specific phospholipase C, as is the case for PrP^{Sc} (see above) (Lehmann and Harris 1996a,b). In addition, Capellari et al. (2000b) showed that E200K PrP^M expressed in a human neuroblastoma cell line (M-17 BE[2C]) had an aberrant glycan at residue 197 and transport of the unglycosylated PrP to the cell surface was impaired.

Transmissibility

Intracerebral inoculation of CJD^{E200K,129M} brain homogenate has resulted in CJD transmission to apes after an incubation period of 6 years (Chapman et al. 1992). Studies with transgenic mice susceptible to human prions show transmission with specimens from familial CJD^{E200K,129M} patients (Telling et al. 1994). Two of three brain samples transmitted the disease following an incubation period of 170 days. The lack of transmission seen with the third sample may be attributed to codon 129 methionine/valine incompatibility (Telling et al. 1995). The precise risk of transmission of CJD from brain and other tissues, especially blood, from CJD^{E200K,129M} patients and healthy mutation carriers remains to be established.

CJD with the E200K Mutation Coupled to the 129V Codon (CJD^{E200K,129V})

Epidemiology

The E200K, 129V haplotype, originally described in a family from Austria (Hainfellner et al. 1999), has been reported in four additional subjects from apparently unrelated families (Parchi et al. 1999b; Puoti et al. 2000). All but two of these subjects carried valine in position 129 of the normal *PRNP* allele.

Characteristics of the Prion Protein

PrP^{Sc} type 2 has been demonstrated in the five subjects examined (Fig. 2) (Hainfellner et al. 1999; Parchi et al. 1999b; Puoti et al. 2000). The glycoform ratio of PrP^{Sc} was similar to that found in the E200K, 129M haplotype (Hainfellner et al. 1999; Parchi et al. 1999b; Puoti et al. 2000).

Clinical and Pathological Features

The disease phenotype associated with the E200K, 129V haplotype appears to differ from that of the E200K, 129M haplotype and resembles the phenotype of sporadic CJD in subjects that are 129V/V and carry PrP^{Sc} type 2 (Parchi et al. 1999b). This phenotype is characterized by ataxia at presentation, whereas myoclonus and PSW complexes on EEG examination appear at later stages. PrP^{Sc} accumulates predominantly in the cerebellum where it forms plaque-like structures.

CJD with the D178N Mutation Coupled to the 129V Codon (CJD^{D178N,129V})

Epidemiology

Currently, twelve apparently unrelated kindreds are known (Goldfarb et al. 1992c; Kretzschmar et al. 1995; Rosenmann et al. 1998; P. Gambetti et al., unpubl.), of which five are American, three French, two Israeli, one Finnish, and one German. Three of the five American kindreds that are of Hungarian-Romanian, Dutch, and French-Canadian origins, respectively, have been repeatedly published, often with the appellation of Day (Friede and DeJong 1964; May et al. 1968; Masters et al. 1981b; Brown et al. 1991a; Fink et al. 1991; Nieto et al. 1991; Goldfarb et al. 1992b,c; Medori et al. 1992), Kui (Masters et al. 1981b; Brown et al. 1991a; Nieto et al. 1991; Goldfarb et al. 1992b,c), and LaP families (Masters et al. 1981b; Goldfarb et al. 1992b), and one is unpublished (P. Parchi and P. Gambetti, unpubl.). Two of the American families are unpublished (P. Gambetti et al., unpubl.). Of the three French kindreds, two commonly identified as Wui and Bel or AB families have been published (Buge et al. 1978; Guidon 1978; Masters et al. 1981b; Vallat et al. 1983; Brown et al. 1991a,b; Nieto et al. 1991; Goldfarb et al. 1992b,c; Medori et al. 1992), and one is unpublished (L.G. Goldfarb, unpubl.). Two cases have been reported in Israel that are members of families of Russian and Yugoslavian Jewish origins, respectively (Rosenmann et al. 1998). The Finnish kindred has been known for several years (Haltia et al. 1979; Masters et al. 1981b; Kovanen and Haltia 1988; Brown et al. 1991a; Goldfarb et al. 1991a, 1992b,c; Medori et al. 1992), and the German kindred is the above-cited Baker family, which was originally reported in 1930 but has been recently shown to have the CJD^{D178N,129V} haplotype (Meggendorfer 1930; Brown et al. 1994a; Kretzschmar et al. 1995).

PRNP *Genotype and Genetic Linkage*

CJD^{D178N,129V} shares the D178N *PRNP* mutation with FFI. The genotypic difference between the two diseases is specified by the codon 129 located on the mutant allele, which encodes valine in the CJD^{D178N,129V} and methionine in FFI (Goldfarb et al. 1992c). Moreover, the alternative presence of the methionine or valine codons at position 129 on the normal allele results in valine homozygous and valine/methionine heterozygous patients (Goldfarb et al. 1992c). Genetic linkage tested in two informative kindreds (Finnish and Day) provided a LOD score of 5.30, supporting the hypothesis that the D178N, 129V haplotype is the cause of the disease (Goldfarb et al. 1992b).

Clinical Features

The mean age at onset of the disease is 39 ± 8 years (range: 26–47) for the 129V/V homozygous patients, and 49 ± 4 years (p <0.01) (range: 45–56) for the 129M/V heterozygous patients. The mean durations are 14 ± 4 months (range: 9–18) for the homozygous patients and 27 ± 14 months (p <0.05) (range: 7–51) for the heterozygous patients (Kirschbaum 1968; Goldfarb et al. 1992c; Kretzschmar et al. 1995). Clinical signs are fairly consistent and apparently independent of the zygosity at codon 129. Presentation is characterized by cognitive impairment, especially memory decrease, often associated with depression, irritability, and abnormal behavior. Ataxia, speech impairments with dysarthria and aphasia, tremor, and myoclonus appear during the course of the disease. Electroencephalographic examination reveals generalized slow-wave activity without periodic complexes in most cases. Insomnia has not been reported in CJD^{D178N,129V}. Polysomnography carried out in two CJD^{D178N,129V} patients revealed a relatively normal sleep pattern (E. Lugaresi et al., unpubl.).

Histopathology

There is apparently no difference between codon 129 homozygous and heterozygous subjects; however, the number of subjects examined at autopsy in which codon 129 is known is limited. The common changes of this familial CJD subtype are spongiosis associated with prominent gliosis, often in the form of gemistocytic astrocytes, and variable degrees of neuronal loss (Fig. 8) (Parchi et al. 1996b). Enlarged or ballooned neurons may be present, which contain argyrophilic and Lewy body-like inclusions, that immunoreact with antibodies to neurofilaments (Fig. 8) (P. Parchi and P.

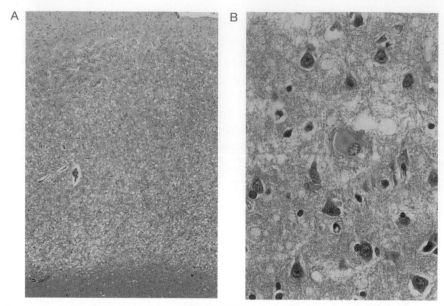

Figure 8. Typical histopathology of CJD^{D178N,129V}. Severe spongiosis with disorganization of the cytoarchitecture (*A*) and occasional neuronal Lewy body-like inclusions or ballooned neurons (*B*) are found in the cerebral cortex. (Hematoxylin and eosin.)

Gambetti, unpubl.). The topography of these lesions is consistent and fairly distinctive. The involvement of the cerebral cortex is widespread, but frontal and temporal cortices are generally more severely affected than the occipital cortex. Within the hippocampal region, these changes are especially prominent in the subiculum and entorhinal cortex, whereas spongiosis is often present in the fascia dentata. Among the subcortical structures, the putamen and the caudate nucleus show severe spongiosis with variable degrees of gliosis; the thalamus is minimally or moderately affected with spongiosis and gliosis, whereas the cerebellum is spared, and minimal or no pathology is seen in the brain stem. The immunostaining pattern is punctate, and its intensity matches the severity of the histopathology. However, the cerebellum shows minimal but definite immunostaining despite the lack of structural changes. No PrP-positive deposits, either in the form of amyloid or non-amyloid plaques, are present.

Characteristics and Allelic Origin of the PrP^{Sc}

All seven cases of CJD^{D178N,129V} examined to date have been associated with PrP^{Sc} that in the unglycosylated form has a gel mobility of 20–21 kD, corresponding to the PrP^{Sc} identified as type 1 (Monari et al. 1994; Parchi et al. 1996a). The glycosylation pattern (Fig. 3) is characterized by a marked underrepresentation of the unglycosylated form, which accounts for ~20% of the total, whereas the intermediate and highly glycosylated forms are similar in amount, and account for the remaining 80% (Parchi et al. 1996a,b).

Determination of the allelic origin has shown that both detergent-insoluble (i.e., aggregated) PrP and PrP^{Sc} derive exclusively from the mutant PrP in brains of CJD^{D178N,129V} subjects either homozygous or heterozygous at *PRNP* codon 129 (Table 3) (Chen et al. 1997). Therefore, there appears to be no direct participation in the disease process of the PrP^C expressed by the normal allele. This finding has implications for the mechanism(s) regulating the disease duration. The role of codon 129 in modulating the disease duration was previously explained as the result of the easier conversion of PrP^C to PrP^{Sc} in 129 homozygotes than in 129 heterozygotes, due to the higher homology of mutant PrP and PrP^C in the 129 homozygotes (Palmer and Collinge 1992). The lack of PrP^{Sc} derived from the wild-type PrP in the brain of subjects with CJD^{D178N,129V} suggests that the mechanism causing a shorter disease duration in the 129 homozygous subjects must be different in this familial CJD subtype.

Animal Transmissibility

CJD^{D178N,129V} has been transmitted to squirrel monkeys with brain tissue of seven out of ten subjects from five of six kindreds (Brown et al. 1992b, 1994b). Squirrel monkeys are homozygous for methionine at codon 129 (Schatzl et al. 1995). Transmission to transgenic mice expressing the human–mouse chimeric PrP has failed (Telling et al. 1996). However, the recipient animals carried the methionine codon at position 128 corresponding to the position 129 in the human *PRNP*, which may be incompatible with the disease transmission with PrP^{Sc} containing valine at position 129.

CJD with the V210I Mutation Coupled to the 129M Codon (CJD^{V210I,129M})

Epidemiology

The V210I,129M haplotype has been reported in 23 affected subjects (Pocchiari et al. 1993; Ripoll et al. 1993; Furukawa et al. 1996; Parchi et al. 1996a; Shyu et al. 1996; Cardone et al. 1999; Mouillet-Richard et al. 1999;

Windl et al. 1999; Huang et al. 2001; Mastrianni et al. 2001; M. Pocchiari and A. Ladogana, pers. comm.). Most of the CJD^{V210I,129M}-affected subjects are either the only affected member of the kindred in which other members carry the V210I mutation but are asymptomatic despite their advanced age, or are from uninformative families. These findings indicate that the V210I mutation has low penetrance, as reported for some of the insertion mutations (Capellari et al. 1997).

Clinical Features

The age at onset in 17 subjects varies between 46 and 80 years with a mean of 59 years (Pocchiari et al. 1993; Ripoll et al. 1993; Furukawa et al. 1996; Shyu et al. 1996; Mouillet-Richard et al. 1999; Piccardo et al. 1999; Huang et al. 2001; Mastrianni et al. 2001; M. Pocchiari and A. Ladogana, pers. comm.). The average duration is 6 months with a 2–24-month range. Although the number of subjects available is small, the polymorphism at codon 129 and a 24-bp deletion polymorphism present on the nonmutant allele in one case do not appear to influence the age at onset or the disease duration. The presentation includes memory, behavioral, and gait disturbances; sudden sensory and motor deficits; clumsiness; dystonic movements; and dysarthria. Subsequent common signs are myoclonus, dysarthria, mutism, and cerebellar signs. The EEG shows the typical PSW complexes in all 7 subjects examined (Pocchiari et al. 1993; Ripoll et al. 1993; Furukawa et al. 1996; Mastrianni et al. 2001). Serial magnetic resonance imaging (MRI) in one case revealed increased signal intensity in the basal ganglia and thalamus in T2-weighted image and proton density, as well as severe brain atrophy. Diffuse white matter degeneration was present in the later stages (Shyu et al. 1996). SPECT scan has demonstrated global or temporal cortical hypoperfusion with intact subcortical structure (Mastrianni et al. 2001).

Histopathology

The six cases examined at autopsy revealed fine spongiform degeneration and gliosis of the gray matter, often more prominent in the cerebral cortex and molecular layer of the cerebellum (Pocchiari et al. 1993; Ripoll et al. 1993; Mastrianni et al. 2001). Occasionally, the spongiosis was composed of large confluent vacuoles (Mastrianni et al. 2001). Ballooned neurons, apparently containing neurofilament proteins, have been reported in the frontotemporal cortex (Mastrianni et al. 2001).

Characteristics and Allelic Origin of the Proteinase-K-resistant PrP (PrP^Sc)

On gel electrophoresis, the unglycosylated PrP^{Sc} migrates at 20–21 kD, corresponding to the PrP^{Sc} type 1 (Parchi et al. 1996b; Cardone et al. 1999; Mastrianni et al. 2001). The ratio of the three PrP^{Sc} glycoforms is similar to that of the PrP^{Sc} associated with sCJDMM1, which is characterized by relative dominance of the intermediate glycoform (Parchi et al. 1999c). Detailed studies have demonstrated that in V210I CJD, PrP^{Sc} is formed by both the mutant and normal PrP, similar to the inherited prion diseases with 5- or 6-repeat insertion mutations (Table 3) (Chen et al. 1997; Silvestrini et al. 1997).

Transmissibility

Intracerebral inoculation of $CJD^{V210I,129M}$ brain homogenates from three subjects has resulted in 100% CJD transmission to transgenic mice carrying a chimeric mouse–human PrP with incubation periods of 182–233 days (Mastrianni et al. 2001). The PrP^{Sc} distribution pattern in transgenic mice was very similar to that produced by sporadic CJD and different from that of other familial prion diseases such as $CJD^{E200K,129M}$ and FFI (Mastrianni et al. 2001).

CJD with the V180I Mutation Coupled to the 129M Codon (CJD^V180I,129M)

Six Japanese cases (one of which was not confirmed by histological examination), and one American case are known (Hitoshi et al. 1993; Kitamoto and Tateishi 1994; Iwaski et al. 1999; Nixon et al. 2000). Four subjects are MM and three M/V heterozygous at codon 129. One has a double mutation, V180I and M232R, on different alleles (described below with the $CJD^{M232R,129M}$). None of the subjects had a family history of CJD. For the Japanese cases, the age at onset of the cases with the V180I mutation only varies between 66 and 81 years of age, and the duration between 1 and 2 years. Of the two subjects examined in detail, one presented with cognitive impairment, especially memory decrease, the other with motor aphasia and hemiparesis, which in both were followed by akinetic mutism, pyramidal and extrapyramidal signs, and myoclonus (Kitamoto and Tateishi 1994; Iwaski et al. 1999). None of the four patients examined demonstrated PSW in the EEG (Kitamoto and Tateishi 1994; Iwaski et al. 1999). Neuropathological examination in three subjects revealed typical spongiform degeneration in the cerebral cortex, basal ganglia, and thalamus.

Neuronal loss and astrocytosis were observed in the cerebral cortex. PrP immunostaining of the grey matter was weak and diffuse. Kuru plaques and plaque-like formations were seen in the two subjects that were heterozygous at codon 129 (Kitamoto and Tateishi 1994; Iwaski et al. 1999). Intracellular perinuclear PrP immunostaining consistent with PrP presence in the Golgi apparatus, rough endoplasmic reticulum, or aggresomal location has been reported in one case (Kopito 2000; Nixon et al. 2000).

CJD with the T183A Mutation Coupled to the 129M Codon (CJD^{T183A,129M})

This haplotype was originally described in one Brazilian kindred of Spanish and Italian origin with 9 and subsequently 12 affected members (Nitrini et al. 1997, 2001). Five additional subjects belonging to families from the US, Germany, and Venezuela have been subsequently reported (Capellari et al. 1999; Windl et al. 1999; Cardozo et al. 2000; Finckh et al. 2000). In the 16 cases examined, the mean age at onset is 44 years (range 37–49 years); mean duration is 4 years (range 14 months–9 years) (Nitrini et al. 1997, 2001; Capellari et al. 1999; Windl et al. 1999; Cardozo et al. 2000). The modifying effect on age at onset and disease duration of codon 129 cannot be assessed since the genotype at codon 129 was determined in only 4 examined subjects. The clinical presentation includes personality changes associated with memory impairment in half of the cases. The predominant subsequent signs include rapidly progressive dementia, aggressive behavior, hyperorality, verbal stereotypes, and often parkinsonian signs. In the 8 cases examined, the EEG failed to demonstrate PSW complexes even in advanced stages of the disease and when the patient exhibited myoclonus. The histologic examination carried out in 6 subjects showed widespread spongiosis and atrophy in the cerebral cortex with more severe involvement of the frontal and temporal lobes. The basal ganglia were variably involved, whereas the hippocampus, thalamus, brain stem, and cerebellum were preserved. Immunostaining for PrP has been reported to be widespread, punctate, and especially strong in the cerebral cortex (Cardozo et al. 2000). A plaque-like pattern of PrP immunostaining has been reported in the putamen (Nitrini et al. 1997). The PrP^{Sc} gel pattern examined in one case is of type 2 (Capellari et al. 1999). The T183A mutation abolishes the N-linked glycosylation site present at codon 181 but, contrary to previous reports, it is not the lack of glycosylation at this site but the T183A mutation itself that causes the retention of the PrP^{M} in the endoplasmic reticulum as observed in cell models (Capellari et al. 2000a).

CJD with the R208H Mutation Coupled to the
129M Codon (CJD^{R208H,129M})

Only two affected subjects (from a kindred carrying the mutation, but negative for history of dementia) have been reported (Mastrianni et al. 1996; Capellari et al. 2001). In both subjects, the disease phenotype appears very similar to that of the typical sporadic CJD or sCJDMM1. Clinically, the patients presented at the age of 58 and 60 years, respectively. In the first patient, the presentation included anorexia, apathy, hallucinations, insomnia, memory loss, and confusion (Capellari et al. 2001). On examination, the patient was agitated and hallucinating with rigidity, tremor, and ataxia. Myoclonus appeared 3 months after the onset, and she died 1 month later. The 14-3-3 CSF test was positive. Sensory and motor difficulties with ataxia of the right leg, which improved after physical therapy, was the presentation in the second patient, but it is not clear whether these signs were part of the main disease (Mastrianni et al. 1996). The patient showed cognitive impairment, including forgetfulness, anomia, and dyscalculia 2 1/2 years later. He became paranoid with hallucinations and delusions, developed an ataxic gait and myoclonus, his speech showed perseverations with minimal output, and he died 7 months after the onset of dementia. The EEG showed PSW complexes in both patients, and the MRI was negative in the first patient and consistent with the atrophy of the left sensory-motor cortex in the second, who also showed hypometabolism of the left hemisphere by positron-emission tomography (PET) imaging. Both patients were M/M at codon 129, and the histopathology, PrP immunohistochemistry, and PrP^{Sc} typing and glycoform ratios were indistinguishable from those associated with typical sporadic CJD or sCJDMM1 (Parchi et al. 1996a,c).

CJD with the M232R Mutation Coupled to the
129M Codon (CJD^{M232R,129M})

The M232R mutation has been observed in eight Japanese subjects (Hoque et al. 1996). It has been found to be associated with the V180I mutation located on the other allele in one subject (also see above) (Hitoshi et al. 1993), whereas another subject was heterozygous at codon 219, the site of a rare glutamine/lysine polymorphism (Barbanti et al. 1996), with the lysine codon on the normal allele (Kitamoto and Tateishi 1994). A third subject was M/V heterozygous at codon 129; the 129 valine codon, which is extremely rare in the Japanese population (Doh-ura et al. 1989), was on the normal allele (Hoque et al. 1996). The remaining mutant subjects were M/M homozygous (Kitamoto and Tateishi 1994;

Hoque et al. 1996). In all of these cases, the family history was negative for neurodegenerative diseases. Clinical and pathologic findings have been reported in detail in three cases (Hoque et al. 1996). Age at onset varies between 55 and 70 years, duration between 4 and 24 months. Common presenting signs are memory and gait disturbances, which progress and are associated with myoclonus and mutism at more advanced stages. EEG is typical, with PSW complexes observed in all cases. CT scan shows widespread cerebral atrophy; PET, carried out in one case, showed decreased cerebral blood flow (Hoque et al. 1996). The histopathology is characterized by widespread spongiosis, gliosis, and neuronal loss of variable degree, regardless of the disease duration. The thalamus shows the most severe spongiosis, especially in the medio-dorsal nucleus. Neuronal loss is severe in the basal ganglia, and the spongiosis is minimal. The cerebral cortex shows moderate spongiosis, and the gliosis and neuronal loss may be severe. All three lesions are present in the brain stem; the cerebellum lacks spongiosis but shows neuronal loss and gliosis (Hoque et al. 1996). PrP immunoreactivity is widespread in the cerebrum and spinal cord with more intense immunostaining in the cerebral cortex, especially the hippocampus, and is of the punctate type with no plaque-like PrP deposits (Hoque et al. 1996). The M232R mutation is of particular interest because it affects a PrP residue that is supposed to be removed posttranslationally when the glycolipid anchor is attached (Stahl et al. 1990). Therefore, it is likely that the arginine that replaces the methionine at codon 232 of PrP^{M232R} is not cleaved by the anchoring process, resulting in a changed conformation of the mutant PrP. The $CJD^{M232R,129M}$ has been transmitted to mice with brain tissue suspensions, but not with suspensions from lymph nodes and spleen, and the resulting numbers of symptomatic animals, incubation times, and pathology were similar to those observed following transmission of sCJD (Hoque et al. 1996). In the subject with the V180I and M232R double mutation who was M/M homozygous at codon 129, the disease started at 84 years of age and lasted 1 year (Hitoshi et al. 1993). The onset was characterized by cognitive impairment with memory deficit and discalculia. At midcourse, the dementia worsened and was accompanied by akinetic mutism. Motor signs were not detected, with the exception of an increase in the deep tendon reflexes and myoclonus. No PSW complexes were detected in several EEG examinations. Serial MRI showed slight atrophy of cerebral white matter and basal ganglia and an abnormally high signal, which was first detected in the temporal cortex and extended into the fronto-parietal-occipital cortex during the first 4 months of illness. PET revealed markedly decreased glucose metabolism involving most of the brain, with relative preservation of sensory-motor and occipital cortices, thalamus, and cerebellum (Hitoshi et al. 1993). Histologic examination was limited to the parietal lobe and demonstrated cortical spongiosis and diffuse immunostaining. The gel pattern of the PrP^{Sc} is unusual because it apparently lacks the high-molecular-weight glycoform (Hitoshi et al. 1993). The M232R, 129M *PRNP* mutation has also been reported in a patient with diffuse Lewy body and progressive dementia, but this patient did not have CJD, based on the absence of spongiform degeneration, abnormal prion aggregates, and PK-resistant PrP (Koide et al. 2002).

CJD with the Two-Repeat Deletion Mutation Coupled to the 129M Codon (CJD^{Del,129M})

A deletion of two 24-bp repeats associated with a CJD phenotype has been identified in two apparently unrelated subjects (Beck et al. 2001; Capellari et al. 2002). The two deletions are distinct and involve the third (R2) and fourth (R3) repeat in the first case, the second (R2) and third (R2) repeats in the second case. There was no family history of neurodegenerative diseases in the first case, suggesting that the deletion mutation has low penetrance, and the family of the second case was uninformative. The ages at onset were 86 and 64 years, and the duration 23 months and 18 months, respectively. In both cases, the symptoms appeared after a head trauma and included memory impairment, dizziness, tremors, and minor ataxia. Myoclonus was observed in one patient (Capellari et al. 2002). EEG recording carried out in one patient was nonspecific (Beck et al. 2001). Histological examination carried out in one case (Capellari et al. 2002) demonstrated fine spongiform degeneration and gliosis with a distribution similar to that described for the typical sporadic CJD (sCJDMM1 subtype) (Parchi et al. 1999c). Immunostaining of PrP revealed a "synaptic" pattern in the cerebral and cerebellar cortices, whereas on immunoblot, PrP^{Sc} was of type 1 with a glycoform ratio similar to that described in sCJDMM1 and some familial prion diseases (Capellari et al. 2002).

Novel Prion Protein Gene Mutations Associated with a CJD-like but Not Fully Characterized Phenotype

P105T

This mutation is reported on a Web site (http://www.madcow.org/prion_point_mutations.html). The father (apparent founder) died of CJD at age 42 with unknown genotype. P105T (129M/V) is confirmed in a son (onset at age 30) and a daughter (still healthy). The codon 129 on the mutant allele was not specified.

R148H, 129M

This haplotype has been reported in a 63-year-old woman. The disease duration was 18 months, and the patient had clinical, pathological, and immunohistochemical features very similar to those of sporadic CJDMV2 (Parchi et al. 1999c; Pastore et al. 2002).

Q160Stop, 129M

This haplotype has been reported in members of an Austrian family with dementia of long duration presenting at 32 and 48 years of age and lasting more than 6–12 years. Repeated EEGs did not show PSW complexes. Brain examination at autopsy revealed a reduced brain weight and diffuse cortical atrophy. No microscopic examination was carried out (Finckh et al. 2000).

T188A, 129M

The only patient who has been shown to carry this haplotype is an 82-year-old woman with a condition of 4-month duration clinically and pathologically indistinguishable from sCJDMM1 subtype (Collins et al. 2000). Immunohistochemistry for PrP was negative. Immunoblot analysis was not carried out. Only a partial family history was available and was negative for neurodegenerative diseases.

T188K and T188R

The T188K mutation has been found in a 59-year-old patient with dysphasia, rapidly progressive dementia, and a negative family history for diseases of the nervous system. Histopathological data are not available. The patient carrying the T188R mutation was alive at the time of publication and affected by an undiagnosed disease (Windl et al. 1999; Finckh et al. 2000).

E196K, 129M

This haplotype was observed in a French family with three likely affected members (Peoc'h et al. 2000). The patients developed rapidly progressive dementia and ataxia with no PSW in EEG recording. The ages of onset were 63–77 years, with a duration of about 1 year. No autopsy data are available.

V203I, 129M

The V203I mutation was reported in a 69-year-old Italian man, with negative family history, admitted for monocular diplopia and dizziness followed by sudden confusion hallucinations, abnormal motor functions, and myoclonus. EEG showed PSW. The patient died 25 days after admission (Peoc'h et al. 2000).

E211Q, 129M

This haplotype has been observed in a French and an Italian family (Peoc'h et al. 2000; Ladogana et al. 2001). The four affected members in each family had clinical signs consistent with classical sporadic CJD, including PSW at EEG. The ages of onset ranged between 42 and 81, and duration ranged from 3 to 32 months (except for one case who died suddenly). No pathological data are available.

P238S

Windl et al. (1999) reported the P238S mutation in a German patient with possible prion disease. The patient was still alive and no further information is available.

Prion Protein Gene Mutations Apparently Not Associated with Prion Disease Phenotypes

The I138M variant has been found in a 57-year-old French male with the diagnosis of "frontal dementia, not CJD." G142S variant has been reported in two patients: a North African man, 69, with multiple sclerosis and a Mali woman, 25, with viral meningoencephalitis. The G142S mutations in these two individuals probably have a common founder, given the geographic proximity of the patients' kindreds and apparent rarity of this variant (http://www.mad-cow.org/prion_point_mutations.html [based on J.L. Laplanche, pers. comm., 2000]).

INHERITED PRION DISEASES WITH FFI PHENOTYPE

FFI with the D178N Mutation Coupled to the 129M Codon (FFI^{D178N,129M})

Epidemiology

Currently, at least 25 pedigrees are known that carry the FFI mutation (Table 4). In addition, there are at least 5 subjects in which the FFI mutation has been found in the absence of a family history of neurodegenera-

Table 4. FFI kindreds

	Kindreds	Origin	Subjects
Goldfarb et al. (1992c)	2[a]	Italian	7
	1	American/British	4
	1	French[e]	2
	1	American/German	2
Montagna et al. (1998)	2[a] (1[b])	Italian	6
	1	French[e]	2
Brown et al. (1998)	1	American/English	2[c]
	1	American/unknown	3
	2	Australian/Irish	3
	1	Japanese	1
Zerr et al. (1998); Harder et al. (1999)	4 (4[b])	German	11
Budka et al. (1998) Almer et al. (1999)	1	Austrian	4
Will et al. (1998)	2	British	2
Pocchiari et al. (1998); Padovani et al. (1998)	1	Italian	2[d]
Rossi et al. (1998)	1	Italian	1
Spacey et al. (2003)	1	Canadian/Chinese	1
Tabernero et al. (2000)	1	Spanish	2
Bär et al. (2002)	1[b]	German	1
Fuchsberger et al. (2003)	1	German	1
Taniwaki et al. (2000)	1	Japanese	–
Marcaud et al. (2003)	1	French[e]	1
Totals	25 (5[b])		58

[a]The same two families in separate reports.
[b]Mutant subjects with no familial history of FFI.
[c]Cases 1 and 2 and III-5; IV-7; IV-12; and V-14.
[d]Case 2 and 4 from FFI 1.
[e]The three FFI might belong to a single pedigree (Marcaud et al. 2003).

tive disease (Table 4). FFI has been reported in several European countries, including Italy, Germany, Austria, Spain, the United Kingdom, and France (Budka et al. 1998; Kretzschmar et al. 1998; Montagna et al. 1998; Padovoni et al. 1998; Pocchiari et al. 1998; Will et al. 1998; Tabernero et al. 2000; Fuchsberger et al. 2003); Australia (Brown et al. 1998); the US (Goldfarb et al. 1992c; Brown et al. 1998); Japan (Brown et al. 1998); and Canada in a family of Chinese descent (Spacey et al. 2003). To date, the FFI haplotype appears to be the third most common after the E200K, 129M and P102L, 129M haplotypes. In Germany, the FFI haplotype was found to be the most common in a survey conducted between 1993 and 1997 (Kretzschmar et al. 1998).

Clinical Phenotype

The age at onset, disease duration, and *PRNP* genotype have been determined in at least 58 FFI patients (Table 4). Several aspects of the phenotype are slightly different in patients who are homozygous or heterozygous at codon 129 of *PRNP* (Montagna et al. 1998). In the 58 subjects in which the genotype at codon 129, age at onset, and duration of the disease are known (Table 4), 40 are methionine homozygous and 18 are methionine/valine heterozygous. The mean age at onset is 49 years (range 20–72 years) and is not significantly different in homozygous and heterozygous subjects. In contrast, the mean disease duration is significantly shorter in the homozygous, 11 ± 4 months, than in the heterozygous subjects, 23 ± 18 (p <0.001), but the ranges, 5–21 and 7–84 months, respectively, overlap considerably. Clinical signs comprise altered sleep–wake cycles with decreased vigilance, dysfunction of the autonomic system, and somatomotor manifestations. Sleep–wake and vigilance abnormalities are characterized by insomnia and oneiric or stuporous episodes with hallucinations and confusion. Autonomic dysfunction includes systemic hypertension, irregular breathing, diaphoresis, pyrexia, and impotence. Diplopia, dysarthria, dysphagia, ataxia/abasia, and dysmetria are common motor signs, but spontaneous and evoked myoclonus, spasticity, and occasional tonic–clonic seizures also are often present. Clinical signs may vary according to the genotype at codon 129. The 129 homozygous patients present prominent sleep–wake and vigilance disturbances as well as spontaneous and evoked myoclonus and more evident autonomic alterations at the onset of the disease. In contrast, these features occur at a more advanced stage in the 129 heterozygous patients who instead present with ataxia and dysarthria. These motor signs worsen and remain prominent in these patients. Tonic–clonic seizures also are more frequent among 129 heterozygous patients.

Routine EEGs do not show specific alterations at onset. With the progression of the disease, especially in the terminal stages, the background EEG activity progressively changes to a monomorphic flat activity. At advanced stages, bursts of repetitive, diffuse 1–2-Hz periodic sharp waves associated with myoclonus may appear in cases of long duration.

Examination by 24-hour polysomnography shows a continuous oscillation between the activity of normal wakefulness and desynchronized theta activity. At more advanced stages, spindles and K complexes, typical of non-rapid eye movement (Nrem) sleep, are lost, and slow-wave sleep may disappear completely. Rem sleep may initially remain normal or display a pathologically preserved muscle tone, with simple gesturing or dream enactment. The total sleep time is drastically reduced, and the cyclic organization of sleep lacks the orderly transition between sleep stages (Sforza et al. 1995).

In 129 heterozygous patients, 24-hour recordings at the onset of the disease are generally characterized by a relative preservation of the cyclic structure of nocturnal sleep and by the persistence of slow (<4 Hz) EEG activity, typical of slow-wave sleep. The lack of physiological muscle atonia in the presence of oneiric activity is less prominent.

Blood pressure, heart rate, and norepinephrine resting plasma levels are higher than in normal controls comparable for age and sex, indicating sympathetic over-activity in all patients. Blood pressure and heart rate circadian rhythms are affected. The nocturnal blood pressure fall is lost early on in the disease, while the physiological bradycardia is still preserved and the rhythmic component persists, although with a reduced amplitude and shifted phase, even in the absence of recorded sleep. Circadian rhythms are abolished only in the terminal stages. Likewise, body core temperature is persistently elevated. Circadian catecholamine rhythms are essentially preserved in the early stages of FFI, but increasing mean plasma levels and decreasing circadian amplitudes mark the progression of the disease, with total loss of rhythms in the terminal stages. Cortisol levels are high, whereas the adrenocorticotropic hormone levels remain normal, suggesting a condition of hypercortisolism added to a functional dysregulation of the hypothalamic-pituitary-adrenal axis. Melatonin levels are gradually decreased in circadian amplitude and often shift in phase to a complete loss of rhythm. Somatotrophin also shows a reduced, or even absent, circadian rhythm, which parallels the loss of deep sleep, whereas prolactin rhythms do not change.

Given the normal cognitive tests, preserved behavior, and the prominent vigilance alterations, FFI does not fulfill standard criteria for dementia.

Neuroradiological and PET Studies

Standard brain CT and MRI disclose cerebral and cerebellar atrophy and ventricular dilatation only in those patients with a prolonged course. PET studies with (18F)-FDG performed initially in four (Cortelli et al. 1997) and then in seven patients for whom histopathological examination was performed, show marked reduction of glucose utilization in the thalamus and, to a lower degree, in the cingular cortex in all cases. Glucose hypometabolism was also found in the basal and lateral frontal and middle and inferior temporal cortices, and the caudate nuclei, in six patients. Generally, cases homozygous at codon 129 with a shorter disease course had less metabolic involvement, restricted to the thalamus, basal frontal and cingular cortex, whereas the subjects heterozygous at codon 129 with longer disease duration also had severe hypometabolism in the hippocampus, putamen, and caudate nucleus. Since the latter were examined at later stages, the more widespread involvement may reflect diffusion of the disease process with time. All brain areas with neuronal loss were hypometabolic on PET, but cerebral hypometabolism was more widespread than the histopathological changes and correlated with the deposition of the abnormal scrapie prion protein (PrP^{Sc}). Remarkably, in the occipital cortex the metabolism remained normal even in late disease stages. Thus, hypometabolism of the thalamus and cingulate cortex is the hallmark of FFI in PET scans. The additional cerebral and cerebellar involvement could be simply the consequence of the more prolonged course in some patients, reflecting the temporal progression of the disease. An alternative explanation, however, could be that the more widespread cerebral and cerebellar hypometabolism represents a metabolic signature of subtle genetic differences related to codon 129 in patients with prolonged disease. The hypometabolism tends to be confined in the homozygous patients and to be more widespread even in relatively early stages in the patients who are heterozygous at codon 129 (Bär et al. 2002). A SPECT using a specific ligand to visualize dopamine and serotonin transporters has shown a reduction of these two transporters by over 50% and 70%, respectively, in the thalamic–hypothalamic region of two FFI patients (Klöppel et al. 2002). The brain serotoninergic system is thought to play an important role in sleep regulation (Jouvet 1999).

Histopathology and Immunohistochemistry

The hallmark is loss of neurons and astrogliosis in the thalamus, which is present in all subjects with detailed histological examination, independent of the disease duration (Fig. 9) (Manetto et al. 1992; Parchi et al. 1995; Almer et al. 1999; Harder et al. 1999; Tabernero et al. 2000; Bär et al. 2002). The medio-dorsal and anterior ventral thalamic nuclei are invariably and severely affected, whereas the involvement of other thalamic nuclei varies (Fig. 9A,B). The inferior olives also show neuronal loss and gliosis in most cases. In contrast, the pathology of the cerebral cortex varies in proportion to the disease duration and is more severe in the limbic lobe than in the neocortex (Fig. 9C) (Manetto et al. 1992; Parchi et al. 1995; Almer et al. 1999; Harder et al 1999; Tabernero et al 2000; Bär et al. 2002). The entorhinal cortex and, to a lesser extent the piriform and paraolfactory cortices, show spongiosis and astrogliosis in most subjects. In contrast, the

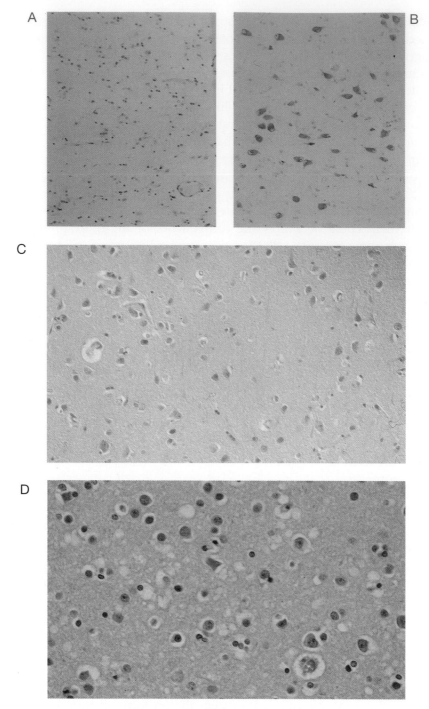

Figure 9. (*See facing page for legend.*)

Table 5. FFI diagnostic criteria

1. Autosomal dominant disease, onset at adult age, and duration of several months to several years
2. Presence of untreatable insomnia, dysautonomia, memory impairment, ataxia and/or myoclonus, pyramidal and extrapyramidal signs
3. Decrease or loss of sleep-related EEG activities
4. Preferential hypometabolism in the thalamic region using [18F]PET
5. Preferential thalamic atrophy by pathologic examination
6. 178N, 129M haplotype
7. PrP^{Sc} of type 2

Any combination of two of the criteria 1–5 makes the diagnosis of FFI highly probable. Criterion 6 in combination with any one of the other criteria, or criterion 7 combined with criteria 2–6, makes the diagnosis definitive.

neocortex is very often spared in the subjects with disease duration of less than 1 year, but focally affected by spongiosis and gliosis in those with a course between 12 and 20 months, and diffusely involved in subjects with duration of more than 20 months (Fig. 9D). In addition, the frontal, temporal, and parietal lobes are affected more severely than the occipital lobe. The presence of Purkinje cell "torpedoes," some neuronal loss especially in the dentate nucleus and vermis, has been reported in the cerebellum (Manetto et al. 1992; Parchi et al. 1995; Almer et al. 1999; Harder et al. 1999; Tabernero et al. 2000; Bär et al. 2002). Other brain stem structures reported to show moderate neuronal loss and gliosis include the periaqueductal grey of the midbrain, and the reticular formation and raphe of most of the brain stem (Almer et al. 1999). Recent findings indicate that neuronal loss in FFI is the result of apoptosis (Dorandeu et al. 1998) and that serotoninergic neurons are selectively increased in median raphe nuclei (Wanschitz et al. 2000). Overall, the thalamus is more severely and consistently involved than any other brain region. Therefore, on the basis of the pathology, FFI can be defined as a preferential thalamic degeneration. Criteria to diagnose FFI are listed in Table 5. The immunohistochemistry of PrP is negative in most cases. However, occasional PrP immunostaining has been reported in the molecular layer of the cerebel-

Figure 9. (*See facing page.*) Histopathology of FFI. (*A*) Severe loss of neurons and gliosis in the dorso-medial thalamic nucleus. (*B*) Control. (*C*) Lack of spongiosis and gliosis in the cerebral cortex of an FFI subject homozygous for methionine at codon 129 with a disease course of 7 months (GFAP immuno-staining). (*D*) Moderate spongiform degeneration in an FFI 129 heterozygous subject with a 25-month disease course. (Hematoxylin and eosin.)

lum with a strip-like pattern similar to that commonly observed in cases with insertional mutations (see below) (Almer et al. 1999). The other brain location containing PrP deposits is the subiculum–entorhinal region (Almer et al. 1999).

Genotype–Phenotype Correlation

The heterogeneity in clinical and histopathologic features that has been noted in FFI-affected subjects is consistent with the existence of two slightly different phenotypes that are determined by the genotype at codon 129 of the normal allele (Goldfarb et al. 1992c). As noted above, 129M/M FFI patients manifest, on average, a more rapid course and prominent sleep and autonomic disturbances, whereas signs of motor and cognitive dysfunction are mild. In contrast, the 129M/V subjects have a more chronic course and manifest motor signs, like ataxia and dysarthria, as a prominent clinical feature at onset, whereas sleep disturbances and autonomic signs are less severe. In addition, signs of cortical involvement appear in the heterozygous subjects, although late in the course of the disease. Furthermore, the histopathology varies as a function of the disease duration, which, in turn, is a function of codon 129 polymorphism. Overall, the thalamus is similarly affected in all subjects, or slightly more in the 129M/M subjects, whereas the cerebral cortex is generally more severely involved in the 129 heterozygous subjects.

Prion Protein in FFI and CJD[D178N]

The PK-resistant fragment of PrP[Sc] associated with FFI has been classified as type 2 and migrates at about 19 kD on gel following deglycosylation (Fig. 3) (Monari et al. 1994; Parchi et al 1995, 2000). Amino acid sequencing showed that the amino terminus of the PK-resistant fragment of the FFI PrP[Sc] is most often located at amino acid 97 (Parchi et al. 2000). The distinctive feature of the FFI-associated PrP[Sc] is the prominent underrepresentation of the unglycosylated form of PrP[Sc]. The ratio of the three glycoforms—highly glycosylated, intermediate, and unglycosylated—expressed as percent of total in the frontal cortex is 58 ± 5, 37 ± 7, and 5 ± 3, respectively (Parchi et al. 1995). The marked underrepresentation of the unglycosylated form is the result of the underrepresentation of the mutant unglycosylated form prior to the conversion into the PrP[Sc] isoform, probably due to the pronounced instability of this D178N mutant glycoform (Petersen et al. 1996). These findings are consistent with the conclusion that the PrP^{Sc} associated with FFI (and CJD^{D178N}) have different protein conformations and/or distinct ligand-binding interactions. Therefore, the FFI and CJD^{D178N} phenotypes are likely to be determined by codon 129 on the mutant allele, which, coupled with the D178N mutation, results in the expression of PrP^{Sc} with distinct biophysical characteristics. Nuclear magnetic resonance (NMR) data have shown that residues 178 and 129 are relatively close in the three-dimensional structure of PrP, suggesting that the perturbation caused by the D178N mutation differs as a function of the residue at position 129. This finding would provide the structural basis for the segregation of FFI and CJD^{D178N} with the two distinct haplotypes D178N, 129M and D178N, 129V, respectively (Goldfarb et al. 1992c; James et al. 1997; Capellari et al. 2000b). However, molecular dynamic simulation analysis does not support this model, showing that the only major change in PrP^{D178N} is the amino-terminal elongation of helix 2 (Gsponer et al. 2001).

Allelic Origin of PrP^{Sc} in Inherited Prion Diseases

As described above for the CJD^{D178N,129V}, both insoluble PrP and PrP^{Sc} derive exclusively from the PrP^{M} in FFI, regardless of whether the affected subjects are homozygous or heterozygous at codon 129 (Table 3) (Chen et al. 1997). Whether the low amount of the PrP^{Sc} found in FFI is due to the monoallelic origin of PrP^{Sc} or to other mechanisms remains to be determined. The transmissibility, to animals not carrying the FFI haplotype, of diseases such as FFI in which only PrP^{M} becomes PrP^{Sc} is puzzling. It is unclear how PrP^{C} can be converted in the recipient animal following inoculation of FFI PrP^{Sc}, while conversion does not occur spontaneously in FFI patients. The high local concentration of PrP^{Sc} achieved by the animal inoculation may overcome the barrier that blocks the conversion of PrP^{C} in FFI.

Transmissibility of FFI

Transgenic mice expressing a chimeric human–mouse PrP^{C} develop a prion disease 200 days after intracerebral inoculation with a homogenate from FFI brains (Telling et al. 1996). At variance with CJD, the pathology as well as the presence of PrP^{Sc} in transmitted FFI is predominantly in the thalamus. These transmission experiments (Collinge et al. 1995; Tateishi et al. 1995; Telling et al. 1996) not only demonstrate that FFI shares transmissibility with other prion diseases, but also provide important clues

Table 6. Known subjects carrying the extra octapeptide repeat(s)

Extrarepeat/Codon 129	No. of families	No. of patients	Age at onset (yr) Mean (range)	Disease duration Mean (range)	References
+1/Met	3	3	65 (58–73)	6 (4–10 mo)	Goldfarb et al. (1993); Laplanche et al. (1995);
+2/Met	1	1	58	3	Pietrini et al. (2003)
+4/Met	4	4	58 (56–65)	4[a] (2–84 mo)	Isozaki et al. (1994); Campbell et al. (1996); Rossi et al. (2000); Yanagihara et al. (2002)
+5/Met	6	10	45 (26–61)	74 (4 mo–19 yr)	Goldfarb et al. (1991b); Cervenáková et al. (1995); Cochran et al. (1996) Skworc et al. (1999); Windl et al. (1999)
+6/Met	6	59	35 (22–53)	86 (3 mo–18 yr)	Collinge et al. (1992); Cervenáková et al. (1995); Nicholl et al. (1995); Oda et al. (1995); Capellari et al. (1997); L. Cervenáková et al. (pers. comm.)
+7/Met	3	5	28 (23–31)	132 (84 mo–16 yr)	Brown et al. (1992c); Mizushima et al. (1994) ; Lewis et al. (2003)
+8/Met	1	11	28 (21–34)	26[b] (12 mo–7 yr)	Laplanche et al. (1999)
+9/Met	2	2	44 (32–55)	>39 (30–>48 mo)	Owen et al. (1992); Duchen et al. (1993); Krasemann et al. (1995)
+2/Val	1	3	59 (52–64)	6[a] (4–84 mo)	van Harten et al. (2000)
+4/Val	1	1	82	4	Laplanche et al. (1995)
+5/Val	1	1	46	48	L. Cervenáková (pers. comm.)
+8/Val	2	12	43 (21–54)	43 (3 mo–13 yr)	Goldfarb et al. (1992a); van Gool et al. (1995)
Total	31	112			

[a]Excluding one case of 7-year duration.
[b]Excluding patients still alive (>2–>12 years).
[c]Counted only in No. of families.

concerning the mechanisms of PrP^C–PrP^{Sc} conversion. The size of the PK-resistant PrP^{Sc} fragment formed in the inoculated transgenic mice is identical in electrophoretic mobility to the PrP^{Sc} fragment present in the inoculum from the donor FFI subjects. Replication of the PrP^{Sc} fragment size was also observed in the transmission of sporadic and inherited CJD associated with PrP^{Sc} type 1 (Telling et al. 1996). Since the recipient mice did not carry any of the donor's *PRNP* mutations and all mice had the same genetic background, this remarkable finding argues that the distinct conformations associated with PrP^{Sc} type 1 and type 2 can be reproduced independently from the original genetic information, likely because the information is carried in the conformation of the donor's PrP^{Sc}. This mechanism explains prion strain diversity by a mechanism that does not require the participation of nucleic acids.

FFI Relationship with Thalamic Dementia and the Sporadic Form of FFI

The concept of thalamic dementia or thalamic degeneration was introduced by Stern in 1939 when he described a subject with severe cognitive impairment of approximately 6 months' duration, which at autopsy showed only severe atrophy of ventral anterior and medial dorsal thalamic nuclei and a variable degree of astrogliosis in the cerebral cortex (Stern 1939). Subsequent classifications (Martin 1970) defined the case by Stern (1939) as preferential thalamic degeneration and added in a separate category the thalamic form of CJD. The discovery of FFI changed the concept of thalamic dementia. After the description of FFI, it was shown that previously reported familial cases of preferential thalamic degeneration or dementia have the same *PRNP* D178N,129M haplotype as FFI (Petersen et al. 1992). Therefore, the condition previously identified as familial thalamic atrophy or dementia, and FFI, are one and the same disease (Kirschbaum 1968; Petersen et al. 1992; Monari et al. 1994).

A disease clinically and pathologically indistinguishable from FFI, but lacking the history of familial incidence and mutations in the PrP gene, was reported in 1999 under the name of sporadic fatal insomnia (sFI) (Mastrianni et al. 1999; Parchi et al. 1999a). The nine proven cases of sFI reported to date have clinical, polysomnographic, and pathological features similar to those of FFI (Mastrianni et al. 1999; Parchi et al. 1999a; Scaravilli et al., 2000). The mean age at onset and the duration vary between 36 and 70 years and 15 and 30 months, respectively. Polysomnographic recordings carried out in one case were indistinguishable from those of FFI (Scaravilli et al. 2000). All the cases examined to date are homozygous for methionine at codon 129. As in FFI, PrP^{Sc} associated with sFI is of type 2 and is present in very low amounts. However, at variance with FFI, PrP^{Sc} of sFI does not show the underrepresentation of the unglycosylated form (Parchi et al. 1999a). The glycoform ratio expressed as percent distribution of the diglycosylated, monoglycosylated, and unglycosylated forms is 26:40:34, similar to that of sporadic CJD (Parchi et al. 1999a). The different PrP^{Sc} glycoform ratio is a consistent finding helpful in distinguishing sFI from FFI on immunoblot. One sFI case has been transmitted to transgenic mice expressing a mouse–human PrP with methionine in position 129. Affected mice developed a disease characterized by the presence of PrP^{Sc} type 2 as in the donor, and a histopathology and PrP^{Sc} distribution similar to that of FFI (Mastrianni et al. 1999).

INHERITED PRION DISEASE WITH VARIABLE PHENOTYPE: THE INSERTIONAL MUTATIONS

Epidemiology and Genetics

The wild-type *PRNP* gene has five repeat sequences between codons 51 and 91. The first sequence, R1, consists of 27 bp, the others, R2–4, comprise four 24-bp repeats. The repeats code for a P(H/Q)GGG(-/G)WGQ amino acid sequence that results in the presence of one nonapeptide (R1) and four octapeptides (R2, R2, R3, R4). The repeat region binds copper with high affinity (Kramer et al. 2001). Owen et al. (1989, 1990) were first to report the insertion of six extra 24-bp repeats in a British family with atypical dementia. Several additional families with one to nine (but not three) 24-bp extra repeat insertions have subsequently been identified in the US, several European countries, and Japan. Altogether, at least 31 families are known to have a hereditary transmissible spongiform encephalopathy (TSE) associated with extra 24-bp repeats in the *PRNP* gene (Table 6) generated most likely through replication slippage or recombination in the area of identical 24-bp repeats. Twenty-eight unique alleles were formed, among which only three alleles with one (Pietrini et al. 2003), four (Isozaki et al. 1994; Campbell et al. 1996), and five (Cochran et al. 1996; L. Cerveánaková, pers. comm.) additional repeats were found twice in independent families. In 26 families the repeat expansion is coupled with methionine at codon 129, whereas in 5 other families the inserts were found on the valine allele (Table 7). One American patient with four extra repeats (R122323234) showed no signs or family history of neurological disease and died of other causes (Goldfarb et al. 1991b).

Table 7. Wild-type and disease-associated *PRNP* alleles with increased and decreased number of 24-bp repeats

Repeats[a] (no.)/ codon 129	Sequence of the repeat area	Ethnic origin	Reference
Nonpathogenic alleles			
5/Met or Val	R12234	~98% of the population	
4/Met or Val	R1234	polymorphism in ~2% of the population	
Pathogenic alleles: Deletions			
3/Met	R124	N. European	Beck et al. (2001)
3/Met	R134	American/NA	Capellari et al. (2002)
Pathogenic alleles: Insertions			
6/Met	R122234	French	Laplanche et al. (1995)
6/Met	R1223**g**34	Italian[b]	Pietrini et al. (2003)
7/Met	R12232**a2a**4	American	Goldfarb et al. (1993)
7/Val	R122**a22a2a**4	Dutch	van Harten et al. (2000)
9/Met	R122222234	British	Campbell et al. (1996)
		Japanese	Isozaki et al. (1994)
9/Val	R1223**2**2234	French	Laplanche et al. (1995)
9/Met	R1223**2**2234	American	Capellari et al. (1997)
9/Met	R1223**g**23**g**234	Italian	Rossi et al. (2000)
9/Met	R122232234	Japanese	Yanagihara et al. (2002)
10/Met	R122**3**23**g**2234	American	Goldfarb et al. (1991b)
10/Met	R1222**a**22**a**2234	American	Cervenáková et al. (1995)
10/Met	R1223222234	Ukranian	Cochran et al. (1996)
		S. African	L. Cervenáková (1995)
10/Met	R1223**g3g3g**2234	German	Skwore et al. (1999)
10/Met	R1223222**a**234	German	Windl et al. (1999)
10/Val	R1223222**a**234	American	L. Cervenáková (pers. comm.)
11/Met	R1222**3**23**g**2234	British	Collinge et al. (1992)
11/Met	R122222223**g**34	American	Cervenáková et al. (1995)
11/Met	R1223**g**223**g**2234	Japanese	Oda et al. (1995)
11/Met	R122**3**23**g**23**g**234	British	Nicholl et al. (1995)
11/Met	R12222222234	Basque	Capellari et al. (1997)
11/Met	R1222**c**2222234	British	L. Cervenáková (pers. comm.)
12/Met	R122**c3**2**3**2**3**2**3g**34	American	Brown et al. (1992c)
12/Met	R12232223**g**2234	Japanese	Mizushima et al. (1994)
12/Met	R122**3**23222234	Australian	Lewis et al. (2003)
13/Val	R1223**2**2222222**a**4	French	Goldfarb et al. (1992a)
13/Val	R1223**g**322222234	Dutch	van Gool et al. (1995)
13/Met	R12232222**a**222**34	French	Laplanche et al. (1999)
14/Met	R122**323g2a2223g**234	British	Owen et al. (1992)
14/Met	R1223**233g22a23**234	German	Krasemann et al. (1995)

[a]Each repeat (R) is a variant 24-bp sequence (R1 is 27 bp), for codes see Goldfarb et al. (1991b). Extra/mutated repeats in bold. (NA) Not available.
[b]Two independent families with the same mutation.

Clinical Features in TSE Associated with Insert Mutations

Clinical data are available in at least 27 families including over 100 affected subjects (Goldfarb et al. 1991b, 1992a, 1993; Brown et al. 1992c; Owen et al. 1992; Isozaki et al. 1994; Mizushima et al. 1994; Krasemann et al. 1995; Laplanche et al. 1995, 1999; Nicholl et al. 1995; Oda et al. 1995; van Gool et al. 1995; Campbell et al. 1996; Cochran et al. 1996; Capellari et al. 1997; Skworc et al. 1999; Rossi et al. 2000; van Harten et al. 2000; Yanagihara et al. 2002; Lewis et al. 2003; Pietrini et al. 2003). If all insertion mutations are considered together, the disease phenotype is highly variable. The age at disease onset ranges between 21 and 82 years (Laplanche et al. 1999), and the disease duration between 2 months and over 19 years (Campbell et al. 1996; Cochran et al. 1996). The clinical and pathological features include either a typical CJD phenotype or a phenotype more consistent with GSS. Some rare cases lack specific histopathology (Collinge et al. 1992; Cochran et al. 1996; Capellari et al. 1997). This phenotypic heterogeneity appears to be at least in part related to the genotype, since the clinical and neuropathological features become more consistent when cases are grouped according to the number of repeats (Table 8). Thus, in patients with four or fewer inserted repeats, the mean age at onset is 62 years (82–52 range) and the duration 6 months (2–14 range) (excluding two cases with an exceptional 7-year duration) (Goldfarb et al. 1991b, 1993; Isozaki et al. 1994; Laplanche et al. 1995; Campbell et al. 1996; Rossi et al. 2000; van Harten et al. 2000; Yanagihara et al. 2002; Pietrini et al. 2003). The majority of the affected subjects have a clinical phenotype with rapidly progressive dementia, ataxia and visual disturbances, myoclonus, and PSW on the EEG (Goldfarb et al. 1991b, 1993; Isozaki et al. 1994; Laplanche et al. 1995; Campbell et al. 1996; Rossi et al. 2000; van Harten et al. 2000; Yanigihara et al. 2002; Pietrini et al. 2003). Disease penetrance rate is low in these families, and the inheritance pattern is undetectable. In contrast, the mean age at disease onset in patients with five or more extra repeats is 32 years (21–61 range), and the mean disease duration is 6 years (3 months to >19 years) (Goldfarb et al. 1991b, 1992a, 1993; Brown et al. 1992c; Owen et al. 1992; Isozaki et al. 1994; Mizushima et al. 1994; Krasemann et al. 1995; Laplanche et al. 1995, 1999; Nicholl et al. 1995; Oda et al. 1995; van Gool et al. 1995; Campbell et al. 1996; Cochran et al. 1996; Capellari et al. 1997; Skworc et al. 1999; Rossi et al. 2000; van Harten et al. 2000; Yanagihara et al. 2002; Lewis et al. 2003; Pietrini et al. 2003). The vast majority of patients present with a slowly progressive syndrome characterized by mental deterioration, cerebellar and extrapyramidal signs, often lacking the PSW complexes on EEG

Table 8. Familial prion diseases with insertions: General phenotypic features

Overall marked heterogeneity

Clinical features:
  Subjects with <4 octapeptide inserts have low penetrance and often the CJD
    phenotype
  Subjects with >4 octapeptide insert have high penetrance and often the GSS
    phenotype; the CJD phenotype is uncommon

Pathological features:
  Are variable and include CJD, GSS, mixed and nonspecific phenotypes
  Presence of non-amyloid (Congo red negative) elongated PrP deposits in the
    cerebellum molecular layer by immunostaining is highly distinctive

examination. A large family with clinical information from eleven affected subjects was characterized by prominent mental signs at presentation and dementia that appeared at later stages (Laplanche et al. 1999). These data show that the clinical phenotype, especially the age at onset and the duration of illness, is affected by the number of repeats. A CJD-like clinical phenotype is associated with four or fewer additional repeats whereas a clinical condition that either does not fit any of the major phenotypes of prion diseases or has GSS-like features is observed in most cases carrying five or more additional repeats. The age at onset correlates inversely, and the duration directly, with the number of repeats. Similar data concerning disease onset and duration have been obtained with the trinucleotide repeat expansions linked to Huntington's disease and spinocerebellar ataxias (Zoghbi and Orr 2000). Huntington's alleles are highly unstable when transmitted from parents to children, especially through male meioses. This is not the case with the *PRNP* repeat expansion. The inserted sequences were exactly the same in the descendants of different lines in the six-generation British family (Poulter et al. 1992), as well as in other smaller families. Anticipation phenomenon that is very characteristic of the trinucleotide repeat expansion disorders is absent or has very little presence (Laplanche et al. 1999) in the repeat-expansion families with spongiform encephalopathy.

As in other inherited prion diseases, the amino acid at codon 129 may influence the phenotypic effects of the repeat-expansion mutations. In the large British family with six extra repeats, the age at death in patients who were homozygous for methionine at codon 129 was significantly lower than in heterozygous subjects, suggesting a protective role of valine at position 129 (Poulter et al. 1992).

Histopathology

The pathological phenotype is also different in affected subjects with low and high numbers of octapeptide repeats, although in a slightly different way. All 5 cases with 1, 2, or 4 extra insertions examined at autopsy have a histopathology indistinguishable from classical CJD (Goldfarb et al. 1993; Campbell et al 1996; Rossi et al. 2000; Pietrini et al. 2003). The neuropathological features in 20 cases with 5 to 7 extra insertions are heterogeneous (Goldfarb et al 1991b; Brown et al. 1992c; Collinge et al. 1992; Mizushima et al. 1994; Nicholl et al. 1995; Oda et al. 1995; Cochran et al. 1996; Capellari et al. 1997; Skworc et al. 1999; Lewis et al. 2003). Thirteen cases exhibited CJD-like changes with widespread spongiform degeneration, astrogliosis, and variable degrees of neuronal loss (Goldfarb et al 1991b; Brown et al. 1992c; Collinge et al. 1992; Mizushima et al. 1994; Nicholl et al. 1995; Oda et al. 1995; Cochran et al. 1996; Capellari et al. 1997; Skworc et al. 1999; Lewis et al. 2003). The other 7 displayed changes which are either consistent with GSS or cannot be easily classified as GSS or CJD (Collinge et al 1992; Skworc et al. 1999). In one case there was virtually no histopathological change (Collinge et al. 1992). A histopathological phenotype fully consistent with GSS was observed in all 7 cases with 8 and 9 extra repeats because they lack specific features (Goldfarb et al. 1991b; Duchen et al. 1993; van Gool et al. 1995; Laplanche et al. 1999). In addition to various degrees of spongiosis, gliosis, and neuronal loss, PrP uni- and multicentric amyloid plaques are present in the molecular layer of the cerebellum (Fig. 10) and often in the cerebral gray matter. In conclusion, the CJD-like histopathological phenotype seems to be the rule in the presence of up to 4 extra insertions. In cases with 5 to 7 extra insertions, the histopathology is heterogeneous and includes CJD, GSS-like, and indefinable phenotypes. Eight and 9 extra insertions are associated with a GSS phenotype.

Characteristics and Allelic Origin of PrP^{Sc}

The gel migration pattern and the glycoform ratios of PrP^{Sc} have been examined in cases with 1 and 4–7 extra repeat insertions (Capellari et al. 1997; Skworc et al. 1999; Rossi et al. 2000; Lewis et al. 2003; Pietrini et al. 2003). The unglycosylated PK-treated PrP^{Sc} migrated at approximately 21 kD, corresponding to PrP^{Sc} type 1 (but occasionally not identical to PrP^{Sc} type 1) in 4 of the 6 subjects in which a detailed immunoblot analysis has been carried out (Capellari et al. 1997; Skworc et al. 1999; Lewis et al. 2003; Pietrini et al. 2003). In one subject, PrP^{Sc} migrated approximately

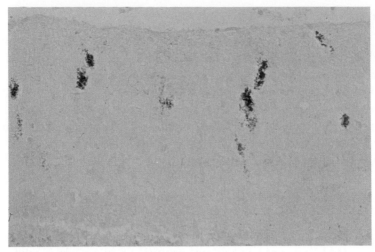

Figure 10. Typical PrP immunostaining of cerebellar cortex in insertion mutation. The molecular layer of the cerebellum shows granular linear aggregates of PrP immunoreactivity. (Immunostaining with the monoclonal antibody 3F4 to PrP.)

but not exactly as PrP^{Sc} type 2 (Pietrini et al. 2003), and in another the PrP^{Sc} molecular weight was considered to be intermediate between those of types 1 and 2 (Rossi et al. 2000). The ratio of the glycoforms is apparently similar in the 3 insertion mutations and is characterized by the predominance of the intermediate glycoform, and the highly glycosylated and unglycosylated forms are similarly represented. This pattern is referred to as type 1A and is similar to that of the sporadic form of CJD (Parchi et al. 1999c). All the cases examined by immunoblotting had the methionine codon on the mutated allele and either 129 methionine or valine in the normal allele.

The amount and physicochemical characteristics of PrP^{M} and PrP^{C} expressed by the mutant and normal alleles have been examined in brain tissues from subjects carrying 4, 5, and 6 extra insertions (Chen et al. 1997). PrP^{M} and PrP^{C} were differentiated because of their different sizes resulting in different gel mobility. The amount of PrP^{M} was ~40% lower than PrP^{C} with either insertion, and ~90% of the PrP^{M} is insoluble in detergents, an indication that it is aggregated. In contrast, only 50% of the total PrP^{C} was detergent-insoluble. Since the detergent-insoluble fraction was resistant to PK digestion, it was concluded that in the insertional mutations with 5 and 6 extra repeats, both PrP^{M} and PrP^{C} are converted to PrP^{Sc}. More recently, a similar conclusion has been reached with cases harboring 1 and 4 extra repeats (Rossi et al. 2000; Pietrini et al. 2003).

Transmissibility and Animal Models

Brain suspensions from 3 of 4 subjects with 5, 7, and 8 extra repeats transmitted the disease to primates after intracerebral inoculation (Brown et al. 1994b). Transgenic mice expressing the mouse homolog of the 9 extra insertion mutation have been obtained (Chiesa et al. 1998). Clinically, the mice developed progressive ataxia that is pathologically associated with cerebellar atrophy due to loss of the granule cells and gliosis. Deposition of PrP was seen in the cerebellum and hippocampus, but the PrP was not resistant to PK treatment in any way comparable to the PK resistance of PrP^{Sc}.

INHERITED PRION DISEASES WITH GSS DISEASE PHENOTYPE

GSS is a chronic hereditary autosomal dominant cerebellar syndrome accompanied by pyramidal signs and cognitive decline, which may evolve into severe dementia (Ghetti et al. 1995). Amyotrophy and parkinsonian signs may be present early or late in the course of the disease. The characteristic pathological feature is the presence of PrP-amyloid plaques in the cerebellar cortex in association with pyramidal tract degeneration. In addition, specific features such as spongiform changes, neurofibrillary tangles (NFT), Lewy bodies, or a combination of these may differentiate the various GSS haplotypes (Ghetti et al. 1995, 1996a,b, 2003). The *PRNP* mutations associated with GSS are shown in Table 1. In view of the consistent presence of PrP-amyloid deposits in GSS and other hereditary prion disease variants, the term hereditary prion protein amyloidosis has been introduced to include forms with extensive PrP-amyloid deposition in disorders that have a phenotype different from GSS. This nomenclature would allow us to include in this group the forms referred to as "inherited prion diseases with variable phenotypes" and the variant characterized clinically by severe dementia and pathologically by the presence of a prion protein cerebral amyloid angiopathy (CAA) and NFT. The variant is a rare entity and is referred to as PrP-CAA (Ghetti et al. 1996c).

In GSS, the symptomatology occurs in the second to eighth decades; the mean duration of illness is five years. To date, at least 56 families affected by GSS have been studied. GSS has been found in Australia, Austria, Canada, Denmark, France, Germany, Hungary, Ireland, Israel, Italy, Japan, Mexico, Poland, United Kingdom, and United States. It is difficult to determine the exact prevalence of GSS for two main reasons: (1) the disease has been reported only in a few countries and (2) the disease may be underreported due to its clinical similarity to olivopontocerebellar atrophy, spinocerebellar ataxia, Parkinson disease, amyotrophic lateral sclerosis, Huntington disease, or Alzheimer disease (AD) (Ghetti et al. 2003).

Although the neuropathologic diagnosis of GSS is based on the presence of PrP-amyloid deposits, their distribution and extent differ widely between families. Amyloid is accompanied by glial proliferation and by loss of neuronal processes and perikarya, leading to variable degrees of atrophy of the affected regions. The clinical phenotypes are associated with mutations of *PRNP*, allelic polymorphism, PrP^{Sc} characteristics, and possibly with environmental and tissue-specific factors.

GSS with the P102L Mutation Coupled to the 129M Codon (GSS^{P102L,129M})

Epidemiology

This mutation is the most common in GSS (Young et al. 1995), and at least 28 families have been reported. The P102L mutation has been found in Austria, Canada, Denmark, France, Germany, Hungary, Israel (Ashkenazi Jew), Italy, Japan, Mexico, United Kingdom, and the US (Rosenthal et al. 1976; Doh-Ura et al. 1989; Goldgaber et al. 1989; Hsiao et al. 1989; Kretzschmar et al. 1992; Goldhammer et al. 1993; Laplanche et al. 1994; Hainfellner et al. 1995; Young et al. 1995; De Michele et al. 2003; E. Alonso et al., unpubl.). The P102L mutation was the first point mutation described in a prion disease, and genetic linkage of the mutation to the disease has been shown (Hsiao et al. 1989; Speer et al. 1991).

Genetics

A CCG to CTG mutation at codon 102 causes a leucine (L) for proline (P) substitution. The mutation has high penetrance; in fact, individuals carrying the mutant allele develop the disease between the fourth and seventh decade of life. The P102L *PRNP* mutation is also shared by two recently described rare GSS phenotypes. The genotypic differences between the three diseases reside in the codon 129 and in the codon 219 located on the mutant allele. The former encodes methionine or valine and the latter encodes glutamic acid or lysine. The P102L mutation has most likely occurred more than once, since it has been seen in families of different ethnic groups.

Clinical Manifestations of GSS^{P102L,129M}

The clinical phenotype is characterized by a progressive cerebellar syndrome with ataxia, dysarthria, and incoordination of saccadic movements. Pyramidal and pseudobulbar signs are also seen. Mental and behavioral deterioration leading to dementia or akinetic mutism occur in the advanced stages of disease. The onset of clinical signs occurs in the fourth to sixth decades of life, and the duration of the disease ranges from a few months to six years. Considerable intrafamilial phenotypic variability may be observed (Adam et al. 1982; Ghetti et al. 1995; Hainfellner et al. 1995; Young et al. 1995). Myoclonus and PSW complexes in the EEG, a finding of diagnostic relevance in CJD, occurs in some of the GSS P102L patients. In some instances, the disease presents a rapid course of 59 months with a clinical picture indistinguishable from that of CJD (Barbanti et al. 1994).

Histopathology

Neuropathologically, deposits of fibrillar and nonfibrillar PrP in the cerebral and cerebellar parenchyma against a background of variable spongiform changes are consistently found (Figs. 11–14) (Masters et al. 1981a; Adam et al. 1982; Hainfellner et al. 1995; Piccardo et al. 1995). Spongiform changes, neuronal loss, and astrocytosis vary in severity even among patients of the same kindred and are most severe when the course of the illness is rapid (Adam et al. 1982; Barbanti et al. 1994; Hainfellner et al. 1995).

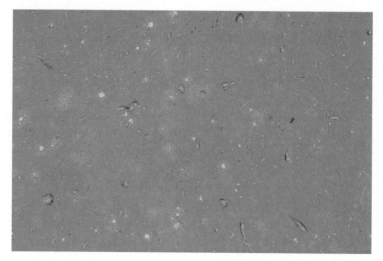

Figure 11. Cerebral cortex of a patient with GSS[P102L] showing multiple amyloid deposits. Note that some deposits show strong fluorescence while others are less bright and not well circumscribed. Also note that some amyloid deposits are multicentric with a larger core at the center and multiple brightly fluorescent or diffuse deposits at the periphery. (Thioflavin S method.)

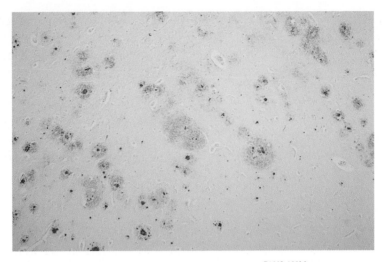

Figure 12. Cerebral cortex of a patient with GSS^{P102L,129M} showing multiple immunopositive deposits. Note that some deposits show strong immunolabeling, whereas others show diffuse immunolabeling. Also note that some of the strongly immunolabeled deposits are multicentric with a larger core at the center and multiple diffuse deposits at the periphery. Immunohistochemistry using anti-PrP serum raised against a synthetic peptide homologous to residue 90-102 of the amino acid sequence deduced from human PrP.

Figure 13. Cerebellar cortex of a patient with GSS^{P102L,129M} showing multiple amyloid deposits. Note that some deposits show strong fluorescence, whereas others are less bright and not well circumscribed. Also note that the amyloid deposits are multicentric in the molecular layer. Deposits in the granule cell layer are small and unicentric. (Thioflavin S method.)

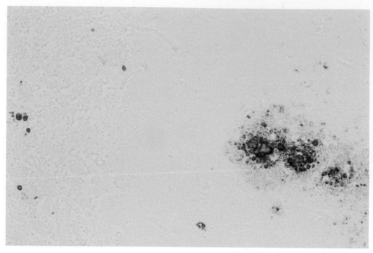

Figure 14. Cerebellar cortex of a patient with GSS[P102L,129M] showing multiple immunopositive deposits in the molecular and granule cell layer. The deposits in the molecular layer are significantly larger than those in the granule cell layer. Immunohistochemistry using anti-PrP serum raised against a synthetic peptide homologous to residue 90-102 of human PrP.

Characteristics and Allelic Origin of PrP^{Sc}

Studies have demonstrated the presence of two major PrP^{Sc} fragments in the brain of individuals with GSS[P102L,129M] (Parchi et al. 1998b; Piccardo et al. 1998b). One has an M_r of ~21 kD and the other of ~8 kD. The 21-kD fragment is similar to the PrP^{Sc} type 1 described in sporadic CJD (Parchi et al. 1996a). However, the ratio of the three major glycoforms of PrP^{Sc} is significantly different and shows, in the GSS[P102L,129M], a pattern in which the diglycosylated form is the most, and the unglycosylated form the least, represented form. The 8-kD fragment is similar to those described in other variants of GSS (Piccardo et al. 1998b). Sequencing and mass spectrometry have shown that the amino terminus of the 21-kD fragment begins at residues 78 and 82, whereas the amino terminus of the 8-kD fragment begins at residues 78, 80, and 82, and the carboxyl terminus ends at a position spanning residues 147–153 (Parchi et al. 1998b). Therefore, the 21-kD and 8-kD fragments differ only in the carboxyl terminus, since the former has an intact carboxyl terminus of PrP^{Sc}, and the latter is truncated at both amino- and carboxy-terminal ends. Both fragments derive exclusively from the mutant allele, as demonstrated by the consistent presence of the mutated L residue at position 102. It was also shown that the presence of the 21-kD form correlates with the presence of spongiform degeneration and "synaptic" pattern of PrP deposition, whereas the 8-kD fragment is found in brain regions showing PrP-positive multicentric amyloid deposits (Parchi et al. 1998b; Piccardo et al. 1998b). These data further indicate that the neuropathology of prion diseases largely depends on the type of PrP^{Sc} fragment that forms in vivo. Because the formation of PrP^{Sc} fragments of 8 kD with ragged amino and carboxyl termini is observed in most GSS subtypes, it may represent a molecular marker for this disorder (Parchi et al. 1998b; Piccardo et al. 1996, 1998b).

Transmissibility

Masters et al. showed that the inoculation into nonhuman primates of brain homogenates obtained from GSS patients induced a spongiform encephalopathy in the recipient animals (Masters et al. 1981a). Some of the patients from whom transmission was obtained had the P102L mutation. One of the patients was the member of the W family reported by Rosenthal et al. (1976). Subsequently, intracerebral inoculation into marmosets from another patient of the W family induced a spongiform encephalopathy indistinguishable from that seen in marmosets inoculated with brain tissue from a case of CJD (Baker et al. 1990). Transmission experiments from P102L GSS patients to mice resulted also in the development of spongiform degeneration (Manuelidis et al. 1987; Tateishi et al. 1996). It is significant that the donor tissue is characterized by the presence of PrP amyloid and severe spongiform changes, whereas the recipient primates and rodents develop a rapidly progressing disease with severe spongiform degeneration but not PrP-amyloid deposition. Transmission is not consistently obtained from P102L patients.

GSS with the P102L Mutation Coupled to the 129M and 219K Codons (GSS$^{P102L, 129M, 219K}$)

Epidemiology

Only one family from Japan has been reported (Furukawa et al. 1995).

Genetics

The P102L mutation was detected in coupling with methionine at residue 129 and lysine at residue 219.

Clinical Manifestations

Symptomatic subjects of this family had either dementia or cerebellar signs (Tanaka et al. 1997).

Histopathology

Neuropathologic studies demonstrated mild PrP deposition in the cerebral and cerebellar cortex and basal ganglia with no amyloid or spongiform changes (Furukawa et al. 1995; Tanaka et al. 1997).

Characteristics and Allele Origin of PrP^Sc

No biochemical data are available.

Transmissibility

No data are available.

GSS with the P102L Mutation Coupled to the 129V Codon (GSS^P102L,129V)

Epidemiology

There are two known cases of GSS^P102L disease that are homozygous for valine at codon 129, most likely from different families (Telling et al. 1995; Young et al. 1997). Both of these cases were found in the US.

Genetics

One patient, homozygous for valine (GTG) at codon 129, had the P102L mutation, CCG to CTG, on one allele. This patient, therefore, had the *PRNP* P102L mutation in coupling with valine at residue 129. This coupling probably arose by a cytosine to thymine mutation occurring at codon 102 on a 129-valine allele, since cytosine to thymine is the most frequent point mutation. However, we cannot rule out the possibility that the methionine to valine mutation (adenine to guanine) occurred on a P102L allele.

Clinical Manifestations

The clinical course in the patient with the P102L, 129V mutation studied clinically was significantly different from that observed in GSS^P102L, 129M in that the presentation was with seizures, the patient had long tract signs, dementia was not part of the clinical phenotype, and the duration of 12 years was unusually long (Young et al. 1997). Typically, in GSS^P102L, 129M the clinical onset is with cerebellar signs, seizures are generally not observed, dementia is frequently seen, and the duration is of ~5 years.

Histopathology

The neuropathologic findings in this patient differ from those frequently seen in P102L, 129M in view of the absence of spongiform changes. The involvement of the corticospinal, spinocerebellar, and gracile tracts, as well as the presence of PrP deposits in the substantia gelatinosa, may be correlated with the severe postural and sensory abnormalities observed in this patient (Young et al. 1997).

Characteristics and Allelic Origin of PrP^{Sc}

No biochemical data are available.

Transmissibility

No data are available.

GSS with the P105L Mutation Coupled to the 129V Codon (GSS$^{P105L, 129V}$)

Epidemiology

Five families have been reported (Nakazato et al. 1991; Amano et al. 1992; Terao et al. 1992; Kitamoto et al. 1993b,c; Yamada et al. 1993; Yamazaki et al. 1997). All five families are from Japan.

Genetics

A proline (CCA) to leucine (CTA) substitution at codon 105 on a 129V allele has been found in patients with hereditary spastic paraparesis from five Japanese families (Nakazato et al. 1991; Amano et al. 1992; Terao et al. 1992; Kitamoto et al. 1993b,c; Yamada et al. 1993).

Clinical Manifestations

Clinically, spastic gait, hyperreflexia, and Babinski sign dominate the picture in the initial stages (Kitamoto et al. 1993c; Yamada et al. 1993).

Extrapyramidal signs such as fine finger tremor and rigidity of limbs may be seen. Paraparesis progresses to tetraparesis and is accompanied by emotional incontinence and dementia. Myoclonus, PSW complexes in EEG, and severe cerebellar signs have not been reported. The onset of the clinical signs is in the fourth and fifth decades of life; the duration of the disease ranges from 6 to 12 years.

Histopathology

Neuropathologically, PrP deposits are found in the neocortex, especially the motor area, striatum, and thalamus. Multicentric PrP-amyloid plaques and diffuse deposits are present in superficial and deep layers of the neocortex, respectively. Spongiform changes are not seen. Neuronal loss and astrocytosis are prominent in the neocortex. NFT are occasionally seen; however, numerous NFT were seen in the cerebral cortex and several subcortical nuclei of a 57-year-old woman with dementia, gait disturbance, but not spastic paraparesis (Yamazaki et al. 1997). In most cases, amyloid plaques are rare in the cerebellum, and axonal losses occur in the pyramidal tracts (Nakazato et al. 1991; Amano et al. 1992; Terao et al. 1992; Kitamoto et al. 1993c; Yamada et al. 1993).

Characteristics and Allelic Origin of PrP^{Sc}

Studies have shown the presence a 6-kD PrP^{Sc} fragment in the brain of individuals with GSS^{P105L,129V}. These studies showed the presence of a large amount of detergent-insoluble, but not PK-resistant, PrP fragments of 27 to 34 kD (Yamada et al. 1999).

Transmissibility

No data are available.

GSS with the A117V Mutation Coupled to the 129V Codon (GSS^{A117V,129V})

Epidemiology

The mutation has been described in nine families: three in France, one in Hungary, one in Ireland, one in the United Kingdom, and three in the US (Warter et al. 1981; Doh-ura et al. 1989; Nochlin et al. 1989; Hsiao et al.

1991a; Tranchant et al. 1991; Mastrianni et al. 1995; Ghetti et al. 1998; Heldt et al. 1998; Mallucci et al. 1999; Kovács et al. 2001; Delisle et al. 2003).

Genetics

The alanine (A) to valine (V) substitution at residue 117 results from a cytosine to thymidine mutation in the second position of the codon. The mutant codon may also contain an adenine to guanine polymorphism in the third position, so that the codon changes from wild-type GCA to GTG. It remains to be determined whether these families share a common founder.

Clinical Manifestations

The GSS$^{A117V, 129V}$ haplotype is associated with a variety of clinical phenotypes, including classic GSS and classic AD (Warter et al. 1981; Nochlin et al. 1989; Hsiao et al. 1991a; Tranchant et al. 1991, 1992; Mohr et al. 1994; Mastrianni et al. 1995; Mallucci et al. 1999). In some individuals, marked extrapyramidal signs with parkinsonian features may occur early or late in the course of the disease (Mohr et al. 1994; Mastrianni et al. 1995; Mallucci et al. 1999). Additional signs that may be seen include pyramidal signs, amyotrophy, myoclonus, emotional lability, and pseudobulbar signs (Mohr et al. 1994; Mastrianni et al. 1995; Mallucci et al. 1999). Behavioral and personality disturbances, such as mood swings, aggressive behavior, and paranoia, frequently present long before neurological signs and symptoms (Mallucci et al. 1999). The phenotypic variability observed among affected individuals even occurs within the same family. EEGs are either normal or nonspecifically abnormal, but no PSW complexes are seen (Mohr et al. 1994; Mallucci et al. 1999). Results from CT scans varied from normality to moderate cerebral atrophy (Mohr et al. 1994; Mallucci et al. 1999). The age at onset of clinical signs is in the second to seventh decade of life; the duration ranges from one to eleven years (Nochlin et al. 1989; Tranchant et al. 1992; Mallucci et al. 1999).

Histopathology

Cerebral atrophy is seen in some cases. Variable amounts of PrP-amyloid plaques and diffuse deposits are widespread throughout the cerebral cortex, hippocampus, basal ganglia, and thalamus; however, in the cerebellum, they may be absent or present in variable amounts (Warter et al. 1981; Nochlin et al. 1989; Tranchant et al. 1991; Mohr et al. 1994; Mastrianni et al. 1995; Mallucci et al. 1999). Pyramidal tract degeneration may be present. Spongiform degeneration may also be seen, but if so, it is focally present in the cerebrum or cerebellum. Neuronal loss, when present, may be severe in the substantia nigra. Neurofibrillary tangles have been seen in individuals who had a long disease duration (Mohr et al. 1994).

Characteristics of PrP^{Sc}

PrP^{Sc} is present in affected individuals with GSS$^{A117V, 129V}$, but was not detected in the samples analyzed from an asymptomatic carrier (Piccardo et al. 2001, 2002). A 14- and 7-kD PrP species were observed in brain extracts and microsome preparations (Piccardo et al. 2001). PK-resistant fragments of 14 and 7 kD were also seen in purified PrP^{Sc} fractions (Piccardo et al. 2001). Sequencing of the 7-kD peptide in non-PK- and PK-digested samples showed a major amino-terminal cleavage site between 87G to 88G and 89W to 90G, respectively (Piccardo et al. 2001). PK digestion changes the stoichiometry of the various PrP peptides observed in the undigested preparations with an increase in the 14- and 7-kD fragments and disappearance of the high-molecular-weight iso-forms of 27 to 35 kD (Piccardo et al. 2001). These findings indicate that patients accumulate larger PrP isoforms that can be cleaved to smaller, insoluble, PK-resistant fragments (Piccardo et al. 2001). These peptides may be the result of the metabolism of PrP in a pathway associated with amyloid formation, because similarly sized fragments have been found in isolated amyloid cores of patients with GSS (Tagliavini et al. 1991, 1994, 2001; Piccardo et al. 2001).

Characteristics and Allelic Origin of PrP Amyloid

Amyloid peptides were purified from the brain tissue of a patient with GSS$^{A117V, 129V}$ who was M/V heterozygous at codon 129 (Tagliavini et al. 2001). The major peptide extracted from amyloid fibrils was a 7-kD PrP fragment. Sequence analysis and mass spectrometry showed that this fragment had ragged amino and carboxyl termini, starting mainly at 88G and 90G and ending with 148R, 152E, or 153N (Tagliavini et al. 2001). Only V was present at positions 117 and 129, indicating that the amyloid protein originated from mutant PrP molecules (Tagliavini et al. 2001). In addition to the 7-kD peptides, the amyloid fraction contained amino- and carboxy-terminal PrP fragments corresponding to residues 23–41, 191–205, and 217–228 (Tagliavini et al. 2001). Fibrillogenesis in vitro with synthetic peptides corresponding to PrP fragments extracted from brain tissue showed that the PrP85–148 peptide readily assembled into amyloid fibrils (Tagliavini et al. 2001). The PrP191–205 peptide also formed fibrillary structures, although with different morphology, whereas the PrP23–41 and PrP217–228 peptides did not (Tagliavini et al. 2001). These findings suggest that the processing of mutant PrP isoforms associated with GSS may occur in the extracellular space (Tagliavini et al. 2001). It is conceivable that full-length PrP and/or large PrP peptides are deposited in the extracellular compartment and then partially degraded by proteases and further digested by tissue endopeptidases. This process may originate a 7-kD protease-resistant core (Tagliavini et al. 2001).

Transmissibility

Transmission from one case was tested in mice and was negative (Tateishi et al. 1996).

GSS with the G131V Mutation Coupled to the 129M Codon (GSS^{G131V,129M})

Epidemiology

Only one family from Australia has been reported (Panegyres et al. 2001).

Genetics

A glycine (GGA) to valine (GTA) substitution at codon 131 on a 129M allele has been found in an individual who was diagnosed at autopsy as having GSS (Panegyres et al. 2001).

Clinical Manifestations

Changes in personality, decrease in cognitive performance, apraxia, tremor, and increased tendon reflexes are the presenting signs (Panegyres et al. 2001). The onset of the disease is early in the fifth decade of life, and the disease has a duration of nine years. MRI images show cerebral and cerebellar atrophy. EEGs do not show PSW. In the late stages, dementia becomes progressively more severe and ataxia develops.

Histopathology

PrP-amyloid plaques and diffuse deposits are seen in the cortex and subcortical nuclei as well as in the cerebellum (Panegyres et al. 2001). NFT are seen in the Ammon's horn and in the entorhinal cortex. No spongiform degeneration is seen.

Characteristics and Allelic Origin of PrP^{Sc}

PK digestion demonstrated evidence of PrP^{Sc} with bands at approximately 8, 18, 26, and 31 kD (Panegyres et al. 2001).

Transmissibility

No data are available.

GSS with the H187R Mutation Coupled to the 129V Codon (GSS^{H187R,129V})

Epidemiology

Only one family from the US has been reported (Butefisch et al. 2000).

Genetics

A histidine (CAC) to arginine (CGC) substitution at codon 187 on a 129V allele has been found in four individuals who presented with cognitive impairment and ataxia (Cervenáková et al. 1999; Butefisch et al. 2000).

Clinical Manifestations

The clinical phenotype is characterized by early progressive cognitive impairment, cerebellar ataxia, and dysarthria followed by myoclonus, seizures, and occasionally pyramidal and extrapyramidal signs (Butefisch et al. 2000). Neuroimaging shows a severe, widespread atrophy of the cerebrum and cerebellum. The age at onset is in the fourth to six decade of life with duration of 7 to 18 years.

Histopathology

PrP deposition is seen in the neocortex, hippocampus, and cerebellum, with the latter two also having PrP-amyloid plaques (Butefisch et al.

2000). The cortical deposits have a round or elongated, "curly" appearance. NFT are also seen in the hippocampus. Atrophy and astrogliosis of the subcortical white matter are present. No spongiform degeneration is seen.

Characteristics and Allelic Origin of PrP^{Sc}

No data are available.

Transmissibility

No data are available.

GSS with the F198S Mutation Coupled to the 129V Codon (GSS$^{F198S, 129V}$)

Epidemiology

Two families have been reported (Farlow et al. 1989; Ghetti et al. 1989, 1995; Hsiao et al. 1992; Mirra et al. 1997). Recently, another family has been identified; however, it has not been reported (D. Dickson et al., pers. comm.). All three families are from the US.

Genetics

A phenylalanine (TTC) to serine (TCC) substitution at codon 198 on a 129V allele has been described in patients from three US families of Caucasian ethnic origin (Dlouhy et al. 1992; Ghetti et al. 1992, 1995; Hsiao et al. 1992; Mirra et al. 1997).

Clinical Manifestations

The clinical phenotype is characterized by cognitive, cerebellar, and pyramidal signs. The main symptoms are gradual loss of short-term memory and progressive clumsiness in walking, bradykinesia, rigidity, dysarthria, and dementia. Signs of cognitive impairment and eye-movement abnormalities may be detected by specific tests before clinical onset of symptoms. Psychotic depression has been seen in several patients; tremor is mild or absent. Symptoms may progress slowly over five years or rapidly over as little as one year. The age at onset of clinical signs is 40–71 years of age; patients homozygous for valine at residue 129 have clinical signs more than ten years earlier, on average, than heterozygous patients

(Dlouhy et al. 1992). The duration of the disease ranges from two to twelve years (Farlow et al. 1989; Ghetti et al. 1992, 1995).

Histopathology

The neuropathologic phenotype is characterized by the presence of massive PrP deposition and amyloid formation in the cerebral and cerebellar parenchyma as well as neurofibrillary lesions in the cerebral gray matter (Ghetti et al. 1989, 1992, 1995; Giaccone et al. 1990, 1992; Tagliavini et al. 1993). In the twelve patients and the two asymptomatic carriers from the Indiana kindred (GSS-IK), unicentric and multicentric PrP-amyloid deposits are distributed throughout the gray structures of the cerebrum, cerebellum, and midbrain. Amyloid deposition is severe in frontal, insular, temporal, and parietal cortices; the highest concentration of deposits is in layers one, four, five, and six. A moderate involvement is seen in the hippocampus, where plaques occur predominantly within the stratum lacunosum-moleculare of the CA1 sector and subiculum. PrP deposits are numerous in the claustrum, caudate nucleus, putamen, thalamic nuclei, cerebellar molecular layer, mesencephalic tegmentum, substantia nigra, and periaqueductal gray matter; however, the degree of amyloid formation in these areas varies. Amyloid deposits are surrounded by astrocytes, astrocytic processes, and microglial cells. In the neocortex, many amyloid cores are associated with abnormal neurites, so that when these lesions are analyzed with classic stains, they are morphologically similar to the neuritic plaques found in Alzheimer's disease (Ghetti et al. 1989, 1992). The neurites are immunoreactive with antibodies to tau, ubiquitin, and amino- and carboxy-terminal domains of the β-amyloid precursor protein (βAPP). The accumulation of βAPP in nerve cell processes is not associated with extracellular deposition of Aβ, except in older patients, where Aβ immunoreactivity also may be observed around PrP-amyloid deposits (Bugiani et al. 1993; Ghetti et al. 1995).

NFT and neuropil threads are found in large numbers in the neocortex and subcortical gray matter. Particularly affected are the frontal, cingulate, parietal, insular, and parahippocampal cortex. In the remaining cortical regions, NFT are present, but less numerous. Of the subcortical gray areas, the caudate nucleus, putamen, nucleus basalis, midbrain and pontine nuclei, including substantia nigra, griseum centrale, and locus coeruleus, show a variable degree of involvement.

Moderate to severe cerebral and cerebellar atrophy, nerve cell loss, and gliosis are found in the neocortex, striatum, red nucleus, substantia nigra, cerebellum, locus coeruleus, and inferior olivary nucleus. Iron deposition is found in the globus pallidus, striatum, red nucleus, and substantia nigra. Spongiform changes are inconspicuous.

Two neurologically asymptomatic individuals carrying the GSS[F198S,129V] mutation died at 42 and 50 years of age. Both individuals were heterozygous at residue 129. PrP-amyloid deposits were numerous in the cerebellar cortex, but rare in the cerebral cortex of the 42-year-old. Numerous PrP-amyloid deposits and numerous NFT were present in the 50-year-old individual (Piccardo et al. 2002).

Characteristics of PrP[Sc]

PK digestion of the brain extracts generated three prominent broad bands of about 27–29, 18–19, and 8 kD and a weaker, but sharp band at 33 kD, as detected with the antibody 3F4 (Piccardo et al. 1996). The latter band (33 kD) may be attributed to PrP and/or to cross-reactivity of antibody 3F4 with residual PK. The stoichiometry among the PrP species differed from that of the undigested peptides by a notable increase in the signal of the low-molecular-weight band.

N-deglycosylation of non PK-treated extracts with PNGase F resulted in disappearance of the 33–35-kD band accompanied by an increased signal of the 28–30-kD band (Piccardo et al. 1996). The 28–30-kD band, seen with antibody 3F4, is consistent with the molecular weight of deglycosylated full-length PrP, as shown by a similar species detected with antibodies raised against synthetic peptides homologous to residues 23–40 and 220–231 of human PrP (PrP23-40 and PrP220-231) (Piccardo et al. 1996). In non-PK-treated brain extracts, the electrophoretic mobility of the 19–20- and 9-kD bands was not modified by deglycosylation.

The combination of PK and enzymatic deglycosylation with PNGase F treatment generated a pattern similar to that of PK treatment alone with prominent fragments at about 27–29, 18–19, and 8 kD. These PrP fragments were immunoreactive with antibody 3F4 and with antisera AS 6800 (raised against a synthetic peptide homologous to residues 89–104 of human PrP), but not with PrP23-40 and PrP220-231.

To investigate the sensitivity of PrP to PK, samples from the cerebellum were exposed to PK under nondenaturing and denaturing conditions. Under nondenaturing conditions, the major PrP isoforms present in GSS-IK retained partial PK resistance, the intensity of the signal after 4 hours being similar to that observed after 1 hour of enzymatic treatment. Conversely, denatured PrP was completely degraded after a 30-minute digestion with PK at 37°C in the presence of sodium dodecyl sulfate (Piccardo et al. 1996).

PK-resistant PrP fragments of similar electrophoretic mobility were seen in all brain regions examined (frontal cortex, caudate nucleus, and cerebellum) in the two GSS-IK patients analyzed. In semiquantitative experiments (similar amounts of total protein loaded), comparable signals were observed in samples from the cerebellum and caudate nucleus, two areas that have a high and a low amount of amyloid, respectively (Piccardo et al. 1996).

In immunoblot studies, the strongest signal was obtained from tissue corresponding to a patient who had the longest clinical course of the disease (12 years). In a follow-up study, PrP^{Sc} obtained from the cerebella of five additional patients of the GSS-IK, whose duration of clinical signs varied from 2 to 7 years, was analyzed. Similar electrophoretic patterns were observed, and high amounts of PrP were present in all cases, regardless of the duration of disease. Nevertheless, in repeated experiments, the patient with a 12-year duration of clinical signs always showed the most intense signal.

PrP was localized in the microsomal fraction. Sarkosyl-soluble and PK-sensitive PrP isoforms from this fraction were seen as prominent bands of ~33–35 kD in both control (familial AD) and GSS-IK (Piccardo et al. 1996). In addition, Sarkosyl-insoluble PrP was present as four major PrP species of approximately 33–35, 28–30, 19–20, and 9 kD (Piccardo et al. 1996). PK digestion of these samples generated three prominent bands of 27–29, 18–19, and 8 kD in GSS-IK, comparable to the PK-resistant species present in brain homogenates (Piccardo et al. 1996). Recent studies have demonstrated that the major amino-terminal cleavage site corresponds to residue 74G, which resides in the octapeptide-repeat region (Piccardo et al. 2001).

Characteristics and Allelic Origin of PrP Amyloid

The biochemical composition of PrP amyloid was first determined in brain tissue samples obtained from patients of the Indiana kindred (Tagliavini et al. 1991, 1994) carrying the F198S mutation in coupling with 129V (Dlouhy et al. 1992; Hsiao et al. 1992). Amyloid cores were isolated by a procedure combining buffer extraction, sieving, collagenase digestion, and sucrose gradient centrifugation. Proteins were extracted from amyloid fibrils with formic acid, purified by gel filtration chromatography and reverse-phase high performance liquid chromatography (HPLC), analyzed by SDS-PAGE and immunoblot, and sequenced. The amyloid preparations contained two major peptides of 11 kD and 7 kD spanning residues 58–150 and 81–150 of PrP, respectively. The amyloid peptides had ragged amino and carboxyl termini.

The finding that the amyloid protein was an amino- and carboxy-terminal truncated fragment of PrP was verified by immunostaining brain sections with antisera raised against synthetic peptides homologous to residues 23–40, 90–102, 127–147, and 220–231 of human PrP. The amyloid cores were strongly immunoreactive with antibodies that recognized epitopes located in the mid-region of the molecule, whereas only the periphery of the cores was immunostained by antibodies to amino- or carboxy-terminal domains. In addition, antisera to the amino and carboxyl termini of PrP labeled large areas of the neuropil that did not possess the tinctorial and optical properties of amyloid. Immunogold electron microscopy showed that antibodies to the mid-region of PrP decorated fibrils of amyloid cores, whereas antisera to amino- and carboxy-terminal epitopes labeled amorphous material at the periphery of the cores or dispersed in the neuropil. These data suggest that amyloid deposition in GSS is accompanied by accumulation of PrP peptides without amyloid characteristics (Giaccone et al. 1992).

In GSS-IK, the amyloid protein does not include the region containing the amino acid substitution. To establish whether amyloid peptides originate from mutant protein alone or from both mutant and wild-type PrP, patients heterozygous M/V for residue 129 were analyzed. 129V was used as a marker for protein from the mutant allele. Amino acid sequencing and electrospray mass spectrometry of peptides generated by digestion of the amyloid protein with endoproteinase Lys-C showed that the samples contained only peptides with 129V, suggesting that only mutant PrP was present in the amyloid (Tagliavini et al. 1994).

Paired Helical Filament Characteristics

Paired helical filament-enriched fractions obtained from the neocortex of GSS-IK patients contained SDS-soluble tau isoforms with electrophoretic mobility and an immunochemical profile corresponding to the isoforms extracted from the brains of patients with AD. These proteins migrate between 60 and 68 kD, immunoreact with antibodies to the amino and carboxyl termini of tau, and require dephosphorylation to be accessible to the antibody Tau-1. Thus, the immunocytochemical findings are consistent with those of the western blot analysis showing that significant similarity exists between GSS-IK and AD as to the Alz50, T46, and Tau-1 immunostaining of NFT (Tagliavini et al. 1993).

Transmissibility

In the case of the Indiana kindred, brain tissue and buffy coat from one affected individual were inoculated into hamsters in two experiments in the laboratory of Drs. E. and L. Manuelidis. No pathologic changes were observed in the primary transmission attempt nor in the second and third serial passage (L. Manuelidis, pers. comm.). Tissue homogenates from another F198S patient have been inoculated into hamsters and mice and no transmission has occurred (Ghetti et al. 1992; Hsiao et al. 1992). Amyloid-enriched fractions and tissue homogenates from GSS-IK patients have been inoculated into marmosets, and no transmission has occurred 30 months after inoculation (H.F. Baker, pers comm.).

GSS with the D202N Mutation Coupled to the 129V Codon (GSS^{D202N,129V})

Epidemiology

One family from the United Kingdom has been reported (Piccardo et al. 1998b). Recently, a second family has been identified (P. Gambetti, pers. comm.). The second family is from Canada.

Genetics

An aspartic acid (GAC) to asparagine (AAC) mutation has been reported in one individual who was diagnosed at autopsy with GSS (Piccardo et al. 1998b).

Clinical Manifestations

Cognitive impairment leading to dementia and cerebellar signs were the main clinical features (Piccardo et al. 1998b). The age at onset is early in the eighth decade and the duration is six years.

Histopathology

Abundant PrP-amyloid deposits are present in the cerebrum and cerebellum, and NFT are seen in the cerebral cortex (Piccardo et al. 1998b). No spongiform degeneration is observed.

Characteristics and Allelic Origin of PrP^{Sc}

PrP^{Sc} was detected in the neocortex of one individual with GSS$^{D202N, 129V}$ (Piccardo et al. 1998b). The sample was characterized by the presence of bands equivalent to ~8 kD, 18–19 kD, and 27–29 kD. This pattern is similar to the pattern seen in GSS$^{F198S, 129V}$.

Transmissibility

No data are available.

GSS with the Q212P Mutation Coupled to the 129M Codon (GSS^{Q212P,129M})

Epidemiology

Only one family from the US with this mutation is known at this time (Young et al. 1998).

Genetics

A glutamine (CAG) to proline (CCG) mutation has been found in one GSS patient who had no family history of neurologic disease (Young et al. 1998).

Clinical Manifestations

Onset occurred at age 60 with gradual development of incoordination and slurring of speech (Young et al. 1998). Three years after onset, the patient was found to have normal mental status, dysarthria, and ataxia. His condition began to progress more rapidly, and approximately 6 years after onset, the patient entered a nursing home where he remained until his death, 8 years after onset. At the time of death, the patient was still mentally competent.

Histopathology

Amyloid deposition was mild throughout the central nervous system. Among GSS variants, this appears to be the form with the least amount of amyloid deposits (Young et al. 1998). The cerebellum was significantly less affected than in all other variants. Immunopositive PrP deposits were present in the cerebellum, where they were more numerous than in the neocortex and striatum. There was axonal degeneration in the anterior and lateral corticospinal tracts throughout the spinal cord.

Characteristics and Allelic Origin of PrP^{Sc}

PrP^{Sc} is seen as a poorly defined smear equivalent to 25–35 kD and two major bands of 18–19 kD and 10 kD (Piccardo et al. 1998a).

Transmissibility

No data are available.

GSS with the Q217R Mutation Coupled to the 129V Codon (GSS^{Q217R,129V})

Epidemiology

Two families from the US have been reported (Ghetti et al. 1995, 2001).

Genetics

A glutamine (CAG) to arginine (CGG) substitution on a 129V allele has been described in two families from the US of Swedish origin (Hsiao et al. 1992; Ghetti et al. 2001).

Clinical Manifestations

The clinical phenotype is characterized by gradual memory loss, progressive gait disturbances, parkinsonism, and dementia. The age at onset of clinical signs is 62–66 years of age. The duration of the disease is five to six years (Ghetti et al. 1994). The neurologic signs may be preceded by episodes of mania or depression that respond to antidepressant medications, lithium, and neuroleptics.

Histopathology

Neuropathologically, there are PrP-amyloid deposits in the cerebrum and cerebellum and abundant NFT in the cerebral cortex and several subcortical gray structures (Ghetti et al. 1994). The neocortex, amygdala, substantia innominata, and thalamus are severely affected.

Characteristics and Allelic Origin of PrP^{Sc}

PrP^{Sc} is essentially similar to that observed in GSS$^{F198S, 129V}$ (Piccardo et al. 1998b).

Transmissibility

Tissue homogenates from one individual with the GSS$^{Q217R, 129V}$ patient have been inoculated into hamsters and mice and no transmission has occurred (Hsiao et al. 1992).

GSS with the M232T Mutation (GSS^{M232T})

Epidemiology

One family from Poland has been reported (Liberski et al. 2000).

Genetics

A methionine (ATG) to threonine (ACG) substitution has been described in one family from Poland (Liberski et al. 2000).

Clinical Manifestations

Cerebellar signs and spastic paraparesis are the initial symptoms followed by dementia (Liberski et al. 2000). The age at onset is in the fifth decade and the duration is six years.

Histopathology

PrP-amyloid plaques and diffuse deposits are seen in the neocortex, subcortical nuclei, and cerebellum (Liberski et al. 2000). It is unclear whether spongiform changes are present.

Characteristics and Allelic Origin of PrP^{Sc}

No data are available.

Transmissibility

No data are available.

Cerebral Amyloid Angiopathy with the Y145Stop Mutation Coupled to the 129M Codon (PrP-CAA$^{Y145Stop, 129M}$)

Epidemiology

Only one family from Japan has been reported in the literature (Ghetti et al. 1996c).

Genetics

A tyrosine (TAT) to stop codon (TAG) substitution on a 129M allele has been found in a Japanese patient with a clinical diagnosis of AD (Kitamoto et al. 1993a; Ghetti et al. 1996c).

Clinical Manifestations

The clinical phenotype is characterized by memory disturbance, disorientation, and a progressive, severe dementia (Ghetti et al. 1996c). The EEG did not show PSW. The age at the onset of the clinical signs was 38 years, and the duration of the disease was 21 years.

Histopathology

Neuropathologically, diffuse atrophy of the cerebrum and dilation of the lateral ventricles were present (Ghetti et al. 1996c). Neuronal loss and gliosis were severe, but no spongiform changes were observed. There were PrP-amyloid deposits in the walls of small and medium-sized parenchymal and leptomeningeal blood vessels and in the perivascular neuropil, as well as neurofibrillary lesions in the cerebral gray matter. The NFT are composed of paired helical filaments with a periodicity of 70–80 nm and were decorated with monoclonal antibodies recognizing abnormal phosphorylation.

Characteristics of PrP^{Sc}

PrP^{Sc} was not detected (Ghetti et al. 1996c).

Characteristics and Allelic Origin of PrP Amyloid

Immunoblot analysis of proteins extracted from amyloid showed that the smallest subunit migrated as a broad band of ~7.5 kD. This band was immunoreactive with antibodies to epitopes located between residues 90 and 147 of PrP (Ghetti et al. 1996c).

Transmissibility

No data are available.

ACKNOWLEDGMENTS

This work was supported by National Institutes of Health grants NS-29822, AG-10133, AG-08992, AG-08155, and AG-14359; the Britton Fund, the Biomed 2 BMH4-CT96-0856 contract, and the Gino Galletti Foundation.

REFERENCES

Adam J., Crow T.J., Duchen L.W., Scaravilli F., and Spokes E. 1982. Familial cerebral amyloidosis and spongiform encephalopathy. *J. Neurol. Neurosurg. Psychiatry* **545:** 37–45.

Almer G., Hainfellner J.A., Brucke T., Jellinger K., Kleinert R., Bayer G., Windl O., Kretzschmar H.A., Hill A., Sidle K., Collinge J., and Budka H. 1999. Fatal familial insomnia: A new Austrian family. *Brain* **122:** 5–16.

Amano N., Yagishita S., Yokoi S., Itoh Y., Kinoshita J., Mizutani T., and Matsuishi T. 1992. Gerstmann-Sträussler-Scheinker syndrome—A variant type: Amyloid plaques and Alzheimer's neurofibrillary tangles in cerebral cortex. *Acta Neuropathol.* **84:** 15–23.

Antonine J.C., Laplanche J.L., Mosnier J.F., Beaudry P., Chatelain J., and Michel D. 1996. Demyelinating peripheral neuropathy with Creutzfeldt-Jakob disease and mutation at codon 200 of the prion protein gene. *Neurology* **46:** 1123–1127.

Apetri A.C. and Surewicz W.K. 2002. Kinetic intermediate in the folding of human prion protein. *J. Biol. Chem.* **277:** 44589–44592.

Baker H.F., Duchen L.W., Jacobs J.M., and Ridley R.M. 1990. Spongiform encephalopathy transmitted experimentally from Creutzfeldt-Jakob and familial Gerstmann-Sträussler-Scheinker diseases. *Brain* **113:** 1891–1909.

Bär K.J., Hager F., Nenadic I., Opfermann T., Brodhun M., Tauber R.F., Patt S., Schulz-Schaeffer W., Gottschild D., and Sauer H. 2002. Serial positron emission tomographic findings in an atypical presentation of fatal familial insomnia. *Arch. Neurol.* **59:** 1815–1818.

Barbanti P., Fabbrini G., Salvatore M., Petraroli R., Pocchiari M., Macchi G., and Lenzi G.L. 1994. No correlation between clinical heterogeneity and codon 129 polymorphism of the prion protein gene (*PRNP*) in Gerstmann-Sträussler-Scheinker syndrome (GSS) with *PRNP* codon 102 mutation. *Neurobiol. Aging* **15:** S156.

Barbanti P., Fabbrini G., Salvatore M., Macchi G., Lenzi G.L., Pocchiari M., Petraroli R., Cardone F., Maras B., and Equestre M. 1996. Polymorphism at codon 129 or codon 219 of PRNP and clinical heterogeneity in a previously unreported family with Gerstmann-Sträussler-Scheinker disease (PrP-P102L mutation). *Neurology* **47:** 734–741.

Barron R.M., Thomson V., Jamieson E., Melton D.W., Ironside J., Will R., and Manson J.C. 2001. Changing a single amino acid in the N terminus of murine PrP alters TSE incubation time across three species barriers. *EMBO J.* **20:** 5070–5078.

Beck J.A., Mead S., Campbell T.A., Dickinson A., Wientjens D.P., Croes E.A., Van Duijn C.M., and Collinge J. 2001. Two-octapeptide repeat deletion of prion protein associated with rapidly progressive dementia. *Neurology* **57:** 354–356.

Bolton D.C., McKinley M.P., and Prusiner S.B. 1982. Identification of a protein that purifies with the scrapie prion. *Science* **218:** 1309–1311.

Booth D.R., Sunde M., Bellotti V., Robinson C.V., Hutchinson W.L., Fraser P.E., Hawkins P.N., Dobson C.M., Radford S.E., Blake C.C.F., and Pepys M.B. 1997. Instability, unfolding and aggregation of human lysozyme variants underlying amyloid fibrillogenesis. *Nature* **385:** 787–793.

Brown P., Cervenáková L., and Powers J.M. 1998. FFI cases from the United States, Australia, and Japan. *Brain Pathol.* **8:** 567–570.

Brown P., Cathala F., Castaigne P., and Gajdusek D.C. 1986. Creutzfeldt-Jakob disease: Clinical analysis of a consecutive series of 230 neuropathological verified cases. *Ann. Neurol.* **20:** 597–602.

Brown P., Goldfarb L.G., Gibbs C.J., Jr., and Gajdusek D.C. 1991a. The phenotypic expression of different mutations in transmissible familial Creutzfeldt-Jakob disease. *Eur. J. Epidemiol.* **7:** 469–476.

Brown P., Cervenáková L., Boellaard J.W., Stavrou D., Goldfarb L.G., and Gajdusek D.C. 1994a. Identification of a PRNP gene mutation in Jakob's original Creutzfeldt-Jakob disease family. *Lancet* **344:** 130–131.

Brown P., Gálvez S., Goldfarb L.G., Nieto A., Cartier L., Gibbs C.J., Jr., and Gajdusek D.C. 1992a. Familial Creutzfeldt-Jakob disease in Chile is associated with the codon 200 mutation of the PRNP amyloid precursor gene on chromosome 20. *J. Neurol. Sci.* **112:** 65–67.

Brown P., Gibbs C.J., Rodgers-Johnson P., Asher D.M., Sulima M.P., Bacote A., Goldfarb L.G., and Gajdusek D.C. 1994b. Human spongiform encephalopathy: The National Institutes of Health series of 300 cases of experimentally transmitted disease. *Ann. Neurol.* **35:** 513–529.

Brown P., Goldfarb L.G., Cathala F., Vrbovska A., Sulima M., Nieto A., Gibbs C.J., and Gajdusek D.C. 1991b. The molecular genetics of familial Creutzfeldt-Jakob disease in France. *J. Neurol. Sci.* **105:** 240–246.

Brown P., Goldfarb L.G., Kovanen J., Haltia M., Cathala F., Sulima M., Gibbs C.J., and Gajdusek D.C. 1992b. Phenotypic characteristics of familial Creutzfeldt-Jakob disease associated with the codon 178 Asn PRNP mutation. *Ann. Neurol.* **31:** 282–285.

Brown P., Goldfarb L.G., McCombie W.R., Nieto A., Squillacote D., Sheremata W., Little B.W., Godec M.S., Gibbs C.J., and Gajdusek D.C. 1992c. Atypical Creutzfeldt-Jakob disease in an American family with an insert mutation in the *PRNP* amyloid precursor gene. *Neurology* **42:** 422–427.

Budka H., Almer G., Hainfellner J.A., Brucke T., and Jellinger K. 1998. The Austrian FFI cases. *Brain Pathol.* **8:** 554.

Buge A., Escourolle R., Brion S., Rancurel G., Hauw J.J., Mehaut M., Gray F., and Gajdusek D.C. 1978. Maladie de Creutzfeldt-Jakob familiale–Étude clinique et anatomique de trois cas sur huit répartis sur trois générations. Transmission au singe écureuil. *Rev. Neurol.* **134:** 165–181.

Bugiani O., Giaccone G., Verga L., Pollo B., Frangione B., Farlow M.R., Tagliavini F., and Ghetti B. 1993. β PP participates in PrP-amyloid plaques of Gerstmann-Sträussler-Scheinker disease, Indiana kindred. *J. Neuropathol. Exp. Neurol.* **52:** 64–70.

Butefisch C.M., Gambetti P., Cervenáková L., Park K.Y., Hallett M., and Goldfarb L.G. 2000. Inherited prion encephalopathy associated with the novel PRNP H187R mutation: A clinical study. *Neurology* **55:** 517–522.

Campbell T.A., Palmer M.S., Will R.G., Gibb W.R.G., Luthert P.J., and Collinge J. 1996. A prion disease with a novel 96-base pair insertional mutation in the prion protein gene. *Neurology* **46:** 761–766.

Canet D., Last A.M., Tito P., Sunde M., Spencer A., Archer D.B., Redfield C., Robinson C.V., and Dobson C.M. 2002. Local cooperativity in the unfolding of an amyloidogenic variant of human lysozyme. *Nat. Struct. Biol.* **9:** 308–315.

Capellari S., Zaidi S.I., Long A.C., Kwon E.E., and Petersen R.B. 2000a. The Thr183Ala mutation, not the loss of the first glycosylation site, alters the physical properties of the prion protein. *J. Alzheimer's Dis.* **2:** 27–35.

Capellari S., Parchi P., Bennett D., Petersen R.B., Gambetti P., and Cochran E. 1999. First North American report of the T183A mutation in the prion protein gene: Clinical, pathological, and biochemical analysis of one case. *Neurology* (suppl. 2) **52:** A324.

Capellari S., Parchi P., Russo C.M., Sanford J., Sy M.S., Gambetti P., and Petersen R.B. 2000b. Effect of the E200K mutation on prion protein metabolism. Comparative study of a cell model and human brain. *Am. J. Pathol.* **157:** 613–622.

Capellari S., Ladogana A., Volpi G., Roncaroli F., Sita D., Baruzzi A., Pocchiari M., and

Parchi P. 2001. First report of the R208H-129MM haplotype in the prion protein gene in an European subject with CJD. *Neurol. Sci.* **22:** S109.

Capellari S., Parchi P., Wolff B.D., Campbell J., Atkinson R., Posey D.M., Petersen R.B., and Gambetti P. 2002. Creutzfeldt-Jakob disease associated with a deletion of two repeats in the prion protein gene. *Neurology* **59:** 1628–1630.

Capellari S., Vital C., Parchi P., Petersen R.B., Ferrer X., Jarnier D., Pegoraro E., Gambetti P., and Julien J. 1997. Familial prion disease with a novel 144-bp insertion in the prion protein gene in a Basque family. *Neurology* **49:** 133–141.

Cardone F., Liu Q.G., Petraroli R., Ladogana A., D'Alessandro M., Arpino C., Di Bari M., Macchi G., and Pocchiari M. 1999. Prion protein glycotype analysis in familial and sporadic Creutzfeldt-Jakob disease patients. *Brain Res. Bull.* **49:** 429–433.

Cardozo J., Caruso G., Molina O., Cardozo D., Chen S., Sirko-Osadsa D.A., Wang W., Xie Z., and Gambetti P. 2000. Familial transmissible spongiform encephalopathy with the T183A mutation on the prion protein gene (PRNP). *J. Neuropathol. Exp. Neurol.* **56:** 433.

Cathala F., Brown P., LeCanuet P., and Gajdusek D.C. 1985. High incidence of Creutzfeldt-Jakob disease in North African immigrants to France. *Neurology* **35:** 894–895.

Cervenáková L., Goldfarb L.G., Brown P., Kenney K., Cochran E.J., Bennett D.A., Roos R., and Gajdusek D.C. 1995. Three new PRNP genotypes associated with familial Creutzfeldt-Jakob disease. *Am. J. Hum. Genet.* **57:** A209. (Abstr.)

Cervenáková L., Buetefisch C., Lee H.S., Taller I., Stone G., Gibbs C.J., Jr., Brown P., Hallett M., Goldfarb L.G. 1999. Novel PRNP sequence variant associated with familial encephalopathy. *Am. J. Med. Genet.* **88:** 653–656.

Cervenáková L., Brown P., Piccardo P., Cummings J.L., Nagle J., Vinters H.V., Kaur P., Ghetti B., Chapman J., Gajdusek D.C., and Goldfarb L. 1996. 24-nucleotide deletion in the PRNP gene: Analysis of associated phenotypes. In *Transmissible subacute spongiform encephalopathies: Prion diseases* (ed. L. Court and B. Dodet), pp. 433–444. Elsevier, Paris.

Chapman J. and Korczyn A.D. 1991. Genetic and environmental factors determining the development of Creutzfeldt-Jakob disease in Libyan Jews. *Neuroepidemiology* **10:** 228–231.

Chapman J., Ben-Israel J., Goldhammer Y., and Korczyn A.D. 1994. The risk of developing Creutzfeldt-Jakob disease in subjects with the *PRNP* gene codon 200 point mutation. *Neurology* **44:** 1683–1686.

Chapman J., Brown P., Goldfarb L.G., Arlazoroff A., Gajdusek D.C., and Korczyn A.D. 1993. Clinical heterogeneity and unusual presentations of Creutzfeldt-Jakob disease in Jewish patients with the PRNP codon 200 mutation. *J. Neurol. Neurosurg. Psychiatry* **56:** 1109–1112.

Chapman J., Brown P., Rabey J.M., Goldfarb L.G., Inzelberg R., Gibbs C.J., Jr., Gajdusek D.C., and Korczyn A.D. 1992. Transmission of spongiform encephalopathies from a familial Creutzfeldt-Jakob disease patient of Jewish Libyan origin carrying the PRNP codon 200 mutation. *Neurology* **42:** 1249–1250.

Chapman J., Arlazoroff A., Goldfarb L.G., Cervenáková L., Neufeld M.Y., Werber E., Herbert M., Brown P., Gajdusek D.C., and Korczyn A.D. 1996. Fatal insomnia in a case of familial Creutzfeldt-Jakob disease with the codon 200(Lys) mutation. *Neurology* **46:** 758-761.

Chen S.G., Teplow D.B., Parchi P., Teller J.K., Gambetti P., and Autilio-Gambetti L. 1995. Truncated forms of the human prion protein in normal brain and in prion diseases. *J. Biol. Chem.* **270:** 19173–19180.

Chen S.G., Parchi P., Brown P., Zou W., Cochran E.J., Vnencak-Jones C.L., Julien J., Vital C., Mikol J., Lugaresi E., Autilio-Gambetti L., and Gambetti P. 1997. Allelic origin of the abnormal prion proteins in familial prion diseases. *Nat. Med.* **3:** 1009–1015.

Chesebro B. 1997. Human TSE disease—Viral or protein only? *Nat. Med.* **3:** 491–492.

Chesebro B., Race R., Wehrly K., Nishlo J., Bloom M., Lechner D., Bergstrom S., Bobbins K., Mayer L., Keith J.M., Garon C., and Haasse A. 1985. Identification of scrapie prion protein-specific mRNA in scrapie-infected and uninfected brain. *Nature* **315:** 331–333.

Chiesa R., Piccardo P., Ghetti B., and Harris D.A. 1998. Neurological illness in transgenic mice expressing a prion protein with an insertional mutation. *Neuron* **21:** 1339–1351.

Cochran E.J., Bennett D.A., Cervenáková L., Kenney K., Bernard B., Foster N.L., Benson D.F., Goldfarb L.G., and Brown P. 1996. Familial Creutzfeldt-Jakob disease with a five-repeat octapeptide insert mutation. *Neurology* **47:** 727–733.

Cohen D., Krausz Y., Lossos A., Ben David E., and Atlan H. 1989. Brain SPECT imaging with Tc-99m HM-PAO in Creutzfeldt-Jakob disease. *Clin. Nucl. Med.* **14:** 808–810.

Cohen F.E., Pan K.M., Huang Z., Baldwin M., Fletterick R.J., and Prusiner S.B. 1994. Structural clues to prion replication. *Science* **264:** 530–531.

Collinge J. 2001. Prion diseases of humans and animals: Their causes and molecular basis. *Annu. Rev. Neurosci.* **24:** 519–550.

Collinge J., Sidle K.C.L., Meads J., Ironside J., and Hill A.F. 1996. Molecular analysis of prion stain variation and the aetiology of "new variant" CJD. *Nature* **383:** 685–690.

Collinge J., Palmer M.S., Sidle K.C., Gowland I., Medori R., Ironside J., and Lantos P. 1995. Transmission of fatal familial insomnia to laboratory animals. *Lancet* **346:** 569–570.

Collinge J., Brown J., Hardy J., Mullan M., Rossor M.N., Baker H., Crow T.J., Lofthouse R., Poulter M., Ridley R., Owen F., Bennett C., Dunn G., Harding A.E., Quinn N., Doshi B., Roberts G.W., Honavar M., Janota I., and Lantos P.L. 1992. Inherited prion disease with 144 base pair gene insertion. *Brain* **115:** 687–710.

Collins S., Boyd A., Fletcher A., Byron K., Harper C., McLean C.A., and Masters C.L. 2000. Novel prion protein gene mutation in an octogenarian with Creutzfeldt- Jakob disease. *Arch. Neurol.* **57:** 1058–1063.

Colombo R. 2000. Age and origin of the PRNP E200K mutation causing familial Creutzfeldt-Jacob disease in Libyan Jews. *Am. J. Hum. Genet.* **67:** 528–531.

Cortelli P., Perani D., Parchi P., Grassi F., Montagna P., De Martin M., Castellani R., Tinuper P., Gambetti P., Lugaresi E., and Fazio F. 1997. Cerebral metabolism in fatal familial insomnia: Relation to duration, neuropathology, and distribution of protease-resistant prion protein. *Neurology* **49:** 126–133.

Creutzfeldt H.G. 1920. Über eine eigenartige herdförmige Erkrankung des Zentralnervensystems. *Z. Gesamte Neurol. Psychiatr.* **57:** 1–18.

———. 1921. Über eine eigenartige herdförmige Erkrankung des Zentralnervensystems. In *Histologische und Histopathologische Arbeiten uber die Grosshirnrinde* (ed. F. Nissl and A. Alzheimer), pp. 1–48. Gustav Fischer, Jena, Germany.

Daude N., Lehmann S., and Harris D.A. 1997. Identification of intermediate steps in the conversion of a mutant prion protein to a scrapie-like form of cultured cells. *J. Biol. Chem.* **272:** 11604–11612.

Delisle M.B., Haik S., Uro-Coste E., Peoc'h K., Puel M., Revesz T., Piccardo P., Hauw J.J., Duyckaerts C., Laplanche J.L., and Ghetti B. 2003. Gerstmann-Sträussler-Scheinker disease presenting as corticobasal degeneration. *J. Neuropathol. Exp. Neurol.* **62:** (in press). (Abstr.)

De Michele G., Pocchiari M., Petraroli R., Manfredi M., Caneve G., Coppola G., Casali C.,

Saccà F., Piccardo P., Salvatore E., Berardelli A., Orio M., Barbieri F., Ghetti B., and Filla A. 2003. Gerstmann-Sträussler-Scheinker disease: A new Italian family with P102L mtuation and variable phenotype. *Can. J. Neurol. Sci.* **30:** 233–236.

Denning G.M., Anderson M.P., Amara J.F., Marshall J., Smith A.E, and Welsh M.J. 1992. Processing of mutant cystic fibrosis transmembrane conductance regulator is temperature-sensitive. *Nature* **358:** 761–764.

Dimitz L. 1913. Bericht des Vereines für Psychiatrie und Neurologie in Wien (Vereinsjahr 1912/13). Sitzung vom 11. Juni 1912. *Jahrb. Psychiatr. Neurol.* **34:** 384.

Dlouhy S.R., Hsiao K., Farlow M.R., Foroud T., Conneally P.M., Johnson P., Prusiner S.B., Hooles M.E., and Ghetti B. 1992. Linkage of the Indiana kindred of Gerstmann-Sträussler-Scheinker disease to the prion protein gene. *Nat. Genet.* **1:** 64–67.

Doh-ura K., Kitamoto T., Sakaki Y., and Tateishi J. 1991. CJD discrepancy. *Nature* **353:** 801–802.

Doh-ura K., Tateishi J., Sasaki H., Kitamoto T., and Sakaki Y. 1989. Pro→Leu change at position 102 of prion protein is the most common but not the sole mutation related to Gerstmann-Sträussler syndrome. *Biochem. Biophys. Res. Commun.* **163:** 974–979.

Dorandeu A., Wingertsmann L., Chretien F., Delisle M.B., Vital C., Parchi P., Montagna P., Lugaresi E., Ironside J.W., Budka H., Gambetti P., and Gray F. 1998. Neuronal apoptosis in fatal familial insomnia. *Brain Pathol.* **8:** 531–536.

Duchen L.W., Poulter M., and Harding A.E. 1993. Dementia associated with a 216 base pair insertion in the prion protein gene. *Brain* **116:** 555–567.

Farlow M.R., Yee R.D., Dlouhy S.R., Conneally P.M., Azzarelli B., and Ghetti B. 1989. Gerstmann-Sträussler-Scheinker disease. I. Extending the clinical spectrum. *Neurology* **39:** 1446–1452.

Finckh U., Muller-Thomsen T., Mann U., Eggers C., Marksteiner J., Meins W., Binetti G., Alberici A., Hock C., Nitsch R.M., and Gal A. 2000. High prevalence of pathogenic mutations in patients with early-onset dementia detected by sequence analyses of four different genes. *Am. J. Hum. Genet.* **66:** 110–117.

Fink J.K., Warren J.T., Drury I., Murman D., and Peacock M.L. 1991. Allele-specific sequencing confirms novel prion gene polymorphism in Creutzfeldt-Jakob disease. *Neurology* **41:** 1647–1650.

Friede R.L. and DeJong R.N. 1964. Neuronal enzymatic failure in Creutzfeldt-Jakob disease. *Arch. Neurol.* **10:** 181–195.

Fuchsberger T., Teipel S.J., Faltraco F., Drezezga A., Hohne C., Schroder M., and Hampel H. 2003. Clinical, neuroimaging and neuropathological findings in a case of fatal familial insomnia. *Neurology* (in press).

Furukawa H., Kitamoto T., Hashiguchi H., and Tateishi J. 1996. A Japanese case of Creutzfeldt-Jakob disease with a point mutation in the prion protein gene at codon 210. *J. Neurol. Sci.* **141:** 120–122.

Furukawa H., Kitamoto T., Tanaka Y., and Tateishi J. 1995. New variant prion protein in a Japanese family with Gerstmann-Sträussler syndrome. *Mol. Brain Res.* **30:** 385–388.

Gabizon R., Kahana E., Hsiao K., Prusiner S.B., and Meiner Z. 1992. Inherited prion disease in Libyan Jews. In *Prion diseases of humans and animals* (ed. S.B. Prusiner et al.), pp. 168–179. Ellis Horwood, London, United Kingdom.

Gabizon R., Telling G., Meiner Z., Halimi M., Kahana I., and Prusiner S.B. 1996. Insoluble wild-type and protease-resistant mutant prion protein in brains of patients with inherited prion diseases. *Nat. Med.* **2:** 59–64.

Gabizon R., Rosenmann H., Meiner Z., Kahana I., Kahana E., Shugart Y., Ott J., and Prusiner S.B. 1993. Mutation and polymorphism of the prion protein gene in Libyan Jews with Creutzfeldt-Jakob disease. *Am. J. Hum. Genet.* **53:** 828–835.

Gajdusek D.C., Gibbs C.J., and Alpers M. 1966. Experimental transmission of a kuru-like syndrome in chimpanzees. *Nature* **209:** 794–796.

Galvez S. and Cartier L. 1984. Computed tomography findings in 15 cases of Creutzfeldt-Jakob disease with histological verification. *J. Neurol. Neurosurg. Psychiatry* **47:** 1244–1246.

Gambetti P., Parchi P., Capellari S., Russo C., Tabaton M., Teller J.K., and Chen S.G. 2001. Mechanisms of phenotypic heterogeneity in prion, Alzheimer and other conformational diseases. *J. Alzheimer's Dis.* **3:** 87–95.

Gerstmann J. 1928. Über ein noch nicht beschriebenes Reflexphanomen bei einer Erkrankung des zerebellaren Systems. *Wien. Med. Wochenschr.* **78:** 906–908.

Gerstmann J., Sträussler E., and Scheinker I. 1936. Über eine eigenartige hereditär-familiäre Erkrankung des Zentralner-vensystems. Zugleich ein Beitrag zur Frage des vorzeitigen lokalen Alterns. *Z. Neurol. Psychiatr.* **154:** 736–762.

Ghetti B., Tagliavini F., Takao M., Bugiani O., and Piccardo P. 2003. Hereditary prion protein amyloidoses. *Clin. Lab. Med.* **23:** 65–85.

Ghetti B., Dlouhy S.R., Giaccone G., Bugiani O., Frangione B., Farlow M.R., and Tagliavini F. 1995. Gerstmann-Sträussler-Scheinker disease and the Indiana kindred. *Brain Pathol.* **5:** 61–75.

Ghetti B., Piccardo P., Parisi J., Takao M., Dlouhy S., Cochran, E.J., and Fox J.H. 2001. Gerstmann-Sträussler-Scheinker (GSSD) disease with neurofibrillary tangles and PrP-amyloid plaques: A new family. *J. Neuropathol. Exp. Neurol.* **60:** 537. (Abstr.)

Ghetti B., Tagliavini F., Giaccone G., Bugiani O., Frangione B., Farlow M.R., and Dlouhy S.R. 1994. Familial Gerstmann-Sträussler-Scheinker disease with neurofibrillary tangles. *Mol. Neurobiol.* **8:** 41–48.

Ghetti B., Piccardo P., Frangione B., Bugiani O., Giaccone G., Young K., Prelli F., Farlow M.R., Dlouhy S.R., and Tagliavini F. 1996a. Prion protein amyloidosis. *Brain Pathol.* **6:** 127–145.

———. 1996b. Prion protein hereditary amyloidosis; Parenchymal and vascular. *Semin. Virol.* **7:** 189–200.

Ghetti B., Young K., Piccardo P., Dlouhy S.R., Pahwa R., Lyons K.E., Koller W.C., Ma M.J., DeCarli C., and Rosenberg R.N. 1998. Gerstmann-Sträussler-Scheinker disease (GSS) with *PRNP* A117V mutation: Studies of a new family. *Soc. Neurosci. Abstr.* **24:** 515.

Ghetti B., Tagliavini F., Masters C.L., Beyreuther K., Giaccone G., Verga L., Farlow M.R., Conneally P.M., Dlouhy S.R., Azzarelli B., and Bugiani O. 1989. Gerstmann-Sträussler-Scheinker disease. II. Neurofibrillary tangles and plaques with PrP-amyloid coexist in an affected family. *Neurology* **39:** 1453–1461.

Ghetti B., Tagliavini F., Hsiao K., Dlouhy S.R., Yee R.D., Giaccone G., Conneally P.M., Hodes M.E., Bugiani O., Prusiner S.B., Frangione B., and Farlow M.R. 1992. Indiana variant of Gerstmann-Sträussler-Scheinker disease. In *Prion diseases of humans and animals* (ed. S. Prusiner et al.), pp. 154–167. Ellis Horwood, New York.

Ghetti B., Piccardo P., Spillantini M.G., Ichimiya Y., Porro M., Perini F., Kitamoto T., Tateishi J., Seiler C., Frangione B., Bugiani O., Giaccone G., Prelli F., Goedert M., Dlouhy S.R., and Tagliavini F. 1996c. Vascular variant of prion protein cerebral amyloidosis with τ-positive neurofibrillary tangles: The phenotype of the stop codon 145 mutation in *PRNP*. *Proc. Natl. Acad. Sci.* **93:** 744–748.

Giaccone G., Tagliavini F., Verga L., Frangione B., Farlow M.R., Bugiani O., and Ghetti B. 1990. Neurofibrillary tangles of the Indiana kindred of Gerstmann-Sträussler-Scheinker disease share antigenic determinants with those of Alzheimer disease. *Brain Res.* **530:** 325–329.

Giaccone G., Verga L., Bugiani O., Frangione B., Serban D., Prusiner S.B., Farlow M.R., Ghetti B., and Tagliavini F. 1992. Prion protein preamyloid and amyloid deposits in Gerstmann-Sträussler-Scheinker disease, Indiana kindred. *Proc. Natl. Acad. Sci.* **89**: 9349–9353.

Gibbs C.J., Jr., Gajdusek D.C., Asher D.M., Alpers M.P., Beck E., Daniel P.M., and Matthews W.B. 1968. Creutzfeldt-Jakob disease (spongiform encephalopathy): Transmission to the chimpanzee. *Science* **161**: 388–389.

Goldfarb L., Korczyn A., Brown P., Chapman J., and Gajdusek D.C. 1990. Mutation in codon 200 of scrapie amyloid precursor gene linked to Creutzfeldt-Jakob disease in Sephardic Jews of Libyan and non-Libyan origin. *Lancet* **336**: 637–638.

Goldfarb L.G., Brown P., Little B.W., Cervenáková L., Kenney K., Gibbs C.J., and Gajdusek D.C. 1993. A new (two-repeat) octapeptide coding insert mutation in Creutzfeldt-Jakob disease. *Neurology* **43**: 2392–2394.

Goldfarb L.G., Brown P., Vrbovska A., Baron H., McCombie W.R., Cathala F., Gibbs C.J., and Gajdusek D.C. 1992a. An insert mutation in the chromosome 20 amyloid precursor gene in a Gerstmann-Sträussler-Scheinker family. *J. Neurol. Sci.* **111**: 189–194.

Goldfarb L.G., Haltia M., Brown P., Nieto A., Kovanen J., McCombie W.R., Trapp S., and Gajdusek D.C. 1991a. New mutation in scrapie amyloid precursor gene (at codon 178) in Finnish Creutzfeldt-Jakob kindred. *Lancet* **337**: 425.

Goldfarb L.G., Brown P., McCombie W.R., Goldgaber D., Swergold G.D., Wills P.R., Cervenáková L., Baron H., Gibbs C.J., and Gajdusek D.C. 1991b. Transmissible familial Creutzfeldt-Jakob disease associated with five, seven and eight extra octapeptide coding repeats in the PRNP gene. *Proc. Natl. Acad. Sci.* **88**: 10926–10930.

Goldfarb L.G., Brown P., Mitrova E., Cervenáková L., Goldin L., Korczyn A.D., Chapman J., Galvez S., Cartier L., Rubenstein R., and Gajdusek D.C. 1991c. Creutzfeldt-Jacob disease associated with the PRNP codon 200^{Lys} mutation: An analysis of 45 families. *Eur. J. Epidemiol.* **7**: 477–486.

Goldfarb L.G., Brown P., Maltia M., Cathala F., McCombie W.R., Kovanen J., Cervenáková L., Goldin L., Nieto A., Godec M.S., Asher D.M., and Gajdusek D.C. 1992b. Creutzfeldt-Jakob disease cosegregates with the codon 178^{Asn} PRNP mutation in families of European origin. *Ann. Neurol.* **31**: 274–281.

Goldfarb L.G., Petersen R.B., Tabaton M., Brown P., LeBlanc A.C., Montagna P., Cortelli P., Julien J., Vital C., Pendelbury W.W., Haltia M., Wills P.R., Hauw J.J., McKeever P.E., Monari L., Schrank B., Swergold G.D., Autilio-Gambetti L., Gajdusek D.C., Lugaresi E., and Gambetti P. 1992c. Fatal familial insomnia and familial Creutzfeldt-Jakob disease: Disease phenotype determined by a DNA polymorphism. *Science* **258**: 806–808.

Goldgaber D., Goldfarb L.G., Brown P., Asher D.M., Brown W.T., Lin S., Teener J.W., Feinstone S.M., Rubenstein R., Kascsak R.J., Boellaard J.W., and Gajdusek D.C. 1989. Mutations in familial Creutzfeldt-Jakob disease and Gerstmann-Sträussler-Scheinker's syndrome. *Exp. Neurol.* **106**: 204–206.

Goldhammer Y., Gabizon R., Meiner Z., and Sadeh M. 1993. An Israeli family with Gerstmann-Sträussler-Scheinker disease manifesting the codon 102 mutation in the prion protein gene. *Neurology* **43**: 2718–2719.

Gsponer J., Ferrara P., and Caflisch A. 2001. Flexibility of the murine prion protein and its Asp178Asn mutant investigated by molecular dynamics simulations. *J. Mol. Graph. Model* **20**: 169–182.

Guidon G. 1978. "Formes familiales de la maladie de Creutzfeldt-Jakob: A propos d'une fratrie de deux." Ph.D. thesis, Université de Rennes, France.

Guiroy D.C., Shankar S.K., Gibbs C.J., Messenheimer J.A., Das S., and Gajdusek D.C. 1989.

Neuronal degeneration and neurofilament accumulation in the trigeminal ganglia in Creutzfeldt-Jakob disease. *Ann. Neurol.* **25:** 102–106.

Hainfellner J.A., Parchi P., Kitamoto T., Jarius C., Gambetti P., and Budka H. 1999. A novel phenotype in familial Creutzfeldt-Jakob disease: Prion protein gene E200K mutation coupled with valine at codon 129 and type 2 protease-resistant prion protein. *Ann. Neurol.* **45:** 812–816.

Hainfellner J.A., Brantner-Inthaler S., Cervenáková L., Brown P., Kitamoto T., Tateishi J., Diringer H., Liberski P.P., Regele H., Feucht M., Mayr N., Wessely P., Summer K., Seitelberger F., and Budka H. 1995. The original Gerstmann-Sträussler-Scheinker (GSS) family of Austria: Divergent clinicopathological phenotypes but constant PrP genotype. *Brain Pathol.* **5:** 201–211.

Haltia M., Kovanen J., Van Crevel H., Bots G.T.A.M., and Stefanko S. 1979. Familial Creutzfeldt-Jakob disease. *J. Neurol. Sci.* **42:** 381–389.

Harder A., Jendroska K., Kreuz F., Wirth T., Schafranka C., Karnatz N., Theallier-Janko A., Dreier J., Lohan K., Emmerich D., Cervos-Navarro J., Windl O., Kretzschmar H.A., Nurnberg P., and Witkowski R. 1999. Novel twelve-generation kindred of fatal familial insomnia from Germany representing the entire spectrum of disease expression. *Am. J. Med. Genet.* **87:** 311–316.

Hegde R.S., Voigt S., and Lingappa V.R. 1998a. Regulation of protein topology by trans-acting factors at the endoplasmic reticulum. *Mol Cell.* **2:** 85–91.

Hegde R.S., Tremblay P., Groth D., DeArmond S.J., Prusiner S.B., and Lingappa V.R. 1999. Transmissible and genetic prion diseases share a common pathway of neurodegeneration. *Nature* **402:** 822–826.

Hegde R.S., Mastrianni J.A., Scott M.R., DeFea K.A., Tremblay P., Torchia M., DeArmond S.J., Prusiner S.B., and Lingappa V.R. 1998b. A transmembrane form of the prion protein in neurodegenerative disease. *Science* **279:** 827–834.

Heldt N., Boellaard J.W., Brown P., Cervenáková L., Doerr-Schott J., Thomas C., Scherer C., and Rohmer F. 1998. Gerstmann-Sträussler-Scheinker disease with A117V mutation in a second French-Alsatian family. *Clin. Neuropathol.* **17:** 229–234.

Hitoshi S., Nagura H., Yamanouchi H., and Kitamoto T. 1993. Double mutations at codon 180 and codon 232 of the PRNP gene in an apparently sporadic case of Creutzfeldt-Jakob disease. *J. Neurol. Sci.* **120:** 208–212.

Hoque M.Z., Kitamoto T., Furukawa H., Muramoto T., and Tateishi J. 1996. Mutation in the prion protein gene at codon 232 in Japanese patients with Creutzfeldt-Jakob disease: A clinicopathological, immunohistochemical and transmission study. *Acta Neuropathol.* **92:** 441–446.

Hosszu L.L.P., Baxter N.J., Jackson G.S., Power A., Clarke A.R., Waltho J.P., Craven C.J., and Collinge J. 1999. Structural mobility of the human prion protein probed by backbone hydrogen exchange. *Nat. Struct. Biol.* **6:** 740–743.

Hsiao K.K., Scott M., Foster D., Groth D.F., DeArmond S.J., and Prusiner S.B. 1990. Spontaneous neurodegeneration in transgenic mice with mutant prion protein. *Science* **250:** 1587–1590.

Hsiao K.K., Cass C., Schellenberg G.D., Bird T., Devine-Gage E., Wisniewski H.M., and Prusiner S.B. 1991a. A prion protein variant in a family with the telencephalic form of Gerstmann-Sträussler-Scheinker syndrome. *Neurology* **41:** 681–684.

Hsiao K., Baker H.F., Crow T.J., Poulter M., Owen F., Terwilliger J., Westaway D., Ott J., and Prusiner S.B. 1989. Linkage of a prion protein missense variant to Gertsmann-Sträussler syndrome. *Nature* **338:** 342–345.

Hsiao K., Dlouhy S.R., Farlow M.R., Cass C., DaCosta M., Conneally P.M., Hodes M.E.,

Ghetti B., and Prusiner S.B. 1992. Mutant prion proteins in Gerstmann-Sträussler-Scheinker disease with neurofibrillary tangles. *Nat. Genet.* **1:** 68–71.

Hsiao K., Meiner Z., Kahana E., Cass C., Kahana I., Avrahami D., Scarlato G., Abramsky O., Prusiner S.B., and Gabizon R. 1991b. Mutation of the prion protein in Libyan Jews with Creutzfeldt-Jakob disease. *N. Engl. J. Med.* **324:** 1091–1097.

Huang Z., Gabriel J.M., Baldwin M.A., Fletterick, R.J., Prusiner S.B., and Cohen F.E. 1994. Proposed three-dimensional structure for the cellular prion protein. *Proc. Natl. Acad. Sci.* **91:** 7139–7143.

Huang N., Marie S.K.N., Kok F., and Nitrini R. 2001. Familial Creutzfeldt-Jakob disease associated with a point mutation at codon 210 of the prion protein gene. *Arq. Neuropsiquiatr.* **59:** 932–935.

Isozaki E., Miyamoto K., Kagamihara Y., Hirose K., Tanabe H., Uchihara T., Oda M., and Nagashima T. 1994. CJD presenting as frontal lobe dementia associated with a 96 base pair insertion in the prion protein gene (Japanese). *Dementia* **8:** 363–371.

Ivanova L., Barmada S., Kummer T., and Harris D.A. 2001. Mutant prion proteins are partially retained in the endoplasmic reticulum. *J. Biol. Chem.* **276:** 42409–42421.

Iwaski Y., Sone M., Kato T., Yoshida E., Indo T., Yoshida M., Hashizume Y., and Yamada M. 1999. Clinicopathological characteristics of Creutzfeldt-Jakob disease with a PrP V180I mutation and M129V polymorphism on different alleles (Japanese). *Rinsho Shinkeigaku* **39:** 800–806.

Jakob H., Pyrkosch W., and Strube H. 1950. Die erbliche Form der Creutzfeldt-Jakobschen Krankheit. (Familie Backer). *Arch. Psychiatr. Nervenkr.* **184:** 653–674.

James T.L., Liu H., Ulyanov N.B., Farr-Jones S., Zhang H., Donne D.G., Kaneko K., Groth D., Mehlhorn I., Prusiner S.B., and Cohen F.E. 1997. Solution structure of a 142-residue recombinant prion protein corresponding to the infectious fragment of the scrapie isoform. *Proc. Natl. Acad. Sci.* **94:** 10086–10091.

Jouvet M. 1999. Sleep and serotonin: An unfinished story. *Neuropsychopharmacology* (suppl. 2) **21:** 24S–27S.

Kahana E., Zilber N., and Abraham M. 1991. Do Creutzfeldt-Jakob disease patients of Jewish Libyan origin have unique clinical features? *Neurology* **41:** 1390–1392.

Kahana E., Milton A., Braham J., and Sofer D. 1974. Creutzfeldt-Jakob disease: Focus among Libyan Jews in Israel. *Science* **183:** 90–91.

Kelly J.W., Colon W., Lai Z., Lashuel H.A., McCulloch J., McCutchen S.L., Miroy G.J., and Peterson S.A. 1997. Transthyretin quaternary and tertiary structural changes facilitate misassembly into amyloid. *Adv. Protein Chem.* **50:** 161–181.

Khurana R., Gillespie J.R., Talapatra A., Minert L.J., Ionescu-Zanetti C., Millett I., and Fink A.L. 2001. Partially folded intermediates as critical precursors of light chain amyloid fibrils and amorphous aggregates. *Biochemistry* **40:** 3525–3535.

Kirschbaum W.R. 1924. Zwei eigenartige Erkrankung des Zentralnervensystems nach Art der spatischen Pseudosklerose (Jakob). *Z. Neurol. Pyschiatr.* **92:** 175–220.

———. 1968. *Jakob-Creutzfeldt disease*, pp. 1–25. Elsevier, New York.

Kitamoto T. and Tateishi J. 1994. Human prion diseases with variant prion protein. *Philos. Trans. R. Soc. Lond. B Biol. Sci.* **343:** 391–398.

Kitamoto T., Iizuda R., and Tateishi J. 1993a. An amber mutation of prion protein in Gerstmann-Sträussler-Scheinker syndrome with mutant PrP plaques. *Biochem. Biophys. Res. Commun.* **192:** 525–531.

Kitamoto T., Ohta M., Doh-ura K., Hitoshi S., Terao Y., and Tateishi J. 1993b. Novel missense variants of prion protein in Creutzfeldt-Jakob disease or Gerstmann-Sträussler syndrome. *Biochem. Biophys. Res. Commun.* **191:** 709–714.

Kitamoto T., Amano N., Terao Y., Nakazato Y., Isshiki T., Mizutani T., and Tateishi J. 1993c. A new inherited prion disease (PrP-P105L mutation) showing spastic paraparesis. *Ann. Neurol.* **34:** 808–813.

Klöppel S., Pirker W., Brucke T., Kovacs G.G., and Almer G. 2002. Beta-CIT SPECT demonstrates reduced availability of serotonin transporters in patients with fatal familial insomnia. *J. Neural Transm.* **109:** 1105–1110.

Koide T., Ohtake H., Nakajima T., Furukawa H., Sakai K., Kamei H., Makifuchi T., and Fukuhara N. 2002. A patient with dementia with Lewy bodies and codon 232 mutation of PRNP. *Neurology* **59:** 1619–1621.

Kopito R.R. 2000. Aggresomes, inclusion bodies and protein aggregation. *Trends Cell Biol.* **10:** 524–530.

Korczyn A.D., Chapman J., Goldfarb L.G., Brown P., and Gajdusek D.C. 1991. A mutation in the prion protein gene in Creutzfeldt-Jakob disease in Jewish patients of Libyan, Greek, and Tunisian origin. *Ann. N.Y. Acad. Sci.* **640:** 171–176.

Kovács G.G., Ertsey C.S., Majtényi C., Jelencsik I., László L., Flicker H., Strain L., Szirmai I., and Budka H. 2001. Inherited prion disease with A117V mutation of the prion protein gene: A novel Hungarian family. *J. Neurol. Neurosurg. Psychiatry* **70:** 802–805.

Kovanen J. and Haltia M. 1988. Descriptive epidemiology of Creutzfeldt-Jakob disease in Finland. *Acta Neurol. Scand.* **77:** 474–480.

Kramer M.L., Kratzin H.D., Schmidt B., Romer A., Windl O., Liemann S., Hornemann S., and Kretzschmar H. 2001. Prion protein binds copper within the physiological concentration range. *J. Biol. Chem.* **276:** 16711–16719.

Krasemann S., Zerr I., Weber T., Poser S., Kretzschmar H., Hunsmann G., and Bodemer W. 1995. Prion disease associated with a novel nine octapeptide repeat insertion in the PRNP gene. *Mol. Brain Res.* **34:** 173–176.

Kretzschmar H.A., Neumann M., and Stavrou D. 1995. Codon 178 mutation of the human prion protein gene in a German family (Backer family): Sequencing data from 72-year old celloidin-embedded brain tissue. *Acta Neuropathol.* **89:** 96–98.

Kretzschmar H.A., Kufer P., Riethmuller G., DeArmond S., Prusiner S.B., and Schiffer D. 1992. Prion protein mutation at codon-102 in an Italian family with Gerstmann-Sträussler-Scheinker syndrome. *Neurology* **42:** 809–810.

Kretzschmar H.A., Giese A., Zerr I., Windl O., Schulz-Schaeffer W., Skworc K., and Poser S. 1998. The German FFI cases. *Brain Pathol.* **8:** 559–561.

Kuwata K., Li H., Yamada H., Legname G., Prusiner S.B., Akasaka K., and James T.L. 2002. Locally disordered conformer of the hamster prion protein: A crucial intermediate of PrP^{Sc}? *Biochemistry.* **41:** 12277–12283.

Ladogana A., Almonti S., Petraroli R., Giaccaglini E., Ciarmatori C., Liu Q.G., Bevivino S., Squitieri F., and Pocchiari M. 2001. Mutation of the PRNP gene at codon 211 in familial Creutzfeldt-Jakob disease. *Am. J. Med. Genet.* **103:** 133–137.

Laplanche J.-L., Chatelain J., Launay J.M., Gazengel C., and Vidaud M. 1990. Deletion in prion protein gene in a Moroccan family. *Nucleic Acids Res.* **18:** 6745.

Laplanche J.-L., Delasnerie-Lauprêtre N., Brandel J.P., Dussaucy M., Chatelain J., and Launay J.M. 1995. Two novel insertions in the prion protein gene in patients with late-onset dementia. *Hum. Mol. Genet.* **4:** 1109–1111.

Laplanche J.-L., Delasnerie-Lauprêtre N., Brandel J.P., Chatelain J., Beaudry P., Alpérovitch A., and Launay J.-M. 1994. Molecular genetics of prion diseases in France. *Neurology* **44:** 2347–2351.

Laplanche J.-L., El Hachimi K.H., Durieux I., Thuillet P., Defebvre L., Delasnerie-

Laupretre N., Peoc'h K., Foncin J.-F., and Destee A. 1999. Prominent psychiatric features and early onset in an inherited prion disease with a new insertional mutation in the prion protein gene. *Brain* **122:** 2375–2386.

Lathrop G.M., Laluel J.M., Julier C., and Ott J. 1984. Strategies for multilocus linkage analysis in humans. *Proc. Natl. Acad. Sci.* **81:** 3443–3446.

Lee H.S., Sambuughin N., Cervenáková L., Chapman J., Pocchiari M., Litvak S., Qi H.Y., Budka H., del Ser T., Furukawa H., Brown P., Gajdusek D.C., Long J.C., Korczyn A.D., and Goldfarb L.G. 1999. Ancestral origins and worldwide distribution of the PRNP 200K mutation causing familial Creutzfeldt-Jakob disease. *Am. J. Hum. Genet.* **64:** 1063–1070.

Lehmann S. and Harris D.A. 1995. A mutant prion protein displays an aberrant membrane association when expressed in cultured cells. *J. Biol. Chem.* **270:** 24589–24597.

———. 1996a. Mutant and infectious prion proteins display common biochemical properties in cultured cells. *J. Biol. Chem.* **271:** 1633–1637.

———. 1996b. Two mutant prion proteins expressed in cultured cells acquire biochemical properties reminiscent of the scrapie isoform. *Proc. Natl. Acad. Sci.* **93:** 5610–5614.

———. 1997. Blockade of glycosylation promotes acquisition of scrapie-like properties by the prion protein in cultured cells. *J. Biol. Chem.* **272:** 21479–21487.

Lewis V., Collins S., Hill A.F., Boyd A., McLean C.A., Smith M., and Masters C.L. 2003. Novel prion protein insert mutation associated with prolonged neurodegenerative illness. *Neurology* **60:** 1620–1624.

Liberski P., Bratosiewicz J., Barvikowska M., Cervenáková L., and Brown P. 2000. A case of Gerstmann-Sträussler-Scheinker disease (GSS) with Met to Thr mutation at codon 232 of the PRNP gene. *Brain Pathol.* **10:** 669. (Abstr.)

Liemann S. and Glockshuber R. 1999. Influence of amino acid substitutions related to inherited human prion diseases on the thermodynamic stability of the cellular prion protein. *Biochemistry* **38:** 3258–3267.

Lope E.S., Junquera S.R., Martinez A.M., and Berenguel A.B. 1977. Acute ascending polyradiculoneuritis in a case of Creutzfeldt-Jakob disease. *J. Neurol. Neurosurg. Psychiatry* **40:** 149–155.

Lorenz H., Windl O., and Kretzschmar H.A. 2002. Cellular phenotyping of secretory and nuclear prion proteins associated with inherited prion diseases. *J. Biol. Chem.* **277:** 8508–8516.

Ma J. and Lindquist S. 2001. Wild-type PrP and a mutant associated with prion disease are subject to retrograde transport and proteasome degradation. *Proc. Natl. Acad. Sci.* **98:** 14955–14960.

———. 2002. Conversion of PrP to a self-perpetuating PrPSc-like conformation in the cytosol. *Science* **298:** 1785–1788.

Ma J., Wollmann R., and Lindquist S. 2002. Neurotoxicity and neurodegeneration when PrP accumulates in the cytosol. *Science* **298:** 1781–1785.

Mallucci G.R., Campbell T.A., Dickinson A., Beck J., Holt M., Plant G., de Pauw K.W., Hakin R.N., Clarke C.E., Howell S., Davies-Jones G.A., Lawden M., Smith C.M., Ince P., Ironside J.W., Bridges L.R., Dean A., Weeks I., and Collinge J. 1999. Inherited prion disease with an alanine to valine mutation at codon 117 in the prion protein gene. *Brain* **122:** 1823–1837.

Manetto V., Medori R., Cortelli P., Montagna P., Baruzzi A., Hauw J., Rancruel G., Vanderhaeghen J.J., Mailleux P., Bugiani O., Tagliavini F., Bouras C., Rizzuto N., Lugaresi E., and Gambetti P. 1992. Fatal familial insomnia: Clinical and pathological study of five new cases. *Neurology* **42:** 312–319.

Manuelidis E.E., Fritch W.W., Kim J.H., and Manuelidis L. 1987. Immortality of cell cultures derived from brains of mice and hamsters infected with Creutzfeldt-Jakob disease agent. *Proc. Natl. Acad. Sci.* **84:** 871–875.

Marcaud V., Laplanche J.L., Defontaines B., Beaudry P., vital A., Vincent D., Sazdovitch V., Hauw J.J., Latinville D., Jung P., Vecchierini F., and Degos C.F. 2003. Usefulness of molecular genetic analysis of the PRNP gene in patients with cerebellar ataxia: A new case of fatal familial incomnia. *Rev. Neurol.* (Paris) **159:** 199–202. (French)

Martin J.J. 1970. Contribution a l'étude de l'anatomie du thalamus et de sa pathologie au cours des maladies degeneratives dites abiotrophiques. *Acta Neurol. Belg.* **70:** 1–211.

Masters C.L., Gajdusek D.C., and Gibbs C.J., Jr. 1981a. Creutzfeldt-Jakob disease virus isolations from the Gerstmann-Sträussler syndrome with an analysis of the various forms of amyloid plaque deposition in the virus-induced spongiform encephalopathies. *Brain* **104:** 559–588.

———. 1981b. The familial occurrence of Creutzfeldt-Jakob disease and Alzheimer's disease. *Brain* **104:** 535–558.

Mastrianni J.A., Iannicola C., Myers R.M., DeArmond S., and Prusiner S.B. 1996. Mutation of the prion protein gene at codon 208 in familial Creutzfeldt-Jakob disease. *Neurology* **47:** 1305–1312.

Mastrianni J.A., Capellari S., Telling G.C., Han D., Bosque P., Prusiner S.B., and DeArmond S.J. 2001. Inherited prion disease caused by the V210I mutation: Transmission to transgenic mice. *Neurology* **57:** 2198–2205.

Mastrianni J.A., Curtis M.T., Oberholtzer J.C., Da Costa M.M., DeArmond S., Prusiner S.B., and Garbern J.Y. 1995. Prion disease (PrP-A117V) presenting with ataxia instead of dementia. *Neurology* **45:** 2042–2050.

Mastrianni J.A., Nixon R., Layzer R., Telling G.C., Han D., DeArmond S.J., and Prusiner S.B. 1999. Prion protein conformation in a patient with sporadic fatal insomnia. *N. Engl. J. Med.* **340:** 1630–1638.

May W.W., Itabashi H.H., and DeJong R.N. 1968. Creutzfeldt-Jakob disease. II. Clinical, pathologic and genetic study of a family. *Arch. Neurol.* **19:** 137–149.

Medori R., Tritschler H.J., LeBlanc A., Villare F., Manetto V., Chen H.Y., Xue R., Leal S., Montagna P., Cortelli P., Tinuper P., Avoni P., Mochi M., Baruzzi A., Hauw J.J., Ott J., Lugaresi E., Autilio-Gambetti L., and Gambetti P. 1992. Fatal familial insomnia is a prion disease with a mutation at codon 178 of the prion gene. *N. Engl. J. Med.* **326:** 444–449.

Meggendorfer F. 1930. Klinische und genealogische Beobachtungen bein einem Fall von spastischen Pseudokosklerose Jakobs. *Z. Neurol. Psychiatr.* **128:** 337–341.

Meiner Z., Halimi M., Polakiewicz R.D., Prusiner S.B., and Gabizon R. 1992. Presence of the prion protein in peripheral tissues of Libyan Jews with Creutzfeldt-Jakob disease. *Neurology* **42:** 1355–1360.

Meyer R.K., McKinley M.P., Bowman K.A., Braunfeld M.B., Barry R.A., and Prusiner S.B. 1986. Separation and properties of cellular and scrapie prion proteins. *Proc. Natl. Acad. Sci.* **83:** 2310–2314.

Mirra S.S., Young K., Gearing M., Jones R., Evatt M.L., Piccardo P., and Ghetti B. 1997. Coexistence of prion protein (PrP) amyloid, neurofibrillary tangles, and Lewy bodies in Gerstmann-Sträussler-Scheinker disease with prion gene (*PRNP*) mutation F198S. *Brain Pathol.* **7:** 1379.

Mishra R.S., Gu Y., Bose S., Verghese S., Kalepu S., and Singh N. 2002. Cell surface accumulation of a truncated transmembrane prion protein in Gerstmann-Sträussler-Scheinker disease P102L. *J. Biol. Chem.* **277:** 24554–24561.

Mitrova E. and Belay G. 2002. Creutzfeldt-Jakob disease with E200K mutation in Slovakia: Characterization and development. *Acta Virol.* **46:** 31–39.

Mizushima S., Ishii K., and Nishimaru T. 1994. A case of presenile with a 168 base pair insertion in prion protein gene (Japanese). *Dementia* **8:** 380–390.

Mohr M., Tranchant C., Heldt N., and Warter J.M. 1994. Alsatian variant of codon 117 form of Gerstmann-Sträussler-Scheinker syndrome: Autopsic study of 3 cases. *Brain Pathol.* **4:** 524.

Monari L., Chen S.C., Brown P., Parchi P., Petersen R.B., Mikol J., Gray F., Cortelli P., Montagna P., Ghetti B., Goldfarb L.G., Gajdusek D.C., Lugaresi E., Gambetti P., and Autilio-Gambetti L. 1994. Fatal familial insomnia and familial Creutzfeldt-Jakob disease: Different prion proteins determined by a DNA polymorphism. *Proc. Natl. Acad. Sci.* **91:** 2839–2842.

Montagna P., Cortelli P., Avoni P., Tinuper P., Plazzi G., Gallassi R., Portaluppi F., Julien J., Vital C., Delisle M.B., Gambetti P., and Lugaresi E. 1998. Clinical features of fatal familial insomnia: Phenotypic variability in relation to a polymorphism at codon 129 of the prion protein gene. *Brain Pathol.* **8:** 515–520.

Morillas M., Vanik D.L., and Surewicz W.K. 2001. On the mechanism of α-helix to β-sheet transition in the recombinant prion protein. *Biochemistry* **40:** 6982-6987.

Mouillet-Richard S., Teil C., Lenne M., Hugon S., Taleb O., and Laplanche J.L. 1999. Mutation at codon 210 (V210I) of the prion protein gene in a North African patient with Creutzfeldt-Jakob disease. *J. Neurol. Sci.* **168:** 141–144.

Nakazato Y., Ohno R., Negishi T., Hamaguchi K., and Arai E. 1991. An autopsy case of Gerstmann-Sträussler-Scheinker's disease with spastic paraplegia as its principal feature. *Clin. Neurol.* **31:** 987–992.

Neufeld M.Y., Josiphov J., and Korczyn A.D. 1992. Demyelinating peripheral neuropathy in Creutzfeldt-Jakob disease. *Muscle Nerve* **15:** 1234–1239.

Neugut R.H., Neugut A.I., Kahana E., Stein Z., and Alter M. 1979. Creutzfeldt-Jakob disease: Familial clustering among Libyan-born Israelis. *Neurology* **29:** 225–231.

Nicholl D., Windl O., de Silva R., Sawcer S., Dempster M., Ironside J.W., Estibeiro J.P., Yuill G.M., Lathe R., and Will R.G. 1995. Inherited Creutzfeldt-Jakob disease in a British family associated with a novel 144 base pair insertion of the prion protein gene. *J. Neurol. Neurosurg. Psychiatry* **58:** 65–69.

Nicholson E.M., Mo H., Prusiner S.B., Cohen F.E., and Marqusee S. 2002. Differences between the prion protein and its homolog doppel: A partially structured state with implications for scrapie formation. *J. Mol. Biol.* **316:** 807–815.

Nieto A., Goldfarb L.G., Brown P., McCombie W.R., Trapp S., Asher D.M., and Gajdusek D.C. 1991. Codon 178 mutation in ethnically diverse Creutzfeldt-Jakob disease families. *Lancet* **337:** 622–623.

Nisipeanu P., El Ad B., and Korczyn A.D. 1990. Spongiform encephalopathy in an Israeli born to immigrants from Libya (letter). *Lancet* **336:** 686.

Nitrini R., Rosemberg S., Passos-Bueno M.R., Gambetti P., Papadopoulos M., Carrilho P.M., Caramelli P., Albrecht S., Zatz M., and LeBlanc A. 1997. Familial spongiform encephalopathy with distinct clinico-pathological features associated with a novel prion gene mutation at codon 183. *Ann. Neurol.* **42:** 138–146.

Nitrini R., Teixeira da Silva L.S., Rosemberg S., Caramelli P., Carrilho P.E., Iughetti P., Passos-Bueno M.R., Zatz M., Albrecht S., and LeBlanc A. 2001. Prion disease resembling frontotemporal dementia and parkinsonism linked to chromosome 17. *Arq. Neuropsiquiatr.* **59:** 161–164.

Nixon R., Camicioli R., Jamison K., Cervenáková L., and Mastrianni J.A. 2000. The PRNP-

V180I mutation is associated with abnormally glycosylated PrPCJD and intracellular PrP accumulations. *Brain Pathol.* **10:** 670. (Abstr.)

Nochlin D., Sumi S.M., Bird T.D., Snow A.D., Leventhal C.M., Beyreuther K., and Masters C.L. 1989. Familial dementia with PrP-positive amyloid plaques: A variant of Gerstmann-Sträussler syndrome. *Neurology* **39:** 910–918.

Oberndorfer S., Urbanits S., Lahrmann H., Jarius C., Albrecht G., and Grisold W. 2002. Familial Creutzfeldt-Jakob disease initially presenting with alien hand syndrome. *J. Neurol.* **249:** 631–632.

Oda T., Kitamoto T., Tateishi J., Mitsuhashi T., Iwabuchi K., Haga C., Oguni E., Kato Y., Tominaga I., Yanai K., Kashima H., Kogure T., Hori K., and Ogino K. 1995. Prion disease with 144 base pair insertion in a Japanese family line. *Acta Neuropathol.* **90:** 80–86.

Oesch B., Westaway D., Walchli M., McKinley M.P., Kent S.B.H., Aebersold R., Barry R.A., Tempst P., Teplow D.B., Hood L.E., Prusiner S.B., and Weissmann C. 1985. A cellular gene encodes scrapie PrP 27–30 protein. *Cell* **40:** 735–746.

Owen F., Poulter M., Collinge J., Leach M., Lofthouse R., Crow T.J., and Harding A.E. 1992. A dementing illness associated with a novel insertion in the prion protein gene. *Mol. Brain Res.* **13:** 155–157.

Owen F., Poulter M., Shah T., Collinge J., Lofthouse R., Baker H., Ridley R., McVey J., and Crow T.J. 1990. An in-frame insertion in the prion protein gene in familial Creutzfeldt-Jakob disease. *Mol. Brain Res.* **7:** 273–276.

Owen F., Poulter M., Lofthouse R., Collinge J., Crow T.J., Risby D., Baker H.F., Ridley R.M., Hsiao K., and Prusiner S.B. 1989. Insertion in prion protein gene in familial Creutzfeldt-Jakob disease. *Lancet* **I:** 51–52.

Padovani A., D'Alessandro M., Parchi P., Cortelli P., Anzola G.P., Montagna P., Vignolo L.A., Petraroli R., Pocchiari M., Lugaresi E., and Gambetti P. 1998. Fatal familial insomnia: A novel Italian kindred. *Neurology* **51:** 1491-1494.

Palmer M.S. and Collinge J. 1992. Human prion diseases. *Curr. Opin. Neurol. Neurosurg.* **5:** 895–901.

Palmer M.S., Mahal S.P., Campbell T.A., Hill A.F., Sidle K.C.L., Laplanche J.L., and Collinge J. 1993. Deletions in the prion protein gene are not associated with CJD. *Hum. Mol. Genet.* **2:** 541–544.

Pan T., Colucci M., Wong B.S., Li R., Liu T., Petersen R.B., Chen S., Gambetti P., and Sy M.S. 2001. Novel differences between two human prion strains revealed by two-dimensional gel electrophoresis. *J. Biol. Chem.* **276:** 37284–37288.

Panegyres P.K., Toufexis K., Kakulas B.A., Cerneváková L., Brown P., Ghetti B., Piccardo P., and Dlouhy S.R. 2001. A new PRNP mutation (G131V) associated with Gerstmann-Sträussler-Scheinker disease. *Arch. Neurol.* **58:** 1899–1902.

Parchi P., Capellari S., Chen S., and Gambetti P. 2003. Familial Creutzfeldt-Jakob disease. In *Neurodegeneration: The molecular pathology of dementia and movement disorders* (ed. D. Dickson et al.), pp. 298–306. Neuropath Press, Pegnitz, Germany.

Parchi P., Castellani R., Cortelli P., Montagna P., Chen S.G., Petersen R.B., Lugaresi E., Autilio-Gambetti L., and Gambetti P. 1995. Regional distribution of protease-resistant prion protein in fatal familial insomnia. *Ann. Neurol.* **38:** 21–29.

Parchi P., Capellari S., Chin S., Schwarz H.B., Schecter N.P., Butts J.D., Hudkins P., Burns D.K., Powers J.M., and Gambetti P. 1999a. A subtype of sporadic prion disease mimicking fatal familial insomnia. *Neurology* **52:** 1757–1763.

Parchi P., Petersen R.B., Chen S.G., Autilio-Gambetti L., Capellari S., Monari L., Cortelli P., Montagna P., Lugaresi E., and Gambetti P. 1998a. Molecular pathology of fatal familial insomnia. *Brain Pathol.* **8:** 539–548.

Parchi P., Capellari S., Chen S.G., Ghetti B., Mikol J., Vital C., Cochran E., Trojanowski J.Q., Dickson D.W, Petersen R.B., and Gambetti P. 1996a. Similar posttranslational modifications of the prion protein in familial, sporadic and iatrogenic Creutzfeldt-Jakob disease. *Soc. Neurosci. Abstr.* **22:** 711.

Parchi P., Capellari S., Sima A.A.F., D'Amato C., McKeever P., Mikol J., Brion S., Brown P., Chen S.G., Petersen R.B., and Gambetti P. 1996b. Creutzfeldt-Jakob disease (CJD) with 178^Asn mutation in the prion protein gene: Neuropathological and molecular features. *J. Neuropathol. Exp. Neurol.* **55:** 635.

Parchi P., Castellani R., Capellari S., Ghetti B., Young K., Chen S.G., Farlow M., Dickson D.W., Sima A.A.F., Trojanowski J.Q., Petersen R.B., and Gambetti P. 1996c. Molecular basis of phenotypic variability in sporadic Creutzfeldt-Jakob disease. *Ann. Neurol.* **39:** 767–778.

Parchi P., Chen S.G., Brown P., Zou W., Capellari S., Budka H., Hainfellner J., Reyes P.F., Golden G.T., Hauw J.J., Gajdusek D.C., and Gambetti P. 1998b. Different patterns of truncated prion protein fragments correlate with distinct phenotypes in P102L Gerstmann-Sträussler-Scheinker disease. *Proc. Natl. Acad. Sci.* **95:** 8322–8327.

Parchi P., Capellari S., Brown P., Sima A.A.F., Mikol J., Gray F., Frosch M.P., Trojanowski J.Q., Vital C., Ghetti B., Giese A., Kretzschmar H.A., and Gambetti P. 1999b. Molecular and clinico-pathologic phenotypic variability in genetic Creutzfeldt-Jakob disease. *Neurology* (suppl. 2) **52:** A323. (Abstr.)

Parchi P., Zou W., Wang W., Brown P., Capellari S., Ghetti B., Kopp N., Schulz-Schaeffer W.J., Kretzschmar H.A., Head M.W., Ironside J.W., Gambetti P., and Chen S.G. 2000. Genetic influence on the structural variations of the abnormal prion protein. *Proc. Natl. Acad. Sci.* **97:** 10168–10672.

Parchi P., Giese A., Capellari S., Brown P., Schulz-Schaeffer W., Windl O., Zerr I., Budka H., Kopp N., Piccardo P., Poser S., Rojiani A., Streichemberger N., Julien J., Vital C., Ghetti B., Gambetti P., and Kretzschmar H. 1999c. Classification of sporadic Creutzfeldt-Jakob disease based on molecular and phenotypic analysis of 300 subjects. *Ann. Neurol.* **46:** 224–233.

Pastore M., Castellani R.J., Chin S., Hua Z., Bell K., Chin S.S., and Gambetti P. 2002. CJD-associated with the novel R148H prion protein gene mutation. *J. Neuropathol. Exp. Neurol.* **61:** 491.

Peoc'h K., Serres C., Frobert Y., Martin C., Lehmann S., Chasseigneaux S., Sazdovitch V., Grassi J., Jouannet P., Launay J.M., and Laplanche J.L. 2000. Identification of three novel mutations (E196K, V203I, E211Q) in the prion protein gene (PRNP) in inherited prion diseases with Creutzfeldt- Jakob disease phenotype. *Hum. Mutat.* **15:** 482.

Peretz D., Williamson R.A., Legname G., Matsunaga Y., Vergara J., Burton D.R., DeArmond S.J., Prusiner S.B., and Scott M.R. 2002. A change in the conformation of prions accompanies the emergence of a new prion strain. *Neuron* **34:** 921–932.

Petersen R.B., Parchi P., Richardson S.L., Urig C.B., and Gambetti P. 1996. Effect of the D178N mutation and the codon 129 polymorphism on the metabolism of the prion protein. *J. Biol. Chem.* **271:** 12661–12668.

Petersen R.B., Tabaton M., Berg L., Schrank B., Torack R.M., Leal S., Julien J., Vital C., Deleplanque B., Pendlebury W.W., Drachman D., Smith T.W., Martin J.J., Oda M., Montagna P., Ott J., Autilio-Gambetti L., Lugaresi E., and Gambetti P. 1992. Analysis of the prion gene in thalamic dementia. *Neurology* **42:** 1859–1863.

Petraroli R. and Pocchiari M. 1996. Codon 219 polymorphism of PRNP in healthy Caucasians and Creutzfeldt-Jakob disease patients. *Am. J. Hum. Genet.* **58:** 888–889.

Piccardo P., Kish S.J., Ang L.C., Young K., Bugiani O., Tagliavini F., Dlouhy S.R., and Ghetti

B. 1998a. Prion protein isoforms in the new variant of Gerstmann-Sträussler-Scheinker disease Q212P. *J. Neuropathol. Exp. Neurol.* **57:** 518.

Piccardo P., Dlouhy S.R., Young K., William A., Feng Y., Quinn B., Dal Canto M., Sufit R., and Ghetti B. 1999. Creutzfeldt Jakob disease (CJD) with prion protein gene (PRNP) V210I mutation. *J. Neuropathol. Exp. Neurol.* **58:** 550. (Abstr. 166)

Piccardo P., Unverzagt F., William A., Takao M., Glazier B., Dlouhy S., DeCarli C., Farlow M., and Ghetti B. 2002. Pathologic prion protein is present in the brain of asymptomatic carriers of *PRNP* mutations associated with Gerstmann-Sträussler-Scheinker disease. *Neurobiol. Aging* **23:** S131.

Piccardo P., Seiler C., Dlouhy S.R., Young K., Farlow M.R., Prelli F., Frangione B., Bugiani O., Tagliavini F., and Ghetti B. 1996. Proteinase-K-resistant prion protein isoforms in Gerstmann-Sträussler-Scheinker disease (Indiana kindred). *J. Neuropathol. Exp. Neurol.* **55:** 1157–1163.

Piccardo P., Liepnieks J.J., William A., Dlouhy S.R., Farlow M.R., Young K., Nochlin D., Bird T.D., Nixon R.R., Ball M.J., DeCarli C., Bugiani O., Tagliavini F., Benson M.D., and Ghetti B. 2001. Prion proteins with different conformations accumulate in Gerstmann-Sträussler-Scheinker disease caused by A117V and F198S mutations. *Am. J. Pathol.* **158:** 2201–2207.

Piccardo P., Ghetti B., Dickson D.W., Vinters H.V., Giaccone G., Bugiani O., Tagliavini F., Young K., Dlouhy S.R., Seiler C., Jones C.K., Lazzarini A., Golbe L.I., Zimmerman T.R., Perlman S.L., McLachlan D.C., St. George-Hyslop P.H., and Lennox A. 1995. Gerstmann-Sträussler-Scheinker disease (*PRNP* P102L): Amyloid deposits are best recognized by antibodies directed to epitopes in PrP region 90-165. *J. Neuropathol. Exp. Neurol.* **54:** 790–801.

Piccardo P., Dlouhy S.R., Lievens P.M.J., Young K., Bird T.D., Nochlin D., Dickson D.W., Vinters H.V., Zimmerman T.R., Mackenzie I.R.A., Kish S.J., Ang L.-C., De Carli C., Pocchiari M., Brown P., Gibbs C.J., Gajdusek D.C., Bugiani O., Ironside J., Tagliavini F., and Ghetti B. 1998b. Phenotypic variability of Gerstman-Sträussler-Scheinker disease is associated with prion protein heterogeneity. *J. Neuropathol. Exp. Neurol.* **57:** 979–988.

Pietrini V., Puoti G., Limido L., Rossi G., Di Fede G., Giaccone G., Mangeri M., Tedeschi F., Bondavalli A., Mancia D., Bugiani O., and Tagliavini F. 2003. Creutzfeldt-Jakob disease with a novel extra-repeat insertional mutation in the *PRNP* gene. *Neurology* (in press).

Pocchiari M., Ladogana A., Petraroli R., Cardone F., and D'Alessandro M. 1998. Recent Italian FFI cases. *Brain Pathol.* **8:** 564–566.

Pocchiari M., Salvatore M., Cutruzzola F., Genuardi M., Allocatelli C.T., Masullo C., Macchi G., Alema G., Galgani S., Xi Y.G., Petraroli R., Silvestrini M.C., and Brunori M. 1993. A new point mutation of the prion protein gene in Creutzfeldt-Jakob disease. *Ann. Neurol.* **34:** 802–807.

Poulter M., Baker H.F., Frith C.D., Leach M., Lofthouse R., Ridley R.M., Shah T., Owen F., Collinge J., Brown G., Hardy J., Mullan M.J., Harding A.E., Bennett C., Doshi R., and Crow T.H. 1992. Inherited prion disease with 144 base pair gene insertion. 1. Genealogical and molecular studies. *Brain* **115:** 675–685.

Priola S.A. and Chesebro B. 1998. Abnormal properties of prion protein with insertional mutations in different cell types. *J. Biol. Chem.* **273:** 11980–11985.

Prusiner S.B. 1982. Novel proteinaceous infectious particles causing scrapie. *Science* **216:** 136–144.

———. 1989. Scrapie prions. *Annu. Rev. Microbiol.* **43:** 345–374.

———. 1997. Prion diseases and the BSE crisis. *Science* **278:** 245–251.

Prusiner S.B., Scott M.R., DeArmond S.J., and Cohen F.E. 1998. Prion protein biology. *Cell* **93:** 337–348.

Puckett C., Concannon P., Casey C., and Hood L. 1991. Genomic structure of human prion protein gene. *Am. J. Hum. Genet.* **49:** 320–329.

Puoti G., Rossi G., Giaccone G., Awan T., Lievens P.M., Defanti C.A., Tagliavini F., and Bugiani O. 2000. Polymorphism at codon 129 of PRNP affects the phenotypic expression of Creutzfeldt-Jakob disease linked to the E200K mutation. *Ann. Neurol.* **48:** 269–270.

Radhakrishnan K. and Mousa E.M. 1988. Creutzfeldt-Jakob disease in Benghazi, Libya. *Neuroepidemiology* **7:** 42–43.

Riek R., Wider G., Billeter M., Hornemann S., Glockshuber R., and Wüthrich K. 1998. Prion protein NMR structure and familial human spongiform encephalopathies. *Proc. Natl. Acad. Sci.* **95:** 11667–11672.

Ripoll L., Laplanche J.L., Salzmann M., Jouvet A., Planques B., Dussaucy M., Chatelain J., Beaudry P., and Launay J.M. 1993. A new point mutation in the prion protein gene at codon 210 in Creutzfeldt-Jakob disease. *Neurology* **43:** 1934–1938.

Rogers M., Taraboulos A., Scott M., Groth D., and Prusiner S.B. 1990. Intracellular accumulation of the cellular prion protein after mutagenesis of its Asn-linked glycosylation sites. *Glycobiology* **1:** 101–109.

Roos R., Gajdusek D.C., and Gibbs C.J. 1973. The clinical characteristics of transmissible Creutzfeldt-Jacob disease. *Brain* **96:** 1–20.

Rosenmann H., Vardi J., Finkelstein Y., Chapman J., and Gabizon R. 1998. Identification in Israel of 2 Jewish Creutzfeldt-Jakob disease patients with a 178 mutation at their PrP gene. *Acta Neurol. Scand.* **97:** 184–187.

Rosenmann H., Talmor G., Halimi M., Yanai A., Gabizon R., and Meiner Z. 2001. Prion protein with an E200K mutation displays properties similar to those of the cellular isoform PrP(C). *J. Neurochem.* **76:** 1654–1662.

Rosenmann H., Meiner Z., Kahana E., Halimi M., Lenetsky E., Abramsky O., and Gabizon R. 1997. Detection of 14-3-3 protein in the CSF of genetic Creutzfeldt-Jakob disease. *Neurology* **49:** 593–595.

Rosenthal N.P., Keesey J., Crandall B., and Brown W.J. 1976. Familial neurological disease associated with spongiform encephalopathy. *Arch. Neurol.* **33:** 252–259.

Rossi G., Giaccone G., Giampaolo L., Iussich S., Puoti G., Frigo M., Cavaletti G., Frattola L., Bugiani O., and Tagliavini F. 2000. Creutzfeldt-Jakob disease with a novel four extra-repeat insertional mutation in the PrP gene. *Neurology* **55:** 405-410.

Rossi G., Macchi G., Porro M., Giaccone G., Bugiani M., Scarpini E., Scarlato G., Molini G.E., Sasanelli F., Bugiani O., and Tagliavini F. 1998. Fatal familial insomnia. Genetic, neuropathologic, and biochemical study of a patient for a new Italian kindred. *Neurology* **50:** 688–692.

Sadeh M., Chagnac Y., and Goldhammer Y. 1990. Creutzfeldt-Jakob disease associated with peripheral neuropathy. *Isr. J. Med. Sci.* **26:** 220–222.

Safar J., Wille H., Itri V., Groth D., Serban H., Torchia M., Cohen F.E., and Prusiner S.B. 1998. Eight prion strains have PrP(Sc) molecules with different conformations. *Nat. Med.* **4:** 1157–1165.

Samaia H.B., Mari J.D.J., Vallada H.P., Moura R.P., Simpson A.J.G., and Brentani R.R. 1997. A prion-linked psychiatric disorder. *Nature* **390:** 241.

Scaravilli F., Cordery R.J., Kretzschmar H., Gambetti P., Brink B., Fritz V., Temlett J., Kaplan C., Fish D., An S.F., Schulz-Schaeffer W.J., and Rossor M.N. 2000. Sporadic fatal insomnia: A case study. *Ann. Neurol.* **48:** 665–668.

Schatzl H.M., Da Costa M., Taylor L., Cohen F.E., and Prusiner S.B. 1995. Prion protein gene variation among primates. *J. Mol. Biol.* **245:** 362–374.

Schoene W.C., Masters C.L., Gibbs C.J., Jr., Gajdusek D.C., Tyler H.R., Moore F.D., and Dammin G.J. 1981. Transmissible spongiform encephalopathy (Creutzfeldt-Jakob disease). *Arch. Neurol.* **38:** 473–477.

Seno H., Tashiro H., Ishino H., Inagaki T., Nagasaki M., and Morikawa S. 2000. New haplotype of familial Creutzfeldt-Jakob disease with a codon 200 mutation and a codon 219 polymorphism of the prion protein gene in a Japanese family. *Acta Neuropathol.* **99:** 125–130.

Sforza E., Montagna P., Tinuper P., Cortelli P., Avoni P., Ferrillo F., Petersen R., Gambetti P., and Lugaresi E. 1995. Sleep-wake cycle abnormalities in fatal familial insomnia. Evidence of the role of the thalamus in sleep regulation. *Electroencephalogr. Clin. Neurophysiol.* **94:** 398–405.

Shyu W.C., Hsu Y.D., Kao M.C., and Tsao W.L. 1996. Panencephalitic Creutzfeldt-Jakob disease in a Chinese family. Unusual presentation with PrP codon 210 mutation and identification by PCR-SSCP. *J. Neurol. Sci.* **143:** 176–180.

Silvestrini M.C., Cardone F., Maras B., Pucci P., Barra D., Brunori M., and Pocchiari M. 1997. Identification of the prion protein allotypes which accumulate in the brain of sporadic and familial Creutzfeldt-Jakob disease patients. *Nat. Med.* **3:** 521–525.

Simon E.S., Kahana E., Chapman J., Treves T.A., Gabizon R., Rosenmann H., Zilber N., and Korczyn A.D. 2000. Creutzfeldt-Jakob disease profile in patients homozygous for the PRNP E200K mutation. *Ann. Neurol.* **47:** 257–260.

Singh N., Zanusso G., Chen S.G., Fujioka H., Richardson S., Gambetti P., and Petersen R.B. 1997. Prion protein aggregation reverted by low temperature in transfected cells carrying a prion protein gene mutation. *J. Biol. Chem.* **272:** 28461–28470.

Skworc K.H., Windl O., Schulz-Schaeffer W.J., Giese A., Bergk J., Nagele A., Vieregge P., Zerr I.I., Poser S., and Kretzschmar H.A. 1999. Familial Creutzfeldt-Jakob disease with a novel 120-bp insertion in the prion protein gene. *Ann. Neurol.* **46:** 693–700.

Spacey S.D., Pastore M., McGillivray B., Fleming J., Gambetti P., and Feldman H. 2003. Fatal familial insomnia: The first account of a family of a Chinese descent. *Arch. Neurol.* (in press).

Speer M.C., Goldgaber D., Goldfarb L.G., Roses A.D., and Pericak-Vance M.A. 1991. Support of linkage of Gerstmann-Sträussler-Scheinker syndrome to the prion protein gene on chromosome 20p12-pter. *Genomics* **9:** 366–368.

Spudich S., Mastrianni J.A., Wrensch M., Gabizon R., Meiner Z., Kahana I., Rosenmann H., Kahana E., and Prusiner S.B. 1995. Complete penetrance of Creutzfeldt-Jakob disease in Libyan Jews carrying the E200K mutation in the prion protein gene. *Mol. Med.* **1:** 607–613.

Stahl N., Baldwin M.A., Burlingame A.L., and Prusiner S.B. 1990. Identification of glycoinositol phospholipid linked and truncated forms of the scrapie prion protein. *Biochemistry* **29:** 8879–8884.

Stender A. 1930. Weitere Beitrage zum Kapitel "Spastische Pseudosklerose Jakobs." *Z. Neurol. Psychiatr.* **128:** 528–543.

Stern K. 1939. Severe dementia associated with bilateral symmetrical degeneration of the thalamus. *Brain* **62:** 157–171.

Stewart R.S. and Harris D.A. 2001. Most pathogenic mutations do not alter the membrane topology of the prion protein. *J. Biol. Chem.* **276:** 2212–2220.

Supattapone S., Bouzamondo E., Ball H.L., Wille H., Nguyen H.O., Cohen F.E., DeArmond S.J., Prusiner S.B., and Scott M. 2001. A protease-resistant 61-residue prion peptide causes neurodegeneration in transgenic mice. *Mol. Cell. Biol.* **21:** 2608–2016.

Swietnicki W., Petersen S.B., Gambetti P., and Surewicz W.K. 1998. Familial mutations and the thermodynamic stability of the recombinant human prion protein. *J. Biol. Chem.* **273:** 31048–31052.

Tabernero C., Polo J.M., Sevillano M.D., Munoz R., Berciano J., Cabello A., Baez B., Ricoy J.R., Carpizo R., Figols J., Cuadrado N., and Claveria L.E. 2000. Fatal familial insomnia: Clinical, neuropathological, and genetic description of a Spanish family. *J. Neurol. Neurosurg. Psychiatry* **68:** 774–777.

Tagliavini F., Prelli F., Ghiso J., Bugiani O., Serban D., Prusiner S.B., Farlow M.R., Ghetti B., and Frangione B. 1991. Amyloid protein of Gerstmann-Sträussler-Scheinker disease (Indiana kindred) is an 11 kd fragment of prion protein with an N-terminal glycine at codon 58. *EMBO J.* **10:** 513–519.

Tagliavini F., Giaccone G., Prelli F., Verga L., Porro M., Trojanowski J., Farlow M., Frangione B., Ghetti B., and Bugiani O. 1993. A68 is a component of paired helical filaments of Gerstmann-Sträussler-Scheinker disease, Indiana kindred. *Brain Res.* **616:** 325–328.

Tagliavini F., Prelli F., Porro M., Rossi G., Giaccone G., Farlow M.R., Dlouhy S.R., Ghetti B., Bugiani O., and Frangione B. 1994. Amyloid fibrils in Gerstmann-Sträussler-Scheinker disease (Indiana and Swedish kindreds) express only PrP peptides encoded by the mutant allele. *Cell* **79:** 695–703.

Tagliavini F., Lievens P.M.J., Tranchant C., Warter J.M., Mohr M., Giaccone G., Perini F., Rossi G., Salmona M., Piccardo P., Ghetti B., Beavis R.C., Bugiani O., Frangione B., and Prelli F. 2001. A 7 kDa prion protein fragment required for infectivity is the mayor amyloid protein in Gerstmann-Sträussler-Scheinker disease A117V. *J. Biol. Chem.* **276:** 6009–6015.

Tanaka Y., Minematsu K., Moriyasu H., Yamaguchi T., Yutani C., Kitamoto T., and Furukawa H. 1997. A Japanese family with a variant of Gerstmann-Sträussler-Scheinker disease. *J. Neurol. Neurosurg. Psychiatry* **62:** 454–457.

Taniwaki Y., Hara H., Doh-ura K., Murakami I., Tashiro H., Yamasaki T., Shigeto H., Arakawa K., Araki E., Yamada T., Iwaki T., and Kira J. 2000. Familial Creutzfeld-Jacob disease with D178N-129M mutation of PRNP presenting as cerebellar ataxia without insomnia. *J. Neurol. Neurosurg. Psychiatry* **68:** 388.

Taratuto A.L., Piccardo P., Reich E.G., Chen S.G., Sevlever G., Schultz M., Luzzi A.A., Rugiero M., Abecasis G., Endelman M., Garcia A.M., Capellari S., Xie Z., Lugaresi E., Gambetti P., Dlouhy S.R., and Ghetti B. 2002. Insomnia associated with thalamic involvement in E200K Creutzfeldt-Jakob disease. *Neurology* **58:** 362–367.

Tateishi J., Kitamoto T., Hoque M.Z., and Furukawa H. 1996. Experimental transmission of Creutzfeldt-Jakob disease and related diseases to rodents. *Neurology* **46:** 532–537.

Tateishi J., Brown P., Kitamoto T., Hoque M.Z., Roos R., Wollman R., Cervenáková L., and Gajdusek D.C. 1995. First experimental transmission of fatal familial insomnia. *Nature* **376:** 434–435.

Telling G.C., Scott M., Mastrianni J., Gabizon R., Torchia M., Cohen F.E., DeArmond S.J., and Prusiner S.B. 1995. Prion propagation in mice expressing human and chimeric PrP transgenes implicates the interaction of cellular PrP with another protein. *Cell* **83:** 79–90.

Telling G.C., Parchi P., DeArmond S.J., Cortelli P., Montagna P., Gabizon R., Mastrianni J.A., Lugaresi E., Gambetti P., and Prusiner S.B. 1996. Evidence for the conformation of the pathologic isoform of the prion protein enciphering and propagating prion diversity. *Science* **274:** 2079–2082.

Telling G.C., Scott M., Hsiao K.K., Foster D., Yang S.-L., Torchia M., Sidle K.C.L., Collinge

J., DeArmond S.J., and Prusiner S.B. 1994. Transmission of Creutzfeldt-Jakob disease from humans to transgenic mice expressing chimeric human-mouse prion protein. *Proc. Natl. Acad. Sci.* **91:** 9936–9940.

Terao Y., Hitoshi S., Shimizu J., Sakuta M., and Kitamoto T. 1992. Gerstmann-Sträussler-Scheinker disease with heterozygous codon change at prion protein codon 129. *Clin. Neurol.* **3:** 880–883.

Tranchant C., Doh-ura K., Steinmetz G., Chevalier Y., Kitamoto T., Tateishi J., and Warter J.M. 1991. Mutation du codon 117 du géne du prion dans une maladie de Gerstmann-Sträussler-Scheinker. *Rev. Neurol.* **147:** 274–278.

Tranchant C., Doh-ura K., Warter J.M., Steinmetz G., Chevalier Y., Hanauer A., Kitamoto T., and Tateishi J. 1992. Gerstmann-Sträussler-Scheinker disease in an Alsatian family—Clinical and genetic studies. *J. Neurol. Neurosurg. Psychiatry* **55:** 185–187.

Tsai M.T., Su Y.C., Chen Y.H., and Chen C.H. 2001. Lack of evidence to support the association of the human prion gene with schizophrenia. *Mol. Psychiatry* **6:** 74–78.

Vallat J.M., Dumas M., Corvisier N., Leboutet M.J., Loubet A., Dumas P., and Cathala F. 1983. Familial Creutzfeldt-Jakob disease with extensive degeneration of white matter. *J. Neurol. Sci.* **61:** 261–275.

van Gool W.A., Hensels G.W., Hoogerwaard E.M., Wiezer J.H.A., Wesseling P., and Bolhuis P.A. 1995. Hypokinesia and presenile dementia in a Dutch family with a novel insertion in the prion protein gene. *Brain* **118:** 1565–1571.

van Harten B., van Gool W.A., van Langen I.M., Deekman J.M., Meijerink P.H.S., and Weinstein H.C. 2000. A new mutation in the prion protein gene: A patient with dementia and white matter changes. *Neurology* **55:** 1055–1057.

Vanik D.L. and Surewicz W.K. 2002. Disease-associated F198S mutation increases the propensity of the recombinant prion protein for conformational conversion to scrapie-like form. *J. Biol. Chem.* **277:** 49065–49070.

Vnencak-Jones C.L. and Phillips J.A., III. 1992. Identification of heterogeneous PrP gene deletions in controls by detection of allele-specific heteroduplexes (DASH). *Am. J. Hum. Genet.* **50:** 871–872.

Wanschitz J., Kloppel S., Jarius C., Birner P., Flicker H., Hainfellner J.A., Gambetti P., Guentchev M., and Budka H. 2000. Alteration of the serotonergic nervous system in fatal familial insomnia. *Ann. Neurol.* **48:** 788–791.

Warter J.M., Steinmetz G., Heldt N., Rumbach L., Marescaux C.H., Eber A.M., Collard M., Rohmer F., Floquet J., Guedenet J.C., Gehin P., and Weber M. 1981. Demence pre-senile familiale syndrome de Gerstmann-Sträussler-Scheinker. *Rev. Neurol.* **138:** 107–121.

Wildegger G., Liemann S., and Glockshuber R. 1999. Extremely rapid folding of the C-terminal domain of the prion protein without kinetic intermediates. *Nat. Struct. Biol.* **6:** 550–553.

Will R.G., Campbell M.J., Moss T.H., Bell J.E., and Ironside J.W. 1998. FFI Cases from the United Kingdom. *Brain Pathol.* **8:** 562–563.

Windl O., Giese A., Schulz-Schaeffer W., Zerr I., Skworc K., Arendt S., Oberdieck C., Bodemer M., Poser S., and Kretzschmar H.A. 1999. Molecular genetics of human prion diseases in Germany. *Hum. Genet.* **105:** 244–252.

Worrall B.B., Rowland L.P., Chin S.S., and Mastrianni J.A. 2000. Amyotrophy in prion diseases. *Arch. Neurol.* **57:** 33–38.

Yamada M., Itoh Y., Fujigasaki H., Naruse S., Kaneko K., Kitamoto T., Tateishi J., Otomo E., Hayakawa M., Tanaka J., Matsushita M., and Miyatake T. 1993. A missense mutation at codon 105 with codon 129 polymorphism of the prion protein gene in a new variant of Gerstmann-Sträussler-Scheinker disease. *Neurology* **43:** 2723–2724.

Yamada M., Itoh Y., Inaba A., Wada Y., Takashima M., Satoh S., Kamata T., Okeda R., Kayano T., Suematsu N., Kitamoto T., Otomo E., Matsushita M., and Mizusawa H. 1999. An inherited prion disease with a PrP P105L mutation: Clinicopathologic and PrP heterogeneity. *Neurology* **53:** 181–188.

Yamazaki M., Oyanagi K., Mori O., Kitamura S., Ohyama M., and Terashi A. 1997. An autopsy case of variant Gerstmann-Sträussler-Scheinker syndrome with codon 105 mutation of the prion protein gene, showing degeneration of the pallidum, thalamus and substantia nigra, and widely distributed neurofibrillary tangles. *Brain Pathol.* **7:** 113–115.

Yanagihara C., Yasuda M., Maeda K., Miyoshi K., and Nishimura Y. 2002. Rapidly progressive dementia syndrome associated with a novel four extra repeat mutation in the prion protein gene. *J. Neurol. Neurosurg. Psychiatry* **72:** 788–791.

Yedidia Y., Horonchik L., Tzaban S., Yanai A., and Taraboulos A. 2001. Proteasomes and ubiquitin are involved in the turnover of the wild-type prion protein. *EMBO J.* **20:** 5383–5391.

Young K., Clark H.B., Piccardo P., Dlouhy S.R., and Ghetti B. 1997. Gerstmann-Sträussler-Scheinker disease with the PRNP P102L mutation and valine at codon 129. *Mol. Brain Res.* **44:** 147–150.

Young K., Piccardo P., Kish S.J., Ang L.C., Dlouhy S., and Ghetti B. 1998. Gerstmann-Sträussler-Scheinker disease (GSS) with a mutation at prion protein (PrP) residue 212. *J. Neuropathol. Exp. Neurol.* **57:** 518.

Young K., Jones C.K., Piccardo P., Lazzarini A., Golbe L.I., Zimmerman T.R., Dickson D.W., McLachlan D.C., St. George-Hyslop P., Lennox A., Perlman S., Vinters H.V., Hodes M.E., Dlouhy S., and Ghetti B. 1995. Gerstmann-Sträussler-Scheinker disease with mutation at codon 102 and methionine at codon 129 of PRNP in previously unreported patients. *Neurology* **45:** 1127–1134.

Zahn R., Liu A., Luhrs T., Riek R., Von Schroetter C., Garcia F. L., Billeter M., Calzolai L., Wider G., and Wüthrich K. 2000. NMR solution structure of the human prion protein. *Proc. Natl. Acad. Sci.* **97:** 145–150.

Zanusso G., Petersen R.B., Jin T., Jing Y., Kanoush R., Ferrari S., Gambetti P., and Singh N. 1999. Proteasomal degradation and N-terminal protease resistance of the codon 145 mutant prion protein. *J. Biol. Chem.* **274:** 23396–23404.

Zhang Y., Swietnicki W., Zagorski M.G., Surewicz W.K., and Sonnichsen F.D. 2000. Solution structure of the E200K variant of human prion protein. *J. Biol. Chem.* **275:** 33650–33654.

Zilber N., Kahana E., and Abraham M.P.H. 1991. The Libyan Creutzfeldt-Jakob disease focus in Israel: An epidemiologic evaluation. *Neurology* **41:** 1385–1389.

Zimmermann K., Turecek P.L., and Schwarz H.P. 1999. Genotyping of the prion protein gene at codon 129. *Acta Neuropathol.* **97:** 355–358.

Zoghbi H.Y. and Orr H.T. 2000. Glutamine repeats and neurodegeneration. *Annu. Rev. Neurosci.* **23:** 217–247.

Zou W., Capellari S., Parchi P., Sy M., and Gambetti P. 2003. Identification of novel proteinase K-resistant C-terminal fragments of PrP in Creutzfeldt-Jakob disease. *J. Biol. Chem.* **278:** (in press.)

15

Neuropathology of Prion Diseases

Stephen J. DeArmond,[1,2,3] James W. Ironside,[3]
Essia Bouzamondo-Bernstein,[1,2,3] and David Peretz,[1,3]
[1]Institute for Neurodegenerative Diseases and
Departments of [2]Pathology (Neuropathology) and [3]Neurology
University of California, San Francisco, California 94143 and
[3]Department of Pathology (Neuropathology Laboratory)
Western General Hospital
Edinburgh EH4 2XU, United Kingdom

Jan R. Fraser
Institute for Animal Health
Edinburgh EH9 3JF, United Kingdom

THE DISCOVERY OF MECHANISMS OF NERVE CELL DYSFUNCTION, degeneration, and death in prion diseases acquired by infection, in dominantly inherited forms, and in sporadic (idiopathic) forms has gone hand in hand with the discovery of the prion and how prions are propagated. The generation of PrP-specific antibodies led to the first immunohistochemical study, in which it was discovered that the amyloid plaques in experimental scrapie in Syrian hamsters contain protease-resistant PrP (Fig. 1A) (Bendheim et al. 1984; DeArmond et al. 1985). That finding convinced us that we had a unique opportunity to obtain a better understanding of the pathogenesis of scrapie in animals and the related human disorders that include sporadic, iatrogenic, and familial Creutzfeldt-Jakob disease (CJD) and the rare familial disorder Gerstmann-Sträussler-Scheinker syndrome (GSS). The overall objective was to test the hypothesis that PrP^Sc accumulation in the brain causes the clinically relevant neuronal dysfunction, vacuolation, and death that are the characteristics of prion diseases. The results of many studies have led to the unifying hypothesis that neuronal degeneration in spontaneous, infectious, and genetic prion diseases and the propagation of prions in those diseases are both related exclusively to abnormalities of PrP.

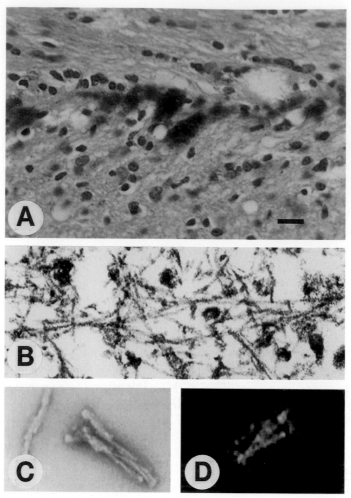

Figure 1. Comparison of PrP amyloid plaques deposited in situ with prion rods formed in vitro from Syrian hamsters inoculated with the Sc237 scrapie prion strain. (*A*) PrP amyloid plaques in the hippocampus are immunopositive with the first generation of PrP-specific antibodies in formalin-fixed, paraffin-embedded tissue sections (Bendheim et al. 1984). (*B*) Transmission electron microscopy of the PrP plaques shows loose aggregates of straight, unbranched 8- to 16-nm-wide amyloid filaments (DeArmond et al. 1985). (*C*) Electron microscopy of negatively stained, whole-mounted PrP 27-30 prion rods: individual rods of the aggregate are 10–16 nm wide (Barry et al. 1985; DeArmond et al. 1985). (*D*) Light microscopy of prion rods show they bind Congo red dye, which displays green–gold birefringence in polarized light. Bar in *A*, 25 μm.

Multiple investigators in the UK, continental Europe, Japan, and the US have contributed to our understanding of the mechanisms of CNS degeneration peculiar to prion diseases. We have attempted to identify and acknowledge their many contributions; however, the number of prion disease investigators is increasing logarithmically and, therefore, inevitably we will have missed some and apologize to them for that oversight. In this revision, we have asked Dr. Jan Fraser, whose laboratory is in Edinburgh, to contribute the outstanding quantitative neuropathologic data from her group to this chapter. Their detailed electron and confocal microscopy approaches delineate for the first time the stepwise progression of neurodegenerative changes in rodent models of scrapie. Their results fill in many of the details of prion disease progression that was relatively superficially described by neuropathologists who relied on classical neurohistopathological techniques. The results of the new studies emphasize that neurodegeneration in prion diseases progresses in a relatively stereotypical set of stages that begins with accumulation of abnormal prion protein that is followed by degeneration of presynaptic nerve terminals, then by atrophy of dendritic trees, and finally by nerve cell degeneration. Why is it important to review the details of disease progression? We have entered a new era of prion disease research heralded by drug trials of human Creutzfeldt-Jakob disease that are designed specifically to clear PrP^{Sc} from the brain. The working hypothesis is that clearance of PrP^{Sc} will prevent further neuronal degeneration and, perhaps, will stimulate synaptic regeneration. The new data suggest that the successful therapy will only be possible if it halts disease during that window of time between synapse loss and nerve cell death.

In the previous edition of this book, the abnormal prion protein found in human prion diseases was designated PrP^{CJD}. However, most investigators use PrP^{Sc} in a generic fashion to designate the pathogenic prion protein conformer found in both animal and human diseases: PrP^{Sc} will be used in this revision. Additionally, our concept of PrP^{Sc} is also evolving. It was originally believed that all of the PrP^{Sc} formed in the brain is protease-resistant; however, recent data indicate that PrP^{Sc} comprising a prion particle is organized into different duplicatable multimers that give prion strains their properties and results in prion strain-specific differences in PrP^{Sc}'s resistance to protease digestion (Peretz et al. 2001, 2002; Wille et al. 2002). Finally, we also are learning that abnormal PrP molecules need not only be β-conformation-rich to be pathogenic: Abnormal concentrations of transmembrane forms of the prion protein, ^{tm}PrP, also cause neurodegeneration. A short review of our current knowledge of different pathogenic types of PrP has been added to this revision of the chapter. Finally, it has been, and remains, the main objective of this chapter to emphasize the molecular and cellular mechanisms of neurodegeneration in prion diseases from the perspective of neuropathologic changes.

ACCUMULATION OF PrP^{Sc} CAUSES THE NEUROPATHOLOGIC CHANGES CHARACTERISTIC OF PRION DISEASES

The neuropathologic features characteristic of prion diseases include spongiform (vacuolar) degeneration of the brain parenchyma, nerve cell degeneration and death, variable reactive astrocytic gliosis, and variable amyloid plaque formation. The discoveries that there is a single-copy PrP gene, designated *Prnp* in animals and PRNP in humans, and that there are two PrP conformers (Cohen et al. 1994; James et al. 1997), a constitutively expressed, nonpathogenic conformer (PrP^C) and a pathogenic conformer in prion diseases (PrP^{Sc}), argued that PrP^{Sc} accumulation, rather than a putative slow virus, was the cause of neuronal degeneration.

Amyloid Plaques

Kuru, CJD, and GSS are, along with Alzheimer's disease, Down syndrome, pathological aging, and cerebral amyloid angiopathy, included among the cerebral amyloidoses. Amyloid is defined as an extracellular mass of protein that has the tinctorial and histochemical properties of starch (e.g., amyloid means starch-like) (Glenner 1980; Glenner et al. 1986). The starch-like characteristics include strong staining by the periodic acid Schiff (PAS) reaction that is most probably due to the high glycosaminoglycan content of all amyloids (Snow et al. 1987, 1989, 1990). In hematoxylin-and-eosin-stained tissue sections, it often appears as homogeneous, glass-like masses that stain pale pink. However, the characteristic of amyloids that distinguishes them from other extracellular masses is their affinity for Congo red dye, which displays green-gold birefringence when viewed with polarized light. The birefringence of the Congo red reaction indicates that the protein component of amyloids has a high β-sheet content. Electron microscopy reveals that the major protein component of amyloid is polymerized into aggregates of straight unbranched filaments. Although all amyloids have similar light and electron microscopic appearances, there are in reality multiple forms, each defined by its main protein component. The most common form of amyloidosis is caused by polymerization of the Aβ42 peptide fragment that forms as the result of aberrant degradation of the β-amyloid precursor protein during pathologic aging, Alzheimer's disease, and Down syndrome. The main protein com-

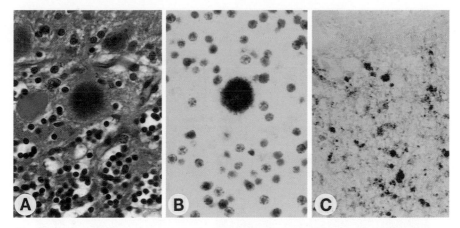

Figure 2. Typical kuru-type amyloid plaques in the cerebellar granule cell layer from a case of kuru can also be found in 5–10% of CJD cases, are abundant in nvCJD, and are mixed with other morphologic amyloid plaque subtypes in GSS. (A) A single kuru plaque stained by the PAS method is about the same size as the Purkinje cells at the top of the micrograph, 15–20 μm in diameter. (B) The kuru plaque is selectively stained by PrP immunohistochemistry following pretreatment of the section with formic acid. (C) PrP immunohistochemistry by the hydrolytic autoclaving method reveals numerous, non-amyloid PrP plaques and smaller punctate deposits in kuru as well as in some CJD cases. Magnification of C is one-half that of A and B.

ponent of the amyloid in prion diseases is composed of protease-resistant PrP peptides.

The first evidence that PrP^{Sc} causes the neuropathologic changes in prion diseases was finding that amyloid plaques in scrapie-infected animals and in human prion diseases (CJD, kuru, and GSS) are composed of protease-resistant PrP (Figs. 1–3) and not of the Aβ42 peptide (Bendheim et al. 1984; DeArmond et al. 1985; Kitamoto et al. 1986; Snow et al. 1989; DeArmond and Prusiner 1995). Amyloid plaques in prion diseases were relatively easy to immunostain with PrP-specific antibodies by routine immunohistochemical techniques. The intensity of immunostaining of PrP amyloid, like Aβ amyloid of Alzheimer's disease, could be increased by pretreatment of the tissue section with formic acid (Fig. 2B) (Kitamoto et al. 1987). PrP immunohistochemistry for electron microscopy indicated that the amyloid plaques in scrapie-infected Syrian hamsters have a filamentous ultrastructure like other amyloids (Fig. 1B) and that the filaments contain protease-resistant PrP (DeArmond et al. 1985). These discoveries were the first to link the neuropathologic features of prion diseases with the prion particle, because Prusiner and colleagues had found earlier that purified prions tend to form into amyloid

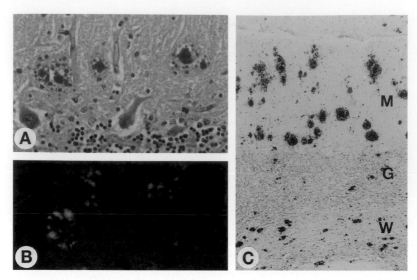

Figure 3. Multicentric GSS-type amyloid plaques in the cerebellum from a case of GSS(P102L). (*A*) PAS stain of multicentric plaques shows a central core of amyloid 15–30 μm in diameter surrounded by multiple smaller satellite amyloid deposits. (*B*) The GSS plaques bind Congo red dye, which displays green-gold birefringence in polarized light. (*C*) PrP immunohistochemistry by the hydrolytic autoclaving method results in strong immunostaining of the multicentric plaques and also reveals numerous large and small primitive plaques that do not stain by the PAS method or bind Congo red dye. The largest of these plaques are located in the molecular layer of the cerebellar cortex and can be 100 μm in length. Smaller PrP plaques are located in the granule cell layer and in the white matter. Magnification in *C* is half that in *A* and *B*. (W) Cerebellar white matter; (M) molecular layer; (G) granular layer.

rods in vitro (Fig. 1C) (Prusiner et al. 1983; Prusiner 1984). During the process of purifying prions from scrapie-infected brain, it was found that exposure of 33- to 35-kD PrP^{Sc} to proteinase K (PK) digested 87 amino acids from its amino-terminal region, leaving PrP 27-30 (Fig. 4), which is as infectious as PrP^{Sc} and has a propensity to polymerize into amyloid rod-like structures. These so-called "prion rods" were subsequently found to be an "artifact" of the procedure used to purify prions, because both PK digestion and detergents were required for their formation (McKinley et al. 1991a). Prion rods bind Congo red dye and display green-gold birefringence in polarized light like purified filaments from tissue amyloids (Fig. 1D) (Prusiner et al. 1983). The greater propensity of PrP 27-30 to form into amyloid rods compared to full-length PrP^{Sc} may be due to its higher β-sheet content or because digestion of a large portion of the amino terminus eliminated any steric hindrance to polymerization.

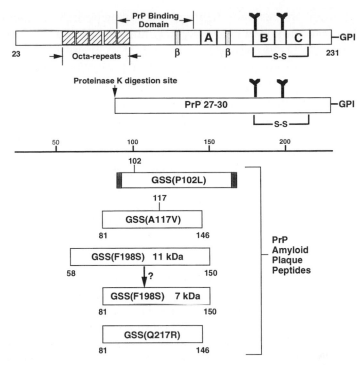

Figure 4. Amyloidosis in prion diseases requires truncation of the PrP molecule to highly amyloidogenic peptides. (Adapted, in part, from Ghetti et al. 1996a.)

Several groups have found that PrP^{Sc} is 43% β-sheet and 30% α-helix, whereas PrP^C is 3% β-sheet and 42% α-helix (Pan et al. 1993; Safar et al. 1993). PrP 27-30 has an even higher β-sheet content (54%) and lower α-helix content (21%) than PrP^{Sc} (Caughey et al. 1991; Gasset et al. 1993).

Amyloidosis In Vivo Involves Truncation of Pathogenic PrP

Amino-terminal truncation of PrP^{Sc} by limited digestion with PK yields PrP 27-30 that is amyloidogenic in vitro (Fig. 1C,D) (McKinley et al. 1991a). Truncation of the PrP also features in GSS syndromes, which, by definition, require dominant inheritance and the presence of large numbers of PrP amyloid plaques in the brain. The amino acid sequence of PrP in GSS amyloid was first determined for the Indiana kindred-type of GSS (Tagliavini and Pilleri 1983; Tagliavini et al. 1994) in which a mutation at codon 198 results in a substitution of a serine for phenylalanine, GSS(F198S) (Dlouhy et al. 1992; Hsiao et al. 1992). Amyloid cores were purified from brain, proteins were extracted from amyloid fibrils with formic acid, and the resulting peptides were sequenced. Two major peptides with ragged amino and carboxyl termini were obtained of about 11 kD and about 7 kD comprising PrP residues 58–150 and 81–150, respectively (Fig. 4). Immunohistochemistry of the GSS(F198S) with antisera to synthetic PrP peptides verified the presence of the ~11-kD and ~7-kD peptides in the plaques in situ. Immunohistochemistry with antibodies specific for the amino and carboxyl termini of PrP immunostained amorphous material at the periphery of the plaques and in the neuropil away from the plaques, suggesting that full-length PrP exists in those locations. To determine whether or not amyloidogenic peptides were derived from wild-type PrP^C or mutant PrP, Val-129 was used as a marker for mutant PrP peptides in patients whose wild-type PrP^C was Met-129. Amino acid sequencing and electrospray mass spectrometry of PrP amyloid peptides indicated they contained only 129V, arguing they were derived exclusively from mutant PrP(F198S). Ghetti et al. (1996a) used the same biochemical techniques to isolate and determine the size of peptide fragments from the amyloid in GSS(A117V) and GSS(Q217R). In both cases, mutated PrP was truncated to about 7-kD peptides spanning residues 81 to 146 (Fig. 4). Limited information was obtained about the amyloid of GSS(P102L). Western analysis of amyloid-enriched fractions showed two PrP immunoreactive bands at 25–30 kD and 15–20 kD. Amino acid sequence analysis of the high-performance liquid chromatography (HPLC)-purified peptides showed they contained the leucine substitution at residue 102, indicating that the amyloid was composed of mutant PrP(P102L). The residues included in the shorter PrP(P102L) peptide have been estimated by immunohistochemistry with antibodies to synthetic PrP peptides. It appears to extend from near residue 90 to near residue 165 (Fig. 4). A fifth mutation leading to a Gerstmann-Sträussler-Scheinker syndrome, GSS(Y145Stop), results in formation of a stop-codon at codon 145 and the synthesis of a truncated form of amyloidogenic PrP (Kitamoto et al. 1993a).

The mutant amino acid substitutions that cause GSS-like cerebral amyloidosis result in both amino- and carboxy-terminal truncation of PrP that yields highly amyloidogenic peptides. All of these peptides contain the putative PrP^{Sc}-to-PrP^C binding domain that studies of the host species barrier to prion transmission in transgenic (Tg) mice suggest resides between residues 90 and 140 (Scott et al. 1993). The propensity of those peptides to polymerize into amyloid filaments testifies to this domain's ability to bind with peptides containing the homologous sequence and also to the tendency for this binding to form highly β-sheeted polymers.

Non-amyloid PrP^{Sc} Accumulation Causes Neuronal Dysfunction and Degeneration

Except for the several genetically distinct forms of GSS in which large amounts of PrP amyloid plaques are deposited throughout the brain, most prion diseases in animals and humans are associated with few or no PrP amyloid plaque deposits. For example, only 5–10% of cases of the most common form of human prion disease, sporadic CJD, contain mature amyloid plaques. It seems very unlikely, therefore, that cerebral amyloidosis in prion diseases accounts for nerve cell dysfunction and degeneration.

The location of a non-amyloid form of PrP^{Sc} or, for that matter, the location of nonpathogenic PrP^{C} in the brain parenchyma, and their association with spongiform degeneration and reactive astrocytic gliosis has been a more complex problem than localizing PrP in amyloid plaques. There are several reasons for this difficulty: First, no PrP-specific antibodies can distinguish PrP^{C} and PrP^{Sc}. Second, PrP^{Sc} is poorly antigenic in its native conformation: It requires denaturation to make its antigenic sites available to antibodies (Serban et al. 1990). Third, because both PrP^{C} and PrP^{Sc} exist together in the brain in prion diseases, PrP^{C} must be eliminated before PrP^{Sc} can be localized by immunohistochemical techniques. Because of these problems, in our first attempts to determine whether or not there is a correlation between sites of PrP^{Sc} accumulation and sites of spongiform degeneration and reactive astrocytic gliosis, we compared our neuropathologic and immunohistochemical results with biochemical measurements of PrP^{Sc} in homogenates of unfixed, dissected brain regions (DeArmond et al. 1987). The results showed for the first time that both spongiform degeneration and reactive astrocytic gliosis colocalize with sites of PrP^{Sc} accumulation in the brain.

Progress became possible with the development of two sensitive immunohistologic techniques that differentiate PrP^{C} and PrP^{Sc} in tissue sections. The histoblot technique combines the morphologic advantages of histologic sectioning with the neurochemical advantages of dot-blotting to distinguish PrP^{C} from PrP^{Sc} on the basis of their differential susceptibility to PK digestion and their relative reactivity with PrP antibodies in their natural and denatured states (Taraboulos et al. 1992b). The second advance, developed in Tatieshi's laboratory, has been named hydrolytic autoclaving (Muramoto et al. 1992). Aldehyde-fixed, paraffin-embedded tissue sections are autoclaved in dilute hydrochloric acid prior to immunohistochemical staining. This technique takes advantage of the relative resistance of PrP^{Sc} to acid hydrolysis compared to PrP^{C}. There are

advantages and disadvantages to both techniques. Histoblots allow quanti-
tation and subregional localization of both PrP^C and PrP^{Sc} in serial sections
of the CNS but do not show cellular localization. Hydrolytic autoclaving is
less sensitive and appears to be selective for PrP^{Sc} based on comparison
with histoblot analysis but, importantly, reveals the cellular localization of
PrP^{Sc} (DeArmond and Prusiner 1995).

Histoblots are made by transferring (blotting) frozen sections of
unfixed brain tissue to nitrocellulose paper where PrP^C can be completely eliminated with certainty by PK digestion (Taraboulos et al. 1992b).
The undigested portion of PrP^{Sc}, PrP 27-30, which is poorly antigenic in
its native configuration, is then denatured with 4 M guanidinium (Gdn),
which greatly enhances binding of PrP-specific antibodies. The result is an
intense immunohistochemical signal with little or no background that
reveals the neuroanatomic location of PrP^{Sc} and is quantifiable. Because
PrP^{Sc} in its native state shows little or no reaction with the current PrP-
specific antibodies, the location of PrP^C can be determined in histoblots
not treated with PK or Gdn. Histoblotting has verified the earlier biochemical and immunohistochemical studies of PrP^{Sc}, which showed that
PrP^{Sc} deposition precedes the development of histopathology and that
spongiform degeneration and reactive astrocytic gliosis colocalize precisely with PrP^{Sc}. An example of the precision of the relationship between sites
of PrP^{Sc} deposition and pathologic changes was seen in the cerebral cortex of Syrian hamsters inoculated with two different scrapie strains (Fig.
5) (Hecker et al. 1992). With the Sc237 prion strain, PrP^{Sc} accumulation
tends to remain localized to the deeper layers of the cerebral cortex;
spongiform degeneration and reactive astrocytic gliosis are localized to
the same region. In contrast, with the 139H strain, PrP^{Sc} is distributed to
all layers of the cerebral cortex as are the pathologic changes. Similar
results were obtained by Bruce and her colleagues (Bruce et al. 1989) using
standard immunohistochemical methods.

Subsequently, we examined the temporal pattern of PrP^{Sc} accumulation by measuring its concentration in dissected brain regions throughout
the course of scrapie in Syrian hamsters (Jendroska et al. 1991). Animals
were inoculated intrathalamically with the Sc237 strain of scrapie agent.
Four relevant observations were made: First, PrP^{Sc} accumulates exponentially in many, but not all, brain regions, beginning with the site of inoculation in the thalamus (Fig. 6). Second, the pattern of spread of PrP^{Sc} from
brain region to brain region was stereotypic and suggested that it occurred
largely by slow axonal transport; however, spread to septum also appeared
to occur by diffusion of PrP^{Sc} in the extracellular space of the CNS. The latter form of intracerebral spread has not been identified in other animal
species or with other strains of scrapie agent. Third, spongiform degener-

Sc237

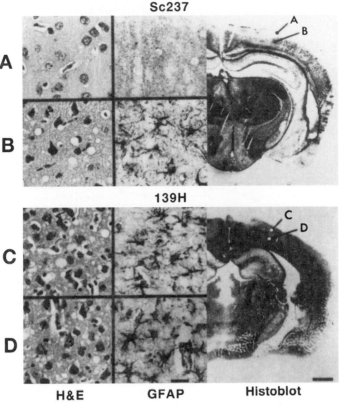

139H

H&E   GFAP   Histoblot

Figure 5. There is a precise colocalization of vacuolar degeneration of the brain parenchyma and reactive astrocytic gliosis with sites of PrP^{Sc} deposition. Syrian hamsters were inoculated intracerebrally with either the Sc237 or 139H strain of scrapie prions. Incubation times (postinoculation interval to clinical signs) were about 65 and 160 days, respectively. Histoblot analysis showed that PrP^{Sc} was more widely distributed and more intensely immunostained with 139H than Sc237. Correlation of the neuropathologic changes with sites of PrP^{Sc} deposition are made for the cerebral cortex. (*A*) With Sc237, there is little or no PrP^{Sc} immunoreactivity in the outer half of the cerebral cortex and (*B*) moderate PrP^{Sc} immunoreactivity in the inner half. No vacuolation (H&E stain) or reactive astrocytic gliosis (GFAP immunostain) is located in the outer cortex, whereas mild to moderate changes are present in the inner cortex. With 139H, there is intense PrP^{Sc} immunoreactivity in both the outer and inner neocortex that is associated with vacuolation and reactive astrocytic gliosis in both locations. Bar on histoblot, 1 mm; bar on histological section, 50 μm. (Reprinted, with permission, from Hecker et al. 1992.)

ation and reactive astrocytic gliosis developed 1–2 weeks after the start of PrP^{Sc} accumulation in a region. Fourth, total brain PrP^{Sc}, based on the sum of PrP^{Sc} levels in each brain region, was directly proportional to the whole-

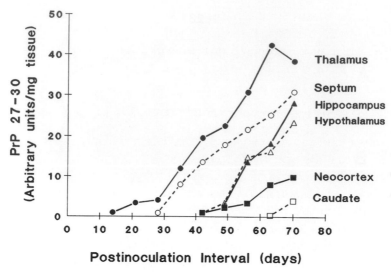

Figure 6. Kinetics of PrP 27-30 accumulation during the course of scrapie in Syrian hamsters inoculated intrathalamically with Sc237 prions. PrP 27-30 is the proteinase K digestion product of PrP^{Sc}. Homogenates of dissected brain regions were treated with proteinase K, and the relative amount of PrP 27-30 was determined by western analysis. Clinical signs of scrapie presented about 65 days postinoculation. (Reprinted, with permission, from Jendroska et al. 1991.)

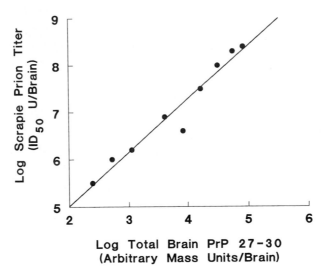

Figure 7. Whole-brain Sc237 strain infectivity titer is directly proportional to the brain concentration of PrP 27-30 throughout the course of scrapie in Syrian hamsters. (Reprinted, with permission, from Jendroska et al. 1991.)

brain scrapie infectivity titer measured throughout the course of scrapie (Fig. 7). From these initial studies, a unifying prion disease hypothesis resulted, which proposed that propagation of prions and the development of neuropathologic changes are both linked to the formation of PrP^{Sc}.

NEUROPATHOLOGIC CHARACTERISTICS OF PRION DISEASES

Prion diseases in animals and humans can be divided into three broad categories based on their neuropathological features: (1) those characterized by vacuolar (spongiform) degeneration of the brain, little or no PrP amyloid plaque formation, and accumulation of protease-resistant PrP; (2) those with abundant PrP amyloid plaque formation, variable amounts of vacuolar degeneration, and abnormal PrP accumulation that is relatively protease-sensitive; and (3) those that are characterized by intense vacuolar degeneration, abundant PrP amyloid plaque formation, and protease-resistant PrP.

The first category includes the vast majority of prion diseases, including scrapie in sheep and rodents; bovine spongiform encephalopathy (BSE); kuru; sporadic, familial, and iatrogenic CJD; and familial and sporadic fatal insomnia (FFI and SFI, respectively). The striking similarity of the neuropathologic changes in these disorders prompted the veterinary neuropathologist, William Hadlow, to propose that scrapie of sheep and kuru of the Fore people of New Guinea were homologous diseases and that kuru could be transmitted to animals by inoculation of brain homogenates into laboratory animals (Hadlow 1959) in the same way that sheep scrapie had been experimentally transmitted to goats 20 years earlier by inoculation of spinal cord homogenates (Cuillé and Chelle 1939). The striking resemblance of the neuropathologic changes in kuru and in CJD was recognized by Klatzo (Klatzo et al. 1959). By 1966 and 1968, respectively, brain homogenates from kuru and CJD patients were successfully transmitted to nonhuman primates (Gajdusek et al. 1966; Gibbs et al. 1968). Successful transmission of familial CJD to nonhuman primates was first reported in 1973 (Roos et al. 1973; Masters et al. 1981b). A mutation of the PRNP gene was genetically linked to FFI in 1992 (Medori et al. 1992a,b): It was successfully transmitted to rodents in 1995 (Tateishi et al. 1995). SFI, which has virtually identical clinical and neuropathological features as FFI, but not PRNP gene mutation, was transmitted to rodents in 1999 (Mastrianni et al. 1999). A summary of the successful transmissions of human neurodegenerative diseases to primates by the National Institutes of Health was published (Brown et al. 1994).

The second neuropathologic category of prion disease appears to be confined largely to humans. The only prion diseases included in this category are the six dominantly inherited syndromes designated GSS. GSS was first transmitted to nonhuman primates in 1981 (Masters et al. 1981a), which was the first evidence that it might be related to CJD. Although PrP^{Sc} is resistant to PK digestion in sporadic, iatrogenic, and familial CJD and in FFI and SFI, mutated PrP in GSS cases is relatively sensitive to protease digestion (Hsiao et al. 1989b, 1991b; Hegde et al. 1998).

In 1996, a new variant of CJD, designated nvCJD, was discovered in the UK that is characterized by abundant PrP amyloid, like GSS, and intense vacuolation and accumulation of protease-resistant PrP^{Sc}, but no PRNP gene mutation, like sporadic and iatrogenic CJD (Will et al. 1996). In GSS syndromes, nonconservative amino acid substitutions in the PrP molecule appear to favor abnormal degradation into highly amyloidogenic PrP peptides (Fig. 4). In contrast, nvCJD is caused by infection with a new strain of prions derived from BSE during the recent epidemic in the UK and mainland Europe (Scott et al. 1999). This new prion strain is highly amyloidogenic. In the context of our current understanding of the propagation of prion strains in a host and of PrP amyloidosis, it appears that the new BSE strain of prions induces a β-sheeted conformation in the host's PrP that results in degradation to highly amyloidogenic PrP peptides.

The first two categories of prion diseases also differ in the potency of their prions. In the National Institutes of Health series, laboratory transmission rates were highest for iatrogenic CJD (100%), less for sporadic CJD (90%), and least for familial prion diseases (68%) (Brown et al. 1994). Of the familial prion diseases, the highest transmission rate occurred with familial CJD(E200K) (85%), whereas GSS(P102L) had a relatively low rate of 38%.

The following clinical and neuropathologic descriptions of human prion diseases have been condensed from other more detailed accounts (see DeArmond and Prusiner 1996, 1998; DeArmond et al. 2002).

Kuru

Kuru is a neurologic disorder confined to the aboriginal Fore people of New Guinea and was one of several CNS diseases discovered in the western Pacific after World War II. The combined anthropologic–epidemiologic–neurologic–neuropathologic investigation of kuru by Australians and Americans and the subsequent laboratory transmission, first of kuru and then of CJD to nonhuman primates, was begun in 1957 when Zigas and Gajdusek began investigations of kuru among the Fore people of the Eastern Highlands of New Guinea (Zigas and Gajdusek 1957). By 1959, the clinical and pathologic features of kuru were published (Gajdusek and Zigas 1959; Klatzo et al. 1959). The physicians who examined victims of kuru noted progressive cerebellar degeneration with ataxia and tremor, and dementia in some cases during the later stages. Investigations of the Fore people by anthropologists suggested that kuru had begun between 1900 and 1920 (Glasse and Lindenbaum 1992) as a case of sporadic CJD. Alpers has convincing epidemiologic evidence that ritualistic cannibalism was the mode of transmission (Alpers 1992). His conclusion was based on three observations: (1) According to the rules of cannibalism among the Fore, women and children consumed all body parts, including brain, whereas the men only consumed skeletal muscle; (2) kuru mainly affected adult women, children, and adolescents of both sexes; and (3) the incidence of kuru among children declined with the end of cannibalism. According to the anthropologists (Glasse and Lindenbaum 1992), little of the corpse was discarded: Even bones were pulverized, cooked, and eaten with green vegetables.

Neuropathology

Except for atrophy of the vermis and flocculonodular lobes of the cerebellum, the brain was grossly normal. Microscopically, the most severe pathologic changes were located in the cerebellum with loss of granule cells, loss of Purkinje cells, fusiform swelling of the proximal portion of many remaining Purkinje cell axons (torpedoes), and intense Bergmann radial gliosis consistent with the clinical features dominated by ataxia. The changes in the cerebral cortex consisted of slight spongiform degeneration in the neuropil between nerve cell bodies in most regions. These lesions were most severe in paramedian cerebral cortical region (cingulate gyrus, subiculum, entorhinal area, parahippocampal gyrus, and superior frontal gyrus), the putamen and caudate nucleus, and the anterior and medial nuclei of the thalamus (Kakulas et al. 1967). Beck and Daniel (1979) were particularly impressed by the severe degeneration in the olivo-ponto-cerebellar system in addition to marked cerebellar vermal atrophy. In the caudate nucleus and putamen, many of the remaining neurons contained intracytoplasmic vacuoles. Finally, amyloid plaques consisting of spherical deposits with radiating spicules at their periphery (spiked-ball plaque) were found in 75% of cases (see Fig. 2). The largest number of plaques was located in the granule cell layer of the cerebellar cortex. They are PAS positive (Fig. 2A) and bind Congo red dye, which shows green birefringence in polarized light, as do all amyloids. Similar amyloid plaques are found in some cases of CJD, GSS, and scrapie, where most neuropathologists continue to refer to them as kuru plaques.

Creutzfeldt-Jakob Disease

CJD is a neurodegenerative syndrome caused by accumulation of pro-tease-resistant PrP^{Sc} in the brain, similar to the accumulation of PrP^{Sc} in scrapie, and is distinguished by a specific set of clinical and neuropatho-logic features. About 10% of CJD cases are caused by one of the several inherited PRNP mutations. In these cases, accumulation of mutated PrP, which is most probably expressed in most if not all nerve cells, causes spongiform degeneration and the spontaneous formation of prions. The great majority of CJD cases, accounting for 85–90% of all human prion diseases, are sporadic with no evidence of an infectious or genetic etiolo-gy. There are reasons to believe that sporadic CJD, in some cases, may be caused by an age-related spontaneous PRNP mutation in a single cell and, in others, by aberrant metabolism of PrP^{C} and its spontaneous conversion to PrP^{Sc} in a "thermodynamically unlucky" victim. Sporadic CJD has a worldwide incidence of about 1 per million per year (Masters et al. 1979; Brown et al. 1987), which may represent the probability that sufficient numbers of PrP molecules spontaneously cross the energy barrier that separates the normal α-helical conformation and the pathogenic β-sheet-ed conformation. Males and females are affected in equal numbers. The peak age of onset is around 60 years with a wide range of up to 90 years; a small number of sporadic CJD cases have been reported in individuals under the age of 20 years (Monreal et al. 1981; Brown 1985; Packer et al. 1980; Berman et al. 1988). Finally, a small proportion of CJD cases are acquired by infection. The vast majority of the latter were transmitted through medical procedures, thus the designation of iatrogenic CJD. Contaminated intracerebral electroencephalogram (EEG) electrodes, human growth hormone (HGH) preparations, human pituitary gonadotropin preparations, dura mater grafts, and corneas have caused iatrogenic CJD (for review, see DeArmond and Prusiner 1996; DeArmond et al. 2002). The epidemiologic evidence indicates that nvCJD is a new form of CJD also acquired by infection; however, the origin of the infec-tious prions is still being debated (see below) (Will et al. 1996). Iatrogenic and nvCJD can often be distinguished from sporadic and familial CJD cases by the age of onset. Therefore, the occurrence of CJD in more than 55 patients ranging in age from 10 to 41 years who had received HGH sug-gested an iatrogenic etiology (Brown 1985; Buchanan et al. 1991; Fradkin et al. 1991; Brown et al. 1992). The early onset of CJD in these patients clearly differentiates them from cases of sporadic CJD and most forms of familial CJD. The risk of CJD among those who have received HGH is estimated to be about 1 per 200, whereas the risk of CJD in the general population under 40 years of age is about 1 per 20 million (Fradkin et al. 1991). All of these patients received injections of HGH every 2–4 days for 4–12 years (Gibbs et al. 1985; Koch et al. 1985; Powell-Jackson et al. 1985; Titner et al. 1986; Croxson et al. 1988; Marzewski et al. 1988; New et al. 1988; Anderson et al. 1990; Billette de Villemeur et al. 1991; Macario et al. 1991; Ellis et al. 1992). Interestingly, most of the patients presented with cerebellar syndromes that progressed over time periods varying from 6 to 18 months. Some patients became demented during the terminal phase of their illnesses. The clinical courses of some patients with dementia occurring late resemble kuru more than ataxic CJD in some respects (Prusiner et al. 1982). Assuming these patients developed CJD from injections of prion-contaminated HGH preparations, the possible incubation periods range from 4 to 30 years (Brown et al. 1992). These incubation times are not surprising, given that some cases of kuru have incubation periods of two to three decades (Gajdusek et al. 1977; Prusiner et al. 1982; Klitzman et al. 1984). Many patients received several common lots of HGH at various times during their prolonged therapies, but no single lot was administered to all the American patients. An aliquot of one lot of HGH has been reported to transmit CNS disease to a squirrel monkey after a prolonged incubation period (Gibbs et al. 1993).

As a general rule, all forms of CJD present with a relatively rapid course leading to death within 4–12 months from start of signs and symptoms; however, cases lasting 2–5 years are not uncommon (Masters and Richardson 1978; Malamud 1979). Familial CJD cases tend to present at an earlier age than sporadic CJD. For example, CJD(E200K) has an onset at 55 ± 8 years of age, which progresses to death in an average of 8 months, and CJD(D178N, V129) has on onset at 46 ± 7 years of age with death occurring in an average of 22 months (Brown et al. 1991). CJD linked to octapeptide inserts presents at 23–35 years of age and is followed by a particularly long progression to death of 4–13 years. Multiple mutations of the PRNP gene have been genetically linked to familial CJD syndromes. The penetrance is close to 100% in familial prion diseases, but it is age dependent. For CJD(E200K), penetrance is as low as 1% for carriers of the mutation at age 40 and close to 100% at age 80 (Rosenmann et al. 1997). Penetrance was found to be related to expression levels of both mutant PrP(E200K)mRNA and mutant PrP(E200K) protein. They were lower than expression levels of the wild-type PrPmRNA and PrP^C in healthy carriers, whereas most of the clinically ill E200K patients expressed equal levels of mutant and wild-type proteins.

Clinical features that suggest the diagnosis of CJD are often preceded by a prodromal period in which nonspecific clinical signs occur, including fatigue, sleep disturbances, memory disturbances, behavioral changes, vertigo, and ataxia. However, the most characteristic clinical features include rapid progression of mental deterioration with dementia, myoclonus, a broad spectrum of motor disturbances (extrapyramidal, cerebellar, pyramidal, and/or anterior horn cell), and an EEG showing periodic short-wave activity. When the quartet of dementia, myoclonus, periodic EEG activity, and rapid progression is seen in a patient, the diagnosis of CJD is relatively certain. In contrast to kuru, in which the clinical and neuropathologic features were relatively uniform, sporadic CJD is characterized by diversity of clinical features and an equally diverse distribution of neuropathology. About 15% of sporadic CJD cases are similar to kuru, with the development of ataxia as an early sign, followed by dementia (Brown et al. 1984). The majority of the patients who developed CJD after intramuscular pituitary growth hormone injections presented with cerebellar syndromes that progressed over 6–18 months. CJD subtypes with visual disturbances and severe occipital cortex pathology have been designated the Heidenhain variant.

Neuropathology

The gross appearance of the brain in CJD is variable and not diagnostic. In some cases, no recognizable abnormalities are seen, whereas others show varying degrees of cerebral cortical, striatal, and/or cerebellar atrophy with brain weights as low as 850 g (Malamud 1979).

Spongiform degeneration and reactive astrocytic gliosis in CJD exhibit a wide spectrum of intensity and distribution (Beck and Daniel 1979). Some investigators divided the neuropathology of CJD into 16 subtypes (Beck et al. 1969), whereas others (Malamud 1979) preferred to subclassify CJD into six clinico-neuropathologic categories: cortical, corticostriatal with or without visual loss, corticostriatocerebellar, corticospinal, and corticonigral. Richardson and Masters (1995) recently reviewed the clinical spectrum of human prion diseases and emphasized the Heidenhain variants with visual symptoms, striatal variants resembling Huntington's disease, thalamic variants including FFI, cerebellar variants that resemble kuru, variants with oculomotor disturbances that can resemble progressive supranuclear palsy, panencephalopathic variants with involvement of white matter, and variants with demyelinative peripheral neuropathy.

The hallmarks of CJD are spongiform degeneration of neurons and their processes, neuronal loss, intense reactive astrocytosis, and amyloid plaque formation; however, these vary considerably from case to case. In some patients, the only histologic clue is delicate vacuolization of the gray matter with minimal or no detectable nerve cell loss or reactive astrocytosis (Fig. 8A). The vacuoles that comprise spongiform degeneration are intracellular and located in the neuropil between nerve cell bodies. They should not be confused with spaces around nerve cell bodies and blood vessels, which are artifacts. The vacuoles tend to be round to oval and vary in diameter from 5 to 25 µm in most cases, but can be as large as 50–60 µm in others (Fig. 8B). Masters and Richardson (1978) designated the late stage of disease "status spongiosis" to distinguish it from the former patterns, which they designated "spongiform degeneration." Status spongiosis is characterized by massive loss of neurons and gray matter neuropil, development of numerous extracellular microcysts as large as 100 µm in the parenchyma, and a dense meshwork of reactive astrocytic processes surrounding the spaces (Fig. 9).

In CJD, spongiform degeneration may be found in the neocortex, subiculum of the hippocampus, putamen, caudate nucleus, thalamus, and molecular layer of the cerebellar cortex. It is usually minimal or absent from the globus pallidus, Ammon's horn, and dentate gyrus of the hippocampus, brain stem, and spinal cord (Masters and Richardson 1978; Beck and Daniel 1979; Malamud 1979); however, we have found cases of CJD with extensive spongiform degeneration of the hippocampus. Spongiform degeneration of the cerebral cortex occurs in virtually all cases, regardless of the clinical presentation. In the cerebral cortex, the amount of vacuolation can vary considerably from region to region even within the same cortical section. It can be diffusely distributed to all cortical layers or may have a pseudolaminar appearance, the latter primarily in the deep cortical layers. Vacuolation of nerve cell bodies has also been reported in some cases of CJD; however, it is more characteristic of natural scrapie in sheep and of kuru.

As shown by electron microscopy, spongiform degeneration consists of focal swelling of neuritic processes, both axonal and dendritic, and synapses with loss of internal organelles and accumulation of lacy abnormal membranes (Lampert et al. 1972; Chou et al. 1980; Beck et al. 1982). The relative disappearance of spongiform degeneration reported by Masters and Richardson (1978) with progressive nerve cell loss is consistent with the hypothesis that vacuolation is mostly confined to nerve cell processes. In status spongiosis, "vacuolation" is primarily extracellular in the form of microcysts.

Reactive astrocytic gliosis in CJD (Figs. 8C and 9B) and in scrapie has been described as more intense than one would expect based on the amount of nerve cell loss (Dormont et al. 1981; Mackenzie 1983).

Nevertheless, Masters and Richardson (1978) found that the degree of reactive astrocytosis correlated very well with the degree of nerve cell loss. PrP 27-30 in the form of prion rods has been found to stimulate astrocyte

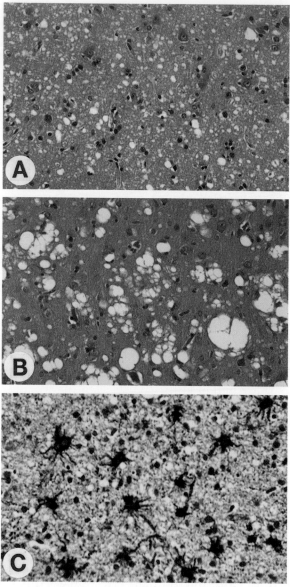

Figure 8. (*See facing page for legend.*)

proliferation in vitro (DeArmond et al. 1992) and may contribute to the very intense astrocytic reaction in prion diseases.

Generally, lesions in the white matter in CJD, as well as in other prion diseases, are secondary to neuronal loss. However, a vacuolar myelopathy is found in some CJD cases, particularly in Japan (Tateishi et al. 1981). Vacuolar myelopathy is a characteristic of prion disease in mice inoculated with Mo(RML) or the Chandler scrapie prion isolates (Carlson et al. 1986; Prusiner et al. 1990) and with human CJD prions (Liberski et al. 1989). Ultrastructurally, the vacuoles are within the myelin sheath, and occasionally intra-axonal.

Kuru-type amyloid plaques are found in only 5–10% of CJD cases, in association with MV genotype at codon 129 in the PrP genes (Parchi et al. 1996). They are immunopositive with PrP antibodies, but are negative with antibodies to β-amyloid (Snow et al. 1989, 1990). The plaques in CJD consist of discrete eosinophilic, periodic-acid Schiff (PAS)-positive spherical masses with radiating amyloid spicules at their periphery (Fig. 2A). In kuru and CJD, they are found most often in the granule cell layer of the cerebellar cortex but can occasionally be seen in other brain regions, including the cerebral cortex, basal ganglia, and thalamus (Klatzo et al. 1959). Most, but not all, patients in whom ataxia is prominent have kuru-type plaques in their cerebellum (Pearlman et al. 1988). These patients exhibit a protracted clinical course that may last up to 3 years. Amyloid angiopathy has also been found in some cases of CJD; however, this vascular amyloid is composed of the Aβ peptide and not of PrP (Tateishi et al. 1992).

It is possible that more cases of CJD with PrP plaques will be found as a result of an improved technique, using hydrolytic autoclaving, for their detection by immunohistochemistry (Muramoto et al. 1992). Prior to immunostaining, glass-mounted histologic sections of brain are immersed in 3 mM HCl (effective range, 1–30 mM) and autoclaved for 10 minutes. This not only results in intense staining of kuru plaques, but it also reveals numerous primitive PrP plaques (Fig. 2C). Primitive PrP

Figure 8. Vacuolar degeneration of the gray matter is characteristic of scrapie and CJD syndromes and less so of GSS syndromes. These vacuoles are in neuronal processes including dendrites, distal axons, and presynaptic boutons. Reactive astrocytic gliosis is variable. (A) The most common form of vacuolation in the cerebral cortex in a case of CJD consists of vacuoles 5–20 μm in diameter between nerve cell bodies (H&E stain). (B) A less common form of vacuolation in a CJD case includes larger vacuoles up to 50–70 μm in diameter (H&E stain). (C) Intense reactive astrocytic gliosis in a different case of CJD (GFAP immunostain). All photomicrographs are at the same magnification.

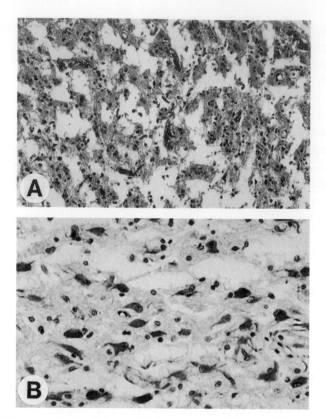

Figure 9. Status spongiosis of the cerebral cortex in a CJD case follows severe loss of neurons. It consists of microcysts in the extracellular space that are surrounded by reactive astrocytes and their processes. (*A*) Low-magnification view of the cerebral cortex showing microcysts (H&E). (*B*) Higher-magnification view of the large reactive astrocytes in the walls of the microcysts in the same case (GFAP immunohistochemistry with hematoxylin counterstain). The magnification in *A* is half that in *B*.

plaques stain poorly by the PAS reaction and do not generally show green birefringence with Congo red dye.

New Variant CJD

There is a substantial degree of concern in the UK that some humans have contracted a prion disease as the result of ingesting meats infected with BSE. There are three main reasons for suspecting an infectious etiology and BSE in particular: (1) nvCJD appears to be a new form of prion disease with unique clinical and neuropathological features and does not resemble cases of sporadic or iatrogenic CJD found in young or old patients. (2) The majority of the victims were from the UK and a couple from France; therefore, they were all geographically in the same region as the BSE epidemic. (3) Known incubation periods for other forms of CJD acquired by infection are well within a range that would link nvCJD with BSE. To date (April 1999), there have been 40 cases of nvCJD identified (22 female, 18 male) with a mean age of death of 30 years (range 18–53 years) and median duration of illness of 13 months (range 7–38 months). All cases that have been tested are methionine homozygotes at codon 129 in the PRNP. Sporadic CJD cases are extremely rare under the age of 40 years and have a mean age of presentation of around 63 years, with a duration of illness of less than 6 months in the majority of cases (Will et al. 1996; Chapter 12). None of the patients were related or had known pathogenic mutations of the PRNP gene. The clinical presentations were atypical for sporadic CJD, since the disease began with sensory or psychiatric disturbances rather than dementia or motor abnormalities. In addition, the neuropathologic features in all of the cases were identical. Moreover, they were markedly different from other cases of sporadic, iatrogenic, or familial CJD in elderly or young patients (Monreal et al. 1981; Koch et al. 1985; Kulczycki et al. 1991; Kitamoto et al. 1992; Billette de Villemeur et al. 1994).

James Ironside and Jeanne Bell, the neuropathologists in the UK National CJD Surveillance Unit who reviewed all of the nvCJD cases (Ironside 1996; Will et al. 1996), emphasize that the most remarkable feature of nvCJD is the massive amount of protease-resistant PrP in the form of innumerable mature, PAS-reactive, PrP amyloid plaques, many with the features of kuru plaques, and the massive number of PrP primitive plaque-like deposits that are PAS-nonreactive (Fig. 10). The number of PrP plaque deposits rivals that seen in the many GSS cases; but, it should be emphasized, this disease is classified as CJD because it is not dominantly inherited, because of the intense spongiform degeneration, and because of the presence of protease-resistant PrP. Many of the mature PrP amyloid plaques are distinctive because they are located at the center of vacuoles, particularly in the cerebral cortex (Fig. 10A). These have been designated "florid plaques." Another distinctive feature is the clustering of mature and immature plaques, particularly well seen by PrP immunohistochemistry using the hydrolytic autoclaving technique (Fig. 10B). There were multiple less well-defined protease-resistant PrP deposits scattered throughout the gray matter. Some of the latter surround nerve cell bodies and proximal portions of neuritic processes (Fig. 10B). Although all regions of the cerebral cortex contain PrP plaques, the greatest concentra-

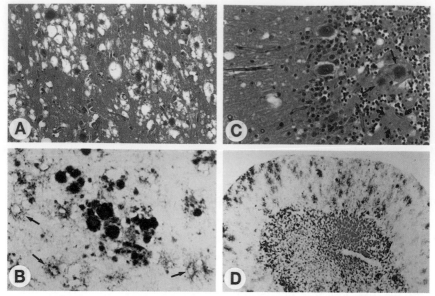

Figure 10. Main neuropathologic features of nvCJD. (*A*) Numerous 15–30-µm florid amyloid plaques and severe vacuolar degeneration in the cerebral cortex (PAS stain, 20x objective). (*B*) Some of the florid plaques in the cerebral cortex are arranged in clusters; in addition, some neurons and the proximal portions of their processes (*arrows*) are encrusted with PrP^Sc (PrP immunohistochemistry, hydrolytic autoclaving method, 40x objective). (*C*) Kuru plaques (*straight arrows*) and primitive plaques (*curved arrows*) in the granule cell layer and vacuolation of the molecular layer of the cerebellum (H&E, 40x objective). (*D*) Abundant PrP^Sc deposition in the granule cell and molecular layer of the cerebellar cortex (PrP immunohistochemistry, hydrolytic autoclaving method, 10x objective).

tions appear to occur in the occipital lobe and cerebellum. In the cerebellum, typical kuru-type plaques can be found in the granule cell layer (Fig. 10C) and occasionally in the molecular layer of the cerebellar cortex. However, the number of kuru plaques seen in H&E and PAS stains is relatively small compared to the uniquely massive deposits of PrP^Sc in the molecular layer and granule cell layer (Fig. 10D). Many of these are PAS-nonreactive and do not have the classic appearance of amyloid, particularly in the molecular layer. The amount of PrP plaque deposition in nvCJD is far more extensive than in typical sporadic CJD or HGH iatrogenic CJD (Ironside 1996). Furthermore, the nvCJD plaques do not resemble plaques found in GSS.

The most intense spongiform degeneration in nvCJD occurs in the basal ganglia and thalamus, with particularly marked astrocytic gliosis in the latter. The massive accumulation of protease-resistant PrP in the cerebellum is associated with severe spongiform degeneration, neuronal loss (Fig. 10C), and reactive astrocytic gliosis.

Another major difference between nvCJD and sporadic CJD is the presence of disease-associated PrP outside the central nervous system. This has been detected both by immunoctyochemistry and by western blotting in lymphoid tissues including the tonsils, lymph nodes, spleen, and appendix (Hilton et al. 1998; Hill et al. 1999). In one case, PrP immunoreactivity was detected in follicular dendritic cells in the appendix, which had been removed 8 months before the onset of neurological disease (Hilton et al. 1998). This observation has led to concerns that lymphocytes may be involved in the transport of disease-associated PrP, which in turn has led to concerns regarding the potential safety of blood and blood product donations from residents of the UK and mainland Europe and from visitors who have spent a significant amount of time in those countries.

The hypothesis that nvCJD is causally related to BSE has received support from observations of experimental BSE transmission in macaque monkeys (Lasmézas et al. 1996), which showed neuropathological changes similar to those in nvCJD (Fig. 11). Subsequent strain typing studies (Bruce et al. 1997) also argued that the BSE and nvCJD agents are indistinguishable on primary transmission into syngeneic mice. More recently, BSE and nvCJD brain extracts resulted in identical incubation times, neuropathological features, and pathogenic PrP isoforms in transgenic mice expressing bovine PrP^C in the absence of mouse PrP^C, Tg(BoPrP)$Prnp^{0/0}$ mice (Hill et al. 1997; Scott et al. 1999). Both BSE and nvCJD were associated with formation of subcallosal PrP amyloid plaques and virtually identical vacuolation distribution profiles with the most intense vacuolation confined to the brain stem and habenula. In contrast, brain extracts from scrapie-infected sheep were associated with low to moderate levels of vacuolation throughout the brain of Tg(BoPrP)$Prnp^{0/0}$ mice and no amyloid plaque formation. These mice were resistant to infection with sporadic CJD. Finding that the disease phenotype in Tg(BoPrP)$Prnp^{0/0}$ mice inoculated with nvCJD prions is identical to that caused by BSE prions is the most compelling evidence that prions from BSE-infected cattle have infected humans.

At present, it is difficult to predict the numbers of future nvCJD cases. One of the major determinants of the size of the nvCJD epidemic is the incubation period, which at present is uncertain; therefore, it is not possible to predict with confidence future disease trends (Cousens et al. 1997). However, based on the magnitude of increasing numbers of nvCJD cases since 1995, it has been calculated that as many as 100,000–200,000 cases

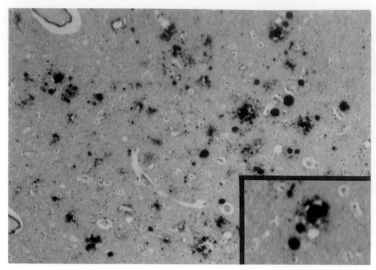

Figure 11. Cerebral cortex of a macaque monkey following inoculation with BSE prions contains numerous protease-resistant PrP plaques that vary in diameter from 5 to 25 μm. *Inset:* Several plaques are within or surrounded by vacuoles, similar to the florid plaques in nvCJD. Magnification of inset is twice that of the main photomicrograph. (This figure was adapted from a 35-mm transparency generously provided by Dr. James Ironside, Neuropathology Laboratory, CJD Surveillance Unit, University of Edinburgh, Edinburgh, United Kingdom.)

of nvCJD may occur if incubation periods are ≥60 years, or less than 700 cases if incubation periods are ≤20 years (Ghani et al. 2000).

Familial Fatal Insomnia (D178N, M129)

This dominantly inherited form of prion disease is an example from nature of the relationship between the amino acid sequence of PrP and selective targeting of neurons for degeneration. Specifically, in FFI, a D178N mutation occurs on an allele that encodes a methionine at position 129, and the major pathology is limited to the thalamus. In contrast, in familial CJD(D178N, V129) in which a valine is encoded at 129 on the mutated allele, the pathologic changes are widespread, but do not include the thalamus (Goldfarb et al. 1992). FFI was recently transmitted to rodents (Tateishi et al. 1995).

The clinical and neuropathologic data have been obtained from studies of two large kindreds with FFI (Manetto et al. 1992; Medori et al. 1992a,b). The age of onset is between 35 and 61 years. FFI progresses relatively rapidly over a period of 7–36 months. The primary sleep/wake disturbances are progressive insomnia followed by complex hallucinations and then stupor and coma. In addition, there are autonomic disturbances including hyperhidrosis, pyrexia, tachycardia, hypertension, and irregular breathing. The principal motor findings are ataxia, spontaneous and evoked myoclonus, dysarthria, and pyramidal signs. In contrast, patients with familial CJD(D178N, V129) present with dementia but not insomnia.

Neuropathology

Pierluigi Gambetti at Case Western Reserve University has made all of the original neuropathologic observations of FFI and reviewed the clinical, neuropathologic, and molecular differences between FFI and CJD(D178N, V129) (Gambetti et al. 1995). In FFI, the most severe neuropathology occurs in the anterior ventral and medio-dorsal nuclei of the thalamus, where there is over 50% loss of neurons with reactive astrocytosis. The cerebral cortex is little affected, with minimal to mild patchy astrocytosis. Mild spongiosis of the cerebral cortex was largely confined to layers 2–4 in one case. In the cerebellum, there is swelling of the proximal axon (torpedoes) of many Purkinje cells and mild loss of Purkinje cells and granule cells. There are no amyloid plaques. There is a greater than 50% loss of neurons from the inferior olives.

FFI patients who are homozygous at codon 129 for methionine have a shorter duration of disease (7–18 months) in contrast to heterozygotes (20–35 months). There is thalamic and olivary atrophy in both. However, there is more widespread spongiosis of the cerebral cortex in homozygotes, whereas spongiosis and gliosis are occasional and focal in the entorhinal cortex and absent from the neocortex in heterozygotes.

Protease-resistant PrP^{Sc} is detected in both FFI and CJD(D178N, V129). In FFI of short duration, significant amounts of PrP^{Sc} are detected in many brain regions that lack pathologic changes, consistent with the hypothesis that abnormal PrP accumulation precedes and causes neuronal dysfunction and neuropathology. There appears to be a greater accumulation of PrP^{Sc} in the brain stem of homozygous patients compared to heterozygous, and this may explain in part the shorter time to death in the former.

Gerstmann-Sträussler-Scheinker Syndromes

GSS is one of the rarest neurodegenerative diseases known, with an incidence of 2–5 individuals per 100 million (Hsiao and Prusiner 1991). The diagnosis of GSS requires dominant inheritance, a mixture of cognitive and motor disturbances, and widespread deposition of PrP amyloid plaques. Like CJD, GSS is not a single disease. There are six dominantly inherited disorders to which this name has been given, each with a different clinical presentation, different neuropathologic features, and linkage to a different mutation of the PRNP gene. A codon-102 mutation resulting in a leucine-for-proline substitution (Hsiao et al. 1989b) is linked to the original GSS family described by Gerstmann, Sträussler, and Scheinker, GSS(P102L) (Kretzschmar et al. 1991). Of the more than 20 known pathogenic PRNP mutations causing familial prion diseases, 5 mutations, in addition to the P102L, are associated with deposition of numerous PrP amyloid plaques: GSS(P105L) (Nakazato et al. 1991; Kitamoto et al. 1993b; Yamada et al. 1993), GSS(A117V) (Doh-ura et al. 1989; Nochlin et al. 1989; Mastrianni et al. 1995), GSS(Y145Stop), GSS(F198S) (Ghetti et al. 1989, 1995; Hsiao et al. 1992), and GSS(Q217R) (Hsiao et al. 1992). Three generalities can be made about GSS syndromes based on combined neurohistological, immunohistochemic, biochemical, and molecular genetic analysis (Ghetti et al. 1996a). (1) The clinical features associated with each mutation are different. (2) Each mutation is associated with different morphologic subtypes of PrP plaques. (3) Some mutations are associated with significant neurofibrillary tangle degeneration of neurons and neuritic plaque formation similar to Alzheimer's disease.

GSS(P102L)

This primarily ataxic form of GSS was the first to be described by the neurologists Gerstmann and Sträussler and the neuropathologist Scheinker in 1936 (Gerstmann et al. 1936). Seitelberger (1981b) defined it as "spinocerebellar ataxia with dementia and plaque-like deposits." The P102L mutation has been found in about 32 families in Austria, the UK, Canada, the US, France, Germany, Italy, Israel, and Japan (Doh-ura et al. 1989; Goldgaber et al. 1989; Hsiao et al. 1989a; Kretzschmar et al. 1992; Goldhammer et al. 1993). Three out of 7 GSS cases have been transmitted to nonhuman primates (Masters et al. 1981a).

Neuropathology. The pathologic hallmark of ataxic GSS is the multicentric amyloid plaque (GSS plaque). These are most numerous in the molecular layer of the cerebellar cortex but are also found in large numbers in the cerebral cortex (Fig. 3A). The GSS plaque consists of a larger central mass of amyloid surrounded by smaller satellite deposits. They bind Congo red dye, which displays green birefringence in polarized light (Fig. 3B). The molecular layer of the cerebellum in some cases contains numerous "amorphous" or "primitive" plaques that can be 150–200 μm in diameter (Fig. 3C) (Kuzuhara et al. 1983). The latter do not fulfill the criteria for mature amyloid, since they are weakly PAS-positive and rarely show green birefringence with the Congo red stain. Unicentric kuru-type plaques can also be found. The GSS plaques, kuru plaques, and primitive plaques immunostain specifically with PrP antibodies (Kitamoto et al. 1986; Roberts et al. 1986; Snow et al. 1989; DeArmond and Prusiner 1995); the reaction can be augmented by the HCl-autoclave technique (Fig. 3C) (Muramoto et al. 1992). White matter degeneration resembling that of other system degenerations, such as Friedreich's ataxia, is a prominent feature in most cases (Seitelberger 1981a). Neuronal loss is scattered throughout the brain and spinal cord. Spongiform degeneration is variable in degree and extent and can be difficult to detect. Neurofibrillary tangles, if present, are usually found in numbers and in locations consistent with the patient's age.

Tg(GSS-P101L) Mice. Tg mouse lines expressing a mutant mouse PrP that carries the codon-102 mutation linked to GSS in humans, designated Tg(MoPrP-P101L), have verified that this PrP amino acid sequence is amyloidogenic and causes a GSS syndrome (Hsiao et al. 1990). Codon 101 in mice is analogous to codon 102 in humans. Tg mouse lines expressing high (H) levels of the transgene product, Tg(MoPrP-P101L)-H mice, developed a spontaneous neurodegenerative disorder with formation of numerous PrP immunopositive amyloid plaques in multiple brain regions similar to those found in human GSS (Fig. 12). Tg mice expressing low levels of MoPrP-P101L, Tg(MoPrP-P101L)-L mice, did not develop a neurodegenerative disease spontaneously. Disease in high expressors of the muPrP resulted in the formation of prions (Hsiao et al. 1994). Brain homogenates from spontaneously ill Tg(MoPrP-P101L)-H mice transmitted to 7% of Syrian hamsters into which it was inoculated and to 40% of Tg(MoPrP-P101L)-L mice. Low rates of transmission of GSS to nonhuman primates were also characteristic of human GSS: In the National Institutes of Health series (Brown et al. 1994), the rate of transmission to primates was highest (85%) for CJD(E200K), 70% for CJD(D178N, V129), and relatively unsuccessful (38%) for GSS(P102L). In a subsequent study of transmission of mutant prions derived from Tg(MoPrP-P101L)-H mice, it was discovered that success rate and incubation time were influenced greatly by wild-type MoPrPC (Telling et al. 1996a). Thus, when low levels of MoPrP-P101L were expressed in combination with wild-type MoPrP^C incubation times with mutant prions were about 350 days. In contrast, when MoPrP-P101L was expressed by itself, in the absence of MoPrP^C, all of the Tg mice expressing low levels of MoPrP-P101L became ill at about 200 days. These

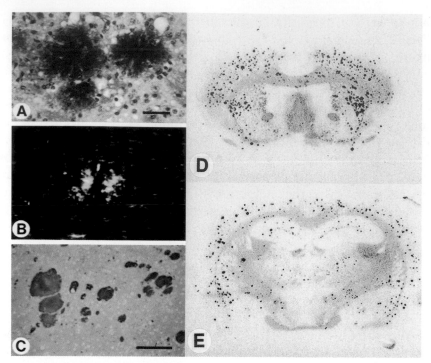

Figure 12. Numerous GSS-type amyloid plaques develop spontaneously throughout the brain in Tg(GSS-P101L) mice. The majority of the plaques are PAS-positive (*A*) and bind Congo red dye that displays green-gold birefringence in polarized light (*B*) and, therefore, fulfill the criteria for amyloid. (*C*) The plaques are PrP immunoposi-tive by the hydrolytic autoclaving method. The histoblot method shows numerous PrP amyloid plaques in both gray and white matter: level of the head of the caudate nucleus (*D*) and level of the hippocampus and thalamus (*E*). Bars, *A*, 50 μm; *C*, 100 μm. (Adapted from Hsiao et al. 1994.)

results argue that MoPrP^C can have an inhibitory effect on the interaction of an infecting prion with a transgene, presumably by competition of MoPrP^C and MoPrP-P101L for protein X. In these studies, successful transmission of MoPrP-P101L prions was 100%, possibly because the Tg mice were constructed differently (modified ILn/J mouse cosmid in this study versus cosSHa.Tet vector in the earlier study).

GSS (A117V)

Originally classified as CJD, it was later reclassified GSS because of the number and location of PrP immunopositive amyloid plaques (Nochlin et al. 1989). The A117V mutation was first described in a French family (Doh-ura et al. 1989) and then in an American family of German descent

(Hsiao et al. 1991a). The main clinical features include progressive dementia usually associated with dysarthria, rigidity, tremor, and hyper-reflexia. Masked facies in several individuals plus tremor and rigidity suggest parkinsonism. Ataxia and myoclonus are uncommon.

Neuropathology. The most striking neuropathologic feature is widespread amyloid plaque formation, of which four morphological types have been described (Nochlin et al. 1989). First, there are multiple multicentric GSS-type plaques consisting of 4 to more than 10 amyloid masses. The entire cluster of amyloid ranges from 150 to 500 μm in diameter. These are located in the neocortex, hippocampus, caudate nucleus, and putamen. Some are also present in the subcortical white matter. Second, there are typical kuru-type amyloid plaques. These are 20–70 μm in diameter and located principally in the white matter. Third, there are 50- to 150-μm unicentric amyloid plaques without radiating spicules at their periphery that resemble the cores of the neuritic plaques of Alzheimer's disease; however, they do not have a halo of dystrophic neurites surrounding them. These are found in the neocortex, hippocampus, caudate nucleus, and putamen. Fourth, there are "amorphous" or "primitive" plaques that are as large as 250 μm in greatest dimension, similar to those seen in some ataxic GSS cases, which are located in the deep cerebral cortical layers and in the molecular layer of the cerebellar cortex (Fig. 13) (Tranchant et al. 1992; Mastrianni et al. 1995). All of the plaques immunostain specifically with PrP antibodies but not with Aβ antibodies. Spongiform degeneration is not found in cortical or subcortical gray matter. Astrocytosis is primarily associated with amyloid plaques in the neocortex. The caudate nucleus, putamen, globus pallidus, and thalamus show severe neuronal loss with astrocytosis. There is mild neuronal loss in the substantia nigra.

GSS(F198S) with Neurofibrillary Tangles

The clinical features of this familial prion disorder are ataxia, parkinsonism, and dementia (Farlow et al. 1989). Individuals carrying the mutation are normal until their mid-30s to early 60s. The presenting symptoms are gradual loss of short-term memory and progressive clumsiness, which may be exaggerated when the individual is under stress or tired. The symptoms can progress rapidly over a period of a year or slowly over 5 years. Rigidity and bradykinesia generally occur late in the disease, at which time dementia also worsens. There is little or no tremor. The rigidity and bradykinesia improves with L-dopa analogs. Without treatment of the parkinsonism, the patient dies within 1 year, and with treatment, suc-

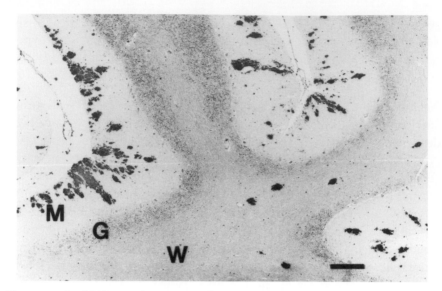

Figure 13. In GSS(A117V), numerous large primitive PrP plaques, 150–200 μm in longest dimension, deposit in the molecular layer of the cerebellar cortex. PrP deposits can also be seen in the white matter. PrP immunohistochemistry by the hydrolytic autoclaving technique with hematoxylin counterstain. (W) White matter; (M) molecular layer; (G) granule cell layer. Bar, 200 μm.

cumbs to intercurrent illness in 2–3 years. The onset of disease is about 10 years earlier in patients who are homozygous for the valine polymorphism at codon 129 compared to patients with Val/Met heterozygosity (Dlouhy et al. 1992).

Neuropathology. Bernardino Ghetti has fully described the neuro-histopathologic features of this unique prion disorder (Ghetti et al. 1989). The neuropathologic changes in GSS(F198S) are remarkable because there are numerous neuritic plaques and intraneuronal neurofibrillary tangles of the type characteristic of Alzheimer's disease. Like Alzheimer's disease, the neuritic plaque consists of an amyloid core surrounded by τ-protein-immunopositive dystrophic neurites; however, the amyloid core in GSS(F198S) is PrP immunopositive and Aβ peptide negative (Fig. 14). Like other PrP amyloids, it is composed of highly truncated PrP peptides (Fig. 4). As with other forms of GSS, large numbers of multicentric GSS-type and unicentric plaques are located throughout the cerebral and cerebellar cortex. In older patients, Aβ immunoreactivity could be found at the periphery of the PrP amyloid deposits (Bugiani et al. 1993). Nerve cell loss occurs in almost all cerebral cortical regions, but is most severe in the

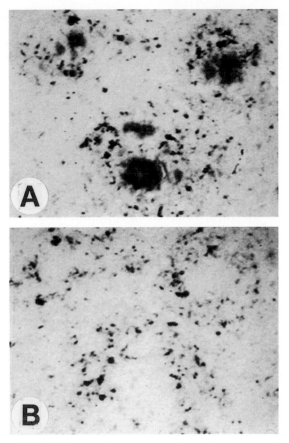

Figure 14. In GSS(F198S), neuritic plaques form that are similar to those in Alzheimer's disease except that the amyloid cores contain PrP peptides and do not contain the Aβ peptide. (*A*) Double immunoperoxidase immunohistochemistry shows that the amyloid cores of neuritic plaques contain PrP (*brown* reaction product) and the dystrophic neurites contain τ-protein (*blue* reaction product). (*B*) The cores of neuritic plaques do not immunostain for the Aβ peptide by double immunostaining. (This figure was adapted from a 35-mm transparency generously provided by Dr. Bernardino Ghetti, Indiana University, Indianapolis, and was previously published [by permission of Oxford University Press] in DeArmond et al. 2002.)

cerebellar cortex, where there is marked loss of Purkinje cells. Neuronal loss is also prominent in many subcortical nuclei, including the substantia nigra, red nucleus, inferior olive, and dentate nucleus. Spongiform degeneration is minimal or focal.

Other GSS Syndromes with Neurofibrillary Tangles

Neurofibrillary degeneration significantly more severe than that consistent with aging occurs with other PRNP mutations causing GSS-like cerebral amyloidosis. Large numbers of neurofibrillary tangles are found in cortical and subcortical regions in patients with the Q217R (Hsiao et al. 1992) and the Y145Stop (Kitamoto et al. 1993a) mutations in addition to the F198S mutation. Smaller numbers are found in some patients with the P105L (Kitamoto et al. 1993b; Yamada et al. 1993) and A117V mutations (Mohr et al. 1994; Ghetti et al. 1995). In Q217R patients, Aβ immunoreactivity was also found in the periphery of PrP amyloid plaques (Ikeda et al. 1992). Numerous neurofibrillary tangles and neuropil threads are found in GSS (Y145Stop); however, no Aβ immunoreactivity was associated with the PrP amyloid plaques. The neurofibrillary tangles in GSS syndromes are composed of phosphorylated τ-protein as they are in Alzheimer's disease (Ghetti et al. 1996a). The mechanisms of neuritic plaque formation and neurofibrillary degeneration of neurons in GSS and Alzheimer's disease are unknown. However, because a single protein abnormality is the cause of GSS, whereas multiple protein abnormalities cause Alzheimer's disease, it is possible that an understanding of the mechanisms of neuritic plaque and neurofibrillary tangle formation will emerge first from investigations of GSS.

GSS(Y145Stop): GSS with Prominent Vascular Amyloid

This mutation at codon 145 results in the formation of a TAG stop codon and the synthesis of a highly truncated form of mutant PrP that has a methionine at residue 129 (Kitamoto et al. 1993a; Ghetti et al. 1996b). A Japanese patient with this form of GSS presented with memory disturbance and disorientation at age 38 that progressed over a 21-year period to severe dementia.

Neuropathology. The most striking characteristic of this disorder is the predominance of 10- to 20-μm spherical PrP amyloid deposits that coat small and medium-sized vessels in the brain parenchyma and in the leptomeninges. Electron microscopy of these plaques shows amyloid fibrils radiating away from the vessel wall (Ghetti et al. 1996a). There was no spongiform degeneration, probably due to the severe neuronal loss with accompanying severe astrogliosis, cerebral atrophy, and hydrocephalus ex vacuo. Some remaining neurons contained τ-protein-immunopositive neurofibrillary tangles.

DIAGNOSTIC PROCEDURES FOR PRION DISEASES

Consensus criteria for the diagnosis of prion diseases have been established (Kretzschmar et al. 1996). A diagnosis of "probable CJD" can be made by biopsy or at autopsy if the quartet of clinical signs and symptoms is present and spongiform degeneration of gray matter is found in the absence of other confounding pathology. In the National Institutes of Health series, 52 of 55 autopsy-verified cases of CJD (95%) showed the characteristic spongiform degeneration of cerebral cortex in surgical biopsy specimens (Brown et al. 1994). The "definitive diagnosis" of a human prion disease requires that one of the four following additional criteria be met: presence of PrP amyloid plaques, transmission of spongiform encephalopathy to animals, presence of PrP^{Sc}, or presence of a pathogenic PrP gene mutation.

PrP Amyloid Plaques

Finding kuru-type "spiked ball" amyloid plaques in the cerebellum is diagnostic of CJD (Fig. 2); however, they are found in only 5–10% of CJD cases. Numerous PrP-immunoreactive amyloid and/or preamyloid plaques in dominantly inherited neurodegenerative diseases are diagnostic of a GSS syndrome (Figs. 3, 10, 14).

Presence of PrP^{Sc}

Proteinase-resistant PrP^{Sc} is only found in animal prion diseases, and PrP^{Sc} is only found in human prion diseases. PrP^{Sc} has been detected by western transfer (Bockman et al. 1985; Brown et al. 1986), dot-blot analysis of homogenates of cerebral cortex from CJD and GSS brains (Serban et al. 1990), peroxidase immunohistochemistry with PrP-specific antibodies on tissues pretreated by the hydrolytic autoclaving technique (Muramoto et al. 1992), and the histoblot procedure (Taraboulos et al. 1992b).

In scrapie, PrP^{Sc} may be found in both gray and white matter; in contrast, in CJD, PrP^{Sc} is confined to the gray matter. PrP^{Sc} is distributed uniformly in all layers of the neocortex in most cases, but it may be confined to deeper cortical layers, particularly in ataxic forms of CJD. In the terminal stages of CJD, PrP^{Sc} is found in most brain regions including the cerebral cortex, basal ganglia, thalamus, brain stem, cerebellum, deep cerebellar nuclei, and spinal cord (Fig. 15). Non-amyloid protease-resistant PrP is not uniformly distributed in the gray matter neuropil in GSS cases as it is in CJD; rather, in GSS, when it occurs, it often deposits focally in primitive plaque-like masses (Fig. 13).

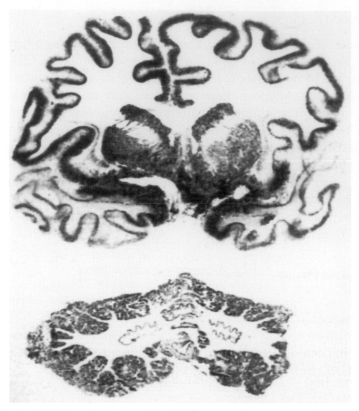

Figure 15. In CJD, histoblot analysis for PrP^{Sc} shows that it accumulates in multiple brain regions. Depicted is accumulation in the cerebral cortex, basal ganglia, and cerebellum. White matter is negative.

Although histoblots reveal the regional distribution of PrP^{Sc}, the cellular compartments in which they are located are not well defined. In retrospect, our earliest modified PrP immunohistochemical techniques accurately revealed the cellular distribution of PrP^{Sc} in scrapie-infected brain tissue (Fig. 16). PrP^{Sc} was localized to presynaptic bouton-like structures in and around nerve cell bodies. It was also located within some nerve cell bodies where it appeared as granular deposits, suggesting localization to endosomes and/or lysosomes. Today, we routinely use the hydrolytic autoclaving technique to localize PrP^{Sc} in animal and PrP^{Sc} in human aldehyde-fixed, paraffin-embedded brain sections (Muramoto et al. 1992). In human brain sections, immunostaining for PrP^{Sc} is confined to the gray matter as it is in histoblots (Fig. 17A). In our experience, the intensity of immunostaining varies considerably in formalin-fixed tissue sections even with hydrolytic autoclaving, whereas the staining in histo-

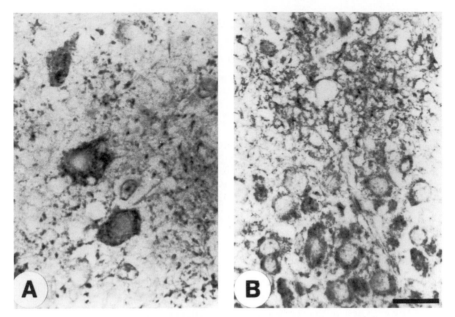

Figure 16. PrP^{Sc} is localized to synapse-like structures in the neuropil adjacent to nerve cell bodies and appears as granular deposits within some nerve cell bodies in Syrian hamsters inoculated with the Sc237 strain of prions. (*A*) Red nucleus. (*B*) CA3 region of the hippocampus. McLean's periodate-lysine-paraformaldehyde fixation and glycomethacrylate embedding. Bar, 20 μm. (Reprinted, with permission, from DeArmond et al. 1987.)

blots is more reproducible. Hydrolytic autoclaving reveals two immuno-staining patterns in the cerebral cortex: first, a diffuse or synaptic pattern that consists of granular immunostaining of the gray matter resembling that seen with synaptophysin antibodies (Fig. 17B) and, second, a coarse or primitive plaque-like pattern that is often particularly intense around vacuoles in the neuropil (Fig. 17C). Direct evidence that PrP^{Sc} accumulates in synaptic boutons is the report that PrP^{Sc} is identified by biochemical analysis of purified synaptosomes from CJD cases (Kitamoto et al. 1992).

Transmission to Animals

In the past, the definitive diagnosis of sporadic, iatrogenic, or familial CJD required transmission to nonhuman primates. The incubation times for transmission of human prion diseases to primates ranges from 17 months for chimpanzees to 64 months for rhesus monkeys (Brown et al. 1994). Tg(MHu2M) mice, which express a chimeric Mo/HuPrP^C, are a more

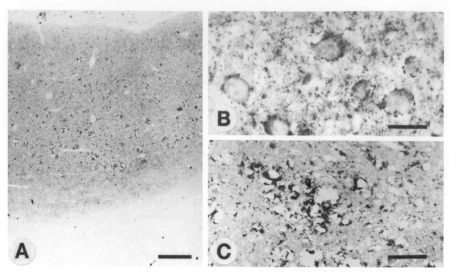

Figure 17. CJD: PrP immunohistochemistry of formalin-fixed, paraffin-embedded tissue sections by hydrolytic autoclaving technique. (*A*) PrP^{Sc} is deposited throughout the entire thickness of the cerebral cortex. (*B*) Higher magnification shows the characteristic synaptic pattern of PrP^{Sc} deposition at the periphery of nerve cell bodies and in the neuropil between nerve cell bodies. (*C*) Coarse masses of PrP^{Sc} surround many vacuoles in some CJD cases, but not all. Bars, *A*, 150 μm; *B*, 25 μm; *C*, 50 μm. (Adapted from DeArmond and Prusiner 1995.)

practical, efficient, and less expensive bioassay for this purpose, since CJD incubation times have been about 200 days (6–7 months) (Telling et al. 1994). In addition to their diagnostic usefulness, results being obtained from these mice are increasing our understanding of human prion strains and the prion disease phenotype.

Different reproducible patterns of PrP^{Sc} accumulation have been found in clinically ill Tg(MHu2M)/*Prnp*$^{0/0}$ mice following intracerebral inoculation with CJD prions from familial CJD(E200K) and FFI(D178N, *cis*-129M) (Telling et al. 1996b). The most intense PrP^{Sc} signal in mice-inoculated FFI prions was localized to thalamic nuclei with smaller amounts deposited in the thalamocortical pathway termination regions of the deeper layers of the frontal cortex (Fig. 18A). There was little or no PrP^{Sc} deposition in the brain stem (not shown). Remarkably, the thalamus is the site of the most severe neuropathologic changes in patients with FFI(D178N, *cis*-129M). In contrast, PrP^{Sc} accumulation in mice inoculated with familial CJD(E200K) prions was deposited in many brain regions including the thalamus, the hypothalamus, the entire thickness of the cerebral cortex, the striatum, and spinal cord, similar to the wide distribution of neuropathologic changes in the human disease counterpart

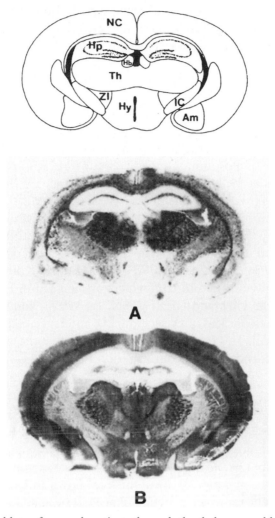

Figure 18. Histoblots of coronal sections through the thalamus and hippocampus of Tg(MHu2M) mouse brains inoculated with brain homogenates from patients with either (*A*) fatal familial insomnia or (*B*) familial CJD(E200K) show markedly different PrP^{Sc} accumulation patterns.

(Fig. 18B). The PrP^{Sc} accumulation pattern following inoculation with sporadic CJD(M/M129) was intermediate with more regions involved than with FFI(D178N, *cis*-129M), but less than with familial CJD(E200K) (not shown). These data support the hypotheses that there are multiple strains of human prions, as there are for scrapie, since their properties remain relatively unchanged following passage from human to the Tg(MHu2M)/*Prnp*$^{0/0}$ mouse. During subsequent passages of prions within Tg(MHu2M)/$Prnp^{0/0}$ mice, the PrP^{Sc} distribution phenotypes of the human strains continued to remain unchanged and different, providing additional support for the existence of multiple human CJD strains (Mastrianni et al. 1999, 2001). Furthermore, the characteristics of the PrP^{Sc}s in human brains and of the respective PrP^{Sc}s from Tg(MHu2M)/$Prnp^{0/0}$ brains remained unchanged after passaging from humans to Tg(MHu2M)/$Prnp^{0/0}$ mice and on subsequent serial passages in Tg(MHu2M)/$Prnp^{0/0}$ mice. Thus, proteinase-K-resistant PrP from multiple sporadic and familial CJD cases, including familial CJD(D178N, cis-129V), migrate as PrP-immunopositive bands at 29, 27, and 21 kD and from FFI cases at 29, 26, and 19 kD (Medori et al. 1992b; Monari et al. 1994). N-deglycosylation of the proteinase-K-digested samples yields single PrP-immunoreactive bands at 19 kD for FFI(D178N, cis-129M) and 21 kD for sporadic CJD and familial CJD(D178N, cis-129V). The respective PrP^{Sc}s had the identical properties after serial transmission in Tg(MHu2M)/$Prnp^{0/0}$ mice. These results argue that distinct protease-resistant PrP glycoforms and conformers are formed in FFI(D178N, cis-129M), sporadic CJD, and CJD(D178N, cis-129V), and that those PrP properties are replicated precisely during interspecies passaging (e.g., humans to mice). This evidence argues strongly that the strain of a prion is most likely encoded in the conformation PrP.

Mutations of the PrP Gene

If patients with a CNS disorder have a mutation in the PRNP gene which results in an amino acid substitution that is not a polymorphism, such as found at codons 129 and 219, then the diagnosis of prion disease is probably secure. If the mutation is one of the five known pathogenic mutations of the PRNP gene for which genetic linkage has been established, then the diagnosis is clearly secure. If the mutation is one of those already described for a prion disease pedigree, but for which genetic linkage has not yet been established, then the likelihood of its being pathogenic is increased. If a new mutation is found, then at least 100 people of similar ethnic background should be screened for the mutation to determine whether it is a polymorphism or a pathogenic change.

MECHANISMS OF NERVE CELL DEGENERATION (PATHOGENESIS)

The most characteristic and clinically relevant neuropathologic feature of prion diseases is vacuolar degeneration of neurons and their processes, particularly in synaptic regions, and its association with failure of synaptic transmission and nerve cell death. A second characteristic is that different populations of neurons are targeted for degeneration as a function of the strain of prion infecting an animal or of a specific PrP gene mutation. The reproducibility of the neuroanatomic distribution PrP^{Sc} and neurodegeneration implies that each neuron population responds to these etiological factors in different ways.

The existence of two markedly different categories of prion diseases, those that resemble scrapie and CJD and those that resemble GSS, has raised the possibility that PrP can cause nerve cell degeneration by two distinctly different mechanisms. Recent discoveries suggest that the two categories of prion disease are associated with two different topologies of the PrP molecule: secreted versus transmembrane.

Secreted Form of PrP

Prusiner's original studies of prion diseases focused on the composition of the scrapie agent and the mechanism of its propagation once it was found to be composed mostly, if not exclusively, of PrP^{Sc}. The results of those studies showed that nascent PrP^{Sc} is formed from PrP^{C} attached to the outer surface of the plasma membrane by a glycolipid anchor. Presumably, although not yet proven, most of the resulting nascent PrP^{Sc} that causes neurodegeneration is also glycolipid-anchored. This mechanism of formation of pathogenic PrP accounts for the propagation of prions and the neuropathologic changes in prion diseases acquired by infection and, most likely, also for prion diseases arising from spontaneous conversion of PrP^{C} to nascent PrP^{Sc}; e.g., cases of sporadic CJD, and in those forms of genetically determined prion diseases that result in CJD-like neuropathologic features. Common to these forms of prion disease are the accumulation of protease-resistant PrP^{Sc} in the brain, little amyloid plaque formation (except for nvCJD), and the propagation of highly infectious prions.

Transmembrane Form of PrP

A second PrP mechanism of nerve cell degeneration was recently identified and may underlie the pathogenesis of some GSS syndromes. The Lingappa laboratory showed that expression of a doubly mutated PrP in transgenic mice, which results in the substitution of a KH to II at residues 110 and 111, causes residues 112–135 of the PrP molecule to adopt a transmembrane topography, with or without glycolipid anchoring of the carboxyl terminus, rather than the usual glycolipid-anchored, "secreted" topography with the entire protein on the outside of cell membranes (Hegde et al. 1998). Animals in which large proportions of ^{tm}PrP(KH-II) are formed develop a spontaneous neurodegenerative disease characterized by spongiform degeneration and reactive astrocytic gliosis that resembles scrapie, but with no PrP plaques. PK digestion of brain homogenates from these animals showed that the mutated PrP molecules are significantly less resistant to PK digestion than PrP^{Sc} and that PK digestion results in the formation of highly truncated PrPs. That study also found that the mutated PrP from a case of GSS(A117V), with abundant PrP plaques, adopted a transmembrane topology and was PK sensitive. In Tg(SHaPrP-A117V) mice, which express Syrian hamster PrP with the A117V mutation, 35% of the mutated PrP adopted a transmembrane topology, and these mice developed amyloid plaques similar to those in human GSS(A117V) (Hegde et al. 1999). Interestingly, GSS(P102L) is also characterized by PK-sensitive PrP (Hsiao et al. 1989b, 1990), raising the possibility that it is also a disease of abnormal PrP transmembrane topology. It is becoming evident that PrP can be made in more than one topographical form, secreted or transmembrane, and that the proportion of PrP molecules that are in one topography or the other depends in part on the PrP amino acid sequence (Hegde et al. 1998).

MECHANISM OF NERVE CELL DEGENERATION BY CONVERSION OF PrP^C TO NASCENT PrP^{Sc}

Conversion of PrP^C to Nascent PrP^{Sc}, Not Just the Presence of PrP^{Sc}, Is a Requirement for Neurodegeneration

The mechanisms of cell dysfunction in prion diseases must be placed into the context of what is known about the conversion of PrP^C to PrP^{Sc} and the subsequent accumulation of PrP^{Sc} in and around cells, because they are the driving forces that result in neuropathologic abnormalities. The evidence that this is the case comes from studies of scrapie-infected cell lines in culture and from studies of PrP knockout mice.

Most of our knowledge of the metabolism and trafficking of PrP^C and PrP^{Sc} is from investigations in the mouse N2a neuroblastoma cell line. Once infected with scrapie prions, the resulting ScN2a cell line retains infectivity chronically and on all subsequent passages in culture (Butler et al. 1988). PrP^C is constitutively expressed in both N2a and ScN2a cells. Ninety percent of the cell's PrP^C becomes attached to the outer surface of the cells by its GPI anchor following its synthesis and passage through the Golgi complex (Borchelt et al. 1990). Like other GPI-anchored proteins, PrP^C reenters the cell through cholesterol-rich caveolae (Ying et al. 1992).

About 10% of the endocytosed PrP^C is degraded with each endocytic cycle. The remainder is recycled to the cell surface along with other pro teins and lipids of the endosome. In pulse-chase experiments, PrP^C in uninfected cells is rapidly labeled by a radioactive amino acid tracer and appears to be completely degraded with loss of the tracer in about 6 hours (Caughey et al. 1989; Borchelt et al. 1990). In scrapie-infected ScN2a cells, labeling of PrP^{Sc} is delayed by about 1 hour after the pulse and increases to a maximum during the time period when the PrP^C pool loses its tracer (Borchelt et al. 1990). Over the next 24 hours, there is little or no loss of tracer from the PrP^{Sc} pool. These results argue that PrP^{Sc} is derived from PrP^C, but, unlike PrP^C, PrP^{Sc} is not degraded and accumulates in the cell. The PrP^C precursor pool from which PrP^{Sc} is formed must reach the cell surface, since blocking PrP^C export from the ER/Golgi complex to the plasma membrane inhibits formation of PrP^{Sc} (Taraboulos et al. 1992a) and since exposure of scrapie-infected cells to phosphatidylinositol-specific phospholipase C (PIPLC), which releases the secretory form of PrP^C from the cell surface, also inhibits formation of PrP^{Sc} (Caughey and Raymond 1991). Details of where PrP^{Sc} becomes distributed in cells are still being determined. Immunoelectron microscopy indicated that PrP^{Sc} accumulates in secondary lysosomes (McKinley et al. 1991b). More recently, we found that a large proportion of nascent PrP^{Sc} accumulates in caveolae-like domains (CLDs) (Vey et al. 1996). Eighty to ninety percent of plasma membrane PrP^C is also localized to CLD regions (Vey et al. 1996; Wu et al. 1997). The only other location in the cell culture definitely known to contain PrP^{Sc} is the conditioned culture medium, which appears to accumulate PrP^{Sc} as a result of excretion from infected cells or as a result of death of infected cells. The site where exogenous PrP^{Sc} comes into contact with PrP^C and where PrP^C is converted to nascent PrP^{Sc} is not known; however, the colocalization of PrP^{Sc} and PrP^C in CLDs suggests that they may be the site of conversion.

Mice in which the PrP gene has been knocked out, $Prnp^{0/0}$ mice (Büeler et al. 1992), have yielded strong support for the concept that PrP^{Sc} is derived from PrP^C, since no PrP^{Sc} is formed, no prions propagated, nor neuropathologic changes developed in these mice. Both acute and chronic exposure of the PrP knockout brains to prions failed to generate new prions or to cause neuropathologic changes (Büeler et al. 1992, 1993; Prusiner et al. 1993; Brandner et al. 1996). These findings indicate that exogenously derived PrP^{Sc} by itself is not pathogenic; rather, in order for PrP^{Sc} to cause neuronal degeneration, it must be derived from PrP^C. It is likely that when nascent PrP^{Sc} is derived from glycolipid-anchored PrP^C, it enters a cellular compartment where it can disrupt functions that PrP^{Sc}

in the extracellular space cannot. It is possible that PrP^{Sc} must be anchored to membranes by a glycolipid anchor and that it can only do so if it is derived from glycolipid-anchored PrP^C.

Prion Disease Phenotype Is a Function of Both the Infecting Prion Strain and the Host's Response to the Prion Strain

One of the most fundamental principles of microbiology is that the infectious disease phenotype is determined both by the infectious agent and by the host's response to it. Thus, conventional agents such as bacteria, viruses, fungi, and parasites attach to specific intracellular and extracellular regions to which they have become adapted for nourishment and proliferation. The host has evolved different immune and inflammatory responses to each pathogenic organism. For example, the host responds to pyogenic bacteria in the subarachnoid space, such as *Neisseria meningitidis*, by an intense neutrophil reaction that, in combination with proliferation of bacteria, can lead to brain swelling, venous infarcts, and cerebral herniation that cause death in hours to days. In contrast, the host response to *Mycobacterium tuberculosis* in the subarachnoid space is a slowly evolving lymphocytic–histiocytic–fibroblastic reaction that leads to obstructive hydrocephalus, arteritis, and arterial strokes that cause death in weeks to months.

The prion/host response complex that generates the clinical and neuropathologic features in prion diseases does not appear to involve a significant host immune response. To date, the only host factor known to influence the prion disease phenotype is the PrP^C expressed by the host. The parameters of the prion disease phenotype most commonly used to identify and classify prion strains are the host species barrier to infection, incubation time, the intensity and distribution of neuropathologic spongiform degeneration, and whether or not amyloid plaques form. Our understanding of the roles played by the prion and by the host in modulating these disease parameters has emerged slowly over the past 17 years in parallel with the discovery of how prions propagate. By viewing prion diseases from the perspective of the neuropathologic changes, we have learned that host-determined variations in PrP^C play as important a role as the conformation of PrP^{Sc} comprising an infecting prion. The role of PrP^C in prion biology cannot be overemphasized, particularly for understanding the mechanism of prion-strain-determined variations in the disease phenotype. Indeed, the critics of the protein-only hypothesis often argue that a protein cannot encode all of the information necessary to account for all of the prion-strain-specific variations in the disease phenotype; in this regard, they may be correct. However, these critics do not differentiate between the coding of strain information in the prion and the combination of prion and host factors that actually generate the disease phenotype.

Our current understanding of the formation of PrP^{Sc} by neurons is that it involves four factors: First, neurons must come into contact with prions, in which strain properties are encoded in the conformation of their PrP^{Sc} (Telling et al. 1996b). Second, a prion's PrP^{Sc} must physically interact with the host's PrP^{C} on the neuron's surface (Scott et al. 1989, 1993; Prusiner et al. 1990; Cohen et al. 1994; Telling et al. 1995). Third, the prion-strain-specific conformation of PrP^{Sc} acts as a template for the refolding of PrP^{C} into an exact duplicate conformation (Cohen et al. 1994). Fourth, an additional cellular mechanism, provisionally designated "protein X," appears to bind to the host's PrP^{C} and facilitate its unfolding and refolding into nascent PrP^{Sc} (Telling et al. 1995; Kaneko et al. 1997b). These events result in an exponential accumulation of PrP^{Sc} in and around neurons (Jendroska et al. 1991). That PrP^{Sc} generated by neurons is sufficient to cause nerve cell degeneration has been verified by neuron-specific expression of SHaPrP^{C} in Tg mice under the control of a neuron-specific enolase promoter and infection with hamster-adapted prions (Race et al. 1995). This is also consistent with the observation that neurons express the highest PrPmRNA at levels significantly higher than glial or endothelial cells in the CNS (Kretzschmar et al. 1986; Jendroska et al. 1991).

The prion attribute that gives it strain properties is the conformation of its PrP^{Sc}, which appears to remain constant during multiple passages in host animals (Telling et al. 1996b). From the perspective of the host animal, when one factors in host-determined variations in the level of expression of PrP^{C}, whether one or two PrP^{C} allotypes (or more in some transgenic mice) are expressed in a host, the level of expression of each PrP^{C} allotype, nerve-cell-specific variations in the conformation of PrP^{C} (see below), and the proportion of ^{tm}PrP and secreted PrP^{C} formed, there are more than sufficient prion and host variations to account for all of the known strain-determined variations in the prion disease phenotype. Other host factors could also influence the disease phenotype, such as the rate of clearance of nascent PrP^{Sc} from the brain, the degree of reactive microgliomatosis and reactive astrocytic gliosis, and possible immune responses (Raeber et al. 1997).

The preeminent roles played by both the prion and the host's PrP^{C} have been highlighted in studies of each component of the prion disease phenotype.

Host Species Barrier Phenotype

The failure of transmission or relatively low success rate of transmission of a prion strain to a host were found to be due in part to a large variation in the amino acid sequence of an infecting prion's PrP^{Sc} and the host's PrP^C, particularly between residues 90 and 120. The failure of transmission of hamster prions to mice could be overcome by expressing hamster PrP^C in Tg mice (Scott et al. 1989; Prusiner et al. 1990) or by expressing chimeric mouse–hamster–mouse PrP^C constructs in which the region between residues 90–120 in the mouse PrP^C molecule was replaced with the hamster amino acid sequence (Scott et al. 1993).

Scrapie Incubation Time Phenotype

Over 25 years ago, Dickinson and his colleagues found that short and long scrapie incubation times in inbred mouse strains are determined by the strain of scrapie agent and by a single host gene (Dickinson et al. 1968; Dickinson and Meikle 1971). Following the discovery of PrP, it was found that the scrapie incubation time gene and the *Prnp* gene are the same (Carlson et al. 1986, 1988). Inbred mouse strains have one of two Prnp alleles that express two $MoPrP^C$ allotypes which differ by two amino acids at codons 108 and 189 (Westaway et al. 1987). These are designated $MoPrP^C$-A and $MoPrP^C$-B. Inoculation of a mouse prion composed of a single $MoPrP^{Sc}$ allotype into congenic and transgenic mice expressing different numbers and combinations of $MoPrP^C$-A and $MoPrP^C$-B showed that short and long incubation times were dependent on whether the amino acid sequence of the PrP^{Sc} of the infecting prion was the same or different from the $MoPrP^C$ allotype expressed and, furthermore, that long and short incubation times were shortened proportional to the number of PrP^C-A and PrP^C-B allotypes expressed (Carlson et al. 1994). Thus, scrapie incubation time is a function of the PrP^{Sc} comprising the prion, the amino acid sequence of the host's PrP^C, and the expression level of PrP^C allotypes.

PrP Amyloid Plaque Phenotype

The two factors that determine whether or not PrP amyloid is formed in prion diseases are the strain of infecting prion and the amino acid sequence of PrP synthesized by the host. That the strain of scrapie agent can determine whether or not amyloid forms was discovered over 25 years ago by neuropathologic analysis of scrapie-infected mice (Fraser and

Bruce 1973, 1983; Bruce et al. 1976). Most mouse-adapted scrapie prion strains are not amyloidogenic; a few strains, such as the 87V, are highly amyloidogenic. Human nvCJD prions, like bovine BSE prions, are also highly amyloidogenic (Fig. 10), unlike other CJD prions (Scott et al. 1999). Our current understanding of this phenomenon is that an amyloidogenic conformation of an infecting PrP^{Sc} is imposed on a host's PrP^{C} and that conformation favors truncation of nascent PrP^{Sc} into peptides that readily polymerize into amyloid filaments (Fig. 4).

Three lines of evidence indicate that the amino acid sequence of PrP^{C} expressed by an animal plays a role in amyloidogenesis. The first is genetic linkage of PRNP mutations in humans to the GSS phenotype for which there are six amyloidogenic mutations at codons 102, 105, 117, 145, 198, and 217. Some of the non-amyloidogenic mutations genetically linked to CJD-like syndromes occur at codons 178, 180, 200, 208, 210, and 232. Intense spontaneous PrP amyloidogenesis in Tg(GSS-P101L) mice expressing MoPrP with the mutation homologous to GSS(P102L) supports the view that some PrP sequences are amyloidogenic (Fig. 12). A second line of evidence comes from transmission of scrapie prion strains to animal species expressing PrP^{C}s with different amino acid sequences. For example, Sc237 prions derived from Syrian hamster brain result in the formation of mature kuru-type PrP amyloid plaques in Chinese hamster (CHa) brain, primitive PrP amyloid plaques in Syrian hamster (SHa) brain, and no amyloid plaques in Armenian hamster (AHa) brain (Lowenstein et al. 1990). AHa and CHa PrP^{C} differ by three amino acids at residues 103, 108, and 112, whereas SHa PrP^{C} differs from CHa by only one amino acid at residue 112. Although those data support the hypothesis that the amino acid sequence of the host's PrP^{C} determines, in part, whether the PrP^{Sc} formed from the host's PrP^{C} is amyloidogenic, they do not rule out other host factors. A third line of evidence more definitively shows a correlation between the amino acid sequence of the host's PrP^{C} and amyloid formation. When Sc237 prions passaged in Syrian hamsters, Sc237(SHa) prions, are inoculated into Tg mice expressing Syrian hamster PrP^{C}, Tg(SHaPrP) mice, small primitive PrP amyloid plaques are formed (Fig. 19A) (Peretz et al. 2002). In contrast, when Sc237(SHa) prions are inoculated into Tg mice expressing a chimeric mouse–hamster–mouse PrP^{C}, Tg(MH2M-PrP) mice, large mature amyloid plaques are formed (Fig. 19B).

Of interest to this discussion, kuru-type amyloid plaques occur in only about 5–10% of sporadic CJD cases. There is a suggestion from genetic studies of CJD patients that these plaques may be linked to a valine polymorphism at codon 129 (de Silva et al. 1994). In contrast, all

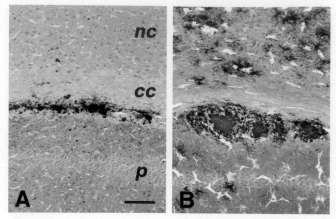

Figure 19. PrP amyloidosis in scrapie is in part a function of the amino acid sequence of the PrP^C converted to nascent PrP^{Sc}. (A) Tg(SHa-PrP) mice inoculated with Sc237 prions. (B) Tg(MH2M-PrP) mice inoculated with Sc237 prions. Bar in A is 50 μm and applies to both A and B. (cc) Corpus callosum; (nc) neocortex; (p) pyramidal cell layer of hippocampal CA1 region. (Adapted from Peretz et al. 2002.)

nvCJD cases, which are characterized by depositon of large numbers of PrP amyloid plaques (Fig. 10), have been homozygous methionine at codon 129 (Collinge et al. 1996).

Vacuolation Histogram Phenotype

The distribution and intensity of spongiform (vacuolar) degeneration, usually displayed in histograms and designated the "lesion profile," has been the most commonly used neuropathologic measure to differentiate one prion strain from another (Fraser and Dickinson 1968, 1973; Fraser 1979; Bruce et al. 1991). The mechanism of differential "targeting" of neurons by prion strains for vacuolar degeneration is not yet completely understood. However, PrP immunohistological data reviewed above revealed a precise correlation between sites of vacuolar degeneration and sites of PrP^{Sc} accumulation. That correlation argues that the neuroanatomic pattern of PrP^{Sc} accumulation, like the pattern of vacuolation, is itself a characteristic of prion strains (Fig. 5).

PrP^{Sc} Distribution Phenotype

With the discovery that the neuropathologic changes in prion diseases colocalize precisely with sites of PrP^{Sc} accumulation (DeArmond et al.

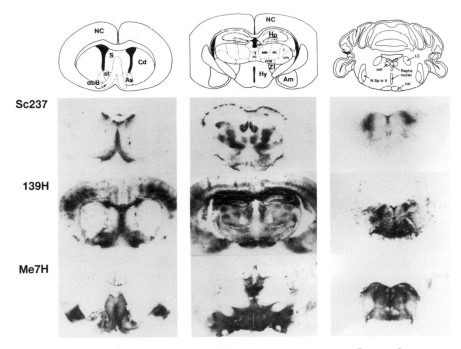

Figure 20. Differential targeting of neurons for conversion of PrP^C to PrP^{Sc} by prion strains. Tg(SHaPrP) mice were inoculated intrathalamically with either the Sc237, 139H, or Me7H prion strains obtained from Syrian hamster brains. (Am) Amygdala; (As) accumbens septi; (Cd) caudate nucleus; (cst) corticospinal tract; (dbB) diagonal band of Broca; (Hp) hippocampus; (Hy) hypothalamus; (LC) locus ceruleus; (mlf) medial longitudinal fasiculus; (NC) neocortex; (N Sp tr V) nucleus of the spinal tract of the trigeminal nerve; (S) septal nuclei; (st) interstitial nucleus of the stria terminalis; (ZI) zona incerta. Thalamic nuclei are in italics.

1987), it was not surprising to find that the distribution of PrP^{Sc} in the brain is also characteristic of prion strains (Figs.18, 20) (Bruce et al. 1989; Hecker et al. 1992; DeArmond et al. 1993). Although vacuolation and reactive astrocytic gliosis occur only at sites where PrP^{Sc} accumulates, the converse is not true: PrP^{Sc} accumulates in some brain regions without obvious neuropathologic changes. One explanation for the latter is that neuropathologic changes require a finite amount of time to develop following local accumulation of PrP^{Sc} (Jendroska et al. 1991). Another possibility is that vacuolation of neuronal processes does not develop in some neuronal populations in response to PrP^{Sc} deposition. A further feature of the PrP^{Sc} deposition phenotype is that, although the distribution of PrP^{Sc} is different for each strain of prions, many of the same brain regions accumulate PrP^{Sc} with each prion strain (Fig. 20). For example, the brain stem tegmentum accumulates PrP^{Sc} with many prion strains. Analysis of this overlap phenomenon is hampered at this time because it is not known whether PrP^{Sc} accumulation in a region, such as the brain stem, is due to local conversion of PrP^{C} to PrP^{Sc} by brain stem nerve cells or whether PrP^{Sc} is transported to the brain stem by anterograde and/or retrograde axonal transport from many distant sites.

Although the mechanism of differential targeting of neurons by prion strains is not understood, analysis of this phenomenon must involve cell-specific mechanisms that influence the interaction of a prion's PrP^{Sc} with the neuron's PrP^{C} or that influence the mechanism that duplicates the conformation of PrP^{Sc} in the cell's PrP^{C}. Analysis must also take into account the early investigations of scrapie within inbred mouse strains, which have shown that the vacuolar degeneration lesion profile varies as a function of the scrapie strain, the route of infection of prions, and the mouse genotype (Fraser and Dickinson 1968, 1973; Fraser 1979, 1993).

PrP^{C} Glycotypes and the Lesion Profile

Where our early studies showed that the distribution of PrP^{Sc} was markedly different in animals of the same genotype inoculated with different prion strains (Fig. 20), more recently, we found that the converse is also true. Thus, the pattern of PrP^{Sc} accumulation in three different inbred mouse strains was different in response to intrathalamic inoculation of a single scrapie prion strain, the RML (Rocky Mountain Laboratory) scrapie isolate (DeArmond et al. 1997). RML prions inoculated into C57BL, CD-1, and FVB mice, all of which carry the $Prnp^a$ gene, exhibited different patterns of PrP^{Sc} deposition (Fig. 21). C57BL mice were distinguished by little or no immunostaining in the hippocampus and neocortex and relatively weak immunostaining in other brain regions. The most intense signals were found in the thalamus and brain stem. In contrast, PrP^{Sc} was uniformly and widely distributed throughout the cerebrum and brain stem in FVB mice. PrP^{Sc} immunostaining in CD-1 mice was intermediate, in that it was more widely distributed in the brain compared to C57BL, but weak or absent in some regions such as the hippocampus, amygdala, and hypothalamus. Common to all three mouse strains was the presence of immunostaining in the brain stem tegmentum and the absence of immunostaining in the cerebellar cortex.

Because each of these mouse strains is homozygous for the $Prnp^a$ gene and, therefore, each expresses PrP^{C}-A with the same amino acid sequence, and because all the mice were inoculated with the RML prion strain pas-

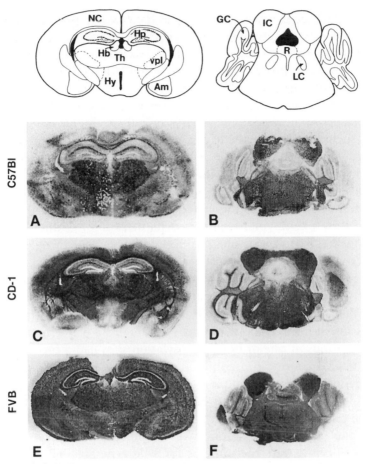

Figure 21. Differential targeting of neurons for conversion of PrP^C to PrP^{Sc}: influence of the strain of animal. A single mouse prion strain, RML(CD1), from a single animal source (CD1 mice) was inoculated intrathalamically into three inbred mouse strains and resulted in three different PrP^{Sc} deposition patterns. (Am) Amygdala; (GC) granule cell layer of the cerebellar cortex; (Hb) habenula; (Hp) hippocampus; (Hy) hypothalamus; (IC) inferior colliculus; (LC) locus ceruleus; (NC) neocortex; (R) dorsal nucleus of the raphe; (vpl) ventral posterior lateral nucleus of the thalamus; (Th) thalamus.

saged through CD-1 mice, finding different patterns of PrP^{Sc} accumulation argues that a non-*Prnp* gene(s) of the host modifies the PrP^{Sc} deposition phenotype. One set of non-Prnp factors that have the potential to influence rate of synthesis, rate of degradation, and posttranslational modification of PrP^C are the series of enzymatic steps in the endoplasmic reticulum and Golgi apparatus that attach and modify the two asparagine-linked oligosaccharides of the PrP molecule at asparagine 181 and 197 (Endo et al. 1989). Neuron-specific glycosylation of PrP^C might underlie selective targeting of neurons because experience with other proteins indicates that asparagine-linked oligosaccharides influence protein conformation and protein/protein interactions (Rademacher et al. 1988; O'Connor and Imperiali 1996). Cell-specific differences in asparagine-linked glycosylation are also well known for other cellular proteins (Rademacher et al. 1988). The fidelity of cell-specific glycosylation patterns is such that it is referred to as a cell's "glycotype." The possibility that each neuron population synthesizes a different PrP^C glycotype is supported by the evidence that the brain contains more than 400 different PrP^C glycoforms (Endo et al. 1989).

To test whether asparagine-linked glycosylation of PrP^C influences the prion disease phenotype, we mutated its threonine residues to alanine within the NXT consensus sites (Taraboulos et al. 1990). Single and double glycosylation-site mutations were made in PrP constructs encoding Syrian hamster PrP, and these were expressed in Tg mice deficient for MoPrP (Prnp$^{0/0}$) (DeArmond et al. 1997). Wild-type SHaPrP^C was uniformly distributed in the neuropil of all gray matter regions in Tg mice, but was absent from cell bodies of neurons and from white matter. In contrast, mutations of one or both glycosylation sites had a profound effect on the neuroanatomical distribution of SHaPrP^C. Mutations of the first glycosylation site alone or in combination with mutation of the second site resulted in low levels of mu-SHaPrP^C, accumulation of mu-SHaPrP^C in nerve cell bodies, and little mu-SHaPrP^C in the gray matter neuropil. When the second site alone was mutated, the levels of mu-SHaPrP^C(T199A) were the same as for wild-type SHaPrP^C; however, it was distributed to all neuronal compartments including the nerve cell body, dendritic tree, and axons of the white matter. These results suggest that the oligosaccharide at residue 181 plays a particularly important role in the trafficking, sorting, and stability of PrP^C.

Two Syrian hamster prion strains, Sc237 and 139H, were inoculated intracerebrally into the Tg mice expressing wild-type and single or doubly deglycosylated SHaPrP^C (DeArmond et al. 1997). No Tg mice expressing mu-SHaPrP^C were affected in which the oligosaccharide at Asn-181 was deleted either because of low levels of mu-SHaPrP^C and/or because these mu-PrPSHa^{C}s did not appear to be transported out of the nerve cell body. Tg mouse lines that expressed mu-SHaPrP^C(T199A) in which the oligosaccharide at Asn-181 was in place but the second at Asn-197 was deleted, Tg(SHaPrP-T199A)/Prnp0/0 mice, did develop scrapie following inoculation with prions; however, the disease phenotype was profoundly different from that in Tg mice expressing wild-type SHaPrP^C,

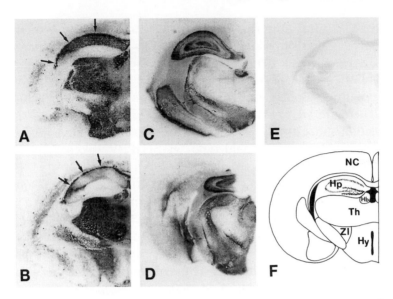

Figure 22. Deletion of the second asparagine-linked oligosaccharide at residue 197 from SHaPrP^C by mutation of the asparagine consensus sequence at codon 199 profoundly alters the PrP^{Sc} accumulation pattern in response to inoculation with Sc237 prions passaged in Syrian hamsters: (*A,B*) Pattern of wild-type SHaPrP^{Sc} accumulation in two Tg mouse lines expressing wild-type SHaPrP^C and (*C,D*) pattern of deglycosylated mutant SHaPrP^{Sc}-T199A accumulation in two Tg mouse lines expressing mutant SHaPrP^C-T199A. (*E*) Inoculation of Tg mouse lines expressing deglycosylation mutant SHaPrP^C-T199A failed to develop scrapie or to accumulate protease-resistant PrP. (*F*) (Hb) Habenula; (Hp) hippocampus; (Hy) hypothalamus; (NC) neocortex; (ZI) zona incerta.

Tg(SHaPrP)7/Prnp$^{0/0}$ mice. First, whereas Tg(SHaPrP)7/Prnp$^{0/0}$ mice became scrapie sick with both Sc237 and 139H prions at 44 and 47 days postinoculation, respectively, Tg(SHaPrP-T199A)/Prnp$^{0/0}$ mice only became scrapie sick with the Sc237 strain and with an incubation time of about 550 days. Second, whereas Sc237 prions resulted in the formation of PrP amyloid plaques in Tg(SHaPrP)7/Prnp$^{0/0}$ mice, no amyloid plaques formed in Tg(SHaPrP-T199A)/Prnp$^{0/0}$ mice (Fig. 22). Third, the neuroanatomic distribution of SHaPrP^{Sc} was markedly different. For example, there was intense immunostaining for SHaPrP^{Sc} in the thalamus of Tg(SHaPrP)7/Prnp$^{0/0}$ mice but little or none in the thalamus of Tg(SHaPrP-T199A)/Prnp$^{0/0}$ mice (Fig. 22).

To test the hypothesis that each neuron population attaches oligosaccharides to its PrP^{C}s that are composed of different carbohydrate and sialic acid moieties, thereby creating brain-region-specific PrP^C glycotypes, we dissected brain regions from Syrian hamsters and subjected their PrP^C to 2D electrophoresis to search for reproducible differences in charge isomers (DeArmond et al. 1999). Earlier studies showed that PrP charge isomers are the result of variable sialation of their asparagine-linked oligosaccharides (Bolton et al. 1985; Endo et al. 1989; Haraguchi et al. 1989). Therefore, we began this study by asking whether or not PrP^C deglycosylation mutants expressed in Tg mice could be distinguished by their charge isomer pattern. Wild-type (wt) SHaPrP^C was characterized by 15–20 charge isomers with isoelectric points ranging from pH 4 to 7. Deletion of both asparagine-linked glycosylation consensus sequences reduced the number of charge isomers to one major spot and two minor spots. An intermediate number of charge isomers characterized SHaPrP^C(T183A) and SHaPrP^C(T199A). This supports the earlier finding that the majority of PrP^C's charge isomers originate from its two asparagine-linked oligosaccharides: The remaining three isomers are likely to originate from PrP^C's glycolipid anchor.

To test for brain region differences in charge isomers, Syrian hamster brains were dissected and equal amounts of SHaPrP^C from each brain region were analyzed. Brain region differences in the number and location of isoelectric points could be identified by visual inspection. The reproducibility of these results was verified by comparing isoelectric points from 2–4 groups of hamsters using two ampholyte ratios. The pH distribution of charge isomers was different for the two ampholyte ratios, and the differences between brain regions were highly reproducible with each ampholyte ratio. For example, with one of the ampholyte ratios, hippocampal PrP^C exhibited 16 constant isoelectric points ranging from pH 5.18 to 7.25 (N = 7 runs comparing 4 groups of animals) whereas the cerebellum exhibited 10 constant isoelectric points ranging from pH 5.51 to 6.75 (N = 11 runs comparing 4 groups of animals).

These results argue that variations in the glycosylation of PrP^C can have a significant influence on the host species barrier, incubation time, amyloidogenesis, and the sites of PrP^{Sc} accumulation, which determines the sites of vacuolar degeneration. How might oligosaccharides attached to PrP^C have such a profound effect on the rate and sites of conversion of PrP^C to PrP^{Sc}? We hypothesize that asparagine-linked oligosaccharides affect the ability of residues 90–140 of the PrP^C molecule to interact with an infecting PrP^{Sc} and, in doing so, influence the rate of PrP^C's conversion into nascent PrP^{Sc}. This kinetic view of PrP^{Sc} formation is consistent with the hypothesis that some PrP^C molecules are more readily converted into PrP^{Sc} than others (Eigen 1996). Moreover, brain region variations in the composition and structure of the two asparagine-linked oligosaccharides might also modulate the amount of convertible PrP^C within the cellular compartment where it interacts with PrP[Sc] by influencing differential trafficking of PrP glycotypes within neurons.

Progression of Molecular and Cellular Events during Prion Diseases

For many years, neuropathologists have recognized a stereotypical stepwise progression of neuropathological changes during CJD that begins with vacuolation of synaptic structures and progresses in later stages to nerve cell degeneration and death (Masters and Richardson 1978). A similar temporal sequence has been reported for scrapie in rodents that begins with axon terminal degeneration and loss, followed by atrophy of the dendritic trees, and finally by neuronal degeneration and death weeks to months later, depending on the animal model (Bouzamondo et al. 2000; Jeffrey et al. 2000). There is considerable evidence from both in vivo (Giese et al. 1995; Lucassen et al. 1995; Gray et al. 1999) and in vitro studies (Schätzl et al. 1997) that the neuron loss in prion diseases has an apoptotic mechanism. In some natural and experimental prion diseases, it can be difficult to detect dying neurons because apoptotic cells are quickly cleared from the CNS and a low-grade loss over the prolonged time course of the diseases is almost impossible to detect. Murine scrapie models in which targeted neuronal loss occurs enable us to identify the sequence of pathological changes that culminate in neuron death and to explore the possibilities for intervention (Fraser 2002).

In two of these models, specific regions of the hippocampal pyramidal cell layer are targeted by different scrapie strains. When C57BL x VM mice are infected with ME7 scrapie, over half of the pyramidal neurons in CA1 hippocampus are lost from day 160 of a 250-day incubation period (Fig. 23a,b) (Jeffrey et al. 2000). When VM mice are infected with 87V scrapie, neurodegeneration is restricted to the small CA2 sector of the hippocampus (Jamieson et al. 2001a,b). Detailed studies of the progression of neurodegenerative changes in these models have shown that the sequence of events at the molecular level depends on the strain of scrapie and the genotype of the mouse.

Neuron Loss in CA1 Region of the Hippocampus during ME7 Scrapie

The ME7 model produces severe pathological changes in the CA1 region, typified at the terminal stages of disease by intense vacuolation (Fig. 23b), reactive astrogliosis, microglial hyperplasia, and dense accumulation of PrP[Sc]. Golgi impregnation of sections from these mice has revealed

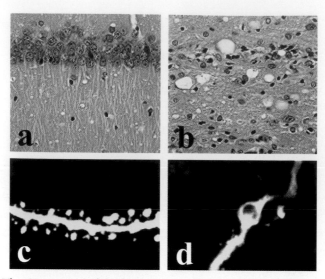

Figure 23. The CA1 region of the hippocampus from C57BL x VM mice: (*a*) normal and (*b*) terminally ill with ME7 scrapie (H&E). Note marked loss of neuronal cell bodies and vacuolation of the neuropil in the latter. Intraneuronal injection of Lucifer yellow reveals the dendrites and their spines in control (*c*) and terminally ill ME7 scrapie-infected mice (*d*). Loss of spines, contortions, and varicosities of dendrites feature in the ME7 scrapie-infected mice. (Photomicrographs *c* and *d* are courtesy of Dr. Debbie Brown.)

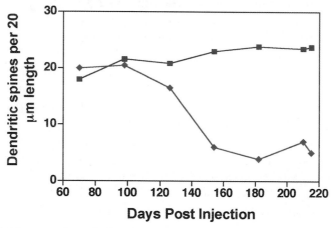

Figure 24. Mean number of spines in a 20-μm length of dendrite visualized by a modified Golgi impregnation technique. Each data point was obtained from between 4 and 9 neurons per group. (*Blue squares*), control mice. (*Red diamonds*), ME7-infected C57BL x VM mice. Standard errors cannot be seen as they are 0.5 or less. A significant loss of spines is apparent from 126 days post injection. (Graph courtesy of Dr. Debbie Brown.)

sequential loss of dendritic spines from 126 days (Fig. 24) (Brown et al. 1997), and subsequent studies using confocal analysis following microinjection of the fluorescent dye Lucifer yellow have confirmed that dendrites of the CA1 neurons become shrunken and contorted by the end of the incubation period (Fig. 23c,d) (Belichenko et al. 2000). Using this technique, significant spine loss can be identified as early as 98 days postinfection (Fig. 25) (D. Brown et al. 2001). Small, focal deposits of PrP^{Sc} can be identified within the pyramidal cell layer from 70 days, preceding all other pathology (Jeffrey et al. 2000). Ultrastructural studies have revealed that synapse loss and axon terminal loss precede spine loss (Jeffrey et al. 2001). Jeffrey et al. have suggested that loss of axon terminals and the associated absence of neurotransmitter release and stimulation of postsynap-

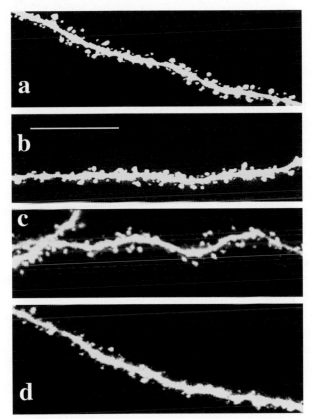

Figure 25. Lucifer-yellow-filled dendrites showing progressive loss of spines in C57BL x VM mice infected with ME7 scrapie. (*a*) Control. (*b–d*) Scrapie-infected: (*b*) 98 days postinoculation (dpi), no significant loss; (*c*) 109 dpi, 18% loss; (*d*) 126 dpi, 51% loss. (Photomicrographs courtesy of Dr. Debbie Brown.)

tic spines result in a decreased excitatory input to pyramidal neurons that triggers apoptosis when those losses are sufficiently extensive across the dendritic tree (Jeffrey et al. 2001).

The sequence of these events suggests that PrP accumulation in CA1 hippocampus is particularly neurotoxic and that its accumulation around axon terminals interferes with synaptic function and leads to an excitatory-like axon terminal degeneration (Jeffrey et al. 1997). Subsequent damage may be part of a cascade of events involving microglial activation (Williams et al. 1997) and astrogliosis, probably stimulated by the presence of fibrillar forms of PrP^{Sc} (Jeffrey et al. 1997).

Neuron Loss in CA2 Region of the Hippocampus during 87V Scrapie

When VM mice are infected with 87V scrapie, neuron loss and abnormal PrP deposition in the hippocampus are restricted to a small sector of pyramidal cells (CA2). These neurons (which do not have spines) show contortions of apical dendrites (Fig. 26a) as early as 70 days into the incubation period of about 320 days. From 100 days onward, evidence of apoptosis of CA1 neurons is found by TUNEL labeling (Fig. 26b) and the characteristic 200-bp DNA laddering (Fig. 26c). In addition, from 100 days onward there is up-regulation of proapoptotic markers c-jun, Fas, and caspase 3 (Fig. 26d) (Jamieson et al. 2001a,b). Although abnormal PrP cannot be identified by immunolabeling in CA2 until 200 days, protoapoptotic markers have been found in cell cultures treated with low concentrations of the neurotoxic peptide PrP 106-126 (White et al. 2001). By the terminal stages of the disease, hyperplasia of both astrocytes and microglia (Fig. 26e) can be found in CA2, and cytoskeletal abnormalities are obvious in sections labeled for tubulin (Fig. 26f,g).

Although there are specific differences between the progression of the degenerative pathology in these two models, both show early evidence of neuronal process damage associated with subsequent apoptosis long before clinical signs of disease become apparent. The precise trigger and mechanisms of the apoptotic pathway operating in prion diseases are not yet clear, although some degree of neuroprotection has been shown by treatment with basic fibroblast growth factor (Fraser et al. 1997) or cell grafts from PrP null mice (K. Brown et al. 2001).

Molecular and Cellular Mechanisms of Vacuolar Degeneration

The growing body of evidence reviewed above indicates that accumulation of PrP^{Sc} triggers a series of events in nerve cells that ultimately result

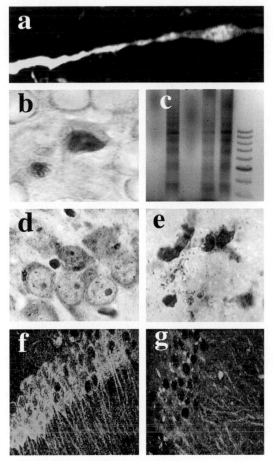

Figure 26. Neurodegenerative changes in the CA2 region of the hippocampus of VM mice infected with 87V scrapie. (*a*) Loss of dendritic spines by 70 days postinoculation (Lucifer yellow injection). (*b*) Apoptosis revealed by TUNEL labeling of CA2 neurons. (*c*) Characteristic DNA laddering in extracts from the CA2 region. (*d*) Activated caspase 3 in CA2 neurons (immunohistochemistry). (*e*) Activated microglia surrounding an amyloid plaque in the CA2 region (immunohistochemistry). (*f,g*) Tubulin immunofluorescence shows disruption of the neuronal cytoskeleton: (*f*) Control. (*g*) Terminally ill, 87V infected. (*a,b,c,d*, modified, with permission, from Jamieson et al. 2001a,b; *e*, courtesy of Dr. Debbie Brown.)

in apoptotic death. Much less is known about the cellular and molecular mechanisms of synapse degeneration and loss. The principal neuropathologic change that best correlates with synapse degeneration is vacuolation of the gray matter neuropil (spongiform degeneration) (Masters and Richardson 1978; Beck and Daniel 1987). Vacuolation occurs primarily in the region of synapses and is characterized by focal neuritic swelling, loss of internal organelles, and accumulation of abnormal membranes (Lampert et al. 1972; Chou et al. 1980). Spongiform degeneration suggests focal abnormalities of the plasma membrane with failure of electrolyte and water homeostasis. Regions of synapses might be particularly vulnerable because of their high density of neurotransmitter receptors and ion channels dedicated to information transfer between neurons.

There are only a few plausible hypotheses backed by data concerning mechanisms of vacuolar degeneration as a result of PrP^{Sc} deposition in the brain. In one scenario, PrP^{Sc} is thought to accumulate in lysosomes and trigger dysfunction as these aberrantly folded molecules are released into the cytosol (Laszlo et al. 1992). The lysosome hypothesis was originally proposed because PrP^{Sc} could only be localized with certainty in lysosomes and in the adjacent cytosol by Immunogold electron microscopy of a scrapie-infected cell line (McKinley et al. 1991b). A more likely scenario is emerging from recent biochemical measurements of PrP^{Sc} in cell fractions derived from scrapie-infected cell lines. These indicate that PrP^{Sc} is formed from PrP^C in CLDs of the plasma membrane where it accumulates, as well as in lysosomes (Gorodinsky and Harris 1995; Taraboulos et al. 1995; Vey et al. 1996; Kaneko et al. 1997a). Accumulation of PrP^{Sc} in CLDs may generate aberrant responses to stimuli mediated through signal transduction systems (Anderson 1993). Accumulation of PrP^{Sc} at or near the surface of a cell may account for a number of plasma membrane abnormalities that have been detected, including decreased plasma membrane fluidity, decreased binding affinity of bradykinin (Bk) receptors, decreased Bk-stimulated inositol triphosphate (IP$_3$) response, and decreased Bk-stimulated intracellular calcium response (Kristensson et al. 1993; DeArmond et al. 1996; Wong et al. 1996). In fact, focal vacuolar degeneration of neurons in prion diseases might be best explained by focal plasma membrane abnormalities controlling transsynaptic signaling and water and electrolyte balance. A third scenario that could result either from abnormal plasma membrane signaling or from accumulation of PrP^{Sc} in several subcellular compartments is a crippled stress response triggered by malfolded, undegradable PrP (Kenward et al. 1994; Tatzelt et al. 1995).

Subtypes of Pathogenic PrP

The failure to find protease-resistant PrP preceding synapse loss, dendritic atrophy, or nerve cell loss in some animal models of scrapie requires commentary by the senior author of this chapter (S. DeA.). Negative correlations such as the one described above for the 87V scrapie strain are the exception rather than the rule. Protease-resistant PrP^{Sc} deposition precedes and colocalizes precisely with neuropathological changes with the great majority of scrapie prion strains. How then can the exceptions be explained? One possibility is the discovery that there is more than one form of pathogenic prion protein. In addition, we also know today that β-sheeted PrP^{Sc} varies greatly in its ability to be denatured and, therefore, in its ability to be detected by immunological techniques.

First, there are transmembrane forms of pathogenic PrP, ^{tm}PrP, that are protease sensitive but, nevertheless, cause gray matter vacuolation. The Lingappa laboratory found that the prion protein molecule has a stop-transfer-effector (STE) sequence between residues 104 and 112 that can determine whether or not the glycolipid-anchored PrP molecule will reside entirely on the outside of the plasma membrane (e.g., secreted PrP^{C}) or will acquire a transmembrane topology (e.g., ^{tm}PrP) from where it has the possibility of being transported into the cytosol (Hegde et al. 1998, 1999). In this regard, it was recently reported that 10% of wild-type PrP^{C} is transported through the ER membrane into the cytosol in vitro where it is ubiquinated and degraded by proteasomes (Yedidia et al. 2001), which is consistent with earlier immunohistochemical studies that found PrP^{C} normally in the cytosol of small numbers of neurons in vivo (DeArmond et al. 1987). The importance of the STE domain was verified when amino acid substitutions within it were found to foster a PrP transmembrane topology (Hegde et al. 1998). Substitution of the K and H to I and I at residues 110 and 111, respectively, caused 50% of the mutated PrP to adopt a transmembrane topography rather than the usual secreted topography. The portion of the PrP molecule that becomes transmembrane is immediately downstream from the STE, residues 112–135. Tg mice in which this PrP construct was expressed developed a spontaneous neurodegenerative disease characterized by spongiform degeneration and reactive astrocytic gliosis that resembled scrapie. PrP (KH-II) in brain homogenates from these Tg mice was significantly less resistant to PK digestion than PrP^{Sc} found in scrapie-infected brains. In addition, PK digestion resulted in the formation of highly truncated, potentially amyloidogenic PrP peptides. The preference for a transmembrane topology was also found for mutated PrPs genetically linked to GSS. Interestingly, GSS is characterized by the formation of numerous amyloid plaques that are also composed of highly truncated PrP peptides that contain residues 112–135 (Fig. 4). GSS is also characterized by protease sensitivity of mutated PrP not located in amyloid plaques. In human GSS(A117V) and in Tg mice expressing PrP(A117V), about 35% of the mutated PrP was found to adopt a transmembrane topology and to be PK sensitive (Hegde et al. 1999). Non-amyloid mutated PrP in the gray matter neuropil in cases of GSS(P102L) is also PK sensitive (Hsiao et al. 1990), raising the possibility that it adopts a transmembrane topology. The one neuropathological feature found in scrapie, CJD, GSS, and Tg(PrP KH-II) mice is the same, vacuolar (spongiform) degeneration of gray matter, suggesting that PrP abnormalities, whether due to abnormal conformations or to abnormal transmembrane topography, alter the same neuronal compartment. ^{tm}PrP has also been detected in scrapie-infected Syrian hamsters (Hegde et al. 1999). Therefore, one cannot exclude the possibility that some scrapie strains, such as 87V, might cause neurodegeneration by preferentially inducing a transmembrane topology to the PrP molecule that would make it protease sensitive, less likely to be detected by standard immunocyto-chemical methods, and pathogenic. Because 87V is an amyloidogenic scrapie strain (Fraser and Bruce 1973), it can be hypothesized that the highly amyloidogenic 90–140 PrP peptide that forms into amyloid in GSS is likely to be protected from degradation by having a transmembrane location. The 87V scrapie strain might be amyloidogenic because it causes PrP to adopt a transmembrane topology, which protects a similar amyloidogenic portion of PrP from degradation by cellular proteases.

A second possible reason for the failure to detect an association between neuropathological changes and PrP^{Sc} accumulation in some scrapie models is the growing evidence that the PrP^{Sc} linked to each prion strain has a different β-sheeted conformation and intermolecular aggregation state and that these differences affect both its ability to be denatured and its rate of degradation by proteases. A number of antibodies directed against different epitopes from the amino to the carboxyl termini of the PrP molecule indicate that the amino-terminal region, specifically from residue 90 to 120, is critical for PrP^{Sc} formation (Muramoto et al. 1996; Peretz et al. 1997). Synthetic peptides corresponding to this region have considerable flexibility consistent with an ability to convert from α-helix to β-sheet conformation (Zhang et al. 1995). Moreover, this region appears to have considerable plasticity, a fact emphasized by the reports that two distinct prion strains exhibit different sites of proteolytic cleavage within this region (Bessen and Marsh 1994; Telling et al. 1996b). More is being learned about the steps in polymerization of PrP^{Sc}. Image processing of negative-stain electron microscopy of 2D protein crystals of PrP 27-30, whose asparagine-linked carbohydrates were labeled with 1.4-nm gold particles, suggests that six PrP 27-30 molecules form a hexagonal multimer with the carbohydrates on the outside of the hexamer (Wille et al. 2002). The best computer model for the hexagonal image seen by electron microscopy was predicted to be six PrP 27-30 molecules with their residues 90–170 formed into a stack of four to five parallel β-helical folds. This amino-terminal β-helix has a triangular cross-section when the stack is viewed from above. The computer model further predicted that two of the sides of the triangular β-helix bind to adjacent PrP 27-30 β-helices for a total of six that form a planar hexamer. In this model, the structurally conserved carboxy-terminal α-helices and the asparagine-linked carbohydrate trees attached to them are located at the periphery of the hexagonal arrangement of six PrP 27-30 molecules. The planar shape of this model provides for vertical assembly of other hexamers into filamentous structures. The relevance to the current discussion is that it illustrates the potential complexity of the PrP^{Sc} denaturation process required for antibody recognition of epitopes buried within its β-structure. Denaturation potentially requires breaking stacks of hexamers, separating individual PrP 27-30 molecules, and breaking hydrogen bonds holding multiple layers of β-helices together. It should also be noted that this is likely to be one of many possible multimers of PrP^{Sc}.

At the molecular level, the only known variable that appears to distinguish different prion strains is the conformation of their PrP^{Sc} molecules. The most recent evidence in support of this hypothesis grew out of a comparison of the properties of PrP^{Sc} comprising two Syrian hamster prion strains (Peretz et al. 2002). Sc237-PrP^{Sc} and DY-PrP^{Sc} were exposed to increasing concentrations of the denaturant, guanidinium HCl (GdnHCl), followed by exposure to PK. PrP^{Sc} surviving this step was then fully denatured with a high concentration of guanidium isothiocyanate to further increase its immunoreactivity for ELISA detection. The "apparent fractional appearance" (Fapp) of PrP^{Sc} was calculated to normalize each ELISA optical density measurement, which was then plotted as a function of GdnHCl concentration (Fig. 27). No Sc237-PrP^{Sc} was degraded by PK until a GdnHCl concentration of about 1.25 M was reached. In contrast, PK digestion of DY-PrP^{Sc} began at 0.25 M GdnHCl. Moreover, DY-PrP^{Sc} was fully degraded by 1.25 M GdnHCl, whereas there was little or no effect of the same concentration on Sc237-PrP^{Sc}. When this conformation-dependent degradation assay was used to study eight different prion strains that propagate in Syrian hamsters, strains that cause disease with similar incubation times grouped together according to similar conformational properties of their PrP^{Sc} (Peretz et al. 2001).

A growing body of evidence indicates some prion strains cause a host's PrP^{C} to adopt a protease-sensitive, transmembrane topology, whereas other strains might result in the formation of pathogenic, secreted forms of PrP^{Sc} that are relatively protease sensitive at one extreme or very difficult to denature at the other. In either extreme case, PrP^{Sc} would

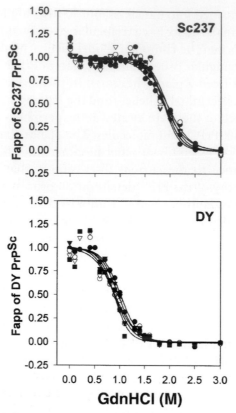

Figure 27. Marked differences in the denaturation transitions of Sc237 and DY PrP[Sc] (see text). (Adapted from Peretz et al. 2002.)

be difficult to detect. In addition to these possibilities, the amount of PrP[Sc] necessary to cause neurodegenerative changes may vary considerably with each strain of agent, with the result that pathogenic levels in some cases may be undetectable by immunohistochemistry. Therefore, the failure to find pathogenic PrP preceding pathological changes is subject to many interpretations (i.e., data are as strong as their detection limits). In some cases, the exception to the rule does not establish a new rule; rather, the exceptions teach us more about how the rule really works.

SUMMARY

The revision of this chapter for the second edition of this book has come during the first year of compassionate treatment of CJD cases with quinacrine, which was once used as an antimalarial drug. It is too early to determine whether or not quinacrine is an effective pharmaceutical agent for human and animal prion diseases; however, therapeutic trials were begun with it because it efficiently cleared PrP^{Sc} from scrapie-infected neuroblastoma cell lines without apparent damage to the cells (Korth et al. 2001). Additional favorable characteristics include its ability to cross the blood-brain barrier and its low toxicity in humans. For those research neuropathologists whose investigations into the molecular and cellular mechanisms of nerve cell dysfunction, degeneration, and death in prion diseases led to the hypothesis that accumulation of PrP^{Sc} causes the clinically relevant neuropathological changes, the therapeutic goal of clearing PrP^{Sc} from the brains of CJD patients to prevent further neurodegeneration is very rewarding. Relevant to the treatment of CJD, the revision of this chapter has focused on the step-wise progression of neurodegenerative changes in prion diseases that begins with dysfunction and degeneration of axon terminals, followed by dendritic atrophy, and last by nerve cell death. There is little hope for patient improvement after a "critical number" of neurons have been lost. The good news from the current neuropathological data is that there is likely to be a significantly prolonged interval between initial synapse degeneration and terminal nerve cell loss during which an effective therapeutic intervention could prevent further clinical deterioration. In this regard, early diagnosis of CJD is essential for any treatment to be effective, because of the rapid rate of PrP^{Sc} formation and accumulation. An eminent neurologist-scientist recently remarked that there will no longer be a need for neuropathologic evaluation of prion diseases because a cure for CJD was imminent! However, our experience indicates an even greater need for neuropathology, specifically for quantitative neuropathologic assessment of treated human and animal prion diseases to determine whether a therapeutic agent effectively clears PrP^{Sc} from the brain and halts progressive neurodegenerative changes. Moreover, quantitative molecular and morphological methods will be required to test whether or not regrowth of dendritic trees, replacement of lost synapses, and improvement of mental function will follow clearance of PrP^{Sc} and, if not, whether additional clearance of activated microglia and/or treatment with trophic factors will also be required.

The discovery of the mechanisms of nerve cell degeneration in prion diseases has gone hand in hand with the discovery of prions and how they replicate. Viewing prion diseases from the perspective of the neuropathologic changes has emphasized the importance of both the conformation of PrP^{Sc} and of host-determined factors, particularly the amino acid and carbohydrate composition of PrP^{C}, that profoundly influence the prion disease phenotype and distinguish each prion strain.

ACKNOWLEDGMENTS

This work was supported by grants from the National Institutes of Health (NS-14069, AG-08967, AG-02132, AG-10770, and NS-22786) as well as by gifts from the Sherman Fairchild Foundation, the Harold and Leila Y. Mathers Foundation, and Centeon. For the first edition, the authors also thank Ms. Juliana Cayetano for the large number of immunohistochemical and neurohistologic stained tissue sections she made, Mr. Thomas Lisse for the many histoblots, and Mr. Ed Shimazu and Dr. Henry Sánchez for their help with image processing. For the revised edition, the authors thank Cynthia Cowdrey for additional neuropathological and immunohistochemical stains, and Helga Thordarson for editorial help.

REFERENCES

Alpers M. 1992. Reflections and highlights: A life with kuru. In *Prion diseases of humans and animals* (ed. S.B. Prusiner et al.), pp. 66–76. Ellis Horwood, New York.

Anderson J.R., Allen C.M.C., and Weller R.O. 1990. Creutzfeldt-Jakob disease following human pituitary-derived growth hormone administration. *Br. Neuropathol. Soc. Proc.* **16:** 543. (Abstr.)

Anderson R.G.W. 1993. Caveolae: Where incoming and outgoing messengers meet. *Proc. Natl. Acad. Sci.* **90:** 10909–10913.

Barry R.A., McKinley M.P., Bendheim P.E., Lewis G.K., DeArmond S.J., and Prusiner S.B. 1985. Antibodies to the scrapie protein decorate prion rods. *J. Immunol.* **135:** 603–613.

Beck E. and Daniel P.M. 1979. Kuru and Creutzfeldt-Jakob disease; neuropathological lesions and their significance. In *Slow transmissible diseases of the nervous system* (ed. S.B. Prusiner and W.J. Hadlow), vol. 1, pp. 253–270. Academic Press, New York.

———. 1987. Neuropathology of slow transmissible encephalopathies. In *Prions—Novel infectious pathogens causing scrapie and Creutzfeldt-Jakob disease* (ed. S.B. Prusiner and M.P. McKinley), pp. 331–385. Academic Press, Orlando, Florida.

Beck E., Daniel P.M., Davey A.J., Gajdusek D.C., and Gibbs C.J., Jr. 1982. The pathogenesis of transmissible spongiform encephalopathy—An ultrastructural study. *Brain* **105:** 755–786.

Beck E., Daniel P.M., Matthews W.B., Stevens D.L., Alpers M.P., Asher D.M., Gajdusek D.C., and Gibbs C.J., Jr. 1969. Creutzfeldt-Jakob disease: The neuropathology of a transmission experiment. *Brain* **92:** 699–716.

Belichenko P.V., Brown D., Jeffrey M., and Fraser J.R. 2000. Dendritic and synaptic alterations of hippocampal pyramidal neurons in scrapie-infected mice. *Neuropathol. Appl. Neurobiol.* **26:** 143–149.

Bendheim P.E., Barry R.A., DeArmond S.J., Stites D.P., and Prusiner S.B. 1984. Antibodies to a scrapie prion protein. *Nature* **310:** 418–421.

Berman P.H., Davidson G.S., and Becker L.E. 1988. Progressive neurological deterioration in a 14-year-old girl. *Pediatr. Neurosci.* **14:** 42–49.

Bessen R.A. and Marsh R.F. 1994. Distinct PrP properties suggest the molecular basis of strain variation in transmissible mink encephalopathy. *J. Virol.* **68:** 7859–7868.

Billette de Villemeur T., Beauvais P., Gourmelon M., and Richardet J.M. 1991. Creutzfeldt-Jakob disease in children treated with growth hormone. *Lancet* **337:** 864–865.

Billette de Villemeur T., Gelot A., Deslys J.P., Dormont D., Duyckaerts C., Jardin L., Denni J., and Robain O. 1994. Iatrogenic Creutzfeldt-Jakob disease in three growth hormone recipients: A neuropathological study. *Neuropathol. Appl. Neurobiol.* **20:** 111–117.

Bolton D.C., Meyer R.K., and Prusiner S.B. 1985. Scrapie PrP 27-30 is a sialoglycoprotein. *J. Virol.* **53:** 596–606.

Bockman J.M., Kingsbury D.T., McKinley M.P., Bendheim P.E., and Prusiner S.B. 1985. Creutzfeldt-Jakob disease prion proteins in human brains. *N. Engl. J. Med.* **312:** 73–78.

Borchelt D.R., Scott M., Taraboulos A., Stahl N., and Prusiner S.B. 1990. Scrapie and cellular prion proteins differ in their kinetics of synthesis and topology in cultured cells. *J. Cell Biol.* **110:** 743–752.

Bouzamondo E., Milroy A.M. Ralston H.J., III, Prusiner S.B., and DeArmond S.J. 2000. Selective neuronal vulnerability of GABAergic neurons during experimental scrapie infection: Insights from an ultrastructural investigation. *Brain Res.* **874:** 210–215.

Brandner S., Isenmann S., Raeber A., Fischer M., Sailer A., Kobayashi Y., Marino S., Weissmann C., and Aguzzi A. 1996. Normal host prion protein necessary for scrapie-induced neurotoxicity. *Nature* **379:** 339–343.

Brown D., Halliday W.G., Jeffrey M., and Fraser J.R. 1997. Visualising scrapie infected and normal neuron in the murine hippocampus using Golgi impregnation and confocal imaging of intracellular dye. *J. Cell Pathol.* **2:** 131–136.

Brown D., Belichenko P.V., Sales, J., Jeffrey M., and Fraser J.R. 2001. Early loss of dendritic spines in murine scrapie revealed by confocal analysis. *Neuroreport* **12:** 179–183.

Brown K., Brown J., Ritchie D., Sales J., and Fraser J.R. 2001. Fetal cell grafts provide long-term protection against scrapie induced neuronal loss. *Neuroreport* **12:** 77–82.

Brown P. 1985. Virus sterility for human growth hormone. *Lancet* **II:** 729–730.

Brown P., Preece M.A., and Will R.G. 1992. "Friendly fire" in medicine: Hormones, homografts, and Creutzfeldt-Jakob disease. *Lancet* **340:** 24–27.

Brown P., Goldfarb L.G., Gibbs C.J., Jr., and Gajdusek D.C. 1991. The phenotypic expression of different mutations in transmissible familial Creutzfeldt-Jakob disease. *Eur. J. Epidemiol.* **7:** 469–476.

Brown P., Cathala F., Raubertas R.F., Gajdusek D.C., and Castaigne P. 1987. The epidemiology of Creutzfeldt-Jakob disease: Conclusion of a 15-year investigation in France and review of the world literature. *Neurology* **37:** 895–904.

Brown P., Rodgers Johnson P., Cathala F., Gibbs C.J., Jr., and Gajdusek D.C. 1984. Creutzfeldt-Jakob disease of long duration: Clinicopathological characteristics, transmissibility, and differential diagnosis. *Ann. Neurol.* **16:** 295–304.

Brown P., Coker-Vann M., Pomeroy K., Franko M., Asher D.M., Gibbs C.J., Jr., and Gajdusek D.C. 1986. Diagnosis of Creutzfeldt-Jakob disease by Western blot identification of marker protein in human brain tissue. *N. Engl. J. Med.* **314:** 547–551.

Brown P., Gibbs C.J., Jr., Rodgers-Johnson P., Asher D.M., Sulima M.P., Bacote A., Goldfarb L.G., and Gajdusek D.C. 1994. Human spongiform encephalopathy: The National Institutes of Health series of 300 cases of experimentally transmitted disease. *Ann. Neurol.* **35:** 513–529.

Bruce M.E., Dickinson A.G., and Fraser H. 1976. Cerebral amyloidosis in scrapie in the mouse: Effect of agent strain and mouse genotype. *Neuropathol. Appl. Neurobiol.* **2:** 471–478.

Bruce M.E., McBride P.A., and Farquhar C.F. 1989. Precise targeting of the pathology of the sialoglycoprotein, PrP, and vacuolar degeneration in mouse scrapie. *Neurosci. Lett.* **102:** 1–6.

Bruce M.E., McConnell I., Fraser H., and Dickinson A.G. 1991. The disease characteristics of different strains of scrapie in *Sinc* congenic mouse lines: Implications for the nature of the agent and host control of pathogenesis. *J. Gen. Virol.* **72:** 595–603.

Bruce M.E., Will R.G., Ironside J.W., McConnell I., Drummond D., Suttie A., McCardle L., Chree A., Hope J., Birkett C., Cousens S., Fraser H., and Bostock C.J. 1997. Transmissions to mice indicate that "new variant" CJD is caused by the BSE agent. *Nature* **389:** 498–501.

Buchanan C.R., Preece M.A., and Milner R.D.G. 1991. Mortality, neoplasia and Creutzfeldt-Jakob disease in patients treated with pituitary growth hormone in the United Kingdom. *Br. Med. J.* **302:** 824–828.

Büeler H., Aguzzi A., Sailer A., Greiner R.-A., Autenried P., Aguet M., and Weissmann C. 1993. Mice devoid of PrP are resistant to scrapie. *Cell* **73:** 1339–1347.

Büeler H., Fischer M., Lang Y., Bluethmann H., Lipp H.-P., DeArmond S.J., Prusiner S.B., Aguet M., and Weissmann C. 1992. Normal development and behaviour of mice lacking the neuronal cell-surface PrP protein. *Nature* **356:** 577–582.

Bugiani O., Giaccone G., Verga L., Pollo B., Frangione B., Farlow M.R., Tagliavini F., and Ghetti B. 1993. βPP participates in PrP-amyloid plaques of Gerstmann-Sträussler-Scheinker disease, Indiana kindred. *J. Neuropathol. Exp. Neurol.* **52:** 64–70.

Butler D.A., Scott M.R.D., Bockman J.M., Borchelt D.R., Taraboulos A., Hsiao K.K., Kingsbury D.T., and Prusiner S.B. 1988. Scrapie-infected murine neuroblastoma cells produce protease-resistant prion proteins. *J. Virol.* **62:** 1558–1564.

Carlson G.A., Goodman P.A., Lovett M., Taylor B.A., Marshall S.T., Peterson-Torchia M., Westaway D., and Prusiner S.B. 1988. Genetics and polymorphism of the mouse prion gene complex: The control of scrapie incubation time. *Mol. Cell. Biol.* **8:** 5528–5540.

Carlson G.A., Kingsbury D.T., Goodman P.A., Coleman S., Marshall S.T., DeArmond S.J., Westaway D., and Prusiner S.B. 1986. Linkage of prion protein and scrapie incubation time genes. *Cell* **46:** 503–511.

Carlson G.A., Ebeling C., Yang S.-L., Telling G., Torchia M., Groth D., Westaway D., DeArmond S.J., and Prusiner S.B. 1994. Prion isolate specified allotypic interactions between the cellular and scrapie prion proteins in congenic and transgenic mice. *Proc. Natl. Acad. Sci.* **91:** 5690–5694.

Caughey B. and Raymond G.J. 1991. The scrapie-associated form of PrP is made from a cell surface precursor that is both protease- and phospholipase-sensitive. *J. Biol. Chem.* **266:** 18217–18223.

Caughey B., Race R.E., Ernst D., Buchmeier M.J., and Chesebro B. 1989. Prion protein biosynthesis in scrapie-infected and uninfected neuroblastoma cells. *J. Virol.* **63:** 175–181.

Caughey B.W., Dong A., Bhat K.S., Ernst D., Hayes S.F., and Caughey W.S. 1991. Secondary structure analysis of the scrapie-associated protein PrP 27-30 in water by infrared spectroscopy. *Biochemistry* **30:** 7672–7680.

Chou S.M., Payne W.N., Gibbs C.J., Jr., and Gajdusek D.C. 1980. Transmission and scanning electron microscopy of spongiform change in Creutzfeldt-Jakob disease. *Brain* **103:** 885–904.

Cohen F.E., Pan K.-M., Huang Z., Baldwin M., Fletterick R.J., and Prusiner S.B. 1994. Structural clues to prion replication. *Science* **264:** 530–531.

Collinge J., Sidle K., Meads J., Ironside J., and Hill A. 1996. Molecular analysis of prion strain variation and the aetiology of 'new variant' CJD. *Nature* **383:** 685–690.

Cousens S.N., Vynnycky E., Zeidler M., Will R.G., and Smith P.G. 1997. Predicting the CJD epidemic in humans. *Nature* **385:** 197–198.

Croxson M., Brown P., Synek B., Harrington M.G., Frith R., Clover G., Wilson J., and Gajdusek D.C. 1988. A new case of Creutzfeldt-Jakob disease associated with human growth hormone therapy in New Zealand. *Neurology* **38:** 1128–1130.

Cuillé J. and Chelle P.L. 1939. Experimental transmission of trembling to the goat. *C.R. Seances Acad. Sci.* **208:** 1058–1060.

DeArmond S.J. and Prusiner S.B. 1995. Etiology and pathogenesis of prion diseases. *Am. J. Pathol.* **146:** 785–811.

———. 1996. Prions. In *Anderson's pathology* (ed. D.T. Purtilo and I. Damjanov), pp. 1042–1060. C.V. Mosby, St. Louis, Missouri.

———. 1998. Prion diseases. In *Neuropathology of dementing disorders* (ed. W.R. Markesbery), pp. 340–376. Arnold, London.

DeArmond S.J., Kretzschmar H.A., and Prusiner S.B. 2002. Prion diseases. In *Greenfield's neuropathology*, 7th edition (ed. D.I. Graham and P.L. Lantos), vol. 2, pp. 273–323. Arnold (Oxford University Press), London.

DeArmond S.J., Kristensson K., and Bowler R.P. 1992. PrP^{Sc} causes nerve cell death and stimulates astrocyte proliferation: A paradox. *Prog. Brain Res.* **94:** 437–446.

DeArmond S.J., McKinley M.P., Barry R.A., Braunfeld M.B., McColloch J.R., and Prusiner S.B. 1985. Identification of prion amyloid filaments in scrapie-infected brain. *Cell* **41:** 221–235.

DeArmond S.J., Mobley W.C., DeMott D.L., Barry R.A., Beckstead J.H., and Prusiner S.B. 1987. Changes in the localization of brain prion proteins during scrapie infection. *Neurology* **37:** 1271–1280.

DeArmond S.J., Qiu Y., Sànchez H., Spilman P.R., Ninchak-Casey A., Alonso D., and Daggett V. 1999. PrP^{C} glycoform heterogeneity as a function of brain region: Implications for selective targeting of neurons by prion strains. *J. Neuropathol. Exp. Neurol.* **58:** 1000–1009.

DeArmond S.J., Qiu Y., Wong K., Nixon R., Hyun W., Prusiner S.B., and Mobley W.C. 1996. Abnormal plasma membrane properties and functions in prion-infected cell lines. *Cold Spring Harbor Symp. Quant. Biol.* **61:** 531–540.

DeArmond S.J., Yang S.-L., Lee A., Bowler R., Taraboulos A., Groth D., and Prusiner S.B. 1993. Three scrapie prion isolates exhibit different accumulation patterns of the prion protein scrapie isoform. *Proc. Natl. Acad. Sci.* **90:** 6449–6453.

DeArmond S.J., Sanchez H., Qiu Y., Ninchak-Casey A., Daggett V., Paminiano-Camerino A., Cayetano J., Yehiely F., Rogers M., Groth D., Torchia M., Tremblay P., Scott M.R., Cohen F.E., and Prusiner S.B. 1997. Selective neuronal targeting in prion diseases. *Neuron* **19:** 1337–1348.

de Silva R., Ironside J.W., McCardle L., Esmonde T., Bell J., Will R., Windl O., Dempster M., Estibeiro P., and Lathe R. 1994. Neuropathological phenotype and "prion protein" genotype correlation in sporadic Creutzfeldt-Jakob disease. *Neurosci. Lett.* **179:** 50–52.

Dickinson A.G. and Meikle V.M.H. 1971. Host-genotype and agent effects in scrapie incubation: Change in allelic interaction with different strains of agent. *Mol. Gen. Genet.* **112:** 73–79.

Dickinson A.G., Meikle V.M.H., and Fraser H. 1968. Identification of a gene which controls the incubation period of some strains of scrapie agent in mice. *J. Comp. Pathol.* **78:** 293–299.

Dlouhy S.R., Hsiao K., Farlow M.R., Foroud T., Conneally P.M., Johnson P., Prusiner S.B., Hodes M.E., and Ghetti B. 1992. Linkage of the Indiana kindred of Gerstmann-Sträussler-Scheinker disease to the prion protein gene. *Nat. Genet.* **1:** 64–67.

Doh-ura K., Tateishi J., Sasaki H., Kitamoto T., and Sakaki Y. 1989. Pro→Leu change at position 102 of prion protein is the most common but not the sole mutation related to Gerstmann-Sträussler syndrome. *Biochem. Biophys. Res. Commun.* **163:** 974–979.

Dormont D., Delpech A., Courcel M.-N., Viret J., Markovitz P., and Court L. 1981.

Hyperproduction de proteine glio-fibrillarie acide (GFA) au cours de l'évolution de la tremblante de la souris. *C.R. Acad. Sci.* **293:** 53–56.

Eigen M. 1996. Prionics or the kinetic basis of prion diseases. *Biophys. Chem.* **63:** A11–A18.

Ellis C.J., Katifi H., and Weller R.O. 1992. A further British case of growth hormone induced Creutzfeldt-Jakob disease. *J. Neurol. Neurosurg. Psychiatry* **55:** 1200–1202.

Endo T., Groth D., Prusiner S.B., and Kobata A. 1989. Diversity of oligosaccharide structures linked to asparagines of the scrapie prion protein. *Biochemistry* **28:** 8380–8388.

Farlow M.R., Yee R.D., Dlouhy S.R., Conneally P.M., Azzarelli B., and Ghetti B. 1989. Gerstmann-Sträussler-Scheinker disease. I. Extending the clinical spectrum. *Neurology* **39:** 1446–1452.

Fradkin J.E., Schonberger L.B., Mills J.L., Gunn W.J., Piper J.M., Wysowski D.K., Thomson R., Durako S., and Brown P. 1991. Creutzfeldt-Jakob disease in pituitary growth hormone recipients in the United States. *J. Am. Med. Assoc.* **265:** 880–884.

Fraser H. 1979. Neuropathology of scrapie: The precision of the lesions and their diversity. In *Slow transmissible diseases of the nervous system* (ed. S.B. Prusiner and W.J. Hadlow), vol. 1, pp. 387–406. Academic Press, New York.

———. 1993. Diversity in the neuropathology of scrapie-like diseases in animals. *Br. Med. Bull.* **49:** 792–809.

Fraser H. and Bruce M.E. 1973. Argyrophilic plaques in mice inoculated with scrapie from particular sources. *Lancet* **I:** 617–618.

———. 1983. Experimental control of cerebral amyloid in scrapie in mice. *Prog. Brain Res.* **59:** 281–290.

Fraser H. and Dickinson A.G. 1968. The sequential development of the brain lesions of scrapie in three strains of mice. *J. Comp. Pathol.* **78:** 301–311.

———. 1973. Scrapie in mice. Agent-strain differences in the distribution and intensity of grey matter vacuolation. *J. Comp. Pathol.* **83:** 29–40.

Fraser J.R. 2002. What is the basis of TSE induced neurodegeneration and can it be repaired? *Neuropathol. Appl. Neurobiol.* **28:** 1–11.

Fraser J.R., Brown J., Bruce M.E., and Jeffrey M. 1997. Scrapie-induced neuron loss is reduced by treatment with basic fibroblast growth factor. *Neuroreport* **8:** 2405–2409.

Gajdusek D.C. and Zigas V. 1959. Clinical, pathological and epidemiological study of an acute progressive degenerative disease of the central nervous system among natives of the eastern highlands of New Guinea. *Am. J. Med.* **26:** 442–469.

Gajdusek D.C., Gibbs C.J., Jr., and Alpers M. 1966. Experimental transmission of a kuru-like syndrome to chimpanzees. *Nature* **209:** 794–796.

Gajdusek D.C., Gibbs C.J., Jr., Asher D.M., Brown P., Diwan A., Hoffman P., Nemo G., Rohwer R., and White L. 1977. Precautions in medical care of and in handling materials from patients with transmissible virus dementia (CJD). *N. Engl. J. Med.* **297:** 1253–1258.

Gambetti P., Parchi P., Petersen R.B., Chen S.G., and Lugaresi E. 1995. Fatal familial insomnia and familial Creutzfeldt-Jakob disease: Clinical, pathological and molecular features. *Brain Pathol.* **5:** 43–51.

Gasset M., Baldwin M.A., Fletterick R.J., and Prusiner S.B. 1993. Perturbation of the secondary structure of the scrapie prion protein under conditions associated with changes in infectivity. *Proc. Natl. Acad. Sci.* **90:** 1–5.

Gerstmann J., Sträussler E., and Scheinker I. 1936. Über eine eigenartige hereditär-familiäre Erkrankung des Zentralnervensystems zugleich ein Beitrag zur frage des vorzeitigen lokalen Alterns. *Z. Neurol. Psychiatrie* **154:** 736–762.

Ghani A.C., Ferguson N.M., Donnelly C.A., and Anderson R.M. 2000. Predicted vCJD mortality in Great Britain. *Nature* **406:** 583–584.

Ghetti B., Dlouhy S.R., Giaccone G., Bugiani O., Frangione B., Farlow M.R., and Tagliavini F. 1995. Gerstmann-Sträussler-Scheinker disease and the Indiana kindred. *Brain Pathol.* **5:** 61–75.

Ghetti B., Piccardo P., Frangione B., Bugiani O., Giaccone G., Young K., Prelli F., Farlow M.R., Dlouhy S.R., and Tagliavini F. 1996a. Prion protein amyloidosis. *Brain Pathol.* **6:** 127–145.

Ghetti B., Tagliavini F., Masters C.L., Beyreuther K., Giaccone G., Verga L., Farlo M.R., Conneally P.M., Dlouhy S.R., Azzarelli B., and Bugiani O. 1989. Gerstmann-Sträussler-Scheinker disease. II. Neurofibrillary tangles and plaques with PrP-amyloid coexist in an affected family. *Neurology* **39:** 1453–1461.

Ghetti B., Piccardo P., Spillantini M.G., Ichimiya Y., Porro M., Perini F., Kitamoto T., Tateishi J., Seiler C., Frangione B., Bugiani O., Giaccone G., Prelli F., Goedert M., Dlouhy S.R., and Tagliavini F. 1996b. Vascular variant of prion protein cerebral amyloidosis with τ-positive neurofibrillary tangles: The phenotype of the stop codon 145 mutation in *PRNP. Proc. Natl. Acad. Sci.* **93:** 744–748.

Gibbs C.J., Jr., Asher D.M., Brown P.W., Fradkin J.E., and Gajdusek D.C. 1993. Creutzfeldt-Jakob disease infectivity of growth hormone derived from human pituitary glands. *N. Engl. J. Med.* **328:** 358–359.

Gibbs C.J., Jr., Gajdusek D.C., Asher D.M., Alpers M.P., Beck E., Daniel P.M., and Matthews W.B. 1968. Creutzfeldt-Jakob disease (spongiform encephalopathy): Transmission to the chimpanzee. *Science* **161:** 388–389.

Gibbs C.J., Jr., Joy A., Heffner R., Franko M., Miyazaki M., Asher D.M., Parisi J.E., Brown P.W., and Gajdusek D.C. 1985. Clinical and pathological features and laboratory confirmation of Creutzfeldt-Jakob disease in a recipient of pituitary-derived human growth hormone. *N. Engl. J. Med.* **313:** 734–738.

Giese A., Groschup M.H., Hess B., and Kretzschmar H.A. 1995. Neuronal cell death in scrapie-infected mice is due to apoptosis. *Brain Pathol.* **5:** 213–221.

Glasse R. and Lindenbaum S. 1992. Fieldwork in the South Fore: The process of ethnographic inquiry. In *Prion diseases of humans and animals* (ed. S.B. Prusiner et al.), pp. 77–91. Ellis Horwood, New York.

Glenner G.G. 1980. Amyloid deposits and amyloidosis. *N. Engl. J. Med.* **302:** 1283–1292.

Glenner G.G., Osserman E.F., Benditt E.P., Calkins E., Cohen A.S., and Zucker-Franklin D. 1986. *Amyloidosis.* Plenum Press, New York.

Goldfarb L.G., Petersen R.B., Tabaton M., Brown P., LeBlanc A.C., Montagna P., Cortelli P., Julien J., Vital C., Pendelbury W.W., Haltia M., Wills P.R., Hauw J.J., McKeever P.F., Monari L., Schrank B., Swergold G.D., Autilio-Gambetti L., Gajdusek D.C., Lugaresi E., and Gambetti P. 1992. Fatal familial insomnia and familial Creutzfeldt-Jakob disease: Disease phenotype determined by a DNA polymorphism. *Science* **258:** 806–808.

Goldgaber D., Goldfarb L.G., Brown P., Asher D.M., Brown W.T., Lin S., Teener J.W., Feinstone S.M., Rubenstein R., Kascsak R.J., Boellaard J.W., and Gajdusek D.C. 1989. Mutations in familial Creutzfeldt-Jakob disease and Gerstmann-Sträussler-Scheinker's syndrome. *Exp. Neurol.* **106:** 204–206.

Goldhammer Y., Gabizon R., Meiner Z., and Sadeh M. 1993. An Israeli family with Gerstmann-Sträussler-Scheinker disease manifesting the codon 102 mutation in the prion protein gene. *Neurology* **43:** 2718–2719.

Gorodinsky A. and Harris D.A. 1995. Glycolipid-anchored proteins in neuroblastoma cells form detergent-resistant complexes without caveolin. *J. Cell Biol.* **129:** 619–627.

Gray F., Chretien F., Adle-Biassette H., Dorandeu A., Ereau T., Delisle M.B., Kopp N., Ironside J.W., and Vital C. 1999. Neuronal apoptosis in Creutzfeldt-Jakob disease. *Neuropathol. Exp. Neurol.* **58:** 321–328.

Hadlow W.J. 1959. Scrapie and kuru. *Lancet* **II:** 289–290.

Haraguchi T., Fisher S., Olofsson S., Endo T., Groth D., Tarantino A., Borchelt D. R., Teplow D., Hood L., Burlingame A., Lycke E., Kobata A., and Prusiner S.B. 1989. Asparagine-linked glycosylation of the scrapie and cellular prion proteins. *Arch. Biochem. Biophys.* **274:** 1–13.

Hecker R., Taraboulos A., Scott M., Pan K.-M., Torchia M., Jendroska K., DeArmond S.J., and Prusiner S.B. 1992. Replication of distinct prion isolates is region specific in brains of transgenic mice and hamsters. *Genes Dev.* **6:** 1213–1228.

Hegde R.S., Tremblay P., Groth D., DeArmond S. J., Prusiner S.B., and Lingappa V.R. 1999. Transmissible and genetic prion diseases share a common pathway of neurodegeneration. *Nature* **402:** 822-826.

Hegde R.S., Mastrianni J.A., Scott M.R., DeFea K.A., Tremblay P., Torchia M., DeArmond S.J., Prusiner S.B., and Lingappa V.R. 1998. A transmembrane form of prion protein in neurodegenerative disease. *Science* **279:** 827–834.

Hill A.F., Desbruslais M., Joiner S., Sidle K.C., Gowland I., Collinge J., Doey L.J., and Lantos P. 1997. The same prion strain causes vCJD and BSE. *Nature* **389:** 448–450, 526.

Hill A.F., Butterworth R.J., Joiner S., Jackson G., Rossor M.N., Thomas D.J. Frosh A., Tolley N., Bell J.E., Spencer M., King A., Al-Sarrag S., Ironside J.W., Lantos P.L., and Collinge J. 1999. Investigation of variant Creutzfeldt-Jakob disease and other human prion diseases with tonsil biopsy samples. *Lancet* **353:** 183–189.

Hilton D.A., Fathers E., Edwards P., Ironside J.W., and Zajicek J. 1998. Prion immunoreactivity in appendix before clinical onset of variant Creutzfeldt-Jakob disease. *Lancet* **352:** 703–704.

Hsiao K. and Prusiner S.B. 1991. Molecular genetics and transgenic model of Gerstmann-Sträussler-Scheinker disease. *Alzheimer Dis. Assoc. Disord.* **5:** 155–162.

Hsiao K.K., Doh-ura K., Kitamoto T., Tateishi J., and Prusiner S.B. 1989a. A prion protein amino acid substitution in ataxic Gerstmann-Sträussler syndrome. *Ann. Neurol.* **26:** 137.

Hsiao K.K., Scott M., Foster D., Groth D.F., DeArmond S.J., and Prusiner S.B. 1990. Spontaneous neurodegeneration in transgenic mice with mutant prion protein. *Science* **250:** 1587–1590.

Hsiao K.K., Cass C., Schellenberg G.D., Bird T., Devine-Gage E., Wisniewski H., and Prusiner S.B. 1991a. A prion protein variant in a family with the telencephalic form of Gerstmann-Sträussler-Scheinker syndrome. *Neurology* **41:** 681–684.

Hsiao K., Scott M., Foster D., DeArmond S.J., Groth D., Serban H., and Prusiner S.B. 1991b. Spontaneous neurodegeneration in transgenic mice with prion protein codon 101 proline ARROW leucine substitution. *Ann. N.Y. Acad. Sci.* **640:** 166–170.

Hsiao K., Baker H.F., Crow T.J., Poulter M., Owen F., Terwilliger J.D., Westaway D., Ott J., and Prusiner S.B. 1989b. Linkage of a prion protein missense variant to Gerstmann-Sträussler syndrome. *Nature* **338:** 342–345.

Hsiao K., Dlouhy S., Farlow M.R., Cass C., Da Costa M., Conneally M., Hodes M.E., Ghetti B., and Prusiner S.B. 1992. Mutant prion proteins in Gerstmann-Sträussler-Scheinker disease with neurofibrillary tangles. *Nat. Genet.* **1:** 68–71.

Hsiao K.K., Groth D., Scott M., Yang S.-L., Serban H., Rapp D., Foster D., Torchia M., DeArmond S.J., and Prusiner S.B. 1994. Serial transmission in rodents of neurodegeneration from transgenic mice expressing mutant prion protein. *Proc. Natl. Acad. Sci.* **91:** 9126–9130.

Ikeda S., Yanagisawa N., Glenner G.G., and Allsop D. 1992. Gerstmann-Sträussler-Scheinker disease showing β-protein amyloid deposits in the peripheral regions of PrP-immunoreactive amyloid plaques. *Neurodegeneration* **1:** 281–288.

Ironside J.W. 1996. Review: Creutzfeldt-Jakob disease. *Brain Pathol.* **6**: 379–388.

James T.L., Liu H., Ulyanov N.B., Farr-Jones S., Zhang H., Donne D.G., Kaneko K., Groth D., Mehlhorn I., Prusiner S.B., and Cohen F.E. 1997. Solution structure of a 142-residue recombinant prion protein corresponding to the infectious fragment of the scrapie isoform. *Proc. Natl. Acad. Sci.* **94**: 10086–10091.

Jamieson E., Jeffrey M., Ironside J.W., and Fraser J.R. 2001a. Apoptosis and dendritic dysfunction precede prion protein accumulation in 87V scrapie. *Neuroreport* **12**: 2147–2153.

———. 2001b. Activation of Fas and caspase 3 precedes PrP accumulation in 87V scrapie. *Neuroreport* **12**: 3567–3572.

Jeffrey M., Goodsir C.M., Bruce M.E., McBride P.A., and Fraser J.R. 1997. In vivo toxicity of prion protein in murine scrapie: Ultrastructural and immunogold studies. *Neuropathol. Appl. Neurobiol.* **23**: 93–101.

Jeffrey M., Martin S., Barr J., Chong A., and Fraser J. R. 2001. Onset of accumulation of PrP^{res} in murine ME7 scrapie in relation to pathological and PrP immunohistochemical changes. *J. Comp. Pathol.* **124**: 20–28.

Jeffrey M., Halliday W.G., Bell J., Johnston A.R., MacLeod N.K., Ingham C., Sayers A.R., Brown D.A., and Fraser J.R. 2000. Synapse loss associated with abnormal PrP precedes neuronal degeneration in the scrapie-infected murine hippocampus. *Neuropathol. Appl. Neurobiol.* **26**: 41–54.

Jendroska K., Heinzel F.P., Torchia M., Stowring L., Kretzschmar H.A., Kon A., Stern A., Prusiner S.B., and DeArmond S.J. 1991. Proteinase-resistant prion protein accumulation in Syrian hamster brain correlates with regional pathology and scrapie infectivity. *Neurology* **41**: 1482–1490.

Kakulas B.A., Lecours A.R., and Gajdusek D.C. 1967. Further observations on the pathology of Kuru (a study of the two cerebra in serial section). *J. Neuropathol. Exp. Neurol.* **26**: 85–97.

Kaneko K., Vey M., Scott M., Pilkuhn S., Cohen F.E., and Prusiner S.B. 1997a. COOH-terminal sequence of the cellular prion protein directs subcellular trafficking and controls conversion into the scrapie isoform. *Proc. Natl. Acad. Sci.* **18**: 2333–2338.

Kaneko K., Zulianello L., Scott M., Cooper C.M., Wallace A.C., James T.L., Cohen F.E., and Prusiner S.B. 1997b. Evidence for protein X binding to a discontinuous epitope on the cellular prion protein during scrapie prion propagation. *Proc. Natl. Acad. Sci.* **94**: 10069–10074.

Kenward N., Hope J., Landon M., and Mayer R.J. 1994. Expression of polyubiquitin and heat-shock protein 70 genes increases in the later stages of disease progression in scrapie-infected mouse brain. *J. Neurochem.* **62**: 1870–1877.

Kitamoto T., Iizuka R., and Tateishi J. 1993a. An amber mutation of prion protein in Gerstmann-Sträussler syndrome with mutant PrP plaques. *Biochem. Biophys. Res. Commun.* **192**: 525–531.

Kitamoto T., Ogomori K., Tateishi J., and Prusiner S.B. 1987. Formic acid pretreatment enhances immunostaining of amyloid deposits in cerebral and systemic amyloids. *Lab. Invest.* **57**: 230–236.

Kitamoto T., Amano N., Terao Y., Nakazato Y., Isshiki T., Mizutani T., and Tateishi J. 1993b. A new inherited prion disease (PrP-P105L mutation) showing spastic paraparesis. *Ann. Neurol.* **34**: 808–813.

Kitamoto T., Shin R.-W., Doh-ura K., Tomokane N., Miyazono M., Muramoto T., and Tateishi J. 1992. Abnormal isoform of prion proteins accumulates in the synaptic structures of the central nervous system in patients with Creutzfeldt-Jakob disease. *Am. J. Pathol.* **140**: 1285–1294.

Kitamoto T., Tateishi J., Tashima I., Takeshita I., Barry R.A., DeArmond S.J., and Prusiner S.B. 1986. Amyloid plaques in Creutzfeldt-Jakob disease stain with prion protein antibodies. *Ann. Neurol.* **20:** 204–208.

Klatzo I., Gajdusek D.C., and Zigas V. 1959. Pathology of kuru. *Lab. Invest.* **8:** 799–847.

Klitzman R.L., Alpers M.P., and Gajdusek D.C. 1984. The natural incubation period of kuru and the episodes of transmission in three clusters of patients. *Neuroepidemiology* **3:** 3–20.

Koch T.K., Berg B.O., DeArmond S.J., and Gravina R.F. 1985. Creutzfeldt-Jakob disease in a young adult with idiopathic hypopituitarism. Possible relation to the administration of cadaveric human growth hormone. *N. Engl. J. Med.* **313:** 731–733.

Korth K., May B.C.H., Cohen F.E., and Prusiner S.B. 2001. Acridine and phenothiazine derivatives as pharmacotherapeutics for prion disease. *Proc. Natl. Acad. Sci.* **98:** 9836–9841.

Kretzschmar H.A., Ironside J.W., DeArmond S.J., and Tateishi J. 1996. Diagnostic criteria for sporadic Creutzfeldt-Jakob disease. *Arch. Neurol.* **53:** 913–920.

Kretzschmar H.A., Prusiner S.B., Stowring L.E., and DeArmond S.J. 1986. Scrapie prion proteins are synthesized in neurons. *Am. J. Pathol.* **122:** 1–5.

Kretzschmar H.A., Kufer P., Riethmüller G., DeArmond S.J., Prusiner S.B., and Schiffer D. 1992. Prion protein mutation at codon 102 in an Italian family with Gerstmann-Sträussler-Scheinker syndrome. *Neurology* **42:** 809–810.

Kretzschmar H.A., Honold G., Seitelberger F., Feucht M., Wessely P., Mehraein P., and Budka H. 1991. Prion protein mutation in family first reported by Gerstmann, Sträussler, and Scheinker. *Lancet* **337:** 1160.

Kristensson K., Feuerstein B., Taraboulos A., Hyun W.C., Prusiner S.B., and DeArmond S.J. 1993. Scrapie prions alter receptor-mediated calcium responses in cultured cells. *Neurology* **43:** 2335–2341.

Kulczycki J., Jedrzejowska H., Gajkowski K., Tarnowska-Dzidousko E., and Lojkowska W. 1991. Creutzfeldt-Jakob disease in young people. *Eur. J. Epidemiol.* **7:** 501–504.

Kuzuhara S., Kanazawa I., Sasaki H., Nakanishi T., and Shimamura K. 1983. Gerstmann-Sträussler-Scheinker's disease. *Ann. Neurol.* **14:** 216–225.

Lampert P.W., Gajdusek D.C., and Gibbs C.J., Jr. 1972. Subacute spongiform virus encephalopathies. Scrapie, kuru and Creutzfeldt-Jakob disease: A review. *Am. J. Pathol.* **68:** 626–652.

Lasmézas C.I., Deslys J.P., Demalmay R., Adjou K.T., Lamoury F., Dormont D., Robain O., Ironside J., and Hauw J.J. 1996. BSE transmission to macaques. *Nature* **381:** 743–744.

Laszlo L., Lowe J., Self T., Kenward N., Landon M., McBride T., Farquhar C., McConnell I., Brown J., Hope J., and Mayer R.J. 1992. Lysosomes as key organelles in the pathogenesis of prion encephalopathies. *J. Pathol.* **166:** 333–341.

Liberski P.P., Yanagihara R., Gibbs C.J., Jr., and Gajdusek D.C. 1989. White matter ultrastructural pathology of experimental Creutzfeldt-Jakob disease in mice. *Acta Neuropathol.* **79:** 1–9.

Lowenstein D.H., Butler D.A., Westaway D., McKinley M.P., DeArmond S.J., and Prusiner S.B. 1990. Three hamster species with different scrapie incubation times and neuropathological features encode distinct prion proteins. *Mol. Cell. Biol.* **10:** 1153–1163.

Lucassen P.J., Williams A., Chung W.C.J., and Fraser H. 1995. Detection of apoptosis in murine scrapie. *Neurosci. Lett.* **198:** 185–188.

Macario M.E., Vaisman M., Buescu A., Neto V.M., Araujo H.M.M., and Chagas C. 1991. Pituitary growth hormone and Creutzfeldt-Jakob disease (letter). *Br. Med. J.* **302:** 1149.

Mackenzie A. 1983. Immunohistochemical demonstration of glial fibrillary acidic protein in scrapie. *J. Comp. Pathol.* **93:** 251–259.

Malamud N. 1979. Creutzfeldt-Jakob's disease: A clincopathologic study. In *Slow transmissible diseases of the nervous system* (ed. S.B. Pruiner and W.J. Hadlow), vol. 1, pp. 271–285. Academic Press, New York.

Manetto V., Medori R., Cortelli P., Montagna P., Tinuper P., Baruzzi A., Rancurel G., Hauw J.-J., Vanderhaeghen J.-J., Mailleux P., Bugiani O., Tagliavini F., Bouras C., Rizzuto N., Lugaresi E., and Gambetti P. 1992. Fatal familial insomnia: Clinical and pathological study of five new cases. *Neurology* **42:** 312–319.

Marzewski D.J., Towfighi J., Harrington M.G., Merril C.R., and Brown P. 1988. Creutzfeldt-Jakob disease following pituitary-derived human growth hormone therapy: A new American case. *Neurology* **38:** 1131–1133.

Masters C.L. and Richardson E.P., Jr. 1978. Subacute spongiform encephalopathy (Creutzfeldt-Jakob disease). The nature and progression of spongiform change. *Brain* **101:** 333–344.

Masters C.L., Gajdusek D.C., and Gibbs C.J., Jr. 1981a. Creutzfeldt-Jakob disease virus isolations from the Gerstmann-Sträussler syndrome. *Brain* **104:** 559–588.

———. 1981b. The familial occurrence of Creutzfeldt-Jakob disease and Alzheimer's disease. *Brain* **104:** 535–558.

Masters C.L., Gajdusek D.C., Gibbs C.J., Jr., Bernouilli C., and Asher D.M. 1979. Familial Creutzfeldt-Jakob disease and other familial dementias: An inquiry into possible models of virus-induced familial diseases. In *Slow transmissible diseases of the nervous system* (ed. S.B. Pruiner and W.J. Hadlow), vol.1, pp. 143–194. Academic Press, New York.

Mastrianni J.A., Capellari S., Telling G.C., Han D., Bosque P., Pruiner S.B., and DeArmond S.J. 2001. Inherited prion disease caused by the V210I mutation: Transmission to transgenic mice. *Neurology* **57:** 2198–2205.

Mastrianni J.A., Curtis M.T., Oberholtzer J.C., Da Costa M.M., DeArmond S., Pruiner S.B., and Garbern J.Y. 1995. Prion disease (PrP-A117V) presenting with ataxia instead of dementia. *Neurology* **45:** 2042–2050.

Mastrianni J.A., Nixon R., Layzer R., Telling G.C., Han D., DeArmond S.J., and Pruiner S.B. 1999. Prion protein conformation in a patient with sporadic fatal insomnia. *N. Engl. J. Med.* **340:** 1630–1638.

McKinley M.P., Meyer R., Kenaga L., Rahbar F., Cotter R., Serban A., and Pruiner S.B. 1991a. Scrapie prion rod formation *in vitro* requires both detergent extraction and limited proteolysis. *J. Virol.* **65:** 1440–1449.

McKinley M.P., Taraboulos A., Kenaga L., Serban D., Stieber A., DeArmond S.J., Pruiner S.B., and Gonatas N. 1991b. Ultrastructural localization of scrapie prion proteins in cytoplasmic vesicles of infected cultured cells. *Lab. Invest.* **65:** 622–630.

Medori R., Montagna P., Tritschler H.J., LeBlanc A., Cortelli P., Tinuper P., Lugaresi E., and Gambetti P. 1992a. Fatal familial insomnia: A second kindred with mutation of prion protein gene at codon 178. *Neurology* **42:** 669–670.

Medori R., Tritschler H.-J., LeBlanc A., Villare F., Manetto V., Chen H.Y., Xue R., Leal S., Montagna P., Cortelli P., Tinuper P., Avoni P., Mochi M., Baruzzi A., Hauw J.J., Ott J., Lugaresi E., Autilio-Gambetti L., and Gambetti P. 1992b. Fatal familial insomnia, a prion disease with a mutation at codon 178 of the prion protein gene. *N. Engl. J. Med.* **326:** 444–449.

Mohr M., Tranchant C., Heldt N., and Warter J.M. 1994. Alsatian variant of codon 117 form of Gerstmann-Sträussler-Scheinker syndrome: Autopsy study of 3 cases. *Brain Pathol.* **4:** 524. (Abstr.)

Monari L., Chen S.G., Brown P., Parchi P., Petersen R.B., Mikol J., Gray F., Cortelli P., Montagna P., Ghetti B., Goldfarb L.G., Gajdusek D.C., Lugaresi E., Gambetti P., and

Autilio-Gambetti L. 1994. Fatal familial insomnia and familial Creutzfeldt-Jakob disease: Different prion proteins determined by a DNA polymorphism. *Proc. Natl. Acad. Sci.* **91:** 2839–2842.

Monreal J., Collinge G.H., Masters C.L., Miller Fisher C., Ronald C.K., Gibbs C.J., Jr., and Gajdusek D.C. 1981. Creutzfeldt-Jakob disease in an adolescent. *J. Neurol. Sci.* **52:** 341–350.

Muramoto T., Kitamoto T., Tateishi J., and Goto I. 1992. The sequential development of abnormal prion protein accumulation in mice with Creutzfeldt-Jakob disease. *Am. J. Pathol.* **140:** 1411–1420.

Muramoto T., Scott M., Cohen F.E., and Prusiner S.B. 1996. Recombinant scrapie-like prion protein of 106 amino acids is soluble. *Proc. Natl. Acad. Sci.* **93:** 15457–15462.

Nakazato Y., Ohno R., Negishi T., Hamaguchi K., and Arai E. 1991. An autopsy case of Gerstmann-Sträussler-Scheinker's disease with spastic paraplegia as its principal feature. *Clin. Neurol.* **31:** 987–992.

New M.I., Brown P., Temeck J.W., Owens C., Hedley-Whyte E.T., and Richardson E.P. 1988. Preclinical Creutzfeldt-Jakob disease discovered at autopsy in a human growth hormone recipient. *Neurology* **38:** 1133–1134.

Nochlin D., Sumi S.M., Bird T.D., Snow A.D., Leventhal C.M., Beyreuther K., and Masters C.L. 1989. Familial dementia with PrP-positive amyloid plaques: A variant of Gerstmann-Sträussler syndrome. *Neurology* **39:** 910–918.

O'Connor S.E. and Imperiali B. 1996. Modulation of protein structure and function by asparagine-linked glycosylation. *Chem. Biol.* **3:** 803–812.

Packer R.J., Cornblath D.R., Gonatas N.K., Bruno L.A., and Asbury A.K. 1980. Creutzfeldt-Jakob disease in a 20-year-old woman. *Neurology* **30:** 492–496.

Pan K.-M., Baldwin M., Nguyen J., Gasset M., Serban A., Groth D., Mehlhorn I., Huang Z., Fletterick R.J., Cohen F.E., and Prusiner S.B. 1993. Conversion of α-helices into β-sheets features in the formation of the scrapie prion proteins. *Proc. Natl. Acad. Sci.* **90:** 10962–10966.

Parchi P., Castellani R., Capellari S., Ghetti B., Young K., Chen S.G., Farlow M., Dickson D.W., Sima A.A., Trojanowski J.Q., Petersen R.B., and Gambetti P. 1996. Molecular basis of phenotypic variability in sporadic Creutzfeldt-Jakob disease. *Ann. Neurol.* **39:** 767–778.

Pearlman R.L., Towfighi J., Pezeshkpour G.H., Tenser R.B., and Turel A.P. 1988. Clinical significance of types of cerebellar amyloid plaques in human spongiform encephalopathies. *Neurology* **38:** 1249–1254.

Peretz D., Scott M., Groth D., Williamson R.A., Burton D.R., Cohen F.E., and Prusiner S.B. 2001. Strain-specified relative conformational stability of the scrapie prion protein. *Protein Sci.* **10:** 854–863.

Peretz D., Williamson R.A., Legname G., Matsunaga Y., Vergara J., Burton D.R., DeArmond S.J., Prusiner S.B., and Scott M.R. 2002. A change in the conformation of prions accompanies the emergence of a new prion strain. *Neuron* **34:** 1–20.

Peretz D., Williamson R.A., Matsunaga Y., Serban H., Pinilla C., Bastidas R.B., Rozenshteyn R., James T.L., Houghten R.A., Cohen F.E., Prusiner S.B., and Burton D.R. 1997. A conformational transition at the N terminus of the prion protein features in formation of the scrapie isoform. *J. Mol. Biol.* **273:** 614–622.

Powell-Jackson J., Weller R.O., Kennedy P., Preece M.A., Whitcombe E.M., and Newsome-Davis J. 1985. Creutzfeldt-Jakob disease after administration of human growth hormone. *Lancet* **II:** 244–246.

Prusiner S.B. 1984. Prions. *Sci. Am.* **251:** 50–59.

Prusiner S.B., Gajdusek D.C., and Alpers M.P. 1982. Kuru with incubation periods exceeding two decades. *Ann. Neurol.* **12:** 1–9.

Prusiner S.B., McKinley M.P., Bowman K.A., Bolton D.C., Bendheim P.E., Groth D.F., and Glenner G.G. 1983. Scrapie prions aggregate to form amyloid-like birefringent rods. *Cell* **35:** 349–358.

Prusiner S.B., Groth D., Serban A., Koehler R., Foster D., Torchia M., Burton D., Yang S.-L., and DeArmond S.J. 1993. Ablation of the prion protein (PrP) gene in mice prevents scrapie and facilitates production of anti-PrP antibodies. *Proc. Natl. Acad. Sci.* **90:** 10608–10612.

Prusiner S.B., Scott M., Foster D., Pan K.-M., Groth D., Mirenda C., Torchia M., Yang S.-L., Serban D., Carlson G.A., Hoppe P.C., Westaway D., and DeArmond S.J. 1990. Transgenetic studies implicate interactions between homologous PrP isoforms in scrapie prion replication. *Cell* **63:** 673–686.

Race R.E., Priola S.A., Bessen R.A., Ernst D., Dockter J., Rall G.F., Mucke L., Chesebro B., and Oldstone M.B.A. 1995. Neuron-specific expression of a hamster prion protein minigene in transgenic mice induces susceptibility to hamster scrapie agent. *Neuron* **15:** 1183–1191.

Rademacher T.W., Parekh R.B., and Dwek R.A. 1988. Glycobiology. *Annu. Rev. Biochem.* **57:** 785–838.

Raeber A.J., Race R.E., Brandner S., Priola S.A., Sailer A., Bessen R.A., Mucke L., Manson J., Aguzzi A., Oldstone M.A., Chesebro B., and Weissmann C. 1997. Astrocyte-specific expression of hamster prion protein (PrP) renders PrP knockout mice susceptible to hamster scrapie. *EMBO J.* **16:** 6057–6065.

Richardson E.P.J. and Masters C.L. 1995. The nosology of Creutzfeldt-Jakob disease and conditions related to the accumulation of PrP^{CJD} in the nervous system. *Brain Pathol.* **5:** 33–41.

Roberts G.W., Lofthouse R., Brown R., Crow T.J., Barry R.A., and Prusiner S.B. 1986. Prion-protein immunoreactivity in human transmissible dementias. *N. Engl. J. Med.* **315:** 1231–1233.

Roos R., Gajdusek D.C., and Gibbs C.J., Jr. 1973. The clinical characteristics of transmissible Creutzfeldt-Jakob disease. *Brain* **96:** 1–20.

Rosenmann H., Halimi M., Kahana I., Biran I., and Gabizon R. 1997. Differential allelic expression of PrP mRNA in carriers of the E200K mutation. *Neurology* **49:** 851–856.

Safar J., Roller P.P., Gajdusek D.C., and Gibbs C.J., Jr. 1993. Conformational transitions, dissociation, and unfolding of scrapie amyloid (prion) protein. *J. Biol. Chem.* **268:** 20276–20284.

Schätzl H.M., Laszlo L., Holtzman D.M., Tatzelt J., DeArmond S.J., Weiner R.I., Mobley W.C., and Prusiner S.B. 1997. A hypothalamic neuronal cell line persistently infected with scrapie prions exhibits apoptosis. *J. Virol.* **71:** 8821–8831.

Scott M., Groth D., Foster D., Torchia M., Yang S.-L., DeArmond S.J., and Prusiner S.B. 1993. Propagation of prions with artificial properties in transgenic mice expressing chimeric PrP genes. *Cell* **73:** 979–988.

Scott M.R., Will R., Ironside J., Nguyen H.O., Tremblay P., DeArmond S.J., and Prusiner S.B. 1999. Compelling transgenetic evidence for transmission of bovine spongiform encephalopathy prions to humans. *Proc. Natl. Acad. Sci.* **96:** 15137–15142.

Scott M., Foster D., Mirenda C., Serban D., Coufal F., Wälchli M., Torchia M., Groth D., Carlson G., DeArmond S.J., Westaway D., and Prusiner S.B. 1989. Transgenic mice expressing hamster prion protein produce species-specific scrapie infectivity and amyloid plaques. *Cell* **59:** 847–857.

Seitelberger F. 1981a. Spinocerebellar ataxia with dementia and plaque-like deposits (Sträussler's disease). *Handb. Clin. Neurol.* **42:** 182–183.

———. 1981b. Straubler's disease. *Acta Neuropathol.* (suppl.) **7:** 341–343.

Serban D., Taraboulos A., DeArmond S.J., and Prusiner S.B. 1990. Rapid detection of Creutzfeldt-Jakob disease and scrapie prion proteins. *Neurology* **40:** 110–117.

Snow A.D., Willmer J., and Kisilevsky R. 1987. Sulfated glycosaminoglycans: A common constituent of all amyloids? *Lab. Invest.* **56:** 120–123.

Snow A.D., Kisilevsky R., Willmer J., Prusiner S.B., and DeArmond S.J. 1989. Sulfated glycosaminoglycans in amyloid plaques of prion diseases. *Acta Neuropathol.* **77:** 337–342.

Snow A.D., Wight T.N., Nochlin D., Koike Y., Kimata K., DeArmond S.J., and Prusiner S.B. 1990. Immunolocalization of heparan-sulfate proteoglycans to the prion protein amyloid plaques of Gerstmann-Sträussler syndrome, Creutzfeldt-Jakob disease and scrapie. *Lab. Invest.* **63:** 601–611.

Tagliavini F. and Pilleri G. 1983. Basal nucleus of Meynert. A neuropathological study in Alzheimer's disease, simple senile dementia, Pick's disease and Huntington's chorea. *J. Neurol. Sci.* **62:** 243–260.

Tagliavini F., Prelli F., Porro M., Rossi G., Giaccone G., Farlow M.R., Dlouhy S.R., Ghetti B., Bugiani O., and Frangione B. 1994. Amyloid fibrils in Gerstmann-Sträussler-Scheinker disease (Indiana and Swedish kindreds) express only PrP peptides encoded by the mutant allele. *Cell* **79:** 695–703.

Taraboulos A., Raeber A.J., Borchelt D.R., Serban D., and Prusiner S.B. 1992a. Synthesis and trafficking of prion proteins in cultured cells. *Mol. Biol. Cell* **3:** 851–863.

Taraboulos A., Jendroska K., Serban D., Yang S.-L., DeArmond S.J., and Prusiner S.B. 1992b. Regional mapping of prion proteins in brains. *Proc. Natl. Acad. Sci.* **89:** 7620–7624.

Taraboulos A., Scott M., Semenov A., Avrahami D., Laszlo L., and Prusiner S.B. 1995. Cholesterol depletion and modification of COOH-terminal targeting sequence of the prion protein inhibit formation of the scrapie isoform. *J. Cell Biol.* **129:** 121–132.

Taraboulos A., Rogers M., Borchelt D.R., McKinley M.P., Scott M., Serban D., and Prusiner S.B. 1990. Acquisition of protease resistance by prion proteins in scrapie-infected cells does not require asparagine-linked glycosylation. *Proc. Natl. Acad. Sci.* **87:** 8262–8266.

Tateishi J., Kitamoto T., Doh-ura K., Boellaard J.W., and Peiffer J. 1992. Creutzfeldt-Jakob disease with amyloid angiopathy: Diagnosis by immunological analyses and transmission experiments. *Acta Neuropathol.* **83:** 559–563.

Tateishi J., Doi H., Sato Y., Suetsugu M., Ishii K., and Kuroiwa Y. 1981. Experimental transmission of human subacute spongiform encephalopathy to small rodents. III. Further transmission from three patients and distribution patterns of lesions in mice. *Acta Neuropathol.* **53:** 161–163.

Tateishi J., Brown P., Kitamoto T., Hoque Z.M., Roos R., Wollman R., Cervenáková L., and Gajdusek D.C. 1995. First experimental transmission of fatal familial insomnia. *Nature* **376:** 434–435.

Tatzelt J., Zuo J., Voellmy R., Scott M., Hartl U., Prusiner S.B., and Welch W.J. 1995. Scrapie prions selectively modify the stress response in neuroblastoma cells. *Proc. Natl. Acad. Sci.* **92:** 2944–2948.

Telling G.C., Haga T., Torchia M., Tremblay P., DeArmond S.J., and Prusiner S.B. 1996a. Interactions between wild-type and mutant prion proteins modulate neurodegeneration in transgenic mice. *Genes Dev.* **10:** 1736–1750.

Telling G.C., Scott M., Mastrianni J., Gabizon R., Torchia M., Cohen F.E., DeArmond S.J., and Prusiner S.B. 1995. Prion propagation in mice expressing human and chimeric PrP transgenes implicates the interaction of cellular PrP with another protein. *Cell* **83:** 79–90.

Telling G.C., Parchi P., DeArmond S.J., Cortelli P., Montagna P., Gabizon R., Lugaresi E., Gambetti P., and Prusiner S.B. 1996b. Evidence for the conformation of the pathologic isoform of the prion protein enciphering and propagating prion diversity. *Science* **274:** 2079–2082.

Telling G.C., Scott M., Hsiao K.K., Foster D., Yang S.-L., Torchia M., Sidle K.C.L., Collinge J., DeArmond S.J., and Prusiner S.B. 1994. Transmission of Creutzfeldt-Jakob disease from humans to transgenic mice expressing chimeric human-mouse prion protein. *Proc. Natl. Acad. Sci.* **91:** 9936–9940.

Titner R., Brown P., Hedley-Whyte E.T., Rappaport E.B., Piccardo C.P., and Gajdusek D.C. 1986. Neuropathologic verification of Creutzfeldt-Jakob disease in the exhumed American recipient of human pituitary growth hormone: epidemiologic and pathogenetic implications. *Neurology* **36:** 932–936.

Tranchant C., Doh-ura K., Warter J.M., Steinmetz G., Chevalier Y., Hanauer A., Kitamoto T., and Tateishi J. 1992. Gerstmann-Sträussler-Scheinker disease in an Alsatian family: Clinical and genetic studies. *J. Neurol. Neurosurg. Psychiatry* **55:** 185–187.

Vey M., Pilkuhn S., Wille H., Nixon R., DeArmond S.J., Smart E.J., Anderson R.G.W., Taraboulos A., and Prusiner S.B. 1996. Subcellular colocalization of cellular and scrapie prion proteins in caveolae-like membranous domains. *Proc. Natl. Acad. Sci.* **93:** 14945–14949.

Westaway D., Goodman P.A., Mirenda C.A., McKinley M.P., Carlson G.A., and Prusiner S.B. 1987. Distinct prion proteins in short and long scrapie incubation period mice. *Cell* **51:** 651–662.

White A.R., Guirguis R., Brazier M.W., Jobling M.F., Hill A.F., Beyreuther K., Barrow C.J., Masters C.L., Collins S.J., and Cappai R. 2001 Sublethal concentrations of prion peptide PrP106-126 or the amylodi beta peptide of Alzheimer's disease activates expression of proapoptotic markers in primary cortical neurons. *Neurobiol. Dis.* **8:** 229–316.

Will R.G., Ironside J.W., Zeidler M., Cousens S.N., Estibeiro K., Alperovitch A., Poser S., Pocchiari M., Hofman A., and Smith P.G. 1996. A new variant of Creutzfeldt-Jakob disease in the UK. *Lancet* **347.** 921 925.

Wille H., Michelitsch M.D., Guénebaut V., Supattapone S., Serban A., Cohen F.E., Agard D.A., and Prusiner S.B. 2002. Structural studies of the scrapie prion protein by electron crystallography. *Proc. Natl. Acad. Sci.* **99:** 3563–3568.

Williams A., Lucassen P.J., Ritchie D., and Bruce M. 1997. PrP deposition, microglial activation, and neuronal apoptosis in murine scrapie. *Exp. Neurol.* **144:** 433–438.

Wong K., Qiu Y., Hyun W., Nixon R., VanCleff J., Sanchez-Salazar J., Prusiner S., and DeArmond S. 1996. Decreased receptor-mediated calcium response in prion-infected cells correlates with decreased membrane fluidity and IP3 release. *Neurology* **47:** 741–750.

Wu C., Butz S., Ying Y., and Anderson R.G.W. 1997. Tyrosine kinase receptors concentrated in caveolae-like domains from neuronal plasma membrane. *J. Biol. Chem.* **272:** 3554–3559.

Yamada M., Itoh Y., Fujigasaki H., Naruse S., Kaneko K., Kitamoto T., Tateishi J., Otomo E., Hayakawa M., Tanaka J., Matsushita M., and Miyatake T. 1993. A missense mutation at codon 105 with codon 129 polymorphism of the prion protein gene in a new variant of Gerstmann-Sträussler-Scheinker disease. *Neurology* **43:** 2723–2724.

Yedidia Y., Horonchik L., Tzaban S., Yanai A., and Taraboulos A. 2001. Proteasomes and ubiquitin are involved in the turnover of the wild-type prion protein. *EMBO J.* **20:** 5383–5391.

Ying Y.-S., Anderson R.G.W., and Rothberg K.G. 1992. Each caveola contains multiple glycosyl-phosphatidylinositol-anchored membrane proteins. *Cold Spring Harbor Symp. Quant. Biol.* **57:** 593–604.

Zhang H., Kaneko K., Nguyen J.T., Livshits T.L., Baldwin M.A., Cohen F.E., James T.L., and Prusiner S.B. 1995. Conformational transitions in peptides containing two putative α helices of the prion protein. *J. Mol. Biol.* **250:** 514–526.

Zigas V. and Gajdusek D.C. 1957. Kuru: Clinical study of a new syndrome resembling paralysis agitans in natives of the Eastern Highlands of Australian New Guinea. *Med. J. Aust.* **2:** 745–754.

16

Some Strategies and Methods for the Study of Prions

Stanley B. Prusiner,[1,2,3] Giuseppe Legname,[1,2] Stephen J. DeArmond,[1,4]
Fred E. Cohen,[1,3,5] and Jiri Safar[1,2]
[1]Institute for Neurodegenerative Diseases, Departments of [2]Neurology,
[3]Biochemistry and Biophysics, [4]Pathology, [5]Cellular and Molecular Pharmacology
University of California, San Francisco, California 94143

Detlev Riesner

Institut für Physikalische Biologie und Biologisch-Medizinisches Forschungszentrum
Heinrich-Heine Universität Düsseldorf
40225 Düsseldorf, Germany

Kiyotoshi Kaneko

National Center of Neurology and Psychiatry
and Core Research for Evolution Science and Technology
Kodaira, Tokyo 187-8502, Japan

THE DISCOVERY OF THE PRION PROTEIN (PrP) by enriching fractions prepared from scrapie-infected hamster brains transformed research on the prion diseases. Prior to identification of PrP 27-30, the protease-resistant core of the misfolded, disease-causing PrP isoform (PrP^{Sc}), almost all studies of prions required bioassays. With the isolation of PrP 27-30, the tools of molecular cloning, genetics, immunology, cell biology, and structural biology could be applied to study prions.

Many different areas of prion disease can now be investigated using a wide variety of approaches. Recombinant PrP expressed in bacteria or mammalian cells can be isolated in large quantities and used for structural studies (Hornemann and Glockshuber 1996; Mehlhorn et al. 1996; Blochberger et al. 1997). The biology of PrP^{Sc} formation as well as the molecular pathogenesis of prion diseases can be studied in cultured cells and transgenic (Tg) mice. Similarly, potential therapeutics might be evaluated by measuring the inhibition of PrP^{Sc} formation in scrapie-infected cultured cells or Tg mice.

Immunoassays for PrP^{Sc} show great promise as tools to diagnose prion disease rapidly and to screen for subclinical cases in humans and domestic animals (Safar et al. 1998, 2002; Biffiger et al. 2002). Because the particle-to-infectivity (P/I) ratio for prions is ~10^5 PrP^{Sc} molecules per ID$_{50}$ unit (Prusiner et al. 1983), it may be possible to develop an immunoassay that is 10- to 100-fold more sensitive than bioassays in animals.

The biology of prions is sufficiently advanced that there are now several reasonably rational approaches available for the development of effective therapies and preventive measures in humans (Chapter 18). First, drugs that alter the conformation of PrP^{Sc} and allow cells to degrade it might prove to be efficacious (Supattapone et al. 1999, 2001). Second, drugs that block the formation of nascent PrP^{Sc} by interfering with the binding of PrP^{C} to PrP^{Sc} might also prove to be effective (Caughey and Race 1992; Caughey et al. 1994; Ingrosso et al. 1995; Heppner et al. 2001; Peretz et al. 2001b). Third, pharmacotherapeutics that mimic dominant-negative inhibition of prion formation by interfering with binding of PrP^{C} to a conversion cofactor provisionally designated protein X may be the most potent strategy among those currently available (Kaneko et al. 1997; Perrier et al. 2000, 2002). A fourth largely empirical approach includes large-scale drug screening that may lead to the identification of an effective therapeutic (Doh-ura et al. 2000; Korth et al. 2001; Kocisko et al. 2003).

In this chapter, we attempt to provide an overview of the application of some modern technologies that have been adapted to prion research. Although we cover such applications rather selectively, we do try to point out where such studies are described in more detail in this monograph as well as in other sources. In addition, we describe a series of rather extensive studies that were mounted to search for a scrapie-specific nucleic acid. Although these studies were unsuccessful, we believe it is important to summarize the scope and thoroughness of that effort.

PRIONS

The only known essential component of the infectious prion particle is PrP^{Sc}, which has an apparent molecular weight (M_r) of 33 kD to 35 kD. PrP^{Sc} is a sialoglycoprotein that contains two complex-type oligosaccharides and a glycosylphosphatidyl inositol (GPI) anchor (Stahl et al. 1992; Rudd et al. 1999); it has a single disulfide bond. Secondary structure analysis by optical spectroscopy suggests that approximately half of the protein is folded into a β-sheet conformation. Electron crystallography argues that the β-structure is unlikely to be a sheet composed of parallel or antiparallel β-strands but more plausibly a β-helix (Wille et al. 2002).

In homogenates of scrapie-infected tissues, PrP^{Sc} is sedimented by centrifugation at 100,000g for 1 hour at 4°C and thus labeled as insoluble. Many attempts to solubilize PrP^{Sc} with detergents, organic solvents, and various salts under nondenaturing conditions in which prion infectivity is preserved have been unsuccessful (Wille et al. 1996, 2000). PrP^{Sc} seems to complex readily with lipids, but no other proteins are found to be consistently associated with PrP^{Sc} in fractions prepared by repeated detergent and salt extraction followed by differential centrifugation.

By electron microscopy (EM), purified PrP^{Sc} has no recognizable quaternary structure, but PrP 27-30 polymerizes into rod-shaped structures with the ultrastructural, tinctorial, and spectroscopic properties of amyloid (Chapter 2). By EM, both PrP^{C} and PrP^{Sc} appeared as amorphous aggregates when dried onto grids; the presence of these proteins was confirmed by Immunogold labeling with the α-PrP 3F4 monoclonal antibody (mAb). In contrast, polymers of PrP 27-30 were visualized as rod-shaped amyloids (Prusiner et al. 1983). Amyloid polymers composed of PrP 27-30 have a higher proportion of β-sheet than PrP^{Sc}; the β-sheet characteristic of the intermolecular bonding associated with polymer formation is more pronounced for PrP 27-30.

The ability of purified PrP 27-30 to assemble into amyloid suggested that amyloid plaques previously reported in kuru, Creutzfeldt-Jakob disease (CJD), and scrapie were composed of the same protein that constitutes most of the infectious prion particle (Chapter 2). PrP 27-30 is formed from PrP^{Sc} after limited proteinase K (PK) digestion; no loss of infectivity occurs during this period of proteolysis (McKinley et al. 1983; Prusiner et al. 1983; Meyer et al. 1986). In contrast, prolonged proteolysis hydrolyzes PrP 27-30 and diminishes prion titers, as determined by bioassay (Fig. 1). The extreme resistance of both PrP 27-30 and prion infectivity to proteolytic digestion and their concomitant decreases after prolonged proteolysis argued persuasively that PrP 27-30 is a component of the infectious particle.

Once the amino-terminal amino acid sequence of PrP 27-30 was determined (Prusiner et al. 1984a), isocoding sets of oligonucleotides were synthesized and used to retrieve cognate cDNA clones from banks prepared from the poly(A)$^{+}$ RNA of hamster and mouse brains (Chesebro et al. 1985; Oesch et al. 1985). The finding of mRNA encoding PrP in the brains of uninfected, control animals led to two different conclusions: In one case, PrP was dismissed as irrelevant to scrapie (Chesebro et al. 1985), and in the other, it prompted the discovery of PrP^{C} (Oesch et al. 1985).

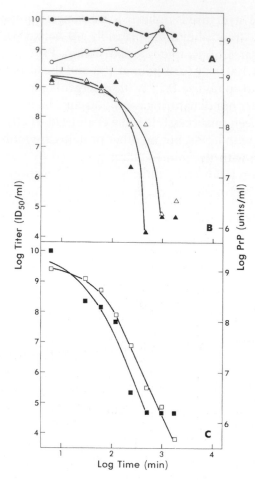

Figure 1. Kinetics of PrP 27-30 degradation and prion activation by proteolytic digestion. Concentration of PrP 27-30 was determined by densitometry and prion titer by
bioassay. (*A*) Control. (*B, C*) Samples were digested with 100 μg/ml proteinase K for
times denoted on the *x*-axis. (*Filled symbols*) Log PrP 27-30; (*open symbols*) log prion
titer. (Reprinted, with permission, from McKinley et al. 1983 [copyright Cell Press].)

PURIFICATION OF PRIONS

Purification of PrP^{Sc} and PrP 27-30 is accomplished by taking advantage
of the insolubility of these proteins in nondenaturing detergents
(Prusiner et al. 1982a, 1983; Diringer et al. 1983; Hope et al. 1986; Bolton
et al. 1987; Turk et al. 1988). Repeated detergent extractions of prion-
infected brain tissue in the presence or absence of salt is frequently used
to isolate both PrP^{Sc} and PrP 27-30. Preparations of PrP 27-30 are gener-

ally more homogeneous than those of PrP^{Sc} because limited digestion with PK removes most protein contaminants. Alternatively, PrP^{Sc} has been isolated by immunoaffinity chromatography after dispersion into detergent–lipid–protein complexes (DLPC) (Gabizon et al. 1988b), but such a procedure has not been adapted to large-scale preparations due to the large amount of purified phospholipid required.

Enriching Fractions for Prion Infectivity

The development of an effective protocol for enriching scrapie prion infectivity in rodent brain fractions led to the discovery of PrP 27-30. Several general principles proved to be useful in developing a purification scheme for prions. First, prions are stable in detergent solutions that do not denature proteins but unstable in detergent solutions that denature proteins, such as SDS (Millson and Manning 1979; Prusiner et al. 1980b). Second, prion infectivity can be separated from cellular components by detergent extractions followed by differential centrifugation or precipitation (Prusiner et al. 1977, 1980b). Although ultracentrifugation is useful in sedimenting prions, it is cumbersome on a large scale; and thus, at an early step in our large-scale purification, we substituted precipitation by polyethylene glycol (PEG) for ultracentrifugation (Prusiner et al. 1982a, 1983). Third, prion infectivity in crude extracts was found to be resistant to nuclease and protease digestions (Hunter 1979; Prusiner et al. 1980b), both of which proved to be valuable in purification. Omission of either micrococcal nuclease or PK resulted in a diminished level of purification. Specifically, subsequent purification by the sucrose gradient procedure was reduced by a factor of 10 when PK digestion was omitted (Prusiner et al. 1982b). The enzyme digestions were carried out at 4°C because warming prions to 37°C caused increased aggregation (Prusiner et al. 1978).

After proteolytic digestion, some of the small peptides and oligonucleotides could be removed by ammonium sulfate precipitation (Prusiner et al. 1980b) in the presence of cholate, because other investigators studying membrane-bound enzymes showed that the effective precipitation of hydrophobic proteins with ammonium sulfate required cholate (Tzagoloff and Penefsky 1971). In one study on the scrapie agent, ammonium sulfate was used in the absence of cholate, but precipitation of the agent seemed to be nonselective (Malone et al. 1979).

Using sodium dodecyl sarcosinate (Sarkosyl) gel electrophoresis, digested peptides and polynucleotides separated from aggregates (Prusiner et al. 1980a). Although the method using protease digestion and gel electrophoresis was cumbersome, it provided sufficient purification to allow us to show convincingly that infectivity in fractions enriched for prions depended on at least one protein (McKinley et al. 1981; Prusiner et al. 1981). Without the purification provided by Sarkosyl gel electrophoresis, neither protease digestion nor chemical modification with diethylpyrocarbonate (DEP) was found to decrease prion titers (Prusiner et al. 1981).

The major drawback of gel electrophoresis for purifying prions was the limited amount of relatively crude material that could be processed. Only 3–4 ml of detergent-extracted, enzyme-digested material could be loaded onto a preparative electrophoresis apparatus at any one time; moreover, electrophoresis required 6–8 hours (Prusiner et al. 1981), followed by a cumbersome procedure using electroelution of prions from the top of the gel.

Because the volumes were severely limited using Sarkosyl gel electrophoresis, an alternative purification scheme using sucrose gradient sedimentation was developed (Prusiner et al. 1982b). This procedure takes advantage of the hydrophobicity of scrapie prions. An enzyme-digested, ammonium sulfate-precipitated fraction was mixed with Triton X-100 and SDS prior to rate-zonal sedimentation through a sucrose step gradient containing no detergent. Presumably, large forms of prions are present either as aggregates that were not dissociated in Triton X-100–SDS or as aggregates that are formed as prions enter the gradient. Both of these scenarios appear to occur. The behavior of prions under these conditions seems to be analogous to that of calcium ATPase, for which this procedure was first used (Warren et al. 1974a,b). Substituting the Triton X-100–SDS mixture with either octylglucoside or Triton X-100 alone, or omitting the Triton X-100–SDS mixture altogether, resulted in virtually no purification of prions (Prusiner et al. 1982a).

The development of the discontinuous sucrose gradient method provided another important step in the level of purification as well as characterization of prions. Preparations were of sufficient purity to allow identification of a unique protein (PrP) within these preparations (Bolton et al. 1982; Prusiner et al. 1982a). Even though the quality of purification was substantially improved and the quantity of material processed in a single centrifugation was 10-fold greater compared to Sarkosyl gel electrophoresis, the purified fractions were still insufficient to extend purification and to provide sufficient amounts for characterization.

To overcome this problem, a large-scale purification procedure was developed (Prusiner et al. 1983). Not only did this protocol increase the amount of purified prions, but equally important, it yielded fractions that contained significantly fewer contaminants compared to preparations using smaller-scale protocols. The zonal rotor sucrose gradient centrifugation used in the large-scale protocol gave increased resolution of the particles being separated due to its configuration and long sedimentation path length, especially when compared to gradient centrifugation in a reorienting vertical rotor (Anderson 1962). In addition, dynamic loading and edge-unloading of the rotor probably also helped to maximize the resolution of purification procedure.

In our early studies on scrapie, we reported that prions could exist in multiple molecular forms, with small forms having a sedimentation coefficient of 40S or less, as well as a succession of larger forms; in fact, some larger aggregates had sedimentation coefficients of more than 10,000S (Prusiner et al. 1978, 1980b). All these data were derived from studies in which the infectivity of gradient fractions was determined by endpoint titrations in mice. Subsequent studies showed that scrapie prions aggregate into rods after exposure to limited proteolysis and that these rods form large arrays or clusters of varying size and shape (Prusiner et al. 1983; McKinley et al. 1991). The aggregation of prions into polymeric structures of varying sizes and shapes made purification extraordinarily difficult.

Large-scale Purification of PrP 27-30

To prepare sufficient amounts of purified fractions containing scrapie prions for a variety of studies, it was necessary to scale up the purification protocol described above to a large-scale procedure (Fig. 2). We also adapted this large-scale protocol for purification of PrP^{Sc} or PrP 27-30 for as few as three scrapie-infected brains from Syrian hamsters.

Typically, brains are taken from Syrian hamsters that were inoculated intracerebrally 75 days earlier with ~10^7 ID$_{50}$ units of Sc237 prions. Hamsters are anesthetized with CO_2 and sacrificed by cervical dislocation. The brains are frozen in liquid nitrogen and stored at −70°C. For each preparation, 900–1000 brains are thawed. Ten liters of a 10% (w/v) homogenate are made by suspending the brains in 0.32 M sucrose and disrupting the tissue with a Brinkmann Polytron equipped with a PT45 generator set at 8 for 2 minutes in a Baker Sterilgard biosafety hood. The homogenization vessel, a stainless steel cylinder with a capacity of 13.5 liters, measures 31.8 cm high and 26.7 cm in diameter. Three screws with wing nuts fasten the lid to the cylinder, and an O-ring around the edge of the cylinder provides a gas-tight seal. The center of the lid has a 4.45-cm diameter opening to allow the PT45 probe to be inserted. This opening is

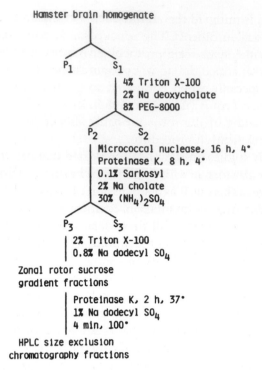

Figure 2. Scheme for large-scale purification of PrP 27-30. (P_n) Pellet; (S_n) supernatant. (Reprinted, with permission, from Prusiner et al. 1984a [copyright Cell Press].)

made nearly gas-tight with a gasket. The homogenization vessel is kept in an ice bath, and the preparation is maintained at 4°C throughout the purification procedure. All solutions are purged with argon to remove any oxygen remaining after evacuation.

The homogenate is clarified by continuous-flow centrifugation at 10,000 rpm using a Beckman JCF-Z rotor equipped with a large pellet core, in a Beckman J21-C centrifuge housed in a Baker biosafety cabinet. The homogenate is pumped through the rotor at 180 ml/minute. To the supernatant (S_1), 1 mM EDTA and 1 mM DTT are added, to which Triton X-100 and doxycholate (DOC) are then added at detergent:protein (w/w) ratios of 4:1 and 2:1, respectively. Once the volume of the S_1 with the detergent is determined, a PEG-8000 cocktail is added to the extract to a final concentration of 0.03 M Tris-OAc, 0.1 M KCl, 20% glycerol, and 8% PEG-8000, followed by stirring at 4°C for 15 minutes to precipitate the PrP^{Sc}.

The pellet (P_2) is collected by continuous-flow centrifugation at 15,000 rpm using the JCF-Z rotor equipped with a standard pellet core.

The fraction S_1 is pumped through the rotor at 160 ml/minute. Fraction P_2 is resuspended in 20 mM Tris-OAc (pH 8.3) containing 0.2% (v/v) Triton X-100 and 1 mM DTT to give a protein concentration of 10 mg/ml. It is digested with 12.5 U/ml of micrococcal nuclease in the presence of 2 mM $CaCl_2$ for 12 hours at 4°C followed by 100 µg/ml PK for 8 hours at 4°C in the presence of 2 mM EDTA and 0.2% (w/v) Sarkosyl. The digestion is terminated with 0.1 mM phenylmethylsulfonyl fluoride (PMSF) and precipitated with $(NH_4)_2SO_4$ (ultrapure) at 30% saturation in the presence of 2% (w/v) sodium cholate. This fraction is centrifuged in a Beckman Type 19 rotor in a Beckman L5-65 ultracentrifuge housed in a Baker biosafety cabinet at 53,7000g for 1 hour at 4°C. The pellet (P_3) is adjusted to 0.825 mg/ml, and 20% Triton X-100 and 8% SDS (specially pure) are added to a final concentration of 2% Triton X-100 and 0.8% SDS. The protein concentration is reduced to 0.75 mg/ml. Typical preparations yield ~500 ml of the P_3 fraction.

A Beckman Ti-15 zonal rotor with a B-29 liner is filled through the edge at a pump speed of 25 ml/minute while rotating at 2000 rpm. The centrifugation is performed in a Beckman L5-65 ultracentrifuge housed in a Baker biosafety cabinet. The rotor is loaded with 600 ml of 25% sucrose in 20 mM Tris-OAc, 1 mM EDTA (pH 8.3) at 4°C, followed by 800 ml of 56% sucrose in 20 mM Tris-OAc, 1 mM EDTA (pH 8.3) at 4°C. Each zonal centrifugation uses 140 ml of the P_3 sample, which is 10% of the total rotor volume. Prior to loading P_3 to the center of the rotor, sucrose is added to 6% at a pump speed of 15 ml/minute, while displacing 56% sucrose through the outer edge. An overlay of 320 ml of 20 mM Tris-OAc, 1 mM EDTA (pH 8.3) followed the sample through the center line. The rotor is set to 5000 rpm and then capped. The speed is increased to 32,000 rpm for a period of 14.5 hours.

Fractions of 40 ml each are collected from the edge by replacing the rotor volume with 20 mM Tris-OAc, 1 mM EDTA (pH 8.3) at 4°C, then kept on ice. To each fraction, sulfobetaine 3-14 is added at a final concentration of 0.05%. All fractions were frozen in a dry ice and ethanol bath, and stored at −70°C.

Each zonal gradient yields two or three fractions of 40 ml each that have specific infectivities of $10^{10.5}$ to 10^{11} ID_{50} U/mg protein, representing a purification of 3,000- to 10,000-fold over the homogenate. As much as 95% of the infectivity loaded onto the gradient is recovered in the two or three fractions with the highest prion titers (Fig. 3). The overall recovery of infectivity using this protocol is generally greater than 20%. Disaggregation of prions during the purification procedure may contribute to this high percentage of recovery.

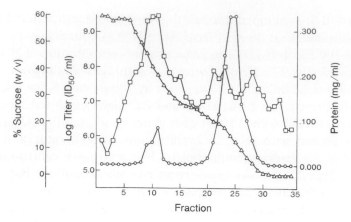

Figure 3. Discontinuous sucrose gradient fractionation of prion infectivity using a zonal rotor. The gradient was centrifuged at 32,000 rpm for 14.5 hours at 4°C. The gradient was fractionated from the outer edge of the Ti15 zonal rotor equipped with a B-29 liner. (*Triangles*) Sucrose concentration; (*circles*) protein concentration; (*squares*) scrapie prion titer. Fraction 1 was from the outer edge of the rotor. Each fraction was 40 ml. (Reprinted, with permission, from Prusiner et al. 1983 [copyright Cell Press].)

PrP^C FROM BRAIN AND RECOMBINANT PrP FROM *E. COLI*

In order to characterize it, PrP^C had to be purified under nondenaturing conditions. This was particularly important because a functional assay for PrP^C does not exist, which prevents a simple and direct assessment of the native state of purified PrP^C. Moreover, measurements of the secondary structure of PrP^C purified from brain provided a baseline for refolding recombinant PrP produced in *E. coli*.

Purification of PrP^C

Purification of PrP^C was modified to avoid using low pH, urea, and SDS-PAGE, all of which might denature PrP^C (Pan et al. 1993; Pergami et al. 1996). Such procedures had been used in earlier studies of PrP^C (Turk et al. 1988; Pan et al. 1992). Typically, a crude microsomal fraction is prepared from 100 normal Syrian hamster brains, extracted with Zwittergent 3-12 (ZW) and centrifuged at 100,000g for 1 hour. The supernatant is applied to an IMAC-Cu^{++} column that is washed with 5 volumes of 0.015 M imidazole; 0.15 M NaCl; 10 mM sodium phosphate (pH 7.0); 0.2% ZW. PrP^C is eluted by increasing the imidazole to 0.15 M. The pH of the eluate was adjusted to 6.4 with 2 M HCl prior to loading onto an SP cation-exchange column (1.5 cm × 3 cm), equilibrated with 0.15 M NaCl; 20 mM MES (pH 6.4); 0.2% ZW. The column is washed with 20 mM MOPS (pH 7.0); 0.2% ZW (buffer A), containing initially 0.15 M NaCl and then 0.25 M NaCl. PrP^C is eluted with buffer A containing 0.5 M NaCl. The eluate is concentrated and desalted on an IMAC-Cu^{++} column (1 cm × 2.5 cm), from which PrP^C is eluted with 0.1 M imidazole in buffer B (0.15 M NaCl; 20 mM MOPS [pH 7.5]; 0.2% ZW) and applied to a wheat germ agglutinin (WGA)–Sepharose column (1 cm × 6.5 cm) equilibrated in the same buffer. During loading, the flow rate is 0.3 ml/minute, then increased to 0.75 ml/minute. After washing with 10 volumes of buffer B, PrP^C is eluted by 0.05 M N-acetyl glucosamine in buffer B. Purified PrP^C fractions are pooled and concentrated with a Centricon-30 filter. N-acetyl glucosamine is removed by 3 washes in 0.15 M NaCl; 10 mM sodium phosphate (pH 7.5); 0.12% ZW (PBSZ); and samples are stored at –20°C. In the final step in PrP^C purification, lectin chromatography is performed, based on earlier observations that PrP 27-30 binds to WGA (Haraguchi et al. 1989). The extent of enrichment relative to other proteins is evaluated by silver-stained SDS-PAGE (Pan et al. 1993).

The amide I′ band of the spectrum from Fourier transform infrared (FTIR) spectroscopy of purified PrP^C showed a symmetrical peak with a maximum at 1653 cm^{-1} (Fig. 4, solid line). Such spectra are characteristic of proteins with high α-helical content. In contrast, the FTIR spectra of PrP^{Sc} (Fig. 4, dashed line) and PrP 27-30 (Fig. 4, dotted line) showed patterns that are characteristic of proteins with high β-sheet content. Deconvolution of the PrP^C spectrum gave an estimate of 42% α-helical content and only 3% β-sheet structure, whereas the PrP^{Sc} and PrP 27-30 spectra revealed 43% and 54% β-sheet structure, respectively (Table 1). The secondary structure of PrP 27-30 determined by transmission FTIR is in very good agreement with data from attenuated total reflection of thin films (Gasset et al. 1993).

Although the FTIR spectrum for PrP^C with a peak at 1653 cm^{-1} is indicative of a secondary structure with a high α-helical content, circular dichroism (CD) spectroscopy, a technique that gives an unambiguous signal for proteins with a large α-helical component, confirmed this interaction. The CD spectrum of PrP^C shows a minimum at 208 nm and a shoulder at 222 nm, clearly indicating that the protein contains one or more α-helices (Fig. 5). To estimate the α-helical content of proteins, far UV data are preferable; however, α-helical content can be determined from ellipticity at 222 nm. At the 222-nm minimum, the rotational strength of an amino acid polymer composed completely of α-helices varies depending on the helix length (Yang et al. 1986). Assuming an average helix length of 14 residues, the α-helical content for PrP^C is ~36%.

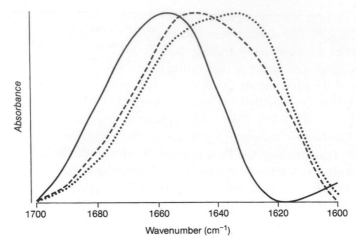

Figure 4. Fourier transform infrared spectroscopy of prion proteins. The amide I′ band (1700–1600 cm^{-1}) of transmission FTIR spectra of PrP^C (*solid line*), PrP^{Sc} (*dashed line*), and PrP 27-30 (*dotted line*). These proteins were suspended in a buffer in D$_2$O containing 0.15 M NaCl; 10 mM sodium phosphate, pD 7.5 (uncorrected); 0.12% ZW. The spectra are scaled independently to be full scale on the ordinate axis (absorbance). (Reprinted, with permission, from Pan et al. 1993 [copyright National Academy of Sciences].)

Expression and Purification of Recombinant PrP

DNA constructs encoding SHaPrP of varying lengths are inserted into expression vectors that are transfected into *E. coli* deficient of proteases (Mehlhorn et al. 1996; Donne et al. 1997). Maximum expression is obtained with SHaPrP containing residues 90–231 corresponding to the sequence of PrP 27-30 and residues 29–231 corresponding to full-length

Table 1. Secondary structural elements of PrP^C, PrP^{Sc}, and PrP 27-30

			Content of secondary structure, %							
		PrP					PrP 27-30			
	FTIR		predicted[a]				predicted[a]			
Structure	PrP^C	PrP^{Sc}	native	α/α	α/β	β/β	FTIR	native	α/α	α/β	β/β
α-Helix	42	30	14	31	27	0	21	20	45	39	0
β-Sheet	3	43	6	0	1	24	54	8	0	1	30
Turn	32	11					9				
Coil	23	16					16				

[a]Predicted values were calculated by using a neural network program (Presnell et al. 1993). For PrP^C and PrP^{Sc}, residues 23–231 were used in the calculation; for PrP 27-30, residues 90–231 were used. (Reprinted, with permission, from Pan et al. 1993 [copyright National Academy of Sciences].)

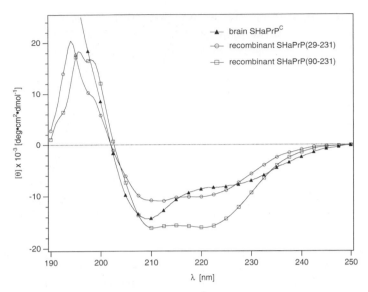

Figure 5. Circular dichroism spectra of PrP^C, recSHaPrP(90–231), and recSHaPrP (29–231). Purified PrP^C was prepared in a solution of 0.15 M NaCl; 10 mM sodium phosphate (pH 7.5); 0.12% ZW. The ordinate axis (ellipticity θ) was calibrated from amino acid analyses (Pan et al. 1993; Mehlhorn et al. 1996; Donne et al. 1997).

PrP except for six amino-terminal residues. Approximately 30 mg of recombinant (rec) SHaPrP(90–231) and SHaPrP(29–231) could be obtained from each liter of *E. coli* culture. Disruption of the bacteria using a microfluidizer gives the highest yields of recSHaPrP that is targeted to the periplasmic space, where it is deposited into refractile bodies. After solubilization of recPrP in 8 M GdnHCl, recSHaPrP is purified by size exclusion chromatography (SEC) and reverse-phase chromatography (RPC). The primary structure of recPrP is determined by Edman sequencing and mass spectrometry; the secondary structure is ascertained by CD and FTIR spectroscopy. When recSHaPrP was purified under reducing conditions, it had, similar to PrP^{Sc}, a high β-sheet content and relatively low solubility, particularly at pH values >7. Refolding of recSHaPrP by oxidation to form a disulfide bond between the two Cys residues of this polypeptide produces a soluble protein with a high α-helical content similar to PrP^C. These multiple conformations of recSHaPrP are reminiscent of the structural plurality that characterizes naturally occurring PrP isoforms. The high levels of purified recSHaPrPs have enabled the determination of their molecular structures by solution nuclear magnetic resonance (NMR) spectroscopy (Donne et al. 1997; James et al. 1997).

Mouse (Mo) PrP has also been expressed in *E. coli* and purified for structural studies (Hornemann et al. 1997; Riek et al. 1997). Full-length MoPrP(23–231) has been recovered from intracellular inclusion bodies and refolded by procedures similar to those described above. A carboxy-terminal fragment of MoPrP comprising 111 residues is expressed in *E. coli* and targeted to the periplasmic space, where it remains soluble and can be purified without denaturation (Hornemann and Glockshuber 1996).

Refolding of Recombinant PrPs

Several related strategies are employed for refolding the lyophilized recSHaPrP obtained from the RPC fractions (Mehlhorn et al. 1996). Recombinant SHaPrP in oxidized form is solubilized at 3–10 mg/ml in 6 M GdnHCl; 50 mM Tris (pH 8), then rapidly diluted into either (1) 20 mM Tris (pH 8); (2) 20 mM sodium phosphate (pH 7.2); or (3) 20 mM sodium acetate (pH 5) to a final protein concentration of 100 µg/ml. The resulting solutions are incubated at 10–25°C for 2–12 hours, then the insoluble material is separated by centrifugation, leaving α-helical recSHaPrP in solution, as determined by CD. The yield of soluble α-recPrP as measured by the bicinchoninic acid (BCA) assay (Pierce) is ≥85%. For further phys-ical studies, this α-recSHaPrP (Fig. 5) is dialyzed against a particular buffer and concentrated by the Centriplus 10 filtration system. Over time, the α-recPrP refolded by this protocol would undergo proteolysis.

In another protocol that also yields α-recPrP, the lyophilized oxidized fractions are dissolved in a small volume of water, assayed for protein con-tent, and then adjusted to 1 mg/ml in 8 M GdnHCl (Zhang et al. 1997). This fraction is then diluted into 10 volumes of 20 mM Tris acetate (pH 8.0) and 5 mM EDTA. The resulting solution is dialyzed against 50 mM HEPES (pH 7.0); 0.005% thimerosal, and the protein is further purified by cation exchange chromatography with a HiLoad SP Sepharose Fast Flow 26/10 column using a gradient of 0–0.4 M LiCl at 4 ml/minute, mon-itoring UV absorption at 280 nm. Fractions containing the protein are dialyzed against 20 mM sodium acetate (pH 5.5); 0.005% thimerosal. In fractions refolded by this method, protein degradation was inhibited by the presence of thimerosal, even when the recPrP solution was maintained at 37°C for 10 days.

To refold recSHaPrP into β-sheet conformation, lyophilized RPC fractions containing freshly purified recSHaPrP obtained under reducing conditions are dissolved in 20 mM MES (pH 6.5). β-sheet structure is con-firmed by CD and FTIR spectroscopy (Zhang et al. 1997). These fractions could be concentrated to 10 mg/ml and appeared to remain in solution over an extended period of time without precipitating. At higher pH (7.5), solubility is reduced substantially. The solubility at pH 6.5 was investigated by ultracentrifugation at 100,000g for 1 hour. The oxidation state of the cysteines was determined by carboxymethylation with iodoacetic acid, before and after reduction with DTT (Mehlhorn et al. 1996).

DISPERSION OF PRIONS INTO LIPOSOMES

The formation of PrP amyloid in vitro requires limited proteolysis in the presence of detergent (McKinley et al. 1991); these structures are often referred to as prion rods due to their long, filamentous shape. Thus, PrP amyloid composed of PrP 27-30 in fractions enriched for infectivity are largely, if not entirely, artifacts of the purification protocol. Although the insolubility of prion rods facilitated purification of infectious prions, it prevented many studies of the molecular structure of prions. The discovery of conditions for solubilization of nondenatured PrP 27-30 and PrP^{Sc} with lipids and detergents to form DLPCs and liposomes therefore represents an important advance (Fig. 6) (Gabizon et al. 1987, 1988c). DLPCs can be centrifuged at 170,000g for 30 minutes while retaining most of PrP 27-30 and infectivity in the supernatant fraction. In fact, infectivity generally increases 10- to 100-fold after dissociation of the prion rods into DLPCs or liposomes.

Subsequently, a protocol was developed to transfer PrP^{Sc} from microsomal/synaptosomal membranes into DLPCs, preserving infectivity and allowing fractionation of PrP^{Sc} as a soluble molecule (Gabizon et al. 1988a,b). The microsomes are solubilized by a combination of 2% (w/v) Sarkosyl and 5 mg/ml phosphatidylcholine. The resulting DLPCs are subjected to ultracentrifugation, and the supernatant is applied to PrP mAb affinity matrix. Monoclonal antibodies raised against purified PrP 27-30 were crosslinked to protein-A–Sepharose in order to minimize the leakage of the antibodies (Barry and Prusiner 1986). After overnight incubation at 4°C, the immunoaffinity matrix is washed with buffers containing increasing concentrations of salt, followed by progressively increasing pH values. Fractions eluted from the column are screened for the presence of PrP and infectivity. Fractions that contained PrP^{Sc} were found to carry infectivity whereas fractions with no detectable PrP^{Sc} harbored little or no infectivity. Moreover, the amount of PrP^{Sc} recovered from the column was loosely correlated with infectivity. These results demonstrated a clear correlation between PrP^{Sc} and infectivity, as reported earlier using other procedures.

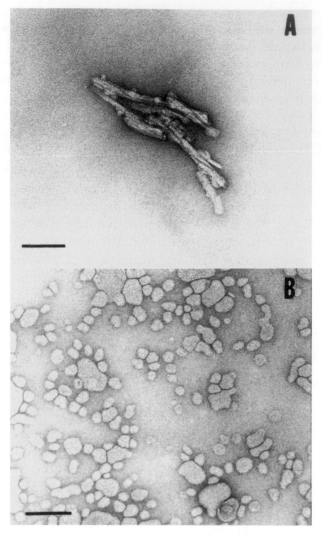

Figure 6. Electron micrographs of negatively stained and Immunogold-labeled prion rods containing PrP 27-30 dispersed into DLPCs. (*A*) Typical prion rods negatively stained with uranyl formate. (Bar, 100 nm.) (*B*) Negatively stained liposomes containing PrP 27-30. (Reprinted, with permission, from Gabizon et al. 1987 [copyright National Academy of Sciences].)

In addition to permitting the correlation between PrP^{Sc} and infectivity to be shown, the transfer of infectivity from prion rods to liposomes also provided a new method by which to search for a cryptic nucleic acid within the prion. Treatment of DLPCs and liposomes with either nucleases or Zn^{++} failed to alter infectivity (Gabizon et al. 1987). In subsequent studies, nucleases were added during the formation of DLPCs. As demonstrated earlier, no change in infectivity was observed after prolonged incubations.

Radiobiological studies were performed on liposomes and DLPCs to characterize the inactivation of prion infectivity. Early radiobiological investigations on the resistance of the scrapie agent to inactivation by ionizing radiation suggested a target size of ~150 kD (Alper et al. 1966, 1978). This small target size supported the proposition that the scrapie pathogen might not possess a nucleic acid, a hypothesis that was generally ignored. Prion rods and liposomes in 14 different preparations were irradiated with increasing doses up to 400 Mrad. The results yielded a mean target size for the prion of 55 ± 9 kD (Fig. 7) (Bellinger-Kawahara et al. 1988). Samples containing prion rods, microsomes, crude homogenates, or liposomes all had the same target size. This finding

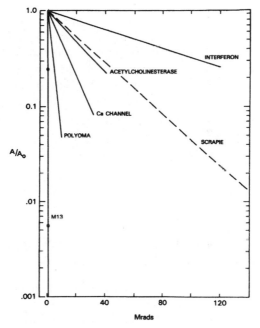

Figure 7. Ionizing radiation target size of prions. Radiation inactivation curves of several different biological activities: enzyme reactions, receptors, and viral infectivity. Data are adjusted to a common irradiation temperature of –135°C (Kempner and Haigler 1982). The double-stranded DNA polyomavirus data are from Latarjet et al. (1967). Diltiazem binding to muscle Ca channel gave a target size of 130 kD (Goll et al. 1984b); acetylcholinesterase size was 66 kD (Goll et al. 1984a); leukocyte interferon displayed a 20-kD target (Pestka et al. 1983). M13 target size >1000 kD. M13 was prepared and assayed as described previously (Bellinger-Kawahara et al. 1987a). (Reprinted, with permission, from Bellinger-Kawahara et al. 1988 [copyright Elsevier Science].)

argues that the target size of infectivity is independent of the physical form of the prion particle. Irradiation of DLPC-incorporated prions with UV light at 254 nm produced an inactivation curve virtually identical to that observed for prion rods (Gabizon et al. 1988c). The kinetics of inactivation of infectivity by UV irradiation yielded exponential survival curves characteristic of a single-hit process. The D_{37} values of 17 J/m^2 to 24 J/m^2 for purified prion rods and DLPCs are in good agreement with those reported two decades earlier for mouse prions in brain homogenates (Alper et al. 1967).

The solubilization of prions with phospholipids helped to clarify issues concerning heterogeneity of the infectious particle with respect to size, density, and charge. The interconversion of prion rods, DLPCs, and liposomes, all with high levels of infectivity, argues that prions are likely to be composed of a single hydrophobic macromolecule, i.e., PrP^{Sc}. All attempts to reveal a putative nucleic acid upon transfer of prion infectivity in DLPCs were unsuccessful.

Liposomes proved to be useful in immunization studies in which prion rods had been ineffective. After immunization of $Prnp^{0/0}$ mice with purified mouse prion rods (Prusiner et al. 1993b), a panel of recombinant antibody (Fab) fragments was obtained (Williamson et al. 1996). Disappointingly, PrP 27-30 was not detected by any of these antibodies prior to treatment with guanidinium thiocyanate (GdnSCN). One plausible explanation for this finding is the lack of immunogenicity of PrP 27-30 when polymerized into prion rods. Purified PrP 27-30 assembles into rod-like structures of 100–200 nm in length that are composed of as many as 1000 PrP 27-30 molecules (Prusiner et al. 1983; McKinley et al. 1991). This aggregation apparently reduces the effective epitope concentrations, thereby hindering efficient immunization and selection of specific antibody phage. Epitope concentration is a critical factor in effective selection from antibody phage display libraries (Parren et al. 1996). To overcome the possible problem of epitopes in PrP 27-30 being buried within prion rods, $Prnp^{0/0}$ mice were immunized with PrP 27-30 incorporated into liposomes. Some of the recombinant Fabs generated by immunization with PrP 27-30 in liposomes reacted with nondenatured PrP^{Sc} and PrP 27-30 (Peretz et al. 1997).

IMMUNOASSAYS OF PrP ISOFORMS

For many years, investigators thought that scrapie, kuru, and CJD were caused by slow viruses (Gajdusek 1977; Parry 1983; Pattison 1988). Among the most puzzling features of these transmissible diseases is the lack of a detectable immune response to the inoculated prions (Chandler 1959; Clarke and Haig 1966; Tsukamoto et al. 1985). Neither scrapie-specific antibodies nor cellular immune responses were detected. In addition, prion-mediated neurodegeneration does not activate the viral surveillance system because no evidence for induction of interferon was found (Katz et al. 1968; Field et al. 1969; Worthington 1972).

Once PrP 27-30 was discovered in fractions highly enriched for prion infectivity, antisera were generated in rabbits (Bendheim et al. 1984; Bode et al. 1985). High-titer, specific polyclonal antibodies are important tools for studying prions. Initial attempts to raise such sera were formidable tasks (Bendheim et al. 1984; Bode et al. 1985), in large part because it was difficult to purify sufficient quantities of PrP 27-30 from scrapie-infected hamster brains for immunization. The poor antigenicity of prion rods also contributed to the difficulty in raising antibodies.

Immunoassay Formats

Because of the lack of antibodies that distinguish PrP^{Sc} from PrP^{C}, current detection systems employ PK resistance, insolubility, or cryptic epitopes of PrP^{Sc} to distinguish between these isoforms (Tatzelt et al. 1999). Different immunoassays currently employed include western blot, ELISA, fluorescence correlation spectroscopy, insolubility, and the conformation-dependent immunoassay (CDI) (Table 2). After limited digestion with PK, PrP 27-30 has been detected by western blots (Bendheim et al. 1984; Oesch et al. 1985; Schaller et al. 1999), dot blots (Serban et al. 1990), and ELISA (Moynagh et al. 1999; Deslys et al. 2001; Grassi et al. 2001; Biffiger et al. 2002), as well as by histoblots (Taraboulos et al. 1992) and immunohistochemistry (Kitamoto et al. 1987; Miller et al. 1994). The sensitivity of western blotting increased substantially by precipitating PrP^{Sc} first with

Table 2. Immunoassays for PrP^{Sc}

Assay procedure	Limited proteolysis	Dynamic range
Western blot	yes	10^2
Ultrasensitive western blot	yes	10^4
ELISA	yes	10^3
Histoblots	yes	10^0
Immunohistochemistry	yes	10^0
Fluorescence correlation spectroscopy	no	10^3
Insolubility	no	10^2
Conformation-dependent immunoassay	no	10^6

phosphotungstic acid (NaPTA) (Wadsworth et al. 2001), a protocol that was developed initially for the CDI (Safar et al. 1998). Increased sensitivity of immunohistochemistry was achieved by denaturing PrP^{Sc} either with chemicals such as formic acid (Kitamoto et al. 1987) or with elevated temperatures by brief autoclaving (Muramoto et al. 1992; Miller et al. 1994). The foregoing immunoassays are relatively insensitive compared to bioassays. Additionally, their application in some transgenic models of prion diseases is limited because the presence of intermediate conformations of PrP can lead to seemingly ambiguous results (Hsiao et al. 1990; Scott et al. 1997b).

Superior assays for PrP^{Sc} are based on conformationally sensitive antibodies that bind to specific epitopes that have differential exposure in PrP^{C} and PrP^{Sc} (Safar et al. 1998, 2002; Barnard et al. 2000). Antibodies that react with native PrP^{C} and denatured PrP^{Sc} but not native PrP^{Sc} are used in the CDI. One study reports an IgM mAb that distinguishes PrP^{Sc} from PrP^{C} (Korth et al. 1997), but this work has not been confirmed. In another study, antibodies raised against the tripeptide motif YYR seem to distinguish PrP^{Sc} from PrP^{C} in immunoprecipitation experiments (Paramithiotis et al. 2003). These findings are unexpected since the PrP molecule has three YYR sequences. In other investigations, fluorescence correlation spectroscopy was used to measure aggregates of PrP in the CSF of patients dying of CJD (Bieschke et al. 2000). Although fluorescence correlation spectroscopy can be extremely sensitive (Walter et al. 1996; Pitschke et al. 1998), the results of this investigation were disappointing because only 50% of the CJD patients tested positively for aggregated PrP (Bieschke et al. 2000).

Conformation-dependent Immunoassay for PrP^{Sc}

Prior to the development of the CDI, all immunoassays for PrP^{Sc} depended on the complete removal of PrP^{C} by limited proteolysis. These immunoassays measured only amino-terminally truncated PrP 27-30 that remains from the digestion. Earlier dot-blotting studies showed that the immunoreactivity of PrP^{Sc}, in contrast to PrP^{C}, was greatly enhanced by denaturation (Serban et al. 1990). Subsequently, studies with anti-PrP mAbs and recombinant Fabs showed that transformation of PrP^{C} into PrP^{Sc} is accompanied by the burying of epitopes near the amino terminus, while the carboxy-terminal epitopes remain exposed (Peretz et al. 1997).

On the basis of the foregoing observations, we set out to develop an immunoassay that did not depend on the proteolysis of PrP^{C}. We chose to build an immunoassay that is instead based on measuring the reactivity of an antibody that binds to native PrP^{C} but does not bind to native PrP^{Sc}. After measuring the binding of the antibody to native PrP^{C}, the binding to denatured PrP^{Sc} can be determined. We thought that such an assay would return somewhat higher values for PrP^{Sc} because we believed that PrP^{Sc} is not uniformly protease-resistant and thus, some PrP^{Sc} molecules are hydrolyzed while PrP^{C} is degraded. To our surprise, a large fraction of PrP^{Sc} was sensitive to proteolytic degradation. To distinguish between the protease-sensitive and the protease-resistant forms of PrP^{Sc}, we labeled them sPrP^{Sc} and rPrP^{Sc}, respectively.

In developing the CDI, we took advantage of the conformational plasticity of recombinant SHaPrP(90–231), which could be refolded into proteins with predominantly α-helical or β-sheet structures. The α-helical form of SHaPrP(90–231) shares some structural features with PrP^{C}, and the β-sheet form has characteristics similar to PrP^{Sc} (Mehlhorn et al. 1996; Zhang et al. 1997). To measure the binding of antibodies to PrP, we adapted a direct ELISA-formatted, dissociation-enhanced time-resolved fluorescence (TRF) detection system (Hemmilä and Harju 1995). The sensitivity limit of detecting denatured SHaPrP(90–231) with Eu-labeled anti-PrP 3F4 mAb IgG (Kascsak et al. 1987) was less than or equal to 5 pg/ml, with a dynamic range of over 10^5 (data not shown).

The 3F4 mAb was ideal for developing the CDI because it reacts weakly with native PrP^{Sc} but extremely well with both native PrP^{C} and denatured PrP^{Sc}. The immunoreactivity of 3F4 was investigated using 5% brain homogenates obtained from $Prnp^{0/0}$ mice (Büeler et al. 1992) and spiked with purified recombinant SHaPrP(90–231) in either α-helical or β-sheet conformation, which was verified by CD spectroscopy (Fig. 8A). SHaPrP(90–231) was unfolded by exposure to 4 M GdnHCl for 5 minutes at 80°C. Following denaturation, the samples were diluted 20-fold, and stored for 16 hours at 5°C before the conformation was determined by CD spectroscopy. The results (Fig. 8A) confirmed that the protein did not refold under the conditions used for crosslinking to the activated plastic ELISA plate.

When Eu-labeled 3F4 IgG was used in the presence of 5% $Prnp^{0/0}$ mouse brain homogenate, the detection limit of denatured SHaPrP(90–231) was ≤ 2 ng/ml and the dynamic range over 10^3. The input/output calibration for denatured SHaPrP(90–231) in both α-helical and β-sheet conformations and with purified infectious PrP^{Sc} resulted in sensitivity and linearity within the same range (data not shown).

A relatively small difference was observed between data obtained from SHaPrP(90–231) in α-helical conformation (Fig. 8B) and from denatured PrP. In contrast, the reactivity with β-SHaPrP(90–231) only marginally exceeded the background (Fig. 8C). When the results were expressed as a ratio between binding of 3F4 mAb to denatured (D) versus native (N) conformations of PrP, the ratio was ≤ 5 for α-helical

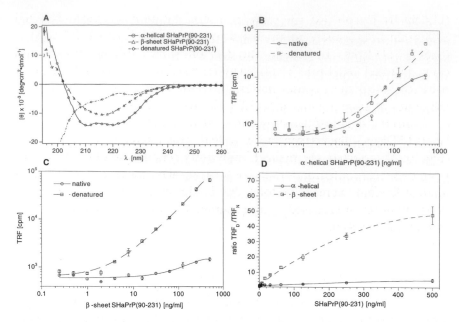

Figure 8. Development of a conformation-dependent immunoassay (CDI) for recombinant SHaPrP(90–231) purified from *E. coli* and folded into different conformations. (*A*) Conformation of recSHaPrP(90–231) as determined by CD spectroscopy. (*B*) Calibration of the CDI with recSHaPrP(90–231) in α-helical conformation and (*C*) in β-sheet conformation. (*D*) Ratios between signals of denatured/native recSHaPrP(90–231) in α-helical and β-sheet conformations. Two major bands in the CD spectrum with minima at 208 nm and 222 nm indicate an α-helical conformation; a single negative band with minimum at 217 nm is characteristic of predominantly β-sheet conformation; a negative trough toward 197 nm documents a random-coil conformation. The calibration of the CDI was performed in the presence of 5% (w/v) *Prnp*[0/0] mouse brain homogenate. The data points and bars represent average ± s.e.m. obtained from four independent measurements. (Reprinted, with permission, from Safar et al. 1998 [copyright Nature Publishing Group].)

SHaPrP(90–231) and >5 for the β-sheet conformation (Fig. 8D). Thus, the ratio of 3F4 mAb binding to denatured and native PrP is a sensitive indicator of the original, native PrP conformation, in which a value >5 in the presence of *Prnp*[0/0] brain homogenate indicates the presence of β-sheet conformers.

Measurement of SHaPrP[Sc] in Brain Homogenates

Fluorescence signals for native and denatured samples of normal SHa brain homogenate containing PrP[C] were similar, and the D/N ratio ≤1.8

(Fig. 9A,B). In contrast, the ratio for a serial dilution of scrapie-infected SHa brain homogenate containing a mixture of PrP^C and PrP^{Sc} was >2 (Safar et al. 1998). In general, the data on serially diluted normal and prion-infected SHa brain homogenate followed the pattern described for α-helical and β-sheet conformations of recombinant SHaPrP(90–231). A D/N ratio of antibody binding >1.8 indicated the presence of $SHaPrP^{Sc}$.

From the signal of denatured and native samples developed with Eu-labeled 3F4 IgG, the concentration of PrP^{Sc} was calculated by Equation 1 (Fig. 9). As expected, the normal brain homogenate contained only PrP^C in an α-helical conformation. In contrast, the data show that the approximately fivefold increase of total PrP in the brains of scrapie-infected Syrian hamsters was caused by the accumulation of PrP^{Sc} in a β-sheet conformation (Fig. 9D).

Evidence for Different Conformations of PrP^{Sc} in Eight Prion Strains

Using the CDI to measure PrP^{Sc} in brain homogenates, we examined eight different prion strains passaged in Syrian hamsters (Safar et al. 1998). Brains from Syrian hamsters were collected when the animals displayed signs of neurologic dysfunction; the incubation times for the prion strains varied from 70 to 320 days (Fig. 10A). Most of the PrP in the brains of Syrian hamsters with signs of neurologic disease was PrP^{Sc}, as defined by the β-sheet conformation. The level of PrP^{Sc} in the brains of these clinically ill animals exceeded that of PrP^C by 3- to 10-fold. The highest levels of PrP^{Sc} were found in the brains of Syrian hamsters infected with the Me7H strain; in contrast, the lowest levels were found in the brains of Syrian hamsters inoculated with the SHa(Me7) strain (Fig. 10C). Interestingly, the Me7H and SHa(Me7) strains, which were both derived from Me7 passaged in mice, possessed similar denatured/native PrP ratios, but they accumulated PrP^{Sc} to quite different levels (Fig. 10A,C). The highest denatured/native PrP ratio of all tested strains was from SHa(RML) prions. Interestingly, no relationship between incubation time and either the concentration of PrP^{Sc} or the ratio of denatured/native PrP seems to exist (Fig. 10A,C).

The apparent independence of the ratio of denatured/native PrP from the PrP^{Sc} concentration became apparent after plotting both parameters in a single plot (Fig. 10B). Each strain occupied a unique position, indicating differences in the conformations of accumulated PrP^{Sc}. Because the PrP^C concentration in each strain was ≤5 μg/ml and the denatured/native PrP^C ratio was ≤1.8, the expected impact of the presence of PrP^C on the final PrP ratio is ≤15%.

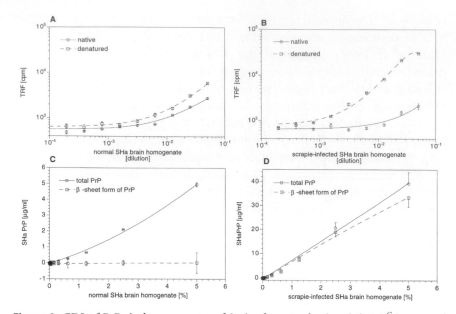

Figure 9. CDI of PrPs in homogenates of Syrian hamster brains. (*A*) PrP^C in normal SHa brain homogenates. (*B*) PrP^C and PrP^{Sc} in homogenates from the brains of Syrian hamsters exhibiting signs of CNS dysfunction ~70 days after inoculation with Sc237 prions. (*C*) Total amount of PrP in normal SHa brain homogenates and the amount of PrP in β-sheet conformation. (*D*) Total amount of PrP in homogenates from the brains of Syrian hamsters inoculated with Sc237 prions and the amount of PrP in β-sheet. Samples of 5% brain homogenates obtained from normal or scrapie-infected Syrian hamsters were serially diluted into 5% *Prnp*$^{0/0}$ mouse homogenate. Data points and bars represent average ± s.e.m. obtained from four independent measurements. The concentration of SHaPrP on the ordinate was calculated from Equation 1 listed below. The ratio between binding of Eu-labeled IgG to PrP in native and denatured conformations measured by TRF was designated TRF$_D$/TRF$_N$. A mathematical model was developed to calculate the content of β-sheet-structured PrP in an unknown sample directly:

$$[PrP_\beta \square \sim \Delta\,TRF_\beta = TRF_D - (TRF_N * f_\alpha) \qquad\qquad Eq.\ 1$$

In Equation 1, an excess of antibody binding to PrP in the transition from the native to denatured state over that expected for the α-helical conformation is directly proportional to the amount of PrP in β-sheet conformation. Symbols in the equations are as follows: Δ TRF$_\beta$, change in the time-resolved fluorescence of PrP in β-sheet conformation during the transition from the native to denatured state; TRF$_D$, time-resolved fluorescence of PrP in the denatured state; TRF$_N$, time-resolved fluorescence of PrP in the native conformation; and f$_\alpha$, the correlation coefficient for dependency of TRF$_D$ on TRF$_N$ obtained with either standard, α-helical recSHaPrP(90–231) or purified SHaPrP^C. The coefficient was obtained by fitting the plot of antibody binding to denatured/native protein by the nonlinear least-squares regression program. (Reprinted, with permission, from Safar et al. 1998 [copyright Nature Publishing Group].)

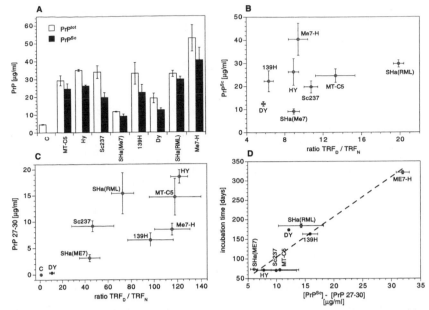

Figure 10. Eight prion strains distinguished by the CDI. (*A*) Concentration of total PrP and PrP^{Sc}. Data represent the average ± S.E.M. obtained from three different brains of LVG/LAK Syrian hamsters infected with different prion strains and measured in three independent experiments. (*B*) Ratio of antibody binding to denatured/native PrP and a function of PrP^{Sc} concentration in the brains of Syrian hamsters infected with different prion strains. Concentration of PrP^{Sc} (Fig. 9, Eq. 1) and the ratio of antibody binding to denatured/native PrP were measured by the CDI. (*C*) Brain homogenates of Syrian hamsters inoculated with different prion strains and uninoculated controls (C) were digested with proteinase K before measurement by the CDI. (*D*) Incubation time plotted as a function of the concentration of the proteinase K–sensitive fraction of PrP^{Sc} ([PrP^{Sc}] – [PrP 27-30]). (Reprinted, with permission, from Safar et al. 1998 [copyright Nature Publishing Group].)

Because only the most tightly folded conformers of PrP^{Sc} are likely to be protease-resistant, each of the brain homogenates was digested with PK prior to measuring the ratio of denatured/native PrP. The positions of many of the strains changed when sPrP^{Sc} was enzymatically hydrolyzed (Fig. 10D). The change was most notable with the DY strain, which was readily detectable before limited proteolysis (Fig. 10B) but became almost undetectable after digestion (Fig. 10D), in accord with earlier western blot studies (Bessen and Marsh 1992, 1994). Equally important, the Sc237 and HY strains were marginally separated prior to PK digestion (Fig. 10B) but became quite distinct afterwards (Fig. 10D). These findings argue that Sc237 and HY are distinct strains even though they exhibit similar incubation times of ~70 days when passaged in Syrian hamsters (Scott et al. 1997a). It is noteworthy that limited proteolysis of PrP^{Sc} from Sc237- and HY-infected brains produced PrP 27-30 that were indistinguishable by migration in SDS-PAGE (Scott et al. 1997a).

Equilibrium Dissociation and Unfolding of PrP by GdnHCl

To extend the foregoing studies on the Sc237 and HY strains, equilibrium dissociation and unfolding of the PrP^{Sc} molecules in GdnHCl was monitored by Eu-labeled 3F4 IgG (Fig. 11) (Safar et al. 1998). After equilibrium unfolding, the proteins were rapidly diluted to the same protein and GdnHCl concentrations, and then crosslinked to the glutaraldehyde-activated plate. The binding of 3F4 IgG was expressed as the D/N ratio (Fig. 11A), which increased as the GdnHCl concentrations increased, indicating continued unfolding. The unfolding patterns of the D/N ratio for PrP^C in normal brain are clearly different from the unfolding patterns of the PrP isoforms in prion-infected brains. Each strain possesses a unique D/N ratio and, consequently, a unique unfolding pattern. The peaks indicate the presence of PrP conformers with a higher affinity than at the next higher or lower GdnHCl concentrations. The largest interstrain differences were at ~1 M, ~3 M, and 6.5 M GdnHCl (Fig. 11A). The D/N ratios for different prion strains at 6.5 M GdnHCl, at which all PrP is unfolded (Safar et al. 1993; Hornemann et al. 1997; Zhang et al. 1997), were similar to those obtained with samples exposed to 4 M GdnHCl and 80°C for 5 minutes (compare Figs. 10B and 11A).

In addition to D/N ratios, antibody binding was also expressed as the fractional change (F_{app}) from the native, folded state (~0) to the denatured, unfolded state (~1). Data expressed as F_{app} enables the comparison of samples with different PrP concentrations. Comparison of the relative abundance of different PrP conformers in Sc237-, DY-, and HY-infected brains showed the largest interstrain differences at ~1 M GdnHCl (Fig. 11B). The highest fraction of conformers that were sensitive to such low concentrations of GdnHCl was found in SHa brains infected with the DY strain.

Selective Precipitation of PrP^{Sc} by NaPTA

To increase the sensitivity of the CDI, a precipitation procedure was developed to enrich samples for PrP^{Sc} prior to immunodetection. Although PrP^{Sc} could be readily detected when the concentration of PrP^{Sc} was equal to or exceeded that of PrP^C, measuring PrP^{Sc} was difficult when samples

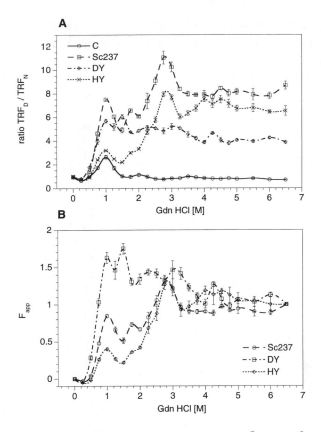

Figure 11. Equilibrium dissociation and unfolding of PrP^C and PrP^{Sc} in three prion strains. Ratio of antibody binding to denatured/native PrP and apparent fractional change of PrP unfolding (F_{app}) during equilibrium dissociation and unfolding. (A) The brains of uninoculated controls (C) and Syrian hamsters infected with Sc237, DY, and HY prion strains were analyzed by the CDI; the ratio of antibody binding to denatured/native PrP is plotted as a function of the GdnHCl concentration. (B) The F_{app} of PrP^{Sc} from the Sc237, DY, and HY strains. Data points and bars represent the average ± S.E.M. obtained from four different measurements. (Reprinted, with permission, from Safar et al. 1998 [copyright Nature Publishing Group].)

contained ≤1% PrP^{Sc} relative to PrP^C. At neutral pH in the presence of Mg^{++}, NaPTA forms complexes with oligomers and polymers of infectious PrP^{Sc} and PrP 27-30 but not with PrP^C (Safar et al. 1998). The resulting dense aggregates are then collected by low-speed centrifugation for 15 minutes. The pellet contains ~99% of the prions but less than 1% of other proteins. With NaPTA precipitation, PrP 27-30 forms rod-like structures that could be detected directly by electron microscopy without further processing. In most respects, the morphology of these rod-like structures resembled the rods found in highly purified preparations of PrP 27-30 (Prusiner et al. 1983).

In a typical experiment, a 5% (w/v) brain homogenate containing 2% Sarkosyl is mixed with stock solution containing 4% NaPTA and 170 mM $MgCl_2$ (pH 7.4) to obtain a final concentration of 0.2–0.3% NaPTA. Generally, 1-ml samples are incubated for 30–60 minutes at 37°C on a rocking platform and centrifuged at 14,000g in a tabletop centrifuge (Eppendorf) for 15 minutes at room temperature. Treatment with 25 µg/ml PK for 1 hour at 37°C is performed before or after precipitation if desired. The pellet is resuspended in H_2O containing protease inhibitors (0.5 mM PMSF; aprotinin and leupeptin, 2 µg/ml each) and tested by the CDI.

After the procedure to precipitate selectively PrP^{Sc} by NaPTA was incorporated into the CDI, it became possible to measure the low concentrations of PrP^{Sc} in brain homogenates from prion-infected animals. Scrapie-infected SHa brain homogenate is serially diluted into normal, uninfected SHa brain homogenate in order to simulate in vivo conditions after infection, in which the levels of PrP^{Sc} are low. The resulting samples contain different levels of infectivity and PrP^{Sc} in the presence of a constant level of PrP^C. The level of PrP^{Sc} can be measured as a function of the D/N ratio of antibody binding over an approximately 10^5-fold range. After scrapie-infected homogenates were diluted $\sim10^{-5}$, the D/N ratio approached 1.8, which is the ratio determined when PrP^C alone was measured (Fig. 12A). Above a value of 1.8, the D/N ratio reliably indicates the presence of increasing quantities of PrP^{Sc}.

The application of Equation 1 (Fig. 9) allowed direct calculation of the concentration of PrP^{Sc} in a mixture of scrapie-infected and normal brain homogenates (Fig. 12B). The sensitivity limit of NaPTA precipitation and CDI is ≤1 ng/ml of PrP^{Sc} in the presence of a 3,000-fold excess of PrP^C. Assuming an average prion titer of 10^8 ID_{50} U/ml for Sc237-inoculated SHa brain homogenate, the sensitivity limit of PrP^{Sc} detection corresponds to an infectivity titer of $\sim10^3$ ID_{50} U/ml without the use of a capture antibody. With a sandwich ELISA format, the sensitivity of the CDI is similar to that of the bioassay (Safar et al. 2002).

Applications of the Conformation-dependent Immunoasssy

The CDI has been adapted in two major ways to measure bovine (Bo) PrP^{Sc} in the brains of cattle with bovine spongiform encephalopathy (BSE) (Safar et al. 2002). First, the CDI was adapted to a robotic system, making the assay completely automated. Second, instead of binding the antigen to the sides of the well, a capture antibody is employed, which

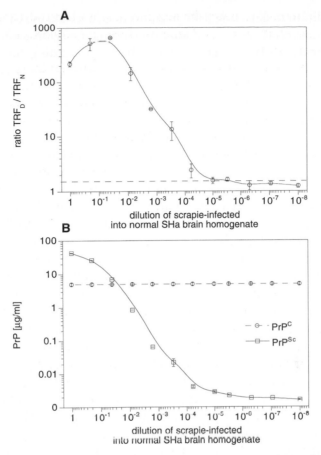

Figure 12. Dynamic range and sensitivity of the ratio of antibody binding to PrP in denatured and native states. Homogenates (5%) prepared from the brains of Syrian hamsters exhibiting signs of CNS dysfunction ~70 days after inoculation with Sc237 prions were serially diluted into 5% normal SHa brain homogenate, and the presence of PrP^{Sc} was measured after NaPTA precipitation by the CDI. (*A*) The ratio of the fluorescence signals of denatured and native aliquots plotted as a function of the dilution of scrapie-infected brain homogenate. (*B*) The absolute amount of PrP^C and PrP^{Sc} was calculated from Equation 1 (legend to Fig. 9) and plotted as a function of the dilution of scrapie-infected brain homogenate. Data points and bars represent average ± s.e.m. obtained from four independent measurements. (Reprinted, with permission, from Safar et al. 1998 [copyright Nature Publishing Group].)

increases the sensitivity between 10- and 100-fold. In fact, the sensitivity of the CDI for bovine prions is similar to that achieved with bioassay in Tg(BoPrP)*Prnp*$^{0/0}$ mice (Chapter 3). The sandwich CDI also detected PrP^{Sc} from deer and elk with CWD, at a sensitivity similar to that for bovine prions (Safar et al. 2002).

In addition to detecting PrP^{Sc} in cattle, deer, and elk, the CDI has been used to measure MoPrP^{Sc} in studies of dominant-negative inhibition of prion formation (Perrier et al. 2002). In those studies, low levels of MoPrP^{Sc} with the dominant-negative mutation were sought but none could be found. The CDI has been applied in studies to detect prions in the muscle of prion-infected mice. MoPrP^{Sc} has been readily detected in the hind limbs of infected mice by the CDI (C. Ryou, in prep.) as well as by bioassay, western blotting, and ELISA (Bosque et al. 2002).

The sandwich CDI has also been employed to detect HuPrP^{Sc} (Bellon et al. 2003), with a sensitivity for HuPrP^{Sc} similar to that obtained by bioassay in Tg(MHu2M)$Prnp^{0/0}$ mice (J. Safar et al., unpubl.). The dynamic range of the CDI for human prions in the brains of sCJD patients is >10^6-fold (see Chapter 3, Fig. 10). Whether or not the CDI can be used to develop an antemortem test for prions in humans remains to be established.

Conformation Stability Immunoassays

Varieties or "strains" of prions are hypothesized to be determined by changes in PrP^{Sc} conformation. The conformational stabilities of several SHa prion isolates were measured using resistance to protease digestion as a marker for undenatured PrP^{Sc}. With GdnHCl treatment, eight prion strains showed sigmoidal patterns of transition from native to denatured PrP^{Sc} as a function of increasing GdnHCl concentration. Half-maximal denaturation occurred at a mean value of 1.48 M GdnHCl for the Sc237, HY, SHa(Me7), and MT-C5 isolates, which all have 75-day incubation periods; 1.08 M for the DY strain (165-day incubation period); 1.25 M for the SHa(RML) and 139H isolates (175-day incubation periods); and 1.38 M for the Me7H strain (320-day incubation period) (Peretz et al. 2001a). In contrast to the CDI, in which levels of sPrP^{Sc} correlate with the length of the incubation time, the half-maximal concentration of GdnHCl for denaturation of PrP^{Sc} does not correlate with incubation times.

SEARCH FOR NUCLEIC ACIDS IN PRIONS

Three approaches have been used to search for a nucleic acid component within the prion particle causing scrapie and CJD (Alper et al. 1966, 1967). These are (1) inactivation of prions by procedures that modify or hydrolyze nucleic acids, (2) characterization of nucleic acids in fractions enriched for prion infectivity, and (3) search for a nucleic acid unique to scrapie-infected preparations.

Inactivation of Prions

Prions are resistant to inactivation by physical, chemical, and enzymatic procedures that hydrolyze, modify, or shear nucleic acids, which supports the proposition that prions are devoid of nucleic acid. However, prions are inactivated by procedures that denature or modify proteins (Prusiner 1982). Similar results were obtained with crude brain homogenates (Alper et al. 1966, 1967), purified prion preparations (Prusiner 1982; Bellinger-Kawahara et al. 1987b), prion liposomes (Gabizon et al. 1988c), and prions from cultured neuroblastoma cells (Neary et al. 1991). The search for a scrapie-specific nucleic acid utilized gel electrophoresis, ultracentrifugation, and differential hybridization employing cDNA libraries, as well as molecular cloning and PCR to search for foreign sequences. Some aspects of the search for a scrapie-specific nucleic acid have assumed that such a polynucleotide would be protected by a protein coat, as is the case for viruses. Such a hypothetical nucleic acid was assumed to possess physical properties similar to those exhibited by known nucleic acids.

Nucleic Acids in Fractions Enriched for Prion Infectivity

In some studies, nucleic acids were reported to be found only in fractions prepared from scrapie-infected animals. By end-labeling and gel electrophoresis (Dees et al. 1985; German et al. 1985; Castle et al. 1987), and by differential hybridization of a cDNA library (Duguid et al. 1988; Aiken et al. 1989; Duguid and Dinauer 1989), differences in the patterns of nucleic acids from infected and noninfected tissue were found. In other studies, copurification of nucleic acids and scrapie infectivity was described. Those investigations include analysis of nucleic acids by ultracentrifugation (Sklaviadis et al. 1989), detection of retroviral and polyadenylated RNA by the PCR technique (Akowitz et al. 1990; Murdoch et al. 1990), identification of mitochondrial DNA by hybridization (Aiken et al. 1990), and reports on ssDNA by electron microscopy (Narang et al. 1988; Narang 1990), all of which were interpreted as evidence for a prion-specific nucleic acid.

The detection of a nucleic acid under these circumstances does not imply that it is an essential component of the infectious prion particle. In published studies, a convincing correlation between a prion-specific nucleic acid and infectivity has not been demonstrated. Ultracentrifugation studies were carried out after disaggregation of CJD-infected brain fractions by detergent treatment and yielded a peak sedimenting at 6–9S containing PrP that was well separated from the infectious fractions. This observation was interpreted as an argument against PrP being the major component of infectivity. However, in similar disaggregation experiments (Riesner et al. 1996), PrP separated from infectivity possessed a drastically changed conformation; thus, the findings support the prion hypothesis rather than falsify it. When residual nucleic acids in highly purified prions were analyzed by molecular cloning, no evidence for a scrapie-specific nucleic acid could be found (Oesch et al. 1988).

Search for a Nucleic Acid Unique to Purified Prion Preparations

Because of the many unsuccessful attempts to identify a nucleic acid component required for prion infectivity, an experimental paradigm was developed to exclude a nucleic acid by demonstrating that prions devoid of such molecules harbor infectivity (Meyer et al. 1991). One infectious unit must contain at least one nucleic acid molecule if it is essential for infectivity. The P/I ratio must be at least unity. Nucleic acids with a P/I <1 would be excluded as prion-related. Because properties of the hypothetical prion-specific nucleic acid are unknown, such a molecule was considered to be either DNA or RNA; single or double stranded; circular or linear; as well as capped, chemically modified, or covalently bound to proteins. In addition, the possibility that a prion-specific polynucleotide might be heterogeneous in size was entertained. Because high levels of prion infectivity have been found only in animals, incorporating radiolabeled nucleotides into the putative prion nucleic acid was not possible. Cell-culture systems producing high levels of prions would obviate this radiolabeling problem, but such cultures are not available.

PAGE combined with silver staining is employed as a general method to analyze all types of nucleic acids. Depending on the size of the polynucleotide, 20–200 pg of nucleic acid per gel band can be detected. For example, 170 pg would be contained in 3×10^9 units of infectious prions if the hypothetical nucleic acid was 100 nucleotides in length and the P/I ratio was unity. Obviously, studies on a smaller prion-specific nucleic acid would require higher titers, whereas studies on a larger nucleic acid would need lower titers to yield the same mass of nucleic acid. The sensitivity of silver staining after conventional PAGE decreases substantially if the nucleic acid is heterogeneous in size.

Search for Homogeneous Nucleic Acids by PAGE

Small nucleic acids up to several hundred nucleotides in length withstand harsh procedures that hydrolyze larger nucleic acids in purified prion preparations. Scrapie prions were purified from brains of Syrian hamsters using a discontinuous sucrose gradient (Prusiner et al. 1983). The infectious fractions contained primarily rod-shaped aggregates composed largely of PrP 27-30. The prion rods were precipitated from the sucrose gradient fractions with ethanol and submitted to a nucleic acid degradation protocol using DNase and Zn^{++}. No significant changes in prion titer were detected after digestion with DNase and exposure to Zn^{++} ions, in agreement with earlier observations (Prusiner 1982). The samples were disaggregated and deproteinized by boiling in 2% SDS, digested with PK, either extracted once with phenol and once with phenol:chloroform (1:1) or, in later experiments, twice with phenol:chloroform:isoamyl alcohol (50:48:2), and finally precipitated with ethanol.

PAGE analysis of the DNase-digested and Zn^{++}-treated prions showed some background smearing as well as distinct bands migrating near the dye front (Fig. 13A, lane 2). Omission of DNase digestion and Zn^{++} hydrolysis resulted in a prion fraction with a large number of silver-stained bands throughout the lane, in addition to rapidly migrating molecules (Fig. 13A, lane 1). The size of the rapidly migrating molecules was estimated to range between 8 and 15 bases as judged by the mobility of control oligonucleotides (Fig. 13A, lane 3). From analyses of samples after additional treatments with DNase, Zn^{++} ions, and NaOH prior to electrophoresis as well as phenol–sulfuric acid measurements, the rapidly migrating bands observed on PAGE are likely to be either complex Asn-linked oligosaccharides released from PrP 27-30 during PK digestion or noncovalently bound sugar polymers that purified with the prion rods. From an analysis of control nucleic acids, 3×10^{10} molecules of each control were readily detected (Fig. 13B, lane 1) and could be recovered (Fig. 13B, lane 2) (Meyer et al. 1991). If the control nucleic acids were hydrolyzed with DNase I and Zn^{++}, no silver-stained bands were detected (data not shown). Silver staining of prions after PAGE showed no signal above 20 nucleotides (Fig. 13B, lane 3). If the control nucleic acid molecules were added to the prion sample after DNase and Zn^{++} treatment but before destroying infectivity with SDS and boiling (Fig. 13B, lane 4), all of them were visible except those of 10–11 nucleotides in length because they were hidden by the non-nucleic acid molecules purifying with the prions.

Prion infectivity was determined to be 1.2×10^{10} ID_{50} U/ml immediately prior to deproteinization. Because 170 pg of nucleic acid in a single band was assumed to be the limit of detection, 1.2×10^{10} nucleic acid molecules of at least 25 nucleotides in length would have been detected. The imprecision of the bioassay required qualification because prion titer determinations frequently vary by a factor of 10. Because the analysis utilized 20% polyacrylamide gels, only nucleic acids shorter than 300 nucleotides in length could be detected. Larger nucleic acids were assessed using a modified procedure.

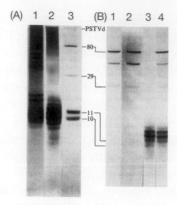

Figure 13. Analysis of nucleic acids in prion rod fractions by 20% PAGE before and after DNase and Zn^{++} treatment. (*A*) Nucleic acids without and with DNase and Zn^{++} treatment. (Lane *1*) Prion rods not treated with DNase and Zn^{++}; (lane *2*) prion rods treated with DNase and Zn^{++}; (lane *3*) marker nucleic acids: circular potato spindle tuber viroid (PSTVd, 300 pg), tRNA (80 nt, 1 ng), oligo DNA (29 nt, 200 pg), oligo RNA (11 nt, 3 ng), oligo RNA (10 nt, 2 ng). (*B*) Comparable amounts of control nucleic acids (number of molecules) and prions (ID_{50} units), 3×10^{10} molecules of control nucleic acids and 1.2×10^{10} ID_{50} units of prions recovered from sucrose gradients were analyzed. (Lane *1*) Control nucleic acid: tRNA (80 nt, 1.4 ng), oligo RNA (11 nt, 0.19 ng), oligo RNA (10 nt, 0.17 ng); (lane *2*) control nucleic acids as in lane *1* but treated by the deproteinization procedure; (lane *3*) prions treated with DNase and Zn^{++} and by deproteinization; (lane *4*) control nucleic acids as in lane *1* added to prions after DNase and Zn^{++} treatment but before deproteinization. The bands of 10 and 11 nt are weak in the photographic reproduction but were clearly seen in the original gel. (Reprinted, with permission, from Meyer et al. 1991 [copyright Society for General Microbiology].)

Analysis of Heterogeneous Nucleic Acids

Although remote and unprecedented, the possibility that prions contain nucleic acid molecules of nonuniform length was considered. In such a case, the nucleic acids would migrate during PAGE as many bands; some bands might be below the threshold for detection or not resolved from neighboring bands, resulting in a smear of staining. The smear could either represent a prion-specific nucleic acid of nonuniform length or be unspecific background hiding a weak specific band.

With RRGE, nucleic acids are separated from other molecules staining with silver and focused into one sharp band. Using RRGE (Fig. 14), heterogeneous nucleic acids were detected with a sensitivity close to that attained with homogeneous nucleic acids.

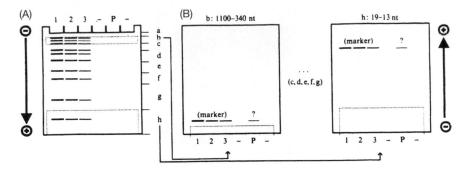

Figure 14. Scheme of return refocusing gel electrophoresis (RRGE). After conventional PAGE (e.g., 100 min, 150 V) as a first step, heterogeneous nucleic acids are dispersed over the whole length of the lane (lane *P* in panel *A*). The lane is cut into a few segments (*a–h*), each corresponding to a well-defined range of M_r. The segments are repolymerized into the bottom of new gel matrices (*B*), and second electrophoresis (250 V) is performed with reversed polarity so that the nucleic acids migrate into the new gel matrix. Because all nucleic acids in a gel segment begin migration from the same position at the beginning of the first PAGE, they meet again after reversal of the polarity of the second electrophoresis if the second run is stopped at a definite time. By adding SDS to the second PAGE, the focusing effect still works for nucleic acids while other substances such as proteins and polysaccharides remain dispersed. This is a significant advantage because proteins, like nucleic acids, stain with silver. The times of refocusing of different gel segments are chosen to be optimal for the different segments (between 42 and 48 min). The unknown nucleic acid of the prion sample is determined by comparison with the nucleic acid markers of known concentrations (markers 1, 2, 3). Only the two gel segments *b* and *h* are given as an example for the refocusing step; gel segment *a* is not used for refocusing. (Reprinted, with permission, from Kellings et al. 1992 [copyright Society for General Microbiology].)

In an initial set of experiments, a 15% polyacrylamide gel matrix was used for RRGE but restricted analyses to polynucleotides smaller than 200 bases. In a subsequent study (Kellings et al. 1992) using a 9% polyacrylamide gel matrix, the size of the polynucleotides analyzed was extended to 1100 bases. In order to obtain a quantitative estimate of the amount of nucleic acid present in a band, known amounts of reference nucleic acids were analyzed simultaneously in adjacent lanes.

Prion samples that were prepared by a similar protocol using DNase and Zn^{++} as described above were evaluated by RRGE (Fig. 15). After the first electrophoresis, only a faintly stained smear in the gel and a ladder of silver-stained bands comigrating with oligonucleotides ranging from 4 to 12 bases were visible (Fig. 15, sections A and B). After RRGE, distinct silver-stained bands were visualized (Fig. 15, sections A and B) and weaker

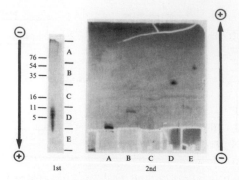

Figure 15. Detection of heterogeneous nucleic acids in prion rods by RRGE. Prion rods were treated with DNase and Zn^{++} and deproteinized prior to RRGE. In this RRGE, the sample contained $10^{7.3}$ ID_{50} units as measured by bioassay prior to boiling in SDS. The separation run is indicated as "1st." Positions of marker oligonucleotides are indicated at the left. An identical gel lane but without staining was cut into gel segments (A–E) and analyzed in the refocusing run, indicated "2nd." Marker nucleic acids as depicted in the scheme of Fig. 14 were not analyzed in this experiment. The total amount of nucleic acid was estimated as 20 ng. (Reprinted, with permission, from Meyer et al. 1991 [copyright Society for General Microbiology].)

bands could be seen (Fig. 15, sections C–E). The total amount of heterogeneous nucleic acid in the prion sample was estimated to be ~20 ng. With the method of normal PAGE (Fig. 13), nucleic acids could not be detected in these samples.

Nuclease digestion studies were carried out prior to RRGE in order to confirm the nucleic acid nature of the bands and to differentiate between RNA and DNA. The RNA and DNA detected differed in size distribution: nucleic acids shorter than 50 nucleotides were mainly RNA, and longer molecules were primarily DNA. The ladder in the range of 4–12 bases was not composed of nucleic acids (Meyer et al. 1991).

During the course of investigations by RRGE, procedures were developed for dispersing prion rods into DLPCs and liposomes while retaining infectivity (Gabizon et al. 1987, 1988c). We thought that nucleic acids, which were possibly protected from degradation by inclusion within the rod-shaped aggregates, might become accessible to nucleases upon formation of the DLPCs.

To purify the prion samples further with retention of infectivity, they were dispersed into DLPCs composed of sodium cholate and egg lecithin (Gabizon et al. 1988c; Meyer et al. 1991). The DLPCs were digested with either a mixture of DNase, Ba313, and RNase A or a mixture of micrococcal nuclease, alkaline phosphatase, RNase A, and phosphodiesterase. A

10-fold reduction in nucleic acid content of the prion preparations was observed after DLPC formation followed by nuclease digestion.

Ratio of Nucleic Acid Molecules Per Infectious Unit

Prion infectivity was monitored by incubation-time assays (Prusiner et al. 1982b) at each step in the preparation. Separation of the prion rods from the sucrose by ethanol precipitation used for discontinuous gradient centrifugation resulted in a loss of infectivity of 1 to 3 orders of magnitude; this was probably due to aggregation as well as denaturation (Meyer et al. 1991). Dispersion of ethanol-precipitated prion rods into DLPCs frequently increased the titer more than 10-fold.

Based on the amount of nucleic acid estimated from RRGE and the titers of the prion fractions prior to boiling in SDS, the P/I ratio of nucleic acid molecules to ID_{50} units was calculated. If the nucleic acids that were detected are related to infectivity, then one of two alternative paradigms must apply. First, a putative scrapie-specific nucleic acid of uniform length might be hidden among an ensemble of background nucleic acids. Such a scrapie-specific nucleic acid would not have been detected by PAGE (Fig. 13), even if it was present in sufficient amounts. Second, a scrapie-specific polynucleotide might be heterogeneous in length. The numbers of nucleic acid molecules per 1 ID_{50} unit are plotted as a function of their length as estimated from the individual gel sections (Fig. 16). In these plots, the calculation was based on the first paradigm; i.e., a well-defined, prion-specific nucleic acid among the heterogeneous background nucleic acids. If the scrapie-specific nucleic acid was longer than 76 nucleotides, the P/I ratio would fall below unity. In Figure 16B, data are presented in which sucrose gradient centrifugation and ethanol precipitation of the prion rods, i.e., steps in which infectivity was lost occasionally, were replaced by ultrafiltration and high-speed centrifugation (Kellings 1995; J. Safar et al., in prep.). This improved protocol gave an even shorter estimate of the maximum length of a hypothetical prion-specific nucleic acid; at a P/I ratio of 1, such a polynucleotide would contain 50 nucleotides.

Estimating the size of a putative prion-specific polynucleotide can be greatly influenced by the method used to determine the prion titer because such measurements must be related to the P/I ratio. The number of ID_{50} units in a sample of SHa prions is measured by determining the incubation time for an aliquot of the sample injected intracerebrally into Syrian hamsters or Tg(SHaPrP)$Prnp^{0/0}$ mice (Chapter 3). From standard curves relating incubation time to prion dose, the prion titer is calculated.

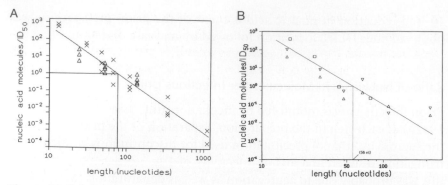

Figure 16. Relationship of the ratio of nucleic acid molecules to ID_{50} units (P/I) to average length of nucleic acid species from determinations of the kind illustrated in Fig. 14 on five (*A*) and three (*B*) independent prion samples. The relationship is linear over the size range of 10–1000 nt with an intercept of ~76 nt (*A*) and 50 nt (*B*) for a P/I of unity. Only smaller nucleic acids have P/I >1. In panel *A*, data shown by open triangles were taken from Meyer et al. (1991) and data depicted by x were taken from Kellings et al. (1992). The data in panel *B* are unpublished (K. Kellings et al., see text). The relationships were calculated as follows, using a fragment containing 450 pg of nucleic acids in the size range of 54–79 nt as an example. Assuming a continuous distribution of different sizes, the 26 species in this class will have an average MW of 22 × 10^3 and there will be:

$$450 \times 10^{-12} \text{ g} \times \frac{6 \times 10^{23} \text{ mole}}{26 \times 22 \times 10^3 \text{ g}} \quad 47 \times 10^8 \text{ mole}$$

of a particular size in this ensemble. Because the starting sample contained $10^{8.7}$ (5 × 10^8) ID_{50} units, the P/I is ~1 for a hypothetical discrete scrapie-specific nucleic acid in the ensemble. (*A*, Modified, with permission, from Riesner 1991.)

From studies with India ink and bacteriophage, it is well documented that 99% or more of the intracerebral inoculum injected into the CNS of mice rapidly exits into the circulation. In preparations of purified prions, there are ~10^5 PrP^{Sc} molecules per ID_{50} unit as measured by intracerebral inoculation; this corresponds to a P/I ratio of ~10^5 (Prusiner et al. 1983). Assuming that 99% of the inoculated PrP^{Sc} molecules exit the CNS rapidly after intracerebral injection, then 1% of PrP^{Sc} is responsible for the infectivity that is measured by bioassay. The P/I ratio is therefore ~10^3 and not 10^5. Applying such an assumption to a putative prion-specific polynucleotide would reduce the size to less than 25 nucleotides.

To determine the applicability of the foregoing arguments, we measured the levels of prions and PrP^{Sc} in SHa brain immediately after inoculation of ~10^7 ID_{50} units of Sc237 prions. The titer of prions in the inoculum was determined by endpoint titration. Hamsters were sacrificed at specified intervals after inoculation, and their brains were removed and homogenized to final 10% (w/v) suspension in PBS. The homogenate obtained at each time point was bioassayed in Syrian hamsters to measure the prion titer, and the concentration of rPrP^{Sc} was measured in parallel by the sandwich CDI.

The results of the time course (Fig. 17A) indicate that both infectivity and rPrP^{Sc} undergo massive, rapid, and parallel clearance within the first 24 hours after inoculation. By 96 hours, the increase in concentration indicates that accumulation of nascent rPrP^{Sc} has already begun. The residual infectivity at that time was only 0.7% of that in the original inoculum. To help to understand this observation, we developed different models for the observed changes in rPrP^{Sc} concentration and tested their fit with experimental data. The best fit (R = 0.99) was with the model assuming a modified exponential decay function for clearance coupled with a simple exponential function for PrP$^{C}\rightarrow$PrP^{Sc} conversion and accumulation of newly formed PrP^{Sc} (Fig. 17B), shown as:

$$[rPrP^{Sc}] = -4.72 + 2.24*e(0.008*t) + 4.95*e(0.072/(0.266*t))$$

The lowest concentration of rPrP^{Sc} detected 24 hours after inoculation was 2.9 ng/g of brain tissue. However, the PrP^{Sc} detected at 24 hours is apparently a mixture of newly formed and residual rPrP^{Sc} molecules. To discriminate both components, we deconvoluted the original data into two components, assuming either no accumulation or no clearance (Fig. 17B). Using this approach, we estimated that the concentration of rPrP^{Sc} remaining in the brain of the recipient animal from the original inoculum after 24 hours is ≤0.26 ng/g (Fig. 17B).

To calculate the number of PrP^{Sc} monomers actually initiating nascent prion formation, we used both the experimentally determined and an extrapolated concentration of PrP^{Sc} in the following formula:

$$m/IU = ([PrP^{Sc}] * 6.022E+23)/(M_r*IU)$$

in which m is the number of PrP^{Sc} molecules, IU is the number of infectious units, and M_r is the apparent molecular weight, measured in kD. The results show that the minimum number of PrP^{Sc} molecules per infectious unit is 582 when the concentration is extrapolated from the model. The upper limit is 5815 when the base for the calculation is the lowest experimental value with no correction. The 90% confidence interval is from 886 to 17,674 and from 89 to 1767 monomers of rPrP^{Sc} per ID$_{50}$ unit for the experimental and model-extrapolated values, respectively. The broad range reflects the standard error of the titration. Using these ranges of values for the number of PrP^{Sc} molecules per ID$_{50}$ unit, we conclude that

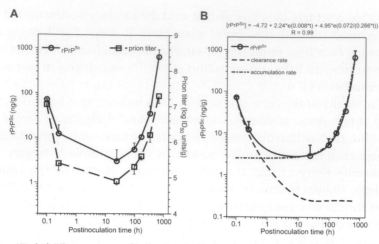

Figure 17. (*A*) Time course of infection in Syrian hamsters after intracerebral inoculation of $10^{7.1}$ ID_{50} units of Sc237 prions. Data points are mean ± S.E.M. from three independent experiments. (*B*) The clearance and accumulation rate of rPrP^{Sc} in the brains of Syrian hamsters. Both mechanisms regulating rPrP^{Sc} concentration were derived from the best fit of the experimental data followed by deconvolution into individual functions.

there are between 500 and 5000 PrP^{Sc} molecules per ID_{50} unit, and the corresponding putative, prion-specific nucleic acid could only consist of 25 nucleotides.

Conclusions about a hypothetical, prion-specific nucleic acid must be considered with respect to the molecular characteristics of the putative prion-specific polynucleotide. As described above, the first case envisions a hypothetical prion-specific nucleic acid that is a well-defined molecular species hidden in the smear of heterogeneous nucleic acids, which represent preparative impurities in the sample. In this case, molecules longer than ~50 nucleotides were not present in concentrations greater than one molecule per ID_{50} unit. If we assume an order of magnitude error in the bioassay and a factor of 2 in the nucleic acid determination, the limit would be 50 nucleotides instead of 25 nucleotides in chain length. Although larger nucleic acids can be excluded as prion-specific candidates, the smaller ones cannot, and no functional significance can be assigned to any of these oligonucleotides.

The second case considers a hypothetical prion-specific nucleic acid that is heterogeneous in length in which more of the detected nucleic acids might function in scrapie agent propagation. Under these circumstances, either all of the polynucleotides or a particular subset must be considered. If all nucleic acids of chain length longer than 240 nucleotides are considered, there is less than one such molecule per infectious unit. We emphasize that a hypothetical prion-specific polynucleotide of heterogeneous length is an extremely remote assumption that is unprecedented in biology. Thus, whatever the structure of a hypothetical prion genome might be, it would have to be constructed of molecules with fewer than 240 nucleotides.

Because nucleic acids smaller than 50 nucleotides with a P/I ratio greater than 1 have been detected in highly purified prion preparations, such oligonucleotides cannot be excluded as an essential component of the prion based on these measurements. However, we have no corroborating evidence to suggest that such molecules have any functional importance in the maintenance or propagation of prions. Such oligonucleotides may simply reflect the decreasing efficiency of nuclease degradation with the decreasing size of the nucleic acid. In this scenario, such nucleic acids are merely impurities in the highly purified preparations without any functional relevance for prion infectivity. It is worth noting that purified prion preparations tend to protect nucleic acids against degradation. In the absence of any evidence to suggest a scrapie-specific nucleic acid, further attempts to evaluate such oligonucleotides seem unlikely to be productive.

If a nucleic acid of 50 nucleotides or fewer were essential for infectivity, it could not function as an mRNA encoding either PrP or another polypeptide, but it might possess some regulatory function. Small regulatory nucleic acids, in most cases RNAs, might act as ribozymes, primers for replication or transcription, antisense RNA for inhibition of transcription, guide RNA for editing, or RNA that forms splicing sites. In any of these scenarios, the genetic information would exert a regulatory influence on the host cell. Whether such a hypothetical oligonucleotide is responsible for prion strains has been considered, but no data for such molecules exist, and the diversity of prion strains is now known to be enciphered in the conformation of PrP^{Sc} (Chapter 9).

PrP BINDING PARTNERS

Structural studies employing bacterially expressed PrPs have shown that a large portion of the amino terminus is unstructured (Donne et al. 1997). The amino terminus of PrP contains octarepeat sequences (PHGGG-WGQ) that coordinate copper ions (Stöckel et al. 1998; Viles et al. 1999), and the binding of copper has been reported to stimulate endocytosis of PrP (Pauly and Harris 1998). PrP-deficient mice have been reported to have diminished levels of copper in their brains and altered levels of superoxide dismutase (SOD) activity (Brown et al. 1997, 1999), but these findings could not be confirmed (Waggoner et al. 2000). Among other suggested functions for PrP^C are roles in neuritogenesis (Graner et al. 2000) and cell adhesion (Schmitt-Ulms et al. 2001).

In mammals examined to date, the protein sequences of PrP and doppel (Dpl) are only ~25% identical although the structures of the two proteins are quite similar (Moore et al. 1999; Mo et al. 2001). Both PrP^C and Dpl are glycoproteins that are bound to the cell surface through a GPI anchor (Stahl et al. 1987; Endo et al. 1989; Rudd et al. 1999; Silverman et al. 2000). The cellular functions of PrP^C and Dpl are not known, but both proteins cause CNS degeneration through different mechanisms. Before causing neurologic disease, PrP^C undergoes a profound conformational change as it is converted into PrP^{Sc} (Pan et al. 1993). The accumulation of PrP^{Sc} to sufficiently high levels leads to CNS degeneration (DeArmond et al. 1987). In contrast, the expression of Dpl is restricted to the testes; when Dpl is expressed in the CNS, it is neurotoxic (Moore et al. 2001).

Initial studies with two lines of $Prnp^{0/0}$ mice revealed no abnormalities (Büeler et al. 1992; Manson et al. 1994), but a third line developed late-onset ataxia caused by Purkinje cell loss (Sakaguchi et al. 1996). The ataxia and Purkinje cell degeneration were prevented in the third $Prnp^{0/0}$ line by expressing a transgene encoding wt MoPrP in these mice (Nishida et al. 1999). Subsequent studies suggested that the ataxia and Purkinje cell degeneration were likely due to expression of Dpl in the cerebellum of these mice (Moore et al. 1999; Li et al. 2000). To test the hypothesis that expression of Dpl in the CNS causes neuronal degeneration, Tg mice were constructed expressing Dpl in the brain (Moore et al. 2001). The expression level of Dpl in the CNS was inversely proportional to the age at which ataxia and Purkinje and granule cell degeneration were observed. The transgenic offspring from crossing Tg(Dpl)$Prnp^{0/0}$ with Tg(SHaPrP)$Prnp^{0/0}$ mice were rescued from the ataxic phenotype, demonstrating that expression of PrP alone was sufficient to prevent disease (Moore et al. 2001).

PrP and Dpl Fusion Proteins

The finding that Tg(Dpl:PrP)FVB/$Prnp^{0/0}$ mice did not develop ataxia and cerebellar degeneration argues that PrP neutralized the toxic effect of Dpl either by directly interacting with Dpl or through another protein. To explore these possibilities, we constructed PrP and Dpl fusion proteins. To the carboxyl termini of MoPrP and MoDpl, the Fc fragment of human immunoglobulin G was fused. In order for these fusion proteins to be glycosylated and for the proper disulfide bonds to be formed, we expressed the proteins in mouse neuroblastoma (N2a) cells (Legname et al. 2002).

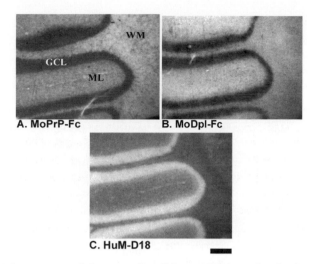

Figure 18. Enlargements of the granule cell layer (GCL), molecular layer (ML), and white matter (WM) region of the coronal section of the cerebellum/pons area of FVB mice. Sections were stained using wt MoPrP-Fc (*A*), wt MoDpl-Fc (*B*), and HuM-D18 (*C*). Bar, 200 μm. (Reprinted, with permission, from Legname et al. 2002 [copyright National Academy of Sciences].)

To assess the conformation of the PrP portion of MoPrP-Fc, we employed a panel of recombinant chimeric human–mouse (HuM) Fabs. Once we found that the immunoreactivity of MoPrP-Fc was similar to that of MoPrP^C, we utilized the fusion proteins as probes in histoblots of cryostat sections of mouse brain (Taraboulos et al. 1992). Unexpectedly, we found intense staining in the granule cell layer of the cerebellum with either wt MoPrP-Fc or wt MoDpl-Fc (Fig. 18). This finding argues that both proteins bind either to different ligands on granule cells or to the same ligand on these cells. Because granule cells do not stain with anti-PrP antibodies, the ligand(s) is not PrP.

To extend these findings, a PrP-Fc fusion protein with a dominant-negative mutation was created (Kaneko et al. 1997; Perrier et al. 2002). The Q218K mutation prevents MoPrP(Q218K) from being converted into PrP^{Sc}; moreover, MoPrP(Q218K) acts as a dominant negative in preventing wt MoPrP^C from being converted into PrP^{Sc} when the two proteins are coexpressed. This dominant-negative mutation was first reported as a common polymorphism in the Japanese population, who seem to be protected from developing CJD (Shibuya et al. 1998).

Because wt MoPrP-Fc bound to granule cells of the cerebellum, which do not express PrP^C, it seems unlikely that wt MoPrP-Fc binding reflects a ligand that is involved in the conversion of PrP^C into PrP^{Sc}. In contrast,

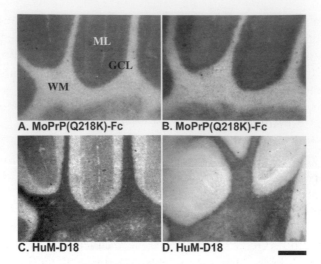

Figure 19. Enlargements of the granule cell layer (GCL), molecular layer (ML), and white matter (WM) region of the coronal section of the cerebellum/pons area of FVB mice (*A, C*), an age-matched FVB/*Prnp*$^{0/0}$ mouse (*B*), and an RML-infected FVB mouse at 132 days after inoculation (*D*). Sections were stained with MoPrP(Q218K)-Fc and HuM-D18 as indicated. Bar, 250 μm. (Reprinted, with permission, from Legname et al. 2002 [copyright National Academy of Sciences].)

dominant-negative MoPrP(Q218K)-Fc binds not only to granule cells, but also to neurons of the molecular layer where PrP^C is expressed (Fig. 19). These findings raise the possibility that the cells of the molecular layer express protein X, which is likely to be distinct from a protein that may mediate Dpl-induced degeneration. Although the binding of the dominant-negative MoPrP(Q218K)-Fc to cells in the molecular layer where PrP^C is expressed builds a case for the possibility that MoPrP(Q218K)-Fc is binding to protein X, the absence of PrP^{Sc} deposition in the molecular layer requires us to hypothesize that PrP^{Sc} formed in the molecular layer is readily transported to the cerebellar white matter where PrP^{Sc} is found.

Protein X

Protein X was proposed in order to explain a series of puzzling results on the transmission of human (Hu) prions to Tg mice. Hu prions transmitted to Tg(HuPrP) mice but not to Tg(MHu2M) mice. In addition, we found that Tg(HuPrP) mice became susceptible to Hu prions when they were crossed with *Prnp*$^{0/0}$ mice. Taken together, these data argue that it is likely that a molecule other than PrP is involved in the formation of PrP^{Sc}

(Telling et al. 1995). Based on the results with the MHu2M transgene and earlier studies showing that the amino terminus of PrP is not required for PrP^{Sc} formation (Rogers et al. 1993), we surmised that the binding of PrP^C to protein X is likely to occur through specific side chains of amino acids located at the carboxyl terminus of PrP^C. An alternative interpretation of these results is that the carboxyl terminus of MoPrP^C bound more avidly than HuPrP^C to HuPrP^{Sc}. In this scenario, PrP^C and PrP^{Sc} with different sequences bound more avidly than PrP^C and PrP^{Sc} with identical sequences. Yet when this had been studied in the past, homotypic interactions seemed to govern conversion of PrP^C to PrP^{Sc} (Prusiner et al. 1990; Scott et al. 1993).

The search for protein X has been frustrating because many proteins are known to bind to PrP^C, but none has been shown to participate in PrP^{Sc} formation (Oesch et al. 1990; Kurschner and Morgan 1995; Edenhofer et al. 1996; Tatzelt et al. 1996; Martins et al. 1997; Rieger et al. 1997; Yehiely et al. 1997; Graner et al. 2000; Keshet et al. 2000; Mouillet-Richard et al. 2000; Gauczynski et al. 2001; Hundt et al. 2001; Schmitt-Ulms et al. 2001).

Studies in ScN2a Cells in Culture

In the studies with ScN2a cells, some results also seemed most readily interpreted in terms of the binding of PrP^C to protein X. Three chimeric PrP constructs, denoted as MHMHuA (Mo residues 214, 218, and 219 replaced with Hu), MHMHuB (Mo residues 226, 227, 228, and 230 replaced with Hu), and MHMHu(A/B) (combined replacements), were transfected transiently into ScN2a cells (Table 3) and were expressed at approximately the same level (Fig. 20A,D). Neither MHMHu(A/B) nor MHMHuA was converted into PrP^{Sc} as judged by the acquisition of protease resistance (Fig. 20B). In contrast, MHMHuB was converted into PrP^{Sc} as efficiently as the control MHM2. We interpreted these results as indicating that Mo protein X did not bind to either MHMHu(A/B) or MHMHuA but recognized MHMHuB and MHM2, both of which were converted into PrP^{Sc}. No inhibition of wt MoPrP^{Sc} formation could be detected (Fig. 20C,F). To explain the results in terms of PrP^C binding to PrP^{Sc}, it can be postulated that the carboxyl terminus of MoPrP^C binds more avidly than MHMHuA PrP^C to MoPrP^{Sc}. This seems antithetical to the interpretation offered above in which PrP^C and PrP^{Sc} with different sequences bind more avidly than PrP^C and PrP^{Sc} of identical sequences (Telling et al. 1995). These data, in concert with earlier results, build a convincing edifice for the existence of protein X.

Table 3. Mutations in epitope-tagged MHM2 PrP inhibit PrP^{Sc} formation in ScN2a cells

Mo codon number	PrP residue[a]				Mutant MHM2	Type of inhibition of PrP^{Sc} formation[b]
	mouse	human	Syr hamster	sheep		
167	Q	E	Q	Q/R	R	2
					E	1
169	S	S	N	S	N	none
170	N	N/S	N	N	S	none
171	Q	Q	Q	Q	R	2
209	V	V	V	V	K	1
210	E	E	E	E	K	none
211	Q	Q	Q	Q	K	none
214	V	I	T	I	I	2
	K	1			E	2
					A	1
					W	2
					P	1
215	T	T	T	T	Q	none
216	Q	Q	Q	Q	R	3
218	Q	E/K	Q	Q	E	1
					K	2
					I	1
					A	2
					W	2
					P	1
					F	1
					R	2
					H	2
219	K	R	K	R	R	none
221	S	S	S	S	A	none
222	Q	Q	Q	Q	K	none

[a]Multiple residues at a particular position indicate naturally occurring polymorphisms.
[b]See Table 4 for explanations of inhibition.

Having identified the HuA region that prevents conversion of mutant PrP^C into PrP^{Sc}, we produced additional constructs to test the replacement of Hu residues at positions 214, 218, and 219 either alone or in combination, and expressed them in ScN2a cells. Substitution of Hu residue 218 alone abolished PrP^{Sc} (Fig. 20E, lane 9), whereas substitution of Hu residue 219 alone did not inhibit PrP^{Sc} formation (Fig. 20E, lane 10). Substitution of Hu residue 214 alone was partially inhibitory (Fig. 20E, lane 7).

Because only a minority of the ScN2a cells expresses the mutant PrPs in these transient transfection experiments, the effect of expressing

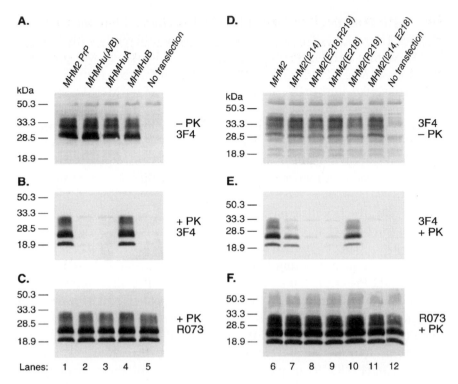

Figure 20. Characterization of the binding site for protein X. Western blot of each chimeric MHM2 construct expressed in ScN2a cells is shown. (*A–C*) (Lane *1*) MHM2 PrP; (lane *2*) MHMHu(A/B); (lane *3*) MHMHuA; (lane *4*) MHMHuB; and (lane *5*) untransfected control ScN2a cells. (*D–F*) (Lane *6*) MHM2 PrP; (lane *7*) MHM2(I214); (lane *8*) MHM2(E218,R219); (lane *9*) MHM2(E218); (lane *10*) MHM2(R219); (lane *11*) MHM2(I214,E218); and (lane *12*) untransfected control ScN2a cells. Panels *A* and *D* demonstrate the expression of each chimeric MHM2 PrP construct: 40 μl of undigested cell lysates was applied to each lane and MHM2 PrP was detected by staining with α-PrP 3F4 mAb. Panels *B* and *E* demonstrate the conversion of chimeric MHM2 PrP^C to PrP^{Sc} and were stained with α-PrP 3F4 mAb. Panels *C* and *F* show endogenous MoPrP^{Sc} as well as chimeric constructs detected with α-PrP RO73 rabbit antiserum. In panels *B–C* and *E–F*, 500 μl of cell lysates was digested with proteinase K (20 μl/ml) at 37°C for 1 hour followed by centrifugation at 100,000*g* for 1 hour and the loading of the resuspended pellet onto the gel. (Reprinted, with permission, from Kaneko et al. 1997 [copyright National Academy of Sciences].)

mutant PrP on conversion of wt MoPrP into PrP^{Sc} could not be assessed (Fig. 20C,F). To measure the influence of mutant PrP^C on the conversion of wt PrP into PrP^{Sc}, we performed cotransfection studies. These results argue that the MHM2 PrP mutants that carry Lys, Ala, Trp, Arg, or His at residue 218 bind to protein X with a greater affinity than does wt MHM2 with Gln at 218 (Table 3). These findings also contend that the two polymorphic Hu residues Glu and Lys interact very differently with Mo protein X. Mutant MHM2 PrP(E218) binds more weakly to Mo protein X than does wt MHM2 PrP(Q218), which results in MHM2 PrP^C(E218) not being converted into PrP^{Sc}, and the conversion of wt MHM2 PrP^C into PrP^{Sc}. In contrast, mutant MHM2 PrP^C(K218) binds more tightly to Mo protein X than does wt MHM2 PrP^C(Q218), which results in both mutant MHM2 PrP^C(K218) and wt MHM2 PrP^C not being converted into PrP^{Sc}.

A HuPrP polymorphism at codon 219, which corresponds to MoPrP codon 218, has been reported in the Japanese population (Kitamoto and Tateishi 1994); ~12% of the people carry the allele encoding Lys instead of Glu. To date, the Lys allele has not been found in 50 autopsied CJD cases in Japan (Shibuya et al. 1998). This is a highly significant finding (Fisher's exact test, $p = 0.00005$) which suggests that HuPrP^C(K219) acts as a dominant negative in preventing CJD. In view of the results presented here with MHM2 PrP(K219), it seems likely that the K219 allele prevents CJD through the high avidity of HuPrP^C(K219) for protein X. The high-affinity binding of HuPrP^C(K219) to protein X prevents HuPrP^C(K219) from being converted into PrP^{Sc}, and it prevents HuPrP^C(E219) from interacting with protein X. The prevention of protein X–HuPrP^C(E219) interaction by HuPrP^C(K219) in patients heterozygous for the polymorphism explains the dominant-negative effect of the K219 substitution. When we introduced the K218 mutation into MHM2 PrP^C expressed in ScN2a cells, the recombinant protein did not convert to PrP^{Sc} and inhibited the conversion of wt MHM2 PrP^C into PrP^{Sc}.

The NMR structure of recSHaPrP(90–231) shows a loop composed of residues 165–171 immediately adjacent to the protein X–binding site on the carboxy-terminal helix, which raises the possibility that one or more of these residues also participates in the binding to protein X. To explore this possibility, we constructed mutants MHM2 PrP(Q167R), MHM2 PrP(Q167E), MHM2 PrP(S169N), MHM2 PrP(N170S), and MHM2 PrP(Q171R) and transfected the DNAs into ScN2a cells (Table 3). MHM2 PrP(N170S) is analogous to human polymorphism N171S (Fink et al. 1994).

In sheep, the substitution of a basic residue at position 171 probably prevents scrapie through a dominant-negative mechanism similar to that postulated for a basic residue at 219 protecting humans from CJD. With few exceptions, only sheep that are Gln/Gln at 171 develop scrapie; sheep that are Gln/Arg or Arg/Arg are resistant to scrapie (Hunter et al. 1993, 1997a,b; Goldmann et al. 1994; Westaway et al. 1994; Belt et al. 1995; Clousard et al. 1995; Ikeda et al. 1995; O'Rourke et al. 1997). These findings suggest that R171 creates a PrP^C molecule in sheep that acts as a dom-

Table 4. Protein X–mediated mechanisms of inhibition of PrP^{Sc} formation

Type of inhibition	Example	Putative mechanism	Relative affinity for protein X[a]
1	HuPrP^C(E219) binding to Mo protein X inhibited by MoPrP^C	competitive	low
2	HuPrP^C(K219) prevents MoPrP^C binding to protein X	noncompetitive	high
3	SHaPrP^C binds to protein X and is not released by MoPrP^{Sc}	noncompetitive	similar

[a]Affinity is relative to that of MoPrP^C for Mo protein X.

inant negative in preventing PrP^{Sc} formation. When we introduced the Q167R or Q171R mutation into MHM2 PrP expressed in ScN2a cells, the recombinant protein was not converted to PrP^{Sc} and inhibited the conversion of wt MHM2 PrP into PrP^{Sc}. Q167R and Q171R in MoPrP correspond to Q171 and Q175 in sheep PrP, respectively. These findings argue that Q167 and Q171 in MoPrP form a discontinuous epitope with V214 and Q218 to which protein X binds.

From these studies, we were able to distinguish three classes of inhibition of PrP^{Sc} formation, designated as types 1, 2, and 3 (Table 4). Type 1 inhibition is illustrated by the competition between MoPrP^C and HuPrP^C(E219) for binding to Mo protein X. In the absence of MoPrP^C, HuPrP^C(E219) is converted into PrP^{Sc} (Telling et al. 1995). Type 2 inhibition appears to be noncompetitive and is depicted by MHM2 PrP(K218), which binds to protein X in ScN2a cells and prevents conversion of wt MHM2 PrP^C to PrP^{Sc}. The binding is sufficiently tight that MHM2 PrP(K218) is also not converted to PrP^{Sc}. Like type 2, type 3 inhibition is also noncompetitive with respect to protein X but occurs through a different mechanism. This case is demonstrated by SHaPrP^C, which binds to Mo protein X but is not released by interacting with MoPrP^{Sc}. In Tg(SHaPrP) mice, SHaPrP^C is converted into PrP^{Sc} in the presence or absence of MoPrP^C when the animals are inoculated with SHa prions (Prusiner et al. 1990, 1993b; Büeler et al. 1993).

COMPLEXES OF PrP^C AND PrP^{Sc}

Studies with Tg mice demonstrated that PrP^C must interact with PrP^{Sc} during its conversion to nascent PrP^{Sc} (Prusiner et al. 1990; Scott et al. 1993). Our attempts to reproduce the formation of PrP^{Sc} in vitro using full-length PrP^C and PrP^{Sc} molecules mixed in equimolar amounts were unsuccessful (Raeber et al. 1992). Subsequent studies using a large excess of PrP^{Sc} defined conditions under which PrP^{C} binds to PrP^{Sc}, as demonstrated by the resistance of PrP^{C} to proteolysis (Kocisko et al. 1994, 1995; Bessen et al. 1995). Because nondenaturing conditions for the release of PrP^{C} bound to PrP^{Sc} have not been identified, it has not been possible to determine whether the conformation of PrP^{C} was altered when the PrP^{C}–PrP^{Sc} complex was formed (Kaneko et al. 1995).

Although some investigators have reported the in vitro formation of PrP^{Sc} by mixing a 50- to 700-fold excess of PrP^{Sc} with [^{35}S]PrP^{C}, their conclusions assume that protease-resistant PrP^{C} is equivalent to PrP^{Sc} (Kocisko et al. 1994; Bossers et al. 1997; DebBurman et al. 1997; Raymond et al. 1997). Interestingly, the binding of PrP^{C} to PrP^{Sc} was found to be dependent on the same residues (Kocisko et al. 1995) that render Tg(MH2M) mice susceptible to SHa prions (Scott et al. 1993) and seems to be dependent on the prion strain (Bessen et al. 1995).

Because a small fraction of PrP^{C} acquired protease resistance when incubated alone, and a much larger fraction showed resistance when incubated with synthetic PrP peptides, we revisited the possibility that PrP^{C} mixed with PrP^{Sc} might render it protease-resistant (Kaneko et al. 1995). Other investigators reported that PrP^{Sc} denatured in 3 M GdnHCl undergoes renaturation and renders PrP^{C} resistant to proteolysis within 2 minutes of mixing (Kocisko et al. 1994). Because numerous attempts to renature prion infectivity from both Gdn and urea failed (Prusiner et al. 1993a), we investigated the effect of 3 M GdnHCl on PrP^{Sc}. As before, we were unable to demonstrate renaturation of PrP^{Sc} that had been denatured in 3 M GdnHCl and then diluted 4- to 10-fold prior to limited protease digestion and SDS-PAGE. Of note, when the dilution was carried out in the same tube to which the 3 M GdnHCl had been added, we observed protease-resistant PrP (data not shown). This was not seen when the tubes were changed, and we surmise that the presence of protease-resistant PrP was due to residual, undenatured PrP^{Sc} binding to the walls of the tube.

When PrP^{Sc} that had been denatured in 3 M GdnHCl and then diluted in buffer to give a final concentration of 0.3–2 M GdnHCl was mixed with PrP^{C}, no protease-resistant [^{35}S]PrP^{C} could be detected. However, mixing undenatured PrP^{Sc} with PrP^{C} produced protease-resistant [^{35}S]PrP^{C}. As reported by other investigators (Kocisko et al. 1994), a 10-fold excess of PrP^{Sc} did not yield protease-resistant [^{35}S]PrP^{C}, and a 50-fold excess of PrP^{Sc} was required to produce protease-resistant [^{35}S]PrP^{C}. The presence of 0.3 M GdnHCl in the reaction mixture seemed to be essential because its removal by methanol precipitation prior to mixing prevented the formation of complexes. Although ~50% of [^{35}S]PrP^{C} was recovered in complexes sedimented at 100,000g for 1 hour, only 10–15% was protease-resistant.

The mixing of PrP^C in crude brain homogenates with PrP^{Sc} has been reported to produce nascent PrP^{Sc} in substantial quantities (Saborio et al. 2001). The production of nascent PrP^{Sc} in a test tube was reported to require repeated cycles, each of 15-second duration, of sonication. Considerable effort has been spent trying to reproduce these results, but without any success (H. Ball and S.B. Prusiner, unpubl.). However, others have reported that brain homogenates containing PrP^C mixed with PrP^{Sc} exhibit increased levels of protease-resistant PrP after prolonged incubation without sonication (Lucassen et al. 2003).

CONCLUDING REMARKS

Although knowledge of prions and the diseases that they cause has increased dramatically over the past 15 years since the discovery of PrP 27-30 was reported, much more remains to be learned. Exhaustive searches for a scrapie-specific nucleic acid have failed to identify any candidates, and many new questions have emerged. For example, little information about the tertiary structure of PrP^{Sc} is available (Chapter 5), and much about the cell biology of PrP^{Sc} formation is unknown (Chapter 10). The identification of protein X, which is thought to function as a molecular chaperone in PrP^{Sc} formation, is another important goal.

The CDI for PrP^{Sc} should find wide application in human and veterinary medicine as well as in agriculture (Chapter 3). It seems likely that many features of the pathogenesis of prion diseases will be uncovered by the availability of a sensitive, high-throughput assay for PrP^{Sc}. Because bovine prions are now accepted as the cause of variant (v) CJD, screening of cattle over 30 months of age for PrP 27-30 in the brain stem is now mandatory upon slaughter in the countries of the European Union. Application of the CDI for the screening of cattle in Europe will provide a much-needed added measure of protection of the human food supply from contamination with bovine prions.

As more is learned about the formation of prions, new approaches to the development of effective pharmacotherapeutics are emerging. Drugs that alter the conformation of PrP^{Sc} and allow the cell to degrade it may prove to be efficacious (Supattapone et al. 1999, 2001). Drugs that block the formation of nascent PrP^{Sc} by interfering with the binding of PrP^C to PrP^{Sc} may also prove to be effective (Caughey and Race 1992; Caughey et al. 1994; Ingrosso et al. 1995; Heppner et al. 2001; Peretz et al. 2001b). Of all the approaches currently available, pharmacotherapeutics that mimic dominant-negative inhibition of prion formation by disrupting the binding of PrP^C to protein X may prove to be the most effective (Kaneko et al. 1997; Perrier et al. 2000, 2002). These and other approaches to therapeutics are discussed in Chapter 18.

REFERENCES

Aiken J.M., Williamson J.L., and Marsh R.F. 1989. Evidence of mitochondrial involvement in scrapie infection. *J. Virol.* **63:** 1686–1694.

Aiken J.M., Williamson J.L., Borchardt L.M., and Marsh R.F. 1990. Presence of mitochondrial D-loop DNA in scrapie-infected brain preparations enriched for the prion protein. *J. Virol.* **64:** 3265–3268.

Akowitz A., Sklaviadis T., Manuelidis E.E., and Manuelidis L. 1990. Nuclease-resistant polyadenylated RNAs of significant size are detected by PCR in highly purified Creutzfeldt-Jakob disease preparations. *Microb. Pathog.* **9:** 33–45.

Alper T., Haig D.A., and Clarke M.C. 1966. The exceptionally small size of the scrapie agent. *Biochem. Biophys. Res. Commun.* **22:** 278–284.

———. 1978. The scrapie agent: Evidence against its dependence for replication on intrinsic nucleic acid. *J. Gen. Virol.* **41:** 503–516.

Alper T., Cramp W.A., Haig D.A., and Clarke M.C. 1967. Does the agent of scrapie replicate without nucleic acid? *Nature* **214:** 764–766.

Anderson N.G. 1962. The zonal ultracentrifuge. A new instrument for fractionating mixtures of particles. *J. Phys. Chem.* **66:** 1984–1989.

Barnard G., Helmick B., Madden S., Gilbourne C., and Patel R. 2000. The measurement of prion protein in bovine brain tissue using differential extraction and DELFIA as a diagnostic test for BSE. *Luminescence* **15:** 357–362.

Barry R.A. and Prusiner S.B. 1986. Monoclonal antibodies to the cellular and scrapie prion proteins. *J. Infect. Dis.* **154:** 518–521.

Bellinger-Kawahara C., Cleaver J.E., Diener T.O., and Prusiner S.B. 1987a. Purified scrapie prions resist inactivation by UV irradiation. *J. Virol.* **61:** 159–166.

Bellinger-Kawahara C.G., Kempner E., Groth D.F., Gabizon R., and Prusiner S.B. 1988. Scrapie prion liposomes and rods exhibit target sizes of 55,000 Da. *Virology* **164:** 537–541.

Bellinger-Kawahara C., Diener T.O., McKinley M.P., Groth D.F., Smith D.R., and Prusiner S.B. 1987b. Purified scrapie prions resist inactivation by procedures that hydrolyze, modify, or shear nucleic acids. *Virology* **160:** 271–274.

Bellon A., Seyfert-Brandt W., Lang W., Baron H., Groner A., and Vey M. 2003. Improved conformation-dependent immunoassay: Suitability for human prion detection with enhanced sensitivity. *J. Gen. Virol.* **84:** 1921–1925.

Belt P.B., Muileman I.H., Schreuder B.E.C., Ruijter J.B., Gielkens A.L.J., and Smits M.A. 1995. Identification of five allelic variants of the sheep PrP gene and their association with natural scrapie. *J. Gen. Virol.* **76:** 509–517.

Bendheim P.E., Barry R.A., DeArmond S.J., Stites D.P., and Prusiner S.B. 1984. Antibodies to a scrapie prion protein. *Nature* **310:** 418–421.

Bessen R.A. and Marsh R.F. 1992. Biochemical and physical properties of the prion protein from two strains of the transmissible mink encephalopathy agent. *J. Virol.* **66:** 2096–2101.

————. 1994. Distinct PrP properties suggest the molecular basis of strain variation in transmissible mink encephalopathy. *J. Virol.* **68:** 7859–7868.

Bessen R.A., Kocisko D.A., Raymond G.J., Nandan S., Lansbury P.T., and Caughey B. 1995. Non-genetic propagation of strain-specific properties of scrapie prion protein. *Nature* **375:** 698–700.

Bieschke J., Giese A., Schulz-Schaeffer W., Zerr I., Poser S., Eigen M., and Kretzschmar H. 2000. Ultrasensitive detection of pathological prion protein aggregates by dual-color scanning for intensely fluorescent targets. *Proc. Natl. Acad. Sci.* **97:** 5468–5473.

Biffiger K., Zwald D., Kaufmann L., Briner A., Nayki I., Purro M., Bottcher S., Struckmeyer T., Schaller O., Meyer R., Fatzer R., Zurbriggen A., Stack M., Moser M., Oesch B., and Kubler E. 2002. Validation of a luminescence immunoassay for the detection of PrP[Sc] in brain homogenate. *J. Virol. Methods* **101:** 79–84.

Blochberger T.C., Cooper C., Peretz D., Tatzelt J., Griffith O.H., Baldwin M.A., and Prusiner S.B. 1997. Prion protein expression in Chinese hamster ovary cells using a glutamine synthetase selection and amplification system. *Protein Eng.* **10:** 1465–1473.

Bode L., Pocchiari M., Gelderblom H., and Diringer H. 1985. Characterization of antisera against scrapie-associated fibrils (SAF) from affected hamster and cross-reactivity with SAF from scrapie-affected mice and from patients with Creutzfeldt-Jakob disease. *J. Gen. Virol.* **66:** 2471–2478.

Bolton D.C., McKinley M.P., and Prusiner S.B. 1982. Identification of a protein that purifies with the scrapie prion. *Science* **218:** 1309–1311.

Bolton D.C., Bendheim P.E., Marmorstein A.D., and Potempska A. 1987. Isolation and structural studies of the intact scrapie agent protein. *Arch. Biochem. Biophys.* **258:** 579–590.

Bosque P.J., Ryou C., Telling G., Peretz D., Legname G., DeArmond S.J., and Prusiner S.B. 2002. Prions in skeletal muscle. *Proc. Natl. Acad. Sci.* **99:** 3812–3817.

Bossers A., Belt P.B.G.M., Raymond G.J., Caughey B., de Vries R., and Smits M.A. 1997. Scrapie susceptibility-linked polymorphisms modulate the *in vitro* conversion of sheep prion protein to protease-resistant forms. *Proc. Natl. Acad. Sci.* **94:** 4931–4936.

Brown D.R., Wong B.-S., Hafiz F., Clive C., Haswell S.J., and Jones I.M. 1999. Normal prion protein has an activity like that of superoxide dismutase. *Biochem. J.* **344:** 1–5.

Brown D.R., Qin K., Herms J.W., Madlung A., Manson J., Strome R., Fraser P.E., Kruck T., von Bohlen A., Schulz-Schaeffer W., Giese A., Westaway D., and Kretzschmar H. 1997. The cellular prion protein binds copper *in vivo*. *Nature* **390:** 684–687.

Büeler H., Aguzzi A., Sailer A., Greiner R.-A., Autenried P., Aguet M., and Weissmann C. 1993. Mice devoid of PrP are resistant to scrapie. *Cell* **73:** 1339–1347.

Büeler H., Fisher M., Lang Y., Bluethmann H., Lipp H.-P., DeArmond S.J., Prusiner S.B., Aguet M., and Weissmann C. 1992. Normal development and behaviour of mice lacking the neuronal cell-surface PrP protein. *Nature* **356:** 577–582.

Castle B.E., Dees C., German T.L., and Marsh R.F. 1987. Effects of different methods of purification on aggregation of scrapie infectivity. *J. Gen. Virol.* **68:** 225–231.

Caughey B. and Race R.E. 1992. Potent inhibition of scrapie-associated PrP accumulation by Congo red. *J. Neurochem.* **59:** 768–771.

Caughey B., Brown K., Raymond G.J., Katzenstein G.E., and Thresher W. 1994. Binding of the protease-sensitive form of PrP (prion protein) to sulfated glycosaminoglycan and Congo red. *J. Virol.* **68:** 2135–2141.

Chandler R.L. 1959. Attempts to demonstrate antibodies in scrapie disease. *Vet. Rec.* **71:** 58–59.

Chesebro B., Race R., Wehrly K., Nishio J., Bloom M., Lechner D., Bergstrom S., Robbins

K., Mayer L., Keith J.M., Garon C., and Haase A. 1985. Identification of scrapie prion protein-specific mRNA in scrapie-infected and uninfected brain. *Nature* **315:** 331–333.

Clarke M.C. and Haig D.A. 1966. Attempts to demonstrate neutralising antibodies in the sera of scrapie-infected animals. *Vet. Rec.* **78:** 647–649.

Clousard C., Beaudry P., Elsen J.M., Milan D., Dussaucy M., Bounneau C., Schelcher F., Chatelain J., Launay J.-M., and Laplanche J.-L. 1995. Different allelic effects of the codons 136 and 171 of the prion protein gene in sheep with natural scrapie. *J. Gen. Virol.* **76:** 2097–2101.

DeArmond S.J., Mobley W.C., DeMott D.L., Barry R.A., Beckstead J.H., and Prusiner S.B. 1987. Changes in the localization of brain prion proteins during scrapie infection. *Neurology* **37:** 1271–1280.

DebBurman S.K., Raymond G.J., Caughey B., and Lindquist S. 1997. Chaperone-supervised conversion of prion protein to its protease-resistant form. *Proc. Natl. Acad. Sci.* **94:** 13938–13943.

Dees C., McMillan B.C., Wade W.F., German T.L., and Marsh R.F. 1985. Characterization of nucleic acids in membrane vesicles from scrapie-infected hamster brain. *J. Virol.* **55:** 126–132.

Deslys J.P., Comoy E., Hawkins S., Simon S., Schimmel H., Wells G., Grassi J., and Moynagh J. 2001. Screening slaughtered cattle for BSE. *Nature* **409:** 476–478.

Diringer H., Gelderblom H., Hilmert H., Ozel M., Edelbluth C., and Kimberlin R.H. 1983. Scrapie infectivity, fibrils and low molecular weight protein. *Nature* **306:** 476–478.

Doh-ura K., Iwaki T., and Caughey B. 2000. Lysosomotropic agents and cysteine protease inhibitors inhibit scrapie-associated prion protein accumulation. *J. Virol.* **74:** 4894–4897.

Donne D.G., Viles J.H., Groth D., Mehlhorn I., James T.L., Cohen F.E., Prusiner S.B., Wright P.E., and Dyson H.J. 1997. Structure of the recombinant full-length hamster prion protein PrP(29–231): The N terminus is highly flexible. *Proc. Natl. Acad. Sci.* **94:** 13452–13457.

Duguid J.R. and Dinauer C.M. 1989. Library subtraction of *in vitro* cDNA library. *Proc. Natl. Acad. Sci.* **85:** 2789–2792.

Duguid J.R., Rohwer R.G., and Seed B. 1988. Isolation of cDNAs of scrapie-modulated RNAs by subtractive hybridization of a cDNA library. *Proc. Natl. Acad. Sci.* **85:** 5738–5742.

Edenhofer F., Rieger R., Famulok M., Wendler W., Weiss S., and Winnacker E.-L. 1996. Prion protein PrP^C interacts with molecular chaperones of the Hsp60 family. *J. Virol.* **70:** 4724–4728.

Endo T., Groth D., Prusiner S.B., and Kobata A. 1989. Diversity of oligosaccharide structures linked to asparagines of the scrapie prion protein. *Biochemistry* **28:** 8380–8388.

Field E.J., Farmer F., Caspary E.A., and Joyce G. 1969. Susceptibility of scrapie agent to ionizing radiation. *Nature* **222:** 90–91.

Fink J.K., Peacock M.L., Warren J.T., Roses A.D., and Prusiner S.B. 1994. Detecting prion protein gene mutations by denaturing gradient gel electrophoresis. *Hum. Mutat.* **4:** 42–50.

Gabizon R., McKinley M.P., and Prusiner S.B. 1987. Purified prion proteins and scrapie infectivity copartition into liposomes. *Proc. Natl. Acad. Sci.* **84:** 4017–4021.

———. 1988a. Properties of scrapie prion proteins in liposomes and amyloid rods. *Ciba Found. Symp.* **135:** 182–196.

Gabizon R., McKinley M.P., Groth D., and Prusiner S.B. 1988b. Immunoaffinity purification and neutralization of scrapie prion infectivity. *Proc. Natl. Acad. Sci.* **85:** 6617–6621.

Gabizon R., McKinley M.P., Groth D.F., Kenaga L., and Prusiner S.B. 1988c. Properties of scrapie prion protein liposomes. *J. Biol. Chem.* **263:** 4950–4955.

Gajdusek D.C. 1977. Unconventional viruses and the origin and disappearance of kuru. *Science* **197:** 943–960.

Gasset M., Baldwin M.A., Fletterick R.J., and Prusiner S.B. 1993. Perturbation of the secondary structure of the scrapie prion protein under conditions that alter infectivity. *Proc. Natl. Acad. Sci.* **90:** 1–5.

Gauczynski S., Peyrin J.M., Haik S., Leucht C., Hundt C., Rieger R., Krasemann S., Deslys J.P., Dormont D., Lasmézas C.I., and Weiss S. 2001. The 37-kDa/67-kDa laminin receptor acts as the cell-surface receptor for the cellular prion protein. *EMBO J.* **20:** 5863–5875.

German T.L., McMillan B.C., Castle B.E., Dees C., Wade W.F., and Marsh R.F. 1985. Comparison of RNA from healthy and scrapie-infected hamster brain. *J. Gen. Virol.* **66:** 839–844.

Goldmann W., Hunter N., Smith G., Foster J., and Hope J. 1994. PrP genotype and agent effects in scrapie: Change in allelic interaction with different isolates of agent in sheep, a natural host of scrapie. *J. Gen. Virol.* **75:** 989–995.

Goll A., Ferry D.R., and Grossman H. 1984a. Target size analysis and molecular properties of Ca^{2+} channels labelled with [^{3}H]verapamil. *Eur. J. Biochem.* **141:** 177–186.

Goll A., Ferry D.R., Striessnig J., Schober M., and Glossmann H. 1984b. (-)-[3H]Desmethoxyverapamil, a novel Ca^{2+} channel probe. Binding characteristics and target site analysis of its receptor in skeletal muscle. *FEBS Lett.* **176:** 371–377.

Graner E., Mercadante A.F., Zanata S.M., Forlenza O.V., Cabral A.L.B., Veiga S.S., Juliano M.A., Roesler R., Walz R., Mineti A., Izquierdo I., Martins V.R., and Brentani R.R. 2000. Cellular prion protein binds laminin and mediates neuritogenesis. *Mol. Brain Res.* **76:** 85–92.

Grassi J., Comoy E., Simon S., Creminon C., Frobert Y., Trapmann S., Schimmel H., Hawkins S.A., Moynagh J., Deslys J.P., and Wells G.A. 2001. Rapid test for the preclinical postmortem diagnosis of BSE in central nervous system tissue. *Vet. Rec.* **149:** 577–582.

Haraguchi T., Fisher S., Olofsson S., Endo T., Groth D., Tarantino A., Borchelt D.R., Teplow D., Hood L., Burlingame A., Lycke E., Kobata A., and Prusiner S.B. 1989. Asparagine-linked glycosylation of the scrapie and cellular prion proteins. *Arch. Biochem. Biophys.* **274:** 1–13.

Hemmilä I. and Harju R. 1995. Time-resolved fluorometry. In *Bioanalytical applications of labelling technologies* (ed. P. Mottram), pp. 113–119. Wallac Oy, Turku, Finland.

Heppner F.L., Musahl C., Arrighi I., Klein M.A., Rülicke T., Oesch B., Zinkernagel R.M., Kalinke U., and Aguzzi A. 2001. Prevention of scrapie pathogenesis by transgenic expression of anti-prion protein antibodies. *Science* **294:** 178–182.

Hope J., Morton L.J.D., Farquhar C.F., Multhaup G., Beyreuther K., and Kimberlin R.H. 1986. The major polypeptide of scrapie-associated fibrils (SAF) has the same size, charge distribution and N-terminal protein sequence as predicted for the normal brain protein (PrP). *EMBO J.* **5:** 2591–2597.

Hornemann S. and Glockshuber R. 1996. Autonomous and reversible folding of a soluble amino-terminally truncated segment of the mouse prion protein. *J. Mol. Biol.* **262:** 614–619.

Hornemann S., Korth C., Oesch B., Riek R., Wide G., Wüthrich K., and Glockshuber R. 1997. Recombinant full-length murine prion protein, mPrP(23–231): Purification and spectroscopic characterization. *FEBS Lett.* **413:** 277–281.

Hsiao K.K., Scott M., Foster D., Groth D.F., DeArmond S.J., and Prusiner S.B. 1990. Spontaneous neurodegeneration in transgenic mice with mutant prion protein. *Science* **250:** 1587–1590.

Hundt C., Peyrin J.M., Haik S., Gauczynski S., Leucht C., Rieger R., Riley M.L., Deslys J.P., Dormont D., Lasmézas C.I., and Weiss S. 2001. Identification of interaction domains of the prion protein with its 37-kDa/67-kDa laminin receptor. *EMBO J.* **20:** 5876–5886.

Hunter G.D. 1979. The enigma of the scrapie agent: Biochemical approaches and the involvement of membranes and nucleic acids. In *Slow transmissible diseases of the nervous system* (ed. S.B. Prusiner and W.J. Hadlow), vol. 2, pp. 365–385. Academic Press, New York.

Hunter N., Goldmann W., Benson G., Foster J.D., and Hope J. 1993. Swaledale sheep affected by natural scrapie differ significantly in PrP genotype frequencies from healthy sheep and those selected for reduced incidence of scrapie. *J. Gen. Virol.* **74:** 1025–1031.

Hunter N., Moore L., Hosie B.D., Dingwall W.S., and Greig A. 1997a. Association between natural scrapie and PrP genotype in a flock of Suffolk sheep in Scotland. *Vet. Rec.* **140:** 59–63.

Hunter N., Cairns D., Foster J.D., Smith G., Goldmann W., and Donnelly K. 1997b. Is scrapie solely a genetic disease? *Nature* **386:** 137.

Ikeda T., Horiuchi M., Ishiguro N., Muramatsu Y., Kai-Uwe G.D., and Shinagawa M. 1995. Amino acid polymorphisms of PrP with reference to onset of scrapie in Suffolk and Corriedale sheep in Japan. *J. Gen. Virol.* **76:** 2577–2581.

Ingrosso L., Ladogana A., and Pocchiari M. 1995. Congo red prolongs the incubation period in scrapie-infected hamsters. *J. Virol.* **69:** 506–508.

James T.L., Liu H., Ulyanov N.B., Farr-Jones S., Zhang H., Donne D.G., Kaneko K., Groth D., Mehlhorn I., Prusiner S.B., and Cohen F.E. 1997. Solution structure of a 142-residue recombinant prion protein corresponding to the infectious fragment of the scrapie isoform. *Proc. Natl. Acad. Sci.* **94:** 10086–10091.

Kaneko K., Zulianello L., Scott M., Cooper C.M., Wallace A.C., James T.L., Cohen F.E., and Prusiner S.B. 1997. Evidence for protein X binding to a discontinuous epitope on the cellular prion protein during scrapie prion propagation. *Proc. Natl. Acad. Sci.* **94:** 10069–10074.

Kaneko K., Peretz D., Pan K.-M., Blochberger T., Wille H., Gabizon R., Griffith O.H., Cohen F.E., Baldwin M.A., and Prusiner S.B. 1995. Prion protein (PrP) synthetic peptides induce cellular PrP to acquire properties of the scrapie isoform. *Proc. Natl. Acad. Sci.* **32:** 11160–11164.

Kascsak R.J., Rubenstein R., Merz P.A., Tonna-DeMasi M., Fersko R., Carp R.I., Wisniewski H.M., and Diringer H. 1987. Mouse polyclonal and monoclonal antibody to scrapie-associated fibril proteins. *J. Virol.* **61:** 3688–3693.

Katz M., Balian Rorke L., Masland W.S., Koprowski H., and Tucker S.H. 1968. Transmission of an encephalitogenic agent from brains of patients with subacute sclerosing panencephalitis to ferrets. *N. Engl. J. Med.* **279:** 793–798.

Kellings K. 1995. "Analysis of residual nucleic acids in scrapie prions with differing degrees of aggregation." Ph. D. thesis, Heinrich-Heine-Universität, Düsseldorf. Verlag Shaker, Aachen, Germany.

Kellings K., Meyer N., Mirenda C., Prusiner S.B., and Riesner D. 1992. Further analysis of nucleic acids in purified scrapie prion preparations by improved return refocussing gel electrophoresis (RRGE). *J. Gen. Virol.* **73:** 1025–1029.

Kempner E.S. and Haigler H.T. 1982. The influence of low temperature on the radiation sensitivity of enzymes. *J. Biol. Chem.* **257:** 13297–13299.

Keshet G.I., Bar-Peled O., Yaffe D., Nudel U., and Gabizon R. 2000. The cellular prion protein colocalizes with the dystroglycan complex in the brain. *J. Neurochem.* **75:** 1889–1897.

Kitamoto T. and Tateishi J. 1994. Human prion diseases with variant prion protein. *Philos. Trans. R. Soc. Lond. B Biol. Sci.* **343:** 391–398.

Kitamoto T., Ogomori K., Tateishi J., and Prusiner S.B. 1987. Formic acid pretreatment enhances immunostaining of cerebral and systemic amyloids. *Lab. Invest.* **57:** 230–236.

Kocisko D.A., Baron G.S., Rubenstein R., Chen J., Kuizon S., and Caughey B. 2003. New inhibitors of scrapie-associated prion protein formation in a library of 2000 drugs and natural products. *J. Virol.* **77:** 10288–10294.

Kocisko D.A., Priola S.A., Raymond G.J., Chesebro B., Lansbury P.T., Jr., and Caughey B. 1995. Species specificity in the cell-free conversion of prion protein to protease-resistant forms: A model for the scrapie species barrier. *Proc. Natl. Acad. Sci.* **92:** 3923–3927.

Kocisko D.A., Come J.H., Priola S.A., Chesebro B., Raymond G.J., Lansbury P.T., Jr., and Caughey B. 1994. Cell-free formation of protease-resistant prion protein. *Nature* **370:** 471–474.

Korth C., May B.C.H., Cohen F.E., and Prusiner S.B. 2001. Acridine and phenothiazine derivatives as pharmacotherapeutics for prion disease. *Proc. Natl. Acad. Sci.* **98:** 9836–9841.

Korth C., Stierli B., Streit P., Moser M., Schaller O., Fischer R., Schulz-Schaeffer W., Kretzschmar H., Raeber A., Braun U., Ehrensperger F., Hornemann S., Glockshuber R., Riek R., Billeter M., Wüthrich K., and Oesch B. 1997. Prion (PrP^{Sc})-specific epitope defined by a monoclonal antibody. *Nature* **389:** 74–77.

Kurschner C. and Morgan J.I. 1995. The cellular prion protein (PrP) selectively binds to Bcl-2 in the yeast two-hybrid system. *Brain Res. Mol. Brain Res.* **30:** 165–168.

Latarjet R., Cramer R., and Montagnier L. 1967. Inactivation by UV-, X- and gamma-radiations, of the infecting and transforming capacities of polyoma virus. *Virology* **33:** 104–111.

Legname G., Nelken P., Guan Z., Kanyo Z.F., DeArmond S.J., and Prusiner S.B. 2002. Prion and doppel proteins bind to granule cells of the cerebellum. *Proc. Natl. Acad. Sci.* **99:** 16285–16290.

Li A., Sakaguchi S., Shigematsu K., Atarashi R., Roy B.C., Nakaoke R., Arima K., Okimura N., Kopacek J., and Katamine S. 2000. Physiological expression of the gene for PrP-like protein, PrPLP/Dpl, by brain endothelial cells and its ectopic expression in neurons of PrP-deficient mice ataxic due to Purkinje cell degeneration. *Am. J. Pathol.* **157:** 1447–1452.

Lucassen R., Nishina K., and Supattapone S. 2003. In vitro amplification of protease-resistant prion protein requires free sulfhydryl groups. *Biochemistry* **42:** 4127–4135.

Malone T.G., Marsh R.F., Hanson R.P., and Semancik J.S. 1979. Evidence for the low molecular weight nature of the scrapie agent. *Nature* **278:** 575–576.

Manson J.C., Clarke A.R., Hooper M.L., Aitchison L., McConnell I., and Hope J. 1994. 129/Ola mice carrying a null mutation in PrP that abolishes mRNA production are developmentally normal. *Mol. Neurobiol.* **8:** 121–127.

Martins V.R., Graner E., Garcia-Abreu J., de Souza S.J., Mercadante A.F., Veiga S.S., Zanata S.M., Neto V.M., and Brentani R.R. 1997. Complementary hydropathy identifies a cellular prion protein receptor. *Nat. Med.* **3:** 1376–1382.

McKinley M.P., Bolton D.C., and Prusiner S.B. 1983. A protease-resistant protein is a structural component of the scrapie prion. *Cell* **35:** 57–62.

McKinley M.P., Masiarz F.R., and Prusiner S.B. 1981. Reversible chemical modification of the scrapie agent. *Science* **214:** 1259–1261.

McKinley M.P., Meyer R.K., Kenaga L., Rahbar F., Cotter R., Serban A., and Prusiner S.B. 1991. Scrapie prion rod formation *in vitro* requires both detergent extraction and limited proteolysis. *J. Virol.* **65:** 1340–1351.

Mehlhorn I., Groth D., Stöckel J., Moffat B., Reilly D., Yansura D., Willett W.S., Baldwin M., Fletterick R., Cohen F.E., Vandlen R., Henner D., and Prusiner S.B. 1996. High-level expression and characterization of a purified 142-residue polypeptide of the prion protein. *Biochemistry* **35:** 5528–5537.

Meyer N., Rosenbaum V., Schmidt B., Gilles K., Mirenda C., Groth D., Prusiner S.B., and Riesner D. 1991. Search for a putative scrapie genome in purified prion fractions reveals a paucity of nucleic acids. *J. Gen. Virol.* **72:** 37–49.

Meyer R.K., McKinley M.P., Bowman K.A., Braunfeld M.B., Barry R.A., and Prusiner S.B. 1986. Separation and properties of cellular and scrapie prion proteins. *Proc. Natl. Acad. Sci.* **83:** 2310–2314.

Miller J.M., Jenny A.L., Taylor W.D., Race R.E., Ernst D.R., Kotz J.B., and Rubenstein R. 1994. Detection of prion protein in formalin-fixed brain by hydrated autoclaving immunohistochemistry for the diagnosis of scrapie in sheep. *J. Vet. Diagn. Invest.* **6:** 366–368.

Millson G.C. and Manning E.J. 1979. The effect of selected detergents on scrapie infectivity. In *Slow transmissible diseases of the nervous system* (ed. S.B. Prusiner and W.J. Hadlow), vol. 2, pp. 409–424. Academic Press, New York.

Mo H., Moore R.C., Cohen F.E., Westaway D., Prusiner S.B., Wright P.E., and Dyson H.J. 2001. Two different neurodegenerative diseases caused by proteins with similar structures. *Proc. Natl. Acad. Sci.* **98:** 2352–2357.

Moore R.C., Mastrangelo P., Bouzamondo E., Heinrich C., Legname G., Prusiner S.B., Hood L., Westaway D., DeArmond S.J., and Tremblay P. 2001. Doppel-induced cerebellar degeneration in transgenic mice. *Proc. Natl. Acad. Sci.* **98:** 15288–15293.

Moore R.C., Lee I.Y., Silverman G.L., Harrison P.M., Strome R., Heinrich C., Karunaratne A., Pasternak S.H., Chishti M.A., Liang Y., Mastrangelo P., Wang K., Smit A.F.A., Katamine S., Carlson G.A., Cohen F.E., Prusiner S.B., Melton D.W., Tremblay P., Hood L.E., and Westaway D. 1999. Ataxia in prion protein (PrP) deficient mice is associated with upregulation of the novel PrP-like protein doppel. *J. Mol. Biol.* **292:** 797–817.

Mouillet-Richard S., Ermonval M., Chebassier C., Laplanche J.L., Lehmann S., Launay J.M., and Kellermann O. 2000. Signal transduction through prion protein. *Science* **289:** 1925–1928.

Moynagh J., Schimmel H., and Kramer G.N. 1999. The evaluation of tests for the diagnosis of transmissible spongiform encephalopathy in bovines. European Commission report (July 1999), pp. 1–36.

Muramoto T., Kitamoto T., Tateishi J., and Goto I. 1992. The sequential development of abnormal prion protein accumulation in mice with Creutzfeldt-Jakob disease. *Am. J. Pathol.* **140:** 1411–1420.

Murdoch G.H., Sklaviadis T., Manuelidis E.E., and Manuelidis L. 1990. Potential retroviral RNAs in Creutzfeldt-Jakob disease. *J. Virol.* **64:** 1477–1486.

Narang H.K. 1990. Detection of single-stranded DNA in scrapie-infected brain by electron microscopy. *J. Mol. Biol.* **216:** 469–473.

Narang H.K., Asher D.M., and Gajdusek D.C. 1988. Evidence that DNA is present in abnormal tubulofilamentous structures found in scrapie. *Proc. Natl. Acad. Sci.* **85:** 3575–3579.

Neary K., Caughey B., Ernst D., Race R.E., and Chesebro B. 1991. Protease sensitivity and nuclease resistance of the scrapie agent propagated in vitro in neuroblastoma cells. *J. Virol.* **65:** 1031–1034.

Nishida N., Tremblay P., Sugimoto T., Shigematsu K., Shirabe S., Petromilli C., Erpel S.P., Nakaoke R., Atarashi R., Houtani T., Torchia M., Sakaguchi S., DeArmond S.J., Prusiner S.B., and Katamine S. 1999. A mouse prion protein transgene rescues mice deficient for the prion protein gene from Purkinje cell degeneration and demyelination. *Lab. Invest.* **79:** 689–697.

O'Rourke K.I., Holyoak G.R., Clark W.W., Mickelson J.R., Wang S., Melco R.P., Besser T.E., and Foote W.C. 1997. PrP genotypes and experimental scrapie in orally inoculated Suffolk sheep in the United States. *J. Gen. Virol.* **78:** 975–978.

Oesch B., Groth D.F., Prusiner S.B., and Weissmann C. 1988. Search for a scrapie-specific nucleic acid: A progress report. *Ciba Found. Symp.* **135:** 209–223.

Oesch B., Teplow D.B., Stahl N., Serban D., Hood L.E., and Prusiner S.B. 1990. Identification of cellular proteins binding to the scrapie prion protein. *Biochemistry* **29:** 5848–5855.

Oesch B., Westaway D., Wälchli M., McKinley M.P., Kent S.B.H., Aebersold R., Barry R.A., Tempst P., Teplow D.B., Hood L.E., Prusiner S.B., and Weissmann C. 1985. A cellular gene encodes scrapie PrP 27-30 protein. *Cell* **40:** 735–746.

Pan K.-M., Stahl N., and Prusiner S.B. 1992. Purification and properties of the cellular prion protein from Syrian hamster brain. *Protein Sci.* **1:** 1343–1352.

Pan K.-M., Baldwin M., Nguyen J., Gasset M., Serban A., Groth D., Mehlhorn I., Huang Z., Fletterick R.J., Cohen F.E., and Prusiner S.B. 1993. Conversion of α-helices into β-sheets features in the formation of the scrapie prion proteins. *Proc. Natl. Acad. Sci.* **90:** 10962–10966.

Paramithiotis E., Pinard M., Lawton T., LaBoissiere S., Leathers V.L., Zou W.Q., Estey L.A., Lamontagne J., Lehto M.T., Kondejewski L.H., Francoeur G.P., Papadopoulos M., Haghighat A., Spatz S.J., Head M., Will R., Ironside J., O'Rourke K., Tonelli Q., Ledebur H.C., Chakrabartty A., and Cashman N.R. 2003. A prion protein epitope selectively for the pathologically misfolded conformation. *Nat. Med.* **9:** 893–899.

Parren P.W., Fisicaro P., Labrijn A.F., Binley J.M., Yang W.-P., Ditzel H.J., Barbas C.F., III, and Burton D.R. 1996. In vitro antigen challenge of human antibody libraries for vaccine evaluation: The human immunodeficiency virus type I envelope. *J. Virol.* **70:** 9046–9050.

Parry H.B. 1983. *Scrapie disease in sheep.* Academic Press, New York.

Pattison I.H. 1988. Fifty years with scrapie: A personal reminiscence. *Vet. Rec.* **123:** 661–666.

Pauly P.C. and Harris D.A. 1998. Copper stimulates endocytosis of the prion protein. *J. Biol. Chem.* **273:** 33107–33110.

Peretz D., Scott M., Groth D., Williamson A., Burton D., Cohen F.E., and Prusiner S.B. 2001a. Strain-specified relative conformational stability of the scrapie prion protein. *Protein Sci.* **10:** 854–863.

Peretz D., Williamson R.A., Matsunaga Y., Serban H., Pinilla C., Bastidas R.B., Rozenshteyn R., James T.L., Houghten R.A., Cohen F.E., Prusiner S.B., and Burton D.R. 1997. A conformational transition at the N-terminus of the prion protein features in formation of the scrapie isoform. *J. Mol. Biol.* **273:** 614–622.

Peretz D., Williamson R.A., Kaneko K., Vergara J., Leclerc E., Schmitt-Ulms G., Mehlhorn I.R., Legname G., Wormald M.R., Rudd P.M., Dwek R.A., Burton D.R., and Prusiner S.B. 2001b. Antibodies inhibit prion propagation and clear cell cultures of prion infectivity. *Nature* **412:** 739–743.

Pergami P., Jaffe H., and Safar J. 1996. Semipreparative chromatographic method to purify the normal cellular isoform of the prion protein in nondenatured form. *Anal. Biochem.* **236:** 63–73.

Perrier V., Wallace A.C., Kaneko K., Safar J., Prusiner S.B., and Cohen F.E. 2000. Mimicking dominant negative inhibition of prion replication through structure-based drug design. *Proc. Natl. Acad. Sci.* **97:** 6073–6078.

Perrier V., Kaneko K., Safar J., Vergara J., Tremblay P., DeArmond S.J., Cohen F.E., Prusiner S.B., and Wallace A.C. 2002. Dominant-negative inhibition of prion replication in transgenic mice. *Proc. Natl. Acad. Sci.* **99:** 13079–13084.

Pestka S., Kelder B., Familletti P.C., Moschera J.A., Crowl R., and Kempner E.S. 1983. Molecular weight of the functional unit of human leukocyte, fibroblast, and immune interferon. *J. Biol. Chem.* **258:** 9706–9709.

Pitschke M., Prior R., Haupt M., and Riesner D. 1998. Detection of single amyloid beta-protein aggregates in the cerebrospinal fluid of Alzheimer's patients by fluorescence correlation spectroscopy. *Nat. Med.* **4:** 832–834.

Presnell S.R., Cohen B.I., and Cohen F.E. 1993. MacMatch: A tool for pattern-based protein secondary structure prediction. *CABIOS* **9:** 373–374.

Prusiner S.B. 1982. Novel proteinaceous infectious particles cause scrapie. *Science* **216:** 136–144.

Prusiner S.B., Hadlow W.J., Eklund C.M., and Race R.E. 1977. Sedimentation properties of the scrapie agent. *Proc. Natl. Acad. Sci.* **74:** 4656–4660.

Prusiner S.B., Groth D.F., Bolton D.C., Kent S.B., and Hood L.E. 1984a. Purification and structural studies of a major scrapie prion protein. *Cell* **38:** 127–134.

Prusiner S.B., Groth D.F., Cochran S.P., McKinley M.P., and Masiarz F.R. 1980a. Gel electrophoresis and glass permeation chromatography of the hamster scrapie agent after enzymatic digestion and detergent extraction. *Biochemistry* **19:** 4892–4898.

Prusiner S.B., Groth D., Serban A., Stahl N., and Gabizon R. 1993a. Attempts to restore scrapie prion infectivity after exposure to protein denaturants. *Proc. Natl. Acad. Sci.* **90:** 2793–2797.

Prusiner S.B., Bolton D.C., Groth D.F., Bowman K.A., Cochran S.P., and McKinley M.P. 1982a. Further purification and characterization of scrapie prions. *Biochemistry* **21:** 6942–6950.

Prusiner S.B., Cochran S.P., Groth D.F., Downey D.E., Bowman K.A., and Martinez H.M. 1982b. Measurement of the scrapie agent using an incubation time interval assay. *Ann. Neurol.* **11:** 353–358.

Prusiner S.B., Groth D.F., McKinley M.P., Cochran S.P., Bowman K.A., and Kasper K.C. 1981. Thiocyanate and hydroxyl ions inactivate the scrapie agent. *Proc. Natl. Acad. Sci.* **78:** 4606–4610.

Prusiner S.B., Hadlow W.J., Garfin D.E., Cochran S.P., Baringer J.R., Race R.E., and Eklund C.M. 1978. Partial purification and evidence for multiple molecular forms of the scrapie agent. *Biochemistry* **17:** 4993–4997.

Prusiner S.B., McKinley M.P., Bowman K.A., Bolton D.C., Bendheim P.E., Groth D.F., and Glenner G.G. 1983. Scrapie prions aggregate to form amyloid-like birefringent rods. *Cell* **35:** 349–358.

Prusiner S.B., Garfin D.E., Cochran S.P., McKinley M.P., Groth D.F., Hadlow W.J., Race R.E., and Eklund C.M. 1980b. Experimental scrapie in the mouse: Electrophoretic and sedimentation properties of the partially purified agent. *J. Neurochem.* **35:** 574–582.

Prusiner S.B., Groth D., Serban A., Koehler R., Foster D., Torchia M., Burton D., Yang S.-L., and DeArmond S.J. 1993b. Ablation of the prion protein (PrP) gene in mice prevents scrapie and facilitates production of anti-PrP antibodies. *Proc. Natl. Acad. Sci.* **90:** 10608–10612.

Prusiner S.B., McKinley M.P., Bolton D.C., Bowman K.A., Groth D.F., Cochran S.P., Hennessey E.M., Braunfeld M.B., Baringer J.R., and Chatigny M.A. 1984b. Prions: Methods for assay, purification and characterization. In *Methods in Virology* (ed. K. Maramorosch and H. Koprowski), vol. 8, pp. 293–345. Academic Press, New York.

Prusiner S.B., Scott M., Foster D., Pan K.-M., Groth D., Mirenda C., Torchia M., Yang S.-L., Serban D., Carlson G.A., Hoppe P.C., Westaway D., and DeArmond S.J. 1990. Transgenetic studies implicate interactions between homologous PrP isoforms in scrapie prion replication. *Cell* **63**: 673–686.

Raeber A.J., Borchelt D.R., Scott M., and Prusiner S.B. 1992. Attempts to convert the cellular prion protein into the scrapie isoform in cell-free systems. *J. Virol.* **66**: 6155–6163.

Raymond G.J., Hope J., Kocisko D.A., Priola S.A., Raymond L.D., Bossers A., Ironside J., Will R.G., Chen S.G., Petersen R.B., Gambetti P., Rubenstein R., Smits M.A., Lansbury P.T., Jr., and Caughey B. 1997. Molecular assessment of the potential transmissibilities of BSE and scrapie to humans. *Nature* **388**: 285–288.

Rieger R., Edenhofer F., Lasmézas C.I., and Weiss S. 1997. The human 37-kDa laminin receptor precursor interacts with the prion protein in eukaryotic cells. *Nat. Med.* **3**: 1383–1388.

Riek R., Hornemann S., Wider G., Glockshuber R., and Wüthrich K. 1997. NMR characterization of the full-length recombinant murine prion protein, mPrP(23–231). *FEBS Lett.* **413**: 282–288.

Riesner D. 1991. The search for a nucleic acid component to scrapie infectivity. *Semin. Virol.* **2**: 215–226.

Riesner D., Kellings K., Post K., Wille H., Serban H., Groth D., Baldwin M.A., and Prusiner S.B. 1996. Disruption of prion rods generates 10-nm spherical particles having high α-helical content and lacking scrapie infectivity. *J. Virol.* **70**: 1714–1722.

Rogers M., Yehiely F., Scott M., and Prusiner S.B. 1993. Conversion of truncated and elongated prion proteins into the scrapie isoform in cultured cells. *Proc. Natl. Acad. Sci.* **90**: 3182–3186.

Rudd P.M., Endo T., Colominas C., Groth D., Wheeler S.F., Harvey D.J., Wormald M.R., Serban H., Prusiner S.B., Kobata A., and Dwek R.A. 1999. Glycosylation differences between the normal and pathogenic prion protein isoforms. *Proc. Natl. Acad. Sci.* **96**: 13044–13049.

Saborio G.P., Permanne B., and Soto C. 2001. Sensitive detection of pathological prion protein by cyclic amplification of protein misfolding. *Nature* **411**: 810–813.

Safar J., Roller P.P., Gajdusek D.C., and Gibbs C.J., Jr. 1993. Conformational transitions, dissociation, and unfolding of scrapie amyloid (prion) protein. *J. Biol. Chem.* **268**: 20276–20284.

Safar J., Wille H., Itri V., Groth D., Serban H., Torchia M., Cohen F.E., and Prusiner S.B. 1998. Eight prion strains have PrP^{Sc} molecules with different conformations. *Nat. Med.* **4**: 1157–1165.

Safar J.G., Scott M., Monaghan J., Deering C., Didorenko S., Vergara J., Ball H., Legname G., Leclerc E., Solforosi L., Serban H., Groth D., Burton D.R., Prusiner S.B., and Williamson R.A. 2002. Measuring prions causing bovine spongiform encephalopathy or chronic wasting disease by immunoassays and transgenic mice. *Nat. Biotechnol.* **20**: 1147–1150.

Sakaguchi S., Katamine S., Nishida N., Moriuchi R., Shigematsu K., Sugimoto T., Nakatani A., Kataoka Y., Houtani T., Shirabe S., Okada H., Hasegawa S., Miyamoto T., and Noda T. 1996. Loss of cerebellar Purkinje cells in aged mice homozygous for a disrupted PrP gene. *Nature* **380**: 528–531.

Schaller O., Fatzer R., Stack M., Clark J., Cooley W., Biffiger K., Egli S., Doherr M., Vandevelde M., Heim D., Oesch B., and Moser M. 1999. Validation of a Western immunoblotting procedure for bovine PrP^{Sc} detection and its use as a rapid surveillance method for the diagnosis of bovine spongiform encephalopathy (BSE). *Acta Neuropathol.* **98:** 437–443.

Schmitt-Ulms G., Legname G., Baldwin M.A., Ball H.L., Bradon N., Bosque P.J., Crossin K.L., Edelman G.M., DeArmond S.J., Cohen F.E., and Prusiner S.B. 2001. Binding of neural cell adhesion molecules (N-CAMs) to the cellular prion protein. *J. Mol. Biol.* **314:** 1209–1225.

Scott M., Groth D., Foster D., Torchia M., Yang S.-L., DeArmond S.J., and Prusiner S.B. 1993. Propagation of prions with artificial properties in transgenic mice expressing chimeric PrP genes. *Cell* **73:** 979–988.

Scott M.R., Groth D., Tatzelt J., Torchia M., Tremblay P., DeArmond S.J., and Prusiner S.B. 1997a. Propagation of prion strains through specific conformers of the prion protein. *J. Virol.* **71:** 9032–9044.

Scott M.R., Nguyen O., Stöckel J., Tatzelt J., DeArmond S.J., Cohen F.E., and Prusiner S.B. 1997b. Designer mutations in the prion protein promote β-sheet formation *in vitro* and cause neurodegeneration in transgenic mice. *Protein Sci.* (suppl. 1) **6:** 84. (Abstr.)

Serban D., Taraboulos A., DeArmond S.J., and Prusiner S.B. 1990. Rapid detection of Creutzfeldt-Jakob disease and scrapie prion proteins. *Neurology* **40:** 110–117.

Shibuya S., Higuchi J., Shin R.-W., Tateishi J., and Kitamoto T. 1998. Codon 219 Lys allele of *PRNP* is not found in sporadic Creutzfeldt-Jakob disease. *Ann. Neurol.* **43:** 826–828.

Silverman G.L., Qin K., Moore R.C., Yang Y., Mastrangelo P., Tremblay P., Prusiner S.B., Cohen F.E., and Westaway D. 2000. Doppel is an *N*-glycosylated, glycosylphosphatidylinositol-anchored protein. *J. Biol. Chem.* **275:** 26834–26841.

Sklaviadis T.K., Manuelidis L., and Manuelidis E.E. 1989. Physical properties of the Creutzfeldt-Jakob disease agent. *J. Virol.* **63:** 1212–1222.

Stahl N., Borchelt D.R., Hsiao K., and Prusiner S.B. 1987. Scrapie prion protein contains a phosphatidylinositol glycolipid. *Cell* **51:** 229–240.

Stahl N., Baldwin M.A., Hecker R., Pan K.-M., Burlingame A.L., and Prusiner S.B. 1992. Glycosylinositol phospholipid anchors of the scrapie and cellular prion proteins contain sialic acid. *Biochemistry* **31:** 5043–5053.

Stöckel J., Safar J., Wallace A.C., Cohen F.E., and Prusiner S.B. 1998. Prion protein selectively binds copper (II) ions. *Biochemistry* **37:** 7185–7193.

Supattapone S., Nguyen H.-O.B., Cohen F.E., Prusiner S.B., and Scott M.R. 1999. Elimination of prions by branched polyamines and implications for therapeutics. *Proc. Natl. Acad. Sci.* **96:** 14529–14534.

Supattapone S., Wille H., Uyechi L., Safar J., Tremblay P., Szoka F.C., Cohen F.E., Prusiner S.B., and Scott M.R. 2001. Branched polyamines cure prion-infected neuroblastoma cells. *J. Virol.* **75:** 3453–3461.

Taraboulos A., Jendroska K., Serban D., Yang S.-L., DeArmond S.J., and Prusiner S.B. 1992. Regional mapping of prion proteins in brains. *Proc. Natl. Acad. Sci.* **89:** 7620–7624.

Tatzelt J., Groth D.F., Torchia M., Prusiner S.B., and DeArmond S.J. 1999. Kinetics of prion protein accumulation in the CNS of mice with experimental scrapie. *J. Neuropathol. Exp. Neurol.* **58:** 1244–1249.

Tatzelt J., Maeda N., Pekny M., Yang S.-L., Betsholtz C., Eliasson C., Cayetano J., Camerino A.P., DeArmond S.J., and Prusiner S.B. 1996. Scrapie in mice deficient in apolipoprotein E or glial fibrillary acidic protein. *Neurology* **47:** 449–453.

Telling G.C., Scott M., Mastrianni J., Gabizon R., Torchia M., Cohen F.E., DeArmond S.J., and Prusiner S.B. 1995. Prion propagation in mice expressing human and chimeric PrP transgenes implicates the interaction of cellular PrP with another protein. *Cell* **83:** 79–90.

Tsukamoto T., Diringer H., and Ludwig M. 1985. Absence of autoantibodies against neurofilament proteins in the sera of scrapie infected mice. *Tohoku J. Exp. Med.* **146:** 483–484.

Turk E., Teplow D.B., Hood L.E., and Prusiner S.B. 1988. Purification and properties of the cellular and scrapie hamster prion proteins. *Eur. J. Biochem.* **176:** 21–30.

Tzagoloff A. and Penefsky H.S. 1971. Extraction and purification of lipoprotein complexes from membranes. *Methods Enzymol.* **22:** 219–231.

Viles J.H., Cohen F.E., Prusiner S.B., Goodin D.B., Wright P.E., and Dyson H.J. 1999. Copper binding to the prion protein: Structural implications of four identical cooperative binding sites. *Proc. Natl. Acad. Sci.* **96:** 2042–2047.

Wadsworth J.D., Joiner S., Hill A.F., Campbell T.A., Desbruslais M., Luthert P.J., and Collinge J. 2001. Tissue distribution of protease resistant prion protein in variant Creutzfeldt-Jakob disease using a highly sensitive immunoblotting assay. *Lancet* **358:** 171–180.

Waggoner D.J., Drisaldi B., Bartnikas T.B., Casareno R.L., Prohaska J.R., Gitlin J.D., and Harris D.A. 2000. Brain copper content and cuproenzyme activity do not vary with prion protein expression level. *J. Biol. Chem.* **275:** 7455–7458.

Walter N.G., Schwille P., and Eigen M. 1996. Fluorescence correlation analysis of probe diffusion simplifies quantitative pathogen detection by PCR. *Proc. Natl. Acad. Sci.* **93:** 12805–12810.

Warren G.B., Toon P.A., Birdsall N.J.M., Lee A.G., and Metcalfe J.C. 1974a. Reversible lipid titrations of the activity of pure adenosine triphosphatase-lipid complexes. *Biochemistry* **13:** 5501–5507.

———. 1974b. Reconstitution of a calcium pump using defined membrane components. *Proc. Natl. Acad. Sci.* **71:** 622–626.

Westaway D., Zuliani V., Cooper C.M., Da Costa M., Neuman S., Jenny A.L., Detwiler L., and Prusiner S.B. 1994. Homozygosity for prion protein alleles encoding glutamine-171 renders sheep susceptible to natural scrapie. *Genes Dev.* **8:** 959–969.

Wille H., Prusiner S.B., and Cohen F.E. 2000. Scrapie infectivity is independent of amyloid staining properties of the N-terminally truncated prion protein. *J. Struct. Biol.* **130:** 323–338.

Wille H., Zhang G.-F., Baldwin M.A., Cohen F.E., and Prusiner S.B. 1996. Separation of scrapie prion infectivity from PrP amyloid polymers. *J. Mol. Biol.* **259:** 608–621.

Wille H., Michelitsch M.D., Guénebaut V., Supattapone S., Serban A., Cohen F.E., Agard D.A., and Prusiner S.B. 2002. Structural studies of the scrapie prion protein by electron crystallography. *Proc. Natl. Acad. Sci.* **99:** 3563–3568.

Williamson R.A., Peretz D., Smorodinsky N., Bastidas R., Serban H., Mehlhorn I., DeArmond S.J., Prusiner S.B., and Burton D.R. 1996. Circumventing tolerance to generate autologous monoclonal antibodies to the prion protein. *Proc. Natl. Acad. Sci.* **93:** 7279–7282.

Worthington M. 1972. Interferon system in mice infected with the scrapie agent. *Infect. Immun.* **6:** 643–645.

Yang J.T., Wu C.-S.C., and Martinez H.M. 1986. Calculation of protein conformation from circular dichroism. *Methods Enzymol.* **130:** 208–269.

Yehiely F., Bamborough P., Costa M.D., Perry B.J., Thinakaran G., Cohen F.E., Carlson G.A., and Prusiner S.B. 1997. Identification of candidate proteins binding to prion protein. *Neurobiol. Dis.* **3:** 339–355.

Zhang H., Stöckel J., Mehlhorn I., Groth D., Baldwin M.A., Prusiner S.B., James T.L., and Cohen F.E. 1997. Physical studies of conformational plasticity in a recombinant prion protein. *Biochemistry* **36:** 3543–3553.

17

Biosafety Issues in Prion Diseases

Henry Baron
Industry and Health Policy
Aventis Behring S.A.
75601 Paris Cedex, 12, France

Jiri Safar,[1] Darlene Groth,[1] Stephen J. DeArmond,[1,3]
and Stanley B. Prusiner[1,2,4]
[1]Institute for Neurodegenerative Diseases
Departments of [2]Neurology, [3]Pathology, and
[4]Biochemistry and Biophysics
University of California, San Francisco, California 94143

BIOSAFETY RELEVANT TO PRIONS HAS BEEN ADDRESSED in several guidelines and recommendations published by health authorities in an attempt to limit the potential risk associated with prion contamination in laboratory studies as well as in foods and medicinal products. Unfortunately, these issues are not always considered within the context of prion pathobiology and epidemiology. Instead, prion diseases still are treated as viral-like diseases, and this can result in erroneous assumptions and misguided regulations.

Of the many distinctive features that separate prion diseases from viral, bacterial, fungal, and parasitic disorders, the most remarkable is that prion diseases can be manifest as infectious, inherited, and sporadic illnesses. Familial Creutzfeldt-Jakob disease (CJD), Gerstmann-Sträussler-Scheinker syndrome (GSS), and fatal familial insomnia (FFI) are all dominantly inherited prion diseases; five different mutations of the PrP gene have been shown to be genetically linked to the development of inherited prion disease. Prions from many cases of inherited prion disease have been transmitted to apes, monkeys, and mice carrying human PrP transgenes (Brown et al. 1994a; Telling et al. 1995, 1996). In all three manifestations of prion disease, infectious prions are generated in the brains of afflicted individuals, and these prions are composed of the disease-causing isoform (PrP^{Sc}) of the prion protein (PrP) with the amino acid sequence encoded by the PrP gene of the affected host. When prions are passaged into the brain of a different host species, a "species barrier," related primarily to interspecies differences in PrP sequences and, in some instances, to prion strain, is responsible for inefficient infection (Chapter 9) (Pattison 1965; Scott et al. 1989, 1999; Prusiner et al. 1990; Telling et al. 1995; Asante et al. 2002). If interspecies transmission does occur, then the prions generated in the brain of the alternate host carry the amino acid sequence encoded by the PrP gene of that particular species and not the PrP sequence found in the original inoculum (Scott et al. 1989; Prusiner et al. 1990). In other words, in interspecies infection, such as from sheep to cattle or from cattle to humans, the prions that replicate in the host brain are not the same as those that initiated replication. This scenario is profoundly different from what happens during a viral infection. Although it is not the intent of this chapter to explore the molecular pathogenesis of the prion diseases (Chapter 4), a basic understanding of this area of biology and medicine is essential to the formulation of sound, intelligent, and effective risk assessment and management. If health authorities evaluate prion-related risk from the perspective of the virologist and ignore the unique properties peculiar to prions, they are doomed to make uninformed and perhaps harmful decisions.

There has been much public and professional concern over the potential transmission of scrapie, chronic wasting disease (CWD), and especially bovine spongiform encephalopathy (BSE) prions to humans and animals through consumption of contaminated foodstuffs. Prions in meat and bone meal (MBM), most likely derived from scrapied sheep offal, are considered to be the cause of the BSE epidemic in the UK (Chapter 12) (Wilesmith et al. 1991), and the transmission of a similar disease to cats is believed to be due to prion-contaminated cat food (Wyatt et al. 1991). More recently, the establishment of a link between BSE and variant (v) CJD (Will et al. 1996; Scott et al. 1999) initiated a profound reassessment of European Union (EU) policy on prion-associated risks to the human population, including consideration of a further potential hazard to human health, i.e., the administration of biological as well as medicinal products derived from or associated with human or animal tissues potentially contaminated with prions. In France, public and political reaction to the observation there of comparatively high numbers of CJD cases related to administration of human growth hormone (HGH) extracted from the pituitary glands of cadavers with undiagnosed CJD (Chapter 13) has regulatory authorities in a precautionary quandary as to how to deal with products derived from human tissues. These concerns have spread to questions regarding the safety of human blood and blood products as well.

All these fears and anxieties have given rise to a number of scenarios and forecasts, ranging from the cavalier to the apocalyptic, with attendant protective guidelines and measures. In this chapter, we strive to sort out the truth from the myth, the legitimate from the unreasonable, and to provide a factual basis for concerns, as well as an informed rationale for actions to be implemented. We also explore the potential consequences of under-reaction as well as over-concern with respect to these issues.

BIOSAFETY LEVEL CLASSIFICATION

Human prions and those propagated in apes and monkeys are considered Biosafety Level 2 or 3 pathogens, depending on the studies being conducted (Safar et al. 1999). BSE prions are likewise considered Risk Group 2 or 3 pathogens due to the fact that BSE prions have been transmitted to humans in the UK, France, and elsewhere (Chapter 12) (Will et al. 1996).

All other animal prions are considered Biosafety Level 2 pathogens. Thus, based on our current understanding of prion biology described above, once human prions are passaged in mice and mouse PrP^{Sc} is produced, these prions should be considered Biosafety Level 2 prions, although the human prions are Biosafety Level 3 under most experimental conditions. An exception to this statement concerns mice expressing human, bovine, or chimeric (human/mouse or bovine/mouse) transgenes. These animals produce human, bovine, or chimeric prions when infected with human or bovine prions and should be treated as Biosafety Level 2 or 3 in accord with the guidelines described above. The mechanism of prion spread among sheep and goats developing natural scrapie is unknown (Dickinson et al. 1974; Foster et al. 1992), as are the risk and number of sheep potentially infected with BSE prions (Kao et al. 2002). CWD, transmissible mink encephalopathy (TME), BSE, feline spongiform encephalopathy (FSE), and exotic ungulate encephalopathy (EUE) are all thought to occur after the consumption of prion-infected foods (Gajdusek 1991; Marsh 1992; Collinge and Palmer 1997; Prusiner 1997).

Physical Properties of Prions

The smallest infectious prion particle is probably a dimer of PrP^{Sc}. This estimate is consistent with an ionizing radiation target size of 55 ± 9 kD (Bellinger-Kawahara et al. 1988). Therefore, prions may not be retained by most of the filters that efficiently eliminate bacteria and viruses. However, prions aggregate into particles of nonuniform size that can affect their filtration behavior, and prions cannot be solubilized by detergents, except under denaturing conditions in which infectivity is lost (Gabizon and Prusiner 1990; Safar et al. 1990). Prions resist inactivation by nucleases (Bellinger-Kawahara et al. 1987b), UV-irradiation at 254 nm (Alper et al. 1967; Bellinger-Kawahara et al. 1987a), and treatment with psoralens (McKinley et al. 1983), divalent cations, metal ion chelators, acids (between pH 3 and 7), hydroxylamine, formalin, boiling, or proteases (Prusiner 1982; Brown et al. 1990).

Care of Patients

In the care of patients dying of human prion disease, those precautions used for patients with AIDS or hepatitis are certainly adequate. In contrast to those viral illnesses, the human prion diseases are not contagious (Ridley and Baker 1993). There is no evidence of contact or aerosol transmission of prions from one human to another. However, they are infectious under some circumstances, such as in the case of ritualistic cannibalism in New Guinea causing kuru, the administration of prion-contaminated growth hormone causing iatrogenic CJD, and the transplantation of prion-contaminated dura mater grafts (Gajdusek 1977; Centers for Disease Control 1997; Public Health Service 1997).

Surgical Procedures

Surgical procedures on patients carrying the diagnosis of prion disease should be minimized. It is thought that CJD has been spread from a CJD patient who underwent neurosurgical procedures to two other patients shortly thereafter in the same operating theater (Brown et al. 1992). Although there is no documentation of the transmission of prions to humans through droplets of blood or CSF or by exposure to intact skin or gastric and mucous membranes, the theoretical risk of such occurrences cannot be ruled out definitively. Sterilization of the instruments and the operating room should be performed in accord with recommendations described below.

Because it is important to establish a "definitive" diagnosis of a human prion disease and to distinguish between sporadic or familial cases and those acquired by infection as a result of medical procedures or from prion-contaminated food products, unfixed brain tissue should be obtained. For all cases of suspected human prion disease, a minimum of 1 cm^3 of unfixed cerebral cortex should be part of any biopsy. This specimen should be bisected from the cortical surface through to the underlying white matter with one half of the specimen formalin-fixed and the other half frozen.

Autopsies

Routine autopsies and the processing of small amounts of formalin-fixed tissues containing human prions require Biosafety Level 2 precautions. At autopsy, the entire brain is collected and cut into coronal sections about 1.5 inches (~4 cm) thick; small blocks of tissue can easily be removed from each coronal section and placed in fixative for subsequent histopathologic analyses. Each coronal section is immediately heat-sealed in a heavy-duty plastic bag. The outside of this bag is assumed to be contaminated with prions and other pathogens. With fresh gloves or with the help of an assistant with uncontaminated gloves, the bag containing the specimen is placed into another plastic bag which does not have a contaminated outer surface. The samples are then frozen on dry ice or placed directly in a –70°C freezer for storage. At the very minimum, a coronal section of cerebral hemisphere containing the thalamus and one of the cerebellar hemisphere and brain stem should be taken and frozen.

The absence of any known effective treatment for prion disease demands caution in the manipulation of potentially infectious tissues. The highest concentrations of prions are in the central nervous system and its coverings. On the basis of animal studies, it is possible that high concentrations of prions may also be found in spleen, thymus, lymph nodes, and lung. Moreover, in vCJD, prions are routinely detected in extraneural lymphoid tissues such as tonsil, appendix, and spleen (Hill et al. 1997a; Bruce et al. 2001; Wadsworth et al. 2001). The main precaution to be taken when working with prion-infected or contaminated material is to avoid puncture of the skin (Ridley and Baker 1993). The prosector should wear cut-resistant gloves if possible. If accidental contamination of skin occurs, the area is swabbed with 1 N sodium hydroxide for 5 minutes and then washed with copious amounts of water. Tables 1–4 provide guidelines to reduce the chance of skin punctures, contamination from aerosols, and contamination of operating room and morgue surfaces and instruments. Unfixed samples of brain, spinal cord, and other tissues containing human prions should be processed with extreme care at Biosafety Level 3.

Bovine Spongiform Encephalopathy

The risk of infection for humans by BSE prions is established. Therefore, the most prudent approach is to study BSE prions in a Biosafety Level 2 or 3 facility, depending on the studies to be performed as noted above for human prions.

Table 1. Autopsies of patients with suspected prion disease: Standard precautions

1. Attendance is limited to three staff including at least one experienced patholo-
   gist. One of the staff avoids direct contact with the deceased but assists with
   handling of instruments and specimen containers.

2. Standard autopsy attire is mandatory. However, a disposable, waterproof gown
   is worn in place of a cloth gown.

   a. Cut-resistant gloves are worn underneath two pairs of surgical gloves, or
      chain mail gloves are worn between two pairs of surgical gloves.

   b. Aerosols are mainly created during opening of the skull with a Stryker saw.
      The personal protective equipment approved at the University of California
      at San Francisco for protection from aerosols is an AIR-MATE® HEPA 12
      Powered Air-Purifying Respiratory System (RACAL Health and Safety, Inc.,
      Frederick, MD). This unit consists of the following parts:

      i. Polycoated Tyvek head cover

      ii. HEPA filter unit containing battery pack, belt, and HEPA filter

      iii. AC adapter/charger

      iv. Airflow indicator

      v. Breathing tube assembly

   c. The above respirator system can be worn throughout the autopsy; however,
      when no aerosols are being generated, such as when the brain is being
      removed or when organ samples are being removed in situ, switching to a sur-
      gical mask with a wraparound splash guard transparent visor is acceptable.

3. To reduce contamination of the autopsy suite:

   a. The autopsy table is covered with an absorbent sheet that has a waterproof
      backing.

   b. Contaminated instruments are placed on an absorbent pad.

   c. The brain is removed with the head in a plastic bag to reduce aerosolization.

   d. The brain can be placed into a container with a plastic bag liner for weighing.

   e. The brain is placed onto a cutting board, and appropriate samples are dis-
      sected for snap freezing (see Table 3).

   f. The brain or organs to be fixed are immediately placed into a container with
      10% neutral buffered formalin.

   g. In most cases of suspected prion disease, the autopsy can be limited to exam-
      ination of the brain only. In cases requiring a full autopsy, consideration
      should be given to examining and sampling of thoracic and abdominal
      organs in situ.

Table 2. Autopsy suite decontamination procedures

1. Instruments (open box locks and jaws) and saw blades are placed into a large stainless steel dish, soaked for 1 hour in 2 N sodium hydroxide[a] or 2 hours in 1 N sodium hydroxide, and/or autoclaved at 134°C (gravity displacement steam autoclaving for 1 hour; porous load steam autoclaving for one 18-minute cycle at 30 lbs. psi or six 3-minute cycles at 30 lbs. psi).

2. The Stryker saw is cleaned by repeated wiping with 2 N sodium hydroxide solution.

3. The absorbent table cover and instrument pads, disposable clothing, etc. are double-bagged in appropriate infectious waste bags for incineration.

4. Any suspected areas of contamination of the autopsy table or room are decontaminated by repeated wetting over 1 hour with 2 N sodium hydroxide.

[a]5% Sodium hypochlorite solution at least 20,000 ppm free chloride for 2 hours, or 96% formic acid may substitute but will corrode the stainless steel.

Table 3. Brain-cutting procedures

1. After adequate formaldehyde fixation (at least 10–14 days), the brain is examined and cut on a table covered with an absorbent pad with an impermeable backing.

2. Samples for histology are placed in cassettes labeled with "CJD precautions." For laboratories that do not have embedding and staining equipment or microtome dedicated to infectious diseases including CJD, blocks of formalin-fixed tissue can be placed in 95–100% formic acid for 1 hour, followed by fresh 10% neutral buffered formalin solution for at least 48 hours (Brown et al. 1990). The tissue block is then embedded in paraffin as usual. Standard neurohistological or immunohistochemical techniques are not obviously affected by formic acid treatment; however, in our experience, tissue sections are brittle and crack during sectioning.

3. All instruments and surfaces coming in contact with the tissue are decontaminated as described in Table 2.

4. Tissue remnants, cutting debris, and contaminated formaldehyde solution should be discarded within a plastic container as infectious hospital waste for eventual incineration.

Table 4. Tissue preparation

1. Histology technicians should wear gloves, apron, laboratory coat, and eye protection.

2. Adequate fixation of small tissue samples (e.g., biopsies) from a patient with suspected prion disease is followed by postfixation in 96–100% formic acid for 1 hour, followed by 48 hours in fresh 10% formalin (see Table 3).

3. Liquid waste is collected in a 4-liter waste bottle containing 600 ml of 6 N sodium hydroxide and diluted to a final volume of 4 liters to maintain the optimal concentration for disinfection. Gloves, embedding molds, and all handling materials are disposed in a biohazard waste receptacle.

4. Tissue cassettes are processed manually to prevent contamination of tissue processors.

5. Tissues are embedded in a disposable embedding mold. If used, forceps are decontaminated.

6. In preparing sections, gloves are worn and section waste is collected and disposed in a biohazard waste receptacle. The knife stage is wiped with 1–2 N NaOH. The used knife is immediately discarded in a "biohazard sharps" receptacle. Slides are labeled with "CJD Precautions." The sectioned block is sealed with paraffin.

7. Routine staining:

    a. Slides are processed manually.

    b. Reagents are prepared in 100-ml disposable specimen cups.

    c. After placing the coverslip on, slides are decontaminated by soaking them for 1 hour in 2 N NaOH.

    d. Slides are labeled as "Infectious-CJD."

8. Other suggestions:

    a. Use disposable specimen cups or slide mailers for reagents.

    b. Process slides for immunocytochemistry in disposable petri dishes.

    c. Decontaminate equipment as described above.

Experimental Rodent Prion Diseases

Mice and hamsters are the experimental animals of choice for all studies of prion disease. With the development of transgenic mice that are highly susceptible to human prions, the use of apes and monkeys is rarely needed. The highest titers of prions ($\sim 10^9$ ID_{50} units/g) are found in the brain and spinal cord of laboratory rodents infected with adapted strains of prions (Eklund et al. 1967; Prusiner et al. 1980); lower titers ($\sim 10^6$ ID_{50}

units/g) are present in the spleen and lymphoreticular system (Kimberlin 1976; Prusiner et al. 1978).

Inactivation of Prions

Prion infectivity is diminished by prolonged digestion with proteases, but other treatments such as boiling in SDS or alkali (>pH 10) are variable. Sterilization of rodent brain extracts with high titers of prions requires autoclaving at 132°C for 4.5 hours; denaturing organic solvents such as phenol, chaotropic agents such as guanidine isocyanate, or alkali such as NaOH can also be used for sterilization (Prusiner et al. 1984, 1993; Taylor et al. 1995, 1997).

With the exceptions noted above, Biosafety Level 2 practices and facilities are recommended for all activities utilizing known or potentially infectious tissues and fluids containing nonhuman prions from naturally or experimentally infected animals. Although there is no evidence to suggest that aerosol transmission occurs in the natural disease, it is prudent to avoid the generation of aerosols or droplets during the manipulation of tissues or fluids and during the necropsy of experimental animals. It is further strongly recommended that gloves be worn for activities that provide the opportunity for skin contact with infectious tissues and fluids. Formaldehyde-fixed and paraffin-embedded tissues, especially of the brain, remain infectious. Some investigators recommend that formalin-fixed tissues from suspected cases of prion disease be immersed for 30 minutes in 96% formic acid or phenol before histopathologic processing (Brown et al. 1990), but such treatment may severely distort the microscopic neuropathology.

Prions are characterized by extreme resistance to conventional inactivation procedures, including irradiation, boiling, dry heat, and chemicals (formalin, betapropiolactone, alcohols). However, they are inactivated by 1 N NaOH, 4.0 M guanidinium hydrochloride or isocyanate, sodium hypochlorite (= 2% free chlorine concentration), and steam autoclaving at 132°C for 4.5 hours (Prusiner et al. 1984, 1993; Taylor et al. 1995, 1997). It is recommended that dry waste be autoclaved at 132°C for 4.5 hours or incinerated. Large volumes of infectious liquid waste containing high titers of prions can be completely sterilized by treatment with 1 N NaOH (final concentration) followed by autoclaving at 132°C for 4.5 hours. Disposable plasticware, which can be discarded as a dry waste, is highly recommended. Because the paraformaldehyde vaporization procedure does not diminish prion titers, the biosafety hoods must be decontaminated with 1 N NaOH, followed by 1 N HCl, and rinsed with water; hepa filters should be autoclaved and incinerated.

HUMAN-TO-HUMAN TRANSMISSION OF PRIONS

Iatrogenic CJD

Since the extinction of kuru, an infectious human prion disorder transmitted by the now abolished practice of ritualistic cannibalism among the Fore population of Papua New Guinea (Alpers 1987), all currently known cases of infectious, human-to-human transmission of prions are iatrogenic CJD (iCJD). Iatrogenic CJD accounts for less than 1% of total CJD worldwide and always involves transmission of prions originating from the central nervous system (CNS), where there is potential for high levels of infectivity, or from tissues intimately associated with the CNS, such as the cornea, pituitary gland, and dura mater (Brown et al. 1992). No case of iCJD in which the source of infection was established to be peripheral tissue or blood has ever been reported.

The current status of iCJD, with respect to source of contamination, numbers of cases, and mode of transmission is summarized in Table 5. New cases continue to appear, particularly in relation to administration of HGH extracted from cadaver pituitary glands in France (89 cases identified thus far), and to dura mater transplantation in Japan, where more than 80 cases have been reported (Chapter 13).

An intriguing susceptibility factor associated with iCJD appears to involve a polymorphism at codon 129 of the PrP gene (Chapter 13). In the Caucasian population, the three amino acid phenotypes at this codon have the following distribution: heterozygous methionine/valine (Met/Val) in approximately half of all subjects tested, homozygous Met/Met in roughly 40%, and homozygous Val/Val in about 10% (Owen et al. 1990; Collinge et al. 1991; Laplanche 1996). In iCJD, however, only about 10% of all studied cases are heterozygous Met/Val, with an over-representation of homozygous Met/Met (~60%) and Val/Val (~30%), suggesting that homozygosity at this site may predispose an individual to the acquisition of disease following exposure to exogenous human prions

Table 5. Current status of iatrogenic CJD worldwide

Tissue source of contamination	Number of cases	Mode of transmission
Brain	4	neurosurgical procedures
Brain	2	implantation of stereotactic EEG electrodes
Eye	3	corneal transplantation
Dura mater	136	dura mater transplantation
Pituitary gland	162	parenteral growth hormone therapy
Pituitary gland	5	parenteral gonadotrophin therapy

(Collinge et al. 1991; Brown et al. 1994b; Deslys et al. 1996; Laplanche 1996). In sporadic CJD, there is an even greater overrepresentation of homozygous Met/Met (~80%) with a representative proportion of Val/Val (~10%) but an underrepresentation of heterozygous Met/Val (~10%) (Palmer et al. 1991; Laplanche 1996; Parchi et al. 1996). Whether these differences at codon 129 truly predispose to heightened susceptibility to acquired CJD, or whether their role involves modulation of the duration of incubation periods, remains to be determined. The situation bears close watching, however, particularly in France, where recent cases of HGH-related CJD in Met/Val heterozygotes could be of ominous portent (Deslys et al. 1998). If heterozygosity at codon 129 is not truly protective but only confers prolonged incubation periods, the current high numbers of iCJD cases could increase in the coming years.

CJD and Human Blood

The possibility of transmission of prions from human to human by blood transfusion or through administration of blood components or plasma derivatives has been a cause for concern among health professionals throughout the world. Rigorous surveillance efforts and scientific scrutiny have not revealed conclusive evidence of infectious prions in blood or plasma of preclinical or clinical cases of CJD, including vCJD, nor of transmission of CJD/vCJD through blood or plasma-derived products (Chapter 13) (Baron 2001). Despite no confirmed example of the transmission of prions from a human blood donor who later developed CJD to a recipient, some investigators have postulated that such transmissions may be possible though infrequent (Chapter 13) (Manuelidis 1994; Brown 1995; Ricketts et al. 1997). When a recipient of a blood transfusion, blood components, or plasma derivatives has developed CJD, the first question that must be asked is whether this is a case of sporadic or iatrogenic CJD. The difficulty comes in distinguishing iatrogenic from sporadic CJD. In both iatrogenic and sporadic CJD, wild-type PrP^{Sc} is usually found in the brains of the patients, which is a unique feature of prion disease. If CJD were caused by a virus, and if this putative virus were present in blood, then the virus isolated from the blood or blood product that was given to the patient could be compared to that recovered from the patient. In prion diseases, however, only PrP^{Sc} can be compared to that recovered from the patient, but the detection of PrP^{Sc} in blood or even a blood product derived from a CJD patient has never been reliably accomplished. Such an approach may prove feasible in the future using transgenic mice (Chapter 9) and conformation-dependent immunoassays (Chapter 16).

The application of transgenic mice and conformation-dependent immunoassays to the field of prion biology is so new that many "applied studies" have not yet been completed. Previous investigations with non-transgenic rodents often have not employed experimental designs utilizing adequate controls and safeguards against potential contamination. Such controls are essential in the study of prion diseases because of the extraordinary resistance of prions to standard decontamination procedures.

Experimental Data

Although there is no convincing evidence demonstrating the presence of CJD prion infectivity in human blood, prion infectivity in blood or blood constituents was reported in early studies with experimentally infected laboratory rodents (Clarke and Haig 1967; Dickinson and Fraser 1969; Manuelidis et al. 1978; Kuroda et al. 1983; Diringer 1984; Casaccia et al. 1989). These studies must be considered in the context of the experimental design in which the animals were inoculated with exogenous prions, either peripherally or intracerebrally, and part of the inoculum probably found its way across disrupted blood vessels into the bloodstream during the inoculation procedure. Infectivity levels in these rodent blood samples, as estimated by the incubation periods in bioassay animals, were $\sim 10^2$ ID_{50} units per ml, compared to 10^6-10^7 ID_{50} units per gram of brain tissue (Brown 1995).

More recent studies have shown that blood components and plasma of mice experimentally infected by intracerebral inoculation contain prion infectivity in buffy coat and plasma during the clinical stage of disease, but at very low levels: ~ 100 infectious units (IU) per ml by limit dilution titration for buffy coat, and ~ 10 IU per ml for plasma (Brown et al. 1998, 1999; Brown 2001). Even lower levels were found in buffy coat, and virtually no infectivity was found in plasma during the preclinical phase of the incubation period (Brown 2001).

It must be stressed that in sporadic and familial CJD, which account for more than 99% of the total number of CJD cases worldwide, no known inoculation occurs as in experimental prion disease in laboratory animals. Moreover, it is now believed that both sporadic and familial CJD result from a spontaneous conversion of normal, cellular PrP (PrP^C) into PrP^{Sc}. This being the case, it can be argued that there is no reason to suspect passage of infectivity into the bloodstream, except possibly in late stages of disease when there is breakdown of the blood–brain barrier. Iatrogenic CJD, which represents less than 1% of all CJD cases and in which inoculation of exogenous prions via injection or surgical procedure is documented, may pose a greater potential risk of blood-borne prions. Current blood donor selection criteria, as discussed below, exclude virtually all donors at risk for iCJD.

In a recent study designed to explore the potential implications of vCJD on the blood supply, sheep were used as a model to evaluate cross-species transmission of cattle-derived BSE prions by the oral route and then intraspecies transmission from these orally infected sheep to naive sheep by transfusion (Hunter et al. 2002). The investigators reported transmission of prion disease in 2 of 24 sheep transfused with blood from animals orally challenged with BSE prions, as well as in 4 of 21 sheep transfused with blood drawn from sheep with natural scrapie. The design of this study remains questionable due to a paucity of positive controls. Moreover, the incubation-time data are perplexing, not allowing to establish or even estimate the level of putative infectivity in the blood that transmitted disease. Perhaps more clarification will emerge as this ongoing study progresses, but its results have been interpreted as a warning that preclinical, vCJD-infected blood donors could pose a risk to the blood supply. The relevance of vCJD in humans to these experimental transmissions in genetically susceptible sheep is uncertain.

To this date, there have been no systematic, well-controlled studies of the potential for transmission of human prions from blood of CJD patients using highly susceptible assay animals. With the advent of transgenic mice carrying a chimeric human/mouse PrP gene (Telling et al. 1994) or the human PrP gene on an ablated background (Telling et al. 1995), this issue can be addressed for the first time. Preliminary observations from studies with transgenic mice carrying chimeric human/mouse PrP inoculated intracerebrally with blood from 13 (12 sporadic and 1 inherited) CJD cases have not revealed any positive transmissions (S.B. Prusiner, pers. comm.). Regarding published data, we can refer presently only to putative isolations of prion infectivity from the blood of five patients clinically diagnosed with CJD, using less susceptible laboratory rodent hosts (guinea pigs, hamsters, or mice). Two patients were reported by one laboratory (Manuelidis et al. 1985) and the remaining three by three different laboratories (Table 6) (Tateishi 1985; Tamai et al. 1992; Deslys et al. 1994). The work of Manuelidis et al. (1985) deserves special mention here since it is frequently cited as the first demonstration of prion infectivity in human blood. As pointed out by Brown (1995), the low transmission rate and very high non-CJD experimental mortality in these studies is disquieting and invokes other more plausible interpretations, e.g., intercurrent illness or laboratory cross-contamination. These suspicions were confirmed when the same authors later reported trans-

Table 6. Putative isolations of prion infectivity from blood of CJD patients

Source of isolation	Number of CJD cases	Assay animals	Reference	Causes for concern
Buffy coat	2/2 (sporadic)	guinea pigs, hamsters	Manuelidis et al. (1985)	lack of controls low transmission rates high unexplained mortality aberrant results in subsequent experiments irreproducibility in subsequent experiments
Whole blood, urine	1/3 (sporadic)	mice	Tateishi (1985)	low transmission rates improbable incubation times (comparable to those obtained with brain) irreproducibility in subsequent experiments
Concentrated[a] plasma	1/1 (sporadic)	mice	Tamai et al. (1992)	low transmission rate improbable incubation time (comparable to that obtained with brain)
Buffy coat	1/1 (HGH-iatrogenic)	hamsters	Deslys et al. (1994)	low transmission rate very long incubation period very high unexplained mortality in both control and inoculated animals

[a]Unconcentrated plasma and white blood cells from the same patient failed to transmit disease.

mission of a spongiform encephalopathy to hamsters from buffy coats of Alzheimer's disease patients and nonaffected relatives (Manuelidis et al. 1988) and subsequently from 26 of 30 neurologically healthy control subjects (Manuelidis and Manuelidis 1993). When the National Institutes of Health (NIH) attempted to verify the results of these latter two studies in duplicate experiments, they proved to be nonreproducible (Godec et al. 1991, 1994). A review of the data of Tateishi (1985), Tamai et al. (1992), and Deslys et al. (1994) raises similar concerns of artifactual, laboratory cross-contamination as suggested by one or more of the following problems: incubation times in animals inoculated with blood, urine (Tateishi 1985), and plasma (Tamai et al. 1992) comparable to those in animals inoculated with brain from the same patient; irreproducibility in a subsequent experiment (Doi 1991 [re Tateishi 1985]; Brown 1995), very long incubation periods and very high unexplained mortality in both inoculated and control groups (Deslys et al. 1994).

In contrast to these anecdotal, putatively positive but controversial isolations of prion infectivity from human blood using relatively unsusceptible rodent hosts presenting significant species barriers are the uniformly negative results obtained using more susceptible nonhuman primates. Attempts at the NIH to transmit disease from the blood of 13 CJD patients to different primate species, including chimpanzees, by using multiple routes of inoculation were all unsuccessful (Brown et al. 1994a).

With respect to vCJD, two recent publications have reported that prions were undetectable in blood, plasma, or buffy coat of patients with vCJD using two different methodologies: bioassay in wild-type mice (Bruce et al. 2001) and a western blot immunoassay employing phosphotungstate acid (PTA) for enrichment of PrP^{Sc} (Wadsworth et al. 2001). However, both methodologies successfully detected prion infectivity in lymphoid tissues from the same vCJD cases. Although these negative results in blood of vCJD patients are interesting, more sensitive assay systems will be needed to rule out the presence of low levels of infectivity (Chapter 16).

In conclusion, a review of the published experimental data does not unearth any convincing evidence to support the current perception that human blood may contain prions. This does not mean that the notion of potential prion infectivity in human blood can be dismissed, but only that it is not supported by any of the foregoing transmission studies, many of which have been flawed. The consistent detection of prions in rodent lymphoid tissues (Eklund et al. 1967; Dickinson and Fraser 1969; Lavelle et al. 1972; Kimberlin 1976; Prusiner et al. 1978; Kitamoto et al. 1991; Muramoto et al. 1993; O'Rourke et al. 1994; Brown et al. 1996;

Fraser et al. 1996; Klein et al. 1997) and in lymphoid tissues of patients afflicted with vCJD (Hill et al. 1997a; Bruce et al. 2001; Wadsworth et al. 2001) may give cause to suspect the presence of prions in blood carried by circulating lymphoid-tissue-derived cells during some phase of the illness. In any event, the apparent difficulty in demonstrating prions in human blood argues that if they are present, they must exist at low concentrations.

At present, we have little knowledge of the titers and distribution of prions in human extraneural tissues. Moreover, we currently have no means of identifying asymptomatic individuals who are destined to develop CJD. Without such diagnostic tools, even the detection and measurement of prions in blood of symptomatic patients with CJD are problematic. We describe these problems not to paint a hopeless picture but to delineate where additional research needs to focus in order to gain urgently needed information. The application of highly sensitive prion detection technologies such as transgenic mice (Chapter 9) and the conformation-dependent immunoassay (CDI) (Chapter 16) to attempt to resolve these important public health issues appears to be crucial. Regarding the latter, recent results with the CDI adapted to a sandwich format using a capture monoclonal antibody specific to human PrP show sensitivity levels virtually equivalent to those obtained by bioassay in transgenic mice (Bellon et al. 2003).

Epidemiologic Data

Available epidemiologic data do not suggest prion contamination of blood, blood components, or plasma derivatives (Baron 2001). In fact, these data argue against transmission of CJD through blood. The data discussed in this section concern predominantly the classical forms of CJD, i.e., sporadic, inherited, and iatrogenic. The relatively little information that is available for vCJD will be explored at the end of this section.

Epidemiologic analysis of this issue is based on five different types of studies: (1) case reports, (2) case control studies, (3) routine surveillance, (4) cohort investigation of recipients from CJD donors, and (5) special population studies.

A survey of case reports attempting to link CJD to blood transfusion or administration of blood products has failed to reveal any convincing instances of prion transmission via blood (Chapter 13). The utility of case reports in identifying causative links between CJD cases and therapeutic products or procedures should not be underestimated, since case reports provided the initial warnings of the implications of HGH treatment

(Centers for Disease Control 1985) and dura mater grafts (Prichard et al. 1987; Centers for Disease Control 1989, 1997) in CJD transmission.

Case control studies of CJD conducted in the US (Davanipour et al. 1985; Centers for Disease Control 1996; Holman et al. 1996), the UK (Esmonde et al. 1993), Europe (van Duijn et al. 1998), Australia (Collins et al. 1999), and Japan (Kondo and Kuroina 1982) have addressed the issue of whether receiving blood is a risk factor for CJD. In all of these investigations, there is no evidence of a higher proportion of CJD cases having a history of blood transfusion than that of control subjects. The balance of evidence indicates that blood exposure is not a risk factor for CJD. There are no significant differences between CJD and control populations with respect to receiving or donating blood.

Routine surveillance studies of the epidemiology of CJD have been conducted in many countries and even repeated in some (Brown et al. 1987; Holman et al. 1995; Will 1996; Ruffie et al. 1997; Nakamura et al. 1999). With the exception of the UK, where vCJD has slightly altered the traditional epidemiologic pattern of CJD (Chapter 13), the disease continues to have a similar and constant incidence (~1 case per million population per year) and age distribution (peak in the 60s to 70s, rare under 30 years of age), despite increasing use of blood transfusion and plasma proteins.

Perhaps the most compelling evidence against transmission of CJD through blood or blood products is provided from cohort studies of recipients from CJD donors and from special population studies, including patients suffering from hemophilia A, hemophilia B, β-thalassemia, or sickle cell disease, and those receiving large quantities of cryoprecipitate. Retrospective, follow-up studies with the goal of tracing all recipients of blood or blood products from CJD donors have been conducted in Germany (Heye et al. 1994) and are ongoing in the US, involving the American Red Cross, the American Association of Blood Banks, the New York Blood Center, and the Centers for Disease Control and Prevention (CDC) (L.B. Schonberger, pers. comm.), and in the UK (UK CJD Surveillance Unit [http://www.cjd.ac.ed.uk]). Data from all these studies appear to indicate that despite identification of large numbers of recipients of potentially tainted blood products (up to 196 investigated recipients of blood products from 15 CJD donors in the American study), none of the identified recipients has yet developed CJD.

In studies from the CDC, no patients afflicted with hemophilia A, hemophilia B, β-thalassemia, or sickle cell anemia have contracted CJD in the US over a 17-year period (L.B. Schonberger, pers. comm.). Since clotting factor concentrates for hemophiliacs can come from pools of 20,000 to 30,000 donors, hemophiliacs should have been exposed to prions if they were present in blood at appreciable levels. The CDC currently is actively seeking cases of CJD in hemophiliacs; as yet, they have seen no evidence of CJD in the hemophilia community despite increased surveillance since 1994 (Evatt et al. 1998). In another CDC study of 101 patients who received over 238,000 units of cryoprecipitate over a 17-year period, no cases of CJD were found. Moreover, 76 of these cryoprecipitate users were still alive a minimum of 12 years later, and 3 had received at least one unit from a known CJD donor.

Finally, a retrospective review of the brains of 33 hemophilia patients who were treated with clotting factor concentrate of predominantly UK donor sources during the years 1962–1995 was conducted in the UK (Lee et al. 1998). Although no neuropathologic evidence of CJD was found in any of these cases, sensitive techniques such as the CDI were not employed to look for low levels of PrP^{Sc}.

Data from epidemiologic studies conducted in several parts of the world argue that CJD prions are not transmitted from human to human through blood or blood derivatives. Although it has been stated that epidemiologic methods are limited in the detection of rare events such as CJD (Ricketts et al. 1997), the multiple studies currently under way in many countries should continue to give additional information about the theoretical risk of transmitting CJD through blood or blood products.

Of course, this conclusion can pertain only to sCJD, which is sometimes called the "classical" form of CJD, at this time. With vCJD, which has emerged only recently, we do not benefit from the vast epidemiologic perspective that has guided our risk assessment and management for the long-studied, well-characterized sporadic form of CJD. Moreover, vCJD presents considerable differences from sporadic CJD in many aspects (Chapter 13), including perhaps that involving theoretical blood infectivity. As stated earlier, vCJD patients have a high frequency of lymphoid (particularly tonsillar) involvement. In addition, PrP immunoreactivity has been reported in appendyceal lymphoid tissue from a person in a preclinical phase of vCJD (Hilton et al. 1998). Since vCJD is peripherally acquired and not inherited or sporadic, it may be more reasonable to fear a hematogenous phase at some point in the evolution of the disease. For these reasons, and because we have no way of knowing or even estimating the number of people potentially harboring vCJD prions, we cannot make the same assumptions in risk assessment and management that we make with CJD.

Nonetheless, there are a few factual statements we can make about vCJD. None of the 138 cases of vCJD reported worldwide can be linked in any way to transfusion or to administration of plasma-derived products. This is despite the fact that several UK vCJD patients were blood donors, and no case of vCJD has been recorded among identified recipients of products prepared from these donations (Chapter 13) (UK CJD Surveillance Unit [http://www.cjd.ed.ac.uk]). Overall, whole blood, red blood cells, and platelets from UK donors continue to be transfused to UK recipients (an estimated 30–40 million transfusions in the UK over the past 10 years); yet, in the country that accounts for roughly 98% of all BSE and 95% of all vCJD reported in the entire world, there is no evidence to date that vCJD has been transmitted through blood or plasma products. Whether or not this will change over time is unknown.

Regulatory Policies

The prevailing view among regulatory authorities with respect to the risk of transmitting CJD or vCJD prions through blood and blood products is that not enough is known about the potential of individuals in an asymptomatic phase to carry infectious prions in their blood. Thus, in the absence of a specific and sensitive diagnostic test, it is feared that such "carriers" could disseminate infection by blood donation while escaping detection until the onset of symptoms and signs of disease. Therefore, regulatory groups have decided that transmission of CJD/vCJD through blood constitutes a "theoretical risk" and that appropriate measures must be taken to preclude the occurrence of blood-related CJD/vCJD in the future. Current actions toward that objective are reflected in recent documents issued by the Committee for Proprietary Medicinal Products (CPMP) of the European Medicines Evaluation Agency (EMEA) and by the Food and Drug Administration (FDA) of the United States Department of Health and Human Services (USDHHS). These include the requirement of excluding from the blood donor population all individuals considered theoretically at risk of carrying infectious prions, essentially those at risk for inherited (family history of neurodegenerative disease), iatrogenic (history of having received cadaver pituitary-extracted HGH or a dura mater graft), or variant CJD (residence for a defined period of time in the UK, France, and other European countries reporting BSE and/or vCJD) (USDHHS 2002; CPMP 2003).

In addition, evaluation of separation and fractionation procedures for removal of prion infectivity has been recommended (USDHHS 2002; CPMP 2003). Such studies are designed to address two essential questions with respect to the prion safety of plasma-derived products: Where does prion infectivity segregate during fractionation, and what clearance factor or safety margin is achieved in fractionation?

The design of these prion clearance evaluations is difficult in many respects: first is the choice of the prion itself, second is the choice of the tissue in which the prion was replicated, and third is the choice of the prion preparation. Since prions do not behave as ideal particles in either crude or purified preparations, it is difficult to choose an appropriate inoculum as spiking agent intended to mimic the theoretical prion contaminant in blood or plasma. These issues raise a number of perplexing questions. If one chooses to do spiking studies, what does one use as starting material, especially when prion infectivity cannot be found in human blood? Is the correct approach to spiking experiments the addition of prion-infected mouse or hamster brain to human whole blood or plasma? What is the validity of using mouse or hamster prions that could have different separation and removal/inactivation profiles from human prions? What is the relevance of using brain homogenates for spiking blood or plasma? Could the association of infectivity with brain tissue affect partitioning or adsorptions? And what is the appropriate detection methodology for measuring the concentrations of prions in different fractions? Is laboratory animal bioassay using rodent prion strains and wild-type rodents as assay animals an appropriate model system, or should human prions and transgenic mice expressing human PrP be used? Or can in vitro immunoassays for PrP^{Sc}, such as the CDI (Chapter 16) be employed in these studies (Baron 2001; Vey et al. 2002)?

Because many of these questions remain unanswered, persisting uncertainties continue to challenge the design of these prion removal evaluations, and consensus has not been achieved on many of these issues (Baron 2001). Nonetheless, considerable progress has been made in recent years by several laboratories using different approaches and methodologies (Brown et al. 1998, 1999; Foster et al. 2000; Lee et al. 2001; Stenland et al. 2002; Vey et al. 2002). In effect, despite the use of different spiking agents and prion detection methodologies, the data all seem to converge to indicate that the manufacturing processes for plasma-derived medicinal products would reduce prion infectivity if it was present in human plasma. Whether rodent prions relevantly mimic the behavior of human prions (including vCJD prions) when used as spiking agents remains to be established.

Regulatory policies with regard to required actions upon discovery of a blood or plasma donor who subsequently develops CJD or vCJD are relatively uniform across many countries. Retrieval and quarantine of all blood components and recovered plasma emanating from a donor who was diagnosed with CJD or vCJD after giving blood is required on a global basis (USDHHS 2002; CPMP 2003). Withdrawal is not considered justified in the case of plasma-derived medicinal products from pooled source plasma involving a sporadic or familial CJD donor, due to the over-whelming absence of epidemiologic evidence of CJD transmission via blood products and to evidence of prion removal capacity of plasma pro-tein manufacturing processes as described above. Such withdrawal is required, however, when a vCJD donor is implicated in a plasma pool, due to continuing uncertainties regarding vCJD. Regulatory requirements diverge somewhat with respect to donor exclusion criteria. Although UK residence is generally considered a risk factor and thus a criterion for donor deferral, policies differ with respect to potential donors who resided in European countries other than the UK (USDHHS 2002; CPMP 2003). Geography-based donor deferral requirements also are less wide-ranging for plasma derivatives than for blood components, due to the robust prion removal capacity demonstrated for many manufacturing process steps (USDHHS 2002; CPMP 2003).

In conclusion, the "theoretical risk" of human-to-human prion trans-mission through blood is under continuing review by worldwide regula-tory authorities. Efforts have been made to assure the safety of the blood supply and of plasma derivatives through the application of donor exclu-sion criteria, the recommendation to perform prion removal studies, and the implementation of recall policies. Whether any of these measures will be successful in reducing the theoretical risk is unknown. Country-based exclusion policies are largely inefficient, because the vast majority of excluded donors will not develop vCJD. In addition, they could result in shortages of blood components, as well as a lack of spare plasma capacity to make up for the shortfalls that such measures might provoke (USD-HHS 2002; CPMP 2003). The closely intertwined issues of safety and availability must always be considered in terms of a risk/benefit scenario when applying the precautionary principle to a theoretical risk such as that of the transmission of CJD/vCJD through blood products. To prevent compromising the supply of life-sustaining blood, blood components, and plasma derivatives requires judicious analysis of risk as well as an ongoing reassessment of new information as it becomes available.

RISK OF ANIMAL-TO-HUMAN TRANSMISSION OF PRIONS

BSE is the most worrisome of all the animal prion diseases from a biosafe-ty standpoint. BSE is widely thought to be a manmade epidemic, caused by a form of industrial cannibalism in which cattle were fed MBM produced from prion-contaminated cattle and sheep offal (Chapter 12) (Wilesmith et al. 1991; Anderson et al. 1996; Prusiner 1997). Epidemiologically, BSE has shown a disquieting propensity to cross species barriers through oral consumption of prion-contaminated bovine foodstuffs (Kirkwood et al. 1990; Wyatt et al. 1991; Willoughby et al. 1992), and this propensity has been confirmed experimentally in laboratory transmission studies employing several routes of inoculation, including the oral and intravenous routes (Barlow and Middleton 1990; Dawson et al. 1990; Fraser et al. 1992; Baker et al. 1993; Foster et al. 1996; Lasmézas et al. 1996, 2001; Hunter et al. 2002). Most alarmingly, there is now compelling epidemiologic and experimental evidence that BSE is responsible for the emergence of vCJD, which is the only known example of a human prion disease caused by consumption of animal products (Chapter 9, Chapter 12, and Chapter 13) (Collinge et al. 1996; Lasmézas et al. 1996; Will et al. 1996; Bruce et al. 1997; Hill et al. 1997b; Zeidler et al. 1997; Scott et al. 1999).

Against this backdrop, considerable concern has erupted throughout the world over the safety of not only foodstuffs but also pharmaceutical and biological products either derived from bovine source materials (active ingredients) or manufactured with bovine raw materials used as reagents in production. Additionally, there is concern about products that contain bovine components as excipients in final formulations or as constituents in the ingested product covering (capsule material). The major categories of "at risk" products include recombinant proteins, vaccines, and gene therapy products produced in cultured cell systems using bovine-derived factors, as well as drugs that employ tallow or gelatin products as binders. A nonexhaustive list of bovine derivatives in pharmaceuticals is provided in Table 7. The pervasiveness of bovine-derived products on our planet is remarkable.

Regulatory Policies

In 1991, the EU issued CPMP Guideline III/3298/91 (CPMP 1991), which recommended four measures for minimizing the risk of transmitting animal prions via medicinal products: (1) transparent traceability of the origin of source animals from low-risk regions for BSE, (2) preferential use of younger animals, (3) avoidance of tissues from high and medium infectivity categories as source materials (Table 8), and (4) process validation studies with implementation of removal and inactivation procedures within processes where feasible.

The above recommendations invited a number of criticisms. First of all, they did not propose any quantitative method for prion risk assessment with respect to pharmaceuticals. Second, the hierarchy of risk levels

Table 7. Bovine derivatives in pharmaceuticals

Active ingredients		Raw materials in manufacture		Ingested covering		Excipients	
bovine derivative	bovine source material	bovine derivative	bovine source material	bovine derivative	bovine source material	bovine derivative	bovine source material
Aprotinin	lung	albumin	serum	gelatin	bone	gelatin	bone
Gelatin	bone/hide	amicase	milk (casein)			lactose	milk
Glucagon	pancreas	brain-heart infusion	brain and heart serum			Mg stearate	tallow
Heparin	intestine	fetal calf serum				polysorbate	tallow
Insulin	pancreas	glycerol	tallow				
Surfactant	lung	liver infusion	liver				
		meat extract	carcass serum				
		newborn calf serum					
		pepticase	milk (casein)				
		peptone	muscle				
		polysorbate	tallow				
		primatore	blood/spleen				
		trypsin	pancreas				
		tryptone	milk				

Table 8. Categories of infectivity in bovine tissues and body fluids (based on scrapie prion infectivity in tissues and body fluids from naturally infected Suffolk sheep and goats with clinical scrapie)

CATEGORY I High infectivity	→	brain, spinal cord, (eye)[a]
CATEGORY II Medium infectivity	→	spleen, tonsil, lymph nodes, ileum, proximal colon, cerebrospinal fluid, pituitary gland, adrenal gland, (dura mater, pineal gland, placenta, distal colon)
CATEGORY III Low infectivity	→	peripheral nerves, nasal mucosa, thymus, bone marrow, liver, lung, pancreas
CATEGORY IV No detectable infectivity	→	skeletal muscle, heart, mammary gland, milk, blood clot, serum, feces, kidney, thyroid, salivary gland, saliva, ovary, uterus, testis, seminal vesicle, fetal tissue (colostrum, bile, bone, cartilaginous tissue, connective tissue, hair, skin, urine)

[a]Tissues in parentheses were not titrated in the original studies (Hadlow et al. 1980, 1982).

from different tissues contained no numerical range of infectivity titers for each of the four categories. Moreover, the whole concept of high-to-low risk levels from various tissues was based on transmission experiments from scrapied sheep and goats to wild-type mice (Hadlow et al. 1980, 1982) which might bear little relation to the tissue distribution of prions in infected cattle. Third, the guidelines gave no clue as to the potential impact of subclinical disease. Fourth, the models generally employed for process validation and for determination of removal and inactivation capacity of certain procedures involve spiking materials with titered mouse or hamster brain homogenates. In such studies, testing of downstream products is done by intracerebral inoculation into wild-type mice or hamsters; such a system may or may not reflect the behavior of bovine prions under similar conditions.

It is our opinion that the use of brain or other infectious tissues from transgenic mice expressing bovine PrP on the null ($Prnp^{0/0}$) background provide the best source of inocula. Such prions can be readily bioassayed in mice expressing bovine PrP transgenes (Chapter 9) (Scott et al. 1997). Moreover, highly sensitive immunoassays such as the CDI can be used in many aspects of these studies as well (Chapter 16).

In 1997, under intense political pressure from the EU Parliament, the European Commission (EC) issued a decision referred to as EC Decision

97/534/EC (EC 1997) which (1) defined "specified risk materials" (SRM) as the skull and its contents (including the brain and eyes); tonsils and spinal cord from bovine, ovine, and caprine animals aged over 12 months or from ovine and caprine animals with an erupted permanent incisor tooth; as well as the spleen from ovines and caprines; and (2) banned use of said SRM for any purpose and under any circumstances, with the exception of research. This decision also prohibited the use of vertebral column of bovine, ovine, and caprine animals for the production of mechanically recovered meat. It further banned import into the EU of any "medical, pharmaceutical or cosmetic products, or their starting materials or intermediate products," unless accompanied by a declaration signed by the "competent authority of the country of production" stating the product neither contains nor is derived from SRM.

As defined in the decision, SRM as such are used in relatively few pharmaceutical products. However, materials such as tallow and gelatin, derivatives of which are found in most tablets and capsules as excipients and/or ingested coverings, can be considered to be derived from starting materials that may have come into contact with SRM, and thus classified as SRM-derived, would have been affected by EC Decision 97/534/EC. Therefore, had the decision been implemented as originally proposed, it would have forced roughly 80% of pharmaceuticals off the European market, creating shortages in essential medicines and imposing alternative formulations for a host of medicinal products, with unforeseen, potentially deleterious consequences.

Because of the potentially adverse implications of the EC decision, and in the face of serious concerns expressed by the pharmaceutical industry, by the manufacturers of gelatin and tallow derivatives, and by the EU's own pharmaceutical regulatory agency EMEA (CPMP 1997), implementation of the decision, which was to have come into effect in January 1998, was repeatedly deferred. Numerous amendments and revisions were proposed and debated, finally resulting in repeal of the decision in its original form and replacement by a new one redefining SRM and applying only to food and feed materials but exempting cosmetics, medicinal products, and medical devices, together with their starting materials or intermediate products, from its regulatory scope. The latter now must comply with a recently issued note for guidance on minimizing the risk of ruminant-derived prion transmission covering both human and veterinary medicinal products (EMEA 2001). This document provides guidelines for control measures covering several factors including source of animals, nature of animal tissue used in manufacturing, validation of production processes, and animal age. Manufacturers are required by law to comply through a complex certification process documenting traceability of the origin of source animals, tissue type, etc., for any material of animal, primarily ruminant, origin, not only for new applications but also for all previously authorized products.

In the US, the FDA has provided guidance through a series of letters to manufacturers and importers of FDA-regulated products containing or manufactured with bovine derivatives from countries reporting BSE cases. Since 1989, the U.S. Department of Agriculture (USDA) began a series of preventive actions to protect against BSE, including prohibiting the importation of ruminant livestock and most ruminant-derived products for animal use from the UK and (since 1991) from other countries declaring BSE. This ban was extended to Europe in 1997. In addition, the USDA has monitored the US cattle population through a "targeted surveillance" approach, designed to test the highest risk animals, including "downer" cattle (animals that are non-ambulatory at slaughter), animals that die on the farm, older animals, and animals exhibiting signs of neurologic disease, for neuropathologic and immunohistochemical evidence of BSE. Of the more than 30,000 head of cattle examined to date (close to 20,000 in 2002), no cases of BSE have been found (USDA [http://www.usda.gov]). As another measure designed to prevent BSE in the US, the USDA initiated in 1997 a "mammalian"-to-ruminant feed ban, although porcine MBM may still be fed to cattle and cattle MBM to hogs. The ban also does not cover refeeding of bovine material back to cattle through plate waste and chicken feces.

A notice in the Federal Register of August 29, 1994, summarized the FDA's position and recommendations to reduce any potential BSE risk in bovine-derived products (USDHHS 1994). In essence, the FDA requested that bovine-derived materials originating from animals born or living in BSE-affected countries not be used in the manufacture of FDA-regulated products intended for humans. Although the FDA did not object to the use of bovine-derived materials from BSE-infected countries in the manufacture of pharmaceutical-grade gelatin, it did consider it prudent to source from BSE-free countries. That position was based on an assessment that the manufacturing conditions for gelatin were likely to remove/inactivate prion infectivity sufficiently to obviate any risk of BSE prion transmission. Since studies with bovine prions were not used in making such an assessment, it is unclear whether the conclusions are warranted.

The safety of gelatin from BSE-stricken countries was revisited in 1997, following concerns raised by the Transmissible Spongiform Encephalopathy Advisory Committee (TSEAC), resulting in the FDA's first BSE-related "Guidance for Industry" document, announced in the

Federal Register of October 7, 1997 (USDHHS 1997), and approved by the TSEAC in April, 1998. This guidance document recommends that manufacturers determine the tissue, species, and country of origin of gelatin raw materials and exclude gelatin from bones and hides of cattle sourced from BSE countries or from countries of unknown/dubious BSE status for use in injectable, implantable, or ophthalmic products. It allows oral and cosmetic use of gelatin from bones of cattle from BSE-affected countries with the condition that the raw material is sourced from BSE-free herds and that the heads, spines, and spinal cords are removed at slaughter. Bovine gelatin sourced in the US or other non-BSE countries is unconditionally authorized, and pig-skin gelatin is permitted if uncontaminated with bovine materials from BSE-stricken or BSE-unknown countries.

MONITORING ANIMALS FOR PRIONS

The specified offals bans instituted in Great Britain in 1988 and 1989 (Chapter 12) were based largely on data derived from studies using Swiss mice for the bioassay of sheep prions. The legislation made two assumptions: (1) the distribution of prions in cattle with BSE is the same as that in sheep with scrapie and (2) bioassays of non-CNS sheep prions in mice yielded reliable data. Although we are unaware of any completed comparative study of the distribution of prions in sheep and cattle, there are studies that allow a comparsion of prion titers in mice and Syrian hamsters. Such quantitative data show clear differences between these rodent hosts (Kimberlin and Walker 1977). Equally disturbing is the bioassay of tissues with low titers in nontransgenic mice. Typically, Swiss mice require an incubation time of ~500 days before showing signs of scrapie when inoculated with a 1% brain homogenate from a sheep with scrapie (Hadlow et al. 1982). Greatly prolonged incubation times have also been recorded for BSE prions with a variety of mouse strains (Fraser et al. 1992; Bruce et al. 1994, 1997; Lasmézas et al. 1997). Moreover, the titer of BSE prions in bovine brain measured by endpoint titrations in cattle (Wells et al. 1998) is more than 1000-fold higher than that reported in earlier studies using an RIII mouse bioassay (Bruce et al. 1994). The implications for accurate assessment of prions in peripheral tissues, where the titers are likely to be much lower, are disconcerting.

The advent of transgenic mice expressing bovine PrP but not endogenous mouse PrP ($Prnp^{0/0}$), which are highly susceptible to BSE prions with abbreviated incubation times (Scott et al. 1997), should provide considerable information that has been lacking to date. Endpoint titration of cattle-derived BSE brain homogenate in these mice resulted in a titer of bovine prions which was ~10-fold greater than that obtained using the cattle-to-cattle bioassay described above (Chapter 9) (Safar et al. 2002). Endpoint titrations of homogenates derived from numerous tissues of BSE-afflicted cattle using Tg(BoPrP)$Prnp^{0/0}$ mice will be important. Such studies will allow the development of standard, titered bovine inocula as well as incubation-time assays. The results of such studies will not only be important in assessing the safety of the food supply, but they should also pave the way for process validation and prion removal studies in the pharmaceutical industry.

The availability of Tg(BoPrP)$Prnp^{0/0}$ mice also provides the basis for calibrating sensitive immunoassays (Chapter 16). Such immunoassays will find widespread use in certifying that cattle and other domestic animals are free of prions. Since the beginning of 2000, testing of bovine brain stems at slaughterhouse has been mandated in all EU member countries for cattle over 30 months of age. Prior to June 2003, three different immunoassays, one western blot and two ELISAs, were approved for this screening program (Moynagh et al. 1999). However, the most sensitive of these three tests is at best comparable to the relatively inefficient wild-type mouse bioassay described above (Moynagh et al. 1999). In a recently published study, the CDI was shown to be capable of detecting the PrP^{Sc} in bovine brain stems with a sensitivity similar to that of endpoint titrations in Tg(BoPrP)$Prnp^{0/0}$ mice, which have been shown to be ~10-fold more sensitive than cattle and thus ~10,000-fold more sensitive than wild-type mice (Chapter 16) (Safar et al. 2002). It is clear that the use of immunoassays considerably less sensitive than the CDI has already been instrumental in detecting PrP^{Sc} in brain stems of hundreds of asymptomatic cattle at slaughterhouse in Europe. How many more cattle would have been found or may be found in the future using the CDI is of considerable interest and may have important implications for the safety of the food supply. In June of 2003, the CDI was approved by the EC for screening bovine brain stems.

Although USDA surveys of "high-risk cattle" have failed to reveal any cases of BSE in the US as noted above, this surveillance is still too limited in the sense that it is not directed toward identifying asymptomatic cases. In Germany, which currently numbers 273 cases of BSE, only a small minority occurred in cattle that exhibited overt signs of clinical disease; the vast majority were detected in asymptomatic animals at slaughterhouse by immunoassay (Office International des Epizooties [http://www.oie.int]). It is also notable that a TME outbreak in Stetsonville, Wisconsin, in 1985 is thought to have been due to a sporadic case of BSE in a "fallen cow" (Marsh et al. 1991). Since the incubation period for BSE is 3–4 years (Chapter 12), most infected cattle will never show neurologic deficits because the vast majority of cattle in the US are slaughtered by 1 year of age. Such animals might harbor significant titers of prions but not show any signs of CNS dysfunction, histologic evidence of spongiform degeneration, or immunohistochemical demonstration of PrP. We argue that the availability of highly sensitive methods for detection of PrP^{Sc} in brain stems and other tissues of cattle ought to be employed routinely in slaughterhouses worldwide to ensure that humans are not infected by bovine prions. Calibration of the sensitivity of such methods by bioassay in Tg(BoPrP)$Prnp^{0/0}$ mice should be mandatory.

The emphasis given to BSE in this chapter is not intended to underestimate the potential biological safety concerns associated with some of the other animal prion diseases, particularly scrapie in sheep and CWD in farmed and wild deer and elk. The epidemiology and pathogenesis of these diseases are described in Chapter 11. Scrapie has been known for over 200 years, and the bulk of evidence over time does not support the notion that the scrapie prion is a human pathogen. CWD is more recent, first described in the 1960s, and first determined to be a prion disease in 1978 (Chapter 11) (Williams and Young 1980). Its growing numbers and ever-expanding geographical distribution in recent years have caused considerable concern. According to the CDC, there is no evidence to date that CWD has been transmitted to humans, despite instances in which young men who consumed wild game and/or were hunters developed CJD (Belay et al. 2001). In addition, in vitro protection studies of PrP^{C} bound to PrP^{Sc} seem to show evidence of a molecular barrier limiting the susceptibility not only of humans, but also of cattle and sheep to CWD prions (Raymond et al. 2000). Nonetheless, despite the lack of evidence of pathogenicity of scrapie or CWD prions to humans, surveillance and testing for these diseases are crucial to limiting their future prevalence and spread, and to ensuring the safety of the food supply with respect to prions regardless of species origin or strain. To this effect, the reported ability of the CDI to detect PrP^{Sc} from CWD-infected animals with high sensitivity and to discriminate PrP^{Sc} from CWD and BSE prions represents a major advance (Safar et al. 2002).

VARIANT CREUTZFELDT-JAKOB DISEASE

The importance of vCJD (Chapter 13) with respect to biosafety is that it represents the first example of the transmission of animal prions to humans. Once kuru was experimentally transmitted to apes (Gajdusek et al. 1966), the search for an animal reservoir intensified, but none was found. Eventually, ritualistic cannibalism became accepted as the mode of kuru transmission (Gajdusek 1977).

The transmission of sCJD to apes (Gibbs et al. 1968) stimulated a 25-year search for a relationship between scrapie in sheep and CJD in humans, but none was identified. Exhaustive epidemiologic studies were conducted in the hope of identifying scrapie prions as the cause of CJD (Malmgren et al. 1979; Will and Matthews 1984; Brown et al. 1987; Cousens et al. 1990). When vCJD was first reported, the differences in one or more of the seven amino acids that distinguish bovine from sheep PrP were hypothesized to be the reason that bovine but not sheep prions may have been transmitted to humans. Subsequently, the BSE strain of bovine prions has been offered as the most important factor in their permissiveness in humans (Scott et al. 1999).

The Link between BSE and vCJD

It now appears to be clearly established that vCJD is due to the transmission of bovine prions to humans. This conclusion is based on multiple lines of inquiry: (1) the spatial-temporal clustering of vCJD (Will et al. 1996; Zeidler et al. 1997), (2) the successful transmission of BSE to macaques with induction of PrP plaques similar to those seen in vCJD (Lasmézas et al. 1996), (3) the similarity of the glycosylation pattern of PrP^{Sc} in vCJD to that noted in mice, domestic cats, and macaques infected with BSE prions (Collinge et al. 1996), (4) experimental transmission studies in wild-type mice which suggest that vCJD and BSE prions exhibit similar strain-related behavior (Bruce et al. 1997), and (5) transmission studies in transgenic mice providing conclusive evidence that BSE and vCJD are the same prion strain, removing virtually any doubt that vCJD is a direct result of consumption of products from BSE-infected cattle (Chapter 9) (Scott et al. 1999).

Yet, a number of enigmas still surround vCJD and seek an adequate explanation. Why should the disease strike young people predominantly, with a majority in their teens and twenties? Why was there not a particularly evocative dietary history for any of the patients? Why does the disease incidence now appear to be stagnating rather than increasing? What is the rationale behind analyzing PrP glycoforms in attempting to link BSE and vCJD, since PrP^{Sc} is formed after glycosylation of PrP^{C}? And why does it remain consistently difficult to transmit disease to mice expressing human PrP or chimeric human/mouse PrP transgenes when inoculated with BSE or vCJD prions (Chapter 9) (Scott et al. 1999; Asante et al. 2002)?

From a biosafety standpoint, there is little doubt that vCJD is different from sporadic or inherited CJD and must be considered separately with respect to biosafety issues. There is evidence for a higher frequency of lymphoid (particularly tonsillar) involvement in vCJD (Hill et al. 1997a), which raises the potential risk of blood infectivity carried by circulating lymphoid-tissue-derived cells. Since we have no way of knowing or even estimating the number of individuals potentially harboring vCJD prions, we cannot make the same assumptions in risk assessment and management that we do with sCJD. Within the context of these considerations, the European CPMP position on vCJD and plasma derivatives advocating precautionary withdrawal of batches of plasma derivatives if a pool donor was subsequently diagnosed as having vCJD seems prudent (CPMP 2003).

CONCLUDING REMARKS

From this review of some of the biosafety issues relevant to prion-related risk, whether the well-documented cattle-to-human risk via BSE-contaminated bovine products, or the unsubstantiated, theoretical human-to-human risk through CJD-contaminated blood, it becomes readily apparent that existing information is generally insufficient. This situation squeezes regulatory authorities into the uncomfortable position of having to recommend highly conservative, precautionary measures aimed at precluding a "theoretical risk." Some of these measures are eminently reasonable, and would have been taken whatever the state of the knowledge. Others invite scrutiny and court the danger of provoking medical product shortages while pursuing a "theoretical risk."

One of the most frustrating aspects of trying to deal with prion risk assessment and management in the past has been the lack of rapid and sensitive methods for detection of human and animal prions. The need for rapid implementation of such methods on a global basis is imperative if we are to minimize the risk of prion contamination to public health. Animal bioassays have been widely used and have generated much of the data upon which we currently rely, but they are precluded from routine monitoring because of the prolonged incubation times that are required. Most of the currently available immunoassays, including western blotting, lack sufficient sensitivity. At best, they can detect PrP^{Sc} in brain from clinically ill persons or animals when levels of PrP^{Sc} are as high as or higher than those of PrP^C, but the key for a diagnostic test is to be able to detect low levels of PrP^{Sc} not only in brain but also in peripheral tissues or fluids of asymptomatic subjects, in which PrP^{Sc} may represent less than 0.00001% of the total PrP present.

Fortunately, recent advances have been made on two fronts. First, the development of transgenic mice which are highly susceptible to human (Telling et al. 1994, 1995; Korth et al. 2003) and bovine (Scott et al. 1997) prions with abbreviated incubation periods have given us precious new tools for the confirmation of prion disease and a better understanding of species barriers, strain characteristics, tissue distribution, and levels of infectivity at various disease stages. Such transgenic mice should be the models of choice for process validation and prion removal studies in the pharmaceutical industry.

Second, the rapid, highly sensitive CDI that takes advantage of differences between predominantly α-helical PrP^C and its pathogenic isoform PrP^{Sc}, which has a high β-sheet content, is now available (Chapter 16), for the detection of both animal (Safar et al. 2002) and human (Bellon et al. 2003) prions with high sensitivity. Such an immunoassay may, for the first time, provide a highly sensitive and extremely rapid alternative test for prion infectivity, which could find practical application in widespread slaughterhouse and live animal testing, validation of key removal steps in manufacturing processes, and, potentially, human testing.

REFERENCES

Alper T., Cramp W.A., Haig D.A., and Clarke M.C. 1967. Does the agent of scrapie replicate without nucleic acid? *Nature* **214:** 764–766.

Alpers M. 1987. Epidemiology and clinical aspects of kuru. In *Prions—Novel infectious pathogens causing scrapie and Creutzfeldt-Jakob disease* (ed. S.B. Prusiner and M.P. McKinley), pp. 451–465. Academic Press, Orlando, Florida.

Anderson R.M., Donnelly C.A., Ferguson N.M., Woolhouse M.E.J., Watt C.J., Udy H.J., MaWhinney S., Dunstan S.P., Southwood T.R.E., Wilesmith J.W., Ryan J.B.M., Hoinville L.J., Hillerton J.E., Austin A.R., and Wells G.A.H. 1996. Transmission dynamics and epidemiology of BSE in British cattle. *Nature* **382:** 779–788.

Asante E.A., Linehan J.M., Desbruslais M., Joiner S., Gowland I., Wood A.L., Welch J., Hill A.F., Lloyd S.E., Wadsworth J.D.F., and Collinge J. 2002. BSE prions propagate as either variant CJD-like or sporadic CJD-like prion strains in transgenic mice expressing human prion protein. *EMBO J.* **21:** 6358–6366.

Baker H.F., Ridley R.M., and Wells G.A.H. 1993. Experimental transmission of BSE and scrapie to the common marmoset. *Vet. Rec.* **132:** 403–406.

Barlow R.M. and Middleton D.J. 1990. Dietary transmission of bovine spongiform encephalopathy to mice. *Vet. Rec.* **126:** 111–112.

Baron H. 2001. Variant Creutzfeldt-Jakob disease: Predicting the future? *Blood Coagul. Fibrinolysis* (suppl. 1) **12:** S29–S36.

Belay E.D., Gambetti P., Schonberger L.B., Parchi P., Lyon D.R., Capellari S., McQuiston J.H., Bradley K., Dowdle G., Crutcher M., and Nichols C.R. 2001. Creutzfeldt-Jakob disease in unusually young patients who consumed venison. *Arch. Neurol.* **58:** 1673–1678.

Bellon A., Seyfert-Brandt W., Lang W., Baron H., Gröner A., and Vey M. 2003. Improved conformation dependent immunoassay: Suitability for human prion detection with enhanced sensitivity. *J. Gen. Virol.* **84:** 1921–1925.

Bellinger-Kawahara C., Cleaver J.E., Diener T.O., and Prusiner S.B. 1987a. Purified scrapie prions resist inactivation by UV irradiation. *J. Virol.* **61:** 159–166.

Bellinger-Kawahara C.G., Kempner E., Groth D.F., Gabizon R., and Prusiner S.B. 1988. Scrapie prion liposomes and rods exhibit target sizes of 55,000 Da. *Virology* **164:** 537–541.

Bellinger-Kawahara C., Diener T.O., McKinley M.P., Groth D.F., Smith D.R., and Prusiner S.B. 1987b. Purified scrapie prions resist inactivation by procedures that hydrolyze, modify, or shear nucleic acids. *Virology* **160:** 271–274.

Brown K.L., Stewart K., Bruce M.E., and Fraser H. 1996. Scrapie in immunodeficient mice. In *Transmissible subacute spongiform encephalopathies: Prion diseases* (ed. L. Court and B. Dodet), pp. 159–166. Elsevier, Paris.

Brown P. 1995. Can Creutzfeldt-Jakob disease be transmitted by transfusion? *Curr. Opin. Hematol.* **2:** 472–477.

———. 2001. Creutzfeldt-Jakob disease: Blood infectivity and screening tests. *Semin. Hematol.* (suppl. 9) **38:** 2–6.

Brown P., Preece M.A., and Will R.G. 1992. "Friendly fire" in medicine: Hormones, homografts, and Creutzfeldt-Jakob disease. *Lancet* **340:** 24–27.

Brown P., Wolff A., and Gajdusek D.C. 1990. A simple and effective method for inactivating virus infectivity in formalin-fixed samples from patients with Creutzfeldt-Jakob disease. *Neurology* **40:** 887–890.

Brown P., Cathala F., Raubertas R.F., Gajdusek D.C., and Castaigne P. 1987. The epidemiology of Creutzfeldt-Jakob disease: Conclusion of 15-year investigation in France and review of the world literature. *Neurology* **37:** 895–904.

Brown P., Cervenáková L., McShane M., Barber P., Rubenstein R., and Drohan W.N. 1999. Further studies of blood infectivity in an experimental model of transmissible spongiform encephalopathy, with an explanation of why blood components do not transmit Creutzfeldt-Jakob disease in humans. *Transfusion* **39:** 1169–1178.

Brown P., Rohwer R.G., Dunstan B.C., MacAuley C., Gajdusek D.C., and Drohan W.N. 1998. The distribution of infectivity of blood components and plasma derivatives in experimental models of transmissible spongiform encephalopathy. *Transfusion* **38:** 810–816.

Brown P., Gibbs C.J., Jr., Rodgers-Johnson P., Asher D.M., Sulima M.P., Bacote A., Goldfarb L.G., and Gajdusek D.C. 1994a. Human spongiform encephalopathy: The National Institutes of Health series of 300 cases of experimentally transmitted disease. *Ann. Neurol.* **35:** 513–529.

Brown P., Cervenáková L., Goldfarb L.G., McCombie W.R., Rubenstein R., Will R.G., Pocchiari M., Martinez-Lage J.F., Scalici C., Masullo C., Graupera G., Ligan J., and Gajdusek D.C. 1994b. Iatrogenic Creutzfeldt-Jakob disease: An example of the interplay between ancient genes and modern medicine. *Neurology* **44:** 291–293.

Bruce M.E., McConnell I., Will R.G., and Ironside J.W. 2001. Detection of variant Creutzfeldt-Jakob disease infectivity in extraneural tissues. *Lancet* **358:** 208–209.

Bruce M.E., Chee A., McConnell I., Foster J., Pearson G., and Fraser H. 1994. Transmission of bovine spongiform encephalopathy and scrapie to mice: Strain variation and the species barrier. *Philos. Trans. R. Soc. Lond. B Biol. Sci.* **343:** 405–411.

Bruce M.E., Will R.G., Ironside J.W., McConnell I., Drummond D., Suttie A., McCardle L., Chree A., Hope J., Birkett C., Cousens S., Fraser H., and Bostock C.J. 1997. Transmissions to mice indicate that 'new variant' CJD is caused by the BSE agent. *Nature* **389:** 498–501.

Casaccia P., Ladogana A., Xi Y.G., and Pocchiari M. 1989. Levels of infectivity in the blood throughout the incubation period of hamsters peripherally injected with scrapie. *Arch. Virol.* **108:** 145–149.

Centers for Disease Control (CDC). 1985. Fatal degenerative neurologic disease in patients who received pituitary derived human growth hormone. *Morb. Mortal. Wkly. Rep.* **34:** 359–360.

———. 1989. Update: Creutzfeldt-Jakob disease in a second patient who received a cadaveric dura mater graft. *Morb. Mortal. Wkly. Rep.* **38:** 37–38, 43.

———. 1996. Surveillance for Creutzfeldt-Jakob Disease—United States. *Morb. Mortal. Wkly. Rep.* **45:** 665–668.

———. 1997. Creutzfeldt-Jakob disease associated with cadaveric dura mater grafts—Japan, January 1979–May 1996. *Morb. Mortal. Wkly. Rep.* **46:** 1066–1069.

Clarke M.C. and Haig D.A. 1967. Presence of the transmissible agent of scrapie in the serum of affected mice and rats. *Vet. Rec.* **80:** 504.

Collinge J. and Palmer M.S. 1997. Human prion diseases. In *Prion diseases* (ed. J. Collinge and M.S. Palmer), pp. 18–56. Oxford University Press, Oxford, United Kingdom.

Collinge J., Palmer M.S., and Dryden A.J. 1991. Genetic predisposition to iatrogenic Creutzfeldt-Jakob disease. *Lancet* **337:** 1441–1442.

Collinge J., Sidle K.C.L., Meads J., Ironside J., and Hill A.F. 1996. Molecular analysis of prion strain variation and the aetiology of "new variant" CJD. *Nature* **383:** 685–690.

Collins S., Law M.G., Fletcher A., Boyd A., Kaldor J., and Masters C.L. 1999. Surgical treatment and risk of sporadic Creutzfeldt-Jakob disease: A case-control study. *Lancet* **353:** 693–697.

Committee on Proprietary Medicinal Products (CPMP). 1991. Minimizing the risk of transmitting agents causing spongiform encephalopathy via medicinal products (III/3298/91).

———. 1997. Summary of the provisional CPMP report: Medicinal products for human use affected by Commission Decision (97/534/EC) (CPMP/961/97).

———. 2003. CPMP position statement on Creutzfeldt-Jakob disease and plasma-derived and urine-derived medicinal products (EMEA/CPMP/BWP/2879/02).

Cousens S.N., Harries-Jones R., Knight R., Will R.G., Smith P.G., and Matthews W.B. 1990. Geographical distribution of cases of Creutzfeldt-Jakob disease in England and Wales 1970–84. *J. Neurol. Neurosurg. Psychiatry* **53:** 459–465.

Davanipour Z., Alter M., Sobel E., Asher D.M., and Gajdusek D.C. 1985. Creutzfeldt-Jakob disease: Possible medical risk factors. *Neurology* **35:** 1483–1486.

Dawson M., Wells G.A.H., Parker B.N.J., and Scott A.C. 1990. Primary parenteral transmission of bovine spongiform encephalopathy to the pig. *Vet. Rec.* **127:** 338.

Deslys J.-P., Lasmézas C., and Dormont D. 1994. Selection of specific strains in iatrogenic Creutzfeldt-Jakob disease. *Lancet* **343:** 848–849.

Deslys J.P., Lasmézas C.I., Billette de Villemeur T., Jaegly A., and Dormont D. 1996. Creutzfeldt-Jakob disease. *Lancet* **347:** 1332.

Deslys J.P., Jaegly A., D'Aignaux J.H., Mouthon F., Billette de Villemeur T., and Dormont D., 1998. Genotype at codon 129 and susceptibility to Creutzfeldt-Jakob disease. *Lancet* **351:** 1251.

Dickinson A.G. and Fraser H. 1969. Genetical control of the concentration of ME7 scrapie agent in mouse spleen. *J. Comp. Pathol.* **79:** 363–366.

Dickinson A.G., Stamp J.T., and Renwick C.C. 1974. Maternal and lateral transmission of scrapie in sheep. *J. Comp. Pathol.* **84:** 19–25.

Diringer H. 1984. Sustained viremia in experimental hamster scrapie. Brief report. *Arch. Virol.* **82:** 105–109.

Doi T. 1991. Relationship between periodical organ distribution of Creutzfeldt-Jakob disease (CJD) agent infectivity and prion protein gene expression in CJD-agent infected mice. *Nagasaki Igakkai Zasshi* **66:** 104–114.

Eklund C.M., Kennedy R.C., and Hadlow W.J. 1967. Pathogenesis of scrapie virus infection in the mouse. *J. Infect. Dis.* **117:** 15–22.

Esmonde T.F.G., Will R.G., Slattery J.M., Knight R., Harries-Jones R., de Silva R., and Matthews W.B. 1993. Creutzfeldt-Jakob disease and blood transfusion. *Lancet* **341:** 205–207.

European Commission (EC). 1997. Decision 97/534/EC. On the prohibition of the use of material presenting risks as regards transmissible spongiform encephalopathies. *Off. J. Eur. Comm.* **L-216:** 95–98.

European Medicines Evaluation Agency (EMEA). 2001. Committee for Propietary Medicinal Products (CPMP) and Committee for Veterinary Medicinal Products (CVMP). Note for guidance on minimising the risk of transmitting animal spongiform encephalopathy agents via human and veterinary medicinal products (EMEA/410/01).

Evatt B.H., Austin H., Barnhart E., Schonberger L., Sharer L., Jones R., and DeArmond S.J. 1998. Surveillance for Creutzfeldt-Jakob disease among persons with hemophilia. *Transfusion* **38:** 817–820.

Foster J.D., Bruce M., McConnell I., Chree A., and Fraser H. 1996. Detection of BSE infectivity in brain and spleen of experimentally infected sheep. *Vet. Rec.* **138:** 546–548.

Foster J.D., McKelvey W.A.C., Mylne M.J.A., Williams A., Hunter N., Hope J., and Fraser H. 1992. Studies on maternal transmission of scrapie in sheep by embryo transfer. *Vet. Rec.* **130:** 341–343.

Foster P.R., Welch A.G., McLean C., Griffin B.D., Hardy J.C., Bartley A., MacDonald S., and Bailey A.C. 2000. Studies on the removal of abnormal prion protein by processes used in the manufacture of human plasma products. *Vox Sang.* **78:** 86–95.

Fraser H., Bruce M.E., Chree A., McConnell I., and Wells G.A.H. 1992. Transmission of bovine spongiform encephalopathy and scrapie to mice. *J. Gen. Virol.* **73:** 1891–1897.

Fraser H., Brown K.L., Stewart K., McConnell I., McBride P., and Williams A. 1996. Replication of scrapie in spleens of SCID mice follows reconstitution with wild-type mouse bone marrow. *J. Gen. Virol.* **77:** 1935–1940.

Gabizon R. and Prusiner S.B. 1990. Prion liposomes. *Biochem. J.* **266:** 1–14.

Gajdusek D.C. 1977. Unconventional viruses and the origin and disappearance of kuru. *Science* **197:** 943–960.

———. 1991. The transmissible amyloidoses: Genetical control of spontaneous generation of infectious amyloid proteins by nucleation of configurational change in host precursors: kuru-CJD-GSS-scrapie-BSE. *Eur. J. Epidemiol.* **7:** 567–577.

Gajdusek D.C., Gibbs C.J., Jr., and Alpers M. 1966. Experimental transmission of a kuru-like syndrome to chimpanzees. *Nature* **209:** 794–796.

Gibbs C.J., Jr., Gajdusek D.C., Asher D.M., Alpers M.P., Beck E., Daniel P.M., and Matthews W.B. 1968. Creutzfeldt-Jakob disease (spongiform encephalopathy): Transmission to the chimpanzee. *Science* **161:** 388–389.

Godec M.S., Asher D.M., Masters C.L., Kozachuk W.E., Friedland R.P., Gibbs C.J., Jr., Gajdusek D.C., Rapoport S.I., and Schapiro M.B. 1991. Evidence against the transmissibility of Alzheimer's disease. *Neurology* **41:** 1320.

Godec M.S., Asher D.M., Kozachuk W.E., Masters C.L., Rubi J.U., Payne J.A., Rubi-Villa D.J., Wagner E.E., Rapoport S.I., and Schapiro M.B. 1994. Blood buffy coat from Alzheimer's disease patients and their relatives does not transmit spongiform encephalopathy to hamsters. *Neurology* **44:** 1111–1115.

Hadlow W.J., Kennedy R.C., and Race R.E. 1982. Natural infection of Suffolk sheep with scrapie virus. *J. Infect. Dis.* **146:** 657–664.

Hadlow W.J., Kennedy R.C., Race R.E., and Eklund C.M. 1980. Virologic and neurohistologic findings in dairy goats affected with natural scrapie. *Vet. Pathol.* **17:** 187–199.

Heye N., Hensen S., and Müller N. 1994. Creutzfeldt-Jakob disease and blood transfusion. *Lancet* **343:** 298–299.

Hill A.F., Zeidler M., Ironside J., and Collinge J. 1997a. Diagnosis of new variant Creutzfeldt-Jakob disease by tonsil biopsy. *Lancet* **349:** 99–100.

Hill A.F., Desbruslais M., Joiner S., Sidle K.C.L., Gowland I., Collinge J., Doey L.J., and Lantos P. 1997b. The same prion strain causes vCJD and BSE. *Nature* **389:** 448–450.

Hilton D.A., Fathers E., Edwards P., Ironside J.W., and Zajicek J. 1998. Prion immunoreactivity in appendix before clinical onset of variant Creutzfeldt-Jakob disease. *Lancet* **353:** 703–704.

Holman R.C., Khan A.S., Belay E.D., and Schonberger L.B. 1996. Creutzfeldt-Jakob disease in the United States, 1979–1994: Using national mortality data to assess the possible occurrence of variant cases. *Emerg. Infect. Dis.* **2:** 333–337.

Holman R.C., Khan A.S., Kent J., Strine T.W., and Schonberger L.B. 1995. Epidemiology of Creutzfeldt-Jakob disease in the United States, 1979–1990: Analysis of national mortality data. *Neuroepidemiology* **14:** 174–181.

Hunter H., Foster J., Chong A., McCutcheon S., Parnham D., Eaton S., MacKenzie C., and Houston F. 2002. Transmission of prion diseases by blood transfusion. *J. Gen. Virol.* **83:** 2897–2905.

Kao R.R., Gravenor M.B., Baylis M., Bostock C.J., Chihota C.M., Evans J.C., Goldman W., Smith A.J.A., and McLean A.R. 2002. The potential size and duration of an epidemic of bovine spongiform encephalopathy in British sheep. *Science* **295:** 332–335.

Kimberlin R.H. 1976. *Scrapie in the mouse.* Meadowfield Press, Durham, United Kingdom.

Kimberlin R. and Walker C. 1977. Characteristics of a short incubation model of scrapie in the golden hamster. *J. Gen. Virol.* **34:** 295–304.

Kirkwood J.K., Wells G.A.H., Wilesmith J.W., Cunningham A.A., and Jackson S.I. 1990. Spongiform encephalopathy in an arabian oryx (*Oryx leucoryx*) and a greater kudu (*Tragelaphus strepsiceros*). *Vet. Rec.* **127:** 418–420.

Kitamoto T., Muramoto T., Mohri S., Doh-ura K., and Tateishi J. 1991. Abnormal isoform of prion protein accumulates in follicular dendritic cells in mice with Creutzfeldt-Jakob disease. *J. Virol.* **65:** 6292–6295.

Klein M.A., Frigg R., Flechsig E., Raeber A.J., Kalinke U., Bluethmann H., Bootz F., Suter M., Zinkernagel R.M., and Aguzzi A. 1997. A crucial role for B cells in neuroinvasive scrapie. *Nature* **390:** 687–691.

Kondo K. and Kuroina Y. 1982. A case control study of Creutzfeldt-Jakob disease: Association with physical injuries. *Ann. Neurol.* **11:** 377–381.

Korth C., Kaneko K., Groth D., Heye N., Telling G., Mastrianni J., Parchi P., Gambetti P., Will R., Ironside J., Heinrich C., Temblay P., DeArmond S.J., and Prusiner S.B. 2003. Abbreviated incubation times for human prions in mice expressing a chimeric mouse–human prion protein transgene. *Proc. Natl. Acad. Sci.* **100:** 4784–4789.

Kuroda Y., Gibbs C.J., Jr., Amyx H.L., and Gajdusek D.C. 1983. Creutzfeldt-Jakob disease in mice: Persistent viremia and preferential replication of virus in low-density lymphocytes. *Infect. Immun.* **41:** 154–161.

Laplanche J.L. 1996. Génétique moléculaire des formes familiales et sporadiques des maladies à prions humaines. *Méd. Mal. Infect.* **26:** 264–270.

Lasmézas C.I., Deslys J.-P., Demaimay R., Adjou K.T., Lamoury F., Dormont D., Robain O.,

Ironside J., and Hauw J.-J. 1996. BSE transmission to macaques. *Nature* **381:** 743–744.

Lasmézas C.I., Deslys J.-P., Robain O., Jaegly A., Beringue V., Peyrin J.-M., Fournier J.-G., Hauw J.-J., Rossier J., and Dormont D. 1997. Transmission of the BSE agent to mice in the absence of detectable abnormal prion protein. *Science* **275:** 402–405.

Lasmézas C.I., Fournier J.-G., Nouvel V., Boe H., Marcé D., Lamoury F., Kopp N., Hauw J.-J., Ironside J., Bruce M., Dormont D., and Deslys J.-P. 2001. Adaptation of the bovine spongiform encephalopathy agent to primates and comparison with Creutzfeldt-Jakob disease: Implications for public health. *Proc. Natl. Acad. Sci.* **98:** 4142–4147.

Lavelle G.C., Sturman L., and Hadlow W.J. 1972. Isolation from mouse spleen of cell populations with high specific infectivity for scrapie virus. *Infect. Immun.* **5:** 319–323.

Lee C.A., Ironside J.W., Bell J.E., Giangrande P., Ludlam C., Esiri M.-M., and McLaughlin J.E. 1998. Retrospective neuropathological review of prion disease in U.K. haemophiliac patients. *Thromb. Haemostasis* **80:** 909–911.

Lee D.C., Stenland C.J., Miller J.L.C., Cai K., Ford E.K., Gilligan K.J., Hartwell R.C., Terry J.C., Rubenstein R., Fournel M., and Petteway S.R., Jr. 2001. A direct relationship between the partitioning of the pathogenic prion protein and transmissible spongiform encephalopathy infectivity during the purification of plasma proteins. *Transfusion* **41:** 449–455.

Malmgren R., Kurland L., Mokri B., and Kurtzke J. 1979. The epidemiology of Creutzfeldt-Jakob disease. In *Slow transmissible diseases of the nervous system* (ed. S.B. Prusiner and W.J. Hadlow), vol. 1, pp. 93–112. Academic Press, New York.

Manuelidis L. 1994. The dimensions of Creutzfeldt-Jakob disease. *Transfusion* **34:** 915–928.

Manuelidis E.E. and Manuelidis L. 1993. A transmissible Creutzfeldt-Jakob disease-like agent is prevalent in the human population. *Proc. Natl. Acad. Sci.* **90:** 7724–7728.

Manuelidis E.E., Gorgacz E.J., and Manuelidis L. 1978. Viremia in experimental Creutzfeldt-Jakob disease. *Science* **200:** 1069–1071.

Manuelidis E.E., Kim J.H., Mericangas J.R., and Manuelidis L. 1985. Transmission to animals of Creutzfeldt-Jakob disease from human blood (letter). *Lancet* **II:** 896–897.

Manuelidis E.E., de Figueiredo J.M., Kim J.H., Fritch W.W., and Manuelidis L. 1988. Transmission studies from blood of Alzheimer disease patients and healthy relatives. *Proc. Natl. Acad. Sci.* **85:** 4898–4901.

Marsh R.F. 1992. Transmissible mink encephalopathy. In *Prion diseases of humans and animals* (ed. S.B. Prusiner et al.), pp. 300–307. Ellis Horwood, London.

Marsh R.F., Bessen R.A., Lehmann S., and Hartsough G.R. 1991. Epidemiological and experimental studies on a new incident of transmissible mink encephalopathy. *J. Gen. Virol.* **72:** 589–594.

McKinley M.P., Masiarz F.R., Isaacs S.T., Hearst J.E., and Prusiner S.B. 1983. Resistance of the scrapie agent to inactivation by psoralens. *Photochem. Photobiol.* **37:** 539–545.

Moynagh J., Schimmel H., and Kramer G.N. 1999. The evaluation of tests for the diagnosis of transmissible spongiform encephalopathy in bovines. European Commission report (July 1999), pp. 1–39.

Muramoto T., Kitamoto T., Tateishi J., and Goto I. 1993. Accumulation of abnormal prion protein in mice infected with Creutzfeldt-Jakob disease via intraperitoneal route: A sequential study. *Am. J. Pathol.* **143:** 1470–1479.

Nakamura Y., Yanagawa H., Hoshi K., Yoshino H., Urata J., and Sato T. 1999. Incidence rate of Creutzfeldt-Jakob disease in Japan. *Int. J. Epidemiol.* **28:** 130–134.

Office Internationale des Epizooties (OIE) (http://www.oie.int).

O'Rourke K.I., Huff T.P., Leathers C.W., Robinson M.M., and Gorham J.R. 1994. SCID mouse spleen does not support scrapie agent replication. *J. Gen. Virol.* **75:** 1511–1514.

Owen F., Poulter M., Collinge J., and Crow T.J. 1990. Codon 129 changes in the prion protein gene in Caucasians. *Am. J. Hum. Genet.* **46:** 1215–1216.

Palmer M.S., Dryden A.J., Hughes J.T., and Collinge J. 1991. Homozygous prion protein genotype predisposes to sporadic Creutzfeldt-Jakob disease. *Nature* **352:** 340–342.

Parchi P., Castellani R., Capellari S., Ghetti B., Young K., Chen S.G., Farlow M., Dickson D.W., Sima A.A.F., Trojanowski J.Q., Petersen R.B., and Gambetti P. 1996. Molecular basis of phenotypic variability in sporadic Creutzfeldt-Jakob disease. *Ann. Neurol.* **39:** 767–778.

Pattison I.H. 1965. Experiments with scrapie with special reference to the nature of the agent and the pathology of the disease. In *Slow, latent and temperate virus infections* (NINDB Monogr. 2) (ed. D.C. Gajdusek et al.), pp. 249–257. U.S. Government Printing, Washington, D.C.

Prichard J., Thadani V., Kalb R., Manuelidis E., and Holder J. 1987. Rapidly progressive dementia in a patient who received a cadaveric dura mater graft. *Morb. Mortal. Wkly. Rep.* **36:** 46–50, 55.

Prusiner S.B. 1982. Novel proteinaceous infectious particles cause scrapie. *Science* **216:** 136–144.

———. 1997. Prion diseases and the BSE crisis. *Science* **278:** 245–251.

Prusiner S.B., Groth D., Serban A., Stahl N., and Gabizon R. 1993. Attempts to restore scrapie prion infectivity after exposure to protein denaturants. *Proc. Natl. Acad. Sci.* **90:** 2793–2797.

Prusiner S.B., Hadlow W.J., Eklund C.M., Race R.E., and Cochran S.P. 1978. Sedimentation characteristics of the scrapie agent from murine spleen and brain. *Biochemistry* **17:** 4987–4992.

Prusiner S.B., Groth D.F., Cochran S.P., Masiarz F.R., McKinley M.P., and Martinez H.M. 1980. Molecular properties, partial purification, and assay by incubation period measurements of the hamster scrapie agent. *Biochemistry* **19:** 4883–4891.

Prusiner S.B., McKinley M.P., Bolton D.C., Bowman K.A., Groth D.F., Cochran S.P., Hennessey E.M., Braunfeld M.B., Baringer J.R., and Chatigny M.A. 1984. Prions: Methods for assay, purification and characterization. In *Methods in virology* (ed. K. Maramorosch and H. Koprowski), vol. 8, pp. 293–345. Academic Press, New York.

Prusiner S.B., Scott M., Foster D., Pan K.-M., Groth D., Mirenda C., Torchia M., Yang S.-L., Serban D., Carlson G.A., Hoppe P.C., Westaway D., and DeArmond S.J. 1990. Transgenetic studies implicate interactions between homologous PrP isoforms in scrapie prion replication. *Cell* **63:** 673–686.

Public Health Service. 1997. Interagency Coordinating Committee. Report on human growth hormone and Creutzfeldt-Jakob disease. *U.S. Public Health Serv. Rep.* **14:** 1–11.

Raymond G.J., Bossers A., Raymond L.D., O'Rourke K.I., McHolland L.E., Bryant P.K., III, Miller M.W, Williams E.S., and Caughey B. 2000. Evidence of a molecular barrier limiting susceptibility of humans, cattle and sheep to chronic wasting disease. *EMBO J.* **19:** 4425–4430.

Ricketts M.N., Cashman N.R., Stratton E.E., and ElSaadany S. 1997. Is Creutzfeldt-Jakob disease transmitted in blood? *Emerg. Infect. Dis.* **3:** 155–163.

Ridley R.M. and Baker H.F. 1993. Occupational risk of Creutzfeldt-Jakob disease. *Lancet* **341:** 641–642.

Ruffie A., Delasnerie-Laupretre N., Brandel J.P., Jaussent I., Dormont D., Laplanche J.L., Hauw J.J., Richardson S., and Alperovitch A. 1997. Incidence of Creutzfeldt-Jakob disease in France, 1992–1995. *Rev. Epidemiol. Sante Publique* **45:** 448–453.

Safar J., Groth D., DeArmond S.J., and Prusiner S.B. 1999. Prions. In *Biosafety in microbi-ological and biomedical laboratories* (ed. J. Richmond and R.W. McKinney). U.S. Department of Health and Human Services, Public Health Service, Centers for Disease Control and Prevention, and National Institutes of Health, Washington, D.C.

Safar J., Ceroni M., Piccardo P., Liberski P.P., Miyazaki M., Gajdusek D.C., and Gibbs C.J., Jr. 1990. Subcellular distribution and physicochemical properties of scrapie-associated precursor protein and relationship with scrapie agent. *Neurology* **40:** 503–508.

Safar J., Scott M., Monoghan J., Deering C., Didorenko S., Vergara J., Ball H., Legname G., Leclerc E., Solforosi L., Serban H., Groth D., Burton D.R., Prusiner S.B., and Williamson R.A. 2002. Measuring prions causing bovine spongiform encephalopathy or chronic wasting disease by immunoassays and transgenic mice. *Nat. Biotechnol.* **20:** 1147–1150.

Scott M., Will R., Ironside J., Nguyen H.-O., Tremblay P., DeArmond S.J., and Prusiner S.B. 1999. Compelling transgenetic evidence for transmission of bovine spongiform encephalopathy prions to humans. *Proc. Natl. Acad. Sci.* **96:** 15137–15142.

Scott M., Foster D., Mirenda C., Serban D., Coufal F., Wälchli M., Torchia M., Groth D., Carlson G., DeArmond S.J., Westaway D., and Prusiner S.B. 1989. Transgenic mice expressing hamster prion protein produce species-specific scrapie infectivity and amy-loid plaques. *Cell* **59:** 847–857.

Scott M.R., Safar J., Telling G., Nguyen O., Groth D., Torchia M., Koehler R., Tremblay P., Walther D., Cohen F.E., DeArmond S.J., and Prusiner S.B. 1997. Identification of a prion protein epitope modulating transmission of bovine spongiform encephalopathy prions to transgenic mice. *Proc. Natl. Acad. Sci.* **94:** 14279–14284.

Stenland C.J., Lee D.C., Brown P., Petteway S.R., Jr., and Rubenstein R. 2002. Partitioning of human and sheep forms of the pathogenic prion protein during the purification of therapeutic proteins from human plasma. *Transfusion* **42:** 1497–1500.

Tamai Y., Kojima H., Kitajima R., Taguchi F., Ohtani Y., Kawaguchi T., Miura S., Sato M., and Ishihara Y. 1992. Demonstration of the transmissible agent in tissue from a preg-nant woman with Creutzfeldt-Jakob disease. *N. Engl. J. Med.* **327:** 649.

Tateishi J. 1985. Transmission of Creutzfeldt-Jakob disease from human blood and urine into mice. *Lancet* **II:** 1074.

Taylor D.M., Woodgate S.L., and Atkinson M.J. 1995. Inactivation of the bovine spongi-form encephalopathy agent by rendering procedures. *Vet. Rec.* **137:** 605–610.

Taylor D.M., Woodgate S.L., Fleetwood A.J., and Cawthorne R.J.G. 1997. Effect of render-ing procedures on the scrapie agent. *Vet. Rec.* **141:** 643–649.

Telling G.C., Scott M., Mastrianni J., Gabizon R., Torchia M., Cohen F.E., DeArmond S.J., and Prusiner S.B. 1995. Prion propagation in mice expressing human and chimeric PrP transgenes implicates the interaction of cellular PrP with another protein. *Cell* **83:** 79–90.

Telling G.C., Parchi P., DeArmond S.J., Cortelli P., Montagna P., Gabizon R., Mastrianni J., Lugaresi E., Gambetti P., and Prusiner S.B. 1996. Evidence for the conformation of the pathologic isoform of the prion protein enciphering and propagating prion diversity. *Science* **274:** 2079–2082.

Telling G.C., Scott M., Hsiao K.K., Foster D., Yang S.-L., Torchia M., Sidle K.C.L., Collinge J., DeArmond S.J., and Prusiner S.B. 1994. Transmission of Creutzfeldt-Jakob disease from humans to transgenic mice expressing chimeric human-mouse prion protein. *Proc. Natl. Acad. Sci.* **91:** 9936–9940.

UK CJD Surveillance Unit (http://www.cjd.ed.ac.uk).

United States Department of Agriculture (USDA) (http://www.usda.gov).

United States Department of Health and Human Services (USDHHS). 1994. FDA Notice

59 FR 44592. Bovine-derived materials: Agency letters to manufacturers of FDA-regulated products. *Fed. Reg.* (August 29, 1994).

———. 1997. FDA Guidance FR Doc. 97-26501. Guidance for industry. The sourcing and processing of gelatin to reduce the potential risk posed by bovine spongiform encephalopathy (BSE) in FDA-regulated products for human use. *Fed. Reg. Announcement* (October 7, 1997).

———. 2002. FDA Guidance for industry. Revised preventive measures to reduce the possible risk of transmission of Creutzfeldt-Jakob disease (CJD) and variant Creutzfeldt-Jakob disease (vCJD) by blood and blood products (January, 2002). (http://www.fda.gov/cber/guidelines.htm).

van Duijn C.M., Delasnerie-Lauprêtre N., Masullo C., Zerr I., de Silva R., Wientjens D.P.W.M., Brandel J.-P., Weber T., Bonavita V., Zeidler M., Alpérovitch A., Poser S., Granieri I., Hofman A., and Will R.G. 1998. Case-control study of risk factors of Creutzfeldt-Jakob disease in Europe during 1993–1995. European Union (EU) Collaborative Study Group of Creutzfeldt-Jakob Disease (CJD). *Lancet* **351:** 1081–1085.

Vey M., Baron H., Weimer T., and Gröner A. 2002. Purity of spiking agent affects partitioning of prions in plasma protein purification. *Biologicals* **30:** 187–196.

Wadsworth J.D.F., Joiner S., Hill A.F., Campbell T.A., Desbruslais M., Luthert P.J., and Collinge J. 2001. Tissue distribution of protease resistant protein in variant Creutzfeldt-Jakob disease using a highly sensitive immunoblotting assay. *Lancet* **358:** 171–180.

Wells G.A.H., Hawkins S.A.C., Green R.B., Austin A.R., Dexter I., Spencer Y.L., Chaplin M.J., Stack M.J., and Dawson M. 1998. Preliminary observations on the pathogenesis of experimental bovine spongiform encephalopathy (BSE): An update. *Vet. Rec.* **142:** 103–106.

Wilesmith J.W., Ryan J.B.M., and Atkinson M.J. 1991. Bovine spongiform encephalopathy: Epidemiologic studies on the origin. *Vet. Rec.* **128:** 199–203.

Will R.G. 1996. Incidence of Creutzfeldt-Jakob disease in the European Community. In *Bovine spongiform encephalopathy: The BSE dilemma* (ed. C.J. Gibbs, Jr.), pp. 364–374. Springer-Verlag, New York.

Will R.G. and Matthews W.B. 1984. A retrospective study of Creutzfeldt-Jakob disease in England and Wales 1970–79. I. Clinical features. *J. Neurol. Neurosurg. Psychiatry* **47:** 134–140.

Will R.G., Ironside J.W., Zeidler M., Cousens S.N., Estibeiro K., Alperovitch A., Poser S., Pocchiari M., Hofman A., and Smith P.G. 1996. A new variant of Creutzfeldt-Jakob disease in the UK. *Lancet* **347:** 921–925.

Williams E.S. and Young S. 1980. Chronic wasting disease of captive mule deer: A spongiform encephalopathy. *J. Wildl. Dis.* **16:** 89–98.

Willoughby K., Kelly D.F., Lyon D.G., and Wells G.A.H. 1992. Spongiform encephalopathy in a captive puma (*Felis concolor*). *Vet. Rec.* **131:** 431–434.

Wyatt J.M., Pearson G.R., Smerdon T.N., Gruffydd-Jones T.J., Wells G.A.H., and Wilesmith J.W. 1991. Naturally occurring scrapie-like spongiform encephalopathy in five domestic cats. *Vet. Rec.* **129:** 233–236.

Zeidler M., Stewart G.E., Barraclough C.R., Bateman D.E., Bates D., Burn D.J., Colchester A.C., Durward W., Fletcher N.A., Hawkins S.A., Mackenzie J.M., and Will R.G. 1997. New variant Creutzfeldt-Jakob disease: Neurological features and diagnostic tests. *Lancet* **350:** 903–907.

18

Therapeutic Approaches to Prion Diseases

Stanley B. Prusiner,[1,2,3] Barnaby C.H. May,[1,4] and Fred E. Cohen[1,3,4]
[1]Institute for Neurodegenerative Diseases
Departments of [2]Neurology, [3]Biochemistry and Biophysics,
[4]Cellular and Molecular Pharmacology
University of California, San Francisco, California 94143

DEVELOPING EFFECTIVE THERAPIES FOR PRION DISEASES is a wonderful challenge. Once an effective therapy is devised, it will have important implications for many other degenerative diseases, some of which are thought to be disorders of protein processing like the prion diseases (Prusiner 2001). An effective therapy for any neurodegenerative disease would invigorate new efforts directed toward other disorders.

The biology of prions is sufficiently advanced to make this a propitious time to develop therapeutically oriented research that might be extremely fruitful in the near future. Prion replication begins with protein synthesis, whereby the normal, cellular isoform of the prion protein (PrP^C) is translated in the endoplasmic reticulum and translocated through the Golgi as it transits to the cell surface. During this journey, PrP^C is folded into a three helix-bundle protein, which has a single disulfide bond, two Asn-linked oligosaccharides, and a glycosylphosphatidyl inositol (GPI) anchor. Both the complex-type oligosaccharides and the GPI anchor become sialyated, presumably as PrP traverses the Golgi. Throughout this process, the quality control machinery of the cell is operative to assure the PrP^C that reaches the cell surface is properly folded and possesses the appropriate posttranslational modifications. Once PrP^C reaches the cell surface, this protein seems to accumulate primarily in cholesterol-rich microdomains or rafts, where it can be converted into the disease-causing isoform (PrP^{Sc}). In principle, there are many potential targets for antiprion compounds to attack and thereby provide effective intervention.

Superb mouse models of the genetic and infectious forms of prion disease recapitulate virtually every aspect of the disease process. Such models can be used to screen compounds as well as to evaluate their efficacy. But prior to screening potential therapeutics in mice, levels of PrP^{Sc} can be measured in prion-infected cultured cells. The inhibition of PrP^{Sc} formation can be measured and increased clearance of PrP^{Sc} can be determined. What we lack are cell-free systems for prion replication that faithfully recreate the cellular and animal models.

A variety of compounds have been proposed as potential therapeutics for the treatment of prion diseases, including polysulfated anions, dextrans, Congo red dye, oligonucleotides, and cyclic tetrapyrroles (Dickinson et al. 1975; Kimberlin and Walker 1983; Ehlers and Diringer 1984; Diringer 1991; Ingrosso et al. 1995; Priola et al. 2000; Sethi et al. 2002). All of these compounds have been shown to increase survival when given at the time of prion infection in rodents, but not when initially administered a month or more after infection has been established.

In addition to studies in rodents, scrapie-infected neuroblastoma (ScN2a) cell prions have been used to identify candidate antiprion compounds (Caughey and Race 1992; Gorodinsky and Harris 1995; Taraboulos et al. 1995; Supattapone et al. 1999a; Perrier et al. 2000; Korth et al. 2001; Proske et al. 2002), but none of these compounds has been shown to be effective in halting prion diseases in either animals or humans.

Because the CNS is the site of cellular dysfunction, any compounds developed to treat prion diseases must traverse the blood–brain barrier (BBB) or they must be infused intrathecally. One approach to finding a suitable drug is empirical, where large-scale drug screening or serendipity leads to identification of an effective lead compound for subsequent optimization, or a therapeutic directly. This approach is responsible for most of the drugs that are currently available. We took such an approach in looking at existing drugs known to cross the BBB (Korth et al. 2001). Another strategy is to use compounds that alter the conformation of PrP^{Sc} and thus allow the cell to degrade prions (Supattapone et al. 1999a, 2001). A third approach is to look for drugs that block the formation of nascent PrP^{Sc} by interfering with the binding of PrP^{C} to PrP^{Sc} (Caughey and Race 1992; Caughey et al. 1994; Ingrosso et al. 1995; Heppner et al. 2001; Peretz et al. 2001). A fourth strategy is to search for compounds that diminish the supply of PrP^{C}, because diminishing the substrate is known to abolish PrP^{Sc} formation (Büeler et al. 1993; Prusiner et al. 1993; Tremblay et al. 1998). A fifth approach is to develop pharmacotherapeutics that mimic dominant-negative inhibition of prion formation by disrupting the binding of PrP^{C} to auxiliary proteins required for PrP^{Sc} formation (Kaneko et al. 1997b; Perrier et al. 2000, 2002).

PREVENTING PRION DISEASES

Any discussion of therapeutics investigations must also touch on preventive measures. In sheep, breeding animals for the R171 polymorphism is likely to prove quite useful in preventing natural scrapie (Chapter 4). The R171 polymorphism is a dominant negative in prion replication, and PrP(R171) is not converted into PrP^{Sc} (Perrier et al. 2002). Preventing scrapie in sheep will result in decreasing the exposure of people to prions through the food supply. This will occur in two ways: (1) decreased exposure to prions by consuming tainted sheep meat and meat products and (2) decreased exposure of other domestic animals to feeds prepared from the rendered offal of sheep. Animals exposed to such feeds include cattle, pigs, chickens, and farmed fish. Whether or not screening cattle worldwide for polymorphisms of the PrP gene would lead to the discovery of a dominant negative is uncertain. If such a dominant-negative mutation was found, it might be used to breed prion-resistant cattle. The implications of breeding scrapie-resistant sheep for the spontaneous generation of BSE prions are discussed in Chapter 4.

Diagnostic Tests

Preventive measures also include excluding prion-infected products from the food supply by using diagnostic tests that determine whether or not an animal is infected with prions. In the European Union, such diagnostic tests that measure the protease-resistant fragment of PrP^{Sc}, PrP 27-30, are used to assess the brain stems of all slaughtered cattle over 30 months of age. Although some asymptomatic, prion-infected cattle have been identified by measuring PrP 27-30, many unidentified, prion-infected cattle are still likely to be entering the human food supply for two reasons. First, measuring PrP 27-30 in the bovine brain stem can be problematic because the levels of PrP^{Sc} within the brain stem vary widely, and protease-sensitive PrP^{Sc} cannot be measured using procedures that depend on limited proteolysis to eliminate PrP^{C} (Chapter 3) (Safar et al. 2002). The conformation-dependent immunoassay (CDI) measures both the protease-sensitive and -resistant forms of PrP^{Sc} and thus, is likely to improve identification of prion-infected cattle. Second, an experimental pathogenesis study of prion spread in cattle after oral inoculation shows that the Peyer's patches of the distal ileum are positive at 6, 10, and 18 months after inoculation, and the brain stem remains negative until 32 months after inoculation (Wells 2002). The bioassays for this study were performed in cattle, limiting the number of samples studied due to the high cost of cattle. Using Tg(BoPrP) mice that are ~10 times more sensitive than cattle for bioassays and the CDI for rapid measurement of BoPrP^{Sc} are clearly superior approaches (Chapter 3) (Safar et al. 2002). Employing Tg(BoPrP) mice and the CDI would allow a sufficiently large number of studies to be performed to determine the kinetics of prion replication and deposition in many different bovine organs. Without this rudimentary knowledge, no diagnostic test can be applied in a reliable manner.

Inactivation of Prions and Disinfectants

Still another means of preventing prion disease is through the use of disinfectants that destroy prions. Of all infectious pathogens, prions are the most resistant to killing. It is the extreme resistance of the scrapie agent to inactivation by formalin and heat treatment that made the agent very puzzling in the 1940s (Gordon 1946). The peculiar nature of the scrapie agent grew more strange when Tikvah Alper and her colleagues reported the extreme resistance of the scrapie agent to ionizing and UV irradiation (Chapter 2) (Alper et al. 1966, 1967, 1978).

At the time of writing, the most widely used means of inactivating prions are autoclaving at 134°C and treatment with 1 N or 2 N NaOH (Prusiner et al. 1981, 1984; Taylor and Fernie 1996; Taylor 2000). Complete inactivation of prions in a crude homogenate can require autoclaving for up to 5 hours at 134°C (Prusiner et al. 1984). Exposure to base either in the form of 1 N or 2 N NaOH or undiluted bleach is generally performed at room temperature for at least 2 hours.

The discovery that branched polyamines rapidly clear PrP^{Sc} from living cells has given rise to an entirely new approach to inactivating prions (Supattapone et al. 1999a, 2001). Whether branched polyamines can be used as a food additive to kill prions without destroying the meat or meat product remains to be established. Protein-specific denaturants may provide a promising approach for prion disinfectants and, possibly, for treating protein deposition diseases.

TRICYCLIC DERIVATIVES OF ACRIDINES AND PHENOTHIAZINES

Patients with Creutzfeldt-Jakob disease (CJD) or other prion diseases develop progressive neurologic dysfunction. Prion diseases are invariably fatal, and death frequently occurs in less than one year after the first symptoms appear (Will et al. 1999). The need for effective therapeutics for prion diseases is clear and, as with many other fatal disorders, significant side effects would be tolerated in compounds with demonstrated therapeutic impact.

Many compounds have been identified that inhibit prion propagation when administered at the time of inoculation in rodents (Dickinson et al. 1975; Kimberlin and Walker 1983; Ehlers and Diringer 1984; Diringer and Ehlers 1991). Treatment with these same compounds administered immediately prior to or during the onset of neurologic dysfunction has proven ineffective. Other compounds that inhibit PrP^{Sc} formation, including Congo red, have been identified using scrapie-infected cultured cells (Caughey and Race 1992; Caspi et al. 1998; Priola et al. 2000); some of these have been examined in rodents, but none has been effective when given around the time that neurologic signs appear (Ingrosso et al. 1995).

In a search for compounds that might prove effective in treating prion diseases, we used ScN2a cells to screen for inhibition of nascent PrP^{Sc} formation as well as the clearance of preexisting PrP^{Sc}. Because PrP^{Sc} formation occurs in cholesterol-rich microdomains, inhibitors of cholesterol biosynthesis were examined for their ability to inhibit the conversion of PrP^{C} into PrP^{Sc} (Gorodinsky and Harris 1995; Taraboulos et al. 1995). Statin drugs were found to inhibit PrP^{Sc} formation in cultured cells, but the level of cholesterol depletion required does not permit such an approach to be used in animals.

Because the BBB restricts the access of many molecules to the CNS, we decided to screen a variety of drugs that are known to penetrate the BBB, including phenothiazine-based compounds, for their ability to inhibit PrP^{Sc} formation. Phenothiazine-based drugs have been used to treat psychoses for nearly 50 years (Delay et al. 1952).

Quinacrine and Chlorpromazine

Tricyclic acridine and phenothiazine derivatives were studied, and the effective concentrations for half-maximal inhibition (EC$_{50}$) of PrP^{Sc} formation were determined in cultured cells. These tricyclic derivatives exhibit EC$_{50}$ values between 0.3 μM and 3 μM in ScN2a cells (Fig. 1). Chlorpromazine inhibited PrP^{Sc} formation at 3 μM whereas quinacrine, a structural antecedent of chlorpromazine, was 10 times more potent (Korth et al. 2001). Various quinacrine analogs were assayed, and structure–activity analysis emphasized the importance of the dibasic alkyl side chain extending from the ring nitrogen for bioactivity against PrP^{Sc} formation in ScN2a cells (Fig. 2). These findings defined a new class of antiprion compounds characterized by a tricyclic scaffold substituted with an aliphatic side chain from the center ring.

ScN2a cells treated with 0.8 μM quinacrine for 7 days were found to be devoid of prions by bioassay in Tg(MoPrP-A)4053 mice (Table 1). Exposure of ScN2a cells to 0.5 μM quinacrine greatly prolonged the incu-

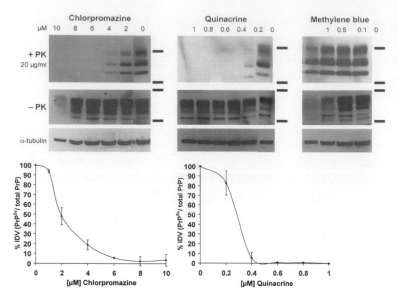

Figure 1. Dose-response relationships for inhibition of PrP^{Sc} formation by chlor-promazine, quinacrine, and methylene blue. Anti-PrP immunoblots of the same ScN2a cell lysates treated with PK (1st row), without PK (2nd row), anti-tubulin immunoblot of the same lysate (3rd row), and average densitometry of three independent immunoblots (*lower panel*; bars represent standard error). ScN2a cells were treated for 6 days with chlorpromazine (*A*), quinacrine (*B*), or methylene blue (*C*). EC$_{50}$ levels are ~2 μM for chlorpromazine and ~300 nM for quinacrine. Methylene blue does not inhibit PrP^{Sc} formation and is cytotoxic at concentrations >500 nM. (Reprinted, with permission, from Korth et al. 2002 [copyright National Academy of Sciences].)

bation time in mice, and only 2 of 5 animals developed disease. 8 μM chlorpromazine failed to cure ScN2a cells of prions, as judged by bioassay.

Three studies preceded the reports mentioned above (Korth et al. 2001) describing the potent inhibition of PrP^{Sc} formation by acridine and phenothiazine derivatives. Chlorpromazine administered repeatedly to mice subcutaneously before and after intracerebral inoculation of scrapie prions resulted in a prolongation of the incubation period by one month, but no change in the incubation period was seen if the prions were inoculated intraperitoneally (Roikhel et al. 1984). In another study, 50 μM chlorpromazine was added to hamster brain fractions enriched for membranes, and the suspension was exposed to UV irradiation (Dees et al. 1985). The prion titers of the unirradiated controls (exposed only to chlorpromazine) were reduced by ~10-fold, whereas the titers of the irradiated samples were reduced >100-fold. In a third study, quinacrine and

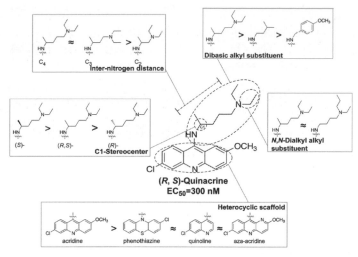

Figure 2. Structure–activity relationship (SAR) of quinacrine and related heterocyclic analogs. SAR analysis revealed a dependence on the tricyclic acridine scaffold and a dibasic alkyl substituent *peri* to the ring nitrogen.

other lysomotrophic reagents added to ScN2a cells resulted in the reduction of PrP^{Sc} (Doh-ura et al. 2000).

Some of the earliest studies of tricyclic compounds in humans are the investigations of Guttmann and Ehrlich, who pioneered the use of methylene blue as an antimalarial (Guttmann and Ehrlich 1891).

Table 1. ScN2a cells treated with quinacrine or chlorpromazine and then inoculated into Tg(MoPrP)4053/FVB mice for bioassay of prion infectivity

Quinacrine (μM)	Incubation time (days ± S.E.M.)	n/n_0[a]
0	76.4 ± 1	4/4
0.2	81.2 ± 5	5/5
0.5	117.5 ± 1	2/5
0.8	—	0/5
Chlorpromazine (μM)	Incubation time (days ± S.E.M.)	n/n_0[a]
0	74.8 ± 4	4/4
2	83.4 ± 6	5/5
5	79.0 ± 5	5/5
8	109.3 ± 7	4/5

After one week of treatment, cells were lysed by freeze–thaw cycles, aspirated several times through a 26-gauge syringe, and inoculated into Tg(MoPrP)4053/FVB mice (C. Korth and S.B. Prusiner, unpubl.). 1% (w/v) RML-infected mouse brain homogenate causes disease in ~50 days after intracerebral inoculation of Tg(MoPrP-A)FVB/4053 (Telling et al. 1995).

[a]n, Number of ill mice; n_0, number of inoculated mice.

Quinacrine was derived from methylene blue and was introduced in the 1930s as a drug with improved antimalarial properties (Green 1932; Schulemann 1932). The phenothiazines were discovered as a by-product of a search for antimalarial substances less toxic than quinacrine (Gilman et al. 1944).

Quinacrine, commonly referred to as atabrine or mepacrine, was used widely during World War II as an antimalarial agent (Goodman and Gilman 1975). Although quinacrine has been replaced by newer antimalarial drugs, it is approved for the treatment of giardiasis, producing cure rates greater than 90%. Quinacrine is known to exert its antitrypanosomal effects by inhibiting trypanothione reductase from *Trypanosoma cruzi* (Bonse et al. 1999), but no mammalian reductase has been shown to bind to quinacrine.

Dosage of quinacrine is recommended to be 100 mg orally for a few days for the treatment of giardiasis, whereas the drug has been administered orally for treatment of malaria in daily doses from 100 mg to 1000 mg for up to many months (Findlay 1951; Goodman and Gilman 1970; Hardman et al. 1996). Quinacrine distributes slowly and achieves steady-state levels only four weeks after the start of therapy. Some of the lowest concentrations in the body are measured in the CNS. At low oral doses, few side effects are described; higher oral doses and parenteral administration increase the occurrence of serious side effects in the cardiovascular system, skin, gastrointestinal system, and CNS. A toxic psychosis from quinacrine has been reported in as many as 0.5% of recipients (Findlay 1951; Goodman and Gilman 1970).

Chlorpromazine has been used in the treatment of schizophrenia and other psychotic conditions since the 1950s (Delay et al. 1952). Daily oral dosage is usually between 100 and 600 mg. Side effects of chlorpromazine at doses used clinically include the tardive dyskinesias and, occasionally, agranulocytosis.

To date, we have found that the bioactivity of phenothiazine- and acridine-based compounds is dependent on the heterocyclic scaffold and a dibasic alkyl substituent (Fig. 2). The structure–activity data define a dependence on a heteroalkyl substituent of modest molecular volume, and even subtle changes to the heterocyclic scaffold are also sufficient to alter dramatically the bioactivity. The structures of chlorpromazine and quinacrine are quite different from those of the small molecules that were designed for dominant-negative inhibition of prion formation described below (Perrier et al. 2000).

These tricylic compounds are also quite different from other known inhibitors of PrP^{Sc} formation, including Congo red, heparins, polysulfated glucosaminoglycans, branched polyamines, and lovastatin. Congo red is a multicyclic, anionic molecule that binds to PrP amyloid (Prusiner et al. 1983) and, more generally, to β-sheet structures of amyloidogenic proteins. It has been suggested that Congo red binds to PrP^{Sc} and thereby sequesters the infectious template (Caughey and Race 1992; Caspi et al. 1998). Other polyanionic molecules, such as the sulfated polyanions, are also thought to act nonspecifically on the PrP^{Sc} template (Ehlers and Diringer 1984; Kimberlin and Walker 1986; Caughey and Raymond 1993). We have no evidence that phenothiazine- or acridine-based compounds bind to PrP^{Sc}, because in applying an assay (Supattapone et al. 1999a) in which purified mouse prion rods were preincubated with these substances for 2 hours, no decrease in the amount of protease-resistant PrP^{Sc} was observed. Lovastatin is thought to act by depleting cells of cholesterol and thereby destroying the raft compartments where the conversion of PrP^{C} to PrP^{Sc} occurs (Taraboulos et al. 1995). The inability of highly charged polyanions to cross the BBB and reach target cells, as well as the toxicity of lovastatin, makes these substances unlikely therapeutics.

The mechanism by which quinacrine and chlorpromazine inhibit PrP^{Sc} formation and possibly enhance clearance is unknown. In contrast to the branched polyamines, both quinacrine and chlorpromazine require a few days of treatment before PrP^{Sc} in ScN2a cells disappears. The branched polyamines require only a few hours of treatment before the levels of preexisting PrP^{Sc} decline (Supattapone et al. 1999a).

Treatment of CJD with Quinacrine

Because quinacrine has been used in humans as an antimalarial drug for over 60 years and is known to cross the BBB, it became an immediate candidate for the treatment of patients dying of CJD and other prion diseases. Humans can tolerate 100 mg to 1,000 mg of quinacrine administered orally on a daily schedule. In an ongoing clinical study, we have been loading patients with 1,000 mg of quinacrine and then giving them 300 mg daily thereafter. In some patients, hepatotoxicity becomes a problem, as evidenced by elevated values of liver function tests (LFTs).

If quinacrine proves to be useful in the treatment of prion disease, the best regimen will be determined in well-controlled clinical studies on humans. For example, it will be critical to determine whether the candidate patients actually have prion disease before administering quinacrine. Because no reliable blood or cerebrospinal fluid (CSF) test for prion disease in humans has been developed, brain biopsies will be critical unless the electroencephalograms and magnetic resonance images show patterns characteristic of prion disease (Chapter 13).

If quinacrine can be shown to cure prion disease in symptomatic humans, prophylactic treatment for carriers of PrP gene mutations and for people previously exposed to prions should be considered. At the same time, it will be reasonable to consider treating animals exposed to prions; this might provide a much-needed rational approach to managing the problems currently plaguing European livestock industries.

Because the practical treatment of people with prion disease requires the drug to cross the BBB, it is unclear whether quinacrine or chlor-prom-azine will be more effective. It is possible that chlorpromazine, with a higher EC_{50} in ScN2a cells than quinacrine, may prove more effi-cacious clinically if it crosses the BBB more readily. It is well known that the phenothiazines exhibit large differences in their distribution between plasma and brain (Svendsen et al. 1988), arguing that drugs, like the per-azines or thioridazine, that could not be reliably examined because of their cytotoxicity to ScN2a cells, might be more potent antiprion drugs than chlorpromazine or even quinacrine. Alternatively, very high doses of phenothiazines assuring high concentrations in the brain might be given to produce a state described at the beginning of the chlorpro-mazine therapy era as "artificial hibernation" (Laborit and Huguenard 1952). Because ScN2a cells can be cured of prions using quinacrine, it is possible that a defined course of therapy might succeed and that lifetime treatment will not be required.

The structure–activity relationship described for the acridine deriva-tives (Fig. 2) provides a platform from which to synthesize many new compounds, some of which will undoubtedly have superior antiprion potencies as well as higher penetration of the BBB. Whether these new compounds will be useful in dissecting the etiologies of such psychotic disorders as the schizophrenias remains unclear. It has been suggested that the schizophrenias might be disorders of protein processing and, thus, are etiologically similar to many other neurodegenerative diseases (Lieberman 1999; Prusiner 2001).

Stereoisomers of Quinacrine

Traditionally, pharmacological potency and selectivity of stereoisomers of various chemicals have been considered for the purpose of drug develop-ment against many diseases (Albert 1960; Brocks and Jamali 1995). Biological activities of stereoisomers can differ enormously (Hutt and Tan 1996). Because the use of a single stereoisomer can be advantageous, there is considerable interest in switching from a racemate to an enantiomeric pure form for many currently available drugs (Tucker 2000).

To measure the antiprion effect of quinacrine enantiomers, we incubated ScN2a cells with individual quinacrine enantiomers (Ryou et al. 2003) or a racemic mixture as described previously (Korth et al. 2001). Chemical synthesis of these enantiomers (Fig. 3A) has been described elsewhere (Craig et al. 1991).

To quantify the antiprion activity of quinacrine enantiomers, we used an ELISA to measure the levels of PrP^{Sc} in ScN2a cells (Fig. 3C). Our ELISA results are similar to those obtained by western blotting (Fig. 3B) (Ryou et al. 2003). The efficiency of PrP^{Sc} reduction by each enantiomer was almost equivalent at concentrations less than 0.2 μM and greater than 0.8 μM. At concentrations between 0.2 and 0.8 μM, (S)-quinacrine was clearly more efficient in reducing PrP^{Sc} levels. At concentrations of 0.3 μM and 0.6 μM, the antiprion activity of (S)-quinacrine was two- to sixfold higher than that of (R)-quinacrine.

We compared the (R,S)-quinacrine (Panorama Compounding) being used for treating patients with prion disease (B. Miller et al., in prep.) to (S)-quinacrine. As expected, (R,S)-quinacrine was less potent than (S)-quinacrine. At a concentration of 0.4 μM, (S)-quinacrine removed almost all PrP^{Sc}, whereas a concentration of 0.5 μM of the (R,S)-quinacrine was required (Fig. 3D). The dose-response curve in ScN2a cells for the (R,S)-quinacrine used for treatment of patients with prion disease was similar but not identical to that of the (R,S)-quinacrine from Sigma Chemical when the western blot in Figure 3B was analyzed by densitometry to quantify the bands.

The EC$_{50}$ values of the quinacrines were calculated from densitometry of western blots and ELISA determinations. Both racemic quinacrines gave similar EC$_{50}$ values for inhibition of PrP^{Sc} that were intermediate of the EC$_{50}$ values of (S)- and (R)-quinacrine. As determined by western blotting followed by densitometry, the EC$_{50}$ values for racemic quinacrine obtained from Sigma and Panorama were 0.36 and 0.35 μM, respectively. EC$_{50}$ values of (S)- and (R)-quinacrine determined by western blotting followed by densitometry were 0.25 and 0.44 μM, respectively; and by ELISA, 0.31 and 0.46 μM, respectively (Fig. 3C).

These findings open a possibility of a more effective quinacrine treatment against prion diseases. Because the (S)-quinacrine enantiomer demonstrates the more potent ability to eliminate PrP^{Sc} in ScN2a cells, it seems reasonable to exclude (R)-quinacrine from treatment in order to minimize the side effects and utilize the maximum tolerated dosage. The stereoselectivity of (S)-quinacrine against prions described in this chapter correlates with results of a pharmacokinetic study, which showed that only (*R*)-quinacrine was found in human urine when racemic quinacrine was administered (Hammick and Chambers 1945). This finding implies that either a receptor for quinacrine binds (*S*)-quinacrine selectively or the pharmacokinetics of each enantiomer are different. One possibility is that the liver more readily metabolizes (*S*)-quinacrine than the (*R*)-enantiomer. In either case, excretion of (*R*)-quinacrine from the body might provide a clue to understanding different antiprion activities of (*S*)- and (*R*)-quinacrine. Not only is the mechanism of quinacrine inhibition of prion propagation not understood, but the mechanisms of quinacrine toxicity are also not known. Whether one or both stereoisomers are responsible for side effects such as liver toxicity, cardiomyopathy, and toxic psychosis remains to be established.

In another plausible but less likely scenario, (*R*)-quinacrine would prove to be superior to (*S*)-quinacrine in treating prion disease if (*R*)-quinacrine is less toxic than (*S*)-quinacrine. In this scenario, inhibition of PrP^{Sc} formation by (*R*)-quinacrine would occur with fewer complications despite the higher dose of drug being administered. To resolve this question, it will be important to develop a mouse model of prion disease in which the quinacrines can be administered orally.

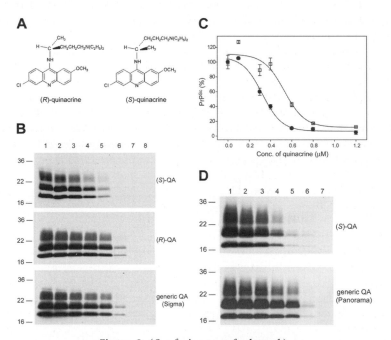

Figure 3. (*See facing page for legend.*)

Bis-acridines

The protein deposition diseases, including the prion diseases, Alzheimer's disease, and Parkinson's disease, involve multimerization and aggregation of a misfolded protein (Bucciantini et al. 2002; Kirkitadze et al. 2002). Oligomerization of PrP seems to be central to the mechanism of PrP^{Sc} formation (Baskakov et al. 2002), suggesting that high local protein concen-

Figure 3. Differential accumulation of PrP^{Sc} in ScN2a cells treated with enantiomers of quinacrine (QA). (*A*) Structures of (*S*)-quinacrine and (*R*)-quinacrine. (*B*) Western blots of PK-digested ScN2a cell lysates incubated with (*S*)-quinacrine, (*R*)-quinacrine, and (*R,S*)-quinacrine from Sigma Chemical. ScN2a cells were cultivated in minimal essential medium (MEM) with 10% fetal calf serum, 1% GlutaMax and 1× penicillin-streptomycin. For treatments, ScN2a cells were plated to ~2–3% confluency. Indicated concentrations (0–0.4 μM) of (*R*)-, (*S*)-, and (*R,S*)-quinacrine were added to the cell cultures. Culture media including quinacrine were changed on alternate days during the 6-day treatment. Cells were harvested in a lysis buffer (20 mM Tris [pH 8.0]; 150 mM NaCl; 0.5% NP-40; 0.5% deoxycholate; sodium salt). Protein contents were quantified using a bicinchoninic acid (BCA) protein quantification kit (Pierce). Molecular weight markers (kD) are shown to the left of each blot. Lanes *1–8* represent quinacrine concentrations of 0, 0.1 μM, 0.2 μM, 0.3 μM, 0.4 μM, 0.6 μM, 0.8 μM, and 1.2 μM, respectively. Cell lysates (100 μg) digested with PK (20 μg/ml) cell for 1 hour at 37°C and then centrifuged at 16,000*g* for 1 hour were separated on a 12% Tris-glycine SDS-PAGE and transferred onto a nitrocellulose membrane (Protran BA-83, Schleicher & Schuell). The membrane was blocked by 5% nonfat milk. The HuM-D13 antibody and goat antihuman Fab antibody–horseradish peroxidase (HRP) conjugate were used as the primary and secondary antibodies, respectively, for the detection of PrP^{Sc}. An enhanced chemiluminescent (ECL, Amersham Pharmacia) substrate for HRP was used for signal development. (*C*) PrP^{Sc} detection by ELISA. PK digestion of cell lysates and centrifugation conditions were the same as described in panel *B*. The pellet was dissolved and denatured for 1 hour using 6 M GdnSCN. The denatured lysates were immobilized on the surface of a 96-well Immulon II ELISA plate (Dynex), and the rest of the surface in the well was blocked by 3% BSA. Detection of PrP^{Sc} was carried out by a 2-hour incubation of HuM-D18 antibody in 1× Tris-buffered saline followed by 1-hour incubation with goat antihuman Fab conjugated to alkaline phosphatase (Pierce). An additional 1-hour incubation utilized 1-step pNPP (Pierce), an alkaline phosphatase substrate for the development of a colored product, which can be measured at 405 nm. SigmaPlot (SPSS Science) was used for mathematical analysis. Mean and standard deviations for individual values were obtained from triplicates of the experiments. (*x*-axis) Concentrations of quinacrine (μM); (*y*-axis) PrP^{Sc} levels (%). (*Filled circles*) (*S*)-quinacrine; (*open squares*) (*R*)-quinacrine. (*D*) Western blots of PK-digested ScN2a cell lysates incubated with (*S*)-quinacrine and (*R,S*)-quinacrine from Panorama Compounding. Lanes *1–7* represent quinacrine concentrations of 0, 0.1 μM, 0.2 μM, 0.3 μM, 0.4 μM, 0.5 μM, and 1.0 μM, respectively. Experimental procedures were the same as described in panel *B*. (Adapted from Ryou et al. 2003.)

trations are achieved in all steps of multimerization leading to an aggregated endpoint. It would be reasonable to expect, therefore, that covalently linked dimers of an agent that reduce PrP^{Sc} load in ScN2a cells may be more potent than its monomeric counterpart. In effect, dimeric molecules could increase the local concentration of the active moiety and hence exploit the multimeric nature of PrP^{Sc}. Although the mechanism by which some acridine compounds (e.g., quinacrine) block PrP^{Sc} replication is unclear, acridine dimers (e.g., 1, 2, or 3, Fig. 4) have been shown to have potent activity in a cell-based model of prion disease. These bis-acridine compounds are characterized by a dimeric motif, of two pendant acridine heterocycles tethered by a linker. Structure–activity data on mono-acridine compounds (e.g., quinacrine) reveal a heavy dependence on the length and composition of the dibasic alkyl substituent *peri* to the ring nitrogen (Fig. 2) (Korth et al. 2001). Bis-acridine compounds have a similar side-chain dependence to their mono-acridine counterparts. In this instance, the side chain serves as a linker to tether the two pendant acridine heterocycles. A study of the antiprion efficacy of these compounds revealed that the bioactivity of this class of compound is dependent on the length, composition, and steric constraints of the linker (May et al. 2003).

A focused library of 21 bis-acridine compounds was prepared according to established literature procedures (Chen et al. 1978). This included compounds tethered via alkyl, polyamine, alkyl-ether, and heterocyclic linkers. A qualitative ScN2a screening assay was used to derive trends in bioactivity across the compound library and to identify potential lead bis-acridines. Additionally, cellular cytotoxicity was used as a secondary screening criterion, as the bis-intercalative cytotoxicity of certain bis-acridine compounds is well characterized. Compounds tethered by polyamine- or alkyl-ether–linked analogs showed improved activity with respect to alkyl-linked bis-acridines. However, certain polyamine-linked analogs were shown to be cytotoxic to uninfected neuroblastoma cells (N2a) at concentrations of 500 nM. To achieve a balance between desirable bioactivity and cytotoxicity, sterically constrained bis-acridine analogs were targeted. It is known that the intercalation of bis-acridines can be mitigated by using rigidified or sterically constrained linkers to tether the acridine heterocycles (Denny et al. 1985). Following this rationale, constrained analogs (e.g., 1, Fig. 4), and *N*-acylated analogs (e.g., 3, Fig. 4), were targeted. In a previous study, compound 1 was shown to be less cytotoxic to MRC-5 fibroblast cells than nonconstrained bis-acridines (Girault et al. 2000). This suggests that inclusion of the constrained linkers disfavors DNA complexation, either through steric hindrance or by restricting the conformational flexibility of the acridine heterocycles, such that bis-intercalation is mitigated. Additionally, the conformational restriction imparted by constrained linkers may be beneficial to activity against PrP^{Sc}. Conformational restriction has been used extensively in drug design to pre-organize ligands into a bioactive conformation. The rationale is that a rigid scaffold can correctly position and orient key structural features without having to undergo an entropically costly conformational rearrangement prior to binding. Thus, restrained ligands can have improved bioactivity with respect to their unrestrained counterparts if the constraints are compatible with the ultimate receptor-relevant conformation. Experimentally it was shown that the conformational constraint imparted by certain heterocyclic linkers (e.g., 1), or bulky linker substituents (e.g., 3), conferred improved activity relative to unconstrained analogs. Additionally, these constrained analogs had reduced cytotoxicity, presumably due to a reduction in the DNA-binding affinity of these compounds.

The initial bioactivity and cytotoxicity screen identified three bis-acridine compounds (1, 2, and 3, Fig. 4) that warranted further characterization. While having efficacious bioactivity against PrP^{Sc}, these compounds were also shown to be nontoxic to uninfected N2a cells at concentrations of 500 nM. Dose–response curves were derived following a 7-day incubation of ScN2a cells with compounds 1, 2, and 3 at concentrations between 5 nM and 500 nM (Fig. 4). Cell lysates were analyzed by ELISA (D. Peretz, pers. comm.; see Fig. 3 legend). The EC$_{50}$ values of 1, 2, and 3 were determined to be 40 nM, 25 nM, and 30 nM, respectively. A ScN2a dose–response curve for compound 1 was also derived by western blot densitometry. Results from both western blot and ELISA methods were consistent. The bioactivity of these compounds is ~10-fold greater relative to the monomeric acridine-based compound, quinacrine (EC$_{50}$ = 300 nM).

The mouse hypothalamic cell line, GT1, can be infected with mouse scrapie and produce stable titers of PrP^{Sc} (ScGT1) (Schatzl et al. 1997). To validate the observed reduction in PrP^{Sc} in ScN2a upon treatment with bis-acridines, ScGT1 cells were also incubated with compound 1. Bis-acridine compound 1 reduced PrP^{Sc} in a dose-dependent manner at concentrations between 100 and 500 nM, as determined by western blot analysis of GT1 cell lysates digested with proteinase K (PK).

Incubation of ScN2a cells with compound 1 at concentrations of 250 nM and 500 nM for 7 days resulted in the complete clearance of protease-resistant PrP^{Sc}, as determined by western blot analysis (Fig. 5). Following this treatment, cells were serially passaged for an additional 21 days in the absence of compound 1. Cell lysates were collected at 7-day intervals and analyzed for PK-resistant PrP^{Sc} by western blot. PrP^{Sc} could not be detect-

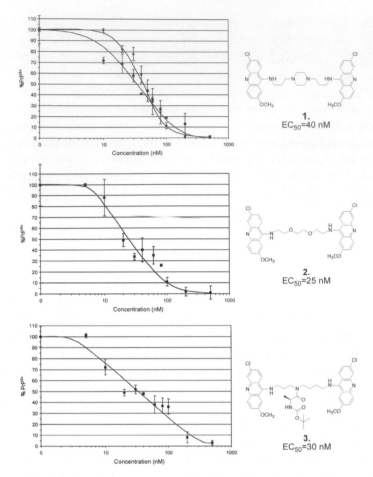

Figure 4. ScN2a dose–response curves for bis-acridine compounds **1**, **2**, and **3**, nanomolar inhibitors of PrP^{Sc} in ScN2a cells. ScN2a cells were incubated for 7 days with compound **1**, **2**, or **3** at concentrations between 5 nM and 500 nM. PK-digested cell lysates were analyzed by ELISA (*squares*) using the procedure of D. Peretz (pers. comm.; see Fig. 3). The EC$_{50}$ values of compounds **1**, **2**, and **3** were 40 nM, 25 nM, and 30 nM, respectively (bars represent standard error from three independent experiments). A dose–response curve for compound **1** was also derived by western blot densitometry of PK-digested cell lysates (*diamond*). Both the western blot and ELISA methods are in good agreement.

ed in cell lysates following the initial treatment with bis-acridine 1, suggesting that the treatment protocol permanently cured ScN2a cells of PrP^{Sc}.

Given the basic nature of bis-acridines, it is possible that these compounds act against PrP^{Sc} by altering lysosomal and/or endosomal com-

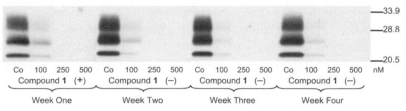

Figure 5. Treatment of ScN2a cells with compound **1** for 7 days clears existing PrP^{Sc}, and cells remain clear of PrP^{Sc} for 21 days after discontinuation of treatment. ScN2a cells were incubated with compound **1** for 7 days at 100 nM, 250 nM, and 500 nM, during which time, media and compound were replaced every 3 days. After this 7-day treatment, the cells were serially passaged for an additional 21 days in the absence of compound **1**. Cell lysates were collected every 7 days and PK-digested prior to western blot analysis (anti-PrP D13). After 21 days treatment-free, protease-resistant PrP^{Sc} could not be detected in cell lysates, suggesting that treatment with bis-acridine **1** for 7 days permanently cured ScN2a cells. Co, control with no compound **1**.

partments. It is thought that the conversion of PrP^{C} to PrP^{Sc} occurs after PrP^{C} has reached the plasma membrane, or along an endocytic pathway to the lysosomes (Caughey et al. 1991a; Taraboulos et al. 1992a). Thus, disrupting the endocytic trafficking of PrP with lysosomotropic agents has been considered as a therapeutic strategy (Doh-ura et al. 2000). The lysosomotropic agent, chloroquine, has previously been shown to reduce PrP^{Sc} load in ScN2a cells (EC$_{50}$ = 2 μM) (Doh-ura et al. 2000; Korth et al. 2001). Additionally, chloroquine has been shown to reduce the protective effects of polyamine dendrimers, presumably by increasing the pH of acidic lysosomal compartments where polyamine dendrimers are thought to exert an effect on PrP^{Sc} (Supattapone et al. 1999a). To determine if the reduction in PrP^{Sc} effected by bis-acridines was synergistic or competitive with lysosomotropic agents, ScN2a cells were co-incubated with compound 1 and chloroquine at 500 nM and 5 μM concentrations, respectively, for 3 days. The cumulative effect of both compounds was observed by western blot analysis of PK-treated cell lysates. PrP^{Sc} was reduced in cells co-incubated with both compounds, versus cells treated with either compound **1** (500 nM) or chloroquine (5 μM) alone.

The challenge in targeting bis-acridine–based therapies is to separate the desirable bioactivity of these compounds from their DNA bis-intercalative cytotoxicity. Therapeutic indices have been used in pharmacology to express the ratio between the efficacious and injurious concentrations of an agent. For agents designed to treat moderate chronic diseases, these ratios overwhelmingly favor efficacy. For most cancer chemotherapeutics, the therapeutic index is unfortunately close to unity. The prion diseases are rapidly and uniformly fatal neurodegenerative diseases, for which no therapy is currently approved. Thus, compounds with modest therapeutic indices, such as bis-acridines, could be considered for clinical use. The data presented, to date, define bis-acridine compounds as a novel class of agents with a potentially acceptable therapeutic index, capable of reducing PrP^{Sc} load in the ScN2a model of prion disease.

BRANCHED POLYAMINES

Branched polyamines can purge ScN2a cells of PrP^{Sc} (Supattapone et al. 1999a). The clearance of PrP^{Sc} was accompanied by a curing of the ScN2a cells of prion infectivity as measured by bioassay in mice (Supattapone et al. 2001).

The ability of branched polyamines to eliminate PrP^{Sc} from ScN2a cells depended on a highly branched structure and a high surface density of primary amino groups (Table 2). The branched polymers investigated included various preparations of polyethyleneimine (PEI), as well as intact polyaminoamide (PAMAM) and polypropylenoimine (PPI) dendrimers and SuperFect, which is a mixture of branched polyamines derived from heat-induced degradation of a PAMAM dendrimer (Tang et al. 1996). Dendrimers are manufactured by a repetitive divergent growth technique, allowing the synthesis of successive, well-defined "generations" of homodisperse structures (Fig. 6). The potency of both PAMAM and PPI dendrimers in eliminating PrP^{Sc} from ScN2a cells increased as the generation level increased: The most potent compounds with respect to eliminating PrP^{Sc} were PAMAM generation 4.0 and PPI generation 2.0, whereas PAMAM generation 1.0 showed little ability to eliminate PrP^{Sc} (Table 2). Similarly, a high-molecular-weight fraction of PEI was more potent than low-molecular-weight PEI.

From the foregoing data, it is clear that for all three branched polyamines tested, increasing molecular size corresponded to an increased potency for eliminating PrP^{Sc}. To determine whether this trend was directly attributable to increased surface density of amino groups on the larger molecules, we also tested PAMAM-OH generation 4.0, a dendrimer that resembles PAMAM generation 4.0 except that hydroxyls replace amino groups on its surface. Unlike PAMAM generation 4.0, PAMAM-OH generation 4.0 did not cause a reduction of PrP^{Sc} levels even at the highest concentration tested (10 mg/ml), establishing that the amino groups are required for the elimination of PrP^{Sc} by PAMAM (Table 2). In an effort to assess the contribution of the branched architecture to the clearing ability of polyamines for PrP^{Sc}, we also tested the linear molecules poly-

Table 2. Clearance and denaturation of PrP^{Sc} by polymer compounds

Compound	Primary NH$_2$ groups	Clearance from ScN2a cells EC$_{50}$ (ng/ml)	PrP^{Sc} in brain homogenates (% control)	PrP^{Sc} in prion rods (% control)
PAMAM 0.0	4	>10,000	100	100
PAMAM 1.0	8	>10,000	100	100
PAMAM 2.0	16	2,000	100	100
PAMAM 3.0	32	400	10	5
PAMAM 4.0	64	80	2	5
PAMAM-OH 4.0	0	>10,000	100	100
PPI 2.0	8	2,000	100	100
PPI 4.0	32	80	5	0
PPI 5.0	64	80	5	0
Low MW PEI		2,000	90	100
Average MW PEI		400	50	20
High MW PEI		80	20	30
Linear PEI		2,000	100	100
Poly-(L)lysine	>500	10,000	100	90
SuperFect		400	50	50

Data from Supattapone et al. (2001). EC$_{50}$, approximate concentration of polymer required to reduce PrP^{Sc} to 50% of control levels in ScN2a cells after exposure for 16 hours. PrP^{Sc} levels were measured by densitometry of western blot signals.

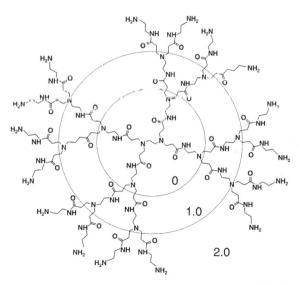

Figure 6. Schematic diagram of PAMAM, generation 2.0 (ethylene diamine core). Successive full generations are indicated by concentric circles. (Reprinted, with permission, from Supattapone et al. 1999a [copyright National Academy of Sciences].)

(L)lysine and linear PEI. Both of these linear compounds were less potent than a preparation of branched PEI with similar average molecular weight (Table 2), establishing that a branched molecular architecture optimizes the ability of polyamines to eliminate PrP^{Sc}, presumably because the branched structures achieve a higher density of surface amino groups.

The most potent compounds identified were generation 4.0 PAMAM and PPI dendrimers. Exposure of ScN2a cells to 3 mg/ml PPI generation 4.0 for 28 days not only reduced PrP^{Sc} to a level undetectable by western blot, but also eradicated prion infectivity as determined by bioassay in Tg(MoPrP-A)4053 mice (Supattapone et al. 2001).

Polyamines Act within Lysosomes and Acidic Endosomes

A series of experiments suggested that branched polyamines cure prion-infected cells by disrupting the structure of PrP^{Sc} within acidic compartments of the cell such as lysosomes and acidic endosomes (Supattapone et al. 2001). The potent activity of branched polyamines in rapidly clearing scrapie prions from cultured ScN2a cells led us to explore the mechanism by which these drugs might act. All of the compounds that effect removal of PrP^{Sc} from ScN2a cells are known to traffic through endosomes (Haensler and Szoka 1993; Boussif et al. 1995), and because PrP^{C} is converted into PrP^{Sc} in caveolae-like domains (CLDs) or rafts (Gorodinsky and Harris 1995; Taraboulos et al. 1995; Vey et al. 1996; Kaneko et al. 1997a), which are then internalized through the endocytic pathway (Caughey et al. 1991a; Borchelt et al. 1992; Peters et al. 2003), we reasoned that polyamines might act upon PrP^{Sc} in endosomes or lysosomes. To test this hypothesis, we first investigated the effect of pretreatment with the lysomotropic agents chloroquine and NH$_4$Cl on the ability of polyamines to eliminate PrP^{Sc}. These lysomotropic agents alkalinize endosomes and have no effect on PrP^{Sc} levels when administered to ScN2a cells (Taraboulos et al. 1992a). In our experiments, 100 μM chloroquine, but not 30 μM NH$_4$Cl, blocked the ability of PEI to eliminate PrP^{Sc} (Fig. 7A). Similar results were obtained with SuperFect (data not shown). Although the failure of NH$_4$Cl to affect PrP^{Sc} levels is not easily explained, the ability of chloroquine to attenuate the ability of branched polyamines to remove PrP^{Sc} is consistent with the notion that these agents act in either endosomes or lysosomes.

Encouraged by these results, we considered the possibility that in an acidic environment, branched polyamines, by indirectly interacting either with PrP^{Sc} or with another cellular component, could cause PrP^{Sc} to become susceptible to hydrolases present in the endosome/lysosome. In an effort to test this hypothesis, we developed an in vitro degradation

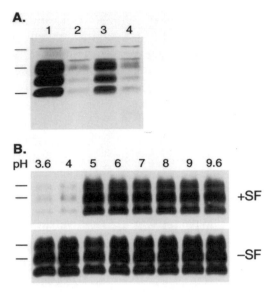

Figure 7. Acidic conditions favor polyamine removal of PrP^{Sc}. (*A*) ScN2a cells were treated with: (lane *1*) control media; (lane *2*) 7.5 µg/ml PEI (average molecular weight ~60,000 D); (lane *3*) PEI plus 100 µM chloroquine; (lane *4*) PEI plus 30 mM NH$_4$Cl. Chloroquine and NH$_4$Cl were added 1 hour prior to addition of PEI. Cells were harvested 16 hours after addition of PEI. All samples shown were subjected to limited proteolysis to measure PrP^{Sc}. Apparent molecular weights based on migration of protein standards are 38 kD, 26 kD, and 15 kD. (*B*) In vitro mixture of crude mouse brain homogenates with SuperFect under a range of pH conditions was performed as described in Supattapone et al. (1999a). pH values are denoted above the lanes. Addition of 60 µg/ml SuperFect denoted as "+SF" and control with no additon as "−SF". All samples shown were subjected to limited proteolysis to measure PrP^{Sc}. Apparent molecular weights based on migration of protein standards are 30 kD and 27 kD. (Reprinted, with permission, from Supattapone et al. 1999a [copyright National Academy of Sciences].)

assay to evaluate the effect of pH on the ability of polyamines to render PrP^{Sc} sensitive to protease. Crude homogenates of scrapie-infected mouse brain were exposed to a broad range of pH values in the presence or absence of SuperFect and then treated with PK prior to western blotting. Whereas PrP^{Sc} remained resistant to protease hydrolysis throughout the pH range (pH 3.6 to pH 9.6) in the absence of SuperFect, addition of the branched polyamine at pH ≤4.0 caused PrP^{Sc} to become almost completely degraded by protease (Fig. 7B).

The dramatic effect of polyamine addition showing that clearance in vitro is optimized below pH 4.0 (Fig. 7B) is consistent with the hypothesis that polyamines act on PrP^{Sc} in an acidic compartment. To establish that the in vitro degradation assay (Fig. 7B) is a valid approximation of the mechanism by which branched polyamines enhance the clearance of PrP^{Sc} from cultured cells, a structure–activity analysis with several of the compounds tested in cultured cells was conducted (Table 2). An excellent correlation between the clearance of PrP^{Sc} in cultured ScN2a cells (Table 2) and the ability to render PrP^{Sc} susceptible to protease at acidic pH in vitro was found. Notably, PAMAM-OH generation 4.0 failed to render PrP^{Sc} susceptible to protease, whereas PAMAM generation 4.0 and PEI exhibited an even stronger activity than SuperFect in vitro, as expected from their observed potency in cultured ScN2a cells (Table 2).

Having established that branched polyamines reduce prion infectivity, we sought to identify the mechanism by which these compounds eliminate PrP^{Sc}. Our first objective was to determine the molecular target of branched polyamines. Previously, we developed an in vitro assay that was used to show that these compounds could render PrP^{Sc} protease susceptible when mixed directly with crude brain homogenates (Fig. 7) (Supattapone et al. 1999a). We performed a similar assay with PrP 27-30 purified from mouse brains infected with RML prions to determine whether or not the molecular target of branched polyamines was present in this highly purified preparation. PrP 27-30 in purified preparations of RML prions was rendered protease-sensitive by branched polyamines with a similar acidic pH optimum and structure–activity profile (Table 2), as previously obtained in crude brain homogenates (Table 2) (Supattapone et al. 1999a). Treatment of purified prions with branched polyamines in vitro also diminished infectivity. We incubated 15 mg/ml RML prion rods in 50 mM sodium acetate (pH 3.0); 1% NP-40; 1 mg/ml bovine serum albumin (BSA) for 2 hours at 37°C with or without 60 mg/ml PPI generation 4.0, and measured prion infectivity using an incubation-time assay in Tg(MoPrP)4053 mice. PPI treatment reduced prion infectivity from 10^7 ID$_{50}$ U/ml to 10^5 ID$_{50}$ U/ml (data not shown).

Branched Polyamines as Therapeutic Agents

Branched polyamines appear to be the first class of compounds to eliminate prion infectivity from living cells that were chronically infected. Polyene antibiotics, anionic dyes, sulfated dextrans, anthracylines, porphyrins, phthalocyanines, dapsone, and a synthetic β-breaker peptide all prolong scrapie incubation times in vivo, but only if administered prophylactically (Farquhar and Dickinson 1986; McKenzie et al. 1994; Ingrosso et al. 1995; Tagliavini et al. 1997; Manuelidis et al. 1998; Adjou et al. 1999; Farquhar et al. 1999; Priola et al. 2000; Soto et al. 2000).

The ability of branched polyamines to cure an established prion infection in cells suggested that these compounds might also reverse disease progression in animals. However, our inability to package branched polyamines in a form that crosses the BBB has prevented us from treating animals with prion disease. When we tried to deliver branched polyamines directly to the CSF through an intraventricular reservoir in mice, the dendrimers were found to not penetrate into the parenchyma of the brain. Instead, the polyamines were found bound to periventricular cells even though continuous intraventricular infusion of PPI generation 4.0 was tolerated by FVB mice up to a total dose of ~0.5 mg/animal (data not shown). Neutralizing the polyamines with glutamic acid did not improve the penetration of dendrimers into the parenchyma, but it did diminish the frequency of seizures.

Mechanism of Dendrimer Action

Several lines of evidence suggest that branched polyamines render PrP^{Sc} molecules protease-sensitive by dissociating PrP^{Sc} aggregates: (1) RML prion rods treated in vitro with PPI become disaggregated, as judged by electron microscopy (Supattapone et al. 2001); (2) prion strains resistant to branched polyamines in vitro appear to be more amyloidogenic than polyamine-susceptible strains, as judged by neuropathology (Fraser and Bruce 1973, 1983; Bruce et al. 1976; Carlson et al. 1994); (3) the ability of branched polyamines to render PrP^{Sc} protease-sensitive in vitro is enhanced by conditions that favor PrP^{Sc} disaggregation. These conditions include acidic pH and the presence of urea (Fig. 7A).

Theoretically, it is possible that the mechanism by which branched polyamines remove PrP^{Sc} and prion infectivity from ScN2a cells does not relate to the ability of these compounds to disaggregate prions in vitro. However, this is unlikely because the relative potency of 14 different polyamines in eliminating PrP^{Sc} from ScN2a cells correlates with the relative ability of these same compounds to render PrP^{Sc} sensitive to proteolysis in crude brain homogenates and purified preparations of RML PrP 27-30 in vitro (Supattapone et al. 1999a). The structure–activity profile obtained from these studies indicates that polyamines become more potent at eliminating PrP^{Sc} as they become more branched and possess more surface primary amines (Table 2). With PPI dendrimers, this effect reached a plateau at the fourth generation; PPI generation 5.0 is no more potent than PPI generation 4.0 at either removing PrP^{Sc} from cells or rendering PrP^{Sc} protease-sensitive in vitro. Homodisperse, uniform PPI and PAMAM dendrimers were more potent than the heterogeneous preparations of PEI or SuperFect, a heat-fractured dendrimer.

We determined that the process by which PPI renders PrP^{Sc} protease-sensitive in vitro was not catalytic. Instead, this process appeared to require a fixed stoichiometric ratio of PPI to PrP^{Sc} of ~1:5. How could PPI denature prion rods stoichiometrically? One possible explanation is that individual amino groups on the surface of PPI might bind to PrP^{Sc} monomers or oligomers that exist in equilibrium with a large aggregate under acidic conditions. The dendrimer might then pry bound PrP^{Sc} molecules apart from the aggregate and/or prevent such molecules from reaggregating.

Another possible mechanism of polyamine-induced prion clearance from ScN2a cells is that branched polyamines facilitate PrP^{Sc} transport from the plasma membrane through the endocytic pathway into secondary lysosomes. Several lines of evidence indicate that the cellular site of action of branched polyamines is secondary lysosomes: First, fluorescein-tagged PPI and PrP^{Sc} both localize to lysosomes (Caughey et al. 1991a; Taraboulos et al. 1992b), and second, the pH optimum of PrP^{Sc} denaturation in vitro is <5.0 (Fig. 7). When cultured cells were studied with fluorescent acidotropic pH measurement dyes, secondary lysosomes were the most acidic cellular compartment detected, with approximate pH values of 4.4 to 4.5 (Anderson and Orci 1988; Diwu et al. 1999). Third, the lysosomotropic agent chloroquine attenuates the ability of branched polyamines to eliminate PrP^{Sc} from ScN2a cells (Supattapone et al. 1999a). Our studies raised the possibility that lysosomal proteases normally degrade PrP^{Sc} in prion-infected cells at a slow rate, and that polyamines accelerate this process by denaturing PrP^{Sc} (Supattapone et al. 2001). As described below, investigations with recombinant (rec) anti-PrP antibody fragments (Fabs) also argue that cells degrade PrP^{Sc} at a slow rate, raising the possibility that PrP^{Sc} is an alternatively folded protein normally present at levels below that detectable by bioassay (Chapter 3) (Peretz et al. 2001).

Other Applications of Branched Polyamines

Beyond their potential use as therapeutic agents and research tools, branched polyamines might also be useful as prion strain-typing reagents and/or prion decontaminants. Presently, typing of prion strains is time-consuming and requires the inoculation of samples into several strains of inbred animals to obtain incubation time and neuropathology profiles (Bruce et al. 1976; Dickinson et al. 1984). Antibody-based PrP^{Sc} conformational stability assays have been developed that are able to distinguish prion strains (Safar et al. 1998; Peretz et al. 2001, 2002). Different species and strains of prions displayed varying susceptibilities to branched polyamine–induced denaturation in vitro (Supattapone et al. 2001). These results suggest that a polyamine-based, in vitro, protease-digestion assay could, in principle, be used as a simple and rapid diagnostic method for prion strain typing. One practical application arising from these results is the possibility that a polyamine-based assay could be used to distinguish between BSE prions and natural scrapie prions in flocks of domestic sheep.

It is very difficult to remove prions from skin, clothes, surgical instruments, foodstuffs, and surfaces (Budka et al. 1995). Standard prion decontamination requires either prolonged autoclaving or exposure to harsh protein denaturants such as 1 N NaOH or 6 M guanidine thiocyanate (GdnSCN) (Prusiner et al. 1981). Dendrimers are nontoxic and relatively inexpensive. These compounds might prove suitable as a disinfecting reagent to limit the commercial and iatrogenic spread of prion diseases.

PrP ANTIBODIES

The difficulty in producing anti-PrP antibodies suggested that animals might be tolerant to PrP because PrP^C is expressed in embryos (Bendheim et al. 1984; Barry et al. 1986; Barry and Prusiner 1986). To test this hypothesis, PrP-ablated ($Prnp^{0/0}$) mice were immunized with purified mouse (Mo) PrP 27-30, the protease-resistant core of PrP^{Sc}, and the antiserum tested; the robust antibody response to PrP gave support to this proposal (Prusiner et al. 1993). From immunized $Prnp^{0/0}$ mice, a panel of recPrP-specific Fabs was obtained corresponding to five major specificities (Williamson et al. 1996). All of the recFabs recognized PrP^{Sc} denatured with 3 M GdnSCN both in vitro and in situ, as well as recPrP spanning residues 90–231, produced in *Escherichia coli* (Mehlhorn et al. 1996). All but one of the recFabs also bound to PrP^C on the cell surface as determined by flow cytometry. Disappointingly, however, native PrP 27-30 was not detected by any of these antibodies prior to treatment with GdnSCN. An explanation for this finding may lie in the physicochemical properties and antigenicity of PrP 27-30 in prion rods. During purification of PrP^{Sc}, brain homogenate was subjected to digestion with PK and extraction with 0.5% Sarkosyl. Purified PrP 27-30 polymerizes into rod-shaped structures of 100 nm to 200 nm in length that are composed of as many as 1000 PrP 27-30 molecules (Prusiner et al. 1983; McKinley et al. 1991). This aggregation may reduce effective epitope concentrations, thereby hindering efficient immunization and selection of specific antibody phage. Epitope concentration is known to be a critical factor in effective selection from antibody phage display libraries (Parren et al. 1996).

Table 3. Epitopes recognized and dissociation constants for the binding of PrP-specific Fabs to recombinant SHaPrP(29–231) refolded into an α-helical conformation

Fab	Epitope	K_d (µg/ml)	K_d (nM)
D13	95–103	0.16 ± 0.03	3.3 ± 0.5
D18	132–156	0.07 ± 0.03	1.6 ± 0.6
R72	151–162	no binding	no binding
R1	220–231	0.09 ± 0.04	1.7 ± 0.8
R2	225–231	0.11 ± 0.02	2.2 ± 0.5
E123	29–37	3.45 ± 0.03	69 ± 0.5
E149	72–86	8.5 ± 0.03	170 ± 0.5

Binding constants were determined by surface plasmon resonance.

To study both the structural transition of PrP^C into PrP^{Sc} and antibody inhibition of prion propagation, a panel of recPrP-specific Fabs were used, consisting of D13, D18, R1, R2, E123, E149, and R72 (Williamson et al. 1998; Peretz et al. 1997; Leclerc et al. 2001). The binding epitopes and affinity constants of the antibodies for recPrP are shown in Table 3. Fab R72 does not recognize PrP in surface plasmon resonance (SPR) or on the cell surface, but binds to PrP^C coated onto the surface of ELISA wells (Peretz et al. 1997).

In studies on the structural transition of PrP^C into PrP^{Sc}, the recFabs were used to identify linear epitopes on PrP^C and PrP 27-30 employing ELISA and immunoprecipitation. An epitope at the carboxyl terminus was accessible in both PrP^C and PrP 27-30; in contrast, epitopes toward the amino-terminal region (residues 90–120) were accessible in PrP^C but largely cryptic in PrP 27-30 (Peretz et al. 1997). Denaturation of PrP 27-30 exposed the epitopes of the amino-terminal domain. These findings argued that the major conformational change underlying PrP^{Sc} formation occurs within the amino-terminal segment of PrP 27-30. This topologic assignment provided localization for spectroscopic measurements, demonstrating that the conversion of PrP^C to PrP^{Sc} involves a major conformational transition. PrP^C has a high α-helical content, some of which is converted into β-sheet when PrP^{Sc} is formed (Pan et al. 1993; Safar et al. 1993; Pergami et al. 1996). PrP 27-30 also has a high β-sheet content (Caughey et al. 1991b; Gasset et al. 1993) but in contrast to PrP^{Sc}, PrP 27-30 assembles into amyloid polymers (Prusiner et al. 1983; McKinley et al. 1991).

The studies with recFabs overcame two problems: First, expression of PrP^C in mammalian cells is low, making purification of sufficient amounts for detailed structural studies impractical, and, second, PrP^{Sc} forms insoluble amorphous aggregates upon purification (Scott et al. 1988; McKinley et al. 1991). Nuclear magnetic resonance (NMR) studies of *E. coli*-derived PrP refolded into PrP^C-like molecules have given a wealth of structural information (Chapter 5) (Mehlhorn et al. 1996; Riek et al. 1996; Donne et al. 1997; James et al. 1997).

To investigate the inhibition of PrP^{Sc} formation by antibodies, a range of concentrations for each of the recFabs (Table 3) was added to ScN2a cultures for a period of 7 days (Peretz et al. 2001). After this time, cells were harvested, and the level of PrP^{Sc} in the culture was analyzed by immunoblotting. The level of PrP^{Sc} in cells treated with Fabs D13, D18, R1, and R2, compared with nontreated cells, was dramatically reduced in a dose-dependent manner. By this analysis, Fabs D18 and D13 appear to be approximately equally effective in blocking PrP^{Sc} formation, having EC$_{50}$ values of 0.45 µg/ml (9 nM) and 0.6 µg/ml (12 nM), respectively. Fabs R1 and R2 were slightly less efficient, with EC$_{50}$ values of 2.5 µg/ml (50 nM) and 2.0 µg/ml (40 nM), respectively. In contrast, 7-day treatment with Fabs E123, E149, or R72, or 5-day treatment with polyclonal IgG recognizing both transmembrane and GPI-anchored forms of mouse N-CAM (Chuong et al. 1982), did not reduce the level of PrP^{Sc} in ScN2a cultures even when these antibodies were used at high concentrations. It is noteworthy that N-CAM has been shown to interact with PrP^C on the cell surface, but the significance of these interactions is uncertain (Schmitt-Ulms et al. 2001). During these experiments, the levels of PrP^C and glyceraldehyde-3-phosphate dehydrogenase in antibody-treated and untreated cells were found to be invariant, indicating that the PrP-specific antibodies used produced no cytotoxic effects that may have indirectly compromised the production of PrP^{Sc}.

To determine whether PrP^{Sc} remained undetectable after removal of the PrP-specific antibody, ScN2a cells were independently passaged for a minimum of 7 days in the presence of 10 µg/ml of each of the recFabs. The recFabs were then removed from the culture, and the cells were passaged for an additional period in Fab-free medium, after which time the level of PrP^{Sc} was measured. PrP^{Sc} concentrations in ScN2a cells passaged for 7 days in the presence of Fab D18 were reduced to undetectable levels but returned to ~50% of the level of an untreated control culture after 7 additional days of growth in the absence of D18. However, if cells were cultured for a 14-day period in the presence of Fab D18, PrP^{Sc} remained at undetectable levels after 28 additional days of culture in antibody-free medium. Similarly, when prion-infected cells were treated with Fab D13 for 21 consecutive days, followed by 7 days of growth in media without antibody, no PrP^{Sc} could be detected. After 7 additional days in culture without Fab, PrP^{Sc} increased back to 5% of the level found in untreated control culture. If, however, cells were subjected to 28 consecutive days of treatment with Fab D13, PrP^{Sc} remained below the level of detection after 28 days of culture without recFab. When employed at a concentration of 10 µg/ml, neither Fab R1 nor Fab R2 was sufficiently potent to prevent the reemergence of PrP^{Sc} in the culture after the antibody was removed. Fab R72 had no impact on the level of PrP^{Sc} in the culture after either 21 consecutive days or 63 consecutive days (data not shown) of treatment. As a second measure of prion titer, we performed bioassays in which Swiss CD-1 mice were inoculated with antibody-treated (10 µg/ml) and untreated ScN2a cells. Mice inoculated intracerebrally with D18-, D13-, or R2-treated cells were disease-free after a period of 350 days, whereas mice inoculated with untreated or R72-treated cells had a mean incubation time to disease of 169 days and 165 days, respectively. The prolonged incubation times correspond to a reduction of $>10^3$ in the infectious prion titer in treated cells (Butler et al. 1988).

Results similar to those described here with Fab D18 were obtained with the monoclonal antibody (mAb) 6H4 (Enari et al. 2001). The 6H4 mAb was added to the culture medium of ScN2a cells, and the levels of PrP^{Sc} were measured. Like Fab D18, 6H4 decreased the levels of PrP^{Sc}.

Clearance of PrP^{Sc} from Cells

An important question posed by the experiments described concerns the removal of preexisting PrP^{Sc} from the ScN2a cells when antibody treatment was initiated. The observations from those experiments might be taken to indicate that PrP^{Sc} levels rapidly diminish when antibodies are added to the cultures. Indeed, data reported in other studies describing the inhibition of prion propagation (Caughey and Raymond 1993; Caughey et al. 1998; Chabry et al. 1999; Supattapone et al. 1999a; Perrier et al. 2000) have been interpreted in this way. However, such an interpretation does not consider the dynamic expansion of ScN2a cell populations in culture. If an antibody or other reagent curtails the formation of nascent PrP^{Sc} molecules, then each successive round of cell division may serve to dilute the effective concentration, but not necessarily the total amount, of residual PrP^{Sc} within the culture, thus creating a potentially erroneous impression of PrP^{Sc} clearance. An accurate calibration of the rate with which PrP^{Sc} is purged from ScN2a cultures needs to account for any increase in cell population and commensurate reduction in PrP^{Sc} concentration that has taken place over the course of the experiment.

To analyze the kinetics of prion clearance, ScN2a cells were grown in the presence of 10 μg/ml PrP-specific Fab (Peretz et al. 2001). Cells were harvested after a period of 1, 2, 3, and 4 days of antibody treatment, and the total mass of cell protein was determined in each case as a measure of cell number. The PrP^{Sc} concentrations in Fab-treated and untreated cells at these time points were determined by immunoblotting (Fig. 8A). Total PrP^{Sc} in the culture at each time point was then calculated by factoring in the total cell mass in each case. When these data were plotted against the duration of antibody treatment, differences in the efficacy with which individual Fabs resolved prion infection became apparent (Fig. 8B).

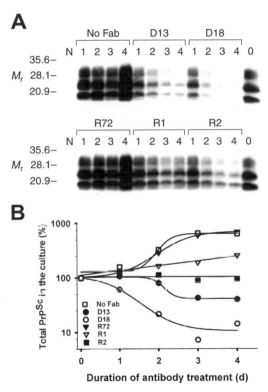

Figure 8. Time course of antibody-mediated PrP^{Sc} clearance. (*A*) The level of PrP^{Sc} in ScN2a cells grown for 1 days, 2 days, 3 days, or 4 days in the presence of PrP-specific Fabs (10 μg/ml) was determined by immunoblotting. Lane *N* indicates N2a cells treated with PK. The apparent molecular weights (*M*$_r$) of the migrated fragments are shown in kD. (*B*) The effect of antibody treatment on the total amount of PrP^{Sc} in ScN2a cell cultures. The data represent the mean of three experiments. (Reprinted, with permission, from Peretz et al. 2001 [copyright Nature Publishing Group].)

Fab D18 was found to be the most effective antibody. The time taken from the initial treatment with D18 to eliminate 50% of PrP^{Sc} from the cells ($t_{1/2}$) was 28 hours (Peretz et al. 2001). The $t_{1/2}$ of PrP^{Sc} in ScN2a cells is thought to exceed 24 hours (Borchelt et al. 1990), suggesting that at a concentration of 10 µg/ml, Fab D18 is able to completely abolish prion propagation and that preexisting PrP^{Sc} is subsequently eliminated from the cells. This finding indicates that a certain amount of PrP^{Sc} is continuously expunged from ScN2a cultures through normal degradation pathways. Fab D13 was the next most potent antibody, also lowering the level of PrP^{Sc} in the culture, but to a lesser extent than Fab D18, indicating that there may be a minimal level of residual PrP^{Sc} synthesis in the presence of this Fab. Fabs R1 and R2, although clearly reducing the rate of prion propagation in ScN2a cells, were not sufficiently effective to yield a reduction in the overall quantity of PrP^{Sc} present in the culture. In untreated cultures, or cultures treated with Fab R72, prion propagation remained unaffected, and PrP^{Sc} levels increased concomitantly with growth in ScN2a cell population.

These findings showed not only that antibody can inhibit prion replication in infected cell cultures, but also that the efficiency of this process varies dramatically between individual antibodies (Peretz et al. 2001). Mechanistically, inhibition of prion replication can be explained by antibody binding to PrP^C molecules on the cell surface, which thereby hinders docking of the PrP^{Sc} template or a cofactor critical for conversion of PrP^C to PrP^{Sc}. In support of this hypothesis, Fab D18, which was the most effective antibody, was distinguished by its capacity to bind a significantly greater number of cell-surface PrP^C molecules than Fabs D13, R1, or R2. In contrast, Fab R72, which had no effect on prion propagation, failed to recognize cell-surface PrP^C even at a concentration of 20 µg/ml. Similarly, Fabs E123 and E149 reacted weakly with MoPrP^C and did not diminish PrP^{Sc} levels. From these findings, it was concluded that the fraction of total cell-surface PrP^C occupied by a given antibody is a key determinant of the inhibitory potency of that antibody. The data imply that discrete PrP^C populations, distinguishable on the basis of immunoreactivity, exist on the surface of ScN2a cells. These populations may be conformationally distinct from one another, may differ in their posttranslational modification, or may result from the interaction of a proportion of PrP^C with other cell-surface components, possibly including other PrP molecules.

To address whether the region of PrP bound by each antibody was of intrinsic importance to its inhibitory potency, we compared inhibition at conditions in which equal amounts of two different Fabs were bound to the ScN2a cell surface. For example, when used at concentrations of 0.6

μg/ml, equivalent amounts of Fabs D18 and D13 were bound to the cell surface, but D18 inhibited prion replication much more efficiently. Similarly, at concentrations of 0.6 μg/ml and 2.5 μg/ml, respectively, Fabs D18 and R1 bound equivalently to ScN2a cells, but D18 was clearly more effective in reducing the level of PrP^{Sc} in the culture. Finally, at a concentration of 2.5 μg/ml, Fabs D13 and R1 bound equivalently to the cell surface, but D13 more actively reduced PrP^{Sc} synthesis.

The demonstration that the region of PrP^C bound by a given antibody is a critical determinant of its inhibitory capacity provided insight into the specific machinery of prion propagation (Peretz et al. 2001). The known binding epitope of Fab D18 spans PrP residues 132–156 (Williamson et al. 1998) and incorporates helix A (residues 145–155) of PrP^C (Fig. 9). Spatially, this sequence is positioned on the opposite face of the protein from residues Q167, Q171, T214, and Q218, which are hypothesized to participate in binding an auxiliary molecule essential to prion propagation (Kaneko et al. 1997b; Zulianello et al. 2000). This finding argues that

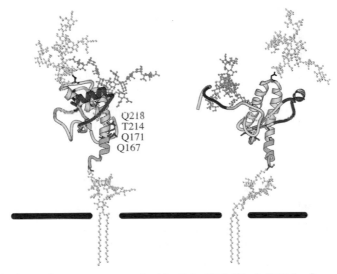

Figure 9. Regions of sequence recognized by Fabs D13 (*blue*), D18 (*red*), and R1 and R2 (*green*) superimposed onto two views of the structure of recPrP(90–231) determined by NMR (James et al. 1997). Carbohydrate moieties linked to Asn-180 and Asn-196 are shown in orange and yellow, respectively (Petrescu et al. 1999; Rudd et al. 2001). The carboxy-terminal GPI anchor is shown in cyan extending into the cell membrane (represented in *black*). Side chains of residues Q167, Q171, T214, and Q218, which are proposed to bind to a cellular cofactor critical to prion propagation, are included (*purple*). (Reprinted, with permission, from Peretz et al. 2001 [copyright Nature Publishing Group].)

D18 operates mechanistically by directly blocking or modifying PrP^C interaction with PrP^{Sc}, rather than by inhibiting the binding of a cofactor. Numerous other reports also identify the 132–156 region of the protein as critical to prion synthesis and interspecies transmission (Scott et al. 1993; Kocisko et al. 1995; Priola and Chesebro 1995; Priola et al. 2001), although prion propagation can proceed in the absence of sequence between residues 140 and 175 (Supattapone et al. 1999b), indicating that helix A of PrP^C is not a prerequisite for PrP^{Sc} binding. Together, these studies identify residues 132–140 of PrP^C as a logical target for antiprion drug development.

In contrast to D18, Fabs R1 and R2 are relatively poor inhibitors of prion propagation. Both these antibodies bind the extreme carboxyl terminus of PrP, much closer to the Q167, Q171, T214, and Q218 cluster and distant from the D18 epitope (Fig. 9). These antibodies may have little effect on PrP^{Sc} binding, but could be in direct competition with the putative cellular cofactor for binding to PrP^C. The lower cell-surface binding we observe for Fabs R1 and R2 compared to D18 may be explained in part by this type of competition, but the proximity of this epitope to the GPI anchor and cell surface may also be of importance in this respect.

Little insight into the mechanism of PrP^{Sc}–PrP^C interactions could be discerned from the inhibitory activity of Fab D13, because its epitope (residues 95–103) is found in the unstructured portion of the protein. The location of this region relative to the structured carboxy-terminal domain has not been accurately defined. Notably, however, this portion of PrP^C is thought to undergo substantial conformational rearrangement as prion infectivity is acquired (Peretz et al. 1997; Leclerc et al. 2001). It is tempting to speculate that some of the inhibitory potency of D13 may derive from its ability to preserve a PrP^C-like conformation in this segment of the protein. Significantly, with the exception of Fabs R1 and R2, none of the antibodies used in this study compete with each other for binding to cell-surface PrP^C, which suggests that the antibodies may be used in combination to achieve a maximum inhibitory effect.

Anti-PrP Antibodies as Therapeutics for Prion Disease

Whether anti-PrP antibodies will show any efficacy therapeutically remains to be established. In one study, Tg mice expressing anti-PrP antibodies were found to have prolonged incubation times when prions were inoculated intraperitoneally (Heppner et al. 2001). Not surprisingly, no prolongation of the incubation time was found in these Tg mice inoculated intracerebrally with prions. Antibodies and antibody fragments do not readily cross the BBB; only when the BBB is disrupted do such large molecules enter the CNS.

In Tg mice producing Aβ deposits in the brain, immunization with the Aβ peptide reduced the levels of Aβ deposition dramatically (Schenk et al. 1999). Aβ deposition was also reduced in the Tg mice by the peripheral administration of antibodies (Bard et al. 2000), but subsequent studies showed that the removal of Aβ probably occurred by lowering the blood levels of this peptide, which induced an efflux of Aβ from the brain to the circulation (DeMattos et al. 2001, 2002). Attempts to treat patients with Alzheimer's disease by Aβ immunization have been unsuccessful, to date (Birmingham and Frantz 2002).

Whether advances in bioengineering in the future will make it practical to deliver antibodies or antibody fragments to the CNS is unknown. Based on the studies described above, the specificity and efficacy of Fab D18 make it an ideal pharmaceutical if it can be delivered throughout the CNS of a human or animal with prion disease.

POLYNUCLEOTIDE THERAPEUTICS

Although infectious prions are devoid of any functional nucleic acid, removal of all nucleic acids from purified preparations of prions was exceedingly difficult (Chapters 2 and 17) (Kellings et al. 1992, 1994). This observation raised the possibility that nucleic acids might preferentially bind to PrP^{Sc}. Subsequent studies demonstrated that DNA and RNA bind to the amino terminus of PrP^{C}, specifically in the region of residues 23–44 (Muramoto et al. 1996; Nandi 1998; Nandi and Leclerc 1999; Barron et al. 2001; Cordeiro et al. 2001; Gabus et al. 2001a,b; Nandi and Sizaret 2001). These experiments used either recombinant protein or peptides from various regions of PrP to demonstrate binding of DNA and RNA and their influence on PrP amyloid formation, which reflects the transformation of this unstructured region (Donne et al. 1997) to one rich in β-structure. Whether this β-structure is composed of typical planar β-sheets or β-helices remains to be determined (Wille et al. 2002).

Phosphothioate Oligonucleotides

Intrigued by this apparent interaction between nucleic acids and PrP, the effect of exogenous nucleic acid on prion formation in ScN2a cells was examined (M.V. Karpuj et al., in prep.). Because both DNA and RNA are readily degraded by nucleases, phosphorothioate oligonucleotides (PS-DNA) that are relatively resistant to nucleases were used (Akhtar et al. 1991; Hoke et al. 1991).

When ScN2a cells were treated with a 22-mer PS-DNA, the levels of both PrP^C and PrP^{Sc} diminished without affecting other proteins in the cells. This phenomenon was not specific to a particular sequence of PS-DNA. The degradation of PrP^C mediated by PS-DNA was similar whether or not the cells were infected with prions. PS-DNA–mediated PrP^C degradation seems likely to occur at the cell surface because recFabs can prevent the destruction of PrP^C. Although the Fab D18 inhibited PrP^{Sc} formation and may have enhanced PrP^{Sc} clearance, there are no data that argue for Fab-induced degradation of PrP^C. Additionally, PS-DNA added to cultures and PrP^C colocalized on the cell surface (Peretz et al. 2001). Thus, the degradation of PrP^C occurs not simply from an interaction with specific regions of PrP, but must be due to some unique property of PS-DNA.

The removal of PrP^{Sc} was concentration-dependent, with an EC_{50} of 0.045 μM. Treating ScN2a cells for 14 days with PS-DNA cured cells of PrP^{Sc} as determined by immunoassay and bioassay (Table 4). The removal of PrP^{Sc} was dependent on the length of the oligonucleotide, dose, and incubation time. As little as 1 μM of the 22-mer oligonucleotide for 48 hours was sufficient to remove PrP^{Sc} in ScN2a cells. Two weeks of incubation with the 1 μM of oligonucleotides was sufficient to remove PrP^{Sc} with no evidence for new PrP^{Sc} formation although newly synthesized PrP^C was present. The fact that the oligonucleotides do not completely abolish the presence of PrP^C in N2a cells even at high doses and even after prolonged treatment indicates that some population of PrP^C is resistant to PS-DNA–mediated degradation. Both PrP^C and PrP^{Sc} were reduced after 14 hours of incubation with 1 μM of oligonucleotides, and no PrP^{Sc}

Table 4. Bioassays of prions in ScN2a cells treated with PS-DNA

Host	Inoculum	Length of experiment	Incubation period (mean days ± S.D.)	n/n_0[a]
CD-1	ScN2a cells		148 ± 6	10/10
CD-1	ScN2a cells treated for 7 weeks with 22-mer PS-DNA	>450	267 ± 85	3/10
Tg(MoPrP-A)FVB-B4053	ScN2a cells		68 ± 2	10/10
Tg(MoPrP-A)FVB-B4053	ScN2a cells treated for 7 weeks with 22-mer PS-DNA	>350	81	1/10[b]

[a]n, Number of mice that developed disease; n_0, number of inoculated mice.
[b]One mouse died at 81 days postinoculation.

remained after 48 hours. That both single- and double-stranded PS-DNA were able to reduce PrP^C and eliminate PrP^{Sc} indicates that the effect occurs through the backbone of the oligonucleotide.

The reduction of PrP^C and PrP^{Sc} by the 22-mer PS-DNA was not specific to a particular sequence of the oligonucleotides. However, the effective removal of PrP^C and PrP^{Sc} seemed to be dependent on the length of the PS-DNA. A 15-mer PS-DNA, bound to either FITC or rhodamine, reduced both PrP^C and PrP^{Sc} in the same manner as the 22-mer PS-DNA, whereas the 15-mer DNA alone did not. This suggests that the lowest effective threshold is between 15- and 22-mers for PS-DNA. It might be that shorter fragments do not form the correct conformation in order to interact with either PrP or another factor necessary for the degradation of PrP.

Decreasing PrP^C Levels

Whereas the foregoing studies with PS-DNA show a dramatic effect on the levels of PrP^{Sc} in cultured ScN2a cells, the mechanism by which PS-DNA rids the cells of prions remains unclear. One possible mechanism that might explain the disappearance of PrP^{Sc} in PS-DNA–treated ScN2a cells involves the reduction in PrP^C levels. The reduction of PrP^C, the substrate for PrP^{Sc} formation, leads to a decrease in prions. Transgenetic studies first showed that increased expression of a SHaPrP gene in mice decreased the incubation time after inoculation with SHa prions (Prusiner et al. 1990). Studies with inducible transgenes showed that repression of the PrP transgene in mice led to decreased formation of PrP^{Sc} (Tremblay et al. 1998). In $Prnp^{0/0}$ mice, the lack of PrP^C prevents any PrP^{Sc} from being formed. Mice hemizygous for PrP deficiency ($Prnp^{+/0}$) exhibited prolonged incubation times (Prusiner et al. 1993) but, inexplicably, were reported to accumulate PrP^{Sc} at approximately the same rate as normal $Prnp^{+/+}$ mice (Büeler et al. 1994). Although no studies have been reported showing antisense PrP RNA inhibition of PrP^{Sc} formation, investigations with small interfering RNAs (RNAi) have shown diminished PrP^{Sc} levels (Daude et al. 2003; Tilly et al. 2003). Whether RNAi can be used as a pharmacotherapeutic remains to be determined.

CpG Deoxyoligonucleotides

Mice inoculated intraperitoneally with RML prions were treated with CpG deoxyoligonucleotides, which consist of a central unmethylated dinucleotide flanked by two 5′ purines and two 3′ pyrimidines (Sethi et al. 2002). Mice received 5 nM CpG deoxyoligonucleotides (H.A. Kretzschmar, pers. comm.) 7 hours after inoculation and daily for the next 4 or 20 days. A third group of mice received the CpG deoxyoligonucleotides at the time of RML inoculation and daily for the next 4 days. The mice that received the CpG deoxyoligonucleotides for 4 days survived ~70 days longer than the untreated controls, which died at ~180 days after inoculation. The mice that received the CpG deoxyoligonucleotides for 20 days survived more than 150 days longer than the untreated controls. None of the mice that received the CpG deoxyoligonucleotides for 20 days had become ill, at 330 days postinoculation. Presumably, the inflammatory response to the CpG deoxyoligonucleotides retarded the transport of intraperitoneally inoculated prions to the CNS.

RNA Aptamers

RNA aptamers were produced against a peptide containing human (Hu) PrP residues 90–129 (Proske et al. 2002). The aptamers, denoted DP7, bound to recMoPrP, recSHaPrP, and chimeric mouse–hamster (MHM2) PrP with a K_d of ~100 nM, whereas the binding to recHuPrP was 10 times less avid. Addition of DP7 at 700 nM to ScN2a cells inhibited the synthesis of newly formed MHM2PrP^{Sc} but had no effect on the preexisting MoPrP^{Sc}. The studies were performed by metabolic labeling of the cultures with [^{35}S]methionine and cysteine. The radiolabeled MHM2PrP^{Sc} was separated from PrP^C by ultracentrifugation after detergent treatment with Sarkosyl. The pellets were resuspended in 1% SDS and boiled for 10 minutes prior to immunoprecipitation with the 3F4 mAb, which reacts with MHM2PrP but not MoPrP.

Whether aptamers such as DP7 have any potential as therapeutic agents is unclear. The inability of the DP7 aptamer to remove existing PrP^{Sc} in ScN2a cells contrasts with quinacrine, dendrimers, anti-PrP recFabs, and PS-DNA, all of which stimulate the clearance of PrP^{Sc}.

PYRIDINE DERIVATIVES MIMICKING
DOMINANT-NEGATIVE INHIBITION

To explain the results on the transmission of Hu prions to Tg mice, the existence of an auxiliary molecule required for PrP^{Sc} formation was postulated and provisionally designated protein X (Chapters 1, 4, and 17) (Telling et al. 1994, 1995). Subsequent investigations with ScN2a cells demonstrated dominant-negative inhibition of PrP^{Sc} formation by amino acid substitutions found in sheep and humans that are resistant to prion disease. The search for protein X has been frustrating because many proteins are known to bind to PrP^C, but none has been shown to be required for PrP^{Sc} formation (Oesch et al. 1990; Kurschner and Morgan 1995; Edenhofer et al. 1996; Tatzelt et al. 1996; Martins et al. 1997; Rieger et al. 1997; Yehiely et al. 1997; Graner et al. 2000; Keshet et al. 2000; Mouillet-Richard et al. 2000; Gauczynski et al. 2001; Hundt et al. 2001; Schmitt-Ulms et al. 2001). Whether protein X is one protein or a complex of proteins remains to be established. It is reasonable to expect that mice deficient for protein X would not replicate prions, or would do so extremely slowly, resulting in prolonged incubation times.

Site-directed mutagenesis combined with the NMR structure of recSHaPrP permitted mapping of the dominant-negative epitope on PrP^C. An expression vector with an insert encoding MHM2PrP was transfected into ScN2a cells; MHM2 carries two Syrian hamster residues, which create an epitope that is recognized by the anti-SHaPrP 3F4 mAb that does not react with MoPrP (Scott et al. 1992). Substitution of a Hu residue at either position 214 or 218 in MHM2PrP prevented formation of MHM2PrP^{Sc} in ScN2a cells (Kaneko et al. 1997b). MHM2(Q218K) was found to be ineligible for conversion into PrP^{Sc}. When MHM2PrP and MHM2PrP(Q218K) were coexpressed, the conversion of MHM2PrP^C into PrP^{Sc} was inhibited, arguing that MHM2PrP(Q218K) was acting as a dominant negative. The side chains of residues 214 and 218 protrude from the same surface of the carboxy-terminal α-helix, forming a discontinuous epitope with residues 167 and 171 in an adjacent loop. Like MHM2PrP(Q218K), substitution of a basic residue at either position 167 or 171 prevented PrP^{Sc} formation. Similar to MHM2PrP(Q218K), MHM2PrP(Q167R) acts as a dominant negative. In Tg mice, neither MHM2PrP(Q218K) nor MHM2PrP(Q167R) was converted into PrP^{Sc} and both inhibited the conversion of wild-type (wt) MoPrP^C into PrP^{Sc} in a dominant-negative manner (Perrier et al. 2002).

Selection of Compounds by a Structure-based Drug Design Strategy

The most suitable therapeutic target within the prion replication cycle was assumed to be the dominant-negative epitope on the surface of PrP^C. Theoretically, three types of ligands might interrupt the binding between the protein that binds the dominant-negative epitope (protein X) and PrP^C: (1) a small molecule that alters the protein X–binding site on the surface of PrP^C by binding to PrP^C; (2) a compound that mimics the protein X–binding site on the surface of PrP^C; and (3) a compound that mimics the PrP^C-binding site on the surface of protein X. Because the protein X–binding site on the surface of PrP lacks suitable cavities, the DOCK algorithm of Shoichet and Kuntz (1991) proved to be ineffective. Because the identity and structure of protein X is still unknown, attention was focused on identifying a mimicker of the protein X–binding site on the surface of PrP^C.

Disruption of protein–protein interactions using a structure-based drug design approach has proved to be difficult; however, this task should be easier if a small subset of residues contributes to a significant fraction of the binding energy and occupies only a small percentage of the overall protein surface (Clackson and Wells 1995; Tilley et al. 1997; Bogan and Thorn 1998). Kaneko et al. (1997b) showed that residues Q168, Q172, T215, and Q219 on the surface of the PrP^C molecule contribute most prominently to the stability of the molecular complex between PrP^C and the hypothetical protein X. Their side-chain coordinates from the PrP(90–231) NMR structure define a plausible pharmacophore target for mimetic design (James et al. 1997; Kaneko et al. 1997b). In the PrP(90–231) NMR structure, residue Q168 is part of a nonhelical loop that is not in close contact with the three other residues involved in the protein X–binding site. The apparent discontinuity of the protein X–binding site side chain is due to NOE restraints between the side chain of V166 and residues Y218, E221, S222, and Y225 in uncomplexed recombinant PrP(90–231). However, it would be quite feasible for Q168 to move adjacent to Q172, T215, and Q219 upon binding to protein X by extending helix B one turn. We therefore placed Q168 in two positions: (1) as it is found in the NMR structure of PrP(90–231) and (2) with helix B extended from residue Q172 to V166 (Fig. 10). As the substitution of basic residues at Q168, Q172, Q219, and acidic or hydrophobic residues at T215 appears to inhibit PrP^{Sc} replication by increasing the affinity of protein X for PrP^C, we modeled Arg, Lys, His, Asp, Glu, and Trp onto the relevant side-chain positions of the PrP(90–231) NMR structure using the program SCWRL (Bower et al. 1997).

Using these amino acid side-chain coordinates, a dataset of 3D pharmacophores was assembled such that all possible combinations of amino acid substitutions at the protein X–binding site were represented (Fig. 10). Emphasis was placed on mimicking the position of the functional atoms in each side chain as it is unlikely that any small molecules will contain all the atoms of each of the side chains that define the dominant-negative epitope. Pharmacophores with more than one basic residue were eliminated as substitution of more than one of the basic residues at the protein X–binding site interferes with the dominant-negative inhibition of prion replication (Zulianello et al. 2000). This analysis resulted in 1,000 poten-

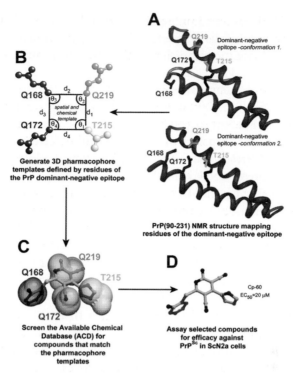

Figure 10. The structure-based design of dominant-negative PrP mimics, which led to the identification of the pyridine-based compound, Cp-60, a micromolar inhibitor of PrP^{Sc} in ScN2a cells. Two conformations of the dominant-negative epitope on the surface of PrP were used to create a dataset of pharmacophores. (*A*) In the first conformation (conformation 1, *top*), residue 168 is positioned as found in the NMR structure of PrP(90–231), whereas in the second (conformation 2, *bottom*), helix B is extended to residue V166 (Liu et al. 1999). In this case, residue Q168 forms a contiguous structural epitope with the other three residues, Q172, T215, and Q219. Using the coordinates of these four residues, datasets of three-dimensional pharmacophores were generated (*B–D*). Pharmacophore templates were used to screen the Available Chemical Database (ACD) for small molecules that spatially and electronically mimicked the dominant-negative epitope. To illustrate the in compuo screen, an overlay of Cp-60 and the dominant-negative pharmacophore is shown. Cp-60 is shown in the conformation predicted by the CONCORD algorithm. Cp-60 is superimposed on the dominant-negative pharmacophore derived from conformation 2 (panel *A, bottom*). The relevant side chains are represented by a van der Waals sphere centered on the side-chain atoms' positions. The color scheme is by residue in the sphere and by atom type in Cp-60: carbon (*green*), oxygen (*red*), nitrogen (*blue*), and sulfur (*yellow*). In total, 63 compounds were selected from the ACD and assayed in ScN2a cells for efficacy against PrP^{Sc}. Of the nontoxic compounds identified, the pyridine-based compound, Cp-60, had an EC$_{50}$ value of 20 μM. (Adapted from Perrier et al. 2000.)

tial templates, which were compared with the 210,000 compounds present in the Available Chemicals Database (ACD) for compounds that mimic both the spatial orientation and chemical properties present in the dataset of pharmacophores. The computer program CONCORD was used to generate the three-dimensional structure of each of the ACD compounds (Ho and Marshall 1995). Our algorithm, called genX, has two main stages (Ullman 1976). First, the covalent connectivity of the pharmacophore functional atoms is compared to the compounds. Then, if at least one match for each of these groups of functional atoms is located within the compound, the mean coordinate position of each group is calculated. This generates a single pseudo-atom representation for each match corresponding to a functional atom cluster. The interatomic distances and bond angles defined by these pseudo-atoms are then compared to the distances and angles previously specified for each of the pharmacophores from an analysis of the relevant side-chain centroids. For each possible match, the small molecule is superimposed upon the optimal collection of side-chain atoms and the root mean-square distance (RMSD) is calculated. A RMSD of less than 2 Å is classified as a match. GenX identified a total of 5000 compounds from the ACD, which were further clustered according to their functional groups and screened visually using MIDAS-PLUS. Of the 80 compounds that were identified for experimental characterization, 63 were available for purchase.

Screening of Compounds

Compounds were screened in a cellular assay for prion replication. To analyze the effect of the compounds on newly synthesized PrP^{Sc}, ScN2a cells were transiently transfected. In this assay, MHM2PrP is used, and $MoPrP^{Sc}$ is immunologically silent. Twenty-four hours after the transfection, each of the 63 compounds was added to the medium and incubated for 3 days. After 72 hours, the cells were lysed, and the crude extract was analyzed on an immunoblot before and after PK digestion. Initial screening of compounds at 10 μM identified 10 potential inhibitors of PrP^{Sc} formation (Perrier et al. 2000). The inhibition observed for 5 of these candidates was not reproducible. Complete dose–response curves confirmed the activity of 4 of the 10 compounds: Cp-18, Cp-32, Cp-60, and Cp-62. Unlike these four compounds, Cp-7 failed to exhibit a classical dose–response curve. As Cp-60 and Cp-62 were solubilized at 1 mM in pure methanol, we verified that the inhibition was not due to the solvent. Immunoblots revealed that 4% methanol, which corresponds to the volume of solvent used with 40 μM of Cp-60, did not inhibit PrP^{Sc} formation.

For all the compounds screened, mmunoblots of the samples not treated with PK provided a control for the transfection efficiency and showed that each of the samples was equally transfected. The EC_{50} value for each compound was determined by densitometry as a ratio of PrP^{Sc}:total PrP. The compounds in descending order of apparent potency are Cp-60 (18 μM), Cp-18 (20 μM), Cp-62 (30 μM), and Cp-32 (60 μM).

Cytotoxicity was evaluated for each of the 63 compounds and carefully analyzed for the four active compounds. The cellular toxicity of the compounds was determined by the quantification of the total protein amounts in the lysates and compared to the untreated control. At concentrations near the EC_{50} values, Cp-18 and Cp-32 showed toxic effects on ScN2a cells. At 30 μM of Cp-18, 5% ($\pm$ 0.1%) of the protein remained, and at 70 μM of Cp-32, 47% ($\pm$ 3%) of the protein was found. Experiments with Cp-18 and Cp-32 were not pursued. In contrast, 95% ($\pm$ 4%) of the protein was found after exposure to 20 μM Cp-60, and 80% ($\pm$ 3%) of the protein was recovered after exposure to 40 μM Cp-62. Among the 63 compounds screened, 3 inhibited PrP^{Sc} formation without significant cellular toxicity (Cp-7, Cp-60, Cp-62). Fifty-eight compounds did not inhibit PrP^{Sc} formation; among these, 5 were toxic to ScN2a cells. Because Cp-60 was not overtly toxic to cells and exhibited the lowest EC_{50} value, we studied its impact on cellular viability using the MTT reagent. At 20 μM of Cp-60, 82% ($\pm$ 0.3%) of ScN2a cells were alive; at 40 μM of Cp-60, 66% ($\pm$ 0.1%) of cells were viable. To assess the toxicity of the methanol used to solubilize Cp-60, the viability of the cells was determined with solvent alone. Methanol was responsible for 10% and 20% of the cell death at 20 μM and 40 μM of Cp-60, respectively.

To obtain an independent verification of the EC_{50} value of the most active compound, we used a more sensitive and quantitative immunoassay using time-resolved fluorescence (TRF) (Safar et al. 1998). Transiently transfected ScN2a cells were incubated with various concentrations of Cp-60 over a period of 3 days. The cell lysates were digested with PK and PrP^{Sc} measured by immunoassay using a Europium-labeled mAb (Eu-mAb 3F4) (Fig. 11A). The EC_{50} for Cp-60 obtained by the TRF technique was 21.5 $\pm$ 0.5 μM. The result of this approach agreed well with the dose–response curve determined by immunoblotting, from which the EC_{50} for Cp-60 was calculated to be 18.0 $\pm$ 1.5 μM (Fig. 11B).

Treatment of ScN2a Cells with Cp-60

The ability of Cp-60 to cure ScN2a cells was studied by incubating cells with solvent alone, Cp-60, or Congo red as a positive control. These cells

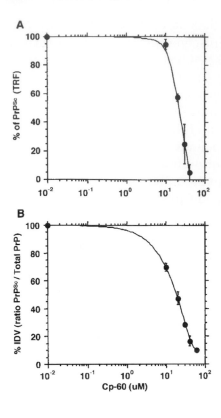

Figure 11. Cp-60 inhibition on MHM2PrP^{Sc} formation. (*A*) Immunoassay using time-resolved fluorescence (TRF). The data represent the mean ± S.E.M. from three independent experiments conducted in duplicate. (*B*) Immunoblots quantified by densitometry. Treatment of transfected ScN2a cells with Cp-60 and immunoblot analysis was described in Perrier et al. (2000). Densitometry was conducted with four immunoblot films from independent experiments. Values represent mean ± S.E.M. (Reprinted, with permission, from Perrier et al. 2000 [copyright National Academy of Sciences].)

were serially passaged every 3 days for a total of 9 days. An immunoblot analysis of the samples showed that after 6 days of incubation with 40 μM Cp-60, half of the PrP^{Sc} molecules disappeared. After 9 days, the cells exhibited little detectable PrP^{Sc}. The magnitude of this effect was comparable to that of Congo red (1 μM), a reagent previously reported to block PrP^{Sc} formation in cells.

Cp-60 was used to search the Available Chemicals Database for related molecules. Five related compounds were analyzed (Fig. 12), and three exhibited activity that is comparable to Cp-60. In the ScN2a cell assay, the EC$_{50}$ values for A3, A4, and A5 were 35 ± 3.0 μM, 18.6 ± 3.1 μM, and 15.5 ± 0.5 μM, respectively. Our enthusiasm for the initial leads described above is tempered by the recognition that existing animal models of scrapie replication are quite time-consuming, and existing cellular data suggest that daily dosing schedules will be required. Mimicking dominant-negative inhibition in the design of drugs that inhibit prion replication may provide a more general approach to developing therapeutics for deleterious protein–protein interactions.

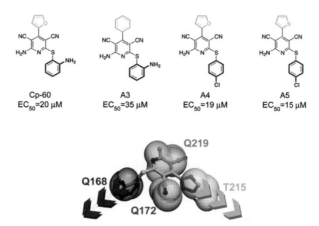

Figure 12. The structure of the pyridine-based Cp-60 and related analogs A3, A4, and A5. Substituents of the pyridinyl compounds are colored to match the corresponding residues of the dominant-negative pharmacophore that these substituents are thought to mimic: T215 (*yellow*) and Q168 (*red*). Arrows on the pharmacophore overlay (*lower*) indicate the direction of subsequent modification to substituents of the pyridine scaffold. By following this pattern of structure–activity analysis, potent pyridine-based compounds (EC$_{50}$ = 300 nM) have been identified (B.C.H. May et al., in prep.).

CONCLUDING REMARKS

Because people at risk for inherited prion diseases can now be identified decades before neurologic dysfunction is evident, the development of an effective therapy for these fully penetrant disorders is imperative (Chapman et al. 1994; Spudich et al. 1995). Although we have no way of predicting the number of individuals who may develop neurologic dysfunction from bovine prions in the future (Cousens et al. 1997; Ghani et al. 2000), seeking an effective therapy now seems most prudent. Interfering with the conversion of PrP^C to PrP^{Sc} would seem to be the most attractive therapeutic target (Cohen et al. 1994).

Defining the pathogenesis of prion disease is an important issue with respect to developing an effective therapy. The issue of whether large aggregates of misprocessed proteins or misfolded monomers (or oligomers) cause CNS degeneration has been addressed in several studies of prion diseases in humans as well as in Tg mice. In humans, the frequency of PrP amyloid plaques varies from 100% in Gerstmann-Sträussler-Scheinker disease and variant CJD (Will et al. 1996) to ~70% in kuru (Klatzo et al. 1959) and ~10% in sporadic CJD, arguing that these plaques are a nonobligatory feature of disease (DeArmond and Prusiner

1997). In Tg mice expressing both MoPrP and SHaPrP, animals inoculated with hamster prions produced hamster prions and developed amyloid plaques composed of SHaPrP (Prusiner et al. 1990). In contrast, Tg mice inoculated with mouse prions did not develop plaques even though they produced mouse prions and died of scrapie.

REFERENCES

Adjou K.T., Demaimay R., Deslys J.-P., Lasmézas C.L., Beringue V., Demart S., Lamoury F., Seman M., and Dormont D. 1999. MS-8209, a water-soluble amphotericin B derivative, affects both scrapie agent replication and PrPres accumulation in Syrian hamster scrapie. *J. Gen. Virol.* **80:** 1079–1085.

Akhtar S., Basu S., Wickstrom E., and Juliano R.L. 1991. Interactions of antisense DNA oligonucleotide analogs with phospholipid membranes (liposomes). *Nucleic Acids Res.* **19:** 5551–5559.

Albert A. 1960. *Selective toxicity.* John Wiley & Sons, New York.

Alper T., Haig D.A., and Clarke M.C. 1966. The exceptionally small size of the scrapie agent. *Biochem. Biophys. Res. Commun.* **22:** 278–284.

———. 1978. The scrapie agent: Evidence against its dependence for replication on intrinsic nucleic acid. *J. Gen. Virol.* **41:** 503–516.

Alper T., Cramp W.A., Haig D.A., and Clarke M.C. 1967. Does the agent of scrapie replicate without nucleic acid? *Nature* **214:** 764–766.

Anderson R.G. and Orci L. 1988. A view of acidic intracellular compartments. *J. Cell Biol.* **106:** 539–543.

Bard F., Cannon C., Barbour R., Burke R.-L., Games D., Grajeda H., Guido T., Hu K., Huang J., Johnson-Wood K., Khan K., Kholodenko D., Lee M., Lieberburg I., Motter R., Nguyen M., Soriano F., Vasquez N., Weiss K., Welch B., Seubert P., Schenk D., and Yednock T. 2000. Peripherally administered antibodies against amyloid β-peptide enter the central nervous system and reduce pathology in a mouse model of Alzheimer disease. *Nat. Med.* **6:** 916–919.

Barron R.M., Thomson V., Jamieson E., Melton D.W., Ironside J., Will R., and Manson J.C. 2001. Changing a single amino acid in the N-terminus of murine PrP alters TSE incubation time across three species barriers. *EMBO J.* **20:** 5070–5078.

Barry R.A. and Prusiner S.B. 1986. Monoclonal antibodies to the cellular and scrapie prion proteins. *J. Infect. Dis.* **154:** 518–521.

Barry R.A., Kent S.B.H., McKinley M.P., Meyer R.K., DeArmond S.J., Hood L.E., and Prusiner S.B. 1986. Scrapie and cellular prion proteins share polypeptide epitopes. *J. Infect. Dis.* **153:** 848–854.

Baskakov I.V., Legname G., Baldwin M.A., Prusiner S.B., and Cohen F.E. 2002. Pathway complexity of prion protein assembly into amyloid. *J. Biol. Chem.* **277:** 21140–21148.

Bendheim P.E., Barry R.A., DeArmond S.J., Stites D.P., and Prusiner S.B. 1984. Antibodies to a scrapie prion protein. *Nature* **310:** 418–421.

Birmingham K. and Frantz S. 2002. Set back to Alzheimer vaccine studies. *Nat. Med.* **8:** 199–200.

Bogan A.A. and Thorn K.S. 1998. Anatomy of hot spots in protein interfaces. *J. Mol. Biol.* **280:** 1–9.

Bonse S., Santelli-Rouvier C., Barbe J., and Krauth-Siegel R.L. 1999. Inhibition of

Trypanosoma cruzi trypanothione reductase by acridines: Kinetic studies and structure-activity relationships. *J. Med. Chem.* **42:** 5448–5454.

Borchelt D.R., Taraboulos A., and Prusiner S.B. 1992. Evidence for synthesis of scrapie prion proteins in the endocytic pathway. *J. Biol. Chem.* **267:** 16188–16199.

Borchelt D.R., Scott M., Taraboulos A., Stahl N., and Prusiner S.B. 1990. Scrapie and cellular prion proteins differ in their kinetics of synthesis and topology in cultured cells. *J. Cell Biol.* **110:** 743–752.

Boussif O., Lezoualc'h F., Zanta M.A., Mergny M.D., Scherman D., Demeneix B., and Behr J.P. 1995. A versatile vector for gene and oligonucleotide transfer into cells in culture and in vivo: Polyethyleneimine. *Proc. Natl. Acad. Sci.* **92:** 7297–7301.

Bower M.J., Cohen F.E., and Dunbrack R.L.J. 1997. Prediction of protein side-chain rotamers from a backbone-dependent rotamer library: A new homology modeling. *J. Mol. Biol.* **267:** 1268–1282.

Brocks D.R. and Jamali F. 1995. Stereochemical aspects of pharmacotherapy. *Pharmacotherapy* **15:** 551–564.

Bruce M.E., Dickinson A.G., and Fraser H. 1976. Cerebral amyloidosis in scrapie in the mouse: Effect of agent strain and mouse genotype. *Neuropathol. Appl. Neurobiol.* **2:** 471–478.

Bucciantini M., Giannoni E., Chiti F., Baroni F., Formigli L., Zurdo J., Taddei N., Ramponi G., Dobson C.M., and Stefani M. 2002. Inherent toxicity of aggregates implies a common mechanism for protein misfolding diseases. *Nature* **416:** 507–511.

Budka H., Aguzzi A., Brown P., Brucher J.-M., Bugiani O., Collinge J., Diringer H., Gullotta F., Haltia M., Hauw J.-J., Ironside J.W., Kretzschmar H.A., Lantos P.L., Masullo C., Pocchiari M., Schlote W., Tateishi J., and Will R.G. 1995. Tissue handling in suspected Creutzfeldt-Jakob disease (CJD) and other human spongiform encephalopathies (prion diseases). *Brain Pathol.* **5:** 319–322.

Büeler H., Raeber A., Sailer A., Fischer M., Aguzzi A., and Weissmann C. 1994. High prion and PrP^{Sc} levels but delayed onset of disease in scrapie-inoculated mice heterozygous for a disrupted PrP gene. *Mol. Med.* **1:** 19–30.

Büeler H., Aguzzi A., Sailer A., Greiner R.-A., Autenried P., Aguet M., and Weissmann C. 1993. Mice devoid of PrP are resistant to scrapie. *Cell* **73:** 1339–1347.

Butler D.A., Scott M.R.D., Bockman J.M., Borchelt D.R., Taraboulos A., Hsiao K.K., Kingsbury D.T., and Prusiner S.B. 1988. Scrapie-infected murine neuroblastoma cells produce protease-resistant prion proteins. *J. Virol.* **62:** 1558–1564

Carlson G.A., Ebeling C., Yang S. L., Telling G., Torchia M., Groth D., Westaway D., DeArmond S.J., and Prusiner S.B. 1994. Prion isolate specified allotypic interactions between the cellular and scrapie prion proteins in congenic and transgenic mice. *Proc. Natl. Acad. Sci.* **91:** 5690–5694.

Caspi S., Sasson S.B., Taraboulos A., and Gabizon R. 1998. The anti-prion activity of Congo red. Putative mechanism. *J. Biol. Chem.* **273:** 3484–3489.

Caughey B. and Race R.E. 1992. Potent inhibition of scrapie-associated PrP accumulation by Congo red. *J. Neurochem.* **59:** 768–771.

Caughey B. and Raymond G.J. 1993. Sulfated polyanion inhibition of scrapie-associated PrP accumulation in cultured cells. *J. Virol.* **67:** 643–650.

Caughey B., Raymond G.J., Ernst D., and Race R.E. 1991a. N-terminal truncation of the scrapie-associated form of PrP by lysosomal protease(s): Implications regarding the site of conversion of PrP to the protease-resistant state. *J. Virol.* **65:** 6597–6603.

Caughey B., Brown K., Raymond G.J., Katzenstein G.E., and Thresher W. 1994. Binding of the protease-sensitive form of PrP (prion protein) to sulfated glycosaminoglycan and Congo red. *J. Virol.* **68:** 2135–2141.

Caughey B.W., Dong A., Bhat K.S., Ernst D., Hayes S.F., and Caughey W.S. 1991b. Secondary structure analysis of the scrapie-associated protein PrP 27-30 in water by infrared spectroscopy. *Biochemistry* **30:** 7672–7680.

Caughey W.S., Raymond L.D., Horiuchi M., and Caughey B. 1998. Inhibition of protease-resistant prion protein formation by porphyrins and phthalocyanines. *Proc. Natl. Acad. Sci.* **95:** 12117–12122.

Chabry J., Priola S.A., Wehrly K., Nishio J., Hope J., and Chesebro B. 1999. Species-independent inhibition of abnormal prion protein (PrP) formation by a peptide containing a conserved PrP sequence. *J. Virol.* **73:** 6245–6250.

Chapman J., Ben-Israel J., Goldhammer Y., and Korczyn A.D. 1994. The risk of developing Creutzfeldt-Jakob disease in subjects with the *PRNP* gene codon 200 point mutation. *Neurology* **44:** 1683–1686.

Chen T.K., Fico R., and Canellakis E.S. 1978. Diacridines, bifunctional intercalators. Chemistry and antitumor activity. *J. Med. Chem.* **21:** 868–874.

Chuong C.M., McClain D.A., Streit P., and Edelman G.M. 1982. Neural cell adhesion molecules in rodent brains isolated by monoclonal antibodies with cross-species reactivity. *Proc. Natl. Acad. Sci.* **79:** 4234–4238.

Clackson T. and Wells J.A. 1995. A hot spot of binding energy in a hormone-receptor. *Science* **267:** 383–386.

Cohen F.E., Pan K.-M., Huang Z., Baldwin M., Fletterick R.J., and Prusiner S.B. 1994. Structural clues to prion replication. *Science* **264:** 530–531.

Cordeiro Y., Machado F., Juliano L., Juliano M.A., Brentani R.R., Foguel D., and Silva J.L. 2001. DNA converts cellular prion protein into the beta-sheet conformation and inhibits prion peptide aggregation. *J. Biol. Chem.* **276:** 49400–49409.

Cousens S.N., Vynnycky E., Zeidler M., Will R.G., and Smith P.G. 1997. Predicting the CJD epidemic in humans. *Nature* **385:** 197–198.

Craig J.C., Labelle B., and Ohnsorge U. 1991. The absolute configuration of the enantiomers of 6-chloro-9-(4′-diethylamino-1′-methylbutyl)-amino-2-methoxyacridine (quinacrine). *Chirality* **3:** 436–437.

Daude N., Marella M., and Chabry J. 2003. Specific inhibition of pathological prion protein accumulation by small interfering RNAs. *J. Cell Sci.* **116:** 2775–2779.

DeArmond S.J. and Prusiner S.B. 1997. Prion diseases. In *Greenfield's Neuropathology*, 6th edition (ed. P. Lantos and D. Graham), pp. 235–280. Edward Arnold, London, United Kingdom.

Dees C., Wade W.F., German T.L., and Marsh R.F. 1985. Inactivation of the scrapie agent by ultraviolet irradiation in the presence of chlorpromazine. *J. Gen. Virol.* **66:** 845–849.

Delay J., Deniker P., and Harl J.-M. 1952. Traitement des états d'excitation et d'agitation par une méthode médicamenteuse dérivée de l'hibernothérapie. *Ann. Méd.-Psychol.* **110:** 267–273.

DeMattos R.B., Bales K.R., Cummins D.J., Paul S.M., and Holtzman D.M. 2002. Brain to plasma amyloid-beta efflux: A measure of brain amyloid burden in a mouse model of Alzheimer's disease. *Science* **295:** 2264–2267.

DeMattos R.B., Bales K.R., Cummins D.J., Dodart J.C., Paul S.M., and Holtzman D.M. 2001. Peripheral anti-A beta antibody alters CNS and plasma A beta clearance and decreases brain A beta burden in a mouse model of Alzheimer's disease. *Proc. Natl. Acad. Sci.* **98:** 8850–8855.

Denny W.A., Atwell G.J., Baguley B.C., and Wakelin L.P. 1985. Potential antitumor agents. 44. Synthesis and antitumor activity of new classes of diacridines: Importance of linker chain rigidity for DNA binding kinetics and biological activity. *J. Med. Chem.* **28:** 1568–1574.

Dickinson A.G., Fraser H., and Outram G.W. 1975. Scrapie incubation time can exceed natural lifespan. *Nature* **256:** 732–733.

Dickinson A.G., Bruce M.E., Outram G.W., and Kimberlin R.H. 1984. Scrapie strain differences: The implications of stability and mutation. In *Proceedings of Workshop on Slow Transmissible Diseases* (ed. J. Tateishi), pp. 105–118. Japanese Ministry of Health and Welfare, Tokyo, Japan.

Diringer H. 1991. Transmissible spongiform encephalopathies (TSE) virus-induced amyloidoses of the central nervous system (CNS). *Eur. J. Epidemiol.* **7:** 562–566.

Diringer H. and Ehlers B. 1991. Chemoprophylaxis of scrapie in mice. *J. Gen. Virol.* **72:** 457–460.

Diwu Z., Chen C.S., Zhang C., Klaubert D.H., and Haugland R.P. 1999. A novel acidotropic pH indicator and its potential application in labeling acidic organelles of live cells. *Chem. Biol.* **6:** 411–418.

Doh-ura K., Iwaki T., and Caughey B. 2000. Lysosomotropic agents and cysteine protease inhibitors inhibit scrapie-associated prion protein accumulation. *J. Virol.* **74:** 4894–4897.

Donne D.G., Viles J.H., Groth D., Mehlhorn I., James T.L., Cohen F.E., Prusiner S.B., Wright P.E., and Dyson H.J. 1997. Structure of the recombinant full-length hamster prion protein PrP(29-231): The N terminus is highly flexible. *Proc. Natl. Acad. Sci.* **94:** 13452–13457.

Edenhofer F., Rieger R., Famulok M., Wendler W., Weiss S., and Winnacker E.-L. 1996. Prion protein PrP^C interacts with molecular chaperones of the Hsp60 family. *J. Virol.* **70:** 4724–4728.

Ehlers B. and Diringer H. 1984. Dextran sulphate 500 delays and prevents mouse scrapie by impairment of agent replication in spleen. *J. Gen. Virol.* **65:** 1325–1330.

Enari M., Flechsig E., and Weissmann C. 2001. Scrapie prion protein accumulation by scrapie-infected neuroblastoma cells abrogated by exposure to a prion protein antibody. *Proc. Natl. Acad. Sci.* **98:** 9295–9299.

Farquhar C.F. and Dickinson A.G. 1986. Prolongation of scrapie incubation period by an injection of dextran sulphate 500 within the month before or after infection. *J. Gen. Virol.* **67:** 463–473.

Farquhar C., Dickinson A., and Bruce M. 1999. Prophylactic potential of pentosan polysulphate in transmissible spongiform encephalopathies. *Lancet* **353:** 117.

Findlay G.M. 1951. *Recent advances in chemotherapy.* 3rd edition, vol. 2. The Blakiston Company, Philadelphia.

Fraser H. and Bruce M.E. 1973. Argyrophilic plaques in mice inoculated with scrapie from particular sources. *Lancet* **1:** 617.

———. 1983. Experimental control of cerebral amyloid in scrapie in mice. *Prog. Brain Res.* **59:** 281–290.

Gabus C., Auxilien S., Pechoux C., Dormont D., Swietnicki W., Morillas M., Surewicz W., Nandi P., and Darlix J.L. 2001a. The prion protein has DNA strand transfer properties similar to retroviral nucleocapsid protein. *J. Mol. Biol.* **307:** 1011–1021.

Gabus C., Derrington E., Leblanc P., Chnaiderman J., Dormont D., Swietnicki W., Morillas M., Surewicz W.K., Marc D., Nandi P., and Darlix J.L. 2001b. The prion protein has RNA binding and chaperoning properties characteristic of nucleocapsid protein NCP7 of HIV-1. *J. Biol. Chem.* **276:** 19301–19309.

Gasset M., Baldwin M.A., Fletterick R.J., and Prusiner S.B. 1993. Perturbation of the secondary structure of the scrapie prion protein under conditions that alter infectivity. *Proc. Natl. Acad. Sci.* **90:** 1–5.

Gauczynski S., Peyrin J.M., Haik S., Leucht C., Hundt C., Rieger R., Krasemann S., Deslys J.P., Dormont D., Lasmézas C.I., and Weiss S. 2001. The 37-kDa/67-kDa laminin receptor acts as the cell-surface receptor for the cellular prion protein. *EMBO J.* **20:** 5863–5875.

Ghani A.C., Ferguson N.M., Donnelly C.A., and Anderson R.M. 2000. Predicted vCJD mortality in Great Britain. *Nature* **406:** 583–584.

Gilman H., van Ess P.R., and Shirley D.A. 1944. The metalation of 10-phenylphenothiazine and of 10-ethlypheothiazine. *J. Am. Chem. Soc.* **66:** 1214–1216.

Girault S., Grellier P., Berecibar A., Maes L., Mouray E., Lemiere P., Debreu M.A., Davioud-Charvet E., and Sergheraert C. 2000. Antimalarial, antitrypanosomal, and antileishmanial activities and cytotoxicity of bis(9-amino-6-chloro-2-methoxy-acridines): Influence of the linker. *J. Med. Chem.* **43:** 2646–2654.

Goodman L.S. and Gilman A., Eds. 1970. *The pharmacological basis of therapeutics; a textbook of pharmacology, toxicology, and therapeutics for physicians and medical students,* 4th edition. Macmillan, New York.

———. 1975. *The pharmacological basis of therapeutics; a textbook of pharmacology, toxicology, and therapeutics for physicians and medical students,* 5th edition. Macmillan, New York.

Gordon W.S. 1946. Advances in veterinary research. *Vet. Res.* **58:** 516–520.

Gorodinsky A. and Harris D.A. 1995. Glycolipid-anchored proteins in neuroblastoma cells form detergent-resistant complexes without caveolin. *J. Cell Biol.* **129:** 619–627.

Graner E., Mercadante A.F., Zanata S.M., Forlenza O.V., Cabral A.L.B., Veiga S.S., Juliano M.A., Roesler R., Walz R., Mineti A., Izquierdo I., Martins V.R., and Brentani R.R. 2000. Cellular prion protein binds laminin and mediates neuritogenesis. *Mol. Brain Res.* **76:** 85–92.

Green R. 1932. A report on fifty cases of malaria treated with atebrin. *Lancet* **I:** 826–829.

Guttmann P. and Ehrlich P. 1891. Ueber die wirkung des methylenblau bei malaria. *Berl. Klin. Wochenschr.* **39:** 953–956.

Haensler J. and Szoka F.C.J. 1993. Polyamidoamine cascade polymers mediate efficient transfection of cells in culture. *Bioconjug. Chem.* **4:** 372–379.

Hammick D.L. and Chambers W.E. 1945. Optical activity of excreted mepacrine. *Nature* **155:** 141.

Hardman J.G., Limbird L.L., Molinoff P.B., Ruddon R.W., and Gilman A.G., Eds. 1996. *Goodman & Gilman's the pharmacological basis of therapeutics,* 9th edition. McGraw-Hill, New York.

Heppner F.L., Musahl C., Arrighi I., Klein M.A., Rülicke T., Oesch B., Zinkernagel R.M., Kalinke U., and Aguzzi A. 2001. Prevention of scrapie pathogenesis by transgenic expression of anti-prion protein antibodies. *Science* **294:** 178–182.

Ho C.M. and Marshall G.R. 1995. DBMAKER: A set of programs to generate three-dimensional databases based upon user-specified criteria. *J. Comput.-Aided Mol. Des.* **9:** 65–86.

Hoke G.D., Draper K., Freier S.M., Gonzalez C., Driver V.B., Zounes M.C., and Ecker D.J. 1991. Effects of phosphorothioate capping on antisense oligonucleotide stability, hybridization and antiviral efficacy versus herpes simplex virus infection. *Nucleic Acids Res.* **19:** 5743–5748.

Hundt C., Peyrin J.M., Haik S., Gauczynski S., Leucht C., Rieger R., Riley M.L., Deslys J.P., Dormont D., Lasmézas C.I., and Weiss S. 2001. Identification of interaction domains of the prion protein with its 37-kDa/67-kDa laminin receptor. *EMBO J.* **20:** 5876–5886.

Hutt A.J. and Tan S.C. 1996. Drug chirality and its clinical significance. *Drugs* **52:** 1–12.

Ingrosso L., Ladogana A., and Pocchiari M. 1995. Congo red prolongs the incubation period in scrapie-infected hamsters. *J. Virol.* **69:** 506–508.

James T.L., Liu H., Ulyanov N.B., Farr-Jones S., Zhang H., Donne D.G., Kaneko K., Groth D., Mehlhorn I., Prusiner S.B., and Cohen F.E. 1997. Solution structure of a 142-residue recombinant prion protein corresponding to the infectious fragment of the scrapie isoform. *Proc. Natl. Acad. Sci.* **94:** 10086–10091.

Kaneko K., Vey M., Scott M., Pilkuhn S., Cohen F.E., and Prusiner S.B. 1997a. COOH-terminal sequence of the cellular prion protein directs subcellular trafficking and controls conversion into the scrapie isoform. *Proc. Natl. Acad. Sci.* **94:** 2333–2338.

Kaneko K., Zulianello L., Scott M., Cooper C.M., Wallace A.C., James T.L., Cohen F.E., and Prusiner S.B. 1997b. Evidence for protein X binding to a discontinuous epitope on the cellular prion protein during scrapie prion propagation. *Proc. Natl. Acad. Sci.* **94:** 10069–10074.

Kellings K., Prusiner S.B., and Riesner D. 1994. Nucleic acids in prion preparations: Unspecific background or essential component? *Philos. Trans. R. Soc. Lond. B Biol. Sci.* **343:** 425–430.

Kellings K., Meyer N., Mirenda C., Prusiner S.B., and Riesner D. 1992. Further analysis of nucleic acids in purified scrapie prion preparations by improved return refocussing gel electrophoresis (RRGE). *J. Gen. Virol.* **73:** 1025–1029.

Keshet G.I., Bar-Peled O., Yaffe D., Nudel U., and Gabizon R. 2000. The cellular prion protein colocalizes with the dystroglycan complex in the brain. *J. Neurochem.* **75:** 1889–1897.

Kimberlin R.H. and Walker C.A. 1983. The antiviral compound HPA-23 can prevent scrapie when administered at the time of infection. *Arch. Virol.* **78:** 9–18.

———. 1986. Suppression of scrapie infection in mice by heteropolyanion 23, dextran sulfate, and some other polyanions. *Antimicrob. Agents Chemother.* **30:** 409–413.

Kirkitadze M.D., Bitan G., and Teplow D.B. 2002. Paradigm shifts in Alzheimer's disease and other neurodegenerative disorders: The emerging role of oligomeric assemblies. *J. Neurosci. Res.* **69:** 567–577.

Klatzo I., Gajdusek D.C., and Zigas V. 1959. Pathology of kuru. *Lab. Invest.* **8:** 799–847.

Kocisko D.A., Priola S.A., Raymond G.J., Chesebro B., Lansbury P.T., Jr., and Caughey B. 1995. Species specificity in the cell-free conversion of prion protein to protease-resistant forms: A model for the scrapie species barrier. *Proc. Natl. Acad. Sci.* **92:** 3923–3927.

Korth C., May B.C.H., Cohen F.E., and Prusiner S.B. 2001. Acridine and phenothiazine derivatives as pharmacotherapeutics for prion disease. *Proc. Natl. Acad. Sci.* **98:** 9836–9841.

Kurschner C. and Morgan J.I. 1995. The cellular prion protein (PrP) selectively binds to Bcl-2 in the yeast two-hybrid system. *Brain Res. Mol. Brain Res.* **30:** 165–168.

Laborit H. and Huguenard P. 1952. Technique actuelle de l'hibernation artificielle. *Presse Med.* **60:** 1455–1456.

Leclerc E., Peretz D., Ball H., Sakurai H., Legname G., Serban A., Prusiner S.B., Burton D.R., and Williamson R.A. 2001. Immobilized prion protein undergoes spontaneous rearrangement to a conformation having features in common with the infectious form. *EMBO J.* **20:** 1547–1554.

Lieberman J.A. 1999. Is schizophrenia a neurodegenerative disorder? A clinical and neurobiological perspective. *Biol. Psychiatry* **46:** 729–739.

Liu H., Farr-Jones S., Ulyanov N.B., Llinas M., Marqusee S., Groth D., Cohen F.E., Prusiner S.B., and James T.L. 1999. Solution structure of Syrian hamster prion protein rPrP(90–231). *Biochemistry* **38:** 5362–5377.

Manuelidis L., Fritch W., and Zaitsev I. 1998. Dapsone to delay symptoms in Creutzfeldt-Jakob disease. *Lancet* **352:** 456.

Martins V.R., Graner E., Garcia-Abreu J., de Souza S.J., Mercadante A.F., Veiga S.S., Zanata S.M., Neto V.M., and Brentani R.R. 1997. Complementary hydropathy identifies a cellular prion protein receptor. *Nat. Med.* **3:** 1376–1382.

May B.C.H., Fafarman A.T., Hong S.B., Rogers M., Deady L.W., Prusiner S.B., and Cohen F.E. 2003. Potent inhibition of scrapie prion replication in cultured cells by bisacridines. *Proc. Natl. Acad. Sci.* **100:** 3416–3421.

McKenzie D., Kaczkowski J., Marsh R., and Aiken J. 1994. Amphotericin B delays both scrapie agent replication and PrP-res accumulation early in infection. *J. Virol.* **68:** 7534–7536.

McKinley M.P., Meyer R.K., Kenaga L., Rahbar F., Cotter R., Serban A., and Prusiner S.B. 1991. Scrapie prion rod formation *in vitro* requires both detergent extraction and limited proteolysis. *J. Virol.* **65:** 1340–1351.

Mehlhorn I., Groth D., Stöckel J., Moffat B., Reilly D., Yansura D., Willett W.S., Baldwin M., Fletterick R., Cohen F.E., Vandlen R., Henner D., and Prusiner S.B. 1996. High-level expression and characterization of a purified 142-residue polypeptide of the prion protein. *Biochemistry* **35:** 5528–5537.

Mouillet-Richard S., Ermonval M., Chebassier C., Laplanche J.L., Lehmann S., Launay J.M., and Kellermann O. 2000. Signal transduction through prion protein. *Science* **289:** 1925–1928.

Muramoto T., Scott M., Cohen F.E., and Prusiner S.B. 1996. Recombinant scrapie-like prion protein of 106 amino acids is soluble. *Proc. Natl. Acad. Sci.* **93:** 15457–15462.

Nandi P.K. 1998. Polymerization of human prion peptide HuPrP 106–126 to amyloid in nucleic acid solution. *Arch. Virol.* **143:** 1251–1263.

Nandi P.K. and Leclerc E. 1999. Polymerization of murine recombinant prion protein in nucleic acid solution. *Arch. Virol.* **144:** 1751–1763.

Nandi P.K. and Sizaret P.Y. 2001. Murine recombinant prion protein induces ordered aggregation of linear nucleic acids to condensed globular structures. *Arch. Virol.* **146:** 327–345.

Oesch B., Teplow D.B., Stahl N., Serban D., Hood L.E., and Prusiner S.B. 1990. Identification of cellular proteins binding to the scrapie prion protein. *Biochemistry* **29:** 5848–5855.

Pan K.-M., Baldwin M., Nguyen J., Gasset M., Serban A., Groth D., Mehlhorn I., Huang Z., Fletterick R.J., Cohen F.E., and Prusiner S.B. 1993. Conversion of α-helices into β-sheets features in the formation of the scrapie prion proteins. *Proc. Natl. Acad. Sci.* **90:** 10962–10966.

Parren P.W.H.I., Fisicaro P., Labrijn A.F., Binley J.M., Yang W.-P., Ditzel H.J., Barbas C.F., III, and Burton D.R. 1996. In vitro antigen challenge of human antibody libraries for vaccine evaluation: The human immunodeficiency virus type I envelope. *J. Virol.* **70:** 9046–9050.

Peretz D., Williamson R.A., Legname G., Matsunaga Y., Vergara J., Burton D., DeArmond S.J., Prusiner S.B., and Scott M.R. 2002. A change in the conformation of prions accompanies the emergence of a new prion strain. *Neuron* **34:** 921–932.

Peretz D., Williamson R.A., Matsunaga Y., Serban H., Pinilla C., Bastidas R.B., Rozenshteyn R., James T.L., Houghten R.A., Cohen F.E., Prusiner S.B., and Burton D.R. 1997. A conformational transition at the N-terminus of the prion protein features in formation of the scrapie isoform. *J. Mol. Biol.* **273:** 614–622.

Peretz D., Williamson R.A., Kaneko K., Vergara J., Leclerc E., Schmitt-Ulms G., Mehlhorn

I.R., Legname G., Wormald M.R., Rudd P.M., Dwek R.A., Burton D.R., and Prusiner S.B. 2001. Antibodies inhibit prion propagation and clear cell cultures of prion infectivity. *Nature* **412:** 739–743.

Pergami P., Jaffe H., and Safar J. 1996. Semipreparative chromatographic method to purify the normal cellular isoform of the prion protein in nondenatured form. *Anal. Biochem.* **236:** 63–73.

Perrier V., Wallace A.C., Kaneko K., Safar J., Prusiner S.B., and Cohen F.E. 2000. Mimicking dominant negative inhibition of prion replication through structure-based drug design. *Proc. Natl. Acad. Sci.* **97:** 6073–6078.

Perrier V., Kaneko K., Safar J., Vergara J., Tremblay P., DeArmond S.J., Cohen F.E., Prusiner S.B., and Wallace A.C. 2002. Dominant-negative inhibition of prion replication in transgenic mice. *Proc. Natl. Acad. Sci.* **99:** 13079–13084.

Peters P.J., Mironov A., Peretz D., van Donselaar E., Leclerc E., Erpel S., DeArmond S.J., Burton D.R., Williamson R.A., Vey M., and Prusiner S.B. 2003. Trafficking of prion proteins through a caveolae-mediated endosomal pathway. *J. Cell Biol.* **162:** 703–717.

Petrescu A.J., Petrescu S.M., Dwek R.A., and Wormald M.R. 1999. A statistical analysis of N- and O-glycan linkage conformations from crystallographic data. *Glycobiology* **9:** 343–352.

Priola S.A. and Chesebro B. 1995. A single hamster PrP amino acid blocks conversion to protease-resistant PrP in scrapie-infected mouse neuroblastoma cells. *J. Virol.* **69:** 7754–7758.

Priola S.A., Chabry J., and Chan K. 2001. Efficient conversion of normal prion protein (PrP) by abnormal hamster PrP is determined by homology at amino acid residue 155. *J. Virol.* **75:** 4673–4680.

Priola S.A., Raines A., and Caughey W.S. 2000. Porphyrin and phthalocyanine antiscrapie compounds. *Science* **287:** 1503–1506.

Proske D., Gilch S., Wopfner F., Schätzl H.M., Winnacker E.L., and Famulok M. 2002. Prion-protein-specific aptamer reduces PrP^{Sc} formation. *Chembiochem.* **3:** 717–725.

Prusiner S.B. 2001. Shattuck Lecture—Neurodegenerative diseases and prions. *N. Engl. J. Med.* **344:** 1516–1526.

Prusiner S.B., Groth D.F., McKinley M.P., Cochran S.P., Bowman K.A., and Kasper K.C. 1981. Thiocyanate and hydroxyl ions inactivate the scrapie agent. *Proc. Natl. Acad. Sci.* **78:** 4606–4610.

Prusiner S.B., McKinley M.P., Bowman K.A., Bolton D.C., Bendheim P.E., Groth D.F., and Glenner G.G. 1983. Scrapie prions aggregate to form amyloid-like birefringent rods. *Cell* **35:** 349–358.

Prusiner S.B., Groth D., Serban A., Koehler R., Foster D., Torchia M., Burton D., Yang S.-L., and DeArmond S.J. 1993. Ablation of the prion protein (PrP) gene in mice prevents scrapie and facilitates production of anti-PrP antibodies. *Proc. Natl. Acad. Sci.* **90:** 10608–10612.

Prusiner S.B., McKinley M.P., Bolton D.C., Bowman K.A., Groth D.F., Cochran S.P., Hennessey E.M., Braunfeld M.B., Baringer J.R., and Chatigny M.A. 1984. Prions: Methods for assay, purification and characterization. In *Methods in Virology* (ed. K. Maramorosch and H. Koprowski), vol. 8, pp. 293–345. Academic Press, New York.

Prusiner S.B., Scott M., Foster D., Pan K.-M., Groth D., Mirenda C., Torchia M., Yang S.-L., Serban D., Carlson G.A., Hoppe P.C., Westaway D., and DeArmond S.J. 1990. Transgenetic studies implicate interactions between homologous PrP isoforms in scrapie prion replication. *Cell* **63:** 673–686.

Rieger R., Edenhofer F., Lasmézas C.I., and Weiss S. 1997. The human 37-kDa laminin receptor precursor interacts with the prion protein in eukaryotic cells. *Nat. Med.* **3:** 1383–1388.

Riek R., Hornemann S., Wider G., Billeter M., Glockshuber R., and Wüthrich K. 1996. NMR structure of the mouse prion protein domain PrP(121–231). *Nature* **382:** 180–182.

Roikhel V.M., Fokina G.I., and Pogodina V.V. 1984. Influence of aminasine on experimental scrapie in mice. *Acta Virol.* **28:** 321–324.

Rudd P.M., Wormald M.R., Wing D.R., Prusiner S.B., and Dwek R.A. 2001. Prion glycoprotein: Structure, dynamics, and roles for the sugars. *Biochemistry* **40:** 3759–3766.

Ryou C., Legname G., Peretz D., Craig J.C., Baldwin M.A., and Prusiner S.B. 2003. Differential inhibition of prion propagation by enantiomers of quinacrine. *Lab. Invest.* **83:** 837–843.

Safar J., Roller P.P., Gajdusek D.C., and Gibbs C.J.J. 1993. Thermal-stability and conformational transitions of scrapie amyloid (prion) protein correlate with infectivity. *Protein Sci.* **2:** 2206–2216.

Safar J., Wille H., Itri V., Groth D., Serban H., Torchia M., Cohen F.E., and Prusiner S.B. 1998. Eight prion strains have PrP^{Sc} molecules with different conformations. *Nat. Med.* **4:** 1157–1165.

Safar J.G., Scott M., Monaghan J., Deering C., Didorenko S., Vergara J., Ball H., Legname G., Leclerc E., Solforosi L., Serban H., Groth D., Burton D.R., Prusiner S.B., and Williamson R.A. 2002. Measuring prions causing bovine spongiform encephalopathy or chronic wasting disease by immunoassays and transgenic mice. *Nat. Biotechnol.* **20:** 1147–1150.

Schätzl H.M., Laszlo L., Holtzman D.M., Tatzelt J., DeArmond S.J., Weiner R.I., Mobley W.C., and Prusiner S.B. 1997. A hypothalamic neuronal cell line persistently infected with scrapie prions exhibits apoptosis. *J. Virol.* **71:** 8821–8831.

Schenk D., Barbour R., Dunn W., Gordon G., Grajeda H., Guido T., Hu K., Huang J., Johnson-Wood K., Khan K., Kholodenko D., Lee M., Liao Z., Lieberburg I., Motter R., Mutter L., Soriano F., Shopp G., Vasquez N., Vandevert C., Walker S., Wogulis M., Yednock T., Games D., and Seubert P. 1999. Immunization with amyloid-beta attenuates Alzheimer-disease-like pathology in the PDAPP mouse. *Nature* **400:** 173–177.

Schmitt-Ulms G., Legname G., Baldwin M.A., Ball H.L., Bradon N., Bosque P.J., Crossin K.L., Edelman G.M., DeArmond S.J., Cohen F.E., and Prusiner S.B. 2001. Binding of neural cell adhesion molecules (N-CAMs) to the cellular prion protein. *J. Mol. Biol.* **314:** 1209–1225.

Schulemann W. 1932. Synthetic antimalarial drugs. *Br. Med. J.* **1:** 100–101.

Scott M.R., Köhler R., Foster D., and Prusiner S.B. 1992. Chimeric prion protein expression in cultured cells and transgenic mice. *Protein Sci.* **1:** 986–997.

Scott M.R., Butler D.A., Bredesen D.E., Wälchli M., Hsiao K.K., and Prusiner S.B. 1988. Prion protein gene expression in cultured cells. *Protein Eng.* **2:** 69–76.

Scott M., Groth D., Foster D., Torchia M., Yang S.-L., DeArmond S.J., and Prusiner S.B. 1993. Propagation of prions with artificial properties in transgenic mice expressing chimeric PrP genes. *Cell* **73:** 979–988.

Sethi S., Lipford G., Wagner H., and Kretzschmar H. 2002. Postexposure prophylaxis against prion disease with a stimulator of innate immunity. *Lancet* **360:** 229–230.

Shoichet B.K. and Kuntz I.D. 1991. Protein docking and complementarity. *J. Mol. Biol.* **221:** 327–346.

Soto C., Kascsak R.J., Saborío G.P., Aucouturier P., Wisniewski T., Prelli F., Kascsak R., Mendez E., Harris D.A., Ironside J., Tagliavini F., Carp R.I., and Frangione B. 2000.

Reversion of prion protein conformational changes by synthetic beta-sheet breaker peptides. *Lancet* **355:** 192–197.

Spudich S., Mastrianni J.A., Wrensch M., Gabizon R., Meiner Z., Kahana I., Rosenmann H., Kahana E., and Prusiner S.B. 1995. Complete penetrance of Creutzfeldt-Jakob disease in Libyan Jews carrying the E200K mutation in the prion protein gene. *Mol. Med.* **1:** 607–613.

Supattapone S., Nguyen H.-O.B., Cohen F.E., Prusiner S.B., and Scott M.R. 1999a. Elimination of prions by branched polyamines and implications for therapeutics. *Proc. Natl. Acad. Sci.* **96:** 14529–14534.

Supattapone S., Wille H., Uyechi L., Safar J., Tremblay P., Szoka F.C., Cohen F.E., Prusiner S.B., and Scott M.R. 2001. Branched polyamines cure prion-infected neuroblastoma cells. *J. Virol.* **75:** 3453–3461.

Supattapone S., Bosque P., Muramoto T., Wille H., Aagaard C., Peretz D., Nguyen H.-O.B., Heinrich C., Torchia M., Safar J., Cohen F.E., DeArmond S.J., Prusiner S.B., and Scott M. 1999b. Prion protein of 106 residues creates an artificial transmission barrier for prion replication in transgenic mice. *Cell* **96:** 869–878.

Svendsen C.N., Hrbek C.C., Casendino M., Nichols R.D., and Bird E.D. 1988. Concentration and distribution of thioridazine and metabolites in schizophrenic post-mortem brain tissue. *Psychiatry Res.* **23:** 1–10.

Tagliavini F., McArthur R.A., Canciani B., Giaccone G., Porro M., Bugiani M., Lievens P.M.-J., Bugiani O., Peri E., Dall'Ara P., Rocchi M., Poli G., Forloni G., Bandiera T., Varasi M., Suarato A., Cassutti P., Cervini M.A., Lansen J., Salmona M., and Post C. 1997. Effectiveness of anthracycline against experimental prion disease in Syrian hamsters. *Science* **276:** 1119–1122.

Tang M.X., Redemann C.T., and Szoka F.C.J. 1996. In vitro gene delivery by degraded polyamidoamine dendrimers. *Bioconjug. Chem.* **7:** 703–714.

Taraboulos A., Raeber A.J., Borchelt D.R., Serban D., and Prusiner S.B. 1992a. Synthesis and trafficking of prion proteins in cultured cells. *Mol. Biol. Cell* **3:** 851–863.

Taraboulos A., Scott M., Semenov A., Avrahami D., Laszlo L., and Prusiner S.B. 1995. Cholesterol depletion and modification of COOH-terminal targeting sequence of the prion protein inhibits formation of the scrapie isoform. *J. Cell Biol.* **129:** 121–132.

Taraboulos A., Borchelt D.R., McKinley M.P., Raeber A., Serban D., DeArmond S.J., and Prusiner S.B. 1992b. Dissecting the pathway of scrapie prion synthesis in cultured cells. In *Prion diseases of humans and animals* (ed. S.B. Prusiner et al.), pp. 434–444. Ellis Horwood, London.

Tatzelt J., Maeda N., Pekny M., Yang S.-L., Betsholtz C., Eliasson C., Cayetano J., Camerino A.P., DeArmond S.J., and Prusiner S.B. 1996. Scrapie in mice deficient in apolipoprotein E or glial fibrillary acidic protein. *Neurology* **47:** 449–453.

Taylor D.M. 2000. Inactivation of transmissible degenerative encephalopathy agents: A review. *Vet. J.* **159:** 10–17.

Taylor D.M. and Fernie K. 1996. Exposure to autoclaving or sodium hydroxide extends the dose-response curve of the 263K strain of scrapie agent in hamsters. *J. Gen. Virol.* **77:** 811–813.

Telling G.C., Scott M., Mastrianni J., Gabizon R., Torchia M., Cohen F.E., DeArmond S.J., and Prusiner S.B. 1995. Prion propagation in mice expressing human and chimeric PrP transgenes implicates the interaction of cellular PrP with another protein. *Cell* **83:** 79–90.

Telling G.C., Scott M., Hsiao K.K., Foster D., Yang S.-L., Torchia M., Sidle K.C.L., Collinge J., DeArmond S.J., and Prusiner S.B. 1994. Transmission of Creutzfeldt-Jakob disease from humans to transgenic mice expressing chimeric human-mouse prion protein. *Proc. Natl. Acad. Sci.* **91:** 9936–9940.

Tilley G., Chapuis J., Vilette D., Laude H., and Vilotte J.L. 2003. Efficient and specific downregulation of prion protein expression by RNAi. *Biochem. Biophys. Res. Commun.* **305:** 548–551.

Tilley J.W., Chen L., Fry D.C., Emerson S.D., Powers G.D., Biondi D., Varnell T., Trilles R., Guthrie R., Mennona F., Kaplan G., LeMahieu R.A., Carson M., Han R.-J., Liu C.-M., Palermo R., and Ju G. 1997. Identification of a small molecule inhibitor of the IL-2/IL-2R receptor interaction which binds to IL-2. *J. Am. Chem. Soc.* **119:** 7589–7590.

Tremblay P., Meiner Z., Galou M., Heinrich C., Petromilli C., Lisse T., Cayetano J., Torchia M., Mobley W., Bujard H., DeArmond S.J., and Prusiner S.B. 1998. Doxycyline control of prion protein transgene expression modulates prion disease in mice. *Proc. Natl. Acad. Sci.* **95:** 12580–12585.

Tucker G.T. 2000. Chiral switches. *Lancet* **355:** 1085–1087.

Ullman J.R. 1976. An algorithm for subgraph isomorphism. *J. Assoc. Comput. Machinery* **23:** 31–42.

Vey M., Pilkuhn S., Wille H., Nixon R., DeArmond S.J., Smart E.J., Anderson R.G., Taraboulos A., and Prusiner S.B. 1996. Subcellular colocalization of the cellular and scrapie prion proteins in caveolae-like membranous domains. *Proc. Natl. Acad. Sci.* **93:** 14945–14949.

Wells G.A.H. 2002. European Commission report on TSE infectivity distribution in ruminant tissues (*State of Knowledge,* December 2001), pp. 10–37.

Will R.G., Alpers M.P., Dormont D., Schonberger L.B., and Tateishi J. 1999. Infectious and sporadic prion diseases. In *Prion biology and diseases* (ed. S.B. Prusiner), pp. 465–507. Cold Spring Harbor Laboratory Press, Cold Spring Harbor, New York.

Will R.G., Ironside J.W., Zeidler M., Cousens S.N., Estibeiro K., Alperovitch A., Poser S., Pocchiari M., Hofman A., and Smith P.G. 1996. A new variant of Creutzfeldt-Jakob disease in the UK. *Lancet* **347:** 921–925.

Wille H., Michelitsch M.D., Guénebaut V., Supattapone S., Serban A., Cohen F.E., Agard D.A., and Prusiner S.B. 2002. Structural studies of the scrapie prion protein by electron crystallography. *Proc. Natl. Acad. Sci.* **99:** 3563–3568.

Williamson R.A., Peretz D., Pinilla C., Ball H., Bastidas R.B., Rozenshteyn R., Houghten R.A., Prusiner S.B., and Burton D.R. 1998. Mapping the prion protein using recombinant antibodies. *J. Virol.* **72:** 9413–9418.

Williamson R.A., Peretz D., Smorodinsky N., Bastidas R., Serban H., Mehlhorn I., DeArmond S.J., Prusiner S.B., and Burton D.R. 1996. Circumventing tolerance to generate autologous monoclonal antibodies to the prion protein. *Proc. Natl. Acad. Sci.* **93:** 7279–7282.

Yehiely F., Bamborough P., Costa M.D., Perry B.J., Thinakaran G., Cohen F.E., Carlson G.A., and Prusiner S.B. 1997. Identification of candidate proteins binding to prion protein. *Neurobiol. Dis.* **3:** 339–355.

Zulianello L., Kaneko K., Scott M., Erpel S., Han D., Cohen F.E., and Prusiner S.B. 2000. Dominant-negative inhibition of prion formation diminished by deletion mutagenesis of the prion protein. *J. Virol.* **74:** 4351–4360.

Glossary

Words or phrases in bold have their own Glossary entry.

α-Helix: Structural element of a protein in which the polypeptide backbone follows a right-handed helical path approximating a cylinder, with a periodicity of 3.6 amino acids per turn.

Amyloid: Proteinaceous fibrillar polymers, identified by their tinctorial, ultrastructural, and spectroscopic properties. Amyloid deposition is a feature of some neurodegenerative diseases; the amyloid of each disease is composed of a different protein. The amyloid in **prion diseases** is composed of **PrP 27-30** or small fragments of **PrP^{Sc}**.

β-Helix: Structural element of a protein in which **β-strands** follow a right- or left-handed helical path forming three faces, each a parallel **β-sheet**.

β-Sheet: Structural element of a protein in which adjacent **β-strands** lie approximately flat in either a parallel or antiparallel orientation to one another.

β-Strand: An extended section of polypeptide backbone, incorporated into **β-helix** or **β-sheet**, in which successive amino acid side chains point in opposite directions.

Bo: Bovine (cattle).

Bovine spongiform encephalopathy: A **prion disease** in cattle. Cases of **BSE** have been reported in Austria, Britain, Canada, Czech Republic, Denmark, France, Germany, Ireland, Israel, Italy, Japan, Poland, Portugal, Spain, and Switzerland.

BSE: Bovine spongiform encephalopathy.

CDI: Conformation-dependent immunoassay.

Cervid: The family including deer and elk.

Chronic wasting disease: A **prion disease** in **cervids**, including elk, mule deer, and white-tailed deer. Presents predominantly as a loss of muscle mass. **CWD** has been reported in Canada, Korea, and the United States.

CJD: Creutzfeldt-Jakob disease.

Conformation-dependent immunoassay: A rapid procedure to measure **PrP^{Sc}** by using antibodies that react with an epitope that is exposed in native **PrP^{C}** but buried in native, infectious **PrP^{Sc}**. Because this assay does not require limited proteolysis to hydrolyze **PrP^C**, it can be used to measure both protease-sensitive (**sPrP^{Sc}**) and protease-resistant (**rPrP^{Sc}**) **PrP^{Sc}**.

Creutzfeldt-Jakob disease: A prion disease in humans. **CJD** can be manifest as a genetic (familial CJD [**fCJD**]), infectious (iatrogenic CJD [**iCJD**] and variant CJD [**vCJD**]) or sporadic (**sCJD**) disorder. Approximately 85% of **CJD** cases occur sporadically and 10–15% are inherited (fCJD); less than 1% are infectious.

^{Ctm}PrP: A transmembrane form of **PrP** in which the carboxyl terminus protrudes into the lumen of the cell. The accumulation of ^{Ctm}PrP seems to induce neurodegeneration. Whether nervous degeneration caused by the accumulation of **PrP^{Sc}** is mediated by ^{Ctm}PrP remains to be established.

CWD: Chronic wasting disease.

Dominant-negative inhibition: A phenomenon in which a polymorphic or mutant protein interferes with the function of the wild-type protein. Dominant-negative inhibition of **PrP^{Sc}** formation has been demonstrated for two different residues in **PrP**, both of which are believed to be part of a discontinuous epitope that forms the binding site for **protein X**. The mechanism of dominant-negative inhibition is thought to be due to the more avid binding of mutant **PrP** to **protein X** compared to that of **wt** PrP. In such a scenario, mutant **PrP** would sequester **protein X** and, thus, prevent it from facilitating the conversion of **wt PrP^C** to **PrP^{Sc}**.

Dpl: Doppel.

Doppel: A paralog of **PrP**, the term is derived from "downstream prion protein-like." Dpl shares only 25% sequence identity with **PrP**, but possesses a quite similar structure. Dpl is anchored to the cell surface by a **GPI** moiety. In mammals, Dpl expression is restricted to the testis; when it is expressed in brain, it causes cerebellar dysfunction and neurodegeneration.

Endpoint titration: Method for determining infectivity of a prion sample by bioassay with a serially diluted inoculum. The endpoint is defined as the dilution at which the inoculum is no longer infectious.

Epitope: A site on a large molecule to which a specific antibody binds. In proteins, most epitopes are composed of 6–8 amino acids.

Familial Creutzfeldt-Jakob disease: An inherited form of human **prion disease** caused by a mutation in the *PRNP* gene that results in the substitution of an amino acid or the insertion of additional amino acids. Whether mutant truncated **PrP** causes familial prion disease is uncertain. The most common example of **fCJD** is caused by a mutation resulting in the substitution of lysine for glutamic acid at position 200.

Fatal insomnia: A human prion disease due to selective degeneration of the dorsal medial nuclei of the thalamus, resulting in reduced sleep. Both sporadic and genetic forms of FI have been described. The inherited form of FI, referred to as fatal familial insomnia (FFI), is caused by a mutation in *PRNP* that results in the substitution of asparagine for aspartic acid at position 178. A methionine at the polymorphic position 129 is also required for the FI phenotype. A valine at position 129 prevents the FI phenotype but produces neurodegeneration with clinical signs typical of **fCJD**.

fCJD: Familial Creutzfeldt-Jakob disease.

Feline spongiform encephalopathy: A prion disease in great cats and small domestic cats. Most cases reported from Britain were in great cats housed in zoos and in pet cats from domestic households. It is presumed that these cats developed **FSE** after consuming food contaminated with **BSE** prions.

FSE: Feline spongiform encephalopathy.

Gerstmann-Sträussler-Scheinker disease: An inherited human prion disease caused by a point mutation in **PRNP** and involving the deposition of **PrP amyloid** plaques. Several different point mutations causing **GSS** have been identified, including the substitution of leucine for proline at position 102, or valine for alanine at position 117. Although the clinical characteristics seem frequently to be determined by the particular mutation, there are many exceptions. Generally, **GSS** has a longer clinical course compared to other prion diseases, often lasting several years.

Glycoforms: Variations of a particular protein differing in quantity as well as the composition of the covalently attached sugar residues. Generally, three different forms of **PrP^C** can be discriminated by polyacrylamide gel electrophoresis: (1) unglycosylated, (2) monoglycosylated, and (3) diglycosylated.

GPI: Glycosylphosphatidylinositol, a glycolipid anchor joined posttranslationally to certain proteins upon cleavage of a carboxy-terminal signal sequence. The lipid portion of the anchor attaches the protein to the external surface of the cellular membrane.

GSS: Gerstmann-Sträussler-Scheinker disease.

Haplotype: A mutation in a gene in combination with a specific **polymorphism**. For example, in human **PrP**, the **fatal insomnia** haplotype D178N,M129 signifies a mutation at amino acid residue 178, where asparagine (N) is substituted for aspartic acid (D), in combination with a methionine (M) at residue 129. A different haplotype, with the D178N mutation and a valine at residue 129, produces fCJD(D178N,V129).

[Het-s]: Prion state of yeast protein HET-s.

Hu: Human.

HuM: Chimeric human–mouse protein.

Iatrogenic Creutzfeldt-Jakob disease: A form of **CJD** that is transmitted accidentally through medical procedures, such as the transplantation of prion-contaminated dura mater or other tissue, injection of prion-contaminated human growth hormone from pituitary extracts, or the use of improperly decontaminated surgical instruments as well as devices.

iCJD: Iatrogenic Creutzfeldt-Jakob disease.

ID_{50} unit: The dose of an infectious agent sufficient to infect 50% of exposed cells, animals, or people. The number of infectious particles comprising one ID_{50} unit for a particular infectious agent may vary depending on the species of animal and the route of exposure. For example, many more PrP^{Sc} molecules are required for one ID_{50} unit when delivered orally compared to intracerebral injection; i.e., the **P/I ratio** for prions is much higher in oral infection than by intracerebral inoculation.

Incubation period: Interval of time between initial prion infection and the presentation of clinical signs of disease.

Kuru: A prion disease afflicting the Fore people of Papua New Guinea transmitted orally by ritualistic cannibalism.

Mad cow disease: Common name for **bovine spongiform encephalopathy.**

MH2M: Chimera of mouse **PrP** and Syrian hamster **PrP**, which harbors the Syrian hamster sequence at amino acids 108, 111, 138, 154, and 169 (mouse numbering) and the mouse sequence at all other positions.

MHM2: Chimera of mouse **PrP** and Syrian hamster **PrP**, which harbors the Syrian hamster sequence at amino acids 108 and 111 (mouse numbering) and the mouse sequence at all other positions. These Syrian hamster residues create an **epitope** that enables the chimera to be differentially recognized from mouse **PrP** by antibodies.

MHu2M: Chimera of mouse **PrP** and human **PrP**, which harbors the human sequence at nine amino acids within the "Hu2" insert and the mouse sequence at all other positions. These nine human residues render transgenic mice susceptible to human prions.

MHu2M(M165V,E167Q): Chimera of mouse **PrP** and human **PrP**, which harbors the human sequence at seven amino acids and the mouse sequence at all other positions. The human residues at positions 165 and 167 were reverted to mouse residues within the "Hu2" insert. The seven human residues within the modified "Hu2" insert render mice more susceptible to some human prions than do the nine residues of the unmodified "Hu2" insert that is found in the **MHu2M** transgene.

Mo: Mouse.

NMR: Nuclear magnetic resonance spectroscopy.

^{Ntm}PrP: A transmembrane form of **PrP** in which the NH_2-terminus protrudes into the lumen of the cell.

Octapeptide repeat: Region in the amino terminus of PrP in which the conserved peptide sequence PHGGGWGQ is repeated, usually five times. Two to nine additional repeats produce **fCJD**. One additional octarepeat in humans or cattle seems to have no effect on the frequency of prion disease.

Ov: Ovine (sheep).

Particle-to-infectivity ratio: The number of infectious particles required to produce one ID_{50} **unit**. The lowest possible **P/I ratio** is 1. The **P/I ratio** for prions is generally $10^5–10^6$ **PrP^{Sc}** molecules per ID_{50} **unit** as determined in rodents.

P/I ratio: Particle-to-infectivity ratio.

PK: Proteinase K.

Polymorphism: Occurrence of two or more common residues at a single position in a sequence. The expression of methionine and valine at position 129 in human **PrP** represents a polymorphism.

Prion: An infectious protein. In mammals, prions are composed solely of **PrP^{Sc}** molecules and cause neurodegenerative diseases. In contrast to all other infectious pathogens, prions do not contain nucleic acid. In yeast, prions are infectious proteins distinct from **PrP** that are transferred in a non-Mendelian manner. To date, four yeast prions have been identified.

Prion disease: An invariably fatal, neurodegenerative disease caused by prions that can be manifest as a genetic, infectious, or sporadic disorder. The term "**transmissible spongiform encephalopathy,**" or **TSE**, is also used to denote prion disease.

Prion protein: A protein of ~250 amino acids encoded by a chromosomal gene in mammals and birds. The normal, cellular form (**PrP^C**) is converted into the disease-causing form (**PrP^{Sc}**) by an as yet undefined process whereby PrP undergoes a profound conformational change.

Prion rod: An **amyloid** polymer composed of **PrP 27-30** molecules. Created by detergent extraction and limited proteolysis of **PrP^{Sc}**.

Prion strain: A distinct variety of prions that results in a characteristic phenotype, as measured by the length of the incubation period and the distribution of PrP^{Sc} in brain as well as the pattern of neuropathological lesions. Considerable evidence argues that the biological information of a particular strain is enciphered in the conformation of **PrP^{Sc}**.

Prnd: The gene that codes for the **doppel** protein.

PRNP: The gene that codes for the human **prion protein,** located near the end of the short arm of chromosome 20.

Prnp: The gene that codes for the **prion protein,** located on syntenic chromosome 2 in mice. *Prnp* controls the length of the prion incubation time and is congruent with the incubation-time genes *Sinc* and *Prni*.

Prnp$^{0/0}$: Genotype in which both alleles of *Prnp* are disrupted, resulting in the absence of **PrP** expression. *Prnp* "knockout" mice are resistant to prion disease and do not replicate prions.

Protein X: A hypothetical macromolecule that is thought to act as a molecular chaperone in facilitating the conversion of **PrP^C** into **PrP^{Sc}**.

Proteinase K: An enzyme (protease) that is used to catalyze the hydrolysis of a protein. **PK** is obtained from the fungus *Tritirachium album*, which derives its carbon and nitrogen from keratin, to which the letter K refers.

PrP: Prion protein. Lack of a superscript usually refers to **PrP^C**, but may also refer more generally to both **PrP^C** and **PrP^{Sc}**.

PrP 27-30: A fragment of **PrP^{Sc}** of ~140 amino acids, generated by amino-terminal truncation through digestion with **proteinase K**. The numbers refer to the molecular mass range (in kilodaltons) of this PrP fragment. PrP 27-30 retains prion infectivity and polymerizes into **amyloid**.

PrP amyloid: Polymers of **PrP 27-30** that exhibit the fibrous ultrastructure and tinctorial properties of **amyloid**. The amyloid plaques in the brains of patients with **GSS** or **kuru** contain a variety of **PrP** peptides presumably derived from **PrP 27-30**. Plaques containing PrP amyloid are found in the brains of some mammals with prion disease.

PrP106: Redacted **prion protein** with two large deletions: residues 23–88 and 141–176 (mouse numbering), resulting in a protein of ~106 residues (depending on the species). Also written as MoPrP(Δ23–88,Δ141–176).

PrP*: The conformation of **PrP** when it is bound to **protein X** and before it is converted to **PrP^{Sc}**.

PrP^{BSE}: PrP^{Sc} in **BSE**.

PrP^C: Normal, cellular isoform of the **prion protein**. PrP^C has three α-**helices** and two small β-**strands**. PrP^C is readily digested by proteases.

PrP^{CJD}: PrP^{Sc} in **CJD**.

PrP^{CWD}: PrP^{Sc} in **CWD**.

PrP^{res}: A form of **PrP** that is highly resistant to digestion by **proteinase K**. PrP^{res} is commonly used to denote **PrP** molecules that have acquired protease resistance by in vitro manipulations. This term has been used as an alternative name for **PrP^{Sc}**, but it became ambiguous with the discovery of protease-sensitive forms of **PrP^{Sc}**, denoted **sPrP^{Sc}**. PrP^{res} is also used as a synonym for **PrP 27-30**.

PrP^{Sc}: Misfolded or alternatively folded isoform of the **prion protein** that is the sole component of the infectious **prion** particle. This protein is the only identifiable macromolecule in purified preparations of prions. PrP^{Sc} has a high β-**sheet** content. Limited digestion of PrP^{Sc} leads to amino-terminal truncation, producing **PrP 27-30**. The "Sc" superscript was originally derived from "scrapie" but is now used generically to indicate all disease-causing isoforms of PrP.

PrP^{sen}: A form of the **prion protein** that demonstrates sensitivity to protease-catalyzed digestion, denoted by the "sen" superscript. This term has been used as an alternative name for **PrP^C**, but it became ambiguous with the discovery of protease-sensitive forms of **PrP^{Sc}**, denoted **sPrP^{Sc}**.

[*PSI*$^+$]: Prion state of yeast protein Sup35p.

rec: Recombinant.

rPrP^{Sc}: Protease-resistant form of **PrP^{Sc}**.

sCJD: Sporadic Creutzfeldt-Jakob disease.

ScN2a: Neuroblastoma cell line chronically infected with prions.

Scrapie: A prion disease in sheep and goats.

sPrP^{Sc}: Protease-sensitive form of **PrP^{Sc}**.

^{Sec}PrP: A secretory form of **PrP** that is protected from proteolysis.

SHa: Syrian hamster.

Species barrier: A prolongation of the incubation period when prions are transferred from one species to another. On second passage of prions in the same species, the incubation period decreases, then remains constant on subsequent passages.

Sporadic Creutzfeldt-Jakob disease: The most common form of **CJD**, whose mechanism is unknown. **sCJD** appears to occur worldwide at a rate of approximately one case per million population. Most cases involve older adults.

Strain: See **Prion strain**.

Strain barrier: The abbreviation or prolongation of the **incubation period** resulting from transmission of a particular **prion strain** from one species to another. Generally, a prolongation of the **incubation period** has been seen due to the **species barrier**; however, abbreviated **incubation periods** have been observed for a particular **prion strain** upon transfer from one species to another compared to passage in the same species. In such cases, the strain barrier is apparent, whereas in other instances, this barrier may be less clear. On second passage of a particular **prion strain** in the same species, the **incubation period** may show an additional diminution, but then it remains constant on subsequent passages. It should be noted that transmission of a particular **prion strain** from one species to another can give rise to a new **prion strain**. Sometimes passage of a particular **prion strain** into multiple species may help decipher the nature of the **transmission barrier**.

Tg: Transgenic.

TME: Transmissible mink encephalopathy.

Titer: Concentration of infectious particles, usually presented as the maximum dilution at which it is still infectious.

Transgene: A gene from one organism that has been transferred and integrated into the DNA of another organism of the same or different species, such that the transferred gene is expressed in the new host organism. Investigators conducting prion transmission studies use transgenes to convey unnatural molecular characteristics to experimental animals so as to circumvent the **species barrier**. For example, transmission of hamster prions to transgenic mice expressing hamster **PrP** abrogates the species barrier. When the mouse *Prnp* is disrupted

(*Prnp*$^{0/0}$) in transgenic mice expressing hamster **PrP**, the **incubation period** for hamster prions decreases. In transgenic mice expressing human **PrP**, disruption of the mouse **PrP** gene is required for transmission of human prions.

Transmissible mink encephalopathy: A **prion disease** in mink.

Transmissible spongiform encephalopathy: A fatal, neurodegenerative disease caused by prions. Although the term **TSE** predates the term **prion disease**, the term **TSE** is ambiguous both with respect to etiology and neuropathology: **TSE** emphasizes the infectious form of **prion disease** and ignores the much more common genetic and sporadic forms. Moreover, the spongiform changes, or vacuolation, found in the central nervous system of mammals with prion disease is quite variable; some cases of prion disease show little, if any, spongiform degeneration.

Transmission barrier: An abbreviated or prolonged **incubation period** on the first passage of prions from one species to another, followed by a shortening of the incubation time on second passage. An abbreviated **incubation period** can be seen due to a prion **strain barrier** in contrast to a prolonged incubation time that is due to a **species barrier**. Deciphering whether the **strain barrier** or **species barrier** is responsible for the change in incubation times may require passage of a particular **prion strain** into multiple hosts.

TSE: Transmissible spongiform encephalopathy.

[URE3]: Prion state of yeast protein Ure2p.

Variant Creutzfeldt-Jakob disease: A new phenotype of **CJD** believed to be caused by the consumption of beef products tainted by infectious **BSE** prions. The majority of vCJD cases occur in young adults. Transmission of vCJD prions by the transfusion of a blood product from a donor who has vCJD has been postulated but is, to date, unproven.

vCJD: Variant Creutzfeldt-Jakob disease.

wt: wild-type.

Appendix

Prion Protein (PrP) and Doppel (Dpl) Sequences

Kurt Giles[1] and Stanley B. Prusiner[1,2]

[1]Institute for Neurodegenerative Diseases and
Departments of Neurology and [2]Biochemistry and Biophysics
University of California, San Francisco, California 94143

THE FIRST COMPLETE PrP SEQUENCES WERE REPORTED in 1986 (Basler et al. 1986; Locht et al. 1986). Since then, the PrP genes from more than 100 species have been sequenced. PrP orthologs have been observed in a wide variety of terrestrial vertebrates, representatives of which are shown in Table 1. Full and partial sequences from many other species are available, particularly for primates, ungulates, and rodents (Schätzl et al. 1995; Wopfner et al. 1999; van Rheede et al. 2003).

The discovery of a PrP paralog, termed "doppel" for "downstream, prion protein-like," which has a tertiary structure similar to that of PrP (Mo et al. 2001) but only ~25% sequence identity (Moore et al. 1999), promises to advance the understanding of PrP function and its role in neurodegeneration. To date, Dpl has been reported only in placental mammals (Table 2).

In the annotated sequence alignment of PrP and Dpl (Fig. 1), the conservation of secondary structure (above sequences) and disulfide bonds (below sequences) between the two proteins is highlighted. Rectangles delineate the predicted signal sequences for translocation (amino termini) and GPI anchor addition (carboxyl termini). Secondary structure, (cylinders) α-helices; (arrows) β-strands; (solid lines) no regular secondary structure; (dashed lines) undetermined structure, is based on a consensus of wild-type structures determined by NMR spectroscopy of human (Zahn et al. 2000), mouse (Rick et al. 1996), Syrian hamster (James et al. 1997), and cow PrP (López García 2000); mouse (Mo et al. 2001) and human Dpl (Lührs et al. 2003). The predominant polymorphic forms of the respective PrP and Dpl sequences are shown.

Table 1. PrP sequences of representative terrestrial vertebrates

Class	Subclass	Order	Common name	Scientific name	Accession number[a]	Identity (%)[b]	Reference
Mammalia	Boreoeutheria	Primates	human	*Homo sapiens*	P04156	100	Kretzschmar et al. (1986)
			chimpanzee	*Pan troglodytes*	P40253	99	Cervenakova et al. (1994)
			Rhesus macaque	*Macaca mulatta*	P40254	95	Schätzl et al. (1995)
			marmoset	*Callithrix jacchus*	P40247	95	Schätzl et al. (1995)
		Dermoptera	flying lemur	*Cynocephalus variegatus*	Q866W9	92	van Rheede et al. (2003)
		Scandentia	tree shrew	*Tupaia tana*	Q866W8	85	van Rheede et al. (2003)
		Rodentia	mouse	*Mus musculus*	P04925	90	Locht et al. (1986)
			rat	*Rattus norvegicus*	P13852	90	Gomi et al. (1994)
			hamster	*Mesocricetus auratus*	P04273	91	Basler et al. (1986)
			squirrel	*Sciurus vulgaris*	Q811W7	90	van Rheede et al. (2003)
		Caviomorpha	guinea pig	*Cavia porcellus*	Q811W5	80	van Rheede et al. (2003)
		Lagomorpha	rabbit	*Oryctolagus cuniculus*	Q95211	89	Loftus and Rogers (1997)
			pika	*Ochotona princeps*	Q866W7	89	van Rheede et al. (2003)
		Eulipotyphia	mole	*Talpa europaea*	Q866W6	84	van Rheede et al. (2003)
			hedgehog	*Erinaceus europaeus*	Q866W5	81	van Rheede et al. (2003)
			gymnure	*Hylomys suillus*	Q866W4	78	van Rheede et al. (2003)
		Chiroptera	fruit bat	*Cynopterus sphinx*	Q866W2	90	van Rheede et al. (2003)
			Daubenton's bat	*Myotis daubentoni*	Q866W0	83	van Rheede et al. (2003)
			leaf-nosed bat	*Macrotus californicus*	Q866V9	82	van Rheede et al. (2003)
		Carniovora	mink	*Mustela vison*	P40244	91	Kretzschmar et al. (1992)
			ferret	*Mustela putorius furo*	P52114	90	Bartz et al. (1994)
			dog	*Canis familiaris*	O46592	90	Doyle D. and Rogers M.S.[c]
			cat	*Felis silvestris catus*	O18754	89	Rohwer R.G., Edelman D., and Protzman J.L.[c]
		Pholidota	pangolin	*Manis tetradactyla*	Q866V8	89	van Rheede et al. (2003)
		Perissodactyla	horse	*Equus equus*	Q866V7	90	van Rheede et al. (2003)
			black rhinoceros	*Diceros bicornis*	Q866V6	91	van Rheede et al. (2003)
		Certartiodactyla	pig	*Sus scrofa*	P49927	91	Martin et al. (1995)
			camel	*Camelus dromedarius*	P79141	91	Kaluz et al. (1997)
			sheep	*Ovis aries*	P23907	92	Goldmann et al. (1990)

Class	Order	Common name	Species	Accession no.[a]	% identity[b]	Reference
		goat	*Capra hircus*	P52113	92	Goldmann et al. (1996)
		cow	*Bos taurus*	P10279	90	Goldmann et al. (1991)
		kudu	*Tragelaphus strepsiceros*	P40242	89	Martin T.C., Hughes S.L., Hughes K.J., and Dawson M.[c]
		antelope	*Antilocapra americana*	Q9MZU6	90	Raymond et al. (2000)
		elk	*Cervus elaphus nelsoni*	P79142	92	O'Rourke et al. (1998)
		giraffe	*Giraffa camelopardalis*	O97695	89	Wopfner et al. (1999)
		hippopotamus	*Hippopotamus amphibius*	Q866V5	91	van Rheede et al. (2003)
		sperm whale	*Physeter catodon*	Q866V4	91	van Rheede et al. (2003)
Xenartha	Pilosa	anteater	*Cyclopes didactylus*	Q866U6	89	van Rheede et al. (2003)
Afrothera	Afrosoricida	tenrec	*Tenrec ecaudatus*	Q866U8	80	van Rheede et al. (2003)
		golden mole	*Amblysomus hottentotus*	Q866U7	84	van Rheede et al. (2003)
	Macroscelidea	elephant shrew	*Macroscelides proboscideus*	Q866U9	85	van Rheede et al. (2003)
	Tubulidentata	aardvark	*Orycteropus afer*	Q866V0	89	van Rheede et al. (2003)
	Hyracoidea	hyrax	*Procavia capensis*	Q866V1	88	van Rheede et al. (2003)
	Sirenia	manatee	*Trichechus manatus*	Q866V2	89	van Rheede et al. (2003)
	Proboscidea	elephant	*Elephas maximus*	Q866V3	87	van Rheede et al. (2003)
Marsupialia	Polyprotodonta	possum	*Trichosurus vulpecula*	P51780	75	Windl O., Dempster M., Estibeiro P., and Lathe R.[c]
Reptilia	Strigiformes	owl	*Tyto alba*	Q9PW93	41	Wopfner et al. (1999)
Aves	Anseriformes	duck	*Anas platyrhynchos*	Q9PW95	43	Wopfner et al. (1999)
	Ciconiiformes	condor	*Vultur gryphus*	Q9PW97	42	Wopfner et al. (1999)
	Gruiformes	crane	*Balearica pavonina gibbericeps*	Q9PW96	42	Wopfner et al. (1999)
	Columbiformes	pigeon	*Columba rupestris*	Q9I9F3	41	Zhang et al. (2002)
	Galliformes	chicken	*Gallus gallus*	P27177	41	Gabriel et al. (1992)
Testudines	Cryptodira	slider turtle	*Trachemys scripta*	Q9I9C0	39	Simonic et al. (2000)
		softshell turtle	*Trionyx sinensis*	Q801Y1	40	Fukuoka S., Yanagihara K., and Goto E.[c]
Amphibia	Salientia	frog	*Xenopus laevis*	Q8QFR0	36	Strumbo et al. (2001)
Lissamphibia						

[a] Accession number from Swiss-Prot database (http://us.expasy.org/sprot/).
[b] Amino acid identity based on a global alignment (Needleman and Wunsch 1970) of the mature protein (signal peptides removed) with mature, full-length human PrP.
[c] Direct submission to database.

Dpl

Human	MRKHLSWWWLATVCMLLFSHLSA VQTRGIKHR------	32
Mouse	.KNR.GT.V.IL....A...T.KA.......	32
Cow	...GGC...I.I...Q.CS.KA.......	32
Sheep	...GGC...I.V...Q..S.KA.......	32

PrP

Human	MANLGCWMLVLFVATWSDLGLC KKRPKPGG-WNTGGSRYPGQGSPGGNRYPPQGGGWGQPHGGGWGQ------PHGGGWGQPHGGG	80
Mouse	.....Y.L.A..TM.TV............-T......S.	79
Hamster	....SY.L.A..M.TV..........-T	80
Mink	MVKSHI.S.L....I.F.....G......	83
Cow	MVKSHI.S.I....M..V.....G......PHGGGWGQ.	91
Sheep	MVKSHI.S.I....M..V.....G......	83
Elk	MVKSHI.S.I....M..V.....G......	83

Dpl   β1   αA   β2

Human	----IKWNRKALPSTA-QITEAQVAE-----NRPGAFIKQGRKLDIFG-AEGNRYYEANYWQFPDGIHYNGCSEAN	98
Mouse	----F.....V..SGG....R.......A....Y.E......	99
Cow	----......V..S-V.RT...-I.......-V....K.....	98
Sheep	----......V..S-V.HT...-I.......N..-V.......	98

PrP   β1   αA   β2

Human	WGQPHGGG-WGQGGGTHSQWNKPSKPKTNMKHMAGAAAAGAVVGGLGGYMLGSAMSRPIIHFGSDYEDRYYRENMHRYPNQVYYRPMDEYS	170
Mouse	.......N..............L.V.........M..N.W...Y...............V.Q.	169
Hamster	......-.....N............MM..N.W...N.............V.Q.N	170
Mink	....G.....S.G..G....V.........L..N...Y..............K.V.Q.	174
Cow	....G..---.G........V.........L.............Y....V.Q.	181
Sheep	....G..-S............V.........L..N.....Y..........V.Q.	173
Elk	....G..---............V.........L..N.....Y..........V.Q.N	173

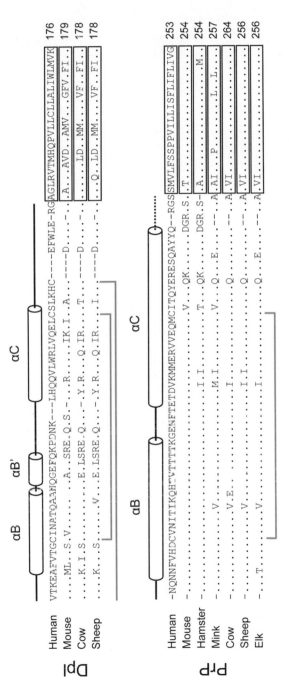

Figure 1. Annotated sequence alignment of PrP and Dpl from selected mammals.

Table 2. Dpl sequences in mammals

Subclass	Order	Common name	Scientific name	Accession number[a]	Identity (%)[b]	Reference
Boreoeutheria	Primates	human	*Homo sapiens*	Q9UKY0	100	Moore et al. (1999)
	Rodentia	mouse	*Mus musculus*	Q9QUG3	78	Moore et al. (1999)
	Caviomorpha	guinea pig	*Cavia porcellus*	Q8K3E6	61[c]	van Rheede T.[d]
	Chiroptera	fruit bat	*Cynopterus sphinx*	Q8MIJ1	55[c]	van Rheede T.[d]
	Carniovora	cat	*Felis silvestris catus*	Q8MIJ3	63[c]	van Rheede T.[d]
	Perissodactyla	tapir	*Tapirus terrestris*	Q8MIJ4	62[c]	van Rheede T.[d]
	Certartiodactyla	sheep	*Ovis aries*	Q9GIY2	79	Tranulis et al. (2001)
		cow	*Bos taurus*	Q9GK16	79	Tranulis et al. (2001)
		sperm whale	*Physeter catodon*	Q8MIJ2	65[c]	van Rheede T.[d]
Afrothera	Hyracoidea	hyrax	*Procavia capensis*	Q8MIJ5	56[c]	van Rheede T.[d]
	Sirenia	manatee	*Trichechus manatus*	Q8MIJ6	61[c]	van Rheede T.[d]
	Probososcidae	elephant	*Elephas maximus*	Q8MIJ7	62[c]	van Rheede T.[d]

[a]Accession number from Swiss-Prot database (http://us.expasy.org/sprot/).
[b]Amino acid identity based on a global alignment (Needleman and Wunsch 1970) of the mature protein (signal peptides removed) with mature, full-length human Dpl.
[c]Fragment of sequence slightly shorter than mature, full-length protein, resulting in a slightly lower value than expected.
[d]Direct submission to database.

Distant homologs of PrP and Dpl have recently been reported in fish (Table 3). These proteins can be split into two groups of orthologs: (1) Japanese and freshwater pufferfish PrP-1 and PrP-2, salmon PrP-1, and zebrafish PrP-1 and PrP-2 (Oidtmann et al. 2003; Rivera-Milla et al. 2003 and in prep.) and (2) PrP-like from Japanese and freshwater pufferfish, and zebrafish (Suzuki et al. 2002; E. Rivera-Milla et al., in prep.). The proteins of group 1 are significantly longer than PrP but contain the cysteine residues of the disulfide bond, and glycosylation sites in similar positions, implying that they may have comparable tertiary structures to PrP and Dpl (Oidtmann et al. 2003; Rivera-Milla et al. 2003). The proteins of group 2 are shorter than PrP and lack the structural protein landmarks, although they share a limited sequence similarity. Genomic synteny of PrP-1, PrP-like, and neighboring proteins with mammalian PrPs provides additional evidence for orthology (Suzuki et al. 2002; Oidtmann et al. 2003; E. Rivera-Milla et al., in prep.). Initial studies on PrP-1 and PrP-2 have revealed that their overexpression affects brain morphology in developing zebrafish embryos (E. Rivera-Milla et al., in prep.). However, further experiments are required to confirm a common functionality with mammalian PrPs.

Table 3. Remote homologs of PrP and Dpl identified in bony fish (class Osteichthyes)

Subclass	Order	Common name	Scientific name	Protein name[a]	Accession number[b]	Reference
Protacanthopterygii	Salmoniformes	salmon	*Salmo salar*	PrP-1	Q80IJ8	Oidtmann et al. (2003)
Neoteleostei	Tetraodontiformes	Japanese pufferfish	*Takifugu rubripes*	PrP-1	Q80IJ9, Q8AX89	Oidtmann et al. (2003); Rivera-Milla et al. (2003)
				PrP-2	Q80OZ8	Oidtmann et al. (2003)
				PrP-like	Q8IJJ1	Suzuki et al. (2002)

[a]Terminolgy for these proteins has not yet been standardized.
[b]Accession number from Swiss-Prot database (http://us.expasy.org/sprot/).

REFERENCES

Bartz J.C., McKenzie D.I., Bessen R.A., Marsh R.F., and Aiken J.M. 1994. Transmissible mink encephalopathy species barrier effect between ferret and mink: PrP gene and protein analysis. *J. Gen. Virol.* **75:** 2947–2953.

Basler K., Oesch B., Scott M., Westaway D., Wälchli M., Groth D.F., McKinley M.P., Prusiner S.B., and Weissmann C. 1986. Scrapie and cellular PrP isoforms are encoded by the same chromosomal gene. *Cell* **46:** 417–428.

Cervenakova L., Brown P., Goldfarb L.G., Nagle J., Pettrone K., Rubenstein R., Dubnick M., Gibbs C.J., Jr., and Gajdusek D.C. 1994. Infectious amyloid precursor gene sequences in primates used for experimental transmission of human spongiform encephalopathy. *Proc. Natl. Acad. Sci.* **91:** 12159–12162.

Gabriel J.-M., Oesch B., Kretzschmar H., Scott M., and Prusiner S.B. 1992. Molecular cloning of a candidate chicken prion protein. *Proc. Natl. Acad. Sci.* **89:** 9097–9101.

Goldmann W., Hunter N., Martin T., Dawson M., and Hope J. 1991. Different forms of the bovine PrP gene have five or six copies of a short, G-C-rich element within the protein-coding exon. *J. Gen. Virol.* **72:** 201–204.

Goldmann W., Hunter N., Foster J.D., Salbaum J.M., Beyreuther K., and Hope J. 1990. Two alleles of a neural protein gene linked to scrapie in sheep. *Proc. Natl. Acad. Sci.* **87:** 2476–2480.

Goldmann W., Martin T., Foster J., Hughes S., Smith G., Hughes K., Dawson M., and Hunter N. 1996. Novel polymorphisms in the caprine PrP gene: A codon 142 mutation associated with scrapie incubation period. *J. Gen. Virol.* **77:** 2885–2891.

Gomi H., Ikeda T., Kunieda T., Itohara S., Prusiner S.B., and Yamanouchi K. 1994. Prion protein (PrP) is not involved in the pathogenesis of spongiform encephalopathy in zitter rats. *Neurosci. Lett.* **166:** 171–174.

James T.L., Liu H., Ulyanov N.B., Farr-Jones S., Zhang H., Donne D.G., Kaneko K., Groth D., Mehlhorn I., Prusiner S.B., and Cohen F.E. 1997. Solution structure of a 142-residue recombinant prion protein corresponding to the infectious fragment of the scrapie isoform. *Proc. Natl. Acad. Sci.* **94:** 10086–10091.

Kaluz S., Kaluzova M., and Flint A.P. 1997. Sequencing analysis of prion genes from red deer and camel. *Gene* **199:** 283–286.

Kretzschmar H.A., Neumann M., Riethmüller G., and Prusiner S.B. 1992. Molecular cloning of a mink prion protein gene. *J. Gen. Virol.* **73:** 2757–2761.

Kretzschmar H.A., Stowring L.E., Westaway D., Stubblebine W.II., Prusiner S.B., and DeArmond S.J. 1986. Molecular cloning of a human prion protein cDNA. *DNA* **5:** 315–324.

Locht C., Chesebro B., Race R., and Keith J.M. 1986. Molecular cloning and complete sequence of prion protein cDNA from mouse brain infected with the scrapie agent. *Proc. Natl. Acad. Sci.* **83:** 6372–6376.

Loftus B. and Rogers M. 1997. Characterization of a prion protein (PrP) gene from rabbit; a species with apparent resistance to infection by prions. *Gene* **184:** 215–219.

López García F., Zahn R., Riek R., and Wüthrich K. 2000. NMR structure of the bovine prion protein. *Proc. Natl. Acad. Sci.* **97:** 8334–8339.

Lührs T., Riek R., Guntert P., and Wüthrich K. 2003. NMR structure of the human doppel protein. *J. Mol. Biol.* **326:** 1549–1557.

Martin T., Hughes S., Hughes K., and Dawson M. 1995. Direct sequencing of PCR amplified pig PrP genes. *Biochim. Biophys. Acta* **1270:** 211–214.

Mo H., Moore R.C., Cohen F.E., Westaway D., Prusiner S.B., Wright P.E., and Dyson H.J. 2001. Two different neurodegenerative diseases caused by proteins with similar structures. *Proc. Natl. Acad. Sci.* **98:** 2352–2357.

Moore R.C., Lee I.Y., Silverman G.L., Harrison P.M., Strome R., Heinrich C., Karunaratne A., Pasternak S.H., Chishti M.A., Liang Y., Mastrangelo P., Wang K., Smit A.F.A., Katamine S., Carlson G.A., Cohen F.E., Prusiner S.B., Melton D.W., Tremblay P., Hood L.E., and Westaway D. 1999. Ataxia in prion protein (PrP)-deficient mice is associated with upregulation of the novel PrP-like protein doppel. *J. Mol. Biol.* **292:** 797–817.

Needleman S.B. and Wunsch C.D. 1970. A general method applicable to the search for similarities in the amino acid sequence of two proteins. *J. Mol. Biol.* **48:** 443–453.

Oidtmann B., Simon D., Holtkamp N., Hoffmann R., and Baier M. 2003. Identification of cDNAs from Japanese pufferfish (*Fugu rubripes*) and Atlantic salmon (*Salmo salar*) coding for homologues to tetrapod prion proteins. *FEBS Lett.* **538:** 96–100.

O'Rourke K.I., Baszler T.V., Miller J.M., Spraker T.R., Sadler-Riggleman I., and Knowles D.P. 1998. Monoclonal antibody F89/160.1.5 defines a conserved epitope on the ruminant prion protein. *J. Clin. Microbiol.* **36:** 1750–1755.

Raymond G.J., Bossers A., Raymond L.D., O'Rourke K.I., McHolland L.E., Bryant P.K., 3rd, Miller M.W., Williams E.S., Smits M., and Caughey B. 2000. Evidence of a molecular barrier limiting susceptibility of humans, cattle and sheep to chronic wasting disease. *EMBO J.* **19:** 4425–4430.

Riek R., Hornemann S., Wider G., Billeter M., Glockshuber R., and Wüthrich K. 1996. NMR structure of the mouse prion protein domain PrP(121–231). *Nature* **382:** 180–182.

Rivera-Milla E., Stuermer C.A., and Malaga-Trillo E. 2003. An evolutionary basis for scrapie disease: Identification of a fish prion mRNA. *Trends Genet.* **19:** 72–75.

Schätzl H.M., Da Costa M., Taylor L., Cohen F.E., and Prusiner S.B. 1995. Prion protein gene variation among primates. *J. Mol. Biol.* **245:** 362–374.

Simonic T., Duga S., Strumbo B., Asselta R., Ceciliani F, and Ronchi S. 2000. cDNA cloning of turtle prion protein. *FEBS Lett.* **469:** 33–38.

Strumbo B., Ronchi S., Bolis L.C., and Simonic T. 2001. Molecular cloning of the cDNA coding for *Xenopus laevis* prion protein. *FEBS Lett.* **508:** 170–174.

Suzuki T., Kurokawa T., Hashimoto H., and Sugiyama M. 2002. cDNA sequence and tissue expression of *Fugu rubripes* prion protein-like: A candidate for the teleost orthologue of tetrapod PrPs. *Biochem. Biophys. Res. Commun.* **294:** 912–917.

Tranulis M.A., Espenes A., Comincini S., Skretting G., and Harbitz I. 2001. The PrP-like protein Doppel gene in sheep and cattle: cDNA sequence and expression. *Mamm. Genome* **12:** 376–379.

van Rheede T., Smolenaars M.M., Madsen O., and De Jong W.W. 2003. Molecular evolution of the mammalian prion protein. *Mol. Biol. Evol.* **20:** 111–121.

Wopfner F., Weidenhofer G., Schneider R., von Brunn A., Gilch S., Schwarz T.F., Werner T., and Schätzl H.M. 1999. Analysis of 27 mammalian and 9 avian PrPs reveals high conservation of flexible regions of the prion protein. *J. Mol. Biol.* **289:** 1163–1178.

Zahn R., Liu A., Lührs T., Riek R., von Schroetter C., López García F., Billeter M., Calzolai L., Wider G., and Wüthrich K. 2000. NMR solution structure of the human prion protein. *Proc. Natl. Acad. Sci.* **97:** 145–150.

Zhang L., Li N., Wang Q.G., Fan B.L., Meng Q.Y., and Wu C.X. 2002. Cloning and sequencing of quail and pigeon prion genes. *Anim. Biotechnol.* **13:** 159–162.

Index

Page numbers followed by an f or t indicate a figure or table, respectively.